2018 IEEE 30th International Symposium on Power Semiconductor Devices and ICs (ISPSD 2018)

Chicago, Illinois, USA
13 – 17 May 2018

IEEE Catalog Number: CFP18ISP-POD
ISBN: 978-1-5386-2928-4

**Copyright © 2018 by the Institute of Electrical and Electronics Engineers, Inc.
All Rights Reserved**

Copyright and Reprint Permissions: Abstracting is permitted with credit to the source. Libraries are permitted to photocopy beyond the limit of U.S. copyright law for private use of patrons those articles in this volume that carry a code at the bottom of the first page, provided the per-copy fee indicated in the code is paid through Copyright Clearance Center, 222 Rosewood Drive, Danvers, MA 01923.

For other copying, reprint or republication permission, write to IEEE Copyrights Manager, IEEE Service Center, 445 Hoes Lane, Piscataway, NJ 08854. All rights reserved.

*** This is a print representation of what appears in the IEEE Digital Library. Some format issues inherent in the e-media version may also appear in this print version.

IEEE Catalog Number:	CFP18ISP-POD
ISBN (Print-On-Demand):	978-1-5386-2928-4
ISBN (Online):	978-1-5386-2927-7
ISSN:	1063-6854

Additional Copies of This Publication Are Available From:

Curran Associates, Inc
57 Morehouse Lane
Red Hook, NY 12571 USA
Phone: (845) 758-0400
Fax: (845) 758-2633
E-mail: curran@proceedings.com
Web: www.proceedings.com

TABLE OF CONTENTS

Monday - May 14, 2018

8:50 – 10:20
Plenary 1
Chairs: John Shen, *Illinois Institute of Technology, USA*
Kuang Sheng, *Zhejiang University, China*

8:50
PL1-1 ISPSD: 30 Year Journey in Advancing Power Semiconductor Technology 1
Ayman Shibib, *Vishay Siliconix, United States;* Leo Lorenz, *ECPE/Infineon, Germany;*
Hiromichi Ohashi, *NEPRC-J, Japan*

9:35
PL1-2 Silicon, GaN and SiC: There's Room for All – An Application Space Overview of Device Considerations 8
Larry Spaziani, Lucas Lu, *GaN Systems, Canada*

10:40 – 12:10
Plenary 2
Chairs: Wai Tung Ng, *Universoty of Toronto, Canada*
Kevin Chen, *Hong Kong University of Science and Technology, China*

10:40
PL2-1 Si Wafer Technology for Power Devices: A Review and Future Directions 12
Norihisa Machida, *SUMCO Corporation, Japan*

11:25
PL2-2 The Future of Power Semiconductors: An EU Perspecitve 15
Bert De Colvenaer, *ECSEL JU, Belgium*

14:00 – 15:40
Session 1 – Superjunction MOS, Diodes and IGBTs
Chairs: Young Chul Choi, *ON Semiconductor, Korea*
Marina Antoniou, *University of Cambridge, UK*

14:00
1-1 IGBT with Superior Long-Term Switching Behavior by Asymmetric Trench Oxide 24
C. Sandow, P. Brandt, H.-P. Felsl, F.-J. Niedernostheide, F. Pfirsch, H.-J. Schulze, A. Stegner, F. Umbach, *Infineon Technologies AG, Germany;* F. Santos, W. Wagner, *Infineon Technologies Austria AG, Austria*

14:25
1-2 6.5 kV Field Shielded Anode (FSA) Diode Concept with 150C Maximum Operational Temperature Capability 28
B.K. Boksteen, C. Papadopoulos, D. Prindle, A. Kopta, C. Corvasce, *ABB Switzerland Ltd., Switzerland*

14:50

1-3 **Low Noise Superjunction MOSFET with Integrated Snubber Structure** 32
Hiroaki Yamashita, Syotaro Ono, Hisao Ichijo, Masataka Tsuji, *Toshiba Electronic Devices & Storage Corporation, Japan;* Takenori Yasuzumi, *Toshiba Corporation, Japan;* Masaru Izumisawa, Wataru Saito, *Toshiba Electronic Devices & Storage Corporation, Japan*

15:15

1-4 **Breakthrough of Drain Current Capability and On-Resistance Limits by Gate-Connected Superjunction MOSFET** 36
Wataru Saito, *Toshiba Electronic Devices & Storage Corporation, Japan*

16:00 – 18:05
Session 2 – SiC Power MOSFETs
Chairs: Peter Losee, *General Electric, USA*
Andrei Petru Mihaila, *ABB, Switzerland*

16:00

2-1 **Investigation of Threshold Voltage Stability of SiC MOSFETs** 40
Dethard Peters, *Infineon Technologies AG, Germany,* Thomas Aichinger, *Infineon Technologies Austria AG, Austria,* Thomas Basler, *Infineon Technologies AG, Germany,* Gerald Rescher, *Infineon Technologies Austria AG, Austria,* Katja Puschkarsky, Hans Reisinger, *Infineon Technologies AG, Germany*

16:25

2-2 **Deep-P Encapsulated 4H-SiC Trench MOSFETs with Ultra Low $R_{on}Q_{gd}$** 44
Yasuhiro Ebihara, Aiko Ichimura, Shuhei Mitani, Masato Noborio, Yuichi Takeuchi, Shoji Mizuno, Toshimasa Yamamoto, Kazuhiro Tsuruta, *Denso Corporation, Japan*

16:50

2-3 **Influence of the Off-State Gate-Source Voltage on the Transient Drain Current Response of SiC MOSFETs** 48
Christian Unger, Martin Pfost, *Technische Universität Dortmund, Germany*

17:15

2-4 **Reduction of RonA Retaining High Threshold Voltage in SiC DioMOS by Improved Channel Design** 52
Atsushi Ohoka, Masao Uchida, Tsutomu Kiyosawa, Nobuyuki Horikawa, Kouichi Saitou, Yoshihiko Kanzawa, Haruyuki Sorada, Kazuyuki Sawada, Tetsuzo Ueda, *Panasonic Corporation, Japan*

17:40

2-5 **Avalanche Ruggedness and Reverse-Bias Reliability of SiC MOSFET with Integrated Junction Barrier Controlled Schottky Rectifier** 56
Cheng-Tyng Yen, Fu-Jen Hsu, Chien-Chung Hung, Chwan-Ying Lee, Lurng-Shehng Lee, Ya-Fang Li, Kuo-Ting Chu, *Hestia Power Inc., Taiwan*

Tuesday - May 15, 2018

8:30 – 10:10
Session 3 – Lateral Devices: Reliability
Chairs: Phil Rutter, *Nexperia, UK*
Jun Cai, *Texas Instruments, USA*

8:30

3-1 **Comprehensive Investigation on Mechanical Strain Induced Performance Boosts in LDMOS** 60
Wangran Wu, Siyang Liu, Jing Zhu, Weifeng Sun, *Southeast University, China*

8:55

3-2 **Investigation on Total-Ionizing-Dose Radiation Response for High Voltage Ultra-Thin Layer SOI LDMOS** 64

Xin Zhou, Lingfang Zhang, Ming Qiao, Zhangyi'an Yuan, Ping Luo, *University of Electronic Science and Technology of China, China;* Lei Shu, *Harbin Institute of Technology, China;* Zhaoji Li, Bo Zhang, *University of Electronic Science and Technology of China, China*

9:20

3-3 **Electromigration Current Limit Relaxation for Power Device Interconnects** 68

Jungwoo Joh, Young-Joon Park, Srikanth Krishnan, Kim Christensen, Jayhoon Chung, *Texas Instruments, United States*

9:45

3-4 **Performance and Reliability Insights of Drain Extended FinFET Devices for High Voltage SoC Applications** 72

B. Sampath Kumar, Milova Paul, Mayank Shrivastava, *Indian Institute of Science, India;* Harald Gossner, *Intel Deutschland GmbH, Germany*

10:30 – 12:10
Session 4 – Smart Power ICs
Chairs: Nicolas Rouger, *CNRS, France*
Budong (Albert) You, *Silergy Corp., China*

10:30

4-1 **High-Speed, High-Reliability GaN Power Device with Integrated Gate Driver** 76

Gaofei Tang, *Hong Kong University of Science and Technology, China;* M.-H. Kwan, *Taiwan Semiconductor Manufacturing Company Limited, Taiwan;* Zhaofu Zhang, Jiabei He, Jiacheng Lei, *Hong Kong University of Science and Technology, China;* R.-Y. Su, F.-W. Yao, Y.-M. Lin, J.-L. Yu, Thomas Yang, Chan-Hong Chern, Tom Tsai, H.C. Tuan, Alexander Kalnitsky, *Taiwan Semiconductor Manufacturing Company Limited, Taiwan;* Kevin J. Chen, *Hong Kong University of Science and Technology, China*

10:55

4-2 **A 600V High-Side Gate Drive Circuit with Ultra-Low Propagation Delay for Enhancement Mode GaN Devices** 80

Yangyang Lu, Jing Zhu, Weifeng Sun, *Southeast University, China;* Yunwu Zhang, *Southeast University and China Resources Microelectronics Limited, China;* Kongsheng Hu, Zhicheng Yu, Jing Leng, *Southeast University, China;* Shikang Cheng, Sen Zhang, *CSMC Technologies Corporation, China*

11:20

4-3 **A Smart Gate Driver IC for GaN Power Transistors** 84

Jingshu Yu, Wei Jia Zhang, Andrew Shorten, Rophina Li, Wai Tung Ng, *University of Toronto, Canada*

11:45

4-4 **[Late News] CMOS Bi-Directional Ultra-Wideband Galvanically Isolated Die-to-Die Communication Utilizing a Double-Isolated Transformer** 88

Mahdi Javid, *Arizona State University, United States;* Karel Ptacek, *ON Semiconductor, Czech Republic;* Richard Burton, *Atomera Inc., United States;* Jennifer Kitchen, *Arizona State University, United States*

13:30 – 15:10
Session 5 – GaN Power Devices - 1
Chairs: Kevin Chen, *Hong Kong University of Science and Technology, China*
Oliver Haeberlen, *Infineon Technologies, Austria*

13:30

5-1   Dynamic-Ron Control via Proton Irradiation in AlGaN/GaN Transistors 92
A. Tajalli, *Università degli Studi di Padova, Italy;* A. Stockman, *ON Semiconductor, CMST imec/Ghent University, Belgium;* M. Meneghini, *Università degli Studi di Padova, Italy;* S. Mouhoubi, A. Banerjee, *ON Semiconductor, Belgium;* S. Gerardin, M. Bagatin, A. Paccagnella, E. Zanoni, *Università degli Studi di Padova, Italy;* M. Tack, *ON Semiconductor, Belgium;* B. Bakeroot, *CMST imec/Ghent University, Belgium;* P. Moens, *ON Semiconductor, Belgium;* G. Meneghesso, *Università degli Studi di Padova, Italy*

13:55

5-2   Bidirectional Threshold Voltage Shift and Gate Leakage in 650 V p-GaN AlGaN/GaN HEMTs: The Role of Electron-Trapping and Hole-Injection 96
Yuanyuan Shi, Qi Zhou, Qian Cheng, P. Wei, L. Zhu, D. Wei, A. Zhang, Wanjun Chen, Bo Zhang, *University of Electronic Science and Technology of China, China*

14:20

5-3   GaN-on-Si Lateral Power Devices with Symmetric Vertical Leakage: The Impact of Floating Substrate 100
Hanyuan Zhang, Shu Yang, Kuang Sheng, *Zhejiang University, China*

14:45

5-4   Short Circuit Robustness Analysis of New Generation Enhancement-Mode p-GaN Power HEMTs 104
M. Riccio, G. Romano, L. Maresca, G. Breglio, A. Irace, *Università degli Studi di Napoli Federico II, Italy;* G. Longobardi, *University of Cambridge, United Kingdom*

15:30 – 17:30
Poster Session 6 – High Voltage
Chairs: Tadaharu Minato, *Mitsubishi, Japan*
Tom Tsai, *TSMC, Taiwan*

6-1   Influence of Doping Profiles and Chip Temperature on Short-Circuit Oscillations of IGBTs 108
Vera van Treek, Hans-Joachim Schulze, Franz-Josef Niedernostheide, Christian Sandow, Roman Baburske, Frank Pfirsch, *Infineon Technologies AG, Germany*

6-2   A 750V Recessed-Emitter-Trench IGBT with Recessed-Dummy-Trench Structure Featuring Low Switching Losses 112
Yao Yao, Haihui Luo, Qiang Xiao, *Zhuzhou CRRC Times Electric Co., Ltd., China;* Chunlin Zhu, *Dynex Semiconductor Ltd., United Kingdom;* Haibo Xiao, Rongzhen Qin, *Zhuzhou CRRC Times Electric Co., Ltd., China;* Luther-King Ngwendson, *Dynex Semiconductor Ltd., United Kingdom;* Xubin Ning, Canjian Tan, *Zhuzhou CRRC Times Electric Co., Ltd., China;* Ian Deviny, *Dynex Semiconductor Ltd., United Kingdom;* Xiaoping Dai, *Zhuzhou CRRC Times Electric Co., Ltd., China*

6-3   Small Current Unclamped Inductive Switching (UIS) to Detect Fabrication Defect for Mass-Production Phase IGBT 116
Kazuya Sano, Yukio Matsushita, Megumi Yachi, Yoji Nakata, *Mitsubishi Electric Corporation, Japan;* Keiichiro Ide, Daisaku Yoshida, *Kyokuyo Semiconductors Corporation, Japan;* Tomohito Kudo, Yasuo Ata, Hideki Haruguchi, Shinya Soneda, Tadaharu Minato, *Mitsubishi Electric Corporation, Japan*

6-4   Tailoring the Performance of Silicon Power Diodes by Predictive TCAD Simulation of Platinum 120
Moritz Hauf, Christian Sandow, Franz-Josef Niedernostheide, *Infineon Technologies AG, Germany;* Gerhard Schmidt, *Infineon Technologies Austria AG, Austria*

6-5   Novel 3D Narrow Mesa IGBT Suppressing CIBL 124
Masahiro Tanaka, *Nihon Synopsys G.K., Japan;* Akio Nakagawa, *Nakagawa Consulting Office, LLC., Japan*

6-6   **N-Buffer Design Optimization for Short Circuit SOA Ruggedness in 1200V Class IGBT** 128
Kenji Suzuki, Koichi Nishi, Mitsuru Kaneda, Akihiko Furukawa, *Mitsubishi Electric Corporation, Japan*

6-7   **High Avalanche Capability Specific Diode Part Structure of RC-IGBT based upon CSTBT™** 132
Shinya Soneda, Shinya Akao, Tetsuo Takahashi, Akihiko Furukawa, *Mitsubishi Electric Corporation, Japan*

6-8   **C_{OSS} Losses in Silicon Superjunction MOSFETs across Constructions and Generations** 136
Grayson Zulauf, Juan M. Rivas-Davila, *Stanford University, United States*

6-9   **Extending the RET-IGBT (Recessed Emitter Trench IGBT) Concept to High Voltages:**
Experimental Demonstration of 3.3kV RET IGBT 140
L. Ngwendson, I. Deviny, C. Zhu, I. Saddiqui, C. Kong, A. Islam, J. Hutchings, J. Thompson, M. Briggs, O. Basset,
Dynex Semiconductor Ltd., United Kingdom; H. Luo, Y. Wang, Y. Yao, *Zhuzhou CRRC Times Electric Co., Ltd., China*

6-10   **Temperature Dependence of the On-State Voltage Drop in Field-Stop IGBTs** 144
L. Maresca, M. Riccio, G. Breglio, A. Irace, *Università degli Studi di Napoli Federico II, Italy;*
P. Mirone, C. Sanfilippo, L. Merlin, *Vishay Semiconductor Italiana, Italy*

6-11   **A High-Voltage p-LDMOS with Enhanced Current Capability Comparable to**
Double RESURF n-LDMOS 148
Bo Yi, Junji Cheng, Moufu Kong, Bingke Zhang, Xing Bi Chen, *University of Electronic
Science and Technology of China, China*

6-12   **Self Terminating Lateral-Vertical Hybrid Super-Junction FET that Breaks $R_{DS}.A$ –**
Charge Balance Trade-Off Window 152
Karthik Padmanabhan, Lingpeng Guan, Madhur Bobde, Sik Lui, *Alpha and Omega Semiconductor, Inc., United
States;* Anup Bhalla, *United Silicon Carbide Inc., United States;* Hamza Yilmaz, *Computime Limited, China;*
Lei Zhang, *Jireh Semiconductor, Inc., United States*

6-13   **Local Lifetime Control for Enhanced Ruggedness of HVDC Thyristors** 156
J. Vobecky, V. Botan, K.U. Meier, K. Tugan, M. Bellini, *ABB Switzerland Ltd., Switzerland*

6-14   **Low Injection Anode as Positive Spiral Improvement for 650V RC-IGBT** 160
Ryu Kamibaba, Mitsuru Kaneda, Tetsuo Takahashi, Akihiko Furukawa, *Mitsubishi Electric Corporation, Japan*

6-15   **Observation of Current Filaments in IGBTs with Thermoreflectance Microscopy** 164
Riteshkumar Bhojani, Jens Kowalsky, Josef Lutz, *Technische Universität Chemnitz, Germany;* Dustin Kendig,
Microsanj, United States; Roman Baburske, Hans-Joachim Schulze, Franz-Josef Niedernostheide, *Infineon
Technologies AG, Germany*

6-16   **IGBT Structure with Electrically Separated Floating-P Region improving**
Turn-On dVak/dt Controllability 168
Yoshihiro Ikura, Yuichi Onozawa, *Fuji Electric Co., Ltd., Japan;* Akio Nakagawa, *Nakagawa
Consulting Office, LLC., Japan*

6-17   **Optimization of Trench Sidewall for Low Leakage Current of the**
Sloped Field Plate Trench Edge Termination 172
Wentao Yang, *Hong Kong University of Science and Technology, China;* Xianda Zhou, *Hong Kong University of
Science and Technology and Sun Yat-sen University, China;* Chao Xiao, Hao Feng, Yong Liu, Xiangming Fang,
Hong Kong University of Science and Technology, China; Yuichi Onozawa, Hiroyuki Tanaka, Kaname Mitsuzuka,
Fuji Electric Co., Ltd., Japan; Johnny K.O. Sin, *Hong Kong University of Science and Technology, China*

6-18   **Analysis of Reverse Temperature Dependent Switching-Off Behavior of Ultra-Thin Fieldstop IGBTs** 176
So-Youn Kim, Euntaek Kim, Jiho Jeon, Jinyoung Jung, Soo-Seong Kim, Kwang-Hoon Oh,
Chongman Yun, *TRinno Technology, Korea*

6-19   **Effect of Charge Imbalance and Edge Structure on the Reverse Recovery Waveform in**
Superjunction Body Diode 180
Daisuke Arai, Mizue Yamaji, Koichi Murakami, Masaaki Honda, Shinji Kunori, *Shindengen Electric
Manufacturing Co., Ltd., Japan*

6-20 **Tight Relationship among Field Failure Rate, Single Event Burn-Out (SEB) and Cold Bias Stability (CBS) as a Cosmic Ray Endurance for IGBT and Diode** 184

K. Suzuki, Y. Yoshiura, K. Uryu, T. Minato, M. Tarutani, Y. Miyazaki, H. Uemura, T. Hagihara, S. Momii, Y. Kusakabe, M. Nakamura, *Mitsubishi Electric Corporation, Japan;* Y. Fujita, K. Takakura, *MELCO Semiconductor Engineering Corporation, Japan*

15:30 – 17:30
Poster Session 7 – GaN
Chairs: Tadaharu Minato, *Mitsubishi, Japan*
Tom Tsai, *TSMC, Taiwan*

7-1 **Gate Architecture Design for Enhancement Mode p-GaN Gate HEMTs for 200 and 650V Applications** 188

N.E. Posthuma, S. You, S. Stoffels, H. Liang, M. Zhao, S. Decoutere, *imec, Belgium*

7-2 **Uni-Directional GaN-on-Si MOSHEMTs with High Reverse-Blocking Voltage based on Nanostructured Schottky Drain** 192

Jun Ma, Elison Matioli, *École Polytechnique Fédérale de Lausanne, Switzerland*

7-3 **Characterization of GaN-HEMT in Cascode Topology and Comparison with State of the Art-Power Devices** 196

Sven Buetow, Reinhard Herzer, *Semikron Elektronik GmbH & Co. KG, Germany*

7-4 **Performance Enhancement of CMOS Compatible 600V Rated AlGaN/GaN Schottky Diodes on 200mm Silicon Wafers** 200

J. Biscarrat, R. Gwoziecki, Y. Baines, J. Buckley, C. Gillot, W. Vandendaele, G. Garnier, M. Charles, M. Plissonnier, *Université Grenoble Alpes, CEA-LETI, France*

7-5 **Novel AlGaN/GaN SBDs with Nanoscale Multi-Channel for Gradient 2DEG Modulation** 204

Anbang Zhang, Qi Zhou, Chao Yang, Yuanyuan Shi, Changxu Dong, *University of Electronic Science and Technology of China, China;* Tong Liu, *Chinese Academy of Sciences, China;* Yijun Shi, Wanjun Chen, Zhaoji Li, Bo Zhang, *University of Electronic Science and Technology of China, China*

7-6 **Switching Performance Analysis of GaN OG-FET using TCAD Device-Circuit-Integrated Model** 208

Dong Ji, Wenwen Li, Srabanti Chowdhury, *University of California, Davis, United States*

7-7 **A Split Gate Vertical GaN Power Transistor with Intrinsic Reverse Conduction Capability and Low Gate Charge** 212

Ruopu Zhu, Qi Zhou, Hong Tao, Yi Yang, Kai Hu, Dong Wei, Liyang Zhu, Yu Shi, Wanjun Chen, Xiaorong Luo, Bo Zhang, *University of Electronic Science and Technology of China, China*

7-8 **Experimental Characterization of the Fully Integrated Si-GaN Cascoded FET** 216

Jie Ren, Chak Wah Tang, Hao Feng, Huaxing Jiang, Wentao Yang, *Hong Kong University of Science and Technology, China;* Xianda Zhou, *Hong Kong University of Science and Technology and Sun Yat-Sen University, China;* Kei May Lau, Johnny K.O. Sin, *Hong Kong University of Science and Technology, China*

7-9 **Effect of Device Layout on the Switching of Enhancement Mode GaN HEMTs** 220

Loizos Efthymiou, Gianluca Camuso, Giorgia Longobardi, Florin Udrea, *University of Cambridge, United Kingdom;* Terry Chien, Max Chen, *Vishay General Semiconductor, Taiwan;* Ayman Shibib, Kyle Terrill, *Vishay Siliconix, United States*

7-10 **A Balancing Method for Low R_{on} and High V_{th} Normally-Off GaN MISFET by Preserving a Damage-Free Thin AlGaN Barrier Layer** 224

Jialin Zhang, Liang He, Liuan Li, Yiqiang Ni, Taotao Que, Zhenxin Liu, Wenjing Wang, Jiexin Zheng, Yanfen Huang, Jia Chen, Xin Gu, Yawen Zhao, Lei He, Zhisheng Wu, Yang Liu, *Sun Yat-sen University, China*

7-11 **Enhancement of Punch-Through Voltage in GaN with Buried P-Type Layer Utilizing Polarization-Induced Doping** 228
Wenshen Li, Mingda Zhu, Kazuki Nomoto, Zongyang Hu, *Cornell University, United States;*
Xiang Gao, *IQE RF LLC, United States;* Manyam Pilla, *Qorvo Inc., United States;* Debdeep Jena,
Huili Grace Xing, *Cornell University, United States*

7-12 **P-Gate GaN HEMT Gate-Driver Design for Joint Optimization of Switching Performance, Freewheeling Conduction and Short-Circuit Robustness** 232
Han Wu, Asad Fayyaz, Alberto Castellazzi, *University of Nottingham, United Kingdom*

7-13 **Monolithic Integration of GaN-Based NMOS Digital Logic Gate Circuits with E-Mode Power GaN MOSHEMTs** 236
Minghua Zhu, Elison Matioli, *École Polytechnique Fédérale de Lausanne, Switzerland*

7-14 **645 V Quasi-Vertical GaN Power Transistors on Silicon Substrates** 240
Chao Liu, Riyaz Abdul Khadar, Elison Matioli, *École Polytechnique Fédérale de Lausanne, Switzerland*

15:30 – 17:30
Poster Session 8 – Packaging
Chairs: Tadaharu Minato, *Mitsubishi, Japan*
Tom Tsai, *TSMC, Taiwan*

8-1 **Effects of Inorganic Encapsulation on Power Cycling Lifetime of Aluminum Bond Wires** 244
Nan Jiang, *Technische Universität Chemnitz, Germany;* Markus G. Scheibel, Benjamin Fabian,
Marko Kalajica, Anton-Zoran Miric, *Heraeus Deutschland GmbH & Co. KG, Germany;*
Josef Lutz, *Technische Universität Chemnitz, Germany*

8-2 **Sn- and Cu-Oxide Reduction by Formic Acid and its Application to Power Module Soldering** 248
Naoto Ozawa, Tatsuo Okubo, Jun Matsuda, Tatsuo Sakai, *Origin Electric Co., Ltd., Japan*

8-3 **Dynamic Characterisation and Optimisation of Multiply Contacted Power Busbars** 252
Vanessa Basler, Andreas Wagner, Wolfgang Hölzl, Gerhard Wachutka, *Technische Universität München, Germany*

8-4 **Development of a Highly Integrated 10 kV SiC MOSFET Power Module with a Direct Jet Impingement Cooling System** 256
Bassem Mouawad, Robert Skuriat, Jianfeng Li, C. Mark Johnson, *University of Nottingham, United Kingdom;*
Christina DiMarino, *Virginia Polytechnic Institute and State University, United States*

8-5 **A More Accurate Electromagnetic Modeling of WBG Power Modules** 260
Ivana Kovačević-Badstübner, Ulrike Grossner, *ETH Zürich, Switzerland;* Daniele Romano, Giulio Antonini,
Università degli Studi dell'Aquila, Italy; Jonas Ekman, *Luleå University of Technology, Sweden*

8-6 **Accelerated Thermal Fatigue Test of Metallized Ceramic Substrates for SiC Power Modules by Repeated Four-Point Bending** 264
Hiroyuki Miyazaki, Hideki Hyuga, Kiyoshi Hirao, Hiroshi Sato, Hiroshi Yamaguchi, *National Institute of Advanced Industrial Science and Technology, Japan;* Shoji Iwakiri, Hideki Hirotsuru, *Denka Co., Ltd., Japan*

8-7 **Dynamic Stability Analysis based on State-Space Model and Lyapunov's Stability Criterion for SiC-MOS and Si-IGBT Switching** 268
Xiao Zeng, Zehong Li, Yuzhou Wu, Wei Gao, Jinping Zhang, Min Ren, Bo Zhang, *University of Electronic Science and Technology of China, China*

Wednesday - May 16, 2018

8:30 – 10:10
Session 9 – GaN Power Devices - 2
Chairs: Peter Moens, *ON Semiconductor, Belgium*
Yang Liu, *Sun Yat-sen University, China*

8:30
9-1 1 kV/1.3 mΩ cm² Vertical GaN-on-GaN Schottky Barrier Diodes with High Switching Performance 272
Shu Yang, Shaowen Han, Rui Li, Kuang Sheng, *Zhejiang University, China*

8:55
9-2 Reverse-Blocking AlGaN/GaN Normally-Off MIS-HEMT with Double-Recessed Gated Schottky Drain 276
Jiacheng Lei, Jin Wei, Gaofei Tang, Kevin J. Chen, *Hong Kong University of Science and Technology, China*

9:20
9-3 Recess-Free AlGaN/GaN Lateral Schottky Barrier Controlled Schottky Rectifier with
Low Turn-On Voltage and High Reverse Blocking 280
Xuanwu Kang, Xinhua Wang, Sen Huang, Jinhan Zhang, Jie Fan, Shuo Yang, Yuankun Wang, Yingkui Zheng,
Ke Wei, Jin Zhi, Xinyu Liu, *Institute of Microelectronics of Chinese Academy of Sciences, China*

9:45
9-4 [Late News] An Industry-Ready 200 mm p-GaN E-Mode GaN-on-Si Power Technology 284
N.E. Posthuma, S. You, S. Stoffels, D. Wellekens, H. Liang, M. Zhao, B. De Jaeger, K. Geens, N. Ronchi,
S. Decoutere, *imec, Belgium;* P. Moens, A. Banerjee, H. Ziad, M. Tack, *ON Semiconductor, Belgium*

10:10 – 12:10
Poster Session 10 – Low Voltage Technology
Chairs: David Tsung-Yi Huang, *Richtek, Taiwan*
Sameer Pendharkar, *Texas Instruments, USA*

10-1 Application-Driven Device/Circuit Co-Simulation Framework for Power MOSFET Design and
Technology Development 288
Tirthajyoti Sarkar, Ashok Challa, Kirk Huang, Prasad Venkatraman, Dean Probst, *ON Semiconductor, United States*

10-2 A Novel High Performance Medium-Voltage DEnMOS in 40nm CMOS Technology 292
Wei Lin, Upinder Singh, Jeoung Mo Koo, Huihua Jiang, *Globalfoundries, Singapore*

10-3 Novel Current Re-Distribution Structure for Improved and Easy-to-Manufacturing 24V LDMOS 295
Cheng-Hua Lin, Yan-Liang Ji, C.H. Jan, C.W. Hu, Keven Chang, H.W. Kao, *MediaTek Inc., Taiwan*

10-4 A Novel Divided STI-Based nLDMOSFET for Suppressing HCI Degradation under
High Gate Bias Stress 299
Takahiro Mori, Shunji Kubo, Takashi Ipposhi, *Renesas Semiconductor Manufacturing Co., Ltd., Japan*

10-5 Hot-Carrier Induced Off-State Leakage Current Increase of LDMOS and
Approach to Overcome the Phenomenon 303
Keita Takahashi, Kanako Komatsu, Toshihiro Sakamoto, Koji Kimura, Fumitomo Matsuoka, *Toshiba Electronic
Devices & Storage Corporation, Japan;* Yoshiaki Ishii, Katsumi Egashira, Masaki Sakai, *Japan Semiconductor
Corporation, Japan*

10-6 Novel Approach for NLDMOS Performance Enhancement by Critical Electric Field Engineering 307
Jaroslav Pjenčák, *ON Semiconductor, Czech Republic;* Moshe Agam, *ON Semiconductor, United States;*
Ladislav Šeliga, *ON Semiconductor, Czech Republic;* Thierry Yao, Agajan Suwhanov, *ON Semiconductor, United States*

10-7 **A 0.35μm 600V Ultra-Thin Epitaxial BCD Technology for High Voltage Gate Driver IC** 311

Huihui Wang, *Shanghai Huahong Grace Semiconductor Manufacturing Corporation, China;* Ming Qiao, *University of Electronic Science and Technology of China, China;* Feng Jin, *Shanghai Huahong Grace Semiconductor Manufacturing Corporation, China;* Yang Yu, ZhangYi'an Yuan, *University of Electronic Science and Technology of China, China;* Binbin Miao, Wenqing Yang, Jie Wu, Wensheng Qian, Tong Deng, Donghua Liu, Ziquan Fang, Wenting Duan, Jiye Yang, Weiran Kong, *Shanghai Huahong Grace Semiconductor Manufacturing Corporation, China;* Bo Zhang, *University of Electronic Science and Technology of China, China*

10-8 **Impact of Self-Heating Effect in Hot Carrier Injection Modeling** 315

Dong Seup Lee, Dhanoop Varghese, Arif Sonnet, Jungwoo Joh, Archana Venugopal, Srikanth Krishnan, *Texas Instruments, United States*

10-9 **Duty-Cycle-Accelerated Hot-Carrier Degradation and Lifetime Evaluation for 700V Lateral DMOS** 319

Siyang Liu, Zhichao Li, Wangran Wu, Weifeng Sun, *Southeast University, China;* Shulang Ma, Yuwei Liu, Wei Su, Xiaohong Liu, *CSMC Technologies Corporation, China*

10-10 **A High-Speed SOI-LIGBT with Electric Potential Modulation Trench and Low-Doped Buried Layer** 323

Shaohong Li, Long Zhang, Jing Zhu, Weifeng Sun, Qingxi Tang, Hao Wang, Ling Sun, *Southeast University, China;* Yan Gu, Shikang Cheng, Sen Zhang, *CSMC Technologies Corporation, China;* Yangbo Yi, *Wuxi Chipown Microelectronics Ltd., China*

10-11 **A Comparison of Close-Cell, Stripe-Cell, and Orthogonal-Cell Low Voltage Superjunction Trench Power MOSFETs for Linear Mode Application** 327

Yi Su, Madhur Bobde, Sik Lui, Hong Chang, *Alpha and Omega Semiconductor, Inc., United States;* Qinhai Jin, Lei Zhang, *Jireh Semiconductor, Inc., United States*

10-12 **A 150V Novel High-Voltage LDMOS in a 0.18um BCD Plug-In Process** 331

Yen-Ming Chen, Chiu-Ling Lee, Min-Hsuan Tsai, Chiu-Te Lee, Chih-Chong Wang, *United Microelectronics Corporation, Taiwan*

10-13 **Application of CS-MCT in DC Solid State Circuit Breaker (SSCB)** 335

Wanjun Chen, Hong Tao, Chao Liu, Yawei Liu, Chengfang Liu, Jie Liu, Yijun Shi, Qi Zhou, Zhaoji Li, Bo Zhang, *University of Electronic Science and Technology of China, China*

10-14 **ESD Failure Analysis and Robustness Improvement for Multi-STI-Finger LDMOS used as Output Device** 339

Ran Ye, Siyang Liu, Zhigang Dai, Hongting Chen, Wangran Wu, Weifeng Sun, Shengli Lu, *Southeast University, China;* Wei Su, Feng Lin, Guipeng Sun, *CSMC Technologies Corporation, China*

10:10 – 12:10
Poster Session 11 – IC Design

Chairs: David Tsung-Yi Huang, *Richtek, Taiwan*
Sameer Pendharkar, *Texas Instruments, USA*

11-1 **Integrated Symmetrical High Voltage Inverter for the Excitation of Touch Sensitive Electroluminescent Devices** 343

Katrin Hirmer, Muhammad Bilal Saif, Klaus Hofmann, *Technische Universität Darmstadt, Germany*

11-2 **A Power Inductor Integration Technology using a Silicon Interposer for DC-DC Converter Applications** 347

Yixiao Ding, *Hong Kong University of Science and Technology, China;* Xiangming Fang, *Shenzhen CoilEasy Technologies Limited, China;* Yuan Gao, Yuefei Cai, Xing Qiu, Philip K.T. Mok, S.W. Ricky Lee, Kei May Lau, Johnny K.O. Sin, *Hong Kong University of Science and Technology, China*

11-3 **A New 1200 V HVIC with High Side Edge Trigger in Order to Solve the Latch on Failure by the Negative VS Surge** 351

Kinam Song, Wonhi Oh, Jinkyu Choi, Seunghyun Hong, Sangmin Park, *ON Semiconductor, Korea*

11-4 A High-Voltage Half-Bridge Gate Drive Circuit for GaN Devices with High-Speed Low-Power and High-Noise-Immunity Level Shifter 355
Xin Ming, Xuan Zhang, Zhi-wen Zhang, Xu-dong Feng, Li Hu, Xia Wang, Gang Wu, Bo Zhang, *University of Electronic Science and Technology of China, China*

11-5 AC/DC Flyback Controller with UHV Integrated Start-Up Current Source in 180nm HVIC Technology 359
Hing Kit Kwan, Bai Yen Nguyen, Wen-Cheng Lin, Xiaoxin Liu, Swapnil Pandey, Saikat Chakraborty, Jongjib Kim, Don Disney, *Globalfoundries, Singapore*

10:10 – 12:10
Poster Session 12 – SiC
Chairs: David Tsung-Yi Huang, *Richtek, Taiwan*
Sameer Pendharkar, *Texas Instruments, USA*

12-1 Evaluation of Gate Oxide Reliability in 3.3kV 4H-SiC DMOSFET with J-Ramp TDDB Methods 363
Masakazu Sagawa, Hiroshi Miki, Yuki Mori, Haruka Shimizu, Akio Shima, *Hitachi Ltd., Japan*

12-2 Repetitive Surge Current Test of SiC MPS Diode with Load in Bipolar Regime 367
Shanmuganathan Palanisamy, Jens Kowalsky, Josef Lutz, *Technische Universität Chemnitz, Germany;*
Thomas Basler, Roland Rupp, Jasmin Moazzami-Fallah, *Infineon Technologies AG, Germany*

12-3 Accumulation Channel vs. Inversion Channel 1.2 kV Rated 4H-SiC Buffered-Gate (BG) MOSFETs: Analysis and Experimental Results 371
Kijeong Han, B. Jayant Baliga, *North Carolina State University, United States;* Woongje Sung, *State University of New York Polytechnic Institute, United States*

12-4 Characterization of 1.2 kV SiC Super-Junction SBD Implemented by Trench and Implantation Technique 375
Baozhu Wang, Hengyu Wang, Xueqian Zhong, Shu Yang, Qing Guo, Kuang Sheng, *Zhejiang University, China*

12-5 Normally-OFF Dual-Gate Ga_2O_3 Planar MOSFET and FinFET with High I_{ON} and BV 379
H.Y. Wong, N. Braga, R.V. Mickevicius, *Synopsys, Inc., United States;* F. Ding, *University of California, Berkeley, United States*

12-6 Analysis of Short-Circuit Break-Down Point in 3.3 kV SiC-MOSFETs 383
Kazuki Tani, Jun-ichi Sakano, Akio Shima, *Hitachi Ltd., Japan*

12-7 Electrical Characterization of 1.2kV SiC MOSFET at Extremely High Junction Temperature 387
Jiahui Sun, Hongyi Xu, Shu Yang, Kuang Sheng, *Zhejiang University, China*

12-8 Methodology for Enhanced Short-Circuit Capability of SiC MOSFETs 391
Junjie An, Masaki Namai, Hiroshi Yano, Noriyuki Iwamuro, *University of Tsukuba, Japan;* Yusuke Kobayashi, Shinsuke Harada, *National Institute of Advanced Industrial Science and Technology, Japan*

12-9 27.5 kV 4H-SiC PiN Diode with Space-Modulated JTE and Carrier Injection Control 395
Koji Nakayama, *National Institute of Advanced Industrial Science and Technology, Japan;* Tomonori Mizushima, Kensuke Takenaka, *National Institute of Advanced Industrial Science and Technology and Fuji Electric Co., Ltd., Japan;* Akihiro Koyama, *National Institute of Advanced Industrial Science and Technology and Mitsubishi Electric Corporation, Japan;* Yuji Kiuchi, *National Institute of Advanced Industrial Science and Technology and New Japan Radio Co., Ltd., Japan;* Shinichiro Matsunaga, Hiroyuki Fujisawa, *National Institute of Advanced Industrial Science and Technology and Fuji Electric Co., Ltd., Japan;* Tetsuo Hatakeyama, *National Institute of Advanced Industrial Science and Technology, Japan;* Manabu Takei, *National Institute of Advanced Industrial Science and Technology and Fuji Electric Co., Ltd., Japan;* Yoshiyuki Yonezawa, *National Institute of Advanced Industrial Science and Technology, Japan;* Tsunenobu Kimoto, *Kyoto University, Japan;* Hajime Okumura, *National Institute of Advanced Industrial Science and Technology, Japan*

12-10 Investigation on Degradation Mechanism and Optimization for SiC Power MOSFETs under Long-Term Short-Circuit Stress 399
Jiaxing Wei, Siyang Liu, Jiong Fang, Sheng Li, Ting Li, Weifeng Sun, *Southeast University, China*

12-11 High Accuracy Large-Signal SPICE Model for Silicon Carbide MOSFET 403
Fu-Jen Hsu, Cheng-Tyng Yen, Chien-Chung Hung, Chwan-Ying Lee, Lurng-Shehng Lee, Kuo-Ting Chu, Ya-Fang Li, *Hestia Power Inc., Taiwan*

12-12 Analysis of Parameters Determining Nominal Dynamic Performance of 1.2 kV SiC Power MOSFETs 407
Roger Stark, Ivana Kovačević-Badstübner, Alexander Tsibizov, Bhagyalakshmi Kakarla, Yanrui Ju, Beat Jaeger, Thomas Ziemann, Ulrike Grossner, *ETH Zürich, Switzerland*

12-13 SiC Trench IGBT with Diode-Clamped p-Shield for Oxide Protection and Enhanced Conductivity Modulation 411
Jin Wei, *Innoscience Technology Co., Ltd., China;* Meng Zhang, *Hong Kong Polytechnic University, China;* Huaping Jiang, *Dynex Semiconductor Ltd., United Kingdom;* Suet To, *Hong Kong Polytechnic University, China;* SungHan Kim, Jun-Youn Kim, *Innoscience Technology Co., Ltd., China;* Kevin J. Chen, *Hong Kong University of Science and Technology, China*

12-14 Surge Current Failure Mechanisms in 4H-SiC JBS Rectifiers 415
Edward Van Brunt, Thomas Barbieri, Adam Barkley, James Solovey, Jim Richmond, Brett Hull, *Wolfspeed, A Cree Company, United States*

12-15 Surge Capability of 1.2kV SiC Diodes with High-Temperature Implantation 419
Hongyi Xu, Jiahui Sun, Jingjing Cui, Jiupeng Wu, Hengyu Wang, Shu Yang, Na Ren, Kuang Sheng, *Zhejiang University, China*

12-16 Ruggedness of 6.5 kV, 30 A SiC MOSFETs in Extreme Transient Conditions 423
Ashish Kumar, Sanket Parashar, *North Carolina State University, United States;* Shadi Sabri, Edward Van Brunt, *Wolfspeed, A Cree Company, United States;* Subhashish Bhattacharya, Victor Veliadis, *North Carolina State University, United States*

12-17 Next Generation 1200V, 3.5mΩ.cm² SiC Planar Gate MOSFET with Excellent HTRB Reliability 427
Sauvik Chowdhury, Kevin Matocha, Blake Powell, Gin Sheh, Sujit Banerjee, *Monolith Semiconductor, Inc., United States*

12-18 Investigation on Single Pulse Avalanche Failure of 900V SiC MOSFETs 431
Na Ren, Hao Hu, Kang L. Wang, *University of California, Los Angeles, United States;* Zheng Zuo, Ruigang Li, *AZ Power Inc., United States;* Kuang Sheng, *Zhejiang University, China*

12-19 Long Term High Temperature Reverse Bias (HTRB) Test on High Voltage SiC-JBS-Diodes 435
Felix Hoffmann, *Universität Bremen, Germany;* Andrei Mihaila, Lukas Kranz, *ABB Switzerland Ltd., Switzerland;* Philippe Godignon, *CNM-CSIC, Spain;* Nando Kaminski, *Universität Bremen, Germany*

13:30 – 15:10
Session 13 – SiC Reliability and Ruggedness
Chairs: Kevin Matocha, *Monolith Semiconductor, USA*
Yoshiyuki Yonezawa, *AIST, Japan*

13:30

13-1 Robustness Improvement of Short-Circuit Capability by SiC Trench-Etched Double-Diffused MOS (TED MOS) 439
Naoki Tega, Kazuki Tani, Digh Hisamoto, Akio Shima, *Hitachi Ltd., Japan*

13:55

13-2 High-Temperature Validated SiC Power MOSFET Model for Flexible Robustness Analysis of Multi-Chip Structures 443
M. Riccio, V. d'Alessandro, G. Romano, L. Maresca, G. Breglio, A. Irace, *Università degli Studi di Napoli Federico II, Italy;* A. Castellazzi, *University of Nottingham, United Kingdom*

14:20

13-3   Reliability Investigation with Accelerated Body Diode Current Stress for 3.3 kV 4H-SiC MOSFETs with Various Buffer Epilayer Thickness 447

Yuji Ebiike, Takeshi Murakami, Eisuke Suekawa, Shigehisa Yamamoto, Hiroaki Sumitani, Masayuki Imaizumi, Masayoshi Tarutani, *Mitsubishi Electric Corporation, Japan*

14:45

13-4   Dynamic Switching and Short Circuit Capability of 6.5kV Silicon Carbide MOSFETs 451

L. Knoll, A. Mihaila, L. Kranz, M. Bellini, S. Wirths, E. Bianda, C. Papadopoulos, M. Rahimo, *ABB Switzerland Ltd., Switzerland*

15:30 – 17:35
Session 14 – Packaging and Enabling Technologies
Chairs: Tomoyuki Miyoshi, *Hitachi, Japan*
            Alberto Castellazzi, *Nottingham University, UK*

15:30

14-1   Improvement of Power Cycling Reliability of 3.3kV Full-SiC Power Modules with Sintered Copper Technology for $T_{j,max}$=175°C 455

Kan Yasui, Seiichi Hayakawa, Masato Nakamura, Daisuke Kawase, Takashi Ishigaki, Koji Sasaki, Toshihito Tabata, Toshiaki Morita, *Hitachi Power Semiconductor Device Ltd., Japan;* Masakazu Sagawa, Hiroyuki Matsushima, Toshiyuki Kobayashi, *Hitachi Ltd., Japan*

15:55

14-2   Enhanced Breakdown Voltage and Low Inductance of All-SiC Module 459

Motohito Hori, Yuichiro Hinata, Katsumi Taniguchi, Yoshinari Ikeda, Tomoyuki Yamazaki, *Fuji Electric Co., Ltd., Japan*

16:20

14-3   Dynamic Performance Analysis of a 3.3 kV SiC MOSFET Half-Bridge Module with Parallel Chips and Body-Diode Freewheeling 463

Abdallah Hussein, Bassem Mouawad, Alberto Castellazzi, *University of Nottingham, United Kingdom*

16:45

14-4   Power Cycling Reliability Results of GaN HEMT Devices 467

Jörg Franke, Guang Zeng, Tom Winkler, Josef Lutz, *Technische Universität Chemnitz, Germany*

17:10

14-5   Individual Device Active Cooling for Enhanced System-Level Power Density and More Uniform Temperature Distribution 471

Y. Zeng, A. Hussein, A. Castellazzi, *University of Nottingham, United Kingdom*

Thursday - May 17, 2018

8:30 – 10:10
Session 15 – Novel Device Structures
Chairs: Dev Alok Girdhar, *Intersil, USA*
            Yoshiyuki Yonezawa, *AIST, Japan*

8:30

15-1   Non-Full Depletion Mode and its Experimental Realization of the Lateral Superjunction 475

Wentong Zhang, *University of Electronic Science and Technology of China and CSMC Technologies Corporation, China;* Song Pu, Chunlan Lai, Li Ye, *University of Electronic Science and Technology of China, China;* Shikang Cheng, Sen Zhang, Boyong He, *CSMC Technologies Corporation, China;* Zhuo Wang, Xiaorong Luo, Ming Qiao, Zhaoji Li, Bo Zhang, *University of Electronic Science and Technology of China, China*

8:55

15-2 Cathode Short Structure to Enhance the Robustness of Bidirectional Power MOSFETs 479

Tanuj Saxena, Vishnu Khemka, Moaniss Zitouni, Raghu Gupta, Ganming Qin, *NXP Semiconductors Inc., United States;* Philippe Dupuy, *NXP Semiconductors Inc., France;* Mark Gibson, *NXP Semiconductors Inc., United States*

9:20

15-3 40V to 100V NLDMOS Built on Thin BOX SOI with High Energy Capability, State of the Art Rdson/BVdss and Robust Performance 483

Yang Hao, Sim Poh Ching, Madelyn Liew, Alexander Hölke, *X-FAB Sarawak Sdn. Bhd., Malaysia;* Uwe Eckoldt, *X-FAB Semiconductor Foundries, Germany;* Martin Pfost, *Technische Universität Dortmund, Germany*

9:45

15-4 Novel Integration Techniques of "Recessed" High Voltage Field-Drift MOSFET with HK/MG RMG Technology 487

C.P. Hsiung, P.H. Chiang, S.C. Pu, C.L. Wang, C.W. Lu, K.L. Liu, K.K. Chang, C.C. Yang, N.C. Lee, S.Y. Hsiao, W.F. Lee, C.C. Wang, *United Microelectronics Corporation, Taiwan*

10:30 – 12:10
Session 16 – IGBTs

Chairs: Thomas Laska, *Infineon Technologies, Germany*
Jan Vobecky, *ABB, Switzerland*

10:30

16-1 A Novel Carrier Accumulating Structure for 1200V IGBTs without Negative Capacitance and Decreasing Breakdown-Voltage 491

Md Tasbir Rahman, Keisuke Kimura, Takeshi Fukami, Masaki Konishi, Tsuyoshi Nishiwaki, Jun Saito, Kimimori Hamada, *Toyota Motor Corporation, Japan*

10:55

16-2 Study on the Improved Short-Circuit Behavior of Narrow Mesa Si-IGBTs with Emitter Connected Trenches 495

K. Eikyu, A. Sakai, *Renesas Electronics Corp., Japan;* H. Matsuura, Y. Nakazawa, *Renesas Semiconductor Manufacturing Co., Ltd., Japan;* Y. Akiyama, Y. Yamaguchi, *Renesas Electronics Corp., Japan*

11:20

16-3 An Advanced Soft Punch through Buffer Design for Thin Wafer IGBTs Targeting Lower Losses and Higher Operating Temperatures up to 200°C 499

Elizabeth Buitrago, Athanassios Mesemanolis, Charalampos Papadopoulos, Chiara Corvasce, Jan Vobecky, Munaf Rahimo, *ABB Switzerland Ltd., Switzerland*

11:45

16-4 Investigation of the Mechanism of Gate Voltage Oscillation in 1.2kV IGBT under Short Circuit Condition 503

Takuo Kikuchi, *Toshiba Corporation, Japan;* Kazutoshi Nakamura, *Toshiba Electronic Devices & Storage Corporation, Japan;* Kazuto Takao, *Toshiba Corporation, Japan*

13:30 – 14:45
Session 17 – Invited and Late News Papers

Chairs: Olivier Trescases, *University of Toronto, Canada*
Alberto Castellazzi, *Nottingham University, UK*

13:30

17-1 [Invited] Design of LED Driver ICs for High-Performance Miniaturized Lighting Systems 508

Yuan Gao, Lisong Li, Philip K.T. Mok, *Hong Kong University of Science and Technology, China*

13:55

17-2   **[Invited] High Voltage Capacitive Voltage Conversion** 512

Randall L. Sandusky, *Helix Semiconductors, United States;* Alexander Hölke, *X-FAB Sarawak Sdn. Bhd., Malaysia*

14:20

17-3   **[Late News] Chip-Scale Cooling of Power Semiconductor Devices: Fabrication of Jet Impingement Design** 516

Feng Zhou, Ki Wook Jung, *Toyota Research Institute of North America, United States;* Yuji Fukuoka, *Toyota Motor Corporation, Japan;* Ercan M. Dede, *Toyota Research Institute of North America, United States*

14:45

17-4   **[Late News] An Innovative Silicon Power Device (i-Si) through Time and Space Control of a Stored Carrier (TASC)** 520

Mutsuhiro Mori, Tomoyuki Miyoshi, Tomoyasu Furukawa, Yujiro Takeuchi, Yusuke Hotta, Masaki Shiraishi, *Hitachi Ltd., Japan*

Proceedings of the 30th International Symposium on Power Semiconductor Devices and ICs

May 13-17, 2018, Chicago, USA

Celebrating **30 YEARS** of Excellence in
Advancing Power Semiconductor Technologies

Sponsored by

Technically Co-sponsored by

GENERAL CHAIR'S MESSAGE

It is my great honor and pleasure to welcome you on behalf of the Organizing Committee to the 30th IEEE International Symposium on Power Semiconductor Devices and ICs (ISPSD 2018) in the beautiful city of Chicago.

ISPSD2018 marks the 30th anniversary of ISPSD. Since its first meeting in Japan in 1988, ISPSD has become the world's premier forum for technical discussions in all areas of power semiconductor devices and power integrated circuits. The global power semiconductor industry has steadily grown into a $30 billion sector over the past three decades, enabling energy-efficient applications such as solar power, wind power, electric vehicles, ICT infrastructures, lighting and industrial drives. Over 1600 technical papers have been presented at ISPSD conferences. Most of the breakthrough power device technologies were first reported at ISPSD before they became phenomenal commercial successes. ISPSD2018 will celebrate 30 years of excellence in advancing power semiconductor technologies with a series of technical and social events during the conference. In particular, we will induct 32 distinguished colleagues into the newly established ISPSD Hall of Fame during the Wednesday 30th Anniversary Celebration Banquet. As a special celebration gift, we present you a complete collection of ISPSD proceedings 1988--2018 (including the two Electrochemical Society Workshops on High Voltage and Smart Power Devices and ICs in 1987 and 1989), thanks to the support of IEEE, IEEJ, ECS, and a group of dedicated volunteers.

Our plenary session on Monday will start with an opening keynote speech on "ISPSD: A 30 Year Journey in Advancing Power Semiconductor Technology" from three founding members of the conference, Drs. Ayman Shibib (USA), Leo Lorentz (Germany), and Hiromichi Ohashi (Japan), and followed by three other plenary speeches on "Silicon, GaN and SiC: There's Room for All" from Mr. Larry Spaziani, GaN Systems Inc., Canada, "Si Wafer Technology for Power Devices: A Review and Future Directions" from Mr. Norihisa Machida, SUMCO, Japan, and "The Future of Power Semiconductors: an EU Perspective" from Dr. Bert De Colvenaer, ECSEL, Belgium, respectively. We are extremely fortunate to have these distinguished leaders from industry to share their visions and wisdoms with us.

ISPSD2018 features 50 oral and 79 poster presentations on both silicon and WBG power devices which are selected from 245 abstracts from 23 countries/regions. On Sunday, we offer 6 short courses on a series of current and practical topics in the field.

Chicago is the third largest city in the United States, and an international hub for finance, commerce, industry, technology, telecommunications, and transportation. The city of Chicago is the birth place of modern skyscrapers and a living museum of modern architecture. ISPSD 2018 is held in the historical Palmer House Hilton Hotel in the beautiful and safe downtown district ("Loop") of Chicago with numerous museums, parks, theaters, restaurants, and shops within a walking distance. We trust you and your family will enjoy your stay in Chicago. On Monday, we will host our Welcome Reception in the splendid Empire Room of the Palmer House. On Wednesday, we expect to see you all at the 30th Anniversary Celebration Banquet to enjoy Chicago jazz/blues music, wines and delicious meals. Our family and companion programs offer exciting city tours on Monday, and Tuesday, and Wednesday.

We are pleased to acknowledge the support of our Gold Partners PowerAmerica and Sinopower Semiconductor, and Silver Partners Applied Materials, ShinDengen, Synopsys, and Tektronix. Their support and participation have created a very strong industrial relevance. We will have 17 exhibitors in the exhibition/coffee area (Salon 4-9 on the 3rd Floor) Monday through Thursday. The exhibitors will showcase their state-of-the-art technologies, products, and solutions, creating a highly interactive networking environment when mixing with the poster sessions and coffee breaks in the same space.

I would like to express my utmost gratitude to the members of the organizing committee, the technical program committee, and the advisory committee, who with hard work and selfless dedication have made this conference possible. I wanted to thank each and every one of you as a presenter, an attendee, an exhibitor, a volunteer, or any combined role of the above for your contribution and participation.

Once again I welcome you to ISPSD2018. Together we help deliver a more sustainable future.

Z. John Shen, General Chair

ORGANIZATION

ORGANIZING COMMITTEE

General Chair
John Shen, *Illinois Institute of Technology, USA*

Vice Chairs
Kuang Sheng, *Zhejiang University, China*
Oliver Haeberlen, *Infineon, Austria*

Technical Program Committee Chair
Wai Tung Ng, *University of Toronto, Canada*

Finance Chair
Sujit Banerjee, *Monolith Semiconductor, USA*

Publicity Chair
David Sheridan, *AOS Semiconductor, USA*

Publication Chair
Olivier Trescases, *University of Toronto, Canada*

Short Course Chair
Alex Huang, *University of Texas at Austin, USA*

Industrial Liaison and Expo Chair
Victor Veliadis, *PowerAmerica, USA*

Local Arrangements Chair
Anup Bhalla, *USCi, USA*

Webmaster
Mengqi Wang, *University of Toronto, Canada*

ADVISORY COMMITTEE

Gehan Amaratunga, *Cambridge University, UK*
Tat-Sing Paul Chow, *Rensselaer Polytechnic Institute, USA*
Mohamed Darwish, *MaxPower Semiconductor, USA*
Don Disney, *GlobalFoundries, USA*
Dan Kinzer, *Navitas Semiconductor, USA*
Leo Lorenz, *ECPE, Germany*
Gourab Majumdar, *Mitsubishi Electric, Japan*
Peter Moens, *ON Semiconductors, USA*
Mutsuhiro Mori, *Hitachi, Ltd., Japan*
Hiromichi Ohashi, *NPERC-J, Japan*
Yasukazu Seki, *Fuji Electric Co., Ltd., Japan*
M. Ayman Shibib, *Vishay Siliconix, USA*
Johnny Sin, *Hong Kong Univ. of Science and Technology, China*
Jan Šonský, *NXP Semiconductors, Belgium*
Yoshitaka Sugawara, *Ibaraki University, Japan*
Richard K. Williams, *Adventive Technology, USA*
Toshiaki Yachi, *Tokyo University of Science, Japan*

TECHNICAL PROGRAM COMMITTEE

Chair
Wai Tung Ng, *University of Toronto, Canada*

Category 1: High Voltage Power Devices (HV)
Category Chair
Anup Bhalla, *United Silicon Carbide, USA*

Members
Marina Antoniou, *University of Cambridge, UK*
Giovanni Breglio, *University of Naples Federico II, Italy*
Young Chul Choi, *ON Semiconductor, Korea (in USA)*
Thomas Laska, *Infineon Technologies, Germany*
Xiaorong Luo, *UESTC, China*
Tadaharu Minato, *Mitsubishi, Japan*
Yasuhiko Onishi, *Fuji Electric, Japan*
Wataru Saito, *Toshiba Corporation, Japan*
Jan Vobecky, *ABB, Switzerland*
Chongman Yun, *Trinno Technology, Korea*
Shuai Zhang, *TSMC, China*

Category 2: Low Voltage Devices and Power IC Device Technology (LVT)

Category Chair
Phil Rutter, *Nexperia, UK*

Members
Jun Cai, *Texas Instruments, USA*
Naoto Fujishima, *Fuji Electric, Japan*
Dev Alok Girdhar, *Intersil, USA*
Alexander Hölke, *XFAB, Malaysia*
Kenya Kobayashi, *Toshiba Corporation, Japan*
Yoshinao Miura, *Renesas Electronics, Japan*
Purakh Raj Verma, *UMC, Taiwan*
Ronghua Zhu, *NXP Semiconductors, USA*

Category 3: Power IC Design (ICD)

Category Chair
Olivier Trescases, *University of Toronto, Canada*

Members
David Tsung-Yi Huang, *Richtek, Taiwan*
Hoi Lee, *UT Dallas, USA*
Takahiro Mori, *Renesas, Japan*
Shuichi Nagai, *Panasonic, Japan*
Nicolas Rouger, *CNRS, France*
Junichi Sakano, *Hitachi, Japan*
Weifeng Sun, *Southeast University, China*
Maarten Swanenberg, *NXP Semiconductors, Holland*
Budong (Albert) You, *Silergy Corp., China*
Alessandro Zafarana, *STMicroelectronics, Italy*

Category 4: GaN and Nitride Base Compound Materials (GaN)

Category Chair
Tom Tsai, *TSMC, Taiwan*

Members
Kevin Chen, *Hong Kong University of Science and Technology, China*
Oliver Haeberlen, *Infineon Technologies, Austria*
Alex Huang, *North Carolina State University, USA*
Yang Liu, *Sun Yat-sen University, China*
Peter Moens, *ON Semiconductor, Belgium*
Sameer Pendharkar, *Texas Instruments, USA*
Jun Suda, *Nagoya University, Japan*
Tom Tsai, *TSMC, Taiwan*
Yasuhiro Uemoto, *Panasonic, Japan*

Category 5: SiC and Other Materials (SiC)
Category Chair
Peter Losee, *GE, USA*

Members
Philippe Godignon, *CNM institute, Spain*
Chih-Fang Huang, *National Tsing Hua University, Taiwan*
Takeharu Kuroiwa, *Mitsubishi Electric, Japan*
Chwan Ying Lee, *Hestia-Power Inc., Taiwan*
Kung-Yen Lee, *National Taiwan University, Taiwan*
Kevin Matocha, *Monolith Semiconductor, USA*
Andrei Petru Mihaila, *ABB, Switzerland*
David Sheridan, *Alpha & Omega Semiconductor, USA*
Ranbir Singh, *GeneSiC, USA*
Jun Suda, *Nagoya University, Japan*
Victor Veliadis, *Power America, USA*
Yoshiyuki Yonezawa, *AIST, Japan*
Jon Zhang, *Wolfspeed, USA*

Category 6: Module and Package Technologies (PK)
Category Chair
Alberto Castellazzi, *Nottingham University, UK*

Members
Sven Berberich, *Semikron, Germany*
Josef Lutz, *Technical University of Chemnitz, Germany*
Tomoyuki Miyoshi, *Hitachi, Japan*
Hiroshi Tadano, *University of Tsukuba, Japan*

AWARDS

THE OHMI BEST PAPER AWARD

The Best Paper Award was renamed to "The Ohmi Best Paper Award" in honor of the late Prof. Ohmi's outstanding contributions to the ISPSD. The Ohmi Best Paper Award will be granted to the author(s) of a paper determined to be the best overall in the ISPSD2018.

ISPSD2017 THE OHMI BEST PAPER AWARD
A Novel Hybrid Power Module with Dual Side-Gate HiGT and SiC-SBD

Abstract: In this paper, a novel hybrid power module using a new combination of dual side-gate HiGTs (high-conductivity IGBT) and SiC-SBDs is proposed. This combination achieves drastic switching loss reductions at a turn-off loss of -43%, a turn-on loss of -71%, and a reverse recovery loss of -98% compared with a conventional combination of trench gate HiGTs and U-SFDs (ultra soft & fast recovery diode). As a result, the proposed DuSH module (dual side-gate HiGT hybrid module) has an extremely low inverter loss of -50%, similar to SiC-MOSFETs.

Yujiro Takeuchi received the B.S. and M.S. degrees in maritime science from Kobe University, Hyogo-ken, Japan in 2010 and 2012. He joined Hitachi, Ltd., Ibaraki-ken, Japan, in 2012, where he has been engaged in research on power semiconductor devices.

Tomoyuki Miyoshi received the B.S., M.S., and Ph.D. in Tohoku University, Miyagi, Japan, in 2005, 2007, and 2015. In 2007, he joined Hitachi, Ltd., Japan. He is currently engaged in research and development of power device technologies.

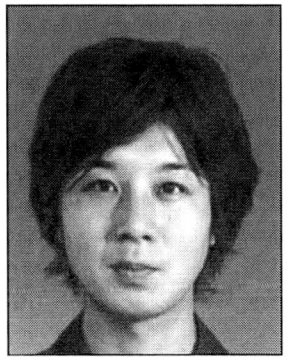

Tomoyasu Furukawa received the B.S. and M.S. degrees in Hiroshima University in 2000 and 2002. He joined Hitachi, Ltd, Japan, in 2002. He is currently engaged in research and development of power device technologies.

Masaki Shiraishi received the B.S. and M.S. degrees in Tokyo Institute of Technology in 1996 and 1998. He joined Hitachi, Ltd, Japan, in 1998. He has been engaged in research and development of power semiconductor devices.

Mutsuhiro Mori received the B.E., M.E. and Ph. D. degrees in science and engineering from Waseda University, Tokyo, Japan. Since 1979, he has been with the Hitachi Research Laboratory, Hitachi, Ltd. He has been engaged in the research and development of optimal energy saving control systems with IT and power semiconductor devices, such as the light-triggered thyristors, GaAs power static induction transistors (SIT), one chip inverter ICs, high-voltage driver ICs, soft and fast recovery diodes (SFD), and high-conductivity insulated gate bipolar transistors (HiGT). He is a member of the Institute of Electrical Engineers of Japan (IEEJ) and a senior member of the Institute of Electrical and Electronics Engineers (IEEE).

CHARITAT AWARD (YOUNG RESEARCHER AWARD)

A young researcher (age less than 30 at the time of the conference) who is both first author and presenter of a paper will be nominated to the award.

The Charitat award will be presented during the closing ceremony of the conference. The Ohmi best paper award will be presented during the ISPSD2019 opening session.

ISPSD 2017 CHARITAT AWARD
High Performance Fully-Recessed Enhancement-Mode GaN MIS-FET with Crystalline Oxide Interlayer

Abstract: In this work, we developed an effective technique to form a sharp and stable crystalline oxidation interlayer (COIL) between the reliable LPCVD (low pressure chemical vapor deposition)-SiN$_x$ gate dielectric and recess-etched GaN channel. The COIL was formed using oxygen-plasma treatment, followed by *in-situ* annealing prior to the LPCVD-SiN$_x$ deposition. The COIL plays the critical role of protecting the etched GaN surface from degradation during high-temperature (i.e. at ~ 780 °C) process, which is essential for fabricating enhancement-mode GaN MIS-FETs with highly reliable LPCVD-SiN$_x$ gate dielectric and fully recessed gate structure. The LPCVD-SiN$_x$/GaN MIS-FETs with COIL deliver normally-off operation with a V_{TH} of 1.15 V, small on resistance, thermally stable V_{TH} and low positive-bias temperature instability (PBTI).

Mengyuan Hua received the B.S. degree in Physics from Tsinghua University, Beijing, China, in 2013. She then joined the Hong Kong University of Science and Technology (HKUST), Hong Kong, China, where she received the Ph.D. degree in Electronic and Computer Engineering in 2017 under the supervision of Prof. Kevin J. Chen. Currently, she is a research associate at HKUST. Her research interests include GaN-based power device technology and device reliability.

ISPSD HALL OF FAME

The purpose of the ISPSD Hall of Fame (IHF) is to honor individuals who have made high impact contributions in advancing power semiconductor technology and/or in sustaining the success of ISPSD. Starting this year, the IHF replaces the traditional "Contributory Awards" and "Pioneer Awards". The AdCom has selected the following 32 distinguished colleagues as the first inductees into the ISPSD Hall of Fame:

Michael S. Adler	for contributions to modern power semiconductor technology, and his leadership role in organizing ISPSD conferences
Gehan Anil Joseph Amaratunga	for contributions to modern power semiconductor technology, and his leadership role in organizing ISPSD conferences
B. Jayant Baliga	for contributions to modern power semiconductor technology, and his leadership role in organizing ISPSD conferences
Xingbi Chen	for contributions to superjunction power semiconductor devices
Tat-Sing Paul Chow	for contributions to silicon and wide bandgap power semiconductor devices, and his leadership role in organizing ISPSD conferences
Mohamed Darwish	for contributions to the advancement of power semiconductor technology, and his leadership role in organizing ISPSD conferences
Taylor R. Efland	for contributions to power IC technology, and his leadership role in organizing ISPSD conferences
Wolfgang Fichtner	for contributions to MOS gated thyristors and TCAD modeling tools, and his leadership role in organizing ISPSD conferences
Min-Koo Han	for contributions to modern power semiconductor technology, and his leadership role in organizing ISPSD conferences
Phil Hower	for contributions to power device safe operating area study and power IC technology
A. A. Jaecklin	for contributions to modern power semiconductor technology, and his leadership role in organizing ISPSD conferences
Daniel Kinzer	for contributions to power MOSFET technology, and his leadership role in organizing ISPSD conferences
Leo Lorenz	for contributions to modern power semiconductor technology, and his leadership role in organizing ISPSD conferences
Gourab Majumdar	for contributions to IGBT and intelligent power module technology, and his leadership role in organizing ISPSD conferences
Jose Millan	for contributions to modern power semiconductor technology, and his leadership role in organizing ISPSD conferences
Peter Moens	for contributions to integrated power technology and GaN power device and reliability, and his leadership role in organizing ISPSD conferences
Akio Nakagawa	for contributions to IGBT and power IC technology
Hiromichi Ohashi	for contributions to modern power semiconductor technology, and his leadership role in organizing ISPSD conferences
Tadahiro Ohmi	for contributions to modern power semiconductor technology, and his leadership role in organizing ISPSD conferences

Masahiro Okamura	for contributions to modern power semiconductor technology, and his leadership role in organizing ISPSD conferences
James Plummer	for contributions to MOS-bipolar power devices and power ICs, and for inspiring and training a new generation of device researchers
C. A. T. Salama	for contributions to power IC technology, and his leadership role in organizing ISPSD conferences
Yasukazu Seki	for contributions to IGBT technology, and his leadership role in organizing ISPSD conferences
M. A. Shibib	for contributions to modern power semiconductor technology, and his leadership role in organizing ISPSD conferences
Dieter Silber	for contributions to modern power semiconductor technology, and his leadership role in organizing ISPSD conferences
Paolo Spirito	for contributions to modern power semiconductor technology, and his leadership role in organizing ISPSD conferences
Yoshitaka Sugawara	for contributions to modern power semiconductor technology, and his leadership role in organizing ISPSD conferences
Yoshiyuki Uchida	for contributions to modern power semiconductor technology, and his leadership role in organizing ISPSD conferences
Harry Vaes	for contributions to RESURF technology
Carl Frank Wheatley	for contributions to IGBT and radiation-hard power device technology
Richard K. Williams	for contributions to trench power MOSFET and power IC technology, and his leadership role in organizing ISPSD conferences
Toshiaki Yachi	for contributions to modern power semiconductor technology, and his leadership role in organizing ISPSD conferences

Deceased members in BOLD

PARTNERSHIP ORGANIZATIONS

GOLD PARTNERS

SILVER PARTNERS

EXHIBITORS

PowerAmerica Institute
930 Main Campus Drive, Suite 200
Raleigh, NC 27606, USA
www.poweramericainstitute.org

The PowerAmerica consortium brings together the brightest minds in the wide bandgap (WBG) power semiconductor world. Semiconductor manufacturers and the companies that use power semiconductors in their products are working together to accelerate the adoption of next generation silicon carbide (SiC) and gallium nitride (GaN) power electronics. Our objective is to educate the workforce and reduce the cost and the perceived risk inherent with this new technology. We fulfill our mission with the backing of the U.S. Department of Energy and the engagement of top researchers from industry, academia, and national laboratories.

Applied Materials, Inc.
3050 Bowers Avenue
P.O. Box 58039
Santa Clara, CA 95054, USA
www.appliedmaterials.com

Applied Materials is a leading equipment supplier to the semiconductor, display and PV industries. Our innovations make possible the technology shaping the future by leveraging our core competencies in:

- Materials Engineering
  We work at the frontiers of physics and materials science, designing systems with exacting precision and unparalleled production volume capacity. We make constant performance improvements to current materials, while researching and developing the next radical materials advancement.
- Customer Engagement
  We are 100% committed to working closely with our customers to identify and solve their high-value problems. We develop differentiated products that improve our customers' device performance and yield.

Shindengen Electric
New-Ohtemachi Bldg.
2-2-1 Ohtemachi, Chiyoda-ku
Tokyo, 100-0004, Japan
www.shindengen.com

Shindengen has developed and provided innumerable power electronics components for over 70 years. Utilizing our three core technologies – Power Devices, Power Supplies, and Module technology we offer vertically integrated power conversion solutions with emphasis on transportation, clean energy, & motor control innovations.

Synopsys, Inc.
690 East Middlefield Road
Mountain View, CA 94043, USA
www.synopsys.com

Synopsys, a leader in EDA and semiconductor IP, is also growing its leadership in software security and quality solutions. Whether you're an SoC designer creating advanced semiconductors, or a developer writing applications that require the highest security and quality, Synopsys has the solutions needed to deliver innovative, high-quality, secure products.

Tektronix, Inc.
14150 SW Karl Braun Drive
P.O. Box 500
Beaverton, OR 97077, USA
www.tekcom

Tektronix delivers innovative, precise and easy-to-operate test and measurement solutions that unlock insights and drive discovery. For wide bandgap applications these include both the Keithley 4200A-SCS parameter analyzer and Series 2400 and 2600B Source Measure Unit instruments as well as the unique Tektronix IsoVu system for isolated differential voltage measurements.

Crosslight Software, Inc.

230-3410 Lougheed Hwy
Vancouver, BC, V5M 2A4 Canada
www.crosslight.com

Crosslight has been a leading provider of TCAD simulation tools since 1995. With a special focus on compound semiconductor devices, such as GaN HEMTs, Crosslight simulation tools have built a solid reputation in the semiconductor industry for their easy convergence and high accuracy.

Everbeing Int'l Corp.

No. 1, Jinshan 2nd Street, East District
Hsinchu City, Taiwan 30080
www.probestation.tw

Everbeing is a world leading manufacturer of probe stations and micropositioners based in Taiwan. With 25 years of history, we strive in producing reliable, precise, user-friendly products with affordable prices. Our solutions cater to vast range of applications, which can be tailored to your specific needs.

FormFactor, Inc.

7005 Southfront Road
Livermore, CA 94551, USA
www.formfactor.com

FormFactor is a leading provider of essential test and measurement technologies along the full IC life cycle – from characterization, modeling, reliability and design debug, to qualification and production test. Our products and services enable our customers to accelerate profitability by optimizing device performance and advancing yield knowledge.

Jedat, Inc.

HSB Teppozu, 1-1-12, Minato, Chuo-ku
Tokyo 104-0043 Japan
www.jedat.co.jp/eng/

Jedat is the leading company for EDA software in Japan. Jedat exhibits PowerVolt, a software product which provide DC analysis, Resistance map analysis, Transient Analysis and Electro-thermal analysis. It works layout data only, no netlist required. User optimizes layout pattern at any phase of design.

Mitsui Bussan Electronics Ltd.

Shiba Park Building, A-10F,
4-1 Shibakoen 2-chome, Minato-ku
Tokyo 105-0011, Japan
www.mbel.co.jp/english/

MPI Corporation

No. 155, Chung-Ho St.
Chu-pei City, Hsinchu County 302, Taiwan
www.mpi-corporation.com

MPI Corporation offers complete Test Solutions based on a variety of engineering probe systems, RF probes from 26 to 110 GHz, and RF calibration software QAlibria®. Major products focus on RF and mmW, Device Characterization for modeling and process development, High Power, and many other complex semiconductor test applications.

Signatone Corporation

393-J Tomkins Court
Gilroy, CA 95020, USA
www.signatone.com

Signatone has been providing manual and semi-automatic wafer probing solutions to semiconductor and materials research community worldwide for 50 years. Signatone PowerPro product line encompass ambient and hot/cold probing of SiC, GaN, and silicon power devices up to 20kV, 500A and in the thermal range of -60°C to 300°C.

Silicon Frontline Technology, Inc.

4030 Moorpark Avenue, Suite 249
San Jose, CA 95117, USA
www.siliconfrontline.com

Silicon Frontline's R3D is the leading power device analysis tool.
- Create distributed model for transient simulations
- Perform transient electro-thermal simulation
- Reduce Rdson, meet EM specifications
- Optimize sense device locations
- Analyze impact of package and PCB layout

With over 70 customers, R3D is the proven choice!

Sinopower Semiconductor Co., Ltd.

Block 6, Innovation and Technology Park
558 Fenhu Road, FOHO
Suzhou City, Jiangsu Province, China
www.sinopowers.cn

TESEC, Inc.

1225 W. 190th Street, Suite 325
Gardena, CA 90248, USA
www.tesecinc.com

TowerJazz

20 Shaul Amor Avenue
P.O. Box 619
Migdal Haemek 2310502, Israel
www.towerjazz.com

Sinopower Semiconductor was established by leading Chinese research and industrial organizations in the field of wide-bandgap semiconductor, including Peking University, Sun Yat-sen University, Epilight Technology and Sino Nitride in 2016. The newly-founded company is endeavoring to become a leading GaN epitaxial wafer and GaN device processing technology provider in China.

TESEC has been a global leader in Power Device Testing Technologies for half a century. The TESEC "SpeKtra" series has become the industry benchmark for Power Device production test. The world focus on energy-saving/low-carbon societies continue to drive TESEC to lead advancements in test systems for new high power device technologies such as SiC/GaN.

TowerJazz specializes in manufacturing analog ICs for more than 300 customers worldwide in growing markets, offering SiGe, BiCMOS, Power and Mixed-Signal/CMOS, RF CMOS, CMOS image sensor, and MEMS technologies. TowerJazz also offers design enablement and process transfer services and operates seven manufacturing facilities in Israel, the US and Japan.

PLENARY TALKS

PL1-1: ISPSD: A 30 Year Journey in Advancing Power Semiconductor Technology
Ayman Shibib, *Vishay Siliconix, United States;* Leo Lorenz, *ECPE/Infineon, Germany;*
Hiromichi Ohashi, *NEPRC-J, Japan*

Abstract: Celebrating the 30th Anniversary of ISPSD is a very special occasion to reflect on the origin and roots of the conference and how it came about to be the premier international conference on Power Semiconductor Devices and ICs. A review of the events that led to its formation and development is presented. Another aspect of this celebration is the review of the contributions of ISPSD to the Power Device technical community covering wide and diverse power device areas and applications throughout the 30 years history of ISPSD. The future prospects of Power Devices and ISPSD in the next years is briefly mentioned.

PL1-2: Silicon, GaN and SiC: There's Room for All –
An Application Space Overview of Device Considerations
Larry Spaziani, Lucas Lu, *GaN Systems, Canada*

Abstract: The discrete power device marketplace is estimated between 15 and 22 billion dollars and is comprised primarily of transistors and diodes in a variety of voltage, current, packaging and power ratings. It is an area of intense focus as new technologies such as wide bandgap emerge and new applications such as electric vehicles emerge. Decision makers from Engineers to CEO's are faced with the same decisions they have always faced, comparing power, efficiency and size, yet the decisions are more difficult given the fast-moving pace of these emerging technologies. In this paper, several application spaces ranging from consumer to vehicle to motors will be reviewed, comparing the most critical aspects of the applications against semiconductor choices these decision makes have available. Considerations of the appropriate technologies will be reviewed comparing where the technologies have been, are today, and where they will be in the next 5 years.

PL2-1: Si Wafer Technology for Power Devices: A Review and Future Directions
Norihisa Machida, *SUMCO Corporation, Japan*

Abstract: Silicon wafers have been widely used in semiconductor devices for years. Their characteristics have been improved by untiring development efforts to meet power device manufacturers' requirements such as lowering substrate resistivity for Power MOSFET and reducing resistivity variation for IGBT. As future directions, by utilizing advantages of silicon wafers, adoption of MCZ grown bulk silicon wafers for low and middle voltage IGBT and introduction of 300mm size silicon wafers will proceed.

PL2-2: The Future of Power Semiconductors: An EU Perspecitve
Bert De Colvenaer, *ECSEL JU, Belgium*

Abstract: With the integration of more renewable energy sources (RES) in the European landscape, with more stringent demands on supply and with cost conscious customers, and also environmental conscious, Europe calls for a smarter energy landscape where power semiconductors will play a major role in the years to come.

SHORT COURSE

The short course this year is supported by Sinopower Semiconductor and other industrial partners.

Time	Title	Speaker
09:00-10:00	Advances in Silicon Power Technology	Madhur Bobde
10:00-10:15	Coffee Break	
10:15-11:15	Perspective of Loss Mechanisms in Silicon and Wide Bandgap Power Devices	Gerald Deboy
11:15-11:30	Coffee Break	
11:30-12:30	Silicon Carbide Power Device Design and Fabrication: Making the Transition from Silicon	Victor Veliadis
12:30-14:00	Lunch Break	
14:00-15:00	Vertical Power Electronic Devices based on Bulk GaN Substrates	Isik C. Kizilyalli
15:00-15:15	Coffee Break	
15:15-16:15	AlGaN/GaN Power Device Reliability	Peter Moens
16:15-16:30	Coffee Break	
16:30-17:30	Multi-Chip Semiconductor Power Module Design and Assembly: Rethinking Established Packaging Solutions for Improved Performance, Robustness and Reliability	Alberto Castellazzi

SC1. Advances in Silicon Power Technology

Madhur Bodge, *Alpha & Omega Semiconductor*

Abstract: Silicon power MOSFETs have made tremendous advancements in the past decade. The key concept that has led to this is that of charge balance. In conventional power MOSFET device the maximum doping level and the thickness of the drift region is limited by blocking voltage constraints and a triangular electric field results in its sub-optimal utilization.

The concept of charge balance involves adding an opposite polarity of charge in the drift region compared to default doping to modify the shape of electric field from triangular to trapezoidal for better utilization of drift region for voltage blocking and allow significantly higher doping concentration for lower conduction losses. For low voltages (below 400V) the popular device structure to achieve this is the Split Gate Transistor (SGT). This device utilizes trench MOS charge balance with a shield electrode under the gate. In addition to significantly improving the On Resistance per unit area (~3x for 100V blocking), the shield electrode also significantly reduced the gate to drain miller capacitance (Crss) and Crss/Ciss ratio to allow for high frequency switching.

For high voltages above 400V, the depth of trench and thickness of liner oxide make SGT device impractical to fabricate. As a result, the Super-Junction transistor has emerged as the most successful MOSFET for high voltages. This device utilizes alternating P and N columns in the drift region thereby creating a charge balance. Methods such as multi epi, deep trench and fill have been demonstrated and are commercially successful for making superjunction transistors. These can achieve an On-Resistance reduction of up to 8x compared to planar DMOS transistor. However, presence of alternating P and N columns also results in peculiar Capacitance curves, particularly the Crss which drops sharply at low drain biases and then increases at higher drain biases. It also exhibits snappy diode reverse recovery.

Charge balanced structure is also finding use in bipolar devices such as IGBT and Fast recovery diodes. In these devices, charge balance is used for various performance enhancements such as improving turn-off losses, injection enhancement, and controlling injection efficiency for faster switching.

The goal of this seminar is to understand the device physics and electrical characteristics of charge balanced devices in unipolar and bipolar power devices. This seminar is intended for intermediate level audience.

SC2. Perspective of Loss Mechanisms in Silicon and Wide Bandgap Power Devices

Dr. Gerald Deboy, *Infineon Technologies Austria AG*

Abstract: This short course will discuss switching losses for power semiconductor devices from a physical device point of view. The focus will be laid on power MOSFETs based on Superjunction technology and GaN high electron mobility transistors as two prominent examples of the silicon and the wide bandgap world. We will give a perspective of loss mechanisms in the light of recent developments of the two fundamental device concepts.

Based on these loss mechanisms appropriate circuits and control methods are discussed yielding best efficiency for both device concepts respectively.

The short course addresses researchers interested in a deep understanding of the device properties as well as users of modern power semiconductor devices seeking best matching between topology, control and power device.

SC3. Silicon Carbide Power Device Design and Fabrication: Making the Transition from Silicon

Victor Veliadis, PhD, *Deputy Executive Director and CTO, PowerAmerica*

Abstract: The tutorial will outline the advantages of SiC over other power electronic materials, and will introduce SiC devices currently developed for power electronic applications. ESD, high-voltage testing, and packaging aspects will be covered. The design and properties of SiC JFETs, MOSFETs, BJTs, IGBTs, Thyristors, and Junction Barrier Schottky and PiN diodes will be discussed, with an emphasis on their performance advantages over those of their Si counterparts. Common SiC Edge Termination techniques, which allow SiC devices to exploit their full high-voltage potential, will be rigorously treated and their impact on device performance will be highlighted. Aspects of device fabrication will be taught with an emphasis on the processes that do not carry over from the mature Si manufacturing world and are thus tailored to SiC. In particular, the tutorial will stress in more detail the design and fabrication of SiC MOSFETs, which are being inserted in the majority of SiC based power electronic systems. Device reliability will be reported through exemplary hard switching results. Exemplary SiC-based power electronics systems like hybrid loaders, fast chargers, PV inverters, EV traction, and circuit breakers will highlight the significant advantages of these systems over their Silicon based counterparts.

This tutorial is intended for intermediate level audiences.

SC4. Vertical Power Electronic Devices based on Bulk GaN Substrates

Dr. Isik C. Kizilyalli, *Program Director at the Advanced Research Projects Agency – Energy (ARPA-E), Department of Energy*

Abstract: Silicon (Si) has been the semiconductor material of choice for power devices for quite some time due to cost, ease of processing, and the vast amount of information available about its material properties. Si devices are, however, reaching their operational limits in blocking voltage capability, operation temperature, and switching frequency due to the intrinsic material properties of Si. Wide bandgap (WBG) power semiconductors, such as gallium nitride (GaN) and silicon carbide (SiC), are an attractive emerging alternative to Si in many applications. Power converters based on WBG devices can achieve both higher efficiency and higher gravimetric and volumetric power conversion densities than the equivalent Si based converters. The power figure of merit (PFOM), which captures the trade-off between the device specific resistance (Rsp) versus the device BV clearly illustrates the advantage of GaN over Si and SiC devices. This arises from the cubed dependence of the figure-of-merit on the critical electric field where the critical electric field for GaN is 10 times that of Si and 1.6 times that of SiC. To date, the majority of GaN power device development has been directed toward lateral architectures, such as high-electron mobility transistors (HEMTs), fabricated in thin layers of GaN grown on foreign substrates (including Si or SiC). Such lateral devices suffer from well-known issues such as current-collapse, dynamic on-resistance, inability to support avalanche breakdown, and inefficient thermal management. Many of these shortcomings arise from defects originating in the very large lattice and coefficient of thermal expansion (CTE) mismatch between GaN and the substrate. Furthermore, most power electronics semiconductor and diodes are vertical architectures. Fabricating vertical semiconductor device structures on lattice and CTE matched bulk GaN substrates possible to realize the material-limited potential of GaN including true avalanche-limited breakdown and more efficient thermal management, leading to large device currents (> 100A) without resorting to device parallelization, high breakdown voltages (1.2 to 5kV), and increased number of die on a wafer. Recent availability of both 2- and 4-inch bulk GaN substrates is enabling breakthroughs in GaN device performance with vertical diode structures. In this tutorial recent advances in bulk GaN substrates and vertical architecture GaN power electronic devices (diodes, transistors, and application circuits) is surveyed with emphasis on the ARPA-E (Department of Energy) funded projects in the SWITCHES and PNDIODES Programs along with recent significant advances made in Japan. The SWITCHES Program (launched 2013) aimed to catalyze the development of vertical GaN devices using innovations in materials and/or device architectures that drive the costs of the devices. The goal was to enable the development of high voltage (>1200V), high current (100A) single die power semiconductor devices that, upon ultimately reaching scale, would have the potential to reach functional cost parity with Si power transistors while also offering breakthrough relative circuit performance (low losses, high switching frequencies, and high temperature operation). The PNDIODES (Power Nitride Doping Innovation Offers Devices Enabling SWITCHES, launched 2017) Program funds transformational advances and mechanistic understanding in the process of selective area doping in the III-Nitride wide band gap (WBG) semiconductor material system and the demonstration of arbitrarily placed, reliable, contactable, and generally useable p-n junction regions that addresses a major obstacle, enables high- performance and reliable GaN vertical power electronic semiconductor devices.

SC5. AlGaN/GaN Power Device Reliability
Dr. Peter Moens, *ON Semiconductor Belgium*

Abstract: AlGaN/GaN power devices have made tremendous progress over the past few years, and first commercial products have entered the market. The quality of MOCVD has reached a level that allows the fabrication of large area transistors with high yield and good reproducibility. Although GaN power devices achieve substantial higher system efficiency compared to their Si counterparts, the widespread adoption of GaN power devices in the market is still hampered by the unknown field reliability.

The tutorial will focus on the current understanding of the different intrinsic and extrinsic reliability mechanisms of AlGaN GaN power devices, and will cover following aspects:

- A methodology on how to extract important information on the conduction mechanisms in the GaN buffer structure out of relatively simple measurements on TLM structures (back-gating or substrate ramp) and transistors (current DLTS).
- Overview of the main intrinsic reliability mechanisms: gate reliability (both MISHEMT and pGaN gate), NBTI/PBTI of MISHEMTs, accelerated drain stress and hot carrier stress (semi-on-state stress). Recoverable versus permanent degradation.
- GaN-specific failure and degradation modes such as the inverse piezo-electric effect and dynamic Ron.
- Acceleration models and statistical distribution models (Weibull, lognormal) applied to GaN.
- Extrinsic reliability (HTRB, HTGB, thermal cycling etc).
- Switching reliability (double pulse testing, boost converter, ...)
- Introduction to the new upcoming JEDEC standard for AlGaN/GaN power devices (JC 70.1).

The topic will be treated in-depth and is for an intermediate-advanced audience.

SC6. Multi-Chip Semiconductor Power Module Design and Assembly: Rethinking Established Packaging Solutions for Improved Performance, Robustness and Reliability
Dr. Alberto Castellazzi, *Associate Professor of Power Electronics, Power Electronics, Machines and Control Group University of Nottingham*

Abstract: This short-course analyses the typical structure and assembly process of commercial power modules. Based on real application examples, it goes on to illustrate key operational electro- thermal and thermo-mechanical effects which prevent the achievement of disruptive efficiency, power density, robustness and reliability. It then presents innovative concepts and design approaches enabling progress beyond state-of-the-art and discusses the transfer of technology to new and upcoming wide-band-gap semiconductor technologies. In closing, package bespoke design methodologies and tools are addressed, with a focus on future virtual prototyping needs to support competitive development of increasingly integrated solutions.

The course targets an audience with entry to intermediate level knowledge of power device packaging; the topic is treated in general at the survey level, with some punctual aspects only dealt more in depth.

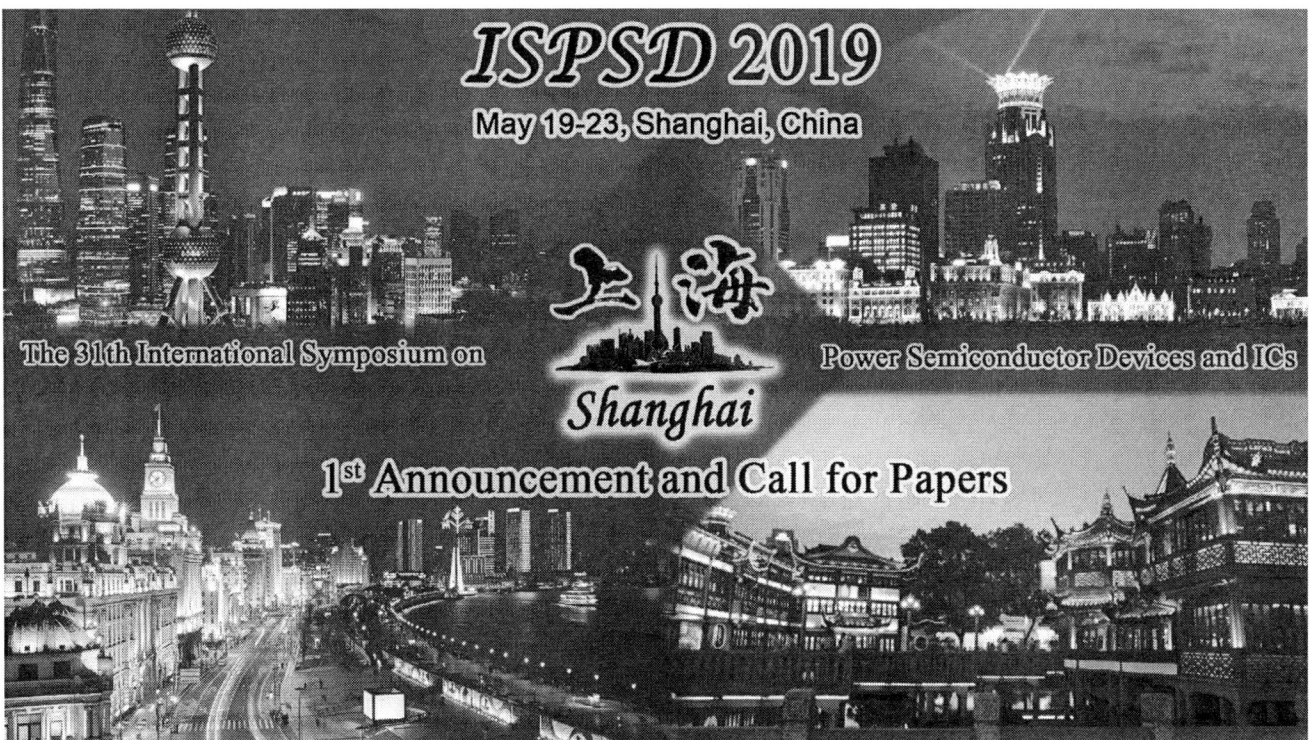

ISPSD is the premier forum for technical discussions in all areas of power semiconductor devices and power integrated circuits. ISPSD'19 will be held in the city of Shanghai, one of the most cosmopolitan, diverse, dynamic, vibrant and fascinating cities in China.

Topics of interest include but are not limited to:

- **High Voltage Power Devices**: Medium and high voltage Si bipolar devices such as IGBT, thyristors, pn diodes, etc.

- **Low Voltage Power Devices**: Low and medium voltage Si unipolar devices such as discrete or integrated power MOSFETs, SJ type devices, etc.

- **SiC Power Devices**: SiC-based devices, materials, and processing.

- **GaN and other WBG Power Devices**: GaN and other WBG-based devices, materials, and processing.

- **Power ICs**: Process integration, power IC design, Power Supply-on-a-Chip, etc.

- **Packaging and Module Technologies**: Integrated power modules and packaging technologies (functionality, power density, isolation, reliability, device cooling, temperature endurance, manufacturing, materials, etc.)

Submission requirement:

- **Abstract submission deadline:** November 12, 2018 (www.ispsd2019.com)

- **Author notification:** January 21, 2019

- **Late news submission (limited acceptance):** March 1, 2019

- **Final manuscript submission deadline:** March 15, 2019

- A PDF abstract should be submitted through the website including a single-page text summary in English (500 words maximum) and up to two additional pages of supporting figures.

General Chair: Prof. Kuang Sheng, Zhejiang University, Email: shengk@zju.edu.cn
Technical Program Chair: Prof. Kevin J. Chen, Hong Kong University of Science and Technology
Email: eekjchen@ust.hk

Technically co-sponsored by

Proceedings of the 30th International Symposium on Power Semiconductor Devices & ICs
May 13-17, 2018, Chicago, USA

ISPSD: 30 Year Journey in Advancing Power Semiconductor Technology

Invited Plenary Paper

Ayman Shibib, IEEE Fellow
Vishay - Siliconix
2201 Laurelwood Road, Santa Clara, CA 95054 USA
email: ayman.shibib@vishay.com

Leo Lorenz, IEEE Fellow
ECPE/Infineon, Germany

Hiromichi Ohashi,
NEPRC-J, Japan
email: oh@nperc-j.or.jp

Abstract—Celebrating the 30th Anniversary of ISPSD is a very special occasion to reflect on the origin and roots of the conference and how it came about to be the premier international conference on Power Semiconductor Devices and ICs. A review of the events that led to its formation and development is presented. Another aspect of this celebration is the review of the contributions of ISPSD to the Power Device technical community covering wide and diverse power device areas and applications throughout the 30 years history of ISPSD. The future prospects of Power Devices and ISPSD in the next years is briefly mentioned.

Keywords—component; ISPSD, ISPSD 2018, Power Devices, Power Semiconductor, High Voltage ICs, Power ICs, BCDMOS, SOI, dielectric isolation, IGBT, MCT, Thyristors, SiC, GaN.

I. INTRODUCTION

After this brief introduction, Section II covers the events that led to the ISPSD as an annually held conference, the opportunity to fulfil a needed function and the realization of that need by a dedicated worldwide group of individuals that made it happen. Section III reviews highlights of technologies, devices, concepts and applications that made this conference the premier conference on Power Semiconductor Devices and ICs. It is important to note that because the focus is the contributions of ISPSD, all references are from ISPSD published papers and the authors avoided relating to outside references made by the ISPSD authors or other authors. To be consistent with this approach the authors of this paper avoided any claims of "firsts" and relied on reflecting what was published in the Proceedings of ISPSD as technical contributions.

The coverage of the span of 30 years is divided into three decades each decade reviewed separately and is covered as an entity to make it easier to point out the three stages; origin, development and current status.

II. ORIGIN AND FOUNDATION OF ISPSD

Power devices main coverage up till the mid-1980s was by having sessions at the IEEE Electron Devices Meeting (IEDM), Sponsored by the IEEE Electron Devices Society, that used to alternate annually between Washington, DC, and San Francisco. During the Washington DC session, a Power Device workshop was held one day after the IEDM at the National Institute of Standards (NIST) for the power community which included reviews and workshops on current power device topics. In the Power Devices Workshop, in the mid-1980s, the need for an international conference on Power Devices was discussed by global volunteer members as the IEDM had limited coverage of Power Device topics with limited papers and sessions coverage.

In 1987, the Electrochemical Society sponsored a session, among its many parallel sessions, on "High Voltage and Smart Power Devices" chaired by Peter Shackle, Philips Laboratories, during the Electrochemical Society 1987 Spring Meeting in Philadelphia, Pa. Worldwide participation ensured good technical coverage and the meeting was quite successful.

In 1988, the first International Symposium on Power Semiconductor Devices (ISPSD) conference was held in Tokyo, Japan, with good worldwide participation but with Japanese Technical, Steering and Organizing Committees. In 1989, The Electrochemical Society sponsored another symposium, Chaired by Ayman Shibib, AT&T Bell Labs, on "High Voltage and Smart Power Devices and ICs". The addition of "ICs" to the title of the conference was important to make sure that the symposium is not just about discrete high power devices but it also includes integrated power devices and applications. The Organizing and Technical committees included members from around the world. After the successful conclusion of that symposium, talks started in earnest to combine the Japanese and American symposia into one international symposium to be held annually and alternate between Japan and the US and to be fully sponsored by the IEE of Japan (IEEJ) when it is held in Japan, and by the IEEE Electron Devices Society when it is held in the US. Working with Mike Adler, GE, we were able to have the IEEE Electron Devices Society's AdCom accept the proposal of full

978-1-5386-2928-4/18 $31.00 © 2018 IEEE

sponsorship of ISPSD, but there was the issue of the copyright material that both the IEEJ and the IEEE had strict guidelines for the copyright. Finally, it was agreed that the copyright of ISPSD, when held in Japan, belongs to IEEJ following the VLSI symposium precedent, and IEEE holds the copyright everywhere else. Along with this agreement, the IEEJ gave the IEEE the ability to create an IEEE catalog number for all ISPSD Proceedings regardless of where it is held, and thus allow its distribution through IEEE which was essentially the best resolution for ISPSD since its Proceedings would be available and distributed worldwide through the large IEEE network that included research institutes and libraries.

While this was being worked out, the second ISPSD symposium was held in Tokyo, Japan, with two noticeable changes. First the addition of "ICs" to the title of ISPSD becoming International Symposium on Power Semiconductor Devices and ICs" and also the addition of "overseas members" to the Technical Program Committee. With the successful conclusion of ISPSD 1990, ISPSD 1991 was scheduled to be held for the first time outside of Japan and in the US chaired by A. Shibib in Baltimore, Maryland, April 22nd-24th.

The successful completion of ISPSD 1991 and the ability for the symposium to meet its objectives, that included a return of surplus to the IEEE Electron Device Society, made the agreements reached earlier very solid and set the stage for years to come in the association and support of ISPSD by the IEEE Electron Devices Society. In 1994, The European part of the technical committee proposed adding Europe to the two year rotation between Japan and US making it a three-way rotation with Japan, US and Europe. Bylaws and guidelines for ISPSD were established including an Ad Com (Advisory) committee made of past General Chairpersons of ISPDS.

The ability to unify the Power Device community spanning a wide range of discrete high voltage high power devices to discrete and integrated low voltage lower current devices and ICs to attend the same conference was among the biggest achievements of ISPSD. Another significant achievement is the formation of a truly international group of Power Device technical experts that can meet annually to discuss technical and developmental aspects of the field and have an open exchange of ideas. A third major contribution of ISPSD is to share common improved device performance among the widely different Power Device types and applications.

The continued success of the conference is supported by the increasing number of papers published in the Proceedings. Figure 1 shows the total number of papers presented at ISPSD per year. Leading members that participated in establishing ISPSD were awarded the first "Contributory Award" at 2001 ISPSD. Those are M. Adler, A. Shibib, T. Ohmi, H. Okamura, H. Ohashi and A. Jacklin.

Some of the Technical Committee members of ISPSD appear in the photographs shown in Figures (3) 1988 ISPSD, (4) 1991 ISPSD and (5) 1997 ISPSD.

III. TECHNICAL CONTRIBUTIONS OF ISPSD

It is important to point out that we have restricted coverage and references to ISPSD papers **only** since we are addressing the technical contributions of ISPSD and the paper is not meant to be a general review and coverage of Power Semiconductor topics outside of ISPSD.

A. *The first decade of ISPSD 1988 to 1998*

In the first decade of ISPSD, the focus of the symposium was a combination of telecom technology applications and high power mainly bipolar switches. The telecom applications required mainly higher voltage ICs to perform the subscriber line interface circuit (SLIC) of opening the channel, ringing access and other supervisory functions in addition to discrete MOSFET applications for RF transmitter from MHz to GHz for digital cellular base stations.

The high voltage or power ICs requirements were fulfilled with a form of silicon-on-insulator (SOI) known as dielectric isolation (DI) that basically isolated each power component of the circuit and some CMOS circuits with a thick oxide layer by depositing a very thick polysilicon layer of about 500 um, to act as a substrate, on an etched silicon wafer, and allowed for the full and reliable integration of Bipolar-CMOS-DMOS (BCDMOS) devices for high reliability telecom applications.

It is interesting to note here that lateral IGBTs were integrated into the BCDMOS technology platform reported in 1988 ISPSD with DI isolation (Gammel et al, p117) and in 1990 ISPSD (Sakurai et al, p66) and Junction Isolated (JI) integrated LIGBTs were reported in 1990 ISPSD (Tsuchiya et al, p60). Other higher voltage, above 1000V, ICs relied on junction isolation for the control circuit with MOSFET output, 1992 ISPSD (Rumennik, p322) and vertical bipolar output devices, 1988 ISPSD (Cini et al, p88).

High voltage ICs that followed in that decade expanded into other than telecom applications like automotive and consumer electronics and continued the SOI developments and applications and added junction isolation and RESURF principles, both in junction and by extending the concept to dielectric isolation 1991 ISPSD (Huang et al, p27), that covered voltage ranges from less than 100 volts to above a thousand volts.

Along with the technology aspect of the development was the focus on optimizing and increasing the performance and reliability of power devices such as reducing the on resistance for the same voltage rating, increasing the switching speed, increasing the efficiency and power handling capability of the device, and improving its reliability and Safe Operating Area (SOA). In particular, reducing the on resistance by junction and dielectric RESURF, and going to vertical MOS channel structures that reduce the surface area consumption. Increasing the frequency in ICs focused on BiCMOS technology integrating a GHz NPN with CMOS, 1990 ISPSD (Yamada et al, p86).

For the discrete devices, two categories have to be distinguished; low power, predominantly MOSFETs and high-

978-1-5386-2928-4/18 $31.00 © 2018 IEEE

power devices that include IGBTs, GTOs and SI devices and modules spanned a wide range of technologies. The lower power discrete devices, with low to medium voltages, were mainly MOSFETs and focused on reducing the specific on resistance to devices using Super Junction structures for higher voltages and trench MOSFETs for lower voltages. For lower voltage devices, a double gate trench MOSFETs was reported for a 70V device with a specific on resistance of 120 mOhm.mm2, 1992 ISPSD (Baba et al, p300), and for a drain trench technology a 100 milli-Ohm.mm2 for 50V device, 1992 ISPSD (Vera et al, p294) and for a trench MOSFET 90 mOhm.mm2 for 70V, 1994 ISPSD (Matsumoto et al, p365) and 90 mOhm.mm2 for 30V rated P-channel device at low gate voltage drive, 1996 ISPSD (Williams et al, p53), furthermore, reduced gate charge Qg were reported in 1997 ISPSD (Narazaki et al, p285). For higher voltage devices, charge balanced, Super Junction, devices were reported in 1999 ISPSD (Lorenz et al, p3) demonstrating a significant improvement of on resistance as a function of voltage for high voltages compared to the classic power law dependence of on resistance on breakdown voltage.

For high power discrete devices that started with bipolar devices like Diodes (D), Bipolar Transistor (BT), Gate-Turn-Off (GTO), Thyristors, Static Induction Thyristors (SIT) as well as MOS-controlled devices like MOS-Controlled Thyristor (MCT) and the Insulated Gate Bipolar Transistor (IGBT). Reported improvements focused on higher voltage, larger currents, high frequency, self-protection features such as over voltage and short circuit sustenance were key areas. Extension of voltages for diodes up to 4 KV, 1992 ISPSD (Kitagawa et al, p60), and 5 KV, 1995 ISPSD (Tornblad, p380), and current levels up to 3000 A, 1988 ISPSD (Murakami et al, p172). For Thyristors, voltage levels of 12 KV with 1 KA of current, 1990 ISPSD (Iwamoto et al, p283), and light triggered 8kV/3kA Thyristors with integrated breakover diode and integrated dv/dt – protection 1990 ISPSD (Schulze et al, p289), and 6 KV and 5.5 KA, 1997 ISPSD (Kato et al, p73), were reported. For Thyristors, GTOs and SITs, the focus was to increase the frequency and to reduce the turn-off losses of the 4.5 to 8 KV devices by structure optimization and proton irradiation, 1990 ISPSD (Shimizu et al, p231) and electron irradiation for SIT, 1997 ISPSD (Morikawa et al, p61). Consequently, power conversion efficiency of high power applications is now more than 99%.

The MCT had a flurry of activities in the first decade of ISPSD from 1990 to 1997. Most of the papers focused on new structures that can be utilized in the MCT improvement, high frequency application, safe operating area, failure mechanisms and comparison of MCTs to IGBTs. The IGBT in the first decade of ISPSD had by far the most attention spanning continuous coverage from 1988 to 1998. In Figure (2) a plot of the number of papers of IGBT and MCT throughout the first decade of ISPSD is shown. Non-latch-up and Injection Enhancement principles of IGBT decided the outcome.

The wide interest and applications of the IGBT required continuous improvement and optimization from device structure, Punch-Through (NP) and Non-Punch-Through (NPT), to reducing the carrier lifetime for higher frequency operation to a wider SOA. The new IGBT concepts with trench cell technology were published in the ISPSD, first for 600V in 1994 (M. Harada, et al p411), then also as a first short circuit robust trench version for 1200V in 1998 (Laska, et al, p433). Also, High Voltage IGBTs in the range up to 6.5kV came into focus and also the injection enhancement principle was discussed intensively in the ISPSD meetings (e.g. 1995 ISPSD, Kitagawa et al, p486).

At the 1994 ISPSD, a new failure mechanism occurring during the operation of power devices at high voltages was presented which could be explained by cosmic radiation-induced streamers (Kabza et al, p9). Comparing the failure rates obtained in a salt mine and in the research laboratories, the failure cause could be clearly identified. Effective measures to improve the cosmic radiation hardness by an optimized vertical distribution of the electrical field strength were proposed and are still in use today.

From the early stage of WBG semiconductors development, ISPSD played an important role. The first decade of ISPSD reported mostly Silicon Carbide (SiC) devices covering advantages of SiC compared to Si and device structures of pn-junctions, Schottky, MOSFET, UMOSFET, SIT and GTO.

Coverage of SiC development started in 1990 ISPSD with an invited paper reviewing expectations for SiC in Power Electronics (Matsunami, p13), followed by device performance advantages, 1991 ISPSD (Bhatnagar et al, p176), and another review in 1993 ISPSD (Helbig, p6). Also in 1993 ISPSD comparison of several compound semiconductors for superior performance using different figures of merit (FOM) was reported (Chow et al, p84), and numerical analysis of SiC Schottky diode and MOSFET (Funaki et al, p212).

The first full session dedicated to SiC was established in 1995 ISPSD that included papers on SiC substrate growth (Lebedev et al, p90), nearly ideal edge termination for 6H-SiC (Alok et al, p96), efficient 4H-SiC power Schottky rectifiers (Itoh et al, p101), Al/Ti Schottky barrier for 6H-SiC (Ueno et al, p107). Another SiC session appeared in 1996 ISPSD focusing on SiC technology and characterization aspects; Nitrogen implantation for edge termination (Alok et al, p107), EBIC investigation of edge termination (Raghunathan et al, p111), and edge termination for 3C-SiC on Si (Tyagi et al, p115). Also in 1996 ISPSD, Agrawal et al, p119, reported on the performance advantages of 4H-SiC UMOSFET but anticipated a limitation due to the Time-Dependent-Dielctric-Breakdown (TDDB) of oxides grown on SiC.

A review paper of SiC for Power Devices was presented in 1997 ISPSD (Palmour et al, p25), outlining advantages of SiC MOSFETs over Si for high power, 4.2 KW, and low trr, 105 ns, and maximum frequency of 250 KHz. Also in 1997 ISPSD, papers covered use of 4H-SiC for 700V to 5KV Static Induction Transistors for power grid applications (Iwasaki et al, p149), analysis of gate dielectrics of 6H and 4H-SiC for UMOSFETS (Sridevan et al, p153), low leakage SiC diodes using different termination schemes (Singh et al, p157), fast 4H-SiC diodes (Mitlehner et al, p165), and measurements of

978-1-5386-2928-4/18 $31.00 © 2018 IEEE

electron and hole impact ionization coefficients (Raghunathan et al, p173).

In 1998 ISPSD SiC session, Sugawara et al, p119, reported on the feasibility of a 5KV 3KA SiC MOSFET that can replace a 6KV 6KA Si GTO in the same application, and Seshadri et al, p131, on the turn off characteristics of a SiC GTO for power switching applications.

Packaging related topics were covered that included thermal cycling, 1988 ISPSD, Yasukawa et al, p36, design and management of Plastic Surface Mount packages for Power Electronics, 1993 ISPSD (Kasem et al, p316), as well as Module developments, 1994 ISPSD (Anderson et al, p21), in the first decade of ISPSD that included chip-on-chip, low inductance module construction for high speed applications.

B. The second decade of ISPSD 1999 to 2008

In the second decade of ISPSD papers covered different aspects of Forward Bias Safe Operating Area (FBSOA), 1999 ISPSD (Hower p 59) and using adaptive RESURF to improve SOA, 2000 ISPSD (Hower p345), avalanche induced thermal instability, 2001 ISPSD (Hower p153) electrical and thermal SOA and tradeoffs in device design with breakdown voltage and specific on resistance, 2002 ISPSD (Hower p1), and rugged integrated LDMOS, 2005 ISPSD (Hower p327), and an anomalous temperature behavior in LDMOS current sensing, 2007 ISPSD (Lin et al p65).

In addition, some of the widely referenced papers of ISPSD that became known as the "Spirito Effect" which changed the way many Low Voltage Power MOSFETs data sheets specify SOA. It used to be that the FBSOA was specified by the power dissipation level and thermal considerations of the package, but the demonstration that there is another restriction which has to do with the whether the temperature coefficient of the drain current is positive or negative in the region of operation of the device. The papers were published on electrothermal instability in low voltage MOSFETs, 1999 ISPSD (Breglio et al, p233), analytical model for thermal instability and SOA, 2002 ISPSD (Spirito et al, p269) and modeling the onset of thermal instability with experimental validation, 2005 ISPSD (Spirito et al, p183).

In the area of IGBTs new vertical concepts came up. The field stop IGBT which was presented in 2000 ISPSD with simulation as well as very first experimental results (T. Laska et al, p355) was an important IGBT milestone for this decade. Later IGBT cell concept considerations with respect to ultimate on state performance were published in 2006 ISPSD (Nakagawa, p5).

For the low voltage trench MOSFETs, a review paper outlined the progression of low voltage (typically 30V) of trench cell increase requiring deep sub-micron-capabilities, 2000 ISPSD (Williams, p19). Other high density low on resistance trench devices achieved an 18 mOhm.mm2 specific on resistance at Vgs=10V, 2001 ISPSD (Zeng et al p147). Another 26V trench MOSFET with self-aligned source contact and a 0.2 um trench width was reported, 2002 ISPSD (Peake et al, p29). Further reduction in on resistance for a 33V

and achieving a 10 mOhm.mm2 was reported in 2003 ISPSD (Ono et al, p28), and a 4 mOhm.mm2 on resistance for a 20V device was reported in 2003 ISPSD (Zandt et al, p32). The reduction of the gate-drain charge to improve the switching efficiency of the trench MOSFET by increasing the bottom gate oxide was reported in 2003 ISPSD (Darwish et al, p24).

A Super Junction, with implant energies up to 2 MeV, 55V trench MOSFET was reported for the range of 55 to 100V achieving 51 mOhm.mm2 at 78V, 2004 ISPSD (Ninomiya et al, p177). Resurf Stepped Oxide (RSO) trench MOSFETs were reported in 2004 ISPSD (Koops at al, p185) for the 80V class with a specific on resistance of 58 m$\square$mm² followed by split gate RSO trench MOSFETs with 46 m$\square$mm² in 2006 ISPSD (Gajda et al p109). A split gate 25V RSO was reported with a 3.8 mOhm.mm2 and a gate-drain charge of 0.9 nC.mm2 2007 ISPSD (Goarin et al, p61).

The RFLDMOS development to higher frequencies and better reliability was demonstrated in the second decade of ISPSD. Dedicated sessions for RFLDMOS in 2001 (Matsumoto et al, p99, Cai et al, p103, Shindo et al, p107) and 2003 ISPSD (Xu et al, p190, Roger et al p270, Letavic et al, p274, Pathirana et al, p278) covering the frequency range of 2 to 5 GHz and improving the power gain, efficiency and linearity. Improvement by design of on resistance stability by control of Hot Carrier Injection was experimentally demonstrated in 2004 ISPSD (Shibib et al, p233).

For Wide Band Gap semiconductors, SiC continued to gain more interest from development and application sides. Starting with 2001 ISPSD, WBG had 2 dedicated sessions and with emergence of GaN, more GaN papers shared the WBG sessions with SiC. Many SiC papers focused on the Schottky diode performance improvements 1999 ISPSD (Chilukur et al, p161) and reduction in leakage for trench MOS barrier, TMBS (Khemka et al, p165). Higher voltage diodes, 2001 ISPSD (Sugawara et al, p27), 4H-SiC diode (Lendenmann et al, p31, and Singh et al, p45) as well as BJT (Ryu et al, p37) and JFET (Takayama et al, p41).

A review of WBG semiconductors for RF Power applications was reported, 2002 ISPSD (Ueda, p17) , a 1000V 130 mOhm.cm2 recessed MOSFET in 6H-SiC (Banerjee et al, p69), and a 1.6KV 4H-SiC with 27 mOhm.cm2 MOSFET. In 2003 ISPSD, a SiC trench JFET achieved 1726V with a specific Rds of 3.6 mOhm.cm2, and accumulation mode GaN devices were reported (Matocha et al, p54, and Yoshida et al, p58). In 2004 ISPSD, the WBG session was shared between SiC and GaN. Three papers covered SiC pin diode (Nakagawa et al, p357), BJT (Agarwal et al, p361), a 12.7KV commuted GTO, and a Normally-off AlGaN/GaN Carbon doped GaN layer. In 2005 ISPSD, there were 3 WBG sessions, 2 SiC and one for GaN and Diamond.

The SiC 2005 ISPSD papers focused on 4H-SiC MOSFETs with 1600V (Zhang et al, p211), 2KV with 10.3 mOhm.cm2 (Ryu et al, p275) and 1200V Resurf MOSFET with 62 mOhm.cm2 (Kimoto et al, p279). 2006 ISPSD saw the introduction of an extremely robust SiC Schottky diode

978-1-5386-2928-4/18 $31.00 © 2018 IEEE

with surge currents up to 35 times the rated current and high dV/dt robustness up to 90V/ns (Rupp et al, p269), and a 4H-SiC P-channel IGBT of 5.8KV and 570 mOhm.cm2 (Zhang et al, p285). In 2008, a WBG Technical Sub-committee of ISPSD was formed to handle the increased submission of WBG papers. A GaN-On-Si paper of 1.8KV on 100 mm substrate was presented with the use of Carbon doped buffer to reduce the current collapse ratio to 1.1 to 1.7 up to Vds of a 1000V.

C. The third decade of ISPSD 2009 to 2018

The most prominent reported development during the last decade is the development of WBG semiconductors, especially GaN-on-Silicon enhancement mode (e-mode) High-Electron-Mobility-Transistor (HEMT) devices and simple IC circuits. Full sessions of GaN appeared in every ISPSD since 2009, and in 2016 ISPSD, the WBG sub-committee was split into 2 sub-committees, for SiC and the other for GaN and other Nitride materials.

Most noted publications in ISPSD focused on P−GaN gate structure HEMT. GaN smart power chips were reported building a voltage reference generator and a voltage comparator with a wide temperature range of up to 250 C, 2009 ISPSD (Wong et al, p57). For HEMT devices, optimizing the Carbon doped buffer allowed achieving a 0.62 mOhm.cm2 for a BV above 1000V, 2011 ISPSD (Hilt et al, p239). In the following year, reducing the trap density in the C-doped buffer with a lower buffer BV was shown to reduce the current collapse effect, 2012 ISPSD (Hilt et al, p345). In 2013 ISPSD, a 2.8V high Threshold voltage was reported for P-GaN gate on 200 mm Si wafer achieving 670V and 20 Ohm.mm (Kim et al, p315). In 2015 ISPSD a co-packaged 0.14 automotive BCD IC with a GaN HEMT for programmable slew rate and negative gate drive and digital current mode control was reported (Rose, et al, p361). Also in 2015 ISPSD, an experimental Solid State circuit breaker using 650V GaN based monolithic bidirectional switch was reported (Shen et al, p79) and a paper on the impact of Mg out-diffusion and activation on P-GaN gate HEMT device performance was presented (Posthuma et al, p95).

A current-collapse-free 850V GaN HEMT structure that utilizes hole injection from the drain was reported in 2016 ISPSD (Kaneko et al, p41). An improvement in high temperature gate bias stress stability MIS-HFET for a depletion mode device was reported, 2014 ISPSD (Wong p55), and a cascade depletion mode GaN and low voltage Silicon MOSFET were reported with an extra Zener protecting the MOSFET, 2016 ISPSD (Wu et al, p255). It is to be noted the contribution of a semiconductor foundry to the GaN development, 2014 ISPSD (Wong et al, p55) and 2016 ISPSD (Wu et al, p255). 2015 ISPSD had a review of GaN transistors status and future prospects (Lidow, p1) and 2017 ISPSD had a review of GaN Power ICs (Kinzer, p19).

For SiC, the third ISPSD decade had continued interest and development for high voltage switching and high power applications. 2010 ISPSD covered 1200V 35A SiC SIT with improved blocking gain of 480 (Tanakaet al, p357), 1400V 5

mOhm.cm2 MOSFET, and high surge current ruggedness 5KV 4H-SiC commuted GTO (Ogata et al, p369). In 2013 ISPSD, SiC papers presented optimized MOSFET performance; a 1400V trench MOSFET with thick bottom oxide and 4.4 mOhm.cm2 (Takaya et al, p43), and a breakthrough in the trade-off between Vth (5.1V) and Rsp (5.2 mOhm.cm2) for 600V SiC MOSFET (Furuhashi et al, p55). 2015 ISPSD had continued SiC MOSFET development achieving a 70% reduction in RonAxQgd FOM over conventional SiC MOSFET (Tegat ey al, p81), a 1200V 2.0 mOhm.cm2 V-groove trench SiC MOSFET (Uchida et al, p85), and 1200V 4H-SiC with 2.7 mOhm.cm2 (Zhang et al, p89). In 2016 ISPSD, a manufacturable and rugged 1.2 KV SiC MOSFET fabricated on 150 mm wafer CMOS foundry fab was reported (Banerjee et al, p279). 2017 ISPSD saw the introduction of the asymmetric cell 1200V trench MOSFET with high ruggedness and gate oxide reliability enabling first time less than 1 fit life time failure rate (Peters et al, p239).

For Si trench power MOSFETs, introduction of dual channel 25V MOSFETs enabling 95% DC-DC conversion efficiency at 600kHz in 2015 ISPSD (Häberlen et al, p65), and a 60V split gate trench MOSFET achieving a specific on resistance of 10.9 mOhm.mm2 and a Figure of Merit of 85 mOhm.nC in 2016 ISPSD (Park et al, p387). The reduction in cell pitch for high performance MOSFETs necessitated the use of Deep-Ultra-Violet (DUV) lithography for critical dimensions.

With respect to Si IGBTs also in this decade still had many innovations and publications were seen. Besides the trend in cell pitches down to the sub-micron range, 2016 ISPSD (Eikyu, et al, p211) and a lot of activities with reverse conducting IGBTs were shown, 2009 ISPSD (Rahimo et al, p283) as well as first approaches with large wafer diameter IGBTs of 300mm, 2016 ISPSD (Schulze et al, p355).

As Power Devices continue to gain better performance and higher efficiencies, well above 90% in power conversion applications, and having wider and ubiquitous use in all areas including automotive, telecom, industrial, consumer and power grid we look to the prospect of new requirements corresponding to electrified future society, and standards of performance, miniaturization and integration with other electrical devices and networks. New areas that are evolving now such as the Internet-Of-Things (IOT), autonomous driving vehicles and connectivity are applications that will define the new applications of Power Devices and the developments that are needed to address those applications. ISPSD, we believe u et al,pwill expand into these areas by investigating new materials, technologies and integrated device concepts unified with network connected CMOS circuitries, pushing many limits beyond what have been accepted as limits.

IV. Conclusions

A review of the origin and foundation of ISPSD is presented underlying how separate efforts started and different continents were unified into a worldwide synchronized effort to promote mutual collaboration to the benefit of the whole Power Device community. A review of the technical contributions of ISPSD symposia and its coverage of the wide devices and technologies of Power Devices to the technical community throughout 3 decades is presented. Finally, the prospects of the future role of Power Devices and their importance and spread into more fields and applications unified future power electronics promise to have an even wider impact of ISPSD in evolving to new materials, technologies and integrated device concepts.

Acknowledgment

To all authors of papers published in ISPSD Proceedings over the past 30 years and members of Technical, Organizing

and Advisory Committees worldwide who volunteered their time, effort and technical expertise to ISPSD, thank you!

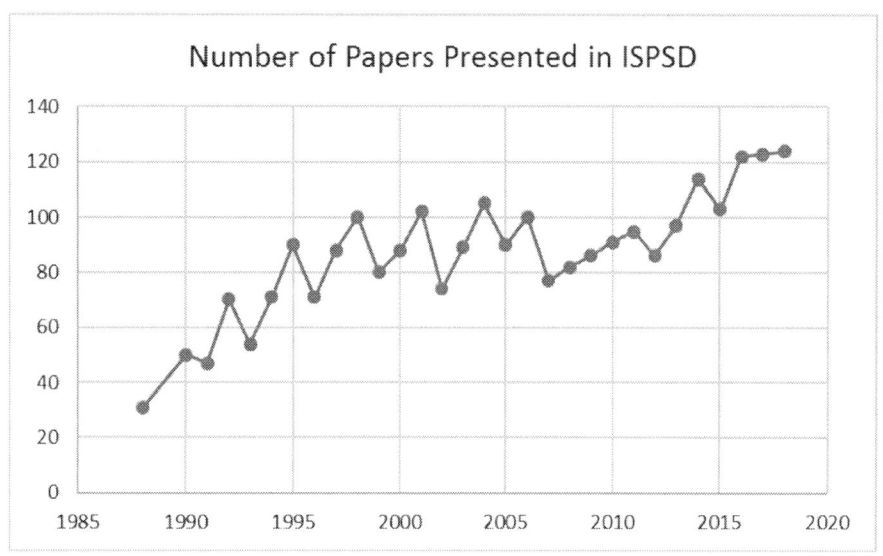

Figure 1: Total number of papers presented at ISPSD per year

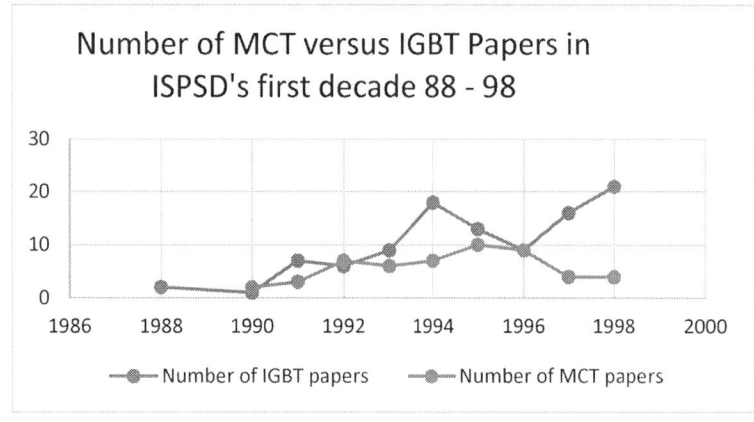

Figure 2: Number of IGBT and MCT papers presented in the first decade of ISPSD

Figure 3: General Chair Prof. Tadahiro Ohmi at the very first ISPSD in 1988.

Figure 4: Committee members at ISPSD1991 (the first ISPSD to be held in North America).

Figure 5: Committee members at ISPSD1997.

Proceedings of the 30th International Symposium on Power Semiconductor Devices & ICs
May 13-17, 2018, Chicago, USA

Silicon, GaN and SiC: There's Room for All

An application space overview of device considerations

Larry Spaziani
GaN Systems, General Manager
Ottawa, Canada
lspaziani@gansystems.com

Lucas Lu
GaN Systems, Principal Application Engineer
Ottawa, Canada
llu@gansystems.com

Abstract— The discrete power device marketplace is estimated between 15 and 22 billion dollars and is comprised primarily of transistors and diodes in a variety of voltage, current, packaging and power ratings. It is an area of intense focus as new technologies such as wide bandgap emerge and new applications such as electric vehicles emerge. Decision makers from Engineers to CEO's are faced with the same decisions they have always faced, comparing power, efficiency and size, yet the decisions are more difficult given the fast-moving pace of these emerging technologies. In this paper, several application spaces ranging from consumer to vehicle to motors will be reviewed, comparing the most critical aspects of the applications against semiconductor choices these decision makes have available. Considerations of the appropriate technologies will be reviewed comparing where the technologies have been, are today, and where they will be in the next 5 years.

Keywords—Wide bandgap, GaN, SiC, IGBT, MOSFET, HEMT, Motor, Electric Vehicle, power supply, wireless power

I. INTRODUCTION

Wide bandgap materials such as Silicon Carbide and Gallium nitride have seen wide adoption in the past five years because of the intrinsic benefits of these materials, when compared to Silicon. Table 1 is a commonly used in comparing wide bandgap materials to silicon.

Table 1 – Material comparison of wide bandgap materials to silicon

Material	Bandgap (eV)	Mobility (cm²/V*s)	Permittivity	Vsat (cm/s)	Critical field (V/cm)
Si	1.1	1400	11.8	1×10^7	3×10^5
GaAs	1.42	8500	12.8	2×10^7	4×10^5
4H-SiC	3.23	260	9.7	2×10^7	2.9×10^6
GaN (Bulk)	3.4	900	9	2.5×10^7	3.3×10^6
GaN (HEMT)	3.4	1800	9	2.5×10^7	3.3×10^6

key market trends in Automotive, Consumer and Industrial spaces, critical parameters required in these spaces, and detailed technology comparisons with regard to semiconductor solutions. First, a series of common applications are reviewed, with a look at the architectures and those parameters which are critical to those topologies. Secondly, additional criteria for deciding on technologies are reviewed which focus on the business aspects of decision making, including cost, capacity, reliability and future development predictions. Including these very realistic factors into decision making now will help any company to decide on their product's development roadmap. Finally, a look at market trends for each technology will be reviewed in the context of additional feature sets such as integration capabilities, packaging and power limitations, process improvements and capacity predictions. This holistic overview will show that all technologies can and should be considered for inclusion into the next generation of power electronics.

II. PERFORMANCE CRITERIA OVERVIEW

In this section, semiconductor device considerations are presented not as a function of their material characteristics as shown in Table 1, but instead, the materials are considered in terms of characteristics the system designer will consider, as shown in Table 2. In Table 2, the primary electrical considerations of voltage, current, and Figure of Merit, as well as other performance factors, are compared, with Figure of Merit being the most universally accepted general comparison between technologies, and between different offerings from various semiconductor suppliers.

Table 2 – Application performance comparison of wide bandgap materials to silicon

Viewing all semiconductor offerings from a simple table such as shown in Figure 1, is, however, presenting the technologies in a very simplistic view. This paper will review

978-1-5386-2928-4/18 $31.00 © 2018 IEEE

Parameter	MOSFET	IGBT	SiC MOSFET	SiC MOSFET	GaN EHEMT	GaN Cascode	GaN GIT
BV$_{ds}$ [V]	650	600	1200	650	650	650	600
I$_{ds}$ [A]	29	30	40	29	31	12	15
R$_{ds,on}$ [mΩ]	45	50*	52	120	41	150	65
Q$_g$ [mΩ]	93	167	115	61	8	6.2	11
Q$_g$× R$_{ds(on)}$ [nC×mΩ]	4185	8350	5980	7320	328	930	715
Q$_{rr}$[nC]	13000	920	283	53	0	54	0
C$_{oss}$[pF]	70	107	150	90	67	133	-

(*) IGBT equivalent on-state resistance $R_{ds,equ} = V_{ce}/I_{c,nom}$

The performance criteria of Table 2 can function as a guide to which technology might be a best fit for a system, but requires a detailed understanding of the application, the critical system requirements and more often than not, the impact of other components such as magnetics in the system. This section will therefore look at a diverse family of applications, in detail, such that a better understanding of the value proposition of any given semiconductor material can be understood.

III. APPLICATIONS REVIEW

Semiconductor decision making applications worldwide, in market segments such as consumer, enterprise, industrial and automotive. In order to understand performance-based decision processes when considering silicon IGBT's and MOSFETs, Gallium Nitride devices and Silicon Carbide devices, the following applications will be reviewed in this section.

- Consumer – Adapters
- Consumer / Enterprise – LLC converters
- Consumer – Wireless Power Transfer
- Enterprise / Automotive   Phase Shifted Full Bridge
- Industrial – Low Voltage Motor drives
- Industrial – Energy Storage systems
- Automotive – On Board Charger
- Automotive – Traction Inverters

In each case, general system priorities such as power density, energy efficiency and system cost will be considered and weighed against the properties of the semiconductor devices which apply to these systems.

A. Consumer Adapters

The adapter space from 25W to 250W, covering commercial fast phone chargers to high end gaming adapters will be presented. Power density and cost will be considered the primary design criteria. Solutions utilizing standard flyback, quasi-resonant flyback and Active Clamp flyback will compared with critical semiconductor parameters considered.

B. Consumer / Enterprise – LLC converters

The LLC converter is considered one of the world's most often-used converter technologies, primarily because it delivers very high efficiency at a low overall cost. It will be shown that the choice of semiconductor device strongly effects the overall performance when combined with the design of the main transformer.

C. Consumer Wireless Power

Wireless power is a fast-emerging technology, ranging from 100kHz to 6.78MHz and higher. Since the driving market is consumer based, the comparison of semiconductor choices is typically between MOSFETs and GaN devices. A comparison of several leading architectural choices, and the appropriate semiconductor device choices will be made.

D. Enterprise / Automotive Phase Shifted Full Bridge

With the onset of electric vehicles, and the wide possible ranges of batteries that can be considered, the Phase Shifted Full Bridge topology can be seen as an excellent choice for vehicle DC/DC converters to convert high voltage batteries to lower system voltages such as 12V or 48V. Design considerations and semiconductor choices will be shown to optimize this architecture.

E. Industrial Low Voltage Motor Drives

Low voltage motor drives can consider tradeoffs in bandwidth, noise, size, efficiency, cost and motor-level-integration to choose from the lowest cost silicon devices to high performance wide bandgap devices. The use of GaN devices have been shown to offer significant system level advantages but requires a relatively new look at the overall architecture. The system benefits of higher frequency and higher efficiency operation will be reviewed against semiconductor device choices.

F. Industrial Energy Storage Systems

Energy Storage Systems (ESS) present an emerging market which highly values energy efficiency. When full power considerations dominate, IGBT silicon solutions are often deemed the best to meet ESS price/performance levels. Since ESS typically run at very low load however, the resistive nature of SiC and GaN offer advantages. An ESS system is

978-1-5386-2928-4/18 $31.00 © 2018 IEEE

presented, with system level benefits studied as a function of semiconductor costs.

G. Automotive – On Board Charger

The On-Board charger (OBC) operates between 3.3 and 22kW in automobiles worldwide, with higher levels for commercial vehicles. In an automobile, size and power density are the most critical parameter, next to system costs. Comparisons of Si, SiC and GaN based OBC architectures will be highlighted. The Totem Pole Bridgeless Power Factor Correction circuit, a highly efficient PFC technology, will also be shown to be a key part of any OBC

H. Automotive – Traction inverter

The high-power levels of the Traction Inverters, ranging from 10's of kW to 100's of kW, have traditionally required high power IGBT solutions, with some SiC and GaN technologies emerging. It will be shown that in normal drive mission profiles, while the IGBT solution offers high efficiency at a low cost, a combination IGBT/Wide bandgap solution may offer the best cost performance. Figure 1 shows a comparison at various drive mission profile levels, of a Silicon, a SiC and a Hybrid GaN/Silicon solution, showing that that semiconductor choice is not automatic.

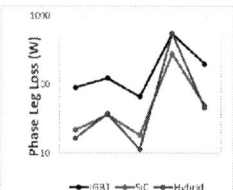

Figure 1 – Historical development of Transistors vs performance

IV. Past, Present and Future Developments

The fact that semiconductors have always evolved should lead us all to realize that semiconductors are now and will continue to evolve. In the power electronics arena, history has demonstrated, as shown in Figure 2, that approximately every 12 years, new technology emerges, evolves and performance is maximized, before new technology emerges. What is not shown in Figure x is that the older technologies do survive long into the marketplace, by either evolving highly specialized versions or by being commoditized.

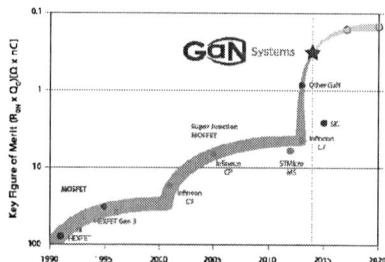

Figure 2 – Historical development of Transistors vs performance

According to IHS [2], the market drivers and inhibitors for the growth of wide bandgap materials is shown in Figure x3. This section will address the state of the semiconductor technologies today and predict where these technologies will go in the next few years. In general, it is assumed that silicon based MOSFET and IGBT technology has fully matured and, as shown by the curves in Figure x, may improve somewhat but dramatically. This section will therefore focus on looking at Silicon Carbide and Gallium Nitride. This analysis will review technology trends that limit or expand the market acceptance of wide bandgap technologies.

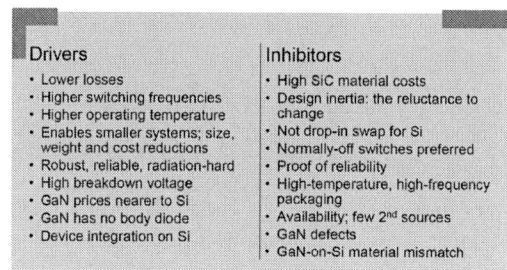

Figure 3 – IHS Market Drivers and Inhibitors

A. Voltage and current range

A comparison of voltage ranges for Silicon IGBT's, MOSFETs, SiC MOSFETs and GaN transistors will be presented showing the breadth of market acceptance by each technology. It will be shown that Silicon MOSFETs, ranging from tens of volts to low kilovolt ranges, compliments Silicon IGBT technologies, that in turn ranges from mid-hundreds of volts to high kilovolt levels.

In a similar way, Gallium Nitride and Silicon Carbine cover the same voltage ranges, with GaN devices dominating from tens to hundreds of volts and Silicon Carbide dominating from approximately one kilovolt to many kilovolts.

Future voltages for GaN devices should range to commercially available 1200V devices to experimental devices to 3300V, while Silicon Carbide devices are now at, and will expand down to 600V.

978-1-5386-2928-4/18 $31.00 © 2018 IEEE

Current values today show silicon devices reaching into ratings of several hundred amperes, with SiC and GaN devices today limited to just above one hundred Amperes. Both devices show the ability to reach higher currents. Paralleling Silicon, SiC and GaN devices will be quickly compared to show the upside capabilities of each technology.

B. Operating Temperature

Silicon devices have comfortably operated between -55°C to +150°C for some generations now, with high end automotive devices stretching to +175°C SiC and GaN devices have also been able to operate in this same range, but with less proven reliability above +150°C. Theoretical limits of wide bandgap devices show that these devices should be able to be operated at significantly higher junction temperatures but are still limited by yet-to-be-proven reliability and packaging limitations that prevent general acceptance to operating at higher temperatures.

SiC devices in the next several years will continue deliver +175°C rated parts, but will continue to have to prove reliability, and GaN device will remain at 150C for the near time and will lag SiC devices by approximately two to three years.

C. Reliability

Reliability, or the requirement to prove a technology is reliable, has been one of the limiting factors of wide bandgap materials. A summary review of state-of-the-art published reliability data will be summarized, and an overview of the activities of the JEDEC's JC-70 committee will be reviewed, revealing the emerging trends in specifying and proving the reliability of both SiC and GaN Devices.

D. Figure of Merit and Rsp improvements

Performance is generally summarized by the Figure of Merit rating of a technology, while the cost/performance ratio is often compared by studying a technology's specific on resistance, or *Rsp*.

GaN and SiC commercial devices will be summarized with relative performance over the past ten years, with a look to how many more generations are likely to emerge in the next few years.

E. Commercial adoption considerations

The following technology considerations will be summarized and predicted, showing the expected growth of the SiC and GaN technologies.

- Production Capacity and wafer sizes
- Material Availability and manufacturability
- Second Sourcing capabilities
- Market pricing
- Integrated circuit functions

V. CONCLUSION

A wide overview of semiconductor considerations, system architectures, and commercial considerations has been presented to show the complex and ever-changing view of engineers and CEOs alike when choosing a semiconductor technology for their systems. A collection of mainstream high-volume systems has been reviewed with a particular scope of choosing Silicon, SiC or GaN technologies. It has been shown that many factors, both commercial and technical, govern the choice of the perfect device, and the rapid adoption of GaN and SiC devices, the choice is made even harder.

REFERENCES

[1] "Market Forecsasts for SiC and GaN Power semiconductors," IHS Markit, 2018

Proceedings of the 30th International Symposium on Power Semiconductor Devices & ICs
May 13-17, 2018, Chicago, USA

Si Wafer Technology for Power Devices

A Review and Future Directions

Norihisa Machida

Customer Product Engineering Department, Technology Division
SUMCO Corporation
Kouhoku, Saga, Japan
machida@sumcosi.com

Abstract—Silicon wafers have been widely used in semiconductor devices for years. Their characteristics have been improved by untiring development efforts to meet power device manufacturers' requirements such as lowering substrate resistivity for Power MOSFET and reducing resistivity variation for IGBT. As future directions, by utilizing advantages of silicon wafers, adoption of MCZ grown bulk silicon wafers for low and middle voltage IGBT and introduction of 300mm size silicon wafers will proceed.

Keywords— silicon wafer, Power MOSFET, IGBT, MCZ, 300mm

I. INTRODUCTION

Silicon wafers have been widely used for semiconductor devices including memory, logic, image sensors, and power devices for a long time. In 2017, worldwide annual consumption of silicon wafers reached 11,810 million square inches in accumulated area [1] or equivalently more than 235 million pieces of 200mm size wafers. SUMCO is a leading silicon wafer manufacturer which has been supplying wide range products to major device manufacturers based on advanced R&D activities [2].

For today's power devices, various silicon wafers are used on the basis of required breakdown voltage for specific power devices. For Power MOSFET devices, epitaxial wafers are mainly used to secure uniform quality active layer. For high breakdown voltage required IGBT, bulk silicon wafers by Floating zone (FZ) crystal growing method are used to secure defect-free active layer. And for lower breakdown voltage IGBT, bulk silicon wafers by Magnetic field applied Czochralski (MCZ) crystal growing method are recently adopted by various device manufactures for stable, large volume material procurement (Fig. 1). In addition, MCZ bulk silicon wafers can provide the device manufacturers with an option to enlarge the wafer size to 300mm. Technical issues of silicon wafers for power devices are basically improvement of switching characteristics: reducing on-resistance for Power MOSFET and saturation voltage between collector and emitter (Vce(sat)) for IGBT.

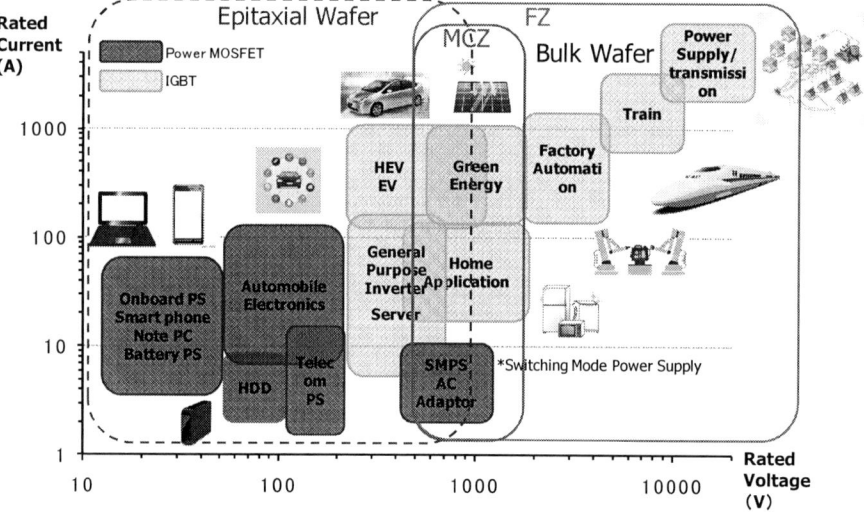

Fig. 1. Silicon wafers used for power devices by rated voltage and rated current

II. POWER MOSFET

For Power MOSFET, specific on-resistance Rsp could be divided to three components: channel on-resistance (Rch), epitaxial layer on-resistance (Repi), and substrate on-resistance (Rsub) (1).

$$Rsp = Rch + Repi + Rsub \qquad (1)$$

Substrate resistivity, Rsub has higher contribution ratio to reducing the specific on-resistance in case of thinner epitaxial layer low voltage Power MOSFET [3].

To lower resistivity of negative conductivity type bulk silicon wafers, arsenic has been used as a dopant until facing a physical limitation of stable single crystal growth around 1.6 mΩ cm level due to the cellular growth. In case of heavy doping, crystal growth would face a constitutional supercooling that could induce the cellular growth and consequent structure loss of single crystal. The constitutional supercooling condition can be described by (2)[4].

$$G_L/V \leqq mC_0/D \cdot (1-k_0)/k_0 \qquad (2)$$

where G_L is temperature gradient of silicon melt; V is silicon crystal growth rate; m is depression of freezing point; C_0 is initial dopant concentration; D is diffusion coefficient of dopant in the silicon melt; and, k_0 is segregation coefficient of dopant.

Instead of arsenic ($k_0 = 0.3$), phosphorous doping which has larger segregation coefficient ($k_0 = 0.35$) has been utilized to grow even lower resistivity bulk silicon crystals for low voltage Power MOSFET devices (Fig. 2). Today, by optimizing the factors described in (2), phosphorous doped silicon wafers with lower than 0.7 mΩ cm are available in market. Further lower resistivity bulk silicon wafers are under development.

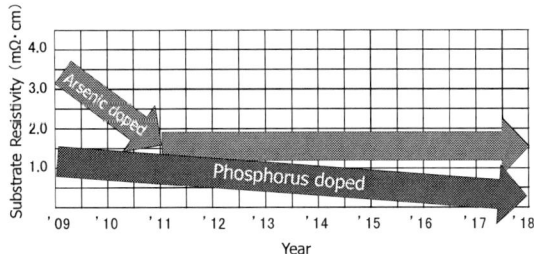

Fig. 2. Trend of substrate resistivity for low voltage Power MOSFET

III. IGBT

Since IGBT modules used for industry, automotive, and electric train consist of a set of IGBT chips, individual IGBT chip's characteristics such as Vce(sat) should be controlled within desired variation range. The resistivity variation that affects Vce(sat) needs to be controlled in a single wafer and among silicon wafers. The neutron transmutation doping (NTD) has been used to obtain controlled variation of resistivity. Meanwhile, long leading time and unstable nuclear facility availability promoted resistivity variation improvement with so-called normal doping techniques for bulk silicon wafers. The capabilities of resistivity variation using gas doping for FZ silicon wafers and phosphorous doping for MCZ silicon wafers are getting to the same or better level comparing with NTD silicon wafers today (Fig. 3).

Although MCZ bulk silicon wafers certainly contain some level of oxygen and crystal originated defects, collaborating efforts with device manufacturers and process tuning enabled adoption of MCZ bulk silicon wafers for low and middle breakdown voltage IGBT today.

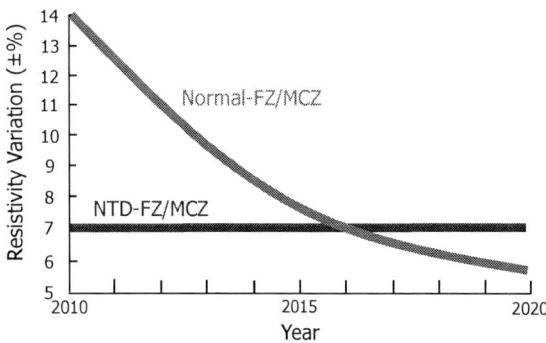

Fig. 3. Trends of resistivity variation of FZ and MCZ silicon wafers by doping method

IV. FUTURE DIRECTIONS

As future directions, silicon wafer advantages of higher productivity and lower cost for high volume market segments will be moreover utilized. Specifically, further adoption of MCZ bulk silicon wafers and 300mm wafers will continue and expand for years. Fig. 4 shows our image of silicon wafers transition for IGBT in near future. More MCZ bulk silicon wafers including 300mm size will support the vigorously growing market. Likewise, there could be a good chance to adopt more 300mm MCZ substrate for Power MOSFET devices in future.

978-1-5386-2928-4/18 $31.00 © 2018 IEEE

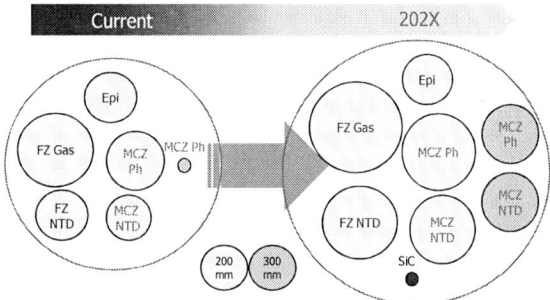

Fig. 4. An image of silicon wafers transition for IGBT

REFERENCES

[1] SEMI press release of February 6, 2018,
http://www.semi.org/jp/jp/node/81221

[2] Hisashi Furuya, Silicon Wafers for Advanced Devices,The 7th International Symposium on Advanced Science and Technology of Silicon Materials, Nov. 21-25 , 2016

[3] International Rectifier Application Note AN-1084, Power MOSFET Basics, Vrej Barkhordarian,
https://www.infineon.com/dgdl/mosfet.pdf?fileId=5546d462533600a4015357444e913f4f

[4] W.A.Tiller, K.A.Jackson, J.W.Rutter, B.Chalmers, Acta Metall. 1, 428 (1953)

Proceedings of the 30th International Symposium on Power Semiconductor Devices & ICs
May 13-17, 2018, Chicago, USA

The Future of Power Semiconductors:
an EU Perspecitve

Bert De Colvenaer
Executive Director
ECSEL JU
Brussels, Belgium
ecsel-office@ecsel.europa.eu

Abstract - With the integration of more renewable energy sources (RES) in the European landscape, with more stringent demands on supply and with cost conscious customers, and also environmental conscious, Europe calls for a smarter energy landscape where power semiconductors will play a major role in the years to come.

Keywords : Policy and technology, ECSEL JU, Europe

I. INTRODUCTION

In the last years, it has become apparent that semiconductor-based innovative technologies have enabled more savings of electrical energy than the growth of demand has been in the same period. The core of the European competitive advantage is within the system knowledge and provision of holistic system solutions. Saving energy is equivalent to reducing the costs and being more competitive. Energy efficiency levels in IEA member countries improved, on average, by 14% between 2000 and 2015. This generated energy savings of 19 exajoules (EJ) or 450 million tonnes of oil equivalent (Mtoe) in 2015. These savings also reduced total energy expenditure by 540 billion USD in 2015, mostly in buildings and industry. While GDP grew by 2% in IEA countries, the efficiency gains led to flattening of the growth in the primary energy demand. In parallel, the global CO2 emissions stalled since 2013 with only 2% growth, in 2014 with 1,1% and in 2015 with -0,1%.

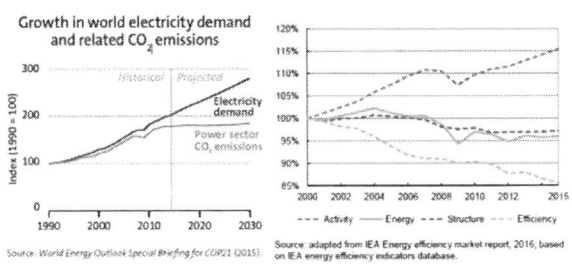

Figure 1 – Growth in world electricity demand and related CO2 emissions

Energy saving is also an opportunity. In fact, by reducing power dissipation and corresponding heat production, energy is available for other uses and equipment.

According to IEA, the analysis of factors driving energy consumption trends for IEA member countries indicates that in IEA the decoupling was mainly due to efficiency improvements (figure right above). Structural changes (mostly shift to less intensive industries and services) also assisted efficiency improvements in reducing the total energy consumption. Cumulative savings over the period 2000 – 2015 were 159 EJ, equivalent to more than one year of final energy consumption in Europe, China and India altogether.

Examples of the most important Electronic Components and Systems (ECS) applications having high impact on the efficient use or generation of energy are power management products with a steady growing market 20 B US$, see Figure 2.

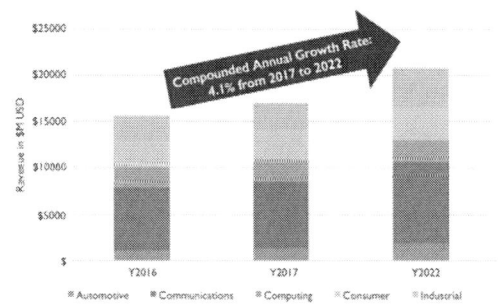

Figure 2 – Annual power management IC market revenue.

Another example of ECS market contributing to the efficient use of energy is the wireless infrastructure RF power device market, with around US$1 billion Total Addressable Market (TAM). The share of GaN based devices increases from 10% in 2015 to expected 40% in 2022 (source ABI research, 2017) which demonstrates how fast new techniques can be deployed

978-1-5386-2928-4/18 $31.00 © 2018 IEEE

if the business added value is achieved. Driven by new developments, such as the electro mobility and Industry 4.0, new energy supply chains and consumption patterns come up. Powering the electro mobility is a major challenge in the coming years with the implementation of a reliable and sustainable charging infrastructure.

European ECS companies are amongst the leaders in smart energy related markets, which is largely driven by political decisions as well as by the move to renewing energies and to added costs on carbon dioxide emissions. Leading market positions are achieved for electrical drives, grid technology and decentralized renewable energy sources. This position will be strengthened and further employment secured by innovative research on European level. Competitive advantages can be gained by research in the following areas:

- significant reduction and recovery of losses (application and SoA related);
- power density increase and decreased size of the systems by miniaturization and integration, on system and power electronics level;
- increased functionality, reliability and lifetime (incl. sensors & actuators, ECS HW/SW, monitoring systems,...);
- manufacturing and supply of energy relevant components, modules and systems;
- the game change to renewable energy sources and decentralized networks, including intermediate storage;
- energy supply infrastructure for e-mobility, digital live and industry 4.0;
- "plug and play integration" of ECS into self-organized grids and multi-modal systems;
- safety and security issues of self-organized grids and multi-modal systems;
- optimization of applications and exploitation of achieved technology advances in all areas where electrical energy is consumed;
- ECS for storage solutions.

The ECS for energy (incl. components, modules, CPS, service solutions), which support the EU and national energy targets, will have a huge impact on the job generation and education if based on the complete supply chain and fully developed in Europe. The key will be the capability to maintain complete systems understanding and competence for small-scale solutions up to balanced energy supply solutions for regions. It is mandatory to have plug- and-play components that are enabled by broad research contributions from SMEs and service providers including EU champions in the energy domain. Thanks to the expected wider proliferation of energy storage devices in the smart city context, new distributed forms of energy storages will become available, to be exploited by smart control systems.

Societal benefits include access to knowledge, development of modern lifestyle and availability of energy all the time everywhere – with a minimum of wasted energy and a minimum of greenhouse gas emissions. Applications having a huge energy demand and therefore a large saving potential are in the areas of High Performance Data centers serving the mobile connected world, the implementation of Smart Cities and the future implementation of e-mobility with widely distributed charging stations, demanding a higher density of energy distribution points with – as a key point - local intermediate storage systems.

Smart grid delivers smart energy from suppliers to consumers using digital technology to control appliances at consumer's homes to save energy, reduce cost and increase reliability and transparency.

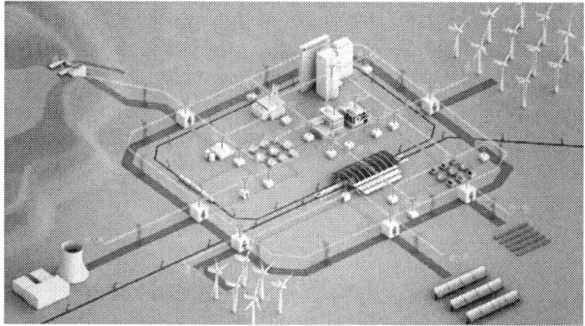

Figure 3 - Smart Energy landscape – from centralized to distributed (PV, wind, biogas, ...) generation and conversion, consisting of High/Medium Voltage grid (orange), Low Voltage grid (yellow) including Communication Network (aquamarine) linking producers and consumers down to regional and community level (source ECSEL JU MASP[1])

Figure 4 - The Prosumer concept. The Power transmission lines turn from one-way to two-way ones.

---

[1] https://www.ecsel.eu/sites/default/files/2017-12/ECSEL%20GB%202017.94%20-MASP%202018_0.pdf

II. ENERGY IN AND FOR EUROPE

EU energy policies aim to ensure that European citizens can access secure, affordable and sustainable energy supplies. The EU is working in a number of areas to make this happen: the Energy Union[2] strategy is focused on boosting energy security, creating a fully integrated internal energy market, improving energy efficiency, decarbonizing the economy (not least by using more renewable energy), and supporting research, innovation and competitiveness, the Energy Security Strategy presents short and long-term measures to shore up the EU's security of energy supply, EU funding and other support is helping to build a modern, interconnected energy grid across Europe, the 'Clean Energy for All Europeans' package, published in November 2016, has three main objectives: putting energy efficiency first, achieving global leadership in renewable energies, and providing a fair deal for consumers, Safety across the EU's energy sectors, with strict rules on issues such as the disposal of nuclear waste and the operation of offshore oil and gas platforms.

As part of its long-term energy strategy, the EU has set targets for 2020 and 2030. These cover emissions reduction, improved energy efficiency, and an increased share of renewables in the EU's energy mix. It has also created an Energy Roadmap for 2050, in order to achieve its goal of reducing greenhouse gas emissions by 80-95%, when compared to 1990 levels, by 2050. Together, these goals provide the EU with a stable policy framework on greenhouse gas emissions, renewables and energy efficiency, which gives investors more certainty and confirms the EU's lead in these fields on a global scale.

The Energy Union's 5 dimensions:

- security, solidarity and trust: diversifying Europe's sources of energy and ensuring energy security through solidarity and cooperation between EU countries,
- a fully integrated internal energy market: enabling the free flow of energy through the EU through adequate infrastructure and without technical or regulatory barriers,
- energy efficiency: improved energy efficiency will reduce dependence on energy imports, lower emissions, and drive jobs and growth
- decarbonizing the economy: the EU is committed to a quick ratification of the Paris Agreement[3] and to retaining its leadership in the area of renewable energy

- research, innovation and competitiveness: supporting breakthroughs in low-carbon and clean energy technologies by prioritizing research and innovation to drive the energy transition and improve competitiveness.

III. TECHNOLOGY OPTIONS IN AND FOR EUROPE

The European Strategic Energy Technology Plan (SET-Plan)[4][5] aims to accelerate the development and deployment of low-carbon technologies. It seeks to improve new technologies and bring down costs by coordinating national research efforts and helping to finance projects. The SET-Plan promotes research and innovation efforts across Europe by supporting the most impactful technologies in the EU's transformation to a low-carbon energy system. It promotes cooperation amongst EU countries, companies, research institutions, and the EU itself. Research, innovation and competitiveness are one of the five dimensions of the Commission's Energy Union strategy.

The integrated SET-Plan is part of a new European energy Research & Innovation (R&I) approach designed to accelerate the transformation of the EU's energy system and to bring promising new zero-emissions energy technologies to market.

- Identifies 10 actions for research and innovation, based on an assessment of the energy system's needs and on their importance for the energy system transformation and their potential to create growth and jobs in the EU,

- Addresses the whole innovation chain, from research to market uptake, and tackles both financing and the regulatory framework,

- Adapts the governance structures under the umbrella of the SET-Plan to ensure a more effective interaction with EU countries and stakeholders,

- Proposes to measure progress via overall Key Performance Indicators (KPIs), such as the level of investment in research and innovation, or cost reductions

The European Technology and Innovation Platforms[6] (ETIPs) were created to support the implementation of the SET Plan by bringing together EU countries, industry, and researchers in key areas. They promote the market uptake of key energy technologies by pooling funding, skills, and research facilities.

---

[2] https://ec.europa.eu/commission/priorities/energy-union-and-climate_en
[3] https://ec.europa.eu/clima/policies/international/negotiations/paris_en
[4] https://publications.europa.eu/portal2012-portlet/html/downloadHandler.jsp?identifier=771918e8-d3ee-11e7-a5b9-

01aa75ed71a1&format=pdf&language=en&productionSystem=cellar&part=
[5] https://setis.ec.europa.eu/sites/default/files/setis%20reports/set-plan_brochure.pdf
[6] https://ec.europa.eu/research/innovation-union/index.cfm?pg=etp

978-1-5386-2928-4/18 $31.00 © 2018 IEEE

The European Energy Research Alliance[7] (EERA) aims to accelerate new energy technology development by cooperation on pan-European programs. It brings together more than 175 research organizations from 27 countries, involved in 17 joint programs. It plays an important role in promoting coordination among energy researchers along the SET Plan objectives and in the technology transfer to the industry.

IV.   The ECSEL JU: Work in Progress

On July 10th 2013, the European Commission issued their proposal for the "Innovation Investment Package" to the budgetary authority. This included the establishment of Joint Undertakings (JU's), implementing Joint Technology Initiatives, among which the JU on "Electronic Components and Systems for European Leadership ECSEL". The Council adopted the Regulation on May 6th 2014 and published it in the Official Journal on June 7th 2014. The Regulation entered into force twenty days later, on June 27th 2014, and ECSEL JU came into being, as the merger of two pre-existing JU's, ENIAC and ARTEMIS, also taking up activities of the European Technology Platform on Smart Systems Integration "EPoSS".

The ECSEL JU, as the acronym indicates, addresses Electronics Components and Systems: a capability of essential importance for each citizen, company and nation in the world. Information and communication technology and its applications all run on this fabric: no industrial product, no entertainment, no transport system is conceivable today without them. Already, the physical and economic well-being of every citizen and society is supported by electronics applications, from healthcare and personal safety to entertainment and safer transport. They are also the main drivers for innovation which, in itself, is the foundation for job creation and economic growth. The trend will become stronger in the future, creating increasingly interconnected devices with unprecedented capabilities, enabling the Internet of Things and the incorporation of ICT in all industrial branches: the very essence of the contemporary industrial revolution described as "Industry 4.0".

ECSEL JU – a Public-Private Partnership – is targeted at re-establishing European leadership in an area of systemic importance for the European economy, and of strategic importance for Europe's security and long-term well-being. It is organized as a Joint Undertaking established on the basis of Article 187 of the Treaty on the Functioning of the European Union (TFEU). ECSEL JU supports this goal by supporting collaborative, industrially- relevant Research, Development and Innovation projects financed by the participating industrial and academic partners, and also by the EU (through the "Horizon2020" program of the European Commission) and by the National/Regional funding authorities of the ECSEL Participating States – a so-called "Tri-partite funding model".
The ECSEL Members are the Union (represented by the European Commission) and the ECSEL Participating States

(EPS) collectively forming the Public Authorities Board (PAB), and the Private Members (the industrial associations AENEAS, ARTEMISIA and EPoSS) grouped to form the Private Members Board (PMB). At of today, the ECSEL JU has the following EPS as members: Austria, Belgium, Bulgaria, Czech Republic, Denmark, Estonia, Finland, France, Germany, Greece, Hungary, Ireland, Israel, Italy, Latvia, Malta, Netherlands, Norway, Poland, Portugal, Romania, Slovakia, Spain, Sweden, Switzerland, Turkey and the United Kingdom.

ECSEL JU's highest governing authority is the Governing Board, comprising representations from the Commission, the EPS and the three industry associations ("industry", including SME's and research institutes). The Governing Board is responsible for all operational aspects of the Joint Undertaking, in particular for the strategic RD&I work to be done (via the Multi-Annual Strategic Plan) and the annual implementation of that plan (Work Plan). Funding decisions of projects selected from Calls is solely the responsibility of the Public Authorities Board, comprising the Commission and the EPS.

As defined in the MASP the identified major challenges to be addressed in the activities of the ECSEL JU are described below. They are to take in account the SWOT analysis as described in the MASP document with following main characteristics : As Europe's strength, one can indicate the leading position : four European based power semiconductor suppliers amongst the top 12 having together a market share of over 24% in 2014. Three power modules suppliers in the top ten with a market share of over 33%. The overall share of European suppliers is increasing in this growing market underlining their competitiveness. To the contrary, weak points are the ability to follow the very fast changing environment and the speed of introduction of regulations: "100 years old" established infrastructure to be converted into a highly flexible and dynamic energy supply infrastructure. There are still many opportunities: the high energy conversion efficiencies (93% - 99% or more) allowing better use of renewable energy resources, further exploiting new materials, new device architectures, innovative new circuit topologies, architectures and algorithms lowering the total energy system cost. And as new business opportunities there are: the new infrastructure for EV charging, implementation of the energy highway through Europe, emission free (or more electric) cities, decentralized smart storage, distributed DC network & grid technology and of course efficient management of data. And as threats or rather main technology challenges : the availability of renewable energies in sufficient and appropriate amounts, the oversupply and peak supply of renewable/variable energy sources, the availability of batteries/energy storage, the complexity of the current distribution grid and missing acceptance of and investments in new HV and DC grid connection and of course the fragmented legislations.

---

[7] https://www.eera-set.eu

978-1-5386-2928-4/18 $31.00 © 2018 IEEE

A. A first major challenge is to ensure sustainable power generation and energy conversion in order to have a loss free energy conversion and generation with approx. 99% efficiency by 2020.

The topic of energy generation can be divided historically into two main fields: (1) traditional energy generation (e.g. fossil or nuclear power plants) and (2) energy generation based on renewable sources (e.g. wind, solar, hydropower, geo-thermal). In both cases, "raw energy" is produced in a form, which cannot be transmitted or used without conversion. A new upcoming application in the field of EV is the need of new batteries for energy storage to manage over-capacities and undersupply. Examples are non-continuous energy sources like wind-mills and solar cells. Using old-fashioned electronics for rectifying, transforming or converting (AC/DC or DC/AC) the currents, only about half of the energy can be used. New, much more dedicated and efficient components have to be used, which partially will be based on new materials. In general, everything must be done to reduce the lifetime capital and operational expenses (CAPEX and OPEX) of renewable energy generation below those of the traditional energy generation.

The need for energy is a fact in the modern society. The question is how to provide the energy in a resource efficient way and at a cost accepted by the society. Nano-electronics is playing an important role in the generation of renewable energies. Highly efficient conversion leads to fewer investments and therefore lower cost for the renewable energies. CAPEX and OPEX reduction per generated power unit is the only way to compete with traditional energy sources. In terms of power semiconductors, which are the fuel for energy efficient systems, Europe has a leading position with four European based suppliers amongst the top 12 having together a market share of over 24% in 2014 for power semiconductors and three in the top ten with a market share of over 33% for power modules. Overall, the share of European suppliers is increasing in this growing market underlining their competitiveness.

The high priority R&D&I areas are:

- Affordable energy conversion efficiencies of 93% to 99% or more allowing better use of renewable energy resources, exploiting new materials, new devices architectures, innovative new circuit topologies, architectures and algorithms lowering the total system cost;
- Enhanced device and system lifetime and reliability with effective thermal management ensuring life expectancy for renewable energy systems to be 20 to 30 years;
- Developing semiconductors-based solar energy technologies including photovoltaic technologies and integrating them with solid-state lighting applications;
- Reduced physical size and weight of individual transformer stations with equivalent power ratings by

the development of solid-state transformers. These actuators will provide new functions for the operation of power systems and avoid infrastructure extensions caused by increasing share of distributed generation;
- Innovative devices exploiting new materials to dramatically increase their power density capabilities to be used in efficient converters, supported by passive elements, new interconnect technologies and packaging techniques to achieve further miniaturization and further reduce losses;
- New nano-materials, devices and systems for improving energy efficiency of the growing worldwide renewable energy technologies, such as photovoltaic, wind and water;
- System EMI research to cope with higher switching frequencies and further miniaturization;
- System reliability enhancement with focus on thermo-mechanical and thermo-electro-mechanical reliability;
- Resilient control strategies, and self-healing systems technologies that enable better use of renewable energy sources, their real-time monitoring, performance prediction, proactive coordination and integration with smart urban systems;
- Smart sensor networks able to measure all internal and external physical parameters that influence energy conversion efficiency and thus help to enable an efficient smart energy landscape. This also includes sensors that support intelligent predictive maintenance concepts resulting in reduced maintenance costs and increased lifetime for equipment and infrastructure;
- Self-powering systems for small IoT nodes have to be developed. The target is that local energy harvesting will substitute battery powered devices and eliminate the high demand of energy for the battery manufacturing and distribution logistics.

It can be expected that new highly efficient technologies (e.g. wide band gap materials, disruptive innovations based on new processing approaches and architectures) are introduced and new competitive solutions lead to a further growth of market share in the supply of power semiconductors. On the system level, it is expected that European suppliers are established in the field of resilient control strategies that enable better use of renewable energy sources, their real-time monitoring, performance prediction, proactive coordination and integration with smart urban systems. For the energy supply of the IoT nodes, harvesters and intermediate storages have to be developed to substitute and minimize batteries.

B. As a second major challenge, the MASP defines the achieving of efficient community energy management.

The decentralization of energy sources, the opportunities with networked systems, the limitations in peak electricity supply, new demand for electric energy supply for the urban mobility and the introduction of storage systems will lead to new challenges in energy management and distribution for communities and cities. To illustrate the change and challenges in the distribution of energy a PV and wind energy example is given: Over the last 6 years, electricity demand in the UCTE countries grids have slowly decreased, from 2 600 to 2 500 TWh. In the same time period, wind and solar PV production increased by 79% and 338% respectively, reaching 226 TWh and 94 TWh in 2015. This development has led to variable renewable energy (VRE) accounting for 12.8% of total electricity production in 2015. The share of VRE for 2015 and projected for 2021 is shown in the following figure of selected UCTE countries.

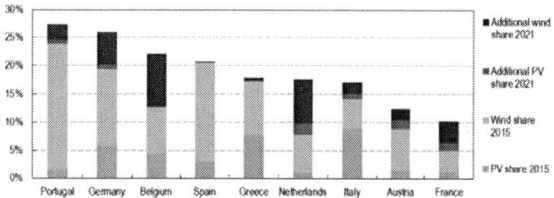

Figure 5 - Share of VRE generation in 2015 and 2021 for selected UCTE countries Source: Adapted from IEA (2016a), Medium Term Renewable Energy Market Report

The scope and ambition is therefore, through the technologies supported, to reach the highest efficiency and most economic energy supply and management solutions for the communities and smart cities, including the distribution of energy to them. Advanced control and monitoring systems are already deployed at the transmission network level (high DC voltage). Broad inclusion of small and medium size renewable energy sources into the grid and their coordination requires adoption of control and monitoring systems at the medium voltage levels as well. In the medium voltage grids where small and medium size energy sources represent a significant part of installed energy production potential, real-time monitoring of energy flows is needed to enable demand/response management (DRM).

The following high priority R&D&I areas have been identified:
- Smart Grid applications that exploit demand/response technology in a robust and secure way, negotiating the trade-off between different levels of urgency in energy need with a varying price of that energy at any given time and accommodating variable renewable electricity;
- Self-organizing grids and multi-modal energy systems;
- Improved grid visibility through advanced grid monitoring, including medium and low voltage levels;

- A highly resilient power grid through the introduction of proactive control algorithms (that go beyond demand/response), significantly improving the grid's self-healing and self-protection capabilities;
- Full implementation of Smart Grid technologies, resulting in the massive deployment of the necessary control options for the complete realization of the Agile Fractal Grid also including smart agriculture (e.g. greenhouse energy efficiency);
- Smart E-Mobility grid for optimized charging, storage and distribution of electric power for light, medium and heavy vehicle transportation;
- Technological solutions for efficient and smart buildings (indoor) and outdoor subsystems including heating, ventilation, air conditioning and lighting, as well as traffic access, to achieve optimal energy-efficient performance, connectivity and adaptive intelligent management while ensuring scalability and security;
- Fog / cloud computing to offer sufficient and cost-effective processing power and to ease maintenance and update of control software - edge computing to support low latency applications, such as real-time grid control.

Through all these R&D&I activities, following achievements are expected: The medium voltage level management (DRM) helps to adjust consumption to the production (presently the production is adjusted to match the consumption), promotes dynamic pricing tariffs that are needed to increase market share of small energy producers and at the same time enables the reduction of energy losses by better matching of production and consumption. Improved energy management at the MV level enables risk-free integration of additional renewable energy sources into the grid without any negative impacts on grid stability of the MV an LV micro-grids. Real-time monitoring at the MV enables the deployment of self-healing MV grid strategies. The impact of electrification will help e-mobility trough a migration versus demanding technologies: Trend towards decentralization of energy & energy storage, new applications for energy distribution: Mandatory to sell EV and fast charging >20kW: High Electronic content / Moving to SiC Based Modules.

C. The third major challenge is to reduce the energy consumption by 2030 to achieve the current EU policy targeted of 30% savings utilizing innovative nano-electronics based solutions.

Therefore three prominent and fast growing areas are addressed: (1) the reduction of power consumption by the electronic components and systems themselves, (2) the systems built upon them and (3) the application level in several areas.

Some examples in the field of electronic components are the following: One of the most challenging issues in High-

978-1-5386-2928-4/18 $31.00 © 2018 IEEE

Performance Computing is energy consumption. It is a well-known fact that the energy consumption of HPC data centers will grow by a significant factor in the next four to five years. Hence the costs of associated cooling infrastructures (with 50%-70% of the overall power dedicated to the cooling task of the current generation data centers) already exceed the costs of the HPC systems themselves. Therefore reduction of energy consumption is becoming mandatory. Otherwise the consumption of exascale systems will reach up to the 100 mega-watt ranges; the demand for mobile electronic equipment: the scaling is tremendous since billions of mobile electronic devices are deployed and connected to the grid each year. Even low percentage improvements have a high impact on energy consumption; The demands for communication networks: increasing data volumes (1000 fold increase in mobile data volume), always-on availability, instant messaging – they all demand a permanently active infrastructure avoiding any inefficient operation. In order to avoid explosion of energy consumption of the communication networks, energy per transmitted data unit needs to be cut drastically. In the 5G development, the target is set to limit the energy per transmitted bit to 1/10th of today's level. To reach the target several measures needs be applied, e.g., electronic beam-forming techniques, efficient communication algorithms and highest efficiency components. Regarding system configurations: The energy efficiency of the system is achieved by using sensors, actuators, drives, controls and innovative components where the loss of energy can be reduced by innovative or even destructive approaches. The ambition is to reach a wider implementation of adaptive and controlled systems to meet the needs through monitoring and the ability to reduce energy losses. For example, intelligent building management systems can guarantee minimal energy use for heating and lighting (also providing safety and security). And at application level: Under the EU Ecodesign Directive, the European Commission sets Minimum Energy performance Standards (MEPS) for 23 categories of products sold in Europe. The Commission is currently considering revising or developing standards for the following product groups: air heating products, cooling products and process chillers, enterprise servers and data storage products, machine tools and welding equipment, smart appliances, taps and showers, lighting products, household refrigeration, household washing machines and dishwashers, computers, standby power consumption, water heaters, pumps and vacuum cleaners. Further, under the Energy Performance of Building Directive, there is a continuous tightening of national minimum energy performance requirements in line with the cost-optimal methodology. The growing number of computing components within the hardware architecture of both HPC and embedded systems requires larger efforts for the parallelization of algorithms. In fact, optimization of parallel applications are still far behind the possibilities offered by today's HPC hardware, resulting in sub-optimal exploitation of system and hence a significant waste of energy consumption.

Having the whole value chain represented and being in world-wide leading positions, Europe has a rather good chance to build up a healthy "green industry" around tools and goods to reduce energy consumption. European companies have acknowledged strengths in power electronics and in nearly all of its applications. Market studies show leading positions of Europe in the field of power electronics and advanced LED lighting and even dominance in power semiconductor modules for renewable energies. Activities inspired, founded and led by European stakeholders such as the GreenTouch® initiative or a number of ETSI and ITU standardization initiatives and focus groups exert worldwide influence. By resorting to latest micro-/nano-electronic technologies and most advanced system concepts, European companies defined and set new standards and raised the bars in performance and energy efficiency. Also the related R&D is very active in all of those domains.

Therefore the high priority R&D&I areas can be indicated as follows:

- Intelligent drive control: technology, components and miniaturized (sub) systems, new system architectures and circuit designs, innovative module, interconnect and assembly techniques addressing the challenges at system, sub-system and device level for efficiently controlled engines and electrical actuation in industrial applications;
- Technologies and control systems to improve energy performance of lighting system;
- Highly efficient and controlled power trains for e-mobility and transportation;
- Efficient (in-situ) power supplies and power management solutions supported by efficient voltage conversion and ultra-low power standby, based on new system architectures, innovative circuit and packaging concepts, specific power components for lighting and industrial equipment serving portable computers and mobile phones, and standby switches for TVs, recorders and computers. Power management solutions in industrial, municipal and private facilities;
- Low weight/low power electronics, with advanced thermal management solutions, based on novel materials and innovative devices particularly benefiting, among other areas, medical applications, where improved energy management is one of the keys to cost-effective solutions (for example, medical imaging equipment);
- Immediate issue to be solved on the way towards exascale computing is power consumption: The root cause of this impending crisis is that the needs for ever increasing performances require larger amount of devices (and associated memory) while the chip power efficiency is no longer improving at previous rates. Therefore, improvements in system architecture (e.g. clock switching, etc.) and computing technologies (i.e. usage of low-power processors and accelerators like GPU, FPGA, etc.) are mandatory to progress further;
- Related issue of heat dissipation in computing system requires sophisticated air or liquid cooling units (e.g.

chilled water doors, refrigerated racks, heat exchangers, etc.) further adding to the costs;

- Together with computing technologies (CPU, GPU, DSP, etc.) interconnect technologies add their own energy consumption, thus requiring further efforts to optimize routing strategies and switching policies in order to minimize the traffic. Usage of 3D nano-electronics based integrated devices and photonics can be envisioned for such improvement;
- Energy efficient sensor networks, including hardware and software application layers;
- Optimal parallelization of traditional sequential algorithms and efficient mapping on parallel and heterogeneous architectures will not only provide necessary performance but help to reduce energy consumption;
- Energy efficient communication networks with highest efficient ECS, beam forming and embedded algorithms;
- Efficient adaptive power management for 5G wireless network.

The expected achievements are directly linked to the R&D&I priorities. Worth of highlighting is that in several applications a huge price pressure, neglecting the benefits via reduced operational costs over the lifetime, asks for severe achievements in cost reduction of technologies. The achievement of exascale high-performance computing capability by 2020 requires a reduction by at least a factor of 5 of the current consumption in order to stay in the domain of technical and economic feasibility.

Following is a list of potential implementations to support the objective of energy consumption reduction: added increased share of intelligent drive control, electrical actuators for robotics, enterprise servers and data storage products, lighting products, household refrigeration, washing machines and dishwashers, computers, standby power consumption overall, water heaters, pumps and vacuum cleaners. Further potential is seen in highly efficient Industry 4.0 improvements based on sensor data and new control for actuators.

V. RELATED ECSEL JU PROJECTS

Many of the above indicated challenges are addressed in the portfolio ECSEL JU projects. In what follows is a short summary of the project main characteristics. More details can be found in the related website.

The EPT300[8] project, running from Apr 2012 to Sep 2015, was based on the concept of 300mm wafers in a 1:1 transfer approach from 200mm to fully prove compliance with application requirements. So the old device in 200mm should

have the same performance of the transferred device in 300mm. The challenges in EPT300 were related to the processes, equipment, automation and handling, substrates that are needed for such a "challenging" transfer to a higher area of wafers. It set up the first pilot line and high volume power 300mm production with the first demonstrators in CoolMOSTM, IGBT and SFET technologies. In Oct 2011 Infineon Technologies AG has produced the first chips ("first silicon") on a 300-millimeter thin wafer for power semiconductors at the Villach site in Austria, making Infineon the first company in the world to succeed in taking this step forward. It introduced unique 300mm diameter substrates thinner than paper, processing equipment, handling and automation concepts creating world's most efficient and most affordable devices.

Second generation power semiconductor devices fabricated in European leading 300 mm pilot lines were at the heart of the EPPL project[9], running from Apr 2013 to Sep 2016, which main goal was to extend the leading position of power semiconductors "Made in Europe" and for which manufacturing excellence, cost competitiveness and challenging applications were critical boundary conditions. With this, to leverage the technical characteristics of power devices and foster the trend towards system-in-package integration, advances in packaging technologies became of prime importance. Work performed included developing next generation power semiconductors based on 300 mm wafers, setting up the required technologies as pilot line manufacturing, and demonstrating the thus achieved reliable and advantageous solutions for a wide range of application fields. It enabled energy efficiency improvements for applications that did not (fully) benefit from advanced power electronics so far by providing power semiconductors with less ON-resistance, faster switching characteristics and better feature/cost performance. It ensured compatibility of enhanced power semiconductors with new, advanced packaging technologies like 3D integration and set up an advanced power semiconductor manufacturing pilot line across European countries, helping to secure important work places for highly skilled European work force.

One of the most important European Research projects on energy efficiency eRAMP[10], running from Apr 2014 to Mar 2017, focused on the rapid introduction of new production technologies, such as packaging technologies for energy-saving chips. The eRamp project covered the entire power electronics value chain, from generation and transmission all the way to consumption. The project contributed to strengthen Germany and Europe as centers of expertise for power electronics.

---

[8] http://www.ept300.eu
[9] http://www.eppl-project.eu

[10] https://www.ecsel.eu/news/eramp-project-strengthens-europe-center-expertise-power-electronics

R3-POWERUP[11], running as from 07 July 2017, will establish the first 300mm Pilot Line in Europe for Smart Power and discrete power devices featuring 90nm lithography for high-density logic, analogue and power devices and embedded Non Volatile Memories for the realization of complex Systems-on-Chip. As such, it will fill the existing gap in the availability of 300mm Pilot Lines in Europe, which covers only Logic CMOS and discrete power devices. It aims at developing and demonstrating a brand new 300mm advanced manufacturing facility for 90nm Smart Power technology, configured as a multi-KET Pilot Line (i.e. Nanoelectronics, Nanotechnology, Advanced Manufacturing), improving productivity and competitiveness of integrated IC solutions for Smart Power and power discrete technologies by supporting a variety of applications in the Automotive and industrial domains.

The PowerBase project[12], running from May 2015 to Jun 2018, is looking into the future: using innovative silicon and gallium nitride substrates combined with embedded chip assembly technologies to achieve unparalleled efficiency in compact power applications. 39 Partners from nine European countries have joined the project to develop the next-generation energy-saving chips (or power devices) based on materials such as gallium nitride (GaN). They will prepare these semiconductors for mass industrial use in smartphones, laptops, servers and many other applications. The PowerBase research focus includes intensive material and reliability research to improve the quality and lengthen the service life of GaN-based semiconductors. Plans also foresee the establishment of pilot lines for 200mm wafers to manufacture GaN-based power components in a high-volume industrial production environment. The project involves the entire value chain, covering expertise in raw materials research, process innovation, assembly innovation, pilot lines up to various application domains.

SemI40[13] running as from May 2016, covers smart manufacturing of power electronics in Europe. The project responds to the urgent need of increasing the competitiveness of the semiconductor manufacturing industry in Europe through establishing smart, sustainable and integrated Enterprise Collaboration Systems manufacturing. It focusses on four highly challenging aspects of utmost importance: 1) Data Safety and security in manufacturing environment with special attention on legacy equipment; 2) Agility in ECS production for fast adaptability to changes; 3) Tools and methodologies for automated decision making in manufacturing shop floor, based on big data analysis methods; 4) Virtualization and digitalization for advanced simulation in fab environment. By the sustainability measures envisioned SemI40 project contributes to safeguard more than 20.000 jobs of people directly employed in the participating facilities and in total more than 300.000 jobs of people employed at all industry partners facilities worldwide. SemI40 targets to contribute to cost competitiveness in manufacturing "Made in Europe". The activities are essentially supported by SME's, research institutes, and academic partners. Particularly the leading edge research performed by the involved academic partners will further enhance manufacturing science in Europe resulting in a solid base for further I4.0 advancements. The Education of the younger generation and the efficient use of natural resources such as energy and materials are in the focus. Through implementing I4.0 and enabling the connected manufacturing, the Semi40 project ensures high skilled and demanding jobs in manufacturing in Europe and maintains the competitiveness of European industry and research in a truly European cooperative consortium.

VI. CONCLUSION : THE NEED OF THINKING TOGETHER, WORKING TOGETHER, INVESTING TOGETHER.

In order to implement the European ambitions on secure, affordable and sustainable energy supplies, there is a need for even more and advanced power semiconductors, fully and smartly integrated in the whole new energy landscape and its distribution network in order to match supply and demand (cost) efficiently. Europe and ECSEL JU will keep on investing in power semiconductors as key enabling technology for mobile, industry, energy, applications. The conditions for success are on three sides - regulation and standards, technology availability, reliability and seamless integration & acceptance by the users. Only when industry, RTO's and public authorities align priorities, cooperate on implementation and share cost, this will/may happen.

---

[11] https://www.ecsel.eu/projects/r3-powerup
[12] http://www.powerbase-project.eu
[13] http://www.semi40.eu

Proceedings of the 30th International Symposium on Power Semiconductor Devices & ICs
May 13-17, 2018, Chicago, USA

IGBT with superior long-term switching behavior by asymmetric trench oxide

C. Sandow, P. Brandt, H.-P. Felsl, F.-J. Niedernostheide,
F. Pfirsch, H.-J. Schulze, A. Stegner, F. Umbach
Infineon Technologies AG
Am Campeon 1-12, D-85579 Neubiberg, Germany

F. Santos, W. Wagner
Infineon Technologies Austria AG
Siemensstr. 2, A-9500 Villach, Austria

Abstract—**The continued shrinking of IGBT chips calls for new design approaches to ensure reliable and stable switching operation during the chip lifetime. We demonstrate a new asymmetric gate oxide concept with a designed variable thickness that leads to stable long-term operation in trench IGBTs and reduces the switching delay and the gate charge without sacrificing electrical performance. These claims are supported by longer-term repetitive switching experiments as well as TCAD simulations on a calibrated model.**

I. INTRODUCTION

The long-term stability of electrical properties during repetitive switching is an important feature of MOS-switches which is required to guarantee safe operation over the lifetime of such devices. The relevant degradation mechanism is based on the injection of highly energetic (hot) charge carriers into the gate oxide of the MOS-switch [1]. A fraction of the injected hot charge carriers will become trapped in the oxide and act as fixed charge which, depending on its position in the oxide, can effect the device characteristics in different ways. The primary position at which fixed charges are created depends on the design of the device. In conventional low-voltage MOS-Field-Effect-Transistors, usually the largest amount of hot carriers hits the oxide at pinch-off condition in the high-field region near the drain side of the channel; this also holds for most DMOS-type devices. If the fixed charges are located near the MOS channel interface, the threshold voltage of the device will change depending on the polarity of the hot carriers.

In Trench-IGBTs ([2], [3]), charge carriers will gain the highest energy when they are accelerated by the high electric fields which are present in the space charge region of IGBTs during the over-voltage phase present in hard switching turn-off events [4]. In IGBTs, the electric field peak close to the top of the space charge region is maximal because for fast turn-off most of the current is carried by positively charged holes in the space charge region. The positive charge of the holes adds to the charge of ionized donors of n-type IGBTs and leads to a steep electric field slope which results in a large absolute value of the electric field. Consequently, the hot carriers are mainly injected into the oxide at the trench bottom. The effects of charge at the bottom of the trenches can be observed as a change of the switching slope dI/dt during turn-on and a change of the switching delays during turn-on and turn-off.

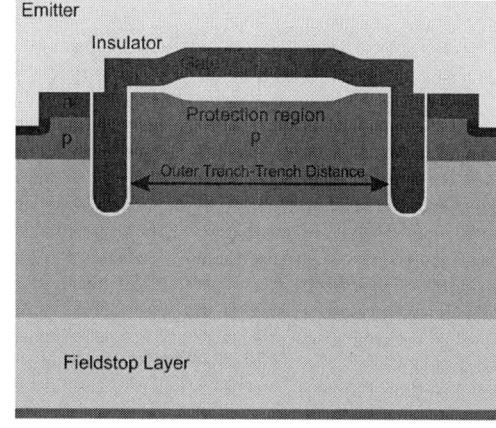

Fig. 1. Conventional IGBT cell with a large trench-trench distance and a p-type protection region to protect the trenches from long-term degradation.

Both changes of the switching characteristics are linked to the shift of the flat-band-voltage at the trench-bottom.

As demonstrated in [4], the long-term stability of trench IGBT devices can be ensured by placing a p-doped protection region at the outer side of the trench (Fig. 1). However, placing such a p-doped region imposes technological constraints for the optimization of the IGBT cell design. The main constraint is that the p-region is usually manufactured by implantation and subsequent annealing. As this annealing step also leads to a significant lateral diffusion of dopants, a further reduction of the outer trench-trench-distance is limited by the diffusion length. In the worst case, if the p-doped region merges with the IGBT p-body implant, the output characteristics of the IGBT starts to show unwanted snap-back effects.

II. MOTIVATION AND FABRICATION PROCESS

The conduction and switching losses of IGBTs strongly depend on two aspects of the IGBT cell design. First, the channel width which impacts the on-state losses and the short-circuit current of the IGBT and second, the ratio between the active cell area (area between two neighboring MOS-channels below the emitter contact) and the total cell area of the chip which changes the on-state losses and the turn-off losses. To improve the performance of the 5th generation IGBT cell, a new choice of channel width and area ratio has

978-1-5386-2928-4/18 $31.00 © 2018 IEEE   24

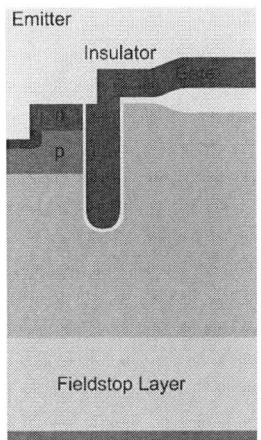 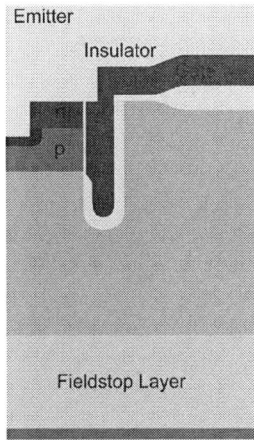

Fig. 2. (Left) Thin-Oxide IGBT with reduced cell pitch which cannot accommodate the protective p-region because of lateral p-diffusion. (Right) Asymmetric thick oxide IGBT with reduced cell pitch and asymmetric trench oxide.

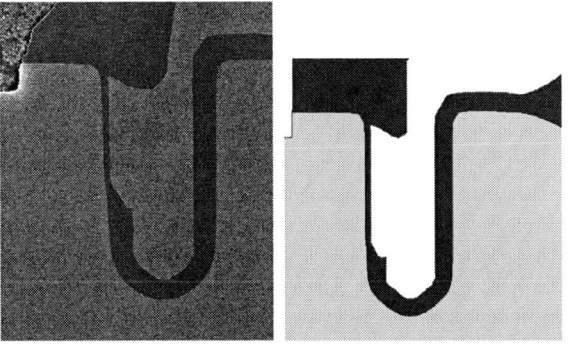

Fig. 3. (Left) Scanning electron micrograph of the 5th generation IGBT with an asymmetric trench oxide. (Right) Result of a calibrated process simulation of the same structure.

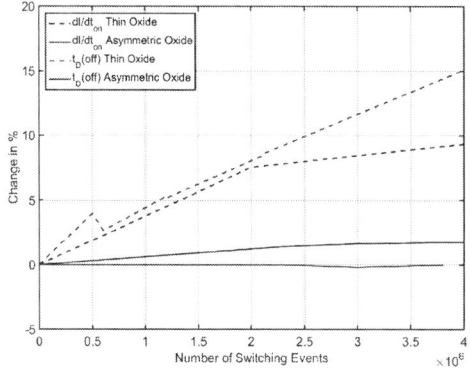

Fig. 4. Relative change of the switching parameters dI/dt during turn-on and turn-off delay $t_d(\text{off})$ as a function of the number of switching pulses. The asymmetric variant is stable.

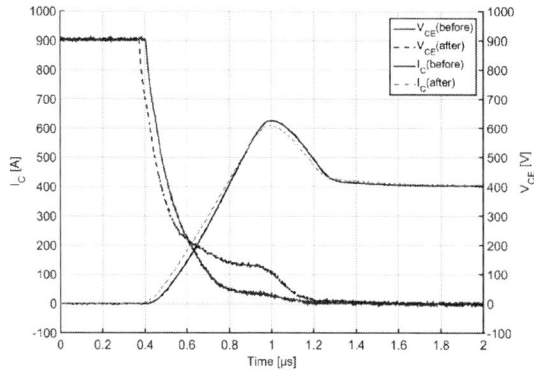

Fig. 5. Thin oxide variant 1700 V - Turn-on of a small pitch thin oxide variant before stress and after $0.8 \cdot 10^7$ switching events. A small drift in $t_d(\text{on})$, dI/dt and a voltage tail becomes visible.

been implemented, which reduces the outer trench-to-trench distance. Therefore, the p-doped region at the off-site of the trench which is present in previous IGBT generations could not be placed and a novel fabrication process which yields an asymmetric trench oxide thickness was introduced (Fig. 2).

The fabrication process starts with a homogenous thick oxide which is covered by photoresist and structured by photolithography. During the subsequent etching, this initial thick oxide is removed by etching in the channel region up to a defined position below the channel. As a last step for the formation of the gate insulator, a thermal gate oxide with a standard thickness is grown. The thin oxide which is present in the channel is needed to maintain a low channel voltage drop during the on-state. The etching of the initial thick oxide is monitored by dedicated test structures.

Figure 3 shows the resulting geometry in an SEM analysis and the result of a calibrated process simulation, which reproduces all relevant structural aspects very well. This process simulation model is the basis for the calibration of an electrical TCAD device model which is discussed in section IV.

III. EXPERIMENTAL INVESTIATION ON LONG-TERM SWITCHING STABILITY

To verify the proposed stabilizing effect of the asymmetric thick gate oxide, long term switching experiments with up to 12 million switching events were performed. For comparison, a variant was studied which as only difference uses a standard thin oxide without any protective p-implant instead of the asymmetric thick oxide. Therefore, the thin oxide structure is expected to be unstable during long-term switching. Both structures have a nominal voltage rating of $V_{\text{nom}} = 1700\,\text{V}$. The stress was performed under hard switching conditions with a current of $2 \cdot I_{\text{nom}} = 400\,\text{A}$ and a voltage of $V_{\text{DC}}(\text{link}) = 900\,\text{V}$ at room temperature which results in over-voltages exceeding the rated RBSOA (for a representitive switching waveform Fig. 8). Under these conditions, a strong dynamic avalanche is present, which is an indication for an accelerated degradation as discussed [2].

In Figure 4, the time evolution of the switching slope (dI/dt) at turn-on and the turn-off delay ($t_d(\text{off})$) are compared for the IGBT with asymmetric thick gate oxide and the

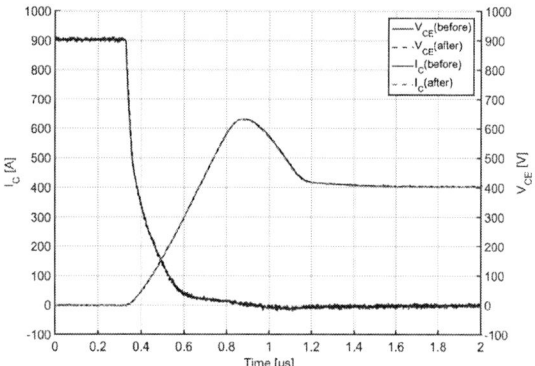

Fig. 6. 5th generation IGBT 1700 V - Turn-on of a small pitch asymmetric oxide variant before stress and after $1.2 \cdot 10^7$ switching events. No changes in the characteristics are visible.

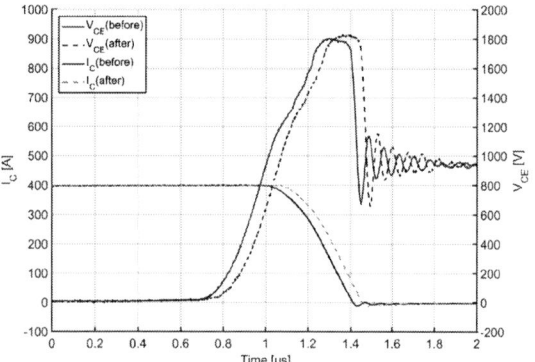

Fig. 7. Thin oxide variant 1700 V - Turn-off of a small pitch thin oxide variant before stress and after $0.8 \cdot 10^7$ switching events. A notable drift in $t_d(\text{off})$ is visible. These hard switching conditions ($2 \cdot I_{\text{nom}}$) are also applied during the stress phase.

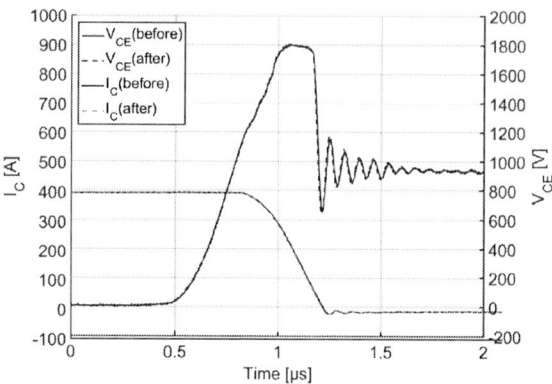

Fig. 8. 5th generation IGBT 1700 V - Turn-off of a small pitch asymmetric oxide variant before stress and after $1.2 \cdot 10^7$ switching events. No changes in the characteristics are visible.

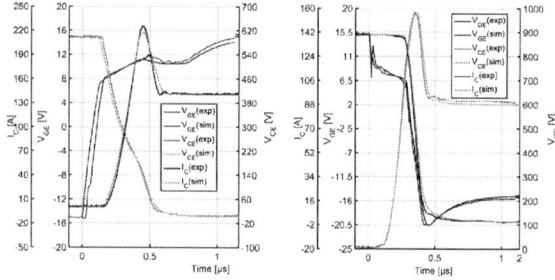

Fig. 9. 5th generation IGBT 1200 V - Calibration of TCAD simulation for turn-off (left) and turn-on (right) of an IGBT5 cell. The deviation between simulation and experiment for switching energies is $< 10\%$ in all cases.

thin oxide variant. While the relative change of the thin oxide variant is significantly larger than 5%, the variant with thick oxide shows only relative drifts smaller than 2.5%. The final readouts for the thin oxide variant (Figs. 5, 7) at turn-on show a drift of $t_d(\text{on})$, dI/dt and a pronounced voltage tail. For turn-off, the main effect is a drift of $t_d(\text{off})$. For the optimized variant with asymmetric thick oxide (Figs. 6, 8), even after 12 million pulses with strong dynamic avalanche present during the test, no drift of electrical properties is apparent for either turn-on or turn-off.

IV. SIMULATION: EFFECT OF INTERFACE CHARGE AT THE TRENCH BOTTOM

To understand the effect of the asymmetric thick gate oxide, a TCAD model was developed, which is based on the presented process model (Fig. 2). To demonstrate the good quality of the model calibration, experimental and simulation data for a 1200 V 5th generation IGBT are plotted in Figure 9. The relative deviation of the switching energies for a wide range of currents, voltages, temperatures and gate resistors is below

10%. As an example for modelling the long-term degradation of the asymmetric thick oxide and the thin-oxide variant, the current waveform during turn-on is shown in Fig. 10 for three different interface charge densities $(0, 5 \cdot 10^{11}/\text{cm}^2, 1 \cdot 10^{12}/\text{cm}^2)$ which are placed at the trench bottom. For the IGBT cell with a thin homogenous gate oxide, the turn-on delay changes up to 40% of the initial value. The IGBT cell with thick asymmetric gate oxide shows a change of $< 2\%$ which is negligible. This shows that the asymmetric thick oxide suppresses the effect of potentially injected charge into the trench-bottom effectively.

V. SIMULATION: EFFECTS OF ASYMMETRIC THICK OXIDE ON SWITCHING BEHAVIOR

Besides the stable long-term switching behavior, the asymmetric thick oxide has another major effect on the switching behavior: the reduction of the gate charge, which becomes apparent as a reduction of switching delay times ($t_d(\text{on})$ and $t_d(\text{off})$). The gate charge reduces from $Q_G = 1020\,\text{nC}$ in the thin-oxide variant structure to $Q_G = 580\,\text{nC}$ in the structure with the asymmetric thick oxide. In Fig. 11, voltage waveforms for turn-off are compared between the asymmetric thick oxide chip and the thin oxide variant for different gate resistor values ($R_G = 2\,\Omega, 11\,\Omega, 20\,\Omega, 40\,\Omega, 80\,\Omega$). All electrical parameters

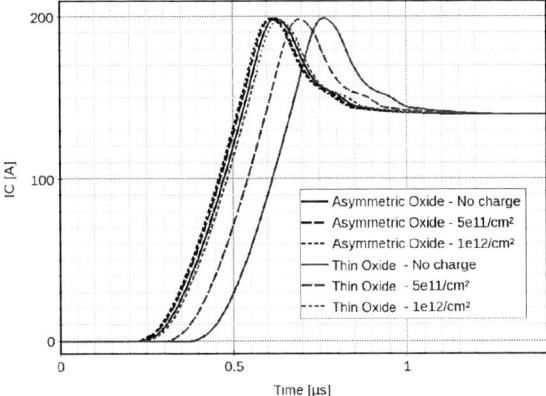

Fig. 10. 5th generation IGBT 1200 V - Simulated current during turn-on for different amounts of fixed charges at the trench bottom for thin (red) and thick asymmetric (black) trench oxides. The effect of charge on the asymmetric oxide variant is negligible.

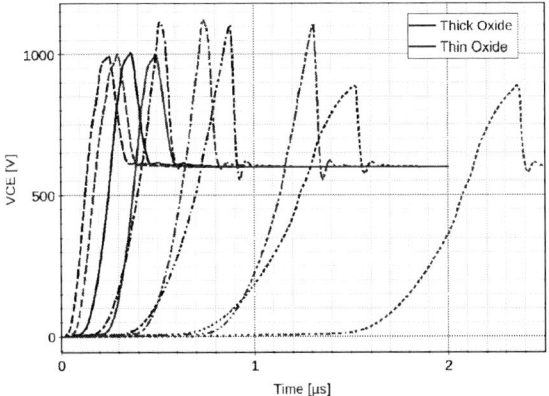

Fig. 11. 5th generation IGBT 1200 V - Comparison of simulated turn-off VCE curves for the asymmetric thick oxide and the thin oxide variants for $R_G = 2\,\Omega, 11\,\Omega, 20\,\Omega, 40\,\Omega, 80\,\Omega$ (same R_G same line style). The curves mainly differ in the switching delay $t_d(\text{off})$.

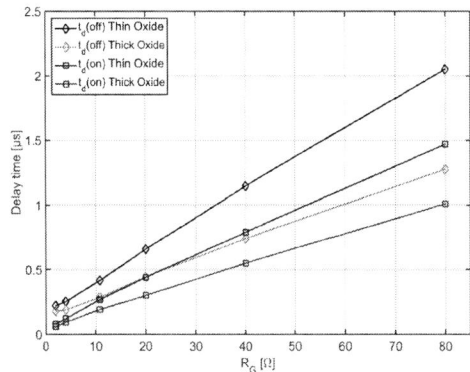

Fig. 12. 5th generation IGBT 1200 V - Simulated comparison of switching delays $t_d(\text{off})$ and $t_d(\text{on})$ for the asymmetric thick oxide and the thin oxide variants for different R_G. Both delays reduce up to 35% for the asymmetric oxide variant.

and switching delays are reduced, which is providing significant benefits in the application.

REFERENCES

[1] C. E. Blat, E. H. Nicollian, E. H. Poindexter, "Mechanism of negativebiastemperature instability", *Journal of Applied Physics*, vol. 69(3), pp. 1712-1720, 1991.

[2] S. Geissmann, L. De Michielis, C. Corvasce, M. Rahimo, M. Andenna, "Extraction of dynamic avalanche during IGBT turn off", *Microelectronics Reliability*, vol. 76, pp. 495-499, 2017.

[3] P. Muenster, D. Wigger, H.G. Eckel, "Impact of the dynamic avalanche on the electrical behavior of HV-IGBTs", *Proc. of PCIM Europe 2015; International Exhibition and Conference for Power Electronics, Intelligent Motion, Renewable Energy and Energy Management* pp. 1-10, 2015.

[4] T. Laska, F. Hille, F. Pfirsch, R. Jereb, M. Bassler, "Long Term Stability and Drift Phenomena of different Trench IGBT Structures under Repetitive Switching Tests", *Proc ISPSD 2007* pp. 1-4, 2007.

like dU/dt and maximum over-voltage do not change for a given gate resistor value between the two variants. Only the turn-off delay time decreases up to 35 %. Fig. 12 shows a comparison of the extracted turn-on and turn-off delays for both variants and different gate resistors. The plot demonstrates that the turn-on delay is reduced similar to the turn-off delay, which is a result of the reduced input capacitance due to the partially thick oxide of the novel structure.

VI. CONCLUSION

Long-term switching experiments show that an asymmetric thick trench oxide enables a stable operation even when stressed under hard conditions with strong dynamic avalanche occuring during switching. This result enables new cell design options which omit the previously needed protective p-doped region for medium-range cell pitches. The new design options enable the further optimization of the trade-off between on-state losses and switching losses. Additionally, the gate charge

978-1-5386-2928-4/18 $31.00 © 2018 IEEE

Proceedings of the 30th International Symposium on Power Semiconductor Devices & ICs
May 13-17, 2018, Chicago, USA

6.5 kV Field Shielded Anode (FSA) Diode Concept with 150C Maximum Operational Temperature Capability

B.K. Boksteen, C. Papadopoulos, D. Prindle, A. Kopta, C. Corvasce
ABB Switzerland Ltd, Semiconductors
Lenzburg, Switzerland
boni.boksteen@ch.abb.com

Abstract— In this paper we present a low leakage current 6.5 kV field shielded anode (FSA) diode with high forward bias safe operating area (FBSOA) ruggedness capable of reliable operation up to 150 °C. This is achieved through optimization of the junction termination, the resistive zone (RZ) between this area and the active region and selective shallow ion irradiation for local lifetime control. As a result, the diode maintains or exceeds the softness, surge current and FBSOA capabilities set by the 6.5 kV carrier axial lifetime (CAL) diode, while also reducing its (125 °C) leakage current by more than 4 times achieving magnitudes typically associated with low leakage emitter efficiency control (EMCON) based concepts.

Keywords—Diode, FSA, Resistive zone, Termination, Leakage, FBSOA, Lifetime control

I. INTRODUCTION

Fast diode concepts like the emitter efficiency control (EMCON [1]) and field shielded anode (FSA [2,3], Fig. 1) have enabled robust operation up to 150 °C for 3.3 kV and up to 175 °C for 1.2 kV and 1.7 kV devices. Together with increased thermal stability, soft switching characteristics and substantial safe operating area (SOA) [4-8] margins have to be provided. For voltages higher than 3.3 kV, the EMCON diode and its inherently low leakage current (Table. 1), is presently the concept of choice for 150 °C operation. However, to fulfill the SOA requirements set for high-voltage devices, this concept requires additional complex hence costly processes such as SPEED and CIBH [9,10]. This work experimentally demonstrates a highly robust 6.5 kV FSA diode capable of reliable operation up to 150 °C with minimal design and process complexity (Table. 1). The paper is outlined as follows. First an overview of the FSA anode is given, followed by the optimization requirements of the termination and resistive zone (RZ). Subsequently measurement results are presented and finally conclusions are drawn.

II. THE FIELD SHIELDED ANODE

In the FSA diode, when compared to the carrier axial lifetime (CAL) [11] concept, the anode side irradiation defects are shifted away (shielded) from the space charge region (SCR) (light green, Fig. 1). Separating the local lifetime control from the SCR reduces the off-state leakage current generation [2] while also removing the need for specific low leakage irradiation species to achieve this

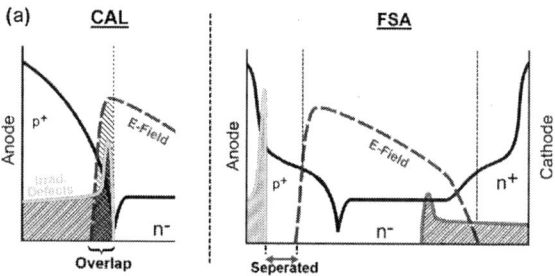

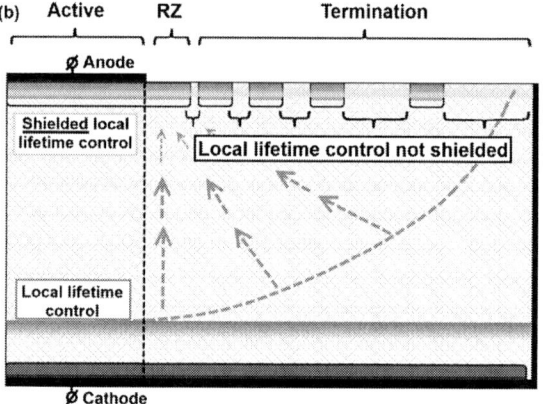

Fig. 1. a) 1D cross-section of the CAL (anode side only) and FSA concept showing overlapping (CAL) and separated (FSA) SCR and local lifetime control (light green). b) 2D cross-sections of a FSA diode including the Active, RZ and Termination. Grey arrows indicate direction of h+ current during reverse recovery.

Figure 2 shows that the transition from CAL to FSA anodes resulted in up to a three times leakage current reduction for the 1.7 kV and 3.3 kV voltage ranges [2,3]. However, for voltages above 3.3 kV this anode change alone resulted in significantly less leakage current reduction. The reason being that for higher voltage devices the size of the termination forms a significant part of the total chip area (Fig. 2b). For such large ring based terminations the non-shielded inter-ring irradiation (Fig. 1b) significantly contributes to the total leakage current and hence, reduces the advantage of the FSA concept regarding leakage.

TABLE I. ANODE'S STRENGTHS AND WEAKNESSESS FOR DESIGNS WITH
EQUAL TERMINATION, CATHODE AND SIMILAR ON-STATE BEHAVIOUR
COMPARED TO OUR OPTIMZED FSA DESIGN (THIS WORK)

Diode (6.5kV)	Leakage Current	FBSOA
CAL	High	High
EMCON	Low	Low
FSA	Medium	Medium
This Work	Low	High

A. Leakage current reduction

Selective masking of irradiation in the inter-ring area of the termination is beneficial for leakage reduction [12]. For the 6.5 kV FSA diode, the leakage can be reduced 2.5 to 3.5 times depending on the masking layout (1, 2, 3) shown in Fig. 2a. These results match or exceed levels reached in non-irradiation masked (smaller termination) FSA devices rated below 3.3 kV (Fig. 2b).

In addition to masking irradiation in the termination, other more conventional methods that have been used to further reduce ('Term. Optimization' in Fig. 2b) the leakage current are:

1) Reduction of the total PN junction area i.e. leakage originating from thermal generation near the termination rings can be decreased by minimizing the ring width and number of rings.

2) Reduction of the effective termination width i.e. optimization of the termination ring placement to reduce the depletion volume in which thermal generation occurs.

B. Resistive zone & FBSOA

For each leakage reduction method that can be implemented one must weigh in the performance trade offs. For instance, preventing irradiation throughout the device except for the active region, as indicated by position (1) in Fig. 2a, will significantly reduce the forward bias safe operating area (FBSOA). To understand why this is the case the charge distribution near the surface in the RZ during reverse recovery should be studied. The charge density along the RZ during reverse recovery can be (qualitatively) described by:

$$\rho (x,t) = q (- N_A(x) + p_{rr} (x,t) + p_{av} (x,t) - n_{av} (x,t)) \quad (1)$$

where N_A is the RZ ring doping, p_{rr} is the reverse recovery hole density and p_{av}, n_{av} the hole and electron density generated by avalanche at local field peaks. Inter-ring RZ depletion that causes local field variations occurs when during the reverse recovery process, the increased flow of holes (grey arrows

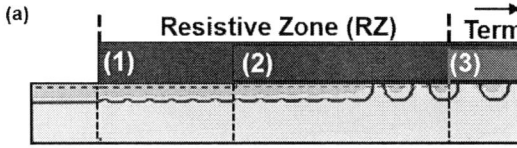

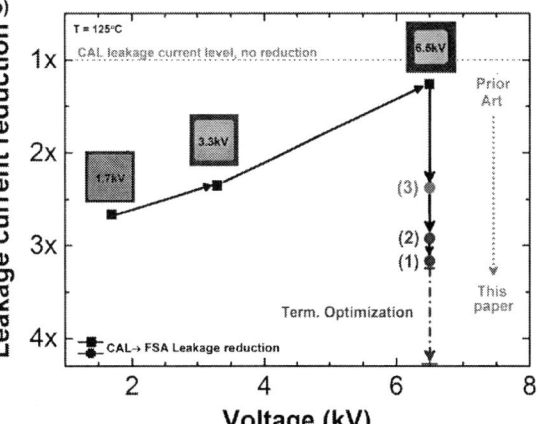

Fig. 2. a) 2D electric field simulation (static) of the RZ during blocking, where also the irradiation masking locations (1, 2, 3) are illustrated. The dashed lines indicate local irradiation. Note that the majority of the irradiated RZ surface does not deplete (grey area) during blocking. b) Measured CAL to FSA leakage current reduction for different voltage classes (squares), and the positive effect of masking the irradiation (1, 2, 3) in the large termination (black border) 6.5 kV chip (circles).

Fig. 1b) towards the lower doped parts between the RZ rings is in the order of N_A ($p_{rr} \sim N_A$). This dynamic effect caused by the reverse recovery hole current is therefore not observed during static blocking. This difference in electric field distribution and depletion is visualized through TCAD in Fig. 2a (blocking state) which has negligible surface depletion, and the EMCON and FSA designs of Fig. 3a (reverse recovery) with clear interring ring depletions. The electric fields in the non-irradiated and irradiated FSA designs of Fig. 3 show that (locally) lowering the reverse recovery (hole) current through irradiation yields a reduction in the RZ electric fields. It is however the combination of both high fields and high (hole) currents in the RZ that results in increased Joule heating (Fig. 3b), and consequently, decreasing FBSOA ruggedness. It is therefore key to have an irradiated RZ (positions (2, 3), Fig. 2a) and optimize its ring placement such that the reverse recovery hole current and peak fields are both minimized and sufficiently separated [4]. The EMCON and FSA TCAD simulations of Fig. 3 show that implementing a wide RZ with equidistant placement of its rings results in sub-optimal distribution of Joule heating with a local energy peak close to the active region (Fig. 3b). The optimized RZ shown in Fig. 3b highlights the positive effect of optimized RZ rings placement on the lateral surface field (inset) and resultant Joule heating. Finally, Fig. 4 shows both a 2D Joule heating simulation at peak power and chip failure location of a device with and without the optimized RZ.

978-1-5386-2928-4/18 $31.00 © 2018 IEEE

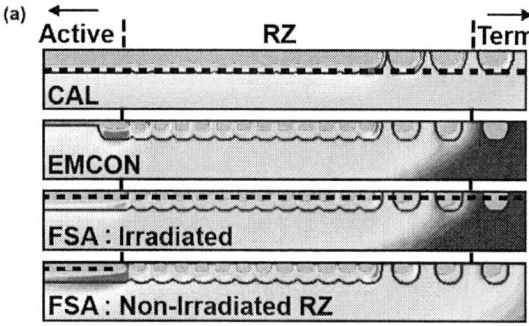

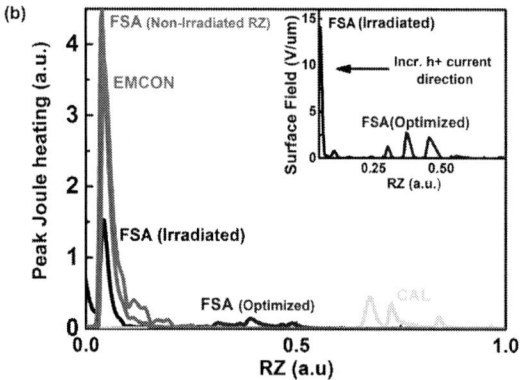

Fig. 3. 2D electric field simulation (dynamic) at peak power during the reverse recovery of a non-optimized RZ applied to the different anode types. The dashed lines indicate local irradiation. Note the inter-ring depletion and fields at the RZ surface of the EMCON and FSA designs b) Lateral cross-section through peak Joule heating (energy) hotspots of the RZ`s in (a) and that of an optimized RZ. Inset shows improvement of the lateral surface (dynamic) field distribution with the point of highest hole current and peak field clearly separated for the optimized design.

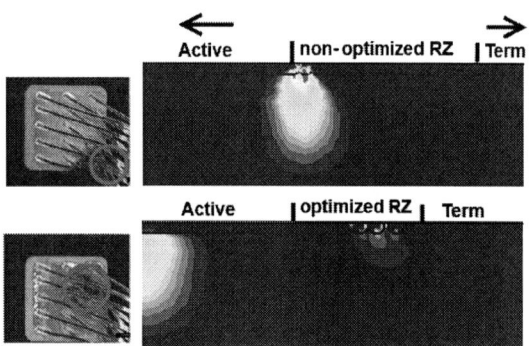

Fig. 4. Chip failure (left) and Joule heating simulation at peak power for a chip without and with the optimized RZ. It is shown that the failure shifts from the termination to the active region.

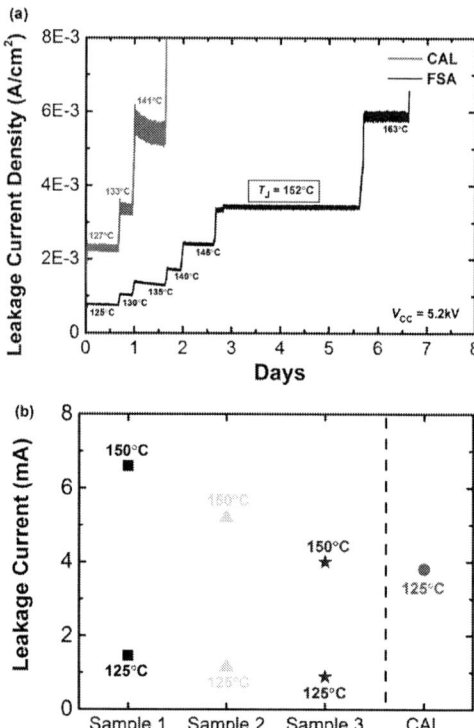

Fig. 5. HTRB stability test showing thermal runaway for the CAL based 6.5 kV diode and stable operation for the irradiation masked FSA diode (Sample 1). b) Measured leakage current at 150 ˚C for devices using the optimization methods discussed in section II. Sample 3 represents our final optimized FSA diode.

III. EXPERIMENTAL RESULTS

In this section we discuss experimental results on the leakage current and FBSOA of our 6.5 kV FSA diode optimized based on the methods treated in section II.

A. Leakage current

Figure 5a shows the high temperature reverse bias (HTRB) test which is the standard method used to assess thermal stability. Thermal runaway is observed for the CAL based 6.5 kV diode. The optimized FSA diode, using only masked irradiation, shows stable operation at $T_J \geq 150$ ˚C. Figure 5b shows leakage current levels of devices of equal chip size using the optimization methods discussed in section II. In sample 1 only irradiation masking in the termination was employed, sample 2 shows how the leakage current was further reduced by minimizing the termination ring widths. Lastly sample 3, our final optimized FSA diode, shows further leakage current reduction achieved by removing the access termination rings and shrinking the effective termination width. The leakage level of sample 3 provides additional margin to the already thermally stable sample 1 (Fig. 5a) and is such that its 150 ˚C leakage is approximately equal to that of the state-of-the-art CAL 6.5 kV diode at 125 ˚C

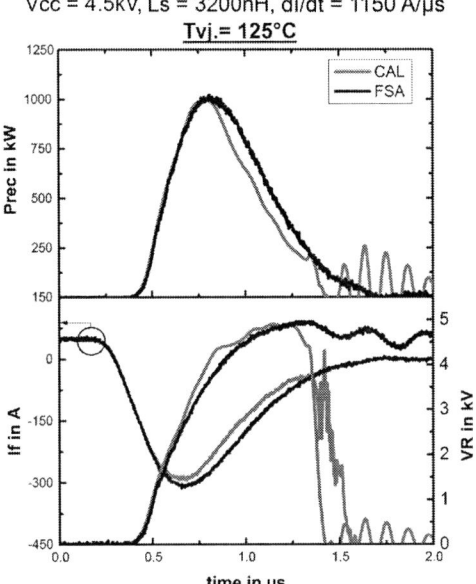

Vcc = 4.5kV, Ls = 3200nH, dI/dt = 1150 A/µs
Tvj.= 125°C

Fig. 6. Optimized 6.5 kV FSA diode passing reverse recovery failure point of state-of-the art robust 6.5 kV CAL diode (single chip).

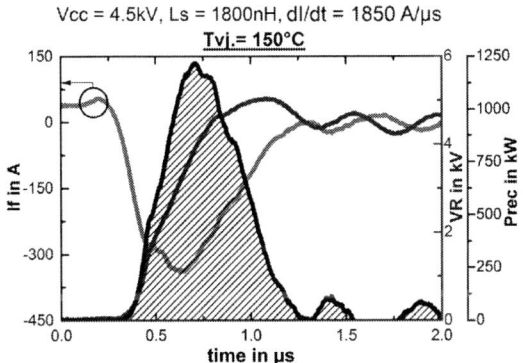

Vcc = 4.5kV, Ls = 1800nH, dI/dt = 1850 A/µs
Tj.= 150°C

Fig. 7. Last pass reverse recovery at $T_J = 150\ °C$ of the optimized FSA chip (single chip)

B. FBSOA

When FBSOA failures in a device are limited by the active region rather than the termination, this indicates that the RZ & termination are sufficiently robust for the device. In Fig. 4 for instance it is shown that the point of failure for devices employing the optimized RZ indeed shifted from the termination to the active region. Figure 6 shows that the FSA chip with optimized RZ surpasses the single chip failure FBSOA set by the robust 6.5 kV CAL diode. In Fig. 7 the last pass reverse recovery at $T_J = 150\ °C$ is shown. The increased single chip peak P_{rec} and the possibility to switch at higher dI/dt conditions shows that the optimized FSA diode outperforms the already substantial FBSOA of the CAL diode platform, allowing save operation under extreme switching conditions. The FBSOA has been performed from low currents (10% nominal) up to two time nominal current.

IV. CONCLUSION

In this paper we discussed different anode concepts (CAL/FSA/EMCON), RZ and termination optimizations, with a focus on leakage current and FBSOA, using TCAD simulations. This resulted in an optimized RZ design for the FSA concept. With reduced size, reduced peak dynamic fields and minimal Joule heating this RZ combined with well-defined masking of irradiation and a termination optimized to reduce leakage current, provides a robust 150 °C capable 6.5 kV FSA diode. Finally the FBSOA and dI/dt ruggedness of the device surpasses that of the already robust 125 °C CAL platform.

ACKNOWLEDGMENT

The authors would like to thank Rachid Jabrany, Chantal Toker-Bieri, Snjezana Ninkovic and Maria Amongero for their support throughout the development phase of this product [13].

REFERENCES

[1] A. Porst, F. Auerbach, H. Brunner, G. Deboy, and F. Hille, "Improvement of the diode characteristics using emitter-controlled principles (EMCON-diode), " in *Proc. IEEE ISPSD*, Weimar, Germany, 1997, pp. 213-216.

[2] S. Matthias, J. Vobecky, C. Corvasce, A. Kopta and M. Cammarata, "Field Shielded Anode (FSA) concept enabling higher temperature operation of fast recovery diodes" in *Proc. IEEE ISPSD*, San Diego, CA, 2011, pp. 88-91.

[3] C. Corvasce *et al*, "New 1700V SPT+ IGBT and diode chip set with 175°C operating junction temperature" in *Proc. EPE*, Birmingham, England, 2011, pp. 1-10.

[4] Y. Tomomatsu *et al*, "An analysis and improvement of destruction immunity during reverse recovery for high voltage planar diodes under high dIrr/dt condition" in *Proc. IEEE ISPSD*, Maui, HI, 1996, pp. 353-356.

[5] K. I. Matsushita *et al*, "4.5kV high-speed and rugged planar diode with novel carrier distribution control" in *Proc. IEEE ISPSD*, Kyoto, Japan, 1998, pp. 191-194.

[6] J. Lutz and M. Domeij, "Dynamic avalanche and reliability of high voltage diodes" in *Microelectronics Reliability,* vol. 43, no. 4, pp.529-536, April 2003.

[7] H. J. Schulze, F. J. Niedernostheide, F. Pfirsch and R. Baburske, "Limiting Factors of the Safe Operating Area for Power Devices," in *IEEE TED*, vol. 60, no. 2, pp. 551-562, Feb. 2013.

[8] C. Papadopoulou *et al*, "Fast Diode Anode Design Concepts for IGBT Applications` in *Proc ISPS*, Prague, Czech Republic, 2016.

[9] H. Schlangenotto, J. Serafin, F. Sawitzki and H. Maeder, "Improved recovery of fast power diodes with self-adjusting p emitter efficiency," in *IEEE EDL*, vol. 10, no. 7, pp. 322-324, July 1989

[10] M. Chen, J. Lutz, M. Domeij, H. P. Felsl and H. J. Schulze, "A novel diode structure with controlled injection of backside holes (CIBH)" in *Proc. IEEE ISPSD*, Naples, 2006, pp. 1-4.

[11] J. Lutz and U. Scheuermann, "Advantage of the new controlled axial lifetime diode" in *Proc. PCIM*, Nuremberg, Germany, 1994, pp. 163-169

[12] S. Matthias and A. Kopta, "Power semiconductor device with new guard ring termination design and method for producing same." U.S. Patent 8384186B2, Feb. 26, 2013.

[13] C. Papadopoulous *et al*, "The third generation 6.5kV HiPak2 module rated at 1000A and 150°C" in *Proc. PCIM*, Nuremberg, Germany, 2018, in press.

Proceedings of the 30th International Symposium on Power Semiconductor Devices & ICs
May 13-17, 2018, Chicago, USA

Low Noise Superjunction MOSFET
with Integrated Snubber Structure

Hiroaki Yamashita, Syotaro Ono, Hisao Ichijo, Masataka Tsuji, Takenori Yasuzumi[+], Masaru Izumisawa and Wataru Saito
Discrete Semiconductor Div., Toshiba Electronic Devices & Storage Corporation
1-1, Iwauchi-machi, Nomi-shi, Ishikawa, Japan E-mail: hiroaki7.yamashita@toshiba.co.jp
[+]Corporate Manufacturing Engineering Center, Toshiba Corporation
33, Shin-Isogo-cho, Isogo-ku, Kanagawa, Japan

Abstract— **Novel superjunction (SJ)-MOSFET structure with distributed internal snubber area is proposed. RC network composed by gate electrode, oxide and P-type pillar provides frequency-dependent capacitive coupling between each terminal. The snubber area acts as gate-drain capacitance (C_{gd}) only at high-frequency band. Switching noise generated during turn-off transient is absorbed by the snubber without increasing switching loss. We verified the concept by simulation and experiments, and confirmed better trade-off between EMI (Electromagnetic interference) and efficiency.**

Keywords— superjunction; noise; switching loss

I. INTRODUCTION

High voltage MOSFETs are widely used in switching applications handling commercial power supply. Superjunction (SJ) is a main stream structure for 300-900V voltage ratings achieving superior specific on-resistance ($R_{on}A$) lower than Si-limit [1]. Compared to advanced compound semiconductor devices such as GaN HEMT and SiC MOSFET, SJ-MOSFET is easy to handle for designers using conventional MOSFETs in terms of noise controllability. Therefore the improvement of the performance has still been demanded. The $R_{on}A$ of SJ-MOSFET has been improved by pitch shrinking and optimization of doping profile of SJ structure which enable reduction of conduction and switching loss [2]-[4]. However, advanced SJ-MOSFETs are sometimes difficult to use because of their high speed switching characteristics. In order to control the switching speed, several techniques are proposed; mesh gate structure to keep gate-drain capacitance (C_{gd}) large [5] or controlling switching speed by optimizing internal gate resistance [6]. However, these strategies increase switching loss by reducing switching speed. Breakthrough of the trade-off between the switching noise and loss further accelerates the application field of SJ-MOSFETs.

In this report, novel SJ-MOSFET with internal snubber is presented. With this structure, switching noise is effectively suppressed by increasing C_{gd} only at high-frequency operation. The trade-off between radiated noise level and efficiency is improved.

II. SWITCHING NOISE GENERATION MECHANISM

The process of switching noise generation can be explained by analyzing small signal equivalent circuit. Fig.1 shows turn-off transient simulation waveform of inductive load switching by SJ-MOSFET in case of parasitic inductances in the circuit. After gate-source voltage (V_{gs}) falls below gate plateau voltage,

the channel current (I_{ch}) starts to drop. The drop-off of I_{ch} is compensated by displacement current (I_{cds} and I_{cgd}) generated by discharging C_{ds} and C_{gd} in order to keep the inductive load current constant. The interval of this process, so-called "miller interval", highly depends on C_{gd} characteristic controlling the switching speed. After C_{ds} is discharged, C_{ds} rapidly decreases by the depletion of SJ. Then V_{ds} starts to rise and I_{cds} falls sharply, because the displacement current ($C_{ds}*dV_{ds}/dt$), also becomes small. Affected by the decrease of I_{cds}, source current (I_s) suddenly drops, and the high dI_s/dt generates voltage drop by parasitic inductance at the end of the miller interval. As a result, source voltage falls down and the gate voltage is also lowered by capacitive coupling via C_{gs}. These fast transients will be the cause of EMI or self turn-on of the MOSFET, resulting noisy switching behavior with low efficiency. Particularly high-frequency noise (e.g. >100MHz) is difficult to suppress by the circuit design, because even package leads (~10nH) become large parasitic impedance at high-frequency band and modification of the layout is not always effective.

Therefore reduction of noisy transients by the device itself is essential for next generation SJ-MOSFETs with ease of use. Especially C_{gd} behavior during miller interval at turn-off process should be designed carefully, while considering the effect on switching loss.

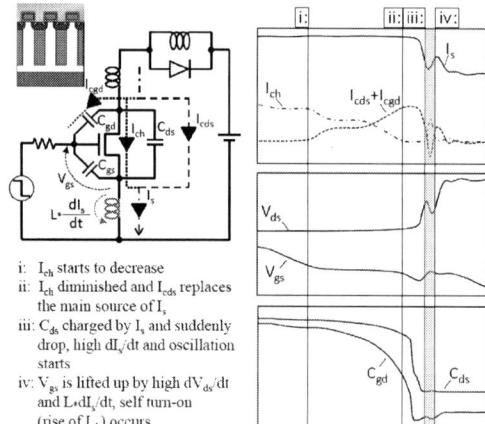

i: I_{ch} starts to decrease
ii: I_{ch} diminished and I_{cds} replaces the main source of I_s
iii: C_{ds} charged by I_s and suddenly drop, high dI_s/dt and oscillation starts
iv: V_{gs} is lifted up by high dV_{ds}/dt and $L*dI_s/dt$, self turn-on (rise of I_{ch}) occurs

Fig.1 Turn-off waveform of inductive load switching simulation by SJ-MOSFET with parasitic inductances. Static capacitance curves (C_{gd}, C_{ds}) estimated from V_{ds} are also shown.

978-1-5386-2928-4/18 $31.00 © 2018 IEEE

III. STRUCTURE, MECHANISM AND SIMULATIONS

To manage both noise characteristic and switching loss, we developed a device having C_{gd} characteristic depending on the operation condition. Fig.2 illustrates the proposed structure composed by conventional MOSFET (Cut plane A) and snubber area (Cut plane B). In the conventional MOSFET area, gate oxide and electrode are only formed on each N-pillar. In the snubber region, the gate structures are also formed on alternate P-pillars. These P-pillars are connected to source electrode through the pillars themselves and base region in the conventional area along the pillar stripe direction. No additional process step is required since the snubber is composed by fundamental SJ and MOS structures.

The effect of the snubber area is explained by an equivalent circuit described by RC network connected to the conventional cell region (Fig.3). The additional area is composed by gate electrode, resistance of P-pillar (R_{sp}), oxide capacitance between gate electrode and P/N-pillar (C_{gp}/C_{gn}), and SJ pillar capacitance (C_{pn}). By the existence of R_{sp} in the circuit, electronic coupling between each electrode depends on signal frequency and V_{ds}. At low frequency, the gate potential mainly couples to source potential through C_{gp} and R_{sp}, because the impedance of R_{sp} is smaller than C_{pn} ($R_{sp} \ll 1/2\pi f C_{pn}$). On the other hand, the gate electrode couples to drain electrode via C_{gp} and C_{pn} at high frequency. Namely, effective gate-drain capacitance is increased by the additional capacitance of the snubber area. The frequency, at which the snubber acts as added C_{gd}, can be controlled with R_{sp} by changing snubber layout. Therefore, C_{gd} can be increased at only high-frequency band (e.g. >100MHz), while keeping C_{gd} and switching loss small at switching frequency (e.g. ~100kHz). Furthermore, the snubber effect is enhanced at the turn-off timing of V_{ds} rise. P-pillar is depleted as V_{ds} increases, then R_{sp}, resistance composed by the P-pillar, becomes large by the extraction of holes in the pillar (Fig.4(a)). The increase of R_{sp} accelerates the coupling between gate and drain potentials. As Fig.4(b) indicates, C_{gd}-V_{ds} curve with the snubber structure has clear frequency dependence. The effect of the snubber on switching characteristic is evaluated by simulation (Fig.5). By modifying the capacitance curve by the snubber, switching noise is effectively suppressed. In addition, self turn-on behavior is also suppressed, which will contribute to the improvement of the set efficiency.

Consequently, the snubber acts as added C_{gd} only at noise-concerned frequency. The proposed structure will realize low switching loss and noise level at the same time.

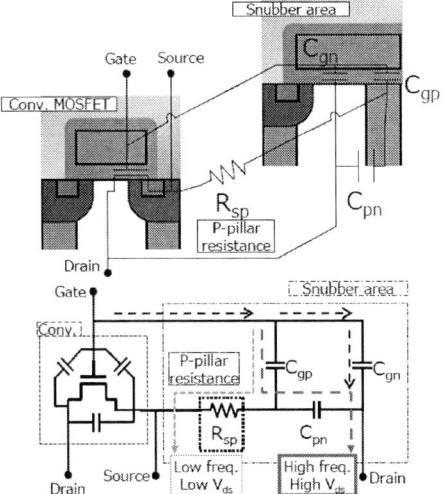

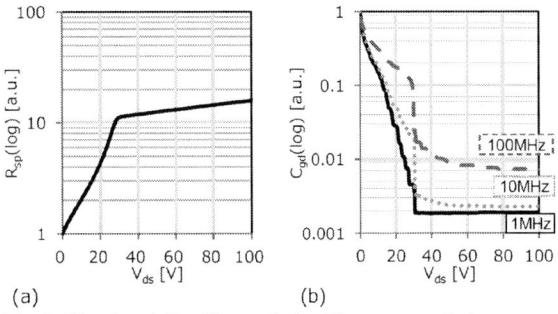

Fig.3 Equivalent small signal circuit of the proposed structure and schematic signal flow dependence on frequency and V_{ds}.

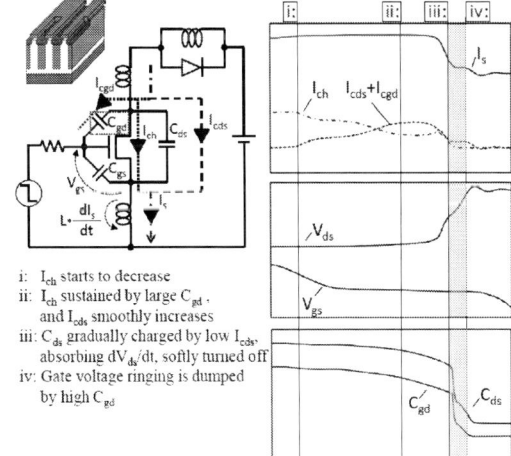

Fig.4 Simulated R_{sp}-V_{ds} and C_{gd}-V_{ds} curve of the proposed structure.

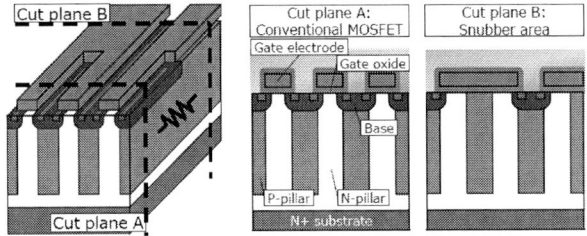

Fig.2 Proposed SJ MOSFET structure with internal snubber.

i: I_{ch} starts to decrease
ii: I_{ch} sustained by large C_{gd}, and I_{cds} smoothly increases
iii: C_{ds} gradually charged by low I_{cds}, absorbing dV_{ds}/dt, softly turned off
iv: Gate voltage ringing is dumped by high C_{gd}

Fig.5 Inductive load switching simulation waveform of the proposed structure with static capacitance curves at 100MHz.

IV. EXPERIMENTS

We fabricated test samples and evaluated static and dynamic characteristics. Three gate layouts are considered: stripe, mesh and snubber. Fig.6 shows schematic figures of each structure. Voltage and current ratings are 600V/12A. Each device is on the same wafer and only the plane layout is changed. Samples are assembled with TO-220FP package.

Table.1 summarizes DC characteristics for each structure normalized by the result of the stripe structure. The main differences between each structure are $R_{on}A$ and V_{th}, resulting from the difference in channel width indicated in the table. Focusing on the capacitive behavior, C_{gd} and gate charge (Q_g) characteristics are evaluated. Fig.7 shows the C_{gd}-V_{ds} curve at low frequency regime (100kHz). Stripe structure has the smallest capacitance because of small gate electrode area. Mesh and snubber samples having similar gate area show almost the same C_{gd} curve at all V_{ds} range. Reflecting the differences of C_{gd}, Q_g characteristic has structure dependence (Fig.8). Same as C_{gd}, stripe has smallest Q_{gd}, and mesh and snubber have almost the same Q_{gd}. Note that Q_g measurement is conducted at quite slow switching of a few tens of microseconds, which is relatively slow compared to the actual usage at switching mode power supply.

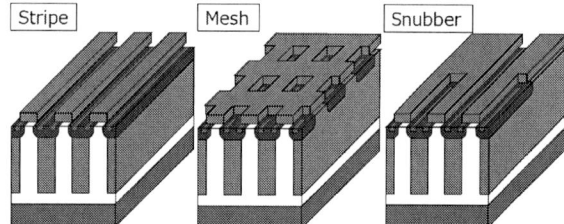

Fig.6 Gate layout structure of fabricated samples.

Table.1 DC characteristics

Gate layout	Channel width	Gate electrode area	BV	$R_{on}A$	Vth
Stripe	100%	100%	100%	100%	100%
Mesh	30%	156%	101%	109%	113%
Snubber	51%	158%	101%	108%	112%

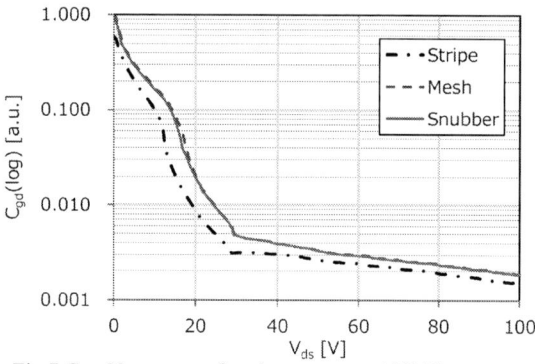

Fig.7 C_{gd} -V_{ds} curves of each structure at 100kHz.

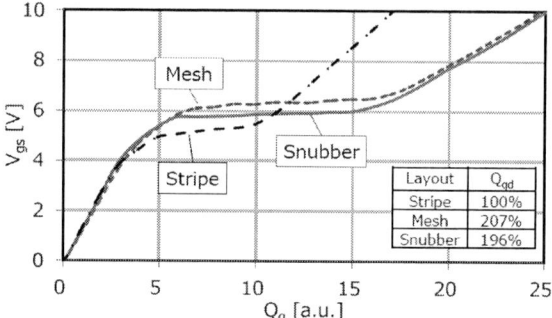

Fig.8 Gate charging characteristics (V_{gs} -Q_g).

In slow transient tests described above, the proposed structure shows almost the same characteristic as mesh structure. By contrast, the snubber structure effectively suppresses the noise at actual switching operation. Fig.9 illustrates inductive load turn-off switching waveforms of each sample at almost equivalent turn-off loss. As predicted from small C_{gd} and Q_{gd}, stripe has the fastest and noisy switching characteristic (Fig.9(a)). On the other hand, snubber device (Fig.9(c)) shows small switching noise compared to mesh sample having almost the same C_{gd} and Q_{gd} (Fig.9(b)). Especially, gate voltage oscillation is effectively suppressed. The oscillation frequency is around 130 MHz which is high enough for activating the snubber effect of increasing C_{gd}. Owing to the added C_{gd} in high-frequency band, smooth switching is achieved with the proposed structure.

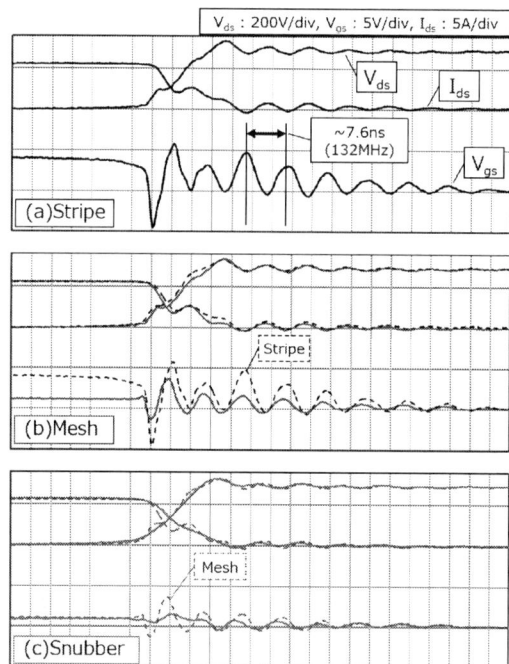

Fig.9 Inductive load turn-off waveform of each structure.

Dotted lines are for comparison among samples.

978-1-5386-2928-4/18 $31.00 © 2018 IEEE

For the purpose of clarifying the improvement in actual usage, the devices are tested in power factor collection (PFC) circuit of AC-DC converter with output of 200W (Fig.10). Efficiency and radiated noise emission at full load condition are evaluated. The switching speed was varied by changing gate drive resistance in order to investigate the trade-off between the efficiency and noise level.

Fig.11 shows radiated noise spectrum of each device. The noise level is reduced by the adoption of the snubber structure at the entire frequency band. Especially, the peak level around 200MHz and 400MHz are clearly suppressed with the proposed structure. The relationship between efficiency and the maximum noise level is shown in Fig.12. As external gate resistance (R_g) is increased, the switching speed is slowing down, having trade-off between these characteristics. The proposed snubber structure shows better trade-off compared to conventional stripe and mesh devices. That is, reduction of radiation noise level is expected at targeting efficiency.

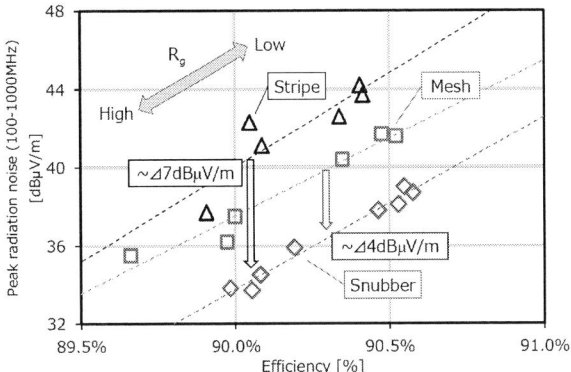

Fig.12 Relationship between radiated noise and efficiency.

V. CONCLUSION

The novel SJ-MOSFET structure with internal snubber is proposed. By adding the snubber area, effective gate-drain capacitance is modulated and increased at high-frequency, resulting in EMI noise suppression. High efficiency with low noise emission level is achieved.

ACKNOWLEDGMENT

The authors would like to thank S. Sato and N. Yokoyama for supporting the device fabrication. We also have had the great support and encouragement of N. Kurihara.

REFERENCES

[1] G. Deboy, M. März, J. –P. Stengl, H. Strack, J. Tihanyi, and H. Weber "A new generation of high voltage MOSFETs breaks the limit of silicon", in Technical Digests of IEDM'98, pp.683-685, 1998.

[2] T. Fujihira, "Theory of semiconductor superjunction devices," Jpn. J. of Appl. Phys., vol. 36, pp.6254-6262, 1997.

[3] A. G. M. Strollo and E. Napoli, "Optimul on-resistance versus breakdown voltage tradeoff in superjunction power devices: a novel analytical model," IEEE Trans. on Electron Devices, vol. 48, pp. 2161-2167, 2001.

[4] W. Saito, "Theoretical limits of superjunction considering with charge imbalance margin," in Proceedings of ISPSD'15, pp. 125-128, 2015.

[5] S. Ono, H. Ohta, H. Yamashita, M. Izumisawa, W. Saito, S. Sato, N. Matsuda, Y. Ohishi, M. Tsuji, J. Onodera, and G. Tchouangue, "Improvement of gate controllability for new generation Superjunction MOSFETs : DTMOS-III series", in Proceedings of PCIM Europe 2011, pp. 320-323.

[6] T. Sakata, Y. Niimura, S. Takenoiri, S. Watanabe, T. Shimatoh, H. Minazawa, Y. Arita, Y. Kobayashi, and J.Li, "A Low-Switching Noise and High-Efficiency Superjunction MOSFET, Super J MOSR S2", in Proceedings of PCIM Asia 2015, pp.419-426.

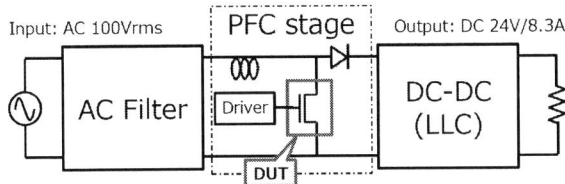

Fig.10 Schematic circuit of AC-DC converter.

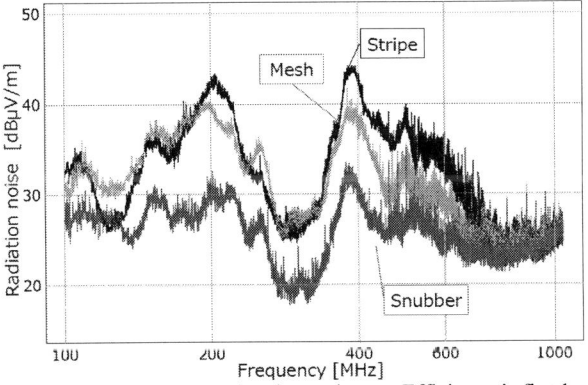

Fig.11 Noise spectrum of each gate layout. Efficiency is fixed at almost the same level (90.4~90.5%) by modifying external gate resistance of the gate driver.

Proceedings of the 30th International Symposium on Power Semiconductor Devices & ICs
May 13-17, 2018, Chicago, USA

Breakthrough of Drain Current Capability and On-Resistance Limits by Gate-Connected Superjunction MOSFET

Wataru Saito

Toshiba Electronic Devices & Storage Corporation
Kawasaki, Japan
wataru3.saito@toshiba.co.jp

Abstract—**This paper reports a new structure of Gate-connected Superjunction (GS) MOSFET to cope with both high drain current density and low on-resistance. The conventional superjunction (SJ) structure is attractive to reduce the specific on-resistance dramatically due to the charge compensation concept. The drain saturation current density, however, is limited by JFET depletion at the bottom region of the SJ structure even if the on-resistance can be reduced by the lateral SJ pitch narrowing. The accumulation-mode operation is effective not only for low on-resistance but also for suppressing the depletion at the SJ bottom due to the accumulation carriers. This paper reports the potential of the GS-MOSFET for high drain current density and low on-resistance based on the simulation results. Dynamic characteristics are also compared with the conventional SJ-MOSFET.**

Keywords—superjunction; on-resistance; saturation current

I. INTRODUCTION

The power-MOSFET is a key component in switching mode power supply circuits and inverter systems. In these applications, low on-resistance of the MOSFETs is desired to reduce power losses in the system. A superjunction (SJ) MOSFET has been commercialized with ultra-low on-resistance below the Si-limit [1]. The SJ structure is consisted with multiple p- and n-columns to allow higher drift region (n-columns) doping concentration than conventional power-MOSFETs. The highly doped n-column directly reduces the on-resistance. As a charge compensation concept, the excess charge in the n-column is counterbalanced by the adjacent charges in the p-column, and thus high breakdown voltage is maintained due to high vertical electric field distribution.

To reduce the specific on-resistance ($R_{on}A$) in the SJ MOSFETs, the lateral pitch of the SJ structure must be narrowed in principle due to maintaining high vertical electric field even with the increase of the n-column doping concentration [2]-[4]. According to this design, the $R_{on}A$ reduction with lateral pitch narrowing has been demonstrated, and also, at the 600-650 V-class power MOSFET products, the $R_{on}A$ has been reduced continuously [5]-[8].

The drain saturation current density (J_{dsat}) is also an important characteristic in the SJ MOSFET product design. Low drain current capability is an obstacle to shrink the chip area, even if the on-resistance can be reduced by the lateral SJ pitch narrowing. The products trend shows the J_{dsat} has been increased with inverse proportional to the $R_{on}A$, because the J_{dsat} is limited by JFET depletion at bottom region of the SJ structure [9], [10]. In addition, the high column doping

concentration induces the breakdown voltage lowering due to the charge imbalance by the process variation [11], [12]. Therefore, the compatibility of high J_{dsat} and low $R_{on}A$ requires not only the SJ pitch narrowing but also the process margin cut and the thermal budget suppression, and it was estimated that the 600 V-class limits of the $R_{on}A$-J_{dsat} characteristics maintained the product trend were J_{dsat} = 900-1300 A/cm² and $R_{on}A$ = 5.5-7.3 mΩcm² in the previous work [9].

This paper reports a new structure of Gate-connected Superjunction (GS) MOSFET to cope with both high drain current density and low on-resistance in the SJ MOSFET. The simulation results show the potential of the GS-MOSFET for the breakthrough of high J_{dsat} and low $R_{on}A$ limits in the SJ MOSFET and the dynamic characteristics are also discussed.

II. DESIGN CONCEPT AND DEVICE STRUCTURE

To break through the limits of high J_{dsat} and low $R_{on}A$, the JFET depletion at the SJ bottom region must be suppressed. The accumulation-mode operation is attractive for suppressing the depletion and the on-resistance can be reduced due to the increase of the drift carrier density [13]-[15]. Although the gate electrode induces the accumulation layer at the MOS interface, it is difficult to realize 600 V-class devices, because an ultra-thick oxide film is necessary for sustaining the drain voltage and the thick oxide weakens the accumulation. A thick oxide film is also an obstacle for the lateral pitch narrowing and the process integration [4], [16].

Therefore the GS structure was chosen to realize the

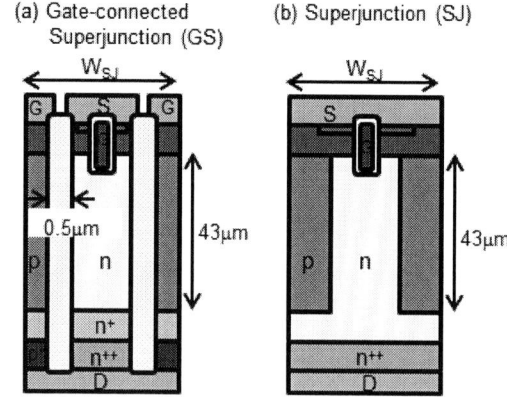

Fig. 1 Cross sectional structure of 600 V-class (a) GS-MOS and (b) SJ-MOS.

978-1-5386-2928-4/18 $31.00 © 2018 IEEE

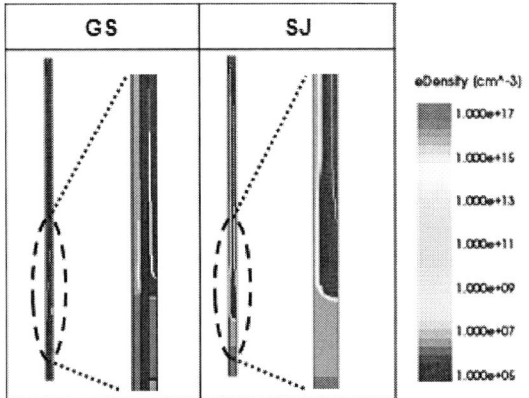

Fig. 2 Suppressing the pinch-off at n-column bottom region at saturation condition ($V_{ds} = 20$ V) in GS-MOS with $W_{SJ} = 3\mu$m.

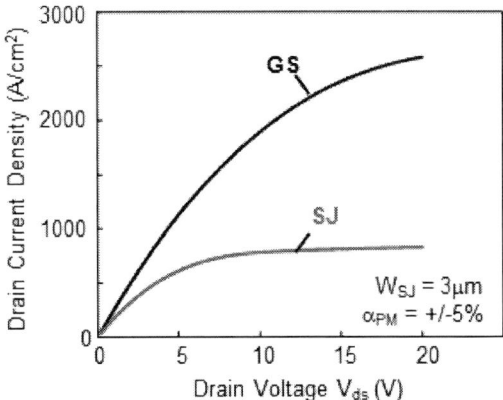

Fig. 3 On-state I_d-V_{ds} characteristics for GS-MOSFET and SJ-MOSFET.

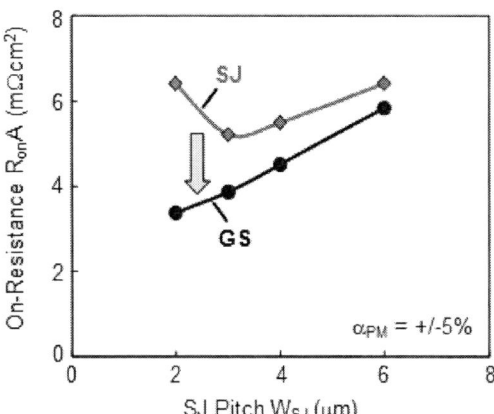

Fig. 4 On-resistance reduction by SJ pitch narrowing in GS-MOSFET.

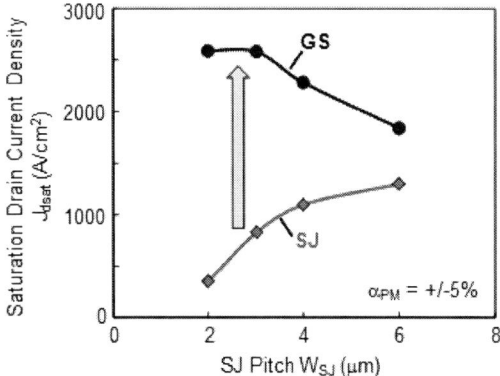

Fig. 5 Maintaining saturation current density increase with SJ pitch narrowing in GS-MOSFET.

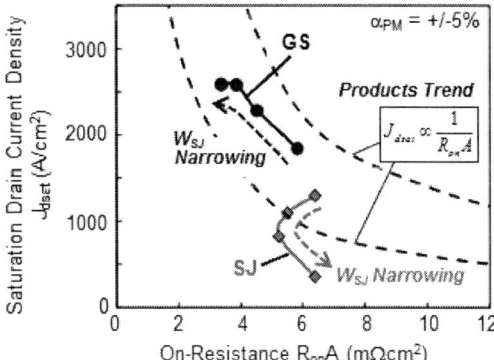

Fig. 6 Fig. 6 Maintaining products trend of drain current density and on-resistance by GS-MOSFET with lateral pitch narrowing.

accumulation-mode operation by the SJ structure without the thick oxide film. In the GS-MOSFET, p-columns are connected to the gate and the accumulation layer is generated at the interface between the n-column and the oxide film at the on-state as shown in Fig. 1. Since the p-columns are depleted at the off-state under low applied voltage and sustains the drain voltage, the oxide film thickness between the p- and n-columns can be designed independently from the breakdown voltage. The ON/OFF switching operation is obtained by the MOS gate structure as same as the conventional SJ-MOSFET. In addition, to avoid the G-D short at the on-state, the p-n-p structure, which is the anti-series connection of diodes, is formed between the drain and the p-column [15]. To avoid the turn-on of the parasitic p-ch MOSFET at the on-state, the doping concentration of the n-layer under the p-column must be optimized from the view point of the threshold voltage for the parasitic p-ch MOSFET.

The saturation current density J_{dsat} and the on-resistance $R_{on}A$ for a 600 V-class device were estimated by the device simulation Sentaurus Device of Synopsys. In this work, the oxide film thickness and the SJ thickness were constant of 0.5 μm and 43 μm, respectively as shown in Fig. 1. The J_{dsat} was

defined as the drain current density at $V_{ds} = 20$ V and $V_{gs} = 10$V and the $R_{on}A$ was calculated from the on-state drain voltage at the drain current density of 100 A/cm^2 [9]. In the actual device, the doping concentration is varied and so the

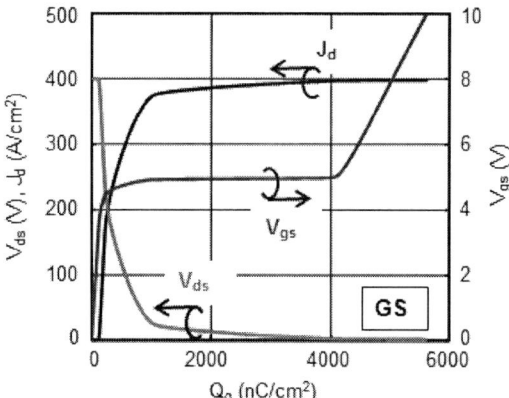

Fig. 7 Gate charge waveform of GS-MOSFET with $W_{SJ} = 3\mu m$.

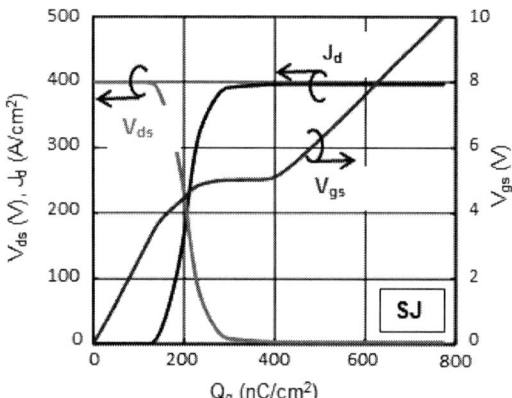

Fig. 8 Gate charge waveform of SJ-MOSFET with $W_{SJ} = 3\mu m$.

process margin α_{PM} was considered in this simulation. In this work, the α_{PM} was a constant of +/-5%. The gate charge was also simulated to discuss the switching characteristics.

III.   DEVICE CHATACTERISITICS

The GS-MOSFET achieved not only lower $R_{on}A$ but also higher J_{dsat} compared with the SJ-MOSFET, because the accumulation mode operation by the gate-connected p-column induces high carrier concentration in the drift layer and suppresses the SJ bottom depletion even under high drain voltage as shown in Fig. 2. At the same conditions of the lateral SJ-pitch W_{SJ} and the α_{PM}, the I_d-V_{ds} curve was improved clearly by the GS-MOSFET compared with the SJ-MOSFET as shown in Fig. 3. The W_{SJ} narrowing improves $R_{on}A$ and J_{dsat} characteristics for the GS-MOSFET, although the characteristics for the SJ-MOSFET are degraded by the W_{SJ} narrowing as show in Figs. 4 and 5. The characteristics for SJ-MOSFETs depart from the products trend at the $W_{SJ} < 4\mu m$ [9]. In contrast, the GS-MOSFET achieves to maintain the products trend even with $W_{SJ} = 2\mu m$ as shown in Fig. 6.

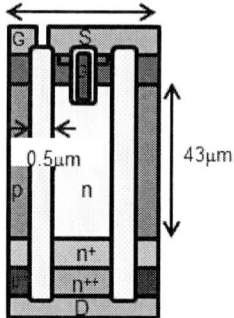

Fig. 9 Cross sectional structure of 600 V-class HGS (Half Gate-connected Superjunction) -MOSFET for gate charge reduction from GS-MOSFET.

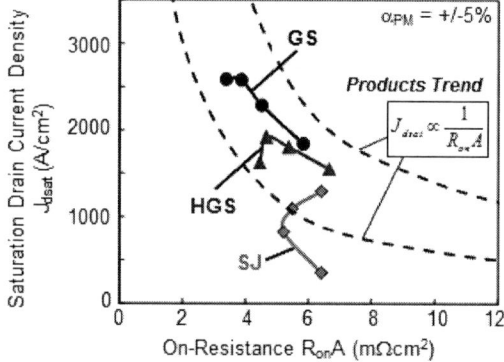

Fig. 10 Trade-off characteristics between saturation drain current density and on-resistance for HGS-MOSFET.

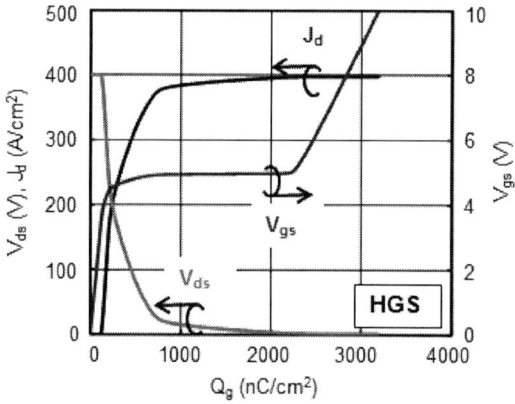

Fig. 11 Gate charge waveform of HGS-MOSFET with $W_{SJ} = 3 \mu m$.

The gate-connected p-column induces the accumulation layer and obtains good on-state characteristics as shown above. However, the gate charge is increased dramatically. This is the same manner at the field plate type device as reported in the previous work [14]. The switching gate charge Q_{SW} of the GS-MOSFET was 14 times larger than that of the SJ-MOSFET as shown in Figs. 7 and 8.

978-1-5386-2928-4/18 $31.00 © 2018 IEEE   38

Table I Static and dynamic characteristics comparison between GS-MOSFET, HGS-MOSFET and SJ-MOSFET.

Device	GS	HGS	SJ
SJ Pitch	3μm	3μm	4μm
$R_{on}A$	3.9mΩcm²	4.7mΩcm²	5.5mΩcm²
J_{dsat}	2600A/cm²	1900A/cm²	1100A/cm²
$R_{on}Q_{sw}$	15.5ΩnC	9.8ΩnC	1.4ΩnC
$R_{on}Q_g$	21.7ΩnC	14.9ΩnC	3.3ΩnC

IV. ARRANGEMENT OF CHARACTERISTICS

The proposed GS-MOSFET has a potential for the breakthrough of the SJ-MOSFET limit. The switching characteristics, however, are degraded by the large gate capacitance. As a middle state structure, Half Gate-connected SJ (HGS)-MOSFET is also proposed to arrange the characteristics as shown in Fig. 9. In the HGS-MOSFET, one p-column is connected to the gate and another one is connected to the source. The HGS-MOSFET improves the $R_{on}A$ and J_{dsat} characteristics compared with the SJ-MOSFET and the characteristics correspond to the middle between those of the GS-MOSFET and the SJ-MOSFET as shown in Fig. 10. The Q_{SW} for the HGS-MOSFET becomes a half of that for the GS-MOSFET as shown Fig. 11. From these results, the static and switching characteristics can be adjusted by the number of p-columns connected to the gate.

The characteristics for the proposed devices in this work are summarized in Table I. It was estimated that the GS-MOSFET breaks through the SJ-MOSFET limits and obtains J_{dsat} = 2600 A/cm² and $R_{on}A$ = 3.9 mΩcm² for 600 V-class. Although 11 times larger $R_{on}Q_{sw}$ compared with the SJ-MOSFET is disadvantage for the power supply application, high J_{dsat} and low $R_{on}A$ are attractive for inverter applications, such as motor drive systems, power conditioning systems and so on. The HGS-MOSFET has a potential for the replacement of the old generation SJ-MOSFET, because $R_{on}Q_{SW}$ and $R_{on}Q_g$ values are almost the same. From these results, the GS-MOSFET and the HGS-MOSFET have a potential for breakthrough the SJ-MOSFET limits.

V. CONCLUSIONS

A new structure of Gate-connected Superjunction (GS) MOSFET was proposed to cope with both high drain current density and low on-resistance. The gate-connected p-column induces accumulation layer at the interface between the n-column and the oxide film. The accumulation-mode operation is effective not only for low on-resistance but also for suppressing the depletion at the SJ bottom due to the accumulation carriers. The GS-MOSFET breaks through the SJ-MOSFET limits and obtains J_{dsat} = 2600 A/cm² and $R_{on}A$ = 3.9 mΩcm² for 600 V-class. Since the gate-connected p-column increases the gate charge, the $R_{on}Q_{sw}$ is 11 times larger compared with the SJ-MOSFET. As a middle state structure, Half Gate-connected SJ (HGS)-MOSFET, in which a half of the p-columns are connected to the gate, was also proposed. The HGS-MOSFET also improves the $R_{on}A$ and J_{dsat} characteristics compared with the SJ-MOSFET and the characteristics can be adjusted by the number of p-columns connected to the gate. From these results, the GS-MOSFET and the HGS-MOSFET have a potential for breakthrough the SJ-MOSFET limits.

REFERENCES

[1] G. Deboy, M. März, J. –P. Stengl, H. Strack, J. Tihanyi and H. Weber "A new generation of high voltage MOSFETs breaks the limit of silicon", in Technical Digests of IEDM'98, pp.683-685, 1998.

[2] T. Fujihira, "Theory of semiconductor superjunction devices," Jpn. J. of Appl. Phys., vol. 36, pp.6254-6262, 1997.

[3] A. G. M. Strollo and E. Napoli, "Optimul on-resistance versus breakdown voltage tradeoff in superjunction power devices: a novel analytical model," IEEE Trans. on Electron Devices, vol. 48, pp. 2161-2167, 2001.

[4] W. Saito, "Comparison of theoretical limits between superjunction and field plate structures," in Proceedings of ISPSD'13, pp. 241-244, 2013.

[5] W. Saito, I. Omura, S. Aida, S. Koduki, M. Izumisawa, H. Yoshioka, H. Okumura, M. Yamaguchi and T. Ogura, "A 15.5mΩcm²-680V Superjunction MOSFET reduced on-resistance by lateral pitch narrowing," in Proceedings of ISPSD'06, pp.293-296, 2006.

[6] J. Sakakibara, Y. Noda, T. Shibata, S. Nogami, T. Yamaoka and H. Yamaguchi, "600V-class super junction MOSFET with high aspect ratio p/n columns structure," in Proceedings of ISPSD'08, pp.299-302, 2008.

[7] W. Saito, "Power Device Trends for High- Power Density Operation of Power Electronics System," Jpn. J. of Appl. Phys., vol. 53, 04EP02, 2014.

[8] F. Udrea, G. Deboy and T. Fujihira, "Superjunction power devices, history, development, and futre prospets," IEEE Trans. on Electron Devices, vol. 64, pp. 713-727, 2017.

[9] W. Saito, "Process design of superjunction MOSFETs for high drain current capability and low on-resistance," in Proceedings of ISPSD'17, pp. 475-478, 2017.

[10] D. Disney and G. Dolny, "JFET depletion in superjunction devices," in Proceedings of ISPSD'08, pp. 157-160, 2008.

[11] P. M. Shenoy, A. Bhalla and G. M. Dolny, "Analysis of the effect of charge imbalance on the static and dynamic characteristics of the super junction MOSFET," in Proceedings of ISPSD'99, pp. 99-102, 1999.

[12] W. Saito, "Theoretical limits of superjunction considering with charge imbalance margin," in Proceedings of ISPSD'15, pp. 125-128, 2015.

[13] B. Jayant Baliga, Tsengyou Syau, and Prasad Venkatraman, "The accumulation-mode field-effect transistor: a new ultralow on-resistance MOSFET," IEEE Electron Device Letters, vol. 13, pp. 427-429, 1992.

[14] M. A. Gajda, S. W. Hodgskiss, L. A. Mounfield, N. T. Irwin, G. E. J. Koops and R. van Dalen, "Industriation of resurf stepped oxide technology for power transistors," in Proceedings of ISPSD'06, pp. 109-112, 2006.

[15] J. Wei, X. Luo, Y. Zhang, P. Li, K. Zhou, Z. Li D. Lei, F. He and B. Zhang, "Accumulation-mode high voltage SOI LDMOS with ultralow specific on-resistance," in Proceedings of ISPSD'15, pp. 185-188, 2015.

[16] R. Siemieniec, C. Braz and O. Blank, "Design considerations for charge-compensated power MOSFET in the medium-voltage range, in Proceedings of ISPS'16, pp. 107-114, 2016.

Proceedings of the 30th International Symposium on Power Semiconductor Devices & ICs
May 13-17, 2018, Chicago, USA

Investigation of threshold voltage stability of SiC MOSFETs

Dethard Peters[*], Thomas Aichinger[†], Thomas Basler[‡],
Gerald Rescher[†], Katja Puschkarsky[‡], Hans Reisinger[‡]

[*] Infineon Technologies AG, Schottkystrasse 10, D-91052 Erlangen
[†] Infineon Technologies Austria AG, Siemensstrasse 2, A-9500 Villach, Austria

[‡] Infineon Technologies AG, Am Campeon 1-12, D-85579 Neubiberg, Germany
dethard.peters@infineon.com

Abstract— Silicon carbide (SiC) based metal-oxide semiconductor-field-effect-transistors (MOSFETs) show excellent switching performance and reliability. However, compared to silicon devices the more complex properties of the semiconductor-dielectric interface imply some natural peculiarities in threshold voltage variation. This paper analyzes threshold voltage hysteresis effects, bias temperature instability effects (BTI) and their relevance for the switching behavior. Most of the effects can be understood by means of simple physical models and do not harm reliability and performance of the device. It turns out that the standard norm test and readout procedures typically used to characterize threshold voltage and threshold voltage drifts for Si devices are insufficient and need to be adapted for SiC MOSFETs in order to get reproducible and solid results.

Keywords—SiC MOSFET, BTI, threshold voltage stability, norm test, threshold hysteresis

I. Introduction

Silicon carbide (SiC) based metal-oxide semiconductor-field-effect-transistors (MOSFETs) show excellent switching performance and reliability [1]. Due to the presence of carbon at the interface between dielectric and SiC the physical properties of the interface is different to silicon. In order to achieve low on-resistances the inversion channel mobility has been improved significantly e.g. by nitridation techniques using nitric oxide. In first order these gate oxide techniques succeeded in lowering the density of interface states close to the conduction band. But due to the physics of the interface states there are some natural peculiarities in threshold voltage variation.

These threshold hysteresis effects and their relevance for the switching behavior are investigated in detail in this paper. Since the hysteresis is totally reversible, it has to be clearly separated from bias temperature instability effects (BTI). The standard norm tests do not define any preconditioning procedures since this is not relevant for Si MOS devices. But for SiC MOSFETs it turns out that an appropriate preconditioning is required in order to measure threshold voltage and on-resistance in a reproducible way.

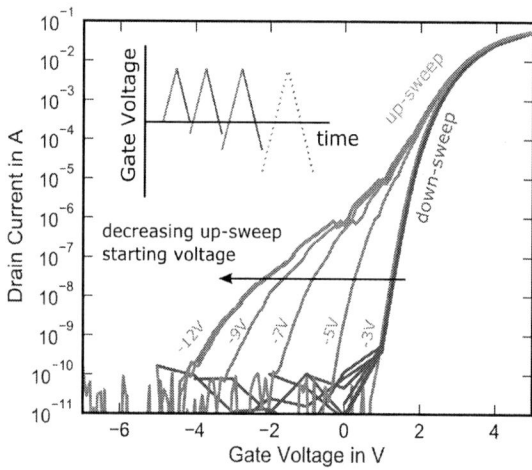

Fig. 1. Transfer characteristics of a CoolSiC™ MOSFET in subthreshold regime (conditions: $V_{DS} = 0.1$ V, V_{GS} ramp 1 V/s, 25°C). The inset shows how the gate voltage is swept with varying low starting voltage. The up-sweep curves (green) depend on the starting voltage. The down-sweep curves (blue) start always from +4 V and do not depend on starting voltage.

II. Hysteresis Phenomena

A. Hysteresis of transfer characteristics

Hysteresis effects can be observed e.g. by measuring the transfer characteristics of a SiC MOSFET, in particular in the subthreshold regime [3]. Fig. 1 shows the result of a CoolSiC™ MOSFET (conditions: $V_{DS} = 0.1$ V, V_{GS} ramp 1 V/s, 25°C). The inset shows how the gate voltage is swept with varying low level starting voltage. The lower the starting voltage of the up-sweep (green curves) and the lower the drain current the higher is the hysteresis between up-sweep and down-sweep and the lower is the subthreshold swing. On the other hand, all curves of the down-sweep (blue) are congruent. Therefore, the down-sweep starts always from +4 V (blue). Please note that this phenomenon is an intrinsic and reversible effect. All devices behave in the same manner.

978-1-5386-2928-4/18 $31.00 © 2018 IEEE   40

The effect can be explained by interface states which are distributed over the entire bandgap of SiC [4, 5]. The interface states are positively charged if a negative gate source voltage is applied and are neutralized at a positive gate bias. The amount of positive interface charges which have to be neutralized during the up-sweep depends on the gate voltage the up-sweep is starting with. Therefore the drain current onset has a lower level if the starting voltage is lower. During the down-sweep the interface remains inverted as long as a drain current is maintained. Thus the Fermi level position changes only slightly. The interface charge state is still nearly neutral. This explains why the down-sweep has a steeper subthreshold slope and does not depend on the sweep ramp. If the gate voltage is always above -3 V the interface is either depleted or inverted and thus there is nearly no hysteresis visible. That means in order to avoid large hysteresis it is beneficial to operate the device between -3V and +15V.

B. Hysteresis effects measured with AC gate signals

However, as power devices are used in switch mode it is important to understand how this hysteresis observed in DC mode affects the behavior in a real power device application. Therefore, the devices were investigated by measurements with very high time resolution and with a rapid technique to determine the threshold voltage within 1 μs [6]. A rectangular gate signal is applied with a frequency of 50 kHz in order to test close to application. Fig. 2 shows how the threshold voltage V_{TH} follows the gate signal. The gate is switched between V_{GSon} = +15 V and an off-level V_{GSoff} which is varied between 0 and −10 V (see color code in Fig. 2). There are two observations to notice:

(i) in *off-state* (see Fig. 2, time slots 0 … 10 μs and 20 … 30 μs, resp.) the threshold voltage V_{TH} is the lower the lower V_{GSoff} is and the longer the devices is kept in off-state.

(ii) in *on-state* (i.e. V_{GS} = +15V, time slots 10 … 20 μs and 30 … 40 μs, resp.) V_{TH} recovers very fast with a recovery time of about 1 μs. The signals are congruent for all V_{GSlow} < -4 V.

This behavior is in fact identical to the previously discussed hysteresis and can again be described by the physical model of positive interface charges. During turn-on the interface states are neutralized by capture of electrons from the conduction band. The higher V_{GSon} the more free electrons exist in the inversion channel and the faster the interface is neutralized. The situation is different in off-state (i.e. V_{GSlow} < V_{th}). Then the Fermi level drops well below the conduction band. Its position depends on V_{GSlow}. Those interface states located above the Fermi level are charged by capturing holes from the valence band. When the Fermi level drops the number of positively charged interface states increases. Thus V_{TH} drops with V_{GSlow}. The recovery time of the off-state scales with V_{GSlow} since the concentration of holes rules the recovery mechanism. For instance if the interface is only weakly accumulated with holes (V_{GS} ≈ -4 V) the off-time of 10μs is too short to charge all interface states above the Fermi level.

In any case the threshold voltage in on-state is the relevant one since it rules the saturation current of the channel which scales with $(V_{GSon} - V_{TH})^2$. Independent of the biasing the hysteresis observed in this AC test is fully reversible.

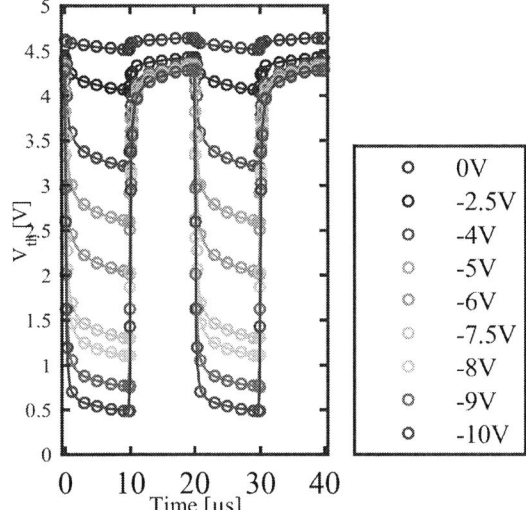

Fig. 2. Rapid threshold voltage measurement under AC gate stress: The gate is switched between +15 V and different off level voltages V_{GSoff} with a frequency of 50 kHz. The threshold voltage V_{TH} is determined within 1 μs with a special readout technique. Conditions: V_{DS} = 0.5 V, I_{Don} = 2 mA, T_j=175°C, V_{GSoff} varied from 0V … -10 V (see color code of legend).

C. Hysteresis effects during turn-on and turn-off

The hysteresis effect described above is as well visible in the turn-on transient if different off-levels of the gate signal are directly compared, e.g. V_{GSoff} = -10 V and -5 V, resp. Fig. 3 shows two corresponding turn-on transients of a CoolSiC™ 11 mΩ/1200V half bridge (type FF11R12W1M1_B11, conditions see subtitle of Fig. 3). Starting from a gate voltage of V_{GSoff} = -10 V (green) the corresponding load current (black) rises at a lower gate voltage compared to the case which starts from V_{GSoff} = -5 V (red). The reason is that the threshold voltage is depending on the preconditioning by a negative gate bias. Nevertheless, the influence of the hysteresis on the current turn-on signal is seen as harmless. There is nearly no difference in the voltage slope since this depends on the product of gate resistor and Miller capacity. Fig. 4 shows the turn-off transients of these 2 cases with different off levels of the gate voltage. As the device turns-off from +15 V there is no difference regarding the hysteresis in both cases. The difference is driven by the modified gate currents which vary by $\Delta I_G = \Delta V_{GSoff}/R_G$. Hence, the hysteresis effect is almost irrelevant for turn-off.

In sum the result of the switching behavior analysis is that the lower threshold voltage during off-state induces a faster turn-on, since the overdrive $V_{GS} - V_{TH}$ is higher and causes a slightly higher drain current. Vice versa, the turn-off is also faster since V_{TH} shifts upward during the on-state. The hysteresis effect of V_{TH} can plausibly be explained by positive interface states and follows the Schottky-Read-Hall mechanism. Charging and discharging time depends on the density of free electrons and holes at the interface and therefore on the gate bias level in on- and off-state.

978-1-5386-2928-4/18 $31.00 © 2018 IEEE

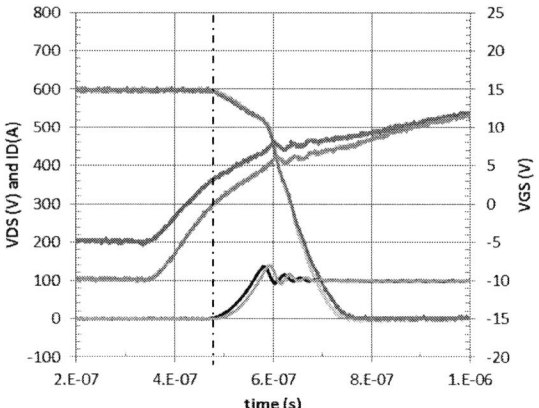

Fig. 3. Turn-on transient showing the influence of hysteresis: the current starts to rise (black) already at a lower gate voltage if the off-state level of the gate signal is -10 V (green) instead of -5 V (red). Conditions: half bridge type FF11R12W1M1_B11, $T_j = 150°C$, $R_G = 20 \, \Omega$, $V_{bus} = 600 \, V$, $I_{load} = 100 \, A$. Color code see Fig. 4.

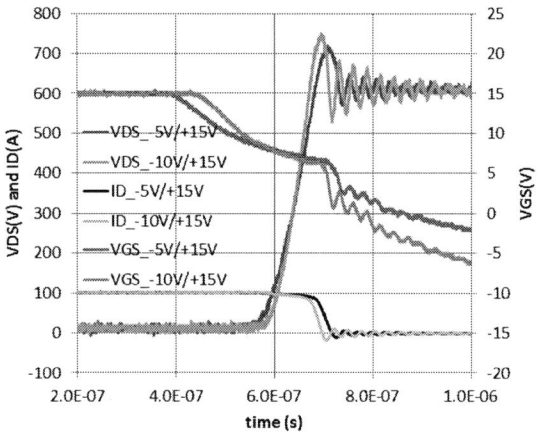

Fig. 4. Turn-off transient which does not show large differences between -5 V and -10 V V_{GSoff} since the down-sweep is stable. At the end of the Miller phase the current starts to switch-off from same gate level of ~6.5 V. Conditions see Fig.3.

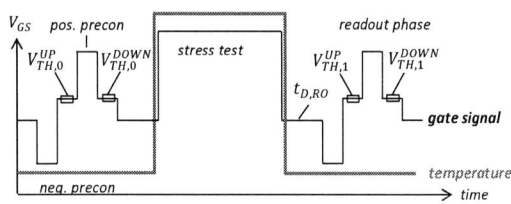

Fig. 5. Measure-stress-measure procedure suggested in [5] to evaluate BTI tests for SiC MOSFETs. Before threshold voltage and on-resistance readout a defined charge state at the interface is generated by negative and positive preconditioning.

III. THRESHOLD VOLTAGE READOUT PROCEDURE

The problem for qualification tests of SiC MOSFETS is now to make a difference between reversible hysteresis effects on the one hand side and long term drift on the other hand which is well known as *bias temperature instability* (BTI). It is more permanent and only detectable for much longer stress times compared to the switching period of some µs. BTI is in general caused by capture of charges of point defects in the gate oxide e.g. dimers, hydroxyl E' and others. The hysteresis effect can be excluded if the interface is forced into a well-defined charge state prior to V_{TH} readout. For this purpose a new *measure-stress-measure procedure* is proposed for SiC MOSFETs [5].

A *readout procedure* is proposed to be done in a sequence with negative and positive preconditioning pulse (see Fig. 5).

i. A negative gate pulse of e.g. -20 V generates accumulation of holes at the MOS interface and charges the interface states completely positive.

ii. Then the threshold voltage V_{TH}^{UP} is measured in MOS diode configuration (i.e. $V_G = V_D$) by forcing a current in mA range. V_{TH}^{UP} corresponds to the up-sweep curve in Fig. 1 (green) labeled with -12V.

iii. A positive preconditioning is done by applying a gate pulse of +20V. An inversion channel is formed which neutralizes the interface quickly.

iv. The threshold voltage V_{TH}^{DOWN} is measured in the same way as V_{TH}^{UP} but now corresponds to the down-sweep curve in Fig. 1 (blue). V_{TH}^{DOWN} is the more relevant threshold voltage since the voltage difference $V_{GSon} - V_{TH}^{DOWN}$ determines directly the on-state loss of the MOSFET. The hysteresis can now be defined by $\Delta V_{TH}^{HYST} = V_{TH}^{DOWN} - V_{TH}^{UP}$.

This procedure keeping the sequence (i)-(iv) has to be done for each readout, at least before and after the stress test. It has been shown that this method makes the readout much more invariant to the time delay after termination of stress. Thus, hysteresis effects can be correctly separated from permanent degradation [6].

IV. BTI TEST RESULTS

The measurement pattern of Fig. 5 is used to assess V_{TH} stability of CoolSiC™ SiC MOSFETs (data from [5]. The result is shown in Fig. 6, with a stress temperature of 150°C and a stress voltage of +35V. The stress voltage is 20V above the recommended use gate voltage of +15V. The time delay between stress and readout $t_{D,RO}$ was in the range of 1h ±15min. This time was needed to cool down the device and perform the readout sequence. Fig. 6 shows that the BTI drift of the tested SiC MOSFET is about 350 mV after 160.000s (= 45 h) under accelerated DC stress. There is only a very small drift in the hysteresis $\Delta\Delta V_{TH}^{HYST}$ indicating that the defects generated by BTI have a negligible impact on the hysteresis. The threshold voltage drifts follow a power law of the form $\Delta V_{TH} = a*t^b$. In case of ΔV_{TH}^{DOWN} the constants are a = 82 mV, b = 0.128, t is the stress time in s.

978-1-5386-2928-4/18 $31.00 © 2018 IEEE

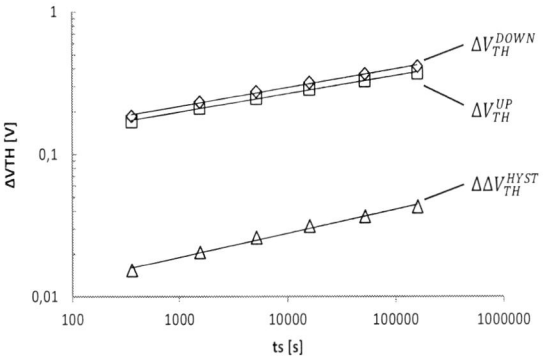

Fig. 6. BTI induced threshold voltage drifts as function of stress time using the measure-stress-measure pattern illustrated in Fig. 5. The stress bias was +35V and the stress temperature was 150°C. Please note that the change $\Delta\Delta V_{TH}^{HYST}$ with time is marginal.

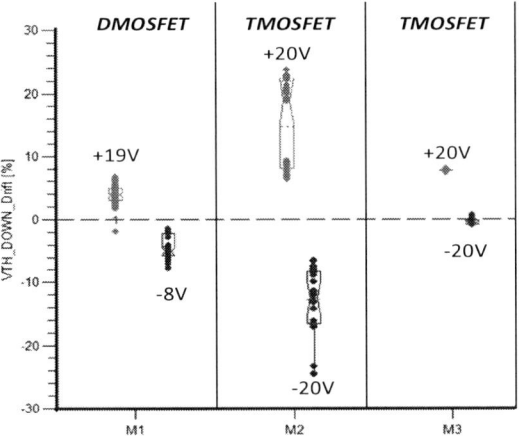

Fig. 7. Relative threshold voltage drift of SiC MOSFET devices of 3 vendors M1, M2, M3. Up to 200 devices per manufacturer were stressed for 500 h at 150°C using positive and negative DC stress between +20 V and -20 V. The stress biases are indicated in the figure. M1 is a DMOSFET device, M2 and M3 are TMOSFET devices. M3 is own hardware.

This BTI result allows a reliable life time extrapolation similar as for Si MOSFETs.

Different SiC MOSFET parts from three different manufacturers were tested in this way (Fig. 7). The tested devices differ considerably in BTI drift amplitude and drift variation. The differences are attributed to variations in device processing and device design. Own devices (M3) show a very narrow drift range and less than 10% increase in V_{TH} after 500h positive BTS at +20V. Applying negative gate bias stress (NBTI) only manufacturer M3 showed a very narrow drift range and almost no drift.

Regarding BTI tests no immanent risks are seen as long as

i. the MOS process is optimized to minimum threshold drift,
ii. a tight distribution of V_{TH} is achieved and
iii. changes of V_{TH} are predictable and homogeneous.

Conclusio

It was pointed out that the hysteresis of SiC MOSFETs can be attributed to interface states which are distributed over the entire bandgap of SiC. The interface states are positively charged if a negative gate source voltage is applied and are neutralized at a positive gate bias. The time needed for charging and neutralization depends on the free carrier densities in off-state and on-state and therefore on the gate bias in off- and on-state. The good news is that this hysteresis (i) is fully reversible, (ii) harmless for turn-on and (iii) almost irrelevant for turn-off. A lower threshold voltage during off-state induces a faster turn-on since the overdrive $V_{GS} - V_{TH}$ is higher causing a higher drain current. Vice versa, the turn-off is also faster since V_{TH} shifts upward during the on-state.

A test sequence is proposed to clearly separate hysteresis effects from permanent threshold voltage drift by gate bias and temperature (BTI). The readout sequence uses both a negative and a positive preconditioning gate pulse which puts the interface states into a well-defined state.

The BTI tests reveal a drift which is higher as compared to silicon devices but is predictable and can be anticipated within the data sheet limitations.

References

[1] D. Peters et al., "Performance and ruggedness of 1200V SiC Trench MOSFET," 2017 29th International Symposium on Power Semiconductor Devices and IC's (ISPSD), Sapporo, 2017, pp. 239-242.

[2] D. B. Habersat, R. Green, A.J. Lelis, "Feasibility of SiC Threshold Voltage Drift Characterization for Reliability Assessment in Production Environments", Materials Science Forum, ISSN: 1662-9760, Vol. 897, pp 509-512.

[3] G. Rescher et al., "On the subthreshold drain current sweep hysteresis of 4H-SiC nMOSFETs" 2016 IEEE International Electron Devices Meeting (IEDM), San Francisco, CA, 2016, pp. 10.8.1-10.8.4.

[4] G. Rescher et al., "Preconditioned BTI on 4H-SiC: Proposal for a Nearly Delay Time-Independent Measurement Technique", IEEE Transactions on Electron Devices (2018).

[5] T. Aichinger, G. Rescher, G. Pobegen, "Threshold voltage peculiarities and bias temperature instabilities of SiC MOSFETs", Microelectronics Reliability 80 (2018): 68-78.

[6] K. Puschkarsky, H. Reisinger, T. Aichinger, W. Gustin, T. Grasser, "Threshold voltage hysteresis in SiC MOSFETs and its impact on circuit operation", IIRW 2017, Lake Tahoe.

Proceedings of the 30th International Symposium on Power Semiconductor Devices & ICs
May 13-17, 2018, Chicago, USA

Deep-P Encapsulated 4H-SiC Trench MOSFETs With Ultra Low $R_{on}Q_{gd}$

Yasuhiro Ebihara, Aiko Ichimura, Shuhei Mitani, Masato Noborio, Yuichi Takeuchi, Shoji Mizuno, Toshimasa Yamamoto, and Kazuhiro Tsuruta

DENSO CORPORATION
Nisshin, Aichi, 470-0111, Japan
Yasuhiro_ebihara@denso.co.jp

Abstract— Deep-P encapsulated 4H-SiC trench MOSFET was proposed. The fabricated MOSFET with a blocking voltage of 1800V demonstrated a ultra low $R_{on}Q_{gd}$ of 133 nCmΩ. The structure optimization was carried out for switching-loss reduction. The improved switching characteristics were obtained by the balanced JFET resistance and the gate-drain capacitance.

Keywords; SiC-MOSFET, Figure of Merit, Miller charge, on-resistance, gate-drain charge

I. INTRODUCTION

SiC MOSFETs are attractive for low loss, voltage-controlled and normally-off power devices. Year by year, performances and reliabilities have been enhanced by reducing specific on-resistance and using a relaxation technique of electric-field crowding at the gate oxide. Among the many types of SiC MOSFETs, 4H-SiC trench MOSFETs have a great potential owing to high MOS-channel density and high channel mobility on the trench sidewall [1,2]. In contrast to the planar-type SiC MOSFETs [3,4], however, there are two major issues to be solved in the trench-type SiC MOSFETs to make the best use of their potentials.

First problem is a trade-off between the JFET resistance and the channel resistance. The P-type layer is commonly formed under or beneath the gate trench to avoid the electric field crowding at gate oxide of the trench bottoms. The P-type layer, whose pitch is similar to the gate pitch, defines the conduction path of the JFET. If the gate pitch is decreased in order to lower the overall channel resistance, it shortens the conduction path. Therefore, it causes unwanted increase in the JFET resistance. Consequently, the cell pitch is limited to the point where the JFET resistance does not accounts for the majority of the specific on-resistance. In order to maximize the channel density, the trade-off should be improved.

Second problem is a high gate-drain charge(Q_{gd}) which deteriorates the switching behavior. As the drift thickness of SiC MOSFET is about ten times thinner than that of the Si, the gate-drain charge of SiC MOSFET tends to be larger than that of the Si. Moreover, the more the gate cell is packed, the more the gate area expands, which results in the larger gate-drain charge than that of the planar-type SiC-MOSFETs. Therefore, efforts must be made to decrease the gate drain charge.

In the previous work, the new type of MOSFET called Deep-P encapsulated 4H-SiC trench MOSFET was presented and demonstrated the improvement of the trade-off between the JFET resistance and the channel resistance while the relaxation of the electric field crowding was retained[5]. In the Deep-P encapsulated 4H-SiC trench MOSFETs, the gate trench pitch and the JFET pitch was independently designed to minimize their resistance to the limit of the frontend fabrication tolerance. As a result, Deep-P encapsulated 4H-SiC trench MOSFET exhibited a ultra-low specific on-resistance of $2m\Omega cm^2$ and a high avalanche breakdown voltage of 1800V.

In this paper, the authors report the drastic reduction of gate-drain capacitance(C_{gd}), charges, and improvement of the Figure of Merit which is defined as the product of specific on-resistance and the gate-drain charge. Furthermore, the optimization of the cell structure for low switching losses is also presented.

II. DEVICE CONCEPT AND FABRICATION

Figures 1 (a) and (b) show the 3D schematic view of the conventional 4H-SiC trench MOSFET [6] and the Deep-P encapsulated 4H-SiC trench MOSFETs, respectively. The p-type layer, named Deep-P region, is formed under the p-base region in order to relax the electric field crowding at the bottom corner of the trench gate oxide. The differences of these structure are the geometrical structure of the Deep-P region and trench gate cell pitch. Although the Deep-P region of conventional structure is parallel to the trench gate sidewall, that of the presented structure is orthogonal to the trench. The depth of each Deep-P region is almost the same. For the on-resistance reduction, the trench gate cell pitch and JFET cell pitch, which equals the Deep-P pitch, must be reduced. The presented structure enables independent design of these pitches. As a result, trench gate pitch and JFET cell pitch of the presented structure are 34% and 64% smaller than those of the conventional structure, respectively. For switching-loss reduction, it is important to reduce the gate-drain capacitance and gate-drain charge. The electric potential and the depletion region of the conventional and presented MOSFETs were calculated by using a device simulator. Figures 2 (a) and (b) show the electric potentials of conventional and presented

978-1-5386-2928-4/18 $31.00 © 2018 IEEE

structure, respectively, under a drain voltage of 100V and a gate voltage of 0V. The depletion region for each structure expands toward the drift region to almost the same depth. Compared with the conventional structure, the presented structure has the non-depleted Deep-P region under the trench-gate oxide. The non-depleted region acts as the shield against the drain bias, which can shrink the effective gate-drain capacitance area. Hence, low gate-drain capacitance and low gate-drain charge are achieved in the presented structure. This cell construction inherently has small ratio of the gate-drain charge with respect to the gate-source charge. This feature is essential to suppress the parasitic turn-on phenomena which may cause the fatal failure in the application using the half bridges. In addition, the proposed structure can avoid the switching delay of the Deep-P charge by connecting Deep-P region closely to each P-base region [7].

Deep-P gap should be selected so as to retain the low JFET resistance and the low gate-drain capacitance. Figure 3 shows the simulated results of Deep-P gap dependences of the gate-drain capacitance and the JFET resistance. From fig. 3, while the gate-drain capacitance increases by increasing the Deep-P gap, the JFET resistance, which directly affects the conduction loss, decreases. For DC/DC converter or high-speed applications, it is more important to reduce the switching losses rather than the conduction losses, and the low gate-drain charge is generally a best choice for minimizing the switching losses. Therefore, it is considered that the narrow Deep-P gap is preferable. On the other hand, the wide Deep-P gap is suitable for low-speed applications because the reduction of conduction losses are more important than that of the switching losses. In that case, the Deep-P gap is limited up to the acceptable level of electric field crowding at the gate oxide. Thus, the device design can be optimized by selecting the appropriate Deep-P gap for each application without modification of the channel layout.

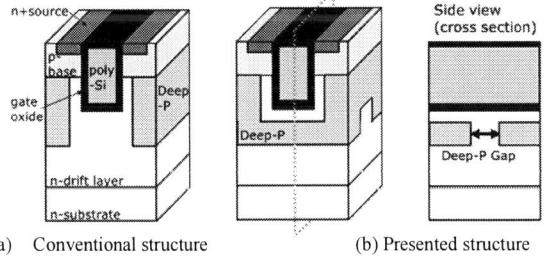

(a) Conventional structure (b) Presented structure

Fig.1 3D sketch of (a) conventional structure and (b) presented structure (Deep-P encapsulated 4H-SiC trench MOSFET).

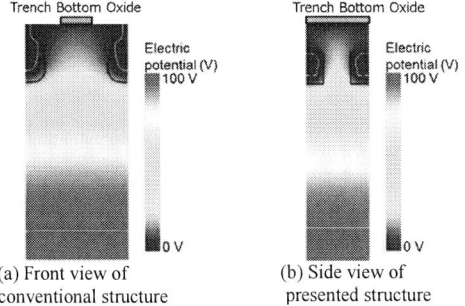

(a) Front view of (b) Side view of
conventional structure presented structure

Fig. 2 Electric potential of (a) conventional and (b) presented structure at V_d=100V and V_g=0V.

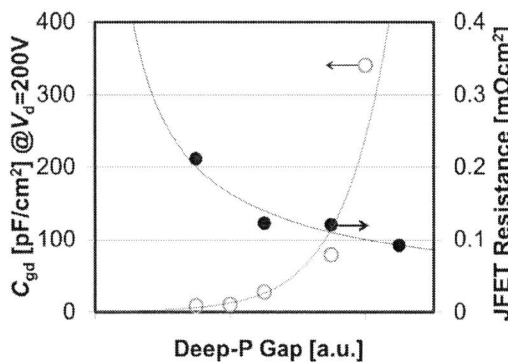

Fig. 3 Deep-P gap dependences of C_{gd} and JFET resistance.

The MOSFETs were fabricated on an n-type 4°-off-axis 4H-SiC epitaxial layer. The concentration and the thickness of the epitaxial layer was chosen to obtain a breakdown voltage of over 1700V. The gate oxide was deposited by using CVD, followed by nitridation and post oxidation annealing (POA). The oxide thickness is designed for the commonly used on-state gate-source voltage of V_{gs}=+20V. The MOSFETs with narrow, intermediate, and wide Deep-P gap were fabricated to optimize the Deep-P gap.

III. RESULTS AND DISCUSSION

A. Deep-P Gap optimization for the switching behavior

Figure 4 shows the measured gate-source capacitance (C_{gs}), gate-drain capacitance, and drain-source capacitance (C_{ds}) of the fabricated MOSFETs with narrow, intermediate, and wide Deep-P gap. The drain-source capacitances strongly depend on the applied drain voltages. The gate-drain capacitances fall drastically at low drain voltages and slightly increases as the drain voltage increases. The increase of the gate-drain capacitance at high drain voltage is caused by weakening of the shield effect with extending the depletion layer towards inside the Deep-P region. The gate-drain capacitance at high drain voltage decreases as the Deep-P gap became narrower, which agrees with the simulated results. Regardless of the Deep-P gap, the gate-source capacitance was almost the same in each MOSFET.

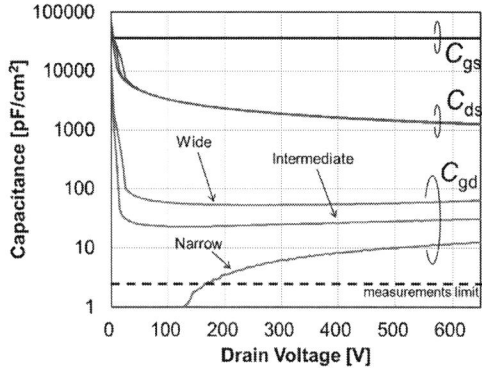

Fig. 4 Measured C_{gd}, C_{ds}, and C_{gs} of fabricated MOSFETs with wide, intermediate, and narrow Deep-P gap.

978-1-5386-2928-4/18 $31.00 © 2018 IEEE

In the case of large gate-source capacitance, low gate-drain capacitance and charge do not necessarily mean short Miller period and low switching losses because of the non-flat Miller effects. It is well known that non-flat Miller effect occurs in the SiC-MOSFET because the transconductance(g_m) of the SiC MOSFETs is relatively lower than that of the Si MOSFETs. In addition, the drain current of the SiC MOSFETs at the high drain voltage shows the non-saturated characteristics. During operation in the non-flat Miller region, not only the gate-drain charge but also gate-source charges are needed, which causes the switching delay [8].

Figure 5 shows the turn-off waveforms of the fabricated MOSFETs with narrow, intermediate, and wide Deep-P gap. The dI/dt of each structure was almost the same because the gate-source capacitance and charge are almost the same. The differences are the dV/dt of each structure and the gate voltages during the Miller periods. As is clear from this figure, the gate voltages of the fabricated MOSFETs with the narrow Deep-P gap was higher than those of the other structures at the beginning of the Miller periods while the gate voltage became close at the end of the Miller periods.

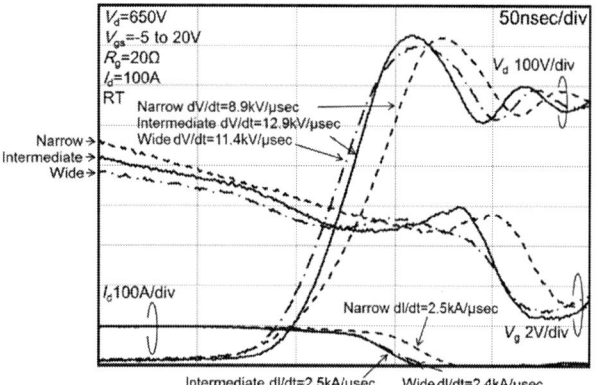

Fig. 5 Turn-off waveforms of fabricated MOSFETs with narrow, intermediate and wide Deep-P gap. The dI/dts and the dV/dts were calculated by the slopes from 30% to 70% of V_d and I_d.

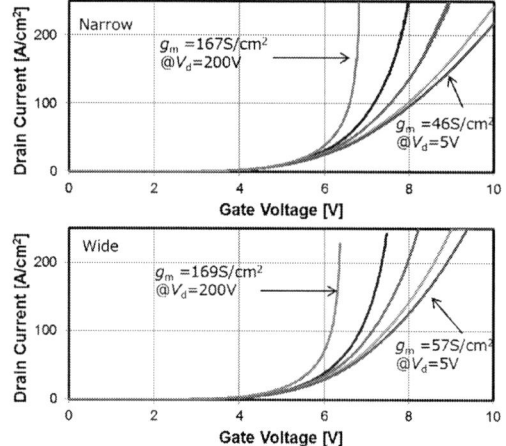

Fig. 6 Measured V_g-I_d curve of 0.6mm□ TEG MOSFETs with narrow and wide Deep-P gap.(V_d=5, 10, 50, 100, 200V, RT). G_m was calculated by the slope from 40A/cm² to 160A/cm².

Thereby, contrary to expectation, the dV/dt of narrow Deep-P gap structure was smallest of all though the gate-drain capacitance was smallest of all. The small dV/dt is caused by the large JFET resistance (see Fig. 3) which deteriorates the transconductance at the beginning of the Miller periods or at a low drain voltage as shown in Fig. 6. At the end of the Miller period or at a high drain voltage, the current was mainly governed by the channel resistance. Hence, the trans-conductance is not affected by the JFET resistance as shown in Fig. 6. On the other hand, although the wide Deep-P gap structure has the lowest JFET resistance (see Fig. 3), the large gate-drain capacitance causes the long Miller period. The fastest dV/dt is achieved at the MOSFET with intermediate Deep-P gap that has the balanced JFET resistance and gate-drain capacitance. As a result, the Deep-P gap was optimized for low switching losses by considering the delay of the gate-source capacitance and the gate-drain capacitance.

B. Static Characteristics of optimized structure

Figures 7 (a) and (b) show the on-state forward I_d-V_{ds} characteristics at room temperatures and 150°C. The measured specific on-resistance was 2.04mΩcm² at room temperature and 3.47 mΩcm² at 150°C. The specific on-resistances were calculated at a gate voltage of 20V and a drain current of 300A/cm². The breakdown voltage at room temperature was above 1800V as shown in Fig. 8 (c). The specific on-resistance of conventional structure was 3.5mΩcm² (not shown). Compared with the conventional MOSFET, the on-resistance was reduced approximately by 43%. The threshold voltage was about 2.7V at room temperature. The threshold voltage was measured at a drain current of 100 mA/cm² and a drain voltage of 10V.

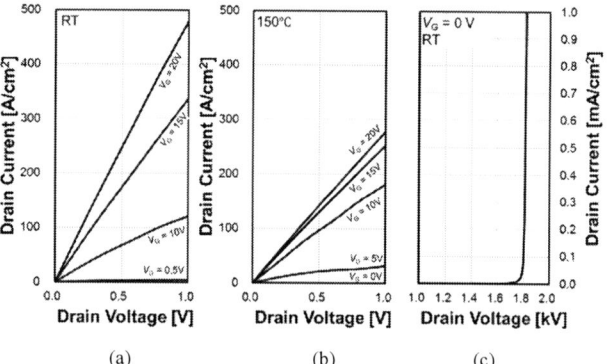

Fig. 7 On- and off- state characteristics of fabricated MOSFET with optimized Deep-P gap at RT and 150°C.

C. Dynamic behavior of optimized structure

Capacitance-Voltage measurements, gate charge tests, and double pulse tests of the presented structure with optimized Deep-P gap and conventional structure were performed to confirm the shielding effect. Figure 8 shows the measured gate-drain capacitance and drain-source capacitance. The gate-drain capacitance of the presented structure at a drain voltage of 200V was about 30 pF/cm², while that of the conventional structure was about 200 pF/cm².

978-1-5386-2928-4/18 $31.00 © 2018 IEEE

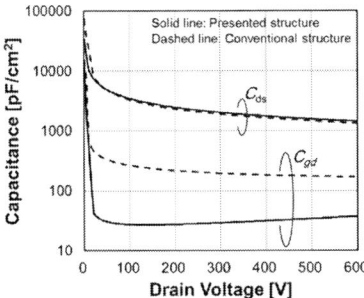

Fig. 8 Measured results C_{gd} and C_{ds} of conventional and presented structure.

Figures 9 (a) and (b) show the tested results of the gate charge (Q_g) of the presented structure and the conventional structure, respectively. The presented structure achieved a gate drain charge of 65nC/cm², which is lower than that of conventional structure (143nC/cm²). Gate-drain charges were calculated within the 10 to 90% range of the drain voltage. In order to estimate only gate-drain charge, the measurements were performed at low load current. The Turn-on and Turn-off waveforms were depicted in Fig. 10. Owing to the low gate-drain capacitance and charge, the presented structure could reduce the E_{on} by 32%, and E_{off} by 40%. Figures 12 (a) and (b) show the drain current dependence of $E_{on} + E_{off}$ and the dI/dt dependence of E_{off}, respectively, in the fabricated MOSFETs with optimized Deep-P gap and conventional structure. Compared with the conventional structure, the total losses were effectively reduced in the presented structure. It is an important issue whether the trade-off between the dI/dt and E_{off} is improved or not to minimize the E_{off}, given that the drain voltage surge is well suppressed during the turn-off period. As shown in Fig. 11 (b), E_{off} was effectively reduced at the same dI/dt, indicating that the high dV/dt of presented structure was attributed to the improvement of the trade-off.

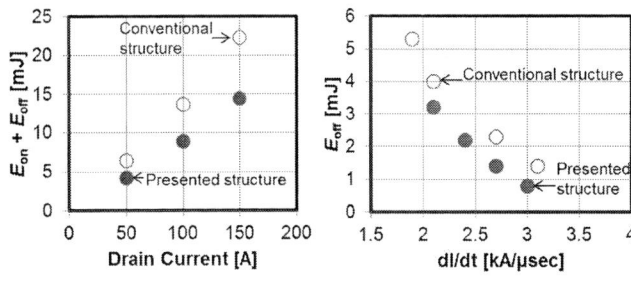

(a) Drain current dependence (b) dI/dt dependence

Fig. 11 (a) Drain current dependence of $E_{on} + E_{off}$ (V_d=650V, V_g=-5/20V, I_d=50, 100,150 A, R_g=30Ω, RT) and (b) dI/dt dependence of E_{off} (V_d=650V, V_g=-5/20V, I_d=100A, R_g=5, 10, 20, 30Ω, RT), respectively.

IV. SUMMARY

The Deep-P encapsulated 4H-SiC trench MOSFET was proposed. The fabricated MOSFET with a blocking voltage of 1800V demonstrated a ultra low $R_{on}Q_{gd}$ of 133 nCmΩ. Furthermore, taking into account of the non-flat Miller effects, the optimization of the gate-drain capacitance and the JFET resistance was carried out to minimize the switching losses. Consequently, the fabricated MOSFET demonstrated not only the ultra low $R_{on}Q_{gd}$ but also superior switching performance.

ACKNOWLEDGMENT

This work is supported by Toyota Motor Corporation and Toyota Central R&D Labs., Inc..

REFERENCES

[1] J.A. Cooper Jr., M.R. Melloch, R. Singh, A. Agarwal, J.W. Palmour, "Status and prospects for SiC power MOSFETs" IEEE Trans. Electron Devices vol.49, p.658-64, 2002.

[2] H. Yano, H. Nakao, T. Hatayama, Y. Uraoka, and T. Fuyuki, "Increased Channel Mobility in 4H-SiC UMOSFETs Using On-Axis Substrates" Mater. Sci. Forum, Vols.556-557, pp.807-810, 2007.

[3] Q. Zhang, G. Wang, H. Doan, S-H. Ryu, B. Hull, J. Young, S. Allen, J. Palmour, "Latest Results on 1200V 4H-SiC CIMOSFETs With R$_{sp,on}$ of 3.9 mΩcm² at 150°C", Proceedings of ISPSD, pp. 89-92, 2015.

[4] P. Losee, A. Bolotnikov, L. Yu, R. Beaupre, Z. Stum, S. Kennerly, G. Dunne et al, "1.2kV Class SiC MOSFETs with Improved Performance over Wide Operating Temperature" Proceedings of ISPSD, pp. 89-92, 2015

[5] A. Ichimura, Y. Ebihara, S. Mitani, M. Noborio, Y. Takeuchi, S. Mizuno, T. Yamamoto, and K. Tsuruta, "4H-SiC Trench MOSFET with Ultra-Low On-Resistance by using Miniaturization Technology", ICSCRM2017, to be published in Mater. Sci. Forum in 2018.

[6] https://www.denso.com/jp/ja/products-and-services/industrial-products/sic/

[7] S. Kyogoku, K. Ariyoshi, R. Iijima, Y. Kobayashi and S. Harada, "Role of trench bottom shielding region on switching characteristics in 4H-SiC Double-trench MOSFETs", ICSCRM2017, to be published in Mater. Sci. Forum in 2018.

[8] B. Agrawal, M. Preindly, B. Bilgin, and A. Emad, "Estimating Switching Losses for SiC MOSFETs with Non-Flat Miller Plateau Region", Applied Power Electronics Conference and Exposition (APEC), pp2664-2670, 2017.

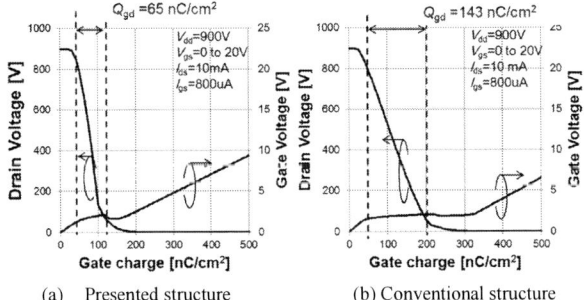

(a) Presented structure (b) Conventional structure

Fig. 9 Measured gate charge of (a) presented and (b) conventional structure.

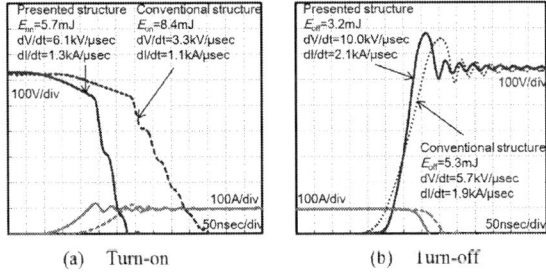

(a) Turn-on (b) Turn-off

Fig. 10 Double pulse waveforms of (a) turn-on and (b) turn-off of conventional and presented structure. (V_d=650V, V_g=-5/20V, I_d=100A, R_g=30Ω, RT)

978-1-5386-2928-4/18 $31.00 © 2018 IEEE

Proceedings of the 30th International Symposium on Power Semiconductor Devices & ICs
May 13-17, 2018, Chicago, USA

Influence of the Off-state Gate-Source Voltage on the Transient Drain Current Response of SiC MOSFETs

Christian Unger, Martin Pfost
Chair of Energy Conversion, TU Dortmund
Emil-Figge-Straße 68, Dortmund, Germany - email: christian.unger@tu-dortmund.de

Abstract—In this work we investigate the effect of negative off-state gate-source voltages on SiC MOSFETs. With increasingly negative $V_{\mathrm{GS,off}}$ voltages, a more pronounced drain current overshoot immediately after turn-on is observed. This effect is most noticeable in saturation, where the drain current is determined primarily by the channel. The phenomenon is attributed to positively charged oxide- and interface-traps that temporarily enhance the inversion charge in the channel before they are gradually neutralized. The amount of charged traps depends on the position of the valence band edge in accumulation, hence the $V_{\mathrm{GS,off}}$ dependence. Two distinct components with very different time constants are observed.

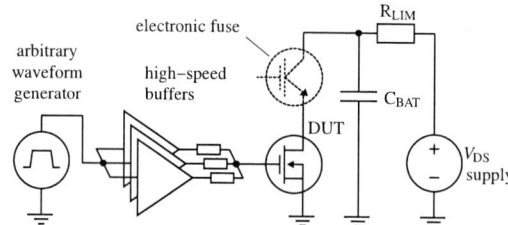

Fig. 1. Setup used for the pulsed measurements.

I. INTRODUCTION

Power modules equipped with SiC MOSFETs are now becoming more readily available. To fully exploit the superior capabilities of SiC-MOSFETs, suitable gate-drive concepts are required. Unfortunately, fast-switching devices and packages that have relatively high gate-loop impedances are particularly susceptible to parasitic turn-on. Thus, gate drivers must manage fast switching while preventing parasitic turn-on. This is usually achieved by using a negative off-state gate-source voltage $V_{\mathrm{GS,off}}$ [1].

However, SiC MOSFETs are known to exhibit threshold voltage instabilities after positive and negative gate bias conditions due to charge trapping in the oxide and at the SiC/SiO_2 interface [2], [3]. This effect is mostly investigated in the context of gate reliability and is often measured via the transfer characteristics after high gate bias stress at high temperature and characterized via a threshold voltage shift ΔV_{th}. With this method, the measurement accuracy is strongly affected by the time it takes to perform a sweep since the various trap states can exhibit very different time constants that range from a few µs [4] to several seconds [5].

In this work we perform a more application-oriented investigation of the impact of negative off-state gate-source voltages on SiC MOSFETs. An emphasis is put on the influence of the trap states and the subsequent transient drain current change over time. Furthermore, the implications on device characterization and performance under short-circuit conditions are analyzed.

II. GATE DRIVER & MEASUREMENT SETUP

The device under test (DUT) is a commercially available 600 V-class device with an $R_{\mathrm{DS,on}}$ of 125 mΩ.

The measurement setup (see Fig. 1) is optimized for low inductance and features a high-speed analog gate-drive that allows to deploy complex waveforms to the DUT. For this purpose,

an arbitrary waveform generator (AWG) in conjunction with several high-speed, high-current buffers is used. To balance the buffers and allow parallel operation, the outputs are coupled via resistors. In case of a device failure, the short-circuit is safely terminated by an IGBT acting as an electronic fuse.

III. MEASUREMENTS

To illustrate the phenomenon, the transient drain current response of a SiC-MOSFET is compared to a conventional Si-Superjunction-MOSFET in Fig. 2. The devices are operated in the saturation region and pulsed to an on-state gate-source voltage $V_{\mathrm{GS,on}}$ that results in a comparable drain current I_{D}.

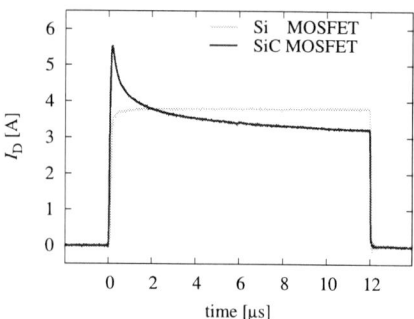

Fig. 2. Comparison between a Si-Superjunction- and a SiC-MOSFET. The devices are operated in saturation at $V_{\mathrm{DS}} = 10$ V. The SiC-device is pulsed from $V_{\mathrm{GS,off}} = -6$ V to $V_{\mathrm{GS,on}} = 7$ V while the Si-device is pulsed from $V_{\mathrm{GS,off}} = -6$ V to $V_{\mathrm{GS,on}} = 4.2$ V to obtain a comparable current.

For the Si device I_{D} reaches a steady state, while the current continuously decreases for the SiC device.

978-1-5386-2928-4/18 $31.00 © 2018 IEEE 48

To analyze the $V_{GS,off}$-dependence of this phenomenon, pulses with different $V_{GS,off}$ levels but same $V_{GS,on}$ are measured in Fig. 3. In addition, short interruptions of $100\,\text{ns}$ are introduced.

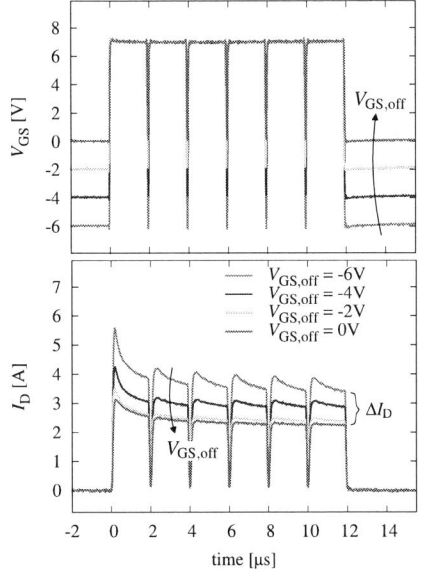

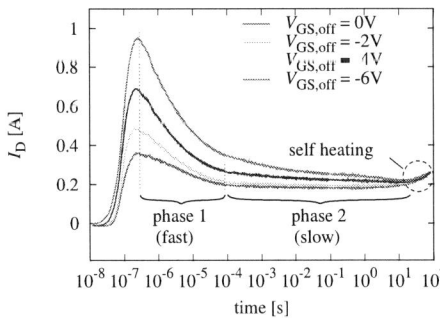

Fig. 3. Influence of the $V_{GS,off}$ voltage on the transient drain current change in saturation at $V_{DS} = 10\,\text{V}$. The device is pulsed from $V_{GS,off} = -6\,\text{V}, -4\,\text{V}, -2\,\text{V},$ and $0\,\text{V}$, respectively, to $V_{GS,on} = 7\,\text{V}$.

A significant increase of the drain current with more negative $V_{GS,off}$ values is observed. Like in Fig. 2, the drain current diminishes with time. However, the I_D reduction extends across the short interruptions. This is a clear indication that the phenomenon possesses a memory effect. At the end of the $12\,\mu\text{s}$ pulse, a current difference ΔI_D remains. To further investigate this phenomenon, longer pulses are measured and displayed in the semilogarithmic plot in Fig. 4.

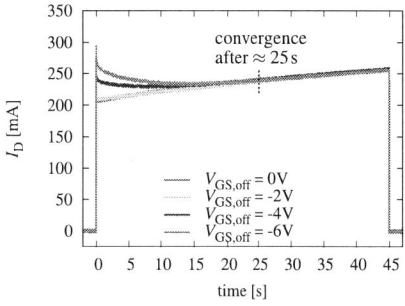

Fig. 4. Time-logarithmic plot of a $100\,\text{s}$ pulse at $V_{DS} = 5\,\text{V}$ and $V_{GS,on} = 5\,\text{V}$ for different $V_{GS,off}$. To maintain a comparable sample rate across ten orders of magnitude, the pulse is constructed from several measurements. By optical inspection two distinct phases can be identified.

The variation of I_D over time can be separated into two phases by using $V_{GS,off} = 0\,\text{V}$ as a reference. During phase 1 a sig-

nificant and fast reduction of the initial I_D peak value over the first $100\,\mu\text{s}$ is observed for all $V_{GS,off}$ voltages. Then, a lower but longer decrease occurs during phase 2 ($100\,\mu\text{s} - 100\,\text{s}$). This however, is only measureable for $V_{GS,off} \leq -2\,\text{V}$.

The I_D increase towards the end of the pulse ($10\,\text{s} - 100\,\text{s}$) can be explained by self-heating. This is even more apparent on the linear time scale in Fig. 5.

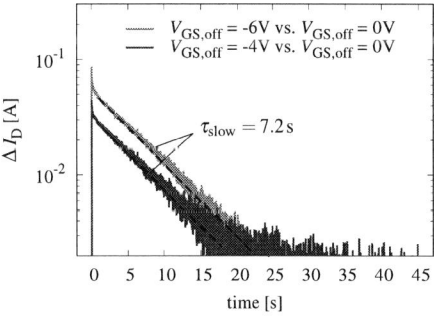

Fig. 5. Long pulses to characterize the slow component of the drain current decrease during phase 2, cf. Fig. 4. The traces for $V_{GS,off}=0\,\text{V}$ and $V_{GS,off}=-2\,\text{V}$ overlap almost completely.

To limit self-heating, low values of $V_{DS} = 5\,\text{V}$ and $V_{GS,on} = 5\,\text{V}$ are used for the measurements in Fig. 4 and Fig. 5. Nevertheless, I_D increases due to operation below the temperature compensation point where $\partial I_D / \partial T$ is positive. In order to extract a time constant for the drain current decay, the superimposed drain current change due to self-heating needs to be compensated. Therefore, the drain current at $V_{GS,off} = 0\,\text{V}$ is used as a basis to compute the drain current difference ΔI_D in Fig. 6. On a first order approximation this compensates the influence of self-heating and allows for the extraction of a time constant.

Fig. 6. Logarithmic plot of the drain current differences ΔI_D between $V_{GS,off}=-6\,\text{V}$ and $V_{GS,off}=0\,\text{V}$ ($V_{GS,off}=-4\,\text{V}$ and $V_{GS,off}=0\,\text{V}$ respectively) from Fig. 5.

Assuming a standard decay process $\Delta I_D(t) \propto \cdot \exp(-t/\tau)$, a time constant of $\tau_{slow} = 7.2\,\text{s}$ is determined for the slow I_D decrease during phase 2 in Fig. 4.

The fast component during phase 1, however, can not be described by a singular time constant. Therefore, a multi-exponential approach $\Delta I_D(t) = \sum_i A_i \cdot \exp(-t/\tau_i)$ is used to extract a spectrum of time constants τ_i and the corresponding amplitudes A_i via the software tool MERA [6]. For this, the drain current difference calculated from Fig. 4 is used. The resulting spectrum is displayed in Fig. 7.

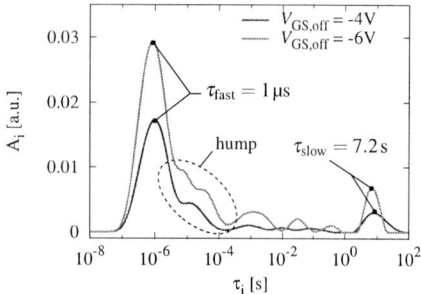

Fig. 7. Spectra extracted from the drain current differences between $V_{GS,off} = -6$ V and $V_{GS,off} = 0$ V ($V_{GS,off} = -4$ V and $V_{GS,off} = 0$ V respectively) from Fig. 4. Two predominant peaks at $\tau_{fast} = 1\,\mu s$ and $\tau_{slow} = 7.2\,s$ are observed.

For both $V_{GS,off}$ voltages, similar spectra with almost identical maxima are observed. The time constant $\tau_{slow} = 7.2\,s$, determined previously by the conventional approach in Fig. 6, is also present in the spectra. In case of the fast components, a peak at $1\,\mu s$ and a hump spanning from approximately $5\,\mu s$ to $150\,\mu s$ is found.

IV. PHYSICAL BACKGROUND & IMPLICATIONS

The reduction of I_D over time is attributed to the high trap density at the SiC/SiO$_2$ interface and within the oxide (cf. Fig. 8).

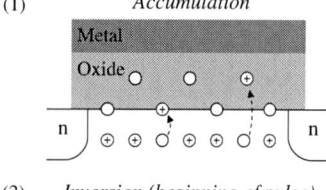

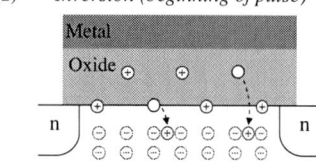

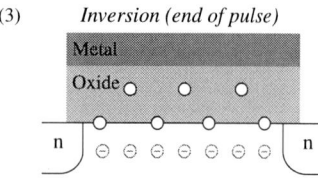

Fig. 8. Schematic representation of the charging and neutralization of traps in a metal-oxide-semiconductor system.

In accumulation, these interface- and oxide traps are charged by holes from the p-body. After turning on the device, the positively charged traps enhance the inversion charge in the channel. However, the positively charged traps are gradually neutralized by electrons from the channel. In accordance, the inversion charge and therefore the drain current decreases over time. The total amount of the trapped charge (and thus the extent of the I_D change) depends on the valence band edge in accumulation which is determined by $V_{GS,off}$.

The temporary increase of the inversion charge can be described via a shift of the threshold voltage which becomes apparent in the pulsed transfer characteristics for different $V_{GS,off}$ in Fig. 9.

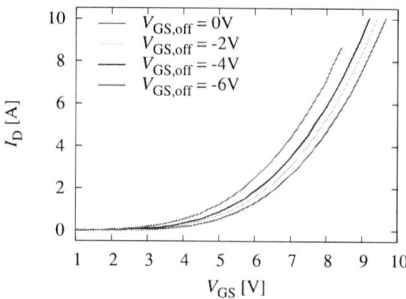

Fig. 9. Transfer characteristics obtained by fast pulsed measurements with a pulse width of $t_{pulse} = 1\,\mu s$ at $V_{DS} = 50$ V for $V_{GS,off} = -6$ V, -4 V, -2 V and 0 V.

As a consequence, the off-state gate biasing must be taken into account during device characterization, especially when measuring the device in the sub-threshold or weak inversion region. However, the impact of the trapping phenomenon is also observed in strong inversion, e.g. during the short-circuit conditions in Fig. 10.

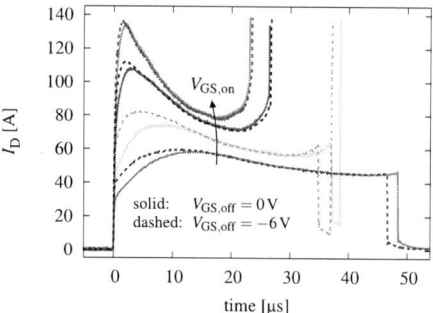

Fig. 10. Drain currents during a short-circuit at $V_{DS} = 400$ V and $V_{GS,on} = 12$ V, 14 V, 16 V, 18 V for $V_{GS,off} = -6$ V and 0 V, respectively.

Switching into a short from $V_{GS,off} = -6$ V leads to higher peak currents compared with $V_{GS,off} = 0$ V. This increases the power dissipation and in consequence causes the device to fail earlier for $V_{GS,on} = 12$ V and 14 V. For higher V_{GS}, the effect is negligible probably due to the already strongly inverted channel. In contrast to the long time constants in Fig. 4, the drain currents converge much quicker during a short-circuit. This is due to the high temperature which accelerates the neutralization of the traps by hot electrons.

During the high power pulse in Fig. 11, this electron current through the oxide becomes large enough to cause a significant voltage drop over the gate resistor.

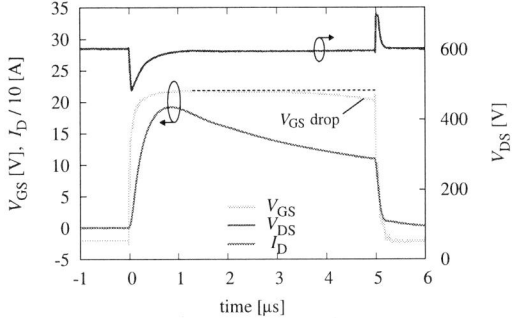

Fig. 11. Pulse with very high power at $V_{DS} = 600\,V$ and $V_{GS} = 21.5\,V$. The pulse is short enough not to damage or degrade the device. The V_{GS} drop towards the end of the pulse is caused by the resistance in series to the gate and indicates the onset of significant gate current.

This is the reason of the often-observed drop of the gate-source voltage in SiC-MOSFETs during short-circuit operation [7] and can be used to directly measure the gate current I_G. The logarithmic plot of I_G in Fig. 12 shows a linear increase towards the end of the pulse which indicates an temperature induced process that can be described by $I_G(T) \propto \exp(-E_A/k_BT)$. To determine the activation energy E_A, the temperature corresponding to the gate current is needed.

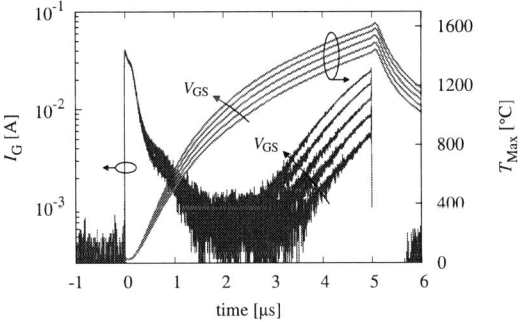

Fig. 12. Simulated maximum temperature and measured gate currents I_G at $V_{DS} = 600\,V$ for $V_{GS,on} = 19.5\,V, 20.0\,V, 20.5\,V, 21.0\,V$ and $21.5\,V$. The gate current shows a linear increase at the end of the pulse.

Hence, the temperature displayed in Fig. 12 is simulated using the approach presented in [8]. The simulation uses a physics-based thermal Cauer network, considering the on-chip metallization as well as the strong temperature-dependency of the thermal conductivity. The resulting Arrhenius plot is displayed in Fig. 13.

An activation energy of $E_A \approx 2.35\,eV$ is extracted. It should be noted that this is only a indicative value since the

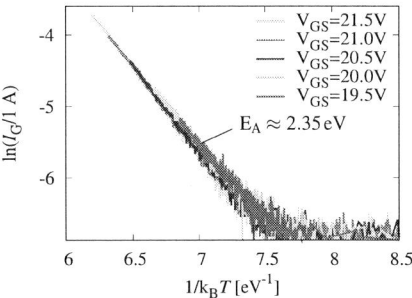

Fig. 13. Arrhenius graph based on the measured gate currents and simulated temperatures displayed in Fig. 12.

temperature data is simulated and only the peak temperature is used. A possible mechanism responsible for the gate currents at high V_{GS} voltages is Fowler-Nordheim tunneling which also has a temperature dependence [9].

V. Conclusions

In this work we demonstrate that negative $V_{GS,off}$ values severely impact the transient drain current of a SiC MOS-FETs. The observed temporary increase of I_D is explained by positively charged oxide and interface traps that temporarily strengthens the inversion charge in the channel. The neutralization by electrons from the channel causes the drain current to decay over time. The decay can be separated into a fast and a slow component. This phenomenon especially affects the accurate determination of the threshold voltage and transfer characteristics but also leads to a reduced short-circuit withstand duration in some operating points.

References

[1] A. Maerz, T. Bertelshofer, M. M. Bakran, and M. Helsper, "A novel gate drive concept to eliminate parasitic turn-on of SiC MOSFETs in low inductance power modules," in *Proc. PCIM Europe*, 2017, pp. 1–7.

[2] A. J. Lelis, D. Habersat, R. Green, A. Ogunniyi, M. Gurfinkel, J. Suehle, and N. Goldsman, "Time dependence of bias-stress-induced SiC MOSFET threshold-voltage instability measurements," *IEEE Trans. Electron Devices*, vol. 55, no. 8, pp. 1835–1840, 2008.

[3] G. Rescher, G. Pobegen, T. Aichinger, and T. Grasser, "On the sub-threshold drain current sweep hysteresis of 4H-SiC nMOSFETs," in *Proc. IEDM*, 2016, pp. 10.8.1–10.8.4.

[4] S. Potbhare, N. Goldsman, A. Akturk, M. Gurfinkel, A. Lelis, and J. S. Suehle, "Energy- and time-dependent dynamics of trap occupation in 4H-SiC MOSFETs," *IEEE Trans. Electron Devices*, vol. 55, no. 8, pp. 2061–2070, 2008.

[5] G. Pobegen and A. Krassnig, "Instabilities of SiC MOSFETs during use conditions and following bias temperature stress," in *Proc. IRPS*, 2015, pp. 6C.6.1–6C.6.6.

[6] *Multi-Exponential Relaxation Analysis (MERA) Toolbox*, (accessed June 1, 2016). [Online]. Available: http://www.vuiis.vanderbilt.edu/~doesmd/MERA/MERA_Toolbox.html.

[7] T.-T. Nguyen, A. Ahmed, T. E. Chang, and J.-H. Park, "Gate oxide reliability issues of SiC MOSFETs under short-circuit operation," *IEEE Trans. Power Electronics*, vol. 30, no. 5, pp. 2445–2455, 2015.

[8] M. Pfost, C. Boianceanu, H. Lohmeyer, and M. Stecher, "Electrothermal simulation of self-heating in DMOS transistors up to thermal runaway," *IEEE Trans. Electron Devices*, vol. 60, no. 2, pp. 699–707, 2013.

[9] A. K. Agarwal, S. Seshadri, and L. B. Rowland, "Temperature dependence of Fowler-Nordheim current in 6H- and 4H-SiC MOS capacitors," *IEEE Electron Device Letters*, vol. 18, no. 12, pp. 592–594, 1997.

Proceedings of the 30th International Symposium on Power Semiconductor Devices & ICs
May 13-17, 2018, Chicago, USA

Reduction of RonA Retaining High Threshold Voltage in SiC DioMOS by Improved Channel Design

Atsushi Ohoka, Masao Uchida, Tsutomu Kiyosawa, Nobuyuki Horikawa, Kouichi Saitou, Yoshihiko Kanzawa,
Haruyuki Sorada, Kazuyuki Sawada, Tetsuzo Ueda
Power Electronics Business Development Office, Automotive & Industrial Systems Company
Panasonic Corporation, Moriguchi-City, Osaka, Japan
E-mail: ohoka.atsushi@jp.panasonic.com

Abstract—Trade-off between threshold voltage and specific on-resistance is successfully overcome in a diode-integrated SiC MOSFET by improving the design of n-type epitaxial channel layer and p-type body region. This new design features enhanced transconductance, hence low on-state resistance, while retaining high threshold voltage. Obtained specific on-resistance of the fabricated 1200V SiC DioMOS is among the lowest achieved for SiC MOSFETs including trench devices. The transconductance enhancement is also demonstrated to be effective in increasing the turn-on switching speed, thus contributing to higher efficiency in power switching systems with reduced conduction and switching losses.

Keywords—SiC MOSFET, diode integration, epitaxial channel

I. INTRODUCTION

Silicon carbide is a promising candidate to replace silicon in high power switching systems owing to its excellent material properties such as high breakdown field, high thermal conductivity, and high electron saturation velocity. Although SiC MOSFETs have been successfully demonstrated for commercial use, the high wafer cost has long hindered its widespread use in the market. In order to minimize the chip cost, much research effort has been focused towards lowering the specific on-resistance (RonA) of the MOSFET.

For inverter applications, however, SiC MOSFETs cannot be used alone but rather must be paralleled with external Schottky barrier diodes, because intrinsic body diodes have a high forward voltage and are unstable from the reliability point of view. Addition of external diodes results in unfavorable consequences such as increased material cost and larger footprint for the modules. To avoid these problems, we have reported a novel device concept, so called the SiC Diode-integrated MOSFET (DioMOS), which integrates a low-V_f diode into a unit cell of the MOSFET. Forward and reverse current conduction in a single unit cell is realized by utilizing thin and highly-doped epitaxial channel [1]. Design of this channel layer as well as its vicinity is therefore the key to controlling forward and reverse characteristics in DioMOS.

In this paper, channel design of DioMOS is further improved to minimize RonA while ensuring high threshold

voltage V_{th} and low forward voltage V_f of the reverse diode are retained. Both hand analysis and experimental results suggest the possibility of independently optimizing RonA, V_{th}, and V_f by taking full advantage of the n-type epitaxial channel. The newly-designed DioMOS exhibits excellent electrical properties such as high V_{th} of 4.5V and low RonA of 3.5mΩcm^2, which help improve the robustness and efficiency of high power switching systems.

II. DEVICE STRUCTURE DESIGN

Figure 1 is a schematic cross-section of DioMOS which shows the channel layer grown directly on top of the drift layer by epitaxial growth. Doping the channel layer highly n-type can lower the potential barrier in the channel, allowing electrons to flow from drain to source during reverse conduction. Since only one type of carrier is involved in the reverse conduction process, this diode-like function named the channel diode constitutes a unipolar current and is free from bipolar degradation, which is a commonly encountered issue in SiC when body diodes are actively used [2]. Presence of this channel diode eliminates the need for externally-connected Schottky barrier diode as summarized in Fig. 2.

Reduction of RonA without lowering of V_{th} can be accomplished by enhancing the transconductance g_m. One way

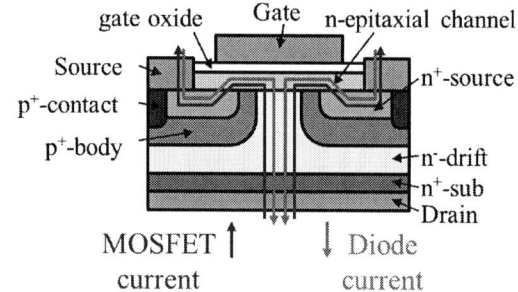

Fig. 1: Device structure of Diode-Integrated MOSFET (DioMOS), with MOSFET and diode current flow directions indicated with blue and red arrows, respectively. Current flows in both directions occur through the n-type channel.

978-1-5386-2928-4/18 $31.00 © 2018 IEEE

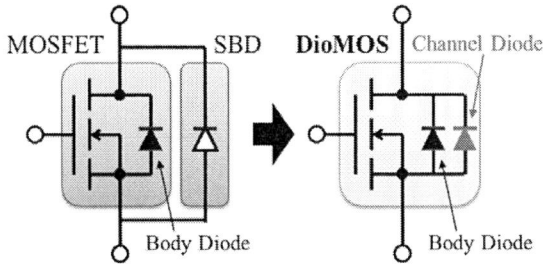

Fig. 2: Concept of diode integration in DioMOS, where channel diode functions as the low-V_f diode in place of the external SiC Schottky barrier diode.

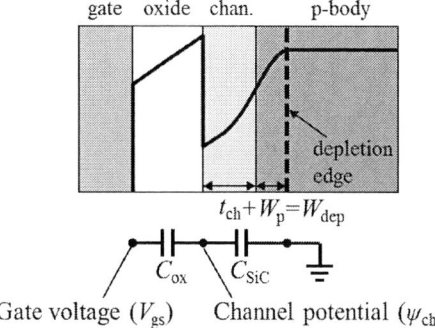

Fig. 3: Conduction band diagram in the vicinity of channel with associated regions overlapped. Equivalent circuit is also shown, where channel potential is determined by the capacitive divider of oxide capacitance and depletion capacitance.

of enhancing g_m is to improve the efficiency of the gate in modulating the channel potential ψ_{ch}. This efficiency is of particular interest in power MOSFETs because they generally have thick gate oxide to mitigate the high electric field, which then results in less control of the channel potential by the gate. Based on a simple capacitive divider circuit as shown in Fig. 3, this modulation efficiency by the gate, defined as $\Delta\psi_{ch}/\Delta V_{gs}$, can be derived as

$$\frac{\Delta\psi_{ch}}{\Delta V_{gs}} = \frac{C_{ox}}{C_{ox}+C_{SiC}} = \frac{1}{1+\frac{t_{ox}}{\varepsilon_{ox}}\frac{\varepsilon_{SiC}}{W_{dep}}} = \frac{1}{1+\frac{\varepsilon_{SiC}}{\varepsilon_{ox}}\frac{t_{ox}}{t_{ch}+W_p}}, \quad (1)$$

where C_{ox} and C_{SiC} are oxide capacitance and depletion capacitance within the SiC layer, respectively. Introduction of n-type channel beneath the gate oxide results in slight modification to the expression of total depletion width W_{dep}. Namely, it is written as the sum of channel thickness t_{ch} and depletion width W_p in the p-type body. From the expression of drain current (2) in the parabolic region of MOSFET [3], g_m is calculated by differentiating it with respect to V_{gs} as in (3),

$$I_{ds} = \mu_{eff} C_{ox} \frac{W}{L}\left(V_{gs} - V_{th} - \frac{1}{\frac{\Delta\psi_{ch}}{\Delta V_{gs}}}\frac{V_{ds}}{2}\right)V_{ds}, \quad (2)$$

$$g_m = C_{ox}\frac{W}{L}\left[\left(V_{gs} - V_{th} - \frac{1}{\frac{\Delta\psi_{ch}}{\Delta V_{gs}}}\frac{V_{ds}}{2}\right)\frac{d\mu_{eff}}{dV_{gs}} + \mu_{eff}\right]V_{ds}, \quad (3)$$

which suggests g_m enhancement is indeed possible by enlarging $\Delta\psi_{ch}/\Delta V_{gs}$ as described earlier. According to (1), the modulation efficiency is improved by either reducing the gate oxide thickness t_{ox} or increasing the total depletion width in the SiC layer.

Large $\Delta\psi_{ch}/\Delta V_{gs}$, however, is undesirable in terms of achieving high V_{th} and low V_f of the channel diode simultaneously. To be more specific, the value of $\Delta\psi_{ch}/\Delta V_{gs}$ has been deliberately set low in the previously-reported DioMOS to address this V_{th}-V_f trade-off [4]. V_f of the channel diode is directly related to the potential barrier height in the channel. V_{th}, in contrast, can be thought as the gate voltage required to completely lower this barrier. Consequently, if the efficiency of the gate to modulate the barrier height is large, V_{th} easily drops to a lower value. To avoid such situation and achieve high V_{th}, small $\Delta\psi_{ch}/\Delta V_{gs}$ is preferred. The challenge in designing DioMOS is therefore to have as large $\Delta\psi_{ch}/\Delta V_{gs}$ as possible for high g_m and as small $\Delta\psi_{ch}/\Delta V_{gs}$ as possible for overcoming the V_{th}-V_f trade-off.

In order to satisfy both criteria with a single parameter $\Delta\psi_{ch}/\Delta V_{gs}$, one must take into account the gate voltage dependence of this modulation efficiency. Examining (1), it is clear that the depletion width W_p is the only parameter with V_{gs} dependence, as both t_{ox} and t_{ch} are physically-defined dimensions. In addition, the relevant gate voltages at which large $\Delta\psi_{ch}/\Delta V_{gs}$ is favored and small $\Delta\psi_{ch}/\Delta V_{gs}$ is favored are recognized as above threshold (on-state) and below threshold (off-state), respectively. At this point, the design problem simplifies to merely making W_p large for the on-state and small for the off-state.

During the on-state, the p-n junction formed between n-type channel and p-type body is reverse-biased, allowing the depletion layer to extend deep into the bulk of p-type body depending on its doping level. On the other hand, the p-n junction during the off-state is weakly reverse-biased, or even forward-biased for negative V_{gs}. As a result, extension of depletion layer into the p-type body is rather limited compared to the case of on-state. This behavior of depletion layer is consistent with the desired tendency for $\Delta\psi_{ch}/\Delta V_{gs}$. Thus, further elaboration of this behavior by means of p-type body doping concentration N_{body} is expected to be beneficial for separately improving the on-state and off-state characteristics. The value of N_{body} is selected such that the increase in depletion width W_p in the on-state is larger than that in the off-state. This ensures larger change in the on-state $\Delta\psi_{ch}/\Delta V_{gs}$ than that in the off-state $\Delta\psi_{ch}/\Delta V_{gs}$. To compensate for the expected increase in W_p in the off-state, channel thickness is also reduced by a minimal amount so as to not affect the on-state W_{dep}.

Concept of the new channel design with lower N_{body} and thinner t_{ch} is validated by numerical simulations using Sentaurus TCAD. As a reference, previously-reported device with high p-type body doping of 1.5×10^{19} cm^{-3} and channel thickness of 60 nm is used. N_{body} and t_{ch} of the improved design are 0.2×10^{19} cm^{-3} and 52 nm, respectively. Figure 4 summarizes the simulated W_{dep} for reference and improved channel designs during on-state and off-state. Comparison of the reference and improved designs reveals that the total depletion width W_{dep} is successfully made larger in the on-state, while the off-state

978-1-5386-2928-4/18 $31.00 © 2018 IEEE

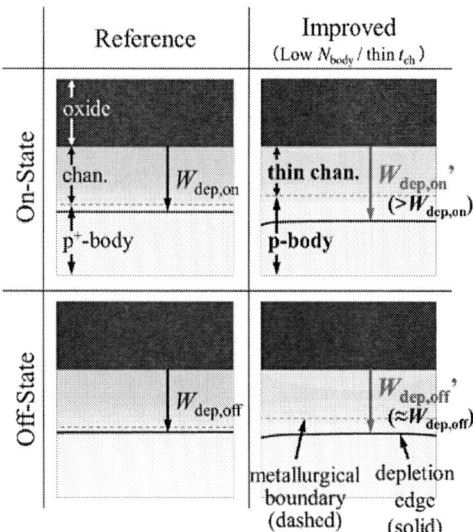

Fig. 4: Simulated total depletion width W_{dep} in the vicinity of channel region for reference and improved channel designs in on- and off-states. On-state W_{dep} is successfully increased, without affecting off-state W_{dep}.

W_{dep} is nearly unaffected owing to the thinned channel. The channel thickness is appropriately selected in the new design, since even thinner channel may result in less g_m enhancement, while thicker channel may result in worse V_{th}-V_f trade-off.

III. RESULTS AND DISCUSSIONS

SiC DioMOS based on the improved channel design is fabricated on a drift layer designed for a blocking voltage of 1200V. With the exception of reduced p-type body doping and thinned n-type channel layer, all processing conditions are exactly identical between the reference and improved DioMOS.

A. Static Characteristics

Figure 5 shows the measured transconductance as a function of V_{gs} for the reference and improved channel designs at room temperature. Nearly double the enhancement in g_m is observed at moderate V_{gs}, though the effect becomes slightly less pronounced at higher V_{gs}. As the gate voltage increases, it becomes more difficult for electrons to resist the transverse electric field, thus resulting in lower channel mobility. Nonetheless, reduction of p-type body doping concentration is shown to be an effective way to enhance the transconductance. The resulting output I-V characteristics are shown in Fig. 6, where the gate voltages used for forward and reverse conduction are +20V and 0V, respectively. The on-state resistance R_{on} is successfully reduced, consistent with the result of enhanced g_m, while the reverse channel diode characteristics are nearly unaffected. No kinks are observed in the diode characteristics, indicating that most of the reverse current does flow through the channel, thus deactivating the body diodes. Figure 7 is the measured specific on-resistance as a function of temperature, comparing the reference and improved channel designs. It is interesting to observe that the temperature

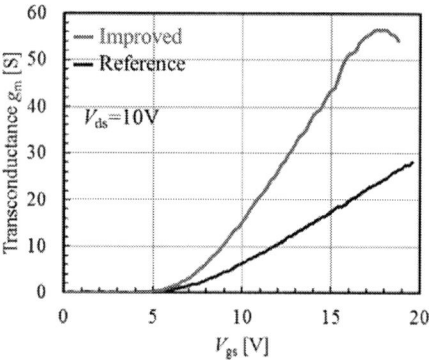

Fig. 5: Measured transconductance versus V_{gs} at V_{ds}=10V. Nearly double the enhancement in g_m is observed with the improved channel design.

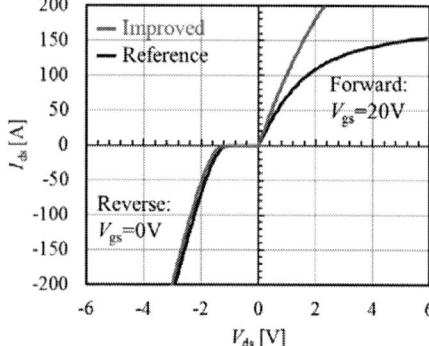

Fig. 6: Measured I_{ds}-V_{ds} characteristics in 1st- and 3rd-quadrants. Improved channel design improves R_{on} while retaining the same low V_f.

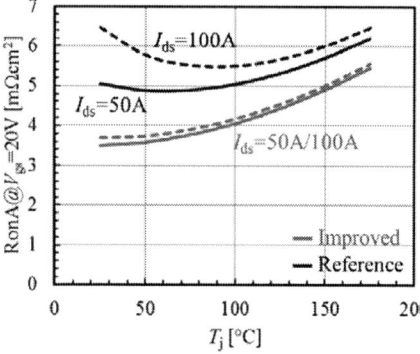

Fig. 7: Measured V_{th} versus temperature at I_{ds}=50A (solid) and I_{ds}=100A (dashed).

behavior is quite different between the two designs, despite the fact that the processing conditions for oxidation and post-oxidation anneal are unchanged. Temperature dependence of RonA is largely determined by that of its dominant components, the channel resistance and drift resistance. The channel resistance has a negative temperature coefficient because

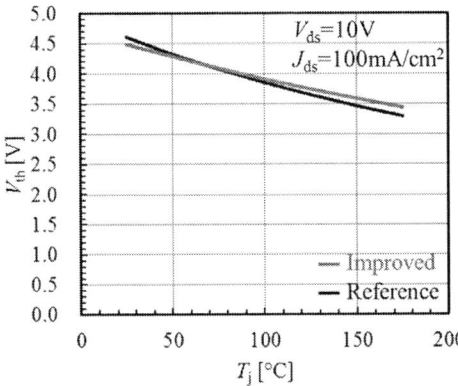

Fig. 8: Measured V_{th} versus temperature. High V_{th} of 4.5V is successfully retained with smaller temperature dependence.

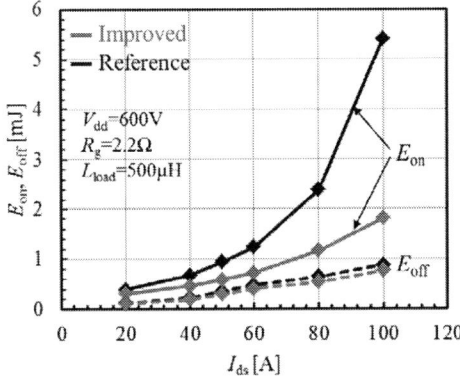

Fig. 9: Measured E_{on} (solid) and E_{off} (dashed) as a function of I_{ds} obtained using TO-247-3 package.

electrons trapped at shallow interface states can get de-trapped as temperature rises, resulting in reduced coulomb scattering [4]. Drift resistance, on the other hand, has a positive temperature coefficient. Since drift resistance is independent of the channel design, the change in the overall temperature behavior can be attributed solely to the reduction of channel resistance. This monotonically increasing RonA can help prevent thermal runaway when paralleling the devices.

Lastly, high threshold voltage is confirmed in Fig. 8. V_{th}, in this case, is defined as the gate voltage at J_{ds}=100mA/cm^2 and V_{ds}=10V. This high V_{th}, obtained across a wide range of temperature improves the robustness during high-speed switching by preventing false turn-on.

B. Dynamic Characteristics

Enhancement of g_m is effective not only for reducing forward conduction losses but also switching losses. Figure 9 shows the switching losses measured under double pulse test setup with chopper configuration. Switching losses are

Table 1: Comparison of the device performances among the presented SiC MOSFET reported in the literature and commercially-available SiC MOSFET.

		DioMOS		SiC MOSFET A	SiC MOSFET B	SiC MOSFET C
		This work	Previous work			
Device Structure		Planar	Planar	Trench	Trench	Planar
V_{th}	[V]	4.5	4.5	4.5	4.0	2.5
RonA	(25°C) [mΩcm^2]	3.5	5.0	3.5	5.2	2.0
	(175°C) [mΩcm^2]	5.5	6.2	5.8	10.9	3.9
Qg·Ron	(25°C) [mΩ·nC]	2650	4130	2790	5270	3220
	(175°C) [mΩ·nC]	4130	5080	4650	10940	6440
(V_{gs} range)		(0/20V)	(0/20V)	(-5/15V)	(0/20V)	(-4V/15V)
Diode Integration		Yes	Yes	No	No	No

(Qg: Gate Charge)

obtained while varying the load current from 20A to 100A. Turn-on loss (E_{on}) of the improved channel design is found to improve greatly, while turn-off loss (E_{off}) is hardly affected. The gate plateau voltage is lowered with enhanced g_m, leading to shorter time for reaching the plateau voltage and shorter time for charging the miller capacitance, which then results in larger dI_{ds}/dt and larger dV_{ds}/dt, respectively. This faster switching speed during turn-on is accountable for the reduction of E_{on}.

C. Performance Benchmark

Table 1 is a summary of device performances including DioMOS, reported SiC MOSFETs in the literature, and commercially-available SiC MOSFETs. The product of on-resistance R_{on} and gate charge Q_g of improved DioMOS is demonstrated to be among the lowest including trench devices.

IV. CONCLUSIONS

Trade-off between threshold voltage and specific on-resistance is successfully overcome in a diode-integrated SiC MOSFET by improving the channel design. Excellent electrical properties of the newly-designed DioMOS simplify the thermal and electrical designs and imply DioMOS to be very promising for the future SiC-based power switching systems.

ACKNOWLEDGMENT

The authors would like to thank Mr. Eiji Fujii as well as members of the development team at Panasonic Corporation for their support and supervision throughout this work.

REFERENCES

[1] M. Uchida et al., "Novel SiC power MOSFET with integrated unipolar internal inverse MOS-channel diode," IEDM 2011, 2011, pp. 26.6.1-26.6.4.

[2] A. Agarwal, H. Fatima, S. Haney and S. H. Ryu, "A new degradation mechanism in high-voltage SiC power MOSFETs," in IEEE Electron Device Letters, vol. 28, no. 7, pp. 587-589, July 2007.

[3] Y. Taur and T. H. Ning, Fundamentals of Modern VLSI Devices, 2nd ed., Cambridge: Cambridge Univ. Press, 2009, p. 157.

[4] S. Potbhare, G. Goldsman, G. Pennington, A. Lelis and J. M. McGarrity, "Numerical and experimental characterization of 4H-silicon carbide lateral metal-oxide-semiconductor field-effect transistor," in Journal of Applied Physics, vol. 100, no. 4, pp. 044545, August 2006.

Proceedings of the 30th International Symposium on Power Semiconductor Devices & ICs
May 13-17, 2018, Chicago, USA

Avalanche Ruggedness and Reverse-Bias Reliability of SiC MOSFET with Integrated Junction Barrier Controlled Schottky Rectifier

Cheng-Tyng Yen, Fu-Jen Hsu, Chien-Chung Hung, Chwan-Ying Lee, Lurng-Shehng Lee,
Ya-Fang Li and Kuo-Ting Chu
Hestia Power Inc.
10F-2, 27 Guanxin Rd, Hsinchu, Taiwan, email: ct.yen@hestia-power.com

Abstract—**A process and a scalable structure were used to implement the SiC MOSFET with integrated junction barrier controlled Schottky diode (JMOS) without area penalty. The JMOS could provide similar on-resistance and drain-source breakdown voltage with the same chip size as the standard double-implanted MOSFET (DMOS). The ideal factor and Schottky barrier height of integrated Schottky diode were 1.13 and 1.22eV for 650V JMOS and 1.11 and 1.27eV for 1200V JMOS. The diode forward voltage drop of JMOS were lower than DMOS when the diode forward current were smaller than 44A for 650V JMOS and 58A for 1200V JMOS. The reverse recovery charge of 650V and 1200V JMOS at 150°C were 22% and 53% lower than corresponding DMOS. The peak reverse recovery current of 650V and 1200V JMOS were 26% and 40% lower than corresponding DMOS. The output capacitance of JMOS were also lower than DMOS. The avalanche energy (E$_{AS}$) of 650V and 1200V JMOS were 1682mJ and 1270mJ, smaller than the corresponding DMOS, but a 17.2 J/cm^2 E$_{AS}$ is still superior to silicon counterparts. The results of diode forward current stress, diode surge current test and 1000 hours high temperature reverse bias test demonstrated that JMOS is reliable.**

Keywords- SiC MOSFET, Schottky, avalanche, bipolar degradation, forward surge, HTRB

I. INTRODUCTION

The adoption of SiC MOSFET has been accelerated recently due to its high efficiency and benefits brought to the system. However, the cut-in voltage of forward conduction in the intrinsic body diode of SiC MOSFET is close to 3V due to a large built-in potential in p-n junction of wide bandgap materials. In applications where the intrinsic body diode is used as the free-wheeling diode, a high cut-in voltage will contribute additional conduction loss and adversely impact the efficiency. Moreover, it has been found that the on-resistance (R$_{DS(on)}$) will be degraded by the bipolar conduction of body diode because the recombination of holes and electrons would induce the basal plane dislocations into large stacking faults [1]. Therefore, it has become a common practice to externally connect an inverse unipolar SiC Schottky barrier diode (SBD) with SiC MOSFET to reduce conduction loss and ensure long-

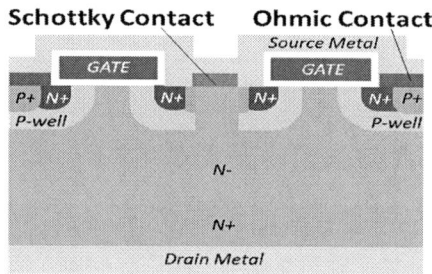

Fig. 1. The cross-section of JMOS.

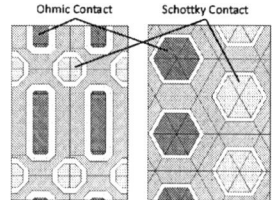

Fig. 2. The layout of first generation JMOS.

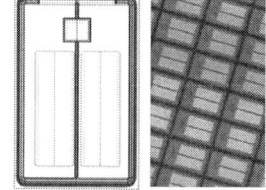

Fig. 3. The layout and fabricated chip of second generation JMOS.

term reliability. Nevertheless, using external SiC SBD is a costly way and requires additional assembly cost and package space which is not suitable to realize highly integrated compact power modules [2]. Hence, several approaches have been proposed to integrate SBD within the MOSFET cells [2-5] to solve the problem. To develop a SBD integrated SiC MOSFET, people may concern two things related to the cost: (1) how much the progresses will be complicated by integrating SBD into the MOSFET since Ohmic contact in SiC typically being formed by annealing to a temperature higher than 950°C, and (2) how much the area will have to be allocated for SBD. And another concern is that if the integrated SBD will impact the performance and reliability of MOSFET. In this work, we demonstrate that a monolithic SiC junction barrier controlled Schottky diode (JBS) integrated MOSFET (JMOS) can be accomplished by identical processes flow as standard SiC double implanted MOSFET (DMOS) without area penalty and shows promising performances.

978-1-5386-2928-4/18 $31.00 © 2018 IEEE

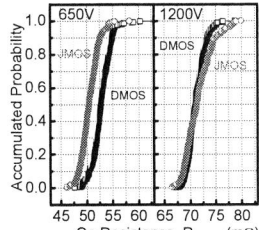

Fig. 4. Accumulated distribution of on-resistance for 650V and 1200V DMOS and JMOS.

Fig. 5. Accumulated distribution of zero gate voltage breakdown voltage BV_{DSS} for 650V and 1200V DMOS and JMOS.

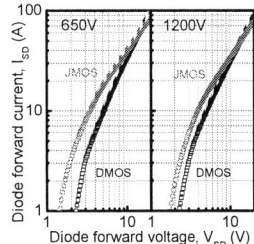

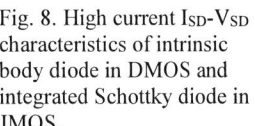

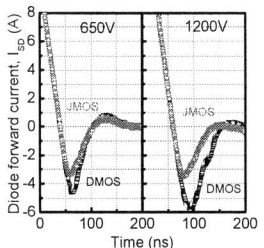

Fig. 8. High current I_{SD}-V_{SD} characteristics of intrinsic body diode in DMOS and integrated Schottky diode in JMOS.

Fig. 9. Diode reverse recovery characteristics of DMOS and JMOS at 150°C.

II. DEVICE CONCEPT AND STRUCTURE

In SiC, Ohmic contact is typically formed by silicidation of Ni on n+ region at a temperature higher than 950°C, but most Schottky contacts come to behave like Ohmic contacts if being annealed up to such high temperature. In our work, to incorporate a good Schottky contact into the MOSFET cell, the Schottky openings were formed together with the openings to the gate, after the Ohmic contacts were formed in the source (body) openings on the n+/p+ regions, and a common Ti/TiN/AlCu metal stack was used to form gate contacts for the gate electrode and Schottky contacts and Ohmic contacts for the source electrode simultaneously as the cross-section shown in Fig.1. In other words, no additional processes were introduced for integrating Schottky contacts into the MOSFET cells. The layout of our first trials for JMOS with this process is shown in Fig.2, where Schottky contacts were formed among MOSFET cells or simply by replacing some MOSFET cells. The first generation JMOS showed good results [2], however, the specific on-resistance ($r_{on,sp}$) of JMOS was about 20% higher than standard DMOS at that time, when the active

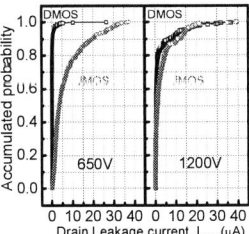

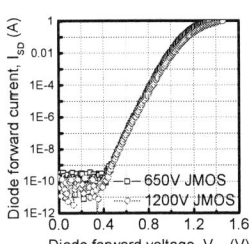

Fig. 6. Accumulated distribution of zero gate voltage drain leakage current I_{DSS} for 650V and 1200V DMOS and JMOS.

Fig. 7. Sub-threshold I_{SD}-V_{SD} characteristics of Schottky diode in 650V and 1200V JMOS.

areas were the same. Although the total chip area could be saved because of a common termination structure and peripheral region were used in JMOS, as compared to separate chips. It will be more attractive if we can get rid of most of area penalties.

To achieve this target, the second generation JMOS was designed and fabricated as shown in Fig. 3. The idea used in this generation of JMOS is to maintain the same effective density of channel width (Wch/cm^2) and to reduce the influence of parasitic resistances as far as possible, as compared to DMOS by the optimization of layout.

III. ELECTRIC CHARACTERISTICS

Fig. 4 shows the distribution of $R_{DS(on)}$ of 650V and 1200V DMOS and JMOS with exact the same active area and chip size at V_{GS}=20V and I_{DS}=20A. The average $R_{DS(on)}$ of 650V

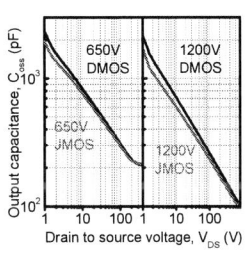

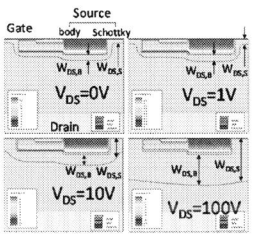

Fig. 10. Output capacitance-voltage characteristics of DMOS and JMOS.

Fig. 11. The variation of depletion boundaries with drain bias in JMOS.

JMOS was 2mΩ lower than 650V DMOS and the average $R_{DS(on)}$ of 1200V JMOS was 1mΩ higher than 1200V DMOS. It should be reasonable to say that the $R_{DS(on)}$ of JMOS and DMOS were essentially the same if the material and process variations such as epi concentration, epi thickness and channel length were taken into account. Fig.5 and Fig.6 shows the distribution of drain-source breakdown voltage (BV_{DSS}) at V_{GS}=0V, I_{DS}=100μA and the drain-source leakage current (I_{DSS}) at V_{GS}=0V, V_{DS}=650V/1200V for JMOS and DMOS. The average BV_{DSS} of 650V and 1200V JMOS were 41V and 30V lower than corresponding DMOS and the average I_{DSS} of JMOS were about one order of magnitude higher than corresponding DMOS. A higher I_{DSS} for both 650V and 1200V JMOS as compared to DMOS can be attributed to the image force induced barrier lowering and tunneling happened in Schottky contact at reverse bias. This situation will be the

978-1-5386-2928-4/18 $31.00 © 2018 IEEE   57

same if people use external parallelly connected MOSFET and SBD, because in most cases, the leakage current of SBD is larger that MOSFET under the same reverse bias. The sub-threshold forward current-voltage (I_{SD}-V_{SD}) characteristics of integrated Schottky diodes in 650V and 1200V JMOS were shown in Fig.7. The ideal factor and Schottky barrier height (Φ_{BN}) were 1.13 and 1.22 eV for 650V JMOS and 1.11 and 1.27eV for 1200V JMOS. These are typical values of Ti based Schottky contacts in SiC, suggesting good Schottky contacts were successfully formed. Fig. 8 compares high current I_{SD}-V_{SD} characteristics of 650V DMOS/JMOS and 1200V DMOS/JMOS. The V_{SD} of 650V JMOS was lower than 650V DMOS when I_{SD} was lower than 58A and the V_{SD} of 1200V JMOS was lower than 1200V DMOS when I_{SD} was lower than 44A, suggesting that within the practical range of I_{SD}, JMOS could provide a lower diode conduction loss. For a very high I_{SD}, the V_{SD} of DMOS will lower because of conductivity modulation. The difference in V_{SD} between JMOS and DMOS was larger for 650 rated devices because I_{SD}-V_{SD} curves of intrinsic body diode in DMOS is not sensitive to the epi concentration and thickness, which is not the case for unipolar Schottky diode. Fig. 9 compares the diode reverse recovery characteristics of JMOS and DMOS at 150°C. The reverse recovery charge (Qrr) of 650V and 1200V JMOS were 22% and 53% lower than their corresponding DMOS. The peak reverse recovery current ($Irmax$) of 650V and 1200V JMOS were also 26% and 40% lower than their corresponding DMOS. The Qrr and $Irmax$ in JMOS is lower as expected because there was no minority carrier injection in the integrated unipolar Schottky diode. Fig.10 shows that JMOS possessed a lower output capacitance C_{oss} as compared to DMOS even though their active area were exactly the same. This can be explained by the change of depletion boundaries with drain biases according to TCAD simulation results as shown in Fig.11. Specific C_{oss} can be expressed as the function of depletion layer widths below gate W_{GD}, below body $W_{DS,B}$ and below Schottky contact $W_{DS,S}$:

$$C_{oss} = C_{GD} + C_{DS} = \frac{\varepsilon_S}{W_{GD}} + \frac{\varepsilon_S}{W_{DS,B}} + \frac{\varepsilon_S}{W_{DS,S}}. \quad (1)$$

As shown in Fig.11, the depletion width $W_{GD,S}$ will become larger than $W_{GD,B}$ almost soon after applying the bias and. We hypothesized that the fast depletion of Schottky was partly

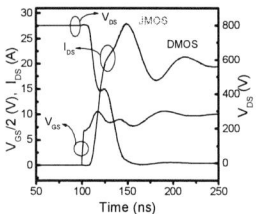

Fig. 12. Simulated 800V/20A turn-on switching waveform of 1200V DMOS and JMOS.

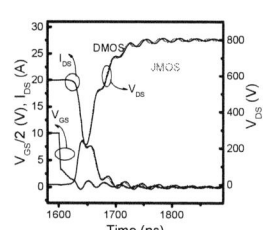

Fig. 13. Simulated 800V/20A turn-off switching waveform of 1200V DMOS and JMOS.

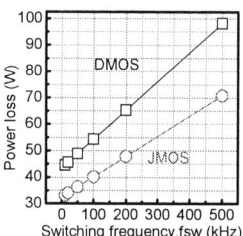

Fig. 14. Calculated 8kW soft-switching converter power loss of DMOS and JMOS when integrated diode conducting currents.

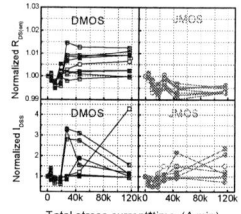

Fig. 15. Variation of $R_{DS(on)}$ and I_{DSS} of DMOS and JMOS over a constant I_{SD} stress.

assisted by the side-depletion of surrounding p-regions around. After the drift layer was fully depleted, $W_{GD,S}$ would still be larger than $W_{DS,B}$, because the distance between Schottky contact and the drain is larger than the distance between the bottom of body pwell and the drain. Since the active area of JMOS and DMOS were the same, a larger $W_{GD,S}$ than $W_{GD,B}$ in JMOS would naturally result in a lower C_{oss} as compared to DMOS.

The turn-on and turn-off switching waveforms of 1200V

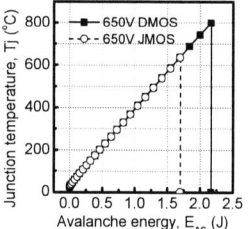

Fig. 16. The corresponding junction temperature with unclamped avalanche energy.

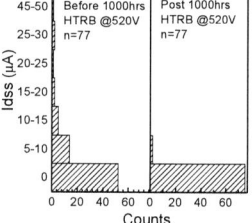

Fig. 17. Histograms of I_{DSS} for n=77 smaller 650V JMOS before and post 1000 hours HTRB at 150°C and V_{DS}=520V.

DMOS and JMOS were simulated by Spice model with a 800V dc link and a 20A load current as shown in Fig.12 and Fig.13. The turn-on transients of DMOS and JMOS were almost the same and only a slight difference in turn-off transients can be distinguished. The energy consumed for each switching transient was 336.81μJ in JMOS, compared to 349.40μJ in DMOS. For a 16kW boost converter with input voltage Vin=480Vdc, output voltage Vout=950Vdc, output current Iout=18A, switching frequency fsw=100kHz, the simulated efficiency was 98.06% with JMOS, higher than 97.31% with DMOS. Since in boost converter, the integrated diode was not conducting, the efficiency gain of JMOS should be come from a lower C_{oss}. To access the benefits of a lower V_{SD} and Qrr in JMOS, topologies such as soft switched full bridge converter where the integrated diode of MOSFET has to conduct currents were used to calculate the power losses by

978-1-5386-2928-4/18 $31.00 © 2018 IEEE   58

Rated Voltage	650V		1200V	
Device	DMOS	JMOS	DMOS	JMOS
Chip Size	4.3 x 2.9 mm^2			
ideal factor	–	1.13	–	1.11
Φ_{BN} (eV)	–	1.22	–	1.27
Rdson (mΩ)	53	51	70	71
Vth (V)	2.4	2.0	2.4	2.2
BVdss (V)	824	783	1582	1552
Idss (μA)	0.4	7.4	0.5	2.6
Qrr @150°C (nC)	148	115	260	121
trr @150°C (ns)	57	59	92	72
Irmax (A)	4.6	3.4	5.8	3.5
E_{AS} (mJ)	2145	1682	1728	1270
I_{FSM} (A)	184	102	168	102
Eon (μJ)	–	–	179.00	178.27
Eoff (μJ)	–	–	170.40	158.54
Esw (μJ)	–	–	349.40	336.81
Efficiency	–	–	97.31%	98.06%

Table 1. Summary of characteristics for 650V and 1200V DMOS and JMOS.

taking conduction losses of MOS channel and diode, output capacitance energy, and reverse recovery energy into considerations [6]. The power losses of 1200V DMOS and JMOS with different operation frequencies were shown in Fig.14. The gains in power losses JMOS compared to DMOS became larger at higher frequencies demonstrated the benefits can be brought by a lower V_{SD}, C_{oss} and Qrr of JMOS.

IV. DEVICE RUGGEDNESS AND RELIABILITY

To investigate the ruggedness of integrated Schottky diodes in JMOS, a constant I_{SD} was used to stress several series connected JMOS and DMOS. The variations of $R_{DS(on)}$ and I_{DSS} of JMOS were lower than DMOS during the 120,000 A.min stress duration as shown in Fig.15. Since no obvious bipolar degradations were found even for the DMOS, the smaller parametric variations in JMOS could be simply because of less self-heating. Unclamped avalanche energy (E_{AS}) and diode forward surge current (I_{FSM}) capabilities were also tested to evaluate the robustness of JMOS. The E_{AS} and I_{FSM} of 650V and 1200V JMOS and DMOS were summarized in Table.1. A lower E_{AS} for JMOS compared to DMOS can be explained by Fig.16, where the junction temperatures Tj, calculated by utilizing the transient thermal resistance, increased with increasing E_{AS}. As shown in Fig.16, the Tj corresponding to the E_{AS} failing JMOS was about 636°C, compared to around 800°C for DMOS. The Tj resulted in the failure of JMOS was the temperature where Ti Schottky

contacts started to behave like Ohmic. A lower I_{FSM} of JMOS compared to DMOS could have two reasons: (1) in the range of I_{FSM}, the V_{SD} of JMOS was higher than DMOS, and (2) the maximum Tj where the catastrophic failures could be avoided was lower in JMOS. Although the E_{AS} and I_{FSM} of JMOS were lower than that of DMOS, the 1682mJ E_{AS} for 650V JMOS, corresponding to a energy density of 17.2J/cm^2, is still superior to silicon power devices, where this value of 650V Si super junction MOSFETs is typical lower than 4J/cm^2. And the around 100A surge capability is comparable to our 1200V/8A rated JBS, suggesting that JMOS is robust enough for real applications. Fig.17 shows the I_{DSS} histograms of smaller 650V JMOS with a sample size of 77 before and after 1000hrs of high temperature reverse bias stress (HTRB) at 150°C and 520V. No failures were found in HTRB suggested JMOS is reliable even its I_{DSS} was relatively higher than DMOS.

V. CONCLUSION

A monolithic SiC JMOS was developed with $R_{DS(on)}$ and BV_{DSS} comparable to DMOS and showing no area penalty. A lower V_{SD}, C_{oss} and Qrr of JMOS is beneficial especially for applications where integrated diode has to conduct currents. The unipolar integrated Schottky in JMOS can get rid of concern over bipolar degradation in SiC and their E_{AS}, I_{FSM} and HTRB results suggest that this device is sufficient robust and reliable for realizing a highly integrated compact module.

REFERENCES

[1] S. H. Ryu, H. Fatima, S. Haney, Q. Zhang, R. Stahlbush, and A. Agarwal, "Effect of recombination-induced stacking faults on majority carrier conduction and reverse leakage current on 10 kV SiC DMOSFETs" *Mater. Sci. Forum*, vol. 740-742, pp. 1127-1130, 2009.

[2] C. T. Yen, C. C. Hung, H. T. Hung, L. S. lee, C. Y. Lee, T. M. Yang, Y. F. Huang, C. Y. Cheng, P. J. Chuang, "1700V/30A 4H-SiC MOSFET with low cut-in voltage embedded diode and room temperature boron implanted termination", *Proc. ISPSD'15*, pp. 265-268, 2015.

[3] S. Hino, T. Hatta, K. Sadamatsu, Y. Nagahisa, S. Yamamoto, T. Iwamatsu, Y. Yamamoto, M. Imaizumi, S. Nakata and S. Yamakawa, "Demonstration of SiC-MOSFET embedding schottky barrier diode for inactivation of parasitic body diode", *Proc. ECSCRM'16*, pp. 129-130, Sep. 2016.

[4] F. J. Hsu, C. T. Yen, C. C. Hung, H. T. Hung, C. Y. Lee, L. S. Lee, Y F Huang, T. Z. Chen, P. J. Chuang, "High efficiency high reliability SiC MOSFET with monolithically integrated Schottky rectifier", *Proc. ISPSD'17*, pp. 45-48, 2017.

[5] W. Sung and B. J. Baliga, "Monolithically integrated 4H-SiC MOSFET and JBS diode (JBSFET) using a single Ohmic/Schottky process scheme", *IEEE Electron Device Lett.*, vol. 37, no. 12, pp.1605-1608, Dec. 2016.

[6] D. Graovac, M. Purschel, A. Kiep, "MOSFET power losses calculation using the datasheet parameters", *Infineon Application Note*, v.1.1, 2006.

Proceedings of the 30th International Symposium on Power Semiconductor Devices & ICs
May 13-17, 2018, Chicago, USA

Comprehensive Investigation on Mechanical Strain Induced Performance Boosts in LDMOS

Wangran Wu, Siyang Liu, Jing Zhu, Weifeng Sun *

National ASIC System Engineering Research Center
Southeast University, Nanjing, China
*E-mail: swffrog@seu.edu.cn

Abstract— In this paper, we have comprehensively investigated the performance of LDMOS under mechanical strain. The electrical properties of nLDMOS under uniaxial tensile (UT) strain along channel direction are examined thoroughly. We find that the nLDMOS with longer gate length (L_g) is more preferred for strain. Both lateral electric field (V_d) and vertical electric field (V_g) play an important role on the strain effects. The piezoresistance coefficients of nLDMOS are evaluated for the first time. Neglectable breakdown voltage (V_{bd}) degradation is observed with the 4.4% drain current (I_D) increase under UT strain. Finally, the biaxial tensile strain and uniaxial compressive strain parallel to channel are proved to be most efficient for nLDMOS and pLDMOS with 8.8% and 14.5% R_{on} reduction, respectively.

Keywords— *mechanical strain, LDMOS, performance enhancement, piezoresistance coefficient*

I. INTRODUCTION

Process-induced strain was first introduced into Si MOSFETs by Intel in 2002 and began to play a major role in the mainstream VLSI semiconductor industry [1, 2]. Device performance benefits from the strain induced carriers' effective mass and scattering rate modulation [3, 4]. Now, strain is induced in various ways to enhance the device performance in almost every modern semiconductor workshop. However, the study and application of strained Si technology in power devices have experienced much slower development. It has been proved that the power MOSFETs with process-induced strain could break through the "Si-limit", i.e. the on-resistance (R_{on}) and breakdown voltage (V_{bd}) curve, which is because of the strain induced carrier mobility enhancement [5, 6]. The strained Si LDMOS with embedded SiGe substrate and deposited poly Si buried layer have shown a prominent R_{on} reduction [7, 8].

The mechanisms of strained LDMOS are complex because the strain effects in the inverted channel and drift region are different. Thus, the comprehensive investigation on strain induced performance boosts in LDMOS is of great importance. The existing reports utilize new structures or techniques unavailable in unstrained devices, which imposes difficulties in identifying the physical mechanisms related to strain. In this paper, the mechanical strain via wafer bending is used to avoid the unexpected variance.

II. DEVICE STRUCTURE AND EXPERIMENTS

Uniaxial and biaxial (tensile or compressive) mechanical strain can be applied using the wafer bending system (Fig. 1). The applied strain can be calculated according to the relative displacement of the top and bottom plates. The LDMOS used in this study is fabricated on the (001) Si substrate with a <110> channel direction through the standard BCD process. LDMOS with the same drift region length of 2.2 μm but different channel lengths of 1.2 μm, 1.4 μm and 20 μm are examined to verify the strain effect on devices with various dimensions. Fig. 2 shows the schematic of nLDMOS used in this study.

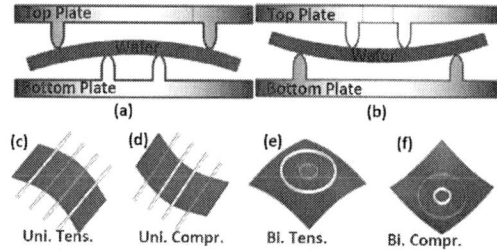

Fig. 1: Illustrations of wafer bending system to apply (a) tensile stress, (b) compressive stress, uniaxial (c) tensile and (d) compressive strain, biaxial (e) tensile and (f) compressive strain.

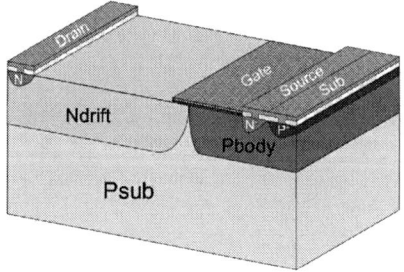

Fig. 2: Schematic of nLDMOS used in this study. Devices with L_g of 20 μm, 1.4 μm, and 1.2 μm are studied.

The amount of strain in Si wafer can be measured by Si Raman peak shifts. Fig. 3 shows the Raman peak shift of the

978-1-5386-2928-4/18 $31.00 © 2018 IEEE

bare Si wafer under uniaxial strain applied by the wafer bending system as a function of the plates' relative displacements. Obviously, the strain is introduced and the amount of strain is calibrated with the one obtained by calculation. Figs. 4 and 5 show the I_d-V_d and I_d-V_g curves of the nLDMOS under UT strain. Clear I_d and transconductance (g_m) enhancements are observed under UT strain. The threshold voltage (V_{th}) shift can be neglected. The change in drain current under the same gate voltage is used to represent the strain effects, i.e. the $\Delta I_d=(I_{dstrain}-I_d)/I_d$. It is found that ΔI_d is strongly correlated to the V_d, V_g and the devices' L_g (Figs. 6, 7 and 8) in LDMOS, which is different from the conventional MOSFET. ΔI_d decreases with increasing V_d in both devices with L_g of 20 μm and 1.4 μm. Higher V_g corresponds to larger ΔI_d in the whole V_d region in 20-μm LDMOS. In 1.4-μm LDMOS, ΔI_d follows the same V_g dependence with that in 20-μm device when V_d is small. While ΔI_d has opposite V_g dependence at large V_d compared with the 20-μm LDMOS. Fig. 8 shows that the strain is more effective in nLDMOS with longer gate.

In LDMOS, I_d in the channel region equals to:

$$I_{d,ch} = \beta[(V_g - V_{th})\,V_m - \frac{1}{2}V_m^2],$$

$$\beta = \mu_{ch}C_{ox}Z / L_{ch} \qquad (1)$$

where V_m is the voltage at the boundary of channel and drift region, μ_{ch} is the electron mobility in the inverted channel, Z is the channel width. Similarly, I_d in the drift region can be represented as:

$$I_{d,drift} = \alpha\{V_d - V_m - \frac{2}{3}(\frac{2\varepsilon_S N_B}{qN_D^2 d^2})^{\frac{1}{2}}[V_d^{\frac{3}{2}} - V_m^{\frac{3}{2}}]\},$$

$$\alpha = \frac{q\mu_n N_D Z d}{L_d} \qquad (2)$$

where μ_n is the electron mobility in drift region, d is the thickness of electron path, N_B and N_D are the substrate doping concentration and implant doping concentration in drift region, respectively. Thus, we have the total on resistance:

$$R_{on} = R_{on,ch} + R_{on,drift}$$

$$= \frac{1}{\frac{\partial I_{d,ch}}{\partial V_m}\big|_{V_g = C}} + \frac{1}{\frac{\partial I_{d,drift}}{\partial V_m}\big|_{V_m = C}} \qquad (3)$$

$$= \frac{1}{\beta(V_g - V_{th})} + \frac{1}{\alpha}\{1 - (\frac{2\varepsilon_s N_B}{qN_D^2 d^2}V_d)^{\frac{1}{2}}\}^{-1}$$

Because the electrons in the inverted channel benefit more from UT strain than majority carriers in drift region in <110>/(001) direction [3], the coefficient β changes more than α under UT strain. When V_d is small (i.e. in the linear region), R_{on} equals to $1/[\beta(V_g-V_{th})]+1/\alpha$. In this case, β affects more in devices with longer gate and the UT strain is more effective in devices with longer gate. When V_d increases, the second term in Eq. (3) increases (the resistance of drift region) and α plays more important role in the total strain effects and the total strain effect decreases.

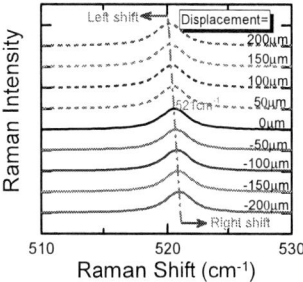

Fig. 3: The Raman peak shift of Si wafer as a function of bending system's top and bottom plates' relative displacement. The amount of strain in Si are calibrated accordingly.

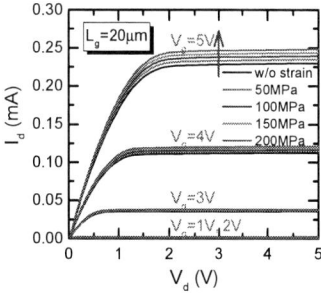

Fig. 4: I_d-V_d curves of nLDMOS under UT strain with V_g from 1 V to 5 V. I_d increases with the increasing strain.

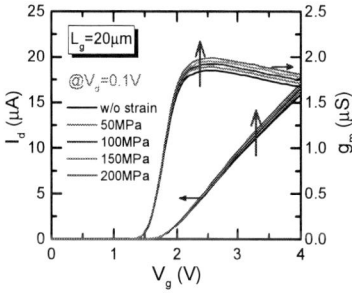

Fig. 5: I_d-V_g curves and g_m-V_g curves of nLDMOS under UT strain at V_d of 0.1 V. I_d and g_m increases with the increasing strain. V_{th} does not shift.

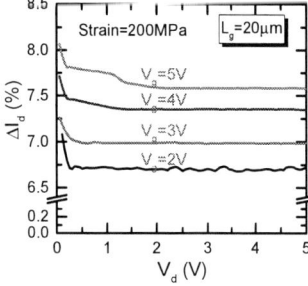

Fig. 6: ΔI_d versus V_d at V_g from 2 V to 5 V under 200 MPa UT strain in nLDMOS with 20-μm L_g. ΔI_d depends on V_g and V_d.

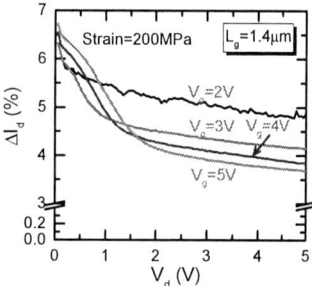

Fig. 7: ΔI_d versus V_d at V_g from 2 V to 5 V under 200 MPa UT strain in nLDMOS with 1.4-μm L_g. ΔI_d depends on V_g and V_d.

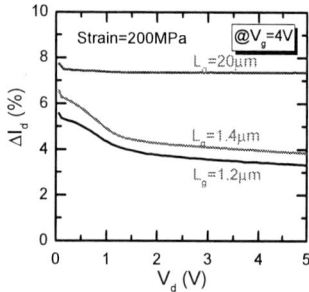

Fig. 8: ΔI_d versus V_d under 200 MPa UT strain in nLDMOS with L_g of 1.2 μm, 1.4 μm and 20 μm at V_g of 4 V. ΔI_d increases with increasing L_g.

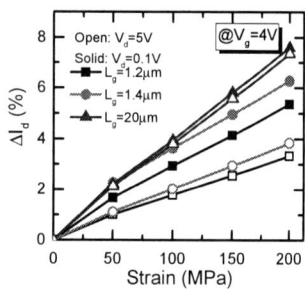

Fig. 9: ΔI_d versus UT strain in nLDMOS with L_g of 1.2 μm, 1.4 μm and 20 μm at V_g of 4 V. ΔI_d increases with increasing L_g.

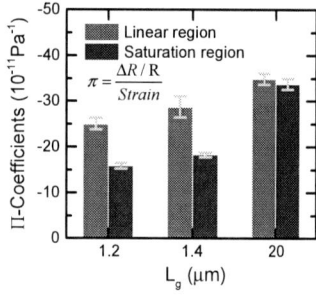

Fig. 10: Piezoresistance (π-) coefficients of LDMOS with various L_g in linear and saturation region. Strain effects in linear region is more prominent.

Fig. 9 shows ΔI_d versus UT strain curves in nLDMOS with L_g of 20 μm, 1.4 μm and 1.2 μm. Then, the piezoresistance (π-) coefficients ($\pi=(\Delta R/R)$/Strain) are obtained in the linear and saturation region in different devices (Fig. 10). nLDMOS with 20-μm L_g has high and similar π-coefficients in the linear and saturation region, while π-coefficients in linear region is much higher than that of saturation region in nLDMOS with 1.4-μm and 1.2-μm L_g. The breakdown characteristics are also studied in nLDMOS under UT strain along channel direction. Neglectable V_{bd} degradation is observed in nLDMOS with 4.4% I_d enhancement (Fig. 11). Thus, strain technology is promising to break through the "Si-limit" in LDMOS.

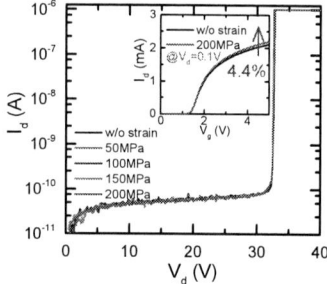

Fig. 11: BV characteristics of nLDMOS under UT strain. Neglectable V_{bd} degradation is observed with 4.4% I_d increase (inset).

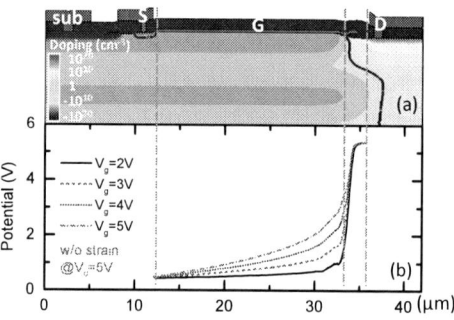

Fig. 12: (a) Cross section view of nLDMOS with L_g of 20 μm; (b) Potential along interface as V_g increases from 2 V to 5 V at V_d of 5 V. Potential drop in channel region domains.

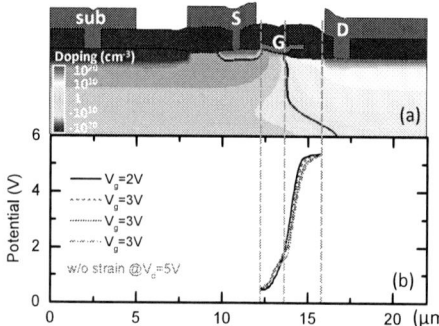

Fig. 13: (a) Cross section view of nLDMOS with L_g of 1.4 μm; (b) Potential along interface as V_g increases from 2 V to 5 V at V_d of 5 V. Potential drop in drift region domains.

Figs. 12 and 13 show the simulation results of nLDMOS with 20-μm L_g and 1.4-μm L_g at V_d of 5 V. The electron mobility increases with UT strain, resulting in better performance. The potential mainly drops in the channel region and drift region in nLDMOS with 20-μm L_g and 1.4-μm L_g respectively and the potential drop varies with V_g. Note that the electrons in the inverted channel benefit more from the UT strain. In this case, more potential drop in channel results in more prominent strain effect, which explains the different V_g dependences on ΔI_d. It also implies that strain engineer in channel region in long-channel nLDMOS and drift region in short-channel nLDMOS are more effective. Figs. 14, 15 and 16 show the R_{on} changes induced by six kinds of strain in nLDMOS and pLDMOS. It is proved that biaxial tensile strain and uniaxial compressive strain parallel to channel are most efficient for nLDMOS and pLDMOS with 8.8% and 14.5% R_{on} reduction, respectively.

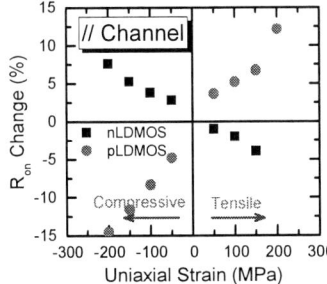

Fig. 14: R_{on} change versus UT strain parallel to channel in nLDMOS and pLDMOS.

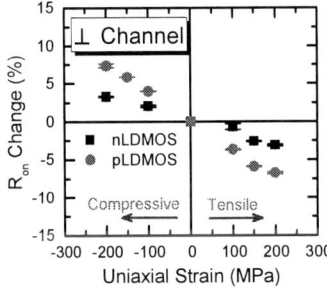

Fig. 15: R_{on} change versus UT strain perpendicular to channel in nLDMOS and pLDMOS.

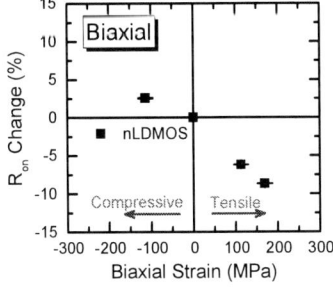

Fig. 16: R_{on} change versus biaxial strain in nLDMOS.

III. CONCLUSIONS

In this work, we investigate the performance of LDMOS under mechanical strain. The mechanisms of strain in LDMOS are more complicated than conventional MOSFETs. Electrical properties of nLDMOS under uniaxial tensile strain along channel direction are examined thoroughly. It is shown that nLDMOS with longer gate benefits more from the UT strain. Strain is more effective in the linear region. The vertical field (V_g) plays an important role on strain effects as well. The piezoresistance coefficients of nLDMOS are evaluated for the first time. Neglectable V_{bd} degradation is observed with the 4.4% drain current increase under UT strain. Finally, the biaxial tensile strain and uniaxial compressive strain parallel to channel are proved to be most efficient for nLDMOS and pLDMOS, respectively.

ACKNOWLEDGMENT

This work was supported by the National Natural Science Foundation of China (61704025, 61674030), the Natural Science Foundation of Jiangsu Province (BK20160691, BK20150627) and the Fundamental Research Funds for the Central Universities.

REFERENCES

[1] S. E. Thompson, M. Armstrong, C. Auth, and S. Cea, "A logic nanotechnology featuring strained-silicon," *IEEE Electron Device Letters*, vol. 25, pp. 191-193, 2004.

[2] H. Iwai, "Roadmap for 22 nm and beyond (Invited Paper)," *Microelectronic Engineering*, vol. 86, pp. 1520-1528, 2009.

[3] M. V. Fischetti and S. E. Laux, "Band structure, deformation potentials, and carrier mobility in strained Si, Ge, and SiGe alloys," *Journal of Applied Physics*, vol. 80, p. 2234, 1996.

[4] W. Wu, C. Liu, J. Sun, W. Yu, X. Wang, Y. Shi, and Y. Zhao, "Experimental study on NBTI degradation behaviors in Si pMOSFETs under compressive and tensile strains," *IEEE Electron Device Letters*, vol. 35, pp. 714-716, 2014.

[5] P. Moens, F. Bauwens, J. Baele, and K. Vershinin, "XtreMOS : The first integrated power transistor breaking the Silicon limit," in *International Electron Devices Meeting*, 2006, 2006, pp. 1-4.

[6] P. Moens, J. Roig, F. Clemente, and I. D. Wolf, "Stress-induced mobility enhancement for integrated power transistors," in *IEEE International Electron Devices Meeting*, 2007, pp. 877-880.

[6] M. Kondo, N. Sugii, Y. Hoshino, W. Hirasawa, Y. Kimura, M. Miyamoto, T. Fujioka, S. Kamohara, Y. Kondo, and S. I. Kimura, "Thick-strained-Si/relaxed-SiGe structure of high-performance RF power LDMOSFETs for cellular handsets," *IEEE Transactions on Electron Devices*, vol. 53, pp. 3136-3145, 2006.

[8] M. Miyamoto, N. Sugii, Y. Kumagai, and Y. Kimura, "Low-on-resistance strain-controlled LDMOS transistors for 0.25-μm power ICs," in *IEEE International Symposium on Power Semiconductor Devices and ICS*, 2011, pp. 168-171.

Proceedings of the 30th International Symposium on Power Semiconductor Devices & ICs
May 13-17, 2018, Chicago, USA

Investigation on Total-Ionizing-Dose Radiation Response for High Voltage Ultra-Thin Layer SOI LDMOS

Xin Zhou[1,*], Lingfang Zhang[1], Ming Qiao[1], Zhangyi'an Yuan[1], Ping Luo[1], Lei Shu[2], Zhaoji Li[1] and Bo Zhang[1]

[1]State Key Laboratory of Electronic Thin Films and Integrated Devices, University of Electronic Science and Technology of China, Chengdu, P. R. China, Email: zhouxin@uestc.edu.cn
[2]Harbin Institute of Technology, Harbin, P. R. China,

Abstract— Total-ionizing-dose radiation response for 600V ultra-thin layer SOI LDMOS transistor is investigated. Radiation conduction modulation model is proposed to reveal the degradation mechanism of conduction current and breakdown voltage. Multi-interface irradiation damage cause positive net trapped charge, which reduces on-resistance equivalently and enhance conduction current. Meanwhile, they suppress the depletion in the drift region at off-state and then decrease breakdown voltage. Based on the model, irradiation induced net trapped charge density are extracted to evaluate the irradiation damage in drift region.

Keywords—Total-ionizing-dose; SOI LDMOS; radiation conduction modulation; multi-interface irradiation damage

I. INTRODUCTION

Due to its ease to integration and fast switching, there is an increasing demand in laterally diffused MOS (LDMOS), which is one of the most widely used in high-voltage power ICs for aeronautics and astronautics applications [1]. However, MOS devices are extremely sensitive to Total-ionizing-dose (TID) radiation [2]-[5]. After irradiation, electrons will rapidly drift toward the gate and holes will drift toward the Si/SiO_2 multi-interface and trapped charges will build up in the insulating oxide layers. Due to large Si/SiO_2 interface area, SOI LDMOS would suffer from more severe and complex interface defect induced reliability issues [6]-[9]. However, compared to VDMOS, little care is devoted to irradiation reliability for SOI LDMOS, especially with breakdown voltage (BV) above 150 V [10]-[13]. In addition, for a given SOI LDMOS, it is difficult to extract directly the trapped charge density in the oxide layer of the drift region by common measurement techniques (such as C–V analysis, DCIV and charge pumping) [14]. And the evaluation for the radiation damage in oxide layer of the drift region is less documented.

In this paper, TID radiation response for 600V ultra-thin layer SOI LDMOS transistor is investigated by modeling, simulation and experiment. An analytical model for the 600V voltage ultra-thin layer SOI LDMOS is proposed to evaluate net oxide trapped charges in the oxide layer of drift region and reveal the mechanism of TID response.

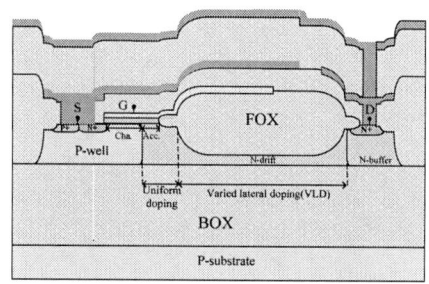

(a)

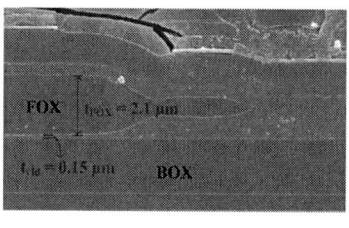

(b)

Fig.1 (a) Schematic cross-sectional view of the high voltage ultra-thin layer SOI LDMOS. (b) SEM photograph for the N-drift ragion.

II. DEVICE STRAUTURE AND MECHANISM

Fig.1 (a) shows the schematic cross-sectional view of the high voltage ultra-thin layer SOI LDMOS. The thickness of SOI layer and buried oxide (BOX) layer are 1.5 μm and 3 μm, respectively. In order to achieve BV above 600V, the ultra-thin drift region is adopted to obtain high critical breakdown electric field of Si, which can enhance the electric field in BOX layer. Meanwhile, varied lateral doping (VLD) technology is utilized to optimize the surface electric field and maximize the lateral BV. For the fresh device, the BV and V_{th} are 618V and 2.9V, respectively. Fig1 (b) shows the SEM photograph of the high voltage ultra-thin layer SOI LDMOS. The thickness of field oxide (FOX) layer and VLD region is 2.1μm and 0.15μm. The N-drift region is thinned by growing FOX layer, which implies the FOX layer is considerable thick as well as BOX layer. The irradiation induced defects density is proportional to the thickness of oxide layer, as a result, the irradiation damage

978-1-5386-2928-4/18 $31.00 © 2018 IEEE

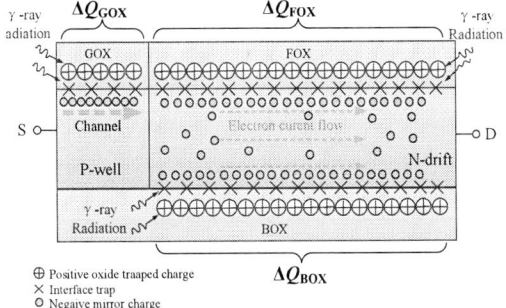

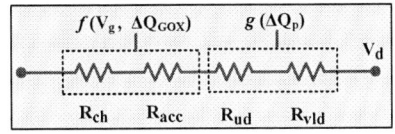

Fig.2 Schematic for multi-interface irradiation damage mechanism.

Fig.3 Equivalent on-resistance schematic for the ultra-thin layer SOI high voltage LDMOS.

in oxide layer of drift region should be evaluated seriously rather than channel region for the SOI LDMOS.

Fig.2 shows the schematic for multi-interface irradiation damage mechanism. Except for recombination, irradiation generated holes transport through SiO_2 by polar on hopping and are trapped by oxide trap near the Si/SiO_2 interface. Positive oxide trapped charge and interface trap are formed, which depends strongly on the electric field and the thickness of oxide layer. The SOI LDMOS has multi-interface with large area including gate oxide layer/SOI layer interface, FOX layer/SOI layer interface and BOX layer/SOI layer interface, of which the irradiation generated net trapped charge density are ΔQ_{GOX}, ΔQ_{FOX} and ΔQ_{BOX}, respectively. These trapped charges introduce equivalently opposite mirror charges in the bulk, which gives rise to carrier concentration and potential field change in the bulk. It is expected that the degradation mechanism would be more complex and severe due to multi-interface irradiation damage.

In order to evaluate positive oxide trapped charges in the multi-interface as well as the degradation of electrical properties induced by radiation, the radiation conduction modulation (RCM) model is proposed. Fig.3 shows the equivalent on-resistance schematic for the ultra-thin layer SOI high voltage LDMOS. The equivalent on-resistance for the high voltage ultra-thin layer SOI LDMOS is considered as four resistances in series, as described:

$$R_{on} = R_{ch}[V_g, \Delta Q_{GOX}] + R_{acc}[V_g, \Delta Q_{GOX}] + R_{ud}[\Delta Q_p] + R_{vld}[\Delta Q_p] \quad (1)$$

$$R_{ch} = L_{ch} / \mu_{ch}(V_g \varepsilon_{ox} / t_{GOX} + \Delta Q_{GOX}) \quad (2)$$

$$R_{acc} = L_{acc} / \mu_{acc}\left(V_g \varepsilon_{ox} / t_{GOX} + \Delta Q_{GOX}\right) \quad (3)$$

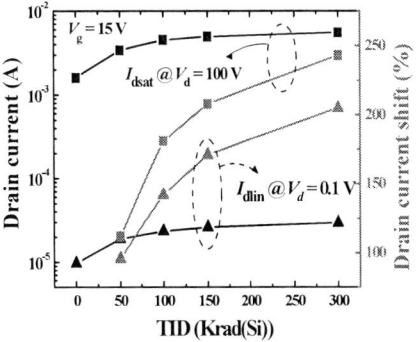

Fig.4 Measured I_{dlin} and I_{dsat} degradation as functions of TID.

$$R_{ud} = (L_{ud} - L_{acc}) / \mu_{ud}(qD_d + \Delta Q_p) \quad (4)$$

$$R_{vld} = \int_0^{L_{vld}} dx / \left[\mu_{vld}\left(N(x)t_{vld}q + \Delta Q_p\right)\right] \quad (5)$$

R_{ch}, R_{acc}, R_{ud} and R_{vld} are the channel resistance, accumulation region resistance, partial uniform doping region resistance and VLD region resistance, respectively. R_{ch} and R_{acc} are modulated by gate voltage V_g and ΔQ_{GOX}, R_{ud} and R_{vld} are modulated by ΔQ_p equal to $\Delta Q_{FOX} + \Delta Q_{BOX}$. Due to the ultra-thin drift region, the effect of ΔQ_{FOX} is assumed to be same as the ΔQ_{BOX}. ΔQ_p is assigned to be the sum of ΔQ_{FOX} and ΔQ_{BOX} with the aim of characterizing the irradiation damage of oxide layer in the drift region. L_{ch}, L_{acc}, L_{ud} L_{vld} represent the length of the channel area, the accumulation zone and the uniform doping drift region and varied lateral doping region respectively. Z is the device width. μ_{ch} μ_{acc} μ_{ud} and μ_{vld} represents the electron mobility of the channel area, the accumulation zone and the uniform doping drift region and lateral linear variation of doping area respectively. D_d is the doping dose in drift region, and $N(x)$ is the doping concentration in varied lateral doping region. ε_{ox} is the dielectric constant of silicon dioxide. t_{GOX} is the thickness of the GOX layer. The trap charge in the oxide layer here is the density of the trapped charge, including the negative interface trap charge. The RCM model also provides a method to evaluate the radiation damage in the oxide layer of the drift region by extracting ΔQ_p. Based on the RCM model, the radiation induced positive oxide trapped charge equivalently reduces R_{on}, resulting in the conduction current increasing.

III. RESULTS AND DISCUSSIONS

Aiming at evaluating the radiation damage, TID irradiation experiment is performed for the high voltage SOI LDMOS. 24 device samples are chosen to be subjected to irradiation with $^{60}C_o$. The TID in experiment are 50, 100, 150 and 300 Krad(Si) with irradiation rate of 100 rad(Si)/s, and all terminal are float during irradiation. I_{dlin} and I_{dsat} are monitored.

Fig.4 shows measured I_{dlin} and I_{dsat} degradation as functions of TID for the ultra-thin layer SOI high voltage LDMOS. I_{dlin} and I_{dsat} are the drain current in linear region and saturation region, extracted at $V_d = 0.1$ and 100V with $V_g = 15V$. It is

978-1-5386-2928-4/18 $31.00 © 2018 IEEE

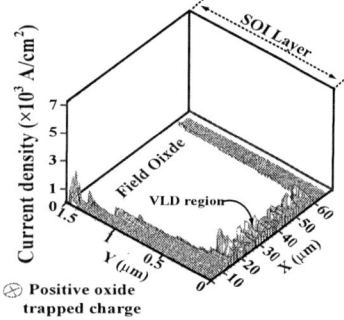

(a)

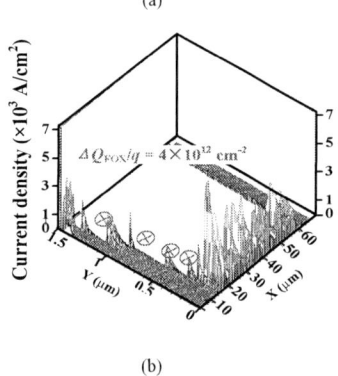

(b)

Fig.5 (a)Simulated current density distribution for $\Delta Q_{FOX}=0cm^{-2}$. (b)Simulated 3D current density distribution for $\Delta Q_{FOX}=4\times10^{12}cm^{-2}$.

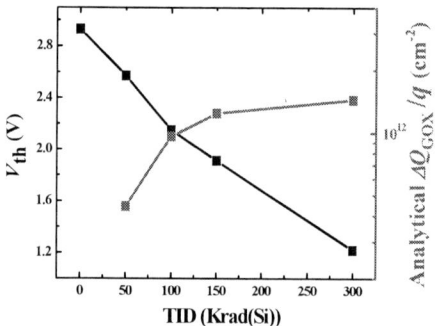

Fig.6 Measured V_{th} and analytical ΔQ_{GOX} as a function of TID.

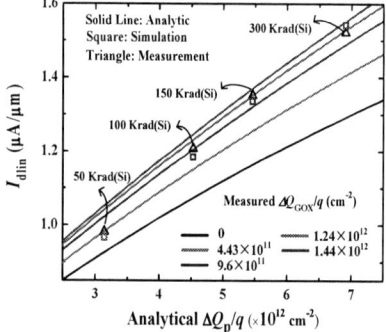

Fig.7 Analytical, measured and simulated I_{dlin} as a function analytical $\Delta Q_p/q$ with different ΔQ_{GOX}.

observed that both the I_{dlin} and I_{dsat} increase after irradiation. When TID = 300 Krad(Si), the I_{dlin} reaches 2.96×10^{-5} A. It is noting that the more severe degradation appears for I_{dsat} than for I_{dlin}. Drain current increasing indicates negative charges in bulk become more which results in on-resistance reduced.

Fig.5 shows simulated 3D current density distribution for different ΔQ_{FOX} when $V_g = 15V$ and $V_d = 0.1$ V. Positive oxide trapped charges are put into the FOX in simulation which will introduce equivalently negative mirror charges into the bulk. It shows the current density in the drift region is enhanced with ΔQ_{FOX} increasing, which is responsible for the measured I_{dlin} and I_{dsat} increasing with TID. Drain current increasing indicates negative charges in bulk become more when total ionizing dose increases which results in on-resistance reduced. It fits well with the RCM model which reveals that the radiation induced positive oxide trapped charge reduces R_{on}, resulting in the conduction current increasing.

Fig.6 shows the measured V_{th} and analytical ΔQ_{GOX} as a function of TID. As TID increases from 50 Krad(Si) to 300 Krad(Si), the V_{th} decreases from nearly 3V to 1.2V. Meanwhile the analytical $\Delta Q_{GOX}/q$ increases from 4.43×10^{11} cm^{-2} to 1.44×10^{12} cm^{-2}. Obviously, the irradiation induced positive oxide charge in GOX layer is responsible for the V_{th} decreasing. According to $\Delta V_{th} = \Delta Q_{GOX}\cdot C_{GOX}$, the positive oxide trapped charge density in GOX layer can be obtained. At threshold, interface traps are predominantly negatively charged for n-

channel transistors and trapped oxide charges are positively charged for n-channel transistors. Both of the interface traps and trapped oxide charges together effect the drift of threshold voltage.

Fig. 7 shows the analytical, measured and simulated I_{dlin} as a function analytical $\Delta Q_p/q$ with different ΔQ_{GOX}. Compared to ΔQ_{GOX}, ΔQ_p dominates in I_{dlin} degradation, that is, oxide layers in the drift region is subjected to more severe radiation damage. It indicates that 3.14×10^{12} cm^{-2}, 4.52×10^{12} cm^{-2}, 5.46×10^{12} cm^{-2} and 6.91×10^{12} cm^{-2} net positive oxide trapped charges are generated in FOX and BOX layer for the ultra-thin layer SOI LDMOS after 50 Krad(Si) to 300 Krad(Si) irradiation. Meanwhile, the corresponding ΔQ_{GOX} and ΔQ_p are putting into oxide layer in I_{dlin} simulation as shown in Fig. 7.The RCM model fits well with the simulation and experimental results.

Fig. 8 (a) shows measured breakdown voltage at off-state with different TID. BV is reduced with TID increasing. As positive oxide trapped charges introduced, the depletion region is collapsed and the electric field peak appears at the source side instead of the drain side, which is responsible for BV reduced. Fig.8 (b) shows the simulated electric field distribution at surface with different $\Delta Q_p/q$. As $\Delta Q_p/q$ increasing from 0 cm^{-2} to 6.91×10^{12} cm^{-2}, the surface electric field peak changes from the drain region to the source region.

978-1-5386-2928-4/18 $31.00 © 2018 IEEE

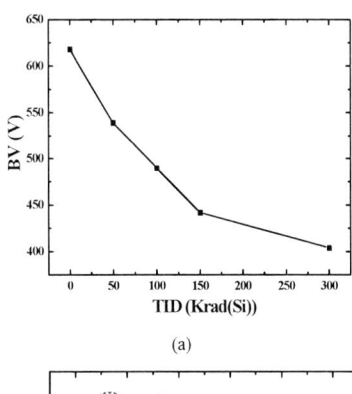

(a)

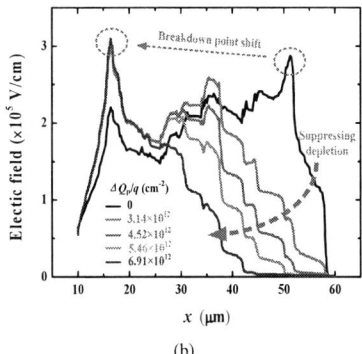

(b)

Fig.8 (a) Measured I_d-V_d curves at off-state with different TID of the SOI LDMOS. (b) Simulated electric field distribution at surface with different $\Delta Q_p/q$.

IV. CONCLUSION

TID radiation response for high voltage ultra-thin layer SOI LDMOS is investigated in this paper. The mechanisms of TID radiation induced I_{dlin} and BV degradation are revealed by modeling, simulation and verified experiment. RCM model is proposed to extract ΔQ_p in the oxide layer of drift region induced by irradiation. The radiation induced positive oxide trapped charges in GOX, FOX and BOX layer equivalently increases charges in the bulk, which results in the I_{dlin} increasing with the on-resistance reduced and BV decreasing with suppressing depletion in the drift region. Evaluated result shows the oxide layers in the drift region are subjected to more severe damage with large ΔQ_p, which dominates the I_{dlin} and BV degradation. Extracted ΔQ_p much more than ΔQ_{GOX} is the main contributor to I_{dlin} and BV, as a result, the drift region should be seriously designed with the consideration to the anticipated charges introduced by radiation.

REFERENCES

[1] P. M. Shea and Z. John Shen, IEEE Trans. on Nuc. Sci., 58, 2739(2011).

[2] M. Qiao, X. Zhou, Y. He, et al., IEEE Electron Dev. Lett., 33, 1438(2012).

[3] K. Hara, S. Wada, J. Sakano, et al., IEEE ISPSD, Waikoloa, USA(2014).

[4] J. F. Chen, S. Y. Chen, K. M. Wu, et al., Appl. Phys. Lett. 93, 223504 (2008).

[5] X. Zhou, M. Qiao, Y. He, et al., Appl. Phys. Lett. 107, 203507(2015).

[6] P. E. Dodd, M. R. Shaneyfelt, B. L. Draper, et al., IEEE Trans. on Nuc. Sci., 56, 3456(2009).

[7] J. M. Lauenstein, A. D. Topper, M. C. Casey, et al., IEEE Radiation Effects Data Workshop, San Francisco, USA,(2013), p. 1.

[8] S. Diez, M. Ullan, G. Pellegrini, M. Lozano, et al., IEEE Trans. on Nuc. Sci., 57, 3322(2010).

[9] F. Faccio, B. Allongue, G. Blanchot, et al., IEEE Trans. on Nuc. Sci., 57, 1790(2010).

[10] J. R. Schwank, M. R.Shaneyfelt, D. M. Fleetwood, et al., IEEE Trans. on Nuc. Sci., 55, 1833(2008).

[11] R. V. Dalen, A. Heringa, P. W. M. Boos, et al., IEEE ISPSD, 2010, pp.89-92.

[12] H. J. Barnaby, IEEE Transactions on Nuclear Science, vol.53, No.6, pp. 3103-3121, 2006.

[13] Patrick M. Shea; Z. John Shen, IEEE ISPSD, 2011, pp.376-379.

[14] Boyi Yang; Jiann-Shiun Yuan; Zheng John Shen, IEEE Transactions on Electron Devices, 2011, pp.4000-4010.

Proceedings of the 30th International Symposium on Power Semiconductor Devices & ICs
May 13-17, 2018, Chicago, USA

Electromigration Current Limit Relaxation for Power Device Interconnects

Jungwoo Joh, Young-Joon Park, Srikanth Krishnan, Kim Christensen, Jayhoon Chung

Analog Technology Development, Texas Instruments, Dallas, TX, U.S.A., jjoh@ti.com

Abstract— **Electromigration (EM) is a key limiting factor for designing lateral power devices. The metal interconnect for power devices features multiple fingers with strapped metal layers to carry large current, which leads to unique EM behaviors. In this paper, we present a new EM methodology for power device interconnects to account for these effects. The new model features circuit performance based failure criteria and allows more EM current limit for power devices with multiple fingers than the conventional rule. This provides relaxed EM rules for more efficient power device design.**

Keywords—electromigration; power device; reliability; failure criteria; multi finger

I. INTRODUCTION

The metal system for lateral power devices is different from typical CMOS interconnects. For example, in order to handle large current in power LDMOS, there are multiple source and drain fingers with wide and thick metal layers, and several metal layers are strapped together to allow high current [1, 2]. These unique characteristics in power device interconnect can result in very different electromigration (EM) degradation behaviors than typical CMOS interconnects [3, 4]. In particular, due to redundancy in conduction path, EM void formation only increases metal resistance gradually, but it may not be fatal as in CMOS interconnects where a single interconnect failure can result in system failure [5, 6].

In this paper, to account for these features and better assess EM reliability, we developed a new model for EM failure criteria for power device metal system based on the circuit performance, namely ON resistance (R_{ON}). Also, we discuss the impact of parallel conduction paths on EM

statistics. We show that the new methodology can allow more current density for power devices than the EM rules based on single lead failure.

II. ELECTROMIGRATION FAILURE BEHAVIOR

In typical CMOS circuits, the interconnect resistance is a very small portion of total R_{ON} of the transistor which it is connected to, and therefore small metal resistance degradation does not generally affect the overall circuit performance. In addition, typical EM failure takes place near VIA areas [7]. Because shunting layers (e.g. TiN) are generally very thin, once full spanned void starts to form in the main metal lead (e.g. Al), the metal resistance increases very sharply with time (**Fig 1**, left). As a result, the EM lifetime is insensitive to failure criteria (FC_{metal}), and it can be defined relatively arbitrarily to represent a large enough resistance increase. The semiconductor industry typically defines 10-20% metal resistance increase as EM failure. Typically, EM current limits (j_{DC}) are based on this methodology.

On the other hand, the characteristics of the metal system for lateral power devices, particularly for source/drain fingers, are quite different. First of all, multiple metal layers are often times strapped together to flow large current as shown in **Fig 1** (right) [2]. As a result, major void formation in one of the metal layers does not result in fatal failure. The metal resistance gradually increases with time because there are alternate current paths. In this case, the EM lifetime is dependent on the definition of failure threshold because the more metal resistance degradation allowed, the longer the device lifetime will be (**Fig 1**). In fact, extracted EM lifetime of VIA-fed single lead structure and strapped S/D finger

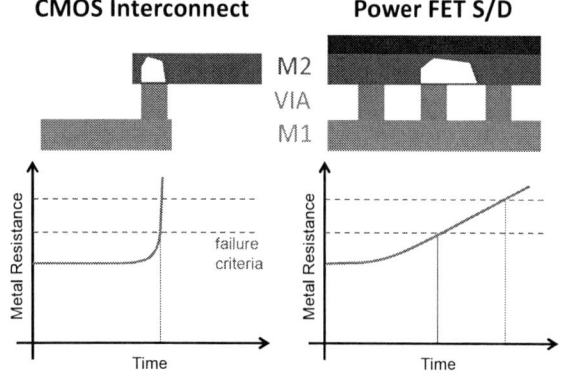

Fig 1. Electromigration behavior of typical CMOS interconnect (left) and power FET (right). For CMOS interconnect, lifetime is failure criteria independent, while lifetime is failure criteria sensitive for Power FET (Fig 2) because the resistance increase is more gradual.

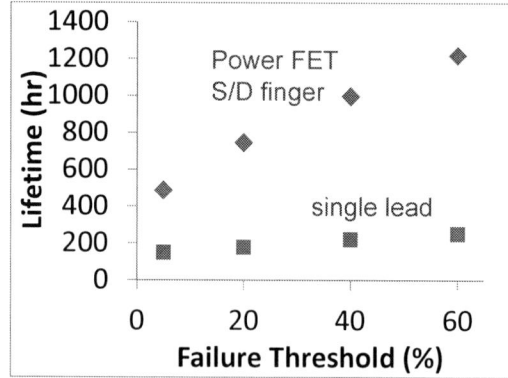

Fig 2. Extracted lifetime as a function of failure threshold from accelerated life tests on power FET S/D finger (MET1 and MET2 strapped) and single lead (MET1 down). As depicted in Fig 1, lifetime depends on failure criteria for S/D fingers while it is nearly FC independent for single lead failure.

978-1-5386-2928-4/18 $31.00 © 2018 IEEE   68

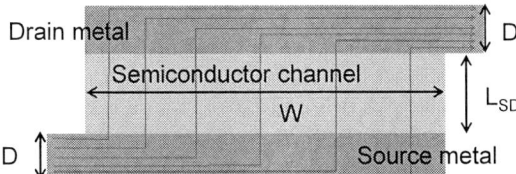

Fig 3. Schematic layout of a unit cell of a lateral FET. The (red) arrows represent current paths. In typical layout, transistor channel width W (=S/D finger length) is much larger than source-drain spacing (L_{SD}) or source/drain finger metal width (D_S and D_D), making metal resistance significant portion of total R_{ON}.

shows this behavior (**Fig 2**): strapped S/D finger shows linear increase in lifetime with FC_{metal} while single lead shows very weak dependence. Then a question arises – where exactly should the failure threshold be defined for power devices? The metal resistance is a significant portion of the total R_{ON} of a typical power transistor due to its large dimension. As a result, change in metal resistance can significantly affect total R_{ON} of the transistor and thus overall circuit performance. Therefore, it is more reasonable to define the failure criteria based on total R_{ON} degradation of a power FET than to use a universal failure threshold (20%). This will result in more accurate prediction of the device and circuit lifetime.

III. NEW EM FAILURE CRITERIA FOR POWER FET

A. Failure criteria based on R_{ON} degradation

While conventional EM failure is universally defined as 20% increase in metal resistance, in our new model we define EM failure as metal resistance degradation that leads to a certain amount of increase in *total* R_{ON} of the power transistor (e.g. 5% of total R_{ON}). For different applications, the maximum allowable degradation in total R_{ON} can be determined according to circuit/product requirements (e.g. datasheet specification or circuit fault). The drift of R_{ON} of a power FET mainly consists of semiconductor ($R_{semiconductor}$) and metal resistance (R_{metal}) degradation:

$$\Delta R_{ON} = \Delta R_{semiconductor} + \Delta R_{metal}$$
$$= \alpha R_{ON}(0) = (\alpha_s + \alpha_m)R_{ON}(0) \qquad (1)$$

where α, α_s, and α_m is the allowable drift (reliability budget) for total, semiconductor-induced (e.g. channel hot-carrier degradation), and metal-induced (EM) degradation (%), respectively. $R_{ON}(0)$ is the initial ON resistance of the fresh device. For example, if maximum R_{ON} degradation budget α is 15% based on circuit simulation, and the end of life R_{ON} degradation due to channel hot carrier is 10% (α_s), maximum allowable ΔR_{metal} from EM is 5% (α_m).

Now we need to evaluate how much metal resistance degradation causes a fatal amount of total R_{ON} drift, $\alpha_m R_{ON}(0)$. A new EM failure criterion, FC_{metal} can be found:

$$\alpha_m R_{ON}(0) = FC_{metal} R_{metal}(0) \qquad (2)$$

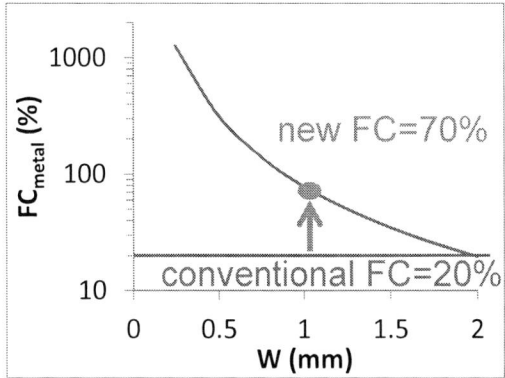

Fig 4. Modeled failure criteria as a function of device unit finger length. The used parameters are from a high voltage device. For comparison, conventional failure criterion (20% - independent of W) is also shown.

Solving for FC_{metal}, we obtain:

$$FC_{metal} = (\alpha - \alpha_s)\left[\frac{R_{semiconductor}}{R_{metal}} + 1\right] \qquad (3)$$

From (3), it can be seen that lower R_{metal} gives rise to higher FC_{metal}. FC_{metal} can be higher than conventional threshold of 20%, which allows longer lifetime or more current density J_{DC}.

B. Model application: HV power FET example

In this section, we demonstrate how our new methodology can be applied to actual power device design. We use a lateral HV power FET as an example. The layout of a unit cell is shown in **Fig 3** (note parameter definitions). The total R_{ON} consists of semiconductor and metal resistance:

$$R_{semiconductor} = R_{SS}\frac{L_{SD}}{W} \qquad (4)$$

$$R_{metal} = R_{SM}\frac{W}{D} \qquad (5)$$

where R_{SS} and R_{SM} are sheet resistance of semiconductor and metal, respectively. Plugging (4) and (5) in (3), one obtains:

$$FC_{metal} = (\alpha - \alpha_s)\left[\frac{L_{SD}D}{W^2}\frac{R_{SS}}{R_{SM}} + 1\right] \qquad (6)$$

Equation (6) shows that the failure threshold is dependent on device layout geometry as well as process parameters (e.g. sheet resistance). This allows designers to optimize performance and EM reliability.

One of the key device layout design parameters in Equation (6) is the device finger length (=channel width), W. In **Fig 4**, we plotted the failure criteria as a function of source/drain finger length for a high voltage power FET. We assumed that R_{ON} degradation budget for metal resistance increase (α_m) is 5%. It can be seen that the device finger length largely impacts failure criteria as expected from Equation (6). For W=1 mm, the new methodology gives higher failure

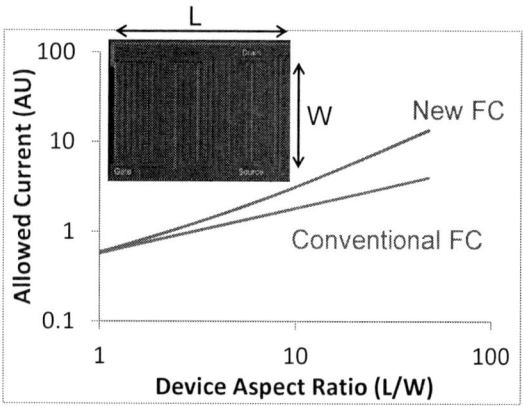

Fig 5. Allowed total current for the same R_{ON} power FET as a function of device aspect ratio (L/W). The current rating is normalized to AR=1. In the inset, power device layout is shown. There is different sensitivity to the aspect ratio for different failure criteria.

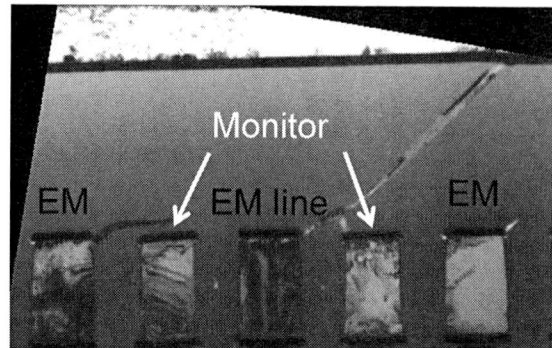

Fig 6. Failure analysis x-section of a failed multi-finger device. This device was stressed to have 4x increase in metal resistance. Hillocking induced dielectric failure is observed in the anode side of the device.

threshold (70%) than conventional 20%. Since the metal resistance degradation occurs almost linearly as shown in [3], this increased failure threshold yields more than 2 times longer lifetime than conventional method for W=1 mm, or can allow >50% more J_{DC} at a given lifetime spec. As W increases, this gain reduces, but the new failure criteria model gives consistently better lifetime than the conventional method for the entire practical range of W (**Fig 4**).

C. Layout example: aspect ratio

One of the options to improve EM J_{DC} rating of a power FET is to reduce the finger length and add more fingers. In order to demonstrate how we can use the new failure criteria model when changing the layout, we have calculated the total current rating of a HV FET with different aspect ratio but with the same area (**Fig 5** inset). The total R_{ON} of the device remains the same because of the same device area (=$L \times W$). With higher aspect ratio ($AR=L/W$), each finger is shorter and thus needs to carry less current, and therefore, the total current rating of the power device increases even with the conventional failure criteria (constant 20%). On top of this, the new methodology gives more benefit because of higher failure criteria for shorter finger length (W) as shown in **Fig 4**. As a result, even higher current can be allowed with the new failure criteria compared with the conventional failure criteria (**Fig 5**). As shown in this example, the new methodology to determine failure criteria can be a very powerful tool to optimize the EM performance of lateral power devices.

D. Impact of increased failure criteria

Because increased failure threshold for R_{metal} could result in more EM voiding, there is a concern in hillock formation on the anode side of the metal where the metal is piled up. In fact, we observed hillocking induced dielectric failure after increasing the metal resistance by 4 times (**Fig 6**). However, hillock induced failure lifetime was found to be 10x higher than the void failure. Although this sets a limit to extend the

failure threshold, there is still a large room to improve EM rating using our new methodology.

IV. FAILURE STATISTICS FOR MULTI FINGER DEVICE

Most power transistors have multiple fingers to carry larger current. Because there are parallel current paths, a significant increase in R_{metal} of one finger has only minor impact on overall R_{ON}. Therefore, failure of the whole device (a certain increase in total R_{ON}) takes place only after R_{metal} of majority of the fingers degrades. Also, as metal resistance starts to degrade in certain fingers, the current flowing through those early failing fingers decreases, and therefore the degradation is slowed down (negative feedback). This means that the variance of degradation behavior for multi-finger devices should be much smaller than single finger cases.

In order to verify this idea, we have performed Monte-Carlo simulations to study the lifetime distribution for different number of fingers. We simulated 100 single finger devices (R_{metal}=1000 Ω) and 100 multi-finger (number of fingers=100, thus R_{metal}=10 Ω) power FETs. In this simulation, we assumed that the metal resistance degradation of each finger is independent and identically distributed (i.i.d.). We have used different degradation models including linear, exponential, and abrupt resistance increase with and without incubation time. Since we found that the detailed degradation model as well as failure threshold do not change the overall behavior, we will use a simple linear model (rate=0.05%/hr) and 50% failure threshold in the following discussion.

Fig 7 shows resistance increase behavior as well as log-normal plot of lifetime for the two groups (1 & 100 fingers). As expected, median time to failure (MTTF or T50) is the same between the two groups. However, it is clearly seen that the distribution is much tighter for the multi-finger device. The standard deviation (σ_{LN}) is 0.14 and 0.0133 for N=1 and 100, respectively. The reduced σ_{LN} is expected from the law of larger numbers ($\sigma \sim 1/\sqrt{N}$) because degradation of each finger is averaged out for multi-finger devices. In fact, we have experimentally confirmed that σ_{LN} scales with $N^{-0.5}$ (**Fig 8**, left). Since device lifetime is determined at low failure fraction, EM lifetime for multi-finger devices is longer due to the small σ_{LN} (**Fig 7**, right). As noted above, this lifetime increase results from better predictability (tighter distribution),

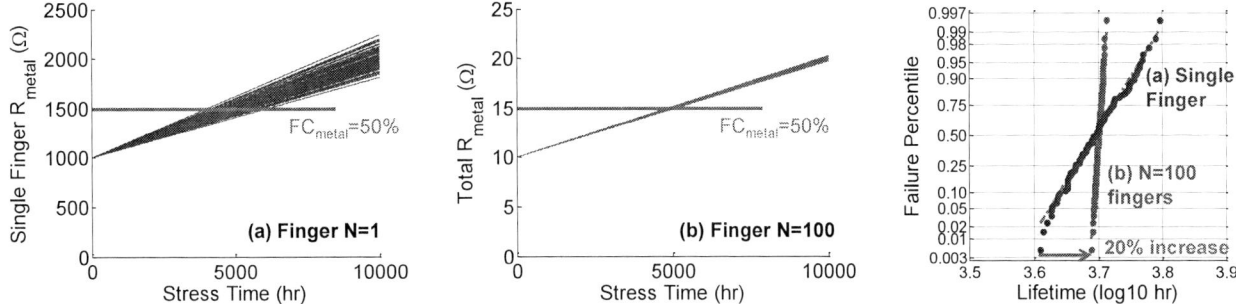

Fig 7. Simulated metal resistance degradation behavior for (a) single finger device and (b) multi-finger (N=100) device. 100 random devices were simulated for each case. Each finger was assumed to have metal resistance of 1000 Ω, and R_{metal} was assumed to increase linearly with time. FC_{metal}=50% was arbitrarily chosen to extract the lifetime. Right: Log-normal distribution of simulated lifetime of single- and multi-finger (N=100) devices. Due to reduction in standard deviation (0.14 to 0.013), 0.5% lifetime (T0.5) increases by 20%.

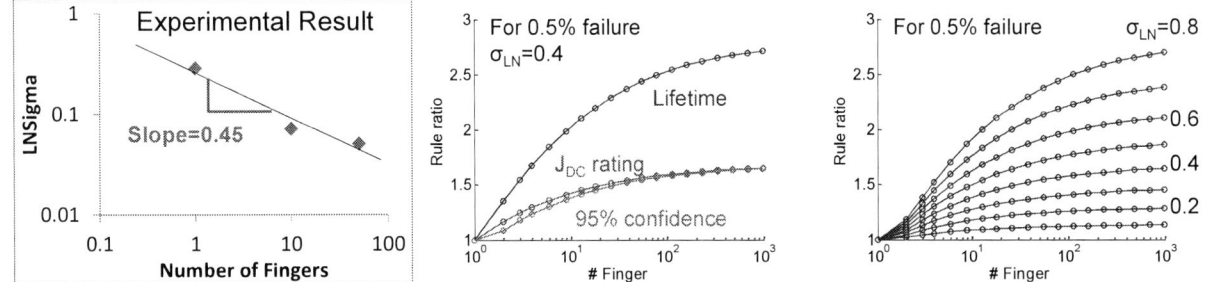

Fig 8. Left: Experimental verification of multi-finger effect. With more number of fingers, the standard deviation decreases as the model predicts (σ_{LN}~1/√N). Middle: Computed lifetime (0.5% failure) and EM current rating as a function of the number of FET fingers. For power FETs with large number of fingers, there is significant EM J_{DC} improvement (>50%). Right: J_{DC} rule ratio (0.5% failure) for different standard deviation.

not from any physical reason (no change in MTTF). The improved lifetime in turn increases the maximum allowed EM limit for a given lifetime:

$$\frac{j_{DC}(N)}{j_{DC}(1)} = exp[\frac{Z\sigma_{LN}}{n}\left(\frac{1}{\sqrt{N}} - 1\right)] \qquad (7)$$

where Z is the inverse of standard normal cdf for the target failure fraction (e.g. Z=-2.58 for 0.5%), and n is the current exponent in Black's equation [4] (we used n=2 for Al).

Based on this model, **Fig 8** (middle) shows computed lifetime and current rating improvement with N (normalized to N=1). Here, we used σ_{LN}=0.4 for the single finger distribution (from single lead test data). It can be seen that 0.5% lifetime for the N=100 finger device is 2.7 times larger than a single finger. This corresponds to 64% increase in EM current rating (J_{DC}) for a given lifetime. Since there is statistical variation in this ratio, we performed additional Monte-Carlo simulations to find 95% confidence level for the current rating (red curve in **Fig 8** middle). It can be seen that even with the small number of fingers (>10) there is >30% improvement in J_{DC} rating.

This gain is expected to be larger for processes with high σ_{LN} and/or when targeting lower failure fraction as shown in Eq (7). **Fig 8** (right) shows the J_{DC} rule improvement ratio for different σ_{LN}. It can be seen that larger standard deviation processes benefit more from the multi-finger rules. Because this J_{DC} gain results from the redundancy that gives better lifetime statistics, this idea can be generalized to other cases where there are multiple parallel paths (e.g. power grid [5, 6]).

V. CONCLUSION

We developed a new EM methodology for power device interconnect that allows relaxed EM rules. This is enabled by power device interconnect structure that features multiple fingers with strapped metal layers. This new EM methodology can provide new insights for designing more optimized power FETs and help reducing manufacturing costs.

REFERENCE

[1] R. K. Williams and M. Kasem, "Lateral Power MOSFET Having Metal Strap Layer to Reduce Distributed Resistance," US Patent 5767546, 1998.

[2] T. Letavic, J. Petruzzello, M. Simpson, J. Curcio, S. Mukherjee, J. Davidson, S. Peake, C. Rogers, P. Rutter, M. Warwick, and R. Grover, "Lateral smart-discrete process and devices based on thin-layer silicon-on-insulator," *ISPSD*, 2001, pp. 407-410.

[3] Y.-J. Park, J. Joh, and K.-S. Ko, "Electromigration in strapped metal layers with large dimensions for lateral power device applications," *IEEE International Reliability Physics Symposium*, 2014, pp. 2A.4.1-2A.4.4.

[4] J. R. Black, "Electromigration - A brief survey and some recent results," *Electron Devices, IEEE Transactions on*, vol. 16, no. 4, pp. 338-347, 1969.

[5] V. Mishra and S. S. Sapatnekar, "The impact of electromigration in copper interconnects on power grid integrity," *Proceedings of the 50th Annual Design Automation Conference*. Austin, Texas: ACM, 2013, pp. 1-6.

[6] S. Chatterjee, M. Fawaz, and F. N. Najm, "Redundancy-aware Electromigration checking for mesh power grids," presented at 2013 IEEE/ACM International Conference on Computer-Aided Design (ICCAD), 2013.

[7] Y.-J. Park, "Interconnect dimension and current waveform effects on electromigration performance," presented at IRPS Tutorial, 2006.

Proceedings of the 30th International Symposium on Power Semiconductor Devices & ICs
May 13-17, 2018, Chicago, USA

Performance and Reliability Insights of Drain Extended FinFET Devices for High Voltage SoC Applications

B Sampath Kumar, Milova Paul and Mayank Shrivastava

Advanced Nano-electronic Device and Circuits Research Group
Department of Electronic Systems Engineering
Indian Institute of Science, Bangalore, India 560012
mayank@iisc.ac.in

Harald Gossner

Intel Deutschland GmbH
Am Campeopn10-12,85576
Neubiberg, Germany

Abstract—In this paper[1], Drain extended FinFET device design and the challenges associated with the performance and reliability are discussed. Physical insights into the performance vs. reliability trade-off for the Fin enabled high voltage designs is elaborated and compared with their planar counterpart (DeMOS). Effect of Fin width discretization over ESD reliability, Safe Operating Area and HCI reliability are discussed.

Keywords—Drain Extended, , ESD , finFET, HCI, reliability, Safe Operating Area

I. INTRODUCTION

Drain Extended MOS (DeMOS) devices were extensively implemented for high voltage (HV), System on Chip (SOC) applications [1]. Wide voltage range of LDMOS devices are used as gate drivers, voltage converters and operational amplifiers. On the other hand, with the aggressive CMOS scaling and with the advent of FinFET technology [2], the novelty of FinFETs served extensively in IC technology beyond 20nm. However, HV integration such as drain extended device technology did not pace up substantially for FinFET technology. Stacking the core/IO transistors can also offer high voltage applications for FinFET, but they are not only limited to few volts of operation and but also adds complexity of design, such as isolation schemes. One step solution for high voltage applications are DeFinFET devices. Where the operating voltage can be tuned with layout and other design parameters. Moreover, thermal resistance, power density, off state dissipation [3][4] threatens the feasibility of the FinFET HV integration. Performance metrics of fin-based HV devices [5]-[7] were discussed so far. However, detailed analysis of performance and reliability were never explored for fin enabled drain extended devices. This work demonstrates the performance and unique features of reliability of drain extended FinFET devices compared to its planar counterpart.

II. DRAIN EXTENDED FINFET: ARCHITECTURE AND DEVICE DESIGN

Fig. 1 shows the various cross sections of drain extended FinFETs. Proposed process FEOL flow for realizing the high

voltage DeFinFET into the fin based technology is designed using 3D TCAD process simulations. Deep Wells for Drift region is implanted and annealed prior to baseline FinFET Process. In FinFET technology since the anneal cycles are shorter to control the depth of implants, high voltage wells suitable for DeFinFETs are implanted prior to the standard process flow the FinFET technology. Gate oxide used is a composure of SiO2 and HfO2 with Effective oxide thickness of 1.1nm, TiN is used as a gate metal and nitride spacers are used for contact isolation. The drain extended part of the device is formed with low doped drift region, unlike having a self-aligned source/drain contact, a selective source/drain contact formation is necessary. Fig. 2 shows the calibrated IV characteristics of FinFET with experimental data [8]. Device simulations and physical investigations are done using the FinFET calibrated setup.

A. On Resistance vs. Breakdown trade-off

Fig.3 shows R_{ON} vs V_{BD} tradeoff of HV-FinFETs ((a)DeFinFET and (b)STI-DeFinFET) in comparison with its

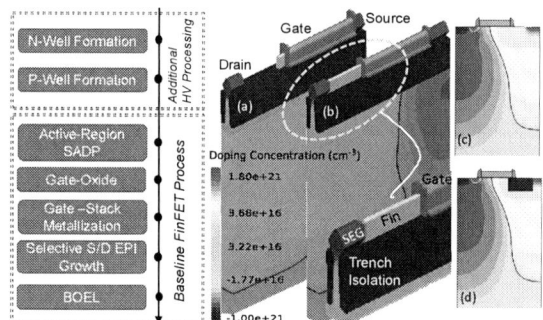

Figure 1: Proposed process FEOL of HV Drain extended devices, (a) STI-DeFinFET: STI isolation in the drift region (b) Drain extended FET: DeFinFET. Inset shows the Fin, Trench Isolation and Gate Metal Stack. (c), (d) are the planar counterpart DeMOS & STI-DeMOS respectively, which resembles the cross section of 3D-DeFinFETs.

---

[1]This work is supported by IMPRINT program of Department of Science and Technology (Project Code: 5360) and Ministry of Human Resource and Development. Government of India.

978-1-5386-2928-4/18 $31.00 © 2018 IEEE

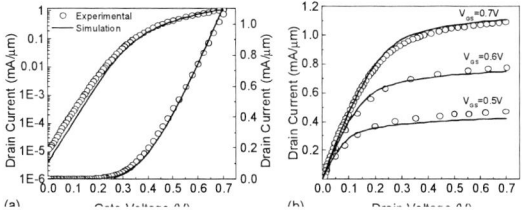

Figure 2: Calibration of TCAD models for drift-diffusion transport with experimental data [7] (a) Transfer characteristics, and (b) Output characteristics [6].

planar counterparts. HV-FinFETs with its architectural limitation of narrow width, shows higher R_{ON} over conventional DeMOS devices. DeFinFETs has an order trade off in RON vs VBD, when compared to planar technology. The inset of Fig. 3 shows the Design of experiments (DOE) used to tune the performance and voltage rating of the device. With the optimized design for RON vs VBD trade off, device with best performance out of DOE is chosen for reliability analysis of DeFinFET in the proceeding sections.

B. Quasi Saturation in DeFinFET

Fig. 4 depicts the quasi saturation behavior of DeFinFETs. With increase in gate bias and when high current is injected in to the drift region, the carrier density exceeds the background doping and causes space charge modulation. Fig. 4 (a) shows the shift in the peak from gate edge towards the drain edge of the device as a result of space charge modulation [9]. Fig. 4(b) shows the current saturation as a function of gate bias. The voltage range between threshold voltage and the onset voltage of quasi-saturation narrows down in DeFinFET, this is attributed to the narrow fin widths. Unlike planar DeMOS device narrow fin width in DeFinFET increases the current crowding and leads to the early quasi saturation effect.

C. DeFinFET : Figure of Merits

Fig. 5(a) shows the I_{ON}/I_{OFF} ratio of HV-FinFETs. Owing to the narrow pitch and fin enabled dual side channel inversion HV-FinFETs are benefited with good gate control over the channel and has low OFF state leakage. However, STI-DeFinFET due to high ON resistance has 2x times lower I_{ON}/I_{OFF} ratio, when compared to DeFinFETs. Current

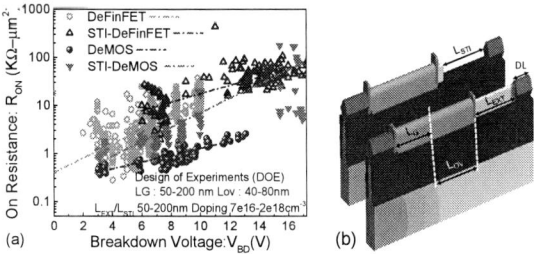

Figure 3: Performance Metric: (a) On Resistance (R_{ON}) vs Breakdown Voltage (V_{BD}) of DeFinFET and STI-DeFinFET devices. (b) 3D view of DeFinFET depicting channel length (L_G), Overlap Length (L_{OV}), Drift Length (L_{EXT}/L_{STI}) and drain contact length (DL).

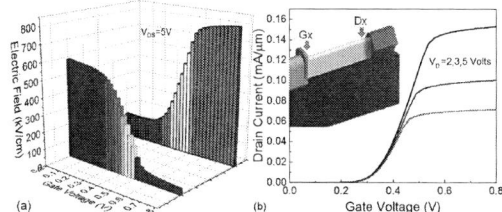

Figure. 4 (a) Quasi Saturation (QS) depicted through a shift in the peak electric field as a function of V_{GS},. Here Gx and Dx are referred to the position at Gate and Drain edge respectively as shown in the inset of (b), (b) transfer characteristics of DeFinFET depicting severe Quasi-Saturation at different Drain biases, is attributed to inevitable narrow fin architecture.

limitation phenomena of HV-FinFETs are described through quasi-onset current, Fig. 5(b) depicts the quasi onset current in DeFinFET and STI-DeFinFETs. Apart from having a narrow fin and device dimensions STI-DeFinFET is effected through severe quasi saturation. And this is attributed to higher current crowding under the drain contact opening. Moreover, space charge modulation is a current driven phenomenon, quasi saturation effect can be substantially lowered by increasing the drain opening (DL). Fig 5(b) inset shows the ~ 0.5x improvement in the quasi onset current.

Moreover, due to pronounced R_{ON} and current crowding in HV-FinFETs, Figure of Merit (FOM) stands 2x times lower (Fig. 6(a)) than Planar DeMOS and STI-DeMOS devices (Fig. 6(b)). The device foot print over the design of experiment of

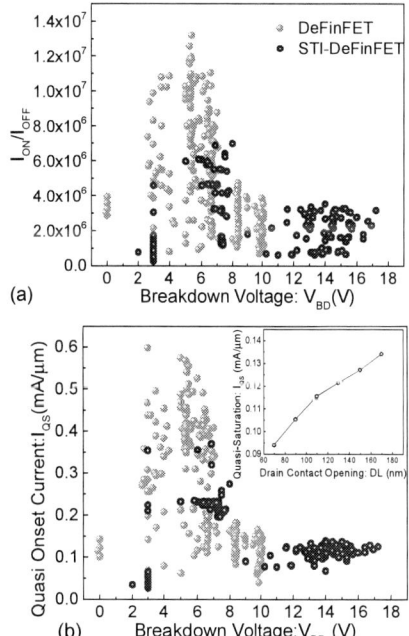

Figure. 5: (a) On Current I_{ON} to I_{OFF} ratio of DeFinFET and STI-DeFinFET devices(b) depicts the current at peak g_m (Quasi-onset) in I_{DS} vs V_{GS} characteristics. This is the current value at which kirk effect takes place. Inset shows the effect of drain opening (DL) on quasi onset current.

978-1-5386-2928-4/18 $31.00 © 2018 IEEE

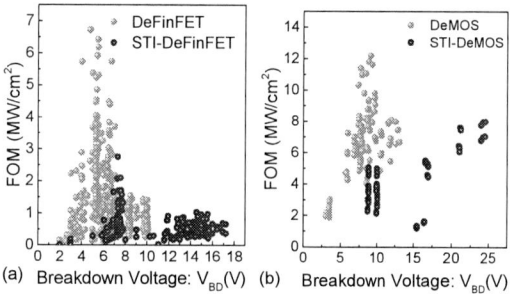

Figure. 6: Power rating of the transistor: Figure of Merit (FOM) of (a) DeFinFET and STI-DeFinFET measured as V_{BD}^2/R_{ON}-specific (b) FOM of Planar DeMOS in comparison with DeFinFET.

both the device types (planar and fin based) are identical. Inclusion of STI along the conducting path improves the V_{BD} by ~2X times by trading off with the FOM. However, Narrow pitch, wrapped gate and high current densities in HV-FinFETs severely degrades the performance when compared to its planar counterpart DeMOS devices

D. Band Tailoring in DeFinFETs

Fig. 7 shows the band diagram of DeFinFET taken under gate of Planar DeMOS whereas, the cutline for band diagram is taken inside the fin region of DeFinFET. Band bending extracted under the same biasing conditions, shows higher gradient being close to gate in Planar DeMOS while it is close to Drain contact in case of DeFinFET, this signature of band diagram reflects DIBL and hot carrier reliability favoring towards the DeFinFETs. Significant reduction of DIBL is seen in DeFinFET attributed to the pinning of band energy which comes from the wrapped gate around the fin. Since, the band bending is more towards the drain extension region in case of the DeFinFETs. Electric field is distributed more inside the drift region.

Due to this significant gate control over the channel is seen in DeFinFETs and owing to the band tailoring in the DeFinFET, the feasibility of channel length scaling is more flexible in fin based HV devices when compared to the planar DeMOS devices. Moreover, Drain induced barrier lowering also least effected in HV-FinFETs. Fig. 7(b) shows the effect of DIBL being more in case of Planar DeMOS when compared to DeFinFETs. Having compared the DIBL and

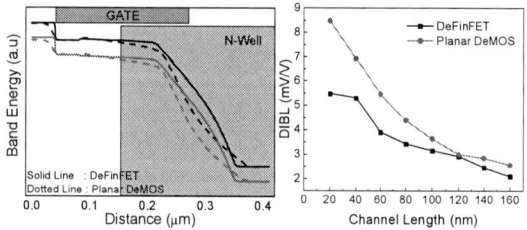

Figure.7: (a) Band Diagram comparison of planar DeMOS and DeFinFET, (b) Drain Induced Barrier Lowering (DIBL) comparison of planar DeMOS and DeFinFET. Significant reduction of DIBL is seen in DeFinFET attributed to the pinning of band energy which comes from the wrapped gate around the fin.

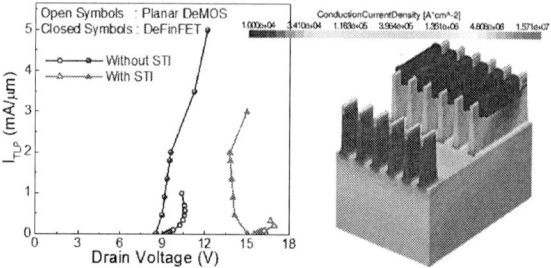

Figure. 8: (a) TLP IV characteristics of Planar DeMOS and DeFinFET (w/ and w/o STI). (b) Uniform current conduction during ESD stress in multifin configuration of DeFinFET.

channel control of DeFinFETs, ON current of fin based HV devices can be improved by scaling down the channel lengths. However, increasing the current in DeFinFET by channel length scaling will lead to the early quasi saturation effect. Therefore, DeFinFET performance gets trade-off between channel length scaling and ON current. Moreover, Quasi saturation in the device will lead to localized hot spot close to the drain edge, and increases the lattice heating and finally cause the device Failure.

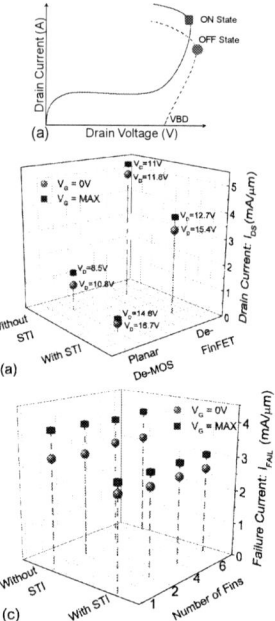

Figure. 9: Comparison of Safe operating area (SOA) of DeFinFETs and Planar DeMOS (a) SOA of drain extended devices are extracted using the failure current: It2. at ON state VGS =VGmax and OFF state VGS=0V. (b) Extraction method used to define SOA. SOA covered by the DeFinFETs is 5x times higher than its planar DeMOS counterpart. (c) SOA is independent of the multifinger arrangement in DeFinFETs, which makes them more suitable for HV integration over planar DeMOS devices

978-1-5386-2928-4/18 $31.00 © 2018 IEEE   74

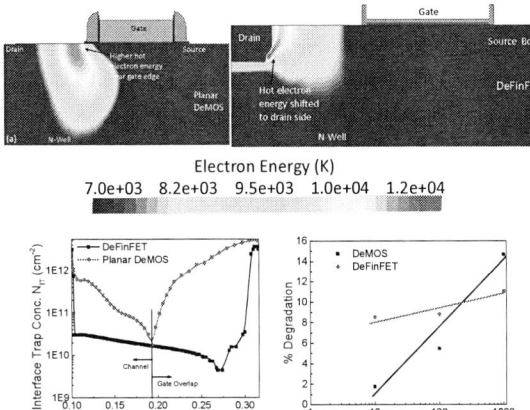

Electron Energy (K)

7.0e+03 8.2e+03 9.5e+03 1.0e+04 1.2e+04

Figure. 10 Hot Carrier Induced (HCI) degradation Drain extended devices (a) Contour showing spatial hot carrier distribution in Planar and FinFET drain extended devices. In Fig. 7 Band bending signifies this phenomenal hot carrier distribution. Therefore, (b) NIT generation at oxide-silicon interface, shows, the degradation is concentrated at gate edge in DeFinFET, while NIT is distributed throughout the interface in Planar DeMOS(c) % Id-sat degradation of Planar and Fin DeMOS w.r.t time, where VDS is biased close to breakdown voltage. With the signature of NIT generation, the rate of degradation in Planar DeMOS is higher than DeFinFET.

III. RELIABILITY

Figure.8 (a) shows the ESD reliability Drain extended devices. Owing to filament formation, failure current It2 stands lower in planar DeMOS, whereas DeFinFET (w/ and w/o STI) can offer 5x times improvement in failure current. This is attributed to the isolation of the current path, coming from the fin to fin isolation. Contour plot in Fig. 8(b) shows the current conduction under TLP stress (a 100ns current pulse). Absence of current filamentation is shown through uniform current conduction. Fin enabled width discretization boosts ESD reliability. (Fig. 8(a)) and SOA (Fig. 9) of HV-FinFETs shows 5x times improvement in ESD reliability and 2x times improvement in SOA, which is attributed to mitigation of localized hotspots, which comes from uniform fin current (Fig. 9(b)), whereas planar devices show early failure signature due to formation of hot spots. Fig. 10 shows the HV-FinFETs having better tolerance towards HCI degradation, due to its nature of band bending. Hot carrier density distribution being away from the gate oxide region, favors in slower rate of degradation, whereas in Planar DeMOS Hot carriers are populated closer to gate edge leading to higher rate of degradation. Due to fin architecture, number of Si-H Bond Concentration will be greater than Planar architecture for a given device area/ foot print. Therefore, initial degradation, which is caused by Si-H bond breaking is significant in DeFinFET. However, the slope of degradation is higher in case

of Planar DeMOS device, as the degradation of Si-O bond breaking is crucial over the overall stress period.

IV. SUMMARY

Fin based architecture and design for High voltage drain extended devices are studied. Attributed to narrow Fin geometry, quasi saturation onset appears early. Moreover, performance FOM of drain extended devices degrade severely as one moves for FinFET technology. However, on the other hand, due to fin enablement gate control over channel is improved. With the fin based architecture and observed band tailoring in DeFinFET are suitable for channel length scaling, where as in planar DeMOS, channel length scaling is often limited due to DIBL.Due to fin width discretization non-uniformity among the fins are suppressed in fin based High voltage devices. Fin width discretization favors in ESD reliability and SOA owing to uniform current conduction. And finally, since hot carrier density is distributed away from the channel region, the life time of the drain extended devices are predicted to increase in FinFET geometry.

REFERENCES

[1] R. Minixhofer et.al, "A 120V 180nm High Voltage CMOS smart power technology for System-on-chip integration," *Proceeding of the 2010 ISPSD*, pp. 75-78, June 2010

[2] C. Auth *et al.*, "A 22nm high performance and low-power CMOS technology featuring fully-depleted tri-gate transistors, self-aligned contacts and high density MIM capacitors," *2012 Symposium on VLSI Technology (VLSIT)*, Honolulu, HI, 2012, pp. 131-132

[3] H. Kawasaki *et al.*, "Demonstration of highly scaled FinFET SRAM cells with high-κ/metal gate and investigation of characteristic variability for the 32 nm node and beyond," *2008 IEEE International Electron Devices Meeting*, San Francisco, CA, 2008, pp. 1-4

[4] S. O. Koswatta *et al.*, "Off-state self-heating, micro-hot-spots, and stress-induced device considerations in scaled technologies," *2015 IEEE International Electron Devices Meeting (IEDM)*, Washington, DC, 2015, pp. 20.2.1-20.2.4

[5] M. Shrivastava, H. Gossner, V. Rao, "A Novel drain-extended FinFET device for high-voltage high-speed applications", *IEEE Electron Device Letters*, vol. 33, no. 10, pp. 1432-1434, October 2012.

[6] B. S. Kumar, M. Paul and M. Shrivastava, "On the design challenges of drain extended FinFETs for advance SoC integration," *2017 International Conference on Simulation of Semiconductor Processes and Devices (SISPAD)*, Kamakura, 2017, pp. 189-192.

[7] Y. T. Wu *et al.*, "Simulation-Based Study of Hybrid Fin/Planar LDMOS Design for FinFET-Based System-on-Chip Technology," in *IEEE Transactions on Electron Devices*, vol. 64, no. 10, pp. 4193-4199, Oct 2017.

[8] S. Natarajan *et al.*, "A 14nm logic technology featuring 2nd-generation FinFET, air-gapped interconnects, self-aligned double patterning and a 0.0588 μm² SRAM cell size," *2014 IEEE International Electron Devices Meeting*, San Francisco, CA, 2014, pp. 3.7.1-3.7.3

[9] B. S. Kumar and M. Shrivastava, "Part I: On the Unification of Physics of Quasi-Saturation in LDMOS Devices," in *IEEE Transactions on Electron Devices*, vol. 65, no. 1, pp. 191-198, Jan. 2018

Proceedings of the 30th International Symposium on Power Semiconductor Devices & ICs
May 13-17, 2018, Chicago, USA

High-Speed, High-Reliability GaN Power Device with Integrated Gate Driver

Gaofei Tang[1], M.-H. Kwan[2], Zhaofu Zhang[1], Jiabei He[1], Jiacheng Lei[1], R.-Y. Su[2], F.-W. Yao[2], Y.-M. Lin[2], J.-L. Yu[2], Thomas Yang[2], Chan-Hong Chern[2], Tom Tsai[2], H. C. Tuan[2], Alexander Kalnitsky[2], and Kevin J. Chen[1]

[1]Department of Electronic and Computer Engineering, The Hong Kong University of Science and Technology, Hong Kong
[2]Analog Power & Specialty Technology Division, Taiwan Semiconductor Manufacturing Company Limited, Hsin-Chu, Taiwan
Phone: +852-23588530, Fax: +852-23581485, Email: eekjchen@ust.hk

Abstract—An enhancement-mode GaN power switch with monolithically integrated gate driver is demonstrated on a 650-V GaN-on-Si power device platform. The integrated GaN-based gate driver features advanced designs such as bootstrapped gate-charging current source that enables high current driving capability during the entire turn-on process and rail-to-rail output. The GaN power transistor with integrated gate driver was characterized up to 300 V/15 A switching operations using a double pulse tester, and exhibits suppressed gate ringing and fast switching speed. The peak drain voltage slew rate dV/dt is above 125 V/ns during turn-on, and 336 V/ns during turn-off.

Keywords—*Double pulse tester; driving capability; gate driver; GaN power switch; rail-to-rail output; slew rate*

I. INTRODUCTION

Regarded as promising candidates for next-generation power electronics applications, GaN power switches are capable of delivering higher efficiency and higher power density for power conversion systems [1]. Although the discrete GaN power devices have already shown superior device performance, majority of the peripheral control/drive functional blocks are implemented with separate chips based on silicon technology. Such a hybrid drive solution (i.e. *"Si-based gate driver + GaN-based power switch"*) will inevitably be challenged by the parasitic inductances from the bonding wires and on-board traces, especially at high operating frequencies. Si and SiC power MOSFETs with a high threshold voltage and robust gate structure (i.e. MOS gate) have a high immunity to the parasitic effects. However, the commercially available enhancement-mode (E-mode) GaN power devices with *p*-type gate barrier structure exhibit relatively lower threshold voltage and lower limit of the maximum gate voltage. Parasitic effects could be more detrimental when these GaN power devices are under high-voltage and high-frequency switching operations. For instance, the narrow gate drive voltage margin of the E-mode GaN power devices makes the gate structures vulnerable to the over-voltage stress by the ringing effect [2]. The high slew rate (dV/dt) of the drain voltage may lead to unexpected false turn-on issue because of the relatively small threshold voltage (< 2 V) in the E-mode GaN power devices [3]. These parasitic effects will result in device performance degradation and even the circuit functionality failure, and raise reliability concerns about the GaN power devices.

To fully leverage the performance advantages of GaN power devices and provide a robust protection, the concept of GaN-based smart power IC has been proposed to accommodate not only the high-voltage GaN power switches but also the peripheral drive/control modules [4-6]. The on-chip integrated gate driver close to the GaN power switch can significantly shorten the gate drive loop, minimize the parasitic inductance and enhance the reliability of GaN power switches. The monolithic integration of *"gate driver + power switch"* can greatly benefit from the high integration level of the planar AlGaN/GaN heterojunction process, and one such implementation has been reported in [7]. Meanwhile, there is still room for improvement in the gate driver design.

In this work, 650-V E-mode GaN power transistors with monolithically integrated gate drivers have been implemented on a GaN-on-Si power device platform. A new gate driver circuit is proposed to address the issue of slower driving speed in the conventional design. Systematic characterization for the fabricated power integrated circuits has been conducted at both the chip and board level.

II. GaN POWER DEVICE WITH INTEGRATED GATE DRIVER

A. Monolithic integration of GaN devices

The monolithic integration of the E-mode GaN power switch and GaN-based gate driver was realized on a commercially available 650-V GaN-on-Si platform as the schematic cross section in Fig. 1 (a). The device fabrication employed the previously reported technology in [5, 8]. The low-voltage rated enhancement-/depletion-mode (E/D-mode) HEMTs are used to form the gate drive circuit. Compared to the discrete E-mode GaN power device fabrication, the GaN power switch with integrated gate driver can be realized without any additional processing steps. The low-voltage rated E/D-mode HEMTs in the gate driver have a gate length L_G of 0.8 μm, a gate-to-source distance L_{GS} of 1 μm, and a gate-to-drain distance L_{GD} of 2 μm. The large-area normally-off GaN power device features a gate width W_G of 120 mm.

The DC characteristics of the low-voltage E/D-mode HEMTs and high-voltage GaN power transistor fabricated on the same wafer are plotted in Fig. 1 (b) and (c). The E-mode HEMTs employing GaN power device technique feature the same threshold voltage V_{th} of +1.6 V as the normally-off GaN power transistor at room temperature. The integrated 650-V

978-1-5386-2928-4/18 $31.00 © 2018 IEEE

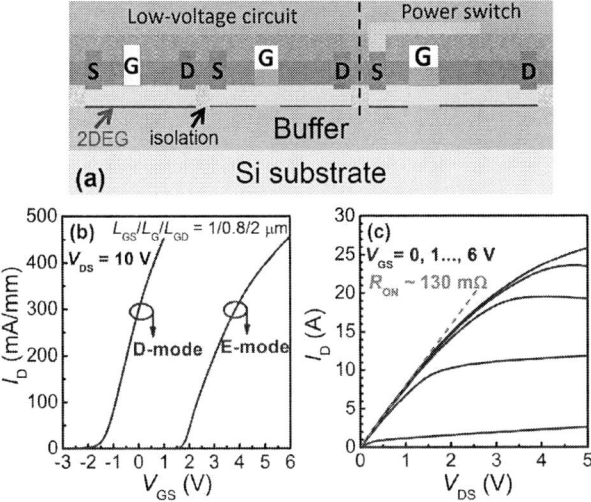

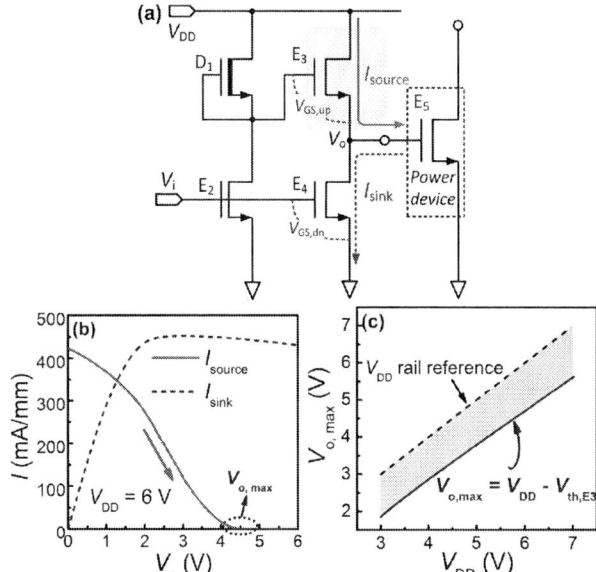

Fig. 1 (a) Schematic platform for monolithic integration of GaN-based gate driver and normally-off GaN power transistor. (b) Transfer characteristics of monolithic integrated E/D-mode GaN HEMTs, and (c) Output characteristic of a 650-V/130-mΩ GaN power transistor at room temperature.

Fig. 2 (a) The schematic circuit of GaN power device integrated with conventional (Gen-I) gate driver. (b) The relationship between driver's output voltage V_o and source/sink current I_{source}/I_{sink}. The I_{source} decreases as V_o rises until $V_{GS,up}$ reaches E3's threshold voltage. (c) The relationship between driver's power supply voltage V_{DD} and output amplitude $V_{o,max}$.

GaN power switch has a static ON-resistance R_{ON} of 130 mΩ. Considering the positive V_{th} shift of the power device under high-voltage switching operations and the long-time stress reliability of the Schottky-contact p-GaN gate [9-10], the recommended gate drive voltage is 6~7 V to ensure the power switch fully turned-on. According to the C-V measurement, the required gate charge Q_G for switching this GaN power transistor from 400-V off-state to on-state with a 6-V gate drive voltage is calculated to be ~2 nC, which is far less than that required for the Si counterparts [1].

B. Chip-level characterization

A conventional design of GaN-based gate driver is illustrated in Fig. 2 (a). The drive circuit consists of two stages, i.e. the logic control stage and the buffer stage. The logic circuit formed by E/D-mode HEMTs can reshape and transfer the input PWM signal, while the buffer stage formed by two push-pull E-mode HEMTs can provide charge/discharge current for the gate capacitance of the integrated power switch. However, such a design results in compromised switching speeds (e.g. a longer turn-on time) because the charge-up process of the power switch slows down as the gate charging current (i.e. source current I_{source}) drops with a rising output voltage (V_o) as depicted in Fig. 2 (b). Meanwhile, the gate driver's output voltage level $V_{o,max}$ cannot maintain the rail-to-rail compatibility as the charging process stops once $V_{GS,up}$ of the pull-up transistor E3 reaches its threshold voltage $V_{th,E3}$ as V_o rises. Then the $V_{o,max}$ will experience a reduction of V_{th} compared to the supply voltage V_{DD} (i.e. $V_{o,max} = V_{DD} - V_{th}$), as described in Fig. 2 (b-c). Although the GaN power device can be fully turned-on by a 4-V gate drive voltage under static measurement as shown in Fig. 1 (c), the gate drive voltage needs to be as high as 6 V under high-speed switching operations, considering the positive shift in the V_{th} caused by the high drain bias stress. To ensure a sufficient gate drive voltage for the power switch, the conventional GaN based gate driver needs to be supplied with a V_{DD} larger than 7.5 V. Then the Schottky-gated E-mode HEMTs in the driver circuit will undergo an aggressive gate stress, leading to accelerated gate junction degradation. Also, the static power consumption of the gate driver will be increased.

To address the issues of charging speed and drive voltage amplitude, the pull-up transistor E3 should be kept fully turned-on until the capacitive load (i.e. the gate of the power switch) is charged up to V_{DD}. In this work, a charge pump circuit is introduced to the gate driver to ensure a constant V_{GS} in the pull-up charging transistor during charging process. The schematic circuit diagram is shown in Fig. 3 (a). The device dimensions in the gate driver are well designed using a compact device model to drive the 650-V/130-mΩ GaN power transistor. The fabricated chip size is 4.6mm×1.1mm. The monolithic integration of the power switch and gate driver can effectively minimize the parasitic inductance in the gate drive loop. The driving capabilities of the non-inverting conventional (Gen-I) and new (Gen-II) gate drivers with the same device dimensions in the buffer stage were characterized with a capacitive load, respectively. Fig. 3 (b-c) show the measured switching waveforms. Compared to the Gen-I gate driver, the Gen-II circuit can deliver both a higher driving capability and rail-to-rail output. The achieved rail-to-rail output capability ensures that the gate driver can well turn on the GaN power device with a power supply of 6~7 V.

The static quiescent current of the Gen-II gate driver is measured to be ~6 mA, meaning that the static power consumption is ~36 mW with a 6-V power supply. The system efficiency will be little affected by this static power consumption because of the large power handling capability of the integrated GaN power switch. Fig. 4 illustrates the propagation delay of the gate driver with a 330-pF capacitive

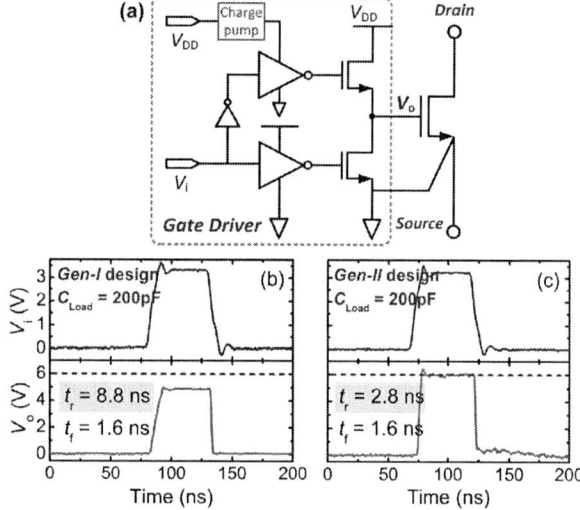

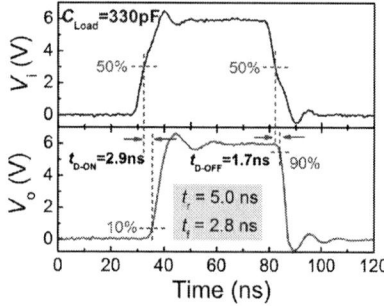

Fig. 3 (a) Schematic circuit of the GaN power device with integrated Gen-II gate driver. Driving capabilities of the gate drivers using (b) Gen-I and (c) Gen-II design with a supply voltage V_{DD} of 6 V.

Fig. 4 Propagation delay of Gen-II GaN-based gate driver with a 330-pF capacitive load.

load, which corresponds to the 2-nC gate charge of the integrated GaN power switch. The turn-on and turn-off propagation delay (t_{D-ON} and t_{D-OFF}) are extracted to be only 2.9 ns and 1.7 ns. The much smaller propagation delay compared to CMOS technology is mostly a result of the larger current density and smaller intrinsic capacitance of the lateral GaN devices.

III. BOARD-LEVEL CHARACTERIZATION

To characterize the power device with integrated gate driver under switching operation, a double pulse tester (DPT) was employed to mimic the hard-switching operation. The printed circuit board (PCB) design refers to the previously reported scheme in Ref. [9]. Fig. 5 shows the circuit schematic and photograph of the designed DPT board. The device under test (DUT) is mounted on the PCB with silver epoxy and electrically connected to the board through the bonding wires. The Kelvin connection of this power integrated circuit allows the gate driver to drive the intrinsic source of the GaN power switch free of the inductive noise generated by the high current fluctuation of the source terminal. A 600-V SiC Schottky barrier diode is used as the freewheeling diode of the inductive load to facilitate fast commutation process.

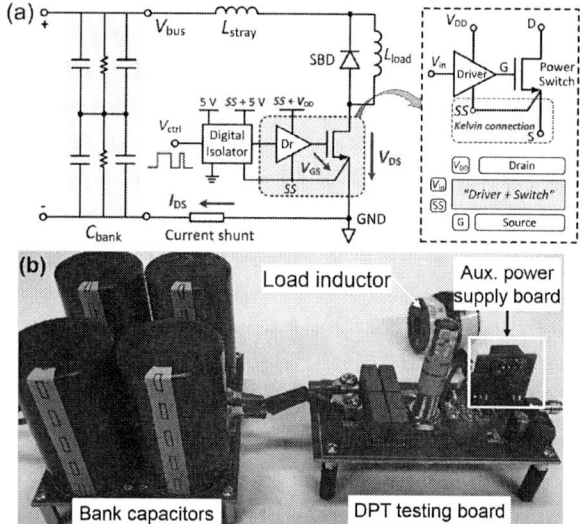

Fig. 5 (a) The schematic circuit of the double pulse tester (DPT) and pin configuration of the power integrated circuit. (b) The photo of the DPT board.

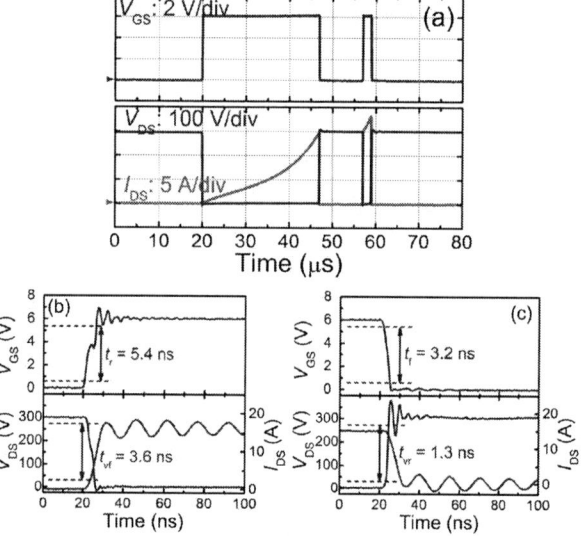

Fig. 6 (a) Double-pulse waveforms of the GaN power device with integrated gate driver. (b) Turn-on, and (c) turn-off waveforms (including: V_{GS}, V_{DS}, I_{DS}). V_{DD} = 6 V, V_{bus} = 300 V, I_{Load} = 15 A.

The double pulse tests have been performed up to 300V/15A without the need of negative power supply for the gate driver, indicating an effective suppression of false turn-on by the monolithic integration of gate driver and GaN power switch. Fig. 6 (a) shows the recorded double pulse waveforms of the DUT switched at 300 V/15A condition. From the turn-on and turn-off switching transients in Fig. 6 (b-c), the gate signal with small oscillation and fast switching transients (e.g. 1.3-ns turn-off time) can ensure the circuit reliability and reduce the switching loss. During the turn-on transient, the suppressed ringing effect as a result of minimized parasitic inductance in the gate drive loop can protect the gate junction of the power switch from the harmful gate overshoot. On the other hand, the GaN power switch with integrated gate driver

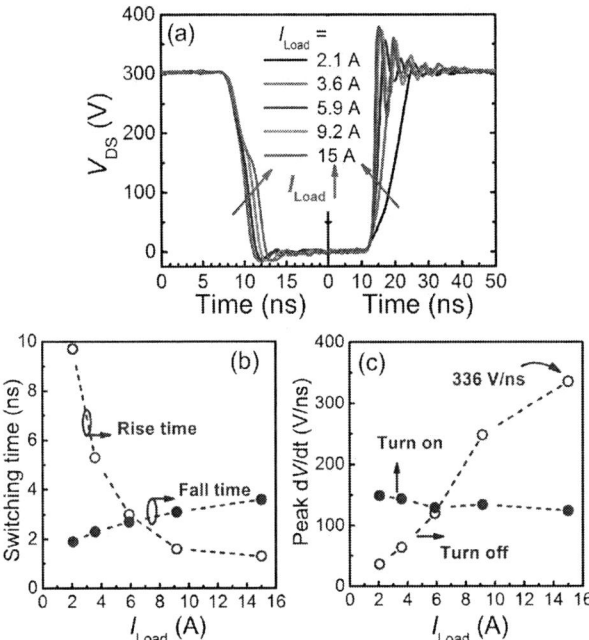

Fig. 7 (a) The switching waveforms of V_{DS} under various load current (V_{bus} = 300 V). (b) The extracted rise time and fall time of V_{DS}. (b) The peak dV/dt extracted from switching V_{DS} waveforms.

exhibits fast turn-off process without the false turn-on issue as shown in Fig. 6 (c). The potential false turn-on issue caused by high slew rate dV/dt of drain voltage is effectively eliminated, indicating that the need for the negative power supply in the monolithic gate driver is relaxed.

The DPT characterization has been performed at different load current I_{Load}, and the corresponding switching transient behavior of V_{DS} are summarized in Fig. 7 (a). With an increasing load current, the turn-on process is moderately increased while the turn-off process is significantly accelerated as the extracted switching time in Fig. 7 (b). The overshoot of V_{DS} during turn-off process is mainly caused by parasitic inductance from the on-board trace and bonding wires to the drain terminal, which can be further optimized by shortening the bonding wires or employing surface-mounting-type package. From the extracted peak slew rate dV/dt of V_{DS} in Fig. 7 (c), the highest slew rate of 336 V/ns is obtained during the turn-off transient under 300V/15A testing conditions. The high switching speed of the GaN power device with integrated gate driver can potentially reduce the switching loss while the GaN power device works in a hard-switching mode at a high frequency.

IV. CONCLUSIONS

In this work, a 650-V GaN power device with integrated gate driver was demonstrated on a commercialized GaN-on-Si platform. The integrated circuits can be fabricated employing E-mode GaN power device technique without any additional processing steps. Compared to the conventional gate driver design, the proposed integrated gate driver with a charge pump can deliver both the high driving capability and rail-to-rail drive signal. In addition to the chip-level characterization, the switching performance of the driver-embedded GaN power device has been evaluated using a double pulse test circuit. The monolithic integration of the gate driver and power switch can effectively eliminate the parasitic inductance in the gate drive loop. The gate signal of the power device exhibits suppressed ringing and alleviated overshoot. The high switching speed of the GaN power switch can dramatically help reduce the switching loss. The high switching speed and high device reliability offered by the driver-embedded GaN power switch can enhance the reliability of the GaN-based power systems. The demonstrated driver-integration scheme paves the way to the reliable GaN-based power switching applications.

REFERENCES

[1] K. J. Chen, O. Häberlen, A. Lidow, C.-L. Tsai, T. Ueda, Y. Uemoto and Y. Wu, "GaN-on-Si Power Technology: Devices and Applications," *IEEE Trans. Electron Devices*, vol. 64, no. 3, pp. 779–795, Mar. 2017.

[2] D. Reusch, and J. Strydom, "Understanding the Effect of PCB Layout on Circuit Performance in a High-Frequency Gallium-Nitride-Based Point of Load Converter," *IEEE Trans. Power Electron.*, vol. 29, no. 4, pp. 2008–2015, Apr. 2014.

[3] R. Xie, H. Wang, G. Tang, X. Yang, and K. J. Chen, "An Analytical Model for False Turn-On Evaluation of High-Voltage Enhancement-Mode GaN Transistor in Bridge-Leg Configuration," *IEEE Trans. Power Electron.*, vol. 32, no. 8 pp. 6416–6433, Aug. 2017.

[4] K.-Y. Wong, W. Chen, K. J. Chen, "Integrated voltage reference and comparator circuits for GaN smart power chip technology," in *Proc. 26th Int. Symp. Power Semiconductor Devices & ICs (ISPSD)*, pp. 57–60, June 2009.

[5] C.-L. Tsai, Y.-H. Wang, M.-H. Kwan, P.-C. Chen, F.-W. Yao, S-C. Liu, J.-L. Yu, C-L Yeh, R.-Y. Su, W. Wang, W.-C. Yang, K.-Y. Wong, Y.-S. Lin, M.-C. Lin, H.-Y. Wu, C.-M. Chen, C.-Y. Yu, C.-B. Wu, M.-H., Chang, J.-S. You, T.-M. Huang, S.-P. Wang, L.Y. Tsai, Chan-Hong Chern, H.C. Tuan and A. Kalnitsky, "Smart GaN Platform: Performance & Challenges," in *63rd International Electron Devices Meeting (IEDM)*, pp. 737–740. Dec. 2017.

[6] D. Kinzer, "GaN Power IC Technology: Past, Present, and Future," in *Proc. 29th Int. Symp. Power Semiconductor Devices & ICs (ISPSD)*, pp. 19 24, June 2017.

[7] S. Ujita, Y. Kinoshita, H. Umeda, T. Morita, S. Tamura, M. Ishida, and T. Ueda, "A compact GaN-based DC-DC converter IC with high-speed gate drivers enabling high efficiencies," in *Proc. 26th Int. Symp. Power Semiconductor Devices & ICs (ISPSD)*, pp. 51–54, June 2014.

[8] G. Tang, M. H. Kwan, K. Y. Wong, J. Lei, R. Y. Su, F. W. Yao, Y. M. Lin, J. L. Yu, Tom Tsai, H. C. Tuan, A. Kalnitsky, and K. J. Chen, "Digital Integrated Circuits on an E-Mode GaN Power HEMT Platform," *IEEE Electron Device Lett.*, vol. 38, no. 9, pp. 1282–1285, Sept. 2017

[9] H. Wang, J. Wei, R. Xie, C. Liu, G. Tang, and K. J. Chen, "Maximizing the performance of 650-V p-GaN gate HEMTs: Dynamic RON degradation and circuit design considerations," *IEEE Trans. Power Electron.*, vol. 32, no. 7, pp. 5539–5549, July 2017.

[10] A N. Tallarico, S. Stoffels, P. Magnone, N. Posthuma, E. Sangiorgi, S. Decoutere, and C. Fiegna, "Investigation of the p-GaN Gate Breakdown in Forward-Biased GaN-Based Power HEMTs," *IEEE Electron Device Lett.*, vol. 38, no. 1, pp. 99–102, Jan. 2017

Proceedings of the 30th International Symposium on Power Semiconductor Devices & ICs
May 13-17, 2018, Chicago, USA

A 600V High-side Gate Drive Circuit with Ultra-low Propagation Delay for Enhancement Mode GaN Devices

Yangyang Lu[1], Jing Zhu[1], <u>Weifeng Sun[1]</u>, Yunwu Zhang[1,2], Kongsheng Hu[1], Zhicheng Yu[1], Jing Leng[1]
Shikang Cheng[3], Sen Zhang[3]
[1] National ASIC System Engineering Research Center, Southeast University, Nanjing, China
[2] China Resources Microelectronics Limited, Wuxi, China
[3] CSMC Technologies Corporation, Wuxi, China
*Email: swffrog@seu.edu.cn

Abstract—Benefit from the presented new Common Mode Dual-Interlock (CMDI) structure, a 600V high-side gate driver base on normal 0.5μm 600V Bipolar-CMOS-DMOS (BCD) technology achieving the high dV_S/dt noise immunity larger than 85V/ns and low propagation delay less than 22ns for enhancement mode gallium nitride (GaN) devices is proposed in this paper.

Keywords—High-side driver; Gallium nitride; Common Mode Dual-Interlock

I. Introduction

High voltage high-side gate drive integrated circuits which always been used to drive high-side power device of the half-bridge application due to its advantages of high efficiency, space-saving, high reliability and etc.[1-3]. Since there are many filter circuits restricting the maximum operating frequency of the conventional high-side gate driver due to the parasitic effect of the LDMOSs in level shifter, developing a low propagation delay gate driver with high noise immunity is crucial [4].

Large propagation delay and dV_S/dt noise are two critical problems for high-side gate driver which limiting the applications in GaN systems. In recent years, lots of attentions have been paid for 600V high-side drivers which are widely used for driving silicon-based power MOSFETs or IGBTs in motor driver system. As we know, in order to improve the dVs/dt noise immunity capability, a filter should be added in the high side circuit of the high-side gate driver utilizing the reset dominant technique [5]. In addition, in fact, the filter is also needed in the high-side gate drivers with the V-I-V (Voltage to Current and Current to Voltage) technique [6] or Cap-loading level shifter [7], because of the process variations and layout mismatch. The above mentioned filter in the high-side gate drive circuit would increase the total propagation delay, as shown in Fig.1, and then limit its maximum operating frequency. Therefore, the high-side gate drivers with the conventional existing high-side circuits are not suitable for driving the GaN power device in some power management systems.

In order to further improve the noise immunity without increasing the propagation delay, a new technique called Common Mode Dual-Interlock (CMDI) for the high side circuit in the 600V high-side gate driver is proposed in this

paper, achieving less than 22ns propagation delay without sacrificing any dV_S/dt immunity capability.

II. Frequency Restriction of Conventional High-Side Gate Drivers With Different Level Shifter Technique

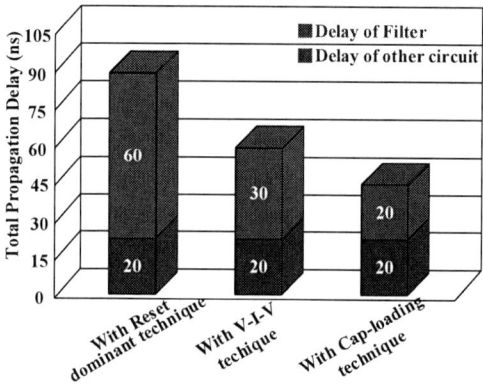

Fig. 1 High side propagation delay contribution with different high side techniques

Three high-side gate drivers with different high side circuits were designed. When utilizing Reset dominant technology, about 60ns filter should be added in the high side circuit to guarantee the dV_S/dt immunity larger than 80V/ns. When utilizing V-I-V technology, about 30ns filter should be added in the high side circuit to guarantee the dV_S/dt immunity larger than 84V/ns. When utilizing Cap-loading technology, about 20ns filter should be added in the high side circuit to guarantee the dV_S/dt immunity larger than 85V/ns. As illustrated in Fig.1, the filter time in the high side channel is the main issue for decreasing the total propagation delay.

III. Dissicussion On The Proposed CMDI Structure

A. Operating machanism of the proposed CMDI

In order to improve the operating frequency, a new CMDI structure for the high side driver is proposed, as shown in Fig.2. The block diagram of the high-side gate driver with the proposed CMDI is shown in Fig. 2(a), it contains input buffer,

---

This work was supported by the National Natural Science Foundation of China (BK20150627 and 61674030), Natural Science Foundation of Jiangsu Province (61504025) and the National key research and development plan (2017YFB0402900).

978-1-5386-2928-4/18 $31.00 © 2018 IEEE

output buffer, level shifter structure and the proposed CMDI structure. The level shifter structure contains two Lateral Double-diffused MOS (LDMOS) LD1 and LD2 are driven directly by the narrow pulse, Diodes D1 and D2 are used to protect the devices P1 and P2 in the high side logic from breakdown.

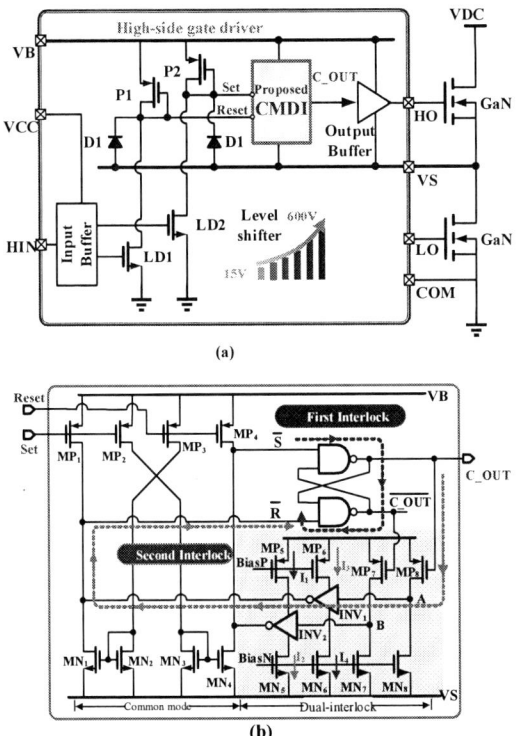

(a)

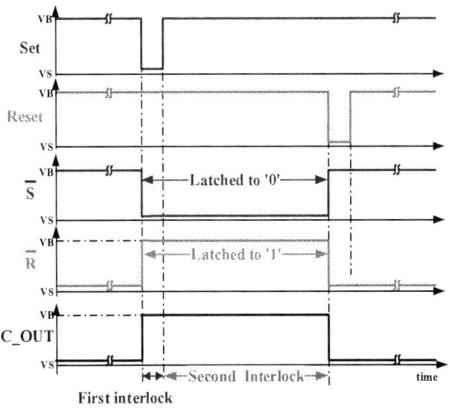

(b)

Fig. 2 (a) the block diagram of the high-side driver with the proposed CMDI circuit and (b) the detail schematic of the proposed CMDI circuit

Fig. 2(b) shows the schematic of the proposed CMDI circuit. Reset and Set are the input and C_OUT is output of the CMDI circuit. Four PMOSFETs MP1~MP4 and Four NMOSFETs MN1~MN4 are included in the common mode structure which will cut off the common-mode noise signal from Set and Reset terminal. The proposed CMDI circuit, including the first interlock and the second interlock structure, do not need additional filter authentically even under the conditions of process variations and layout mismatch. The first interlock structure is a RS trigger, which can transfers the input signal to output rapidly. The second interlock structure is a dynamic current source circuit, which can counteract the differential current caused by the mismatch, then maintain the dV_S/dt noise immunity capacity without any extra filter.

The basic working principle of the proposed CMDI circuit is shown in the Fig. 3 and its working process is illustrated as follow:

(1) When Set='0', Reset='1', $\overline{S}$='0' and $\overline{R}$='1', C_OUT is latched to '1' by the first interlock circuit.

(2) Then Set='1', Reset='1', $\overline{S}$ and $\overline{R}$ terminal are latched to

'0' and '1' by the second interlock circuit, because $I_{MP1}=I_{MN1}=I_{MP4}=I_{MN4}=0$ and $\overline{C\_OUT}$='0', A='1', B='0', the output of INV$_1$ and INV$_2$ are at high level and low level, respectively.

(3) When Set='1', Reset='0', $\overline{S}$='1' and $\overline{R}$='0', C_OUT is latched to '0' by the first interlock circuit,

(4)Then Set='1', Reset='1', $\overline{S}$ and $\overline{R}$ are latched to '1' and '0' by the second interlock circuit, because $I_{MP1}=I_{MN1}=I_{MP4}=I_{MN4}=0$ and $\overline{C\_OUT}$='1', A='0', B='1', the output of INV$_1$ and INV$_2$ are at low level and high level, respectively.

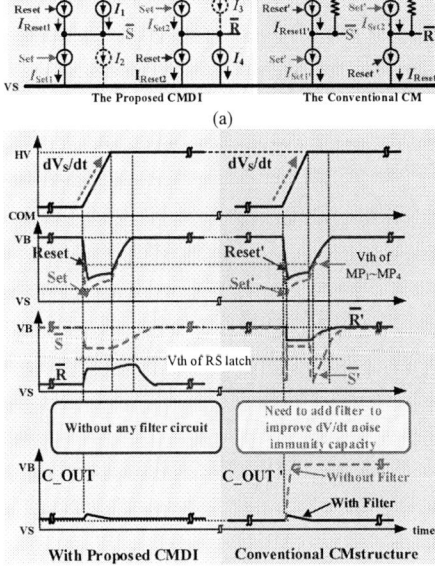

Fig. 3 The basic working principle of the proposed CMDI circuit

B. Noise immuniy machanism

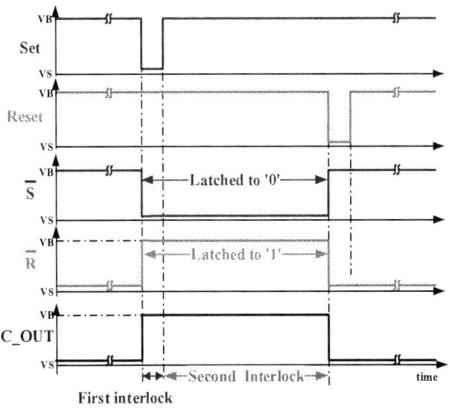

(a)

(b)

Fig. 4 (a) the equivalent model of the input signal of RS latch and the differential-mode noise filtered process of the high-side driver utilizing the proposed CMDI structure and the conventional common mode(CM) structure.

The equivalent model of the input signal of RS latch utilizing the proposed CMDI structure verse the conventional

978-1-5386-2928-4/18 $31.00 © 2018 IEEE   81

common mode structure is shown in Fig. 4 (a). I_{Reset1} and I_{Reset2} is generated by MP_1 and MN_4 I_{Set1}, I_{Set2} is generated by MN_1 and MP_4. For the proposed CMDI structure, assuming in the initial state C_OUT =0, then $\overline{S}$ ='1' and $\overline{R}$ ='0'. I_1 and I_4 is effective, I_2 and I_3 is ineffective. When the dV$_S$/dt is applied, Set and Reset signal would be out of synchronism because of the process or layout mismatch. Differential current will appear between I_{Set} and $I_{Reset,}$ then I_1 and I_4 will compensate or counteract this differential current without any filter circuit. But only a pull-up resistor is applied in the conventional common mode structure which will latch on or latch off without filter circuit in the high *dVs/dt* environment which is shown in the Fig. (4)b.

IV. MENERAMENT RESULTS

A 600V high-side gate driver utilizing the proposed CMDI circuit is implemented in 600V BCD process with a Novel Double-Well Isolation Structure [8]. Its micrograph is shown in Fig. 5. The propagation delay of high side channel (HIN to HO) is measured in the Fig.6. It shows that the turn-off propagation delay is 21.28ns and the turn on propagation delay is only 20.93ns at 25°C when the supply voltage is 15V.

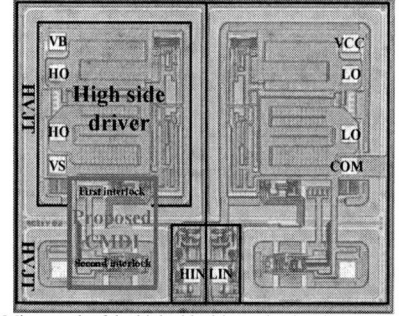

Fig. 5 Micrograph of the high-side driver adopting the proposed CDMI

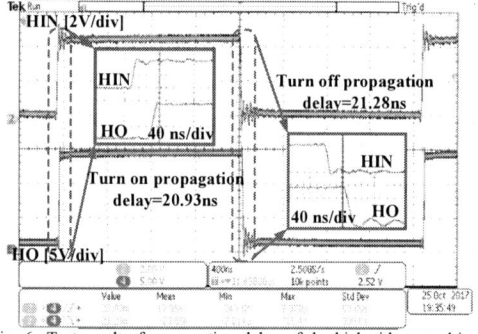

Fig. 6 Test result of propagation delay of the high-side gate driver with proposed CMDI structure

Fig.7 shows temperature characteristics of the propagation delay at different supply voltage. It is indicated that the maximum propagation is smaller than 30ns from -40°C to 125°C at different supply voltage. Fig.8 shows that the *dVs/dt* immunity capability of the high-side driver with proposed CMDI circuit can achieve to 85V/ns without any logic malfunction and latch-up issues.

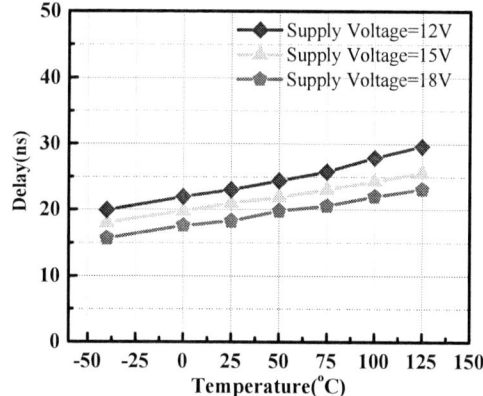

Fig. 7 Propagation delay Vs temperature and supply voltage

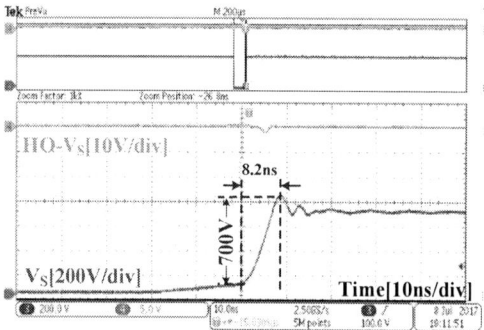

Fig. 8 *dVs/dt* noise immunity capability test

Table. 1 Performance summary and comparison with previous works

Performance	This work	Reset dominant technique[1]	V-I-V technique[2]	Cap-loading level shifter[3]	Company A[9]
Maximum offset supply voltage (V$_S$) [V]	600	600	600	600	100
Low-side supply voltage (V$_{CC}$) [V]	15	15	15	15	5
dV$_S$/dt noise immunity (dV$_S$/dt) [V/ns]	85	80	84	85	50
Typical propagation delay [ns]	21.28	80	50	40	28
Process Platform	600V HV BCD Process				100V BCD process

Finally, the performance is summarized in table 1 by comparing with that of other researches. It can be seen from the table that this work with the proposed CMDI circuit can achieve the smallest propagation delay without sacrificing the *dVs/dt* immunity capability.

V. CONCLUSION

A 600V high-side gate driver implemented in standard 0.5μm 600V Biplor-CMOS-DMOS (BCD) process utilizing a Common Mode Dual Interlock circuit is proposed in this paper. The proposed CMDI circuit can block the noise signal regardless of common mode and differential mode without any RC filter circuit. The measured results show that it achieving less than 22ns propagation delay without sacrificing any *dVs/dt* immunity capability.

978-1-5386-2928-4/18 $31.00 © 2018 IEEE

REFERENCES

[1] S. Ujita, Y. Kinoshita, H. Umeda & T. Morita, "A compact GaN-based DC-DC converter IC with high-speed gate drivers enabling high efficiencies." IEEE International Symposium on Power Semiconductor Devices & IC's, 2014, pp.51-54.

[2] J. Chen, and W. T. Ng. "Design trends in smart gate driver ICs for power MOSFETs and IGBTs." IEEE International Conference on Asic, 2017, pp.112-115.

[3] F. Hong, H. Lei, B. Ji, L. Li & S. Wang, "Buck inverter without shoot through." IET Power Electronics, 2015, 10(13), pp.1740-1750.

[4] L. Zhidong, and H. Lee. "A 100V gate driver with sub-nanosecond-delay capacitive-coupled level shifting and dynamic timing control for ZVS-based synchronous power converters." IEEE Custom Integrated Circuits Conference, 2015, pp:1-4.

[5] David C. Tam and C. C. Choi.. "Reset dominant level-shift circuit for noise immunity," US Patent 5514981, 1996

[6] J. T Hwang., M. S. Jung, S. K. Jin, H. K. Dong, "Noise Immunity Enhanced 625V High-Side Drive." ESSCIRC 2006, pp.572-575.

[7] Yunwu Zhang, Jing Zhu, Weifeng Sun, Yangyang Lu, et al. "A capacitive-loaded level shift circuit for improving the noise immunity of high voltage gate drive IC", IEEE International Symposium on Power Semiconductor Devices & IC's 2015, pp. 173-176.

[8] Weifeng Sun, Jing Zhu, et al. "Electrical Investigation on a Novel Double-Well Isolation Structure in 600V-Class HVIC", IEEE Trans. Electron Devices, 2012, 59(12), pp.3477-3481.

[9] Texas Instruments. "LM5113 Datasheet: LM5113 5A, 100V half bridge gated driver for e-GaN FETs," Apr. 2013. [online]

Proceedings of the 30th International Symposium on Power Semiconductor Devices & ICs
May 13-17, 2018, Chicago, USA

A Smart Gate Driver IC for GaN Power Transistors

Jingshu Yu*, Wei Jia Zhang, Andrew. Shorten, Rophina Li and Wai Tung Ng[+]

The Edward S. Rogers Sr. Department of Electrical and Computer Engineering
University of Toronto
Toronto, Ontario, Canada M5S 3G4
*yujingshu@vrg.utoronto.ca; [+]ngwt@vrg.utoronto.ca

Abstract—In this paper, an integrated smart gate driver IC with segmented output stage topology, programmable sense-FET, current sensing circuits and an on-chip stacked-based CPU for flexible digital control is presented. This IC is fabricated using TSMC's 0.18 μm BCD GEN2 process for driving a d-mode GaN power HEMT in cascode configuration. Using a segmentation technique, this IC can dynamically adjust the gate driving strength during switching transition to achieve slope control and EMI reduction. Programmable sense-FET and current sensing circuit monitor the load current for peak-current regulation. The embedded CPU can update all digital configuration bits on-the-fly. In dynamic driving mode, current spike at turn-on transition is reduced by 83% without sacrificing the switching speed. Current sensing circuit can detect peak current value and response within 5 ns. The pre-stored driving patterns can be loaded to the driving circuit in 1 μs under active driving mode.

Keywords—cascode d-mode GaN driver; segmented output stage; dynamic switching; current sensing; active driving

I. INTRODUCTION

Comparing with silicon-based power MOSFETs, Gallium nitride (GaN) devices have superior performance in higher frequency, higher temperature and higher power applications. There are two types of GaN power transistors, depletion-mode (d-mode) and enhancement-mode (e-mode) devices. Driving e-mode GaN power transistors, which have positive threshold voltage, is similar to driving normally-off n-channel MOSFETs but with many challenges, including low threshold voltage, narrow tolerance between maximum and rated gate voltage, *dv/dt* and *di/dt* issues due to high slew rates during transition [1]. On the other hand, normally-on d-mode GaN transistors require negative gate-source voltage to be turned off. However, providing negative voltage levels is uncommon and impracticable in real applications. Sudden power failure may lead to severe safety issues for circuits with normally-on

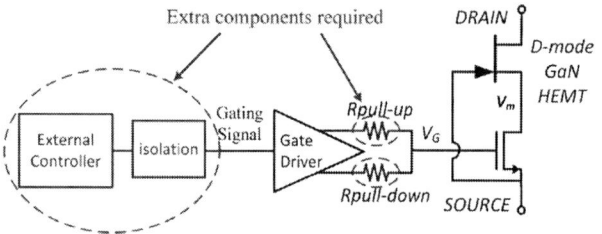

Fig. 1 Extra components required when using conventional gate drivers for d-mode GaN in cascode configuration [1].

devices. Using d-mode GaN HEMTs in cascode structure is a well-accepted solution, as shown in Fig. 1 [1]. The source of a d-mode GaN power transistor is connected to the drain of a low voltage (LV), low on-resistance silicon power MOSFET. The gate of the GaN device is connected to the source of the LV MOSFET. The current path between the drain and the source of this composite device is controlled by the MOS gate of the silicon transistor. The mid-point voltage, V_m, in off state is determined by the ratio between the drain to source capacitance of the GaN power transistor and that of the LV silicon power MOSFET.

Currently, there are a few number of commercially available gate drivers for e-mode GaN transistors. However, there are no dedicated drivers for d-mode GaN devices [2]. Most commercial gate driver ICs require two external resistors for setting the pull-up and pull-down speeds, external controllers to generate gating signals, as shown in Fig. 1. This leads to an increase in PCB space and extra parasitics. Other drawbacks such as fixed output voltage levels, the lack of precise timing control capability also limit applications using d-mode GaN transistors.

In this paper, a smart gate driver IC for d-mode GaN power transistors with built-in current sensing, tunable output resistance [2], tunable current sensing ratio and smart digital control is presented. It is designed to drive an external d-mode GaN HEMT in cascode configuration [1] with a built-in LV silicon sense-FET. Together with an on-chip stack-based CPU. This fully integrated gate driver IC allows flexible internal control, current mode regulation [3], active ringing suppression, and efficiency improvement. Section II describes the circuit implementation of this proposed design. Experimental setup and simulation results are presented in Section III. Section IV contains the summary and future work.

II. PROPOSED DRIVING TECHNIQUES

A. Design Approach and System Diagram

Figure 2 shows the micrograph of the smart date driver IC for d-mode GaN power transistors. The chip was fabricated using TSMC's 0.18 μm BCD GEN2 process. The co-packaging diagram and the expended view of the silicon driver system is shown in Fig. 3. The silicon IC includes an on-chip stacked-based CPU, a hybrid DPWM block, segmented gate drivers, and current sensing modules. There are 3 different voltage inlands on the silicon die, 1.8 V for the CPU, logic

978-1-5386-2928-4/18 $31.00 © 2018 IEEE   84

controller and current sensing blocks, 5 V for the segmented driver circuits and 20 V for a programmable power sense-FET, which are inherently protected from the 600 V high voltage on the power line by the cascode structure.

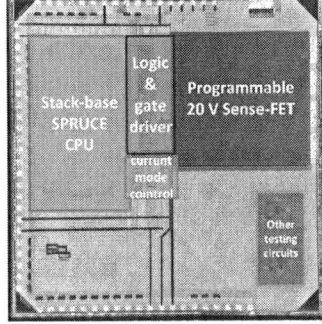

Fig. 2 Micrograph of the smart gate driver IC fabricated using TSMC's 0.18μm BCD GEN2 process (5 mm × 5 mm).

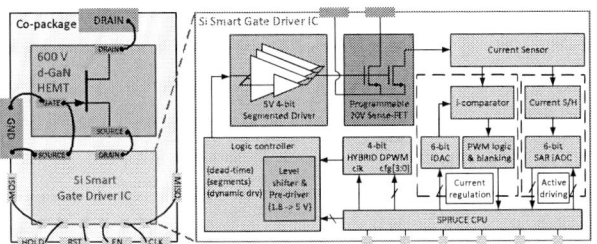

Fig. 3 Co-packaging diagram and the expanded view of the silicon smart gate driver IC.

B. On-chip Digital Control and Hydrid DPWM

In order to achieve smart controls, the proposed driver IC has 280 internal configuration bits. Conventional designs use either a long scan chain or parallel inputs, which incurs die areas while the flexibility is still limited. The proposed IC requires only 6 I/O pins for controlling the entire chip, as shown in Fig. 4. The CPU can generate and updated the digital control signals on-the-fly, including the number of enabled driver segments, optimum dead-time, dynamic switching related parameters, etc. In the active driving mode, the 6-bit current mode ADC monitors the load current and feeds the digital results to CPU. By using a look-up-table (LUT) with pre-optimized driving patterns, the CPU can continuously configure the gate driving IC to provide the best performance.

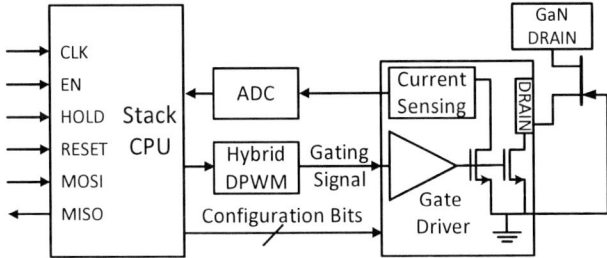

Fig 5 Block diagram showing the interface between the embedded CPU and analog gate driver circuits.

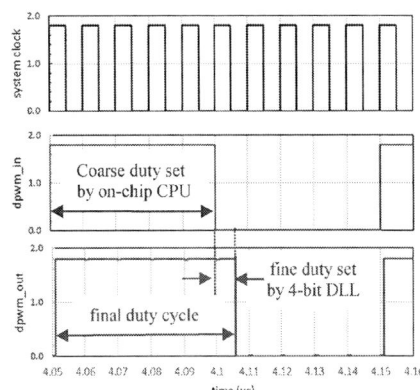

Fig. 4 Timing diagram of the hybrid DPWM block.

Accurate and precise control of the gating signals is required when GaN devices are used in high frequency applications in the MHz range. The embedded CPU can generate PWM signals from 100 kHz to 50 MHz with an internal 100MHz system clock with 10 ns resolution. Together with a 4-bit DLL block, the hybrid DPWM can achieve a resolution of 625 ps, as shown in Fig. 5. When a GaN-based DC/DC converter is operating at 1 MHz, the proposed hybrid DPWM has over 10-bit resolution.

C. Segmented Gate Driver and Dynamic driving

The concept and advantages of the dynamic driving techniques have been verified using IGBT, power MOSFET and e-mode GaN HEMT [4][5][6]. In this paper, the driver IC is dedicated to drive a d-mode GaN HEMT in cascode configuration, as shown in Fig. 6. A 5 V gate driver, consisting of four segments (SEG[0] to SEG[3]) with a size ratio of 8:25:50:100, is designed for the segmented 20 V MOSFET. The transition time can vary from 1 ns to 20 ns for different output resistances.

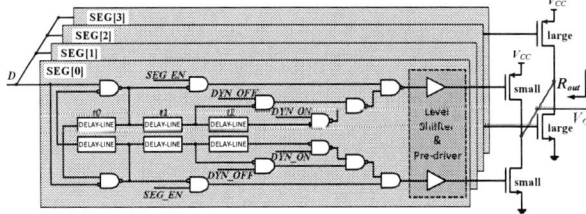

Fig. 6 Circuit implementation for dynamic driving technique.

D. Programmable sense-FET and Current Sensing

The LV power MOSFET of the cascode configuration has a built-in programmable sense-FET, as shown in Fig. 7. The common-source structure eliminates the influence of different gate-source voltage V_{gs} due to the voltage drop across the sense resister R_{SENSE}. The main cell and the sensing cell, sharing the same gate and source connection, are designed with different widths to achieve sensing ratio from 500:1 to 3000:1. This programmable feature allows the driver to be accommodate different d-mode GaN transistors and load currents. The current sensing blocks can also be used to implement a closed loop current regulation system as shown in Fig. 8. The drain current I_{ds} of the d-mode GaN HEMT is monitored by the programmable silicon sense-FET. The sense current I_{SENSE} is compared with a reference current $I_{reference}$. When I_{SENSE} reaches the pre-set limit, the comparator output resets the gating signal, turning off the d-mode GaN HEMT. When a GaN-based DC/DC converter is operated in a closed-loop circuit, the embedded CPU takes signals from external voltage control loop and finish local digital compensation.

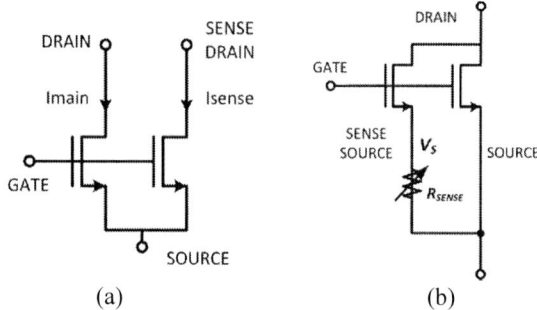

Fig. 7 (a) Common-source structure sense-FET used in this work, (b) conventional common-drain structure sense-FET.

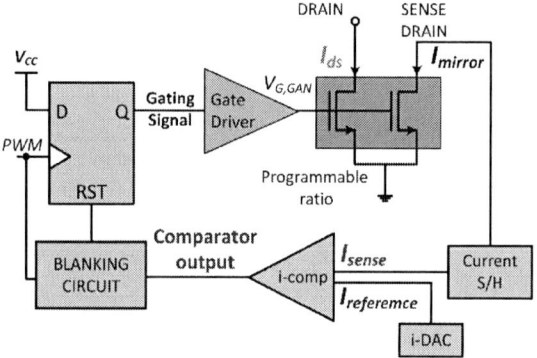

Fig. 8 Current sensing and peak-current regulation circuits.

III. EXPERIMENTAL SETUP

The fabricated gate driver IC and customized 10 A d-mode Gan HEMT are mounted on a printed-circuit-board as shown in Fig. 9. The printed circuit board is made of aluminum substrate to provide good thermal dissipation, and 2-µinch immersion gold surface for chip-on-board (COB) bonding purpose. Epoxy is used to protect the co-packaged chips.

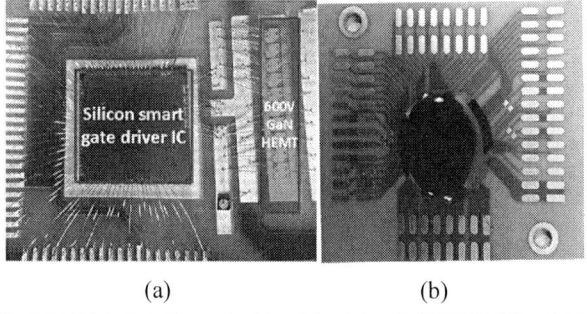

Fig. 9 (a) Fabricated silicon gate driver IC and d-mode GaN HEMT bonded to printed-circuit-board; (b) Co-packaged chips protected by epoxy.

A. Dynamic Driving for Slope Contorl

The equivalent model of the cascode d-mode GaN is shown in Fig. 10. Internal parasitic inductors exist on the bonding wires between silicon gate driver IC and d-mode GaN HEMT. Figure 11 compares the switching behaviors of the cascode d-mode GaN with fixed R_{out} and dynamic R_{out}. Using the dynamic switching technique, ringing at the gate node of the HEMT is reduced by 86%, and current spike during turn-on is reduced by 83% without sacrificing the switching speed. This slope-control approach can effectively reduce the effect of high voltage and current slew rates, solve the dv/dt and di/dt issues and optimize EMI.

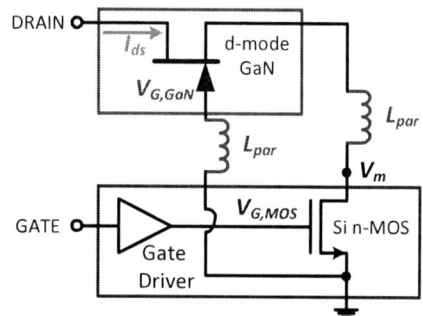

Fig. 10 Equivalent circuit of the co-packaged d-mode GaN.

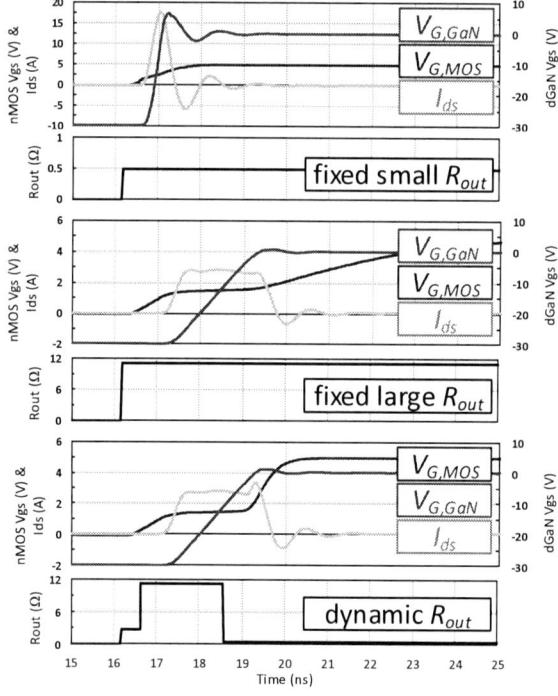

Fig. 11 Simulated waveforms showing switching behaviors with large R_{ou}, small R_{ou} and dynamic R_{ou}.

B. Peak-current Regulation

The peak-current detection feature illustrated in Fig. 8, and is verified by the simulation results as shown in Fig. 12. The peak conduction current is preset to 3.5 A. The programmable sense-FET is configured to provide a sensing ratio of 1000:1. Therefore, the 6-bit current-mode DAC (iDAC) is set to generate a reference current of 350 mA. The response time is less than 5 ns. By configuring the iDAC on-the-fly and generating a ramping-down $I_{reference}$ signal, slope compensation can be achieved when duty cycle is greater than 50%.

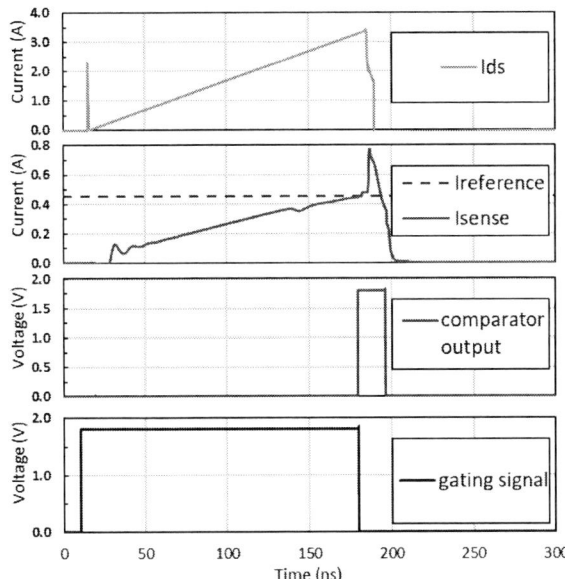

Fig. 12 Peak-current regulation verified by simulations.

C. Active Driving Mode

For a synchronous SMPS converter, dead-time is the time interval during which gating signals for both power transistors are off. An excessively long dead-time leads to body-conduction power loss, while an insufficiently short dead-time will cause shoot-through current. With optimal dead-times, one power device is turned on when the drain-to source current of the other power device just drops to zero. Since the power the transition time of power transistors varies with load current, dead-time needs to be carefully under different conditions. Figure 13 shows the optimal dead-time between two identical 200 V GaN HEMTs in a synchronous buck converter with a 100 V input, 1 MHz switching frequency and a varying load

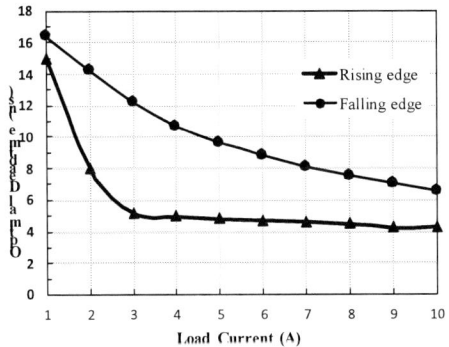

Fig. 13 Optimal dead-time vs. load current of a typical 200 V d-mode GaN.

current between 1 A to 10 A. The plot shows the optimal deadtime decreases when load current changes from light to heavy. The on-chip CPU can store the best driving strategy, which includes optimal dead-time, enabled gate driver segments, dynamic driving patterns for a given load current. By reading the outputs of the integrated 6-bit current-mode ADC, the CPU can search the pre-stored driving patterns in a LUT and update the configuration bits in less than 40 CPU system cycles with a 100 MHz system clock. Power conversion efficiency can be actively and continuously optimized over a wide range of load currents.

IV. CONCLUSIONS

This paper presents a smart driver IC that could dynamically adjust the gate driving strength to achieve slope control during switching, monitor the load current using a built-in sense-FET, provide peak-current regulation and actively adapt the best driving pattern according to the loading conditions. The embedded CPU can control all the internal configuration bits locally, and provide on-chip digital close-loop compensation. Current spike during turn-on transition is reduced by 83% without sacrificing the switching speed. The response time for peak-current regulation is 5 ns. In active driving mode, the best gate driving pattern is updated in less than 1 μs according to the load conditions. Future work will be carried out to obtain the experimental results from the co-packaged chip. The close-loop current-mode regulation and active driving on a GaN DC/DC buck-converter in close-loop will be tested.

ACKNOWLEDGMENT

The author would like to thank NSERC CANADA for financial support, TMSC for providing access to their 0.18 μm BCD GEN2 technology and IC fabrication, Prof. Kui Ma from Guizhou University, Zhenghua Yongguang Electronic Inc. and Guizhou Zhenghua Fengguang Semiconductor Inc. for their generous help in wire-bonding.

REFERENCES

[1] A. Lidow, J. Strydom, M. de Rooij, and D. Reusch, *GaN Transistors for Efficient Power Conversion*, 2nd ed. Wiley, 2015.

[2] E. A. Jones, F. F. Wang, and D. Costinett, "Review of Commercial GaN Power Devices and GaN-Based Converter Design Challenges," *IEEE J. Emerg. Sel. Top. Power Electron.*, vol. 4, no. 3, pp. 707–719, 2016.

[3] M. Rose, Y. Wen, R. Fernandes, R. Van Otten, H. J. Bergveld, and O. Trescases, "A GaN HEMT driver IC with programmable slew rate and monolithic negative gate-drive supply and digital current-mode control," *Proc. Int. Symp. Power Semicond. Devices ICs*, vol. 2015–June, pp. 361–364, 2015.

[4] J. S. Yu, W. J. Zhang, and W. T. Ng, "A segmented output stage H-bridge IC with tunable gate driver," *Proc. Int. Symp. Power Semicond. Devices ICs*, pp. 205–208, 2014.

[5] J. Chen, W. J. Zhang, A. Shorten, J. S. Yu, W. T. Ng, M. Sasaki, T. Kawashima, and H. Nishio, "An IGBT gate driver IC with collector current sensing," in *2017 29th International Symposium on Power Semiconductor Devices and IC's (ISPSD)*, 2017, pp. 275–278.

[6] H. C. P. Dymond, J. Wang, D. Liu, J. J. O. Dalton, N. McNeill, D. Pamunuwa, S. J. Hollis, and B. H. Stark, "A 6.7-GHz Active Gate Driver for GaN FETs to Combat Overshoot, Ringing, and EMI," *IEEE Trans. Power Electron.*, vol. 33, no. 1, pp. 581–594, Jan. 2018.

978-1-5386-2928-4/18 $31.00 © 2018 IEEE

Proceedings of the 30th International Symposium on Power Semiconductor Devices & ICs
May 13-17, 2018, Chicago, USA

CMOS Bi-Directional Ultra-Wideband Galvanically Isolated Die-to-Die Communication Utilizing A Double-Isolated Transformer

Mahdi Javid
Arizona State University
Tempe, AZ, USA
mahdi.javid@asu.edu

Karel Ptacek
ON Semiconductor
Rožnov, Czech Republic

Richard Burton
ON Semiconductor
Phoenix, AZ, USA
(Now with Atomera Inc. USA)

Jennifer Kitchen
Arizona State University
Tempe, AZ, USA

Abstract— In this work, an ultra-wideband (UWB) bi-directional galvanic isolator (BDGI) is reported for the first time. The proposed design methodology uses time-division-duplex (TDD) protocol to merge the functionality of two passive galvanically isolated channels into one magnetically coupled communication channel between two chips, enabling up to 50% form-factor and assembly cost reduction while achieving state-of-art performance. A low-power UWB pulse polarity-modulated transceiver architecture is presented to maximize the channel's capacity to 300 Mb/s and minimize power consumption and propagation delay to 200 pj/b and 5 ns respectively. The communication channel utilizes a double-isolated transformer coupled channel consisting of two transformers connected in series using bondwires and achieves 11 kVpk (7.8 kVrms) high voltage isolation, the highest reported without adding extra steps or alternating the native CMOS fabrication process. The system is realized in a 0.25 um BCD (Bipolar-CMOS-DMOS) process with 0.8 mm² silicon area per channel. The system uses odd-symmetry center-tapped transformers and differential transceivers to increase noise/transient immunity.

Keywords— Isolators, Gate-Driver, System-on-chip, Ultra-Wide-Band, RF, Chip-to-Chip, Integrated-Passive-Device, HVIC

I. INTRODUCTION

In recent years, CMOS galvanically (DC) Isolated chip-to-chip communication has attracted lots of attention as an enabling technology for replacing opto-couplers. Galvanic isolators (GIs) have a wide range of applications including power management, biomedical sensors, and isolated serial links [1-4], where data or control signals must be reliably communicated from a low-voltage chip to a high-voltage chip. GIs in many cases, must support bi-directional communication to either enable read/write functionality for data applications or to enable closed-loop control of power management systems. Fig. 1(a) illustrates an application of BDGI in CMOS power management solutions for isolated gate drive and monitoring of power devices (e.g. IGBTs or GaN FETs). For GIs, key performance concerns are: cost (associated with the silicon fabrication, package assembly and integration), isolation rating, power consumption and common mode transient immunity (CMTI). Control applications require minimal propagation

This work is supported by ON Semiconductor's R&D, Phoenix, AZ, USA

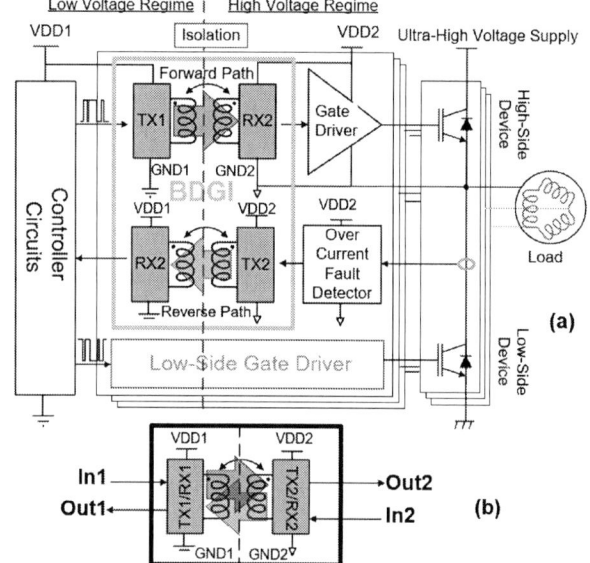

Fig. 1. (a) Galvanically isolated Half-Bridge system using a conventional BDGI (b) Proposed BDGI using a Time-Division Duplex channel

delays to ensure system stability in closed-loop configurations for fault status or voltage/current monitoring.

Currently, BDGI products use two individual uni-directional isolators, with one isolator dedicated to the forward path signal flow and one dedicated to the reverse path (Fig. 1(a)), which provides a duplex solution that consumes significant area [2].

This work proposes time-division-duplexing (TDD) realized with ultra-wideband (UWB) pulse polarity modulation (PPM) transceiver to merge functionality of two inductively coupled channels into one inductively coupled channel between the two chips that can be used in both directions (Fig.1 (b)). The proposed system enables up to 50% form-factor and assembly cost reduction from existing topologies, as juxtaposition of conventional BDGI (Fig. 1(a)) and proposed BDGI in Fig. 1(b) shows. This paper also presents novelties applicable to both uni-directional and BDGIs, including a new high data rate, low

978-1-5386-2928-4/18 $31.00 © 2018 IEEE

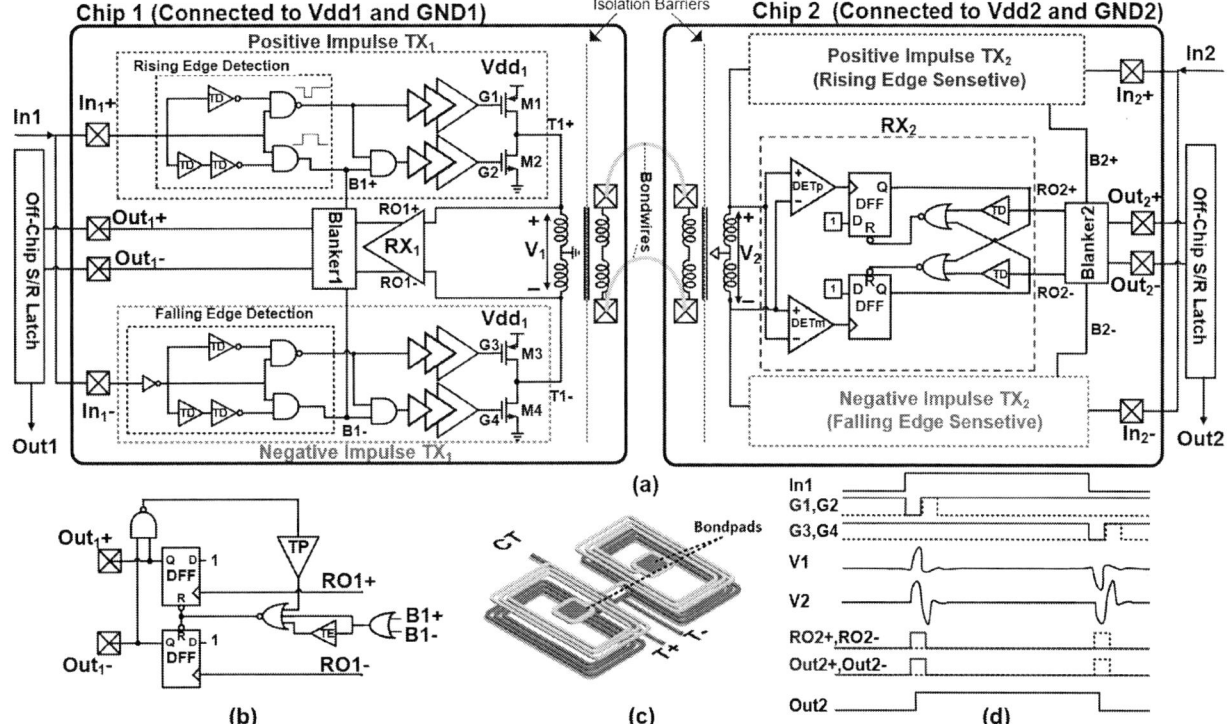

Fig. 2. (a) Implemented UWB BDGI (b) Blanker circuit (c) Center-tapped transformer (d) BDGI waveforms during communication from Chip 1 to Chip 2

power and low propagation delay UWB transceiver architecture and a double isolated transformer coupled channel using two series connected transformers to achieve state-of-art performance .

To the best of authors' knowledge, this is the first time a UWB Bi-Directional Galvanic Isolator is reported.

II. PROPOSED ARCHITECTURE AND IMPLEMENTATION

The implemented BDGI is depicted in Fig. 2(a) and includes two chips connected together through two bondwires. Each chip consists of a differential coreless transformer with a center-tapped primary coil, and a transceiver connected to the primary of the transformer. The two chips have identical architecture. The transmitter structure in Fig. 2(a) is detailed on Chip 1, and the receiver is detailed on Chip2. Fig. 2(d) illustrates the circuit waveforms during a uni-directional communication from Chip 1 to Chip 2. The transmitter encodes the input data (In1) rising edge and falling edge to a differentially positive and negative impulse respectively (V1). The receiver (RX1) translates the received impulses (V2) to a short digital pulse on RO2+ upon detection of a positive impulse and a short digital pulse on RO2- upon detection of a falling edge. Blanker2 and Blanker1 are the self-interference cancellation units, here because Chip 1 is transmitting and Chip 2 is not transmitting Blanker2 passes the data from RX2's outputs to the output pads. An off-chip S/R latch decodes the received impulses (Out2). Since Chip 1 is transmitting in this example, during the transmission of Chip 1,

Blanker1 ensures that Chip 1's outputs (Out1+ and Out1-) remain at zero and doesn't pass RX1's output to Chip1's outputs to cancel self-interference. Encoding the input data rising and falling edges to short impulses (UWB) minimizes the transmit and receive signal time length, and also allows the channel to remain free as long as the data value remains unchanged, which is desirable for TDD. Additionally, UWB modulation reduces the power consumption by shortening the transmission and reception time. Moreover, UWB pulse polarity was chosen over OOK and narrowband (single tone) communication methods because narrow band methods take more time for the transmitted and received voltage to rise above the receiver's sensitivity. Thus UWB PPM minimizes the propagation delay[4].

A. Inductive Coupled Channel: Double Isolated Transformer

The bottleneck for isolation rating in CMOS isolators with transformer coupling is the thickness of inter-metal-dielectric (IMD) between two stacked coils forming a transformer [1]. In this work a double isolated transformer is formed by connecting two stacked transformers in series through bondwires. Since Chip 1 and Chip 2 are connected through bondwires placed between the transformers' secondary coils (the sides that have bondpads), the two transformers are connected in series, realizing effectively two IMD barriers between the two chips and achieve two times the isolation rating of one transformer, without adding extra steps or alternating the native CMOS fabrication process. Fig. 2 (c) shows one of the two transformers. The transformer's coils benefit from an odd-symmetry (s-shaped) design that translates magnetic transients to a common mode electrical signal that is rejected by the differential receiver, thus increasing the CMTI of the circuit from an even-symmetry design. The center-tap of each transformer is also connected to the ground of its respective chip to reject common mode transients coming from high voltage fluctuations of GND2 and eliminate ground loops.

B. Ultra-Wideband Transmitter

Detailed architecture of transmitter is shown in Fig. 2(a) for Chip 1. Chip 2's TX is identical to Chip 1 and is depicted on a block level. The UWB transmitter on each chip consists of two branches: the positive impulse generator (connected to T1+ on Chip 1) that creates an impulse when detecting a rising edge at its input In1+, and the negative impulse generator (connected to T1-) that creates an impulse when detecting a falling edge at its input In1-. The logic circuit at input of each transmitter branch generates two non-overlapping signals upon detection of their respective activating edge (G1-G2 at rising edges and G3-G4 at falling edges). These signals control the PMOS-NMOS output stage (M1-M2 and M3-M4) connected to the transformer. The transformer's inductance acts as a UWB pulse-shaping filter. Since the transformer is center-tapped, either a differentially positive or negative impulse is generated across the transformer input (V1 in Fig. 2 (d)). Each transmitter creates an impulse by turning on and off its corresponding output stage's PMOS switch(e.g. M1), which can cause a slowly decaying ringing accompanying the generated impulse. To eliminate this ringing, the output stage's NMOS device (e.g. M2) is therefore turned on immediately after the PMOS turns off to accelerate the signal damping. This technique releases the channel for reception (high impedance mode) faster than the slow current control introduced in [4] and can therefore attain higher data rates.

C. Receiver and Blanker (Self-interference Cancellation)

Due to use of two series transformers in the presented system, the received signal can be distorted into a doublet waveform (two sequential impulses with opposite polarity), where a positive (negative) generated impulse on Chip1 is received as a positive (negative) impulse accompanied by a negative (positive) impulse, as shown in the transient waveform of V2 in Fig. 2(d). This doublet phenomenon is removed by using the receiver structure of Fig. 2(a) detailed on Chip 2 as RX2, which consists of two rectifying detectors DETp that amplifies and detects only positive impulses of the received doublet and DETm that amplifies and detects only negative impulses of the received doublet. Upon detection of either a positive or negative impulse, the cross-coupled D-type flipflop mutes the other detector's output for a short time period to block reception of the counter pulse. Thus RX only detects the first impulse of every doublet (that has the same polarity with the generated impulse) and ignores the second impulse.

A blanker unit (Fig. 2(b)) is placed between the receiver outputs (RO+ and RO-) and the chip's outputs (Out+ and Out-) to mute the chip's receiver outputs during data transmission and avoid self-interference from the chip's transmitter. The blanker is triggered on the rising edge of the blanking input

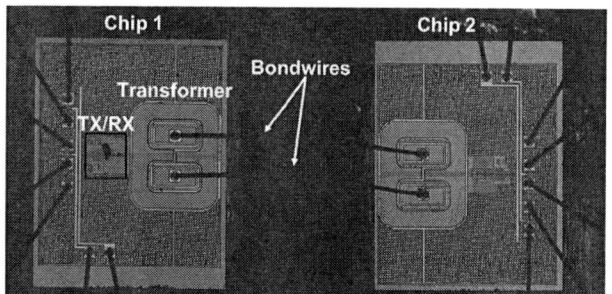

Fig. 3. System module micrograph with chips assembled together in a SOIC-16 Wide Body package

(OR combination of B+ and B-) and mutes the output for a total time of TE. The TPW delay sets the output pulse widths for each received impulse and was designed to be long enough for the pulses to be detected using an external decoder (e.g. SR latch).

D. Independent control of Positive and Negative Impulse Transmitters

The positive and negative impulse transmitters can be controlled independently. In Fig. 2(a) they are shorted together at the input to automatically encode input data edges to PPM impulses. This independent control provide flexibility to the user to implement modulations other than PPM. Moreover, since two consecutive positive or negative impulses will not be interpreted as a data value change, they can be used to transfer protocol/scheduling information. Furthermore, in control applications requiring fault detection, a rising edge can be sent multiple times for higher reliability.

III. FABRICATION AND MEASUREMENT RESULTS

The chip was fabricated in a 0.25 μm BCD process and co-packaged in a 16-pin SOIC Wide Body Package, as shown in Fig. 3. The two chips within the system are identical, thus increasing the yield and reducing development costs when compared with other proposed solutions [2,3]. In the presented measurements, the transmitter's two inputs (In+ and In-) are tied together so that the transmitter operates as a standard edge modulator that translates the input's rising and falling edges to positive and negative impulses respectively. A successful full channel communication at 300 Mb/s with a propagation delay of 5ns can be seen from the measured unidirectional performance in Fig. 4. Fig. 5 shows the input and decoded outputs of the measured BDGI during a 120Mbps bi-directional communication which illustrates the interleaved decoded (Out 2 and Out1) and non-decoded (e.g. Out2+, etc.) receive and transmit signals. The chips' outputs for measurement in Fig. 5 are connected to a board-integrated S/R latch to decode the received signal. A 1-minute dielectric breakdown test has shown a 11 kV (7.8 kVrms) isolation rating, the highest reported to date for a fully CMOS isolator. For a supply voltage of 4 V, the total system shows a low power consumption of 200 pj/b. Table I summarizes the presented system's performance and compares it to other state-of-the-art fully CMOS GI architectures.

978-1-5386-2928-4/18 $31.00 © 2018 IEEE

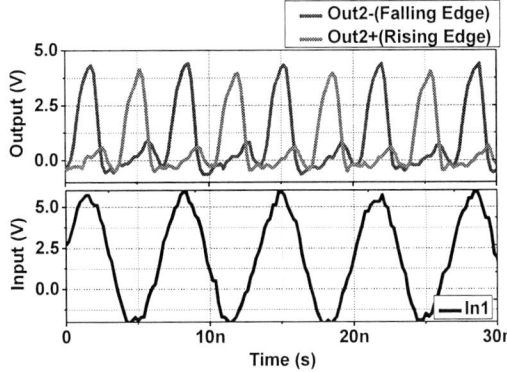

Fig. 4. Measured uni-directional communication from Chip 1 to Chip 2 at 300 Mb/s

IV. CONCLUSION

A UWB CMOS BDGI is fabricated and measured. The presented system achieves smallest total area per channel of 0.8 mm^2. Using the half-duplex topology allowed reducing the number of isolating transformers and bondwires by 50%. The double-isolated transformer enabled a 7.8 kVrms isolation rating, the highest isolation rating reported in fully CMOS, using less than 10% of the chip area of previously highest reported isolation [2]. Furthermore, possibility of reaching an isolation rating two times of the rating allowed by process's inter-metal-dielectric's thickness, without alternating the native process or adding additional steps to fabrication, or packaging is demonstrated. The UWB transceiver presented in this work achieves the smallest propagation delay of 5 ns, with a low-power consumption of 200 pj/b. This power consumption is equal to the lowest reported power consumption to date [2] and a third of power consumption in the UWB GI reported in [4]. This work also demonstrates that UWB TDD can achieve a data rate of 300 Mb/s, 7.5 times that achieved with half duplex through load modulation in [3].

ACKNOWLEDGMENT

Authors want to acknowledge ON Semiconductor R&D, Phoenix, USA for supporting this research.

REFERENCES

[1] M. Javid, R. Burton, K. Ptacek and J. Kitchen, "CMOS integrated galvanically isolated RF chip-to-chip communication utilizing lateral resonant coupling," 2017 IEEE Radio Frequency Integrated Circuits Symposium (RFIC), Honolulu, HI, 2017, pp. 252-255.

[2] S. Mukherjee et al., "A 500Mb/s 200pJ/b die-to-die bidirectional link with 24kV surge isolation and 50kV/μs CMR using resonant inductive coupling in 0.18μm CMOS," 2017 IEEE International Solid-State Circuits Conference (ISSCC), San Francisco, CA, 2017, pp. 434-435.

[3] P. Lombardo, V. Fiore, E. Ragonese and G. Palmisano, "A fully-integrated half-duplex data/power transfer system with up to 40Mb/s data rate, 23mW output power and on-chip 5kV galvanic isolation," 2016 IEEE International Solid-State Circuits Conference (ISSCC), San Francisco, CA, 2016, pp. 300-301.

[4] S. Kaeriyama, S. Uchida, M. Furumiya, M. Okada, T. Maeda and M. Mizuno, "A 2.5 kV Isolation 35 kV/us CMR 250 Mbps Digital Isolator in Standard CMOS With a Small Transformer Driving Technique," in IEEE Journal of Solid-State Circuits, vol. 47, no. 2, pp. 435-443, Feb. 2012

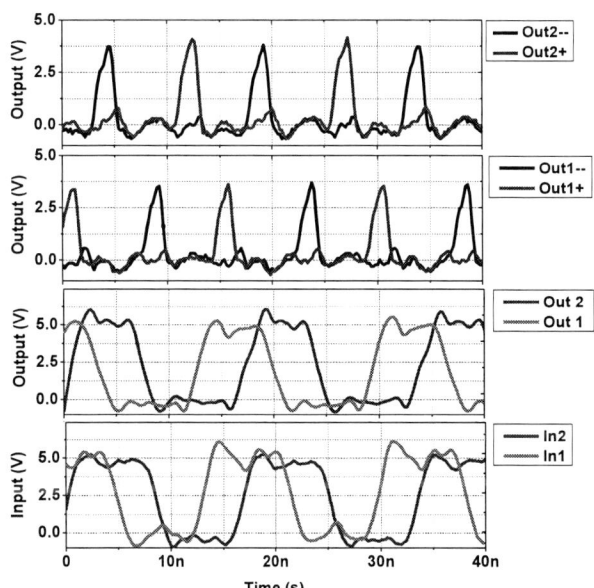

Fig. 5. Measured bi-Directional interleaved performance at 120 Mb/s

TABLE I. PERFORMANCE SUMMARY AND COMPARISON

Parameter	This Work		[2]	[3]	[4]
Technology(nm)	250		180	350	500
Data-Rate(Mb/s)	300		500	40(forward path) 3 (reverse path)	250
Isolation Rating (kVrms)	7.8		7.5	5	2.5
Vdd(V)	5	4	N.R.	3	5
Total Power Consumption(pj/b)	300	200	200	520	650
Modulation Scheme	Pulse Polarity		OOK	OOK	Pulse Polarity
Silicon Area(mm^2 /channel)	0.81		12.5	2.5	1
Prop. Delay(ns)	5		N.R.	N.R.	5.5
Bi-Directivity	Yes Shared GI Link		Yes Two parallel GI Links	Yes Load modulation	No

Proceedings of the 30th International Symposium on Power Semiconductor Devices & ICs
May 13-17, 2018, Chicago, USA

Dynamic-Ron Control via Proton Irradiation in AlGaN/GaN Transistors

A.Tajalli[1], A. Stockman[2,3], M. Meneghini[1], S. Mouhoubi[2], A. Banerjee[2], S. Gerardin[1], M. Bagatin[1], A. Paccagnella[1], E. Zanoni[1], M. Tack[2], B. Bakeroot[3], P. Moens[2]and G. Meneghesso[1]

[1]Univ. of Padova, Dept. of Information Engineering, via Gradenigo 6/B 35131 Padova, Italy
[2]ON Semiconductor, Oudenaarde, Belgium, [3]CMST imec/Ghent University, Ghent, Belgium
email: tajalli@dei.unipd.it

Abstract— **Dynamic-Ron is still a key issue in GaN power HEMTs. Recently [2] we demonstrated that proton irradiation is an effective and controllable way to reduce dynamic-Ron in AlGaN/GaN HEMTs; this beneficial effect is ascribed to the minute increase in the leakage of the uid-GaN layer, promoting charge de-trapping from the buffer. The effect is dependent on L_{GD}, shorter L_{GD} is better. The shorter L_{GD} corresponds to a shorter region for trapping, and therefore the dynamic-Ron is not strong when L_{GD} is short. We demonstrate that samples submitted to proton irradiation at high fluences (1.5×10^{14} p/cm², 3 MeV) show a complete suppression of dynamic-Ron (complete voltage range, 150 °C), without significant modifications in the other device parameters. Combined pulsed measurements, drain current transient (DCT) characterization and electroluminescence (EL) analysis are used to explain the experimental data.**

Keywords—GaN; High Electron Mobility Transistors(HEMT); electroluminescence (EL); proton irradiation; dynamic-Ron.

I. INTRODUCTION

GaN High Electron Mobility Transistors (HEMTs) show a great performance in high power switching applications (650V-1200V) [1], owing to the facts that they have a high breakdown electric filed (3.3 MV/cm), a two-dimensional electron gas (2DEG), and a low on-resistance. However, GaN-based power transistors still suffer from trapping and other degradation mechanisms that affect the dynamic performance and the reliability of the transistor. As a matter of fact, dynamic-Ron is a key issue in GaN power HEMTs. The charge trapped during OFF-state operation has a negative impact on the ON-resistance in GaN-based power HEMTs.

It has been shown in [2] that the dynamic Ron can be decreased to zero by proton irradiation. Moreover, as a result of this proton irradiation, dynamic-Ron is no longer dependent to the trapping voltage applied to the drain of a GaN HEMT.

In this paper, the proton irradiation method is employed to control the dynamic-Ron of GaN-based power HEMTs. The effects of proton irradiation are studied by variety of measurements such as double pulse, drain current transient, and electroluminescence analysis. The results presented in this paper indicate the decrease of the trap level after irradiation and accordingly an increase of the electro luminescence signal.

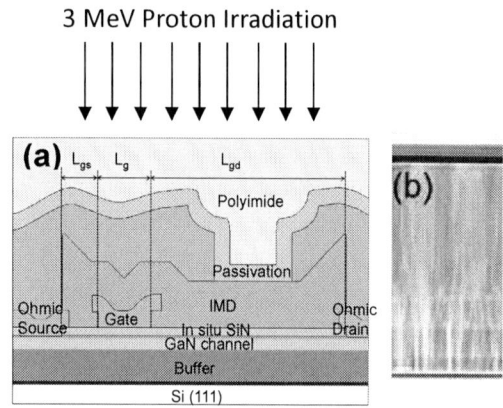

Fig. 1. (a) schematic representation of the structure of the device under analysis and of the proton implantation process used for eliminating dynamic Ron; (b) TEM cross section of the buffer of the analyzed devices.

II. EXPERIMENTAL DETAILS

A. Device and measurement Description

The study is carried out on 6-inch GaN metal insulator-semiconductor (MIS) HEMTs on silicon substrate power transistors. The epitaxial structure was grown by metalorganic chemical vapor deposition (MOCVD) with the gate width of 200µm designed for 650V operation. A schematic of the device is depicted in Fig. 1. (see [3] for details). Proton radiation was performed at Legnaro National Laboratories in Legnaro, Padova, Italy. Devices were submitted to proton irradiation at room temperature with no bias applied at 3MeV. In order to analyze the dynamic ON-resistance, pulsed I-V measurements performed at 150 °C (the worst case for dynamic-R_{ON}) [8] before and after proton irradiation up to $V_{DSQ} = 600$ V (pulse times: ON=20µs, OFF=2ms) are reported in Fig. 2. Fig. 2a shows a current collapse before irradiation on the device at T= 150°C and fig 2b, no current collapse after irradiation, indicating total suppression of the dynamic-R_{ON} post irradiation. This effect is for the first time reported in [2], whereas this paper provides a full in-depth characterization and understanding (through current- DTLS and EL) and

978-1-5386-2928-4/18 $31.00 © 2018 IEEE

verification on large area power transistors. Fig. 3 reports the results of pulsed-IV measurements carried out on 4 devices with different gate drain distance (L_{GD} =15-20 μm) before and after radiation at T=150°C. As can be noticed, the dynamic-R_{ON} is completely superimposed after radiation at 1.5×10^{14} p/cm².

B. Measurement Description and results

The physical origin of the decrease in dynamic-Ron is characterized by means of drain current transient measurements. DCT measurements were performed with (V_{GSQ}, V_{DSQ}) = (-20V, 200V) (corresponding to maximum dynamic- R_{ON} before radiation) and T_{amb} ranging from 110°C to 170°C step of 20°C with time constants in the range 0.1-10 s (see Fig. 4 (a), (b)). One trap level, having activation energy of 1.05-1.1 eV, was detected, correlated to carbon acceptors in the buffer (CN traps, [5]) (Fig. 4c). No trap level state was detected after the proton irradiation (1.5×10^{14} p/cm²), which is in accordance to the total suppression of dynamic Ron.

The impact of proton radiation on the device performance was studied by means of emission microscopy. The electroluminescence (EL) signal was detected for several gate and drain voltage levels. In the discussed analysis the drain voltage levels range from 50V to 200V are used to investigate the presence of hot electrons before/after proton irradiation, see Fig. 5; the gate voltage levels were chosen according to the threshold voltage of the devices and to the dissipated power (not to induce device degradation). An acquisition time of 25s and an EM gain of 200 were used.

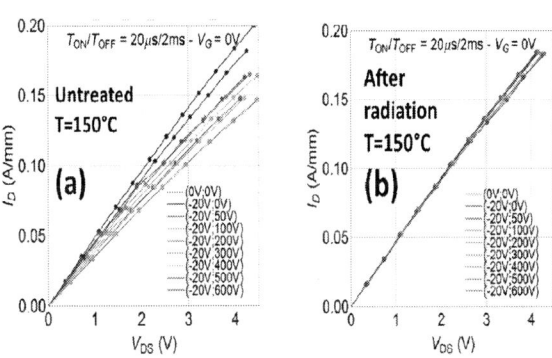

Fig. 2. (a) current collapse before radiation at 150°C; (b) current collapse after radiation at 150°C.

We plot the EL signal as a function the difference between the gate and the threshold voltage (V_{GS}-V_{TH} level). After the radiation the samples irradiated at 1.5×10^{14} p/cm² show a higher EL signal, which is indicative of a higher electron current/energy, also a consequence of the reduction of the trapping phenomena [5], [6]. Irradiated samples have less trapping, i.e. less virtual gate leakage, i.e. higher electric field for the same drain voltage. For this reason, the EL signal is higher in the irradiated samples. On the other hand, having less defects, the irradiated device has less hot electron trapping(or an easier de-trapping). Fig. 5c shows the false color EL maps

showing the distribution of the EL signal before and after radiation. an acquisition time of 25 s and an EM gain of 200. The electroluminescence signal increased on the device after radiation from 8M to 18M for instance at VD = 200V.

III. IMPACT OF PROTON IRRADIATION ON THE DYNAMIC R_{ON} OF ALGAN/GAN HEMTs

The on-resistance variation of large power transistors in the vicinity of the irradiated area is shown in Fig. 6. A circular pattern of increasing dynamic Ron at V_{DS} = 200V is visible spreading outward from the two irradiation zones. The highest proton fluence corresponds to the lowest dynamic Ron.

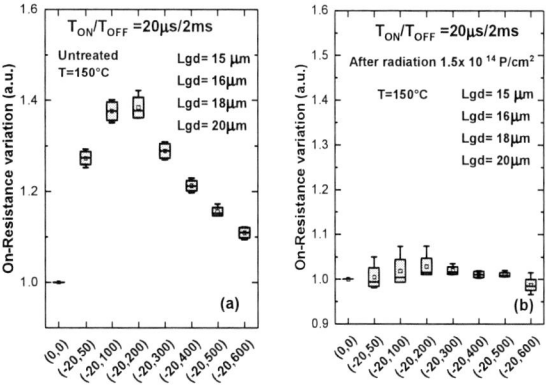

Fig. 3. Variation of dynamic Ron, with the quiescent bias up to 600V for different L_{GD}. (a) untreated devices, (b) after radiation at T= 150°C.

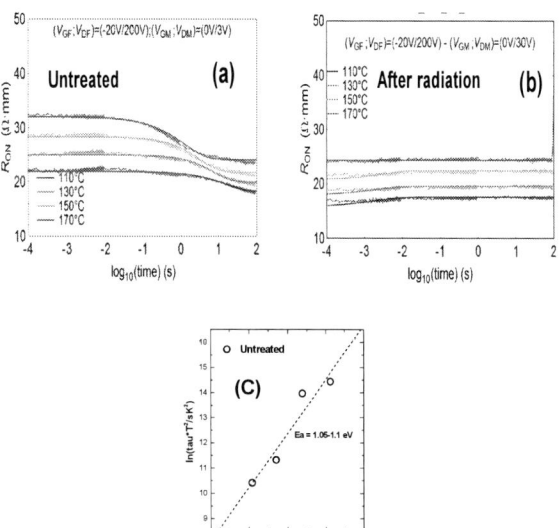

Fig. 4. drain current transient measurement before and after radiation at VDSQ =200V.(b) no detectable trap level states shows after irradiation. (c) Arrhenius plot of the defect responsible for peak dynamic Ron, with activation energy around 1.05 eV

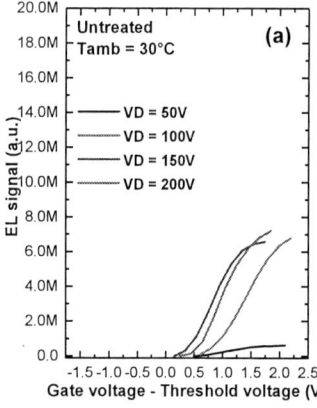 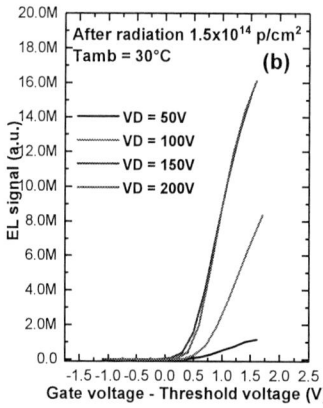 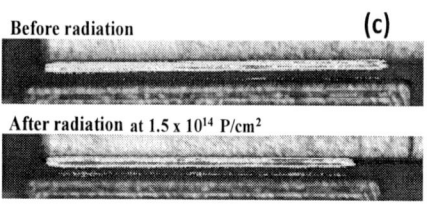

Fig. 5. EL intensity versus gate voltage overdrive at different drain voltage on a GaN HEMT. (a) before radiation at V_D from 50 V up to 200 V (b) the samples after radiation show a higher EL signal. (c) false color EL maps showing the distribution of the EL signal before and after radiation. an acquisition time of 25 s and an EM gain of 200

For die in the irradiated zone, devices with smaller gate/drain spacing (L_{GD}) exhibit a larger R_{on} decrease. This can be seen when comparing devices below ($L_{GD} = 20\mu m$) to those above ($L_{GD} = 14\mu m$) the irradiated cells indicated with arrows. The decrease of dynamic Ron with decreasing L_{GD} is also clearly visible from Fig. 7, which will be discussed later. The inset of Fig. 6 shows the source leakage at $V_{DS} = 200V$ after normalization w.r.t. the median of the leakage distribution at every L_{GD}, to filter out the L_{GD} dependence on the drain leakage. So, the color of each cell represents the relative drain leakage w.r.t. all the other devices measured with the same L_{GD}. A similar pattern arises, corresponding to a higher drain-to-source leakage close to the impact zone, which gradually decreases with distance from the latter.

Fig. 7 shows the off-state drain leakage of large transistors as a function of their measured R_{ON} variation at $V_{DS} = 200V$ and T = 150°C. The distinction is made between devices out of and within the vicinity of the impact zone of both proton beams. Overall, devices irradiated with the highest proton fluence correspond to the cloud with lowest R_{on} combined with a high $I_{D,off}$ at $V_{DS} = 200V$. Shorter L_{GD} devices show the lowest dyn-R_{on} after irradiation. It means that the shorter L_{GD} corresponds to a shorter region for trapping, and therefore the dynamic-Ron is not strong when L_{GD} is short. The effects are reproduced on several batches of proton irradiation, and they are very consistent. We believe that dynamic Ron in the non-treated devices results from the absence of de-trapping process due to the uid acting as an insulator thus preventing the vertical transfer of electrons- at high trapping bias- from buffer to the 2DEG [7]. The irradiated devices exhibit a reduced dynamic Ron thanks to a process of de-trapping induced by well-controlled but minute increase of leakage in the uid layer. In this case, the transfer of electrons is not vertical (from buffer to 2DEG) but lateral (from source to drain).

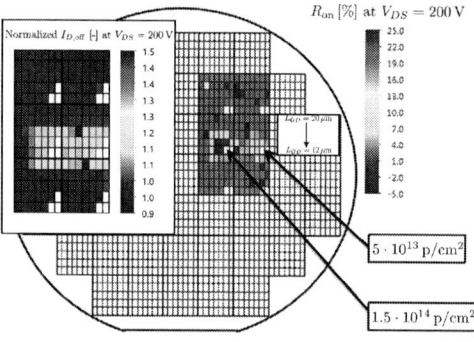

Fig. 6. Wafer map of the on-resistance variation at $V_{DS} = 200V$ and T = 150°C on large power transistors (W=140mm). Inset: source leakage at $V_{DS} = 200V$ normalized w.r.t. the median of all devices with the same L_{GD}.

IV. CONCLUSION

In summary we demonstrate that proton irradiation is an effective and controllable way to reduce dynamic-Ron in AlGaN/GaN HEMTs. The effect is studied by means of combined pulsed characterization, transient measurements and EL on untreated and irradiated devices. We demonstrate the following relevant results : the electroluminescence signal increased on the device after radiation, while dynamic-Ron in AlGaN/GaN HEMTs decreased after irradiation; this beneficial effect is ascribed to the minute increase in the leakage of the uid-GaN layer, promoting charge de-trapping from the buffer. No trap level state was detected after the proton irradiation ($1.5x10^{14}$ p/cm²) by means of drain current transient.

ACKNOWLEDGMENT

This work was partially supported by the Horizon2020 project Innovative Reliable Nitride based Power Devices and Applications (InRel-NPower).

978-1-5386-2928-4/18 $31.00 © 2018 IEEE

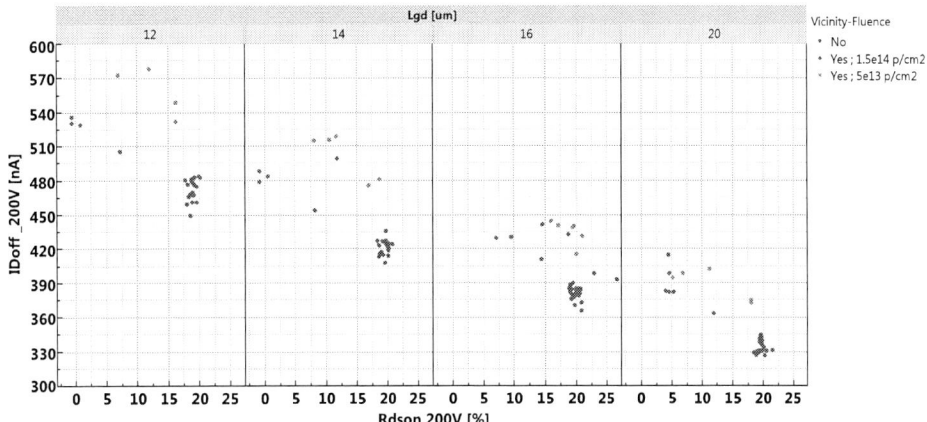

Fig. 7. Off-state drain leakage at V_{DS} = 200V and T = 150°C on large power transistors (W=140mm). Blue dots represent large transistors out of the vicinity of any impact zone, while red and green dots represent those in the direct vicinity of the impact zone with proton fluence of 5 10^{13} p/cm2 and 1.5 10^{14} p/cm2, respectively.

REFERENCES

[1] P. Moens, A. Banerjee, P. Coppens, F. Declercq, and M. Tack, "AlGaN / GaN Power Device Technology for High Current (100 + A) and High Voltage (1 . 2 kV)," in Proceedings of the 28th international symposium on power semiconductor devices and ICs (ISPSD), 2016,, pp. 455–458, Prague, Czeck Republic.

[2] M. Meneghini, A. Tajalli, P. Moens, A. Banerjee, A. Stockman, M. Tack, S. Gerardin, M. Bagatin, A. Paccagnella, E. Zanoni, and G. Meneghesso, Total Suppression of Dynamic-Ron in AlGaN/GaN-HEMTs Through Proton Irradiation, 2017 IEEE International Electron Devices Meeting (IEDM) DOI 10.1109/IEDM.2017.8268492.

[3] P. Moens et al., An Industrial Process for 650V rated GaN-on-Si Power Devices using in-situ SiN as a Gate Dielectric , in Proceedings of the 28th international symposium on power semiconductor devices and ICs ISPSD, 2014, pp374-377, Waikoloa, Hawaii, USA.

[4] M. A. Reshchikov et al., Carbon defects as sources of the green and yellow luminescence bands in undoped GaN, Physical Review B **90**, 235203 (2014)

[5] M. Meneghini, G. Meneghesso, E. Zanoni, Analysis of the Reliability of AlGaN/GaN HEMTs Submitted to On-State Stress Based on Electroluminescence Investigation, IEEE Transactions on Device and Materials Reliability, vol. 13, 2013 357–361 (no. 2).

[6] A. Tajalli et al., Field and hot electron-induced degradation in GaN-based power MIS-HEMTs, Microelectronics Reliability **76–77** (2017) 282–286

[7] M.J. Uren et al., Leaky Dielectric" Model for the Suppression of Dynamic RON in Carbon-Doped AlGaN/GaN HEMTs, Transactions on Electron Devices **64** (7), (2017), DOI 10.1109/TED.2017.2706090.

[8] I Rossetto, M Meneghini, A Tajalli, S Dalcanale, C De Santi, P Moens, A Banerjee, E Zanoni, G Meneghesso, Evidence of Hot-Electron Effects During Hard Switching of AlGaN/GaN HEMTs, IEEE transactions on electron devices **64**(9) 3734-3739 (2017)

Proceedings of the 30th International Symposium on Power Semiconductor Devices & ICs
May 13-17, 2018, Chicago, USA

Bidirectional threshold voltage shift and gate leakage in 650 V p-GaN AlGaN/GaN HEMTs: The role of electron-trapping and hole-injection

Yuanyuan Shi, Qi Zhou*, Qian Cheng, P. Wei, L. Zhu, D. Wei, A. Zhang, Wanjun Chen, and Bo Zhang

State Key Laboratory of Electronic Thin Films and Integrated Devices, University of Electronic Science and Technology of China, Chengdu 610054, P.R. China, Email: zhouqi@uestc.edu.cn,zhangbo@uestc.edu.cn

Abstract— The threshold voltage (V_{TH}) instability of a 650 V p-GaN gate AlGaN/GaN HEMTs and its underlying physical mechanism was investigated by forward gate stress. A uniquely bidirectional shift in the V_{TH} with the critical gate voltage $V_{critical}$ of 6 V was observed in the device after the static and dynamic gate stress. The temperature- and time-dependent gate leakage current revealed that the occurrence of electron-trapping and hole-injection in sequence with the increasing gate bias responsible for the inhomogeneous shift in V_{TH}. At small positive gate bias (V_G<6V), the positive shift in V_{TH} is induced by electron filling of acceptor-like traps in AlGaN barrier, while the gate leakage is accordingly dominated by *trap-dominated SCLC*. At large positive gate bias (V_G>6V), the hole-injection is triggered that results in a negative shift in V_{TH} and the gate leakage exhibits a substantial increase due to the forward turn on of the gate pn junction. Besides, the effective hole-injection also leads to a significant increase in OFF-state drain leakage, which is believed to be the pronounced electron-hole recombination in the channel.

Keywords—threshold voltage instability; gate leakage; gate stress; electron trapping; hole injection

I. INTRODUCTION

AlGaN/GaN-on-Si high-electron-mobility transistors (HEMTs) have been studied extensively for both high frequency and power electronics application over the past two decades [1]. For power electronics application, Normally-OFF devices is required to guarantee the high-desirable fail-safe operation and simple gate drive configuration [2,3]. In order to achieve the intrinsic normally-off operation, several approaches have been proposed such as the recessed gate [4-5], fluorine ion implantation [6], and p-GaN gate [7,8]. Among them the p-GaN Normally-OFF devices are already commercially available to date in recent years [9]. Because the reliable forward gate swing of p-GaN device is relatively small (e.g. ≤+7 V), it is essential to bias the p-GaN device at a stable forward gate voltage for reliable operation and simultaneously obtaining high performance. However, the possible threshold voltage (V_{TH}) instability of the device may induce unintentional operation point drift that could jeopardize the power conversion efficiency of the system or even cause gate degradation and then device failure. Therefore, it is of great interest to evaluate the V_{TH} stability of p-GaN gate AlGaN/GaN HEMTs.

In this work, a commercial 650 V rating p-GaN

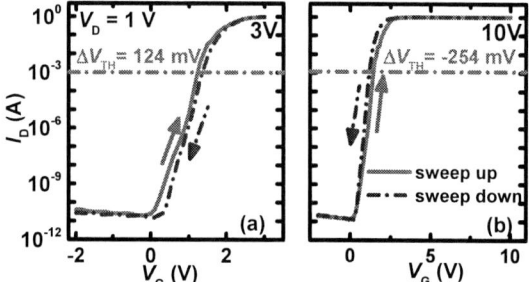

Fig.1. Transfer curves of the 650 V p-GaN gate AlGaN/GaN HEMT in double-sweep with the maximum gate bias at (a) $V_{G,max}$=3 V, and (b) $V_{G,max}$=10 V.

AlGaN/GaN HEMTs was investigated under forward gate stress. A bidirectional V_{TH} shift was observed. The critical voltage for the V_{TH} turned from positive shift into negative shift was found to be +6 V, which is close to the forward turn on voltage of the gate p-GaN/AlGaN junction. A physical model of electron-trapping followed by hole-injection was proposed to responsible for the bidirectional V_{TH} shift. At small positive gate bias, the positive V_{TH} shift is dominated by electron filling of acceptor traps in AlGaN barrier while the gate leakage is accordingly dominated by trap assisted tunneling. At large positive gate bias, hole-injection is observed that induces the negative V_{TH} shift and the gate leakage exhibits a substantial increase due to the forward turn on of the gate p-GaN/AlGaN junction. Besides, a significant increase in OFF-state drain leakage was also observed after hole-injection, which is believed to be the pronounced electron-hole recombination in the channel.

II. THRESHOLD VOLTAGE INSTABILITY

The p-GaN gate HEMTs used in the experiments are the state-of-the-art 650 V rating commercial devices. The devices were characterized both in static and dynamic gate stress by using the Keithley 4200 semiconductor characterization system. Fig. 1 shows the transfer characteristics measured in double-sweep mode with different sweeping ranges. The device exhibits a clockwise hysteresis with ΔV_{TH}=124 mV (defining the V_{TH} at drain current of I_D=1 mA) for V_G sweep up to 3 V, while it tuned into an anticlockwise hysteresis with ΔV_{TH}=254 mV for V_G sweep up to 10 V as shown in Fig. 1(a) &(b), respectively. Fig.2&3 shows the transfer characteristics

This work was supported in part by the National Natural Science Foundation of China under Grant 61674024, and in part by the Assembly Pre-research Project under Grant JZX2017-1643/Y537, and in part by the National Science and Technology Major Project under Grant 2013ZX02308-005, and in part by the opening project of the State Key Laboratory of Electronic Thin Films and Integrated Devices under Grant KFJJ201609, and in part by the Fundamental Research Funds for the Central Universities under grant ZYGX2016J211. corresponding authors: Qi Zhou, Bo Zhang.

978-1-5386-2928-4/18 $31.00 © 2018 IEEE

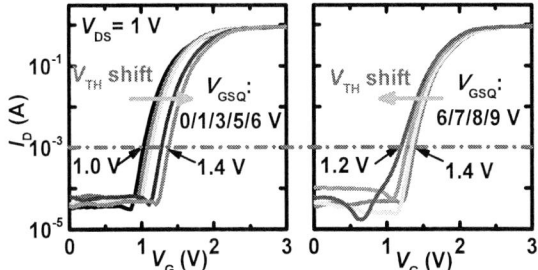

Fig.2. The transfer curves measured by pulsed gate stress with positive quiescent gate bias V_{GSQ} of (a) 0~6 V and (b) 6~9 V. The V_{DS} is fixed at 0 V.

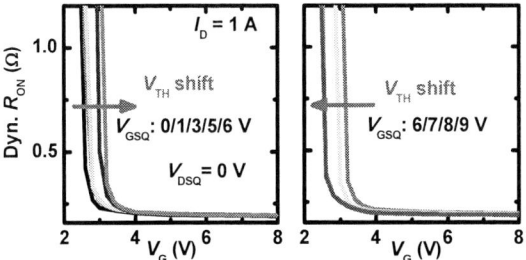

Fig.3. The on-resistance R_{ON} (@ 1 A) versus V_G for 0 V<V_{GSQ}<6 V and 6 V<V_{GSQ}<9 V

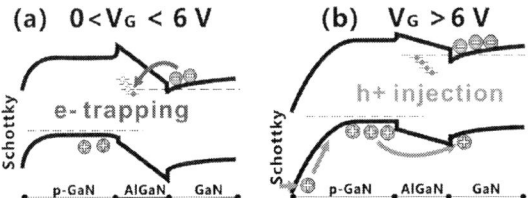

Fig.4. The band diagram and dynamic charge transfer in p-GaN gate region of the device in (a) medium forward gate bias and (b) high forward bias.

I_D-V_G and dynamic R_{ON}-V_G of the device measured under forward gate pulse stress with a pulse width of 5ms and a period of 1 s, while the positive quiescent bias V_{GSQ} ranges from 0 to +9 V. The V_{TH} exhibits a positive shift for $V_{GSQ}\leq6$ V and a negative shift for V_{GSQ}>6 V in Fig.2. Accordingly, the positive V_{TH} shift leads to a distinctly positive shift of the R_{ON}-V_G curve for $V_{GSQ}\leq6$ V, suggesting a remarkable increase in R_{ON} (measured at I_D=1 A) for V_G less than +4 V in Fig.3(a). Whereas the negative V_{TH} shift for V_{GSQ}>6 V results in a respectable reduction in R_{ON} for V_G around +3 V in Fig.3(b). The positive V_{TH} shift attributes to the electron-filling in AlGaN barrier as shown in Fig. 4(a). At higher V_{GSQ} (e.g. >6 V) the effective hole injection that tends to populate electrons in the channel is triggered by the sufficiently large-positive stress voltage (e.g. V_{GSQ}=6-9 V) as shown in Fig. 4(b). Consequently, the V_{TH} turned into negative shift. In order to reinforce the proposed model, the temperature- and time-dependent gate leakage characterization at both of negative and positive gate biases were carried out. The gate leakage mechanism associate with the electron trapping/detrapping was also analyzed.

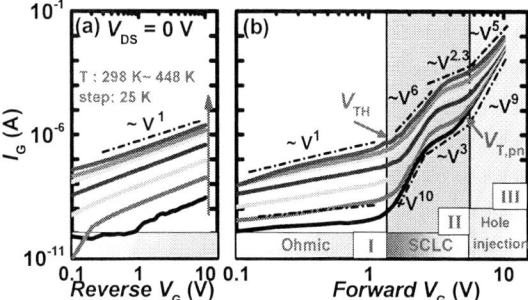

Fig.5. The gate leakage as a function of temperature in (a) reverse and (b) forward bias.

III. GATE LEAKAGE

Figure 5 show the gate leakage current I_G versus gate bias V_G at temperatures of 298~448 K plotted in log-scale. At low voltage for V_{GS}<V_{TH} (Reg.I), the slope of I_G-V_G follow the trend of $I_G \propto V_G$ with the exponent m=1 at all measured temperatures, which reveals that the gate leakage is governed by the *Ohmic conduction*. In Reg.II, I_G starts to exhibit a steeper increase with m=6~10 for V_G>V_{TH} and then followed by a slow increase with m=2~3. Such a I_G-V_G behavior indicates that the gate leakage is dominated by the *space-charge-limited current* (SCLC) mechanism due to the traps in AlGaN barrier [10,11]. The steep increase of I_G is a result of electron-tunneling and subsequent trap filling as the 2DEG is populated in the channel while the device is turned on. Then, after a sufficient trap filling the negatively charged traps in AlGaN barrier in turn to suppress the follow-up electron-tunneling, which leads to a premature saturation of I_G as observed in Region II. Further increase the gate bias (V_G>5.5 V), another steep increase in I_G is observed which is believed to be the effective hole-injection from the p-GaN gate triggered by the turn-on of the gate p-GaN/AlGaN junction at high positive gate bias (in Reg.III).

Fig. 6(a) plots the temperature dependence (T) of gate leakage I_G-T in Reg. I as a function of V_G. The curves can be fitted by Arrhenius equation $I_G = I_0 \exp(-\dfrac{Ea}{kT})$, in which the activation energies (Ea) of the traps are determined by the slope of the I_G. As the channel is biased in OFF-state, most of the applied gate voltage drop across the AlGaN layer. The low I_G is dominated by a moderate de-trapping and trapping (generation- recombination) process of the deep traps (Ea=0.6 eV) for T< 125°C. While for T >125 °C, the slope of I_G-T curve was considerably reduced with the activation energy of 0.25 eV, which may relate to the surface hopping conduction at the interface between the passivation layer and III-Nitride semiconductor[13]. As shown in Fig.6(b), the I_G exhibits a consistently-monotonic decrease with time for the first few milliseconds and then saturate at leakage current level as low as ~10^{-9} A. The schematic cross section of device and energy band diagram of the gate region for V_G<1 V is shown in Fig. 6(c) and illustrates the field-enhanced electron generation-recombination (G-R) process in AlGaN layer in Reg. I.

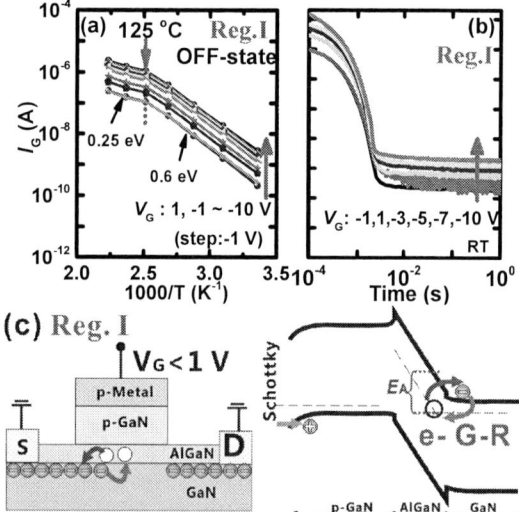

Fig.6 (a) Arrhenius plots of I_G in Reg. I. (b) Time resolved I_G characteristics in Reg. I at room temperature. (c) Schematic cross section of device and energy band diagram of the gate for $V_G < 1$ V to visualize the electron generation-recombination process.

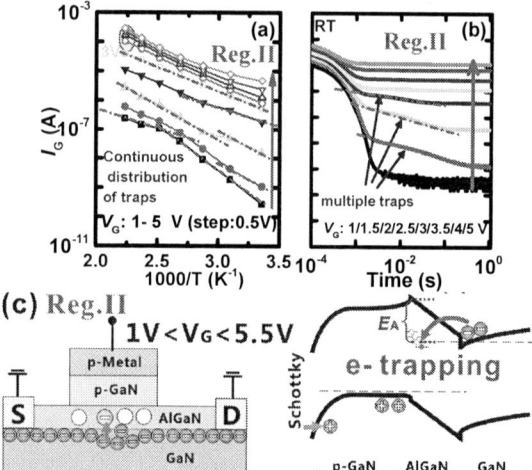

Fig.7. (a) Arrhenius plots of I_G in Reg. II. (b) Time resolved I_G characteristics for 1 V$<V_G<$ 5.5 V at RT. (c) Schematic cross section of the device and energy band diagram of the gate region to illustrate the trap-assisted tunneling process.

The I_G-T curves exhibit multiple slopes for different gate bias voltage between 1V and 3V as marked with dash dot lines in Fig. 7(a). These different slopes result from the dynamic charging process of the traps featuring continuous energies in AlGaN barrier while the Fermi level moves toward the conduction band with the increasing gate voltage shown in Fig. 7(c). Meanwhile due to the pronounced electron tunneling, the overall I_G exhibits a discernible increases as marked in Fig. 7(b), which indicates that the energetically distributed traps take part in the multiple transitions (trap filling) in the SCLC process. Further increase V_G ($\geq$3 V), the uniform slope of I_G-T curves suggest the Fermi level pinning as a result of the populated high 2DEG density. Schematic cross section of the device and energy band diagram shown in Fig. 7(c) illustrates that the enhanced electron-tunneling can be triggered by the reformed 2DEG beneath the gate region with the increased positive gate bias. Accompanied with electron-tunneling is trap-filling in AlGaN barrier. Hence the occupied traps are negatively charged which depletes the 2DEG and causes the positive V_{TH} shift as observed in Fig. 2(a).

As show in Fig.8(a), the gate leakage I_G can be fitted by $I_G = I_0 \exp(-\dfrac{Ea}{kT}) + I_1$ for each V_G, in which the first term (e.g. $I_0 \exp(-\dfrac{Ea}{kT})$) complies with the dynamic process of electron emission and captured by the traps as observed in Reg. I & II. However, the second term I_1 increases exponentially with gate bias ($I_1 =\exp(\eta V_G)$) as shown in the inset of Fig. 8(a). Hence, it can be inferred that another conduction mechanism involves in the gate leakage at $V_G >$5.5V. It is well known that the p-GaN/AlGaN junction will turn on at high positive gate bias. Then the holes in in the p-GaN will be activated and injected

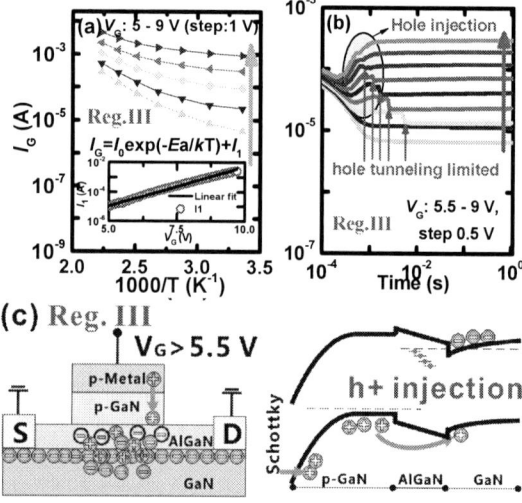

FIG. 8. (a) Arrhenius plots of I_G in Reg.III. The gate leakage can be fitted by $I_G = I_0 \exp(-\dfrac{Ea}{kT}) + I_1$, where $I_1 =\exp(\eta V_G)$ increases exponentially with gate bias as shown in the inset. (b) Time resolved I_G characteristics for $V_G >$ 5.5 V at RT. (c) The band diagram of the gate region to illustrate hole ejecting from the metal/p-GaN, over the P-GaN/AlGaN junction and then injecting into the channel.

into the channel of AlGaN/GaN hetero-structure as depicted in Fig. 8(c).The hole injection process is identified by the humps in the time dependent gate leakage curves in Fig. 8(b). The subsequent drop edge as indicated in Fig. 8(b) originates from the diminished hole injection that is limited by the metal/p-GaN barrier. Because of the effective hole-injection, the V_{TH} of the device exhibits a negative shift as observed in Fig. 2 (b).

978-1-5386-2928-4/18 $31.00 © 2018 IEEE

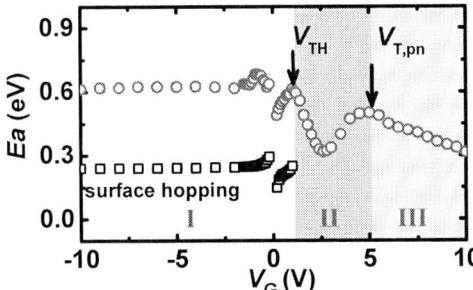

Fig.9. Thermal activation energy (Ea) as a function of gate voltage extracted from the Arrhenius plot of gate leakage.

Fig.9. plots the thermal activation energy (Ea) as a function of gate voltage extracted from the Arrhenius plot of gate leakage. For negative gate voltage, the dominate thermal activation process attributes to the detrapping/trapping of deep-level acceptor traps with Ea of 0.6 eV (T<125 °C) and surface hopping conduction with Ea of 0.25eV (T>125 °C). While for positive gate voltage, the activation energy firstly reaches to peak value at the threshold voltage for the first time as marked with "V_{TH}" in Fig.9. The trapping center moves deeper in the AlGaN bandgap accompanied with the larger positive gate voltage. Then the Femi level starts to move up to the conduction band at the threshold voltage. Accompanied with the upward movement of the Femi level is the reduction of the trapping energy level(i.e. the energy separation between the Fermi level and the conduction band). Then activation energy increases again and reaches to the second peak (marked with "$V_{T,pn}$" in Fig. 9) as hole-injection gets the edge on electron-trapping at the turn on voltage of the gate p-GaN/AlGaN junction.

The effective hole-injection can be also reinforced by the OFF-state leakage current after 5 ms positive gate stress as shown in Fig. 10(a). The OFF-state drain leakage current was measured right after a positive gate stress (e.g. V_{GSQ}=0~9 V) for 5 ms. A significantly increase of the OFF-state drain leakage emerges immediately after 5-ms positive gate stress with V_{GSQ} > 6 V shown in fig.10 (a) , which is accordance in time with hole-injection process shown in fig.8(b). These humps can be clearly attributed to the enhanced electron-hole recombination. After that the injected holes were sufficiently consumed and the drain leakage current reduced close to the current level measured for V_{GSQ} < 6V, which is also supported by the overlap between the second sweeping of OFF-state drain leakage after V_{GSQ}=7 V and the initial sweeping without stress as shown in fig.10 (b).

IV. CONCLUSION

In conclusion, the mechanisms of district V_{TH} stabilities in 650 V pGaN/AlGaN/GaN HEMTs have been investigated. A bidirectional shift of the V_{TH} was observed as the DUTs subjected to forward gate stress. The critical voltage for the V_{TH} turned from positive shift into negative shift was found to be +6 V, which is close to the turn on voltage of the gate pn junction. A physical model of electron-trapping followed by

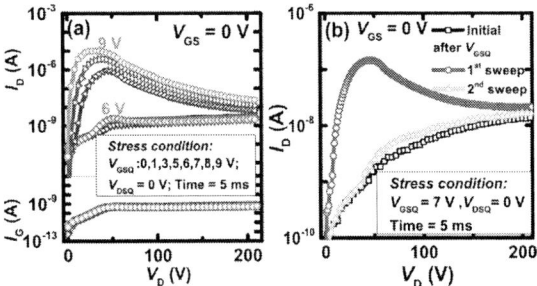

Fig.10. (a) The I_D-V_D and I_G-V_D characteristics measured in OFF-state (V_{GS} = 0 V) immediately after a 5 ms positive gate stresses with V_{GSQ} =0~ 9 V and V_{DSQ} = 0 V. (b) The I_D-V_D characteristics measured in OFF-state (V_{GS} = 0 V) in initial state, 1st and 2nd sweeping after a 5 ms positive gate stress with V_{GSQ} = 7 V.

hole-injection responsible for the bidirectional shift in V_{TH} was proposed. At small positive gate bias, the positive shift in V_{TH} is dominated by electron filling of acceptor traps in AlGaN barrier while the gate leakage is accordingly dominated by *trap-dominated SCLC* mechanism. At higher positive gate bias (V_G>6 V), hole-injection is observed that induces a negative shift in V_{TH} and the gate leakage exhibits a substantial increase due to the forward turn on of the gate p-GaN/AlGaN junction. Besides, a significant increase in OFF-state drain leakage was also observed after hole-injection, which is believed to be the pronounced electron-hole recombination in the channel.

REFERENCES

[1] U. K. Mishra, L. Shen, T. E. Kazior, Y-F. Wu, "GaN-Based RF Power Devices and Amplifiers", Proc. IEEE, vol. 96, no.2, pp.287-305, 2008,

[2] M. Su, C. Chen, S. Rajan, "Prospects for the application of GaN power devices in hybrid electric vehicle drive systems" Semicond. Sci. Technol. vol.28, no.7, pp.074012, 2013

[3] M.J. Scott *et.al.,* "Merits of gallium nitride based power conversion", Semicond. Sci. Technol. vol.28,no.7, pp.074013, 2013.

[4] T. Oka and T. Nozawa, "AlGaN/GaN Recessed MIS-Gate HFET With High-Threshold-Voltage Normally-Off Operation for Power Electronics Applications", IEEE Electron Device Lett., vol 29, no.9, pp.668, 2008.

[5] Q.Zhou *et.al.,* "7.6 V Threshold Voltage High-Performance Normally-Off Al2O3/GaN MOSFET Achieved by Interface Charge Engineering" IEEE Electron Device Lett., vol 37, no.2,pp.165-168, 2016.

[6] Y.Cai, Y. Zhou, K. J. Chen, and K. M. Lau,"High-Performance Enhancement-Mode AlGaN/GaN HEMTs Using Fluoride-Based Plasma Treatment", IEEE Electron Device Lett., vol 26, no.7, pp.435-437,2005.

[7] Y. Uemoto *et.al.,* "Gate Injection Transistor (GIT)—A Normally-Off AlGaN/GaN Power Transistor Using Conductivity Modulation",IEEE Trans. Electron Devices,vol.54,no.12, pp.3393, 2007.

[8] I. Hwang *et.al.,* "p-GaN Gate HEMTs With Tungsten Gate Metal for High Threshold Voltage and Low Gate Current",IEEE Electron Device Lett., vol.34,no.2,pp.202-204, 2013.

[9] EPC, http://epc-co.com/epc; Panasonic, http://www.mouser.hk; GaN Systems, http://www.gansystems.com

[10] F.-C. Chiu, H.-W. Chou, and J. Y. Lee, "Electrical conduction mechanisms of metal/La2O3/Si structure" J. Appl. Phys., vol.97, pp.103503 (2005).

[11] P. W. M. Blom, M. J. M. de Jong, and J. J. M. Vleggaar, "Electron and hole transport in poly(p-phenylene vinylene) devices" Appl. Phys. Lett., Vol 68, 3308 (1996).

[12] S. Arulkumaran *et.al.,,* "Temperature dependence of gate–leakage current in AlGaN/GaN high-electron-mobility transistors" Appl. Phys. Lett. vol.**82**,no.18, pp. 3110 (2003).

Proceedings of the 30th International Symposium on Power Semiconductor Devices & ICs
May 13-17, 2018, Chicago, USA

GaN-on-Si Lateral Power Devices with Symmetric Vertical Leakage: The Impact of Floating Substrate

Hanyuan Zhang, Shu Yang[*], Kuang Sheng

College of Electrical Engineering, Zhejiang University, Hangzhou 310027, China
[*]Email: eesyang@zju.edu.cn

Abstract—In a monolithically integrated GaN-on-Si chip with multiple devices, the conductive Si substrate is shared by all the devices. How to properly terminate the common Si substrate is of critical significance for the voltage blocking and dynamic performance. In this paper, the impact of the floating substrate on the current collapse of the lateral GaN-on-Si devices featuring *symmetric* vertical leakage has been investigated, which facilitates future evaluation of the dynamic performance of a monolithically integrated half-bridge. The floating substrate can provide enhanced breakdown voltage, alleviated buffer-related dynamic R_{ON} degradation, and can possibly suppress the cross-talk particularly for the devices with symmetric vertical leakage.

Keywords—GaN-on-Si, symmetric vertical leakage, buffer trapping, current collapse, floating substrate

I. INTRODUCTION

GaN-on-Si platform is desirable to implement lateral power devices, because of the large-diameter wafer and CMOS-compatible process which lead to low material and manufacturing cost. In particular, monolithically integrated GaN-on-Si half-bridge power module and digital ICs have been demonstrated recently [1, 2]. The monolithically integrated GaN power module can lead to suppressed parasitic inductance, reduced chip size, and enhanced flexibility of the circuit design [3]. In a monolithically integrated GaN-on-Si chip with multiple devices, the conductive Si substrate is shared by all the devices though they are laterally isolated on the surface. How to properly terminate the common Si substrate is of critical significance for optimal breakdown voltage (*BV*) and dynamic performance. It is found that floating substrate can deliver lower dynamic R_{ON} under high drain bias condition but slightly higher dynamic R_{ON} at modest drain bias in lateral GaN-on-Si devices with *asymmetric* vertical top-to-substrate leakage and buffer trapping [4]. However, the substrate cross-talk could also be a concern with floating substrate [5].

In this paper, for GaN-on-Si power transistors featuring *symmetric* vertical leakage, we investigate the floating substrate potential during the switching process, and reveal its impact on the buffer trapping at the source and drain sides, which is valuable for the dynamic performance assessment for the upper/lower transistors in a half-bridge circuit.

II. *BV* ENHANCEMENT

The AlGaN/GaN-on-Si epitaxial structure studied in this work features 25-nm thick $Al_{0.25}Ga_{0.75}N$ barrier layer and 4.2-μm GaN buffer stack grown on a low-resistivity of P-type (111) Si substrate. The D-mode MIS-HEMT was fabricated using a standard process with SiN_x passivation.

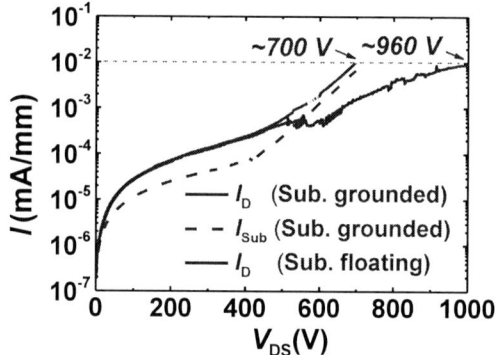

Fig. 1. OFF-state leakage characteristics of the D-mode MIS-HEMT with $L_{GS}/L_G/L_{GD}$ = 2/2/15 μm with substrate grounded and floating, respectively. The top-to-substrate vertical leakage current (I_{Sub}) with substrate grounded is also shown for reference.

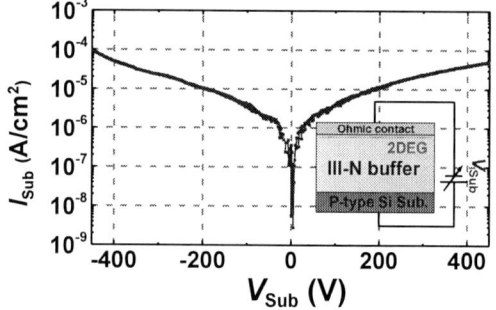

Fig. 2. Vertical *I-V* characteristics measured between a stand-alone ohmic contact and substrate.

978-1-5386-2928-4/18 $31.00 © 2018 IEEE

Fig. 1 shows the OFF-state *I-V* characteristics of the D-mode MIS-HEMT with the substrate grounded and floating, respectively. A floating substrate can yield an enhanced *BV* of ~960 V, in comparison with *BV* ~ 700 V with grounded substrate. When substrate is grounded, the vertical leakage (I_{Sub}) from drain to substrate starts to dominate in the OFF-state leakage at high bias (>500 V). On the other hand, a floating substrate can reduce the vertical electric field, leading to suppressed vertical leakage and enhanced *BV*.

The vertical leakage between a stand-alone top ohmic contact and the substrate under negative and positive substrate biases (V_{Sub}) is shown in Fig. 2. The GaN-on-Si epi-structure and devices studied in this work exhibit nearly *symmetric* top-to-substrate vertical leakage current with opposite substrate bias polarities.

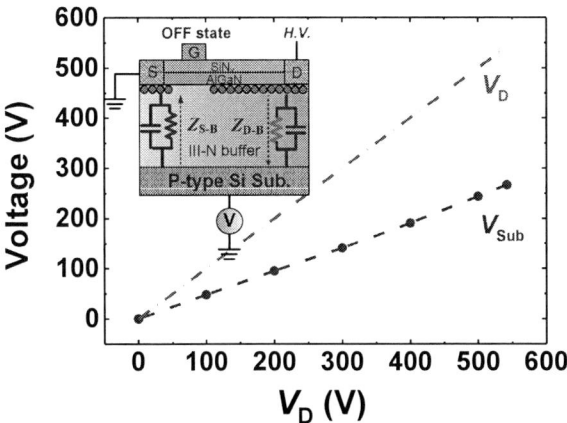

Fig. 3. The induced potential of the floating substrate as a function of the high voltage (V_D) applied to the top ohmic contact. The inset shows the measurement setup.

As shown in Fig. 3, when a high voltage V_D (up to 500 V) was applied to the drain, a positive voltage around $V_D/2$ was induced in the floating substrate, suggesting nearly equal source-to-substrate (Z_{S-B}) and drain-to-substrate (Z_{D-B}) impedance (Fig. 3 inset). The reduced vertical electric field between the top terminal and substrate with floating substrate is also consistent with the suppressed OFF-state leakage in Fig. 1.

III. CROSS-TALK EFFECT

Fig. 4(a) shows the schematic cross section of the test structure for the evaluation of the cross-talk effect [5]. An ohmic contact pad (P), which was fabricated in the adjacent region and laterally isolated from the device under test (DUT), was used to mimic the high voltage (*H.V.*) node. The isolation distance between the DUT and P node is 500 μm, which is expected to be large enough to guarantee negligible lateral leakage current through the buffer stack under *H.V.* stress. In the cross-talk evaluation test, an *H.V.* pulse of 5 s was applied to the P node, whereas the DUT was biased in linear region at V_{GS}/V_{DS} = 0 V/1 V. The induced potential of the floating substrate, as well as the drain current (I_{DS}) of the DUT was monitored.

As shown in Fig. 4(b), when applying an *H.V.* pulse to the P node, a positive voltage ~*H.V.*/2 was induced in the floating substrate. When the *H.V.* stress was withdrawn, the potential of the floating substrate can return to zero. In this way, the cross-talk issue [5] could be possibly suppressed, as a result of symmetric vertical leakage and identical impedance with opposite top-to-substrate bias polarities. By comparison, for GaN-on-Si devices with asymmetrical vertical leakage characteristics, a negative bias could be induced in the floating substrate after the *H.V.* stress was withdrawn, and thus, the negative back-gating effect [7, 8] could result in more severe current collapse.

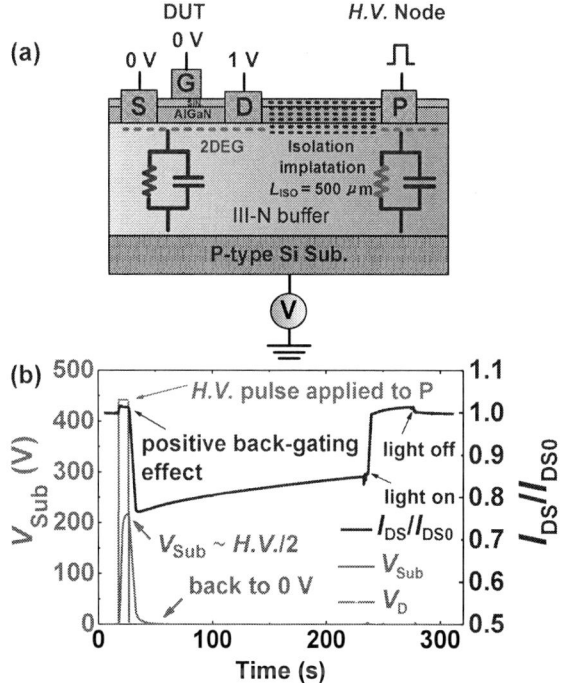

Fig. 4. (a) Schematic cross section of the test structure for cross-talk effect evaluation. The shared substrate is floating. (b) Time-resolved I_{DS} of DUT and the induced potential of the floating substrate with an *H.V.* pulse applied to the P node and then switched back to 0 V.

I_{DS} was normalized with respect to the initial current value (I_{DS0}) without *H.V.* stress. During the *H.V.* stress, the induced positive V_{Sub} and the positive back-gating effect tends to increase the 2DEG density, such that I_{DS} is slightly increased. In the meantime, the induced positive V_{Sub} (or negative top-to-substrate bias, $-V_{Top-to-Sub}$) could cause electron injection from the top ohmic contact or 2DEG, and negative charge trapping in the buffer stack [8]. After the *H.V.* stress was withdrawn, I_{DS} decreases by 25% compared to the initial value, which is due to the negative space charges stored in the buffer stack that can not be released immediately. Afterwards, I_{DS} shows a slow recovery, which is limited by the relatively long emission time constant of the buffer traps as well as the high resistivity of the buffer stack [8]. It is also found that white light illumination can significantly accelerate the I_{DS} recovery.

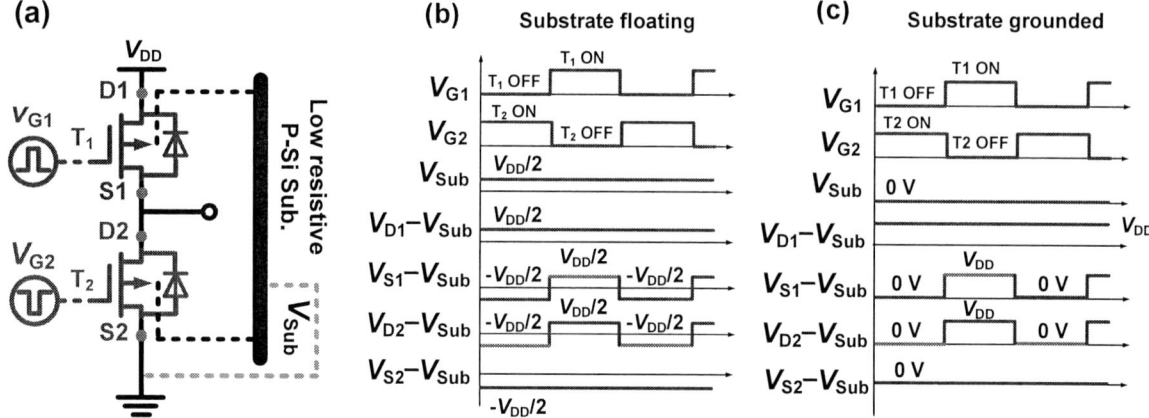

Fig. 5. (a) Schematic diagram of a monolithically integrated half-bridge circuit. V_{DD} is the bus voltage. T_1 and T_2 share the common substrate which is floating or grounded. (b) and (c) show the vertical voltage stress conditions at drain and source terminals of T_1 and T_2 when the substrate is floating or grounded, respectively.

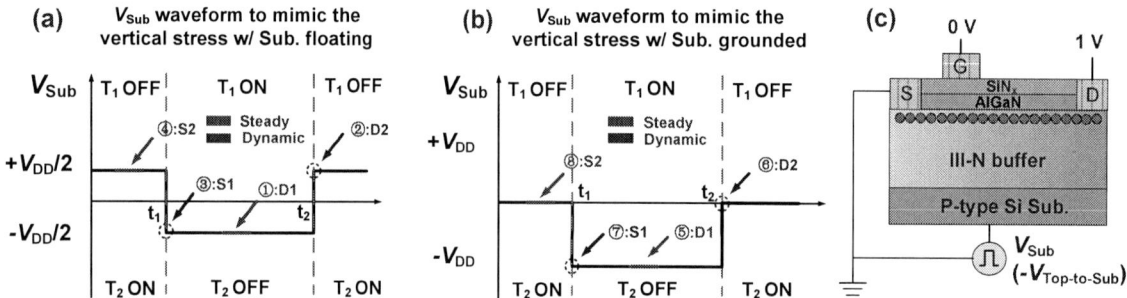

Fig. 6. (a) and (b) are the voltage stress waveforms applied to the substrate to mimic the vertical voltage stress with substrate floating and grounded, respectively. (c) Schematic cross section of the test setup of transient back-gating measurement. $V_{DS} = 1$ V and $V_{GS} = 0$ V.

IV. BUFFER-INDUCED CURRENT COLLAPSE

In a monolithically integrated half-bridge configuration, the switching process can induce distinct top-to-substrate vertical bias stresses ($V_{Top-to-Sub}$) at the source/drain sides for the upper/lower transistors (T_1/T_2) with substrate floating and grounded, as shown in Fig. 5. According to the result shown in Fig. 3, the induced potential of the floating substrate is around $V_{DD}/2$ for the devices with symmetric I_{Sub}. Accordingly, the effective $V_{Top-to-Sub}$ for the drain/source terminals of both upper and lower transistors during high-voltage switching can be determined. After identifying the $V_{Top-to-Sub}$ stress conditions at critical terminals, the impact of $V_{Top-to-Sub}$ on the current collapse for the upper/lower transistors has been evaluated separately with transient back-gating characterizations [8] by applying equivalent V_{Sub} to the Si substrate.

Fig. 6(a) and (b) show the equivalent V_{Sub} waveforms for the critical terminals, which were applied to the substrate to mimic the vertical voltage stress conditions during high-voltage switching. The setup of the transient back-gating measurement is shown in Fig. 6(c), where a small V_{DS} of 1 V was used to minimize the surface trapping effect. During the switching process of a half-bridge configuration, the potentials of D1 and S2 with respect to that of the substrate are constant,

therefore, the dynamic performance of these two terminals are characterized at steady state. On the other hand, S1 and D2 terminals are subjected to varied vertical electric stress during the switching process, and their dynamic performance (or current degradation) was characterized right after switching from OFF-state to ON-state (t_1 and t_2). In the transient back-gating measurement, the equivalent V_{Sub} waveforms corresponding to OFF-state and ON-state voltage stress last for 500 s, respectively.

I_{DS0}/I_{DS} is used to represent the current degradation (or dynamic performance) at various critical terminals. Fig. 7 summarizes the current degradation at critical terminals with substrate floating and grounded, respectively. Compared with the grounded substrate, floating substrate delivers alleviated current collapse up to 500 V, particularly for the drain terminal of the upper switch (D1), which is mainly because of the reduced vertical bias stress. The positive $V_{Top-to-Sub}$ could cause electron injection from the Si substrate and charge trapping in the buffer stack, which could partially deplete the 2DEG.

For the lower switch T_2, floating substrate also leads to better dynamic performance compared with the grounded substrate termination. the mechanisms are illustrated in Fig. 8. For the drain terminal of the lower switch (D2), the positive

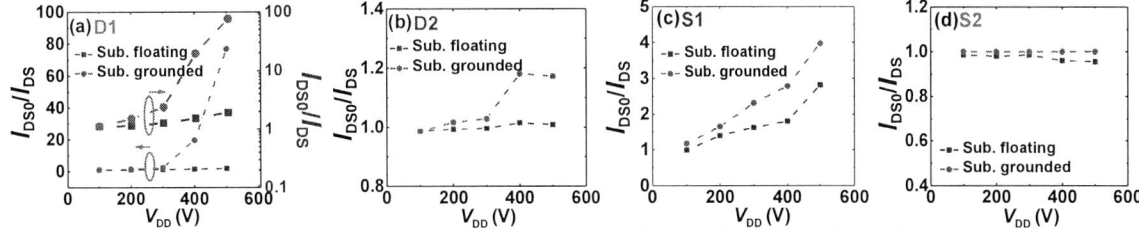

Fig. 7. Current degradation (I_{DS0}/I_{DS}) at (a) D1, (b) D2, (c) S1 and (d) S2 between substrate floating and grounded conditions with varying V_{DD}.

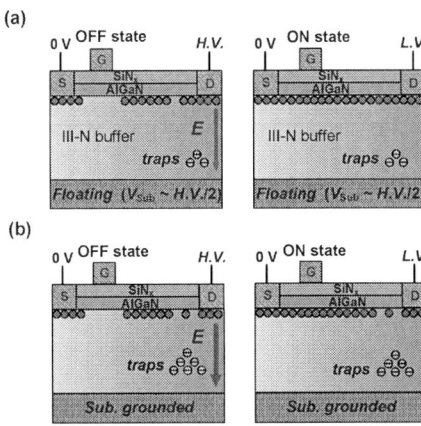

Fig. 8. Schematics showing the buffer trapping in T_2 with substrate (a) floating and (b) grounded, respectively. In a half-bridge circuit with floating substrate, when T_1 is off and T_2 is on, a positive voltage (~$H.V./2$ for GaN-on-Si devices with symmetric I_{Sub}) is induced in the substrate. Such positive back-gating effect tends to enhance the 2DEG density or compensate the adverse influence of the buffer trapping. By comparison, when substrate is grounded, the stored charges in the buffer stack, as a result of the higher vertical electric field during OFF-state, could cause more severe current collapse.

$V_{Top-to-Sub}$ stress at OFF-state with floating substrate is approximately 50% lower than that of the grounded substrate termination, and consequently suppress electron injection from the Si substrate and buffer trapping. Moreover, at ON-state, the floating substrate termination can induce positive potential in the substrate, and such positive back-gating effect could enhance the 2DEG density or compensate the adverse influence of the buffer trapping. Compared with the grounded substrate, the improvement in dynamic performance with floating substrate is more significant at higher voltage.

In summary, the impact of the floating substrate on dynamic performance is twofold. The reduced $V_{Top-to-Sub}$ stress at OFF-state, as well as the presence of a positive back-gate bias during ON-state [8], enables improved dynamic performance.

V. CONCLUSIONS

In conclusion, the impact of the floating substrate on the current collapse of the lateral GaN-on-Si devices featuring *symmetric* vertical leakage has been investigated, which facilitates future evaluation of the dynamic performance of a monolithically integrated half-bridge. The floating substrate

can provide enhanced *BV*, and can possibly suppress the cross-talk particularly for the devices with symmetric vertical leakage. The buffer-related current collapse of the upper/lower transistors in a half-bridge circuit have been studied using transient back-gating measurements. It is shown that the floating substrate termination can alleviate the buffer trapping for both upper and lower switches particularly at higher bus voltage, as a result of the reduced vertical electric field at OFF-state and positive potential induced in the floating substrate at ON-state.

ACKNOWLEDGMENT

This work was supported in part by grants from the Power Electronics Science and Education Development Program of Delta Environmental & Educational Foundation and the Fundamental Research Funds for the Central Universities.

REFERENCES

[1] G. Tang, A. M. H. Kwan, R. K. Y. Wong, J. Lei, R. Y. Su, F. W. Yao, Y. M. Lin, J. L. Yu, T. Tsai, and H. C. Tuan, "Digital integrated circuits on an E-mode GaN power HEMT platform," *IEEE Electron Device Lett*, vol 38. no. 9, pp. 1282-1285, Jul. 2017.

[2] B. Weiss, R. Reiner, P. Waltereit, R. Quay, O. Ambacher, A. Sepahvand, and D. Maksimovic, "Soft-switching 3 MHz converter based on monolithically integrated half-bridge GaN-chip,", in *IEEE Workshop on Wide Bandgap Power Devices and Applications (WiPDA)*, Nov. 2016, pp. 215-219.

[3] D. Reusch, J. Strydom, and J. Glaser, "Improving high grequency DC-DC converter performance with monolithic half bridge GaN ICs," in *IEEE Energy Conversion Congress and Exposition (ECCE)*, Sep. 2015, pp. 381-387.

[4] G. Tang, J. Wei, Z. Zhang, X. Tang, M. Hua, H. Wang, and K. J. Chen, "Dynamic R_{ON} of GaN-on-Si lateral power devices with a floating substrate termination," *IEEE Electron Device Lett.*, vol. 38, no. 7, pp. 937-940, Jul. 2017.

[5] Q. Jiang, Z. Tang, C. Zhou, S. Yang, and K. J. Chen, "Substrate-coupled cross-talk effects on an AlGaN/GaN-on-Si smart Power IC platform," *IEEE Trans. Electron Devices*, vol. 61, no. 11, pp. 3808-3813, Nov. 2014.

[6] X. Li, M. V. Hove, M. Zhao, K. Geens, V. P. Lempinen, J. Sormunen, G. Groeseneken, and S. Decoutere, "200 V enhancement-mode p-GaN HEMTs fabricated on 200 mm GaN-on-SOI with trench isolation for monolithic integration," *IEEE Electron Device Lett.*, vol. 38, no. 7, pp. 918-921, Jul. 2017.

[7] P. Moens, P. Vanmeerbeek, A. Banerjee, J. Guo, C. Liu, P. Coppens, A. Salih, M. Tack, M. Caesar, and M. Uren, "On the impact of carbon-doping on the dynamic R_{ON} and OFF-state leakage current of 650V GaN power devices," in *Proc. Int. Symp. Power Semiconductor Devices IC's (ISPSD)*, May 2015, pp. 37-40.

[8] S. Yang, C. Zhou, S. Han, J Wei, K. Sheng, and K. J. Chen, "Impact of substrate bias polarity on buffer-related current collapse in AlGaN/GaN-on-Si power devices" *IEEE Trans. Electron Devices*, vol. 64, no. 12, pp. 5048-5056, Nov. 2017.

Proceedings of the 30th International Symposium on Power Semiconductor Devices & ICs
May 13-17, 2018, Chicago, USA

Short circuit robustness analysis of new generation Enhancement-mode p-GaN power HEMTs

M. Riccio, G. Romano, L. Maresca, G. Breglio, A. Irace
Dept. of Electrical Engineering and Information Technologies
University of Naples Federico II
Naples Italy
Email: michele.riccio@unina.it

G. Longobardi
Dept. of Engineering
Cambridge University
Cambridge CB30FA, U.K
Email: gl315@cam.ac.uk

Abstract — **This work discusses about the short-circuit capability of new generation Enhancement-mode p-GaN power HEMTs. The electrothermal behavior of two commercially available devices is experimentally verified up to the failure events. Mission profile compact thermal simulations are used to estimate temperature increase during short-circuit tests. The assumption of a temperature-dependent gate current is then investigated by means of 2D electro-thermal TCAD simulations on a reference structure. Finally, a possible trade-off between gate driver resistance and short-circuit capability is discussed.**

Keywords—component; Electro-thermal, failure analysis, gallium nitride, short-circuit, HEMT.

I. INTRODUCTION

An increasing number of device solutions for improving the reliability of GaN transistors have been proposed in literature. However, to be adopted in applications such as motor drive, where devices are subject to harsh electro-thermal (ET) conditions, the Short-Circuit (SC) robustness must be ensured. Nowadays, although the even increasing number of commercially available GaN HEMTs, the SC robustness was not addressed for all the emerging technology solutions. In the last years few papers analyzed the SC capability of GaN-based HEMTs. In 2013, a study on a previous generation 200 V Enhancement-mode (E-mode) HFET showed a threshold voltage (V_{TH}) reduction at high temperature and a potential unstable ET behavior during SC conditions [1]. The SC-SOA of a 650V D-mode HEMT was discussed in 2014, asserting drastic low SC withstand time above 250V [2]. In 2016, a 5µs SC withstand capability was proved for a 600 V E-mode device in cascode configuration [3]. In 2017, two papers analyzed the SC performances of E-mode p-GaN Gate Injunction Transistor (GIT), a current-driven technology, reporting failure events depending on the gate current and for $V_{DC} > 350$ V [4], [5]. All these results are strictly related to the specific DUT technology. This work analyses and discusses the SC capability of a commercial, voltage-driven, 650V p-GaN HEMT with extremely low gate leakage current. The performed analysis consists of a broad set of SC measurements at different electrical conditions (DC supply V_{TEST} and gate-driver resistance R_G). To gain a deep inside understanding of the phenomena involved during SC tests numerical analysis was also performed. To quantify the temperature increase during the SC event, SPICE mission-profile thermal simulations were

carried-out using a compact thermal network provided by the manufacturer [6]. Finally, preliminary 2D electro-thermal TCAD simulations were used to verify the impact of high temperature on the device gate current.

II. ANALYSIS APPROACH

Experiments are conducted on two commercial devices from GaN Systems with breakdown voltage $BV_{DS}=650$V and different on-state resistance (R_{ON}): namely DEV1 ($R_{ON}=200$ mΩ) and DEV2 ($R_{ON}=50$mΩ). The circuit designed and used for the SC test is shown in Fig. 1. Crucial parameters for the analysis were the SC voltage, V_{TEST}, and the gate driver impedance, R_G. The latter play an important role in applications; in fact R_G is used to tune the turn-on slew-rate to avoid critical voltage and current oscillations during device switching transients. An example of high frequency burst oscillations are shown in Fig. 2. SPICE thermal simulations, with a mission-profile approach, were done using the measured power dissipation within the device during SC tests. In Fig. 3 the equivalent RC thermal network for DEV1 is depicted.

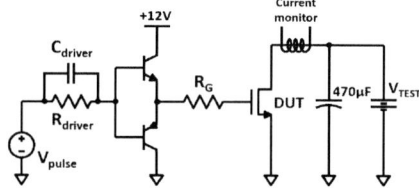

Fig. 1 Circuit used for SC tests. The driving circuit is formed by a bipolar push-pull configuration. The V_{PULSE} amplitude was adjusted to have $V_{GS}=6$V.

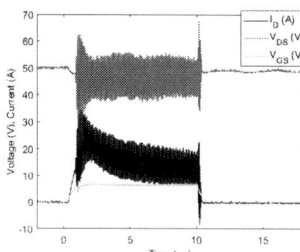

Fig. 2 Measured SC electrical waveforms when high frequency burst oscillations occurs.

978-1-5386-2928-4/18 $31.00 © 2018 IEEE

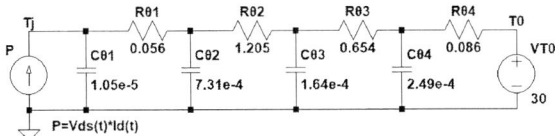

Fig. 3 Cauer RC network for DEV1. Values for R and C elements are extracted from the device datasheet. The current source (P) is a behavioral source which temporal value is obtained by experimental waveforms.

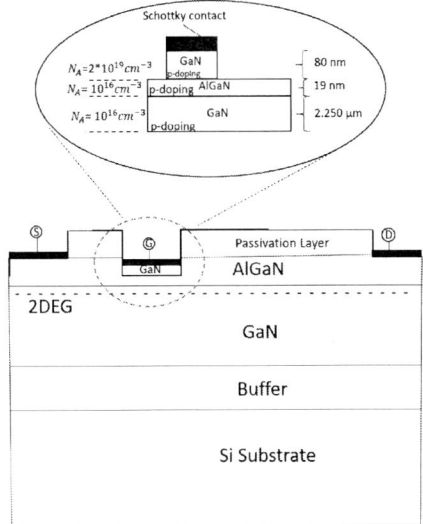

Fig. 4 p-GaN/AlGaN/GaN HEMT device cross-section used for TCAD simulations.

TCAD ET simulations were performed on the structure in Fig. 4. Dimensions and TCAD parameters were extrapolated from a reference work [7]. The device investigated is a lateral three-terminal device with an AlGaN/GaN heterostructure. A buffer layer is used to allow a high quality GaN layer to be grown despite the significant lattice mismatch between GaN and Si. Fixed charges were included in the TCAD simulation deck according to *Ambacher et al.* [8] to consider the piezo-polarization effect observed in GaN devices. A p-type doping of 1×10^{16} cm^{-3} was added in the GaN bottom layer to take into account the carbon doping as reported in the literature. Finally, a thin cap GaN layer was added to form the gate with a Magnesium (Mg) p-type doping density of 2×10^{19} cm^{-3}. This doping value, also reported elsewhere in the literature, matched the experimental results. To simplify the analysis, the p-GaN doping included in the TCAD model is fully ionized.

III. RESULTS DISCUSSION

The measured iso-thermal trans-characteristics (I_D-V_{GS}) and output characteristics (I_D-V_{DS}) for both devices show a V_{TH} almost insensible to temperature up to 425K. For reference, DEV1 curves are reported in Fig. 5. This result proves that these devices are not subject to a positive temperature coefficient for the drain current and a stable electrothermal behavior is expected during SC tests.

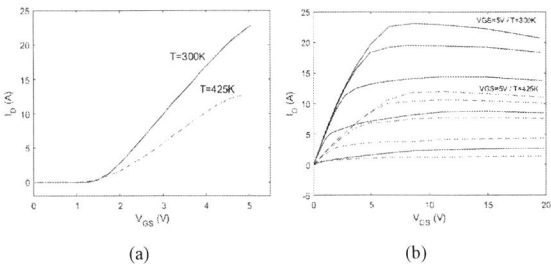

(a)   (b)

Fig. 5 (a) I_D-V_{GS} curve (V_{DS}=10V) at T=300K (solid line) and T=425K (dashed line); (b) I_D-V_{DS} at T=300K (solid lines) and T=425K (dashed lines). V_{GS} linearly spaced from 1.8V to 5V.

A. Short-circuit tests

The impact of V_{TEST} was first analyzed on DEV1 for both low and high gate driver impedance. Fig. 6 reports drain and gate current during SC pulses depending on V_{TEST} ranging from 50V to 260V for the case R_G=46Ω (suggested value in datasheet). Fig. 7 reports drain and gate current for SC tests up to 300V for the case R_G=250Ω. It was found that during the SC pulses a considerable gate current arises, probably due to the fast temperature increase inside the device. In the first case (R_G=46Ω), the device exhibits a gate current up to 9mA, while in the second case (R_G=250Ω) the gate current reaches 6mA at the end of the SC pulse (V_{TEST}=300V). This current, dramatically higher than the value reported in the datasheet (40μA@V_{GS}=6V), imposes a voltage drop across the gate resistance that causes a reduction of the effective V_{GS} applied to the device.

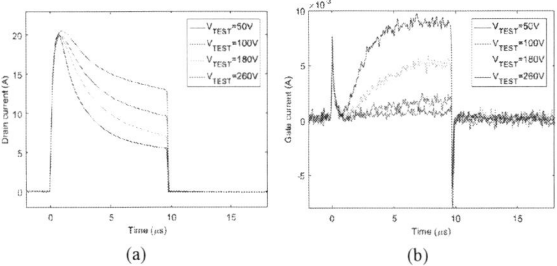

(a)   (b)

Fig. 6 (a) Drain current (a) and gate current (b) during SC pulses at different V_{TEST} (R_G=46Ω). (DEV1).

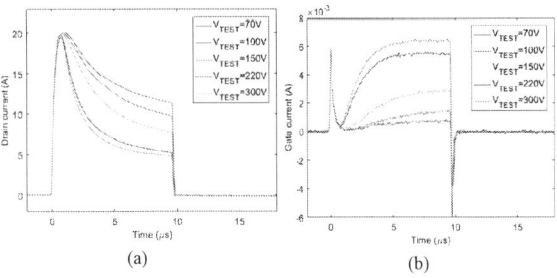

(a)   (b)

Fig. 7 (a) Drain current (a) and gate current (b) during SC pulses at different V_{TEST} (R_G=250Ω). (DEV1).

The combined effect of the self-heating and voltage drop across R_G, due to $I_G(T)$, determine a drastic I_D reduction. Moreover, the experimental results shown in Fig. 6 and Fig. 7 suggest that the SC peak current slightly depend on the V_{TEST}, while remain almost constant for both R_G values. In the case of low driver impedance, despite the voltage drop across the R_G, the device experiences a premature failure for V_{TEST}=300V after a time $t_{failure}\approx$980ns, as shown in Fig. 8. To be sure that the above analysis was not affected by a sensible device ageing effect during repetitive SC tests, the cumulative ET stress was also evaluated after 100 SC pulses. As reported in Fig. 9, output characteristics show a negligible impact on the R_{ON} value that increases of about 8% with respect to the *fresh device*. SC tests were also conducted on DEV2, for two different V_{TEST} values (200V and 450V), varying R_G in a wide range. In Fig. 10 drain current and gate current are shown for the case V_{TEST}=250V. One can note a more pronounced effect of I_D reduction during the short-circuit pulse, while the gate current starts to increase immediately after the device turn-on, reaching about 22mA for the case of low driver impedance (R_G=100Ω). For this device, the datasheet indicates 160µA gate current when no power is dissipated within the device. This large gate current value enables the DEV2 to safely sustain SC events up to V_{TEST}=450V (T_{ON}>10µs). Fig. 11 shows the drain and gate current, depending on the gate resistance, for this test. The gate current reaches 30mA for R_G=100Ω case and the drain current is reduced by more than 70% during the SC pulses.

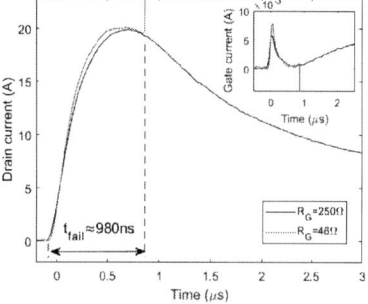

Fig. 8 SC test with V_{TEST}=300V. Failure occurs with low impedance gate driver. (DEV1)

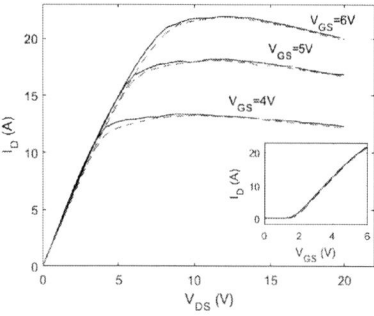

Fig. 9 Ageing effect on the I_D-V_{DS} and I_D-V_{GS} curves. Ron increases of about 8% (DEV1).

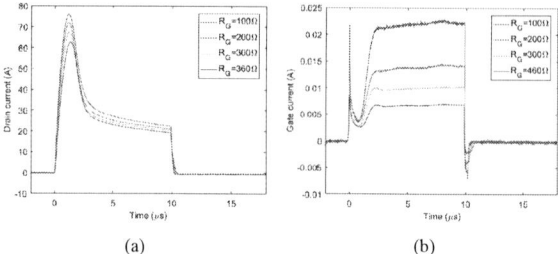

Fig. 10 (a) Drain current (a) and gate current (b) during SC pulses with different R_G (V_{TEST}=200V). (DEV2).

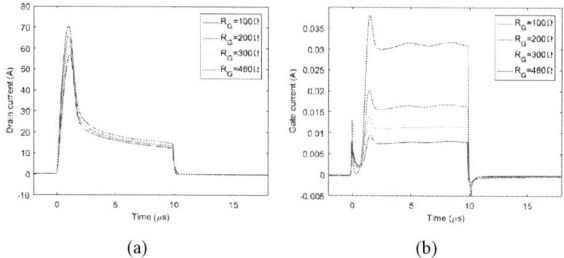

Fig. 11 (a) Drain current (a) and gate current (b) during SC pulses with different R_G (V_{TEST}=450V). (DEV2).

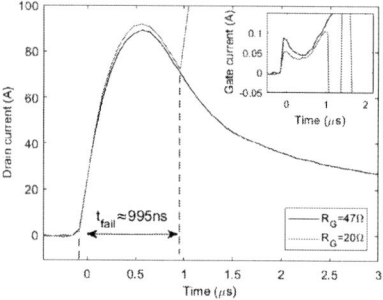

Fig. 12 SC test with V_{TEST}=450V. Failure occurs with low impedance gate driver. (DEV2)

To trigger the failure of DEV2, lower gate resistance values were used. The failure arises for V_{TEST}=400V @ R_G=20Ω after ≈995ns of SC time. The electrical waveforms for the cases R_G=47Ω and R_G=20Ω are reported in Fig. 12. It is visible a weak increase in the drain peak current for the lower gate resistance. The analysis performed suggests that the fast increase of the temperature has a strong impact on the gate current. In conclusion, these experimental results confirm a possible trade-off between R_{G-ON} and SC capability for p-GaN HEMT c technology.

B. SPICE thermal simulation

Using the experimental waveforms acquired during SC tests, thermal simulations are performed to evaluate fast thermal dynamic inside the device. In the following, the results obtained considering low impedance gate driver (R_G=100Ω) is shown for different test voltage. Fig. 13 reports the temperature increase for both devices.

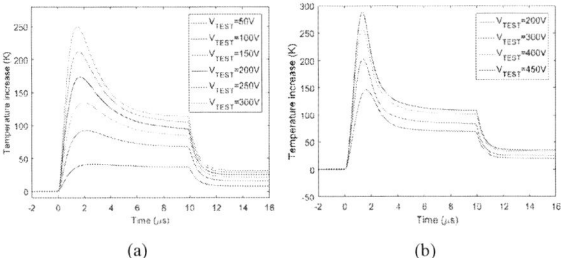

(a)	(b)

Fig. 13 Mission-profile SPICE thermal simulations depending on the test voltage V_{TEST}: case with R_G=100Ω. a) DEV1; b) DEV2.

It is clearly visible that the peak temperature is reached just after 1÷2μs the SC pulse starts. For both DEV1 and DEV2 the estimated temperature increase before the failure is $\Delta T_{max} \approx 250K$. These results also confirm that the ET behavior of p-GaN HEMT during short-circuit is deeply different from the case of Si devices. The HEMT reaches the peak temperature during the first phase of the SC pulse instead of the turn-off phase. This phenomenon is due to the combined effect of drastic drain current reduction during the SC pulse and the low value of the thermal impedance [9].

C. TCAD simulation

2D TCAD analysis was mainly used to evaluate the ET behavior within the device during SC test and support the assumption of a gate current increase due to the fast temperature increase. Fig. 14 shows the temperature distribution during a SC test when the maximum temperature is reached (V_{TEST}=250V). A hot spot is formed close to the gate field plate.

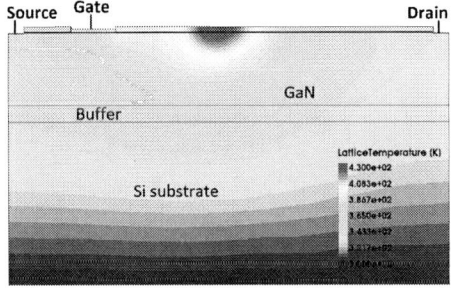

Fig. 14 Temperature distribution during 2D TCAD SC simulation when the maximum temperature is reached (V_{TEST}=250V).

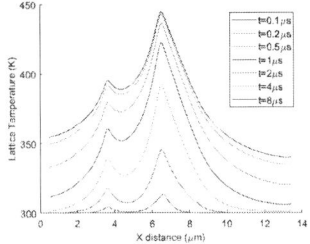

Fig. 15 Temperature profile along x distance at AlGaN/GaN interface at different time instants. (V_{TEST}=250V).

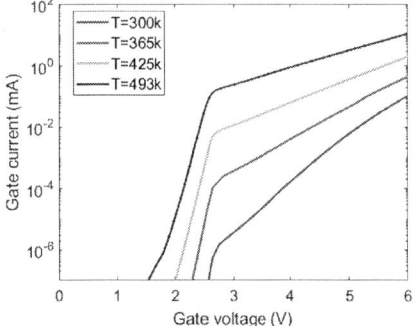

Fig. 16 Simulated I_G-V_{GS} curve for different device steady-state temperature.

Fig. 15 reports temperature profile along x distance at AlGaN/GaN interface at different time instants. It is confirmed the fast temperature dynamic inside the device during the first μs of the SC pulse. Lastly, in Fig. 16 the steady-state gate current for different temperatures was evaluated, confirming a significant increase with the temperature. This is consistent with the explanation of the gate current being formed by carriers reaching the gate electrode via Fowler–Nordheim tunneling and thermionic emission; both strongly temperature dependent phenomena.

IV. CONCLUSION

In this work the short-circuit capability of new generation p-GaN HEMT devices has been analyzed. Experimental and numerical results have reported a strong impact of gate current on the SC withstand time. It has been found that the very fast temperature increase inside the device implies a consistent gate current increase. In turn, this current is responsible of a significant voltage drop across the gate resistance with a following reduction of the effective gate-source voltage V_{GS}. This mechanism act as a negative feedback on the device electrothermal behavior, boosting the SC capability of new p-GaN HEMT structures.

REFERENCES

[1] C. Abbate *et al.*, Microelectronics Reliability, Volume 53, Issues 9–11, 2013, Pages 1481-1485, ISSN 0026-2714.

[2] X. Huang, IEEE International Symposium on Power Semiconductor Devices and ICs, ISPSD 2014, pp. 273-276.

[3] T. Nagahisa *et al.*, Japanese Journal of Applied Physics, 55 04EG01, 2016.

[4] T. Oeder et al., IEEE International Symposium on Power Semiconductor Devices and IC's, ISPSD 2017, pp. 211-214.

[5] M. Fernández *et al.*, IEEE Electron Device Letters, vol. 38, no. 4, pp. 505-508, April 2017.

[6] GN007 Application Note, https://gansystems.com/wp-content/uploads/2018/01/GN007_Modelling-Thermal-Behavior-Using-RC-Thermal-SPICE-Models.pdf

[7] L. Efthymiou *et al.*, On the physical operation and optimization of the p-GaN gate in normally-off GaN HEMT devices, *Applied Physics Letters* 110.12 (2017): 123502.

[8] O.Ambacher *et al.*, J. Appl. Phys. 85,3222 (1999).

[9] M. Riccio *et al.*, Electro-thermal characterization of AlGaN/GaN HEMT on Silicon Microstrip Technology, Microelectronics Reliability, Volume 51, Issues 9–11, 2011, Pages 1725-1729, ISSN 0026-2714.

Proceedings of the 30th International Symposium on Power Semiconductor Devices & ICs
May 13-17, 2018, Chicago, USA

Influence of Doping Profiles and Chip Temperature on Short-Circuit Oscillations of IGBTs

Vera van Treek, Hans-Joachim Schulze, Franz-Josef Niedernostheide,
Christian Sandow, Roman Baburske and Frank Pfirsch

Infineon Technologies AG
Neubiberg, Germany
Vera.vanTreek@infineon.com

Abstract—TCAD simulations of 1200 V IGBT with different *p*-emitter and field-stop doses, and different junction temperatures show that oscillations are likely to occur for short-circuit conditions with relatively low bipolar current gains of the IGBT's collector-side *p-n-p*-transistor and for relative low junction temperatures respectively. Measurement results from different IGBTs are in qualitative agreement with the simulation results. Accordingly, short-circuit oscillations can be controlled by an adjustment of the bipolar current gain by means of an appropriate field-stop and *p*-emitter design.

Keywords—IGBTs, short-circuit oscillations, TCAD simulation

I. Introduction

Short-circuit oscillations (SCOs) can limit the reliability of insulated gate bipolar transistors (IGBTs) [1] and can violate electromagnetic-interference requirements if the oscillation amplitudes are too high and the DC-link-voltage range with SCOs is too broad. Hence, one essential criterion for the optimization of IGBTs is the control of gate-voltage oscillations during short-circuit-operating conditions. For the analysis and the reduction of SCOs, a TCAD-simulation approach is highly desired. This paper presents a two-stage procedure which enables both the analysis of the intrinsic electrical characteristics of IGBTs in different operating points and the analysis of the electrical characteristics of IGBTs during SCOs.

The impact of operating conditions (junction temperature T_J, collector-emitter voltage V_{CE} and gate-emitter voltage V_{GE}) on the occurrence of SCOs was investigated in [2]. However, short-circuit measurements with different IGBTs showed that the IGBT design itself has an impact on SCOs. For similar operating conditions (similar short-circuit currents and collector-emitter voltages), the impact of the cell design is thereby relatively small compared to the impact of the field-stop (FS) and the *p*-emitter design. Different FS and *p*-emitter designs modify the bipolar current gain α_{pnp} of the IGBT's *p-n-p*-transistor part. Therefore, the impact of α_{pnp} on the occurrence of SCOs is discussed in this paper. Compared to previous works [2] - [5], which discussed capacitive effects as cause for SCOs, this work concentrates on the IGBT's back side and the possibility to avoid SCOs with an optimized back-side design.

II. TCAD Simulation of Short-Circuit Oscillations

A. Simulation Approach

For the simulation of high-frequency SCOs, a relatively small simulation-time step (in a range of 0.1 nanoseconds) is necessary because a too high maximum step width results in a too strong damping of oscillations. The simulation of an entire

short-circuit pulse with such a low time step would result in relatively high simulation times and, hence, would limit the screening of SCOs by means of TCAD simulations. Therefore, the following approach was used for the simulation of the investigated 1200 V IGBTs:

- At first, the output characteristic of the considered T_J was quasi-stationary simulated up to the DC-link voltage $V_{DC\text{-}link}$ of interest (Fig. 1). Thereby, the desired gate-driver voltage V_{Dr} was used as V_{GE}. In contrast to the approach in [2], this enabled the analysis of the intrinsic electrical characteristics of the IGBT in the operating point (short-circuit current I_C at $V_{GE} = V_{Dr}$ and $V_{CE} = V_{DC\text{-}link}$). For the last operating points of the output characteristics in Fig. 1, the vertical distributions of the absolute electric-field strength and the charge-carrier densities are shown in Fig. 2 for the middle of the IGBT cell's mesa between 10 µm and 110 µm.[1] Three characteristic regions can be distinguished: a 'quasi-plasma', a space-charge and a plasma region. The high electric-field strength in the FS region is due to a negative effective charge in the drift zone and a necessary but not a sufficient condition for the investigated high-frequency SCOs.

- Secondly, the quasi-stationary simulation result was used as starting point for the isothermal transient simulation of the circuit shown in Fig. 3. Whether the operating point is or is not prone to SCOs was tested by means of a small perturbation of $V_{DC\text{-}link}$ at the beginning of the simulation. If no or only damped oscillations occurred, it was concluded that the operating condition is free of SCOs; if permanent oscillations occurred after a transition time, SCOs were expected during a short-circuit pulse with the applied V_{Dr} and $V_{DC\text{-}link}$. The simulation of time ranges below one microsecond is usually sufficient; near the border of operating conditions with and without SCOs, longer time ranges are necessary. Results of transient simulations of the last operating points of the output characteristics in Fig. 1 are shown in Fig. 4.

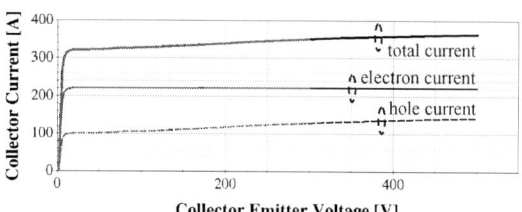

Fig. 1. Output characteristic for $V_{GE} = 15$ V and $T_J = 301$ K of a 1200 V IGBT - simulated up to $V_{CE} = 300$ V (red) and $V_{CE} = 500$ V (black).

---

[1] For the 110 µm thick IGBT structures investigated in this paper, homogenous lateral distributions can be expected in this range.

978-1-5386-2928-4/18 $31.00 © 2018 IEEE

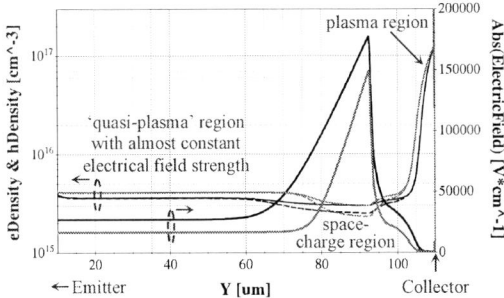

Fig. 2. Vertical distributions of the absolute electric-field strength, the electron (solid) and hole (dashed) density for $V_{CE} = 300$ V (red) and $V_{CE} = 500$ V (black) of the output characteristic in Fig. 1.

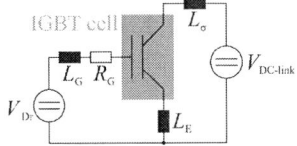

Fig. 3. TCAD-simulation circuit for the transient short-circuit simulations.

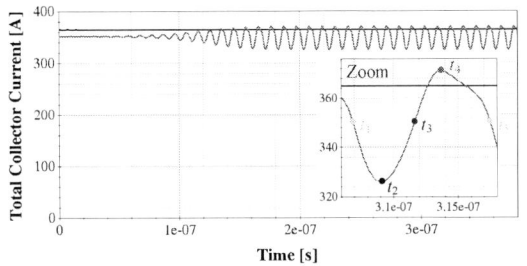

Fig. 4. Transient simulation of the total collector currents with $T_J = 301$ K, $R_G = 0$ Ω, $L_σ = L_G = L_E = 0$ H and $V_{Dr} = 15$ V for $V_{DC\text{-}link} = 300$ V (red) and $V_{DC\text{-}link} = 500$ V (black). The corresponding vertical absolute electric-field strength and charge-carrier-density profiles are shown in Fig. 2.

B. Short-Circuit Oscillation Mechanism

Fig. 5 shows the vertical charge-carrier-density distributions of the five points in time marked in the zoom in Fig. 4 and, as reference, the corresponding quasi-stationary distributions already shown in Fig. 2. According to Fig. 5, electrons and holes are basically stored across the device between t_1 and t_3. Between t_2 and t_3, a charge-carrier-plasma front develops

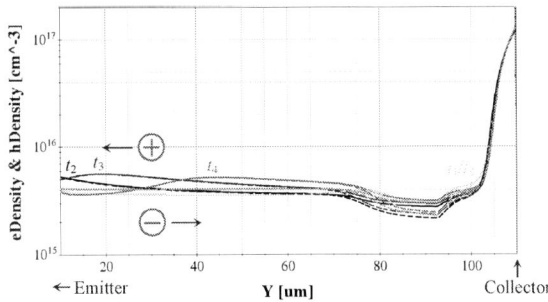

Fig. 5. Comparison of the vertical distributions of the electron (solid) and hole (dashed) density of the points in time in the zoom in Fig. 4 and of the corresponding quasi-stationary distributions (grey).

and propagates approximately until t_5 towards the FS region. Thereby, electrons and holes are released at the front. The transient simulation with $L_σ = L_G = L_E = 0$ H and $R_G = 0$ Ω – and, thus, with a constant electron current at the emitter during the oscillation – verifies that this periodic storage and release of charge-carriers across the device together with the corresponding periodic change of the electric-field-strength distribution is the root cause for the investigated high-frequency SCOs. The reported input capacitance effects in [2] - [5] are solely a result of the charge-carrier storage and release across the device.

III. IMPACT OF THE BIPOLAR CURRENT GAIN ON THE OCCURRENCE OF SHORT-CIRCUIT OSCILLATIONS

As shown in the output characteristics in Fig. 1, the channel current I_{CH} usually saturates at relatively low V_{CE}. The hole current keeps increasing from low to high V_{CE}. This behavior is explained as follows: The expansion of the high-field region at the collector side rises with V_{CE}. As shown in Fig. 2, this results in a smaller plasma region in front of the p-emitter and an increased plasma gradient. Hence, the injected hole current $I_{C,pnp}$ at the collector and $α_{pnp}$ increase. Along an output characteristic, SCOs start at a relatively low V_{CE} and finally disappear with increasing V_{CE} and $α_{pnp}$.

Subsequently, the impact of the p-emitter and FS dose and the impact of T_J on the $V_{DC\text{-}link}$ range with SCOs and the peak-to-peak-collector-current amplitudes are investigated. Thereby, the modification of $α_{pnp}$ and the vertical distributions of the electric-field strength and the charge-carrier densities is regarded. For the $α_{pnp}$ calculation, the following definition is used

$$α_{pnp} = (I_C - I_{CH}) / I_C = I_{C,pnp} / I_C. \quad (1)$$

A. Variation of the p-Emitter Dose

The output characteristics with $V_{GE} = 15$ V and $T_J = 301$ K of different p-emitter doses are shown in Fig. 6. The impact of the emitter dose on the hole injection and $α_{pnp}$ is most distinct up to $V_{CE} = 250$ V and, hence, in a range in which SCOs typically may occur for 1200 V IGBTs. In Fig. 7, the vertical profiles of the electric-field strength and the charge-carrier densities in the drift zone are compared for $V_{CE} = 200$ V. For the raised p-emitter doses, the extent and the maximum level of the remaining plasma in front of the p-emitter is increased; the absolute electric-field strength is slightly increased in the drift zone in front of the FS and decreased in the FS region. The calculated values of $α_{pnp}$ and the peak-to-peak-collector-current amplitudes extracted from the transient simulations are shown in Fig. 8. Both, the $V_{DC\text{-}link}$ range with SCOs and the amplitudes decrease with an increase of the p-emitter dose and $α_{pnp}$.

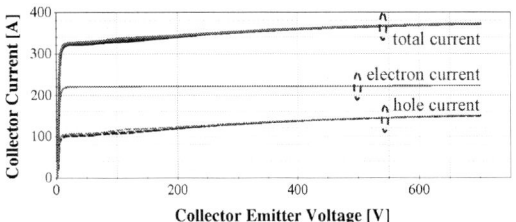

Fig. 6. Output characteristics for $V_{GE} = 15$ V and $T_J = 301$ K for different p-emitter doses *dpem* (*dpem₁* (black), 2·*dpem₁* (blue) and 5·*dpem₁* (red)).

978-1-5386-2928-4/18 $31.00 © 2018 IEEE 109

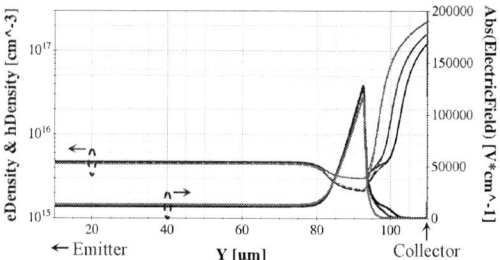

Fig. 7. Vertical distributions of the absolute electric-field strength, the electron (solid) and hole (dashed) density for $V_{CE} = 200$ V, $V_{GE} = 15$ V and $T_J = 301$ K for different p-emitter doses $dpem$ ($dpem_1$ (black), $2 \cdot dpem_1$ (blue) and $5 \cdot dpem_1$ (red)).

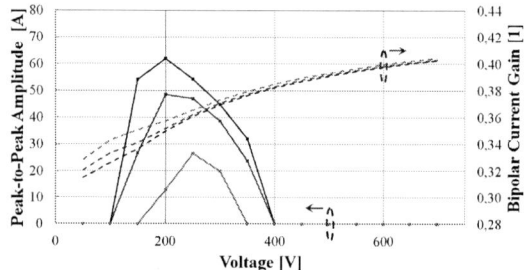

Fig. 8. Calculated α_{pnp} from the quasi-stationary simulation results in Fig. 6 and peak-to-peak-collector-current amplitudes of transient simulations with $V_{DC\text{-link}}$ between 50 V and 700 V, $V_{Dr} = 15$ V and $T_J = 301$ K for different p-emitter doses $dpem$ ($dpem_1$ (black), $2 \cdot dpem_1$ (blue) and $5 \cdot dpem_1$ (red)).

B. Variation of the Field-Stop Dose

The output characteristics with $V_{GE} = 15$ V and $T_J = 301$ K of different FS-stop doses are shown in Fig. 9. The impact of the FS doses on the hole injection and on α_{pnp} becomes significant above $V_{CE} = 50$ V and, hence, in a range in which SCOs typically may occur for 1200 V IGBTs. In Fig. 10, the vertical profiles of the absolute electric-field strength and the charge-carrier densities in the drift zone are compared for $V_{CE} = 200$ V. For the decreased FS doses, the extent of the remaining plasma in front of the p-emitter is reduced; as for the increase of the p-emitter dose, the absolute electric-field strength is slightly increased in the drift zone in front of the FS and decreased in the FS region. The calculated α_{pnp} and the peak-to-peak amplitudes of the collector currents extracted from the transient simulations are presented in Fig. 10. With a decrease of the FS dose and an increase of α_{pnp}, the $V_{DC\text{-link}}$ range with SCOs shifts to lower voltages, and both the $V_{DC\text{-link}}$ range with SCOs and the peak-to-peak-collector-current amplitudes decrease.

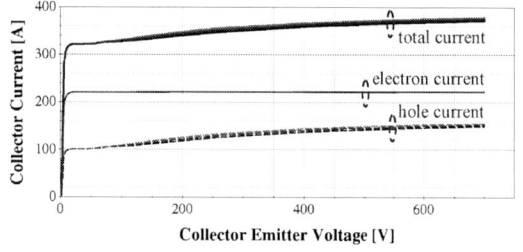

Fig. 9. Output characteristics for $V_{GE} = 15$ V and $T_J = 301$ K for different field-stop doses dfs (dfs_1 (black), $0.8 \cdot dfs_1$ (blue) and $0.6 \cdot dfs_1$ (red)).

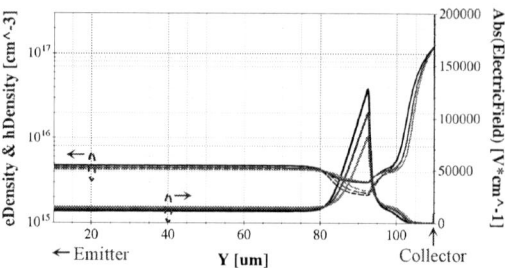

Fig. 10. Vertical distributions of the absolute electric-field strength, the electron (solid) and hole (dashed) density for $V_{CE} = 200$ V, $V_{GE} = 15$ V and $T_J = 301$ K for different field-stop doses dfs (dfs_1 (black), $0.8 \cdot dfs_1$ (blue) and $0.6 \cdot dfs_1$ (red)).

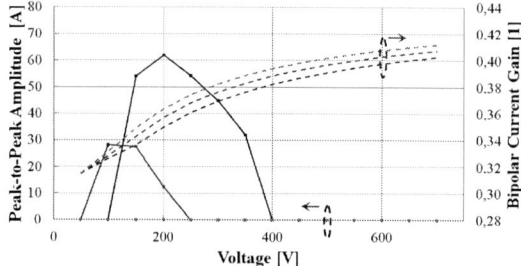

Fig. 11. Calculated α_{pnp} from the quasi-stationary simulation results in Fig. 9 and peak-to-peak-collector-current amplitudes of transient simulations with $V_{DC\text{-link}}$ between 50 V and 700 V, $V_{Dr} = 15$ V and $T_J = 301$ K for different field-stop doses dfs (dfs (dfs_1 (black), $0.8 \cdot dfs_1$ (blue) and $0.6 \cdot dfs_1$ (red)).

C. Variation of the Junction Temperature

The output characteristics with $V_{GE} = 15$ V and different T_J are shown in Fig. 12. The temperature variation modifies both the hole and channel current. In Fig. 13, vertical profiles of the absolute electric-field strength and the charge-carrier densities in the drift zone are compared for $V_{CE} = 200$ V. The extent of the remaining plasma region in front of the p-emitter is reduced for higher junction temperatures; as for the increase of the p-emitter dose and the reduction of the FS doses, the absolute electric-field strength is increased in the drift zone in front of the FS and decreased in the FS region. However, not only the $V_{DC\text{-link}}$ range with SCOs and the amplitudes but also the calculated α_{pnp} at lower V_{CE} decrease with T_J (Fig. 14). This different behavior is most likely caused by the reduction of the charge-carrier mobilities with increasing T_J. For this reason, the increase of the electric-field strength in the drift zone in front of the FS correlates also only for the T_J variation with an increase of the charge-carrier densities.

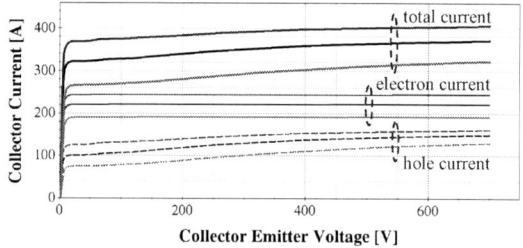

Fig. 12. Output characteristics for $V_{GE} = 15$ V for different junction temperatures T_J ($T_J = 240$ K (blue), $T_J = 301$ K (black) and $T_J = 450$ K (red))

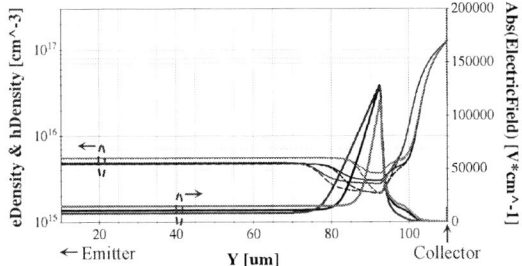

Fig. 13. Vertical distributions of the absolute electric-field strength, the electron (solid) and hole (dashed) density for $V_{CE} = 200$ V and $V_{GE} = 15$ V for different junction temperatures T_J ($T_J = 240$ K (blue), $T_J = 301$ K (black) and $T_J = 450$ K (red)).

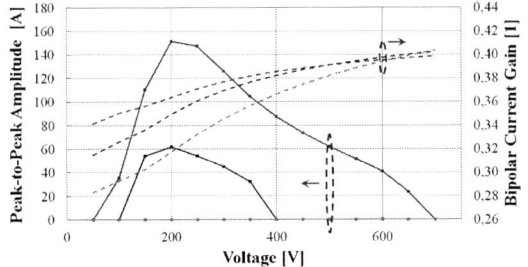

Fig. 14. Calculated α_{pnp} from the quasi-stationary simulation results in Fig. 12 and peak-to-peak-collector-current amplitudes of transient simulations with $V_{DC\text{-link}}$ between 50 V and 700 V and $V_{Dr} = 15$ V for different junction temperatures T_J ($T_J = 240$ K (blue), $T_J = 301$ K (black) and $T_J = 450$ K (red)).

IV. SCO MEASUREMENTS

In Fig. 15, measured short-circuit transients at $T_J = 25$ °C are shown for a 1200 V IGBT with a non-optimized α_{pnp}. Below 200 V and above 450 V, SCOs cannot be observed, but in the $V_{DC\text{-link}}$ range of 300 V to 400 V, SCOs are very pronounced. Experiments at $T_J = 25$ °C with 650 V IGBTs (Fig. 16 and Fig. 17) reveal that the $V_{DC\text{-link}}$ range with SCOs is much broader in case of low p-emitter doses and high FS doses, i.e. if α_{pnp} is low. Typical dependencies of the $V_{DC\text{-link}}$ range with SCOs on the FS and p-emitter dose are shown in TABLE I. Experiments with $T_J = 125$ °C show no SCOs.

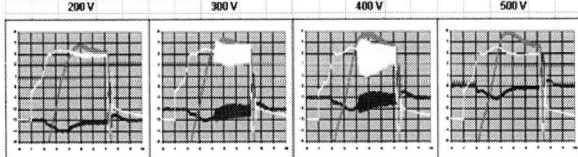

Fig. 15. Short-circuit transients $V_{GE}(t)$ (yellow), $I_C(t)$ (magenta), and $V_{CE}(t)$ (blue) at $V_{Dr} = 15$ V and $T_J = 25$ °C of a non-optimized 1200 V IGBT with SCOs between $V_{DC\text{-link}} = 200$ V and $V_{DC\text{-link}} = 400$ V.

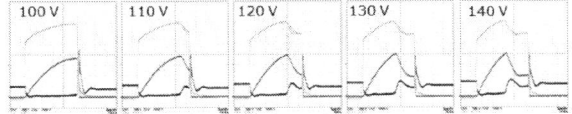

Fig. 16. Short-circuit transients $V_{GE}(t)$ (green), $I_C(t)$ (red), and $V_{CE}(t)$ (blue) at $V_{Dr} = 15$ V and $T_J = 25$ °C of a 650 V IGBT with a relatively low α_{pnp} reveal SCOs between $V_{DC\text{-link}} = 110$ V and $V_{DC\text{-link}} = 130$ V.

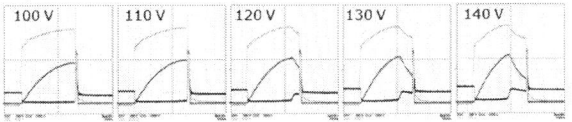

Fig. 17. Short-circuit transients $V_{GE}(t)$ (green), $I_C(t)$ (red), and $V_{CE}(t)$ (blue) at $V_{Dr} = 15$ V and $T_J = 25$ °C of a 650 V IGBT with an optimized α_{pnp} show nearly no SCOs in the entire $V_{DC\text{-link}}$ range.

TABLE I. $V_{DC\text{-LINK}}$ RANGE WITH SCOs OF 650 V IGBTs WITH $V_{Dr} = 15$ V AND $T_J = 25$ °C AS FUNCTION OF THE IMPLANTED FIELD-STOP AND P-EMITTER DOSE

Normalized field-stop dose	Normalized p-emitter dose	$V_{DC\text{-link}}$ range with SCOs
67 %	100 %	15 V
100 %	100 %	43 V
100 %	112 %	30 V

V. CONCLUSIONS

Vertical distributions of the electron and hole density during SCOs revealed that high-frequency SCOs are caused by a periodic storage and release of charge carriers across the IGBT. TCAD simulations of 1200 V IGBTs with different p-emitter and FS doses demonstrated that the intensity of this mechanism can be reduced by back-side-design measures which involve an increase of α_{pnp} during short-circuit-operating conditions. For sufficiently high values of α_{pnp}, SCOs can be avoided completely, however, of course at the expense of higher leakage currents, higher turn-off losses and a reduced thermal short-circuit robustness. TCAD simulations showed also that SCOs are reduced for high junction temperatures.

The introduced two-step simulation approach enabled the analysis of the intrinsic electrical characteristics of the IGBTs in the operating points (I_C at $V_{GE} = V_{Dr}$ and $V_{CE} = V_{DC\text{-link}}$). For the investigated variations of the p-emitter and FS dose and of the junction temperature, the reduction of SCOs came along with an increase of the absolute electric-field strength in the drift zone in front of the FS and a decrease of the absolute electric-field strength in the FS region.

Experimentally observed dependencies of the SCOs on the p-emitter dose, the FS profile and the operating temperature were well reproduced by the presented TCAD simulations.

REFERENCES

[1] P. Reigosa, F. Iannuzzo, M. Rahimo, C. Corvasce, and F. Blaabjerg, "Improving the short-circuit reliability in IGBTs: How to mitigate oscillations," IEEE Transaction on Power Electronics , in press.

[2] P. Reigosa, F. Iannuzzo, and M. Rahimo, "TCAD analysis of short-circuit oscillations in IGBTs," in Proc. of the 29th International Symposium on Power Semiconductor Devices & ICs, pp. 151 - 154, June 2017.

[3] I. Omura, W. Fichtner, and H. Ohashi, "Oscillation effects in IGBT's related to negative capacitance phenomena," IEEE Transaction on Electron Devices, vol. 46, no. 1, pp. 237 – 244, Jan. 1999.

[4] S. Milady, D. Silber, F. Pfirsch, and F.-J. Niedernostheide, "Simulation studies and modeling of short circuit current oscillations in IGBTs", in Proc. of the 21st International Symposium on Power Semiconductor Devices & ICs, pp. 37 - 40, June 2009.

[5] C. Ronisvalle et al."High frequency capacitive behavior of field stop trench gate IGBTs operating in sort circuit," Proc. of the 28th. Annual IEEE Applied Power Electronics Conference and Exposition, pp. 183 – 188, March 2013.

Proceedings of the 30th International Symposium on Power Semiconductor Devices & ICs
May 13-17, 2018, Chicago, USA

A 750V Recessed-Emitter-Trench IGBT with Recessed-Dummy-Trench Structure Featuring Low Switching Losses

Yao Yao[1,2], Haihui Luo*[1,2], Qiang Xiao[1,2], Chunlin Zhu[3], Haibo Xiao[1,2], Rongzhen Qin[1,2],

Luther-King Ngwendson[3], Xubin Ning[1,2], Canjian Tan[1,2], Ian Deviny[3], Xiaoping Dai[1,2,3]

[1]State key Laboratory of Advanced Power Semiconductor Devices, Zhuzhou, P.R. China

[2]Zhuzhou CRRC Times Electric Co.,Ltd., Zhuzhou, P.R. China

[3]Research and Development Centre, Dynex Semiconductor Ltd., Lincoln, UK

*Email: luohh@csrzic.com

Abstract—In this paper, a novel 750V Recessed-Emitter-Trench IGBT (RET-IGBT) featuring Recessed-Dummy-Trench (RDT) structure is proposed. The mesa width is shrunk to 1.2μm with the advantage of RET, which improves the trade-off relationship between on-state voltage drop ($V_{CE(ON)}$) and turn-off energy loss (E_{OFF}) without any performance sacrifice. Furthermore, by applying proper dummy trench and dummy P-well ground scheme through RDT concept, 34.1% lower Miller capacitance, 34.9% lower turn-on loss and 12.3% lower turn-off loss are achieved with better turn-on dI/dt and reverse recovery dV/dt controllability, which are favourable for high frequency operations.

Keywords—*IGBT; Recessed-Emitter-Trench; RET; Recessed-Dummy-Trench; RDT; low switching losses; high turn-on dI/dt controllability*

I. INTRODUCTION

Si-IGBTs are widely used in high power applications, such as electric locomotives, automobiles and so on. The urgent demand to minimize energy consumption strongly requires IGBTs with low $V_{CE(ON)}$, low E_{OFF} and low E_{ON}. "Fine pattern" cell structure is one of the main methods to reduce the $V_{CE(ON)}$ without increasing the E_{OFF}. A new structure, RET-IGBT, was demonstrated to shrink the mesa width without any performance sacrifice [1], which shows that RET-IGBT is a highly promising technology to realize sub-μm mesa width without much reliance on special processes. On the other hand, the design of the dummy region, such as the arrangement of grounded dummy-trench and grounded dummy-P-well, plays an important role in determining capacitance, hence the switching behavior and switching losses [2-4]. However, sub-μm or less contact line width is required to realize grounded dummy-trench and dummy-P-well for fine pattern IGBT, which also places very high demands on the process platform.

In this work, a 750V RET-IGBT with Recessed-Dummy-Trench (RDT) structure (see fig.1a), which grounds the dummy-trench and dummy-P-well at the same time and subsequently decreases the process difficulty, has been

proposed and demonstrated. The mesa width is shrunk to 1.2μm with the advantage of RET, which improves the trade-off relationship between on-state voltage drop ($V_{CE(ON)}$) and turn-off energy loss (E_{OFF}). By applying proper dummy-trench and dummy-P-well ground scheme through RDT concept, lower Miller capacitance and lower switching losses are achieved with better turn-on dI/dt and reverse recovery dV/dt controllability, which are favourable for high frequency operations.

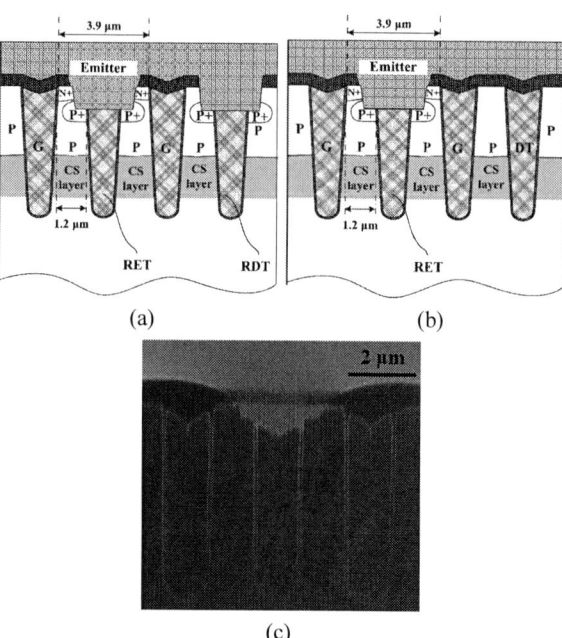

Fig.1 Schematic cross-section of proposed RET-IGBT with RDT structure (a), standard RET-IGBT (b), and SEM cross-section of fabricated RET/RDT structure (c).

This work was supported by Zhuzhou CRRC Times Electric Co., Ltd. under grant K12TZ18A0090.

978-1-5386-2928-4/18 $31.00 © 2018 IEEE

II. DEVICE STRUCTURE

Two kinds of RET-IGBT which named RET-IGBT with RDT (see fig.1a) and standard RET-IGBT (see fig.1b) are fabricated. The RET and RDT structure are fabricated in the same process flow and there is no additional mask. Figure 1c shows the SEM cross-section of fabricated RET/RDT structure. The simulated and experimental results of these two structures are compared.

III. RESULTS AND DISCUSSION

Figure 2 shows the simulated J_C-V_{CE} characteristics. The $V_{CE(ON)}$ (at J_C =250A/cm^2) of RET-IGBT with RDT is slightly higher than the standard RET-IGBT. The reason is that the RDT structure partially grounded the dummy-P-well, resulting in slightly lower carrier injection enhancement effect. At the same time, the RET-IGBT with RDT is able to turn off faster (higher dV/dt) because holes are swept through both the p$^+$-

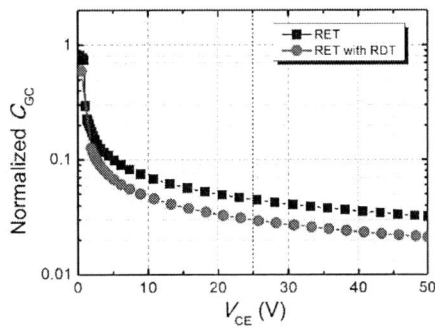

Fig.2 Simulated J_C-V_{CE} characteristics at 150^{0}C.   Fig.3 Simulated C_{GC}-V_{CE} characteristics.

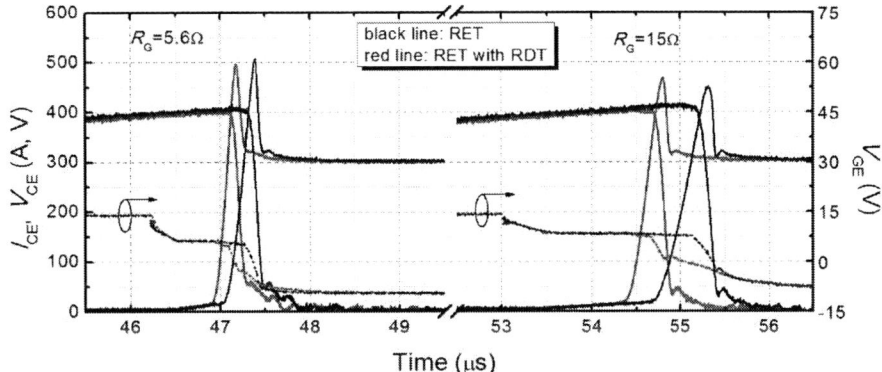

Fig.4 Measured turn-off waveforms. V_{CC}=300V, I=400A, T_J=150°C, V_{GE}=+15/-10V, R_G=5.6/15Ω.

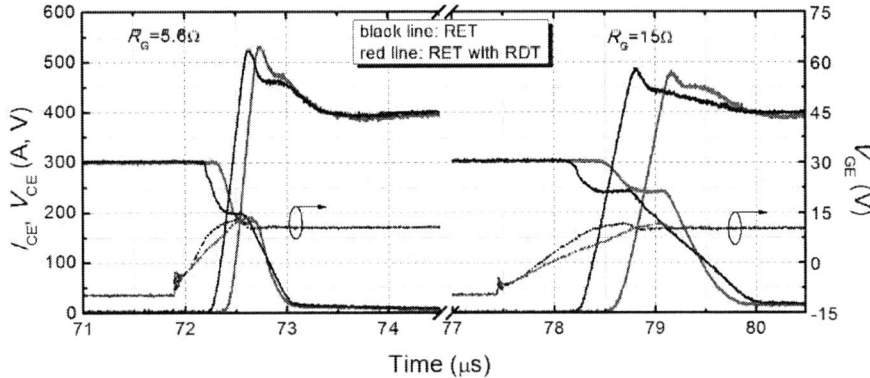

Fig.5 Measured turn-on waveforms. V_{CC}=300V, I=400A, T_J=150°C, V_{GE}= -10/+15V, R_G=5.6/15Ω.

contact and grounded dummy-P-well [3]. The simulated C_{GC}—V_{CE} characteristics (see fig.3) shows that the Miller capacitance of RET-IGBT with RDT is decreased by 34.1% due to the grounded dummy-trench when compared to the standard RET-IGBT at V_{CE} =25V, which is able to reduce the

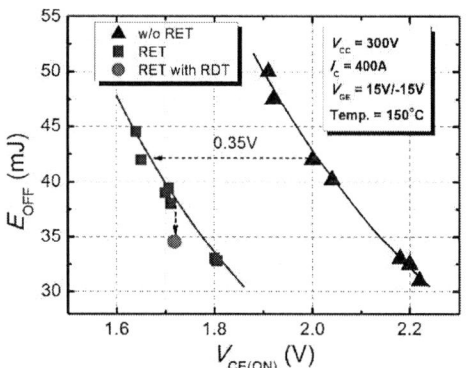

Fig.6 Measured $V_{CE(ON)}$—E_{OFF} trade-off relationship.

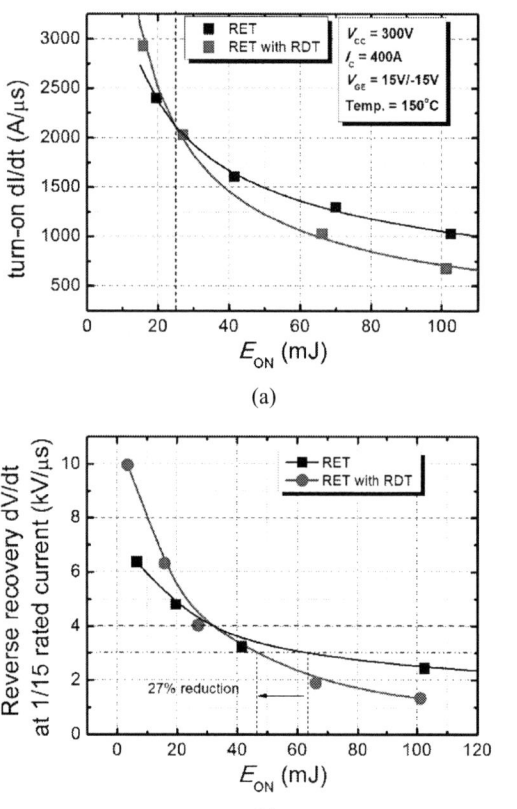

(a)

(b)

Fig.7 Measured trade-off relationship of E_{ON}—turn on dI/dt at rating current (a) and E_{ON}—reverse recovery dV/dt at 1/10th rating current (b).

V_{CE} tail of the RET-IGBT with RDT during turn-on [5]. As a result, both the E_{OFF} and E_{ON} of the RET-IGBT with RDT are decreased, which is evident from the measured turn-on and turn-off waveforms (see fig.4 and fig.5, the rating current is 400A).

Figure 6 shows the $V_{CE(ON)}$—E_{OFF} trade-off relationship. The $V_{CE(ON)}$ of standard RET-IGBT is decreased by 0.35V when compared to conventional IGBT without RET. Furthermore, the E_{OFF} of RET-IGBT with RDT is decreased approximately 10% when compared to the standard RET-IGBT at the same $V_{CE(ON)}$ of 1.72V.

For inverters driving electrical machines, where the highest dV/dt (always during reverse recovery transient at low current, such as 1/10th rating current) is typically limited to below 5kV/μs [2]. Low reverse recovery dV/dt means low turn-on dI/dt, and then high turn-on loss. With optimal dummy-trench and dummy-P-well ground scheme through RDT concept, both the trade-off between turn on dI/dt at rating current and E_{ON}, and trade-off between reverse recovery dV/dt at 1/10th rating current and E_{ON} are improved (see fig.7), which means higher turn on dI/dt controllability and reverse recovery dV/dt controllability. Under the same dI/dt of 1.5kA/μs and reverse recovery dV/dt of 3kV/μs (below 5kV/μs), a reduction of E_{ON} by 23% and 27% is achieved, respectively.

Figure 8 shows a calculation result of the losses of IGBT at a frequency of 10 kHz. The E_{ON} and E_{OFF} is decreased by 34.9% and 12.3% respectively, while on-state conduction loss (E_{CON}) is increased marginally due to the slightly larger $V_{CE(ON)}$. As a result, a reduction of total loss by 11% is achieved. In addition to the low losses, a sufficient short-circuit (duration of 10μs without latch-up even with short circuit current 6.5 times of rating current, see fig.9) and high current turn-off (6.5 times of rating current, see fig.10) capability at 175 °C is also ensured.

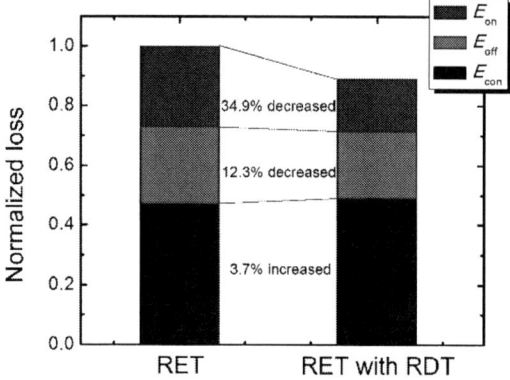

Fig.8 Loss calculation of IGBT at a frequency of 10 kHz

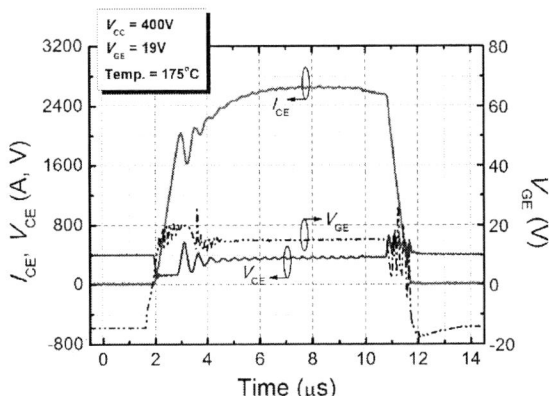

Fig.9: Measured SCSOA for RET-IGBT with RDT at T=175°C and V_{GE}=19V

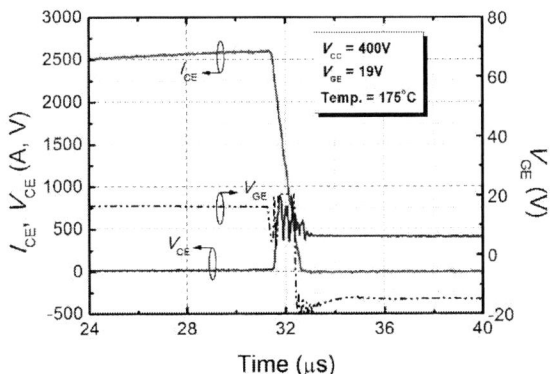

Fig.10: Measured RBSOA for RET-IGBT with RDT at T=175°C and V_{GE}=19V

IV. CONCLUSION

In conclusion, a novel 750V RET-IGBT featuring RDT structure is proposed without much reliance on process platform. The mesa width is shrunk to 1.2μm with the advantage of RET, which improves the $V_{CE(ON)}$—E_{OFF} trade-off relationship with a $V_{CE(ON)}$ reduction of 0.35V. With optimal dummy-trench and dummy-P-well ground scheme through RDT concept, higher turn on dI/dt controllability and reverse recovery dV/dt controllability are realized. At the same time, the E_{ON} and E_{OFF} is decreased by 34.9% and 12.3% at a frequency of 10 kHz respectively, which are favorable for high frequency applications.

REFERENCES

[1] Ian Deviny, Haihui Luo, Qiang Xiao, Yao Yao, Chunlin Zhu, Luther-King Ngwendson, Haibo Xiao, Xiaoping Dai and Guoyou Liu, "A novel 1700V RET-IGBT (Recessed Emitter Trench IGBT) Shows Record Low $V_{CE(ON)}$, Enhanced Current Handling Capability and Short Circuit Robustness", Proc. ISPSD'17, pp. 147-150, 2017

[2] Christian Jaeger, Alexander Philippou, Antonio Vellei, Johannes G. Laven and Andreas Härtl, "A new sub-micron trench cell concept in ultrathin wafer technology for next Generation 1200 V IGBTs" , Proc. ISPSD'17, pp. 69-72, 2017

[3] M. Sawada, Y. Sakurai, K. Ohi, Y. Ikura, Y. Onozawa, T. Yamazaki and Y. Nabetani, "Hole Path Concept for Low Switching Loss and Low EMI Noise with High IE-effect" , Proc. ISPSD'17, pp. 65-68, 2017

[4] Kazuya Konishi, Ryu Kamibaba, Mariko Umeyama, Atsushi Narazaki, Tetsuo Takahashi, Akihiko Furukawa and Masayoshi Tarutani, "Experimental Demonstration of the Active Trench Layout Tuned 1200V CSTBT™ for Lower dV/dt Surge and Turn-on Switching Loss", Proc. ISPSD'16, pp. 363-366, 2016

[5] K.Ohi, Y.Ikura, A.Yoshimoto, K.Sugimura, Y.Onozawa, H.Takahashi and M.Otsuki, "Ultra Low Miller Capacitance Trench-Gate IGBT with the Split Gate Structure", Proc. ISPSD'15, pp. 25-28, 2015

Proceedings of the 30th International Symposium on Power Semiconductor Devices & ICs
May 13-17, 2018, Chicago, USA

SMALL CURRENT UNCLAMPED INDUCTIVE SWITCHING (UIS) TO DETECT FABRICATION DEFECT FOR MASS-PRODUCTION PHASE IGBT

Kazuya Sano[1], Yukio Matsushita[1], Megumi Yachi[1],
Yoji Nakata[1], Keiichiro Ide[2] and Daisaku Yoshida[2],
1: Kumamoto Factory, Wafer Engineering Dept.,
Mitsubishi Electric Corporation
2: Kumamoto Factory, Test Engineering Sect.,
Kyokuyo Semiconductors Corporation
997 Miyoshi, Koushi city, Kumamoto, Japan
E-mail: Sano.Kazuya@dh.MitsubishiElectric.co.jp

Tomohito Kudo[3], Yasuo Ata[3], Hideki Haruguchi[3],
Shinya Soneda[4] and Tadaharu Minato[4]
3: Power Semiconductor Device Dept. A,
4: Power Semiconductor Device Development Dept.,
Power Device Works,
Mitsubishi Electric Corporation
1-1-1   Imajyuku-Higashi, Nishi-ku, Fukuoka, Japan

Abstract—In **U**nclamped **I**nductive **S**witching (UIS) with a small capacitance, the L load current flows into the breakdown point selectively. V_{ava}, which is a peak voltage during this UIS turn-off, is also widely changed by the ON state forward voltage drop V_{on}. Along this V_{on} index, two types of failure modes are detected. One is the weak point of the design and the other is the fabrication defect. Our proposing UIS is one of the most effective approaches to detect the fabrication defects instead of the huge current class module test like an SCSOA. And also V_{on} is one of the major controlling factors of UIS capability other than the index of power loss.

Keywords—IGBT, Unclamped Inductive Switching (UIS)

I.   INTRODUCTION AND METHODOLOGY

To assure IGBT module's **S**afety **O**peration **A**rea (SOA), especially for an automotive and/or high power application, a module level SOA test is necessary. **S**hort **C**ircuit endurance of SOA evaluation (SCSOA) or **U**nclamped **I**nductive **S**witching (UIS) capability reaches the device's destruction, whose energy, order of 1000A and/or 1000V, is extremely high to protect neither the module itself nor the test facilities. On the other hands, a large current UIS test for naked chips, without any wire-bonding or soldering onto a heat-spreader, is also very difficult from almost the same reasons mentioned above.

UIS test has been applied even for IGBT other than MOSFET to simulate the turn-off mode of SCSOA[1,2] or detect the defects originated in missteps during the wafer fabrication process. From this point of view, all of the cell region in the entire chip has the uniform current flowing even in the rated and high current density. It is very difficult for the conventional UIS test to detect the slight fabrication defect as the fatal error in the chip, which could pass the DC test up to the rated breakdown voltage and/or the sensitive gate leak test as described in the TABLE I. For the pnp-bipolar device like an IGBT, since the V_{ava}, which is a peak voltage during UIS turn-off, of a small current UIS is almost the same but a little higher than that of a large current UIS, this small current UIS is able to apply the highest voltage in the entire DC test sequence. Instead of the current probing during the large current UIS[3,4],

the small current UIS is supposed to be effective to detect the process defects. In contrast, a certain large power is necessary for the defects turning into an irreversible damage to be detected. Our proposing "C-UIS" would be the approach applying the small amount of current selectively and efficiently on the defect, which has the relatively low breakdown voltage. Here, the "C-UIS" circuit configuration has an extra-capacitance (C) connected in parallel with the inductance (L) load (Fig. 1). The order of this small capacitance coincidentally corresponds an input or a feedback capacitance of the upper arm IGBT. As the nature of the capacitance C and L, a part of the L load current is charged in (C). Once the avalanche occurs at the defect, V_{CE} rise halts to charge C. At the stop-moment of the current shunting to (C), small but all the current immediately flows into IGBT being in which position of the V-I plane of turning-off locus. As the result, the current being small, it can be possible for this method to detect the silent defect sensitively or slight defect efficiently through the critical current path automatically. We confirmed this hypothesis by the following simulation and experiments as the IGBT structural dependence.

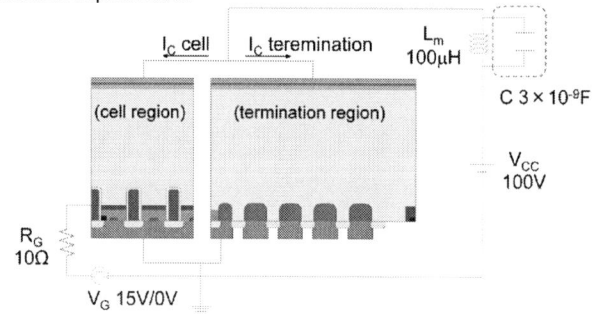

Fig.1. Equivalent circuit of our "C-UIS" includes a small capacitance for both the experiments and simulation. Both cell and the termination structure are schematically described.

II.   Experimetanl Results (I) : EFFECTS OF SMALL CURRENT UIS WITH SMALL CAPACITANCE (C-UIS)

978-1-5386-2928-4/18 $31.00 © 2018 IEEE   116

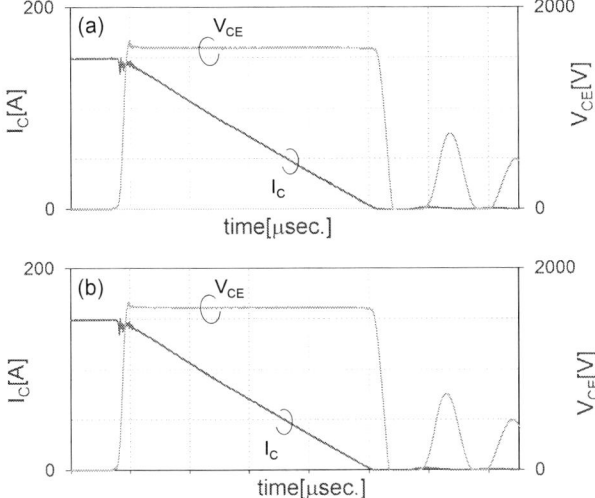

Fig. 2. Experimental waveform in a large current test of 150A at 150degC. (a) "lower" V_{on} (b) "higher" V_{on}

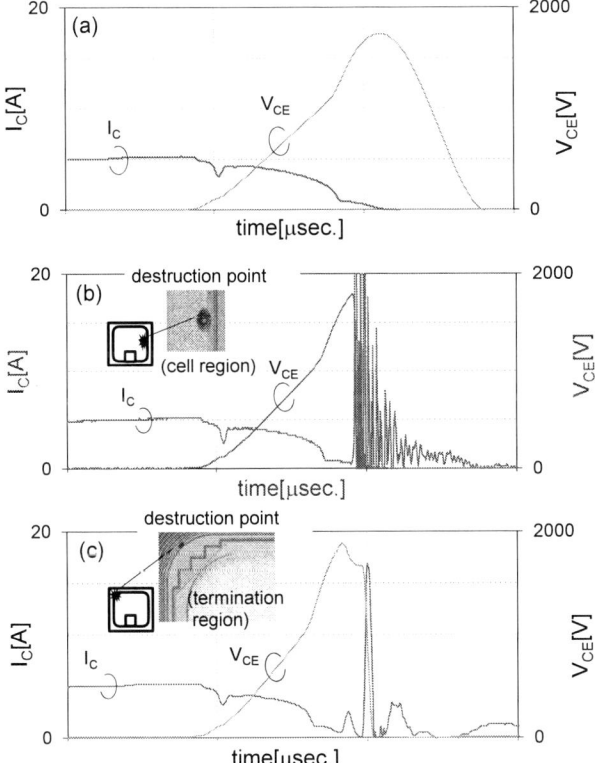

Fig. 3. Experimental waveforms of a small current test of 5A at 150degC. (a) "lower" V_{on} without failure (b) "higher" V_{on} with failure detected in the cell region (c) "higher" V_{on} with failure detected in the termination region

Figure 2 shows a UIS turn-off portion from the collector current I_c of 150A large current, almost the same as rated current. At the moment of turned off, the collector-emitter voltage V_{CE} abruptly reaches the avalanche breakdown voltage V_{ava} as the nature of IGBT device. These Fig. 2 type I_c and V_{CE} waveforms are well known to be the standard UIS characteristic and their shapes are almost identical to be independent of the ON state forward voltage drop V_{on}. But, the UIS of the small current, which is 1/80 of rated current, is completely different (Fig. 3). The time delay of V_{CE} is caused by the combination of L and C mentioned above circuit configuration. As far as a "healthy" chip without any defect or weak-point is concerned, there is no special behaviour (Fig. 3(a)). Among the huge numbers of the mass-production chips, for the higher V_{on} IGBT chip, two types of destruction modes are detected by the UIS of the small current. One is the breakdown in the active cell region and assigned to be the silent "fabrication defect" (Fig. 3(b)). In this case, the breakdown occurred before V_{CE} reaching to the standard cell's V_{ava}. This is the right effect that we expected to detect the silent defects, which could pass the DC test described in TABLE I. The other is the breakdown at the corner of the termination region which is one of the weak points of the design (Fig. 3(c)), and the "current jump" is observed at the end of turn-off time period. This is the most important behaviour to be found out, because this sample had passed the 150A UIS before the small current UIS (Fig. 2(b)). This is the effect of "C-UIS".

As we expected, this small current C-UIS is more sensitive to detect a silent defect even after the standard UIS at the rated current. By the numerical simulation, we also analyzed the origin of "current jump" and the reason why being observed only for the higher V_{on} IGBT chip than that of sample in Fig. 3.

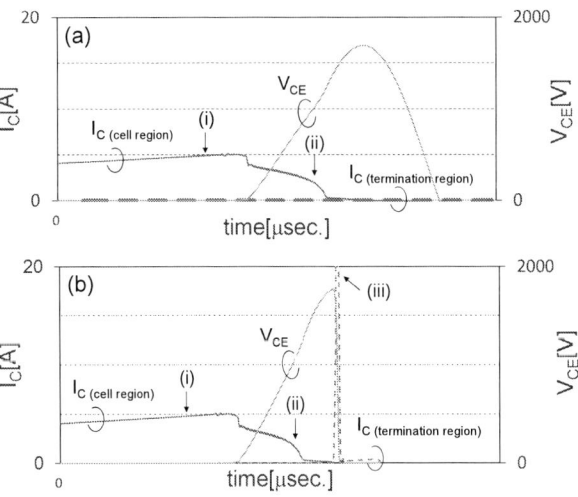

Fig. 4. Numerical simulation result of C-UIS in a small current of 5A at 150degC (a) "lower" V_{on} (b) "higher" V_{on}

III.   SIMULATION ANALYSYS

Focusing on the "current jump" observed in C-UIS test of 5A, the numerical simulation model is set as the parallel connection of the active region and the termination region. The experimental waveforms were sufficiently traced by this numerical simulation as shown in Fig. 4. As the V-I locus (Fig. 5 (a) (b)), the C-UIS turn-off waveforms (red line) are

978-1-5386-2928-4/18 $31.00 © 2018 IEEE   117

superimposed on the breakdown characteristics (black line). The cell region characteristics and the termination region ones are shown separately. We simulated the characteristics up to the secondary breakdown indicating the Negative Differential Resistance (NDR), which is the limitation of the IGBT turn-off [6,7].

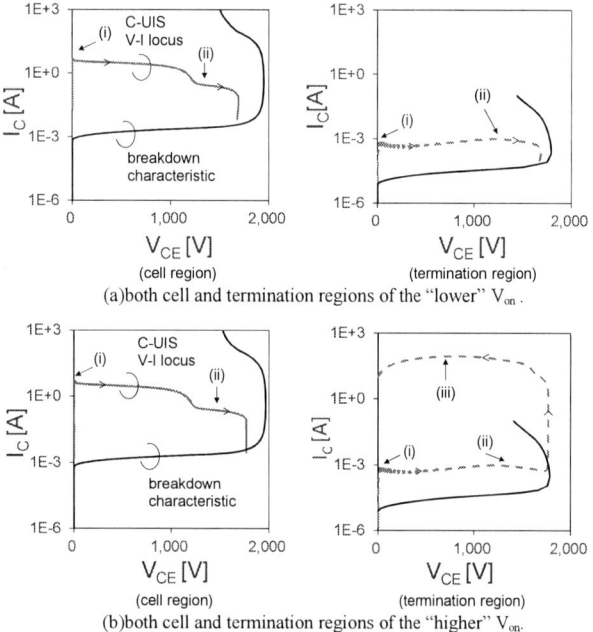

(a)both cell and termination regions of the "lower" V_{on}.

(b)both cell and termination regions of the "higher" V_{on}.

Fig.5. C-UIS V-I locus of 5A is superimposed on the breakdown characteristics.

Since the C-UIS V-I locus reaches the termination breakdown voltage designed as the lower than the cell's one, there is not the current filamentation path shifting from the cell to the termination [5], but simply avalanche breakdown rising. This is shown in Figure 6 schematically as the points (i) to (iii). (i), a current flows through the cell region only. (ii), a small part of the current flows through the termination region. (iii), all the current shifts to the termination region. Finally, (iv), IGBT turns off the current (not illustrated).

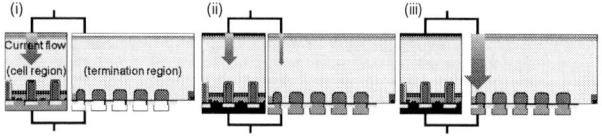

Fig.6. The current path shifting the cell region to the termination region at the point Fig.4(b) (i),(ii) and (iii).

For the higher V_{on} IGBT chip, this small current C-UIS test gives a higher energy stress. It is effective as the screening test for not only chips having a weak point by fabrication defects but also chips of low secondary breakdown voltage capability especially at the corner portion of a termination which we designed 150V lower than the main cell region. In the latter case, "current jump" is caused by avalanche breakdown of the termination region. Since the absolute current value is apparently larger than the one as used in the

DC measurement configuration, it is originated in the termination avalanche breakdown, i.e. impact ionization. The analysis results of the electric field and current density distributions are shown as the evidence of the avalanche not in the cell region but in the termination region in Figure 7 and 8 respectively.

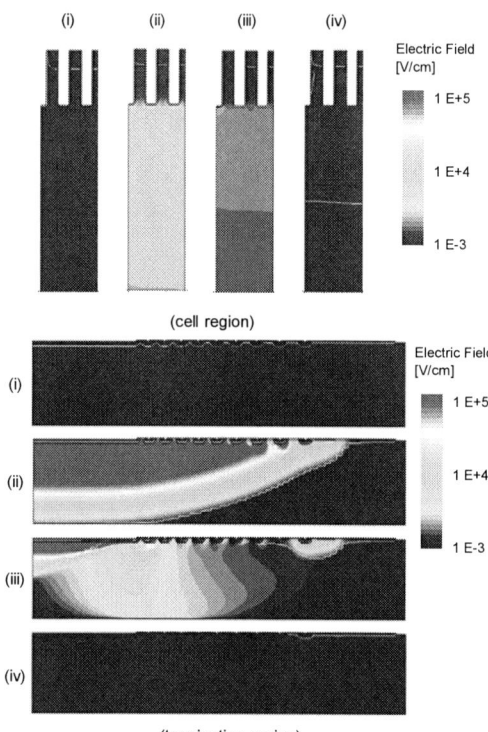

Fig.7. The electric field in C-UIS of "higher" V_{on} IGBT at the point (i), (ii), (iii) and (iv).

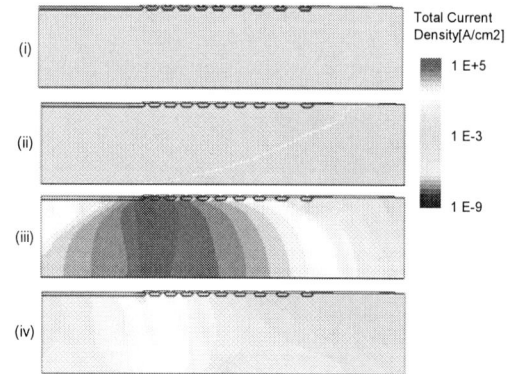

Fig.8. The total current density in C-UIS of "higher" V_{on} IGBT at the point (i), (ii), (iii) and (iv).

IV. EXPERIMENTAL RESULTS (II)

We tested more than 500 chips which belong to 5 groups of V_{on} characteristics from group A to E. Before the C-UIS test, they were carefully tested to reject the defect through the special sequence (TABLE I.) as sensitively as possible. T1 and T2 are the standard characteristics, i.e. V_{th}, V_{on}, I_{GES}, I_{CES}

978-1-5386-2928-4/18 $31.00 © 2018 IEEE

and transient characteristic. The enhanced DC test sequence T3 includes much higher DC voltages than the standard ones and switching items not to miss the fabrication defect, e.g. around the trench gate. Only the chips that cleared from T1 to T3 were tested by "C-UIS".

TABLE I. TEST SEQUENCE FOR THIS EXPERIMENT

test Category		test item	specification (Additional for this experiment)
T1	DC (Additional)	V_{th}	Taget Spec. (leak at a small bias, no kink on V_{GE}-I_C)
		V_{on}	Small current (Rated current)
		I_{GES}	Rated voltage
		I_{CES}	Rated voltage (Lower voltage to detect small leakage current)
T2	Switching	td(ON), tr td(OFF), tf	(Rated current) (rejecting filamentation)
T3	Enhanced DC	E-I_{GES}	(2 times higher than the conventional)
		E-BV_{CES}	greater than 1.2 times the rated voltage
T4	Proposed UIS	C-UIS	(Proposed small current, 1/80 of rate current)
		C-UIS	(Large current(ca. 1/3)C-UIS to ensure the effect of above test)

(Left margin flowchart: T1 — Additional test for this experiment; T2; T3; T4)

TABLE II. THE EXPERIMENTAL C-UIS RESULT AND THE EFFECTIVENESS

5 Group of Von characteristics	A	B	C	D	E
failure rate of a "healthy" destruction	nothing		small but increasing		large
failure rate of a fabrication misstep destruction	nothing		very slight but maintaining a certain value		
screening effectiveness of "C-UIS"	week		suitable		too strong

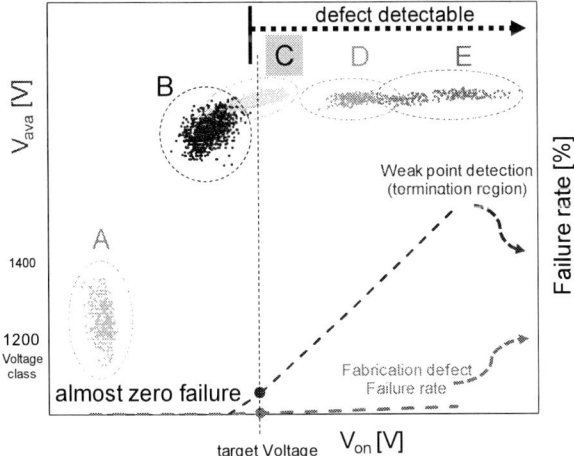

Fig. 9. The experimental C-UIS result of V_{ava} and their failure rates more than 100chips for each group A to E

The higher V_{on}, which stands for the lower hole injection or pnp transistor gain of the parasitic pnp transistor of IGBT, shows the higher V_{ava}. In the group A and B of Figure 9, the failure is not detected. In the group C and D, two types of same failures are detected as the mentioned in the previous section, one is located in the termination region and the other is randomly located inside the active cell region. We assigned the former is "healthy" and the latter is "fabrication defect", because the termination breakdown voltage is designed to be a little lower than the active cell region as the mentioned above. We'd like to emphasize that all the chips applied "C-UIS" have already passed the standard and enhanced DC tests to reject the obvious defect including the termination region. So, in the group C and D, we could identify that the failure rate of the "healthy" destruction is small and the failure rate of the fabrication misstep destruction is very slight. In the group E, the number of "healthy" ones was still small but reached the level not to be acceptable from the mass-production point of view. So, it is also clear that V_{on} is one of the major controlling factor of UIS capability other than the index of power loss. As shown in TABLE II and Fig. 9, even from the silent defect as the enhanced DC test, C-UIS could distinguish three categories, i.e. healthy (no defects), defect detectable and over-killing.

VI. CONCLUSION

Our proposing C-UIS in the small current range is effective to find out the fabrication defects with very high sensitivity. V_{on}, which is translated into the pnp transistor gain, is one of the major controlling factor of UIS capability other than the index of power loss. Along this V_{on} index, sufficiently large number of large current chips were categorized into the healthy, defect detectable and over-killing regions from the mass-production yield point of view. The small current C-UIS is one of the most effective approaches to detect the fabrication defects instead of the huge current class module test like an SCSOA.

REFERENCES

[1] J. Lutz, R. Dobler, J. Mari and M. Menzel, "Short Circuit III in High Power IGBTs", in Proc. EPE2009, pp. 1-8.

[2] H. Suzuki , M. Ciappa, "TCAD simulation of current filamentation in adjacent IGBT cells under turn-on and turn-off short circuit condition", Microelectronics Reliability 55 (2015), pp. 1976-1980.

[3] T. Shoji, M. Ishiko, T. Fukami, T. Ueda, K. Hamada, " Investigations on current filamentation of IGBTs under unclamped inductive switching conditions", in Proc. ISPSD2005, pp. 227-230.

[4] M. Riccio, A. Irace, G. Breglio, P. Spirito, E. Napoli and Y. Mizuno, " Electro-thermal instability in multi-cellular Trench-IGBTs in avalanche condition: experiments and simulations ", in Proc. ISPSD2011, pp. 124-127.

[5] S. Soneda, A. Narazaki, T. Takahashi, K. Takano, S. Kido, Y. Fukada, K. Taguchi and T. Terashima, " Analysis of a Drain-Voltage Oscillation of MOSFET under High dV/dt UIS Condition ", in Proc. ISPSD2012, pp.153-156.

[6] T. Minato, N. Thapar and B. J. Baliga, "Correlation between the static and dynamic characteristics of the 4.5kV Self-aligned trench IGBT", in Proc. ISPSD1997, pp. 89-92.

[7] F. Masuoka, K. Tanaka, T. Kachi, Y. Yoshiura and K. Shimizu, "RFC diode with High Avalanche Stability and UIS Capability", in Proc. ISPSD2017, pp. 131- 134.

[8] M. Tanaka, A. Nakagawa, "Simulation studies for short circuit current crowding of MOSFET-mode IGBT", in Proc. ISPSD2014, pp. 119- 122.

Proceedings of the 30th International Symposium on Power Semiconductor Devices & ICs
May 13-17, 2018, Chicago, USA

Tailoring the Performance of Silicon Power Diodes by Predictive TCAD Simulation of Platinum

Moritz Hauf, Christian Sandow, Franz-Josef Niedernostheide

Infineon Technologies AG
Neubiberg, Germany
moritz.hauf@infineon.com

Gerhard Schmidt

Infineon Technologies Austria AG
Villach, Austria

Abstract— **Today's silicon power diodes typically make use of platinum to adjust the charge carrier lifetime, which influences the device performance. In this work, we present how platinum in silicon can be modelled in terms of process and device simulation in a predictive manner. Furthermore, we demonstrate how this simulation tool can be used to predict performance tradeoffs which allow the tailoring of the electrical device performance.**

Keywords—platinum; silicon power diode; TCAD simulation; optimization; recombination center

I. INTRODUCTION

Since many years, Pt is known to serve as an efficient recombination center in silicon, providing a good balance between efficient charge carrier recombination and, at the same time, low generation rates [1]. As such, Pt is frequently used to adjust the performance tradeoff between the static (V_f) and dynamic losses (E_{rec}) in silicon power diodes, however, avoiding detrimental leakage current levels [2]. Despite being an established process in today's power semiconductor industry, the capability of simulating the influence of Pt on the device properties in a predictive manner has been lacking so far. For this purpose it is important to be able to integrate the diffusion mechanisms related to substitutional platinum recombination centers, Pt_s, in the process simulation. In subsequent device simulation, the Pt_s profile obtained from the process simulation serves as an input to the trap distribution. Furthermore, a predictive simulation of the electrical device behavior requires thoroughly calibrated recombination center properties. Only in combination, reliable simulation results can be obtained, which allow tailoring the performance of silicon power diodes by means of TCAD simulations only.

In the following, we first present the basis for a process simulation of platinum diffusion in silicon. In a second section, the process simulation is combined with the calibration of the parameters of Pt recombination center properties (energy level, electron and hole capture cross sections) to simulate the electrical device behavior in a predictive manner. Finally, we give an example for the use in the development of silicon power diodes.

II. PLATINUM PROCESS SIMULATION

Substitutional platinum Pt_s is acting as a recombination center in silicon [1]. It can be introduced into silicon by depositing a layer of Pt on the silicon surface. This layer serves as an infinite source of Pt for diffusion into the silicon during a subsequent furnace anneal.

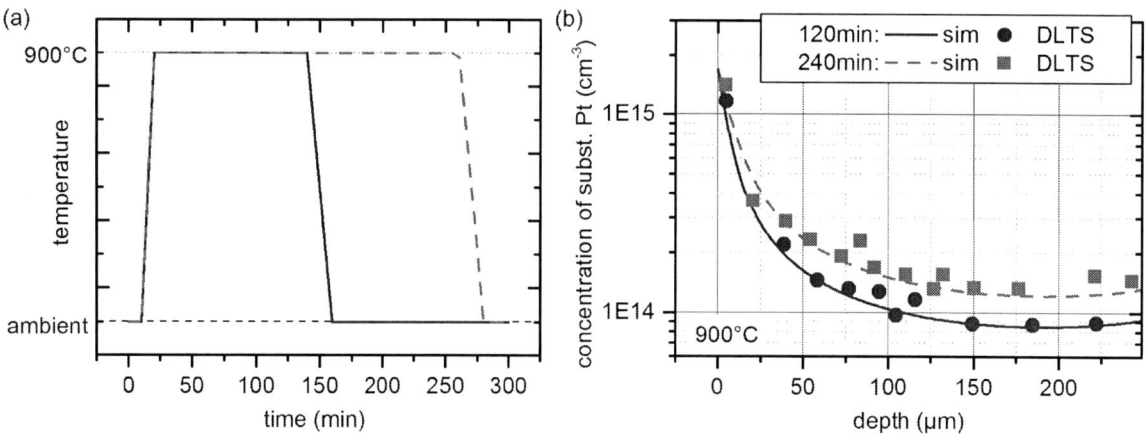

Fig. 1. a) Two different Pt diffusion temperature profiles with 120 min (blue, solid line) and 240 min (red, dashed line) plateau time at 900 °C. b) Corresponding simulated substitutional Pt depth profiles (lines). Experimental DLTS results are shown for comparison (symbols).

978-1-5386-2928-4/18 $31.00 © 2018 IEEE

The following reactions have proven to be sufficient for an accurate description of the diffusion process [3]:

$$Pt_i + V \leftrightarrow Pt_s \qquad (1)$$

The Frank-Turnbull reaction (1) [4] describes an interstitial Pt atom, Pt_i, taking a vacant place in the silicon lattice, V (vacancy), thus, turning into a substitutional Pt atom. The arrow $\leftrightarrow$ should indicate that the reactions in general happen with certain forward and backward reaction rates.

$$Pt_i \leftrightarrow Pt_s + I \qquad (2)$$

The kick-out mechanism (2) describes an interstitial Pt atom, which creates a silicon self-interstitial I and directly takes its place in the silicon lattice.

$$I + V \leftrightarrow 0 \qquad (3)$$

Finally, recombination (3) describes the process of a self-interstitial finding a vacancy and taking a place in the silicon lattice again.

From these three reactions, a set of coupled differential equations can be derived. Implementation of these equations into the TCAD simulator Sentaurus Process allows solving the set of equations as part of the standard process simulation flow of a silicon power diode. Finally, the involved forward- and backward-reaction rates, initial and equilibrium concentrations, as well as the respective temperature dependencies were calibrated to fit the vertical Pt_s profiles from dedicated DLTS (deep-level transient spectroscopy) experiments.

Fig. 1a exemplarily visualizes two different furnace protocols with a plateau time of 120 min and 240 min during a platinum diffusion at 900 °C. Fig. 1b compares the process-simulated profiles to experimental DLTS profiles for a 400 μm thick silicon wafer. The concentration of substitutional platinum decreases from more than 1E15 cm^{-3} at the surface, down to around 1E14 cm^{-3} in the wafer middle (200 μm). The increased plateau time of the Pt diffusion hardly influences the surface concentration. However, the bulk concentration level increases by around 50%. The Pt diffusion model explains this effect the following way: Both, interstitial Pt and silicon self-interstitials show a high diffusivity. However, vacancies are considered to be mainly introduced at the silicon surface and diffuse rather slowly. Therefore, the vacancy concentration in the bulk increases only slowly and shows a dependency on the plateau-time increase from 120 min to 240 min. Via the Frank-Turnbull mechanism (1) the vacancies are occupied by Pt. At the surface, Pt_s can reach the temperature-dependent equilibrium values quite quickly. The concentrations would therefore rather depend on the maximum plateau temperature than on the diffusion time (see also Sec. IV).

III. DEVICE SIMULATION

Device simulations were performed using Sentaurus Device. Here, substitutional platinum was implemented as a recombination center with a certain vertical distribution, which was taken directly from the output of the process simulation.

CALIBRATION

First, the electrical trap properties, i.e. the hole and electron recombination cross-sections, the trap energy level, and the temperature dependencies of these parameters were calibrated. Great care was taken to separate the effect of platinum from other parameters, which influence the charge carrier distribution and extraction during commutation, by using a set

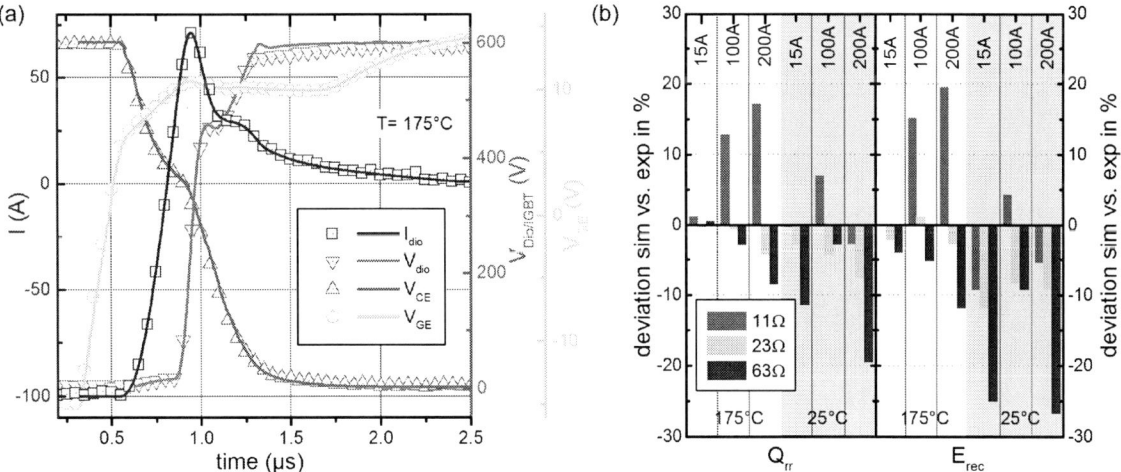

Fig. 2. a) Time transients of the diode current (I_{dio}), diode voltage (V_{dio}), the IGBT collector-emitter voltage (V_{CE}), and the IGBT gate voltage (V_{GE}) for simulation (solid lines) and experiment (open symbols), showing an excellent agreement. 1200 V 100 A free-wheeling diode in a half-bridge configuration with a 1200 V 140 A IGBT. DC-link voltage: 600 V, switching current: 100 A, gate resistor: 23 Ω, temperature: 175 °C. b) Relative deviation between simulation and experiment of the diode recovery charge Q_{rr} (left side) and reverse recovery losses E_{rec} (right side) at different current levels, switching speeds and temperatures.

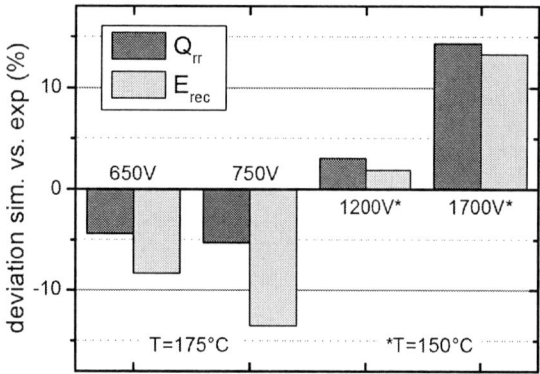

Fig. 3. Relative deviation of the diode recovery charge Q_{rr} (red) and reverse recovery losses E_{rec} (green) for power diodes of different voltage classes from 650 V to 1700 V at their respective nominal currents and operating temperatures for an R_G of 23 Ω.

of dedicated 1200 V diodes, with and without platinum. For the calibration, the simulation results were always compared to dynamic electrical data from the turn-on event of a well-calibrated 1200 V IGBT for various switching conditions on a calibrated test setup. The same IGBT was used for all measurements and simulations in this work. Since literature values for the electron and hole cross-sections of substitutional platinum from DLTS measurements are valid at cryogenic temperatures, these values only served as a starting point for further calibration. Power law temperature dependencies gave good results for electron- and hole recombination cross-sections. Fig. 2a shows the result of calibration for the IGBT turn-on at 175 °C operating temperature, nominal current, using a standard gate resistor R_G of 23 Ω. All features of the switching transients of IGBT collector-emitter voltage V_{CE}, gate voltage V_{GE}, diode voltage V_{dio}, and, most important, the diode current I_{dio} were accurately reproduced. The effect of the diode is mainly seen in the reverse recovery current, so after changing the current direction at 0.8 μs. Via parasitic stray

inductances, the slopes of the diode current (dI/dt) are interlinked with the diode voltage increase as well as the decline of the IGBT collector-emitter voltage V_{CE}.

Fig. 2b gives an overview over the deviation of the two most important diode key figures Q_{rr} (the reverse recovery charge) and E_{rec} (the diode reverse recovery losses) between simulation and experiment for different temperatures (25 °C and 175 °C), current levels (15 A, 100 A, and 200 A), as well as switching speeds (represented by R_{GS} of 11 Ω, 23 Ω, and 63 Ω). Here, it becomes obvious that the calibration focused on the nominal values, where both Q_{rr} and E_{rec} deviate only minimal (<2%). The larger deviations, which are mostly seen for R_G values of 11 Ω and 63 Ω can partially be attributed to a minor mismatch in the dI/dt calibration of the IGBT model. For 23 Ω and operating temperature (calibration condition of the IGBT model), both E_{rec} and Q_{rr} show deviations below ±4% over a wide current range from 15 A up to 200 A. At room temperature, Q_{rr} deviates up to -8% over the whole current range, indicating that not only the current dependency but also the temperature dependency of Pt was calibrated properly.

PREDICTIVE POWER

The predictive power of the combined process and device simulation is demonstrated by using the unified calibrated Pt model to simulate a variety of diodes from different generations and with rated voltages of 650 V, 750 V, 1200 V, and 1700 V. In terms of Pt process simulation, the consideration of these diodes also covers a wide range of Pt diffusion temperatures. Fig. 3 shows the deviation of Q_{rr} and E_{rec}, extracted from the switching transients of the diodes at their nominal currents, an R_G of 23 Ω, and operating temperature of 175 °C, always in combination with the same 1200 V IGBT. None of the investigated diodes show a deviation of more than 14% between simulation and experiment over this diverse selection of silicon power diodes.

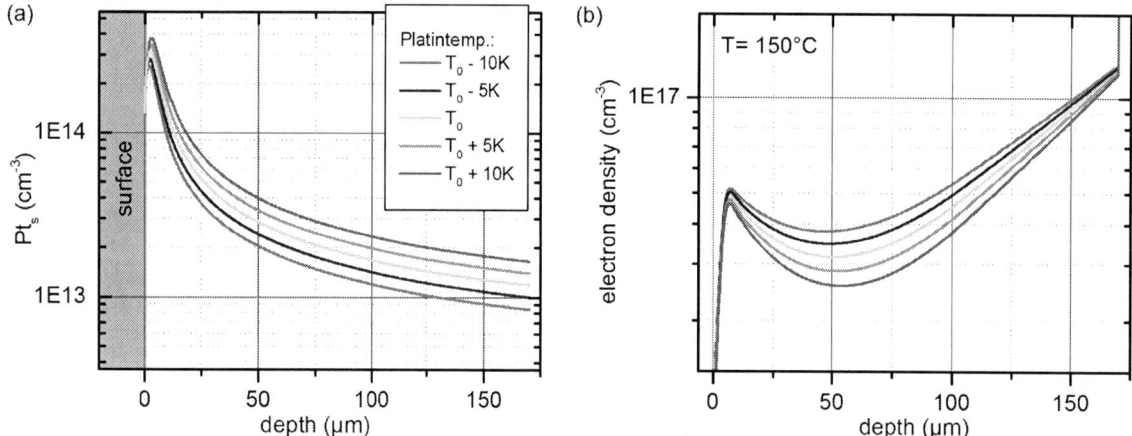

Fig. 4. a) Simulated depth profiles of substitutional platinum (Pt_s) for diffusion temperatures varied in steps of 5 K around a central diffusion temperature T_0 (170 μm thick diode structure). b) Corresponding electron-density profiles at nominal current under forward bias (temperature: 150 °C), with the anode at 0 μm and the cathode at 170 μm.

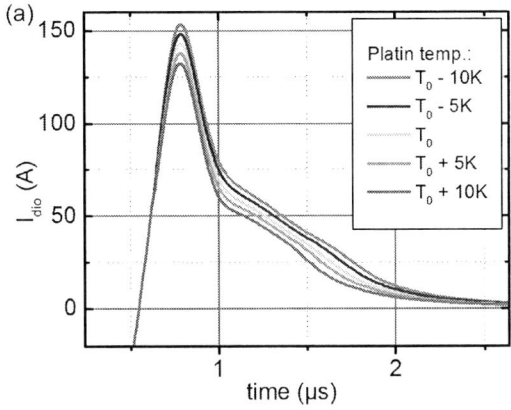 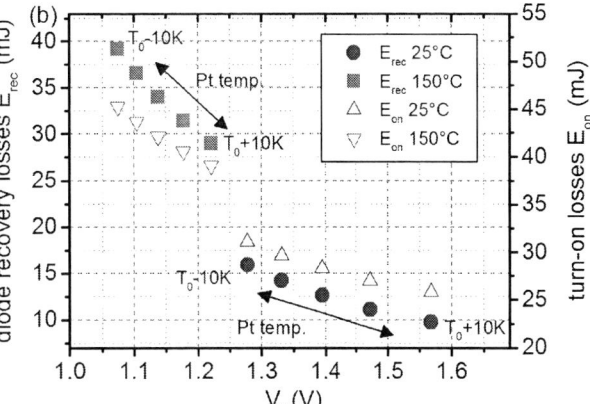

Fig. 5. a) Simulated diode current transients during the reverse recovery for diffusion temperatures varied in steps of 5 K around a central diffusion temperature T_0 (170 μm thick diode structure). High Pt concentrations lead to a decrease in the reverse-recovery peak height and less current tail. b) Simulated trade-off diagram comparing the dynamic (E_{rec}, E_{on}) and static (V_f) losses at room temperature and 150 °C.

IV. DIODE OPTIMIZATION

Finally, we demonstrate how the unified TCAD model of process and device simulation for platinum in silicon power diodes can be used to tailor the device performance. As an example, a test diode structure with a thickness of 170 μm was investigated for a range of Pt diffusion temperatures, varied in steps of 5 K around a central diffusion temperature T_0. Fig. 4a shows how the higher furnace plateau temperature leads to an increase in the local Pt_s concentration both at the surface and in the device's bulk. Unlike above (Sec. II), where the plateau time variation left the surface Pt_s concentration unchanged, the temperature dependency of the equilibrium concentrations is reflected. In device simulation, the quasi-stationary charge-carrier plasma distribution of the diode under forward bias is compared in Fig. 4b. An increased Pt_s concentration profile leads to more recombination throughout the whole depth of the device, resulting in a larger curvature of the electron and hole concentration profiles and, in general, less charge carriers in the device. Furthermore, an impact of the platinum on the effective emitter efficiencies of anode and cathode can be observed.

In Fig. 5a, the effect of increasing Pt concentration on the reverse-recovery behavior is illustrated: the higher recombination rate reduces both, the maximum current I_{RRM}, as well as the tail current. Therefore, also the dynamic losses of the diode (E_{rec}) and the IGBT (E_{on}) during turn-on are affected such that they decrease with increasing platinum temperature. Fig. 5b finally shows that this way also predictive tradeoff diagrams, E_{rec} and E_{on} vs. V_f, can be created. In combination with the Infineon power simulation tool (IPOSIM) [5], a full product optimization can be conducted without requiring any experimental support, tailoring our products to a best fit to the application requirements.

V. CONCLUSION

A unified TCAD model for the consideration of Pt in silicon power diodes was created by implementing a suitable Pt diffusion model into the process simulation and using the resulting Pt_s profile in the device simulation. After thorough calibration, this model demonstrated a remarkable predictive power and, therefore, can be used for the optimization of the device performance. For the first time, now, the complete development process of silicon power diodes can be covered by simulation, which drastically speeds up development times and reduces costs.

ACKNOWLEDGMENT

The authors thank Anna Johnsson and Peter Pichler from the Fraunhofer Institute for Integrated Systems and Device Technology IISB for their support with the DLTS measurements and fruitful discussions.

REFERENCES

[1]. B.J. Baliga and E. Sun, "Comparison of gold, platinum, and electron irradiation for controlling lifetime in power rectifiers", IEEE Transactions on Electron Devices, vol. 24, pp. 685-688, 1977.

[2] F. J. Niedernostheide, F. Hille, and H.-J. Schulze, „Non-uniform platinum profiles in diodes: Experimental extraction and influence on the switching behaviour", Proc. 9th International Seminar on Power Semiconductors (ISPS), Prague 2008, pp. 93 - 97.

[3] E. Badr, P. Pichler, and G. Schmidt, "Modelling platinum diffusion in silicon", J. Appl. Phys. 116, 133508 (2014)

[4] F. Frank and D. Turnbull, "Mechanism of Diffusion of Copper in Germanium", Phys. Rev. 104, 617 (1956)

[5] Infineon Technologies IPOSIM power simulation tool, [online] www.infineon.com/iposim.

Proceedings of the 30th International Symposium on Power Semiconductor Devices & ICs
May 13-17, 2018, Chicago, USA

Novel 3D narrow mesa IGBT suppressing CIBL

Masahiro Tanaka

Nihon Synopsys G.K.

2-21-1 Tamagawa, Setagaya-ku Tokyo, 158-0094, JAPAN.

E-mail: mtanaka@mem.iee.or.jp

Akio Nakagawa

Nakagawa Consulting Office, LLC.

3-8-74 Hamatake, Chigasaki-city, Kanagawa, 253-0021, JAPAN.

E-mail: akio.nakagawa.dr@ieee.org

Abstract—**It was reported that the experimentally fabricated very narrow mesa IGBT has poor short-circuit withstand capability because of CIBL. We propose a novel narrow mesa IGBT, which suppresses CIBL. Additional deep P+ diffusion layer inside the P-base improves CIBL by reducing the enhanced conductivity modulation in the channel inversion layer. The structure achieves good short-circuit withstand capability and superior trade-off relationship between on-state voltage drop and turn-off loss.**

Keywords—Narrow mesa IGBT, CIBL, Short-Circuit, Conductivity Modulation

I. INTRODUCTION

IGBTs are widely used in the middle to high power electronics field, and the market is expanding to Zero Emission Vehicles and renewable energy sources. It was predicted that the IGBTs with extremely narrow mesa realize theoretical silicon limit performance[1]. However, it was reported recently that the actually fabricated very narrow mesa IGBTs[4-8] have poor short-circuit withstand capability because of CIBL (Collector bias Induced Barrier Lowering) related current filamentation[2]. In ISPSD 2017, we showed that the cause of CIBL in the very narrow mesa IGBT is that the enhanced conductivity modulation occurs in the channel inversion layer, which includes the whole mesa region[3]. We discussed that the combination of the CIBL and high transconductance is the cause of short-circuit failure because the electric potential increase in the P-base induces the gate voltage overshoot. We also mentioned that the mesa width should be set above 200nm to prevent the conductivity modulation in the channel inversion layer of the whole mesa and achieve sufficient short-circuit withstand capability.

In this paper, we, for the first time, propose a novel narrow mesa structure with additional deep P+ layers, which suppress CIBL even in 40nm mesa IGBTs. The potential in the deep P+ layer becomes stable even when large current flows in the mesa region. It successfully prevents CIBL and achieves good short-circuit withstand capability. The structure even improves on-state voltage drop.

II. TCAD SIMULATION SETUP

Two types of 40nm narrow mesa 1200V IGBTs have been analyzed by TCAD 3D simulations. First one is conventional narrow mesa structure, which has 0.1um wide N+ source and 0.4um wide P+ contact layers (see Fig.1 and 2.) The other is the proposed structure, which has an additional deep P+ diffusion under the P+ contact layer. The mesa width is varied from 40nm to 100nm. The half unit cell pitch for the width direction is kept at 0.5um, and the pitch for the depth direction is 0.5um. The trench gate depth is 1.0um. The gate oxide thickness and the rated gate voltage are set to be 33nm and 5V, respectively. The N-base thickness is 100um. 8-cell IGBT structure was also used in short-circuit simulation to confirm whether current filament appears. The points: (A), (B), (C), and (D), shown in Fig. 1, are used to probe the values of potential and current density, etc. in the short-circuit simulations in the next Section. The cut-lines: (E), (F), (G) and (H) in Fig. 2 are used to plot the carrier profiles in the next section.

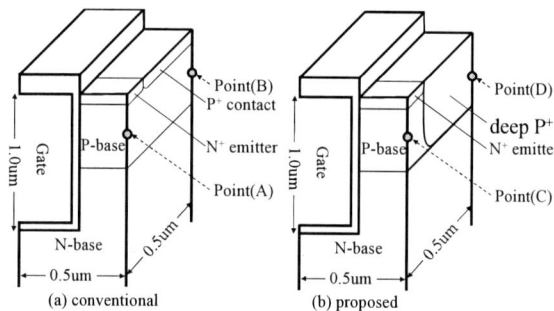

Fig. 1. Schematic view of the simulated structures. The mesa width is varied from 40nm to 100nm. Proposed structure has additional deep P+ diffusion layer. The points: (A), (B), (C), and (D) in the figure are used to plot the values of potential and current density, etc. in Figs. 7.

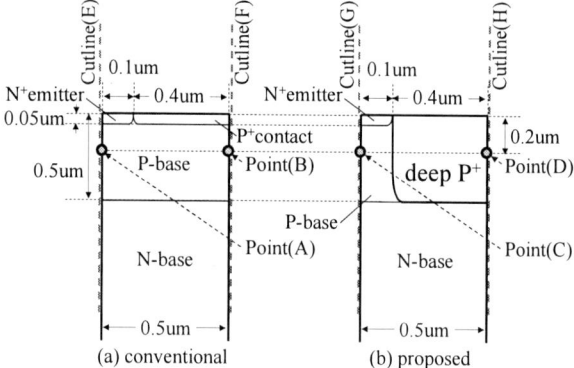

Fig. 2. Cross sectional view in the depth direction of the simulated structures. The cut-lines: (E), (F), (G) and (H) in the figure are used to plot the carrier profiles in Figs. 3, 8 and 10.

978-1-5386-2928-4/18 $31.00 © 2018 IEEE

III. RESULTS AND DISCUSSION

A. CIBL Suppression

We show that CIBL of the conventional narrow mesa IGBT can be successfully suppressed by the proposed structure. The proposed deep P^+ layer even prevents the potential increase inside the N^+ emitter. Fig. 3 shows the potential diagrams of the conduction band edge. In the conventional narrow mesa IGBT, the potential inside the N^+ emitter rises as the applied collector voltage increases. This is because the conductivity modulation occurs in the whole mesa region and the collector voltage directly affects the potential in the P-base and the N^+ emitter. In contrast, in the proposed structure, the potential is stable even when the collector voltage is high. The conductivity modulation in the channel inversion layer is weakened because a large portion of the hole current flows in the deep P^+ layer. In result, the proposed structure improves I_C-V_C saturation characteristics as shown in Fig. 4. And the proposed structure has lower transconductance than the conventional narrow mesa structure as shown in Fig. 5. It stabilizes the current level during the short-circuit operation.

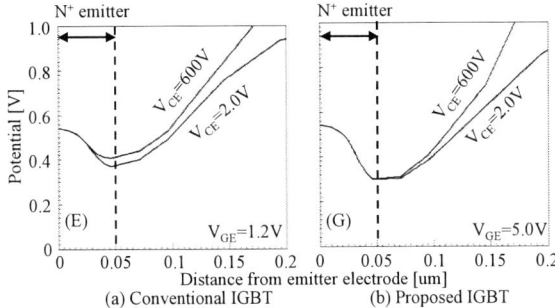

(a) Conventional IGBT   (b) Proposed IGBT

Fig. 3. Potential diagrams in the cutline (E) and (G) of calculated 40nm mesa IGBTs. Proposed structure prevents potential increase inside N^+ emitter and improves CIBL.

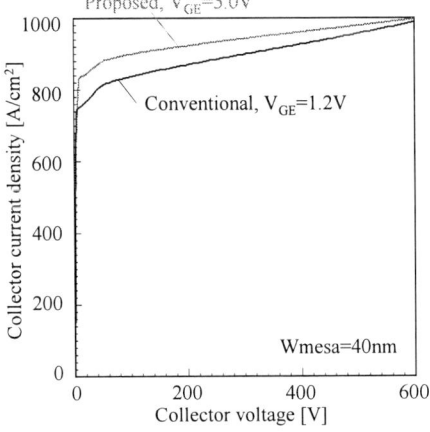

Fig. 4. Current saturation characteristics of two calculated IGBTs. The gate voltage is adjusted so that the saturation current reaches 1000A/cm² at V_{CE}=600V. Proposed structure shows good current saturation characteristics.

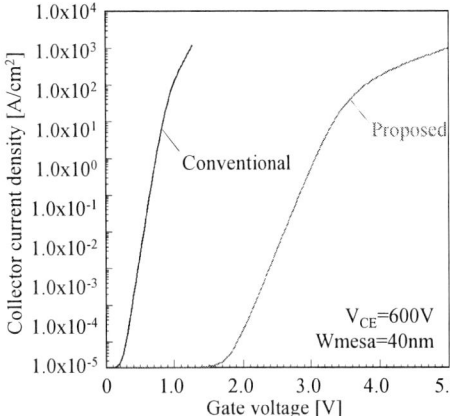

Fig. 5. I_C-V_G characteristics of calculated 40nm mesa IGBTs. Proposed structure reduces the transconductance. It stabilizes the short-circuit current.

B. Short-Circuit Withstand Capability

The proposed 40nm IGBT successfully exhibits a good short-circuit withstand capability, as seen in Fig. 6(b), although conventional 40nm mesa IGBT fails, as seen in Fig. 6(a). It should be noted that in the 40nm mesa IGBT, the whole mesa becomes channel inversion layer when a positive gate voltage is applied.

Conventional narrow mesa IGBT has poor short-circuit withstand capability because of the gate voltage overshoot during the operation. When large hole current flows in the mesa region, the potential at point (A) under the N^+ emitter becomes lower than that of point (B) under the P^+ contact layer as shown in Fig. 7(a). It follows that the hole current in the P-base flows toward the N^+ emitter and enhances the conductivity modulation in the channel inversion layer as shown in Fig. 8(a). It induces the gate voltage overshoot, resulting in the collector current increase far beyond the saturation current because of high transconductance.

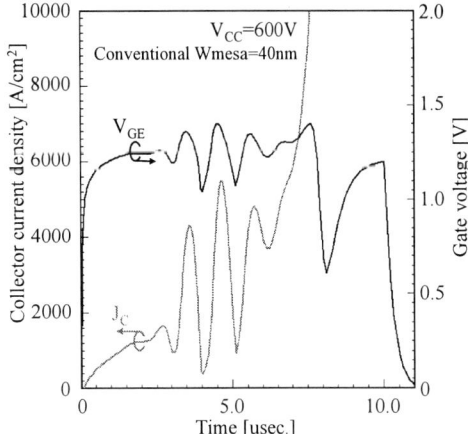

Fig. 6(a). Short-circuit waveforms of conventional 40nm IGBT. Gate voltage overshoot occurs and it increases collector current beyond the saturation current because of the high transconductance.

978-1-5386-2928-4/18 $31.00 © 2018 IEEE   125

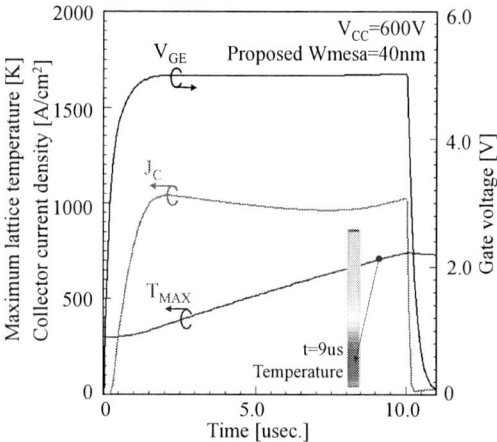

Fig. 6(b). Short-circuit waveforms of proposed 40nm IGBT. The simulated structure is 8-cell. Gate voltage overshoot does not occur. The collector current is saturated. The maximum lattice temperature was observed in the high electric field region in the N-base as discussed in[3]. No current filament was observed.

In contrast, in the proposed structure, the potential in the deep P⁺ layer is stably lower than that of point (C) under the N⁺ emitter during short-circuit operation, as shown in Fig. 7(b). Fig. 7(b) also shows that a large portion of the hole current flows inside the deep P⁺ layer. The hole density under the N⁺ emitter is greatly reduced, as shown in Fig. 8(b). The conductivity modulation in the channel inversion layer is weakened and the gate voltage overshoot is suppressed in the proposed structure. Also, the transconductance of the proposed structure is lower than conventional narrow mesa IGBT as shown in Fig. 5. In result, the proposed structure achieved stable short-circuit operation.

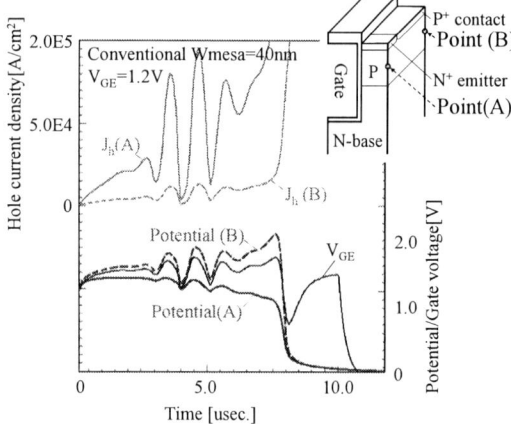

Fig. 7(a). Hole current densities, electric potentials and gate voltage are plotted as a function of time for points (A) and (B) for conventional narrow mesa IGBT during the short-circuit operation. The potential of point (A) under the N⁺ emitter becomes lower than that of point (B) under the P⁺ contact layer. It follows that the hole current in the P-base flows toward the N⁺ emitter and enhances the conductivity modulation. It induces the gate voltage overshoot and rapidly increases the collector current, resulting in the short-circuit failure.

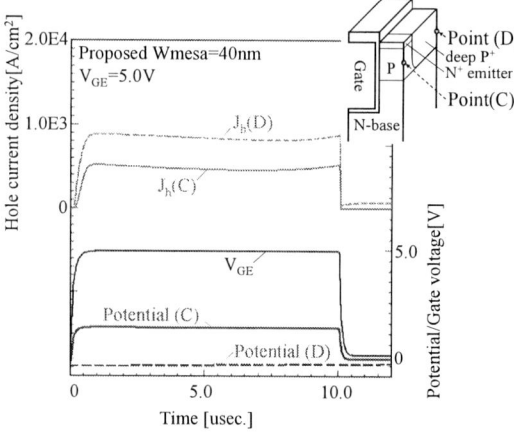

Fig. 7(b). Hole current densities, electric potentials and gate voltage are plotted as a function of time for points (C) and (D) for proposed IGBT during the short-circuit operation. The potential of point (D) in the deep P⁺ layer is lower than that of point (C) under the N⁺ emitter. A large portion of the hole current flows inside the deep P⁺ layer. The conductivity modulation and gate voltage overshoot are suppressed.

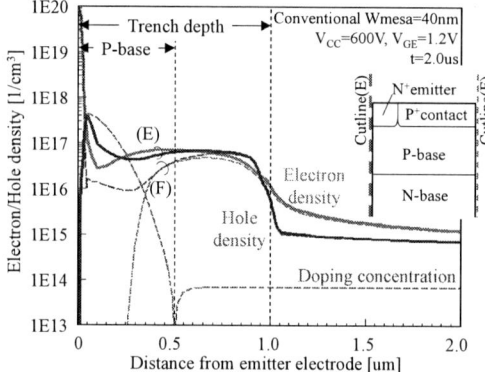

Fig. 8(a). Carrier distributions along the cut-lines (E) and (F) at t=2us are plotted for conventional narrow mesa IGBT. The hole density inside the P-base under the N⁺ emitter along the cut-line (E) is high

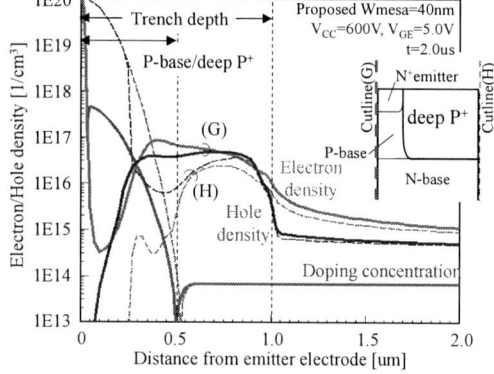

Fig. 8(b). Carrier distributions along the cut-lines (G) and (H) at t=2us are plotted for proposed IGBT. The hole density inside the P-base under the N⁺ emitter along cut-line (G) is greatly reduced.

C. $V_{CE}(sat)$-E_{OFF} Trade-off

It is found that the proposed 40nm IGBT achieves very low on-state voltage drop below 1.1V as shown in Fig. 9, even if the saturation current is set to be 1000A/cm^2 at V_{CE}=600V. The rated gate voltage of 5V can be applied to the proposed IGBT because it has lower transconductance than conventional narrow mesa IGBT. Because of the higher 5V gate voltage, the higher electron density is induced in the mesa region as shown in Fig. 10, resulting in lower on-state voltage drop. Fig. 11 shows the trade-off relationship between on-state voltage drop and turn-off loss. The 40nm mesa structure realizes extremely low on-state voltage drop, which is close to the Silicon limit characteristic. It is found that the 100nm mesa IGBT with deep P$^+$ diffusion shows superior trade-off relationship of $V_{CE}(sat)$=1.28V at 200A/cm^2 and E_{OFF}=10mJ at 200A.

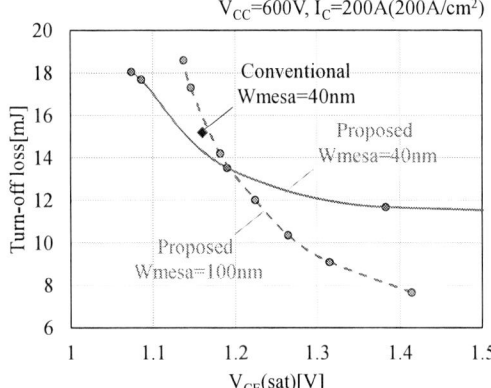

Fig. 11. Trade-off between on-state voltage drop and turn-off loss of calculated IGBTs. The stray inductance between bridge arms was set to 10nH. The gate resistance was set to 1ohm.

IV. CONCLUSION

A novel narrow mesa IGBT structure is proposed. Additional deep P$^+$ layer in the P-base improves CIBL by suppressing the conductivity modulation in the channel inversion layer in the narrow mesa IGBTs of 40nm and 100nm mesa width. The structure achieves sufficient short-circuit withstand capability and superior trade-off relationship between on-state voltage drop and turn-off loss.

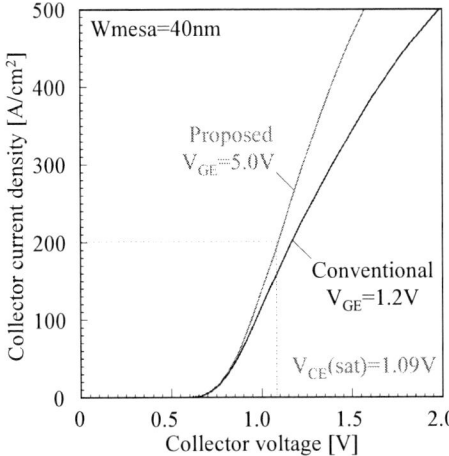

Fig. 9. I_C-V_C characteristics of calculated IGBTs. The gate voltage is adjusted as shown in Fig. 4. Proposed structure improves on-state voltage drop.

REFERENCES

[1] A. Nakagawa, "Theoretical Investigation of Silicon Limit Characteristics of IGBT", Proc. of ISPSD'06, pp. 5-8, 2006.

[2] K. Eikyu, A. Sakai, H. Matsuura, Y. Nakazawa, Y. Akiyama, Y. Yamaguchi and M. Inuishi, "On the Scaling Limit of the Si-IGBTs with Very Narrow Mesa Structure", Proc. of ISPSD'16, pp. 211-214, 2016.

[3] M. Tanaka and A. Nakagawa, "Conductivity modulation in the channel inversion layer of very narrow mesa IGBT", Proc. of ISPSD'17, pp.61-64, 2017.

[4] M. Sumitomo, J. Asai, H. Sakane, K. Arakawa, Y. Higuchi, and M. Matsui, "Low loss IGBT with Partially Narrow Mesa Structure (PNM-IGBT)", Proc. of ISPSD'12, pp.17-20, 2012.

[5] M. Antoniou, N. Lophitis, F. Udrea, F. Bauer, I. Nistor, M. Bellini and M. Rahimo, "Experimental demonstration of the p-ring FS+ Trench IGBT concept: A new design for minimizing the conduction losses", Proc. of ISPSD'15, pp. 21-24, 2015.

[6] K. Kakushima, T. Hoshii, K. Tsutsui, A. Nakajima, S. Nishizawa, H. Wakabayashi, I. Muneta, K. Sato, T. Matsudai, W. Saito, T. Saraya, K. Itou, M. Fukui, S. Suzuki, M. Kobayashi, T. Takakura, T. Hiramoto, A. Ogura, Y. Numasawa, I. Omura, H. Ohashi, and H. Iwai, "Experimental Verification of a 3D Scaling Principle for Low Vce(sat) IGBT", IEDM 2016 Technical Digest, pp. 10. 6. 1-4, 2016.

[7] H. Feng, W. Yang, Y. Onozawa, T. Yoshimura, A. Tamenori and J. K. O. Sin, "A 1200V-class Fin P-body IGBT with Ultra-narrow-mesas for Low Conduction Loss", Proc. of ISPSD'16, pp. 203-206, 2016.

[8] M. Shiraishi, T. Furukawa, S. Watanabe, T. Arai and M. Mori, "Side Gate HiGT with Low dv/dt Noise and Low Loss", Proc. of ISPSD'16, pp. 199-202, 2016.

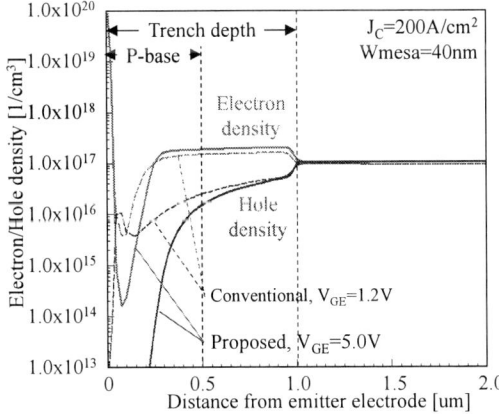

Fig. 10. Comparison of the on-state carrier distribution between the cutline (E) and (G) of calculated 40nm mesa IGBTs. The conductivity modulation in the mesa region is weakened in the proposed IGBT(solid lines). It has higher electron density in most of the mesa region than conventional one because higher gate voltage can be applied. It leads lower on-state voltage drop.

Proceedings of the 30th International Symposium on Power Semiconductor Devices & ICs
May 13-17, 2018, Chicago, USA

N-buffer design optimization for Short Circuit SOA ruggedness in 1200V class IGBT

Kenji Suzuki, Koichi Nishi, Mitsuru Kaneda, Akihiko Furukawa

Power Device Works
Mitsubishi Electric Corporation
1-1-1 Imajyuku-Higashi, Nishi-ku, Fukuoka, 819-0192, Japan
e-mail: Suzuki.Kenji@cs.MitsubishiElectric.co.jp

Abstract— In this paper, a new IGBT with deep n-buffer structure called <u>C</u>ontrolling charge carrier <u>P</u>lasma <u>L</u>ayer (CPL) is proposed. CPL is shown to be effective from the device simulation of the electric field and temperature distribution during the short circuit. A large 230A rated current chip with CPL in 1200V class demonstrated the better turn-off softness and short circuit ruggedness than the conventional structure. On behalf of CPL, even 10% thinner n-drift device shows much higher turn-off capability up to 1670A corresponding to 7.2 times of the rated current, less ringing and higher short circuit endurance energy (Esc), maintaining the better V_{CEsat}-E_{OFF} tradeoff.

Keywords: Trench-gate IGBT, Short Circuit SOA, N-buffer design optimization

I. INTRODUCTION

For the progress of power electronics, it is necessary to enhance the performance of Insulate Gate Bipolar Transistors (IGBTs) and diodes, both mounted on the power modules. We have been developing low-loss structures for IGBTs, such as CSTBT^{TM}, the fine pattern processing and the advanced thin wafer technology [1-3]. CSTBT^{TM} with fine pattern processing has made a drastic improvement of the trade-off relationship between on-state forward voltage drop (V_{CEsat}) and the turn-off energy loss (E_{OFF}) towards the ideal flat carrier distribution [3]. Overcoming the weak point of the first generation CSTBT^{TM}, i.e. relatively narrow SCSOA based upon the short channel like action and the high saturation current, CSTBT^{TM} with dummy trenches could come to be the best solution [4]. To compensate the thin wafer structure like a Light Punch Through (LPT), Field Stop (FS) and Soft Punch Through (SPT) [4-7], the electric field inside the device has been relaxed by optimizing the back side n-buffer structure to keep for the sufficiently wide short-circuit ruggedness, while maintaining the better tradeoff characteristic between power losses and SOAs. As the n-buffer design, thick and/or low concentration are effective to soften the electric field distribution around the backside. Simply maintaining the total thickness of the Si device causes the deterioration of the breakdown voltage. The wafer thicknesses is going to be close to the theoretical limit for a given voltage class and it is careful to design the bulk structure as well as the n-buffer structure. The latest trend of Si bulk design indicates the higher resistivity and the thinner n- drift layer to ensure both the high blocking capability and low total loss [8]. Our proposed deep n-buffer structure called

Controlling charge carrier Plasma Layer (CPL), which is very effective for the diode to improve the tradeoff between the total power loss and RRSOA [9], would be also applicable for IGBT's high ruggedness against short circuit.

II. NEW N-BUFFER STRUCTURE

A. Device Structure

Figure 1 shows the cross-sectional views of the conventional CSTBT^{TM} and the new CSTBT^{TM} with CPL region. New structure has thick but low concentration n-buffer. The CPL is formed by high-energy light ion implantation and annealed under a relatively low temperature process. The two structures have the same MOS structure based on the fine pattern processing.

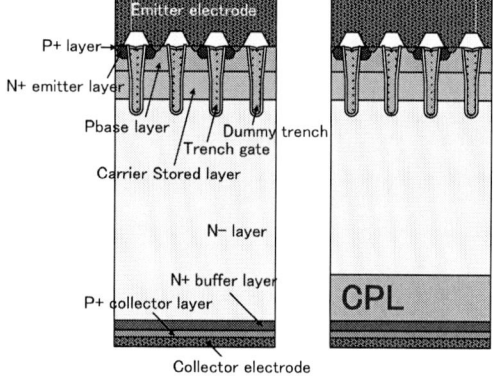

(a) Conventional CSTBT^{TM} (b) New CSTBT^{TM} with CPL

Fig. 1: Cross sectional view of Conventional CSTBT^{TM} and New CSTBT^{TM} with CPL

B. Device Simulation

The device simulations were done by using a simple setup with a two-dimensional IGBT half-cell, and we extracted the time-intersections for the electric field and temperature distribution during an IGBT short-circuit operation to evaluate the effectiveness of CPL. After several V_{CEsat} values obtained by changing the p-collector concentration following the same manner of the fabrication step, the saturation current is adjusted to match the measured value by connecting the small resistance on the emitter electrode. The electric field of the conventional structure with only thin n-buffer is likely to become high at the front side of a chip (Fig.2, 3) during the short circuit. Since the electron current from the individual MOS-channels of the IGBT cells is limited, high hole concentration at the backside hindered the depletion layer sufficiently spreading (Fig.3). The device temperature of the front side of the two structures with the same V_{CEsat} under short circuit is shown in Fig.4. Our CPL region acts not only as the thick and low concentration n-buffer to support the space charge region deeply penetrating into, but also as the hole-filter to regulate hole injection from the relatively highly doped p-collector during the short circuit. So, the preferable electric field distribution shown in Fig. 3 is formed while keeping the preferable distribution of the total carrier concentration of CSTBT™. Even as thinning fashion of the IGBT drift region, CPL region is one of the wild card to maintain the total balance of electrical characteristics including SCSOA. Furthermore, the thin and high concentration n-buffer suppresses the pnp bipolar gain and the extension of the depletion layer until the collector layer at high V_{CC}, so it is effective to reduce leak current at high temperature operation. As a result, the device failure is hard to occur after a few hundred microseconds of the short-circuit turn-off.

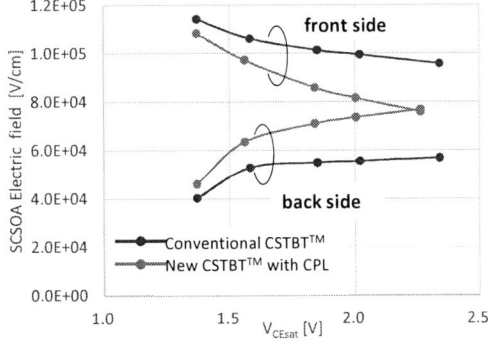

Fig. 2: Simulated relationship between V_{CEsat} and the electric field of the front side and the backside under short circuit

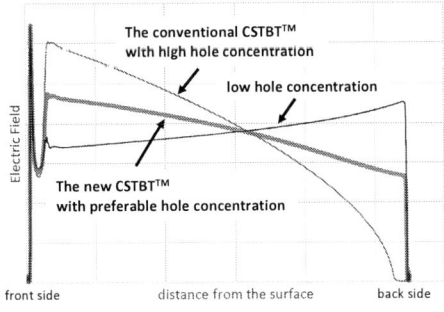

Fig. 3: Preferable electric field distribution under short circuit condition

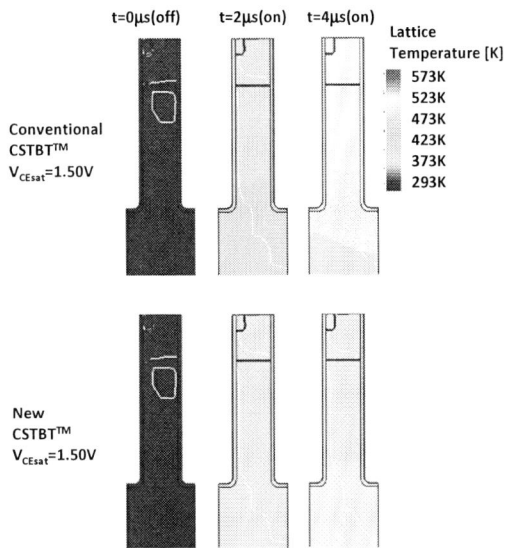

Fig. 4: Simulated device temperature having the same V_{CEsat} at t=0, 2, 4µs under short circuit condition

III. EXPERIMENTAL RESULTS

A. Turn-off Oscillation

A large 230A rated current chip with CPL in 1200V voltage class was fabricated. Turn-off waveforms for conventional CSTBT™ and new CSTBT™ are shown in Fig.5 at the condition of V_{CE}=300 - 600V, I_C=230A, Tj=25 degC with large stray inductance. The conventional structure shows a high and sharp voltage peak (snap-off) in case of V_{CE}=600V where the tail current was sharply shut off in the latest time period. The new structure did not show oscillation until high V_{CC} on behalf of the remaining hole injected from the relatively highly doped p-collector through CPL. As a result, CSTBT™ with CPL is able to lower the collector concentration and be suitable for the high frequency applications.

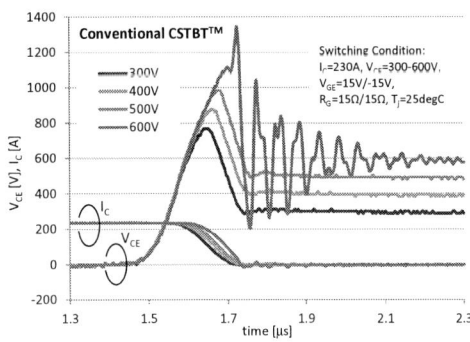

978-1-5386-2928-4/18 $31.00 © 2018 IEEE

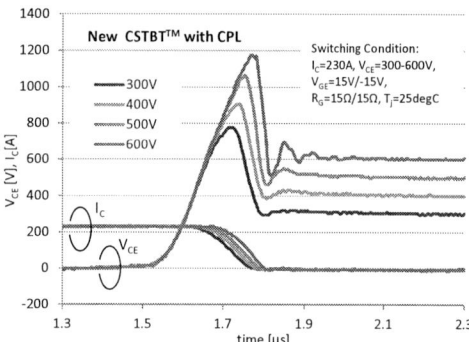

Fig. 5: Turn-off waveforms at the condition of large stray inductance

B. SCSOA

Short Circuit (SC) time endurance t_w of the new structure is 20% longer than the conventional one (Fig. 6), and this longer t_w is translated into 10% higher E_{SC} at the condition of V_{CE}=800V, Tj=150 degC. On behalf of CPL structure as the "hole-filter", I_C slowly raised up, and its temperature elevation is relatively lower in the early stage of SC. And this effect for E_{SC} is more clearly shown as V_{CEsat} dependence in the wide range (Fig, 7). V_{CEsat} is controlled by p-collector concentration. E_{SC} of the new CSTBT™ stays almost constant. E_{SC} of the conventional CSTBT™ becomes low in the case of low V_{CEsat} because the electric field at the front side becomes high as shown in Fig.2, 3. Fig. 8 shows the wafer thickness dependence, i.e. the final chip thickness. New CSTBT™ is able to shrink the wafer thickness, maintaining wide SCSOA.

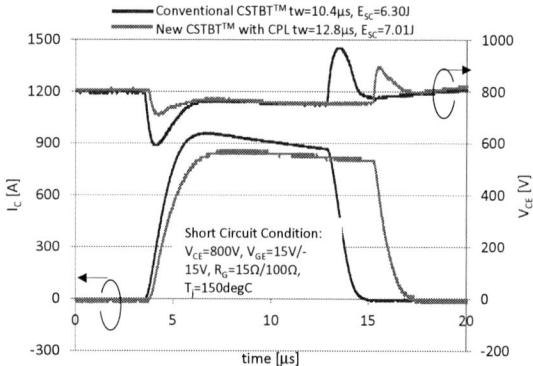

Fig. 6: Short circuit waveform

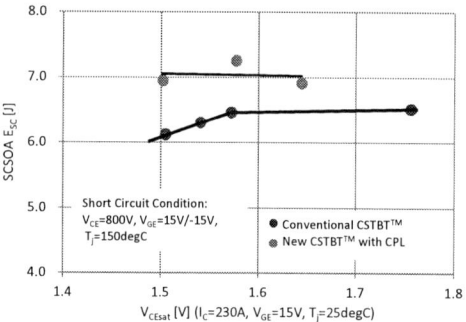

Fig. 7: Short-circuit capability dependence on V_{CEsat}

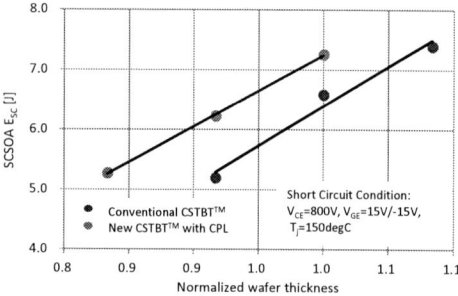

Fig. 8: Short-circuit capability dependence on the wafer thickness

C. Switching Characteristics

CPL effect maintaining both less-ringing and high E_{SC}, about 10% reduction of wafer thickness improved V_{CEsat} and E_{OFF} tradeoff at the condition of V_{CE}=800V, I_C=230A, Tj=150 degC (Fig. 9). And this thin wafer structure also has much higher turn-off capability up to 1670A (Fig. 10) corresponding to 7.2 times of the rated current at the condition of V_{CE}=800V, Tj=175 degC.

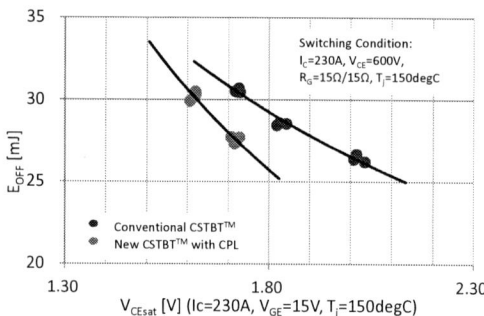

Fig. 9: The trade-off relationship between V_{CEsat} and E_{OFF}

978-1-5386-2928-4/18 $31.00 © 2018 IEEE   130

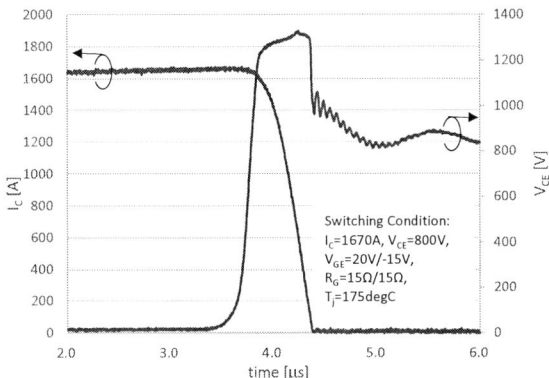

Fig. 10: RBSOA waveform

D. The Summary of the Device Performance

The wafer resistivity, as the n-drift doping concentration, is increased by 25 percent to maintain sufficient breakdown voltage, considering the distribution of the bulk Si material doping. The high wafer resistivity satisfies a softer electric field distribution and a higher breakdown voltage. So the proposed IGBT structure with CPL realizes the good total characteristics by the thinner wafer summarized in Table.1.

Table. 1: Comparison of Conventional CSTBT™ and New CSTBT™

	Conventional CSTBT™	New CSTBT™
Wafer resistivity	1.0	1.25
Wafer thickness	1.0	0.9
n-buffer	lightly and thin n-buffer	lightly and thin n-buffer and **CPL region**
Turn-off oscillation at the condition of high stray inductance	Appeared	Without
FOM = $J_C/(V_{CEsat} \times E_{OFF})$	1.0	1.1
SCSOA Esc	1.0	1.0
RBSOA	1670A (>7.2 times higher than the rated current) at 175degC	

IV. CONCLUSION

Through the SCSOA device simulation, the CPL confirmed to improve the electric field and temperature distribution. A new IGBT structure with CPL realizes the good characteristics by reducing the wafer thickness, maintaining turn-off softness and high SCSOA. That is because CPL seems to act as not only the thick and low concentration n-buffer but also the selective hole injection-recombination filter dynamically corresponding to the current density as the preferable direction of high power bipolar devices.

ACKNOWLEDGMENT

The authors would like to thank Mr. Y. Kawase, Mr. K. Kanada, and Mr. H. Minamitake for fabricating the devices utilized in this study; Mr. K. Nakamura and Mr. T. Takahashi for useful discussions on the CPL concept; and Mr. Y. Takata for his unrelenting encouragement.

REFERENCES

[1] H. Takahashi, H. Haruguchi, H. Hagino and T. Yamada, "Carrier Stored Trench-Gate Bipolar Transistor (CSTBT) – A Novel Power Device for High Voltage Application", Proc. ISPSD '96, pp. 349-352.

[2] Y. Haraguchi, S. Honda, K. Nakata, A. Narazaki and Y. Terasaki, "600V LPT-CSTBTTM on Advanced Thin Wafer Technology", Proc. ISPSD'11, pp.68-71.

[3] T. Takahashi, Y. Tomomatsu and K. Sato, "CSTBT™(III) as the next generation IGBT", Proc. ISPSD'08, pp.72-75.

[4] K. Nakamura, S. Kusunoki, H. Nakamura, Y. Ishimura, Y. Tomomatsu and M. Harada, "Advanced Wide Cell Pitch CSTBTs Having Light Punch Through (LPT) Structures", Proc. ISPSD'02, pp.277-280

[5] M. Otsuki, Y. Onozawa, M. Kirisawa, H. Kanemaru, K. Yoshihara and Y. Seki, "Investigation on the Short-Circuit Capability of 1200V Trench Gate Field-Stop IGBTs", Proc. ISPSD'02, pp.281-284.

[6] S. Dewar, S. Linder, C. von Arx, A. Mukhitinov and G. Debled, "Soft Punch Through (SPT) – Setting new Standards in 1200V IGBT", Proc. PCIM2000. pp. 593-601..

[7] T. Laska, M. Munzer, F. Pfirsch, C. Schaeffer, T. Schmidt, "The Field Stop IGBT (FS IGBT) - A New Power Device Concept with a Great Improvement Potential", Proc. ISPSD'00, pp.355-358.

[8] C. Jaeger, A. Philippou, A. Vellei, J.G. Laven and A. Härtl, "A new sub-micron trench cell concept in ultrathin wafer technology for next Generation 1200 V IGBTs", Proc. ISPSD'17, pp.69-72.

[9] K. Nakamura and K. Shimizu, "Advanced RFC Diode utilizing a Novel Vertical Structure for Softness and High Dynamic Ruggedness", Proc. ISPSD'17, pp.117

Proceedings of the 30th International Symposium on Power Semiconductor Devices & ICs
May 13-17, 2018, Chicago, USA

High avalanche capability specific Diode part structure of RC-IGBT based upon CSTBT™

Shinya Soneda, Shinya Akao, Tetsuo Takahashi and Akihiko Furukawa
Power Device Works, Mitsubishi Electric Corporation
1-1-1 Imajyuku-Higashi, Nishi-ku, Fukuoka, 819-0192, Japan
Phone: +81-92-805-3332 E-mail: Soneda.Shinya@dw.MitsubishiElectric.co.jp

Abstract— A thin RC-IGBT is facing to a difficulty of certifying a stable BV characteristic among the three portions, i.e. IGBT (pnp), Diode (pin) and termination (pnp too). We propose a high avalanche capability specific Diode part structure of the RC-IGBT based upon CSTBT™. This stabilization of the BV characteristic was realized by adjusting the BV balance of three portions described above. One of the successful approach was lowering the Diode's BV by widening Diode's trench pitch (W$_{DTP}$).

Keywords—RC-IGBT; CSTBT; stable blocking voltage; p-n-p; p-n-n+; Avalanche; Capability

I. INTRODUCTION

There is a strong demand for a miniaturization and a high power density of a power module. A Reverse Conducting – Insulated Gate Bipolar Transistor (RC-IGBT) has the advantage for downsizing and a high power density because of the integrating Free Wheel Diode (FWD) into a single IGBT chip. A total power loss of the RC-IGBT has been improved by a progress of a thin wafer process technology, optimizations of the layout and the structure, or using an optimized gate-pulse pattern [1-3]. Applications have been spreading for home appliances [4], industrial motor drives [5], and electric vehicles [6]. A stable Blocking Voltage (BV) characteristic, which means that the avalanche phenomenon occurs without a destruction, is required under the specific usage such as high V$_{CC}$ and/or frequency condition or a high reliability usage. However, it is difficult to certify the stable BV characteristic of the RC-IGBT within the thin wafer because of a Negative Differential Resistance (NDR) characteristic of pnp structure [7]. It has become an obstacle to improve the total power loss of the RC-IGBT within the thin wafer. We tried to solve this problem by proposing high avalanche capability specific Diode part structure of the RC-IGBT based upon CSTBT™.

II. ISSUE OF CONVENTIONAL RC-IGBT STURCTURE

A chip layout and three major components, i.e. IGBT part, Diode part and Edge termination part, of the conventional RC-IGBT are shown in Fig.1 as a schematic overview and cross sectional views. The IGBT part has CSTBT™ structure with an extra n type carrier stored layer (CS-layer) underneath the p-base and narrow pitch deep trench gates. The trench gates act as a filed plate to suppress the electric field rise at the CS-layer

under the high V$_{CE}$ condition. The Diode part has the same CS-layer and trenches which is connected to the emitter electrode in order to avoid the additional processes for the Diode part. And Diode's trench pitch (W$_{DTP}$) is the same as IGBT's one. The back side layers of the IGBT part and the Diode part are p type collector and n type cathode, respectively. These are formed separately by using a backside photo-lithography technology.

The edge termination part is a Filed Limiting Ring (FLR) with p type diffusion layers. From the viewpoint of certifying the stable BV characteristic, the back side layer of the FLR is preferable to use n type or partial n type layer [8]. However, since our RC-IGBT has the CSTBT™ structure, the FLR needs to be deep and high concentration p regions as a guarding for the last trench in the trench array. For this reason, the backside of our FLR have to be p type layer to suppress the strong FLR action of pin Diode.

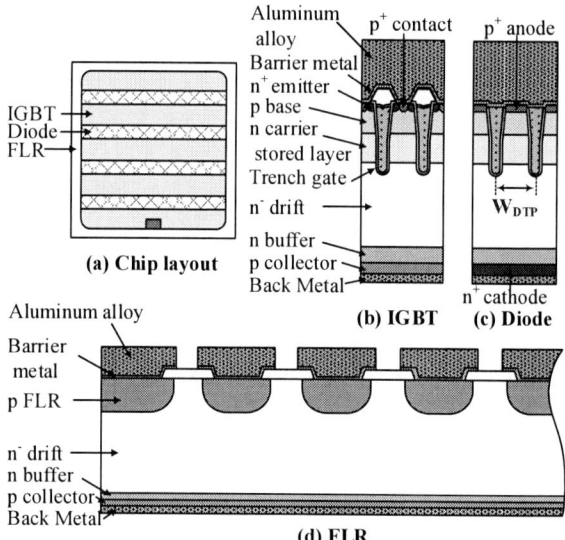

Fig. 1: The chip layout and three major components as a schematic overview and cross sectional views of the conventional RC-IGBT.

978-1-5386-2928-4/18 $31.00 © 2018 IEEE   132

On the other hand, the FLR has three-dimensional complicated structure such as a chip corner, therefore BV of FLR (BV_{FLR}) tends to become the lowest point in the limited area. In addition, the FLR has the NDR characteristic of pnp structure. In case of these worse combination, the destruction of the avalanche phenomenon tends to occur at the FLR. Fig. 2 is the photograph of an emission analysis at the conventional chip after the avalanche destruction. The emission is obtained as the destruction point at the FLR of the chip corner. In order to avoid this destruction, increasing the wafer thickness is effective. But they cause degradation of the total power loss of the RC-IGBT.

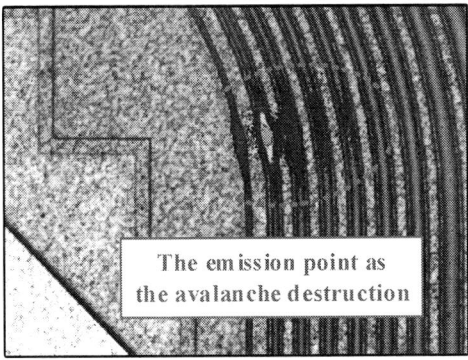

Fig. 2: The photograph of the emission analysis for the conventional RC-IGBT chip after the avalanche destruction.

III. APROACH AND SIMULATION

In case of the conventional RC-IGBT based upon CSTBT™ structure, BV of IGBT (BV_{IGBT}) and Diode (BV_{Di}) are close to the one-dimensional BV on behalf of its narrow trench pitch design, despite having the CS-layer and the deep trenches. In other words, both BV_{IGBT} and BV_{Di} have a high sensitivity for the trench pitch due to the presence of CS-layer and deep trenches. In addition, the Diode part has a Positive Differential Resistance (PDR) characteristic of pin structure [7] [8] and wide active area, so, it has potential for the more stable blocking characteristic than the IGBT part or the termination. It seemed to be better for the BV characteristic of the entire RC-IGBT structure to be designed and regulated by the Diode part structure. We considered to control BV_{Di} by W_{DTP} in order to adjust the preferable balance like Fig. 3.

Fig. 4 is simulated BV waveforms of each parts. BV_{FLR} was lower than BV_{IGBT} and BV_{Di}. Although the NDR characteristic appeared in FLR and IGBT, Diode had the highest BV and the PDR characteristic. The dash line is BV waveform of the proposal Diode with wide W_{DTP}. The BV of proposal Diode was controlled to be the lowest point keeping the PDR characteristic. Fig. 5 is simulated results of the electric field distribution for the conventional and proposal Diodes. In the proposal structure, the electric field around the trench is raised compared with the conventional one as intended.

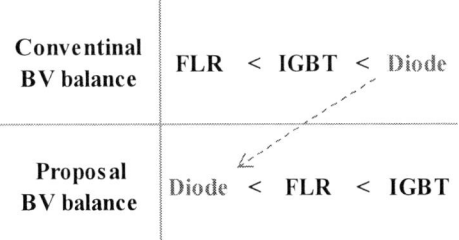

Fig. 3: Target BV relationship at the conventional and proposal RC-IGBTs.

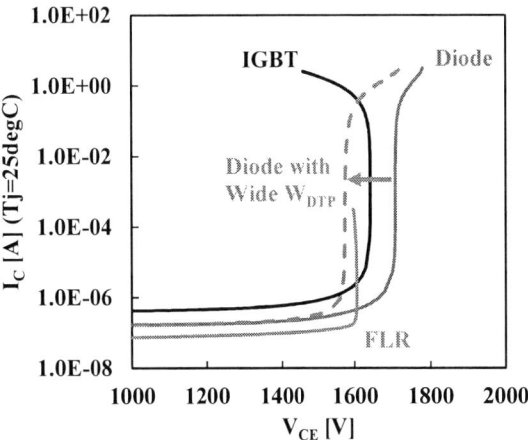

Fig. 4: The simulated results of BV waveforms with each regions.

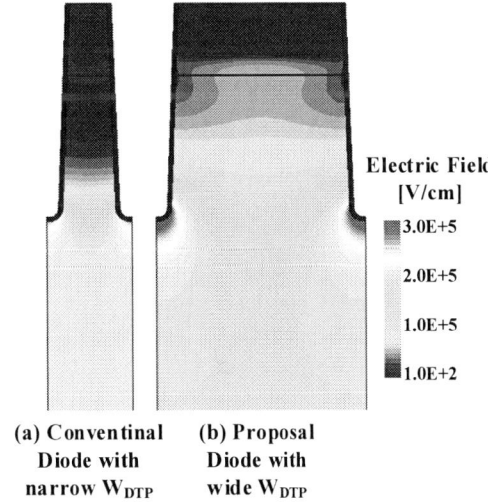

Fig. 5: The simulated electric field distribution with the conventional and proposal Diodes.

IV. EXPERIMENTAL RSULTS

From the simulation results, we found that BV_{Di} could be controlled with W_{DTP}. To adjust BV well balance, we fabricated the RC-IGBTs with variable W_{DTP}. The experimental results of the relationship between W_{DTP} and BV_{CES} is shown in Fig.6. The square mark of the point A is the conventional RC-IGBT. The diamond marks including the point B are the proposal RC-IGBTs with the various W_{DTP}. The circle mark of the point C is the proposal RC-IGBT with thinner wafer than the point A and B. The proposal RC-IGBTs applied a higher resistivity (ρ) on the n drift than the conventional RC-IGBT to compensate a decrease of BV by widening W_{DTP}.

In the narrow W_{DTP} region ($W_{DTP} \leq 1$), BV_{CES} didn't depend on W_{DTP}, because BV_{FLR} was lower than BV_{Di} and the avalanche destruction was occurred at the FLR. In the wide W_{DTP} region ($W_{DTP} > 1$), we could control BV_{CES} by using W_{DTP} and obtain the stable BV waveform. Fig.7 shows the BV waveforms of point A, B and C. The avalanche destruction was observed at the point A only. The point B and C as the proposal structures indicated the stable BV waveforms. The decrease of BV_{CES} due to W_{DTP} expansion was compensated by the increase of resistivity (point B). Fig. 8 shows the V_F-I_F curves, Fig. 9 presents the reverse recovery waveforms of point A and C. There is no adverse effects such as a snapback due to W_{DPT} spreading.

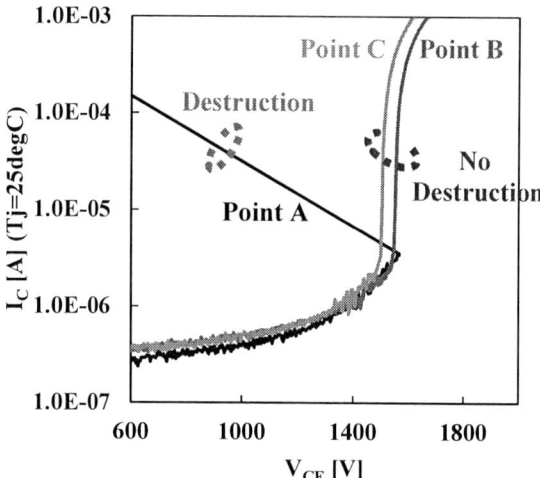

Fig. 7: The experimental BV waveforms of point A (conventional structure), B (wide W_{DTP} + high resistivity) and C (wide W_{DTP} + high resistivity + thin wafer) in Fig.6.

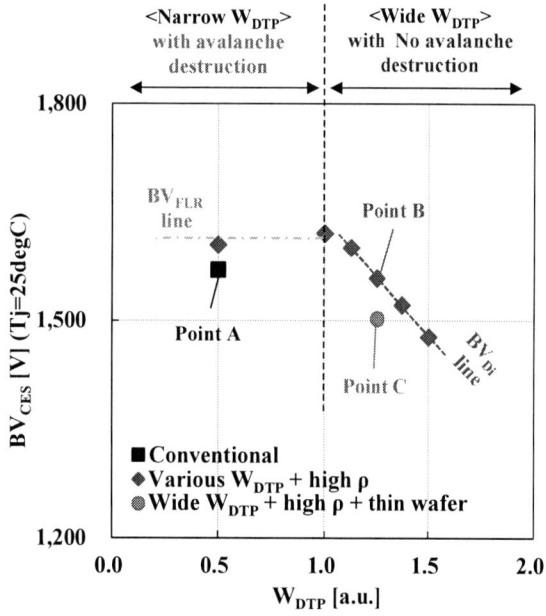

Fig. 6: The experimental results of the relationship between W_{DTP} and BV_{CES} of the fabricated RC-IGBTs. The square mark is the conventional RC-IGBT. The diamond marks are the proposal RC-IGBTs with the various W_{DTP}. The circle mark is the proposal RC-IGBT with thinner wafer than the points A and C. (W_{DTP} is a relative value.)

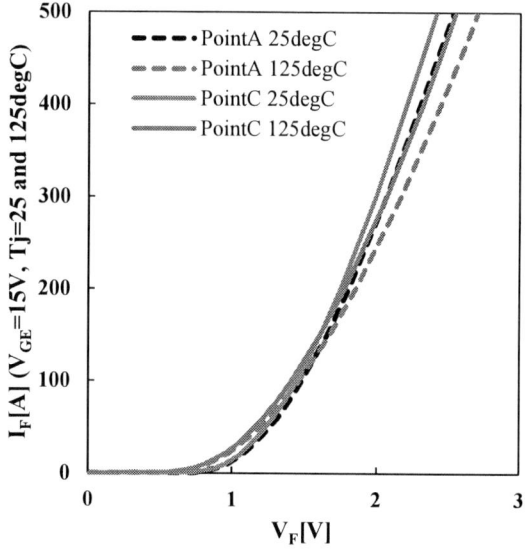

Fig. 8: The experimental V_F-I_F curve of point A (dash line) and C (line) in Fig.6.

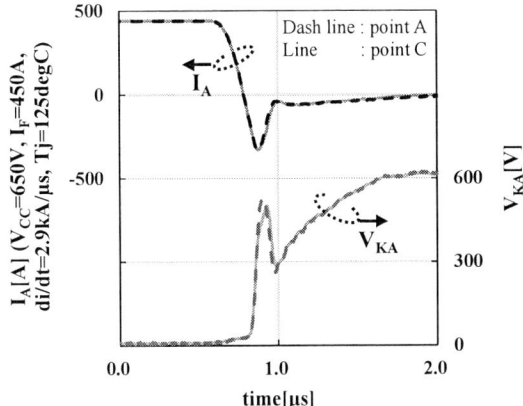

Fig. 9: The experimental reverse recovery waveforms f point A (dash line) and C (line) in Fig.6. The waveforms almost overlapped.

Fig.10 is the V_F-E_{rr} tradeoff of the points A, B and C in Fig.6. Tradeoff lines are described by changing the carrier lifetime. V_F of Point C with proposal Diode structure and thin wafer is improved 7.5% at same E_{rr} (29.2mJ) in comparison with point A as the conventional Diode structure. Our proposed structure improved the V_F-E_{rr} tradeoff and stabilized the BV characteristic at the same time. We have also confirmed there are no degradation effect for the Diode operation such as RRSOA with the spreading W_{DTP}.

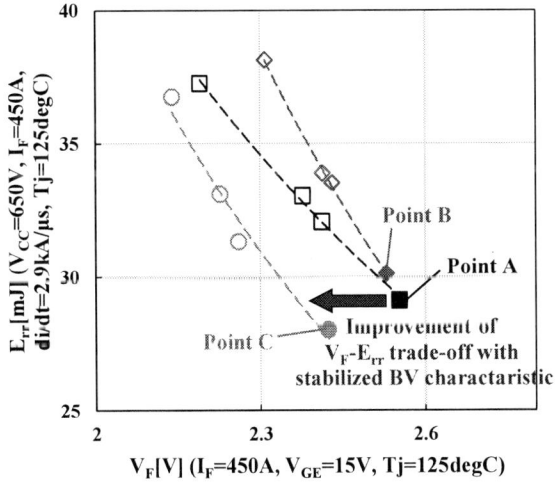

Fig. 10: The V_F-E_{rr} tradeoff at the points A, B and C in Fig.6. The trade-off lines described by the carrier lifetime-control.

V. CONCLUSION

We applied our high avalanche capability specific Diode part structure for RC-IGBT based upon CSTBT™. This proposed Diode structure stabilized BV characteristic, which has no switch back destruction around avalanche breakdown. And the total power loss was improved as the better V_F-E_{rr} tradeoff through the wafer thinning at the same time. This stabilization of the BV characteristic was realized by adjusting the BV balance of three portion of RC-IGBT, i.e. Diode cell, termination region and IGBT cell. One of the certain approach was lowering the Diode's BV by widening Diode's trench pitch (W_{DTP}). We have also confirmed there are no degradation effect for the Diode operation such as RRSOA.

Our proposal RC-IGBT performance is well balance from the entire SOA point of view.

ACKNOWLEDGMENT

The authors are grateful to Mr. H. Nakamura and Mr. K. Kimura for fabricating devices and the test evaluation. Fruitful discussion with Dr. M. Tarutani are greatly appreciated. The authors sincerely thank Dr. T. Minato and Mr. Y. Takata for their unrelenting encouragement.

REFERENCES

[1] H. Rüthing, F. Hille, F.-J. Niedernostheide, H.-J. Schulze, B. Brunner, "600 V Reverse Conducting (RC-)IGBT for Drives Applications in Ultra-Thin Wafer Technology," Proc. ISPSD2007, pp. 89-92, 2007.

[2] T. Yoshida, T. Takahashi, K. Suzuki, M. Tarutani, "The Second-generation 600V RC-IGBT with Optimized FWD", Proc. ISPSD2016, pp. 159-162, 2016.

[3] J. G. Laven, R. Baburske, A. Philippou, H. Itani, M. Dainese, "RCDC-IGBT Study for low-Voltage Applications", Proc. ISPSD2016, pp. 347-350, 2016.

[4] S.Shibata, M. Kato, H. Zhang, "New Transfer-Molded SLIMDIP for white goods using thin RC-IGBT with a CSTBT™ structure", Proc. PCIM2015, pp. 1149-1154, 2015.

[5] M. Takahashi, D. Hofmann, S. Yoshida, A. Tamenori, Y. Kobayashi, O. Ikawa, "Extended Power Rating of 1200V IGBT Module with 7G RC-IGBT Chip Technologies", Proc. PCIM2016, pp. 438-444, 2016.

[6] S. Adachi, S. Yoshida, H. Miyata, T. Kouge, D. Inoue, Y. Takamiya, F. Naganune, H. Kobayashi, T. Heinzel, A. Nishiura, "Automotive power module technologies for high speed switching", Proc. PCIM2016, pp. 1956-1962, 2016.

[7] T. Minato, N. Thapar, B. J. Baliga, "Correlation between the static and dynamic characteristics of the 4.5kV self-aligned trench IGBT", Proc. ISPSD'97, pp. 89-92, 1997.

[8] F. Masuoka, K. Tanaka, T. Kachi, Y. Yoshiura, K. Shimizu, "RFC diode with High Avalanche Stability and UIS Capability", Proc. ISPSD2017, pp. 131-134, 2017.

Proceedings of the 30th International Symposium on Power Semiconductor Devices & ICs
May 13-17, 2018, Chicago, USA

C_{OSS} Losses in Silicon Superjunction MOSFETs across Constructions and Generations

Grayson Zulauf[†] and Juan M. Rivas-Davila
Stanford University, Electrical Engineering
Stanford, CA, USA.
[†] Email: gzulauf@stanford.edu

Abstract—**The superjunction (SJ) structure breaks the unipolar material limit of silicon power MOSFETs, and has achieved widespread adoption in commercial power converters. In resonant applications, these SJ devices experience losses due to charging and discharging the parasitic output capacitor, C_{OSS}, resulting in losses that increase with switching frequency. We document C_{OSS} losses in commercially-available 600 V superjunction devices, showing that even devices grown with the trench-filling epitaxial method have non-negligible losses in MHz converters. Further, progressing in generations within a manufacturer appears to correspond to higher C_{OSS} losses as the cell pitch is reduced and doping is increased. The C_{OSS} losses may exceed conduction losses in many applications for the devices tested here.**

Keywords—*Power semiconductor devices, power transistors, power MOSFET, superjunctions, resonant power conversion*

I. INTRODUCTION

Losses in power semiconductors are typically separated into conduction losses - which occur due to finite resistance during the device on-time - and switching losses, where current-voltage overlap results in energy dissipation at device turn-on or turn-off. Switching losses scale with frequency, and reducing them through resonant techniques has been critical to moving switching frequencies (f_{SW}) into the MHz range, where higher power density can be achieved through the reduction of passive component size and energy storage [1]. In these resonant topologies, the energy stored (E_{OSS}) in C_{OSS} is resonantly cycled through the circuit with timing such that, at turn-on, there is zero V_{DS}, zero E_{OSS}, and therefore zero turn-on energy dissipated. Figure 1 shows two resonant inverter topologies of interest – relative to the half-bridge topologies (Fig. 1b), the ground-referenced single-switch topologies (Fig. 1a) have simpler gate drives but higher device voltage stresses.

A new loss element – interchangeably referred to as C_{OSS} losses, intrinsic energy (E_i), or ZVS energy loss – is non-negligible at MHz switching frequencies in both wide-bandgap [5]–[7] and superjunction (SJ) silicon MOSFETs [8]–[11]. From a circuit perspective, this mechanism appears as a lossy parasitic output capacitance, which, in these resonant circuits, unavoidably forms part of the waveshaping network and undergoes a charge-discharge cycle once per switching cycle during the device off-time. These C_{OSS} losses reintroduce a loss dependence on frequency, with increasing importance with higher f_{SW}.

SJ C_{OSS} losses are relevant at f_{SW} in the hundreds-of-kHz [8], [9] and would significantly impact device selection for

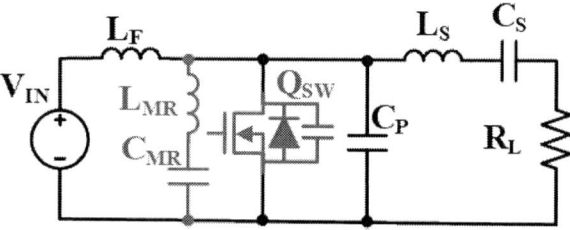

(a) Ground-referenced single-switch example topologies, the Class-E [2] or Class-Φ_2 [3] (added components shown in grey).

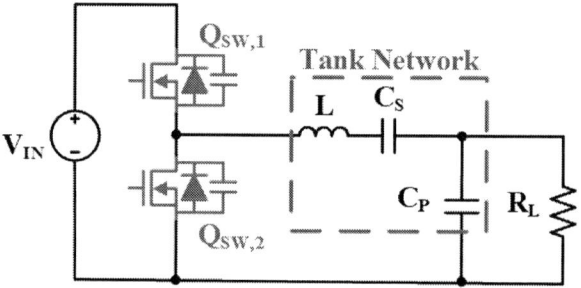

(b) Half-bridge resonant example topology, with a variety of networks capable of replacing the LCC shown above [4].

Figure 1: Two sample classes of topologies where soft-switching can be achieved through resonant waveshaping. The energy in the device output capacitance is typically assumed to be losslessly recycled in both topologies.

many commercial applications. Despite this broad commercial use, very little literature has been published on these losses since their first publication in 2014 [8] outside of work by the same authors [9], a recent seminar [12], and mixed-mode simulations to explain and model the losses [10], [11].

Manufacturers do not publish C_{OSS} loss data or include these losses in simulation models, and despite the widespread adoption of SJ devices in resonant converters operating up to a few MHz (in, for example, power factor correction (PFC) and LLC converters for computer or television power supplies), designers have no way to determine *a priori* which devices will have high C_{OSS} losses and which are candidates for their designs. We show that C_{OSS} losses are present in all SJ devices, even those built with the trench-filling method, and that C_{OSS} losses increase with each generation within a given manufacturer.

978-1-5386-2928-4/18 $31.00 © 2018 IEEE

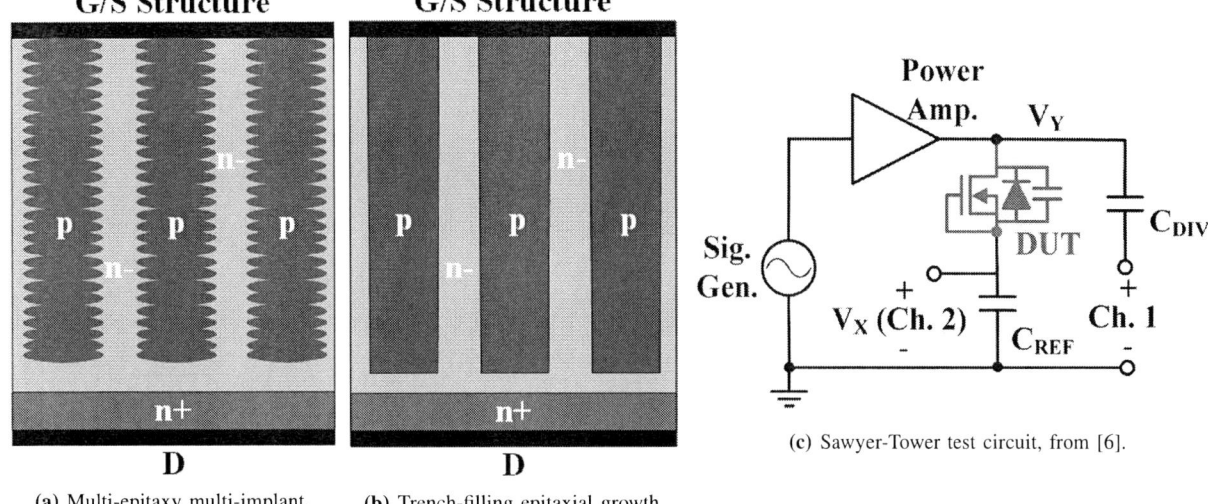

(a) Multi-epitaxy multi-implant.　　**(b)** Trench-filling epitaxial growth.

(c) Sawyer-Tower test circuit, from [6].

Figure 2: Subfigures a) and b) show mockups (not to scale) of the SJ structures for the two predominant methods to grow the p-type columns in SJ devices [10], [13]. Subfigure c) shows the Sawyer-Tower test circuit used to characterize C_{OSS} losses in this paper [6], [8], [9].

II. C_{OSS} LOSS MEASUREMENTS IN SUPERJUNCTIONS

We measure C_{OSS} losses using the Sawyer-Tower circuit shown in Figure 2c, which applies a sine wave across the output capacitor of a V_{GS}-shorted SJ MOSFET while maintaining the body diode in reverse bias (for detailed operation of the circuit, see [5], [6], [8], [9]). In this work, we apply sine waves at 200 kHz, 300 kHz, and 500 kHz with amplitudes from 50 V_{DS} to the device voltage rating. These frequencies are selected as typical operating frequencies for resonant circuits using these SJ devices. We aim to answer two primary questions with the loss characterization to understand the evolution of C_{OSS} losses with device modernization.

We presume that SJ device manufacturers are moving from the multi-epitaxy multi-implant (MEMI) growth process (Fig. 2a) to the trench-filling epitaxial growth (TFEG) process (Fig. 2b) to reduce the number of manufacturing steps [13] and to shrink the cell pitch [14]. Ref. [10] showed that C_{OSS} losses can be reduced by an order-of-magnitude (for a fixed column charge) when moving from the MEMI process to the TFEG process. Firstly, we aim to understand whether C_{OSS} losses are negligible in TFEG-processed devices. Secondly, within a device fabrication technique, we want to understand the scaling of C_{OSS} losses across generations and technology node size. Taken together, these understandings should indicate whether C_{OSS} losses will increase or reduce as SJ technology improves broadly (i.e. moving from MEMI to TFEG processing and reducing $R_{DS,ON}$ through a reduction in cell pitch [13]).

A. TFEG-Manufactured Devices

Manufacturers do not disclose internal device constructions and we have no access to proprietary information, so we assume device constructions only from materials available to the public. We caveat the following work with the understanding that these external materials may lead competitors, customers,

Table I: Toshiba SJ MOSFETs tested in this study. $R_{DS,ON}$ values are nominal, from the respective datasheets. DTMOS[II] E_{OSS} values are measured in the Sawyer-Tower circuit, as datasheets only give C-V curves to 100 V_{DS}.

Gen.	Part Number	Voltage	Current	$R_{DS,ON}$	E_{OSS} at 400 V_{DS}
DTMOS[II]	TK13A65U	650 V	13 A	320 mΩ	4.0 µJ
DTMOS[IV]	TK11P65W5	650 V	11 A	350 mΩ	3.0 µJ
DTMOS[V]	TK290P65Y	650 V	12 A	230 mΩ	3.8 µJ
DTMOS[II]	TK20J60U	600 V	20 A	165 mΩ	6.9 µJ
DTMOS[IV]	TK20V60W5	600 V	20 A	156 mΩ	5.6 µJ
DTMOS[IV]	TK7P60W5	600 V	7 A	540 mΩ	1.9 µJ
DTMOS[V]	TK560P60Y	600 V	7 A	430 mΩ	2.4 µJ

or researchers like us astray; nonetheless, there are strong public indicators from which we will proceed.

For example, Toshiba's DTMOS[IV] and DTMOS[V] generation SJ MOSFETs "use the state-of-the-art single epitaxial process" [15] and, according to the company, "Toshiba's DTMOS-IV technology uses a deep-trench process that narrows the lateral SJ pitch" [14], [16]. Based on these press releases, we assume the DTMOS[IV] and DTMOS[V] series are grown using the TFEG process and previous generations are grown using the MEMI process.

Fig. 3a shows the measured losses at 200 kHz and 300 kHz for 650 V_{DS} devices with similar current ratings in progressing generations, DTMOS[II] (MEMI) to DTMOS[IV] and DTMOS[V] (TFEG). Fig. 3a shows a substantial reduction in C_{OSS} losses when the construction is (presumably) changed from MEMI (DTMOS[II]) to TFEG (DTMOS[IV] and DTMOS[V]), consistent with the simulation predictions in [10]. The two TFEG generations dissipate similar percentages of stored energy over the tested voltage range, with the scaled datasheet E_{OSS} overlaid on the figure. The measurements are consistent with [8], [9], where a large percentage of the C_{OSS} loss occurs at the voltage where the structure becomes fully depleted. This $\approx$ 12%

978-1-5386-2928-4/18 $31.00 © 2018 IEEE　　　137

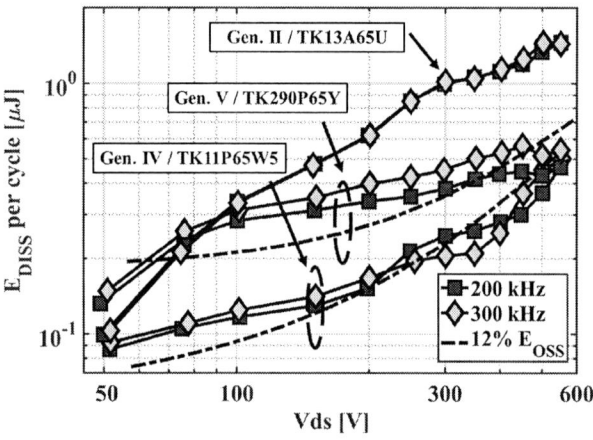 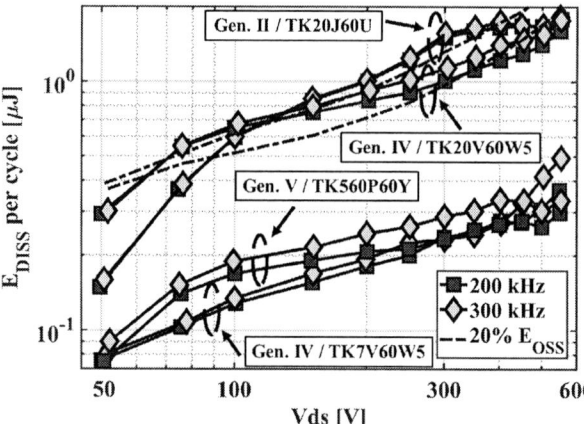

(a) Toshiba 650 V SJ MOSFETs tested in this study (top three devices in Table I). DTMOS^{II} generation datasheets do not include E_{OSS} curves.

(b) Toshiba 600 V SJ MOSFETs tested in this study (bottom four devices in Table I).

Figure 3: Comparison of C_{OSS} losses with a sine wave input for two generations of Toshiba devices, which are presumed to be TFEG-fabricated. V_{DS} refers to the peak voltage of the sine wave applied to charge and discharge the device.

dissipation factor for C_{OSS} in TFEG SJ devices is similar to the 10% dissipation estimated for SiC FETs in [7].

Surprisingly, this improvement in losses between TFEG and MEMI generations does not appear to hold for the 20 A, 600 V_{DS} devices, which are shown in Fig. 3b. While the MEMI device again appears to dissipate $\approx 20\%$ of the stored energy, the TFEG device performs noticeably worse than the characterized 650 V_{DS} devices. Fig. 3b also illuminates that scaling C_{OSS} losses between same-family parts cannot be done as a simple $C_{O,ER}$ scaling – as can be performed with GaN HEMTs [6] – as the 7 A Gen. IV device has 3× smaller $C_{O,ER}$ but $\approx 4.5\times$ smaller losses than the 20 A version.

The C_{OSS} losses in the larger TFEG devices are certainly non-negligible, corresponding to over 1 W per MHz of switching frequency in the 20 A Gen. IV device (TK20V60W5). For example, using this device at f_{SW} = 1 MHz would result in C_{OSS} losses that are larger than conduction losses for current levels up to 3.5 A_{RMS} (ignoring gating losses). In the TFEG devices, C_{OSS} losses appear to increase at low voltages in newer generations as a result of higher low-voltage energy storage, but are similar as a percentage of total E_{OSS}.

B. Progression through Generations

Infineon has clearly defined generations for their high-power SJ MOSFET families. The move from the "CP" generation to the "C6" generation to the "C7" generation corresponds, accordingly to the application note, to a continued reduction in cell pitch [17] and an accompanying reduction in specific $R_{DS,ON}$ [13], [18]. Similarly, we examine the progression from the C6/E6 family to the P6 family to the P7 family. The datasheets do not clearly specify whether the devices are TFEG- or MEMI-fabricated, and we proceed assuming that the cell pitch is reduced between generations but the fabrication method is not changed. The tested devices, which are shown in Table II, were selected to maintain similar nominal $R_{DS,ON}$ values across generations.

Table II: Infineon SJ MOSFETs tested in this study. $R_{DS,ON}$ values are nominal, from the respective datasheets.

Gen.	Part Number	Voltage	Current	$R_{DS,ON}$	E_{OSS} at 480 V_{DS}
CP	IPB60R299CP	600 V	11 A	270 mΩ	5.3 μJ
C6	IPB60R380C6	600 V	11 A	340 mΩ	3.5 μJ
C7	IPL65R230C7	650 V	10 A	204 mΩ	2.9 μJ
P6	IPL60R360P6S	600 V	11 A	320 mΩ	4.4 μJ
C6	IPB60R380C6	600 V	11 A	340 mΩ	3.5 μJ
P7	IPL60R185P7	600 V	19 A	149 mΩ	3.3 μJ

Fig. 4 shows the measured C_{OSS} losses in the tested Infineon devices. For two distinct families, we see the same trend of increasing C_{OSS} losses with the modernization of the devices. From [17], we understand that this modernization is related to both increased doping and finer cell pitch, and the increased loss with reduced cell pitch agrees with the simulations performed in [10]. While manufacturers focus on reducing E_{OSS} to improve performance in hard-switched converters, lower overall E_{OSS} might not correspond to less energy *dissipation* in soft-switched applications.

These measurements indicate that older devices may be preferred for a significant portion of resonant applications [10]. For example, consider a soft-switched converter with 400 V_{DS} voltage swing operating at 500 kHz with 3 A_{RMS} through the device. Selecting the IPL65R230C7 device – which has both the lowest $R_{DS,ON}$ and E_{OSS} in the family – would result, ignoring gating losses, in 2.66 W of device power dissipation (1.84 W P_{COND}, 0.82 W P_{COSS}) while the older IPB60R299CP device would dissipate only 2.44 W (2.43 W P_{COND}, 0.01 W P_{COSS}). While, of course, this example is cherry-picked, the key point is that simply upgrading devices does not necessarily improve efficiency, with more caution necessary at higher operating frequencies or with paralleled devices. Designers must optimize between conduction, C_{OSS}, and gating losses on a per-application basis, with f_{SW}, voltage swing, and current driving the power device selection.

978-1-5386-2928-4/18 $31.00 © 2018 IEEE

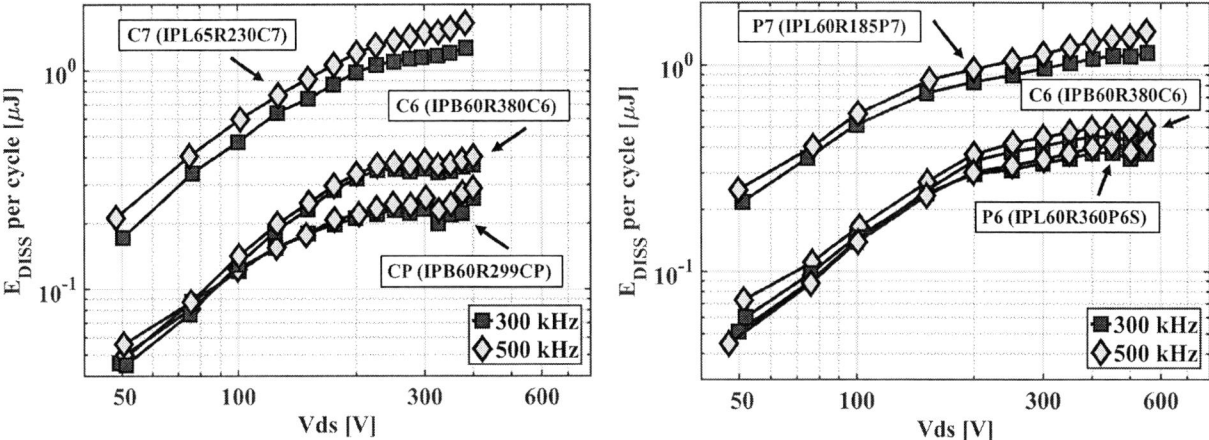

(a) CP, C6, and C7 generations (top three devices in Table II).

(b) P6, C6, and P7 generations (bottom three devices in Table II).

Figure 4: Comparison of C_{OSS} losses with a sine wave input for progressing generations of Infineon devices. V_{DS} refers to the peak voltage of the sine wave applied to charge and discharge the device.

III. CONCLUSION

In this paper, we use the Sawyer-Tower circuit to characterize losses from charging and discharging the parasitic output capacitor, C_{OSS}, of silicon superjunction devices. These losses occur once per cycle in soft-switched converters, including many applications where SJ devices dominate.

As manufacturers continue to focus on reducing $R_{DS,ON}$ in progressive designs, we find that these C_{OSS} losses worsen with the continued reduction in cell pitch. Further, C_{OSS} losses appear to be non-negligible in devices fabricated with trench-filled epitaxial growth (TFEG) processes, indicating that the modernization of the manufacturing processes will not, by itself, eliminate these soft-switching losses.

SJ manufacturers could focus on building both $R_{DS,ON}$-optimized families and C_{OSS}-optimized families, allowing designers to choose on a per-application basis. Further, manufacturers should include an estimation of C_{OSS} losses in both datasheets and simulation models.

ACKNOWLEDGEMENTS

G. Zulauf thanks Dr. Jim Plummer for his insights on the construction of and loss mechanisms in SJ devices.

REFERENCES

[1] D. J. Perreault *et al.*, "Opportunities and challenges in very high frequency power conversion," in *Proc. IEEE Applied Power Elec. Conf.*, 2009, pp. 1–14.

[2] N. O. Sokal, "Class-E RF power amplifiers," *QEX*, vol. 204, pp. 9–20, Jan/Feb 2001.

[3] J. M. Rivas *et al.*, "A high-frequency resonant inverter topology with low voltage stress," in *Proc. IEEE Power Elec. Spec. Conf.*, 2007, pp. 2705–2717.

[4] R. W. Erickson and D. Maksimovic, *Fundamentals of power electronics*. Springer Science & Bus. Media, 2007.

[5] G. Zulauf *et al.*, "Output capacitance losses in 600 V GaN power semiconductors with large voltage swings at high-and very-high-frequencies," in *Proc. IEEE Workshop on Wide Bandgap Power Devices and Applications (WiPDA)*, 2017, pp. 352–359.

[6] G. Zulauf *et al.*, "C_{OSS} losses in 600 V GaN power semiconductors in soft-switched, high- and very-high-frequency power converters," *IEEE Trans. Power Electronics*, 2018.

[7] M. Kasper *et al.*, "ZVS of power MOSFETs revisited," *IEEE Trans. Power Electronics*, vol. 31, no. 12, pp. 8063–8067, December 2016.

[8] J. Fedison *et al.*, "C_{OSS} related energy loss in power MOSFETs used in zero-voltage-switched applications," in *Proc. IEEE Applied Power Electronics Conf.*, 2014, pp. 150–156.

[9] J. Fedison and M. Harrison, "C_{OSS} hysteresis in advanced superjunction MOSFETs," in *Proc. IEEE Power Electronics Conf.*, 2016, pp. 247–252.

[10] J. Roig *et al.*, "Origin of anomalous C_{OSS} hysteresis in resonant converters with superjunction FETs," *IEEE Trans. Electron Devices*, vol. 62, no. 9, pp. 3092–3094, 2015.

[11] J. Roig *et al.*, "High-accuracy modelling of ZVS energy loss in advanced power transistors," in *Proc. IEEE Applied Power Elec. Conf.*, 2018, pp. 263–269.

[12] B. Keogh, *HighVolt Interactive – hysteresis loss in high voltage MOSFETs*, September 2017, Accessed: 03.05.2018. [Online]. Available: https://training.ti.com/sites/default/files/docs/HVI_Coss_ZVS_15-Sept_complete.pdf

[13] F. Udrea *et al.*, "Superjunction power devices, history, development, and future prospects," *IEEE Trans. Electron Devices*, vol. 64, no. 3, pp. 720–734, 2017.

[14] *PCIM: Toshiba super-junction mosfets cut resistance and noise*, May 2016, Accessed: 03.11.2018. [Online]. Available: https://www.electronicsweekly.com/news/products/power-supplies/pcim-toshiba-super-junction-mosfets-cut-resistance-and-noise-2016-05/

[15] *Toshiba Gen-4 Super Junction DTMOS MOSFETs*, Accessed: 03.11.2018. [Online]. Available: https://www.mouser.com/new/Toshiba-Semiconductors/toshiba-dtmos-iv/

[16] *Toshiba debuts 600V Super Junction MOSFET DTMOS IV high-speed diode series at APEC 2013*, March 2013, Accessed: 03.11.2018. [Online]. Available: https://phys.org/news/2013-03-toshiba-debuts-600v-super-junction.html

[17] *AN 2013-04 - CoolMOS C7: Mastering the Art of Quickness*, April 2013, Accessed: 03.05.2018. [Online]. Available: https://www.infineon.com/dgdl/Infineon-ApplicationNote_650V_CoolMOS_C7_Mastering_the_Art_of_Quickness-AN-v01_00-EN.pdf?fileId=db3a30433e5a5024013e6a966779640b

[18] P. M. Shenoy *et al.*, "Analysis of the effect of charge imbalance on the static and dynamic characteristics of the super junction MOSFET," in *Proc. IEEE Power Semiconductor Devices and ICs.*, 1999, pp. 99–102.

Proceedings of the 30th International Symposium on Power Semiconductor Devices & ICs
May 13-17, 2018, Chicago, USA

Extending the RET-IGBT (Recessed Emitter Trench IGBT) Concept to High Voltages: Experimental Demonstration of 3.3kV RET IGBT

L. Ngwendson, I. Deviny, C. Zhu, I. Saddiqui, C. Kong,
A. Islam, J. Hutchings, J. Thompson, M. Briggs, O. Basset
Research and Development Centre,
Dynex Semiconductor Ltd., Lincoln, UK
Email: Luther_Ngwendson@dynexsemi.com

H. Luo, Y. Wang, Y. Yao
Zhuzhou CRRC Times Electric Co., Ltd.,
Zhuzhou, China

Abstract—**In this paper we show simulation and experimental results of new 3.3kV RET-IGBT (Recessed Emitter Trench IGBT). Simulation results show that although the RET concept reduces the active region's trench to trench mesa, it does not show the reported inversion layer modulation phenomenon in conventional "narrow mesa" devices, which can degrade performance with very fine dimensions. It is also shown that High Voltage RET-IGBT can show superior $V_{CE}(sat)$ performance compared to standard Trench IGBT without degrading the dynamic performance such as RBSOA and SCSOA when devices are paralleled.**

Keywords—*IGBT, Narrow mesa, MOS,*

I. INTRODUCTION

Of recent "fine pattern" cell structures, proposed to scale down the silicon mesa in IGBTs to achieved sub-µm dimensions. Narrow Mesa Technology (NMT) generally results in low $V_{CE}(sat)$ in IGBTs and significantly improves the $V_{CE}(sat)$—E_{OFF} trade-off relationship especially when combined with enhanced nwell or carrier storage (CS) layer and thin silicon technology and optimized field-stop design[1-2]. The RET-IGBT concept was experimentally demonstrated in ISPSD'17 with medium voltage 1700V IGBT [3]. It was shown that with a reduced silicon mesa width, record low $V_{CE}(sat)$ can be achieved without compromising the SCSOA, RBSOA and BV performances. The benefits of the RET IGBT concept over conventional narrow mesa includes: (1) inversion layer depletion is on only one side of the active silicon mesa, (2) the contact size is not limited by silicon mesa width (3) photolithographic capability is not limited by trench to trench spac,e and (4) the recessed emitter enhances the chip robustness. Unlike in traditional NMT, in RET IGBT, the approach of the channel depletion in the on-state from only one side of the active mesa delays the conductivity modulation of the channel inversion layer which is beneficial to short circuit performance [4]. In this paper, it is shown that at 3.3kV, up to 20% (0.5V) lower $V_{CE}(sat)$ can be achieved at 150°C with practical 0.5µm trench to trench separation/silicon mesa. Experimental 3.3kV RET IGBT results of 4 chips in parallel also show lower $V_{CE}(sat)$ and excellent SOA. With further optimizations such as much enhanced nwell, optimized field-stop design and thin wafer technology, RET-IGBT is a highly promising technology to realize excellent IGBT performance at high voltages without much reliance on special processes. SEM images of RET and conventional trench gate IGBT are shown in Fig1a and b respectively.

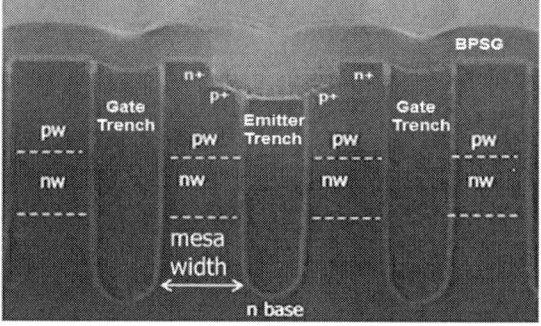

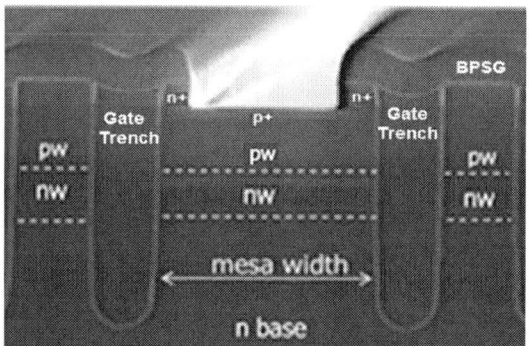

Fig.1 SEM cross-section of fabricated a) RET-IGBT and b) conventional trench IGBT

II. SIMULATION RESULTS

Simulations have been performed to investigate the influence of mesa width (down to sub-µm 0.5µm) on static and dynamic performance. 3.3kV RET IGBT Simulation results in Fig.2a shows RET concept can achieve record low $V_{CE}(sat)$ at 75Acm⁻² current density and more importantly about 40% lower temperature coefficient at 150°C, with reduced silicon mesa. The increasing E_{OFF} with reducing mesa dimension can be improved with thin wafer technology [5]. The reason for lower $V_{CE}(sat)$ and its lower temperature coefficient in RET

978-1-5386-2928-4/18 $31.00 © 2018 IEEE

device is due to the much enhanced IE effect resulting in up to six times higher conductivity modulation in the emitter side compared to the conventional device as shown in Fig.2b. Hence when silicon resistivity increases at high temperatures, RET IGBT shows smaller $V_{CE}(sat)$ increase. The channel depletion in RET IGBT is formed only on one side of the active area silicon mesa (Fig.3). So under short-circuit condition, about half of the Pwell for mesa as small as 0.5µm is still undepleted, even though the Nwell is fully depleted. This means less susceptibility to the channel inversion layer modulation phenomenon [4], which degrades SCSOA compared to conventional NMT where channel is formed on both sides of the silicon mesa. Fig.3's bottom right picture shows that 0.5µm mesa structure has the highest Pwell potential closest to the N+ emitter.

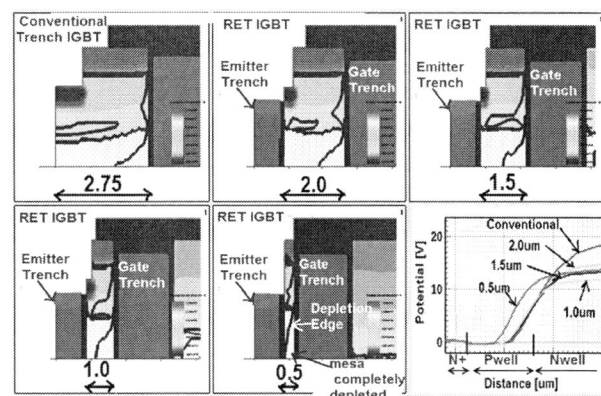

Fig.3: Simulated results showing the depletion layer under short circuit condition (V_{CE}=1800V, V_{GE}=15V). Channel depletion is only along the gate trench side wall. The bottom right picture is the potential distribution within the silicon mesa region

III.   EXPREMINTAL RESULTS

Several experimental batches of 13.5mm x 13.5mm 3.3kV/75A chips with RET and conventional stripe layout design have been manufactured and extensively evaluated to ensure robustness of the technology. Fig.3b shows 4 paralleled 3.3kV/75A IGBT chips + 2FRD on a substrate for testing and characterization.

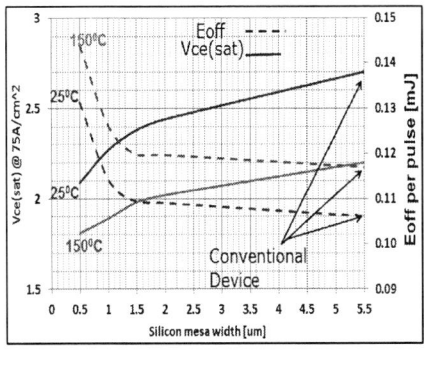

(a)

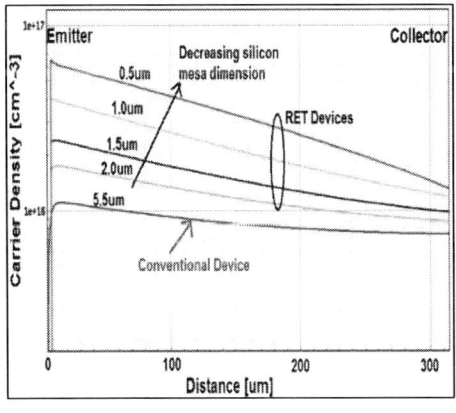

(b)

Fig.2: Simulated influence of mesa width on a) $V_{CE}(sat)$ and E_{off} (b) carrier concentration profile within the n-base region.

Fig.4) 1800V/300A substrate assembly of 4 IGBTs + 2 FRDs in parallel.

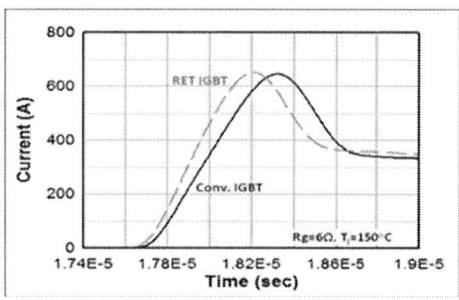

(a)

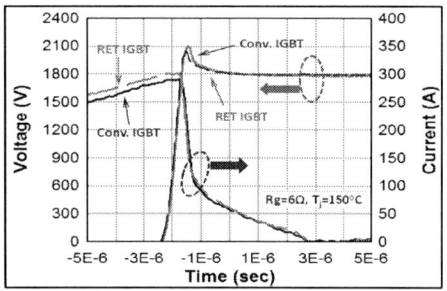

(b)

Fig.5: Measured turn-on (a) and turn-off (b) waveforms. V_{cc}=1800V, I_{ON}=300A (4Chips), T_J=150°C, V_{GE}=+/-15V

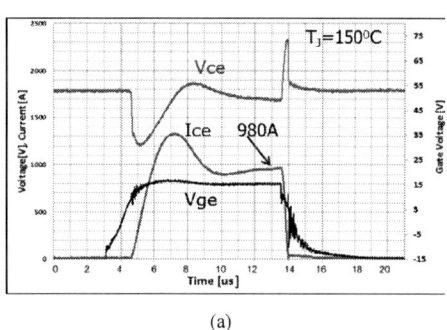

(a)

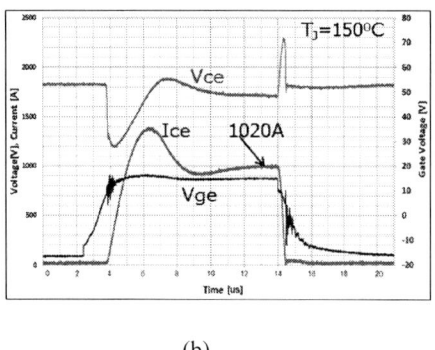

(b)

Fig.6: Measured short circuit performance at V_{GE}=15V for a) Conventional IGBT and b): RET IGBT

Mesa width for standard and RET devices are 5.3μm and 1.6μm respectively. Identical low Nwell have been used for this work and V_{TH} of fabricated samples is 6.5V. At 150°C, typical $V_{CE}(sat)$ is 2.7V and 2.5V respectively, for conventional and RET devices. Standard double pulse dynamic tests have been performed and fig.5 shows the turn-on and turn-off waveforms. Fig.5a shows that RET-IGBT's turn-on dI/dt at 1562A/μs is ~20% faster than conventional device due to quicker build-up of plasma within the drift region hence lower E_{on}. A key design feature to achieve desired dI/dt is the arrangement and configuration of the dummy trenches within the dummy region [2]. Fig.5b shows no significant difference in turn-off current tail between the two devices, however calculated E_{OFF} is 10% higher for RET due to difference in turn-off dV/dt. Fig.6a and b show performances under standard 10us short circuit test. The desired short circuit current of 3 to 4*Inom is a function of V_{TH}, cell pitch and the configuration of the dummy region. In terms of robustness, fig.7a shows RET-IGBT RBSOA performance where more than 4*Inom (1200A) is turned off successfully while fig.7b shows SCSOA at V_{GE} = 17V with the *line voltage* from 1800V to 2200V.

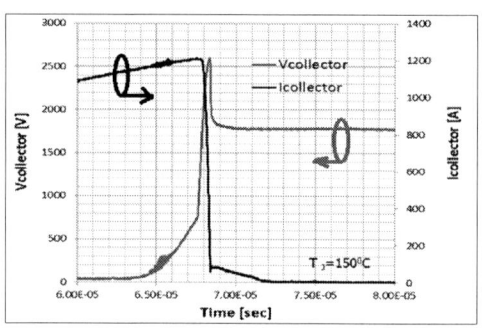

(a)

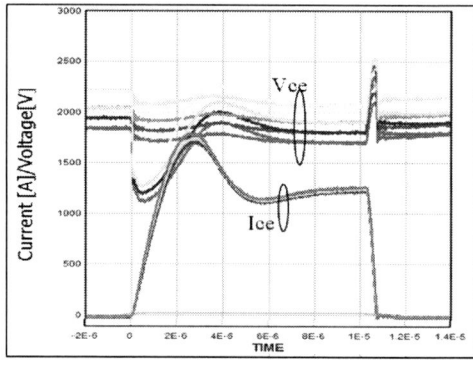

(b)

Fig.7:RE-IGBT Robustness tests: a) RBSOA (Ic > 4*Inom) and (b) SCSOA. T_J=150°C, V_{GE}=+/-17V, V_{LINE}=1800 to 2200V

The emitter trench in RET-IGBT is at the emitter potential (zero volts) during all phases of operation. So under dynamic test conditions, there is hole-accumulation along its side walls which forms a low resistance path for holes flowing to the emitter contact. Hence especially during turn-off, most holes flowing to the emitter contact avoid the n+emitter/pwell junction which enhances RBSOA and SCSOA [3].

IV. CONCLUSION

In this paper we have presented simulation results as well as the initial experimental results of the extension RET-IGBT concept to 3.3kV. The results show that the RET concept which enhances the conductivity modulation in the emitter side, is promising for high voltage trench IGBT where the drift region resistance tend to dominate $V_{CE}(sat)$ *in high p*ower modules. It is also shown that the concept is less prone to inversion layer modulation phenomenon reported in conventional narrow mesa devices. In order to achieve the correct balance between SOA, Eon and dI/dt for high voltage applications, careful optimization of the dummy region configuration is necessary. The next step exploring the full benefits of high voltage RET-IGBTs through much enhanced N-well and thin silicon technology.

REFERENCES

[1] W. Frank et al., "Multi-dimensional trade-off considerations of the 750V micro pattern trench IGBT for electric drive train applications", Proc. ISPSD'15, pp. 105-108, 2017

[2]C. Jaege et.al., "A new sub-micron trench cell concept in ultrathin wafer technology for next Generation 1200 V IGBTs", Proc. ISPSD'17, pp. 69-72, 2017

[3]I. Deviny et. al., "A Novel 1700V RET-IGBT (Recessed Emitter Trench IGBT) Shows Record Low Vce(on), Enhanced Current Handling Capability and Short Circuit Robustness", Proc. ISPSD'117, pp. 147-150, 2017

[4] M. Tanaka1, et al., "Conductivity Modulation in the Channel Inversion Layer of Very Narrow Mesa IGBT", Proc. ISPSD'17, pp. 61-64, 2017

[5]J. Vobecky, et al., "Exploring the Silicon Design Limits of Thin Wafer IGBT Technology: The Controlled Punch Through (CPT) IGBT", ISPSD'08, pp. 76-79, 2008

Proceedings of the 30th International Symposium on Power Semiconductor Devices & ICs
May 13-17, 2018, Chicago, USA

Temperature dependence of the On-State voltage drop in Field-Stop IGBTs

L. Maresca, M. Riccio, G. Breglio, A. Irace

Dept. of Electrical Engineering and Information
Technologies, University Federico II, Naples, Italy
luca.maresca@unina.it

P. Mirone, C. Sanfilippo, L. Merlin

Vishay Semiconductor Italiana,
Borgaro, Torinese (TO), Italy

Abstract—**Insulated Gate Bipolar Transistor (IGBT) is the reference design for power semiconductor switches in the range of the medium-high power applications. Different designs were proposed along the development story of this device, but the actual trend is leaded by the Field-Stop IGBT (FS-IGBT) concept. One of the main advantages of this design is the great accuracy in defining the Emitter injection efficiency of the vertical PNP. However a careful design of the Collector side has to be carried out to avoid unwanted effects such as a negative temperature coefficient for the V_{ON} of the device. In this work we present the two main cause of the aforementioned effect, with a detailed analysis of the effect of the arising of a Schottky Barrier (SB) at the Collector contact for low doping concentrations.**

Keywords—*Field-Stop IGBT; Schottky Barrier; Voltage drop temperature coefficient; TCAD simulations*

I. INTRODUCTION

The IGBT is the most adopted device in the market of the middle-high power semiconductor switches and the state-of-the-art in terms of design is represented by the FS-IGBT [1]. As any IGBT structure, the FS-IGBT design takes the advances of both the MOSFET structure and the power Bipolar Junction Transistor one. Basically, it has the thermal stability typical of the MOSFET structure and the low on-state voltage drop of the BJT structure, thanks to the conductivity modulation occurring in the Drift-layer of the device [2]. However, the advances of the two structures are achieved only if a careful balanced design of them is carried out. As mentioned before, the MOSFET part introduces a positive temperature coefficient of the V_{ON}, that leads to the homogeneous current distribution over the area of the power device. Briefly, if part of the device area has an higher temperature, the local voltage drop is higher compared to the colder part of the device. This leads to a reduced current flowing in the hot part. This is the negative feedback that gives the thermal stability of an IGBT. However, this is true only if the MOSFET component is dominant compared to the BJT one. If the latter component is dominant, a negative temperature coefficient for the V_{ON} arises and, when this happens, the current density increases in the hottest area because of the lower voltage drop. This leads to the destruction of the device, even if in on-state, since a local hot spot arises with the consequent temperature increase. One of

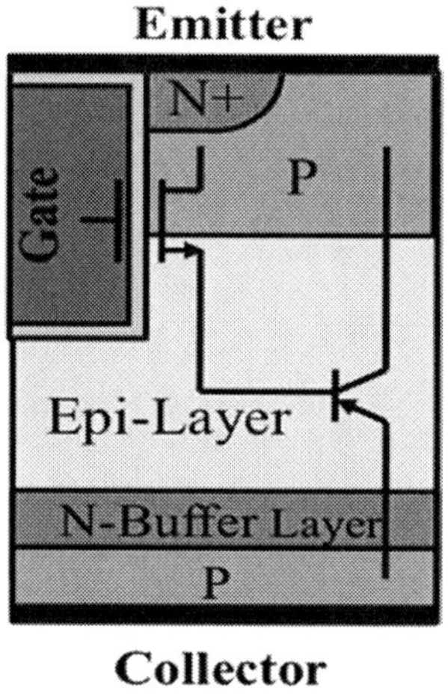

Figure 1. A sketch of a FS-IGBT. The P-Collector layer and the N-Buffer Layer are made by ionic implantation.

the advances of the FS-IGBT design is the vertical PNP emitter (Collector/buffer junction) injection efficiency regulation by means of ionic implantation of the back side of the device (Buffer layer/Collector doping profiles). A careful calibration of the technological steps is mandatory for the Collector side profiling as well as for the Buffer layer. Generally speaking, the design of the collector side affects different aspect of the IGBT performances (e.g. [3]-[6]). In this work we show the temperature dependence of the on-state voltage drop (V_{ON}) for a FS-IGBT, focusing the analysis on the effect of the Collector side doping concentration. In the first section, the analytical model of the three components of the V_{ON} for an IGBT is recalled and a detailed analysis of the temperature effect is reported when the Collector doping concentration is high enough to avoid SB effects. In the last section the arising of the negative temperature coefficient is investigated when the SB occurs for low doping concentrations at the Collector side.

978-1-5386-2928-4/18 $31.00 © 2018 IEEE

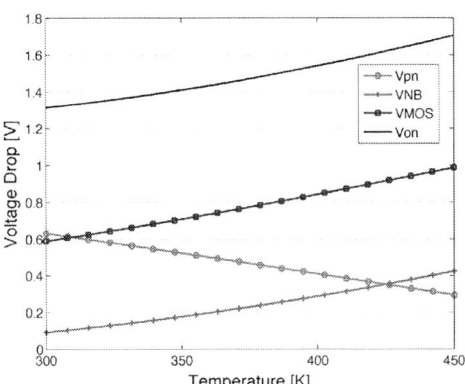

Figure 2. V_{ON} components for a well designed 600V FS-IGBT.

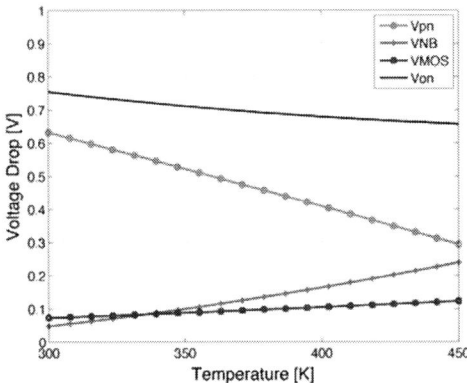

Figure 3. V_{ON} components for a thermally unstable 600V FS-IGBT.

II. V_{ON} TEMPERATURE DEPENDENCE

A sketch of the analyzed FS-IGBT structure is reported in Fig. 1. The IGBT carriers flux can be represented by a PNP, where the Base minority carriers are provided by the MOSFET. The voltage drop between the Collector and the Emitter contacts in forward conduction mode (V_{ON}) can be modeled as the sum of three components: The one due to the MOS channel (V_{MOS}), the one due to the Drift layer (V_{NB}) and the one due to the p/n junction at the collector side (V_{PN}) [7]. Equations (1), (2) and (4) model the three contribute to the voltage drop of an IGBT, starting from the technological features of the structure:

$$V_{PN} = \frac{kT}{q} ln\left(\frac{p_0^2}{n_i^2}\right) \tag{1}$$

$$V_{NB} = \frac{WJ}{(1+\frac{1}{b})\mu_n q n_{eff}} - \frac{D_a}{\mu_n} ln\left(\frac{p_0+N_D}{N_D}\right) \tag{2}$$

where:

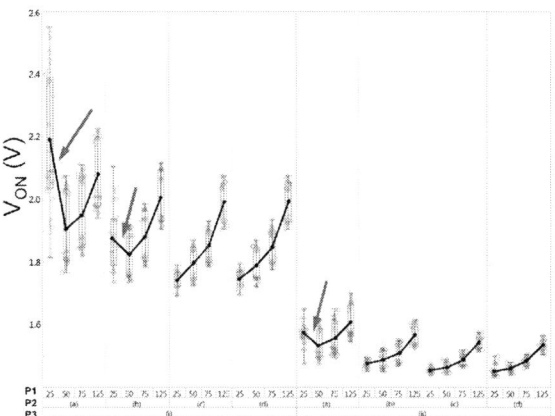

Figure 4. Experimental V_{ON} variation depending on both temperature (P1 [°C]), FS activation process (P2) and Doping profile (P3). The activation process has to be considered for an area that is 30% of the overall area. Red arrows indicates the segments with negative temperature coefficient.

$$n_{eff} = \frac{W}{2L_a} \frac{\sqrt{\{N_D^2+p_0^2 cosech^2(W/L_a)\}}}{tanh^{-1}\left[\frac{\sqrt{\{N_D^2+p_0^2 cosech^2(W/L_a)\}}tanh(W/L_a)}{\{N_D+p_0 cosec(W/L_a)\}tanh(W/L_a)}\right]} \tag{3}$$

$$V_{MOS} = \frac{J_C L_{CH} W_{CELL}}{2\mu_{ni} C_{OX}(V_G-V_{TH})} \tag{4}$$

where k is Boltzman constant, p_0 is the hole concentration at the PNP Emitter junction, D_a is the ambipolar diffusivity, N_D is the Drift-layer doping concentration, L_a is the ambipolar diffusion length, W is the Drift-layer width, $b = \mu_n/\mu_p$, L_{CH} is the MOS channel length, W_{CELL} is the cell pitch and the rest of the symbols belong to the standard notation for semiconductor devices modeling.

Keeping into account the temperature dependence of equations (1)-(4), the three components are reported in Fig. 2 for a 600V rated FS-IGBT. The technological parameters are those of a well designed device, therefore the overall voltage drop is linearly increasing with temperature, to avoid thermal hot spots [8]. From Fig. 2, it is visible as the MOS component (V_{MOS}) and the Drift layer component (V_{NB}) are dominant compared to Vpn and V_{ON} has a linear increasing trend with temperature. Therefore, especially for devices such as the FS-IGBTs, where no lifetime killing techniques are adopted, the VMOS component has to be dominant. To highlight the importance of the balance between the MOS component and the bipolar component, in Fig. 3, as an example, the MOS component is drastically reduced. More in detail, a very short MOS channel was defined to reduce the V_{MOS} component and an high lifetime is defined in the drift region. Therefore, the p/n junction voltage drop (V_{PN}) becomes dominant and the V_{ON} has a linear decreasing trend with temperature, leading to the thermal instability and premature failures due to the hotspots.

978-1-5386-2928-4/18 $31.00 © 2018 IEEE   145

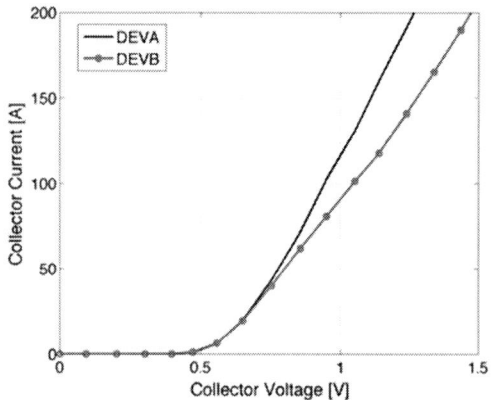

Figure 5. Experimental output curves of DEVA (Standard Collector doping concentration) and DEVB (30% of the active area with low collector doping concentration and the rest with a standard concentration).

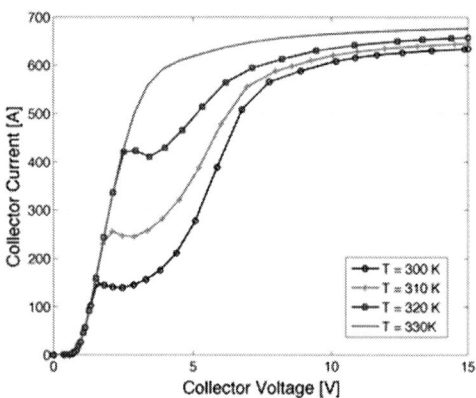

Figure 6. Numerical output curves of a device where 100% of the device has a low doping concentration of the Collector layer. Simulations at T=300K, T=310K, T=320K and T=330 show the high sensitivity of the threshold voltage with the temperature. $\varphi_B = 0.8\ eV$.

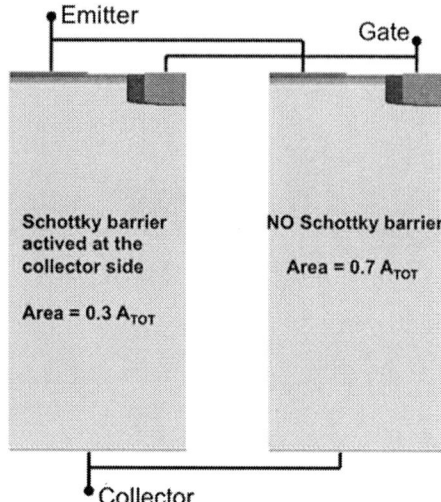

Figure 7. A sketch of the mixed-mode TCAD simulation framework to keep into account for the presence of a Schottky barrier at the collector contact of the device, with a coverage of 30% of the overall area of the device.

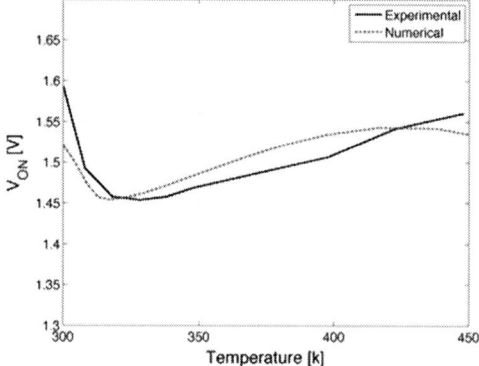

Figure 8. Numerical output curves of a device where 100% of the device has a low doping concentration of the Collector layer. Simulations at T=300K, T=310K, T=320K and T=330 show the high sensitivity of the threshold voltage with the temperature. $\varphi_B = 0.8\ eV$.

III. V_{ON} NEGATIVE TEMPERATURE COEFFICIENT DUE TO THE SCHOTTKY BARRIER EFFECT ACTIVATION

To reduce the bipolar component, the Emitter efficiency of the vertical PNP in a FS-IGBT is reduced by acting on the doping concentration of both the Buffer-layer and the Collector doping profile. To reduce the injection efficiency of the Collector/Buffer-layer junction, the Buffer-layer doping concentration can be increased or the Collector doping concentration can be reduced. The latter solution has a limitation in terms of minimum acceptable doping concentration at the Collector side, since a negative temperature coefficient can also occur because of the arising of a Schottky barrier (SB) at the silicon/metal junction. This aspect becomes relevant when the active doping concentration at the semiconductor/metal junction becomes too low. A possible reason for a too low doping concentration could be a reduced activation percentage of the implanted dopants. When the boron concentration is in the range of 10^{17} - 10^{18} cm^{-3}, the thermionic component is dominant at the metal-semiconductor interface and the SB effect becomes relevant [9]. Since the Collector diffusion in made by p-type dopant, the Schottky junction operates in reverse conducting conditions. Therefore, for low V_{CE} the current is sustained by the leakage current of the Schottky junction. Higher current levels can flow only if the BV condition is achieved. A Schottky diode with a doping concentration in the order of 10^{17} cm^{-3} has a BV in the range of few Volts [10], that adds to the three components evaluated by equation (1)-(4). The experimental observation of this phenomenon is reported in Fig. 4, where the experimental V_{ON} versus both Temperature (P1), activation process (P2) and Collector doping profile (P3) curves are reported for a device with a 30% of the Collector doping profile with a lower concentration and the rest with a standard value. A negative temperature coefficient arises up to T=330K for different designs. It has to be highlighted that this effect does not lead to the thermal instability since the negative temperature coefficient is related to the combination of the high voltage drop due to the area where the SB occurs and the area where it does not occur. A further experimental evidence of the effect is visible when a standard device is compared to a device where the SB effect occurs in terms of output curves. In Fig. 5 the experimental output curves of a standard device (DEVA) and a design "ii, (a)" of Fig. 4 (DEVB) device are reported. It is clear as for low current levels the voltage drop difference between DEVA and DEVB is negligible, since the leakage current of the reverse biased SB sustain the flowing current with a reduced voltage drop contribution. For higher current density (I>20A), DEVB has an higher V_{CE} compared to DEVA. Therefore, the SB area has to work in BV condition to sustain the high current, adding an higher voltage drop. For a better comprehension of this effect TCAD simulations were carried out. Calibrated TCAD simulations were performed keeping into account for the presence of the SB. In Fig. 6 the output curves of a device with a SB of ϕ_B=0.8 eV are reported for four temperatures and it is assumed that the SB occurs for all of device area. The SB introduces an abrupt increase of the V_{CE} after a threshold current and the higher is the temperature, the higher is the threshold current. This is due to the exponential behavior of a SB leakage current to respect to the temperature. For T=330K the effect disappear. To compare TCAD simulations to experimental data, two elementary cells were simulated in parallel to consider a structure equivalent to DEVB (See Fig. 7). More in detail, one cell (70% of coverage) keeps into account the area of the device where the SB does not occurs, while the other cell (30% of coverage) keeps into account for the SB effect. The paralleling of the two cells is compulsory to keep into account that part of the structure has a SB at the Collector contact and the overall voltage drop depends on the current distribution between the two cells. Numerical and experimental V_{ON}-T curves are reported in Fig. 8. A good agreement between the experimental and numerical data is visible. It has to be mentioned that this phenomenon does not lead to thermal current hotspots, since a local temperature increase in the SB area leads to the voltage drop increase (as demonstrated in Fig. 6). On the other hand, if the temperature increase occurs in the area without the presence of the SB effect, the voltage drop increases with temperature. Therefore, even if the overall device exhibits a negative V_{ON}

temperature coefficient, this do not leads to current filamentation effects.

IV. CONCLUSIONS

In this work the arising of a negative temperature coefficient of the V_{ON} of a FS-IGBT, 600 V rated, is analyzed and two main causes are addressed. The first one is related to a dominant contribution of the vertical PNP to respect to the MOS part. The second one is related to the presence of a Schottky barrier at the Collector contact. The former effect leads to the thermal instability and a careful design of the elementary cell is compulsory to avoid it. The latter effect arises when the doping concentration of the Collector diffusion is too low, with a consequent activation of a Schottky barrier effect. Results, coming from both numerical and experimental investigations, demonstrate as the back side of a FS-IGBT has to be carefully made to take advantage of the design features and to avoid the rising of a SB.

REFERENCES

[1] Laska, T., et al. "The Field Stop IGBT (FS IGBT). A new power device concept with a great improvement potential." Power Semiconductor Devices and ICs, 2000. Proceedings. The 12th International Symposium on. IEEE, 2000.

[2] Baliga, B. Jayant. *Fundamentals of power semiconductor devices.* Springer Science & Business Media, 2010.

[3] Spirito, P., Maresca, L., Riccio, M., Breglio, G., Irace, A., & Napoli, E. (2015). Effect of the collector design on the IGBT avalanche ruggedness: A comparative analysis between punch-through and field-stop devices. *IEEE Transactions on Electron Devices, 62*(8), 2535-2541.

[4] Spirito, P., Breglio, G., Irace, A., Maresca, L., Napoli, E., & Riccio, M. (2014). Physics of the negative resistance in the avalanche $I\{-\}V$ curve of field stop IGBTs: Collector design rules for improved ruggedness. *IEEE Transactions on Electron Devices, 61*(5), 1457-1463.

[5] Maresca, L., Breglio, G., & Irace, A. (2014). Automatic TCAD model calibration for multi-cellular Trench-IGBTs. *Solid-State Electronics, 91*, 36-43.

[6] Baburske, R., van Treek, V., Pfirsch, F., Niedernostheide, F. J., Jaeger, C., Schulze, H. J., & Felsl, H. P. (2014, June). Comparison of critical current filaments in IGBT short circuit and during diode turn-off. In *Power Semiconductor Devices & IC's (ISPSD), 2014 IEEE 26th International Symposium on* (pp. 47-50). IEEE.

[7] Khanna, Vinod Kumar. Insulated gate bipolar transistor IGBT theory and design. John Wiley & Sons, 2004.

[8] Spirito, P., Breglio, G., d'Alessandro, V., & Rinaldi, N. (2002). Analytical model for thermal instability of low voltage power MOS and SOA in pulse operation. In *Power Semiconductor Devices and ICs, 2002. Proceedings of the 14th International Symposium on* (pp. 269-272). IEEE.

[9] Pierret, Robert F. Semiconductor device fundamentals. Pearson Education India, 1996.

[10] Sze, Simon M., and Kwok K. Ng. *Physics of semiconductor devices.* John wiley & sons, 2006.

Proceedings of the 30th International Symposium on Power Semiconductor Devices & ICs
May 13-17, 2018, Chicago, USA

A High-Voltage p-LDMOS with Enhanced Current Capability Comparable to Double RESURF n-LDMOS

Bo Yi, Junji Cheng, Moufu Kong, Bingke Zhang, Xing Bi Chen

State Key Laboratory of Electronic Thin Films and Integrated Devices
University of Electronic Science and Technology of China
Chengdu, China
Email: yb@uestc.edu.cn

Abstract—**In this paper, a simple p-LDMOS structure with significantly improved performances based on a novel three dimensional concept is proposed. The hole current in the Ptop region of the *signal region* flows into the floating P⁺, then through the integrated resistor R_P formed by the Pbase region in the z-axis direction with a distance of W_2+W_3, into electrode D. A voltage drop (V_{Gn}) which controls the n-channel will be auto-generated across R_P during the on- and off-state of the hole current. Thus the p-LDMOS applies both hole and electron as majority carriers to conduct current. The proposed p-LDMOS, having only one external controlling signal (G_P), has a current capability comparable to or even larger than that of an optimized double RESURF n-LDMOS implemented through the same process steps. The power loss can be reduced by 74.9% compared with the conventional p-LDMOS.**

Keywords—High-voltage; p-LDMOS; current capability; double RESURF; auto-controlled

I. INTRODUCTION

Lateral Diffused Metal Oxide Semiconductor (LDMOS) is suitable for Smart Power Integrated Circuit (SPIC) [1]-[6]. By using p-LDMOS in SPICs, the circuit design can be simplified and the power efficiency can be much improved [1]-[5],[7]. But compared with an n-LDMOS implemented through the same process, a p-LDMOS usually needs much more chip areas to obtain the same current capability. Many improved ways have been proposed to enhance the current capability of the p-LDMOS [5],[8]-[12]. But the current capabilities of the p-LDMOSs are still limited by the hole mobility. And some even sacrifices the dynamic performance due to large gate charge [12]. LDMOSs applying both hole and electron to conduct current have been proposed and studied [7],[13]-[15]. But extra controlling circuits and components are need to control the n-channel for the p-LDMOS [7]. Or a current source is needed to control the p-channel for n-LDMOS [14]. Or the specific on-resistance is increased to improve the safe

This work is support by the Open Foundation of State Key Laboratory of Electronic Thin Films and Integrated Devices (KFJJ201708) and National Natural Science Foundation of China (No. 51237001).

operation area of n-LDMOS [15]. To overcome the above issues, a new structure of p-LDMOS based on the double RESURF technology is proposed. The proposed p-LDMOS has an ultra-high current capability which is comparable to that of a double RESURF n-LDMOS.

II. STRUCTURE AND RESULTS

A. Static Characteristics

Fig. 1 shows the structure of proposed p-LDMOS and its equivalent circuit. As shown in Fig. 1(a) the p-LDMOS is much more like a double-RESURF n-LDMOS. The only difference is that it has a p-type channel. The external control signal is applied on the gate of p-channel (G_p). The gate of the n-channel (G_n) is auto-controlled by the signal on the floating P⁺ region as shown in Fig. 1(a). The Nwell-iso region is formed by an implantation window gap of 5 μm for both the Nwell and Ptop region. Since the width of W_4 is quite larger than that of W_1, the hole current within W_4 needs to flow out directly to electrode D considering the limited current capability of the metal on the floating P+. Therefore, the Nwell-iso region is used to isolate the two P⁺ regions in the Ptop region due to their different potentials.

When the channel under G_p is turned on, the hole current in the Ptop region which is within the *main current* region flows directly into electrode D. While the hole current in the Ptop region which is within the *signal region* flows into the floating P⁺ region, then across the resistor R_p formed by the Pbase region in the z-axis direction and eventually into electrode D. A voltage drop (V_{Gn}) is caused across the R_p and V_{Gn} is connected to G_n. Then a positive voltage is applied on the G_n to turn on the n-channel automatically. Electron current begins to flow from electrode S to D through the Nwell region and the n-channel. Due to the electron current, the p-LDMOS has a large current capability compatible with that of a double RESURF n-LDMOS. When the p-channel is turned off, the hole current across R_p decreases to zero and the voltage on G_n decreases to zero. Thus, the n-channel is turned off automatically.

978-1-5386-2928-4/18 $31.00 © 2018 IEEE   148

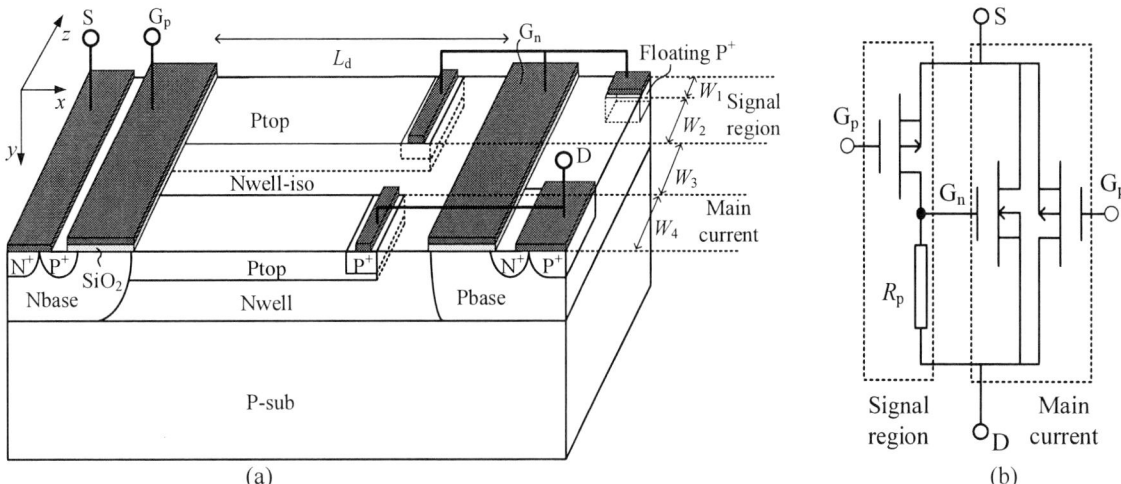

(a)   (b)

Fig. 1. (a) Proposed p-LDMOS and (b) equivalent circuit. S and D represent the source and drain of the p-LDMOS

TCAD simulator is used to verifying the characteristics of the proposed p-LDMOS. Table I shows the key simulation parameters used.

TABLE I

KEY SIMULATION PARAMETERS FOR PROPOSED P-LDMOS IN FIG. 1

Symbol	Description	--	
W_2 (μm)	Main length of integrated resistance R_p	From 35 to 55	
W_4 (μm)	Width of main current region	$2W_2$ to $5W_2$	
L_d (μm)	Length of the drift region	45	
Ptop	P-type region doping for double RESURF	Dose (cm^{-2})	Diffusion Length (μm)
		1.6×10^{12}	0.8
Nwell	N-type region doping for double RESURF	2.2×10^{12}	2.6

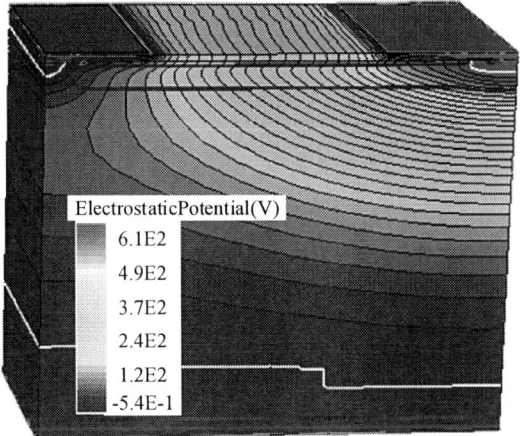

Fig. 2. Potential contour of proposed p-LDMOS at BV = 610 V

Fig. 2 shows the potential contours of the proposed p-LDMOS at breakdown voltage. It shows that although there is an Nwell-iso region formed in the drift region of proposed p-LDMOS, the proposed p-LDMOS can obtain a BV of 610 V, while the BV of a double RESURF n-LDMOS without the Nwell-iso region is 619 V.

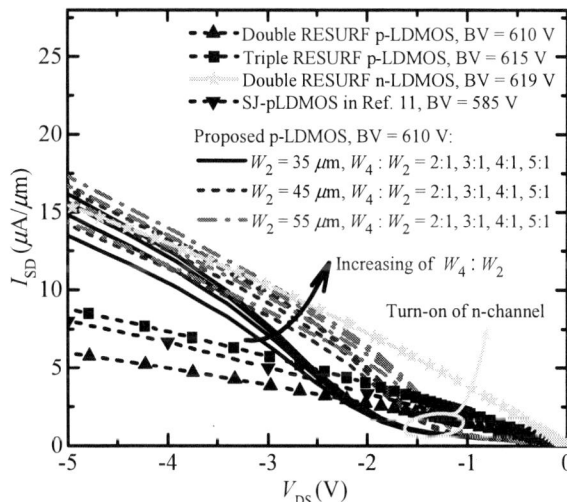

Fig. 3. Comparison of current capabilities for different high-voltage LDMOSs

Fig. 3 compares the output characteristics of different LDMOSs. For the 610 V high-voltage LDMOS, the on-state voltage is set to be 5 V. For the proposed p-LDMOS, when the V_{SD} increases to a certain value, none-linear region is caused in the output waveforms. This is because when the V_{SD} is large enough, the voltage on G_n (V_{Gn}) induced by the hole current across the R_p increases with the increasing of V_{SD}. Therefore, electron current increases with the increasing of V_{SD}. When the n-channel is fully turned on, the total current density increases linearly with the increasing of V_{SD} which means no conductivity modulation effect happens in the proposed p-LDMOS.

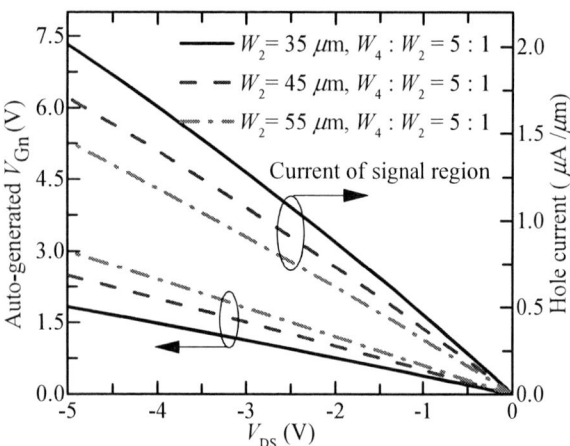

Fig. 4. Relationship of auto-generated V_{Gn} and hole current densities for different designs

Besides, with the increasing of W_2, the n-channel is turned on at a lower V_{SD} resulting a larger current density. This is because the R_p increases with the increasing of W_2 and the V_{Gn} obtains a larger value at a lower V_{SD} as shown in Fig. 4. And for the same W_2, the current density increases with the increasing of $W_4 : W_2$ as shown in Fig. 3. This is because the electron current in the *signal region* is not uniform since the body effect exists in the Pbase region under G_n. To be specific, the electron current close to floating P^+ is zero, while the electron near the main current region reaches the maximum value. Therefore, with the increasing of $W_4 : W_2$, the current ratio in the *signal region* becomes small, and the current density approaches the maximum value.

To be concluded, as shown in Fig. 3, the proposed p-LDMOS has a much larger current capability than that of the p-type silicon limit or SJ-pLDMOS, and is even larger than that of a double RESURF n-LDMOS since the proposed p-LDMOS has one more hole current than that of the double RESURF n-LDMOS.

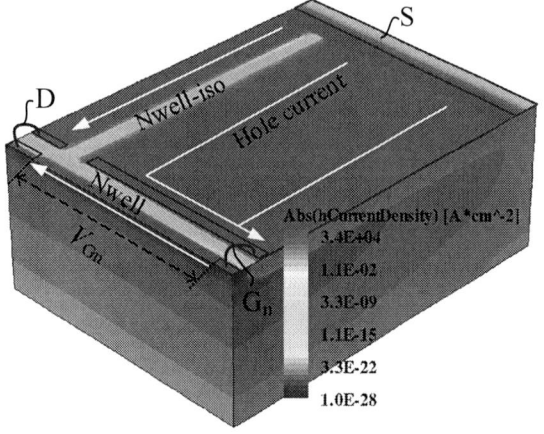

Fig. 5. Hole current density distribution of proposed p-LDMOS at $V_{SD} = 5$ V, W_4 is set to 5 μm to save the simulation time due to its periodicity

Fig. 5 shows the hole current distribution at $V_{SD} = 5$ V. The simulation results verify that no conductivity modulation effect

happens in the proposed p-LDMOS because the hole current increases linearly with V_{SD} shown in Fig. 4 and flows only in the P-type region shown in Fig. 5.

B. Dynamic Characteristics

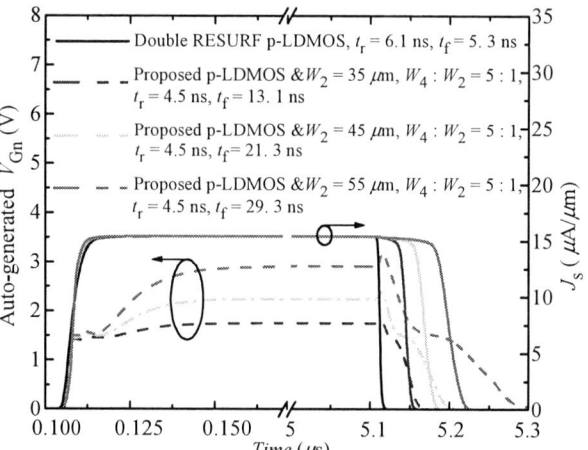

Fig. 6. Switching process of different p-LDMOSs

Fig. 6 compares the switching process of the conventional double RESURF p-LDMOS and the proposed p-LDMOS with different value of W_2. It shows that, the proposed p-LDMOS has a turn-on speed of 4.5 ns which is faster than that of the conventional one due to its larger current capability. While the proposed p-LDMOS has a slower turn-off speed than that of the conventional one. This is because after the p-channel is turn off, more time is still needed to discharge the voltage on G_n. And with the increasing of W_2, the voltage on G_n increases as shown in Fig. 4 and Fig. 6. Therefore, with the increase of W_2, more time is needed to discharge G_n resulting in an increased turn-off time as shown in Fig. 6. However, the proposed p-LDMOS is still a unipolar device. Thus, the turn-off time is still fast which is 13.1 ns, 21.3 ns, and 29.3 ns for different value of W_2, respectively.

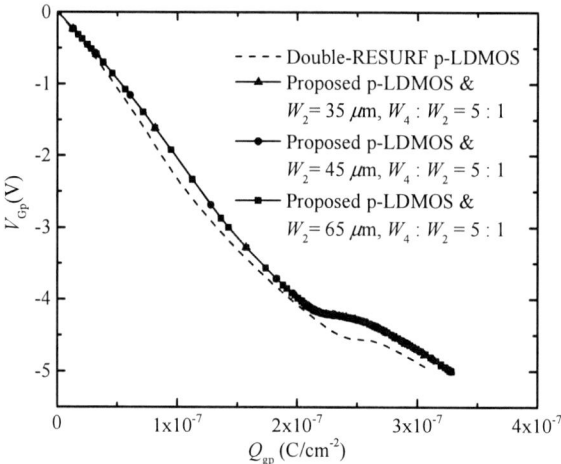

Fig. 7. Comparison of gate charge densities for different p-LDMOSs

Fig. 7 compares the gate charges of conventional double RESURF p-LDMOS and the proposed p-LDMOS with

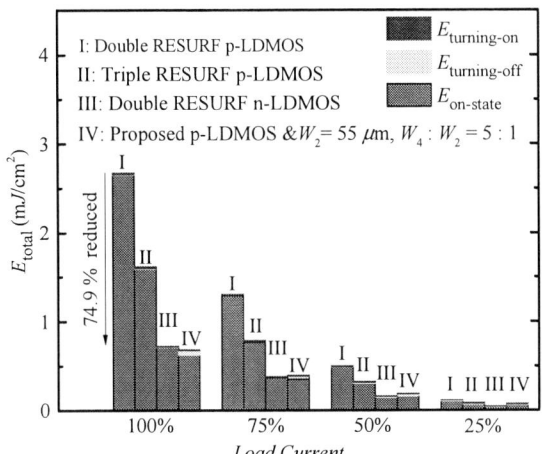

Fig. 8. Comparison of total power loss for different LDMOSs at different load currents at $f = 100$ kHz

different W_2. As it shows, the proposed p-LDMOS has an approximate gate charges with that of the double RESURF p-LDMOS since the n-channel gate is auto-controlled.

Fig. 8 compares the total power loss of different devices including double RESURF, triple RESURF p-LDMOS, double RESURF n-LDMOS and the proposed p-LDMOS with $W_2 = 55$ μm and $W_4 : W_2 = 5:1$ at different load currents. It shows that, although the turn-off power loss of the proposed p-LDMOS is a little larger than that of the double or triple RESURF p-LDMOS, the proposed p-LDMOS has a much smaller power loss than that of the double or triple RESURF p-LDMOS. The power loss at the rated load current, is reduced by about 74.9% and 58.8% compared with that of the double RESURF and triple RESURF p-LDMOS, respectively. The power loss of the proposed p-LDMOS is even compatible with that of the double RESURF n-LDMOS at different load currents.

III. CONCLUSION

In this paper, we proposed a p-LDMOS with auto-controlled n-channel based on double RESURF structure in a novel three dimensional concept. The proposed p-LDMOS obtains a much larger current capability than that of the p-type *"silicon limit"* and has a unipolar switching characteristics. The total power loss of proposed p-LDMOS is much smaller than that of the p-type *"silicon limit"* and is compatible with that of the double RESURF n-LDMOS. By applying the proposed p-LDMOS in HVICs, chip area can be much reduced and system efficiency can be much improved.

REFERENCES

[1] M . Nakano, K. Takahashi, A. Tanaka, A. Osawa and H. Yoshida, "Full-complementary high-voltage driver ICs for flat display panels," in *Proc. VLSI Technology, Systems and Applications*, 1989, pp. 55-58

[2] K. Kobayashi, H. Yanagigawa, K. Mori, S. Yamanaka and A. Fujiwara, "High voltage SOI CMOS IC technology for driving plasma display panels," in *Power Semiconductor Devices and ICs (ISPSD), 1998 IEEE 10rd International Symposium on*, 1998, pp. 141-144.

[3] T. Liang, Y. He, L. Lu, M. Qiao and B. Zhang, "200-V high-side thick layer SOI field PLDMOS for HV switching IC," *Power Electronics and Motion Control Conference (IPEMC-ECCE Asia)*, 2016, pp. 3116-3119.

[4] W. F. Sun, L. Shi, Z. Sun, Y. Yi, H. Li and S. Lu, "High-voltage power IC technology with nVDMOS, RESURF pLDMOS, and novel level-shift circuit for PDP scan-driver IC," *IEEE Trans. Electron Devices*, vol. 53, no. 4, pp. 891-896, 2006.

[5] S. Shimamoto, Y. Yanagida, S. Shirakawa, K. Miyakoshi, T. Imai, T. Oshima, *et al.*, "High performance Pch-LDMOS transistors in wide range voltage from 35V to 200V SOI LDMOS platform technology," in *Power Semiconductor Devices and ICs (ISPSD), 2011 IEEE 23rd International Symposium on*, 2011, pp. 44-47.

[6] D. H. Lu, T. Mizushima, H. Sumida, M. Saito and H. Nakazawa, "High voltage SOI P-channel field MOSFET structures," in *Power Semiconductor Devices and ICs (ISPSD), 2009 IEEE 21rd International Symposium on*, 2009, pp. 17-20.

[7] B. Yi and X. Chen, "A 300 V ultra low specific on-resistance high-side p-LDMOS with auto-biased n-LDMOS for SPIC", *IEEE Transactions on Power Electronics*, vol. 32, no. 1, pp. 551-560, 2017.

[8] M. Sambi, D. Merlini, P. Galbiati, E. Bonera, and F. Belletti, "A novel 0.16um-300 V SOIBCD for ultrasound medical applications," in *Power Semiconductor Devices and ICs (ISPSD), 2011 IEEE 23rd International Symposium on*, 2011, pp. 36-39.

[9] S. Tokumitsu, T. Nitta, T. Shiromoto, T. Kuroi, K. Hatasako, and S. Maegawa, "Enhancement of current drivability in field PMOS by optimized field plate," in *Power Semiconductor Devices and ICs (ISPSD), 2010 IEEE 22rd International Symposium on*, 2010, pp. 253-256.

[10] Z. Xin, Q. Ming, H. Yitao, W. Zhuo, L. Zhaoji, and Z. Bo, "Back-gate effect on Ron,sp and BV for thin layer SOI field p-Channel LDMOS," *Electron Devices, IEEE Transactions on*, vol. 62, pp. 1098-1104, 2015.

[11] L. Bo, Q. Ming, W. Yongchun, K. Mingliang, Y. Jun, Z. Bo, et al., "High voltage SJ-pLDMOS with Variation Lateral Doping drift layer," in *Communications, Circuits and Systems (ICCCAS), 2010 International Conference on*, 2010, pp. 503-506.

[12] X. R. Luo, Q. Tan, J. Wei, K. Zhou, G. Deng, Z. Li and B. Zhang, "Ultralow on-resistance high-voltage p-Channel LDMOS with an accumulation-effect Extended Gate," *Electron Devices, IEEE Transactions on*, vol. 63, pp. 2614-2619, 2016.

[13] X. Chen, "Lateral low-side and high-side high-voltage devices," U.S. Patent 6 998 681 B2, Feb. 14, 2006.

[14] Y. Zhang, S. Pendharkar, P. Hower, S. Giombanco, A. Amoroso and F. Marino, "A RESURF PN bimodal LDMOS suitable for high voltage power switching applications", in *Power Semiconductor Devices and ICs (ISPSD), 2015 IEEE 27rd International Symposium on*, 2015, pp. 61-64.

[15] W. Du, X. B. Chen, "Design of a double-gate power LDMOS with improved SOA by complementary majority carrier conduction paths", *IEEE Transactions on Power Electronics*, vol. 31, no. 7, pp. 5133-5140, 2016.

Proceedings of the 30th International Symposium on Power Semiconductor Devices & ICs
May 13-17, 2018, Chicago, USA

Self Terminating Lateral-Vertical Hybrid Super-Junction FET that Breaks $R_{DS}.A$ –Charge Balance Trade-off Window

Karthik Padmanabhan, Lingpeng Guan, Madhur Bobde, Sik Lui, Anup Bhalla[+] and Hamza Yilmaz[++]

Alpha and Omega Semiconductor, Inc., Sunnyvale, CA,
[+] United Silicon Carbide, Monmouth Junction, NJ
[++] Computime Limited, Queensway, Hong Kong
Email: {karthik.p, lpguan, mbobde, slui}@aosmd.com

Lei Zhang

Jireh Semiconductor, Inc.
Hillsboro, OR, USA
Email: lei.zhang@jfab.aosmd.com

Abstract— **In this paper, we present a novel lateral/vertical hybrid super-junction structure that breaks the fundamental trade-off between the $R_{DS}.A$ and Charge imbalance window for a Super-Junction MOSFET. This device structure can continue the scaling of Superjunction MOSFET well below $10m\Omega.cm^2$ with good manufacturability window. Methods of scaling the Blocking Voltage and $R_{DS}.A$ are also discussed for this hybrid device structure. This device structure also offers an optional drain electrode on the top surface, if needed for co-packaging with other chips.**

Keywords— Superjunction, Charge Balance, High Voltage, Avalanche, Breakdown, Lateral, Vertical, LDMOS, $R_{DS}.A$, Trench Power MOSFETs, Scalability, Cascode, Termination.

I. INTRODUCTION

Superjunction MOSFETs are currently the most popular MOSFETs for many high voltage applications, with increasing viability for higher switching frequency & power density applications. The technology has progressed rapidly towards a low $R_{DS}.A$, an important Figure of Merit (FOM) for high voltage MOSFETs [1],[2],[3]. Scaling of the Super-Junction MOSFET $R_{DS}.A$ has brought it in the ballpark of Wide Band-Gap (WBG) devices [4], while avoiding the higher cost and yield complexity issues associated with the latter. However, a major challenge of scaling Superjunction MOSFET Rds.A is the reduction of charge balance window with lower $R_{DS}.A$[5]. The popularly used methods of manufacturing super junction FETs include using multiple epi layers to form the superjunction columns and a trench filling approach [6] [7]. While the multiple epi approach requires a masked high energy implant which is prone to variation from mask CD and resist profile, the deep trench approach has additional variations from trench CD, angle and epi growth in trench. This is the reason why commercially available superjunction FETs are still significantly higher in $R_{DS}.A$ compared to the ones published in literature.

While a lateral superjunction MOSFET has been proposed before [1] [8], the devices mentioned there still depend on the superjunction columns for the breakdown, and thus are still prone to the same trade-off between $R_{DS}.A$ and Charge imbalance window. The proposed structure [9] addresses this fundamental BV-vs-$R_{DS}.A$ trade-off [10] by decoupling the two parameters, thus greatly improving the scalability factor. The formation of superjunction layers does not require any masking or in-situ doping step, thus achieving precise charge control. The device structure also includes a built in bode diode, that does not include current flow through the super-junction layer, and can be designed independently for excellent robustness.

II. DEVICE STRUCTURE AND CROSS SECTION

A cross section of the device is shown in Fig. 1. The device consists of lateral superjunction channels with N columns at left and right edges to connect to the surface MOSFET and the

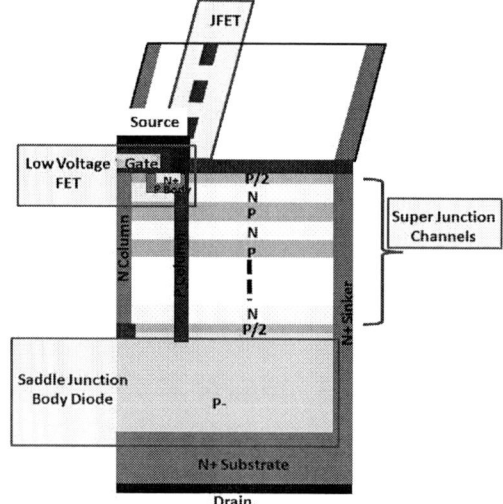

Fig. 1: Cross section of the proposed device.

978-1-5386-2928-4/18 $31.00 © 2018 IEEE

N+ substrate respectively. An additional P column serves the purpose of a vertical JFET to pinch off the N column on the left at high drain biases. A surface low voltage MOSFET is used in cascode configuration to connect the MOSFET source to the vertical JFET through N column on the left. A lightly doped P epi region under the superjunction channels forms a saddle junction body diode, and can be design to bypass avalanche current from the super junction region.

The formation of superjunction region in this structure does not require any lithography or trench etch/epi fill. It can be formed by growing alternately doped P and N type epi regions on a N+substrate/P- epi. Alternately, it can also be formed by growing lightly doped epi and blanket implanting N/P type dopants to form the desired region. Trenches to connect all super junction channels are formed after all the epi growth. These trenches are filled with heavily doped impurities and are completely free of electric field. So there is no need for a perfect epi fill, or charge control. Scaling the $R_{DS}.A$ of this device only requires additional alternating N/P layers without any changes in the width or charge of the superjunction channel as shown in Fig. 2. This in turn means no degradation in charge balance window for lower $R_{DS}.A$ structures.

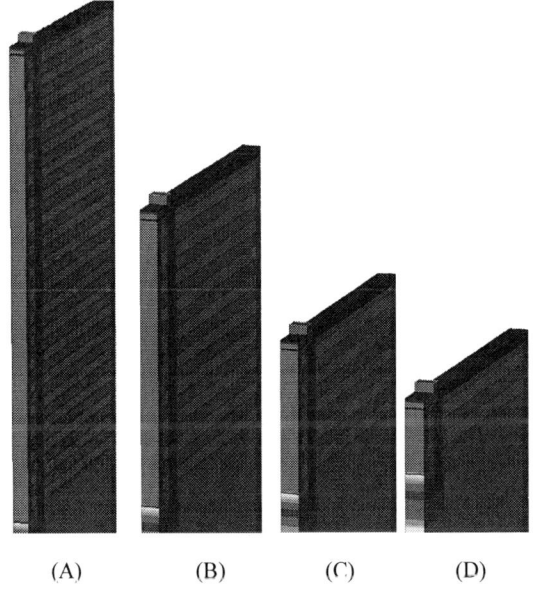

Fig. 2: Simulated 3D Cross sections of proposed super-junction structure for 4 cases: (A) 20 Channels (B) 10 Channels (C) 5 Channels (D) 2 Channels

The avalanche breakdown of the device is determined by two factors: the P-epi saddle junction and superjunction rows. The breakdown voltage of either is unrelated to the other. Whichever of the two has a lower breakdown voltage will clamp the avalanche capability of the entire device. The breakdown voltage of the P-type diode is dictated by the thickness and resistivity of the epi, while the superjunction row breakdown voltage is controlled by the implanted charge and the length of the rows. So, one design consideration is to always have saddle junction clamp the avalanche breakdown.

This is accomplished by making the superjunction rows longer to have a higher breakdown capability. This is beneficial especially with regards to the UIS avalanche capability of the device, and makes it less reliant on the superjunction charge balance. Fig. 3 shows the electrostatic potential lines and the impact ionization distribution in the device.

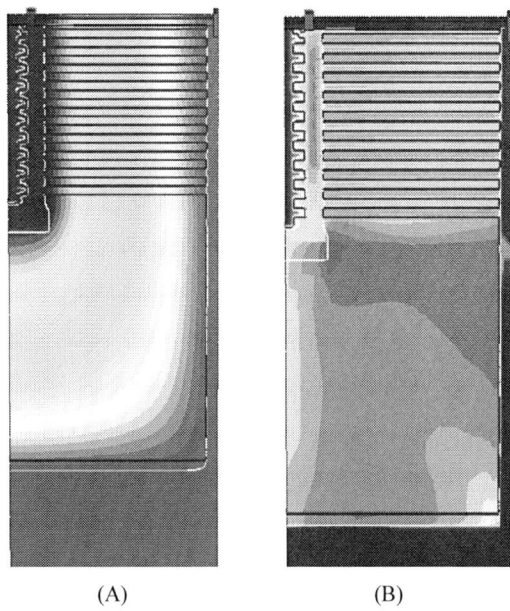

(A)   (B)

Fig. 3: Breakdown Simulations: (A) Electrostatic potential @ BV. (B) Impact ionization @ BV

The MOSFET switching is controlled by a low voltage lateral LDMOS structure. This structure, along with the cross section that shows the conducting channel is shown in Fig. 4.

The LDMOS structure is formed through implants, where the self aligned P+ aligns with the P column trench as the end of the SJ rows, as shown in Fig. 1, while the drain connects to an N column which connects all the conducting superjunction N rows. The LDMOS typically is designed to have a 30V breakdown, similar to the typical breakdown voltage of the gate oxide in a MOSFET. When the LDMOS is turned off, there is no current flowing through the N column, so all the SJ N rows stay disconnected. The vertical P columns shield the high electric field from the drain columns on the right and assist the alternating N/P rows deplete to support high voltage. In the forward blocking mode, the P- epi region also depletes out to support the voltage between the drain at the substrate and P column. As discussed, the breakdown voltage can be clamped by this vertical diode for a robust diode diode. When the low voltage FET on surface is turned on, current flows from the source to the N column on the left and spreads out into the superjunction rows. It is collected by the N column on the right side of the device structure and directed to the N+ substrate at the bottom

978-1-5386-2928-4/18 $31.00 © 2018 IEEE

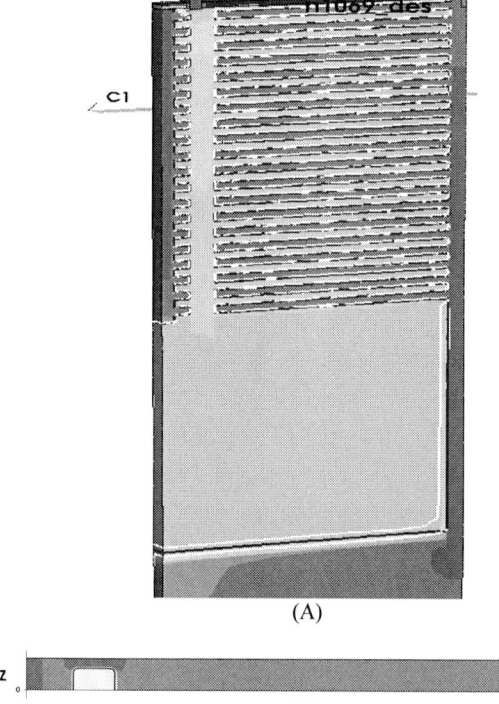

Fig. 4: Conduction Simulations: (A) Current density for the 3D Structure; (B) Current Density for a 2D slice through one of the N Super-Junction channels.

The default device structure terminals are exactly the same as a vertical MOSFET, with source and gate at the surface and drain at the substrate. However, this device also provides the additional option of tapping drain on the surface, if required for co-packaging with other chips/ICs. This structure can also be flipped to make it a bottom source device by using a P+ substrate instead

The superjunction rows are designed using conventional charge balance techniques. Each epitaxial layer has a P and N type implant at different energies, and number of stacked superjunction layers translates to more superjunction layers in parallel. The major advantage of this approach is that it is devoid of any masked implants. This is important because as we target lower $R_{DS}.A$ numbers, the process variations from either the mask CD at smaller feature sizes, or narrower trench CDs in the case of Trench SJ MOSFETs, cause significant variations in both the BV and R_{DS}-on. Since the device is clamped by the vertical saddle junction, it makes it easier to stack multiple superjunction rows to achieve lower R_{DS}-on while maintaining the avalanche breakdown voltage. This significantly helps with the BV-vs-$R_{DS}.A$ trade-off that typically troubles the SJ MOSFET technology.

III. PROPOSED FABRICATION FLOW

The process steps involved in the fabrication of the the device are as shown in Fig. 5. It starts with growing the P-blocking epi on the N+ substrate. The thickness and resistivity, as mentioned earlier, are determined by the desired avalanche rating. After that, we grow the superjunction layers, which are formed by growing a thin layer of intrinsic silicon followed by P and N implants to form the SJ rows. The number of

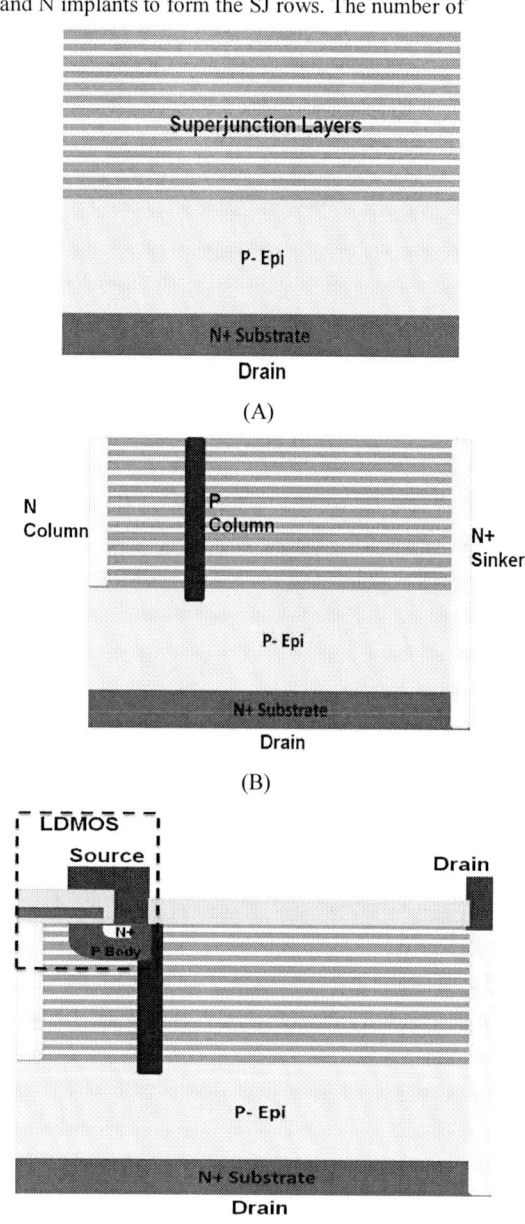

Fig. 5: Device Fabrication Steps: (A). P- epi growth, superjunction layer growth. (B) Vertical P and N row formation. (C) LDMOS formation.

superjunction layers is determined by the target $R_{DS}.A$. Once the SJ layers are grown, the device undergoes a thermal diffusion process to form the SJ rows. The P- layer and the SJ columns are the 2 most basic components of the device.

Once these are done, the deep trench etches to form the vertical N+ sinker to link to the bottom N+ substrate and the vertical P and N columns to link the lateral superjunction rows are formed. These P and N trenches are filled by heavily doped Boron and Phosphorus type Poly-silicon/epi layers respectively. A thermal diffusion of these impurities should be followed to ensure that the deep trenches are field free. The P column needs to extend through the buffer till the P- epi. The left N column connected to the LDMOS needs to extend through all the superjunction layers, while the right N+ sinker needs to connect all the way to the N+ substrate. Once the columns are formed, the topside LDMOS to control the device can be formed using a conventional planar MOSFET structure.

IV. SIMULATION RESULTS AND DISCUSSIONS

Complex numerical 3D simulations of the proposed device structure with more than 200k grid points were performed using Sentaurus. Multi-core & parallel processing tools were used to cut down simulation time. Cell pitch of 50μm and Super-Junction pitch of 3μm was chosen. The resulting technology curve is shown in Fig. 6. A breakdown voltage of >650V was achieved, with the diode supporting the breakdown voltage. Stacking as many as 20 superjunction rows laterally helps achieve a low $R_{DS}.A$ of 7mΩ.cm², and can be easily scaled below 5. The figure clearly shows that the BV-vs-$R_{DS}.A$ trade-off is broken. There eventually will be a point of diminishing returns with respect to how low an $R_{DS}.A$ we can achieve because additional superjunction layers will no longer provide the necessary reduction in $R_{DS}.A$. But it stretches the limits to which Silicon can still the semiconductor of choice to achieving extremely low R_{DS}-on numbers.

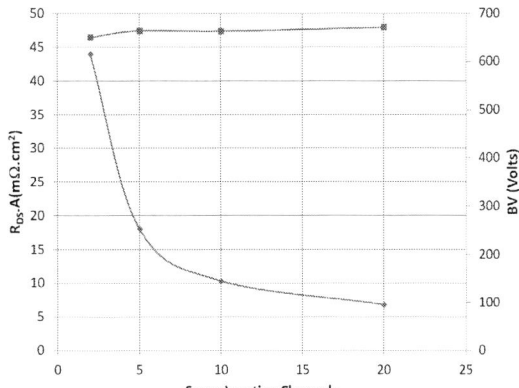

Fig. 6: Simulated $R_{DS}.A$ and BVDss as a function of number of Super-Junction channels.

REFERENCES

[1] Fujihira, Tatsuhiko. (1997). Theory of Semiconductor Superjunction Devices. Japanese Journal of Applied Physics.

[2] T. Fujihira and Y. Miyasaka, "Simulated superior performances of semiconductor superjunction devices," *Power Semiconductor Devices and ICs, 1998. ISPSD 98. Proceedings of the 10th International Symposium on*, Kyoto, 1998, pp. 423-426.

[3] F. Udrea, G. Deboy and T. Fujihira, "Superjunction Power Devices, History, Development, and Future Prospects," in *IEEE Transactions on Electron Devices*, vol. 64, no. 3, pp. 720-734, March 2017.

[4] T. Kobayashi *et al.*, "High-voltage power MOSFETs reached almost to the silicon limit," *Power Semiconductor Devices and ICs, 2001. ISPSD '01. Proceedings of the 13th International Symposium on*, Osaka, 2001, pp. 435-438.

[5] W. Saito, "Theoretical limits of superjunction considering with charge imbalance margin," *2015 IEEE 27th International Symposium on Power Semiconductor Devices & IC's (ISPSD)*, Hong Kong, 2015, pp. 125-128.

[6] H. Kapels, "Superjunction MOS devices — From device development towards system optimization," *2009 13th European Conference on Power Electronics and Applications*, Barcelona, 2009, pp. 1-7.

[7] R. K. Williams, M. N. Darwish, R. A. Blanchard, R. Siemieniec, P. Rutter and Y. Kawaguchi, "The Trench Power MOSFET: Part I—History, Technology, and Prospects," in *IEEE Transactions on Electron Devices*, vol. 64, no. 3, pp. 674-691, March 2017.

[8] M. Rub *et al.*, "A 600V 8.7Ohmmm²Lateral Superjunction Transistor," *2006 IEEE International Symposium on Power Semiconductor Devices and IC's*, Naples, 2006, pp. 1-4.

[9] M. Bobde, Lingpeng Guan, Anup Bhalla and Hamza Yilmaz, Lateral super junction device with high substrate-gate breakdown and built-in avalanche clamp diode, *US20170117386A9*

[10] A. G. M. Strollo and E. Napoli, "Optimal ON-resistance versus breakdown voltage tradeoff in superjunction power devices: a novel analytical model," in *IEEE Transactions on Electron Devices*, vol. 48, no. 9, pp. 2161-2167, Sep 2001.

Proceedings of the 30th International Symposium on Power Semiconductor Devices & ICs
May 13-17, 2018, Chicago, USA

Local Lifetime Control
for Enhanced Ruggedness of HVDC Thyristors

J. Vobecky, V. Botan, K. U. Meier, K. Tugan
ABB Switzerland Ltd, Semiconductors
Lenzburg, Switzerland

M. Bellini
ABB Corporate Research
Dättwil, Switzerland

Abstract— **Proton irradiation is experimentally demonstrated to increase the ruggedness of large area thyristors for HVDC. While maintaining very low On-state losses at V_T below 1.7 V and 1.8 V for 7.2 and 8.5 kV classes (I_T = 6.25 kA, T = 90 ºC), record low leakage current has been achieved at 150 mm silicon wafers together with increased surge current, higher dV/dt capability and lower circuit commutated recovery time t_q.**

Keywords—Thyristors; Silicon devices; Proton irradiation; HVDC transmission;

I. Introduction

UHVDC transmission technology with Load Commutated Converters, which has passed the 10 GW power rating milestone, requires phase control thyristors (PCT) with further lowered losses [1]. Conventionally, achieving the very low on-state losses (V_T), which is the primary parameter of HVDC valves, requires compromising the dynamic parameters (recovery charge Q_{rr}, circuit commutated recovery time t_q, dV/dt capability) and ruggedness (surge current).

This paper shows that when we combine the existing technology for the reduction of wafer thickness [2, 3] with sophisticated local lifetime control, we can obtain thyristors with lower V_T, lower leakage current and massively enhanced ruggedness at the same time. This is demonstrated on the largest existing thyristors of 7.2 and 8.5 kV classes with current rating up to I_T = 6250 A produced on 150 mm Si wafers. All this provides devices superior for the UHVDC, Flexible AC transmission systems (FACTS), pumped-storage hydropower and also for the industry, for which is the achievement of very low leakage current at large area thyristor wafers of special importance because of much higher T_{jmax}.

II. Device Design

A. Existing Design

Thyristor is non-punch through device, in which is the space charge region (SCR) neither at reverse nor at forward blocking allowed to reach the opposite p-n junction. If it happens, the leakage current excessively grows via the *punch-through* effect and breakdown voltage is reached. A sufficient N-base thickness can prevent the *punch-through* from appearing before the blocking junction undergoes the avalanche ionization, which can be set by an appropriate resistivity of the N-base. As this results in a higher ON-state voltage drop V_T, new techniques to reduce the N-base thickness have been developed.

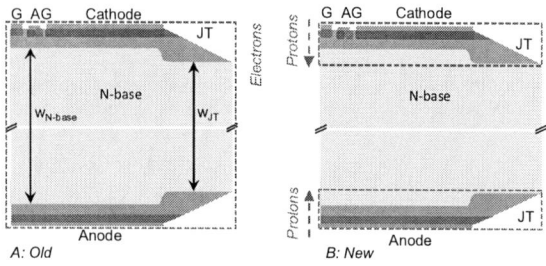

Fig.1: Existing (left) and new concept (right).

The design on Fig.1 (left) eliminates the *punch-through* effect in the active region using a masked diffusion process, which results in a device with $w_{N-base} > w_{JT}$ at about 90% of device area [2, 3]. The benefit of reduced leakage current and increased breakdown voltage can be invested into the reduction of wafer thickness, which can give us a lower V_T for the original blocking capability. The electron irradiation is used to precisely set the recovery charge into a narrow Q_{rr} band, which is necessary for a reliable operation of up to one hundred parts in series.

B. New Design

Further improvement can be achieved by locally modified recombination rate of free carriers in both the OFF and ON-state, if we replace the uniform lifetime control from electron irradiation by the local lifetime control shown in Fig.1 – right.

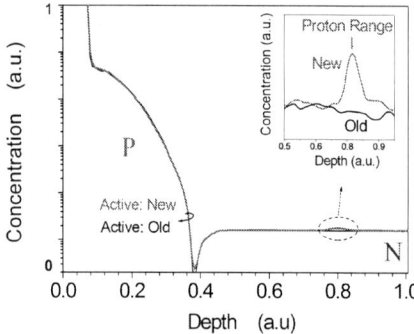

Fig.2: Spreading resistance profiles of electron irradiated (Old) and proton irradiated thyristors (New) in the active region.

978-1-5386-2928-4/18 $31.00 © 2018 IEEE   156

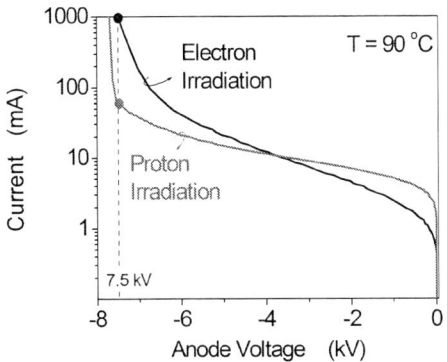

Fig.3: Simulated reverse I-V curves for old and new designs (7.2 kV class).

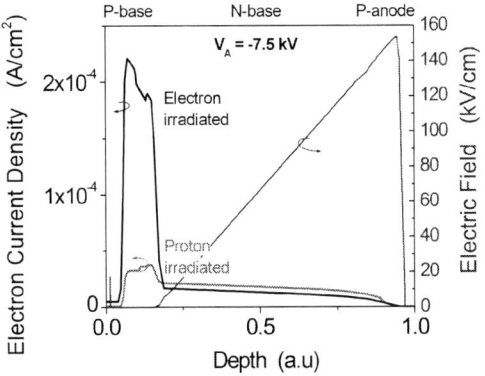

Fig.4: Electron current density for the old and new concept (cathode at left) for the bias points denoted by full circles in Fig.3.

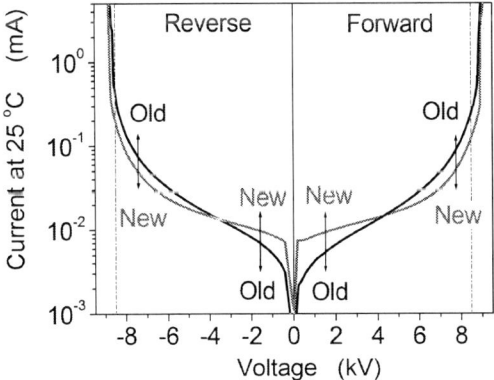

Fig.5: Measured forward and reverse blocking curves of the old and new design concepts at room temperature (8.5 kV class).

The proton irradiation is applied from both anode and cathode side to improve both the forward and reverse blocking capability (OFF-state). Optimized irradiation parameters maintain the original blocking voltage while reducing the leakage current, if we assure a negligible change of doping concentration at the proton range as shown in Fig.2. Contrary to prior art claiming that this approach leads to significantly increased leakage current [4], our work proves the opposite: using an optimal irradiation parameters, it is possible to reduce the leakage current.

The introduced radiation defects with defect peak (end of proton range) placed in the N-base increase the recombination rate of free carriers, their diffusion length reduces accordingly and leakage current decreases. This situation is simulated in Figs.3 and 4 for the case of reverse blocking, where the devices using the new and old irradiation concepts and having the same ON-state voltage drop V_T are compared. In the 2D device simulation using Synopsys simulation package, the whole device cross-section including the junction termination is simulated in cylindrical coordinates.

The proton irradiated device in Fig.3 shows an increased leakage current at reverse voltages up to ≈4 kV due to the increased generation rate caused by the radiation defects created at the anode side. Above 4 kV, the leakage current of devices with protons is lower as a result of the decreased diffusion length of electrons in the N-base close to the P-base (blocking junction) caused by the radiation defects created at the cathode side. For opposite polarity of the anode to cathode voltage, it works in analogous way.

Fig.4 shows that the current density of electrons outside the space charge region is reduced more than fifty times after the proton irradiation. As this reduction of leakage current at high voltages is much stronger than its growth at lower voltages, the final leakage current at nominal breakdown voltage of the new concept is much lower. The measured forward and reverse IV curves shown in Fig.5 confirm the simulations above.

The symmetric proton irradiation results in a qualitatively symmetric effect on the blocking characteristics. However, there can be found a quantitative difference in the suppression of leakage current due to the different amplification mechanisms under the forward and reverse blocking, which can be found in the AC blocking curves shown below (Fig.7).

The protons also modify the uniform ON-state plasma distribution from electron irradiation (see Fig.6). The slight asymmetry of the plasma distribution results from different injection efficiency of the anode and cathode. On top of it, the proton irradiation significantly reduces the plasma concentration in the P-type regions and leaves a higher plasma concentration in the N-base. This has then a positive impact on the plasma dynamics during circuit commutation (turn-off) by making the device less prone to re-triggering not only after a typical commutation (giving a shorter t_q), but also after a giant current pulses under short circuit conditions followed by application of forward or reverse voltage (giving a higher surge current I_{TSM} with re-applied voltage). On the other hand, this feature slightly worsens the V_T-Q_{rr} trade-off curve.

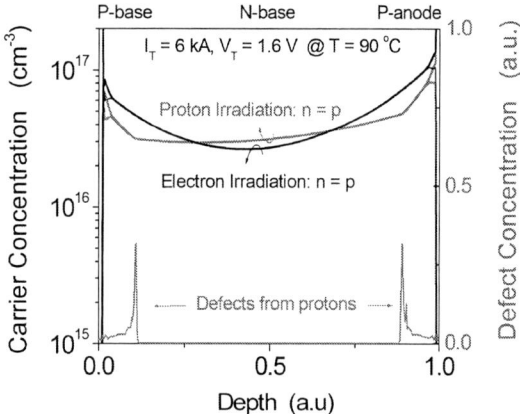

Fig.6: Simulated ON-state carrier distribution for old and new designs (7.2 kV class). Defect peaks resulting from proton irradiation are shown for clarity. The two devices have equal voltage drop V_T.

III. ELECTRICAL CHARACTERISTICS

A. Leakage Current

The achievement of significantly lower leakage current after proton irradiation is evident from the AC forward blocking tests in Figs.7a) and b), which have been performed with the old and new devices having the same V_T and Q_{rr}. As the test was performed up to the nominal breakdown voltage of 8.5 kV and even 300 V higher for the 7.2 kV class, it demonstrates the blocking capability of a phase control thyristor without compromise.

The new approach advances the prior art by achieving more than 50% lower leakage current. For HVDC applications typically used up to 90 - 100 °C, it further increases the safety margins or it allows one to use even a thinner silicon wafer with a lower V_T. For the industrial applications typically used up to 125 °C, it greatly lowers the demands on back-end processing meaning that the free-floating concept is good enough for sufficient cooling and the more robust, but more complicated, more expensive and more active area consuming technique of the low temperature bonding of silicon wafer to molybdenum disk need not be used.

Figs.8a) and b) demonstrate another important feature of the proton irradiation, namely the drop of the forward leakage current to basically same magnitude as for the reverse current. Typically, the forward blocking is attributed by a higher forward leakage current (compare the curves *Old*) due to the current amplification by the internal NPN transistor structure. The local lifetime control efficiently reduces this gain and provides thyristors with 10 °C higher maximal junction temperature or much lower required cooling power.

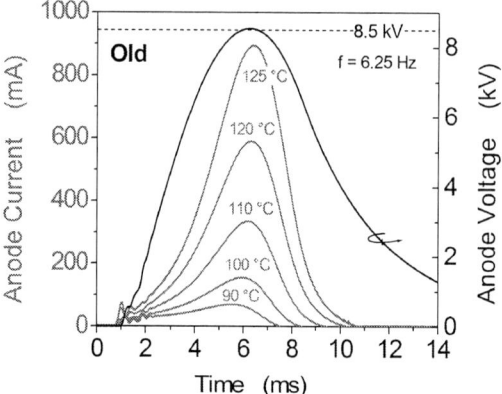

Fig.7a): AC forward blocking test up to full rating voltage of 8.5 kV. **Old** design. 150 mm Si wafer.

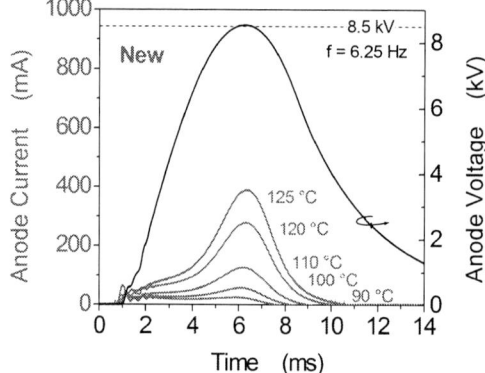

Fig.7b): AC forward blocking test up to full rating voltage of 8.5 kV. **New** design. 150 mm Si wafer.

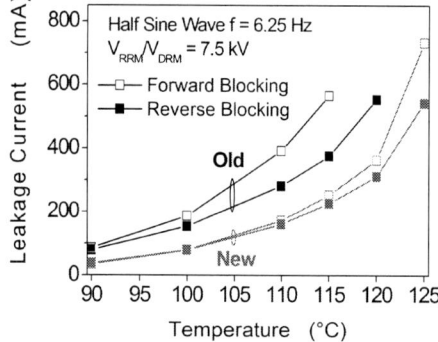

Fig.8a): AC leakage current vs. junction temperature of 7.2 kV class measured up to 7.5 kV. **Old** vs. **New** design.

978-1-5386-2928-4/18 $31.00 © 2018 IEEE

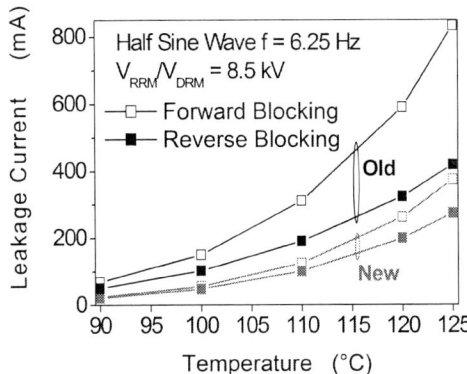

Fig.8b): AC leakage current vs. junction temperature of 8.5 kV class measured up to full rating voltage of 8.5 kV. **Old** vs. New design.

B. Surge Current

The locally reduced recombination lifetime close to the blocking junctions after proton irradiation also increases the magnitude of surge current I_{TSM} with re-applied forward voltage. It is shown for thyristors with equal parameters of the final silicon wafer, V_T and Q_{rr} in Fig.9a). The application of forward voltage after the surge current pulse is given by the application specifics of HVDC valve and represents the most difficult type of surge current parameter to be achieved. The less demanding surge current test with re-applied reverse voltage from Fig.9b) gives higher last pass magnitudes, which are equal to the case without the re-applied voltage. In both Fig.9a) and b) the maximal magnitudes of surge current represent the limit of our tester and the surge current capability of the proton irradiated devices is likely higher.

All the parameter improvements shown above were achieved for other parameters unchanged except for the technology curve V_T - Q_{rr}, which is slightly worse after proton irradiation. As the difference in V_T is typically 10 - 20 mV for a given Q_{rr}, it can be compensated by further reduced wafer thickness not possible in the prior art. The reduction of wafer thickness presented in this paper is a challenging technological task to reduce the losses. An alternative way would be to increase the device area at a thicker silicon. On one hand, this approach would reduce the ON-state losses and thermal resistance. On the other hand, the increase of total volume would lead to a higher recovery charge, hereby reducing the efficiency of V_T reduction. To make an improvement of V_T analogous to the one presented in this paper would then require the usage of Si wafers much more than 150 mm in diameter.

IV. CONCLUSIONS

Proton irradiation of phase control thyristors of 7.2 and 8.5 kV voltage classes respectively was demonstrated to achieve ON-state loss of V_T below 1.7 V and below 1.8 V at $I_T = 6.25$ kA, more than 50% reduced leakage current, about 9% increased surge current with re-applied forward voltage and overall higher dynamic ruggedness. The new devices passed complete electrical qualification in series connection in the HVDC valve and fulfilled all HVDC application specifics. Hereby, the power handling capability of PCTs has further extended to justify the next generation of PCTs for UHVDC, FACTS, pumped-storage hydropower and industry as well.

ACKNOWLEDGMENT

We acknowledge M. Waldmann for running the experimental lots and D. Di-Iorio for reliability testing. We also acknowledge P. Karlsson, A. Blomberg and M. Räms for the HVDC valve testing.

REFERENCES

[1] J. Vobecky, H.-J. Schulze, P. Streit, F.-J. Niedernostheide, V. Botan, J. Przybilla, U. Kellner-Werdehausen, and M. Bellini, "Silicon Thyristors for Ultrahigh Power (GW) Applications", IEEE Trans. El. Dev., 64, 2017, p.760-768.

[2] J. Vobecky, V. Botan, K. Stiegler, U. Meier, M. Bellini, "A Novel Ultra-Low Loss Four Inch Thyristor for UHVDC", in Proc. ISPSD, 2015, pp. 413–416.

[3] J. Vobecky, K. Stiegler, M. Bellini, U. Meier, „New Generation Large Area Thyristor for UHVDC Transmission", in Proc. PCIM 2017, Nuremberg, pp. 761 – 764.

[4] Bartko et al, Reducing the Switching Time of Semic. Devices by Nuclear Irradiation, US Patent 4,056,408, 1977.

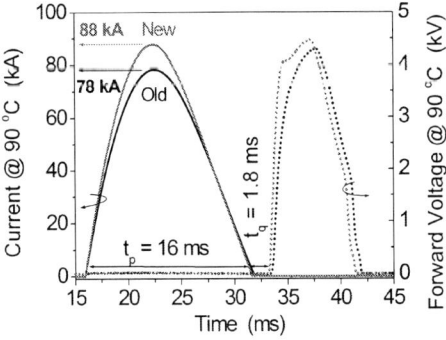

Fig.9a): Surge current (bold) test with re-applied forward voltage (dotted) of 7.2 kV class. Limit of the tester reached for the New device.

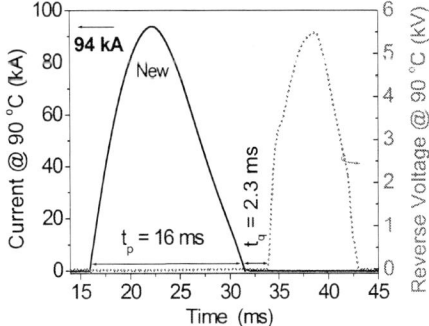

Fig.9b): Surge current test with re-applied reverse voltage of the 7.2 kV class. The limit of the tester was reached.

978-1-5386-2928-4/18 $31.00 © 2018 IEEE

Proceedings of the 30th International Symposium on Power Semiconductor Devices & ICs
May 13-17, 2018, Chicago, USA

Low Injection Anode as Positive Spiral Improvement for 650V RC-IGBT

Ryu Kamibaba, Mitsuru Kaneda, Tetsuo Takahashi and Akihiko Furukawa

Power Device Works, Mitsubishi Electric Corporation

1-1-1 Imajyuku-Higashi, Nishi-ku, Fukuoka, 819-0192, Japan

Phone: +81-92-805-3332 E-mail: Kamibaba.Ryu@cj.MitsubishiElectric.co.jp

Abstract—A novel Reverse Conducting IGBT (RC-IGBT) with a low impurity concentration p⁻ anode is proposed. To improve an embedded anti-parallel diode characteristic, a low hole injection anode structure is effective. The common ohmic contact for both IGBT and diode parts is the strong regulation for the RC-IGBT from the wafer process point of view. One of the solution is utilizing the different metallization independently for IGBT and diode portions. This diode oriented approach has finally played a preferable roll even for IGBT characteristics as if in the positive spiral steps.

Keywords—RC-IGBT, reverse recovery characteristics, low hole injection anode, barrier metal, Lifetime control technique, EB irradiation

I. INTRODUCTION

A Reverse Conducting IGBT (RC-IGBT) which include both IGBT and diode into a single chip [1] [2] is a promising device to reduce a size and cost of the power module. The RC-IGBT has an extremely complex tradeoff relationship among IGBT characteristics, V_{on}-E_{off}-SCSOA, and diode characteristics, V_F-E_{rec}-RRSOA. Therefore, it is difficult to control well-balanced performance between IGBT and diode for the wide frequency range. In the case of a high frequency operation, lifetime control techniques [3] [4] are widely used for a diode. For the conventional RC-IGBT with p⁺/p anode layers fabricated through the common wafer process for both IGBT and diode areas, a heavy lifetime control needs to achieve the lower switching energy loss, especially for diode reverse recovery. On the other hand, an individual IGBT can control its turn-off characteristic only through lowering impurity concentration of the collector layer without the lifetime control technique. Once an IGBT portion of RC-IGBT is fabricated as the low concentration p collector, the IGBT's characteristics are affected by the heavy lifetime control in the not-preferable direction. In contrast, the conventional RC-IGBT with p⁺/p anode layers needs the heavy lifetime control to realize the lower diode's switching energy loss. It largely sacrifices the IGBT's performance. There are many efforts to solve this [5] [6].

In this compromising situation, we tried to find out a solution to improve I_{rr} of diode with the minimum sacrifice of increasing the wafer process fabrication steps.

II. DEVICE STRUCTURE

A "separated type" [6] of RC-IGBT design is one of the major designs. Figure 1 depicts the examples of "separated type" that we called "stripe type" and "island type" [7]. In both types, strongly depending upon the applications, the active area size of RC-IGBT is set forth identical or less size of the summation of individual IGBT's and diode's ones.

Figure 2 shows the cross sectional view of the conventional RC-IGBT. In order to improve the IGBT's tradeoff characteristics, it formed the CSTBT™ utilizing the carrier stored layer and Light Punch Through (LPT) type using a thin wafer process. Diode's area has the p⁺/p anode layers and carrier stored layer. Those concentration are common with IGBT's p⁺ contact layer and p base layer. This is because the diode structure is simultaneously formed through the same wafer process of IGBT's Metal Oxide Semiconductor (MOS) cell portion. The front metal is also formed as the same layers as IGBT.

In the regulation to use the same wafer process for both IGBT and diode regions, the difficulty in RC-IGBT's fabrication process is to unify the front metal for the different doping concentrations of surfaces. Generally, IGBT's front metal uses the three layered structure, Silicide-Barrier metal-Aluminum alloy. Although the Titanium Silicide is the suitable ohmic contact metal for both n⁺ emitter and p⁺ contact region, it is not suitable for low surface concentration contact layer such as p⁻ anode layer. The conventional common front metal using the barrier metal for both IGBT and diode region needs the high impurity concentration anode to maintain the ohmic contact. Consequently, the diode will be suitable for the relatively low frequency operation.

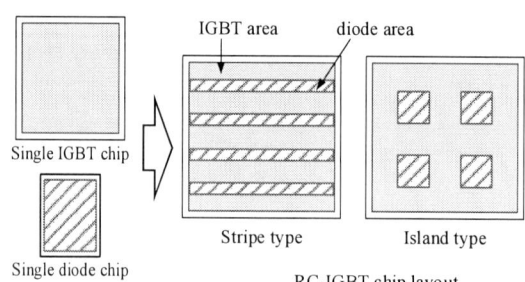

Fig. 1. Top view of "Separated type" RC-IGBTs

978-1-5386-2928-4/18 $31.00 © 2018 IEEE

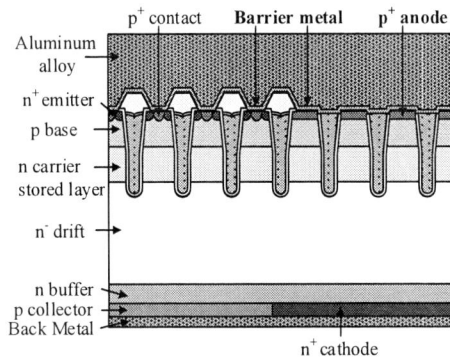

Fig. 2. Cross-sectional view of the conventional RC-IGBT. diode region; High hole injection anode and barrier metal

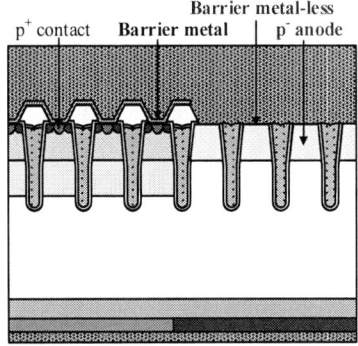

Fig. 3. Cross-sectional view of the proposed RC-IGBT. diode region; Low hole injection anode and barrier metal-less

Figure 3 shows the proposed RC-IGBT with the additional wafer process steps. The proposed one has the low impurity concentration anode to restrain the hole injection and an Al alloy direct contact without the barrier metal only in the diode region.

Table I shows the approaches to obtain the suitable performance for the high frequency operation. "Approach A" shows the conventional RC-IGBT that is applied the lifetime control using Electron Beam (EB) irradiation, which is widely used to reduce I_{rr} of diode and turn-off switching energy loss (E_{off}) of IGBT. High dose EB irradiation is preferable to reduce I_{rr}. But, IGBT's on-state forward voltage drop (V_{on}) - E_{off} tradeoff is worse to be the overcompensation for the I_{rr} improvement and other characteristics. Therefore, RC-IGBT for the high frequency operation should object the lower EB irradiation than the mentioned above case.

"Approach B" is to use the low hole injection anode layer. For obtaining the suitable performance as the high frequency operation, it is effective to the reduction of hole injection from anode layer. However, it is also important to reduce the contact resistance between the front metal and low impurity concentration layer. In case of the conventional front metal, diode's forward voltage drop (V_F) characteristic is not acceptable by formed Schottky like contact. So, we formed the proposed RC-IGBT without the barrier metal in the diode region. In fact, the one has Al alloy direct contact to diode region. One of the solution to realize the high frequency operation suitable for the RC-IGBT is to form the low injection anode layer corresponding the barrier metal-less structure for the front metal. In the next section, we show the electrical characteristics of the fabricated RC-IGBT.

TABLE I. Solution extracted from the approach matrix for high frequency operation

Approach		Condition or detail	High frequency operation suitable		Remarks
			IGBT	diode	
A	Uniform carrier lifetime control (EB irradiation)	**High dose** EB irradiation	Not acceptable	Insufficient	IGBT's V_{on}-E_{off} tradeoff going to be worse Diode's I_{rr} being insuffciently reduced
		Low dose EB irradiation	Suitable	Not acceptable	Diode's E_{rec} being insuffciently reduced
B	Low hole injection anode	**Barrier metal** as the common p/n ohmic contact for IGBT	**No affect**	Not acceptable	Diode's V_F being worse
		Al alloy direct contact as barrier metal-less for diode	**No affect**	Suitable	Diode's V_F-I_{rr} tradeoff being improved

III. EXPERIMENTAL RESULTS

In this section, we explain the electrical characteristics of comparing the proposed 650V RC-IGBT with the conventional one. Figure 4 shows the V_{on}–E_{off} tradeoff of EB dose dependence. The conventional one has the high impurity concentration collector layer, the proposed one has the medium impurity concentration collector layer. The wafer thickness and MOS structure of both IGBT areas are set to be identical.

In spite of the same top side IGBT cell structures, E_{off} as the high frequency range of the proposed RC-IGBT was improved 8% at the same V_{on} (1.82V) in comparison with the conventional RC-IGBT. Figure 5 shows the leakage current characteristics as the EB dose dependence.

The structural change of the diode part pushes up the leakage current is almost two times higher in the medium EB dose condition, but it is almost a half of the leakage of the high dose condition, which indicates the same V_{on} (1.82V). This tendency of the less EB dose leads the better direction of total performances of the RC-IGBT.

In other words, the proposed RC-IGBT is supposed to obtain the positive spiral for the total RC-IGBT's characteristics by reducing EB dose on behalf of the low injection anode layer of diode. Other than the above leakage current characteristics, the details are discussed as the followings especially using Figure 6, which shows the V_F-I_{rr} tradeoff of the diode's reverse recovery as EB dose dependence.

I. As the low and middle frequency operation

On behalf of the low hole injection from anode layer, I_{rr} of the proposed RC-IGBT improved 36% in comparison with the conventional RC-IGBT as the same EB dose, while V_F increase remains 28% from the conventional RC-IGBT. In the low frequency range, the proposed RC-IGBT indicates the almost same tradeoff as the conventional RC-IGBT. In the middle frequency range, I_{rr} of the proposed one is better than the conventional. This is the reason why our proposed approach is getting involved with the positive spiral of improvement of the total RC-IGBT's characteristics. As the experimental results, the conventional one without the additional wafer process steps is more preferable than the proposed one in the low frequency range.

II. As the high frequency operation

The proposed RC-IGBT realizes to improve the controllability of the tradeoff characteristic in the high frequency operation range. In figure 6, at the relatively high V_F (1.55V) value, I_{rr} of the proposed one was improved 14% in comparison with the conventional one.

Figure 7 (a), (b) and (c) show the tradeoff relationship between V_F-t_{rr}, V_F-Q_{rr} and V_F-E_{rec}, respectively. In contrast to V_F-I_{rr} tradeoff (Fig. 6), the difference between the proposed one and conventional one is very slight for these three tradeoff curves characterizing diode's reverse recovery.

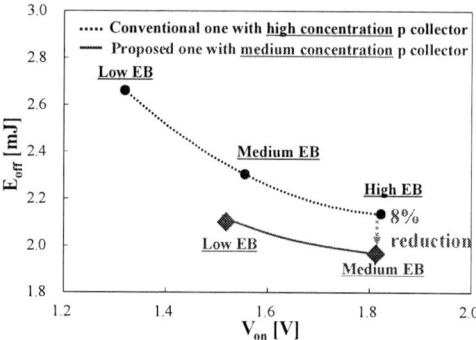

Fig. 4. Tradeoff characteristics V_{on} versus E_{off}.
Tradeoff curve of IGBT's characteristics is controlled the concentration of the collector layer better than for lifetime control technique.
(T_j=125°C, V_{CC}=300V, I_C=75A, V_{GE}=15V/0V)

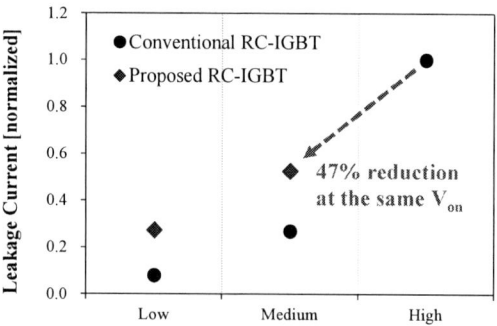

Fig. 5. Leakage current at the EB irradiation dose dependence.
(T_j=125°C)

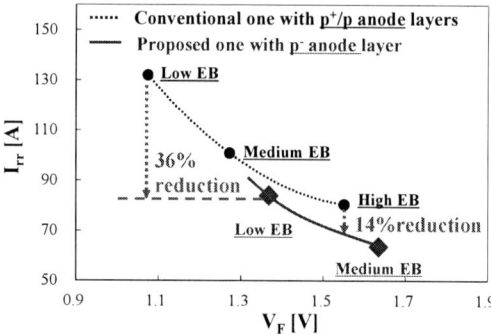

Fig. 6. Tradeoff characteristics V_F versus I_{rr}.
V_F – I_{rr} tradeoff as the high frequency operation, the proposed RC-IGBT is expected to show the better characteristic than the conventional RC-IGBT.
(T_j=125°C, V_{CC}=300V, I_F=75A, V_{GE}=15V/0V)

Figure 8 shows the reverse recovery waveforms of the RC-IGBT having the low impurity concentration p⁻ anode without the barrier metal of the front metal and the conventional one having the high impurity concentration p⁺/p anode layers under the high current condition. The proposed one has the lower peak and softer recovery. It is effective to suppress the hole injection from anode layer, so as to maintain the relatively longer carrier lifetime resulting the lower di/dt, and the smaller surge peak voltage and the less voltage ringing.

As the experimental results, the proposed RC-IGBT based upon the low hole injection and the mild lifetime control achieves the soft reverse recovery waveform even the high frequency range.

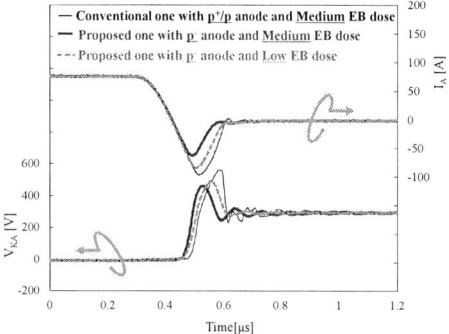

Fig. 8. Reverse recovery waveforms of the proposed RC-IGBT with p⁻ anode layer and the conventional RC-IGBT with p⁺/p anode.
(T_j=125°C, V_{CC}=300V, I_F=75A, V_{GE}=15V/0V)

IV. CONCLUSION

We have successfully fabricated the 650V RC-IGBT having p⁻ anode layer without the barrier metal. Focusing on the high frequency operation with the mild lifetime control condition, it has come to the positive spiral improvement for the RC-IGBT characteristics. One of the remarkable improvement is 14% of I_{rr} maintaining the same V_F (1.55V) as the conventional one. IGBT's characteristics are also getting involved with this positive spiral improvement on behalf of the mild lifetime control of less affecting to V_{on}-E_{off} tradeoff even in the higher frequency operation.

Our solution for future RC-IGBT is the combination of the "IGBT - diode" individually formed the ohmic contact and the medium dose EB irradiation as the mild carrier lifetime control.

REFERENCES

[1] H. Takahashi, A. Yamamoto, S. Aono and T. Minato, "1200V Reverse Conductiong IGBT", *Proc. ISPSD2004*, pp. 133-136, 2004.

[2] H. Ruthing, F.Hille, F. –J. Schulze and B. Brunner, "600V Reverse Conductiong (RC-) IGBT for Drives Applications in Ultra-Thin Wafer Technology", *Proc. ISPSD2007*, pp. 89-92, 2007.

[3] J. Vobecký and P. Hazdra, "Future Trends in Local Lifetime Control", *Proc. ISPSD1996*, pp. 161-164, 1996.

[4] K. Nishiwaki, T. Kushida and A. Kawahashi, "A Fast & Soft Recovery Diode with Ultra Small Qrr (USQ-Diode) Using Local Lifetime Control by He Ion Irradiation", *Proc. ISPSD2001*, pp. 235-238, 2001.

[5] R. Gejo, T. Ogura, S. Misu, Y. Maeda, U. Matsuoka, N. Yasuhara and K. Nakamura "High Switching Speed Trench Diode for 1200V RC-IGBT Based on the Concept of Schottky Controlled Injection (SC)", *Proc. ISPSD2016*, pp. 155-158, 2016.

[6] K. Takahashi, S. Yoshida, S. Noguchi, H. Kuribayashi, N. Nishida, Y. Kobayashi, H. Kobayashi, K. Mochizuki, Y. Ikeda and O. Ikawa "New Reverse-Conducting IGBT (1200V) with Revolutionary Compact Package", *Proc. ISPSD2014*, pp. 131-134, 2014.

[7] T. Yoshida, T. Takahashi, K. Suzuki and M. Tarutani "The second-generation 600V RC-IGBT with Optimized FWD", *Proc. ISPSD2016*, pp. 159-162, 2016.

[8] T. Matsudai, T. Ogura, Y. Oshino, T. Kobayashi, S. Misu, Y. Ikeda and K. Nakamura "Advanced Cathode and Anode Injection Control Concept for 1200V SC(Schottky Controlled Injection)-Diode", *Proc. ISPSD2014*, pp.19-22, 2014.

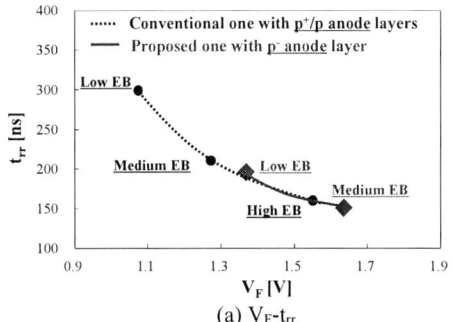

(a) V_F-t_{rr}

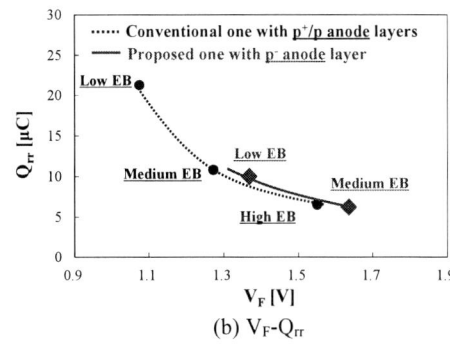

(b) V_F-Q_{rr}

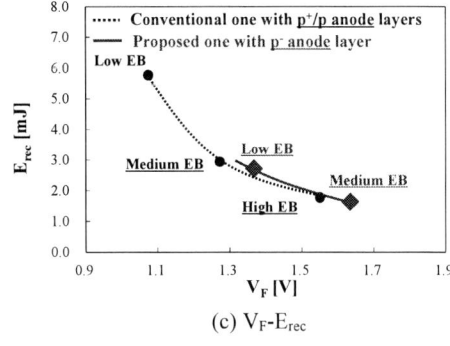

(c) V_F-E_{rec}

Fig. 7. Tradeoff characteristics V_F versus (a) t_{rr}, (b) Q_{rr}, (c) E_{rec}, respectively. Lying between the proposed RC-IGBT and the conventional RC-IGBT, there are no critical differences characteristics for t_{rr}, Q_{rr}, E_{rec}.
(T_j=125°C, V_{CC}=300V, I_F=75A, V_{GE}=15V/0V)

978-1-5386-2928-4/18 $31.00 © 2018 IEEE

Proceedings of the 30th International Symposium on Power Semiconductor Devices & ICs
May 13-17, 2018, Chicago, USA

Observation of Current Filaments in IGBTs with Thermoreflectance Microscopy

Riteshkumar Bhojani[1], Jens Kowalsky[1], Josef Lutz[1], Dustin Kendig[2], Roman Baburske[3], Hans-Joachim Schulze[3], Franz-Josef Niedernostheide[3]

Email: riteshkumar.bhojani@etit.tu-chemnitz.de

[1]Chair of Power Electronics and EMC, Technische Universität Chemnitz, Chemnitz, Germany
[2]Microsanj, Santa Clara, USA
[3]Infineon Technologies AG, Neubiberg, Germany

Abstract—**In this paper, we demonstrate for the first time experimentally measured current filaments in IGBTs under repetitive Short-Circuit (SC) events. These current filaments were discovered with the help of Thermo-Reflectance Microscopy (TRM). The destruction current as function of the applied collector-emitter voltage (V_{CE}) was determined for two differently wire-bonded 15A-1200V IGBT chips. The repetitive SC events in combination with TRM measurement indicate a wide range of non-destructive current filaments at different V_{CE}. Similar filament formation under short-circuit conditions were observed in supporting TCAD device simulations based on multi-cell IGBT structure. These filaments have similar dimensions to the current filaments measured by TRM.**

Keywords—IGBT, short-circuit, current filaments, Thermo-reflectance microscopy

I. INTRODUCTION

In previous works, several authors have shown short-circuit destruction limits and the improvement potential of the SC-SOA for IGBTs of different voltage classes with a minimum current destruction limit at about 50 % of the rated voltage [1-4]. These observations were linked to the effect of current crowding at the collector side of the IGBT. Moreover, different failure modes like failure during the SC pulse and during SC turn-off were also related with the current crowding effect in IGBTs. Up to now, direct experimental evidence for the appearance of the current filaments in IGBTs during short-circuit is missing.

In this work, we investigated the behavior of the current filaments using CCD-based thermo-reflectance microscopy [5]. Under different voltage and current conditions, thermo-reflectance measurements were performed to observe the change in temperature on the top of the IGBT metallization.

II. THERMO-REFLECTANCE MICROSCOPY

In general, the complex refractive index of a material changes with temperature, and the thermo-reflectance coefficient C_{th} simply represents the linear change in optical reflectance of a surface due to a change in temperature (ΔT) [5]:

$$\frac{\Delta R}{R_0} = C_{th}(\lambda)\Delta T \qquad (1)$$

Here, C_{th} is the thermo-reflectance coefficient, R is the reflective intensity in (J/m²), T is the temperature in (K) and λ is the wavelength in (m).

The stroboscopic imaging was used to create time-resolved image data with temporal resolution down to 50 ns. The device is turned ON for a certain period of time at a certain duty cycle, and the CCD exposure time is also set by the user as shown in Fig. 1(a). Nearly all the light reaching the CCD will come from the LED pulses which are offset by a specific time delay relative to the device excitation, and a CCD image is generated as the sum of the LED pulses inside a single CCD exposure window. The timing diagram in Fig. 1(a) shows how the time-resolved image data is collected by pulsing the device excitation and LED probe light with a controllable relative time delay.

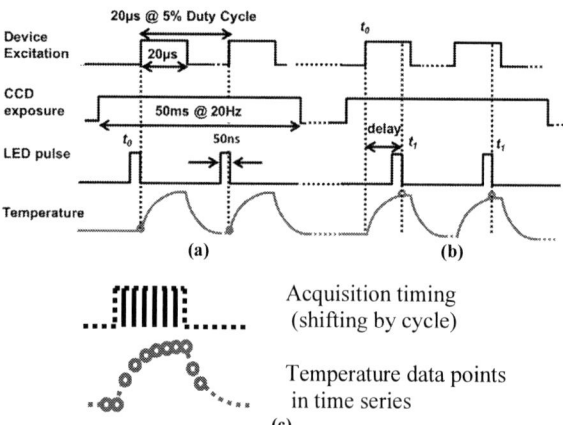

Fig. 1(a) Timing diagram showing how the $t = 0$ data point is collected (b) Timing diagram showing how the $t > 0$ data point is collected for heated device (c) Acquired temperature data points with respect to time.

Fig. 1(b) and (c) show how the temperature data points were acquired by shifting the time delay after each cycle to achieve time resolution, limited only by how tightly the pulse of the LED probe light was set. To create a two-dimensional transient thermo-reflectance image, a reference CCD image with LED pulse delay t = 0 was collected, followed by recording a CCD image at t > 0. Then, a sub-pixel image registration algorithm was applied to a subset of the CCD images to align the t > 0 image to the reference. After this, the

978-1-5386-2928-4/18 $31.00 © 2018 IEEE

registered t > 0 images were subtracted from the t = 0 reference image. This process with recording pairs of t = 0, t > 0 images was continued as a running average to suppress noise to an acceptable level. In this way, the software automatically stepped through user-designated time delays and averaging times to collect a full time-domain image set, from the moment the device was turned on, to the time period after the device switched off and was cooling down.

III. SC THEORY AND FILAMENT BEHAVIOR

A. Simulation results of current filaments during SC

Electro-thermal device simulations were performed to estimate the surface temperature during the SC pulse when current filaments are present in the chip. Fig. 2 shows the transients of the maximum and the average current density together with the course of the maximum temperature under SC conditions for a DC-link voltage $V_{DC} = 500$ V, a collector current $I_C = 240$ A, corresponding to a gate-emitter voltage $V_{GE} = 30$ V at an initial temperature $T_{int} = 300$ K for a multi-cell IGBT structure with a simplified front side structure [4]. The thermal resistance of 3.2 K/W was set at the collector contact and self-heating was considered for SC simulations.

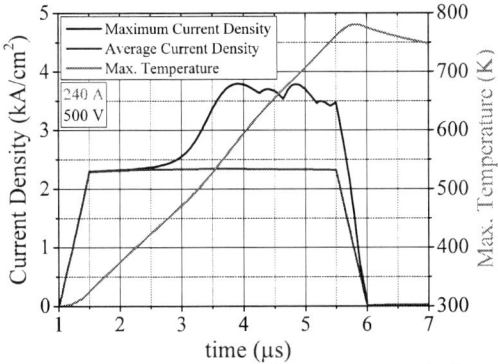

Fig. 2. Transients of the maximum and the average current density and the transient of the maximum temperature from electro-thermal SC simulation at $V_{DC} = 500$ V, $I_C = 240$ A, $T_{int} = 300$ K.

A small window at n⁻-base/n-field-stop junction (Fig. 3) was selected to extract the average and maximum current densities from the device. At 2.5 µs, the two current density transients start to split. At 4 µs, multiple filaments are present in the device (Fig. 3). The typical width of the filament at the emitter-side metallization was between 80 to 120 µm.

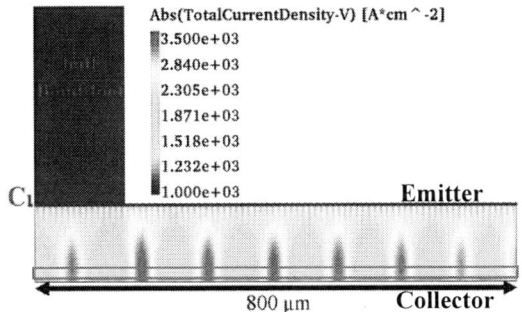

Fig. 3. Current density distribution at 4 µs in IGBT with ref. to Fig. 2.

Fig. 4 shows the surface temperature profile with respect to a horizontal cut C1 in Fig. 3 on top of the Al metallization. At the filament positions, where the simulated temperature close to the collector is about 600 K at 4 µs, the surface temperature shows an increase by 2 to 5 K, which can be resolved using TRM.

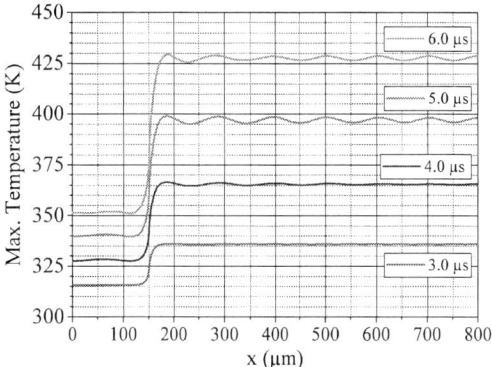

Fig. 4. Surface temperature profile on top of the Al metallization at different time points from Fig. 2.

B. Low Inductive Short-circuit Setup

Fig. 5(left) displays the SC-setup combined with thermo-reflectance microscopy setup. Fig. 5(right) shows a close view of a SC-setup including capacitor bank, measuring probes, TRM lens and a Device Under Test (DUT). The DUT is soldered on a Direct Copper Bonding (DCB) substrate.

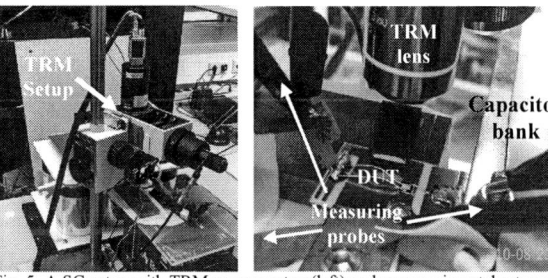

Fig. 5. A SC-setup with TRM camera setup (left) and an experimental setup of TRM lens with a low inductive SC-setup (right).

A very low inductive ($L_{par} \leq 60$ nH) short-circuit setup was designed to investigate the current destruction limit for the 15 A-1200 V IGBTs. The short-circuit destruction limit was measured for two differently wire-bonded IGBT chips shown in Fig. 6. The gate resistors for turn-on and turn-off were set to $R_G = 220$ Ω during all measurements.

The results shown in Fig. 6 indicate that the critical short-circuit current is independent of the number of bond wires on the top of aluminum metallization, allowing us to use chips with only one bond wire for TRM. The temperature on the top surface of the IGBT chip was continuously monitored using a "Microsanj NT210 transient thermo-reflectance camera" by performing repetitive non-destructive SC events. Fig. 7 shows a SC type-1 waveform at $V_{DC} = 400$ V and $V_{GE} = 30$ V. An estimated average temperature swing ΔT_{SC} of the chip during this given pulse can be about 67 K using the whole chip volume given by Equation 2 in [6].

978-1-5386-2928-4/18 $31.00 © 2018 IEEE   165

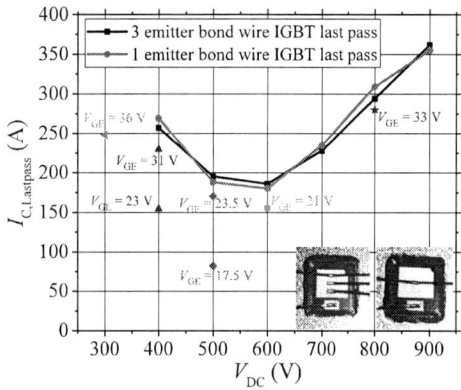

Fig. 6. Destruction limit of 1.2 kV IGBTs for two differently wire-bonded chips at T = 300 K, $L_{par} \leq$ 60 nH, R_G = 220 Ω (red and black points) and SC measurement conditions at which non-destructive filaments were investigated by means of TRM (scattered points).

$$\Delta T_{SC} = \frac{V_{CE} \cdot I_{SC} \cdot t_{SC}}{c \cdot \rho \cdot d \cdot A} \qquad (2)$$

where c is the specific heat capacity of silicon, ρ is the density of silicon, d and A are thickness and area of the chip.

With TRM, the uncovered Al surface was monitored for different V_{DC} by performing repetitive SC events close below the destruction limit. The measurements were performed for gate voltages close to the chip destruction as well as for reduced gate voltages. The SC pulse length was set to 4 μs. The time between each consecutive SC pulse was set to 200 ms which was sufficient to cool down the IGBT top aluminum metallization. The scattered points in Fig. 6 refer to measurement conditions of the IGBT chips at which TRM measurements were performed with different V_{DC}, V_{GE} and I_C.

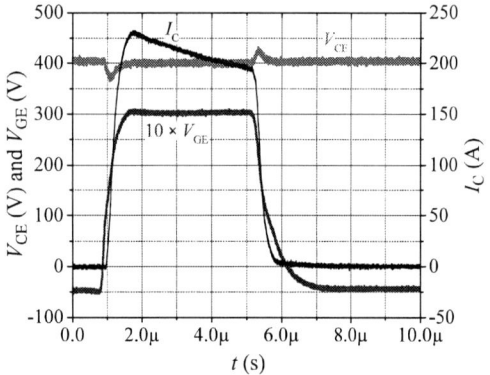

Fig. 7. Transients of the $V_{CE}(t)$, $V_{GE}(t)$ and $I_C(t)$ for a SC type-1 event at V_{CE} = 400 V, V_{GE} = 30 V, T = 300 K, L_{par} = 60 nH and R_G = 220 Ω.

C. Destructive and non-destructive filaments

Fig. 8 schematically illustrates the areas in the I_C-V_{CE} phase space in which either no filaments or non-destructive filaments were observed by TRM. The red-solid line marks the dependency of the destruction current on V_{CE} under SC conditions with the minimum critical SC current lying at about $V_{CE,rated}/2$ for the investigated 1200 V IGBTs. The green line

explicates the current level above which filament appears. The hatched area in Fig. 8 between the red-solid and the green-dashed line covers the non-destructive visible filaments, especially at a low V_{CE} range from 300 V to 600 V. The filament pattern becomes finer from V_{CE} of 300 V to 600 V. For V_{CE} > 600 V, the filament behavior needs to be inquire further. In the area below the green-dashed line, current crowding did not appear and the current flowed uniformly in the IGBT during SC event. The black line indicates the I_C-V_{CE} characteristic for $V_{GE} \leq$ 15 V. For the safe operation of the IGBT device during SC case, it is essential to evaluate this green-dashed boundary line. Therefore, the focus of the TRM was different gate voltages at which the uniform current distribution transformed into a distribution with current filaments, indicated by green-dashed line in Fig. 8, especially at low V_{CE} voltages.

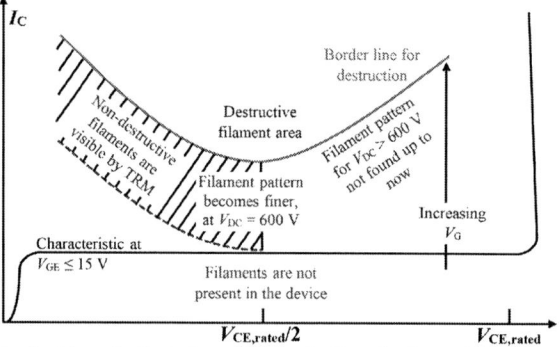

Fig. 8. Schematic illustration of the border lines for destructive and non-destructive filament areas in I_C-V_{CE} phase space.

IV. TRM RESULTS AND DISCUSSION

Fig. 9 displays the TRM images at different DC-link voltages and different gate voltages. At V_{DC} = 400 V and V_{GE} of 30 V (Fig. 9a), after integration over several thousand SC pulses, temperature peaks are visible for the IGBT chip with one emitter bond wire. The TRM images show the temperature difference (ΔT), between t = 0 and t > 0 images, and not the actual temperature. The blue color indicates the coldest region on the IGBT chip. The red spots show high temperature areas indicating filaments in the IGBT. They have similar dimensions as the filaments found in simulation (Fig. 3). The diameter of the current filaments ranges from minimum 40 μm to maximum 140 μm. Their distance varies from 50 μm to 200 μm. The different maximum temperatures at the position of the current filaments indicate that the filaments were carrying different currents. For certain conditions also clustering of filaments was observed.

For V_{DC}/V_{GE} = 300 V/36 V, 400 V/23 V and 500 V/17.5 V (Fig. 9c-e) again the filaments appeared. For V_{DC} = 400 V and 500 V, the gate voltage is 12 V and 7 V lower that the critical gate voltage at which destruction occurs, indicating a wide gate voltage and collector current interval at which non-destructive filaments may appear. The filament pattern was almost regular for V_{DC} = 400 V (Fig. 9d). However, there is a gradient of the maximum filament temperature from the left to the right edge of the TRM image.

978-1-5386-2928-4/18 $31.00 © 2018 IEEE

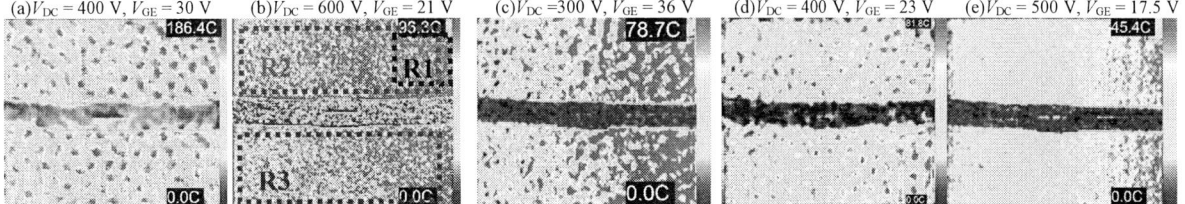

(a)V_{DC} = 400 V, V_{GE} = 30 V (b)V_{DC} = 600 V, V_{GE} = 21 V (c)V_{DC} =300 V, V_{GE} = 36 V (d)V_{DC} = 400 V, V_{GE} = 23 V (e)V_{DC} = 500 V, V_{GE} = 17.5 V

Fig. 9. Surface temperature observation after series of SC pulses using TRM camera at different V_{DC} and V_{GE}. (Scale: temperature difference to coldest point (x: 3.7 mm, y: 2.8 mm))

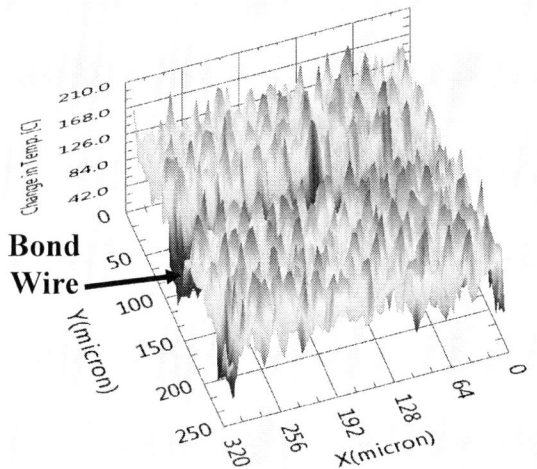

Fig. 10. 3D-TRM image after SC events at V_{DC} = 400 V (ref. to Fig. 9(a), Scale: 0.1× (X, Y)).

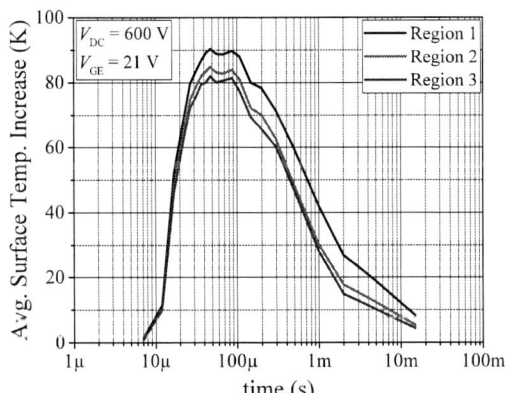

Fig. 11. Measured average surface temperature profile on top of the Al metallization from three different regions with respect to time.

Fig. 10 shows the 3D-TRM picture of the temperature profile with respect to Fig. 9(a) at V_{DC} =400 V. Fig. 11 shows the measured average surface temperature profile on top of aluminum metallization from three different regions with respect to time after repetitive SC events. These three different regions R1, R2 and R3 were selected from Fig. 9(b) after several repetition of the short-circuit events at V_{DC} = 600 V and V_{GE} = 21 V. Since the heat inside the chip takes some time to reach the top metallization, one can initially see a very

small increase in temperature. After 50 µs, the temperature on the chip surface reaches the maximum value and remains the same for another 50 µs. From 100 µs, the temperature of the aluminum metallization starts to decay. Several 10 milli-seconds after the SC pulse, the Al surface metallization reaches its initial temperature.

V. CONCLUSION

Current filaments were observed in IGBTs during repetitive SC events using thermo reflectance microscopy. The diameters of the filaments varied between 40 µm and 140 µm and were comparable to the width of filaments (80 µm to 120 µm) found in supporting TCAD device simulations. The filaments were more pronounced for low DC-link voltages (300 V to 600 V), while at higher DC-link voltages, the filament pattern became finer and transformed to a homogeneous distribution. The filaments were also found at gate voltages lower than the critical gate voltages. The measured results indicate that there is a wide current range where the non-destructive filaments can appear inside the IGBT device during repetitive SC events. In all measurements, irregular patterns of filaments were observed. However, the filaments showed characteristic dimensions and were spaced apart from each other by characteristic distances.

ACKNOWLEDGEMENT

The authors would like to thank Madhu Lakshman Mysore for his help during the measurement and also to Manuela Schulze for critically reading the manuscript.

REFERENCES

[1] A. Kopta, M. Rahimo, et al, "Limitation of the Short-Circuit Ruggedness of High Voltage IGBTs", in Proc.ISPSD, pp. 33-36, 2009.

[2] R. Baburske, V. Treck, et al, "Comparison of Critical Current Filaments in IGBT Short Circuit and during Diode Turn-off", in Proc. ISPSD, pp. 47-50, 2014.

[3] M. Tanaka and A. Nakagawa, "Simulation studies for avalanche induced short-circuit current crowding of MOSFET-Mode IGBT", in Proc. ISPSD, pp. 121-124, 2015.

[4] R. Bhojani, S. Palanisamy, R. Baburske, et al, "Simulation study on collector side filament formation at short-circuit in IGBTs",in Proc. ISPS, pp.70-76, 2016.

[5] M. Farzaneh, K. Maize, D. Lüerßen, et al, "CCD-based thermo-reflectance microscopy: principles and applications", J. Phys. D: Appl. Phys., Vol. 42, pp. 1-20, 2009.

[6] J. Lutz, H. Schlangenotto, et al, Semiconductor Power Devices-Physics, Characteristic, Reliability. Springer, 2011.

978-1-5386-2928-4/18 $31.00 © 2018 IEEE

Proceedings of the 30th International Symposium on Power Semiconductor Devices & ICs
May 13-17, 2018, Chicago, USA

IGBT structure with electrically separated floating-p region improving turn-on dVak/dt controllability

Yoshihiro Ikura, Yuichi Onozawa
Device Development Department, Fuji Electric, Matsumoto, Nagano, Japan, e-mail:ikura-yoshihiro@fujielectric.com

Akio Nakagawa
Nakagawa Consulting Office, Chigasaki, Kanagawa, Japan

Abstract—**IGBT cell structure for improving the turn-on dIc/dt controllability is presented. The difference from the conventional structure is that the floating-p region of the new structure is electrically disconnected from the trench-side-wall region. TCAD simulation suggests that the new device structure achieves better turn-on characteristics and almost the same static and turn-off characteristics compared to the conventional structure. It turned out that the high potential of the trench-side-wall region of the new structure, at the moment just before the collector current start to flow, is the cause of the improved turn-on dIc/dt controllability.**

Keywords— Insulated gate bipolar transistors, Floating P, Turn on, Controllability, Noise, Transient analysis

I. INTRODUCTION

It is well known that the IGBT with floating-p region has a poor controllability over the turn-on dIc/dt (Collector current increase rate). High dIc/dt of IGBT causes high reverse recovery dVak/dt (Anode-Cathode voltage increase rate) of FWD at the opposite arm which leads to EMI noise and voltage oscillation during switching [1]. Previous research has suggested that the rapid increase of the gate voltage during the turn-on period is caused by the displacement current flowing from the floating-p region of which potential show rapid increase induced by the start of the collector current flow [1-3]. In ISPSD 2017, it was suggested that the initial potential of the floating-p region has significant influence on the turn-on dIc/dt controllability [4]. When the potential of the floating-p region is high, the turn-on dIc/dt controllability is better. In this study, a novel IGBT cell structure for improving the turn-on dIc/dt controllability is presented.

II. DEVICE STRUCTURE

Figure 1 shows IGBT device structures used for TCAD simulation analysis. Compared to the conventional structure A (Fig. 1(a)), the floating-p region of the structure B is separately placed from the trench gate (Fig. 1(b)). Compared to the structure B, n+ region is placed on the surface of separation region of the structure C (Fig. 1(c)). The portion "trench-side-wall region" is indicated and defined in the three figures, and is frequently referred, hereafter, in the present paper.

In the next section, TCAD simulation results for these structures are shown.

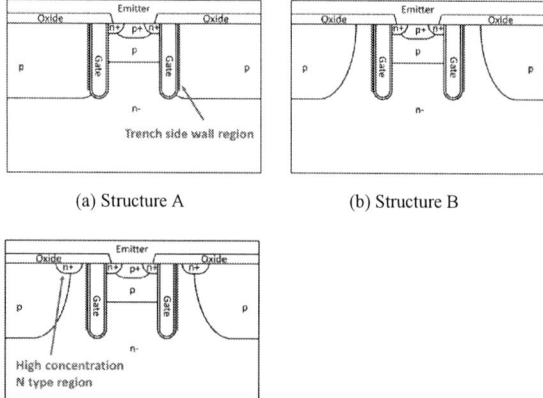

(a) Structure A (b) Structure B

(c) Structure C

Fig. 1. Cross section of IGBT device structures

III. SIMULATION RESULT

Figure 2 shows the simulated distributions of doping concentration, electric potential and hole density of the structure A during the off state. The potential of the floating-p region is not zero but have a certain value. The turn-on dIc/dt controllability improves as the potential value is higher.

Figure 3 shows the simulated distributions of doping concentration, electric potential and hole density of the structure B during the off state. The calculated electric potential of the floating-p region of the structure B is not much higher than that of the structure A. It turned out that hole inversion layer created at the surface of separation region connects the floating-p region and trench-side-wall region. It can be noticed that once the hole inversion layer created, the potential of the floating-p region is fixed at the potential of inversion layer and cannot increase further more.

It is expected that if the creation of hole inversion layer is suppressed, the potential of the floating-p region can take a much higher value. Figure 4 shows simulated distributions of doping concentration, electric potential and hole density of the structure C during the off state. By placing n+ region on the separation region, the creation of the hole inversion layer is successfully suppressed. Instead, hole current flows from the floating-p region to the trench-side-wall region through a path of an electric potential minimum point between the floating-p region and the trench-side-wall region.

978-1-5386-2928-4/18 $31.00 © 2018 IEEE

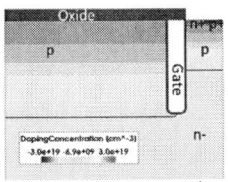
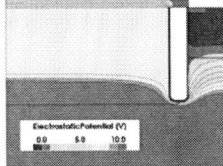

(a) Doping concentration   (b) Electrostatic Potential

(c) Hole Density

Fig. 2. Simulated off state of the structure A

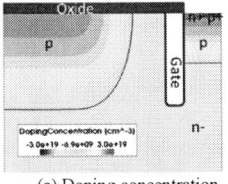
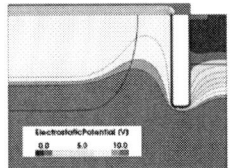

(a) Doping concentration   (b) Electrostatic Potential

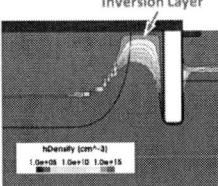

(c) Hole Density

Fig. 3. Simulated off state of the structure B

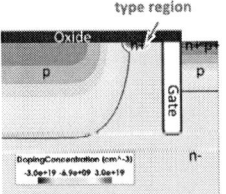
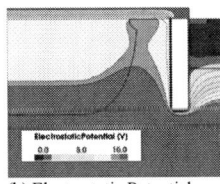

(a) Doping concentration   (b) Electrostatic Potential

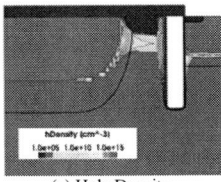

(c) Hole Density

Fig. 4. Simulated off state of the structure C

Figure 5 shows the turn-on waveforms of structure A, B and C for three different gate resistances, Rg. The waveforms of the structure A (solid line) and B (dotted line) show no significant difference. And both gate voltages rapidly increase as the collector current start to flow. On the other hand, the increase of the gate voltage is relatively small in the waveform of the structure C (red line). Figure 6 shows dIc/dt as a function of gate resistance Rg. The dIc/dt controllability by gate resistance of the structure C is better than that of the other structures. Figure 7 shows the turn-on loss Eon as a function of reverse recovery dVak/dt of FWD. The turn-on characteristics of the structure C is better than that of the other structures.

Figure 8 shows the turn-off loss, Eoff, as a function of Collector-Emitter saturation voltage Vce(sat). The turn-off characteristics of the structure C is almost the same as that of the other structures.

Table 1 lists the breakdown voltages. Although the breakdown voltage of the structure C is slightly lower than that of the other structures, the difference is small.

As described above, the structure C has significantly improved turn-on characteristics without any substantial deterioration of other characteristics.

In the next section, the reason why the structure C is effective to improve the turn-on dIc/dt controllability is analyzed.

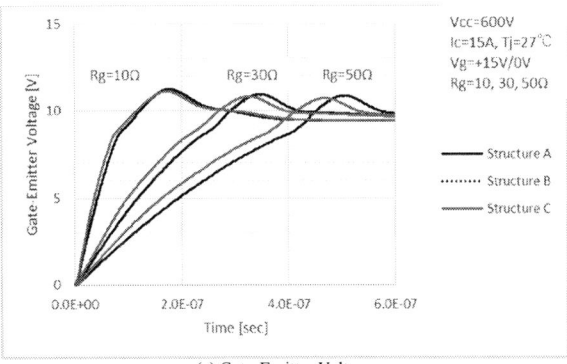

(a) Gate-Emitter Voltage

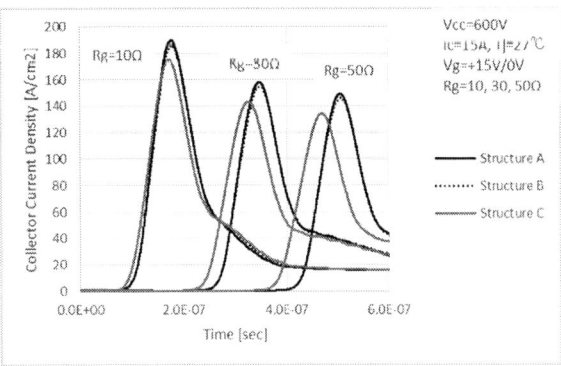

(b) Collector Current Density

Fig. 5. Turn-on waveform with varied gate resistance Rg

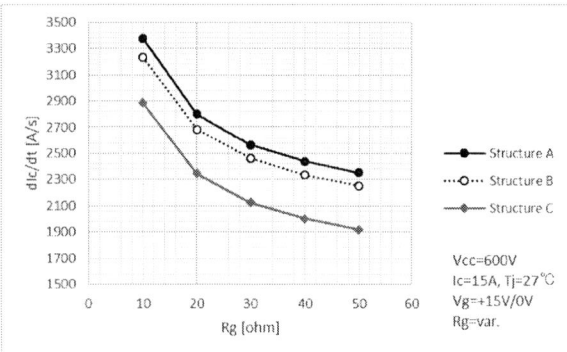

Fig. 6. dIc/dt as a function of gate resistance Rg

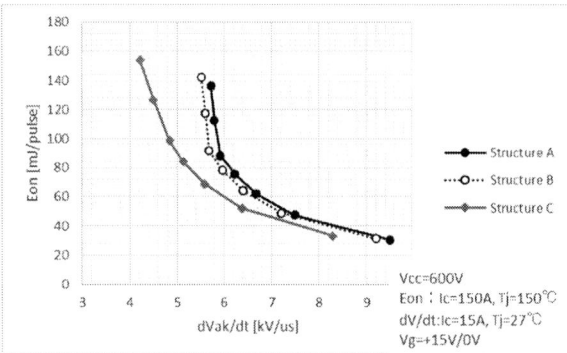

Fig. 7. Turn-on loss Eon as a function of Reverse recovery dVak/dt

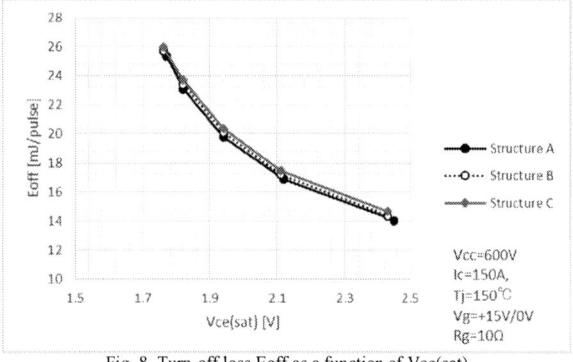

Fig. 8. Turn-off loss Eoff as a function of Vce(sat)

Table 1. Breakdown voltage

	Breakdown Voltage [V]
Strucutre A	1460
Structure B	1428
Structure C	1412

IV. DISCUSSION

When the floating-p region is separately placed from the trench gate, as is the cases of the structure B and C, the potential of the trench-side-wall region, facing the trench gate, is important. Figure 9 shows the turn-on waveform of the potential of the trench-side-wall region, the gate-emitter voltage and the collector current density.

The potential of trench-side-wall region of the structure C increase more rapidly than that of the structure A and B until the rise of collector current. The potential of the trench-side-wall region increases depending on the rate of the gate voltage increase during this period. Because the potential of the trench-side-wall region of the structure A and B is electrically connected to the floating-p region by the inversion layer, the capacitance between the floating p-region and Emitter electrode is relatively large. Thus the increase rate of the potential of trench-side region is relatively small. On the other hand, because the potential of the floating-p region of the structure C is electrically separated from the trench-side-wall region by the n+ region, the capacitance between the trench-side-wall region and the Emitter electrode is relatively small. Thus, the increase rate of the potential of trench-side-wall region is large in the structure C.

The increase rate of the potential of the trench-side-wall region is decided by the hole accumulation rate after the start of the collector current flow. It follows that the potential value of the trench-side-wall region at the very moment just before the collector current start to increase is important. As the potential of trench-side-wall region becomes higher, the hole accumulation and the resulting potential increase rate will be smaller.

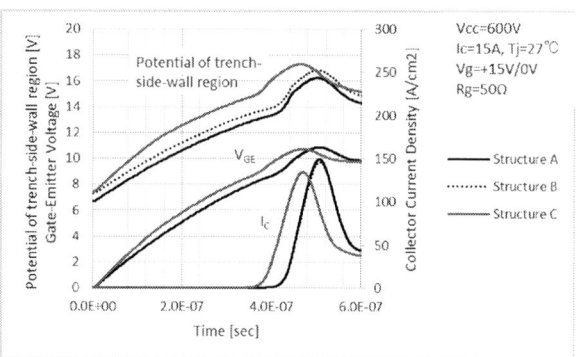

Fig. 9. Turn-on waveform of potential of trench-side-wall region, Gate-Emitter voltage and Collector current density

A TCAD simulation procedure to calculate the turn-on characteristics with arbitrary potential of the floating-p region is described in ISPSD 2017 [4]. In this method, the off state is calculated in which the floating-p region is connected to a fixed voltage source Vfix and the turn-on characteristics are calculated in which a large resistance is connected between the floating-p region and the voltage source Vfix. The simulation suggested that the increase rate of the collector current dIc/dt becomes lower as Vfix increases.

Figure 10 shows the relation between collector current increase rate dIc/dt and the electric potential of the trench-side-wall region at the very moment just before the collector current start to increase for the structure A, B and C when the fixed voltage source value Vfix is changed. All three curves for the structure A, B and C coincide with each other. The electirc potential of trench-side-wall region at the very moment just before the collector current start to increase has the same influence on the increase rate of collector current dIc/dt even for the different device structure A, B and C.

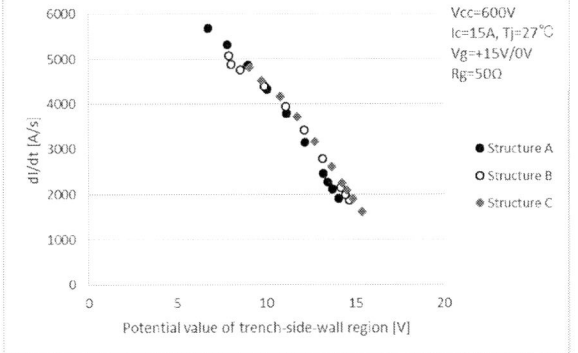

Fig. 10. dIc/dt as a function of the electric potential of the trench-side-wall region at the very moment just before the collector current start to increase

V.   CONCLUSIONS

In this study, IGBT cell structure for improving the turn-on dIc/dt controllability is investigated. Simulation suggested that, even if the floating-p region is separately placed from the trench gate, the influence from increasing potential of the floating-p region is not suppressed because the hole inversion layer at the surface of the separation region connects the floating-p region and the trench-side-wall region. Placing n+ region on the separation region as a countermeasure for the hole inversion layer improves the turn-on dIc/dt controllability. Electrically disconnecting the floating-p region from the trench-side-wall region causes the elevation of the potential of trench-side-wall region after the gate voltage starts to increase. It turned out that the potential of trench-side-wall region at the very moment just before the collector current start to increase rules the turn-on dIc/dt controllability.

REFERENCES

[1]   Y. Onozawa, et al., "Development of the next generation 1200 V trench-gate FS-IGBT featuring lower EMI noise and lower switching loss," in Proc.19th ISPSD, pp. 13-16, May 2007.

[2]   Y. Onozawa, et al., "1200V Super Low Loss IGBT Module with Low Noise Characteristics and High dI/dt Controllability", in Proc. 2005 IEEE-IAS Annual Meeting, pp. 383-387, Oct 2005.

[3]   H. Feng, et al., "Transient turn-on characteristics of the fin p-body IGBT," IEEE Trans. Electron Devices, vol. 62, no. 8, pp. 2555-2561, Aug. 2015.

[4]   Y. Ikura, et al., "Study of the electrostatic potential of the floating-p region during the turn-on period of IGBT" , in Proc. of The 29th International Symposium on Power Semiconductor Devices and IC's, pp. 123-126, May 2017.

[5]   S. Watanabe, et al., "1.7kV Trench IGBT with Deep and Separate Floating p-Layer Designed for Low Loss, Low EMI Noise, and High Reliability" , in Proc. of The 23th International Symposium on Power Semiconductor Devices and IC's, pp. 48-51, May 2011.

Proceedings of the 30th International Symposium on Power Semiconductor Devices & ICs
May 13-17, 2018, Chicago, USA

Optimization of Trench Sidewall for Low Leakage Current of the Sloped Field Plate Trench Edge Termination

Wentao Yang[1], Xianda Zhou[1,2], Chao Xiao[1], Hao Feng[1], Yong Liu[1], Xiangming Fang[1], Yuichi Onozawa[3], Hiroyuki Tanaka[3], Kaname Mitsuzuka[3], and Johnny K.O. Sin[1]

[1]Department of Electronic and Computer Engineering, The Hong Kong University of Science and Technology
Clear Water Bay, Kowloon, Hong Kong, China
[2]School of Electronics and Information Technology, Sun Yat-Sen Univerisity, Guangzhou, P.R. China
[3]Fuji Electric Co., Ltd, 4-18-1 Tsukama, Matsumoto, Nagano 390-0821, Japan
Email: wyangag@connect.ust.hk

Abstract— In this paper, trench sidewall optimization of the sloped field plate trench edge termination structure for low leakage current is experimentally conducted. It is found that the leakage current is highly dependent on the trench sidewall treatment. Rough trench sidewall results in large stress during trench sidewall oxidation, leading to large leakage current. Smooth trench sidewall and a thin layer of thermal oxide should be applied to the trench for achieving a significantly lower leakage current.

Keywords—edge termination, trench, field plate, etching, stress, band-gap narrowing, oxidation, leakage current.

I. Introduction

Edge termination is located at the periphery of the active region in power semiconductor devices for alleviating junction curvature effect and avoiding premature breakdown. Conventional edge termination structures, such as guard rings [1], field plates [2], semi-insulating polycrystalline silicon [3], junction termination extension [4], and variation lateral doping [5], feature long edge termination length, leading to a large non-active region area to total chip area ratio. Deep trench isolation is a very promising technology for dramatically reducing the edge termination length [6, 7]. However, breakdown at the trench region should be avoided to prevent premature and unstable breakdown characteristics and ensure good avalanche robustness [8, 9].

Previously, a sloped field plate trench termination approach has been demonstrated to achieve the ideal planar junction breakdown voltage and good surface charge immunity with record-short lengths for 600 V-class and 1200 V-class devices [10, 11]. On the other hand, leakage current is also a crucial parameter for edge termination. To further validate the structure, special cautions need to be taken during trench fabrication to minimize the leakage current caused due to poor trench sidewall preparation.

In this paper, optimization of the trench sidewall of the structure for over an order of magnitude reduction in leakage

current will be presented in detail experimentally.

II. Mechanisms and sources of leakage current

Figure 1 shows the cross-section of the 600 V-class sloped field plate trench edge termination structure [10]. It features a BenzoCycloButene (BCB) dielectric filled trench for supporting a high voltage within a short length due to the high dielectric strength of the BCB. Besides, a field plate is placed in the trench and connected to the anode electrode. In blocking states, the field plate modulates the electric field distributions along the trench and shifts the peak of the electric field from the trench region to the active region. Furthermore, the field plate can also shield the surface charge from altering the electric field distributions along the trench sidewall. Thus, by using this approach, the ideal breakdown voltage and good surface charge immunity can be achieved with a short edge termination length [12].

As reverse voltage is applied to the device, the depletion region expands from the p-base to the n-drift in the active region. On the other hand, the depletion region also expands along the trench sidewall. Thus, the total leakage current ($I_{R, total}$) consists of two parts: the leakage current flowing in the active region ($I_{R, active}$) and that flowing in the edge region ($I_{R, edge}$). $I_{R, total}$ can be expressed in Eqn. (1) as:

$$I_{R, total} = I_{R, active} + I_{R, edge} \qquad (1)$$

In the active region, the device is with a planar p^+/n^- junction. $I_{R, active}$ consists of two parts: the diffusion current and generation current. $I_{R, active}$ can be kept at a low level if no heavy metals or contaminations are introduced during the device fabrication.

As for the leakage current in the edge termination part ($I_{R, edge}$), it is mainly caused by the current generated at the trench sidewall, and can be expressed in Eqn. (2) as:

$$I_{R, edge} \propto \frac{n_i}{\tau_{eff}} \qquad (2)$$

where n_i and τ_{eff} are the intrinsic carrier density and effective carrier life-time, respectively at the trench sidewall. Both parameters are highly dependent on the trench sidewall quality

This work was supported by Fuji Electric Company, Ltd. under the grant FECL14150550.

978-1-5386-2928-4/18 $31.00 © 2018 IEEE

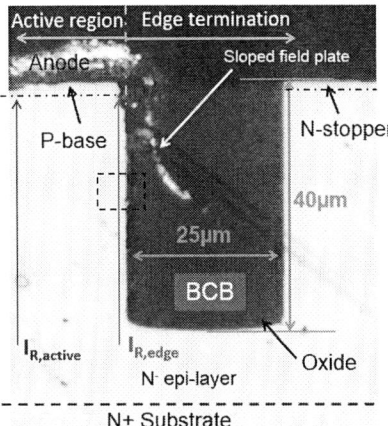

Fig. 1. Optical microscope cross-section of the 600 V-class sloped field plate edge termination structure implemented as a part of a high-voltage diode.

and trench treatment used. Thus, special attention should be paid to the trench implementation process for reducing the leakage current.

III. Experimental results and discussions

The trench sidewall quality is highly dependent on the trench etching and oxidation processes. In this section, experimental results will be presented to show the leakage current dependency on the trench sidewall roughness and oxidation methods.

A. Trench sidewall roughness

The detailed fabrication process of the 600 V-class edge termination structure was presented previously [10]. The deep and wide trench is implemented by using "Bosch" etching method, including etching and passivation steps as illustrated in Fig. 2 [13]. After multiple cycles of etching and passivation, deep and steep trenches can be achieved. However, scallop profile is easy to form along the trench sidewall due to the alternating of the etching and passivation steps. Long etching cycles result in fast etching speed but rough trench sidewall because of lateral etching.

Two "Bosch" etching recipes with different parameters, as listed in Table I, are used in the investigation of the leakage current dependency on trench sidewall roughness. The gas of C_4F_8 is used to deposit the polymer layer during the passivation cycle. SF_6 and O_2 are used in the etching cycle. O_2 plasma is for oxidizing the polymer, and SF_6 acts to etch the silicon. Recipe A has a long cycle time and high etching speed, leading to a rough trench sidewall with a scallop profile as shown in Fig. 3 (a). Recipe B is with a short cycle time and low etching speed, resulting in a smooth trench sidewall as shown in Fig. 3 (b).

The different etching recipes result in different reverse I-V characteristics as shown in Fig. 4. Device A (with Recipe A) features higher leakage current than that of Device B (with Recipe B).

Numerical simulations [14] are used to investigate the above two different processes. The stress model, STRESS.D,

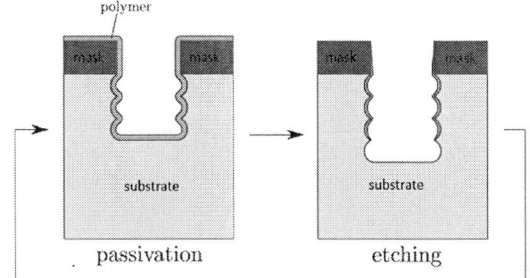

Fig. 2. A schematic illustration of the "Bosch" etching process.

TABLE I
PROCESS PARAMETERS IN "BOSCH" ETCHING PROCESS

Recipe name	A		B	
Gas flow	Etching	Passivation	Etching	Passivation
C_4F_8 (Sccm)	10	75	0	100
SF_6 (Sccm)	160	0	130	0
O_2 (Sccm)	18	0	13	0
Time per cycle(s)	12	10.6	3.5	2.5
Etching speed(μm/min)	2.7		2.2	

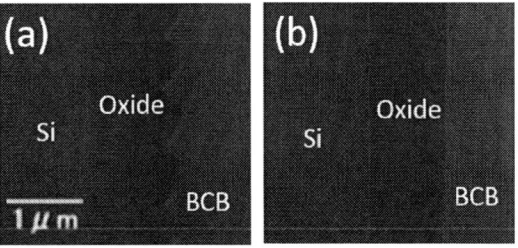

Fig. 3. SEM pictures of trench sidewall profiles (at the region indicated by the dashed rectanglar outline in Fig. 1); (a) Device A (rough trench sidewall), and (b) Device B (smooth trench sidewall).

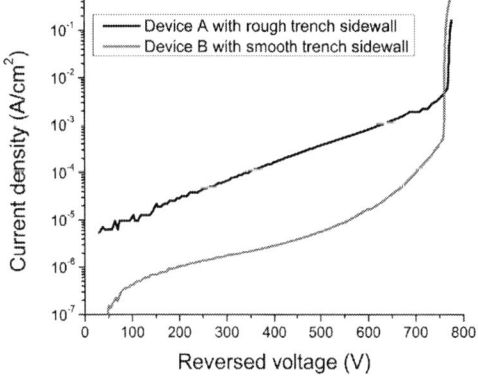

Fig. 4. Reverse I-V characteristics of the fabricated Device A and Device B by using Recipe A and Recipe B, respectively.

is activated to include stress simulation during the oxidation process. The results reveal that during the wet oxidation the rough trench sidewall (in Device A) results in a larger stress than that of the smooth trench sidewall (in Device B) as

978-1-5386-2928-4/18 $31.00 © 2018 IEEE

compared in Fig. 5. The large stress between the oxide and silicon induces Si band-gap narrowing [15-16] as illustrated in Fig. 6.

The narrow band-gap will lead to a high leakage current since intrinsic carrier concentration (n_i) is exponentially proportional to the band-gap (E_g) as expressed in Eqn. (3) as:

$$n_i \propto \exp(-\frac{E_g}{2kT}) \tag{3}$$

where k is the Boltzmann constant and T is the absolute temperature.

Furthermore, the large stress also induces additional generation center [15-16], which reduces the effective carrier life-time at the trench sidewall region and causes the increase in leakage current as previously explained in Eqn. (2).

Thus, the trench sidewall should be smoothened to avoid large stress between the Si/SiO$_2$ interface, and thus to minimize the leakage current.

B. Trench sidewall oxidation process

In addition, different trench sidewall oxidation methods, as listed in Table II, are applied during the 1200 V-class device fabrication to further investigate the leakage current.

The optimized trench etching process (Recipe B) is adopted to achieve a smooth trench sidewall. After the sacrificial oxidation and removal of the sacrificial oxide, different oxidation treatments are applied to the above three devices. Device C and Device D are designed with approximately the same oxide thickness of 0.9 μm but different process conditions. For Device C, only wet oxidation is performed. For Device D, after dry oxidation a layer of LTO is deposited and annealed for LTO densification. Device E is oxidized in N$_2$ and O$_2$ ambient, and has a layer of oxide with a thickness of ~ 0.04 μm. In addition, Device F, a conventional guard ring structure with 17 floating rings having a junction depth of 3 μm, is designed and fabricated together with Device E for comparison.

The corresponding reverse I-V curves are compared in Fig. 7. Device D and Device E have breakdown voltages higher than 1400 V, which are verified as the ideal planar junction breakdown voltage. Device C has a lower breakdown voltage of 1252 V, which is suspected to be due to the improper implementation of the field plates, which leads to a premature breakdown at the trench region. The conventional guard ring structure (Device F) features a premature breakdown of only 980 V, which is far below the designed value of 1300 V. It is suspected that this premature breakdown is due to the poor thermal budget control during device fabrication. However, the leakage current before premature breakdown is reasonable low and mainly dependent on the fabrication conditions. The leakage current of Device C can therefore be taken as the control for comparisons.

Device D features a lower leakage than that of Device C at low voltages because of the thinner thermal oxide used and lower stress along the trench sidewall. However, the leakage current of Device D increases sharply when the reverse biased voltage exceeds ~ 400 V. It is suspected that the sharply increase in leakage is caused by the large stress induced at the

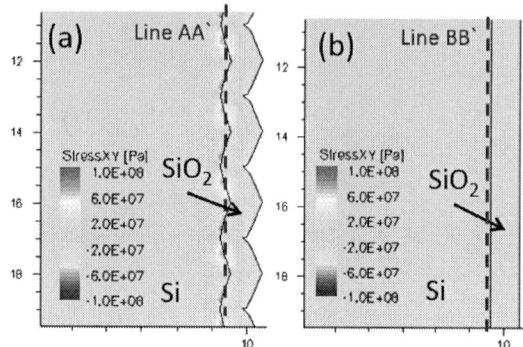

Fig. 5. Simulated stress distributions for different trench sidwall profiles; (a) Device A with rough trench sidewall, and (b) Device B with smooth trench sidewall.

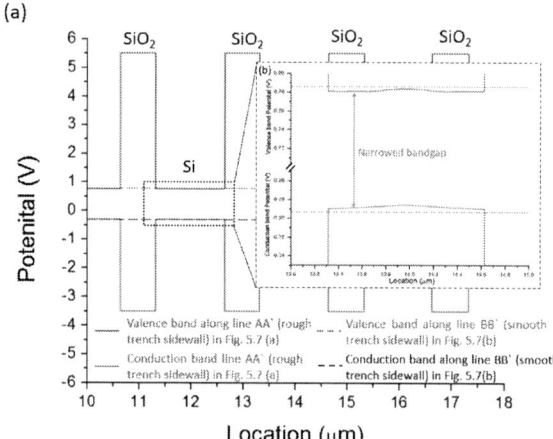

Fig. 6. (a) Comparisons of potential distributions of conduction band and valence band for different trench sidewalls; (b) the zoom-in view of the dotted rectanglar outline in (a).

TABLE II
PROCESS SPLITS IN 1200 V DEVICE FABRICATION

	Trench etching	Oxidation process	Oxide thickness
Device C	Recipe B	Wet (1000 °C, 200 mins)	~ 0.9 μm
Device D	Recipe B	Dry (1000 °C, 50 mins) + LTO depostion (~ 1μm) + LTO densification (900 °C, 30 mins)	~ 0.9 μm
Device E	Recipe B	Dry (1000 °C, 50 mins)	~ 0.04 μm

bottom corner of the trench due to the LTO densification. Figure 8 shows the depletion region distributions at different reverse biased voltages. At a low voltage of 200 V, the depletion region is away from the trench bottom. It reaches the trench bottom when the reverse voltage is above ~ 400 V. The high voltage across the large stress region causes a dramatic increase in leakage current. Thus, the use of LTO should be avoided, and the use of a thin thermal oxide at the trench sidewall is the best for minimizing the leakage current. When doing so as in Device E, over an order of magnitude reduction in leakage current is achieved. The leakage level is compatible

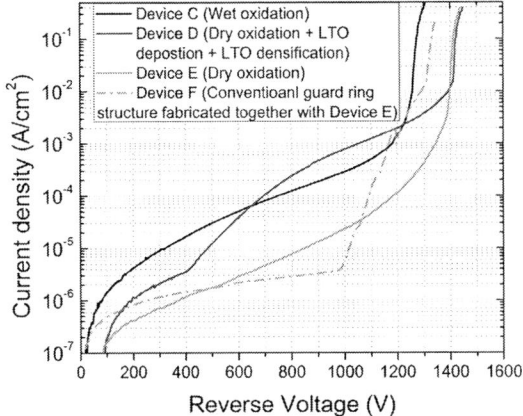

Fig. 7. Reverse I-V characteristics of the 1200 V-class devices with different trench sidewall treatment methods.

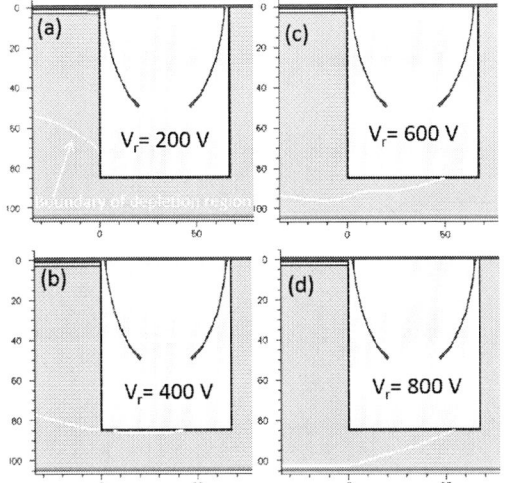

Fig. 8. Depletion region distributions (white line) for 1200 V-class devcie at different reverse biased voltages; (a) Vr = 200 V, (b) Vr = 400 V, (c) Vr = 600 V, and (d) Vr = 800 V.

to that of the conventional guard ring structure in Device F. With a smooth trench sidewall, it is expected that further reduction in leakage current can be achieved with further reducing on the thermal oxide thickness.

IV. Conclusion

In summary, it is found that the leakage current of the sloped field plate trench edge termination structure is highly dependent on the trench sidewall treatment. Smooth trench sidewall and a thin layer of thermal oxide should be applied to the trench for achieving a significantly lower leakage current.

ACKNOWLEDGMENT

The authors would like to thank the staff of Fuji Electric Company Ltd. for their technical advices and support, and the staff of the Nanoelectronics Fabrication Facility and the Semiconductor Product Analysis and Design Enhancement Center at the Hong Kong University of Science and Technology for their help in device fabrication and characterization.

REFERENCES

[1] Y.C. Kao and E.D. Wolley, "High-voltage planar p-n junction," *Proceedings of IEEE*, vol. 55, pp. 1409-1414, Aug. 1967.

[2] F. Conti and M. Conti, "Surface breakdown in silicon planar diodes equipped with field plate," *Solid-State Electron.*, vol. 15, pp. 93-105, 1972.

[3] T. Matsushita, T. Aoki, T. Ohtsu, H. Yamoto, H. Hayashi, M. Okayama, and Y. Kawana, "Highly reliable high-voltage transistors by use of the SIPOS process," *IEEE Trans. Electron Devices*, vol. 23, pp. 826-830, Aug. 1976.

[4] V.A.K. Temple, "Junction termination extension (JTE), a new technique for increasing avalanche breakdown voltage and controlling surface electric fields in P-N junctions," in *IEDM Tech. Dig.*, pp. 423-426, 1977.

[5] R. Stengl, U. Gösele, C. Fellinger, M. Beyer, and S. Walesch, "Variation of lateral doping as a field terminator for high-voltage power devices," *IEEE Trans. Electron Devices*, vol. 33, pp. 426-428, Mar. 1986.

[6] C. Park, N. Hong, D.J. Kim, and K. Lee, "A new junction termination technique using ICP RIE for ideal breakdown voltages," in *Proc. ISPSD*, Santa Fe, USA, pp. 257-260, Jun. 2002.

[7] L. Théolier, H. Mahafoz-Kotb, K. Isoird, F. Morancho, S. Assié-Souleille, and N. Mauran, "A new junction termination using a deep trench filled with BenozCycloButene," *IEEE Electron Devices Letters*, vol. 30, pp. 687-689, Jun. 2009.

[8] R. Kamibaba, K. Takahama, and I. Omura, "Design of trench termination for high voltage devices," in *Proc. ISPSD*, Hiroshima, Japan, pp. 107-110, Jun. 2010.

[9] Y.L. Tsang and J.M. Aitken, "Junction breakdown instabilities in deep trench isolation structures," *IEEE Trans. Electron Devices*, vol. 38, pp. 2134-2138, Sep. 1991.

[10] W.T. Yang, H. Feng, X.M. Fang, Y. Onozawa, H. Tanaka, and J.K.O. Sin, "A Novel Sloped Field Plate Enhanced Ultra-short Edge Termination Structure," *IEEE Electron Device Letters*, vol. 37, 471-473, Apr. 2016.

[11] W.T. Yang, H. Feng, X.M. Fang, Y. Onozawa, H. Tanaka, and J.K.O. Sin, "A new 1200 V-class edge termination structure with trench double field plates for high dV/dt performance," in *Proc. ISPSD*, Sapporo, Japan, pp. 109-112, May 2017.

[12] W.T. Yang, H. Feng, X.M. Fang, Y. Onozawa, H. Tanaka, and J.K.O. Sin, "Design and Characterization of Sloped-field-plate Enhanced Trench Edge Termination," *IEEE Trans. Electron Devices*, vol. 64, pp. 713-719, Mar. 2017.

[13] F. Laermer and A. Schilp of Robert Bosch GmbH, "Method of anisotropically etching silicon", US-Patent, No. 5501893.

[14] *Sentaurus Device User Guide, Version J-2014.09-SP*, Synopsys, Inc., Mountain View, CA, USA, 2014.

[15] J.J. Wortman, J.R. Hauser, and R.M. Burger, "Effect of mechanical stress on pn-junction device characteristics," *J. Appl. Phys.*, vol. 35, p.2122, 1964.

[16] P. Smeys, P.B. Griffin, Z.U. Rek, I.D. Wolf, and K.C. Saraswat, "Influence of process-induced stress on device characteristics and its impact on scaled device performance," *IEEE Trans. Electron Devices*, vol. 46, no. 6, pp. 1245-1252, Jun. 1999.

Proceedings of the 30th International Symposium on Power Semiconductor Devices & ICs
May 13-17, 2018, Chicago, USA

Analysis of Reverse Temperature Dependent Switching-Off Behavior of Ultra-thin Fieldstop IGBTs

So-Youn Kim, Euntaek Kim, Jiho Jeon, Jinyoung Jung, Soo-Seong Kim, Kwang-Hoon Oh, and Chongman Yun
TRinno Technology, Songpa-Gu, Seoul, Republic of Korea
Email-Address: matilda@trinnotech.com

Abstract— **For trench FS IGBTs having different thickness of drift layers, the trade-off performances between conduction loss and switching loss have been evaluated. At switching off mode, it was observed that turn-off loss increases with decrease in temperature, which appears at high current switching with high collector-emitter bias. This temperature dependency of switching–off behavior is unusual and has not been fully investigated before.**

In this research, we analyze underlying physical mechanisms for the reverse temperature dependency of switching-off behavior using device simulation as well as extensive switching measurements. In addition, we propose a new concept of switching SOA of ultra-thin FS IGBTs, identifying switching conditions which lead to abnormal temperature dependency.

Keywords— *IGBT; fieldstop; ultra-thin wafer process; switching-off behavior; dynamic avalanche breakdown*

I. INTRODUCTION

In a battery driven inverter circuit for an EV (electric vehicle) or a HEV (hybrid electric vehicle) IGBT has been employed as a core power switch, the voltage ratings of which need to be optimized depending on the supply voltages of battery systems.

In this regard, we fabricated various medium voltage trench FS (fieldstop) IGBT devices using the ultra-thin wafer process which covers the thickness of IGBT device from 45μm to 70μm. The trade-off performances between conduction and switching losses have been extensively evaluated for those ultra-thin trench FS IGBT devices. As is known, the trade-off performances are dramatically improved with decrease in thickness of the drift layer.

However, unlike what is commonly known, it was observed that switching-off loss becomes even smaller at high temperature under high current density and high field switching operation. This reverse temperature dependent switching–off (RTDS) behavior is unusual and has not been fully explored before.

In this work, we analyze underlying physical mechanisms for the RTDS behavior through TCAD device simulations as well as extensive switching measurements using various test conditions. In addition, we propose a new concept of switching SOA for the ultra-thin FS IGBTs showing reliable switching performances without abnormal temperature dependency.

II. DEVICE EVALUATION

Using a 600V trench FS IGBT platform as well as ultra-thin wafer process, we fabricated ultra-thin trench FS IGBTs having BVs (breakdown voltages) ranging from 350V to 700V. For those devices, the thickness of drift layer is processed from 45 μm to 70 μm and the other design and process parameters for active and termination structures are set to the same. The representative trench FS IGBT is shown in Fig.1.

 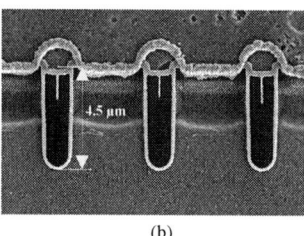

(a)   (b)

Fig. 1. Fabricated trench FS IGBT using the ultra-thin wafer process. IGBTs used in this work are varied with thickness of drift layers.
(a) Top view, and (b) Cross sectional view

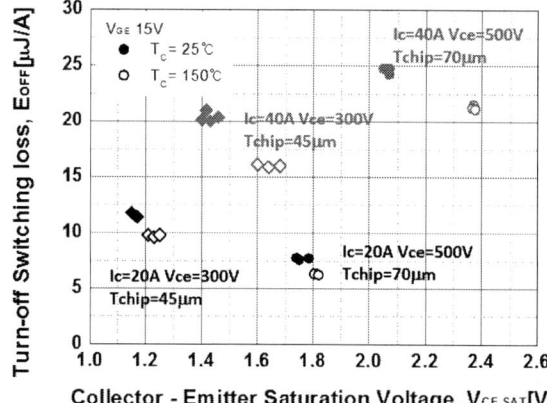

Fig. 2. Trade-off performances between conduction loss and switching loss for the ultra-thin FS IGBTs with measurement temperature. (Ic = 40A is equivalent to Jc = 350A/cm²)

978-1-5386-2928-4/18 $31.00 © 2018 IEEE
176

During the evaluation of trade-off performances, collector current density J_c and collector-emitter voltage V_{ce} are changed. Conducting inductive-load switching test, current and voltage waveforms were observed at 25 °C and 150 °C. When the collector-emitter bias V_{ce} is close to the BV of the FS IGBTs, the RTDS behavior was observed. The switching-off loss, E_{off} at 25 °C is even larger than E_{off} at 150 °C. For the two different 45 µm and 70 µm FS IGBT device groups, test results are plotted in Fig. 2.

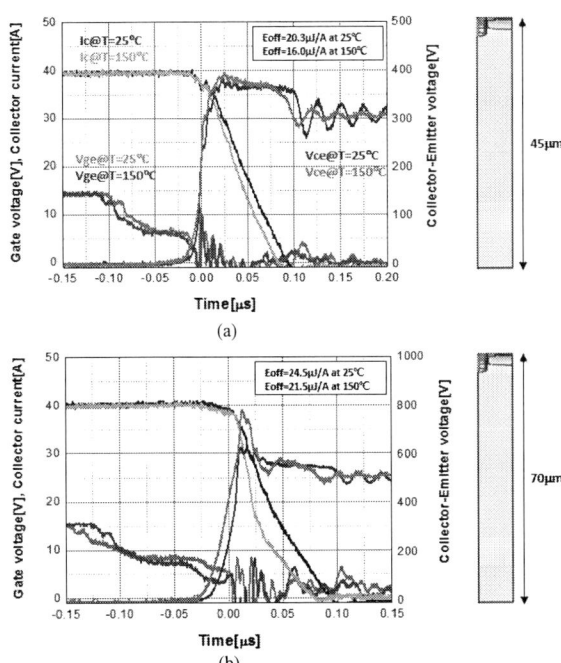

Fig. 3. Switching-off waveforms of the FS IGBTs with inductive load at 25 °C and 150 °C. (a) 45 µm FS IGBT, and (b) 70 µm FS IGBT

To investigate more clearly this RDTS behavior, the actual switching waveforms are shown in Fig. 3. When the collector-emitter bias is high enough to induce dynamic avalanche breakdown, the collector currents start to gradually fall increasing E_{off}. Compared with the switching-off loss at 150 °C, measured E_{off} at 25 °C is higher by 27% to 14% for 45 µm FS IGBT and 70 µm FS IGBT, respectively.

More switching waveforms for FS IGBTs with different thickness of drift layers, which are shown in Fig. 4, demonstrate that the sudden increase in E_{off} of the IGBTs is related to dynamic avalanche breakdown [1, 2]. Using the switching condition where V_{ce} = 300V, I_c = 40A with Rg = 5Ω, turn-off waveforms are examined for the FS IGBTs. As can be seen from waveforms of 45 µm FS IGBT, the RTDS behavior is apparent with strong dynamic breakdown phenomenon.

When the dynamic breakdown occurs at turn-off transition, the collector-emitter voltage V_{ce} is clamped to the constant voltage value and the collector current I_c gradually decreases during that breakdown period.

In order to analyze physical mechanisms underlying this behavior in more detail, TCAD device simulation has also been conducted.

III. RESULTS AND DISCUSSION

Extensive switching tests show that the collector-emitter voltage is clamped under dynamic avalanche breakdown mode which is similar to SSCM [3], and the turn-off process is also governed by the impact ionization process. Since the impact ionization is temperature dependent, this dynamic breakdown phenomenon also becomes weak at higher temperature.

Fig. 5 shows turn-off energy losses with collector voltages for the ultra-thin FS IGBTs. When the V_{ce} increases, the RTDS phenomenon appears above the specific voltage value. Similarly, increase in the collector current leads to the same phenomenon above the specific current value. The higher V_{ce} and higher I_c, the RTDS behavior becomes clearer.

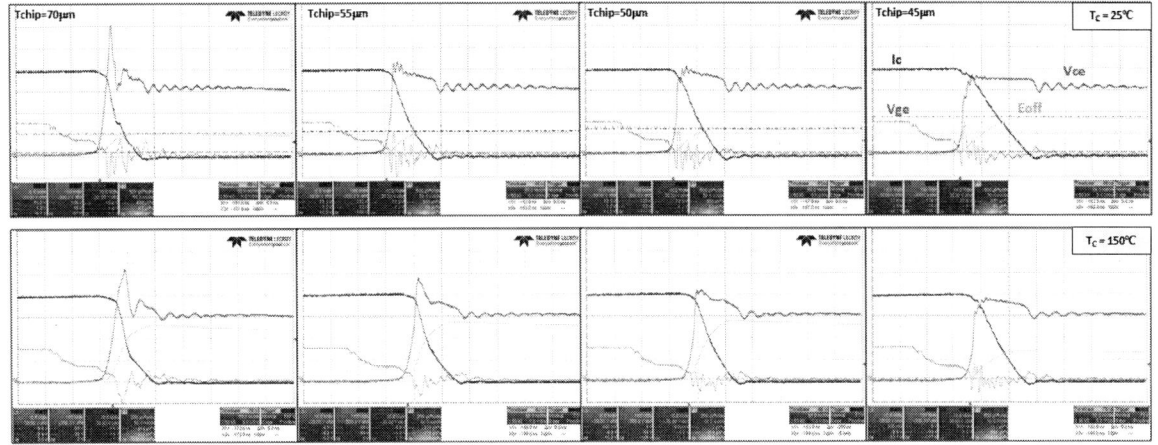

Fig. 4. Switching-off waveforms for FS IGBTs with different thickness of drift layers from 45 µm to 70 µm.

978-1-5386-2928-4/18 $31.00 © 2018 IEEE

Based on the investigations, it seems that for the switching operations with the collector current density over 400 A/cm² or high collector-emitter bias over 70% of BV, the turn-off losses of the FS IGBTs rapidly increase.

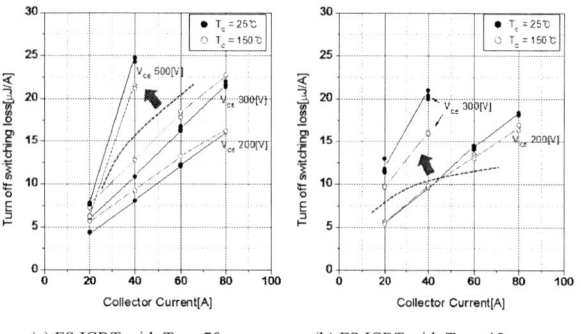

(a) FS IGBT with T$_{chip}$=70μm (b) FS IGBT with T$_{chip}$=45μm

Fig. 5. Turn-off losses vs. collector currents with various collector-emitter voltages for 2 FS IGBT groups. For the switching conditions indicated by the arrow, the turn-off losses at high temperature become even smaller than ones at room temperature.

This switching-off mechanism can also be explained by the TCAD simulations. For the FS IGBT with T$_{chip}$ =70 μm, this RTDS behavior has been reproduced through the simulations. Fig. 6 (a) and (b) show switching off waveforms when the collector-emitter bias is applied by as much as 60% and 100% of static breakdown voltage, respectively. As in Fig. 6 (b), fall time of the collector current at room temperature increases as the collector bias reaches its BV.

In Fig. 6 (c) and (d), electric-field distribution, impact ionization, electron-density, and hole-density are given with the evolution of time. In the normal switching-off process in Fig. 6 (c), the electric field at the overshoot voltage is the largest but less than ~200kV/cm. Since the impact ionization by such electric field is not strong, electron hole carrier density is not significantly influenced by impact ionization and the carriers decay slowly at high temperature.

However, as in Fig. 6 (d), when the collector bias is high enough to induce dynamic breakdown mode, the generated electron-hole carriers are replenished inside the drift layer of IGBT and as a result, turn-off process is abnormally delayed.

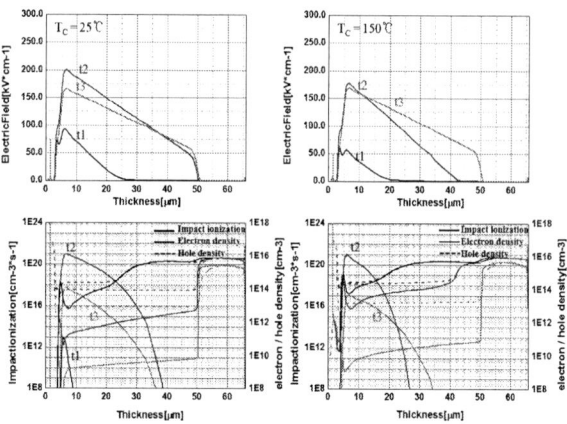

(c) Carrier density and impact ionization in the case of (a)

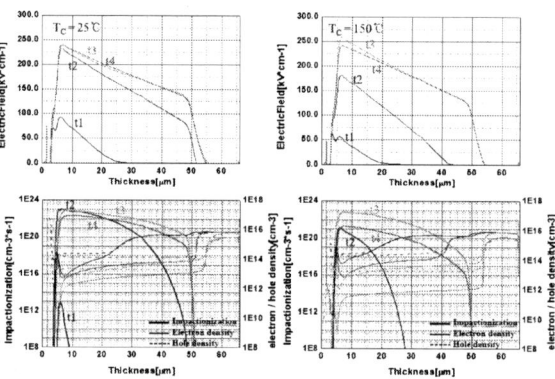

(d) Carrier density and impact ionization in the case of (b)

Fig. 6. Switching-off waveforms for 70μm FS IGBT by TCAD simulation. Collector-emitter bias at switching-off is (a) Vce = 0.6BV, (b) Vce= 1BV, (c) electric field, carrier density, II rate when Vce = 0.6BV, and (d) electric field, carrier density, II rate when Vce = 1BV

At this condition, peak electric field is over 230kV/cm and impact ionization rate is also over ~10²² cm⁻³s⁻¹. As the switching temperature increases, the delayed turn-off switching process recovers to the normal switching mode due to weakening of impact ionization process. Therefore, it can be said that this temperature dependent impact ionization process during the switching-off causes the RTDS behavior.

Varying the switching collector current level, the RTDS phenomenon has also been reproduced in simulation. As can be seen in Fig. 7, the impact ionization rate inside the drift layer increases with collector current so as a result, the RTDS behavior becomes more apparent at higher switching current level. In Fig. 7 (c) and (d), difference in carrier profiles under switching-off conditions at 25 °C and 150 °C can also be noticed.

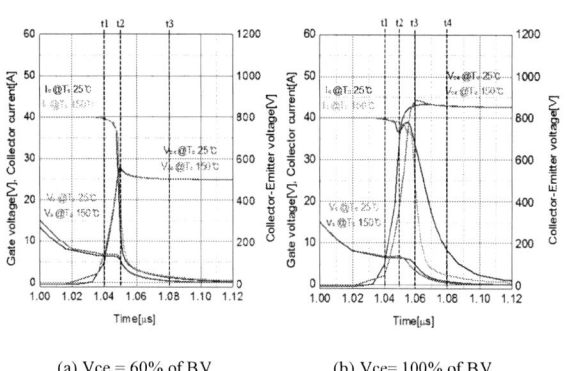

(a) Vce = 60% of BV (b) Vce= 100% of BV

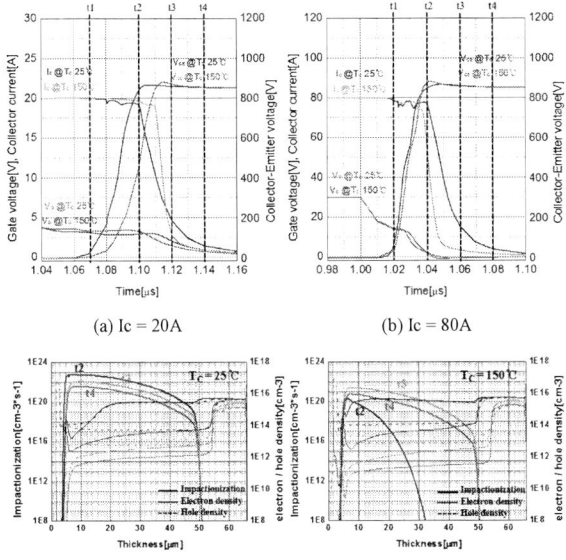

(a) Ic = 20A   (b) Ic = 80A

(c) Carrier density and impact ionization when Ic = 20A

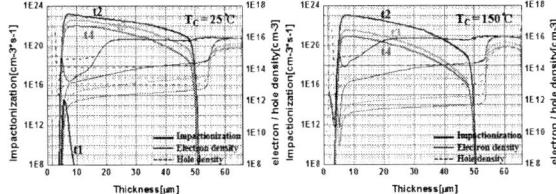

(d) Carrier density and impact ionization when Ic = 80A

Fig. 7. Switching-off waveforms for 70μm FS IGBT by TCAD simulation. Collector-emitter bias at switching-off is 100% of BV. (a) Ic = 20A, (b) Ic = 80A, (c) carrier density, II rate when Ic = 20A, and (d) electric field, carrier density, II rate when Ic = 80A

In general, RBSOA suggests voltage and current conditions where the reliable switching operation is possible. However, as observed in the experimental results, when switching with very high collector-emitter bias, RTDS phenomenon may happen and switching loss can be severely increased.

Considering this RTDS phenomenon, we redefined the switching SOA within RBSOA as shown in Fig. 8. Since RBSOA only informs ruggedness aspect of the device in a switching transition, a new switching SOA is required where the reliable switching operation can be achieved. For the reasonable switching performance, the switching conditions of the IGBT should be set out of the current-voltage dependent dynamic breakdown region.

Moreover, to prevent the reduction of RBSOA resulting from abnormal turn-off delays, the design guideline for ultra-thin FS IGBTs which can minimize the dynamic breakdown region under given switching conditions can be created.

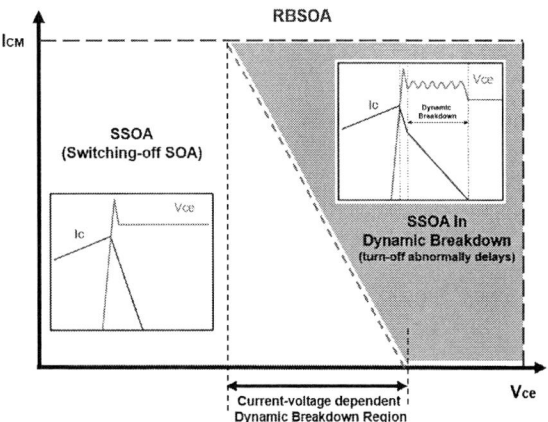

Fig. 8. Switching-off SOA within RBSOA. For a reliable switching operation, dynamic breakdown condition should be avoided.

IV. CONCLUSIONS

For the ultra-thin trench FS IGBTs having different thickness of drift layers, it was observed that turn-off loss increases with decrease in temperature, which appears at high current switching with high internal electric field. This unusual reverse temperature dependency of switching–off behavior is analyzed through extensive device evaluations as well as intense TCAD simulations. The main cause of the RTDS behavior is identified as dynamic breakdown phenomenon during the switching-off period.

Temperature dependent dynamic avalanche breakdown process during switching-off mode generates excessive electron hole carriers, which lead to delay in turn-off process increasing switching-off loss. In addition, we propose a new switching SOA concept for the ultra-thin FS IGBT showing reliable switching characteristics without abnormal temperature dependency.

ACKNOWLEDGMENT

This work was supported by the Ministry of Trade, Industry and Energy (MoTIE) and Korea Evaluation Institute of Industrial Technology (KEIT) through industrial technology innovation project (10080429).

REFERENCES

[1] Josef Lutz, "Semiconductor Power_devices_Physics, Characteristics, Reliability," Springer, 2011, pp. 460-462

[2] S. Machida et. al., "Turn-off switching analysis considering dynamic avalanche effect for low turn-off loss high-voltage IGBTs," Proc. ICSSD'13, pp.1048-1049

[3] M. Rahimo et al., "A Study of Switching-Self-Clamping-Mode "SSCM" as an Over-voltage Protection Feature in High Voltage IGBTs," Proc. ISPSD'05, pp. 67-70, 2005.

978-1-5386-2928-4/18 $31.00 © 2018 IEEE

Proceedings of the 30th International Symposium on Power Semiconductor Devices & ICs
May 13-17, 2018, Chicago, USA

Effect of Charge Imbalance and Edge Structure on the Reverse Recovery Waveform in Superjunction Body Diode

Daisuke Arai, Mizue Yamaji, Koichi Murakami, Masaaki Honda and Shinji Kunori

Shindengen Electric Mfg. Co., Ltd.
10-13, Minami-cho, Hanno-city, Saitama, JAPAN
E-mail : daisuke_arai@si.shindengen.co.jp

Abstract—The reverse recovery characteristics of the body diode in Si superjunction (SJ) have been investigated at the fixed condition of dif/dt = 100A/μsec. It was found that the recovery waveforms are much affected by the charge imbalance (CIB) and by the edge termination structure. N-type rich CIB and the support by the current through the edge structure result in softer recovery. When local lifetime control by $^3He^{2+}$ ion irradiation is performed, the SJ diode with the defect peak at half the depth of the P-column showed softer recovery than the one with the defect peak at the bottom of the P-column. These results indicate that the conductivity modulation carriers remain around the bottom of the P-column during the reverse recovery transient and to control the elimination rate of the remaining carriers is important to achieve soft and fast recovery for SJ diodes. Understanding of these transient phenomena will provide a guideline for realizing general use of SJ-MOSFETs for inverters.

Keywords—*Superjunction; body diode; reverse recovery; CIB; lifetime control; softness*

I. INTRODUCTION

SJ-MOSFETs have improved the trade-off relationship between on-resistance and blocking voltage for high voltage MOSFETs beyond the Si limit [1, 2]. They have excellent static characteristics, whereas the switching characteristics as MOSFETs and the reverse recovery characteristics of the body diode have to be carefully studied and controlled to be more popularly used for inverters [3 - 9].

In this article, we investigated the reverse recovery characteristics of SJ body diodes. First we changed the CIB in SJ diodes experimentally and numerically. Next we compared the recovery waveforms of those SJ diodes with and without the edge termination structure by the calculations. Finally we experimentally performed local lifetime control by $^3He^{2+}$ ion irradiation with the depth of the defect peak changed.

II. EFFECT OF CAHRGE IMBALANCE ON THE RECOVERY

A. Experimental

We fabricated 650V / 280mΩ SJ-MOSFETs with varied CIB and measured reverse recovery characteristics of their body diodes. Lifetime control was not performed. As shown in Fig. 1, N-type 15% rich SJ diode exhibited softer recovery than the one with P-type 15% rich CIB.

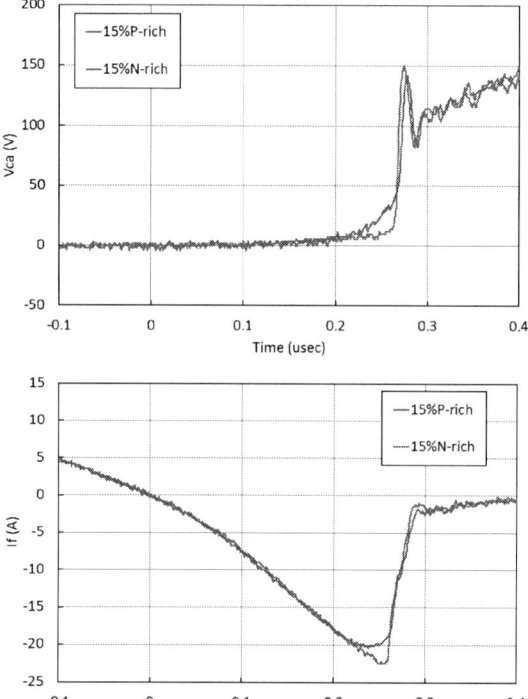

Fig. 1. Measured reverse recovery waveforms for SJ diodes with varied CIB. Lifetime control is not performed.

B. Device Simulation

To investigate the internal physics in the SJ diodes, device simulations were carried out with Sentaurus Device ver. L-2016.03. Fig. 2 shows the circuit model. A conventional Si MOSFET with planar gate was applied to the switch while the Si SJ body diodes were tested as the free wheel diode. In these calculations the edge termination structure was included in each SJ diode model. The structures of the active cell and the edge termination are described in Fig. 3. The calculated recovery waveforms with respect to CIB are shown in Fig. 4. Both experiments and calculations indicated that the SJ diode

978-1-5386-2928-4/18 $31.00 © 2018 IEEE

with N-type rich CIB exhibits softer recovery than the one with P-type rich CIB.

Fig. 5 describes the internal hole density during the reverse recovery transient for the SJ diodes while Fig. 6 describes the hole current density at the same moments. The moments (A) to (F) are indicated in Fig. 4.

According to Fig. 5, the hole escape path is formed through the center of the P-column from the bottom to the source (anode) electrode during the transient. For N-type rich SJ the path is constricted because the P-column is depleted at the early stage of the transient. Hence the conductivity modulation carriers remain longer around the bottom of the P-column which leads to soft recovery. When the CIB is P-type rich, wider hole escape path is formed at the center of P-column and the carriers are swept out of the semiconductor quickly which leads to hard recovery.

These results indicate that to keep the carriers longer around the bottom of the P-column is important to achieve soft recovery. It is consistent with the prior research [4].

III. THE CONTRIBUTION OF THE EDGE TERMINATION TO SOFT RECOVERY

For the calculations shown in Figs. 4 to 6, the device models include both active cell and edge termination structure. The area of the active cell was taken as 0.1cm^2 while the length of the edge termination along the die side was taken as 1.0cm. We confirmed the contribution of the edge structure to the softness. Fig. 7 shows the calculated recovery current waveforms for SJ diodes without edge termination structure. These waveforms exhibit harder recovery compared with those shown in Fig. 4. We split the recovery current in Fig. 4 into the current through the active cell and through the edge termination by calculation. Figs. 8 (a) and (b) show the classified recovery currents for P-type rich and N-type rich CIB respectively. These results indicate that the recovery current through the edge termination has its peak at the tail end of the recovery current through the active area. It helps to keep the softness which is consistent with the prior work [7].

IV. LOCAL LIFTIMECONTROL BY HELIUM ION IRRADIATION

We experimentally performed local lifetime control by using $^3He^{2+}$ ion irradiation on the SJ diodes having just CIB. In this experiment the depth of the defect peak was changed as 20μm and 40μm, while the depth of the P-column was fixed at about 40μm. Fig. 9 shows the recovery waveforms. When the defect peak is formed at 20μm depth, the SJ diode exhibits soft and relatively fast recovery while the sample with the defect peak as 40μm depth shows hard and very fast recovery.

We assume the schematic view of the cross section with the defects induced by the helium irradiation as shown in Fig. 10. It indicates that to achieve soft and fast recovery the defect peak formed by the local lifetime control should avoid the depth of carrier remaining area described in Fig. 4.

V. SUMMARY

It has turned out that the conductivity modulation carriers remain around the bottom of the P-column and their elimination rate determines the softness and fastness of the reverse recovery for SJ diodes. These knowledges will provide a guideline for general use of SJ-MOSFETs for inverters.

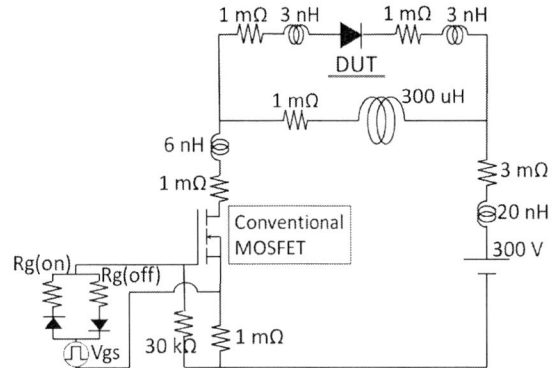

Fig. 2. The circuit model used in the simulation to evaluate the reverse recovery characteristics of the SJ body diodes.

Fig. 3. The SJ device model for (left) active cell and (right) edge termination structure.

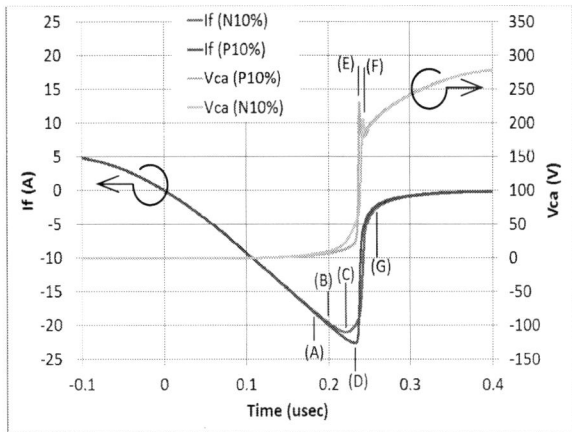

Fig. 4. Calculated reverse recovery waveforms for the SJ body diodes with the CIB changed as (blue) P-type 10% rich and (red) N-type 10% rich.

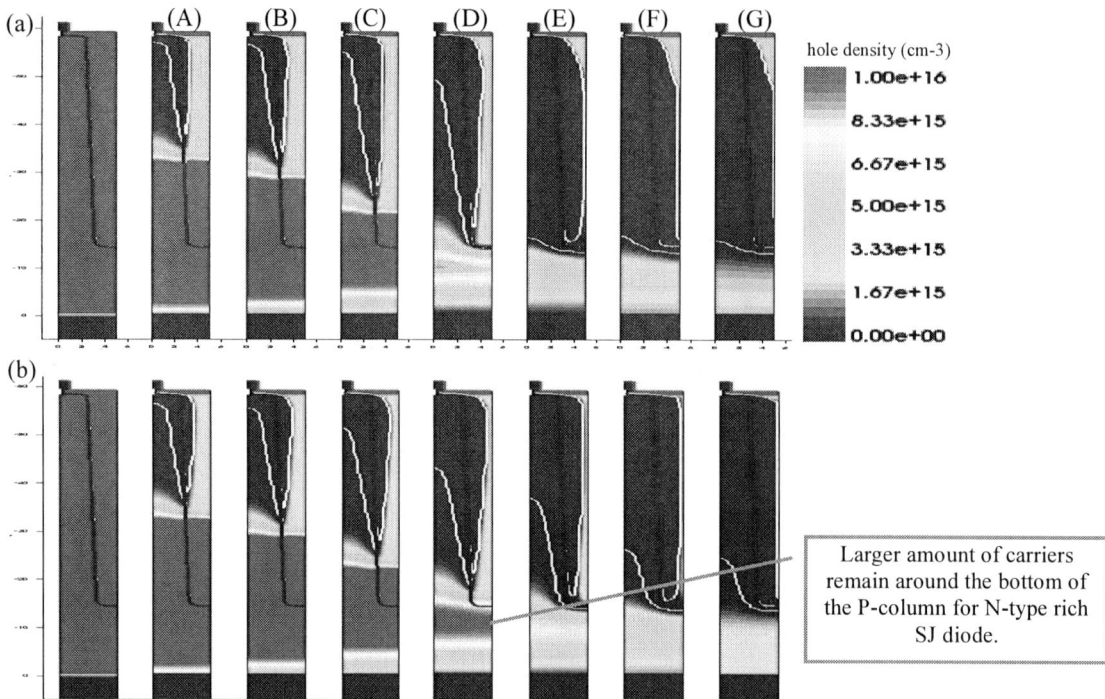

Fig. 5. Calculated internal hole density during the reverse recovery transient for SJ body diodes with CIB as (a) P-type 10% rich (b) N-type 10% rich. Left ones show perfectly forward biased states while the moment (A) to (G) are indicated in Fig. 4.

Larger amount of carriers remain around the bottom of the P-column for N-type rich SJ diode.

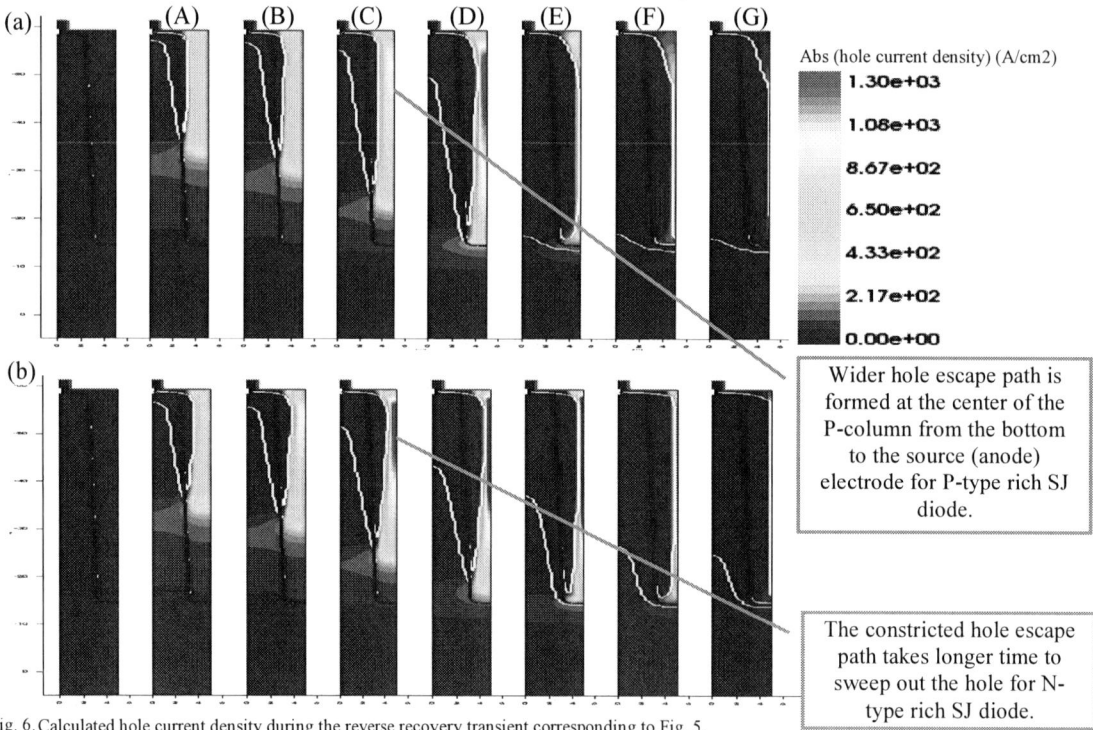

Wider hole escape path is formed at the center of the P-column from the bottom to the source (anode) electrode for P-type rich SJ diode.

The constricted hole escape path takes longer time to sweep out the hole for N-type rich SJ diode.

Fig. 6. Calculated hole current density during the reverse recovery transient corresponding to Fig. 5.

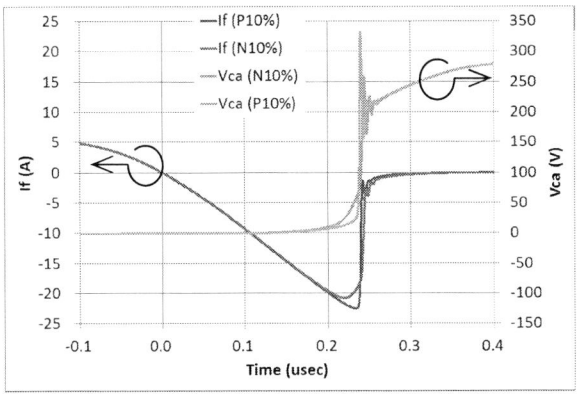

Fig. 7. Calculated reverse recovery waveforms for SJ body diodes without edge termination structure with CIB as (blue) P-type 10% rich and (red) N-type 10% rich.

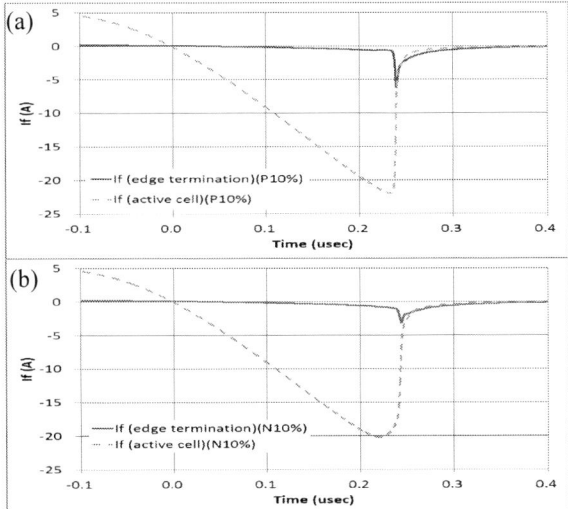

Fig. 8. Calculated reverse recovery current components through the active cell and through the edge termination area for (a) P-type rich (b) n-type rich CIB. The total current corresponds to the current waveform described in Fig. 4.

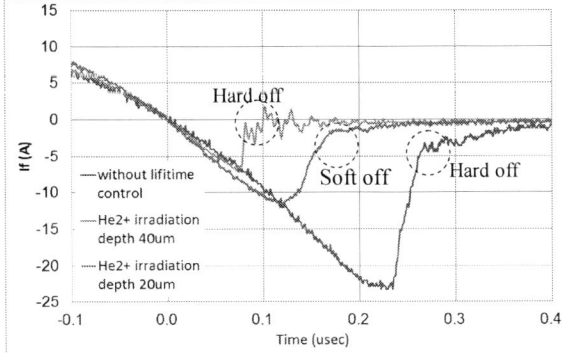

Fig. 9. Measured reverse recovery waveforms for SJ diodes with local lifetime controlled by using $^3\mathrm{He}^{2+}$ ion irradiation. The depth of the defect peak was changed.

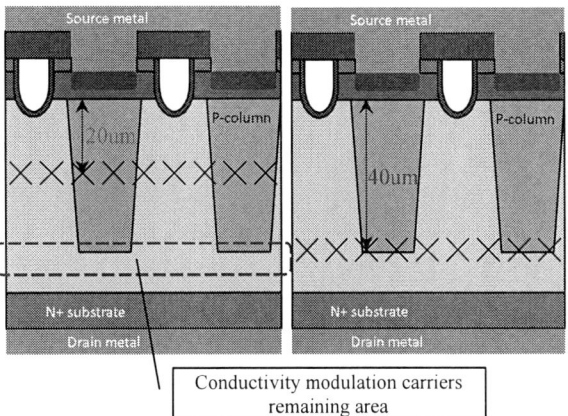

Fig. 10. Schematic view of the defects induced by $^3\mathrm{He}^{2+}$ ion irradiation. Local defects are formed around (left) half the depth of the P-column while (right) the bottom of the p-column.

REFERENCES

[1] T. Fujihira, "Theory of semiconductor superjunction devices," Jpn. J. of Appl. Phys., vol. 36, pp.6254-6262, 1997.

[2] G. Deboy, M. März, J. –P. Stengl, H. Strack, J. Tihanyi and H. Weber "A new generation of high voltage MOSFETs breaks the limit of silicon", Tech. Digest IEEE IEDM 98, pp.683-685, 1998.

[3] P.M. Shenoy, A. Bhalla, and G.M. Dolny, "Analysis of the effect of charge imbalance on the static and dynamic characteristics of the super junction MOSFET," 11th International Symposium on Power Semiconductor Devices and IC's, pp. 99-102, 1999.

[4] S. Ono, Y. Kawaguchi, and A. Nakagawa, "Simulation of the Reverse recovery characteristics of the Super Junction MOSFET," 2001 IEEJ Technical Meeting document, EDD-01-71, pp.19-22, Nov.25, 2001 (in Japanese).

[5] W. Saito, I. Omura, S. Aida, S. Koduki, M. Izumisawa, and T. Ogura, "600V semi-superconjunction MOSFET," 15th International Symposium on Power Semiconductor Devices and IC's, pp. 45-48, 2003.

[6] W. Saito, I. Omura, S. Aida, S. Koduki, M. Izumisawa, and T. Ogura, "Semisuperjunction MOSFETs: new design concept for lower on-resistance and softer reverse-recovery body diode," IEEE Trans. Electron Devices, Vol. 50, No. 8, pp. 1801-1806, Aug. 2003.

[7] Z. Yang, J. Zhu, X. Tong, W. Sun, F. Bian, Y. Tian, Y. Zhu, P. Ye, Z. Li, and B. Hou, "Investigations of inhomogeneous reverse recovery behavior of the body diode in superjunction MOSFET," 29th International Symposium on Power Semiconductor Devices and IC's, pp. 155-158, 2017.

[8] H. Yamashita, H. Ura, S. Ono, M. Nashiki, K. Mii, W. Saito, J. Onodera, and Y. Hokomoto, "Suppression of switching loss dependence on charge imbalance of superjunction MOSFET," 27th International Symposium on Power Semiconductor Devices and IC's, pp. 405-408, 2015.

[9] D. Arai, S. Hisada, M. Yamaji, and S. Kunori, "Dependence of Switching Waveform on Charge Imbalance in Superjunction MOSFET used in Inductive Load Circuit," 29th International Symposium on Power Semiconductor Devices and IC's, pp. 487-490, 2017.

Proceedings of the 30th International Symposium on Power Semiconductor Devices & ICs
May 13-17, 2018, Chicago, USA

Tight relationship among Field Failure Rate, Single Event Burn-out (SEB) and Cold Bias Stability (CBS) as a cosmic ray endurance for IGBT and diode

K. Suzuki, Y. Yoshiura, K. Uryu, T. Minato,
M. Tarutani, Y. Miyazaki, H. Uemura,
T. Hagihara, S. Momii, Y. Kusakabe, M. Nakamura
Power Device Works, Mitsubishi Electric Corporation
1-1-1 Imajyuku-Higashi, Nishi-ku, Fukuoka, Fukuoka,
819-0192, Japan
e-mail: Suzuki.Kenji@cs.MitsubishiElectric.co.jp

Y. Fujita, K. Takakura
MELCO Semiconductor Engineering Corporation
1-1-1 Imajyuku-Higashi, Nishi-ku, Fukuoka, Fukuoka,
819-0192, Japan

Abstract— The applied voltage (Vcc) dependence of SEB characteristics of the Failure In Time (FIT) is generally estimated by the accelerated test, because it takes a long time to cause SEB under the natural condition. So, it is meaningful to confirm the relationship among Field Failure Rate (FFR), SEB, FIT and CBS characteristic. Through physical analysis, the destruction point is confirmed to be located around the electric field peak position during SEB experiment using the neutron irradiation. After both SEB curve fitting and sufficient numbers of analysis for the destruction points, the first major factor to characterize the SEB curve is confirmed to be the electric field strength.

Keywords— SEB, cosmic ray, LTDS, CBS, IGBT, Diode

I. INTRODUCTION

Expanding the renewable energy application like a Photovoltaic or Wind generation, the maintenance-free characteristics including the cosmic ray or SEB endurance, or Long Term DC bias Stability (LTDS), have been important. Basically there is the strong tradeoff relationship between the total electrical power losses and SEB endurance [1]. Which kind of device structure being superior to the others has been intensely discussed from the production point of view [1, 2, 9]. Diodes fabricated in the authentic thick wafer technology are able to survive cosmic radiation induced streamer events, leading to the extraordinary high cosmic radiation hardness. Most likely the highly doped and deeply diffused anode and cathode regions of diode have the high robustness against SEB [1, 9]. In the case of IGBT, it is one of the most important approach to reduce the current amplification factor of the parasitic pnp and npn transistors to suppress destruction by neutron-induced SEB [3, 4, 15]. On the other hand, the measurements show that the failure rate of an IGBT is almost the same as that of a MOSFET or diode, and the parasitic bipolar does not affect the failure rate [4, 14]. To certify the effectiveness of highly accelerated test sequence with neutron for evaluating the cosmic radiation hardness, it is meaningful to confirm the relationship among Field Failure Rate (FFR), SEB characteristics and CBS, in which "Cold" stands for a room temperature. Through physical analysis, the destruction point is confirmed to be located around the electric field peak position during SEB experiment using the neutron irradiation.

II. FIT CURVE OF FFR, SEB AND CBS

From the early stage, the applying voltage [5,6] or peak Electric field (Ep) strength [7] is assumed to be a major factor to characterize SEB phenomena, and there have been same kinds of approaches [8,9]. According to this stream, equation (1), which is the impact ionization coefficient formula along Fulop's approximation [10], is applied as the fitting equation for SEB curve (Fig. 1). Our approximation stands on the hypothesis which SEB in the low electric field is caused by the high energy but low frequency neutron coming across to be observed according to the neutron's differential flux of cosmic-ray-induced neutron as a function of neutron energy [11].

$$\alpha_{eff}(Si) = 1.8 \times 10^{-35} E^7 \qquad (1)$$

In Fig. 1, both FFR and CBS results are simultaneously plotted on the SEB curves' edges for three voltage classes (1200V, 1700V and 4500V) of IGBT modules including IGBT and diode chips. SEB characteristic curves are sandwiched by CBS results on the high voltage side and FFRs on the low voltage side. CBS results at the rated voltage indicate good agreement for SEB results for all the voltage class. Each FFR is carefully estimated as the constant random failure from the return back cases during several years, and their applying voltages are estimated as the relatively higher Vcc application case being clear as using conditions. In spite of the mentioned calculation procedure, FFRs of 1200V and 1700V modules seem to be overestimated and FFR of 4500V modules is much higher than others. This is because all FFRs still include another failure modes other than the cosmic ray origin.

978-1-5386-2928-4/18 $31.00 © 2018 IEEE

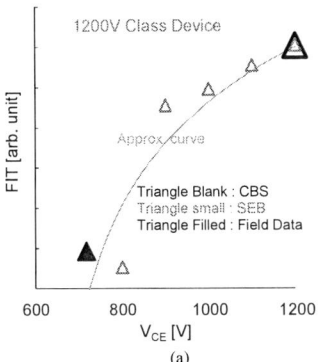

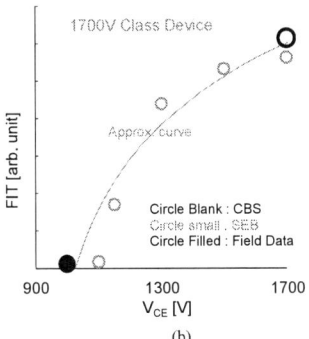

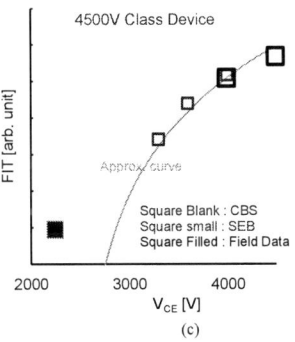

Fig. 1 SEB characteristics sandwiched by Cold Bias Stability (CBS) test on the high voltage side and the Field Failure Rate (FFR) on the low voltage side

The other important topic is whether "threshold" being or not in the low voltage side of the SEB curve. There are several reports for the threshold like a reference 3. Considering the mechanisms of SEB, there are no clear theory or physical evidences of this threshold [1, 5, 6, 7, 8, 11]. As the mathematical curve fitting procedure, some threshold would be one of the convenient parameter to be adapted in the above procedure in this case. One of the possibility is the rapid decrease of neutron flux distribution in the high energy region

[12], but there are another sight of the calibration of the flux distribution for the high energy neutrons [13]. During the experimental procedure for the SEB characterization, certainly there were time-limitation to wait SEB phenomena at the low voltage application, i.e. it was impossible to wait for several hours until SEB observation. This time limitation would more or less affect the FIT calculation, we also think there might be the experimental horizon for this kind of statistic or probability phenomena.

III. SEB FAIURE ANALYSIS

A. Analysis of Destrcution Cases

Among more than hundred chips of the destruction cases during SEB experiments, typical three categories (Fig. 2) are carefully picked up from the accumulation map (Fig. 3). Another voltage class devices had almost same features of the destruction spot pattern, in which almost all the spots were randomly located in the active region and a few spots were at the very inside of the termination region close to the active region, so 1700V class devices are shown as the typical example directly referring our previous study [14].

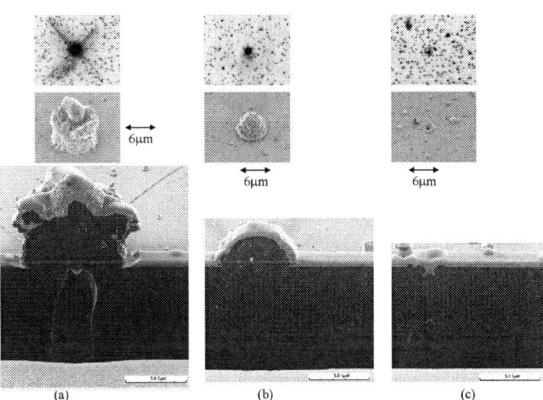

Fig. 2 Three categories of magnitude for SEB failures in the anode region of 1700V class diodes
(a) Relatively large case but further small than ref. 10
(b) Small case
(c) Slight case

Three categories of magnitude for SEB failures in the anode region of 1700V class diode are the followings, (a) relatively large case but further small than ref. 12, (b) small case and (c) slight case being unable to be identified in the surface image shown in the upper row (same as the noise (black spots) level). There are no examples to be pierced the entire chip thickness without exception by using the critical regulation resistor between the DUT and the DC power supply during the SEB experiment. Destruction phenomena were minimized, it was possible to analyze the profiles of destruction points. Those procedures were one of the key approach of our experiments. All the destruction centers are located around the pn main junction depth of several micrometers form the top surface. Upper row of Fig.2 is Superimposed photos of EMS (photo Emission Micro-Scope) image shown as red marker on IR (Infrared) image which can see through the entire Si thickness to detect the defect as shown in thick black circle. Middle row is Tilted SEM images at the Si surface after removing the Al electrode. Bottom row is the cross-sectional SEM images after cut by FIB across the defect position.

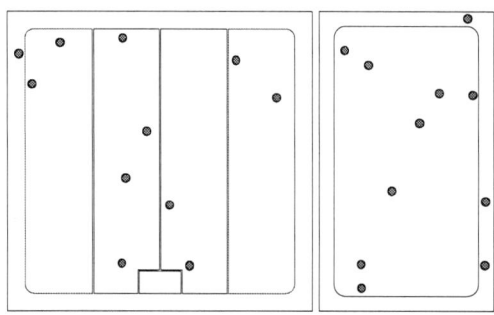

Fig. 3 Accumulation Map of destruction Points for IGBT and diode chips on the SEB curve of Fig.1 (b) 1700V class.

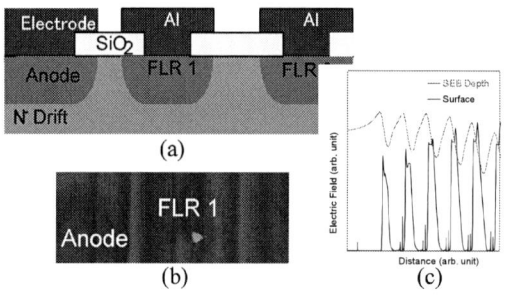

Fig. 4 SEB destruction example in the termination region at $V_{CC}=1200V$ and simulation also corresponding to the same V_{CC}
(a) Schematic illustration of FLR region
(b) EMS + IR image
(c) Simulated electric field distribution

In addition, the destruction point around the termination region is precisely analyzed. Fig.4 shows the SEB destruction example in the termination region at the bias of $V_{CC}=1200V$ (Fig. 4 (b)) with the schematic cross-sectional illustration (Fig. 4 (a)), and the simulation result also corresponding to the same V_{CC} (Fig. 4 (c)). The destruction point agrees reasonably well with the highest electric filed position of the first Field Limiting Ring (FLR 1) as shown in the simulation result (Fig. 4 (c)). This is because the depth of the destruction point of the

termination region is assigned at around the pn main junction as well as the cell region in Fig. 2 and 5. Since it is necessary to arrange the electric field as uniformly as possible in order to prevent the destruction around the termination region during the long term field application, the electric field strength are set forth to be lower in the Si surface than the pn junction depth inside of Si as shown in Fig. 4 (c).

 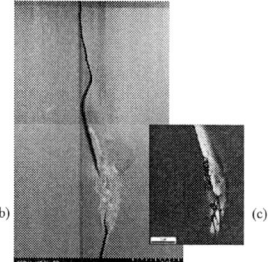

Fig. 5 SEB destruction depth analysis
(a) TEM
(b) ZC
(c) Grain Reference Orientation Deviation (GROD) image as the confirmation where zone melting re-crystallization occurred ZC: the Contrast proportional to the squared atomic number (Z) or HAADF (High-Angle Annular Dark Field)

B. Structural parameter dependence of The failure Rate

As the macroscopic point of view, almost all the experimental results support that SEB results are characterized by Ep or the Ionization Integral (I.I.) [7]. FIT of figure 6 is also normalized by the total chip area including the termination region and the peripheral parts like a gate pad and gate metal street in case of IGBT. Strictly speaking, this FIT normalization contains a little unfair factor for diode. Although a local structural parameter dependences, such as thickness or resistivity of n-drift for a certain device, are also described by Ep or I.I. from the microscopic point of view (Fig. 6), the sights of the medium range order, such as the large structural deference or "device difference" between IGBT and diode, are too complicated to explain along the single parameter of Ep or I.I. (Fig. 6). Considering the absolute thickness of diode being thinner than the IGBT's one, it is quite natural that the diodes' FIT is higher than the IGBTs' from the general point of view. Moreover, along our previous relative Ep [3] normalized by the active BV of device, FIT characteristics of both IGBT and diode is still almost identical. The universality of no clear difference between pin type device and pnp type device has been maintained [4, 14].

Although it has been still unclear to explain the medium range order correlation between structural parameters and SEB characteristics, the first major factor to characterize the SEB curve is confirmed to be the electric field strength through both SEB curve fitting and sufficient numbers of analysis for the destruction points.

978-1-5386-2928-4/18 $31.00 © 2018 IEEE   186

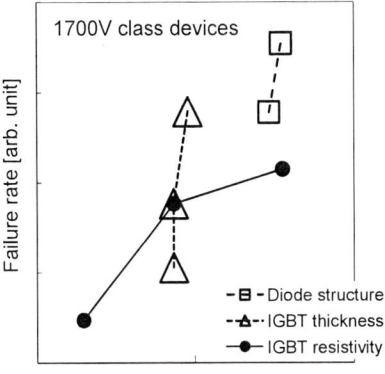

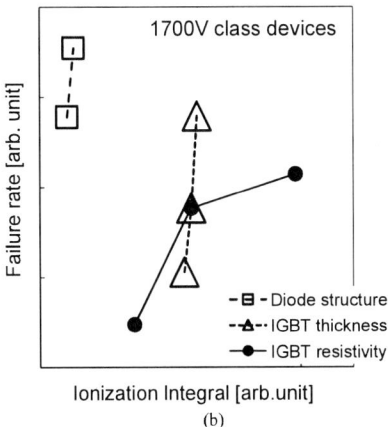

Fig. 6 Structural parameter dependence of the failure rate in the SEB experiment.
(a) Electric field peak (Ep)   (b) Ionization Integral (I.I.)

IV. CONCLUSION

We confirmed the relationship among Field Failure Rate (FFR), SEB characteristics and CBS. Through both SEB curve fitting and sufficient numbers of analysis for the destruction points, the first major factor to characterize the SEB curve is confirmed to be the electric field strength.

ACKNOWLEDGMENT

We thank Mr. H. Muraoka, Mr. M. Tabata, Mr. T. Marumo, Mr. Y. Konishi, Mr. K. Ishimoto, Mr. T. Negishi, Mr. T. Kanenari and Mr. J. Nakashima for their intensive support during the acceleration experiment and data analysis. We also thank Mr. Y. Yasuda, Mr. T. Kuroda, Mr. S. Iura, Mr. T. Shimizu and Mr. A. Furukawa to give us the opportunity and encouragement of this heavy experiment.

REFERENCES

[1] F. Hille, A. Härtl, G. Sölkner, C. Weiß, F. Pfirsch, H. Schulze and S. Aschauer "A Comparative Study of Cosmic Radiati on Hardness and Electrical Performance of Power Diode Concepts", Proc. ISPSD'16, pp.359-362.

[2] F. J. S. Rodriguez, D. Schloegl, F. Hille, P.C. Brandt, M. Pfaffenlehner, A. R. Stegner and A. Härtl, "Novel Emitter Controlled Diode with Copper Metallization in Ultrathin Wafer Technology: Setting a Performance Benchmark", Proc. ISPSD'17, pp.121-122.

[3] S. Nishida, T. Shoji, T. Ohnishi, T. Fujikawa, N. Nose, M. Ishiko and K. Hamada, "Cosmic Ray Ruggedness of IGBTs for Hybrid Vehicles", Proc. ISPSD'10, pp.129-132.

[4] T. Nitta, Y. Sakiyama, R. Kotani, T. Inoue, R. Ohara, K. Sano and M. Yamaguchi, "Cosmic Ray Failure Mechanism and Critical Factors for 3.3kV Hybrid SiC Modules", Proc. PCIM'16, pp.566-572.

[5] H. R. Zeller, "Cosmic Ray Induced Breakdown in High Voltage Semiconductor Devices, Microscopic Model and Phenomenological Lifetime Prediction", Proc. ISPSD'94, pp.339-340.

[6] N. Kaminski and A. Kopta, ABB application note 5SYA 2042-04 " Failure rates of HiPak modules due to cosmic rays".

[7] H. Kabza, H.-J. Schulze, Y. Gerstenmaier, F. Pfirsch and K. Platzoder, P. Voss, J. Wilhelmi and W. Schmid, "Cosmic Radiation as a cause for Power Device Failure and Possible Countemeasures", Proc. ISPSD'94, pp.9-12.

[8] F. Pfirsch and G. Soelkner, "Simulation of Cosmic Ray Failures Rates using Semiempirical Models", Proc. ISPSD'10, pp.125-128.

[9] C. Felgemacher, S. V. Araújo, P. Zacharias, K. Nesemann and A. Gruber, "Cosmic Radiation Ruggedness of Si and SiC Power Semiconductors", Proc. ISPSD'16, pp.51-54.

[10] W. Fulop, "Calculation of avalanche breakdown voltages of silicon p-n junctions", Solid-State Electronics, Volume 10, Issue 1, January 1967, pp.39-43.

[11] Y. Shiba, E. Dashdondog, M. Sudo and I. Omura, "Formulation of Single Event Burnout Failure Rate for High Voltage Devices in Satellite Electrical Power System", Proc. ISPSD'17, pp. 167-170

[12] Measurement and Reporting of Alpha Particle and Terrestrial Cosmic Ray-Induced Soft Errors in Semiconductor Devices, JEDEC Standard JESD89A, August 2001.

[13] M. S. Gordon, P. Goldhagen, K. P. Rodbell, T. H. Zabel, H. H. K. Tang, J. M. Clem and P. Bailey, "Measurement of the flux and energy spectrum of cosmic-ray induced neutrons on the ground," IEEE Trans. Nucl. Sci., Vol. 51, No. 6, pp. 3427-3434, 2004.

[14] Y. Yoshiura, M. Tabata, H. Muraoka, N. Taniguchi, K. Suzuki, S. Aono, M. Tarutani, T. Minato and K. Takakura, "Simple simulation approach for the first trigger step of SEB (Single Event Burn-out) based upon physical analysis for Si high voltage bipolar device", Proc. ISPSD'16, pp.315-318.

[15] T. Shoji, S. Nishida, K. Hamada, and H. Tadano, "Observation and Analysis of Neutron-Induced Single-Event Burnout in Silicon Power Diodes", Special Issue on Robust Design and Reliability in Power Electronics, 2015, pp. 1-6

978-1-5386-2928-4/18 $31.00 © 2018 IEEE

Proceedings of the 30th International Symposium on Power Semiconductor Devices & ICs
May 13-17, 2018, Chicago, USA

Gate architecture design for enhancement mode p-GaN gate HEMTs for 200 and 650V applications

N.E. Posthuma, S. You, S. Stoffels, H. Liang, M. Zhao and S. Decoutere

imec, Leuven, Belgium,

Email: niels.posthuma@imec.be

Abstract—— Enhancement mode p-GaN gate HEMTs with two different gate architectures are compared. The gate is realized by stacked (1-mask) or separate patterning (3-mask) of the p-GaN and gate metal layers. The 3-mask gate architecture, in this work implemented with a novel TiN interlayer, offers the advantage of a low gate resistance, increased flexibility in field plate design and reduced dynamic R_{DS-ON} at high V_{DS}. Both for 200 and 650 V applications excellent device performance is demonstrated on 200 mm substrates using Au-free processing, with a threshold voltage of well above 2 V and a dynamic R_{DS-ON} of below 20%. The 650 V rated device, with a hard breakdown voltage of 1000 V, passes the wafer level HTRB test at 150 °C.

Keywords—Power transistors, gallium-nitride, GaN-on-Si, HEMT, p-GaN gate, gate architecture

I. INTRODUCTION

High electron mobility transistors (HEMTs) based on GaN are typically normally-on devices. In circuits, normally-off power components, also called enhancement mode (e-mode) transistors, are preferred for fail-safe operation. To realize e-mode transistors, the gate region needs to be adapted to fully switch off the transistor at a gate voltage of 0 V. One of the approaches to obtain this is to use the Mg-doped p-type GaN (p-GaN) gate, lifting the conduction band in the channel at equilibrium, therefore realizing e-mode operation.

p-GaN gate HEMTs described in literature can be distinguished in two main groups, namely p-GaN gate HEMTs with a Schottky contact on the p-GaN layer [1,2,3] and the gate-injection transistor (GIT) [4, 5] with an ohmic contact to the p-GaN layer. In this work, we consider the p-GaN gate with a Schottky contact realized by TiN metal deposited on the p-GaN layer. The TiN to p-GaN Schottky diode is reverse biased during on-state operation of the transistor, reducing the gate leakage current compared to an ohmic contact on the p-GaN layer. Having a low gate leakage is important for the gate driver in the final application [6].

In this paper two options to define and fabricate the gate of p-GaN gate HEMTs are explored, highlighting the differences between these approaches. The goal of this work is to compare and screen the better one amongst the two considered gate architectures. In paragraph II, the gate architectures will be explained in detail. Electrical results and detailed discussion are presented in paragraph III. Extension of the p-GaN gate HEMT voltage range for 650 V applications is discussed in section IV.

II. P-GAN GATE ARCHITECTURES

A. Overview of the gate architectures

Fig. 1 shows the schematic illustrations and cross-section transmission electron microscope (TEM) micrographs of the two gate architectures considered in this study. The first, and most straightforward implementation of the p-GaN gate HEMT is the 1-mask approach, see Fig 1(a). In this approach, the TiN gate metal and p-GaN layer patterning are done in a single etch step, using one lithography mask. Here the gate metal and p-GaN layer have similar dimensions. The second architecture, illustrated in Fig. 1(b), is the 3-mask approach, using 3 lithography masks to pattern the p-GaN layer (1), to open the dielectric (2) and finally to pattern the gate metal layer (3). In this architecture, a thick Al containing gate metal stack is applied. In the presented implementation, a TiN interlayer is added on top of the p-GaN layer, below the gate metal stack. This novel interlayer has the advantage of protecting the p-GaN surface in the gate area during the device processing.

B. Device structure and fabrication

The growth of III-N layers has been carried out using metalorganic chemical vapor deposition (MOCVD) on 200 mm Si (111) wafers in a 3x200 mm Veeco Maxbright reactor. The p-GaN layer is grown in sequence with the GaN buffer, channel and barrier.

Transistor processing is done in imec's 200 mm pilot line using an Au-free process flow. In both architectures, a TiN layer is deposited directly on the p-GaN surface. Compared to the 1-mask approach, in the 3-mask approach a thinner TiN layer is applied. First the TiN layer is etched, followed by a selective p-GaN to AlGaN dry etch using a BCl_3/SF_6 chemistry. The access area is passivated by a dielectric stack containing Al_2O_3 and SiO_2. Specific for the 3-mask approach, the dielectric layer is etched open in the gate area followed by the deposition and patterning of the gate metal stack. The 3-mask gate processing is finalized by deposition of a second SiO_2 layer.

The remaining process sequence is similar for the two gate architectures. Ohmic contact formation of the source and drain contacts is done in sequence with the gate processing using a metal stack consisting of Ti, TiN and Al [7]. The process flow is completed by back-end metallization modules, where SiO_2 is used for the inter-metal dielectric layers.

978-1-5386-2928-4/18 $31.00 © 2018 IEEE

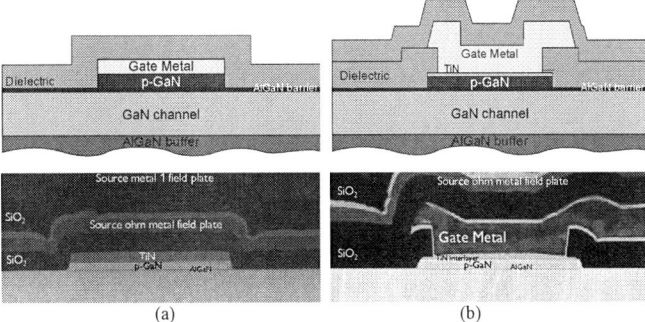

Fig. 1. Schematic illustration (top) and X-TEM micrograph (bottom) of the gate of p-GaN HEMT devices for the 1-mask gate architecture (a) and 3-mask gate architecture with TiN interlayer (b).

C. Comparison of the gate architectures

There are two major differences between the 1- and 3-mask approach that can be critical for device operation.

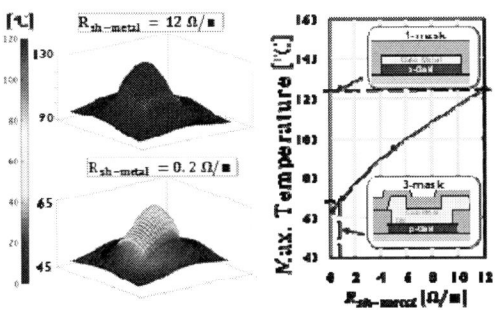

Fig. 2. Electro-thermal simulation of the 2d-temperature profile (left) and maximum operating temperature (right) as function of the gate metal sheet resistance $R_{sh-metal}$. Simulation conditions are for a 18 mm power transistor with 1 mm long gate finger design, real package configuration and 200 μm substrate thickness [8].

First of all, in the 1-mask approach the TiN layer, used for the formation of the Schottky contact, is relatively thin and only contains TiN (i.e. no Al) not to complicate the metal / p-GaN stacked etch sequence. However, as a result, the metal sheet resistance of the gate metal ($R_{sh-metal}$) is high, about 13 Ohm/sq. On the other hand, for the 3-mask approach a thicker, Al-containing gate metal stack is used. The gate metal stack is etched separately from the p-GaN layer, such that there is no impact on the p-GaN layer patterning. For the 3-mask architecture, $R_{sh-metal}$ is significantly lower, about 0.35 Ohm/sq. Depending on the device layout, p-GaN HEMTs with the 1-mask approach will have a high gate resistance (R_{gate}), which is detrimental for the switching speed and operating temperature of the device during switching. The impact of $R_{sh-metal}$ is simulated using electro-thermal modeling [8, 9], an example of the device operating temperature for different values of $R_{sh-metal}$ is shown in Fig. 2 [8]. Here, the worst-case scenario is simulated, where a device layout with long gate fingers is considered. The devices with the 1-mask approach can become twice as hot during operation compared to the 3-mask gate architecture, as indicated in Fig. 2.

A second difference between the two architectures is that a gate metal field plate can be implemented in the 3-mask approach. The gate metal field plate can be of importance to reduce the dynamic R_{DS-ON} and can also have an impact on the device breakdown voltage. In the 3-mask approach in total 4 metal field plates can be implemented, namely gate metal, ohm metal, metal1 and power metal field plates. In the 1-mask approach maximum 3 levels of field plates can be used.

III. RESULTS AND DISCUSSION: SIDE-BY-SIDE COMPARISON OF GATE ARCHITECTURE

A side-by-side comparison of the device performance of the two gate architectures is shown in Table 1 for 36 mm power transistors. The devices are realized on 200 mm substrates and target 200 V applications, with a gate-drain distance of 6 μm. The mean threshold voltage is well above 2 V, the on-resistance is 6 Ohm.mm. The application of the Ti/TiN/Al based contacts [7] results in good ohmic contacts with a contact resistance of 0.35 Ohm.mm. The $I_D V_{GS}$ curves are shown in Fig. 3 and the dynamic R_{on} dispersion is presented in Fig. 4.

TABLE I. OVERVIEW OF DEVICE PARAMETERS OF 36 MM POWER TRANSISTORS FOR THE 1- AND 3-MASK GATE ARCHITECTURE FOR 200 V APPLICATIONS. HIGH THRESHOLD VOLTAGE, LOW CONTACT RESISTANCE, ON-RESISTANCE AND LOW DYNAMIC R_{DS-ON} IS DEMONSTRATED.

Parameter	Unit	1-mask	3-mask
R_{sh}-gate metal	Ohm/sq	13	0.35
$V_{T at max gm}$	V	2.6	2.4
$V_{T at 10\mu A/mm}$	V	1.4	1.3
R_{on}	Ohm.mm	6.8	6.0
R_c	Ohm.mm	0.45	0.35
Dynamic R_{on}	%	< 20	< 20
Breakdown voltage	V	> 200 V	> 200V

As shown in Table 1, the device parameters for power transistors targeting 200 V applications are very similar for the two gate architectures. The main difference is the resistivity of the gate metal, resulting in a significantly lower $R_{sh-metal}$ for the 3-mask approach. No obvious difference is observed in dynamic R_{DS-ON} between the two architectures. Excellent results with a normalized dynamic R_{DS-ON} below 20 percent can be obtained with and without gate metal field plate for both 25 and 150 °C up to 200 V (Fig. 4).

978-1-5386-2928-4/18 $31.00 © 2018 IEEE

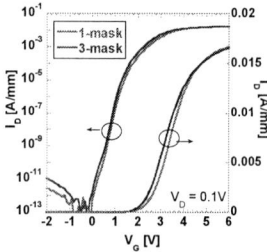

Fig. 3. $I_D V_{GS}$ curves (at $V_{DS} = 0.1$ V, 25 °C) of the realized 36 mm power transistors for the two gate architectures for 200 V applications, showing the high threshold voltage and drive current.

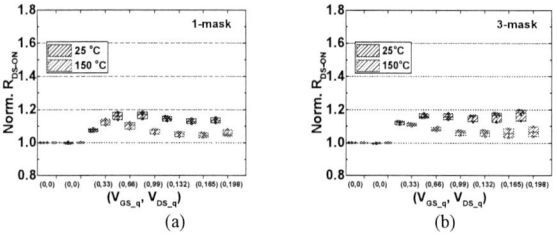

Fig. 4. Normalized dynamic R_{DS_ON} demonstrated below 20% for 25 and 150 °C with excellent uniformity over the full 200 mm wafer for the 1-mask (a) and 3-mask (b) gate architecture, measured using an Auriga AU4850 system for VDs_q from 0 to 200 V.

Both types of devices were subjected to wafer level high temperature reverse bias (HTRB) stress testing for 3 hours, at a drain voltage of 160 V (80 percent of the rated voltage) at a chuck temperature of 150 °C. As shown in Fig. 5, the devices passed HTRB testing, showing minor shift in the IV-curves after stress. Both gate architectures show similar behavior during and after the HTRB test.

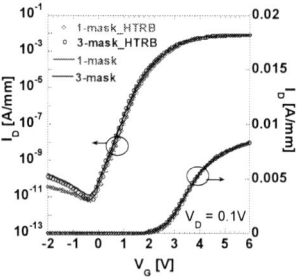

Fig. 5. $I_D V_{GS}$ curves (at $V_{DS} = 0.1$ V, 150 °C) before and after HTRB testing (3 hours, $V_{DS} = 160$ V, 150 °C), showing that no shift in IV behavior is observed after HTRB stress.

IV. P-GAN GATE HEMT FOR 650V APPLICATIONS

Starting from the technology for 200 V applications, the development to realize a power device for a voltage application of 650 V is focused on the GaN buffer stack, AlGaN barrier, the device layout, field plate configuration, and the thickness of the dielectric layers. Maximum 4 levels of metal field plates are applied, namely gate metal, ohm metal, metal1 and power metal field plates.

TABLE II. OVERVIEW OF REALIZED DEVICE PARAMETERS OF 36 MM POWER TRANSISTORS FOR 3-MASK GATE ARCHITECTURE FOR 650 V APPLICATIONS. HIGH THRESHOLD VOLTAGE, LOW CONTACT RESISTANCE, ON-RESISTANCE AND LOW DYNAMIC R_{DS-ON} IS DEMONSTRATED.

Parameter	Unit	3-mask
R_{sh}-gate metal	Ohm/sq	0.35
$V_{T\,at\,max\,gm}$	V	2.9
$V_{T\,at\,10\mu A/mm}$	V	1.8
R_{on}	Ohm.mm	16.1
R_c	Ohm.mm	0.3
Dynamic R_{on}	%	< 20
Breakdown voltage	V	> 650V

An overview of the realized device parameters of a 36 mm power transistor with the 3-mask gate architecture is shown in Table 2. The mean threshold voltage is further improved to 2.8 V and an on-resistance of 16.1 Ohm.mm is obtained for a gate drain distance of 16 µm. The contact resistance is as low as 0.3 Ohm.mm. The $I_D V_{DS}$ and $I_D V_{GS}$ curves of a typical device are shown in Fig. 6 (a) and (b), respectively.

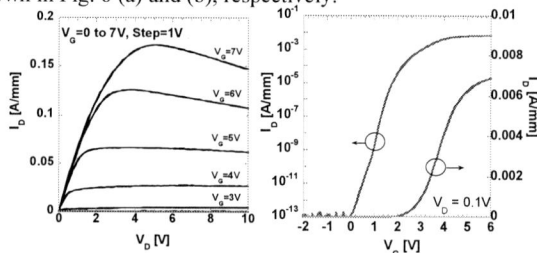

Fig. 6. $I_D V_{DS}$ curve of the realized 36 mm power transistors for 650 V applications, with V_{GS} stepped from 0 to 7 V (a), and $I_D V_{GS}$ curves (at $V_{DS}=0.1$ V, 25 °C), indicating the high threshold voltage and drive current.

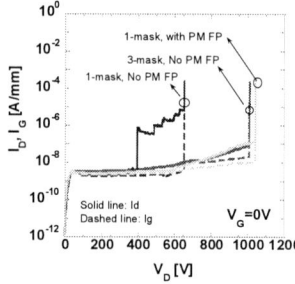

Fig. 7. Device breakdown of a 36 mm power transistor reaching a hard breakdown of 1000 V for the 3-mask gate architecture, both the I_G and I_D current are shown, $V_{GS}=0$ V.

The device breakdown curves of the 36 mm power transistor comparing the 1- and 3-mask gate architecture is shown in Fig 7. The device with the 1-mask approach without gate field plate and without power metal (PM) field plate breaks relatively early at 600 V. Both for the 1-mask approach with an additional PM field plate and the 3-mask approach with a gate metal field plate, a hard device breakdown voltage of 1 kV is obtained. This data shows that the gate field plate present in the 3-mask gate architecture, can offer additional flexibility in the field plate design for high voltage applications.

978-1-5386-2928-4/18 $31.00 © 2018 IEEE

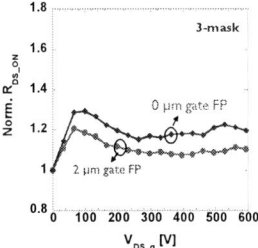

Fig. 8. Normalized dynamic R_{DS-ON} on 1-finger transistors comparing a gate field plate length of 0 and 2 μm for V_{DS-q} from 0 to 600 V.

The impact of the gate field plate length on the dynamic R_{DS-ON} is determined using small, one finger transistors, comparing a gate field plate aligned to the edge of the p-GaN layer (field plate length of 0 μm) and a field plate extending 2 μm over the edge of the p-GaN, see Fig 8. As shown, the measured device dispersion is improved for the device with the long gate field plate. Further optimization of the field plate design of a 36 mm power transistor has resulted in a normalized dynamic R_{DS-ON} of around 10 percent, as shown in Fig. 9.

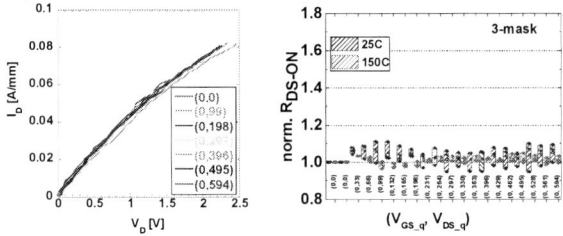

Fig. 9. A normalized R_{DS-ON} of 10% is demonstrated for 25 and 150 °C of 36 mm p-GaN gate HEMTs with a tight distribution over the 200 mm wafer up to 600 V. The shown figures are the pulsed $I_D V_{DS}$ curves (left) and the box plot of normalized dynamic R_{DS-ON} (right) measured using an Auriga AU4850 system for V_{DS-q} from 0 to 600 V.

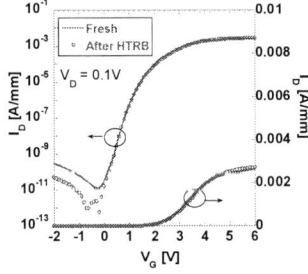

Fig. 10. $I_D V_{GS}$ curves (at V_{DS}=0.1 V, 150 °C) before and after HTRB testing (3 hours, V_{DS}=520V, 150 °C), no shift in IV behavior is observed.

The power transistors designed for 650 V applications are tested under HTRB stress conditions, for 3 hours, at a drain voltage of 520 V (80 percent of the rated voltage) at a chuck temperature of 150 °C. As shown in Fig. 10, the device passed HTRB testing, showing minor shift in the IV-curves after stress.

V. CONCLUSION

p-GaN gate HEMT power transistors with two different gate architectures have been fabricated on 200 mm GaN-on-Si substrates and the electrical performance is compared. The 1-mask approach has the advantage of a shorter process flow, at the cost of a high resistivity of the gate metal. The 3-mask gate architecture is more complicated to realize, but offers the clear advantage of a much lower gate metal sheet resistance. The lower gate metal sheet resistance is important for the final gate resistance and resulting switching behavior of the device. The application of the 3-mask gate architecture offers a larger flexibility in field plate design and reduced dynamic R_{on} device dispersion for 650 V applications.

Both for 200 and 650V applications excellent device performance is demonstrated, with a threshold voltage well above 2 V and a dynamic R_{on} dispersion below 20%. The 650 V rated devices, with a hard breakdown voltage of 1000 V, passes the wafer level HTRB test.

ACKNOWLEDGMENT

The authors would like to thank the imec industrial affiliation GaN Power program partners for their contribution.

This project has received funding from the Electronic Component Systems for European Leadership Joint Undertaking under grant agreement No 662133. This Joint Undertaking receives support from the European Union's Horizon 2020 research and innovation program and Austria, Belgium, Germany, Italy, Netherlands, Norway, Slovakia, Spain, United Kingdom. This article reflects only the authors' view and the JU is not responsible for any use that may be made of the information it contains.

REFERENCES

[1] O Hilt, "Normally-off GaN Transistors for Power Applications" Journal of Physics: Conference Series 494 (2014), 012001

[2] J. Kim, "High Threshold Voltage p-GaN Gate Power Devices on 200 mm Si", ISPSD 2013, pp. 315

[3] N.E.. Posthuma, "Impact of Mg out-diffusion and activation on the p-GaN gate HEMT device performance" ISPSD 2016, pp. 95

[4] Uemoto, "Gate Injection Transistor (GIT) - A Normally-Off AlGaN/GaN Power Transistor Using Conductivity Modulation", IEEE TED (2007), 54(12), pp. 3393–3399.

[5] H. Okita, "Through Recessed and Regrowth Gate Technology for Realizing Process Stability of GaN-GITs", ISPSD 2016, pp. 23–26.

[6] Wang B., "An Efficient High-Frequency Drive Circuit for GaN Power HFETs", IEEE Transactions on Industry Applications (2009), 45(2), pp. 843–853.

[7] Patent US 9,634,107 B2

[8] V. Sodan, "Distributed electro-thermal model based on fast and scalable algorithm for GaN power devices and circuit simulations", ISPSD 2017, pp. 179

[9] Chvála, Aleš, et al. "Effective 3-D Device Electrothermal Simulation Analysis of Influence of Metallization Geometry on Multifinger Power HEMTs Properties." IEEE TED 64.1 (2017): 333-336.

Proceedings of the 30th International Symposium on Power Semiconductor Devices & ICs
May 13-17, 2018, Chicago, USA

Uni-directional GaN-on-Si MOSHEMTs with High Reverse-Blocking Voltage based on Nanostructured Schottky Drain

Jun Ma and Elison Matioli

École polytechnique fédérale de Lausanne (EPFL),
Lausanne, CH-1015, Switzerland.
E-mail: jun.ma@epfl.ch, elison.matioli@epfl.ch

Abstract—**In this work we present uni-directional GaN-on-Si MOSHEMTs with state-of-the-art reverse-blocking performance. We integrated tri-anode Schottky barrier diodes (SBDs) with slanted tri-gate field plates (FPs) as the drain electrode, and achieved a high reverse-blocking voltage (V_{RB}) of -759 ± 37 V at 0.1 µA/mm with grounded substrate. The hybrid Schottky drain did not degrade the ON-state performance when compared with conventional ohmic drain, and the turn-ON voltage (V_{ON}) was as small as 0.64 ± 0.02 V. These results show the potential of GaN-on-Si transistors as high-performance uni-directional power switches, and open enormous opportunities for future highly integrated GaN power devices.**

Keywords—*GaN; HEMTs; SBDs; reverse blocking; slanted tri-gate; tri-anode.*

I. INTRODUCTION

Lateral GaN-on-Si HEMTs are very promising for power applications, yet their bi-directional conduction is not ideal in many topologies of power converters. To achieve high-voltage reverse-blocking (RB) capability, a power diode is added in series with the HEMT, which complicates the circuit design, increases the ON-resistance (R_{ON}) and parasitic elements, and degrades the efficiency of power conversion. HEMTs with integrated RB capability (RB-HEMTs) would therefore be highly desirable to address these issues, however, in the few reports on RB-HEMTs in the literature [1]-[6], the devices exhibited a relatively large V_{ON}, a significant increase in forward voltage (ΔV_F), a small V_{RB} as well as a large reverse leakage current (I_R), limited by the SBDs integrated in their drain electrodes.

Recently we have proposed to reduce the V_{ON} and I_R in GaN SBDs using a tri-anode structure [8], [9], and presented its capabilities for GaN RB transistors [10]. In addition, we demonstrated a slanted tri-gate structure to improve the breakdown voltage of lateral GaN devices [11], [12]. These results paved the path for high-performance GaN RB-HEMTs from both theoretical and experiment aspects. In this work we extend these concepts to demonstrate high-performance GaN MOSHEMTs on Si with state-of-the-art reverse-blocking

This work was supported in part by the European Research Council under the European Union's H2020 program/ERC Grant Agreement 679425 and in part by the Swiss National Science Foundation under Assistant Professor (AP) Energy Grant PYAPP2_166901.

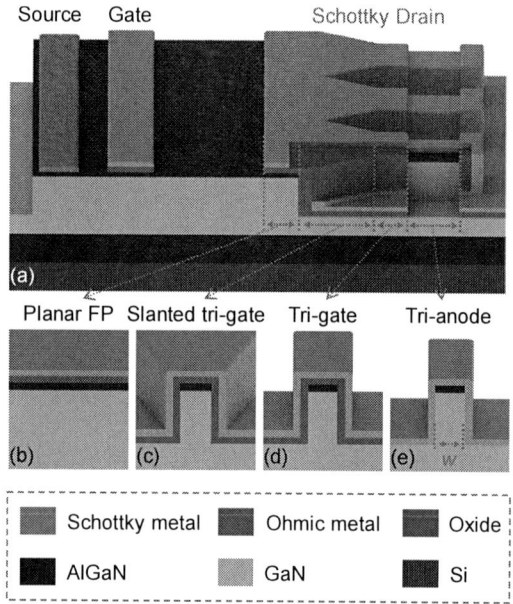

Fig. 1. (a) Schematic of the MOSHEMTs with the slanted tri-gate Schottky drain. Cross-sectional schematics of the (b) planar FP, (c) slanted tri-gate (sTG), (d) tri-gate (TG) and (e) tri-anode (TA) regions.

performance, using a novel tri-anode Schottky drain integrated with slanted tri-gate FPs. The devices presented a high V_{RB} of -759 ± 37 V at 0.1 µA/mm and a small I_R of 65 ± 11 nA/mm at -700 V, measured with grounded substrate. The V_{ON} was 0.64 ± 0.02 V and the ΔV_F was as small as 0.7 V, with little degradation in R_{ON} compared to devices with conventional ohmic contacts.

II. DEVICE FABRICATION

The devices with the tri-anode Schottky drain (tri-SCH) were fabricated on an AlGaN/GaN-on-silicon wafer, and their schematics are shown in Fig. 1(a)-(e). The fabrication process started with e-beam lithography to define the nanowires, which were then etched by inductively coupled plasma with a depth of ~180 nm. The width (w) and spacing (s) for the nanowires in the tri-gate and tri-anode region were both 300 nm, while the w in the slanted tri-gate region increased continuously from 300

978-1-5386-2928-4/18 $31.00 © 2018 IEEE

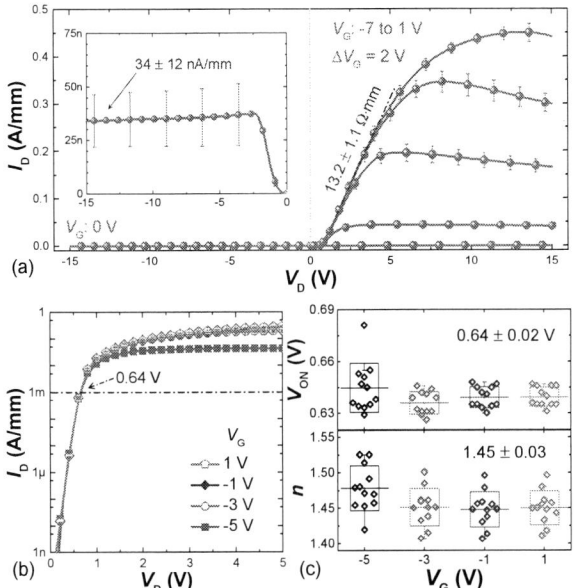

Fig. 2. (a) Output characteristics of devices with the slanted tri-gate SCH drain (tri-SCH), in which the inset shows the I_R in the tri-SCH at $V_G = 0$ V, and (b) their turn-on characteristics. (c) Dependence of the V_{ON} and the ideality factor (n) on V_G.

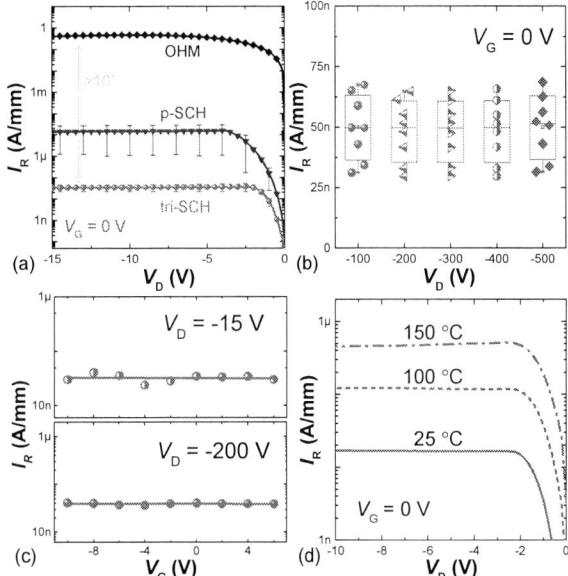

Fig. 3. (a) Comparison of I_R in devices with ohmic (OHM), planar Schottky (p-SCH) and tri-anode Schottky (tri-SCH) drains. Dependence of the I_R on V_D (b) and V_G (c) in the tri-SCH. (d) Dependence of the I_R in tri-SCH on temperature.

nm to 600 nm towards the gate. The devices were isolated from each other by mesa etching with a depth of ~350 nm, followed by deposition and annealing of ohmic metals as source electrodes. A stack of 10 nm SiO_2 and 10 nm Al_2O_3 was deposited by atomic layer deposition as the gate dielectric, and then selectively removed in the tri-anode region. Finally the tri-anode and the gate were formed using Ni/Au, which was later used as the mask to remove the oxide in access/ohmic regions. The gate-to-source length (L_{GS}), gate length (L_G) and gate-to-drain length (L_{GD}) were 1.5 μm, 2.5 μm and 12.5 μm, respectively. The lengths for the planar FP (L_{FP}), slanted tri-gate (L_{sTG}), and tri-gate (L_{TG}) regions were 1.3 μm, 0.7 μm and 0.5 μm, respectively.

The working principle of the tri-SCH can be simply summarized as follows. The hybrid tri-anode SBDs pins the voltage drop at the Schottky junction and hence fixes the I_R at a small level [8],[9], despite the large reverse bias. On the other hand, the slanted tri-gate effectively spreads the electric field [11] and, together with the planar FP region [13], improves the V_{RB}, which is similar to conventional slant FPs but in a more controllable manner. More detailed analysis of this structure can be found in Ref. [12].

MOSHEMTs with the same dimensions but conventional ohmic (OHM) and planar Schottky drain (p-SCH) electrodes were also fabricated on the same chip as references. All devices shared the same dimensions as the tri-SCH except for the different drain electrodes. The ohmic drain was formed by alloying Ti/Al/Ti/Ni/Au, which was the same as the source electrode. The planar Schottky drain was the same as the tri-anode Schottky drain, however without the patterned features. All current values in this work, such as I_R and drain current (I_D), were normalized by the width of the device footprint,

which was 60 μm. Twelve devices of each type were randomly chosen for the investigation, which defined the error bars presented in the results.

III. RESULTS AND DISCUSSION

Figure 2(a) shows the output characteristics of the tri-SCH, presenting its excellent performance as a uni-directional switch. Under forward V_D, the differential R_{ON} and the maximum I_D were 13.2 ± 1.1 Ω·mm and 450 ± 17 mA/mm, respectively. Under reverse biases, the I_R was 34 ± 12 nA/mm at $V_D = -15$ V and $V_G = 0$ V. The V_{ON} was as small as 0.64 ± 0.02 V (Fig. 2(b)) at $I_D = 1$ mA/mm, due to the direct contact of the metal to the 2DEG at the sidewalls of the nanowire [14],[15]. The ideality factor (n) was 1.45 ± 0.03, indicating the high quality of the Schottky contact despite the etching. Figure 2(c) plots the V_{ON} and n at different V_G, revealing very little dependence on V_G.

The I_R in OHM, p-SCH and tri-SCH are compared in Fig. 3(a), all measured at $V_G = 0$ V. While the bi-directional nature of the OHM resulted in large reverse currents, reverse-blocking capability was achieved in p-SCH by replacing the ohmic drain with a planar Schottky drain. However, a large I_R was observed, similarly to other reports in the literature, leading to a small V_{RB} (defined at 0.1 μA/mm) of -0.9 V. This reveals the unsuitability of the planar Schottky drain for practical and efficient power applications. The I_R was dramatically reduced by over two orders of magnitude in tri-SCH. This is because the voltage drop at the Schottky junction (V_{SCH}) was pinned at a smaller value in the tri-anode, compared with the planar Schottky structure. As V_{SCH} was fixed and did not increase with the reverse bias, the I_R was

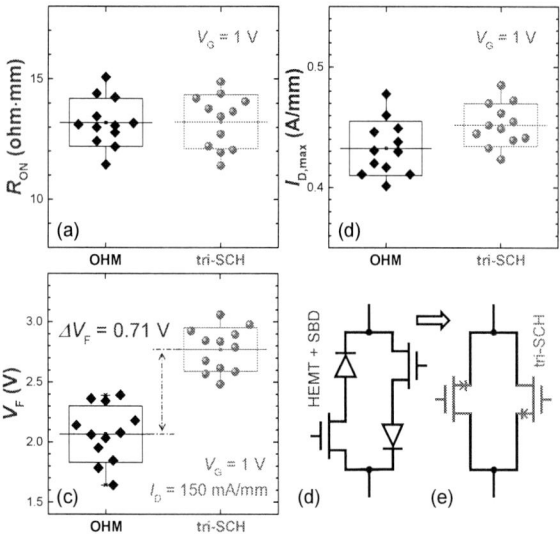

Fig. 4. Comparison of the tri-SCH and OHM in (a) R_{ON}, (b) maximum I_D ($I_{D,max}$) and (c) forward voltage (V_F). Examples of bi-directional power switches using a conventional scheme including two HEMTs and two SBDs (d) and a more advanced scheme using only two reverse-blocking transistors as demonstrated in this work (tri-SCH) (e).

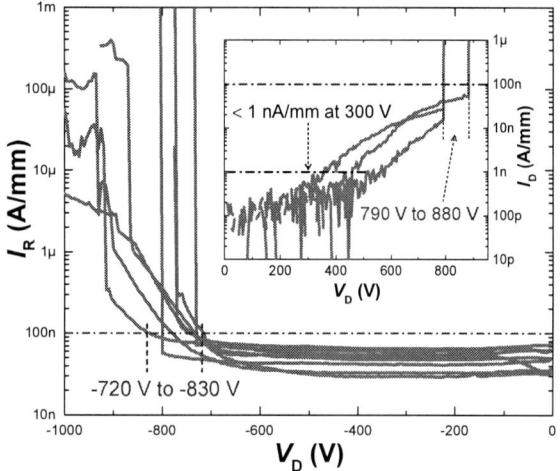

Fig. 5. Room-temperature breakdown characteristics of the tri-SCH, measured with grounded substrate. The V_G for measuring the V_{RB} and V_B was 0 V and -10 V, respectively.

constant even at high reverse biases (Fig. 3(b)), and was not affected by the V_G for a large range of voltages from -10 V to 6 V (Fig. 3(c)). Another improvement with the tri-anode Schottky drain was the better uniformity of the I_R. The variation of the I_R is about three orders of magnitude for the p-SCH, while less than 50 nA/mm for the tri-SCH. As shown in Fig. 3(d), the I_R of the tri-SCH was small and below 1 µA/mm at even 150 °C, revealing the potential of this structure for high-temperature applications.

In addition to its excellent reverse blocking capability, the tri-anode Schottky drain did not degrade the ON-state characteristics of the transistors. The R_{ON} and I_D of the tri-

Table 1. Comparison of the tri-SCH with reverse-blocking GaN (MOS)HEMTs in the literature.

	Substrate	V_{RB}	I_R (µA/mm)	V_{ON} (V)	ΔV_F (V)
This work	Si	-759 ± 37 V at 0.1 µA/mm (grounded sub.)	0.065 ± 0.011 at -700 V (grounded sub.)	0.64 ± 0.02	0.7
[1]	Si	-321 V at 1 mA/mm (floating sub.) -200 V at 1 mA/mm (grounded sub.)	≥ 10 at -75 V (floating sub.)	0.55	1.25
[2]	SiC	-110 V at 10 mA/mm	≥ 1000 at -20 V	--	--
[3]	Si	-685 V at hard breakdown [a]	~ 6 at -100 V [a]	0.4	--
[4]	Al$_2$O$_3$	-49 V at 1 mA/mm	> 100 at -25 V	1.7	≥ 2
[5]	Si	--	~0.4 at -20 V (floating sub.)	1.91	--
[7]	Si	-650 V at ~0.15 mA/mm [a]	--	1.5	--
[6]	Si	-900 V at 1 µA/mm [b]	~0.25 µA/mm at -700 V [b]	0.38	--

[a] Substrate connection not reported
[b] Simulation results

SCH were about the same as those of the OHM (Fig. 4(a) and (b)). The forward voltage (V_F) for the tri-SCH and the OHM was 2.77 ± 0.17 V and 2.07 ± 0.17 V, respectively, extracted at I_D = 150 mA/mm, rendering a small ΔV_F of 0.7 V (Fig. 4(c)) which was very close to the V_{ON} of the tri-anode SBD. This is very important to improve the efficiency while reducing the size and complexity of power converters. For instance, the number of components in a bi-directional power switch can be reduced from four to two using the tri-SCH (Fig. 4(d) and (e)), and the resistive loss from the SBDs can be eliminated as the ΔV_F is so close to the V_{ON} and the R_{ON} of the tri-SCH is about the same as that of the OHM.

Figure 5 shows the breakdown characteristics of the tri-SCH under both forward and reverse drain voltages. The breakdown voltage in this work was defined at a leakage current of 0.1 µA/mm with the Si substrate grounded. The reverse breakdown voltage (V_{RB}) varied from -720 V to -830 V under V_G = 0 V, along with a consistently small I_R of 65 ± 11 nA/mm at -700 V. The forward breakdown voltage (V_B) varied from 790 V to 880 V (the inset of Fig. 5), which was measured under a V_G of -10 V.

The tri-SCH was compared with other reverse-blocking GaN transistors in the literature in Tab. 1. The tri-SCH presented the highest V_{RB}, the lowest I_R and the smallest ΔV_F, despite the grounded substrate and the much stricter definition of V_{RB} in this work, along with a small V_{ON} comparable to the state-of-the-art results.

The forward and reverse performance of the tri-SCH were further benchmarked against state-of-the-art discrete lateral GaN-on-Si power (MOS)HEMTs and SBDs in Fig. 6, respectively. For calculation of the figure-of-merit (FOM) in this work, average R_{ON} (13.2 ± 1.1 Ω·mm), V_{RB} (-759 ± 37 V at 0.1 µA/mm) and V_B (820 ± 42 V at 0.1 µA/mm) were adopted, along with a total transfer length of 3 µm, accounting for both source and drain contacts. The tri-SCH presented high

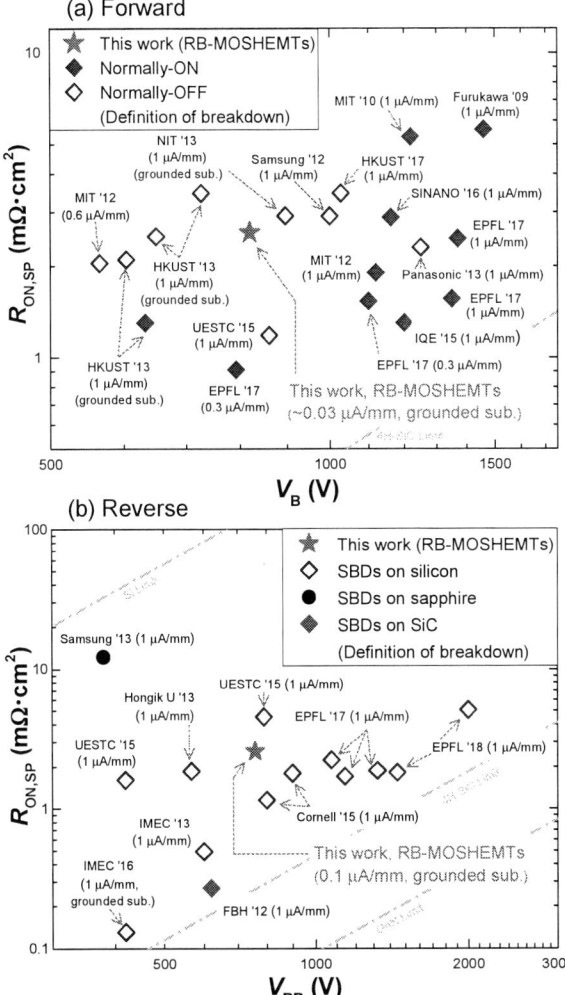

(a) Forward

(b) Reverse

Fig. 6. $R_{ON,SP}$ versus breakdown voltage benchmarks of (a) the forward performance of the tri-SCH (RB-MOSHEMTs) against discrete lateral GaN-on-silicon power (MOS)HEMTs and (b) their reverse performance against discrete lateral GaN SBDs on various substrates. The breakdown voltage for all reference devices was re-calculated based on the reported data following the definition of V_B at $I_{OFF} \leq 1$ μA/mm. For fair comparison, literature results with unspecified R_{ON} or I_R were not included.

V_B and V_{RB}, comparable to state-of-the-art discrete devices measured with grounded substrates, but at a much smaller current, revealing its excellent performance as uni-directional power switches. More importantly, these high blocking voltages under both forward and reverse biases were achieved in a single integrated device, instead of using a discrete transistor in series with an SBD, which can greatly simplify the circuit design, reduce its size, resistance and parasitic components, and improve the efficiency of power converters.

IV. CONCLUSION

In this work we presented high-performance reverse-blocking GaN power MOSHEMTs on Si, by integrating tri-anode SBDs and slanted tri-gate FPs as the drain electrode. The device presented high V_{RB} of -759 ± 37 V at 0.1 μA/mm with grounded substrate, along with small V_{ON} and zero degradation in R_{ON} and I_D. These results can potentially enable GaN transistors as uni-directional power switches for power conversion and open enormous opportunities for highly integrated GaN power devices for future advanced power applications.

REFERENCES

[1] C. Zhou, W. Chen, E. L. Piner and K. J. Chen, "Schottky-ohmic drain AlGaN/GaN normally off HEMT with reverse drain blocking capability," *IEEE Electron Device Lett.*, vol. 31, no. 7, pp. 668-670, Jul. 2010.

[2] E. Bahat-Treidel, R. Lossy, J. Wurfl and G. Trankle, "AlGaN/GaN HEMT with integrated recessed Schottky-drain protection diode," *IEEE Electron Device Lett.*, vol. 30, no. 9, pp. 901-903, Sept. 2009.

[3] J. -G. Lee, S. -W. Han, B. -R. Park, and H. -Y. Cha, "Unidirectional AlGaN/GaN-on-Si HFETs with reverse blocking drain," *Appl. Phys. Express*, vol. 7, pp. 014101-1-014101-4, Dec. 2013.

[4] S. -L. Zhao, M. -H. Mi, B. Hou, J. Luo, Y. Wang, Y. Dai, J. -C. Zhang, X. -H. Ma, and H. Yue, "Mechanism of improving forward and reverse blocking voltages in AlGaN/GaN HEMTs by using Schottky drain," *Chinese Phys. B*, vol. 23, pp. 107303-1-107303-5, Aug. 2014.

[5] A. Taube, J. Kaczmarski, R. Kruszka, J. Grochowski, K. Kosiel, J. Gołaszewska-Malec, M. Sochacki, W. Jung, E. Kamińska, and A. Piotrowska, "Temperature-dependent electrical characterization of high-voltage AlGaN/GaN-on-Si HEMTs with Schottky and ohmic drain contacts," *Solid-State Electron.*, vol. 111, pp. 12-17, Sept. 2015.

[6] Y. Shi, W. Chen, C. Liu, G. Hu, J. Liu, X. Cui, H. Tao, J. Zhang, Y. Shi, A. Zhang, Z. Li, Q. Zhou, and B. Zhang, "A high-performance GaN E-mode reverse blocking MISHEMT with MIS field effect drain for bidirectional switch," *2017 29th International Symposium on Power Semiconductor Devices and IC's (ISPSD)*, Sapporo, 2017, pp. 207-210.

[7] T. Morita, M. Yanagihara, H. Ishida, M. Hikita, K. Kaibara, H. Matsuo, Y. Uemoto, T. Ueda, T. Tanaka, and D. Ueda, "650 V 3.1 mΩ·cm² GaN-based monolithic bidirectional switch using normally-off gate injection transistor," *2007 IEEE International Electron Devices Meeting (IEDM)*, Washington, DC, 2007, pp. 865-868.

[8] J. Ma and E. Matioli, "High-Voltage and Low-Leakage AlGaN/GaN Tri-Anode Schottky Diodes With Integrated Tri-Gate Transistors," *IEEE Electron Device Lett.*, vol. 38, pp. 83-86, Jan. 2017.

[9] J. Ma, D. C. Zanuz and E. Matioli, "Field Plate Design for Low Leakage Current in Lateral GaN Power Schottky Diodes: Role of the Pinch-off Voltage," *IEEE Electron Device Lett.*, vol. 38, pp. 1298-1301, Sept. 2017.

[10] J. Ma, M. Zhu and E. Matioli, "900 V Reverse-Blocking GaN-on-Si MOSHEMTs With a Hybrid Tri-Anode Schottky Drain," *IEEE Electron Device Lett.*, vol. 38, pp. 1704-1707, Dec. 2017.

[11] J. Ma and E. Matioli, "Slanted Tri-Gates for High-Voltage GaN Power Devices," *IEEE Electron Device Lett.*, vol. 38, pp. 1305-1308, Sept. 2017.

[12] J. Ma and E. Matioli, "2 kV slanted tri-gate GaN-on-Si Schottky barrier diodes with ultra-low leakage current," *Appl. Phys. Lett.*, vol. 112, pp. 052101-1-052101-4, Jan. 2018.

[13] J. Ma and E. Matioli, "High Performance Tri-Gate GaN Power MOSHEMTs on Silicon Substrate," *IEEE Electron Device Lett.*, vol. 38, no. 3, pp. 367-370, Mar. 2017.

[14] E. Matioli, B. Lu and T. Palacios, "Ultralow Leakage Current AlGaN/GaN Schottky Diodes With 3-D Anode Structure," *IEEE Trans. Electron Devices*, vol. 60, no. 10, pp. 3365-3370, Oct. 2013.

[15] J. Ma, G. Santoruvo, P. Tandon and E. Matioli, "Enhanced Electrical Performance and Heat Dissipation in AlGaN/GaN Schottky Barrier Diodes Using Hybrid Tri-anode Structure," *IEEE Trans. Electron Devices*, vol. 63, no. 9, pp. 3614-3619, Sept. 2016.

Proceedings of the 30th International Symposium on Power Semiconductor Devices & ICs
May 13-17, 2018, Chicago, USA

Characterization of GaN-HEMT in Cascode Topology and Comparison with State of the Art-Power Devices

Sven Buetow, Reinhard Herzer

SEMIKRON Elektronik GmbH & CoKG, Germany; Sigmundstrasse 200, D-90431 Nuremberg
Tel.: +49-911-6559-406, Email: sven.buetow@semikron.com

Abstract— The paper presents a fully static and dynamic characterization and also reliability investigations of a 650V, 28mΩ GaN-HEMT in Cascode-topology. To show the overall performance of GaN-HEMT the output current per chip area (A per mm^2) versus switching frequency of a three phase inverter is presented for the 650V GaN-HEMT in comparison to 650V Si-IGBT3 (IFX) and Si-CooLMOS C7 (IFX, with fast Si-FWD and SiC-FWD).

Keywords—GaN-HEMT, 650V, Cascode-topology, gate driver, GaN-HEMT comparison to silicon devices

I. INTRODUCTION

The benefit of using high switching frequencies for power inverters is that they allow their output voltage and current to fit more accurately to an ideal sinus waveform. The high ratio between switching-frequency and output-frequency reduces size, losses, cost and weight of filter and DC-link components. But switching losses increase linearly with the switching frequency and using this approach means a significant drop of the achievable output current of the inverter. The solution that overcomes this problem is the use of GaN-HEMTs [1] (≤ 650V) or SiC-MOSFETs [2]-[4] (≥1200V) and to drive it in a way that the switching losses are significantly reduced in comparison to Si-devices (CoolMOS or IGBT respectively). Making use of the outstanding performance of GaN-devices it is extremely important to improve their application and system environment. This imposes the demand for low inductivities and high temperature operation (≥175°C). On the driver side highly integrated gate drivers with driving and monitoring functions for higher frequencies and higher operation temperatures are necessary. But actually industrial customers prefer to compare wide bandgap devices to silicon devices in conventional packages to get quick results and to reduce the assembly and test effort [5]. So the focus of this paper is also the usage of GaN in a conventional package.

II. GaN-DEVICE AND ITS ASSEMBLY

The 650V, 28mΩ GaN-HEMT [4] is assembled in a MiniSKiiP2-module in Cascode Light-topology [6] with back side glueing and front side bonding (see Fig.1). The normally-on GaN-HEMT has a V_{th} of approx. -20V and an operating voltage range of V_{GS}= 0...-30V. To use the normally-on HEMT in inverter operation a low voltage (LV) pMOSFET (V_{br}= 40V, low R_{DSon}) is switched in series and a SiC-freewheeling diode (FWD) in parallel (per switch); see in Fig. 1 and Fig. 2 the DBC-substrate with the devices and the schematic respectively.

III. INTEGRATED GATE DRIVER AND ITS TECHNOLOGY

Fig. 3 presents the block circuit diagram of an integrated gate driver for two switches with GaN-HEMTs in a half bridge. Each switch consists of a Cascode Light- topology with a high voltage (HV) normally-on GaN-HEMT in series with a LV normally-off Si-pMOSFET. Both transistors are directly driven by separate gate driver output stages. In normal operation the LV-pMOSFET is permanently in the on-state and the HV-HEMT is switched directly. In case that the operation voltage fails the LV-pMOSFET is turned off by the monitoring sub-circuit and its V_{DS} drop generates the negative V_{GS} to turn off the HV-HEMT too. To transmit control signals from the primary side to the secondary side 650V (1200V)-level shifters are used for the TOP switch and medium voltage (MV) level shifters (±40V) are implemented for the BOT switch. The optimum gate voltage range for the used devices can be adopted easily by the secondary side operation voltages, e.g. -30V...0V for the HV-GaN-HEMT and -10V...+2V for the LV-p-channel MOSFET. The new developed gate driver IC with its low propagation delay, low dead times and output stages with I_{peak}= ±1.4A is able to switch the devices (up to I_{out}= 100A or even more) with frequencies up to 200kHz

Fig. 1: Photograph of DBC with the devices of one switch and the different commutation paths

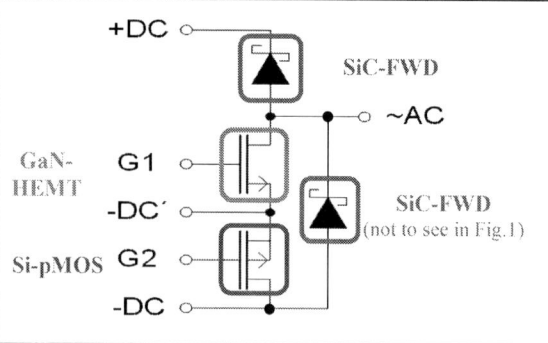

Fig. 2: Schematic of Cascode Light-topology of normally-on GaN-HEMT and normally-off Si-pMOSFET with SiC-FWD

978-1-5386-2928-4/18 $31.00 © 2018 IEEE

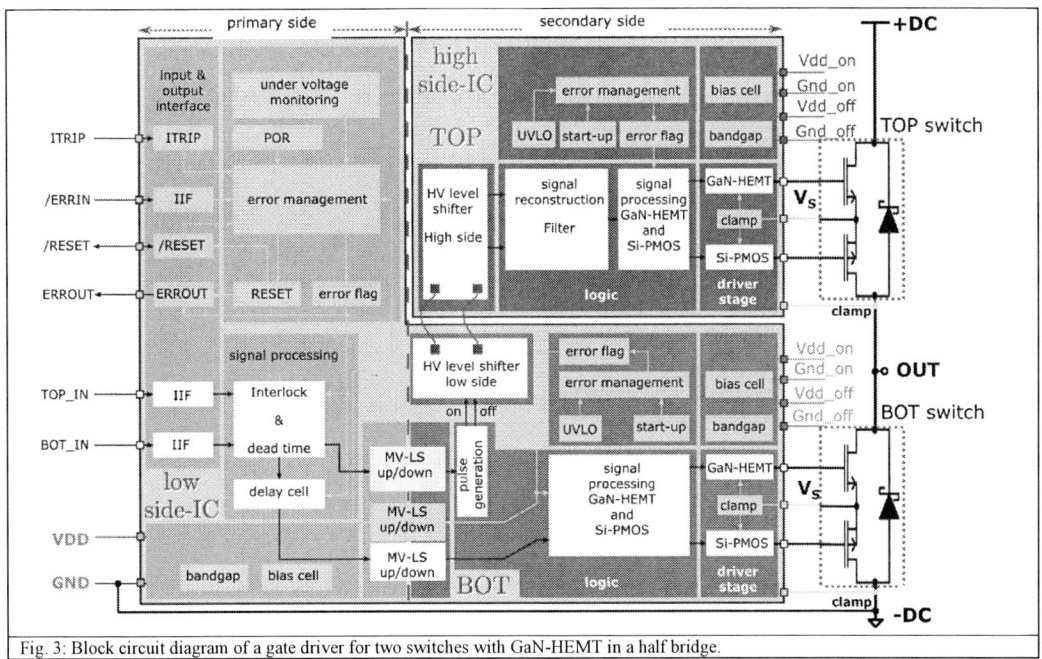

Fig. 3: Block circuit diagram of a gate driver for two switches with GaN-HEMT in a half bridge.

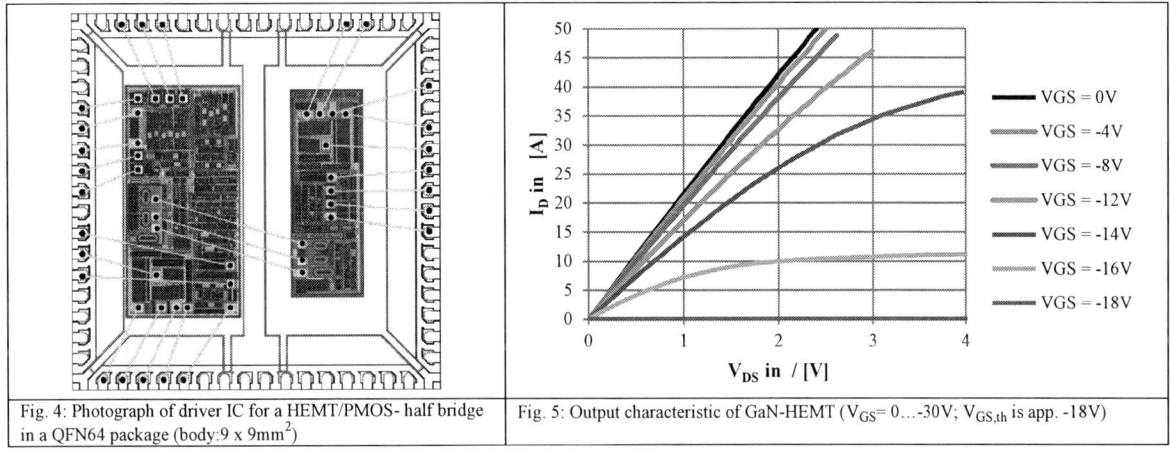

Fig. 4: Photograph of driver IC for a HEMT/PMOS- half bridge in a QFN64 package (body:9 x 9mm^2)

Fig. 5: Output characteristic of GaN-HEMT (V_{GS}= 0...-30V; $V_{GS,th}$ is app. -18V)

with extremely low switching losses. It contains all established subcircuits for monitoring (under voltage monitoring on primary and both secondary sides), safety operation (interlock and dead time generation) and signal and error processing (see Fig. 3).

The half bridge driver IC is realized in a HV-SOI-CMOS-technology [7],[8] with a fully dielectric isolation of every device from each other. It provides a complete latch up immunity and the operational temperature range can be considerably extended up to 200°C- ideal for module integration. Advanced level shifter concepts allow negative secondary side offset voltages down to -30V for the TOP and BOTTOM (BOT) secondary sides. Fig. 4 shows the photograph of the Cascode half bridge driver IC inside a QFN64 package (body: 9 x 9mm^2) with sufficient isolation distances between the low voltage and high voltage circuit parts and external pins for PCB assembly.

IV. MEASUREMENTS

Fig. 5 shows the output characteristic of the 650V, 28mΩ GaN-HEMT-transistor with a threshold voltage of app. -18V. For V_{GS}= 0V the output current is 42A at a V_{DS} drop of 2V. (During the static and dynamic measurement the LV normally-off Si-pMOSFET is turned on and its R_{DSon} is app. 1mΩ.) Fig. 6 and 7 present the turn-on and turn-off switching characteristic of the HEMT at V_{DC}= 400V, I_D= 30A and V_{GS}= -30V...0V. The turn-off behavior is relative soft and only some oscillations can be observed during turn-on with no negative influence on the system performance.

In Fig. 8 and 9 the switching losses (E_{on}, E_{off} and E_{rr}) in dependence of load current and DC-link voltage are summarized. The E_{rr} of SiC FWD ist extremely low and the switching losses of the GaN HEMT are significant higher. Usually the GaN-HEMT has to be switched with very low gate resistances, to reach fast switching and with low switching

978-1-5386-2928-4/18 $31.00 © 2018 IEEE

losses (see Fig. 10, right). But unfortunately high negative voltage spikes below -40V occur at the gate during turn-off of the complementary switch, which are not allowed (minimum transient rating in the data sheet is V_{GS}= -40V) and would lead to a degradation of the gate in continuous operation. This spikes are caused by the high dv/dt, the capacitances of the device and the inductance of the gate circuit (within the package and of the gate driver). Higher gate resistances have to be used to limit the negative gate voltage spikes to -40V, but then the switching losses increase (see Fig. 10, left).

To present the overall performance of GaN-HEMT and to make a comparison to the state of the art Si power devices the output current per chip area (A/mm^2) versus switching frequency of a three phase inverter is presented for 650V GaN-HEMT [1] (with low and higher R_G), Si-IGBT3 (IFX) [10] and Si-CoolMOS C7 [9] (with fast Si-FWD and SiC-FWD) in Fig. 11. For switching frequencies below approx. 30kHz the Si-IGBT3 has the highest output current per mm^2. At higher frequencies (>30kHz) the GaN-HEMT has the best output performance compared to the IGBT and to the CoolMOS, but only by use of a small R_G, which normally leads to unallowed negative gate voltages below -40V during turn-off. To get a higher output performance (A/mm^2) for the GaN-HEMT in comparison to the CoolMOS C7 with SiC-FWD for frequencies >100kHz a further increase of the switching speed

is necessary for the HEMT to reduce the switching losses even more. Keys are lower inductivities inside the package and the commutation circuits. Otherwise a real benefit for the use of the HEMT in comparison to the CoolMOS C7 cannot be demonstrated for frequencies >100kHz.

V. RELIABILITY INVESTIGATIONS

Several reliability investigations (HTRB, HTGB, forward bias test etc.) were be made to characterize the stability and reliability of new GaN-HEMT devices. Fig. 12 shows the curves for the leakage current I_{DSS} during HTRB (at 95% of V_{DSmax}= 618V and T_{Jmax}= 150°C) and for the forward voltages (V_{DS}) of the GaN device at I_D= 10A over 500h and 200h respectively. The leakage current I_{DSS} degreases to approx. 90% within the first 24h of the test and is stable for the rest of the test. The devices are also stable during the forward bias test and HTGB (V_{GS}= -30V, T_J= 150°C, t= 500h; not shown in the figure) over the test time. Fig. 13 shows the threshold voltages of two dies before and after the HTRB. Before the test the threshold voltage of the devices (measured at I_D= 10mA) has a hysteresis (h) of 3 to 4V between turn on and turn off. After the test this hysteresis is almost gone. But the difference between the two devices still remains. It seems that the hysteresis disappears in a kind of burn-in effect which also reduces the blocking current within the first 24h of the HTRB test (see in Fig. 12).

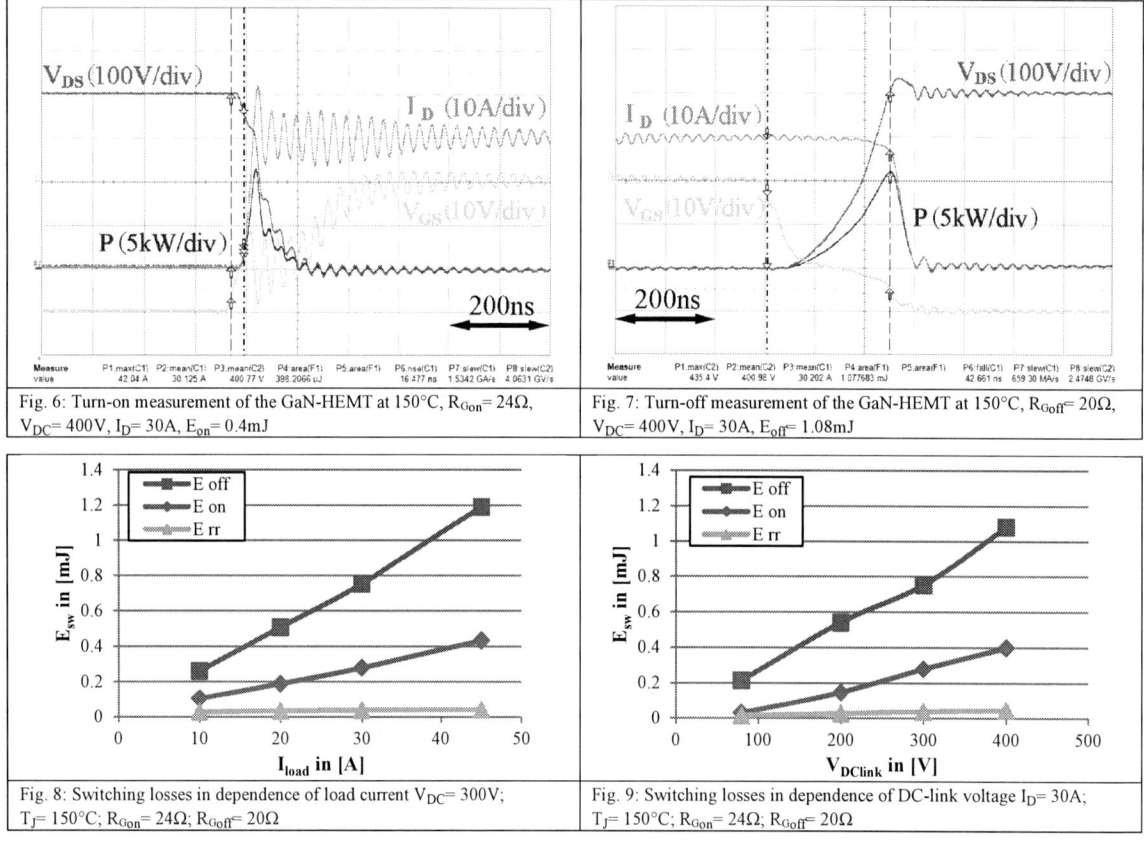

Fig. 6: Turn-on measurement of the GaN-HEMT at 150°C, R_{Gon}= 24Ω, V_{DC}= 400V, I_D= 30A, E_{on}= 0.4mJ

Fig. 7: Turn-off measurement of the GaN-HEMT at 150°C, R_{Goff}= 20Ω, V_{DC}= 400V, I_D= 30A, E_{off}= 1.08mJ

Fig. 8: Switching losses in dependence of load current V_{DC}= 300V; T_J= 150°C; R_{Gon}= 24Ω; R_{Goff}= 20Ω

Fig. 9: Switching losses in dependence of DC-link voltage I_D= 30A; T_J= 150°C; R_{Gon}= 24Ω; R_{Goff}= 20Ω

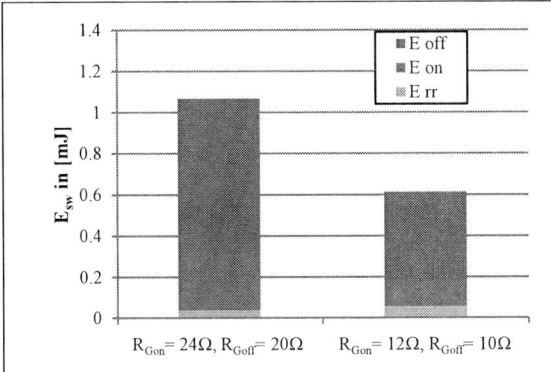

Fig. 10: Comparison of switching losses for nominal and reduced R_G for faster switching (at reduced R_G negative gate voltage spikes during turn-off are out of specification), $V_{DC}= 300V$; $I_D= 30A$; $T_J= 150°C$;

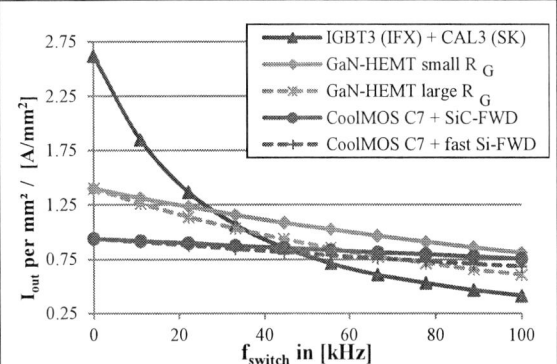

Fig. 11: Comparison of output current per chip area (A/mm²) over switching frequency of a three phase inverter with 650V GaN-HEMT, Si-IGBT3 and Si-CoolMOS at $V_{DC}= 400V$, $T_J= 150°C$, forced air cooling, cosφ= 0,85

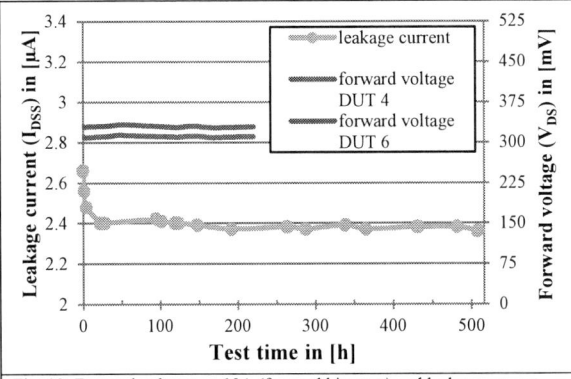

Fig. 12: Forward voltages at 10A (forward bias test) and leakage current during HTRB ($V_{DS}= 618V$ and $T_{Jmax}= 150°C$) versus time

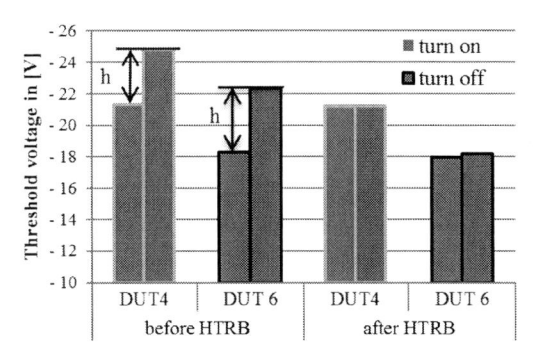

Fig. 13: Threshold voltage (measured at $I_{DS}= 10mA$) of the GaN-HEMT before and after the HTRB test

VI. CONCLUSION

A 650V, 28mΩ GaN-HEMT is assembled in a state of the art MiniSKiiP2-module in Cascode Light-topology with a low voltage (LV) pMOSFET in series and a SiC-freewheeling diode (FWD) in parallel (per switch). The high performance static output capability and the low dynamic losses are presented. During several reliability investigations (HTRB, HTGB, forward bias test, etc.) the GaN-HEMT devices was absolutely stable during the forward bias test and HTGB. In the HTRB test the leakage current I_{DSS} decreases to approx. 90% within the first 24h of the test and is stable for the rest of the test. Only before the HTRB test a hysteresis in the threshold voltage could be observed which disappears after HTRB in a kind of burn in effect which probably reduces the leakage current within the first 24h of the HTRB test too. A presented SOI-gate driver IC with low propagation delay, low dead times and output stages with $I_{peak}= \pm 1.4A$ is able to drive GaN-HEMT switches in a half bridge (up to $I_{out}= 100A$ or even more) with frequencies up to 200kHz with extremely low switching losses. The overall performance of GaN-HEMT in comparison to the state of the art Si power devices (IGBT and CooLMOS) is demonstrated in the simulation of the output current per chip area (A/mm²) versus switching frequency for a three phase inverter.

VII. REFERENCES

[1]   1 Data sheet TPH3207C; Transphorm 650V, 28mΩ GaN-HEMT

[2]   2 Data sheet S2307; Rohm 1200V, 45mΩ MOSFET

[3]   3 Data sheet CPM2-1200-0025B; Wolfspeed 1200V, 25mΩ MOSFET

[4]   4 D. Heer, D. Domes, D. Peters "Switching performance of a 1200V SiC-Trench MOSFET in low-power module" (IFX); Proc. PCIM 2016, pp. 53-59

[5]   C. Schmidt, R. Roeblitz "A Performance Comparison of SiC Power Modules with Schottky and Body Diodes" PCIM 2017

[6]   5 A. Melkonyan, "Elektronische Schalteinrichtung mit mindestens zwei Halbleiterschaltelementen", patent applied for DE102006029928B3, 09 2007.

[7]   R. T. Letavic, E. Arnold, et.al. "High Performance 600V Smart Power Technology Based on Thin Layer Silicon-on-Insulator", Proc. ISPSD 1997, pp. 49-52.

[8]   R.T. Letavic, M. Simpson, E. Arnold, et.al. "600V Power Conversion System-on-a-Chip Based on Thin Layer Silicon–on Insulator", Proc. ISPSD 1999, pp. 325-328.

[9]   Data sheet IPZR019C7; CoolMOS C7 IFX 650V, 19 mΩ

[10] Data sheet SKiiP 26NAB066V1; 600V,50A Si-IGBT3/CAL3-FWD in MiniSKiiP-module (Semikron)

Proceedings of the 30th International Symposium on Power Semiconductor Devices & ICs
May 13-17, 2018, Chicago, USA

Performance enhancement of CMOS compatible 600V rated AlGaN/GaN Schottky diodes on 200mm Silicon wafers

J. Biscarrat, R. Gwoziecki, Y. Baines, J. Buckley, C. Gillot, W. Vandendaele, G. Garnier, M. Charles and M. Plissonnier

CEA, LETI, MINATEC Campus
F-38054 Grenoble, France
jerome.biscarrat@cea.fr

Abstract—**We present significant performance enhancements of AlGaN/GaN power Schottky diodes on 200 mm silicon substrates achieved by optimizing the anode fabrication and the epitaxial layers. 600V rated AlGaN/GaN power diodes using a MIS (Metal Insulator Semiconductor)-gated Schottky anode were processed using a CMOS compatible process flow transferable to mass production environments. Turn-on voltages V_T around 0.6 V at 25°C, forward voltages V_F lower than 1.6 V at 100 mA/mm and 25°C, reverse leakage currents I_{REV} lower than 1 µA/mm at 600 V and 150°C and excellent dynamic performances were achieved and outperform state of the art 600V rated AlGaN/GaN Schottky diodes.**

Keywords— *III-V semiconductor materials, Aluminum gallium nitride, Schottky diodes, HEMTs, Heterostructures.*

I. INTRODUCTION

GaN based heterostructure power devices offer a great opportunity to achieve higher levels of performance and efficiency in power electronic systems [1]. Moreover, the availability of CMOS-compatible fabrication processes on 200mm GaN-on-Si wafers provides a promising solution for mass-production GaN devices at competitive cost [2]. Additionally, the lateral geometry of 2-dimensionnal electron gas (2DEG) devices simplifies the monolithic commuted cell integration, such as the association of Schottky rectifiers and HEMTs, in order to achieve compact power converters and GaN-based integrated circuits (GaN ICs) [3]. Enhancement of static/dynamic performances and reliability of discrete devices is a preliminary and essential work towards the emergence of GaN power integrated circuits.

II. TECHNOLOGICAL FLOW AND DIODE ARCHITECTURE

200mm GaN-on-Si wafers were grown using metalorganic chemical vapor deposition (MOCVD). The complete epitaxial stack is the same as in [4]. The structures grown on 1mm thick p-type silicon substrates consist of an AlN nucleation layer followed by transition layers, then a thick Carbon doped GaN layer and a non-intentionally doped GaN layer which forms the channel [5]. The 2-D electron gas (2DEG) is defined by the

combination of a 1nm AlN spacer and a 24nm AlGaN barrier with 23% Al content, capped by a 10nm in-situ Si_3N_4 layer.

Post growth, a 170nm LPCVD Si_3N_4 is deposited to form the passivation layer under the MIS-gated part of the anode (1st field plate FP1 definition). A 400nm SiO_2 is then added to the passivation stack to form the 2nd field plate level (FP2). In order to define the anode, Titanium nitride (TiN) is used as the contact metal (Schottky contact). Finally, a Ti/Al bilayer annealed at 875°C is used as the cathode metal (Ohmic contact). Typical anode-to-cathode length for this study is 15µm, ensuring a high lateral breakdown voltage (greater than 1500V measured with floating substrate).

The lateral AlGaN/GaN rectifier architecture presented in this work features a MIS-gated Schottky anode that consists of the association of a field-plate termination acting as a MIS-gate and a recessed Schottky contact (Fig. 1-a). This architecture can be considered as a "cascode-diode" (Fig. 1-b) where the MIS-gate allows the voltage drop to be clamped across the Schottky contact in the off-state, thereby limiting the lateral leakage current.

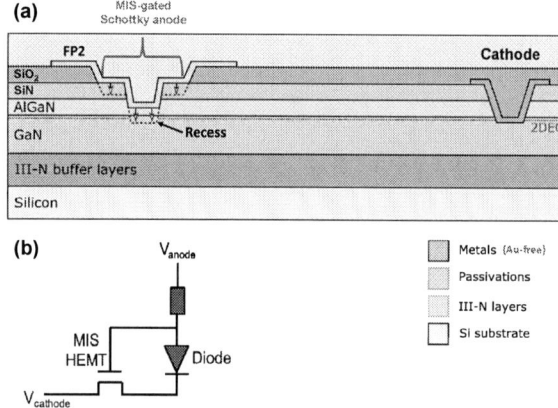

Fig. 1. Schematic cross section (a) of the AlGaN/GaN MIS-gated Schottky diode and (b) its equivalent circuit

Part of this work was performed in the frame of TOURS 2015, project supported by the French "Programme de l'économie numérique des Investissements d'Avenir".

978-1-5386-2928-4/18 $31.00 © 2018 IEEE

III. RESULTS AND DISCUSSION

First focus is set on the static properties of the fabricated diodes, more precisely on the on/off current ratio which is expressed in the following through the turn-on voltage and the reverse leakage current density. Fig. 2 presents the various technological paths explored to improve the overall static performances. We start from a partially recessed anode [5] (12 nm of AlGaN left under the anode) combined to a MIS-gate (FP1) containing a 180nm Si_3N_4 layer. The obtained diodes fall in the range of 1.5V as far as the turn-on voltage is concerned and 1μA/mm as far as the reverse leakage current density at -200V is concerned (Fig. 2 green points). On the one hand, by fully recessing the AlGaN barrier at the anode vicinity, a strong reduction of the turn-on voltage is achieved from 1.5V to 0.7V (Fig. 2 red points). On the other hand by thinning the Si_3N_4 under the MIS-gate from 180nm to 30nm, hence increasing its coupling the underlying 2DEG, the reverse leakage current is reduced by almost two orders of magnitude (Fig. 2 blue points). Both improvements are then combined within a single architecture which allows to improve almost independently the turn-on voltage and the reverse leakage current (Fig. 2 black points), a feature that can be understood in the framework of a cascode diode as illustrated in Fig. 1 (panel b).

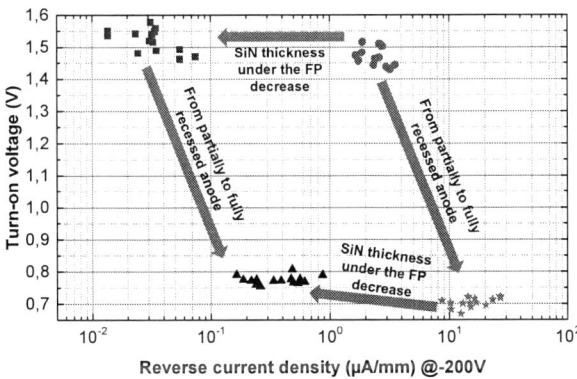

Fig. 2. : Evolution of turn-on voltage (VT) and reverse current density (IREV) with anode recess configuration and SiN thickness under the first field-plate (FP1)

Given the strong dependence of the reverse leakage as a function of the Si_3N_4 thickness under the MIS-gated (FP1), a fine technological control of this parameter appears critical to ensure stable fabrication. To address this key point, a specific etching process was developed and allows a uniform, reproducible and controlled Si_3N_4 layer with thicknesses as thin as 10 nm in the MIS-gate stack. This is illustrated in Fig. 3 which highlights the technological step from a non-uniform Si_3N_4 thickness under the Field plate 1 based on non-optimized etching process (panel a) and a controlled etching process (panel b)

As presented in Fig.4, the leakage in the range (0-400V) (fully controlled by the leakage at the anode side) has been

further reduced by 2 decades (hence an overall reduction of 4 orders of magnitude from the first process point) at 25°C and 1 decade at 150°C (Fig.4-a). It is also shown that even if the anode optimization allows to reduce the leakage current on several hundred of volts, the total reverse current density around 600V is dominated by the vertical leakage, both at 25°C and 150°C in the 3 terminal setup employed (substrate maintained at anode potential). Moreover, it is demonstrated on Fig. 4-b that the important leakage reduction is obtained, again, without degrading the forward conduction properties, with equivalent diode series resistance and only a minor shift in turn-on voltage between the 2 anode processes.

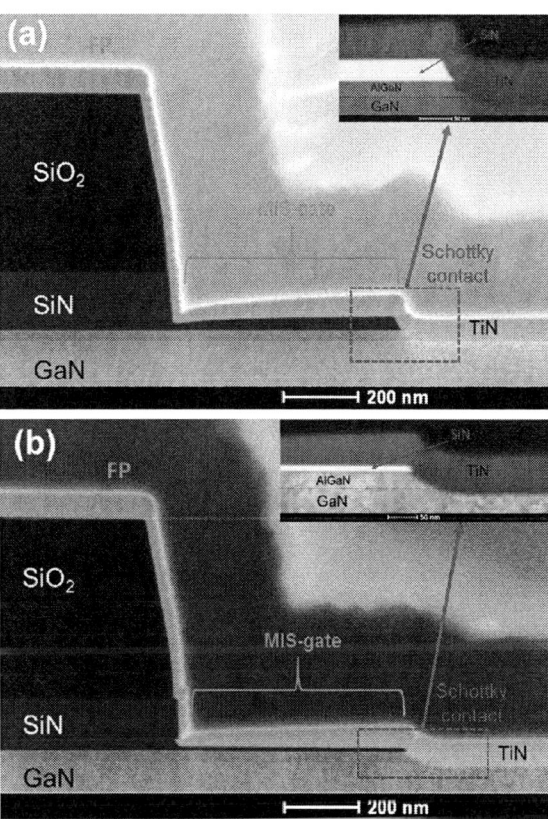

Fig. 3. : TEM cross-section of the MIS-gated hybrid anode with the unoptimized (a) and optimized (b) MIS-gate.

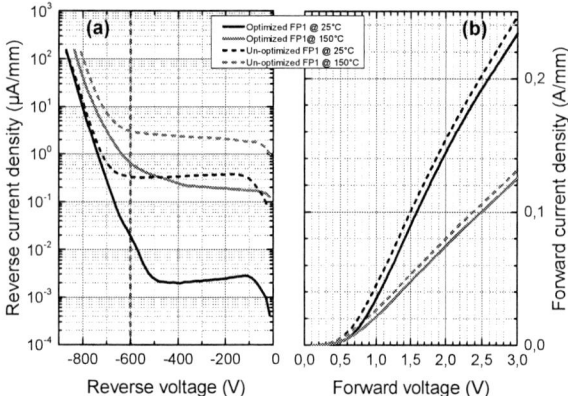

Fig. 4. : Reverse (a) and forward (b) I(V) characteristics at 25°C and 150°C measured on 52mm heterojunction diodes with (full lines) and without (dashed lines) MIS-Gated Schottky anode optimization.

The obtained on/off performances compromise is benchmarked with state-of-the art GaN diodes-on-Si [7-12] in figure 5. The data reveal that from a static point of view, the fabricated diodes outperform the other architectures compatible with 600V rated applications. Note that the two references [8,12] present better tradeoff for forward voltage versus reverse leakage current, but at the expense of reduced breakdown voltage (<600V). Moreover, through work on the ohmic contacts' resistance (R_C<1Ω.mm) and more aggressive diode designs (anode to cathode distance of 12µm instead 15µm), further improvement has been achieved while still maintaining high breakdown voltages (greater than 850V, epi limited in the present case).

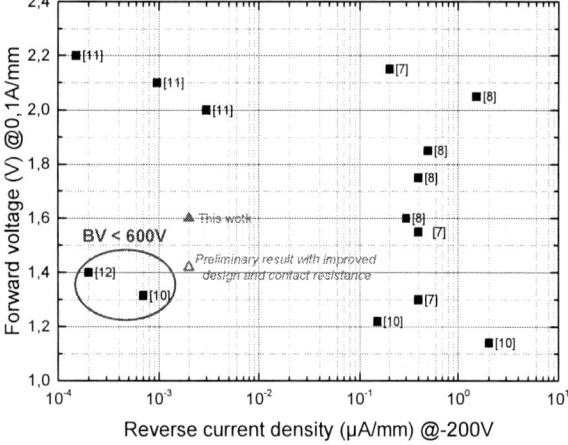

Fig. 5. : Benchmark of the fabricated diodes with the state-of-the art heterojunction diodes. Forward voltage at 100mA/mm versus leakage current density at -200V.

Beyond the static properties, GaN-on-Si devices are known to suffer from current collapse. Therefore, it is crucial to address equally the dynamic properties of the fabricated MIS-gated diodes. Through work on the epitaxial structure, from the buffer to the active layers, it is demonstrated that very promising dynamic behavior is achieved with controlled current collapse after 10s of stress at 600 V at 25°C and 150°C. We emphasize that these characteristics are measured using a microsecond time resolved set up [13], where the dynamic forward voltage (voltage drop V_F when I_F=100mA/mm) after a high voltage reverse stress has been recorded for each epitaxial stack variation (fig.6). As can be seen, a reduction of roughly 3 orders of magnitudes as far as the pre/post stress V_F ratio is obtained when going from epi generation 1 (G1) to generation 4 (G4). By focusing on the last epitaxial generation (G4), one can observe on the one hand that at 25°C the forward voltage ratio remains unchanged on the fabricated MIS-gated diodes. On the other hand, at 150°C, an increase of 50% of the dynamical forward voltage is recorded which the authors believe can be suppressed by further work on the epitaxial structure.

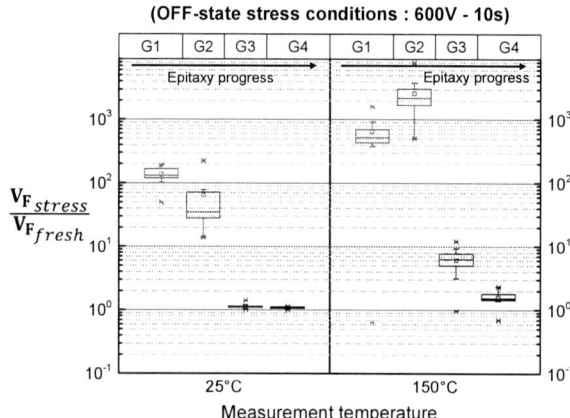

Fig. 6. : Differential forward voltage of four generations of Schottky diodes after {600V, 10s} reverse stress at 25°C and 150°C. Forward current is set at IF = 1A (JF~20mA/mm) with a short relaxation time of 20µs after reverse stress.

IV. CONCLUSION

In conclusion, state of the art 600V rated AlGaN/GaN power Schottky diodes have been fabricated on 200mm Si wafers using a CMOS compatible integration flow. From a static point of view, low turn-on voltages and low forward voltages are obtained in combination with low leakage current densities allowing thereby the best on/off tradeoff reported so far on GaN on Silicon lateral diodes. Moreover, dynamic performances after 10s OFF-state stresses are reported at microsecond time scales for blocking voltages of 600V at 25°C and 150°C. At room temperature no current collapse is observed and an encouraging 50% increase in forward voltage is measured at 150°C. Further work involving in particular epitaxial improvements is under progress to suppress current collapse at high temperature to ensure efficient conversion in harsh applicative environments.

REFERENCES

[1] Chen, Kevin J., et al.," GaN-on-Si Power Technology: Devices and Applications", IEEE Transactions on Electron Devices 64 (3) pp. 779-795, 2017

[2] De Jaeger, Brice, et al., " Au-free CMOS-compatible AlGaN/GaN HEMT processing on 200 mm Si substrates", IEEE 24th ISPSD conference, pp. 49-52, 2012

[3] Chen, Wanjun, et al., "Single-Chip Boost Converter Using Monolithically Integrated AlGaN/GaN Lateral Field-Effect Rectifier and Normally Off HEMT", IEEE electron device letters 30 (5), pp. 430-432, 2009

[4] W.Vandendaele et al., "On the understanding of cathode related trapping effects in GaN-on-Si Schottky diodes", ESSDERC, 2017

[5] M. Charles, et al., "The effect of AlN nucleation temperature on inverted pyramid defects in GaN layers grown on 200 mm silicon wafers", Journal of Crystal Growth **464**, pp. 164-167, 2017

[6] Y. Baines et al, 'Coherent tunneling in an AlGaN/AlN/GaN heterojunction captured through an analogy with a MOS contact", Scientific Reports **7**, 8177, 2017

[7] Zhou, Qi, et al. "High reverse blocking and low onset voltage AlGaN/GaN-on-Si lateral power diode with MIS-gated hybrid anode." IEEE Electron Device Letters 36, 7 (2015), 660-662.

[8] Zhu, Mingda, et al. "1.9-kV AlGaN/GaN lateral Schottky barrier diodes on silicon." IEEE Electron Device Letters 36,4 (2015), 375-377.

[9] Hu, Jie, et al. "Statistical analysis of the impact of anode recess on the electrical characteristics of AlGaN/GaN Schottky diodes with gated edge termination." IEEE Transactions on Electron Devices 63, 9 (2016), 3451-3458.

[10] Hu, Jie, et al. "Performance optimization of Au-free lateral AlGaN/GaN Schottky barrier diode with gated edge termination on 200-mm silicon substrate." IEEE Transactions on Electron Devices 63,3 (2016), 997-1004.

[11] Ma, Jun, and Elison Matioli. "High-Voltage and Low-Leakage AlGaN/GaN Tri-Anode Schottky Diodes With Integrated Tri-Gate Transistors." IEEE Electron Device Letters 38,1 (2017), pp. 83-86.

[12] Matioli, Elison, Bin Lu, and Tomas Palacios. "Ultralow leakage current AlGaN/GaN Schottky diodes with 3-D anode structure." IEEE Transactions on Electron Devices 60, 10 (2013), pp. 3365-3370.

[13] T. Lorin, "A microsecond time resolved current collapse test setup dedicated to GaN-based Schottky diode characterization", ICMTS, 2017 International Conference of. IEEE (2017). pp. 1-4

Proceedings of the 30th International Symposium on Power Semiconductor Devices & ICs
May 13-17, 2018, Chicago, USA

Novel AlGaN/GaN SBDs with Nanoscale Multi-Channel for Gradient 2DEG Modulation

Anbang Zhang[1], Qi Zhou[1]*, Chao Yang[1], Yuanyuan Shi[1], Changxu Dong[1], Tong Liu[2], Yijun Shi[1], Wanjun Chen[1], Zhaoji Li[1], and Bo Zhang[1]

[1]State Key Laboratory of Electronic Thin Films and Integrated Devices, University of Electronic Science and Technology of China, Chengdu 610054, China. (zhouqi@uestc.edu.cn, zhangbo@uestc.edu.cn)
[2]Vacuum Interconnected Nanotech Workstation (NANO-X), Suzhou Institute of Nano-Tech and Nano-Bionics, Chinese Academy of Sciences, Suzhou 215123, China

Abstract—Novel lateral AlGaN/GaN Schottky barrier diodes (SBDs) featuring nanoscale multi-channel for gradient two-dimensional electron gas (2DEG) modulation have been proposed and successfully demonstrated on silicon substrates. The 5 °C low temperature plasma etching with improved resolution is developed to form the nanoscale trenches. Due to the aspect ratio dependent etching, the trenches with different widths and depths can be fabricated in one step etching process. Owing to the small discontinuous etching area of the nanoscale trenches, the lattice strain presented in the original AlGaN/GaN heterostructure is marginally modified. Hence the piezoelectric polarization induced 2DEG can be well maintained and gradually modulated beneath the Schottky contact, which is beneficial for a low turn-on (V_T) and high breakdown voltage (BV). The fabricated SBDs exhibit uniform V_T of 0.61±0.02 V and maximum BV of 1317 V. The proposed nanoscale multi-channel structure can also be applied in the gate structure design for normally-off GaN high electron mobility transistors (HEMTs) as well as edge termination for electric-field distribution optimization of power devices.

Keywords—Schottky barrier diode (SBD); AlGaN/GaN; nanoscale multi-channel; gradient 2DEG modulation; low temperature plasma etching; turn-on voltage; breakdown voltage

I. INTRODUCTION

The Schottky barrier diode (SBD) with low turn-on voltage (V_T) and fast switching capability is essential for most power electronics applications. Owing to the superior properties of wide bandgap material, AlGaN/GaN SBD has attracted tremendous research attention for potential wide implementation and eventual commercialization. The advantage of developing high performance AlGaN/GaN SBD also attributes to the prospective for monolithically integrating AlGaN/GaN SBDs and transistors to realize full-GaN ICs with minimized parasitics toward high performance and fast speed switching applications. However, the conventional AlGaN/GaN SBD (conv.-SBD) exhibits an undesirable high V_T, resulting in a considerable conduction loss. Although the recessed-anode technology [1-8] is an effective method to reduce the V_T, the large area recessed-anode may unavoidably deteriorate the robustness and reliability of the diode. Moreover, the conv.-SBD is usually suffering from peak electric field at the edge of the Schottky junction, which resulting in a low BV. The field-plate (FP) structure is proven to be effective to reshape the electric field distribution. However,

additional fabrication process is necessary to form the FP structure.

In this paper, lateral nanoscale multi-channel (NM) AlGaN/GaN-on-Silicon SBDs featuring gradient 2DEG modulation in the anode have been proposed and successfully demonstrated. The nanoscale trenches are fabricated by 5 °C low temperature plasma etching with improved resolution. The fabricated SBDs exhibit uniform V_T of 0.61±0.02 V and maximum BV of 1317 V.

II. STRUCTURE AND MECHANISM

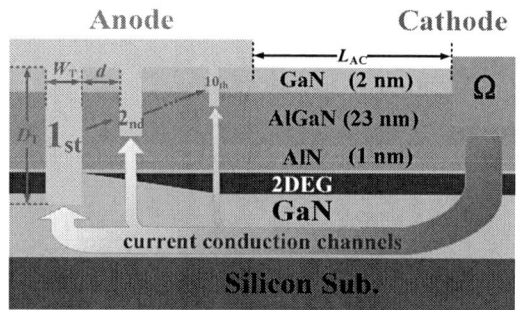

Fig. 1 Schematic cross section of the proposed SBD.

Fig. 1 shows the schematic cross section of the proposed NM-SBD. The epitaxial AlGaN/GaN heterostructures used in the work were grown on 6-inch (111) silicon substrates by metal organic chemical vapor deposition (MOCVD). The epitaxial layers consist of 2 nm GaN cap, 23 nm AlGaN barrier, 1 nm AlN interlayer, and 300 nm GaN channel. The Hall effect measured density and mobility of the two dimensional electron gas (2DEG) are 9.5×10^{12} cm^{-2} and 1500 cm^2/V·s, respectively. The device fabrication process commenced with ten nanoscale trenches formation including electron beam lithography and following low power Cl_2/BCl_3-based inductively coupled plasma (ICP) etching at 5 °C. After trenches formation, mesa isolation was conducted by the former ICP etching system. Then the samples were treated by the tetramethylammonium hydroxide (TMAH) solution (25% concentration) at 85 °C for 30 min to clean the samples and to modify the trench sidewall. After that, the

---

(*Corresponding authors: Qi Zhou, Bo Zhang.*)

978-1-5386-2928-4/18 $31.00 © 2018 IEEE

Ti/Al/Ni/Au metal stacks were deposited by the electron beam evaporation, followed by rapid thermal annealing at 870 °C for 35 s in N_2 ambient. The ohmic contact resistance of 1.5 $\Omega \cdot$mm and sheet resistance of 390 Ω/square were measured by the transmission line method. Finally, the fabrication process ended up with the Ni/Au Schottky metal stacks deposition. The anode-to-cathode distance (L_{AC}) are 4.5, 9.5 and 19.5 μm, and the key structural parameters of the trenches are listed in Table I. All the device widths are fixed as 10 μm.

TABLE I
Key Structural Parameters of the Trenches

Trench No.	Depth D_T (nm)	Width W_T (nm)	Space d (nm)
1	30	700	150
2	29	500	170
3	28	150	200
4	27	120	220
5	25	100	240
6	22	90	260
7	15	70	280
8	11	60	300
9	9	50	300
10	6	40	500

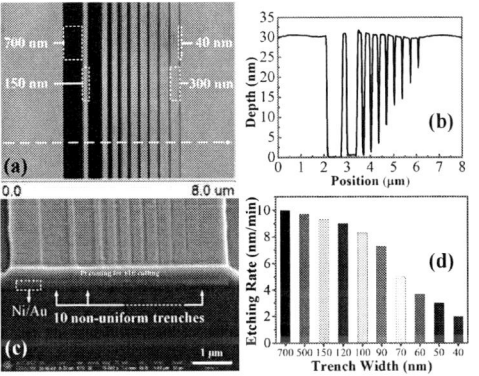

Fig. 2. (a) AFM image of the trenches. (b) The trench depths profile. (c) SEM image of the multi-trench anode. (d) The non-uniform etching rates of the trenches.

Fig. 2(a) shows the atomic force microscope (AFM) image of the trenches and its corresponding depths profile is shown in Fig. 2(b). The non-periodically spaced nanoscale trenches with gradient depths and widths were simultaneously formed during the one-step ICP etching since the etching rate is trench width dependent as summarized in Fig. 2(d). The 2DEG density under the trench can be modulated by adjusting the thickness of the remaining AlGaN barrier in the trench, and the 2DEG density between the two adjacent trenches is modulated by changing the width of the AlGaN/GaN pillar (i.e. 150–300 nm). Fig. 2(c) shows the scanning electron microscope (SEM) image of the fabricated multi-trench anode prepared by focused ion beam (FIB) cutting.

Fig. 3(a) shows the sites where the Raman measurement takes place, and the measurement results from spot A and B are compared in Fig. 3(b). The small peak shift implies that the lattice strain presented in the as-grown AlGaN/GaN heterostructure is marginally modified. Hence, the piezoelectric polarization induced 2DEG can be well maintained beneath the NM-anode owing to the significantly scaled trenches. Such well-maintained 2DEG is beneficial for forward characteristics improvement.

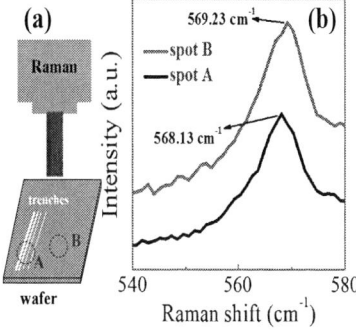

Fig. 3. (a) Schematic of the spots on the fabricated device for Raman measurement. (b) Micro-Raman spectra taken from spot A and B in the active area of the fabricated SBD.

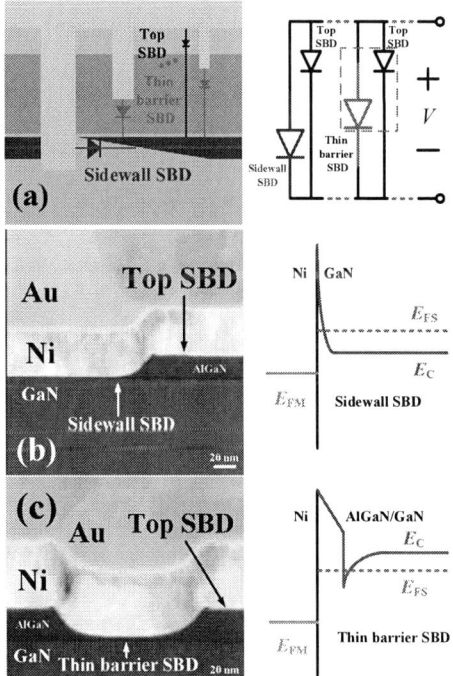

Fig. 4. (a) Equivalent circuit of the NM-SBD under forward bias. (b) and (c) are the STEM images of the Schottky junction area of the over recessed deep trench and partial recessed shallow trench and the corresponding energy band diagrams.

The operation mechanism of the NM-SBD under forward bias can be modeled by the parallel-diode circuit as shown in Fig. 4(a). The over-recessed deep trench, partial-recessed shallow trench and the non-recessed area form the local sidewall SBD, thin barrier SBD and top SBD, respectively. The scanning transmission electron microscopy (STEM) images of the Schottky junction area of the over recessed deep trench and partial recessed shallow trench and their corresponding energy band diagrams are shown in Fig. 4(b) and (c), respectively. Due to the reduced effective Schottky barrier height, the overall low V_T of the device is determined by the sidewall SBD. While the thin barrier SBD and top SBD featuring different intrinsic turn-on voltage can be turned on in a sequence for effective forward current conduction.

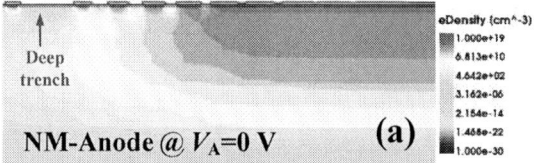

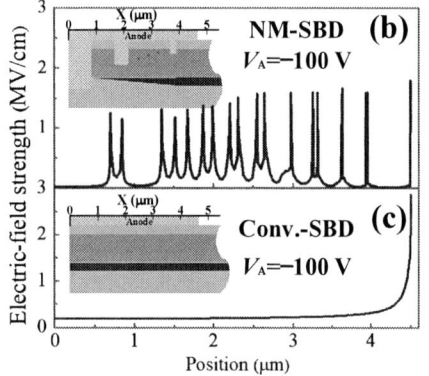

Fig. 5. TCAD simulated (a) 2DEG distribution beneath the NM anode at V_A=0 V. (b) E-field profile in the NM-SBD and (c) the conventional SBD at V_A=−100 V.

TCAD simulation was performed to illustrate the concept of breakdown characteristic enhancement. Fig. 5(a) gives the simulated 2DEG distribution at V_A=0 V showing that a gradient 2DEG modulation effect can be achieved by the NM-anode. Under reverse bias, the electric-field (E-field) profile in the NM-SBD exhibits small multi-peaks while a high single peak E-field appears in the conventional planar Schottky junction as shown in Fig. 5(b) and (c), respectively. The gradient 2DEG modulation is beneficial for obtaining a low and uniform E-field distribution to improve the reverse blocking characteristic of the device.

III. RESULTS AND DISCUSSION

Fig. 6 shows the forward I-V characteristics of the NM-SBDs in linear and semi-log scale. The typical V_T (@ 1 mA/mm) as low as 0.61 V is obtained. The on-resistance R_{ON} of 7.8 Ω·mm is extracted at 2.5 V for device with L_{AC}=4.5 μm. More importantly, the devices exhibit sharp turn-on characteristics with the current density increased by ~6 orders of magnitude

from 0 to 0.6 V. The ideality factor is extracted to be 1.2. Since the forward current below V_T is dominated by the Schottky junction, the overlapping current indicates excellent Schottky junction uniformity, which is regardless of the L_{AC}.

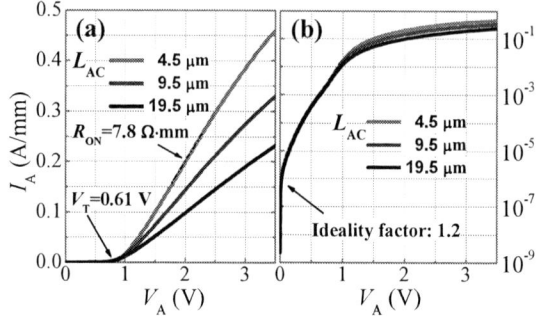

Fig. 6. Forward I-V characteristics of the fabricated NM-SBDs in (a) linear and (b) semi-log scale.

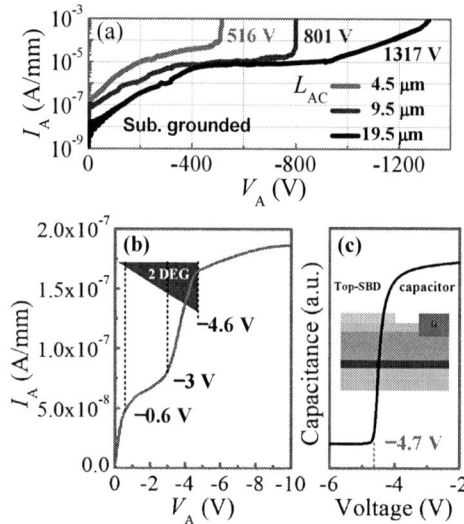

Fig. 7. (a) Reverse I-V characteristics of the fabricated NM-SBDs. (b) Reverse bias I-V curve of the NM-SBD with L_{AC}=4.5 μm under small reverse bias. (c) Top-SBD C-V at 1 MHz.

The breakdown characteristics of the NM-SBDs are shown in Fig. 7. With substrate grounded, the maximum BV of 1317 V is measured for the device with L_{AC}=19.5 μm. As shown in Fig. 7(a), with substrate grounded, the BV (@ 1mA/mm) of 516, 801, and 1317 V is measured for the device with L_{AC} of 4.5, 9.5 and 19.5 μm, respectively. Hence, the calculated critical E-field ranging from 0.7 to 1.1 MV/cm is comparable to that measured from SBDs with FPs. As shown in Fig. 7(b), three turning point at V_A of −0.6, −3 and −4.6 V are observed, suggesting that the depletion process of the 2DEG during reverse blocking in the NM-anode could be divided into three major steps which is different from that of conv.-SBD (i.e. one step or abrupt pinch-off). Such multi-step depletion results from gradient 2DEG modulation effect and is beneficial for achieving a uniform electric-field distribution as studied by the TCAD simulation.

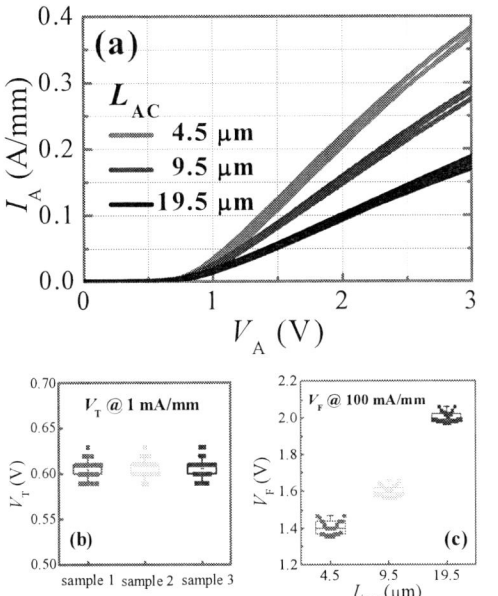

Fig. 8. (a) Typical forward characteristics of NM-SBDs with L_{AC} of 4.5, 9.5, 19.5 μm fabricated on 3 samples in 3 separate process runs. Statistical plots of (b) turn-on voltage (V_T @ 1 mA/mm) and (c) forward voltage (V_F @ 100 mA/mm) extracted from 63 NM-SBDs.

The forward characteristics including V_T and V_F (@ 100 mA/mm) exhibit excellent uniformity in 3 samples fabricated in 3 separate process runs, as shown in Fig. 8. The typical V_T of 0.61 V regardless of the L_{AC} is shown in Fig. 8(b). The typical V_F of 1.4, 1.6 and 2 V for the L_{AC} of 4.5, 9.5 and 19.5 μm is plotted in Fig. 8(c). Such excellent uniformity indicates that the NM structure and the low-temperature ICP etching used for multi-trench formation is quite stable.

IV. CONCLUSION

In summary, a novel lateral AlGaN/GaN-on-Silicon SBD featuring nanoscale multi-channel for gradient 2DEG modulation have been proposed and successfully demonstrated. By the introduction of the nanoscale trenches in the anode, the V_T can be effectively reduced. Benefiting from the uniform E-field distribution resulting from gradient 2DEG modulation,

competitive reverse blocking characteristics are obtained. The developed low temperature ICP etching is appropriate for the NM structure fabrication.

ACKNOWLEDGMENT

This work was supported in part by the National Natural Science Foundation of China (No. 61674024), the Assembly Pre-research Project under Grant JZX2017-1643/Y537, the National Science and Technology Major Project of China (No. 2013ZX02308-005), the Opening project of State Key Laboratory of Electronic Thin Films and Integrated Devices under Grant KFJJ201609, and the Fundamental Research Funds for the Central Universities under Grant ZYGX2016J211.

REFERENCES

[1] E. Bahat-Treidel, O. Hilt, R. Zhytnytska, A. Wentzel, C. Meliani, J. Wurft, and G. Trankle, "Fast-switching GaN-based lateral power Schottky barrier diodes with low onset voltage and strong reverse blocking," *IEEE Electron Device Lett.*, vol. 33, no. 3, pp. 357–359, Mar. 2012.

[2] W. Chen, K.-Y. Wong, W. Huang, and K. J. Chen, "High-performance AlGaN/GaN lateral field-effect rectifiers compatible with high electron mobility transistors," *Appl. Phys. Lett.*, vol. 92, no. 25, pp. 253501-1–253501-3, Jun. 2008.

[3] Y. Yao, J. Zhong, Y. Zheng, F. Yang, Y. Ni, Z. He, Z. Shen, G. Zhou, S. Wang, J. Zhang, J. Li, D. Zhou, Z. Wu, B. Zhang, and Y. Liu, "Current transport mechanism of AlGaN/GaN Schottky barrier diode with fully recessed Schottky anode," *Jpn. J. Appl. Phys.*, vol. 54, pp. 011001-1–011001-6, Jan. 2015.

[4] M. Zhu, B. Song, M. Qi, Z. Hu, K. Nomoto, X. Yan, Y. Cao, W. Johnson, E. Kohn, D. Jena and H. G. Xing, "1.9-kV AlGaN/GaN lateral Schottky barrier diodes on silicon," *IEEE Electron Device Lett.*, vol. 36, no. 4, pp. 375–377, Apr. 2015.

[5] Q. Zhou, Y. Jin, Y. Shi, J. Mou, X. Bao, B. Chen, and B. Zhang, "High reverse blocking and low onset voltage AlGaN/GaN-on-Si lateral power diode with MIS-gated hybrid anode," *IEEE Electron Device Lett.*, vol. 36, no. 7, pp. 660–662, Jul. 2015.

[6] H.-S. Lee, D. Y. Jung, Y. Park, J. Na, H.-G. Jang, H.-S. Lee, C.-H. Jun, J. Park, S.-O. Ryu, S. C. Ko, and E. S. Nam, "0.34 V_T AlGaN/GaN-on-Si Large Schottky Barrier Diode With Recessed Dual Anode Metal," *IEEE Electron Device Lett.*, vol. 36, no. 11, pp. 1132–1134, Nov. 2015.

[7] C. W. Tsou, K. P. Wei, Y. W. Lian, and S. H. Hsu, "2.07-kV AlGaN/GaN Schottky Barrier Diodes on Silicon with High Baliga's Figure-of-Merit," *IEEE Electron Device Lett.*, vol. 37, no. 1, pp. 70–73, Jan. 2016.

[8] Q. Zhou, Y. Yang, K. Hu, R. Zhu, W. Chen, and B. Zhang, "Device technologies of GaN-on-Si for power electronics: enhancement-mode hybrid MOS-HFET and lateral diode," *IEEE Trans. Ind. Electron.*, vol. 64, no. 11, pp. 8971–8979, Nov. 2017.

Proceedings of the 30th International Symposium on Power Semiconductor Devices & ICs
May 13-17, 2018, Chicago, USA

Switching Performance Analysis of GaN OG-FET Using TCAD Device-Circuit-Integrated Model

Dong Ji, Wenwen Li, and Srabanti Chowdhury
Department of Electrical and Computer Engineering
University of California, Davis
California, USA
dongji@ucdavis.edu

Abstract—This paper presents the superior switching performance of the in-situ Oxide-GaN interlayer FET (OG-FET) obtained by modeling a 1.4 kV/2.2 m$\Omega\square$cm^2 device fabricated by the authors [1]. Based on the parameters extracted from fabricated devices, an accurate 2D physics-based device model was developed. Using the device-circuit-integrated model built in Silvaco's MixedMode platform [2], we evaluated the switching performance of the OG-FET in 1) a double-pulse switch circuit, and 2) in a 200 V : 800 V boost converter as well. The OG-FET showed a remarkably low device figure-of-merit ($R_{on}\cdot Q_{GD}$) of 1275 m$\Omega\square$nC. Our results indicate that our recently fabricated large area GaN OG-FET has the potential of attaining higher efficiencies and also enabling MHz range conversions.

Keywords—OG-FET; GaN power transistors; switching performance; TCAD simulation; device-circuit hybrid simulation

I. INTRODUCTION

Power switches are integral to all power conversion systems. Due to the continuous increase in requirements of power density and efficiency in power converters, the state-of-the-art Si-based power devices are reaching their performance bottlenecks, which are limited by the material's physical properties. Wide bandgap (WBG) semiconductors are emerging as the next-generation material for semiconductor power devices. Gallium nitride (GaN), with the merits of wide bandgap, high breakdown electric field, and high electron mobility, is considered as the superior technology for high-power density and high-efficiency power-switching applications [3].

AlGaN/GaN high electron mobility transistors (HEMTs) which take advantage of the high electron mobility two-dimensional electron gas (2DEG) induced by the polarization charges at the AlGaN/GaN interface, have attracted significant research interests and demonstrated excellent performances for power switching applications. However, the lateral topology has limited power handling capability of < 10 kW. In order to obtain higher power-density, a vertical topology is

This work was supported by ARPA-E SWITCHES program

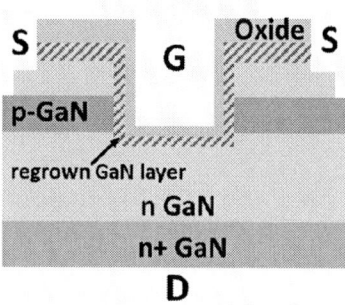

Fig. 1. Schematic of the designed GaN OG-FET.

required. Due to the availability of the high-quality free-standing GaN substrates, the field of GaN vertical power devices is growing exponentially since 2008 [4-16]. Among all the reported vertical structures, only trench MOSFETs have a robust normally-off behavior. However, there are two drawbacks of the trench MOSFET: 1) the channel electron mobility is low, limited by the surface roughness and impurity scatterings [17]; and 2) the gate oxide reliability is an issue, since a robust oxide for GaN is still unavailable. In order to solve the aforementioned issues, a novel in-situ oxide, GaN interlayer based FET (OG-FET) structure was proposed and demonstrated to have superior performance [1, 15]. By introducing a thin unintentional doped (UID) GaN interlayer underneath the gate oxide, the OG-FET is working in an accumulation mode, which distinguishes it from the conventional inversion mode MOSFETs. The low doping density in the UID GaN interlayer enhances the channel electron mobility by reducing the impurity scattering. Moreover, the in-situ grown oxide also enhance the gate dielectric reliability. From our recent study, the channel electron mobility of the OG-FET is 185 cm^2/Vs, which is over 3 times higher than that of the MOSFET [1].

Although the OG-FET has been demonstrated to have excellent DC performance in both on- and off- states [1, 18], the switching performance has not yet been studied. In this paper, a simulation study has been implemented to evaluate the switching performance of the OG-FET. Based on the

978-1-5386-2928-4/18 $31.00 © 2018 IEEE   208

TCAD device-circuit-integrated model reported in our previous study [2, 19], this study analyzes switching performance and power loss in a double-pulse test circuit as well as in a boost converter.

II. SIMULATION METHODOLOGY AND CALIBRATION

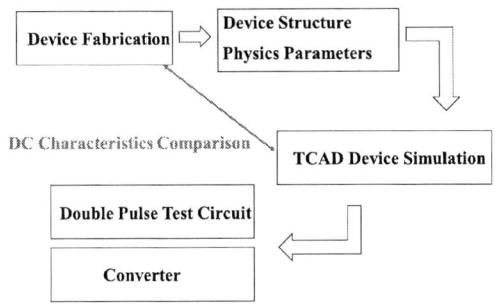

Fig. 2. Simulation methodology used in this study.

TABLE I
PARAMETERS USED IN THE SIMULATION

Parameters	Values
Channel electron mobility (cm^2/Vs)	185
Bulk GaN electron mobility (cm^2/Vs)	1000
Trench length (μm)	2
Trench depth (μm)	0.4
Drift region thickness (μm)	15
Drift region doping density (cm^{-3})	5×10^{15}
Source contact resistance ($\Omega \cdot mm$)	1.682

In this paper, we have assessed a GaN OG-FET using our device-circuit-integrated model that allows a direct evaluation under switching operations. The flow chart of the device-circuit-integrated model is shown in Fig. 2. Before the model development, a high-performance vertical GaN transistor was fabricated. Physical parameters, such as channel electron mobility and ohmic contact resistance, were extracted from the measured result of the fabricated device. A "virtual" device was then built in the TCAD device simulator (Silvaco ATLAS) with the identical device structure dimensions and the physical parameters. A drift-diffusion model was used to simulate the DC performance. A DC output comparison between the fabricated and simulated OG-FETs was performed to calibrate the physical model (drift-diffusion model). Once the TCAD model was calibrated, the 2D device simulator was integrated into a double-pulse test circuit to analysis the switching performance of the "virtual" device, which was identical to the fabricated one. Finally, a boost converter was simulated to evaluate the system performance. Different to other power converters modeling, the circuit-device-integrated simulator used in this study has a more

accurate 2D device model, and provides a chance to observe the physics insight in the transistor during the converter operation.

The parameters used in the simulation are summarized in Table I. Fig. 3 (a) shows the comparison of I_D-V_{DS} characteristics between simulated and fabricated GaN OG-FETs. The accuracy of the 2D device model is indicated by the agreement of the simulated and the experimental data. The measured off-state characteristics of the fabricated OG-FET are shown in Fig. 3 (b), indicating a breakdown voltage of 1.4 kV.

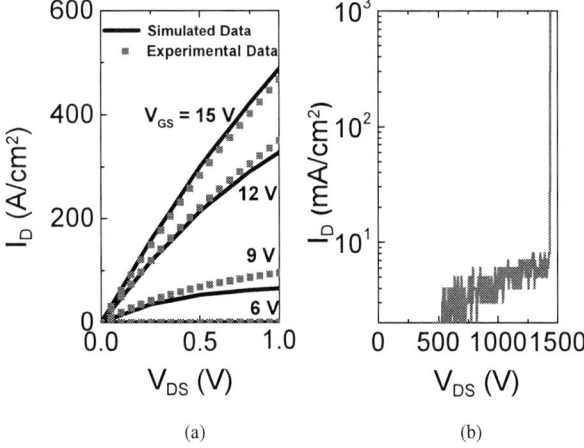

(a)   (b)

Fig. 3. (a) Comparison of I_D-V_{DS} characteristics of the simulated and fabricated OG-FETs. (b) The measured off-state characteristics of the fabricated OG-FET [1].

III. SWITCHING PERFORMANCE ANALYSIS

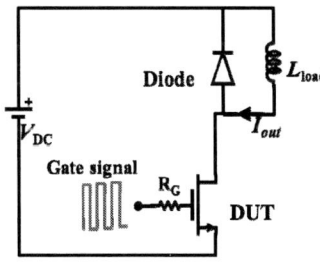

Fig. 4. The double-pulse test circuit used in the simulation.

Fig. 4 shows the double-pulse test circuit used in the simulation. An ideal scaled OG-FET with a total on-state resistance (R_{on}) of 75 mΩ was evaluated as the device under test (DUT). In order to simplify the simulation, a current source was used instead of an inductor as the load.

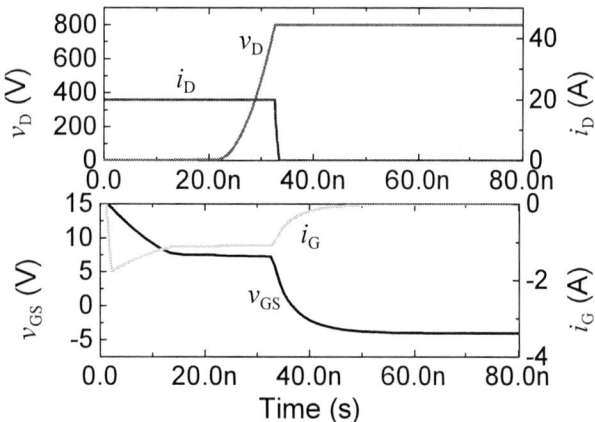

Fig. 5. v_D, i_D, v_{GS}, and i_G waveforms during the turn-off transient.

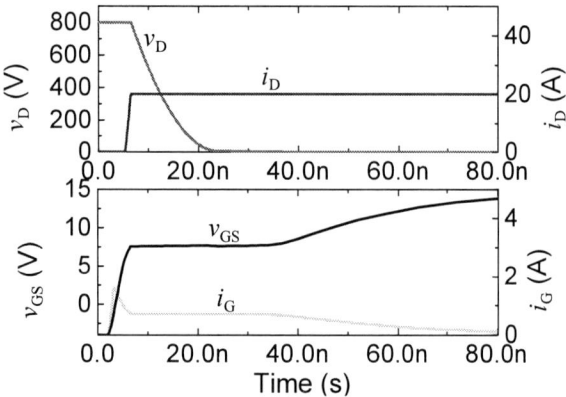

Fig. 6. v_D, i_D, v_{GS}, and i_G waveforms during the turn-on transient.

Fig. 5 shows the switching waveforms during the turn-off transient. The on-state gate voltage was 15 V, and the off-state one was -4 V. The turn-off delay time ($t_{d(off)}$) was 24.75 ns, fall time (t_f) was 7.3 ns, and the turn-off $\Delta v/\Delta t$ was 87.7 kV/μs. The total gate charge is 42.7 nC, including a Q_{GD} of 17 nC.

The switching waveforms of the turn-on transient are shown in Fig. 6. The turn-on delay time ($t_{d(on)}$) was 7.5 ns, rise time (t_r) was 11.3 ns, and the turn-on $\Delta v/\Delta t$ was 56.6 kV/μs.

IV. POWER LOSS ANALYSIS IN BOOST CONVERTERS

A boost converter was used in the simulation to study the power loss of the OG-FET during the switching cycles. Fig. 7 shows the circuit, and the parameters used are shown in Table II.

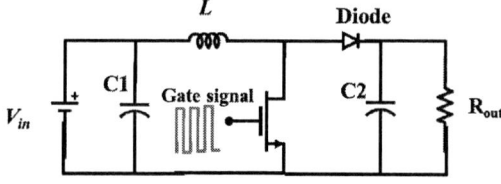

Fig. 7. The boost converter circuit used in simulation.

TABLE II
PARAMETERS USED IN CIRCUIT

Parameters	Values
V_{in} (V)	200
C1 (F)	10^{-5}
L (H)	10^{-3}
C2 (F)	5×10^{-6}
Diode	ideal
D	0.75

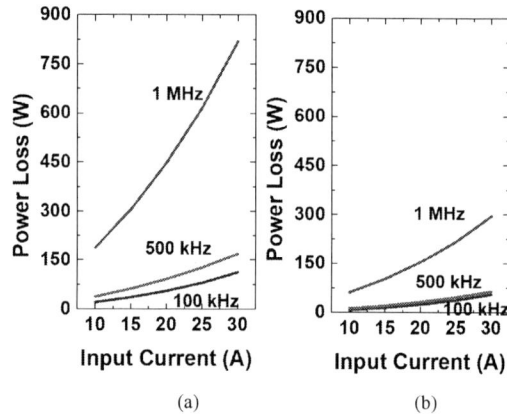

Fig. 8. Power losses as a function of input current for (a) 200 V : 800 V, and (b) 200 V : 400 V converters.

The power losses of GaN OG-FET based boost converters as a function of input current were simulated and shown in Fig. 8. For a 200 V : 800 V conversion, the typical power loss was between 20 W to 110 W under an operation frequency of 100 kHz, while for a frequency of 1 MHz, the typical switching loss was between 180 W to 820 W. For the 200 V : 400 V conversion, the typical power loss was less than 300 W under 1 MHz frequency.

Fig. 9 shows the efficiency of system as a function of input power. Peak efficiencies for different operation frequencies are 99% (100 kHz), 95% (500 kHz) and 90.6% (1 MHz).

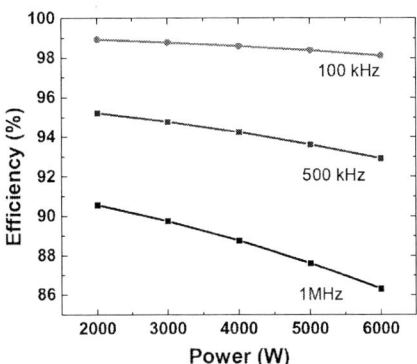

Fig. 9. Efficiency of 200 V : 800 V boost converter as a function of input power. Peak efficiencies for different operation frequencies are 99% (100 kHz), 95% (500 kHz) and 90.6% (1 MHz).

V. CONCLUSION

In summary, we present a switching study of a 1.4 kV GaN OG-FET using Silvaco's device-circuit modeling platform. An accurate physical device model was developed using the extracted parameters from the fabricated devices. The switching performance was analyzed using the device-circuit-integrated model. The OG-FET was found to have a remarkably low device figure-of-merit of 1275 mΩ□nC, which was 19% smaller than commercial SiC MOSFET [20]. The boost converter using the OG-FET as the switch had peak efficiencies of 99% and 90.6% under switching frequencies of 100 kHz and 1 MHz, respectively. Results indicate the OG-FET has the potential for MHz power conversion with substantial high efficiency.

ACKNOWLEDGMENT

The authors would like to acknowledge Dr. Isik Kizilyalli, Dr. Timothy Hiedel, Dr. Eric Carlson and Dr. Daniel Cunnignham for supporting this work under the ARPA-E SWITCHES program.

REFERENCES

[1] D. Ji, C. Gupta, S. H. Chan, A. Agarwal, W. Li, S. Keller, U. K. Mishra, and S. Chowdhury, "Demonstrating > 1.4 kV OG-FET performance with a novel double field-plated geometry and the successful scaling of large-area devices," in Proceedings of IEEE Electron Devices Meeting (IEDM), pp. 223-226, Dec. 2017. DOI: 10.1109/IEDM.2017.8268359.

[2] D. Ji, and S. Chowdhury, "A discussion on the DC and switching performance of a gallium nitride CAVET for 1.2kV application," in Proc. IEEE 3rd Workshop on Wide Bandgap Power Devices and Applications (WiPDA), 2-4 Nov. 2015, pp. 174-179.

[3] P. Kruszewski, J. Jasinski, T. Sochacki, M. Bockowski, R. Jachymek, P. Prystawko, M. Zajac, R. Kucharski, M. Leszczynski, "Vertical Schottky Diodes Grown on Low-Dislocation Density Bulk GaN Substrate," Int. Workshop Nitride Semiconductor, 2014

[4] S. Chowdhury, B. L. Swenson, and U. K. Mishra, "Enhancement and depletion mode AlGaN/GaN CAVET with Mg-ion-implanted GaN as current blocking layer," IEEE Electron Device Letters, vol. 29, no. 6, pp. 543-545, Jun. 2008. DOI: 10.1109/LED.2008.922982.

[5] D. Ji, M. A. Laurent, A. Agarwal, W. Li, S. Mandal, S. Keller, and S. Chowdhury, "Normally OFF Trench CAVET With Active Mg-Doped GaN as Current Blocking Layer," IEEE Transaction on Electron Device, vol. 64, no. 3, pp. 805-808, Dec. 2016. DOI: 10.1109/TED.2016.2632150.

[6] H. Nie, Q. Diduck, B. Alvarez, A. P. Edwards, B. M. Kayes, M. Zhang, G. Ye, T. Prunty, D. Bour, and I. C. Kizilyalli, "1.5-kV and 2.2-mΩ-cm2 vertical GaN transistors on bulkGaN substrates," IEEE Electron Device Letters, vol. 35, no. 9, pp. 939-941, Sept. 2014. DOI: 10.1109/LED.2014.2339197.

[7] D. Shibata, R. Kajitani, M. Ogawa, K. Tanaka, S. Tamura, T. Hatsuda, M. Ishida, T. Ueda, "1.7kV/1.0 mΩ·cm2 normally-off vertical GaN transistor on GaN substrate with regrown p-GaN/AlGaN/GaN semipolar gate structure," in Proceedings of IEEE Electron Devices Meeting (IEDM), Dec. 2016, pp. 248-251. DOI: 10.1109/IEDM.2016.7838385.

[8] D. Ji, and S. Chowdhury, "Design of 1.2kV Power Switches With Low Ron Using GaN-Based Vertical JFET," IEEE Trans. Electron Devices, vol. 62, no. 8, pp. 2571-2578, Aug. 2015. DOI: 10.1109/TED.2015.2446954.

[9] I. C. Kizilyalli, and O. Aktas, "Characterization of vertical GaN p-n diodes and junction field-effect transistors on bulk GaN down to cryogenic temperatures," Semicond. Sci. Technol., vol. 30, no. 12, p. 24001, Nov. 2015. DOI: 10.1088/0268-1242/30/12/124001.

[10] T. Oka, T. Ina, Y. Ueno, and J. Nishii, "1.8 mΩ·cm² vertical GaN-based trench metal-oxide-semiconductor field-effect transistors on a free-standing GaN substrate for 1.2-kV-class operation," Applied Physics Express, vol. 8, p. 054101, Apr. 2015. DOI: 10.7567/APEX.8.054101.

[11] R. Li, Y. Cao, M. Chen, and R. Chu, "600V/1.7Ω Normally-Off GaN Vertical Trench Metal-Oxide-Semiconductor Field-Effect Transistor," IEEE Electron Device Letters, vol. 37, no. 11, pp. 1466-1469, Nov. 2016. DOI: 10.1109/LED.2016.2614515.

[12] W. Li, and S. Chowdhury, "Design and fabrication of a 1.2kV GaN-based MOS vertical transistor for single-chip normally off operation," Phys. Status Solidi A, vol. 213, no. 10, pp. 2714-2720, June 2016. DOI: 10.1002/pssa.201532575.

[13] W. Li, D. Ji, R. Tanaka, S. Mandal, M. A. Laurent, and S. Chowdhury, "Demonstration of GaN Static Induction Transistor (SIT) Using Self-Aligned Process," IEEE Journal of Electron Device Society, vol. 5, no. 6, pp.485-490, Nov. 2017. DOI: 10.1109/JEDS.2017.2751065.

[14] M. Sun, Y. Zhang, X. Gao, T. Palacios, "High-Performance GaN Vertical Fin Power Transistors on Bulk GaN Substrates," IEEE Electron Device Letters, vol. 38, no. 4, pp. 509-512, Feb. 2017. DOI: 10.1109/LED.2017.2670925.

[15] C. Gupta, S. H. Chan, Y. Enatsu, A. Agarwal, S. Keller, and U. K. Mishra, "OG-FET: An in-situ Oxide, GaN interlayer based vertical trench MOSFET," IEEE Electron Device Letters, vol. 37, no. 12, pp. 1601-1604, Dec. 2016. DOI: 10.1109/LED.2016.2616508.

[16] D. Ji, C. Gupta, A. Agarwal, S. H. Chan, C. Lund, W. Li, M. A. Laurent, S. Keller, U. K. Mishra, and S. Chowdhury, "First report of scaling a normally-off In-situ Oxide, GaN interlayer based vertical trench MOSFET (OG-FET)," in the proceedings of Device Research Conference (DRC), June 2017. DOI: 10.1109/DRC.2017.7999442.

[17] C. Gupta, S. H. Chan, C. Lund, A. Agarwal, O. S. Koksaldi, J. Liu, Y. Tnatsu, S. Keller, and U. K. Mishra, "Comparing electrical performance of GaN trench-gate MOSFETs with a-plane (1120) and m-plane (1100) sidewall channels," Applied Physics Express, vol. 9, no. 12, p. 121001, Nov. 2016.

[18] C. Gupta, S. H. Chan, A. Agarwal, N. Hatui, S. Keller, and U. K. Mishra, "First Demonstration of AlSiO as Gate Dielectric in GaN FETs; Applied to a High Performance OG-FET," IEEE Electron Device Letters, vol. 38, no. 11, pp. 1575-1578, Nov. 2017.

[19] D. Ji, Y. Yue, J. Gao, and S. Chowdhury, "Dynamic Modeling and Power Loss Analysis of High-Frequency Power Switches Based on GaN CAVET," IEEE Transactions on Electron Devices, vol. 63, no. 10, pp. 4011-4017, Oct. 2016.

[20] Cree C3M0075120K https://www.wolfspeed.com/c3m0075120k.

Proceedings of the 30th International Symposium on Power Semiconductor Devices & ICs
May 13-17, 2018, Chicago, USA

A Split Gate Vertical GaN Power Transistor with Intrinsic Reverse Conduction Capability and Low Gate Charge

Ruopu Zhu, Qi Zhou†, Hong Tao, Yi Yang, Kai. Hu, Dong. Wei, Liyang. Zhu, Yu. Shi,
Wanjun Chen, Xiaorong Luo and Bo Zhang†
State Key Laboratory of Electronic Thin Films and Integrated Devices, School of Electronic Science and
Engineering, University of Electronic Science and Technology of China, Chengdu, China
Email: zhouqi@uestc.edu.cn, zhangbo@uestc.edu.cn

Abstract—In this work, a vertical normally-off GaN device featuring split-gate with intrinsic reverse conduction (RCVFET) functionality and low gate capacitance is proposed and studied by simulation. Different from the lateral AlGaN/GaN HEMT, the *RC* characteristics of the proposed RCVFET are independent with the threshold voltage of the device, while a low $V_{R,ON}$ of 0.8 V is obtained. Owing to the split-gate design, the gate charge is respectably reduced that is beneficial for improving the switching speed of the RCVFET. The device exhibits a low R_{on} of 0.93 mΩ·cm^2 and a *BV* of 1280V. The reverse recovery time is 13ns. The Q_{GD} is 80 nC that is only one fifth of that obtained in the reference device without split-gate.

Index Terms—GaN, vertical, normally-off, split gate, gate charge, reverse conduction, schottky source, reverse recovery time, turn-on voltage.

I. INTRODUCTION

THE wide band gap semiconductor Gallium Nitride (GaN) is considered to be a robust semiconductor for power switching devices in the high frequency and high power density applications [1]-[3]. To date, by taking the advantage of the high mobility and high density of the two-dimensional electron gas (2DEG) [4]-[6], most of effort has been devoted towards the lateral GaN high electron mobility transistors (HEMTs), in which current flows horizontally. Nevertheless, the performance of lateral GaN HEMTs is still far behind the material limits and there are reluctance in further improving the performance of GaN based lateral devices [7] such as current collapse induced by "virtual gate" effect [8] and premature breakdown [9]. For lateral HEMTs, extension of gate-to-drain spacing is the fundamental approach to enhance the breakdown voltage *BV*, however, it comes at the cost of chip size. On the other hand, the vertical devices exhibit the advantage of delivering higher *BV*, higher power density with smaller chip size. More importantly, in vertical devices, the current collapse

This work was supported in part by the National Natural Science Foundation of China under Grant 61674024, and in part by the Assembly Pre-research Project under Grant JZX2017-1643/Y537, and in part by the National Science and Technology Major Project under Grant 2013ZX02308-005, and in part by the opening project of the State Key Laboratory of Electronic Thin Films and Integrated Devices under Grant KFJJ201609, and in part by the Fundamental Research Funds for the Central Universities under grant ZYGX2016J211. (*Corresponding author: Qi Zhou, Bo Zhang*)

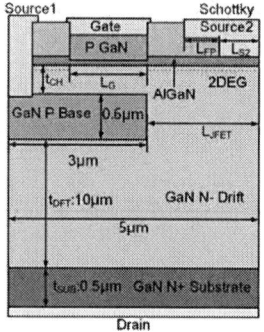

Fig. 1. Schematic cross section of the proposed RCVFET.

induced by surface states can be effectively mitigated because the high electric field transfers from the AlGaN/GaN surface into GaN bulk. Furthermore, the electric field distribution in vertical device is more uniform. Hence, the "virtual gate" effect can be well suppressed, which may lead to less dynamic R_{on} degradation and higher figure-of-merit [10].

Although vertical AlGaN/GaN heterojunction field effect transistors, such as the current aperture vertical electron transistors (CAVETs) [10]-[13], insulated gate AlGaN/GaN heterojunction field-effect transistors [14], trench MOSFETs [15][16] and vertical transistors with p-GaN gate [17] have been successfully demonstrated, GaN vertical transistors with intrinsic reverse conduction (*RC*) functionality are rarely reported. Because the body diode in GaN vertical transistors exhibits a very high turn-on voltage (~3V) and an extremely low mobility of holes (~20 cm^2/V·s), which requires an anti-paralleled SBD performs as a freewheeling diode to sustain the reverse conduction in many power electronic systems for lowering the conduction losses. However, besides the larger volume size, such a "two-chip" solution inevitably increases parasitic capacitance and inductance that would enhance increase switching losses as well as reduces switching speed. Although the lateral AlGaN/GaN HEMT is able to deliver the *RC* functionality, the *RC* voltage drop is determined by the forward V_{TH} of the 2DEG channel which is much larger than the body diode of Si-based power devices (e.g. 0.7V). Not to mention that higher forward V_{TH} is preferred to withstand the

978-1-5386-2928-4/18 $31.00 © 2018 IEEE

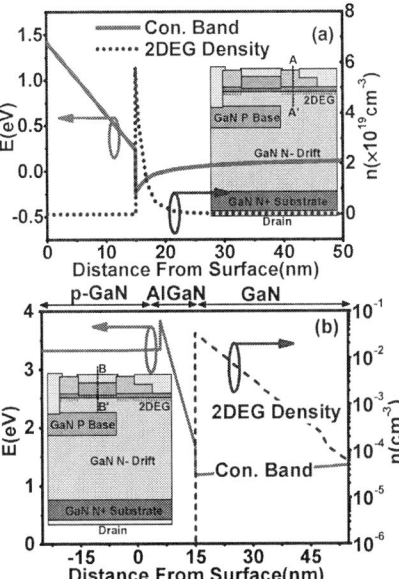

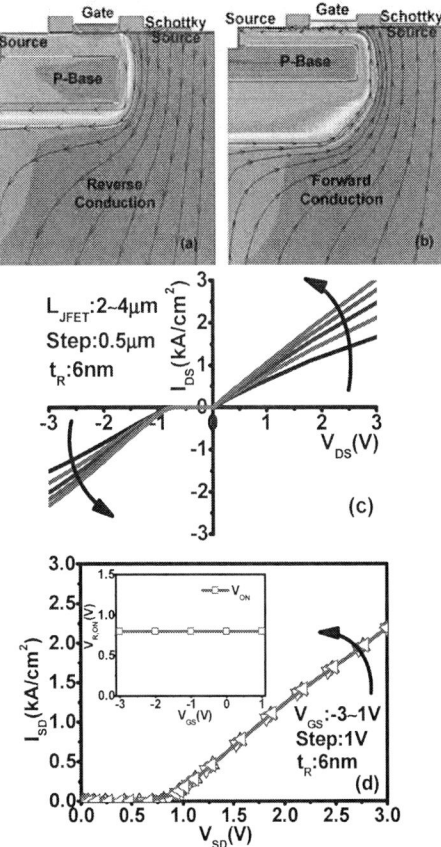

Fig. 2. (a) The simulated 2DEG density and the conduction band diagram along AA'. The heterojunction generates a 2DEG density of 8×10^{12}cm^{-2} and the surface potential is 1.4eV. (b) The density of electrons and the conduction band diagram across the p-GaN gate region along BB'.

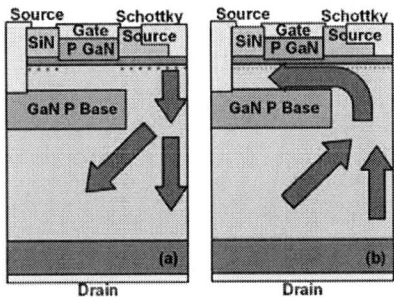

Fig. 3. The current conduction path in (a) reverse conduction and (b) forward conduction.

Fig. 4. Current contours and flowlines for (a) reverse on-state mode and (b) forward on-state mode. Reverse and forward conduction characteristics of the RCVFET (c) with varied L_{JFET}. (d) with varied V_{GS}. V_{GS} is 4V for the forward conduction and 0V for the reverse conduction. The inset shows reverse turn-on voltage maintains the same under varied V_{gs} bias. The gate recess depth of all devices is set as 6nm.

parasitic-induced gate drive noise for improved false turn on immunity, which will impose a severe contraction between the low RC voltage drop and high V_{TH}. Hence, it is highly desired to integrate the Schottky barrier diode with short recovery time into the GaN vertical field effect transistor.

In this work, we proposed a split-gate vertical GaN power transistor with intrinsic reverse conduction capability (RCVFET) by introducing the Schottky source between the two split gates. The central Schottky source performs as an embedded diode with low turn-on voltage, which effectively eliminates the reverse recovery issue that is associated with the GaN PN junction rectifier. Besides, the Schottky source greatly reduces the gate-to-drain charge and consequently the switching losses.

II. DEVICE STRUCTURE AND OPERATION MECHANISM

A schematic cross sectional view of the proposed RCVFET and the key device parameters are shown in Fig. 1. The device simulation is implemented with Sentaurus TCAD. The thickness of p-GaN/Al$_{0.25}$Ga$_{0.75}$N/n$^-$-GaN/n$^+$-GaN is

0.1μm/15nm/10.9μm/0.5μm, respectively. The thickness of channel region t_{CH} is 0.3μm. The doping concentrations of the p-gate and p-base are 3×10^{17} cm^{-3}. The doping concentrations of the N$^-$ and N$^+$ region are 1×10^{16} cm^{-3} and 1×10^{21} cm^{-3} respectively. The SBH of the centered Schottky source is 1 eV. The channel thickness t_{CH} and the Schottky source length L_{S2} are fixed as 0.3 and 0.5 μm. L_{JFET}, L_{FP} and L_{Gate} will be varied to study their impact on device characteristics. The AlGaN/GaN heterojunction features a 2DEG density of 8×10^{12} cm^{-2} with a surface potential of 1.4eV [4] [8] as shown in Fig. 2 (a). The simulated values are comparable to the experimental data [4] [18]. To realize the normally-off operation, the p-GaN gate and p-base are adopted to effectively deplete the 2DEG, leading to the raised up conduction band in the channel (see Fig. 2 (b)).

III. RESULTS AND DISCUSSIONS

Fig. 3 (a) and (b) show the conduction path in reverse bias and forward bias respectively. Fig. 4 (c) shows the simulated RC characteristics dependent with the aperture width between the two P-base (L_{JFET}). At V_{SD} of 3V the maximum RC current density is 2.3 kA/cm^2 for L_{JFET} = 4μm. More importantly, the

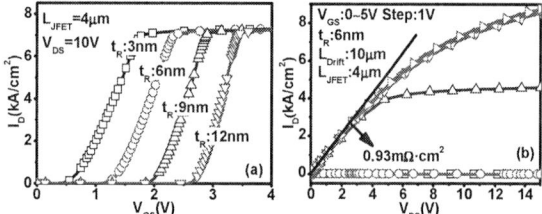

Fig. 5. (a) Transfer characteristics of the RCVFET with varied gate recess depth (t_R=3~12 nm). (b) Output characteristics of the device with t_R=6nm. The saturated output current can exceed 8kA/cm^2 and the on-resistance is 0.93 m$\Omega\cdot$cm^2 at the gate bias of 4V.

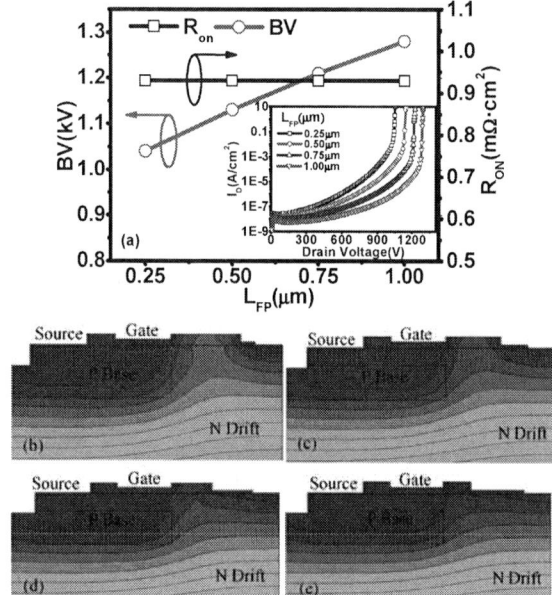

Fig. 6. (a) The impact of the field plate length upon the breakdown voltage and on-resistance. The inset shows off-state characteristics with varied field plate length. The electrostatic potential contours at 1054V on the off-state in the device with L_{FP} of (b) 0.25μm, (c) 0.5μm, (d) 0.75μm, (e) 1μm. The t_R for all devices is 6nm.

RC characteristics are independent with the gate bias and the $V_{R,ON}$ exhibits negligible variation. A stable $V_{R,ON}$ as low as 0.8 V is consistently obtained (see Fig. 4 (d)). It can be seen from Fig. 4 (a), the *RC* current is solely delivered by the centered Schottky source rather than the lateral 2DEG channel, which eliminates the gate modulation effect on *RC*. Fig. 5 shows transfer and output characteristics of RCVFET. By employing a partial gate recess, the V_{TH} can be shifted from 0.6 to 2.6 V. The maximum drain current density is 8.78 kA/cm^2 with a specific R_{ON} of 0.93 m$\Omega\cdot$cm^2.

The impact of length of the source field plate on the breakdown characteristics is studied as shown in Fig. 7. It can be seen that the breakdown voltage deteriorates with the field plate length decreasing and the length of the field plate has almost no impact upon the on-resistance as shown in Fig. 6 (a). Thanks to the existence of high density of 2DEG, the on-resistance maintains almost the same at varied L_{FP}. The breakdown voltage increases and leakage current decreases when the length of the source field plate increases from 0.25μm

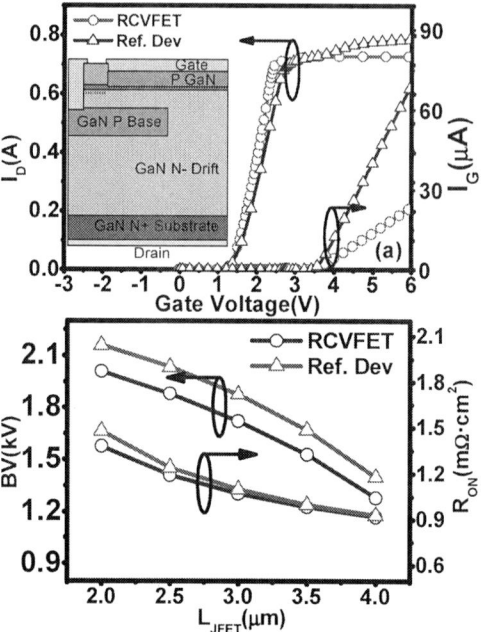

Fig. 7. (a) Transfer characteristics and gate current of RCVFET and Ref. Dev. The length of gate and JFET region for RCVFET are 1μm and 4μm respectively. (b) The impact of L_{JFET} on the breakdown voltage and on-resistance for both devices. The gate recess of all devices is 6nm.

to 1μm. The increased length of the field plate can alleviates the peak electric field at the same drain voltage of 1054V on the off-state leading to the mitigation of potential crowding as depicted in Fig. 6 (b), (c), (d) and (e). The breakdown voltage of the device with L_{FP} of 0.25μm is 1054V. The breakdown voltage of the RCVFET with $L_{Gate}/L_{JFET}/L_{FP}$=1μm/4μm/1μm is 1280V, leading to a FOM of V^2_{BR}/R_{ON} ~1.76×10^9 V$^2\cdot\Omega\cdot$cm^{-2}. All devices have the same dimensions of Schottky source L_{S2} and JFET region L_{JFET}. The t_R for all devices is 6nm.

The impact of L_{JFET} on *BV* and R_{on} for RCVFET and Ref. Dev is depicted in Fig. 7. The breakdown voltage will increase as L_{JFET} declines due to the mitigation of electric field crowding at the corner of the p-base. However a decreased L_{JFET} will cause a more severe JFET effect, leading to R_{on} degradation as depicted in Fig. 7 (b). It can be seen that the values of R_{on} for both devices are exactly the same as each other and the ratio of *BV* for RCVFET to that for Ref. Dev exceeds 90% with varied lengths of JFET region. It should be noted from Fig. 7 (a) that the gate current of RCVFET is obviously lower than that of Ref. Dev due to the split gate design of RCVFET.

The reverse recovery characteristics are studied by mixed mode simulation as shown in Fig. 8 (a). The inset is test circuit as illustrated in [19]. The reverse recovery characteristics of the RCVFET are simulated under the condition of I_F = 16.6A, V_{SD}=2.4 V, V_{RR} = 800 V and di/dt = 2100 A/μs [20]. It can be seen that t_{rr} of the RCVFET is 13ns and the Q_{rr} is 47nC. The active area is 0.01 cm^2. Fig. 8 (b) and (c) show the gate charge characteristics of the RCVFET and Ref. Dev for L_{jfet} being 4μm. Both devices have the same parameters except the gate length. The L_{Gate} for them is 1μm and 3μm respectively. The mixed mode simulation is implemented with Sentaurus TCAD using

978-1-5386-2928-4/18 $31.00 © 2018 IEEE   214

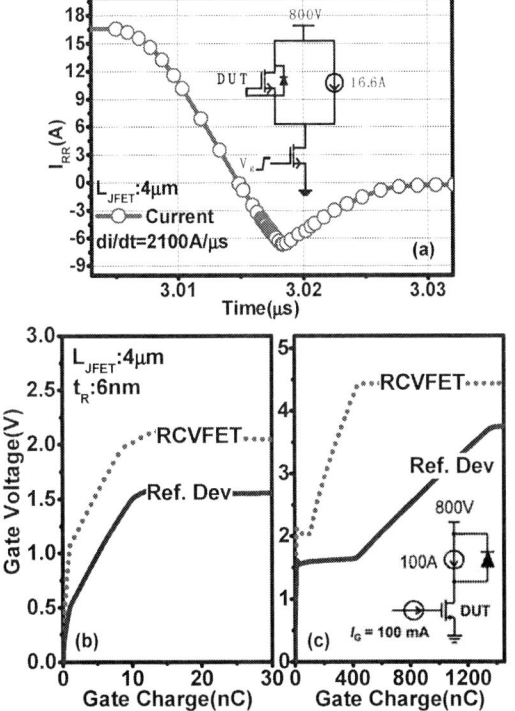

Fig. 8. (a) The reverse recovery characteristics of embedded diode in RCVFET. (b) Gate-to-source charge for RCVFET and Ref. Dev. The inset shows the structure of Ref. Dev. (c) Gate charge for both devices and the active areas are 1cm².

the circuit as illustrated in the inset in Fig. 8 (c). The Q_{GD} for both devices is 80nC and 400nC due to the overlap area between the gate and the drift region of RCVFET is dramatically decreased than that of Ref. Dev.

IV. CONCLUSION

In this work, the split-gate vertical GaN power transistor is proposed with high saturation current, small on-resistance and high breakdown voltage. By introducing a Schottky source between the two split gates, the RCVFET features an intrinsic reverse conduction capability with low turn on voltage of 0.8 V. Owing to the split-gate design, the device also exhibits much smaller gate charge and short reverse recovery time with slightly deteriorated breakdown voltage. Furthermore, the switching loss in a power switching system can reduce thanks to the lower gate charge in RCVFET and elimination of the external diode. Therefore, the proposed RCVFET is of great potential for power conversion system to achieve low power loss and high switching speed.

ACKNOWLEDGMENT

We would like to thank Dr. Ruize Sun with the National University of Singapore and Kai Zhang with State Key Laboratory of Electronic Thin Films and Integrated Devices of UESTC for their helps on device and mixed-mode simulation.

REFERENCES

[1] U. K. Mishra, P. Parikh, Yi-Feng Wu. "AlGaN/GaN HEMTs—An Overview of Device Operation and Applications," *Proceedings of the IEEE*, vol. 90, no. 6, pp. 1022-1031, Jun. 2002.

[2] Tetsu Kachi *et al.*, "Recent progress of GaN power devices for automotive applications," *Japanese Journal of Applied Physics*, vol. 53, no. 10, Sept. 2014.

[3] Nariaki Ikeda *et al.*, "GaN Power Transistors on Si Substrates for Switching Applications," *Proceedings of the IEEE*, vol. 98, no. 7, pp. 1151-1161, Jul. 2010.

[4] O. Ambacher *et al.*, "Two dimensional electron gases induced by spontaneous and piezoelectric polarization in undoped and doped AlGaN/GaN heterostructures," *Journal of Applied Physics*, vol. 87, no. 1, Jan. 2000.

[5] Sten Heikman *et al.*, "Polarization effects in AlGaN/GaN and GaN/AlGaN/GaN heterostructures," *Journal of Applied Physics*, vol. 93, no. 12, Jun. 2003.

[6] Tohru Oka, Tomohiro Nozawa. "AlGaN/GaN Recessed MIS-Gate HFET with High-Threshold-Voltage Normally-Off Operation for Power Electronics Applications," *IEEE Electron Device Letters*, vol. 29, no. 7, pp. 668-670, Jul. 2008.

[7] Gaudenzio Meneghesso *et al.*, "Reliability of GaN High-Electron-Mobility Transistors: State of the Art and Perspectives," *IEEE Transactions on Device and Materials Reliability*, vol. 8, no. 2, pp. 332-343, Jun. 2008.

[8] Ramakrishna Vetury *et al.*, "The Impact of Surface States on the DC and RF Characteristics of AlGaN/GaN HFETs," *IEEE Transactions on Electron Devices*, vol. 48, no. 3, pp. 560-566, Mar. 2001.

[9] Huili Xing *et al.*, "High Breakdown Voltage AlGaN–GaN HEMTs Achieved by Multiple Field Plates," *IEEE Electron Device Letters*, vol. 25, no. 4, pp. 161-163, Apr. 2004.

[10] Yaakov IB *et al.*, "AlGaN/GaN current aperture vertical electron transistors with regrown channels," *Journal of Applied Physics*, vol. 95, no. 4, Jan. 2004.

[11] Srabanti Chowdhury *et al.*, "CAVET on Bulk GaN Substrates Achieved With MBE-Regrown AlGaN/GaN Layers to Suppress Dispersion," *IEEE Electron Device Letters*, vol. 33, no. 1, pp. 41-43, Jan. 2012.

[12] Ramya Yeluri *et al.*, "Design, fabrication, and performance analysis of GaN vertical electron transistors with a buried p/n junction," *Applied Physics Letters*, vol. 106, no. 18, May 2015.

[13] Srabanti Chowdhury *et al.*, "Enhancement and Depletion Mode AlGaN/GaN CAVET with Mg-Ion-Implanted GaN as Current Blocking Layer," *IEEE Electron Device Letters*, vol. 29, no. 6, pp. 543-545, Jun. 2008.

[14] Masakazu Kanechika *et al.*, "A Vertical Insulated Gate AlGaN/GaN Heterojunction Field-Effect Transistor," *Japanese Journal of Applied Physics*, vol. 46, Part 2, no. 20–24, May 2007.

[15] Tohru Oka *et al.*, "1.8 mΩ·cm² vertical GaN-based trench metal–oxide–semiconductor field-effect transistors on a free-standing GaN substrate for 1.2-kV-class operation," *Applied Physics Express*, vol. 8, no. 5, Apr. 2015.

[16] Chirag Gupta *et al.*, "In Situ Oxide, GaN Interlayer-Based Vertical Trench MOSFET (OG-FET) on Bulk GaN substrates," *IEEE Electron Device Letters*, vol. 38, no. 3, pp. 353-355, Mar. 2017.

[17] Hui Nie *et al.*, "1.5-kV and 2.2-mΩ-cm² Vertical GaN Transistors on Bulk-GaN Substrates," *IEEE Electron Device Letters*, vol. 35, no. 9, pp. 939-941, Sept. 2014.

[18] J. P. Ibbetson *et al.*, "Polarization effects, surface states, and the source of electrons in AlGaN/GaN heterostructure field effect transistors," *Applied Physics Letters*, vol. 77, no. 2, Mar. 2000.

[19] Loizos Efthymiou *et al.*, "Zero reverse recovery in SiC and GaN Schottky diodes a comparison," in *Proceedings of the 28th International Symposium on Power Semiconductor Devices and IC's (ISPSD)*, Prague, Czech Republic, Jun. 2016, pp. 71-74.

[20] Masaki Ueno *et al.*, "Fast recovery performance of vertical GaN Schottky barrier diodes on low-dislocation-density GaN substrates," in *Proceedings of the 26th International Symposium on Power Semiconductor Devices and IC's (ISPSD)*, Waikoloa, Hawaii, Jun. 2014, pp. 309-312.

Proceedings of the 30th International Symposium on Power Semiconductor Devices & ICs
May 13-17, 2018, Chicago, USA

Experimental Characterization of the Fully Integrated Si-GaN Cascoded FET

Jie Ren[1], Chak Wah Tang[1], Hao Feng[1], Huaxing Jiang[1], Wentao Yang[1], Xianda Zhou[1, 2], Kei May Lau[1], and Johnny K.O. Sin[1]

[1]Department of Electronic and Computer Engineering, the Hong Kong University of Science and Technology
Clear Water Bay, Kowloon, Hong Kong
[2]School of Electronics and Information Technology, Sun Yat-Sen University, Guangzhou, China
Email: jrenab@connect.ust.hk; eesin@ust.hk

Abstract—In this paper, fabrication process of the fully integrated Si-GaN cascoded field effect transistor is presented first. Device performance of the individual Si MOSFET and AlGaN/GaN MIS-HEMT, which are fabricated along with the Si-GaN cascoded FET, are then characterized. Finally, the device performance of the fully integrated cascoded FET is characterized in detail, including the transconductance compression at high V_{GS} and high V_{DS}, the reverse conduction, the power figure-of-merit (FOM), and the high-temperature characteristics. It is shown that the Si-GaN monolithic cascoded FET is a promising candidate for high performance power switching applications.

Keywords—monolithic integration; cascode configuration; power FETs; GaN; MIS-HEMT; power switching electronics.

I. INTRODUCTION

As the fundamental components of modern power electronic systems, power semiconductor devices are widely used in various electrical energy conversion and control applications. Conventional power semiconductor devices are Si-based. In the past few years, wide bandgap semiconductors, such as SiC and GaN, are attracting extensive research interests. Thanks to the low on-resistance and low junction capacitance, GaN-based high electron mobility transistors (HEMTs) are particularly suitable for high-frequency power switching applications [1].

However, the parasitics introduced in device assembly with the Si-based driver MOSFETs/ICs increase the switching power loss and decrease the system reliability [2-5]. In order to fully utilize the high-speed properties of the GaN-based power transistor, a technology to fully integrate Si MOSFETs and GaN HEMTs on the same substrate in very close proximity is highly demanded. Using metal-line interconnection on IC instead of wire bonding or flip chip bonding in the multichip approach will dramatically reduce the parasitic inductance, and therefore the switching loss, and system reliability will be improved. A monolithically integrated Si-GaN cascoded FET has been demonstrated previously [6]. The fabrication process of this integration technology will be presented in this work. Furthermore,

detailed characterization results of the fabricated Si-GaN cascoded FET will also be presented.

II. DEVICE STRUCTURE AND FABRICATION

As shown in Fig. 1 (a), the Si-GaN cascoded FET is formed by series connection of a low-voltage normally-off Si MOSFET and a high-voltage normally-on GaN MIS-HEMT. A top view image of the fabricated cascoded FET is shown in Fig. 1 (b).

The process begins with a 4-inch n- Si (111) wafer. The active regions of Si devices are isolated by local oxidation of silicon (LOCOS) with 300 nm field oxide. Then, the Si and GaN device regions are isolated by etched narrow trenches (with width of 3 µm and depth of 4 µm into the Si substrate) on the Si device region. Next, 290 nm poly-Si is deposited right after the 30 nm gate oxide. Then, 1.5 µm SiO$_2$ is deposited and it will work as the SEG mask in the following GaN epitaxial growth. 150 nm SiN$_3$ is deposited by LPCVD on top of the SEG mask. The schematic cross-sectional view of the wafer with the above layers is shown in Fig. 2 (a). Next, the SiN$_3$/SiO$_2$/poly-Si/SiO$_2$/Si stack on the GaN device region is removed by dry etching to form recessed windows for SEG of the AlGaN/GaN epitaxial structure. The recessed windows have an etching depth of 3 µm into the Si substrate. After that,

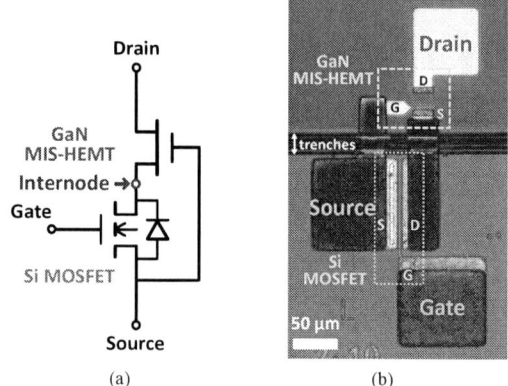

Fig. 1. (a) Circuit schematic of the cascoded FET, and (b) a top view image of the fabricated cascoded FET.

This work was supported by the Research Grants Council of Hong Kong under the General Research Fund (GRF) 16212415 and Grant T23-612/12-R, and also supported by the Introduction of Innovative R&D Team project.

978-1-5386-2928-4/18 $31.00 © 2018 IEEE

4 times of 0.5 μm sacrificial oxide are grown and removed to smoothen the bottom and sidewall Si surfaces in the windows before the subsequent epitaxial growth. The SiN₃ is used as the wet etching mask to protect the underneath SiO₂ during the sacrificial oxide removal step. The final depth of the recessed windows is approximately 4 μm after smoothing. At the end of the substrate preparation, the SiN₃ is removed by hot H₃PO₄, as shown in Fig. 2 (b).

After substrate preparation, approximately 4 μm of AlGaN/GaN epitaxial layers with 8 nm *in-situ* SiN cap are selectively grown in the Si recessed windows by MOCVD [7]. As shown in Fig. 2 (c), a near planar configuration of the GaN and Si device surface is achieved, which is important for fine-pattern lithography. Subsequently, the Si lateral diffused MOSFET processes are carried out, including poly Si gate doping and etching, p-body implantation and drive-in. It needs to be mentioned that SiO₂ is used to cover the GaN epitaxial layers to protect the surface from decomposition during the high-temperature drive-in step of the p-body region. After that, the source/drain implantation of phosphors and the p-body contact implantation of boron are implemented. As shown in Fig. 2 (d), the 330 nm SiO₂ passivation layer is deposited by high-frequency PECVD followed by dopants activation and annealing.

The AlGaN/GaN MIS-HEMT is processed accordingly. The in-situ SiN is selectively etched to open the source/drain windows for the subsequent 20/150/50/80 nm Ti/Al/Ni/Au stack deposition. A 850 °C, 30 sec rapid thermal annealing (RTA) is carried out to form ohmic alloy. This RTA step is the last high-temperature step. After that, the contact holes of Si devices are opened and the Al metal pad are formed by sputtering and lift-off. Forming gas annealing is conducted at 420 ⁰C for 30 min. At this stage, the Si devices are done and ready for measurement (Fig. 2(e)).

Then the GaN devices are isolated by multiple-energy argon implantation. Approximately 19 nm of ALD ZrO₂ is deposited at 200 ⁰C and acts as the gate dielectric together with the in-situ SiN. Then, the gate electrode is formed by 20/200 nm Ni/Au metal evaporation, as shown in Fig. 2 (f). Finally, the Si and GaN devices are interconnected by Al metal-lines (Fig. 2 (g)).

III. DEVICE PERFROMANCE AND DISCUSSION

A. *Characteristics of the Si MOSFET*

Device performance of the individual Si MOSFET and AlGaN/GaN MIS-HEMT, which are fabricated along with the Si-GaN cascoded FET, are characterized. The linear scale transfer characteristics of the Si MOSFET is shown in Fig. 3 (a). The threshold voltage (V_{th}) determined by the linear extrapolation method is extracted to be 2.8 V. The on/off current ratio is measured to be 4×10^{10} as shown in Fig. 3 (b). The subthreshold slop (*SS*) is calculated to be 160 mV/dec. The relatively high *SS* is due to the thick gate oxide, which is used to enable the large gate voltage swing. The dc output characteristics is shown in Fig. 4 (a). The dc specific on-

resistance ($R_{on,sp}$) is extracted as 0.21 mΩ·cm² (on-resistance normalized to Si active area) and 7 Ω·mm (on-resistance normalized to W_{Si}) at a gate bias of 15 V. The off-state I-V characteristics is shown in Fig. 4. (b) which demonstrates the avalanche capability of the Si MOSFET with a breakdown voltage (V_{BD}) of 31 V.

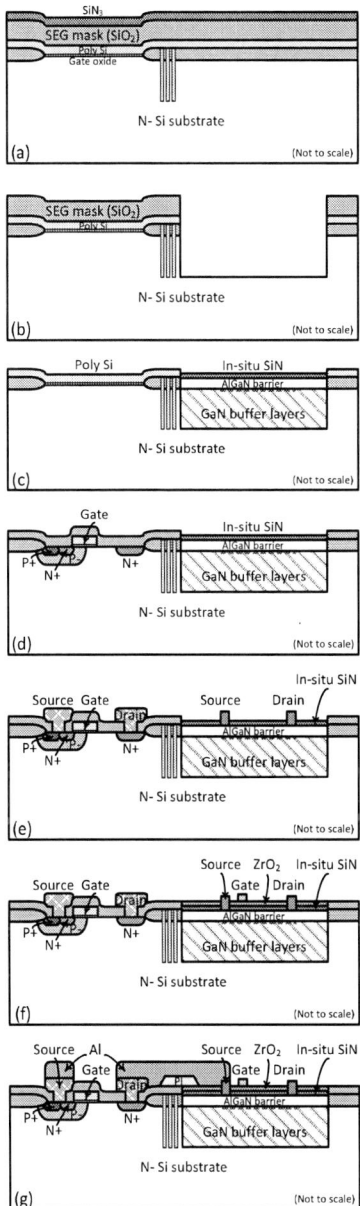

Fig. 2. Schematic cross-sections of (a) after LOCOS, narrow trenches etching and films deposition, (b) after recessed windows formation, (c) after GaN epitaxial structure growth, (d) after Si MOSFET junction formation, (e) after GaN ohmic contacts formation and Si metallization, (f) after GaN MIS-HEMT fabrication, and (g) after interconnection metal-line deposition.

B. Characteristics of the AlGaN/GaN MIS-HEMT

Figure 5 (a) shows the semi-log scale transfer characteristics of the AlGaN/GaN MIS-HEMT, which exhibits an on/off current ratio of 2×10^8 and a steep subthreshold slope (SS) of 78 mV/dec. A V_{th} (determined by the linear extrapolation method) of -6.6 V is measured. The dc output characteristics is shown in Fig. 5 (b). The dc specific on-resistance ($R_{on,sp}$) is extracted as 1.4 mΩ·cm² (on-resistance normalized to GaN active area) and 7.6 Ω·mm (on-resistance normalized to W_{GaN}) at a gate bias of 2 V. V_{BD} of the MIS-HEMT is measured as ~ 700 V at a current density of 1 mA/mm.

C. Characteristics of the Si-GaN Cascoded FET

Figure 6 (a) shows the linear scale dc transfer characteristics of the Si-GaN cascoded FET, which have been normalized to the total active area (the sum of the active area of the Si MOSFET and the GaN MIS-HEMT). The V_{th} determined by the linear extrapolation method is extracted to

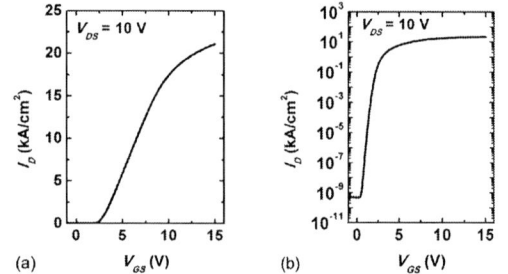

Fig. 3. DC I_D-V_{GS} characteristics in linear scale (a), and semi-log scale (b) of the Si LDMOS FET fabricated with the Si-GaN cascoded FET.

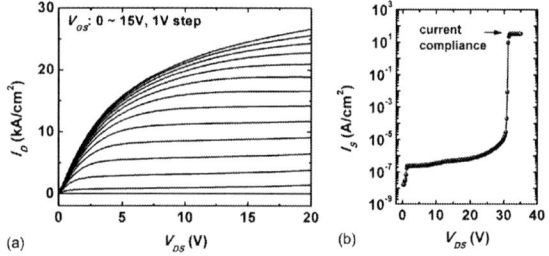

Fig. 4. (a) DC I_D-V_{DS} characteristics, (b) Off-state characteristics in semi-log scale of the Si LDMOS FET fabricated with the Si-GaN cascoded FET.

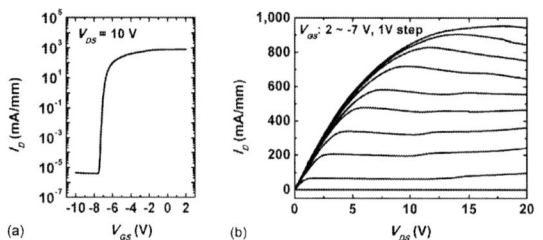

Fig. 5. (a) DC I_D-V_{GS} characteristics in semi-log scale, and (b) DC I_D-V_{DS} characteristics of the AlGaN/GaN MIS-HEMT fabricated with the Si-GaN cascoded FET.

be 3.2 V. There is a transconductance (GM) compression on high V_{GS}, typically for $V_{GS} > 6$ V. It can be seen from Fig. 6 (a) that the voltage dropping across the Si MOSFET (V_{Si}) is around 1 V when $V_{GS} > 6$ V. This means that the Si MOSFET is working in the linear region rather than saturation region. The smaller GM of the Si MOSFET in the linear region compared with that in the saturation region explains the GM compression in the cascoded FET at high V_{GS} and high V_{DS}. It is worth pointing out that the GM compression is just happened on the cascoded FET working in the saturation region. For the device working in the linear region, both the Si MOSFET and GaN MISHEMT are working in the linear region, and there will be no GM compression. For power switching applications, the power switches are mainly working in the linear region. Therefore, the GM compression in the saturation region will not influence much on the performance of the cascoded FET in power switching applications. Figure 6 (b) shows the semi-log scale dc transfer characteristics. The on/off current ratio is measured to be 2×10^8, and the subthreshold slope (SS) is calculated to be 210 mV/dec. The relative high SS is due to the high SS of the Si MOSFET. The transfer curve is obtained using double sweeping mode at V_{DS} of 10 V, and no V_{th} shift is observed when the V_{GS} sweeps from 15 V down to 0 V and from 0 V up to 15 V. A very good interface quality of the Si MOSFET is indicated. It can be seen that the gate leakage current is one order of magnitude smaller than the off-state drain leakage current.

During the reverse conduction period, the gate and source are grounded and the cascoded FET is turned off from forward conduction. But the body diode of the Si MOSFET and the GaN MIS-HEMT formed a cascoded diode which can conduct current in reverse conduction, as shown in Fig. 7 (a). The reverse conduction of a switch is important during dead time operation which cannot be avoided in some power management circuits such as the half-bridge inverter operating with complementary driving pulses. The reverse voltage drop is 1.9 V at 500 A/cm². The ideality factor of the cascoded "body diode" is calculated to be 1.3 from Fig. 7 (b), which indicates the good junction property of the Si p-n junction.

Figure 8 presents the power figure-of-merit (FOM) plot of

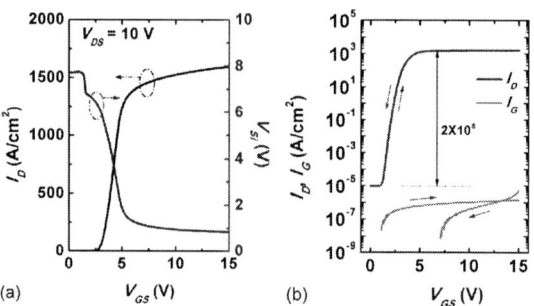

Fig. 6. (a) DC I_D-V_{GS} characteristics of the Si-GaN cascoded FET plotted with the voltage dropping across the Si MOSFET, and (b) DC I_D-V_{GS} and I_G-V_{GS} characteristics of the cascoded FET in semi-log scale with the gate bias negative/positive sweeping at a drain bias of 10V.

$R_{on,sp}$ versus V_{BD} [8]. It shows that the performance of the cascoded FETs is beyond the Si limit and has a power FOM ($V_{BD}^2/R_{on,sp}$) of 162 MW/cm^2. The performance can be further improved by optimizing the size of the Si and GaN devices which constitute the cascoded FET.

The high-temperature measurements are conducted. And the output characteristics of the cascoded FET at room temperature (25 ^{0}C) and 150 ^{0}C are compared in Fig. 9. The $R_{on,sp}$ of the cascoded FET with $L_{GD,GaN}$ of 10 µm increases from 2.3 mΩ·cm^2 at 25 ^{0}C to 3.8 mΩ·cm^2 at 150 ^{0}C. The high-temperature reverse characteristics are measured with a bias of up to 200 V (the voltage is limited by the high-temperature measurement equipment). As shown in Fig. 10, the leakage current of the Si-GaN cascoded FET (with $L_{GD,GaN}$ of 10 µm) increases by 20 times from its room temperature (25 ^{0}C) value to the high-temperature (150 ^{0}C) value. The measurement results demonstrated that the Si-GaN cascoded FET can work properly at 150 ^{0}C.

The switching performance of the monolithic integrated Si-GaN devices has already been demonstrated previously in using the Si-GaN cascoded diode [9]. The switching performance of the cascoded FET was not measured in this work because of the small device size built at the current

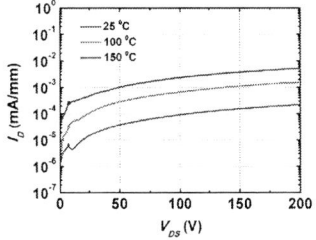

Fig. 10. High temperature measurements of the fully integrated Si-GaN cascoded FET at reverse bias.

stage.

IV. CONCLUSION

In this paper, the fabrication technology of fully integrating the Si MOSFETs and GaN MIS-HEMTs on the same Si substrate is presented. The characterization results of the fabricated cascoded FET demonstrated that the developed technology is compatible with both the Si and GaN fabrication processes, and the Si-GaN monolithic cascoded FET is promising for high-speed power electronic applications.

ACKNOWLEDGMENT

The authors would like to thank the staff of the Nanosystem Fabrication Facility, Photonics Technology Center, and Semiconductor Product Analysis and Design Enhancement Center, the Hong Kong University of Science and Technology, for their help in device fabrication and characterization.

REFERENCES

[1] U.K. Mishra, P. Parikh, and Y.F. Wu, "AlGaN/GaN HEMTs-An Overview of Device Operation and Applications," *Proceedings of the IEEE*, vol. 90, no. 6, pp. 1022-1031, Nov. 2002.

[2] Z. Liu, X. Huang, F. C. Lee, and Q. Li, "Package parasitic inductance extraction and simulation model development for the high-voltage cascode GaN HEMT," *IEEE Trans. on Power Electron.*, vol. 29, no. 4, pp. 1977-1985, Apr. 2014.

[3] X. Huang, Z. Liu, F. C. Lee, and Q. Li, "Characterization and enhancement of high-voltage cascode GaN devices", *IEEE Trans. on Electron Devices*, vol. 62, no. 2, pp. 270-277, Feb. 2015.

[4] Y. Xie, and P. Brohlin, "Optimizing GaN performance with an integrated driver," [Online]. Available: http://www.ti.com/lit/wp/slyy085/slyy085.pdf.

[5] D. Reusch and J. Strydom, "Understanding the effect of PCB layout on circuit performance in a high frequency gallium nitride based point of load converter," *IEEE Trans. Power Electron.*, vol. 29, no. 4, pp. 2008-2015, Apr. 2014.

[6] J. Ren, C.W. Tang, H. Feng, H.X. Jiang, W.T. Yang, X.D. Zhou, K.M. Lau, and J.K.O. Sin, "A Novel 700 V Monolithically Integrated Si-GaN Cascoded Field Effect Transistor," *IEEE Electron Device Lett.*, vol. 39, no. 3, pp. 394-396, Mar. 2018.

[7] J. Ren, C. Liu, C.W. Tang, K.M. Lau, and J.K.O. Sin, "A novel Si-GaN monolithic integration technology for a high-voltage cascoded diode," *IEEE Electron Device Lett.*, vol. 38, no. 4, pp. 501-504, Apr. 2017.

[8] B. J. Baliga, "Gallium nitride devices for power electronic applications," *Semicond. Sci. Technol.*, vol. 28, no. 7, pp. 074011, Jun. 2013.

[9] J. Ren, C. Liu, C.W. Tang, K.M. Lau, and J.K.O. Sin, "Switching Characteristics of Monolithically Integrated Si-GaN Cascoded Rectifiers," in *Proc. 29th Int. Symp. Power Semiconductor Devices ICs (ISPSD)*, Sapporo, Japan, pp. 223-226, May/Jun. 2017.

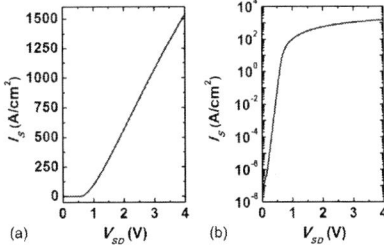

Fig. 7. Characteristics of the reverse conduction in the cascoded FET in linear scale (a), and in semi-log scale (b).

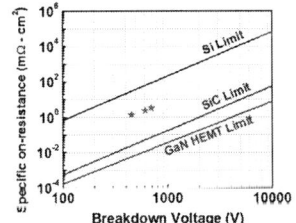

Fig. 8. The power figure-of-merit (FOM) plot of $R_{on,sp}$ versus V_{BD} for the fabricated Si-GaN cascoded FETs.

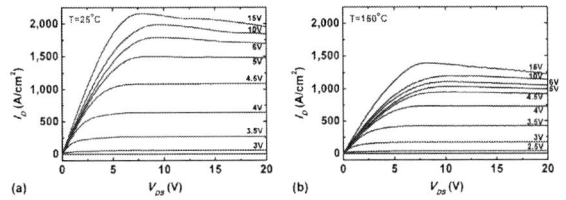

Fig. 9. DC I_D-V_{DS} characteristics of the cascoded FET with $L_{GD,GaN}$ of 10 µm at 25 ^{0}C (a) and 150 ^{0}C (b).

Proceedings of the 30th International Symposium on Power Semiconductor Devices & ICs
May 13-17, 2018, Chicago, USA

Effect of device layout on the switching of enhancement mode GaN HEMTs

Loizos Efthymiou, Gianluca Camuso,
Giorgia Longobardi and Florin Udrea
Department of Engineering,
University of Cambridge
Cambridge, UK
Email: le257@cam.ac.uk

Terry Chien and Max Chen
Vishay General Semiconductor
Taipei, Taiwan

Ayman Shibib
and Kyle Terrill
Vishay Siliconix
Santa Clara, California

Abstract—In this study, a comparison of the on-state and switching performance of different die layout configurations of GaN HEMTs is given. Devices with different layout designs are shown to exhibit varying degrees of susceptibility to oscillatory behaviour when switching. This is demonstrated to be related to the level of source inductance and the unbalances of such inductances within the die rather than the package; this behaviour was quantified using a 3D electromagnetic simulation software. SPICE simulations were used to further support these findings.

I. Introduction

With Gallium Nitride (GaN) products being ramped up for volume production, research has now expanded from the device to the system level. The performance of GaN devices in various power electronic converters such as, boost, buck-boost and half-bridge topologies is being assessed [1][2]. Several publications discuss the challenges circuit design engineers are presented with when using GaN devices due to their fast switching behaviour [1][2]. In particular, attention has been focused on how the parasitic common source inductance (CSI) can create undesirable harmonics and therefore could lead to additional circuit losses [3][4]. The source of CSI could be circuit related or package related [4][5]. More recently, the parallelization of GaN HEMTs has also been attempted [2][6], and the presence of unbalanced circuit-level CSI has again been identified as a major issue. This study shows that the unbalance of inductances at the device design level, from one finger or cell to another is also critical and can induce additional losses and oscillatory behaviour. This puts additional constraints on the layout design and leads to different geometries and configurations than those known in lateral power devices in silicon.

II. Device structure and layout designs

A normally-off AlGaN/GaN HEMT device (650V, 15A) based on p-gate technology was used in this study. The device has a lateral configuration with an AlGaN/GaN heterostructure grown epitaxially on a silicon wafer as shown in Fig. 1. More information on the cross-section of the device used is given in [7]. Devices with different inter-digitated layouts were produced as shown in the schematic designs in Figs. 2, 3 and 4 and were packaged in TO-220 before characterisation. TO-220

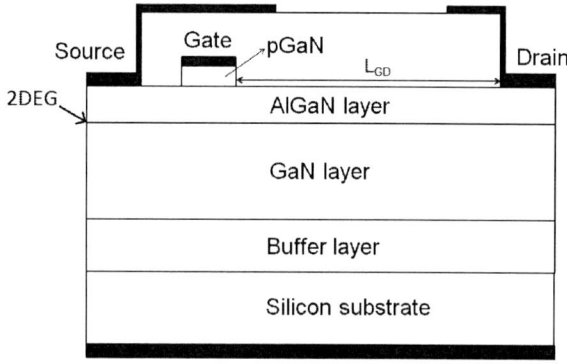

Fig. 1. AlGaN/GaN HEMT cross section

packaging in not very suitable for high frequency applications due to the considerable presence of parasitic components due to the package [8]. In this study, the aim is not to demonstrate state of the art fast switching but rather to compare the switching capability of different layouts thus the use of TO-220 packaging is reasonable. All three layouts (S-D, S-D-S, D-S-D) have an equivalent gate width. The S-D layout has twice the finger length of the other two designs while the S-D-S and D-S-D layout have twice the number of fingers thus maintaining the equivalent gate width. The differences between the three layouts are found in the placement of the contact pads for the different terminals (drain, source, gate) and the metallisation tracks which connect the fingers to the pads. The three different layouts were designed as test structures to investigate the effect of layout on the device performance. During the design, the ability to make good bonding wire contacts to the different contact pads was a priority.

III. Results and analysis

A. On-state performance comparison

A comparison of the specific resistance, $R_{on,spec}$ of the three different layouts was initially made. Measurements were taken using Keithley 2600 source-measurement units and are plotted in Fig. 5. The S-D design (Fig. 2) appears to be

978-1-5386-2928-4/18 $31.00 © 2018 IEEE

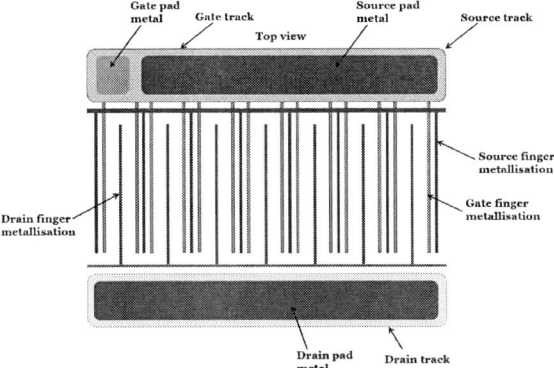

Fig. 2. S-D layout

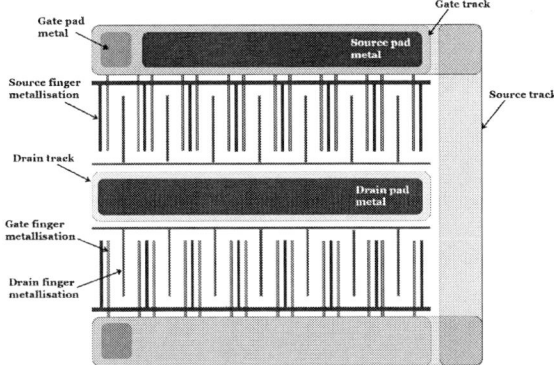

Fig. 3. S-D-S layout

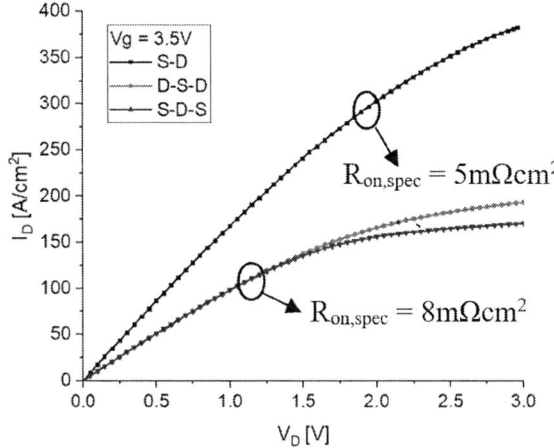

Fig. 5. I_d - V_d characteristic for three different layouts. S-D layout has the lowest $R_{on,spec}$ due to its reduced area for an equivalent gate width and shorter tracks.

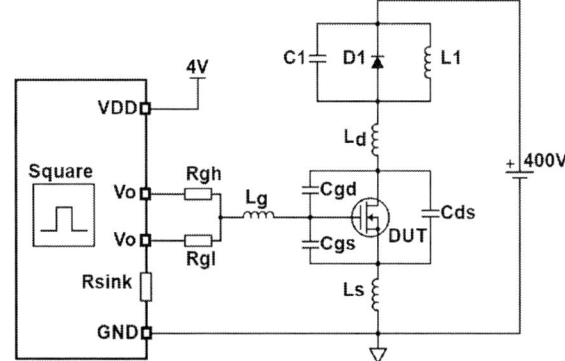

Fig. 6. Clamped inductive switching circuit

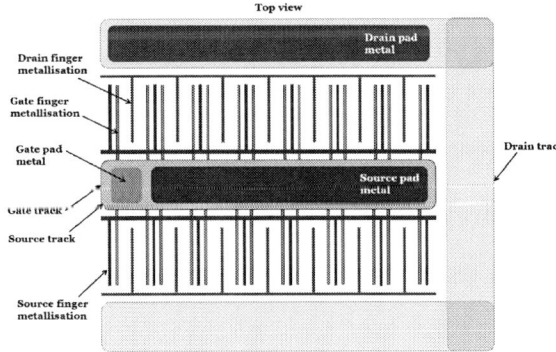

Fig. 4. D-S-D layout

B. Switching performance comparison

The effect that changes in the device layout have on the device switching capability were then investigated. The clamped inductive switching setup shown in Fig. 6 was used to switch the devices using a double pulse test. The switching current was gradually increased until dI/dt related oscillations (see example in Fig. 7) were observed during device turn-off. The current at which oscillations were first observed for the three layouts (S-D, S-D-S, D-S-D) are summarized in Table I. Two scenarios are shown in Table I where different gate drivers and external gate resistances (R_g) were used in the switching circuit. While the maximum current that was switched without oscillations differs in the two scenarios (as expected [4]), the trend in the switching performance for the three different designs remains the same. S-D-S (Fig. 3) and D-S-D (Fig. 4) layouts were found to switch without oscillations at higher currents compared to the S-D design. D-S-D layout was the best performing design.

The difference in the switching capability of the three layouts can be explained when looking at the parasitic source

the best solution when considering just the device on-state current capability. S-D layout has the lowest $R_{on,spec}$ due to its reduced area for an equivalent gate width. Furthermore, the additional track metallisation present in the S-D-S and D-S-D design, compared to the S-D design, leads to an increase in the total resistance of those devices and thus contributes to the increased specific R_{on}.

TABLE I

MAXIMUM SWITCHING CURRENT BEFORE SEVERE OSCILLATIONS ARE OBSERVED FOR THE THREE DIFFERENT LAYOUT DEVICES.

	Driver	R_g	S-D		S-D-S		D-S-D	
1	TC4452V	68Ω	$5A$	$98.7A/cm^2$	$7A$	$110.9A/cm^2$	$13A$	$207.4A/cm^2$
2	TC4432V	10Ω	$2A$	$39.5A/cm^2$	$3A$	$47.5A/cm^2$	$8A$	$127.6A/cm^2$

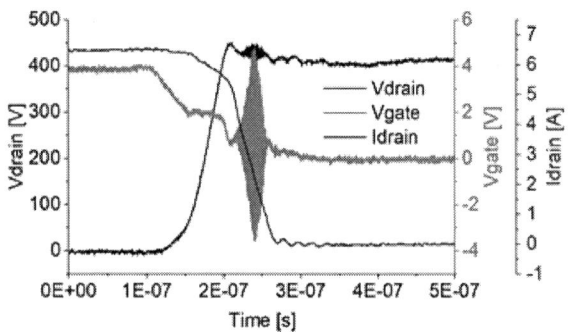

Fig. 7. Example of dI/dt related oscillations observed during HEMT turn-off. Measurement of an S-D-S device shown.

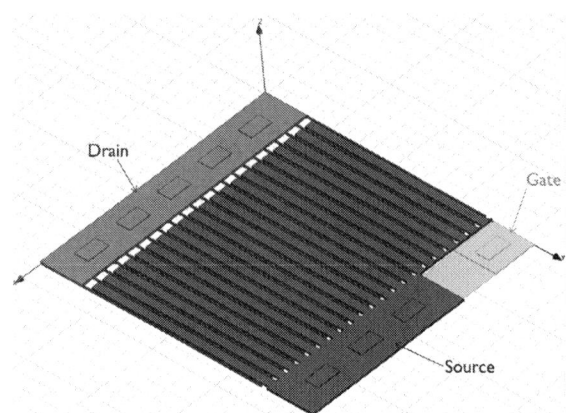

Fig. 8. S-D layout design reproduced in Q3D extractor electromagnetic simulation software for assessment of device inductance.

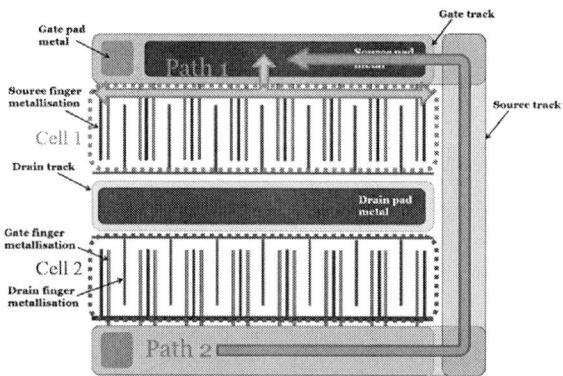

Fig. 9. Example of source inductance imbalance (in the S-D-S layout) at the device level as quantified using Q3D extractor. Additional inductance in path 2 (i.e. Ls2) is approximately 2nH.

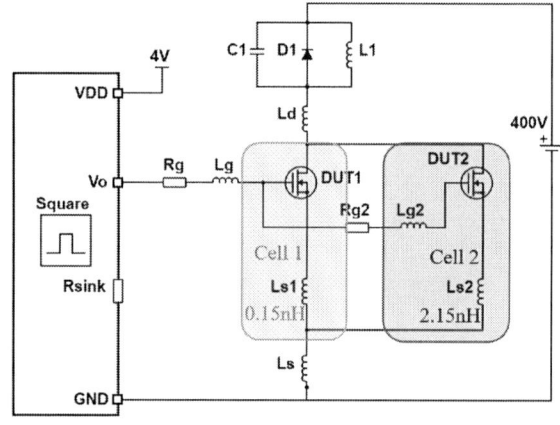

Fig. 10. Two cell equivalent circuit in SPICE to investigate effect of unbalanced inductance (i.e. Ls1 - Ls2 as shown) on switching performance. Rg2, Lg2 set to zero in this example but can also play an important role.

inductance that arises from each design. Electromagnetic simulation software, Q3D extractor, was used to quantify these parasitic parameters (see Fig. 8 for example).

The poor performance of the S-D design can be accounted for when considering that the length of the fingers in this design is twice the size of the fingers in the other two designs which leads to additional inductance to be present per finger (inductance is larger by a factor of 2.25). In the S-D-S design, the source pad is only placed at the top. The group of fingers at the bottom end thus see an additional inductance when conducting current due to the source connection track (approximately 2nH). The D-S-D design which performed the best has both shorter fingers and a shorter source contact pad/finger connection which minimises the source inductance observed. In the D-S-D design the drain pad is only placed on one side and thus the device will have an increased

drain inductance due to the additional connection track. This however does not lead to the same problems that increased source inductance causes.

C. Unbalanced common source inductance (CSI)

It is well known that to achieve high frequency switching the total parasitic CSI in the circuit, and as demonstrated above, the die, has to be minimized [3]. Nevertheless, another very important consideration for high frequency switching is that of unbalanced CSI where attention needs to be paid when parallelizing devices for higher current capabilities. This study investigates whether problems related to unbalanced CSI can

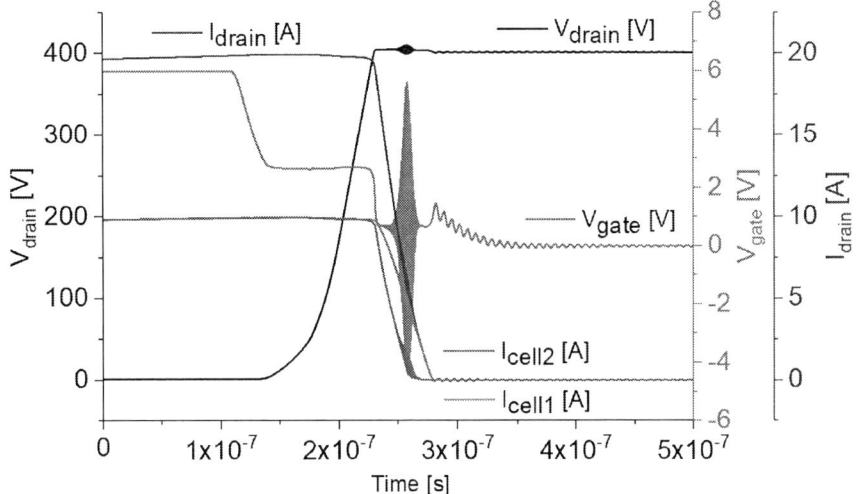

Fig. 11. Example of dI/dt related oscillations during turn-off, observed in SPICE simulations of the circuit shown in Fig. 10.

be an issue at the die level and not just the circuit level for fast switching devices such as GaN HEMTs.

This was done by examining the differences in inductance which arise in areas of the design where asymmetry exists. An example of such asymmetry can be found in the S-D-S design as illustrated in Fig. 9. Cell 2 which includes half the device fingers sees a higher inductive path to the source pad than Cell 1 due to the additional source inductance arising from the source track connection. The value of this additional inductance was estimated using Q3D extractor software to be approximately 2nH.

To understand and to demonstrate the effect of the unbalanced inductance, a representation of the S-D-S device split into two cells was simulated in a SPICE circuit with a clamped inductive switching configuration as seen in Fig. 10. The SPICE models used for the device (DUT1, DUT2) were adjusted versions of the commercially available SPICE models provided by GaN Systems. The adjustments made are related to the device parasitics as described in more detail in [4].

Fig. 11 illustrates some of the issues with unbalanced source inductance during turn-off. The dI/dt rates of the drain current curves (I_{cell1}, I_{cell2}) during turn-off are significantly different in the two cells which can lead to unwanted variations in the energy dissipated at different locations in the die. Furthermore, quite significant oscillations are observed on the gate signal (V_{gate}). Additional simulations under similar conditions were performed to identify whether the unbalance (as opposed to the total level) in CSI was causing the oscillations. It was observed that in an equivalent simulation setup with a single device (of double the size as either DUT1/DUT2) and equivalent parasitics, oscillations were much less severe (not shown here). The unbalanced inductance at die level was thus found to be of sufficient magnitude to harm the switching performance.

IV. CONCLUSION

A comparison of on-state and switching performance of different die layout configurations (S-D, S-D-S, D-S-D) of GaN HEMTs was made. Devices with different designs were found to exhibit varying degrees of susceptibility to oscillatory behaviour when switching. The design which showed increased immunity against oscillations was the D-S-D layout. Nevertheless, this was accompanied by a slight increase in specific on-state resistance ($R_{on,spec}$). The oscillatory behaviour was found to be related to the level of source inductance and the unbalances of such inductances within the die rather than the package; this behaviour was quantified using a 3D electromagnetic simulation software. SPICE simulations were used to further support these findings. The findings of this study are useful in demonstrating that extra care must be taken when designing the layout of GaN HEMT devices for use in high frequency applications.

REFERENCES

[1] R Mitova et al., "Investigations of 600 V GaN HEMT and GaN Diode for Power Converter Applications," in IEEE Transactions on Power Electronics, vol. 29, no. 5, pp. 2441-2452, May 2014.

[2] J. Lu et al., "Design consideration of gate driver circuits and PCB parasitic parameters of paralleled E-mode GaN HEMTs in zero-voltage-switching applications," 2016 IEEE APEC, Long Beach, CA, 2016, pp. 529-535.

[3] A. Lidow et al., "GaN Transistors for efficient power conversion," Power Conversion Publications, El Segundo, California, USA, 2012.

[4] L. Efthymiou et al., "On the source of oscillatory behaviour during switching of power enhancement mode GaN HEMTs," Energies Journal, vol. 10, no. 3, 2017.

[5] Z. Liu et al., "Package Parasitic Inductance Extraction and Simulation Model Development for the High-Voltage Cascode GaN HEMT," in IEEE Transactions on Power Electronics, vol. 29, no. 4, pp. 1977-1985, April 2014.

[6] J. L. Lu et al., "Paralleling GaN E-HEMTs in 10kW100kW systems," 2017 IEEE APEC, Tampa, FL, 2017, pp. 3049-3056.

[7] L. Efthymiou et al., "On the physical operation and optimization of the p-GaN gate in normally-off GaN HEMT devices," Applied Physics Letters, vol. 110, no. 12, 2017, pp. 123502.

[8] AN-9005, "Driving and Layout Design for Fast Switching Super-Junction MOSFETs," Fairchild, 2013.

Proceedings of the 30th International Symposium on Power Semiconductor Devices & ICs
May 13-17, 2018, Chicago, USA

A balancing method for low R_{on} and high V_{th} Normally-off GaN MISFET by preserving a damage-free thin AlGaN barrier layer

Jialin Zhang[1], Liang He[1], Liuan Li[1], Yiqiang Ni[1], Taotao Que[1], Zhenxin Liu[1], Wenjing Wang[1], Jiexin Zheng[1],

Yanfen Huang[1], Jia Chen[1], Xin Gu[1], Yawen Zhao[1], Lei He[1,2], Zhisheng Wu[1,3], and Yang Liu[1,2,3*]

[1] School of Electronics and Information Technology,

[2] Institute of Power Electronics and Control Technology,

[3] State Key Laboratory of Optoelectronic Materials and Technologies,

Sun Yat-Sen University, Guangzhou, People's Republic of China

*email: liuy69@mail.sysu.edu.cn

Abstract—Partially AlGaN recessed scheme based on selective area growth was experimentally demonstrated to improve the performance of normally-off GaN MISFET. The damage-free thin AlGaN barrier layer with lower Al-content in recessed region contributes to a positive V_{th} shift compared with the reference one (from 1.8 V to 2.5 V). At the same time this method realizes a high peak μ_{FE} of 2033 cm^2/V·s and a low gate channel sheet resistance of 519 $\Omega/\square$ (@V_g = 12 V), which is a significant improvement compared with the fully recessed-gate device. The higher Al-contents AlGaN barrier layer regrown in accessed region is adopted to maintain high-conductivity 2DEG transport property. As a result, a maximum drain current of 645 mA/mm and a low on-resistance of 6.8 Ω·mm are obtained. The GaN MISFET also exhibits a low hysteresis, low gate leakage and slight current collapse. This technique could fabricate very promising normally-off GaN devices by designing thickness and Al-content of the controllable-growth thin AlGaN barrier layer.

Keywords—thin AlGaN barrier, GaN MISFET; normally-off; threshold voltage; field-effect mobility.

I. INTRODUCTION

AlGaN/GaN-based heterojunction field-effect transistors (HFETs) on Si substrate have great potentiality and market in the application of high-performance power switching devices, owing to superior figure of merits of the devices [1, 2]. Conventional AlGaN/GaN HFETs exhibit naturally normally-on operation with a negative threshold voltage (V_{th}). For fail-safety, normally-off operation are particularly desirable and indispensable. During the past decades, several techniques

The work was partially supported by the National Key Research and Development Program (Grant No. 2017YFB0402801), National Natural Science Foundation of China (Grant No. U1601210), Guangdong Natural Science Foundation (Grant No. 2015A030312011), Science and Technology Plan of Guangdong Province, China (Grant No. 2017B010112002), International Science and Technology Collaboration Program of Guangzhou City, China (Grant No. 201604030055).

such as recessed-gate structure, fluorine implanting, and p-type gate [1-5] were proposed to achieve normally-off GaN devices.

Among those structures, fully recessed-gate GaN metal-insulator-semiconductor field-effect transistor (MISFET) is a common solution because of its larger tolerance in gate swing voltage and more positive V_{th}, but it suffers from very low channel mobility because of serious MIS interface scattering, leading unsatisfactory on-resistance (R_{on}) issue [6-7]. Compared with the fully recessed-gate, the partially recessed-gate structure can achieve higher gate channel mobility due to the formation of AlGaN/GaN hetero-interface with several nanometers residual AlGaN thin barrier layer (TBL). Some methods have been proposed to develop thin barrier partially recessed-gate GaN devices. By using a etch-stop structure/process and subsequent TMAH treatment, Lu *et al.* demonstrated a normally-off GaN MISFET with a maximum effective channel mobility of 1131 cm^2/V·s [8]. Based on high-temperature ICP etching process, Huang *et al* fabricated partial AlGaN recessed-gate MISFET featuring high maximum drain current ($I_{d,max}$) and low on-resistance [3]. However, it requires a very rigorous etching conditions and precision to control the recess process. A self-terminating wet-etching technique was developed by Wang' group keeping 4 nm AlGaN barrier to obtain high channel mobility of 1400 cm^2/V·s [9]. However, the V_{th} of the thin barrier devices is generally smaller than 1 V.

Alternatively, selective area growth (SAG) technique can provide a powerful method to fabricate recessed-gate structure and suppress MIS interface-related reliability issue. In our previous work, a partial recessed-gate AlGaN/GaN MISFET was successfully fabricated [10]. In this study, we present the performance improvement of normally-off GaN MISFET with partial AlGaN recessed-gate structure by using SAG method. A low Al-contents (20%) Al$_{0.2}$GaN TBL was grown to remain

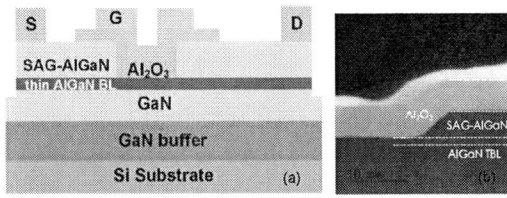

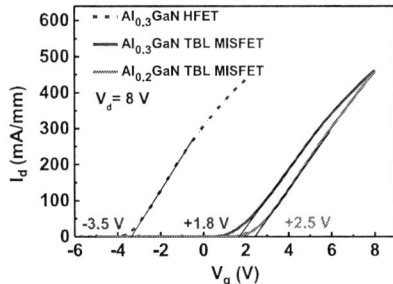

Fig. 1 (a) Schematic cross section of fabricated recessed gate Al_2O_3/AlGaN/GaN MISFET. (b) TEM cross-sectional view of the recessed-gate edge.

Fig. 3 Transfer characteristics of $Al_{0.3}$GaN HFET, $Al_{0.3}$GaN TBL MISFET, and $Al_{0.2}$GaN TBL MISFET (linear scale).

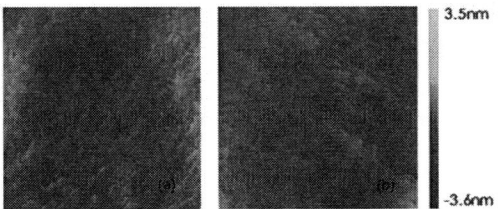

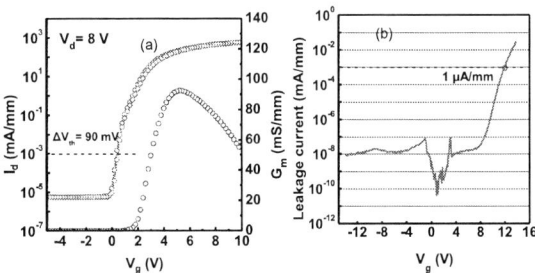

Fig. 2 Surface morphology of (a) gate-recessed region and (b) regrown AlGaN barrier layer with a scanned area of 5×5 μm^2.

Fig. 4 (a) Transfer characteristics of $Al_{0.2}$GaN TBL MISFET (semilog scale) and (b) I_g leakage with V_g sweep from -14 ~ +14V.

an AlGaN/GaN heterojunction at the recessed-gate region, contributing to a more positive V_{th} and achieving a higher gate channel mobility. At the same time, the higher Al-content (30%) $Al_{0.3}$GaN BL was selectively grown on the access area to maintain high conductivity 2DEG transport property and naturally form a recessed gate without lattice damage. Based on the structure, the conductivity property and threshold voltage of fabricated GaN MISFET are both improved greatly.

II. DEVICE PROCESS

Fig. 1 (a) shows the schematic cross-section view of the GaN MISFET. The $Al_{0.2}$GaN/GaN template was grown on a 2-inch diameter Si (111) substrate by MOCVD. The epitaxial layer mainly consists of a GaN buffer layer, a GaN channel layer and an $Al_{0.2}$GaN TBL. Then a thick $Al_{0.3}$GaN barrier was selectively grown on the $Al_{0.2}$GaN/GaN template with the gate region masked to naturally form a recessed-gate structure without lattice damage. Fig. 1 (b) shows the cross-sectional transmission electron microscopic (TEM) image of the device fabricated by SAG. A recessed gate structure could be clearly distinguished. The surface morphology of gate-recessed region and regrown AlGaN barrier layer was measured by AFM with a scanned area of 5×5 μm^2, shown in Fig. 2. The roughness of gate-recessed region and regrown barrier layer are 0.8 nm and 1.1 nm, respectively. The surface morphology could be further optimized by improving epitaxial growth. For comparison, a reference $Al_{0.3}$GaN TBL recessed-gate structure and the corresponding $Al_{0.3}$GaN HFET were also fabricated by regrowing 20 nm $Al_{0.3}$GaN on $Al_{0.3}$GaN/GaN template.

In the device fabrication process [11], the gate-region mask

was removed by a buffered oxide etch (BOE) solution. The device isolation was performed by ICP etching using BCl_3/Cl_2 gas mixture. After that, high-quality Al_2O_3 was deposited at 300 ℃ by atomic layer deposition (ALD) as gate dielectric layer. Ti/Al/Ni/Au-based ohmic contacts are formed by e-beam evaporation and annealed at 830 ℃ in N_2 ambient. Finally, Ni/Au was deposited as gate metal. The ohmic contact resistance (R_c) and sheet resistance (R_{sh}) of the $Al_{0.2}$GaN TBL MISFET are measured to be 1 Ω·mm and 418 $\Omega/\square$ from TLM pattern, respectively.

III. RESULT AND DISCUSSION

Transfer characteristics of $Al_{0.3}$GaN HFET, $Al_{0.3}$GaN TBL MISFET, and $Al_{0.2}$GaN TBL MISFET was shown in Fig. 3. The reference HFET features a normally-on operation with a V_{th} of -3.5 V, which is similar with the as-grown HFET without regrowth process. For the recessed-gate $Al_{0.3}$GaN TBL MISFET, normally-off operation is realized with the V_{th} shifts to approximately 1.8 V. Furthermore, a higher V_{th} of 2.5 V is obtained by employing the $Al_{0.2}$GaN TBL compared with the $Al_{0.3}$GaN TBL sample. Those results demonstrate that the threshold voltage can be effectively controlled by optimizing the Al-content of thin AlGaN barrier layer. Moreover, the $Al_{0.2}$GaN TBL MISFET was further analysis, as

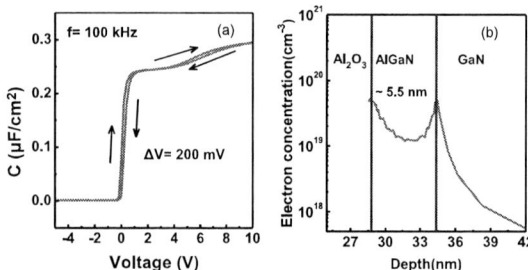

Fig. 5 (a) Double sweep C-V curve of GaN MIS diode at f=100 kHz, (b) Electron distribution extracted by differentiating $1/C^2$ against V.

shown in the transfer curve in semilog scale in Fig. 4 (a), the pinched off characteristics is implemented at $V_g = 0$, which is a true normally off operation for GaN MISFET. The device also exhibits a large I_{on}/I_{off} current ratio of more than 10^8 and a high $g_{m,max}$ of 92 mS/mm. A small V_{th} hysteresis of 90 mV is obtained in double I-V sweep, which indicates an improved interface with lower interface states density by growing the AlGaN TBL. Moreover, gate leakage current with V_g sweep from -14 ~ +14 V is plotted in Fig 4 (b) and the I_g is only 1 µA/mm at $V_g = 12$ V.

In order to investigate the MIS interface characteristics, the capacitance-voltage (C–V) measurements were performed on the recessed $Ni/Al_2O_3/Al_{0.2}GaN/GaN$ MIS diode with a rectangular area of 100 µm×100 µm, as shown in Fig. 5 (a). The CV curve with two rising edge shows steep slope and small hysteresis, which indicates a high-quality MIS interface. Using differentiating $1/C^2$ against V, the electron distribution was extracted in Fig. 5 (b), showing two peaks at the AlGaN/GaN interface and the $AlGaN/Al_2O_3$ interfaces. The thickness of AlGaN TBL can be estimated to approximate 5.5 nm from Fig. 5 (b), which is consistent with the TEM measurement presented in Fig 1 (b). Moreover, the density of interface states are estimated to be 10^{11} cm^{-2} order of magnitude at the $Al_2O_3/AlGaN/GaN$ interface. The small D_{it} value also confirms the high-quality MIS interface between Al_2O_3 and thin AlGaN/GaN.

The output property of the device was shown in Fig.6 (a), in which the maximum drain current is 645 mA/mm and the on resistance (R_{on}) is 6.8 Ω·mm at $V_g = 12$ V. The R_{on} (with the unit of Ω·mm) can be expressed by the following expression:

$$R_{on} = 2R_C + R_{sh} \times (L_{gd} + L_{gs}) + R_{sh\text{-}ch} \times L_g, \quad (1)$$

where R_c, R_{sh}, $R_{sh\text{-}ch}$, L_g are the ohmic contact resistance, sheet resistance of access region, recessed gate channel sheet resistance, and the length of recessed gate, respectively. The extracted $R_{sh\text{-}ch}$ of 519 Ω/□ is slightly larger than the square resistance in the access region R_{sh} (418 Ω/□). For comparison, our previous work [12] shows that the calculated $R_{sh\text{-}ch}$ for Al_2O_3/GaN MISFET without TBL by SAG is approximately 3300 Ω/□ (Fig. 7 (a)), implying that the introduction

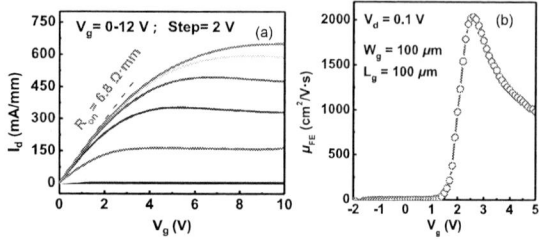

Fig. 6 (a) Output characteristics of $Al_{0.2}GaN$ TBL MISFET. ($L_g/W_g/L_{gs}/L_{gd}$ =2/100/4/5µm) (b) The μ_{FE} of FAT-$Al_2O_3/Al_{0.2}GaN/GaN$ MISFET at $V_d =$ 0.1 V ($L_g/W_g = 100/100$ µm).

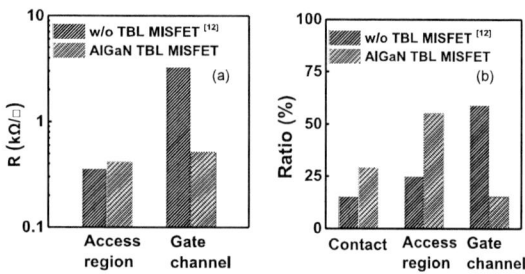

Fig. 7. (a) Sheet resistance in gate region and in access region (b) the proportion of each part resistance for w/o TBL MISFET and AlGaN TBL MISFET

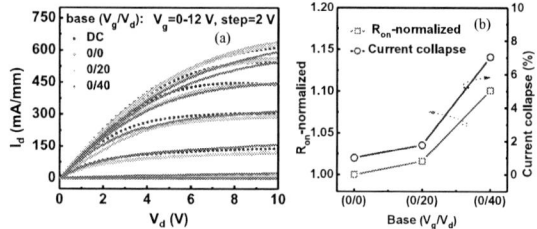

Fig. 8 (a) The output I–V characteristics based on the double pulse-mode with gate/drain quiescent biases ($V_{g, base}$, $V_{d, base}$) of (0 V, 0 V), (0 V, 20 V), and (0 V, 40 V). (b) Dynamic on-resistance and current collapse extracted by double pulse-mode output I–V characteristics. (L_{gd}= 10 µm).

of AlGaN TBL contributes to low on-resistance and high saturation current. As shown in Fig. 7 (b), the percent of each part resistance in the MISFETs contributed to the R_{on} is illustrated. The ratio of $R_{sh\text{-}ch}$ for AlGaN MISFET is four time smaller than w/o TBL MISFET, which is a significant decrease. The corresponding field-effect mobility (μ_{FE}) extracted from FAT-$Al_2O_3/Al_{0.2}GaN/GaN$ MISFET with $W_g/L_g = 100/100$ µm. Using $\mu_{FE} = g_m L_g/(W_g C_{MIS} V_d)$, a high field-effect mobility of 2033 cm^2/V·s was achieved at peak point measured at $V_d = 0.1$ V, indicating the high conductivity channel at $Al_2O_3/AlGaN/GaN$ interface (Fig. 6 (b)).

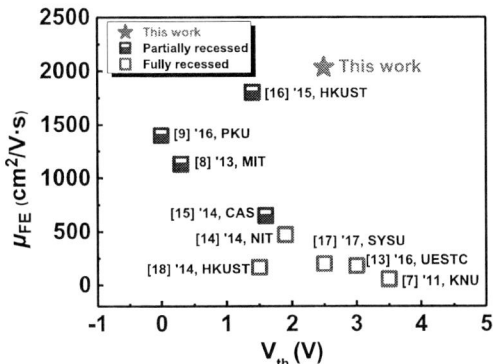

Fig.9. Benchmark of μ_{FE} and V_{th} of normally-off GaN MISFETs.

Double pulse-mode output $I\text{-}V$ measurements were performed to investigate the on-resistance stability of the MISFET, as shown in Fig. 8 (a). The pulse width/period were 1 ms/20 ms and the three gate/drain quiescent off-state biases ($V_{g,\,base}$, $V_{d,\,base}$) were set at (0 V, 0 V), (0 V, 20 V) and (0 V, 40 V), respectively. The percentage of maximum current collapse at quiescent bias of (0 V, 40 V) compared with (0 V, 0 V) reference is only 7%, and the corresponding dynamic R_{on} is increased by10% [Fig. 8 (b)].

Fig. 9 gives a comparison of μ_{FE} and V_{th} for the $Al_{0.2}GaN$ TBL MISFET by SAG technique in this work and some state of-the-art GaN-based normally-off MISFETs on Si substrate with recessed-gate structure (partially recessed and fully recessed) in recent years [7-9, 13-18]. It demonstrates that by employing the low Al-content and thin AlGaN barrier layer, a good balance between higher μ_{FE} (closely relate to R_{on} of the device) and higher V_{th} is realized in our work.

IV. CONCLUSION

Normally-off Al_2O_3/AlGaN/GaN MISFET on Si substrate with good performance was successfully fabricated using SAG technique. A damage-free low Al-contents thin AlGaN barrier layer is employed for remaining an AlGaN/GaN interface at the recessed-gate region and contributed to achieve a more positive V_{th} and high-mobility gate channel. The higher Al-contents AlGaN barrier layer was selectively grown on the template to maintain high-conductivity 2DEG transport property in the access area. The device exhibits improved conductivity property and stable threshold voltage. It is expected that the performance of the device can be further improved by utilizing the controllable-growth thickness and Al-content of thin AlGaN barrier layer.

REFERENCES

[1] K. J. Chen, et al. "GaN-on-Si power technology: Devices and applications." *IEEE Transactions on Electron Devices*, vol. 64, no. 3, pp. 779-795, 2017.

[2] P. Parikh, et al., "650 Volt GaN Commercialization Reaches Automotive Standards." *ECS Transactions.* vol. 80, no. 7, pp. 17-28, 2017.

[3] S. Huang, et al., "High-Temperature Low-Damage Gate Recess Technique and Ozone-Assisted ALD-grown Al_2O_3 Gate Dielectric for High-Performance Normally-Off GaN MIS-HEMTs," in *2014 IEEE International Electron Devices Meeting*, pp. 17.4.1-17.4.4, Dec. 2014.

[4] C. Liu, et al., "Thermally Stable Enhancement-Mode GaN Metal-Isolator-Semiconductor High-Electron-Mobility Transistor with Partially Recessed Fluorine-Implanted Barrier." *IEEE Electron Device Letters.* vol. 36, no. 4, pp. 318-320, 2015.

[5] H. Okita, et al., "Through recessed and regrowth gate technology for realizing process stability of GaN-GITs". *Proc. Int. Symp. Power Semicond. Dev. and ICs.* pp. 23-26, 2016.

[6] T. Hung, et al., "Interface Charge Engineering for Enhancement-Mode GaN MISHEMTs." *IEEE Electron Device Letters.* vol. 35, no. 3, pp. 312-314, 2014.

[7] K. Kim, et al., "Effects of TMAH treatment on device performance of normally off Al_2O_3/GaN MOSFET," *IEEE Electron Device Lett.*, vol. 32, no. 10, pp. 1376-1378, Oct. 2011.

[8] B. Lu, et al., "An etch-stop barrier structure for GaN high-electron-mobility transistors." *IEEE Electron Device Letters.* vol. 34, no. 3, pp. 369-371, 2013.

[9] S. Lin, et al. "A GaN HEMT structure allowing self-terminated, plasma-free etching for high-uniformity, high-mobility enhancement-mode devices." *IEEE Electron Device Letters* vol. 37, no.4, pp. 377-380, 2016.

[10] Y. Wen, Z. He, J. Li, et al., "Enhancement-mode AlGaN/GaN heterostructure field effect transistors fabricated by selective area growth technique," *Appl. Phys. Lett.*, vol. 98, no.7, pp. 072108.1-072108.3, Feb. 2011.

[11] Y. Yao, Z. He, F. Yang, et al., "Normally-off GaN recessed-gate MOSFET fabricated by selective area growth technique," *Appl. Phys. Express.*, vol. 7, no. 1, pp. 016502.1-016502.4, Dec. 2013.

[12] Y. Zheng, F. Yang, L. He, et al., "Selective area growth: A promising way for recessed gate GaN MOSFET with high quality MOS interface." *IEEE Electron Device Letters* vol. 37, no. 9, pp. 1193-1196, 2016.

[13] Y. Shi, et al., "Normally OFF GaN-on-Si MIS-HEMTs Fabricated With LPCVD-SiN_x Passivation and High-Temperature Gate Recess," *IEEE Transactions on Electron Devices,* vol. 63, no. 2, pp. 614-619, 2016.

[14] J. J. Freedsman, et al., "Recessed gate normally-OFF Al_2O_3/InAlN/GaN MOS-HEMT on silicon." *Applied Physics Express* vol. 7, no.10, pp. 104101, 2014.

[15] S. Huang et al., "High RF performance enhancement-mode Al_2O_3/AlGaN/GaN MIS-HEMTs fabricated with high-temperature gate-recess technique," *IEEE Electron Device Letters.* vol. 36, no. 8, pp. 754–756, 2015.

[16] J. Wei, et al., "Low On Resistance Normally-Off GaN Double-Channel Metal–Oxide–Semiconductor High-Electron-Mobility Transistor." *IEEE Electron Device Letters.* vol. 36, no. 12, pp. 1287-1290, 2015.

[17] L. He, F. Yang, et al., "High threshold voltage uniformity and low hysteresis recessed-gate Al_2O_3/AlN/GaN MISFET by selective area growth." *IEEE Transactions on Electron Devices.* vol. 64, no. 4, pp. 1554-1560, 2017.

[18] S. Liu, et al., "Performance enhancement of normally-off Al_2O_3/AlN/GaN MOS channel-HEMTs with an ALD-grown AlN interfacial layer," in *Proc. Int. Symp. Power Semicond. Dev. and ICs.* pp. 362-365, Jun. 2014.

Proceedings of the 30th International Symposium on Power Semiconductor Devices & ICs
May 13-17, 2018, Chicago, USA

Enhancement of Punch-through Voltage in GaN with Buried P-type Layer Utilizing Polarization-induced Doping

Wenshen Li[1], Mingda Zhu[1], Kazuki Nomoto[1], Zongyang Hu[1], Xiang Gao[2], Manyam Pilla[3],
Debdeep Jena[1,4] and Huili Grace Xing[1,4]

[1]School of Electrical and Computer Engineering, Cornell University, Ithaca, NY 14853 USA
[2]IQE RF LLC, Somerset, NJ 08873 USA
[3]Qorvo Inc., Richardson, TX 75080 USA
[4]Department of Material Science and Engineering, Cornell University, Ithaca, NY 14853 USA
Email: wl552@cornell.edu

Abstract—The effect of polarization induced (PI)-doping in GaN buried p-type layer on reverse blocking is studied for the first time. Forward and reverse I-V characteristics is measured on n-p-n diodes. With PI-doping in the buried p-type layer, the reverse punch-through voltage increases from 30 V to 240 V, even with hydrogen passivating the Mg acceptors, indicating the unique advantage of PI-doping on reverse blocking. The enhanced punch-through voltage is attributed to the polarization fixed charge in the p-layer, which is extracted to be ~1.3×10^{17} cm^{-3} and closely matched with the expected value of 1.4×10^{17} cm^{-3}. Vertical trench-MOSFETs with a breakdown voltage of 225 V are also demonstrated on the same sample.

Keywords—GaN; polarization-induced doping; buried p-GaN; punch-through; MOCVD; vertical transistors

I. INTRODUCTION

With the availability of high quality GaN bulk substrate, GaN vertical transistors have seen rapid development in recent years, demonstrating high breakdown voltage and low on-resistance in several device designs [1-9]. GaN vertical power transistors possess unique advantages over the lateral HEMTs including higher power density, better reliability and thermal management. Buried p-type body region is prevalent in GaN vertical power transistors, providing important capabilities including reverse blocking, avalanche and RESURF (reduced surface field). It is well-known that p-GaN gets passivated due to the hydrogen present during epitaxial growth by metal-organic chemical vapor deposition (MOCVD), and requires an post-growth activation process to activate the Mg acceptors. The activation process can be realized at an elevated temperature over 700°C [10], where the Mg-H bonds is broken and H is driven out of the p-GaN surface by diffusion process. The activation of p-GaN is well-established with exposed p-GaN surface, however, it is extremely difficult for buried p-GaN [11], since the hydrogen atoms cannot go through the n-type layers above the buried p-GaN due to much lower diffusivity in n-GaN than in p-GaN. To achieve activation, the surface of the buried p-GaN needs to be exposed during the activation, which posts challenges in the design and fabrication of the vertical GaN transistors.

Polarization-induced (PI) doping is a unique doping scheme in GaN material system, offering advantages such as fully activated dopants independent of temperature [12] and increased breakdown field with the introduction of Al. Different from the p-type doping technique using Mg impurities, PI-doping technique introduces fixed polarization charge to induce mobile holes, thus immune to hydrogen passivation. PI-doped p-type layer in GaN material system has been demonstrated previously in the p-n diode platform using both molecular beam epitaxy (MBE) [12-14] and MOCVD [15] with decent reverse blocking capability. To our knowledge, no study on PI-doping in buried p-type structure has been reported. The utilization of PI-doped p-AlGaN buried body in vertical power transistors was proposed by our group [16] taking advantage of the higher critical field in AlGaN, but the advantage of the PI-doping regarding hydrogen passivation in MOCVD growth was not highlighted. In this work, we conduct the first study on the effect of PI-doping in buried p-type layer on reverse blocking. Much higher punch-though voltage is achieved with PI-doped buried p-type layer compared with the impurity-doped counterpart.

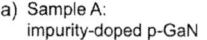

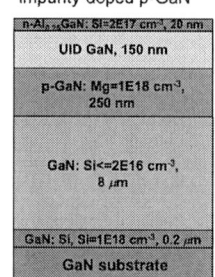
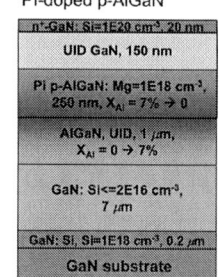

Fig. 1: Epitaxial structures of the buried p-GaN samples grown by MOCVD on bulk GaN substrate. (a) sample A: impurity-doped p-GaN (b) sample B: graded p-AlGaN with polarization-induced (PI) doping.

978-1-5386-2928-4/18 $31.00 © 2018 IEEE   228

II. EXPERIMENTS

Two similar structures are designed without or with PI-doping in the buried p-type layer, as shown in Fig. 1. The epitaxial layers are grown by MOCVD at about 1050°C on two-inch Ga-polar bulk n-GaN substrates. On the impurity-doped sample A, the epitaxial structure has an 8 μm GaN: Si (Si: 1-2×10^{16} cm^{-3}) drift layer, followed by a 250 nm p-type GaN: Mg (Mg: 1×10^{18} cm^{-3}) layer, and then topped by a 150 nm unintentionally-doped GaN layer with a thin n-type cap. In the PI-doped sample B, the buried p-type AlGaN layer has a linearly graded Al content from 7% to 0% and the same Mg concentration as the impurity-doped sample. In order to avoid the formation of a heterojunction, the Al content is first graded linearly from 0% to 7% in the top 1 μm of the drift layer. The rest of the layers are the same as sample A. After growth, secondary ion mass spectroscopy (SIMS) is performed on sample A, as shown in Fig. 2. Low background impurity level is observed alongside with a high H concentration in the buried p-GaN layer, which suggests the passivation of the Mg dopants.

To access the voltage blocking capabilities of the buried p-type layer electrically, vertical n-p-n diodes are fabricated on both samples. The schematic cross-section of the diodes and the fabrication process flow is shown in Fig. 4. The fabrication process starts with mesa isolation by dry etch, followed by anode ohmic contact formation on the thin n-type capping layer and cathode ohmic contact formation on the back side of the substrate. Trench MOSFETs are also fabricated on sample B to further test the reverse blocking capability of the buried PI-doped p-type layer, as shown in Fig. 8.

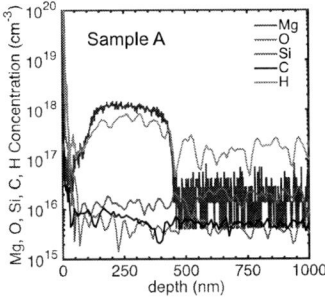

Fig. 2: SIMS profile of the various types of impurities in sample A. High level of H is present in the p-GaN layer, indicative of the passivation of Mg dopants.

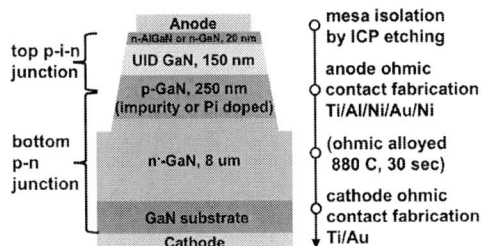

Fig. 3: Schematic cross-section of n-p-n diodes fabricated on both sample A and sample B. The process flow is indicated on the right.

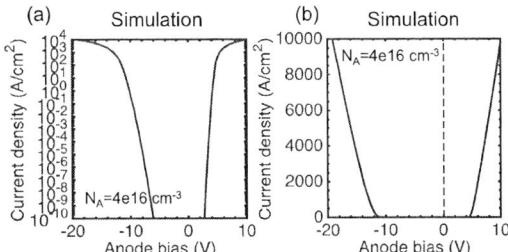

Fig. 4: Simulated I-V characteristics of the n-p-n diode using TCAD Sentaurus. The active N_A in the p-type layer is set to be 4×10^{16} cm^{-3}. Punch-through behavior under forward and backward bias is clearly observed.

III. RESULTS AND DISCUSSION

To better understand the I-V characteristics of the n-p-n diodes at punch-through, the I-V characteristics is first simulated in TCAD Sentaurus as shown in Fig. 4, where the active acceptor concentration (N_A) in the buried p-GaN is set to be 4×10^{16} cm^{-3}. Since the i-GaN thickness is much thinner in the top p-i-n junction than the n-GaN thickness in the bottom p-n junction, the forward and reverse punch-through behavior of the diode is different. At the beginning of punch-through, the current increase exponentially, as can be seen in Fig. 4a. After punch-through, the I-V characteristics is linear, indicating a resistive behavior of the device, as shown in Fig. 4b. The punch-through voltage can be extracted at the onset of the resistive behavior, where the buried p-GaN is fully depleted.

The measured forward and reverse I-V characteristics of the n-p-n diodes on sample A and sample B is shown in Fig. 5 and Fig. 6, respectively. On each plot, the I-V curves of three typical devices are shown. The reverse punch-through voltages of the diodes on sample B are around 240 V, which is much higher than that of the sample A (~30 V), suggesting the effect of the PI-doping in boosting the punch-through voltage. The low punch-through voltages in the sample A indicate that the buried p-GaN is largely passivated. The reverse leakage current of the diodes on sample B does not show a tight distribution, likely due to the leakage through dislocation and process non-uniformity.

To understand the punch-through characteristics of the two samples, the extracted punch-through voltages at forward and

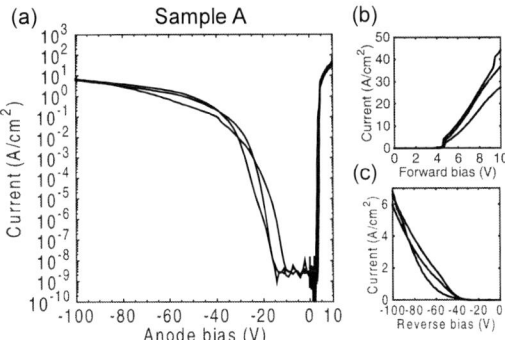

Fig. 5: Measured I-V characteristics of sample A. I-V curves of three typical devices are shown.

978-1-5386-2928-4/18 $31.00 © 2018 IEEE   229

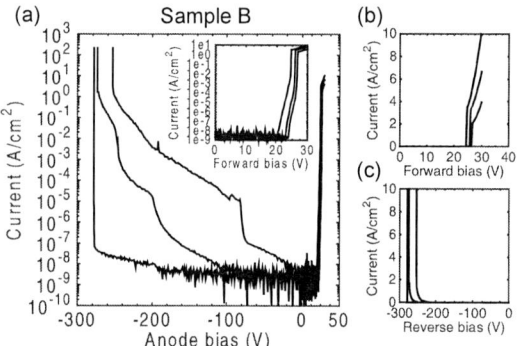

Fig. 6: Measured I-V characteristics of sample B. I-V curves of three typical devices are shown.

reverse bias are extracted on multiple devices and plotted on Fig. 7, alongside the data from simulation. The forward and reverse punch-through voltages are also calculated analytically, based on the electric field distribution when the p-type layer is fully depleted. As shown in Fig. 7, excellent agreement between the analytical calculation and the simulation is observed, validating the analytical calculation. The effective acceptor concentration, i.e. the total negative charge in the p-type layer at punch-through, is extracted for both samples by overlaying the extracted punch-through voltage on the analytical calculation. For the impurity-doped sample A, the active acceptor concentration is extracted to be $\sim 6\times 10^{16}$ cm^{-3}. Since the total Mg concentration is 10^{18} cm^{-3} as determined from SIMS, most of the Mg dopants ($\sim 94\%$) is inactive, indicating strong hydrogen passivation in the buried p-GaN layer grown by MOCVD. In comparison, the total negative charge in the PI-doped sample B is extracted to be $\sim 1.9\times 10^{17}$ cm^{-3}, which is $\sim 1.3\times 10^{17}$ cm^{-3} higher than the sample A. According to the theoretical calculation [12], the polarization charge density in the buried graded p-AlGaN layer in sample is $\sim 1.4\times 10^{17}$ cm^{-3}. This number is very close with the extracted net negative charge difference ($\sim 1.3\times 10^{17}$ cm^{-3}) between the two

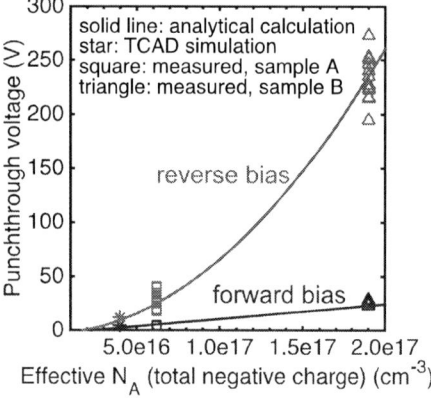

Fig. 7: Measured forward and reverse punch-through voltage of the n-p-n diodes. Solid lines are analytical calculation of the punch-through voltage. Stars shows the extraction from TCAD simulation under $N_A=4\times 10^{16}$ cm^{-3}. Multiple devices on sample A and B are measured. Extracted N_A in sample A: 6×10^{16} cm^{-3}. Extracted total negative charge in sample B: 1.9×10^{17} cm^{-3}.

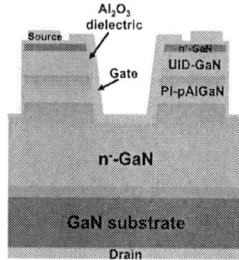

Fig. 8: Schematic cross-section of the vertical trench-MOSFETs fabricated on sample B.

samples. Thus, it is reasonable to attribute the extra negative charge in the PI-doped sample to be the polarization fixed charge, assuming that similar percentage of the Mg dopants in the PI-doped sample B are passivated by hydrogen as in sample A ($\sim 94\%$). This is the first time the polarization charge is extracted from punch-through measurements in GaN.

On the other hand, the vertical trench-MOSFETs fabricated on sample B exhibit decent transistor behavior in the on-state, as shown in the transfer and output characteristics in Fig. 9 and Fig. 10, respectively. The 3-terminal reverse breakdown characteristics at V_{gs}=0 V is shown in Fig. 11. The transistor shows a breakdown voltage (BV) of 225 V, similar with the reverse bias punch-through voltage measured on the n-p-n diodes, indicating that the breakdown mechanism is due to the

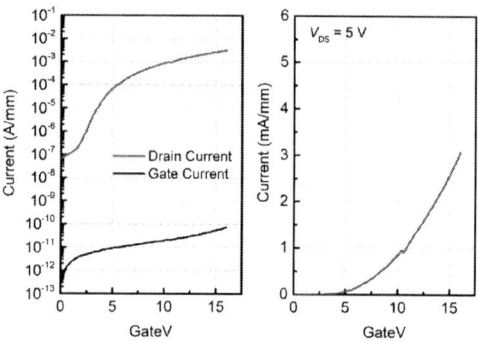

Fig. 9: Transfer characteristics of the trench-MOSFETs on sample B.

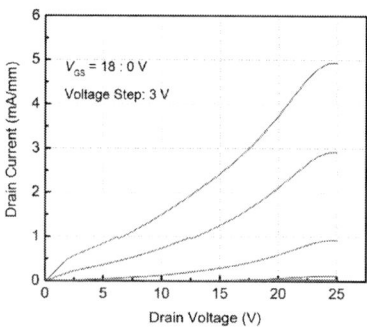

Fig. 10: Output characteristics of the trench-MOSFETs on sample B.

978-1-5386-2928-4/18 $31.00 © 2018 IEEE

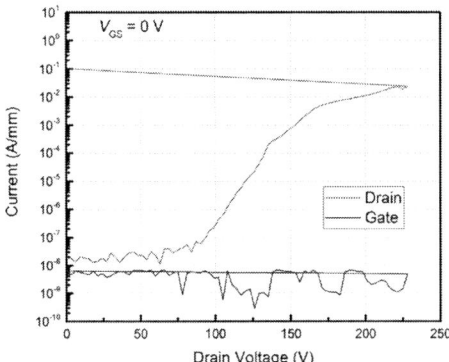

Fig. 11: 3-terminal breakdown characteristics of the trench-MOSFETs measured at V_{gs}=0 V. Measured BV=225 V.

punch-through of the buried PI-doped p-type layer. Although the BV is not very high, it highlights the unique advantage and potential offered by the PI-doping in enhancing the breakdown voltage in buried p-type structure, especially when the passivation of Mg dopants is present.

IV. CONCLUSIONS

The effect of polarization induced (PI)-doping in GaN buried p-type layer on reverse blocking is studied. N-p-n diodes are fabricated on MOCVD-grown buried p-type epitaxial structures with or without PI-doping in the p-type layer. The n-p-n diodes with only impurity doping in the buried p-GaN have a reverse punch-through voltage of ~30 V, correspond to only ~6% active Mg acceptors due to hydrogen passivation. In comparison, the n-p-n diodes with PI-doping have a much higher reverse punch-through voltage of 240 V. Aided by simulation and analytical calculation, an extra negative charge density of ~1.3×10^{17} cm^{-3} is extracted in the PI-doped p-type layer, which is attributed to the polarization fixed charge. Vertical trench-MOSFETs with a breakdown voltage of 225 V are also demonstrated on the PI-doped sample. The results indicate the unique advantage of utilizing PI-doping technique in GaN buried p-type layer: respectable enhancement in punch-through voltage can be achieved irrespective of Mg passivation due to hydrogen, which could benefit the design of a wide variety of GaN vertical power devices.

ACKNOWLEDGMENT

This work was supported in part by the ARPA-E SWITCHES program (DE-AR0000454) monitored by Tim Heidel and Isik Kizilyalli and carried out at the Cornell Nanoscale Science and Technology Facilities (CNF) sponsored by the NSF NNCI program (ECCS-15420819) and New York State.

REFERENCES

[1] S. Chowdhury, M. H. Wong, B. L. Swenson, and U. K. Mishra, "CAVET on bulk GaN substrates achieved with MBE-regrown AlGaN/GaN layers to suppress dispersion", IEEE Electron Device Letters, vol. 33, no. 1, pp. 41-43, 2012.

[2] H. Nie, Q. Diduck, B. Alvarez, A. P. Edwards, B. M. Kayes, M. Zhang, G. Ye, T. Prunty, D. Bour, and I. C. Kizilyalli, "1.5-kV and 2.2-mΩ-cm² Vertical GaN Transistors on Bulk-GaN Substrates", IEEE Electron Device Letters, vol. 35, no. 9, pp. 939-941, 2014.

[3] T. Oka, T. Ina, Y. Ueno, and J. Nishii, "1.8 mΩ·cm² vertical GaN-based trench metal–oxide–semiconductor field-effect transistors on a free-standing GaN substrate for 1.2-kV-class operation", Applied Physics Express, vol. 8, no. 5, p. 054101, 2015.

[4] R. Li, Y. Cao, M. Chen, and R. Chu, "600 V/1.7 Ω Normally-Off GaN Vertical Trench Metal–Oxide–Semiconductor Field-Effect Transistor", IEEE Electron Device Letters, vol. 37, no. 11, pp. 1466-1469, 2016.

[5] D. Shibata, R. Kajitani, M. Ogawa, K. Tanaka, S. Tamura, T. Hatsuda, M. Ishida, and T. Ueda, 2016, December. "1.7 kV/1.0 mΩ·cm² normally-off vertical GaN transistor on GaN substrate with regrown p-GaN/AlGaN/GaN semipolar gate structure", in IEEE International Electron Devices Meeting (IEDM), pp. 10-1, December 2016.

[6] W. Li, K. Nomoto, K. Lee, S. M. Islam, Z. Hu, M. Zhu, X. Gao, M. Pilla, D. Jena, and H. G. Xing, "600 V GaN vertical V-trench MOSFET with MBE regrown channel", in Device Research Conference (DRC), 2017 75th Annual. IEEE, pp. 1-2, June 2017.

[7] Z. Hu, W. Li, K. Nomoto, M. Zhu, X. Gao, M. Pilla, D. Jena, and H. G. Xing, "GaN vertical nanowire and fin power MISFETs", in Device Research Conference (DRC), 2017 75th Annual. IEEE, pp. 1-2, June 2017.

[8] M. Sun, Y. Zhang, X. Gao, and T. Palacios, "High-performance GaN vertical fin power transistors on bulk GaN substrates", IEEE Electron Device Letters, vol. 38, no. 4, pp. 509-512, 2017.

[9] C. Gupta, C. Lund, S. H. Chan, A. Agarwal, J. Liu, Y. Enatsu, S. Keller, and U. K. Mishra, "In situ oxide, GaN interlayer-based vertical trench MOSFET (OG-FET) on bulk GaN substrates", IEEE Electron Device Letters, vol. 38, no. 3, pp. 353-355, 2017.

[10] S. Nakamura, T. Mukai, M. Senoh, and N. Iwasa, "Thermal annealing effects on p-type Mg-doped GaN films", Japanese Journal of Applied Physics, vol. 31, no. 2B, p. L139, 1992.

[11] W. Li, K. Nomoto, K. Lee, S. M. Islam, Z. Hu, M. Zhu, X. Gao, A. Xie, M. Pilla, D. Jena, and H. G. Xing, "Activation of buried p-GaN in MOCVD-regrown vertical structures", unpublished.

[12] J. Simon, V. Protasenko, C. Lian, H. Xing, and D. Jena, "Polarization-induced hole doping in wide–band-gap uniaxial semiconductor heterostructures", Science, vol. 327, no. 5961, pp.60-64, 2010.

[13] S. Li, M. Ware, J. Wu, P. Minor, Z. Wang, Z. Wu, Y. Jiang, and G. J. Salamo, "Polarization induced pn-junction without dopant in graded AlGaN coherently strained on GaN", Applied Physics Letters, vol. 101, no. 12, p. 122103, 2012.

[14] S. Li, T. Zhang, J. Wu, Y. Yang, Z. Wang, Z. Wu, Z. Chen, and Y. Jiang, 2013. "Polarization induced hole doping in graded Al$_x$Ga$_{1-x}$N (x=0.7~1) layer grown by molecular beam epitaxy", Applied Physics Letters, vol. 102, no. 6, p. 062108, 2013.

[15] Y. Enatsu, C. Gupta, S. Keller, S. Nakamura, and U. K. Mishra, "P–n junction diodes with polarization induced p-type graded In$_x$Ga$_{1-x}$N layer", Semiconductor Science and Technology, vol. 32, no. 10, p. 105013, 2017.

[16] H. G. Xing, B. Song, M. Zhu, Z. Hu, M. Qi, K. Nomoto, and D. Jena, "Unique opportunity to harness polarization in GaN to override the conventional power electronics figure-of-merits", in Device Research Conference (DRC), 2015 73rd Annual. IEEE, pp. 51-52, June 2015.

978-1-5386-2928-4/18 $31.00 © 2018 IEEE

Proceedings of the 30th International Symposium on Power Semiconductor Devices & ICs
May 13-17, 2018, Chicago, USA

P-gate GaN HEMT gate-driver design for joint optimization of switching performance, freewheeling conduction and short-circuit robustness

Han Wu, Asad Fayyaz, Alberto Castellazzi
Power Electronics, Machines and Control Group
University of Nottingham
Nottingham, UK
asad.fayyaz@nottingham.ac.uk; alberto.castellazzi@nottingham.ac.uk

Abstract—This paper proposes the design and prototype development and testing of a gate-driver which enables to jointly optimize the performance in application of gate-injection type high electron mobility transistors, taking into account a number of diverse operational conditions, including nominal and abnormal events. The results show that it is possible to optimize the gate-driver parameters in such a way as to ensure optimum switching and free-wheeling performance, while ensuring enhanced short-circuit robustness.

Keywords—GaN HEMTs; GaN GITs; gate-drivers; wide-band-gap semiconductors.

I. INTRODUCTION

GaN transistors are finding rapidly growing interest in power conversion applications, where fully GaN normally-off solutions are oftentimes the preferred choice. In this work, the focus is in particular on devices in the 650 V rating class, of ample interest for automotive and renewable energies applications. Among the few commercially available packaged device options, p-gate type gate injection transistors (GIT) have been demonstrated in previous works to offer excellent performance in application, to show very good promise for high robustness and longer term reliability, while requiring some specific gate-driver design features for best operation [1-4]. Recently, a hybrid-drain (HD) GIT device has been introduced, which significantly advances well known issues related to current collapse in GaN HEMT technology [5]. The specific device under analysis in this work is the PGA26E19BA transistor, with a nominal DC current rating of 13 A [6].

Fig. 1 reports data-sheet information for the gate-source p-n junction type characteristics, together with the device output characteristics. As can be seen, although a threshold-voltage is specified for the transistor, it is actually not fully voltage-driven; in other words, the gate and drain current, I_D, are not independent one of the other. Moreover, as visible in Fig. 1 a), the gate bias current increases with temperature for a given gate-source bias voltage. The investigations of [7] have also shown that for a given drain current value, the on-state resistance does not vary significantly increasing the gate bias current value from a few mA to several tens of mA.

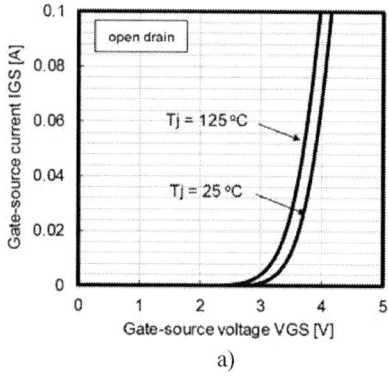

a)

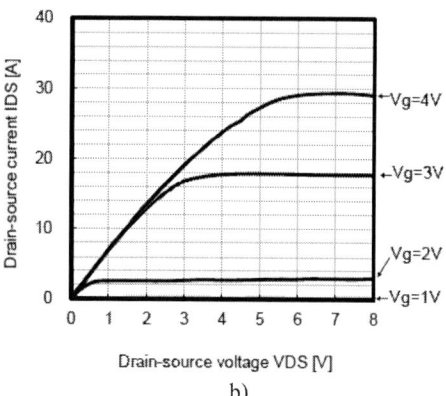

b)

Fig. 1: GaN GIT gate-source *p-n* junction characteristics, a), and output-characteristics at 25 °C, b) (data-sheet information, PGA26E19BA).

Based on these preliminary considerations, this work proposes the development of a bespoke original gate-driver design, which addresses all the main needs for device performance optimization in real operational conditions.

978-1-5386-2928-4/18 $31.00 © 2018 IEEE

II. GATE DRIVER DESIGN

The device-manufacturer recommended design for the terminal stage of the gate driver is shown in Fig. 2: this design optimizes the switching transitions in terms of application of a fast gate voltage and current peak at turn-on and turn-off, controlled by the values of R1 and C, followed by a region of current limitation, according to the on-state characteristics discussed above, controlled by the value of R2, which also controls the discharge rate of C during the off-state along with R1. The corresponding gate-source voltage, V_{GS}, drive waveform is illustrated in Fig. 3 a). In particular, with this design, due to the relatively high value of R2 needed to limit the gate current, the off-state voltage remains negative throughout the off-state period. This solution is not satisfactory in guaranteeing optimum device switching performance in real applications. Indeed, in a real switching converter, the most typical situation is that a half-bridge connection of a high-side and low-side transistors (HST, LST) is switched on and off alternately, with a *dead-time* between commutations to avoid current shoot-through phenomena, as schematically illustrated in Fig. 3 b). During the dead-times, the load current freewheels in one of the two switches, which

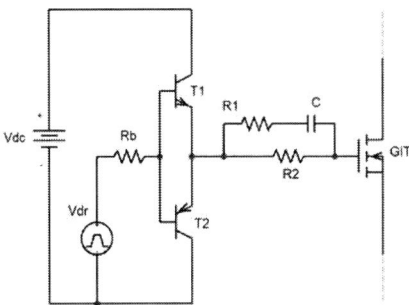

Fig. 2: Recommended gate-driver power stage design.

is then biased in the third quadrant of its output characteristics: device performance in this region of operation is strongly dependent upon the applied V_{GS}: higher losses are associated with more negative V_{GS} values. Experimental results in support of that are presented in Fig. 4 a). Moreover, switching tests reveal that when the device is turned-on again, the transition is much faster and efficient if the off-state V_{GS} value is closer to zero, as clearly visible in the results of Fig. 4 b). in which the curve labelled GaN-HEMT corresponds to a turn-on transition from a negative V_{GS} value; the curve labelled GaN-HEMT-Z corresponds to a turn-on transition from a zero V_{GS} value.

Thus, to improve both free-wheeling and switching performance, the gate-driver design was enhanced with the addition of the diode-resistor network, shown in Fig. 5 a), which effectively by-passes R2 during the off-state: the discharge rate of C is now controlled by R1 and R3 and can be made much faster, while enabling more freedom for the value of R2, that is, of the nominal on-sate drive current. Simulation

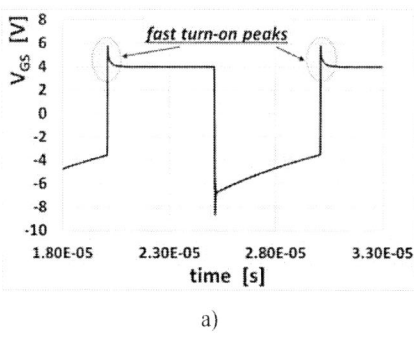

a)

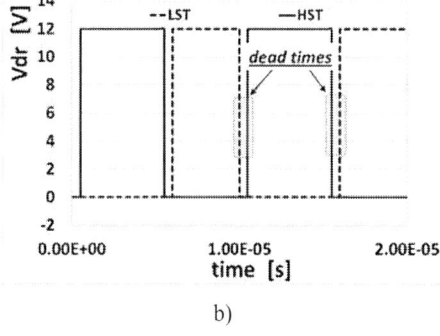

b)

Fig. 3: a); V_{GS} waveform corresponding to the recommended gate-driver design, a); illustration of typical half-bridge switching sequence, including dead-times, b).

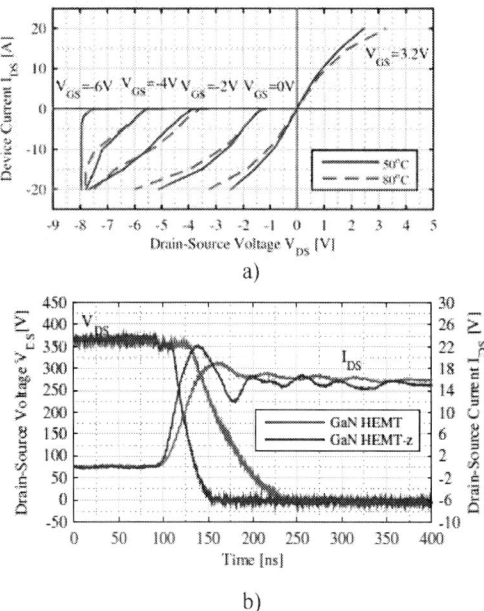

a)

b)

Fig. 4: Measured 1st and 3rd quadrant output characteristics of the GaN GIT for various V_{GS} values, a); measured drain-source voltage and drain current waveforms at turn-on, for two different values of off-state V_{GS}, b)

978-1-5386-2928-4/18 $31.00 © 2018 IEEE

results illustrating the faster discharge to zero of C are shown in Fig. 5 b). The value of C can also be used to the same aim to some extent, but reducing it too much causes a too low on-state V_{GS} value. The solution implemented here allows much more freedom in shaping the driving waveform according to the needs.

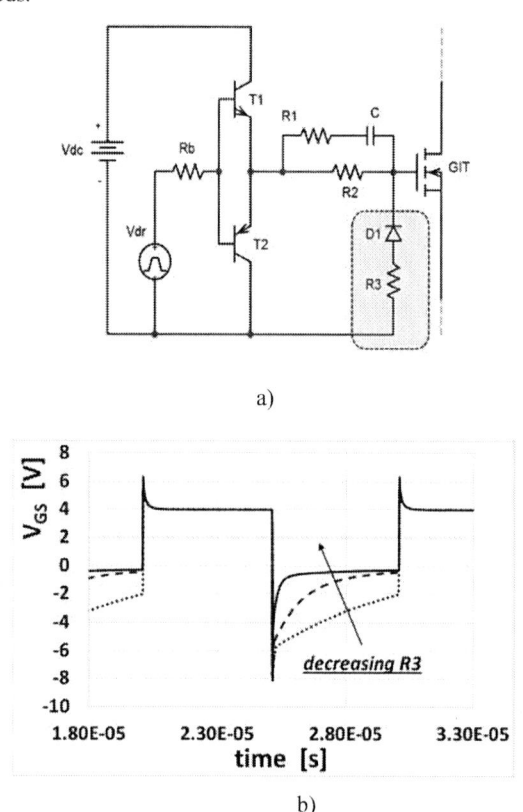

a)

b)

Fig. 5: Gate-driver design improvement, a), to achieve rapid return to zero of V_{GS} during turn-off, b), for improved free-wheeling and turn-on switching performance: the *D1-R3* network bypasses *R2*, enabling a faster discharge of *C*, controlled by the values of *R1* and *R3*.

Finally, recent investigations have shown that GaN GITs have the potential to offer excellent short-circuit robustness, provided that the large gate and drain charge injection taking place during the turn-on transient are duly limited, for example by reducing the V_{GS} peak at turn-on [8]. To this aim, the gate-driver design was further enhanced here with the addition of a clamp circuit for the gate terminal, as shown in Fig. 6 a). This solutions effectively reduces the short-circuit drain and, correspondingly, gate current peak, without compromise on nominal operation: Fig. 6 b) and c) show simulation results for I_D and V_{GS} during two consecutive pulses of operation, the first corresponding to nominal current switching on resistive load and the second corresponding to a short-circuit condition; the results of b) refer to the case when no V_{GS} clamp circuit is inserted, the results in c) correspond to

the use of the clamp circuit; it is visible from these results that the clamp is very effective in reducing the short-circuit current, while nominal operation remains unaffected.

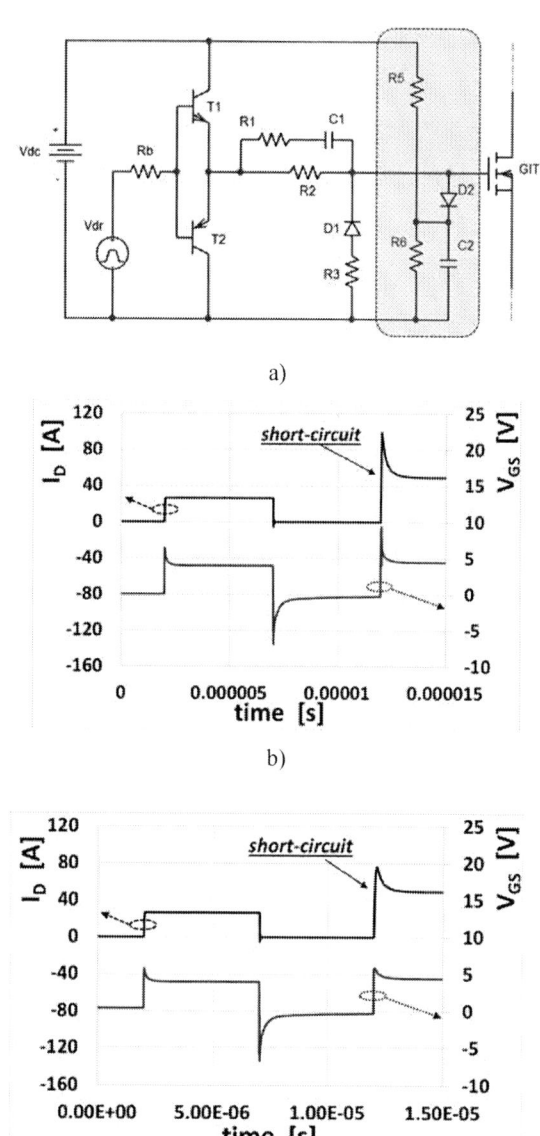

a)

b)

c)

Fig. 6: Gate-driver design improvement to achieve V_{GS} positive peak clamping, a). Illustration of nominal (first pulse, resistive load) I_D and V_{GS} waveforms and their change during short-circuit (second pulse, current limited by device characteristics), a): b): R5 and R6 set the clamp voltage value, whereas *C2* limits the voltage increase when the clamp is active and *D2* is conducting; corresponding reduction of short-circuit peak current (from about 105 A to less than 80 A) for enhanced device robustness, c).

978-1-5386-2928-4/18 $31.00 © 2018 IEEE   234

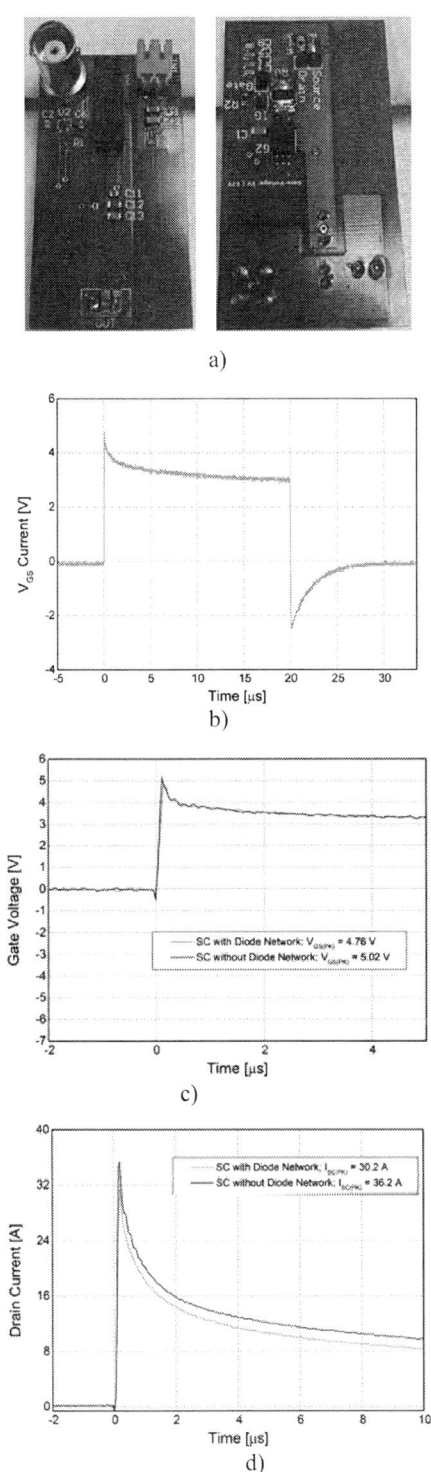

a)

b)

c)

d)

Fig. 7: Gate-driver prototype, a) and b); representative switching waveform showing the ability to control turn-on and turn-off peak amplitudes and rapid return to zero during the turn-off phase.

The clamp network can also serve as a fast short-circuit monitor by careful selection of R6 and C2.

III. PROOF-OF-CONCEPT DEMONSTRATION

A prototype of the gate-driver designed as per above description was implemented. Fig. 7 a) shows the top and bottom side of the gate-driver PCB circuit, which was nearly entirely built with surface mount devices. Fig. 7 b) shows a typical experimental transient profile for V_{GS} and it should be noted that the amplitude of the plateau can be easily adjusted by controlling the value of the bias voltage of the push-pulls driver stage, Vdc in Fig. 6 a), taking into account the decrease with temperature due to the above mentioned increase in the gate bias current. Figs. 7 c) and d) show V_{GS} and I_D during a short-circuit event, with and without the clamp circuit, respectively.

IV. CONCLUSION

This paper has presented the design and prototype demonstration of a gate driver dedicated to optimize the operational performance of gate-injection high electron mobility GaN transistors in real applications. The design enables independent setting of the parameters required to achieve best switching performance, both on and off, reverse current conduction for free-wheeling of inductive load currents, and withstand of abnormal overload short-circuit events to deliver the typical required robustness of drive applications, as required by a number of strategic application domains, such as automotive.

REFERENCES

[1] Y. Uemoto et al., *Gate injection transistor (GIT) - a normally-off AlGaN/GaN power transistor using conductivity modulation*, IEEE Trans. Electron Devices, 54 (12) (2014) 3393–3399.

[2] E. Gurpinar, A. Castellazzi, *Tradeoff Study of Heat Sink and Output Filter Volume in a GaN HEMT Based Single-Phase Inverter*, IEEE Transactions on Power Electronics (Volume: 33, Issue: 6, June 2018), 5226 – 5239.

[3] T. Oeder, A. Castellazzi, M. Pfost, Experimental study of the short-circuit performance for a 600V normally-off p-gate GaN HEMT, in Proc. ISPSD 2017, Sapporo, Japan, 28 May-1 June 2017.

[4] T. Oeder, A. Castellazzi, M. Pfost, Electrical and thermal failure modes of 600 V p-gate GaN HEMTs, Microelectronics Reliability Vol. 76–77, September 2017, 321-326.

[5] K. Tanaka et al., *Reliability of hybrid-drain-embedded gate injection transistor*, 2017 IEEE International Reliability Physics Symposium (IRPS), 2-6 April 2017, Monterey, CA, USA.

[6] https://industrial.panasonic.com/content/data/SC/ds/ds8/c2/FLY000075_EN.pdf

[7] Y. C. Fong, K. W. E. Cheng, *Experimental Study on the Electrical Characteristic of a GaN Hybrid Drain-embedded Gate Injection Transistor (HD-GIT)*, 7th International Conference Power Electronics Systems and Applications - Smart Mobility, Power Transfer & Security (PESA), 2017, 12-14 Dec. 2017, Hong Kong, China.

[8] A. Castellazzi, T. Oeder, A. Fayyaz, S. Zhu, M. Pfost, *Single pulse short-circuit robustness and repetitive stress aging of GaN GITs*, 2018 IEEE International Reliability Physics Symposium, March 11-15, 2018, Burlingame, California, USA.

978-1-5386-2928-4/18 $31.00 © 2018 IEEE   235

Proceedings of the 30th International Symposium on Power Semiconductor Devices & ICs
May 13-17, 2018, Chicago, USA

Monolithic Integration of GaN-Based NMOS Digital Logic Gate Circuits with E-Mode Power GaN MOSHEMTs

Minghua Zhu and Elison Matioli

Power and Wide-band-gap Electronics Research Laboratory (POWERLAB)
École Polytechnique Fédérale de Lausanne (EPFL)
Lausanne, Switzerland
minghua.zhu@epfl.ch, elison.matioli@epfl.ch

Abstract— In this work, we demonstrate high-performance NMOS GaN-based logic gates including NOT, NAND, and NOR by integration of E/D-mode GaN MOSHEMTs on silicon substrates. The load-to-driver resistance ratio was optimized in these logic gates by using a multi-finger gate design of E-mode GaN MOSHEMT to increase the logic swing voltage and noise margins, and reduce the transition periods. State-of-the-art NMOS inverter was achieved with logic swing voltage of 4.93 V at a supply voltage of 5 V, low-input noise margin of 2.13 V and high-input noise margin of 2.2 V at room temperature. Excellent high temperature performance, at 300°C, was observed with a logic swing of 4.85 V, low-input noise margin of 1.85 V and high-output noise margin of 2.2V. In addition, GaN-based NAND and NOR NMOS logic gates are reported for the first time with very good performance. Finally, the logic gates were monolithically integrated with high-voltage E-mode power transistors, which reveals a significant step forward towards monolithic integration of GaN power transistors with gate drivers.

Keywords— *Logic gates, GaN, high temperature, inverters, NAND, NOR, E-mode, D-mode, monolithic integration*

I. INTRODUCTION

The high switching frequency of GaN-based power converters can lead to a significant reduction of the size of passive components, such as capacitors and inductors, and thus, increasing the power density of the overall system [1], [2]. GaN is one of the most promising materials for high frequency power switching due to its exceptional properties such as large saturation velocity, high carrier mobility, and high breakdown field strength. Despite the advantages of GaN power devices, discrete Si-based logic control and gate drivers are still used to control the GaN power devices [3], which limits the switching frequency due to parasitic inductances from external connections of GaN power devices and gate drivers.

A monolithic integration of GaN power devices with GaN-based gate drivers would minimize parasitic components and unveil the full potential of GaN transistors for high frequency power conversion with high efficiency. GaN-based logic gates are essential components to realize level shift, driver control, dead time control and under voltage-lockout (UVLO) for driver circuits. However, high performance CMOS logics in GaN are not feasible today due to the poor transport properties of p-type

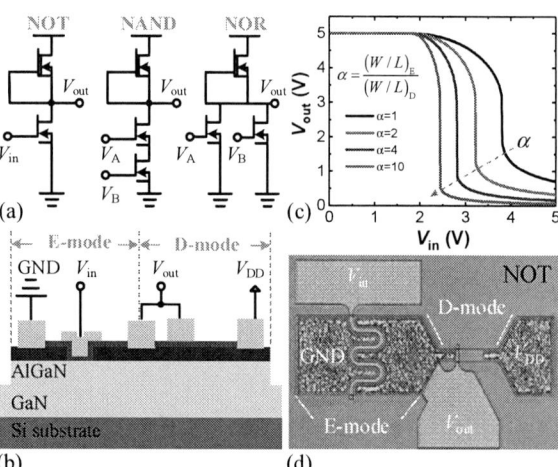

Fig. 1. (a) Equivalent circuits of inverter, NAND and NOR logic gates. (b) Cross-sectional schematic and (c) top view of monolithic integration of E/D-mode MOSHEMTs. (d) Simulated transfer characteristics versus different α ratios.

GaN devices [4], [5]. Ideas to demonstrate purely n-type logic circuits date back of more than 30 years with NMOS logics [6]–[8], also named direct-coupled FET logic (DCFL). GaN DCFLs have been demonstrated [9]–[12], however their performances are not sufficient to satisfy the logic requirements, due to their small noise margins, large logic transition voltages, small logic swing and large low-level output voltages (V_{OL}). Recent research show steady improvements on GaN DCFL [10], however the V_{OL} is still quite high, up to 0.3 V and the maximum voltage swing is 4.66 V, which would lead to high logic losses and safety problems. A very distinct property of GaN compared to Silicon is its high temperature operation, which led to the demonstration of GaN DCFLs operating at high temperatures, up to 375 °C [9], [11], however, it resulted in a much degraded operation compared to room temperature (RT).

In this paper, we demonstrate high-performance NOT, NAND and NOR logic gate units by integrating E/D-mode MOSHEMTs. These logic units were optimized for larger voltage swing, wider noise margin, and smaller transition

---

This work was supported in part by the Technology Research Programme (TRP) of the European Space Agency.

978-1-5386-2928-4/18 $31.00 © 2018 IEEE

periods. High performance was observed even up to 300°C, which could be applied for high temperature applications.

II. INTEGRATED LOGIC DESIGN AND FABRICATION

The NMOS logic gate circuits, shown in Fig. 1(a), were fabricated on AlGaN/GaN epitaxial layer on silicon substrate (Fig. 1(b)) consisting of 3.75 μm buffer, 322 nm of un-doped GaN channel, 23.7 nm of AlGaN barrier and 2.4 nm of GaN cap layer. D-mode and E-mode MOSHEMTs were fabricated at the same time, with the sole difference of one additional gate recess process to achieve E-mode operation. To optimize the design of the logics, we simulated the driver-to-load resistance ratio $\alpha = (W/L)_E / (W/L)_D$, as shown in Fig. 1 (c), where W and L are the width and length of the E- and D-mode transistors respectively. Larger values of α result in sharper transitions, higher logic voltage swings and higher noise margins, since the equivalent resistance of E-mode transistor is much smaller than the overall resistance of the circuit. To achieve large values of α, we designed a multi-finger structure for the E-mode MOSHEMT that results in a much smaller resistance of the E-mode compared to the D-mode device without oversizing too much the E-mode transistor (Fig. 1(d)). The chip fabrication started with the definition of the mesa regions by Cl_2-based inductively coupled plasma (ICP) etching. For the E-mode devices, a 1.5 μm-long gate recess was defined by optical lithography and followed by a Cl_2-based slow-rate dry etching, which leads to a precise control of the etching depth. To improve the surface morphology after gate recess, therefore the electron transport of the e-mode devices, we have combined a slow, low-damage ICP etch with a 5%-TMAH wet treatment performed at 80 °C for 30 minutes (Fig. 2). This step is very critical to obtain a good threshold voltage (V_{th}) control and reasonable on-resistance (R_{dson}) of E-mode transistor. A metal stack of Ti (200 □)/Al (1200 □)/Ti (400 □)/ Ni (600 □)/ Au (500 □) was deposited in the source and drain contact regions by electron-beam evaporation, followed by rapid thermal annealing at 830°C under N_2 atmosphere. The gate dielectric was 25 nm-thick SiO_2 deposited by atomic layer deposition (ALD) at 300°C, immediately after a surface treatment in 37% HCl for 1 min. Finally, gate and contact pads were formed by depositing Ni/Au for both E-mode and D-mode devices. Fig. 1(d) shows the integration of these devices for the NOT logic gate.

III. RESULTS AND DISCUSSION

A. DCFL inverter

Fig. 3(a) shows the measured voltage transfer characteristics (VTC) for DCFL inverters with different α, which is consistent with the simulations shown in Fig. 1(c). The performance of the DCFL inverters was characterized with a V_{DD} of 5 V from RT to 300 °C revealing a proper operation with very little variations for this entire temperature range (Fig. 3(b)). The V_{OL} was nearly unchanged, since the

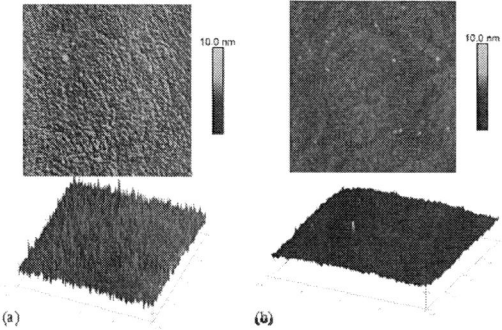

Fig. 2. Surface morphology (a) before and (b) after TMAH treatment.

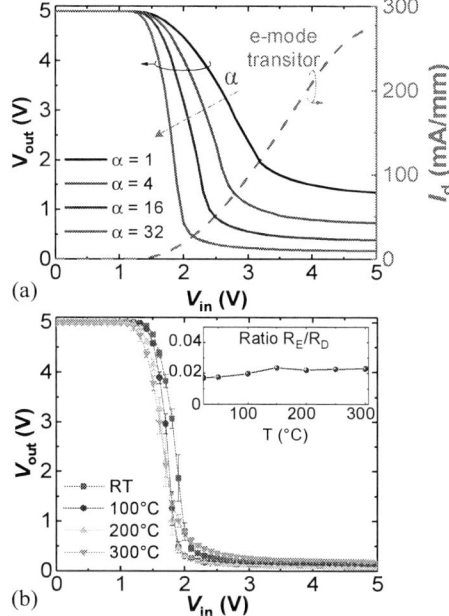

Fig. 3. (a) DCFL VTC versus different α ratio and the transfer characteristics of E-mode transistors (b) DCFL VTC under different temperature varied from RT to 300°C. Inset: resistance ratio of E/D-mode logic versus temperature for output logic equal to 0.

resistance of the E- and D-mode transistors increases at the same rate with temperature (inset of Fig. 3(b)), which results in a nearly constant E/D-mode resistance ratio and thus in a similar V_{OL} since:

$$V_{OL} = \frac{R_E}{R_E + R_D} V_{DD} \qquad (1)$$

As shown in Fig. 3(a), the transition voltage of the DCFL inverter is affected by two main factors: the V_{th} of the E-mode transistor and the ratio α. For high performance logics, the ratio α must be as large as possible and the V_{th} of E-mode transistor must be close to $V_{DD}/2 = 2.5$ V in linear scale, which can be controlled by adjusting the gate recess depth.

978-1-5386-2928-4/18 $31.00 © 2018 IEEE   237

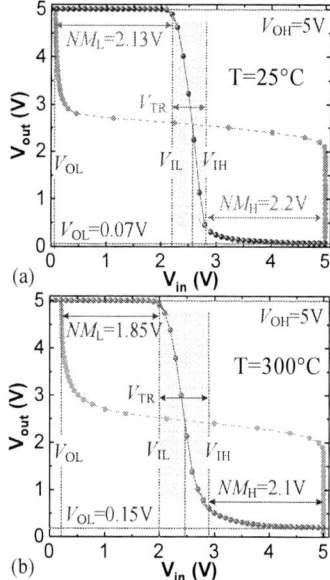

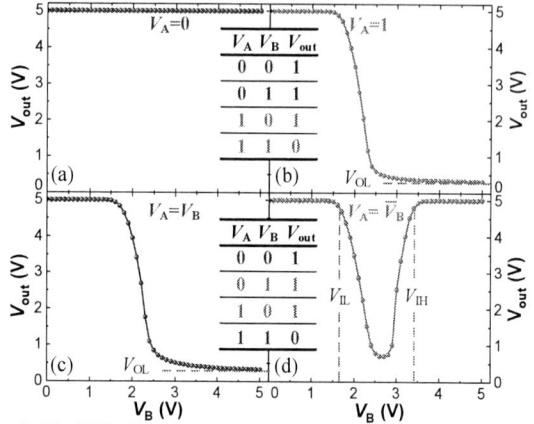

Fig. 5. The VTC of NAND gate logics under different input voltages, where (a) $V_A = 0$, (b) $V_A = 1$, (c) $V_A = V_B$, and (d) $V_A = \bar{V}_B$, V_B is sweeping from 0 to 5V. Insets: corresponding truth table for NAND logic.

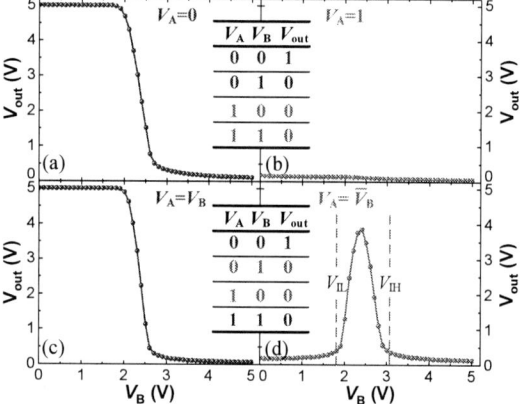

Fig. 6. The VTC of NOR gate logics under different input voltages, where (a) $V_A = 0$, (b) $V_A = 1$, (c) $V_A = V_B$, and (d) $V_A = \bar{V}_B$, V_B is sweeping from 0 to 5V.

Fig. 4. The VTC of the optimized DCFL inverter with $\alpha = 64$ at (a) 25°C and (b) 300°C with supply voltage V_{DD}=5V.

We optimized the design of the DCFL inverter by integrating a D-mode MOSHEMT with $(W/L)_E = 160\ \mu m/8\ \mu m$ acting as the active load and an D-mode MOSHEMT with a $(W/L)_E = 10\ \mu m/32\mu m$ acting as the driver source, which corresponds to a ratio α of 64. By adjusting the gate recess from 25 nm to 27 nm, the V_{th} of the E-mode transistor was increased from 1.8 V (Fig.2 (a)) to 2.5 V in linear scale. The VTCs of the optimized DCFL inverter are shown in Fig. 54. The high-level output voltage (V_{OH}) and V_{OL} were 5 V and 0.07 V, respectively, yielding a much larger voltage swing of 4.93 V/5 V (98.6%) compared to the best values in the literature of 4.66 V/5 V (93.2%) [10] and 6.3 V/7 V (90%) [13]. The low-level input voltage (V_{IL}) and high-level input voltage (V_{IH}), defined at dV_{out}/dV_{in} = -1, were 2.2 V and 2.8 V respectively. The VTC consist of three main regions: the low-input region at $v_{in}<V_{IL}$, the transition region at $V_{IL}\leq v_{in}\leq V_{IH}$, and the high-input region at $v_{in}>V_{IH}$. The width of the transition voltage region ($V_{TR}= V_{IH} - V_{IL}$) is a way of measure the ambiguity of the logic inverter [14], which must be as low as possible. The transition voltage width was 0.6 V/5 V (12%) in our case, which is smaller than the lowest values in the literature 1 V/5 V (20%) [10] and 1.6 V/7 V (22%) [13]. The low-input-logic noise margin ($NM_L=V_{IL} - V_{OL}$) and high-input-logic noise margin ($NM_H=V_{IH} - V_{OH}$) are 2.13 V/ 2.5 V (85.2%) and 2.2 V/2.5 V (88%) respectively, which outperforms the best values of 1.7 V/2.5 V (68%) and 2 V/2.5 V (80%) in [10].

High temperature performance of this inverter is shown in Fig. 54(b), revealing a very good behavior at 300 °C with V_{OH}, V_{OL}, V_{IL}, V_{IH} equal to 5 V, 0.15 V, 2 V, 2.9 V, respectively. The voltage swing at 300 °C was 4.85 V/5 V (97%), with 1.6% degradation compared with the value at RT. This value

is much larger than the best values in the literatures 4.5 V/5 V (90%) in [10] at 200°C and 6.5 V/7 V (93%) in [11] at 300°C. NM_L and NM_H at 300°C were 1.85 V/2.5 V (74%) and 2.1 V/2.5V (84%), respectively, which shows an excellent operation even at 300°C.

B. NAND and NOR logic gate

Fig. 5 shows the VTC of the fabricated NAND logic gates (equivalent circuit in Fig. 1(a)), for which V_{DD} was 5 V, V_B was swept from 0 to 5V and V_A was set at 4 different cases:

case 1: $V_A=0$, thus $V_{out} = \bar{V}_A + \bar{V}_B = 1$ *(Fig. 5 (a));*

case 2: $V_A=1$, thus $V_{out} = \bar{V}_A + \bar{V}_B = \bar{V}_B$ *(Fig. 5 (b)). The* V_{OL} *= $2R_E/(R_D+2R_E)\cdot V_{DD}$ is larger than that of the inverter since the equivalent resistance of E-mode transistor is $2R_E$ instead of R_E.*

case 3: $V_A=V_B$, thus $V_{out} = \bar{V}_A + \bar{V}_B = \bar{V}_B + \bar{V}_B$. *In this case,* V_A *and* V_B *are changing simultaneously, which is quite similar to the case 2 (Fig. 5(c)), except that the transition*

978-1-5386-2928-4/18 $31.00 © 2018 IEEE

region is relatively longer, since the equivalent resistance in the lower side is larger;

case 4: $V_A = \overline{V_B}$, *thus* $V_{out} = \overline{V_A} + \overline{V_B} = V_B + \overline{V_B}$, *(Fig. 5(d)).* When $V_B < V_{IL}$ and $V_B > V_{IH}$, $V_B = 0$ and $V_B = 1$, *thus* $V_{out} = 1$, and $V_{IL} \leq V_B \leq V_{IH}$, *the logic output is ambiguous.*

Fig. 6 shows the VTC of the fabricated NOR logic gates (equivalent circuit in Fig. 1(a)), for which V_A was set at 4 different cases:

case 1: $V_A = 0$, *thus* $V_{out} = \overline{V_A} \cdot \overline{V_B} = \overline{V_B}$ *(Fig.5 (a)).*

case 2: $V_A = 1$, *thus* $V_{out} = \overline{V_A} \cdot \overline{V_B} = 0$ *(Fig.5 (b)).*

case 3: $V_A = V_B$, *thus* $V_{out} = \overline{V_A} \cdot \overline{V_B} = \overline{V_B} \cdot \overline{V_B}$. *In this case,* V_A *and* V_B *are changing simultaneously, which is quite similar to case 1 (Fig. 6(c)), except that the equivalent resistance is* $R_E/2$ *instead of* R_E, *which led to a lower* V_{OL} *and sharper transition.*

case 4: $V_A = \overline{V_B}$, *thus* $V_{out} = \overline{V_A} \cdot \overline{V_B} = V_B \cdot \overline{V_B}$ (Fig. 6(d)). When $V_B < V_{IL}$ and $V_B > V_{IH}$, $V_B = 0$ and $V_B = 1$, thus $V_{out} = 0$. When $V_{IL} \leq V_B \leq V_{IH}$, the logic output is ambiguous.

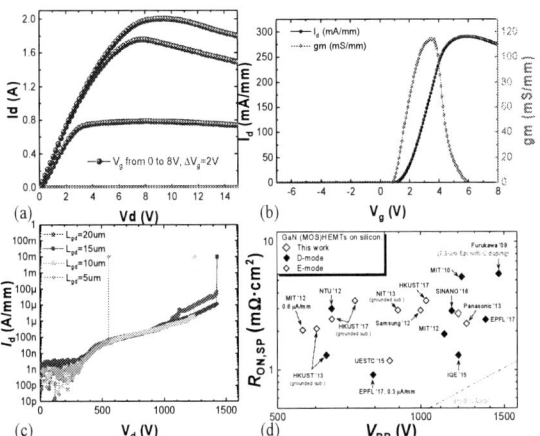

Fig. 7. (a) The output, (b) transfer characteristics, (c) breakdown voltage and (d) benchmark of E-mode power GaN MOSHEMT

C. Logic intergrated with E-Mode power MOSHEMTs

In order to demonstrate the probability of integration with GaN logic with power transistors, we fabricate them on same wafer. The E-mode power MOSHEMTs fabricated on the same chip as the logic gates, presented low leakage current, threshold voltage of 1.5V, and high breakdown voltage (up to 1400V) (Fig. 7). Our integrated E-mode power GaN MOSHEMTs was benchmarked against state-of-the-art GaN MOSHEMTs on Si (Fig. 7(d)).

IV. CONCLUSION

In this paper, we have demonstrated high performance NOT logic with high logic swing voltage, high noise margin, and low transition period both at RT and at 300°C. NAND and NOR GaN logic gates were demonstrated for the first time with very good performance. E-mode power MOSHEMTs with 1kV breakdown and 2A current capability were fabricated on the same chip as the logic gates. These results show the enormous potential of GaN-based logic gates for future monolithic integration in gate drivers as well as for high temperature applications.

ACKNOWLEDGMENT

We would like to thank the staff in CMi and ICMP cleanrooms at EPFL for technical support and advice.

REFERENCES

[1] S. Ujita *et al.*, "A fully integrated GaN-based power IC including gate drivers for high-efficiency DC-DC Converters," in *2016 IEEE Symposium on VLSI Circuits (VLSI-Circuits)*, 2016, pp. 1–2.

[2] E. A. Jones, F. F. Wang, and D. Costinett, "Review of Commercial GaN Power Devices and GaN-Based Converter Design Challenges," *IEEE J. Emerg. Sel. Top. Power Electron.*, vol. 4, no. 3, pp. 707–719, Sep. 2016.

[3] B. Wang, M. Riva, J. D. Bakos, and A. Monti, "Integrated Circuit Implementation for a GaN HFET Driver Circuit," *IEEE Trans. Ind. Appl.*, vol. 46, no. 5, pp. 2056–2067, Sep. 2010.

[4] H. Hahn *et al.*, "First monolithic integration of GaN-based enhancement mode n-channel and p-channel heterostructure field effect transistors," in *72nd Device Research Conference*, 2014, pp. 259–260.

[5] R. Chu, Y. Cao, M. Chen, R. Li, and D. Zehnder, "An Experimental Demonstration of GaN CMOS Technology," *IEEE Electron Device Lett.*, vol. 37, no. 3, pp. 269–271, Mar. 2016.

[6] M. I. Elmasry, "Multidrain NMOS for VLSI logic design," *IEEE Trans. Electron Devices*, vol. 29, no. 4, pp. 779–781, Apr. 1982.

[7] M. I. Elmasry, "Nanosecond NMOS VLSI current mode logic," *IEEE Trans. Electron Devices*, vol. 29, no. 4, pp. 781–784, Apr. 1982.

[8] T. Mimura, S. Hiyamizu, K. Joshin, and K. Hikosaka, "Enhancement-Mode High Electron Mobility Transistors for Logic Applications," *Jpn. J. Appl. Phys.*, vol. 20, no. 5, p. L317, May 1981.

[9] Y. Cai, Z. Cheng, Z. Yang, C. W. Tang, K. M. Lau, and K. J. Chen, "High-Temperature Operation of AlGaN/GaN HEMTs Direct-Coupled FET Logic (DCFL) Integrated Circuits," *IEEE Electron Device Lett.*, vol. 28, no. 5, pp. 328–331, May 2007.

[10] G. Tang *et al.*, "Digital Integrated Circuits on an E-mode GaN Power HEMT Platform," *IEEE Electron Device Lett.*, vol. PP, no. 99, pp. 1–1, 2017.

[11] Z. Xu *et al.*, "High Temperature Characteristics of GaN-Based Inverter Integrated With Enhancement-Mode (E-Mode) MOSFET and Depletion-Mode (D-Mode) HEMT," *IEEE Electron Device Lett.*, vol. 35, no. 1, pp. 33–35, Jan. 2014.

[12] B. Wang, M. Riva, J. D. Bakos, and A. Monti, "Integrated Circuit Implementation for a GaN HFET Driver Circuit," *IEEE Trans. Ind. Appl.*, vol. 46, no. 5, pp. 2056–2067, Sep. 2010.

[13] Z. Xu *et al.*, "High Temperature Characteristics of GaN-Based Inverter Integrated With Enhancement-Mode (E-Mode) MOSFET and Depletion-Mode (D-Mode) HEMT," *IEEE Electron Device Lett.*, vol. 35, no. 1, pp. 33–35, Jan. 2014.

[14] M. H. Rashid, *Microelectronic Circuits: Analysis & Design*, 2 edition. Stamford, CT: CL Engineering, 2010.

Proceedings of the 30th International Symposium on Power Semiconductor Devices & ICs
May 13-17, 2018, Chicago, USA

645 V Quasi-Vertical GaN Power Transistors on silicon substrates

Chao Liu, Riyaz Abdul Khadar, and Elison Matioli

École Polytechnique Fédérale de Lausanne (EPFL)
CH-1015 Lausanne, Switzerland
chao.liu@epfl.ch, elison.matioli@epfl.ch

Abstract—In this paper, we present GaN-on-Si vertical transistors consisting of a 6.7 µm thick n-p-n heterostructure grown on 6-inch silicon substrates by MOCVD. The fabricated vertical trench gate MOSFETs exhibited E-mode operation with a threshold voltage of 3.3 V and an on/off ratio of over 10^8. A specific on-resistance of 6.8 mΩ·cm² and a high off-state breakdown voltage of 645 V were achieved. These results show the great potential of the GaN-on-Si platform for the next generation of cost-effective power electronics.

Keywords—*GaN; vertical transistors; GaN-on-Si; power; semiconductor; devices; MOSFETs*

I. INTRODUCTION

GaN-based semiconductors and devices have progressed rapidly in the past decades. The dominant III-nitride power devices with a high breakdown voltage are lateral AlGaN/GaN high electron mobility transistors (HEMTs), which have the distinct advantages of polarization charges at the heterointerface to produce a high-density and high-mobility two-dimensional electron gas (2DEG) that effectively reduces the on-resistance and increases the switching speeds of high-voltage devices. However, lateral GaN HEMTs have some limitations for power applications. For example, the threshold voltage (V_{TH}) of most GaN HEMTs is not high enough for fail-safe operation in power systems. In addition, the lateral topology is not suitable for high-voltage and high-current applications. Substantial gate-to-drain spacing is required to achieve high breakdown voltages, which increases the chip size and reduces the effective device current density.

Vertical topology, on the other hand, provides a feasible solution for high-power-density devices. By increasing the thickness of the drift layer, the breakdown voltage can be increased without sacrificing the device size. There have been several reports of high performance GaN-based vertical transistors on freestanding GaN substrates [1-13]. However, the high cost and small available size of bulk GaN substrates limit the wide-spread commercial adoption of vertical power devices on bulk GaN. The GaN-on-Si platform offers a cost-effective alternative for vertical GaN power devices, due to its large-scale availability, low cost, and a mature fabrication technology [14-20].

In this work, we demonstrate quasi-vertical GaN-on-Si vertical transistors based on a 6.7 µm-thick n-p-n heterostructure grown on 6-inch silicon substrates. The fabricated vertical trench gate MOSFETs exhibited E-mode

This work was supported by the European Research Council under the European Union's H2020 programme/ERC Grant Agreement 679425.

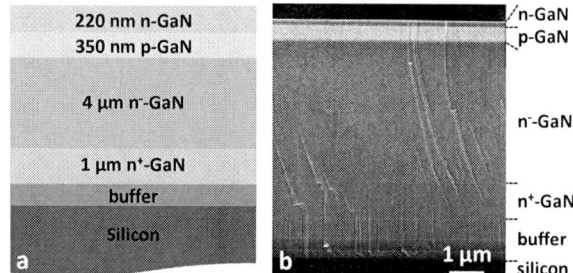

Fig. 1. Cross-sectional (a) schematic, and (b) SEM image of the as-grown n-p-n heterostructure on 6-inch silicon substrates.

operation with a V_{TH} of 3.3 V and an on/off ratio of over 10^8. A specific on-resistance ($R_{ON,SP}$) of 6.8 mΩ·cm² and a high off-state breakdown voltage of 645 V were achieved. The excellent performance can be attributed to the optimized 4 µm drift layer. This excellent performance reveals the great potential of GaN-on-Si to serve as a platform for future power electronic applications.

II. DEVICE STRUCTURE AND FABRICATION

The n-p-n heterostructure used in this work was grown on 6-inch Si (111) substrates by metal organic chemical vapor deposition system. Fig. 1 (a) presents the schematic structure of the n-p-n structure on silicon substrates. From bottom to top, the n-p-n structure consisted of a 1.1 µm-thick buffer layer, a 1 µm-thick n-type GaN layer (Si dopant concentration $\sim 1 \times 10^{19}$ cm^{-3}), a 4 µm-thick n-type GaN layer (Si dopant concentration $\sim 2 \times 10^{16}$ cm^{-3}), a 350 nm-thick p-type GaN layer (Mg dopant concentration $\sim 4 \times 10^{19}$ cm^{-3}), a 200 nm-thick n-type GaN layer (Si dopant concentration $\sim 5 \times 10^{18}$ cm^{-3}), and a 20 nm-thick n-type GaN layer (Si dopant concentration $\sim 1 \times 10^{19}$ cm^{-3}). Cross-sectional SEM image in Fig. 1 (b) shows a total epitaxial thickness of 6.7 µm, with an abrupt junction interface. A thick GaN drift layer with high crystalline quality is required to achieve high breakdown characteristics for GaN vertical devices. A 4 µm-thick continuously grown GaN drift layer on silicon was achieved with a low dislocation density of 3×10^8 cm^{-2}, without the use of any interlayer which would significantly degrade the $R_{ON,SP}$. After growth, the wafer bowing was X ~ -57 µm, Y~ -45 µm, indicating excellent strain management of the thick GaN growth.

A schematic process flow is depicted in Fig. 2. The device fabrication started with a plasma-based dry etching process of

978-1-5386-2928-4/18 $31.00 © 2018 IEEE

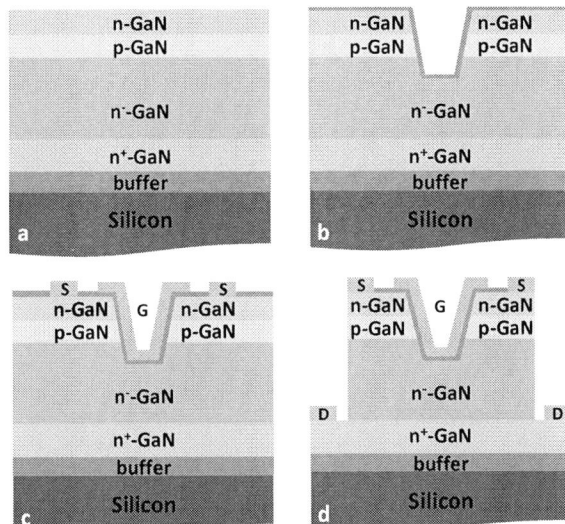

Fig. 2. Schematic process flow for GaN-on-Si vertical MOSFETs, (a) as-grown n-p-n structure on silicon, (b) trench etching and ALD SiO₂ deposition, (c) source contact opening, and gate/source metal deposition, (d) deep etching and drain metal deposition.

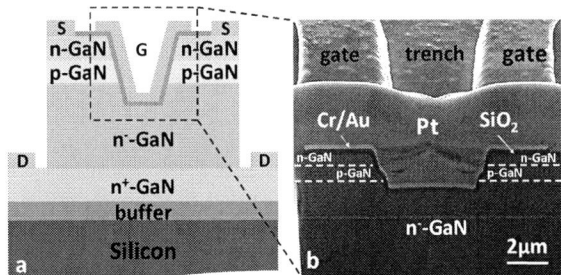

Fig. 3. Cross-sectional (a) schematic, and (b) SEM image of the fabricated quasi-vertical trench gate MOSFETs on Si substrate.

GaN to form the trench structures for the vertical gate. The sample was then treated with a Tetra Methyl Ammonium Hydroxide (TMAH) solution. A rapid thermal annealing was performed at 850°C for 20 min in a N₂ ambient to activate the p-type GaN. Subsequently, a SiO₂ gate dielectric for the vertical MOSFETs was deposited on the top surface and trenches by atomic layer deposition. After opening contact holes, a double-layer metal stack of Cr/Au was evaporated to form both the source and gate electrodes for the vertical MOSFETs. Finally, a 5 μm-deep etching was performed using a ICP etch, followed by the evaporation of drain electrodes. More Details of fabrication can be found in [21].

III. RESULTS AND DISCUSSION

Fig. 3 (a) and (b) presents the cross-sectional schematic and SEM image of the GaN-based vertical trench gate MOSFET on silicon substrates. A 1.6 μm deep gate trench can be observed by FIB-SEM. The trench sidewalls were inclined by 13.2° from the c-axis, resulting in a trench width of 4.0 μm and 5.5 μm at the bottom and the top, respectively. Due to the erosion of the edge of the SiO₂ hard mask during dry etching, a small kink

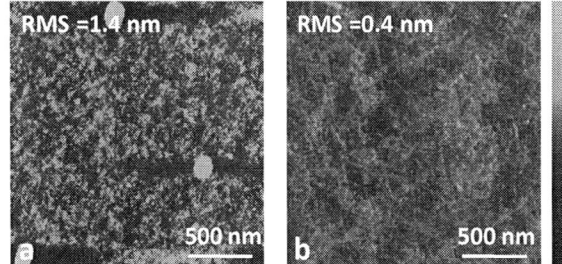

Fig. 4. AFM images of the ICP etched surface (a) without TMAH treatment, and (b) with TMAH treatment.

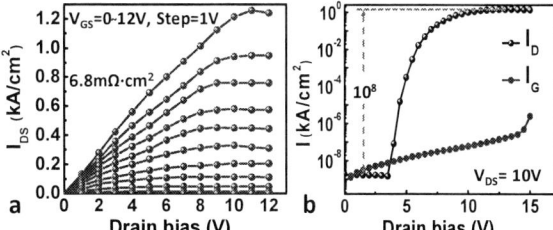

Fig. 5. (a) Output I-V characteristics and (b) semi-log transfer characteristics of the vertical trench gate MOSFETs on silicon substrate.

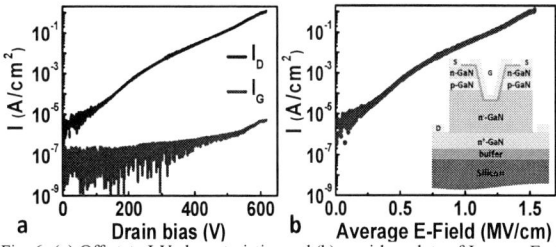

Fig. 6. (a) Off-state I-V characteristics and (b) semi-log plots of I versus E at $V_{GS} = 0$ V of the vertical trench gate MOSFETs on silicon substrate.

was formed at the middle position of the sidewall. The sidewall and bottom of the gate trench were well covered by SiO₂ gate dielectric and Cr/Au gate metal, which is crucial to obtain a low leakage and low on-resistance vertical trench gate MOSFET.

It has been reported that TMAH wet etch is effective in removing damage from the sidewall and bottom of the dry-etched gate trench, which can improve the conductivity of the inversion channel and eliminate the high electric field peaks at the bottom of the gate region [22]. To investigate the effect of TMAH treatment on the dry etched GaN surface, ICP etching was carried out to remove the top 1 μm layer from the as-grown n-p-n structure, followed by TMAH treatment at 85°C for 60 minutes. Fig. 4 compares the AFM images of the dry etched sample (a) before and (b) after TMAH treatment in an area of 2 × 2 μm². Large dots can be observed on the ICP exposed n⁻-GaN surface, which might be caused by the un-optimized ICP etching conditions. The surface morphology of the TMAH treated sample is dramatically improved with a root mean squire (RMS) roughness of 0.4 nm from 1.4 nm before treatment, indicating that TMAH is effective in removing the dry etching damages and smoothing the etched surface.

978-1-5386-2928-4/18 $31.00 © 2018 IEEE   241

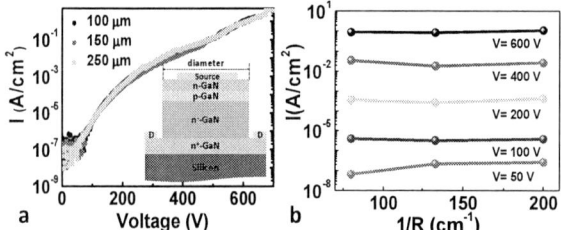

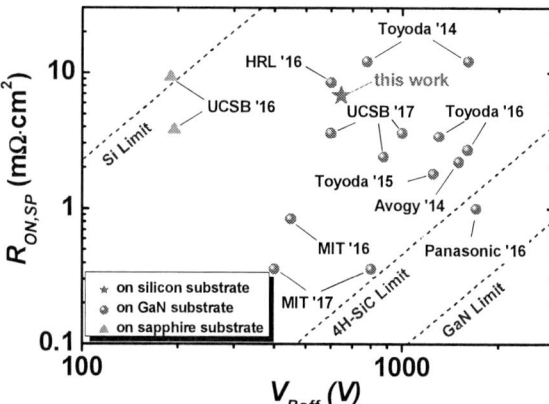

Fig. 7. (a) Two-terminal I-V characteristics of the as-grown n-p-n structure and (b) leakage current density of the two-terminal circular n-p-n test structure with different mesa radius R.

Fig. 5 (a) shows the output I-V characteristics (I_D-V_{DS}) of the fabricated vertical MOSFETs. The $R_{ON,SP}$ calculated from the linear region using the trench area was 6.8 m$\Omega\cdot$cm^2, which could be further reduced by post-trench-etching treatment using a higher concentration of TMAH. At V_{GS} = 12 V and V_{DS} = 11 V, the drain current density of 1.3 kA/cm^2 was obtained, with good saturation behavior. Fig. 5 (b) shows the transfer I-V characteristics (I_D-V_{GS}) of the vertical MOSFETs at a drain-source voltage of 10 V. The fabricated vertical MOSFETs showed a current on/off ratio of over 10^8 and a sub-threshold swing (SS) of 250 mV/dec. A slight increase in gate leakage at V_{GS} = 15 V can be observed, which indicates that the trench etching and gate dielectric deposition can be further improved. A V_{TH} of 3.3 V can be extracted from semi-log scale plots. The high V_{TH} is highly desirable to guarantee a fail-safe operation in high power applications. The off-state leakage current level was below 10^{-8} kA/cm^2, indicating effective current blocking by the n-p-n heterostructure.

Fig. 6 (a) shows the off-state I-V characteristics measured at V_{GS} = 0 V for the fabricated trench gate MOSFETs. A large hard breakdown voltage (V_{Boff}) of 645 V was observed at 2.8 A/cm^2 for the vertical MOSFETs. The breakdown was destructive and mainly observed at the mesa edge regions, which could be improved with edge termination technologies. However, the performance observed even without edge termination is remarkable, which shows the huge potential for GaN on Si vertical transistors. The gate current below 10^{-5} A/cm^2, implying that the drain leakage current originates from a leakage current between the source and the drain contacts. The off-state leakage mechanism was further identified by studying the correlation between leakage current density I and E field. A nearly linear relationship of ln (I) $\propto$ E was found for the fabricated vertical MOSFETs, as plotted in Fig. 6 (b). Referring to [23], the ln (I) $\propto$ E linearity indicates that the leakage mechanism is dominated by variable-range hopping for the vertical MOSFETs in this work.

A two terminal breakdown test for the as-grown n-p-n structure was also performed. As shown in Fig. 7 (a), the breakdown voltage of the n-p-n structure (V_{BR}) is ~ 679 V, which is consistent with the MOSFET breakdown voltage. The measured leakage current mainly flows through the n-p-n heterostructures, instead of the etched surfaces. To test this, we have measured the leakage current from two-terminal circular n-p-n test structures with different mesa radius (R =

Fig. 8. $R_{ON,SP}$ versus V_{Boff} benchmarks of the vertical trench gate MOSFETs on silicon substrates with state-of-the-art E-mode vertical transistors on sapphire and GaN substrates.

50 μm, 75 μm, and 125 μm), and found the exact same leakage current density with different radius (no dependence of leakage current density on the mesa periphery in Fig. 7 (b)). This indicates that the device leakage current is mainly through the heterostructure instead of the etched sidewall.

Our GaN-on-Si vertical transistors were benchmarked in Fig. 8 against state-of-the-art E-mode vertical GaN transistors on GaN, and sapphire substrates as a function of V_{Boff}. With a V_{Boff} of 645 V and $R_{ON,SP}$ of 6.8 m$\Omega\cdot$cm^2, our GaN-on-Si vertical transistors showed a very good Baliga figure-of-merit (FOM) of 61 MW/cm^2. The FOM can be significantly improved by designing field plate and edge termination technologies as well as optimizing the growth and device fabrication.

CONCLUSION

In summary, GaN-on-Si vertical MOSFETs are demonstrated on 6-inch silicon substrates. The fabricated vertical MOSFETs exhibited a threshold voltage of 3.3 V and a specific on-resistance of 6.8 m$\Omega\cdot$cm^2. An off-state breakdown voltage of 645 V was achieved, thanks to the high-quality 4-μm-thick GaN drift layer. These results make this GaN-on-Si vertical MOSFETs very promising for cost-effective and high-performance power electronics applications.

ACKNOWLEDGMENT

We would like to thank the staff at CMi and ICMP cleanrooms at EPFL for technical support and advice. We would also like to thank Dr. Kai Cheng from Enkris Semiconductor, Inc for the collaboration on the growth of high-quality customized wafers.

REFERENCES

[1] H. Otake, S. Egami, H. Ohta, Y. Nanishi, and H. Takasu, "GaN-based trench gate metal oxide semiconductor field effect transistors with over 100 cm^2/(Vs) channel mobility," Jpn. J. Appl. Phys., vol. 46, no. 25, pp. L599–L601, Jul 2007.

[2] H. Otake, K. Chikamatsu, A. Yamaguchi, T. Fujishima, and H. Ohta, "Vertical GaN-based trench gate metal oxide semiconductor field-effect transistors on GaN bulk substrates," Appl. Phys. Exp., vol. 1, no. 1, p. 011105, Jan 2008.

[3] D. Ji, M. A. Laurent, A. Agarwal, W. Li, S. Mandal, S. Keller, and S. Chowdhury, "Normally OFF Trench CAVET With Active Mg-Doped GaN as Current Blocking Layer." IEEE Trans. Electron Devices, vol. 64, no. 3, pp 805–808, Mar 2017.

[4] C. Gupta, S. H. Chan, C. Lund, A. Agarwal, O. Koksaldi, J. Liu, Y. Enatsu, S. Keller, and U. K. Mishra, "Comparing electrical performance of GaN trench-gate MOSFETs with a-plane (1120) and m-plane (1100) sidewall channels," Appl. Phys. Exp., vol. 9, no. 12, pp. 121001-1– 121001-3, Nov 2016.

[5] C. Gupta, A. Agarwal, S. H. Chan, O. S. Koksaldi, S. Keller and U.K. Mishra, "1 kV field plated in-situ Oxide, GaN interlayer based vertical trench MOSFET (OG-FET)", in Proc. 75th Annu. Device Res. Conf. (DRC), Jun 2017, pp. 1–2.

[6] C. Gupta, C. Lund, S. H. Chan, A. Agarwal, J. Liu, Y. Enatsu, S. Keller and U.K. Mishra, "In-situ Oxide, GaN interlayer based vertical trench MOSFET (OG-FET) on bulk GaN substrates", IEEE Electron Device Lett., vol. 38, no. 3, pp. 353–356, Mar 2017.

[7] M. Sun, Y. Zhang, X. Gao and T. Palacios, "High performance GaN vertical fin power transistors on bulk GaN substrate", IEEE Electron Device Lett., vol. 38, no. 4, pp. 509–512, Apr 2017

[8] T. Oka, Y. Ueno, T. Ina, and K. Hasegawa, ''Vertical GaN-based trench metal oxide semiconductor field effect transistors on a free-standing GaN substrate with blocking voltage of 1.6 kV," Appl. Phys. Exp., vol. 7, no. 2, p. 021002, 2014.

[9] T. Oka, T. Ina, Y. Ueno and J. Nishii, "1.8 $m\Omega \cdot cm^2$ vertical GaN-based trench metal–oxide–semiconductor field-effect transistors on a freestanding GaN substrate for 1.2-kV-class operation," Appl. Phys. Exp., vol. 8, no. 5, pp. 054101-1–054101-3, May 2015.

[10] T. Oka, T. Ina, Y. Ueno, and J. Nishii, "Over 10 a operation with switching characteristics of 1.2 kV-class vertical GaN trench MOSFETs on a bulk GaN substrate," in Proc. 28th Int. Symp. Power Semiconductor Devices ICs (ISPSD), Jun 2016, pp. 459–462.

[11] R. Li, Y. Cao, M. Chen and R. Chu, "600V/1.7Ω Normally-Off GaN Vertical Trench Metal-Oxide-Semiconductor Field-Effect Transistor", IEEE Electron Device Lett., vol. 37, no. 11, pp.1466–1469, Nov 2016.

[12] H. Nie, Q. Diduck, B. Alvarez, A. P. Edwards, B. M. Kayes, M. Zhang, G. Ye, T. Prunty, D. Bour and I. C. Kizilyalli, "1.5-kV and 2.2-$m\Omega.cm^2$ vertical GaN transistors on bulk-GaN substrates," IEEE Electron Device Lett., vol. 35, no. 9, pp. 939–941, Sep 2014

[13] D. Shibata, R. Kajitani, M. Ogawa, K. Tanaka, S. Tamura, T. Hatsuda, M. Ishida, and T. Ueda, "1.7 kV/1.0 $m\Omega \cdot cm^2$ normally-off vertical GaN transistor on GaN substrate with regrown p-GaN/AlGaN/GaN semipolar gate structure," in IEDM Tech. Dig., 2016, pp. 10.1.1–10.1.4.

[14] X. Zhang, X. Zou, X. Lu, C. W. Tang, and K. M. Lau, "Fully- and Quasi-Vertical GaN-on-Si p-i-n Diodes: High Performance and Comprehensive Comparison," IEEE Trans. Electron Devices, vol. 64, no. 3, pp. 809–815, Mar 2017.

[15] X. Zou, X. Zhang, X. Lu, C. W. Tang, and K. M. Lau, "Fully Vertical GaN p-i-n Diodes Using GaN-on-Si Epilayers," IEEE Electron Device Lett., vol. 37, no. 5, pp. 636–639, May 2016.

[16] X. Zou, X. Zhang, X. Lu, C. W. Tang, and K. M. Lau, "Breakdown Ruggedness of Quasi-Vertical GaN-Based p-i-n Diodes on Si Substrates," IEEE Electron Device Lett., vol. 37, no. 9, pp. 1158–1161, Sep 2016.

[17] S. Mase, Y. Urayama, T. Hamada, J. J. Freedsman, and T. Egawa, "Novel fully vertical GaN p-n diode on Si substrate grown by metalorganic chemical vapor deposition," Appl. Phys. Exp., vol. 9, no. 11, p. 111005, Nov 2016.

[18] Y. Zhang, M. Sun, D. Piedra, M. Azize, X. Zhang, T. Fujishima, and T. Palacios, "GaN-on-Si Vertical Schottky and p-n Diodes," IEEE Electron Device Lett., vol. 35, no. 6, pp. 618–620, Jun 2014.

[19] Y. Zhang, D. Piedra, M. Sun, J. Hennig, A. Dadgar, L. Yu, and T. Palacios, "High-Performance 500 V Quasi- and Fully- Vertical GaN-on-Si pn Diodes," IEEE Electron Device Lett., vol. 38, no. 2, pp. 248–251, Feb 2017.

[20] R. A. Khadar, C. Liu, and E. Matioli, "820 V GaN-on-Si Quasi-Vertical PiN Diodes with BFOM of 2.0 GW/cm²," IEEE Electron Device Lett., vol. 39, no. 3, pp. 401–404, Mar 2018.

[21] C. Liu, R. A. Khadar, and E. Matioli, "GaN-on-Si Quasi-Vertical Power MOSFETs," IEEE Electron Device Lett., vol. 39, no. 1, pp. 71-74, Jan 2018.

[22] MKodama, M. Sugimoto, E. Hayashi, N. Soejima, O. Ishiguro, M. Kanechika, K. Itoh, H. Ueda, T. Uesugi, and T. Kachi, "GaN-Based Trench Gate Metal Oxide Semiconductor Field-Effect Transistor Fabricated with Novel Wet Etching," Appl. Phys. Exp., vol. 1, no. 2, pp. 021104-1 – 021104-3, Feb 2008.

[23] D. Han, C. Oh, H. Kim, J. Shim, K. Kim, and D. Shin, "Conduction Mechanisms of Leakage Currents in InGaN/GaN-Based Light-Emitting Diodes", IEEE Trans. Electron Devices, vol. 62, no. 2, pp. 587–592, Feb 2015

978-1-5386-2928-4/18 $31.00 © 2018 IEEE

Proceedings of the 30th International Symposium on Power Semiconductor Devices & ICs
May 13-17, 2018, Chicago, USA

Effects of Inorganic Encapsulation on Power Cycling Lifetime of Aluminum Bond Wires

Nan Jiang[*], Markus G. Scheibel[+], Benjamin Fabian[+], Marko Kalajica[+], Anton-Zoran Miric[+], Josef Lutz[*]

Email: nan.jiang@etit.tu-chemnitz.de

[*]Faculty of Electrical Engineering and Information Technology, Chemnitz University of Technology, Chemnitz, Germany

[+]Heraeus Electronics, Heraeus Deutschland GmbH & Co. KG, Hanau, Germany

Abstract—Bond wire lift-offs and heel cracks are regarded as one of the main failure mechanisms of the bond wires in the power modules. In previous studies, the reliability of the bond wires can be improved by applying encapsulations to cover the bond wires. In this work, the investigation was mainly focused on the effect of inorganic encapsulation on the reliability of the aluminum bond wires under the power cycling test. Samples with different encapsulation positions were tested in order to evaluate the dependency of the encapsulation position on the reliability of the bond stitches. The results of power cycling tests and shear tests showed a reliability improvement of the tested samples when the aluminum bond wires was covered by the inorganic encapsulation. Besides, the reliability improvement of the samples was highly related to the covered position of the bond stitches. Thermal-mechanical simulation results indicated the heel cracks had stronger influence on the encapsulated bond wires at the certain position than the other position, which may be the reason of the position dependency of the encapsulation on the reliability improvement of the bond wires.

Keywords—power cycling test; inorganic encapsulation; aluminum bond wires; reliability

I. INTRODUCTION

Aluminum (Al) bond wires with the properties of easy process capability and acceptable high thermal/electrical conductivity as well as low cost, are widely used as the top-side interconnection in the nowadays power module manufacturing. Many research works have been conducted over the decades in order to improve the reliability performance of the Al bond wires. The applied encapsulation within the package is regarded as one of the influence factors on the reliability performance of the Al bond wires. However, silicone gels, as one of the most commonly applied encapsulation materials, was not suitable for the high temperature application due to its limited operation temperature [1]. In this work, the investigation was mainly focused on the effect of inorganic encapsulation on the reliability of the aluminum bond stitches at different positions. A type of newly developed inorganic encapsulation materials based on cements with advantages of low setting temperature, high operation temperature and controllable coefficient of thermal expansion (CTE) was chosen as encapsulation material for power devices in order to investigate the influence of the covered position on the reliability of the aluminum bond wires.

Power cycling test (PC test) is considered as the most important accelerated reliability tests to evaluate the lifetime of power devices. During the switch on operation, the tested sample is heated up by the heat dissipation due to the power loss of the chip, afterwards the sample is cooled down during the switch off operation. The temperature swing during the operation cycles and the mismatch of CTE between the bond wires and the chip introduce the stress concentration on the bond wires. This will finally lead to bond wire failures, such as bond wire lift-offs or heel cracks.

In this work, power cycling tests were performed for diodes bonded with Al bond wires. The test samples were divided into four groups depending on the encapsulated position of the bond stitches. Shear tests for bond stitches were carried out in parallel in order to investigate the interface degradation of the Al bond stitches. Thermal-mechanical simulations were performed in order to figure out the strain distribution in the bond wires.

II. TEST SET-UP

Eight aluminum bond wires with the diameter of 400 μm (Heraeus Al H14 CR) were bonded on the top Al-metallization layer of the diode (Semikron SKCD81C170IHD). The diode was soldered on the direct copper bonded (DCB) Al_2O_3 substrate with SnAg solder. No baseplate was applied for the device under test (DUT). The configuration of DUTs and the encapsulated positions of the bond stitches are illustrated in Fig.1.

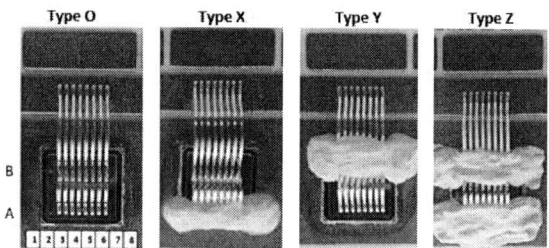

Fig.1. Four types of DUTs encapsulated on the different positions (position A and B).

The presented work received the financial support by the German Federal Ministry of Education and Research BMBF (Contract: 16EMO0226).

A new class of encapsulation materials is under current industry development by Heraeus. Reliability studies on power modules potted with this new class of encapsulation materials reviled migrations and reactions on the chip surface as new failure mechanism after a lifetime of factor 3 as compared with the respective silicone reference [2].

In the power cycling tester, DUTs were mounted on a liquid-cooled heat-sink. The thermal paste was used as a thermal interface material to transport the heat from DUTs to the heat-sink. The load current (I_{load}) was conducted through copper-bars. A thermocouple was placed inside the copper adapter plate in order to measure the heat sink temperature. The junction temperature was measured by using a sense current of 100 mA ($V_{CE}(T)$ method) [3] [4]. The control strategy of constant t_{on} and I_{load} was applied for power cycling test [5]. The control strategy of constant switching time (t_{on}/t_{off}) and load current (I_{load}) was chosen for the PC tests. The chosen test conditions were: $I_{load} = 132$ A, $t_{on} = 1$s, $t_{off} = 2$s, $\Delta T_j = 120$ K, $T_{j,min} = 30°C$. 5% increase of the forward voltage (V_f) and 20% increase of thermal resistance (R_{thjc}) were considered as the criteria of failure.

The shear strength and the sheared area demonstrated a linear correlation in the previous study [6]. In order to investigate the interface degradation of the bond stitches, one of the test devices was taken to the shear tester at certain cycle numbers during the power cycling tests. The remaining DUTs were tested until end of life. The shear height of the shear tool is 10% of the wire diameter above the bonding interface. The shear test for the device started at bond stitch A1, and ended at B8 (as shown in Fig. 1).

III. TEST RESULTS AND ANALYSIS

A. Power cycling test

Results of power cycling tests of the aforementioned types of DUTs are demonstrated in Fig. 2, which were generated by Minitab software in order to fit into the Weibull distribution (95% confidence intervals). DUTs of Type Z (encapsulated at position A and B) showed the best reliability performance among all the types.

TABLE I.   TEST RESULTS

Test Results	Types			
	Type O	*Type X*	*Type Y*	*Type Z*
Average N_f	20270	30443	24074	34522
Failure criteria	5% V_f	5% V_f	5% V_f	5% V_f
Improvement	100 %	150 %	119 %	170 %

In order to evaluate the influence of the encapsulated bond stitches on the power cycling capability of the DUTs, the average lifetime of different types of DUTs are compared in TABLE I. The improvement of Type X (encapsulated at position A) is much larger compared to Type Y (encapsulated at position B).

As shown Fig. 2, the shape factor of DUTs with encapsulation materials covered at position B (Type Y) is smaller compared to others, which means the spreading of lifetime of Type Y is larger than other types. It has to be mentioned, the shape of bond stitches at position B increases the complexity to maintain the identical encapsulation condition at this position, which may be the reason of the larger shape factor.

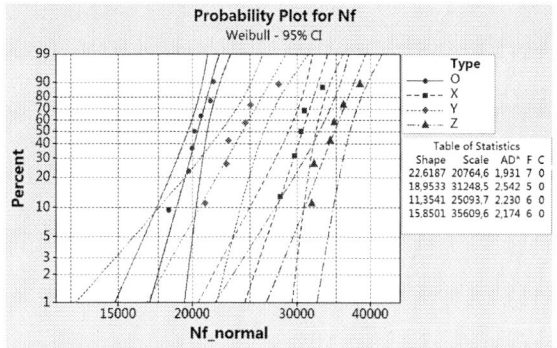

Fig. 2. Weibull distributions of power cycling lifetime of DUTs of different types.

B. Shear test

At certain number of cycles during the power cycling tests, one of the tested samples was taken out to perform the shear test. However, it cannot be guaranteed that the encapsulated bond stitches were not damaged during the de-encapsulating process. Thus all the shear tests were performed for the un-encapsulated bond stitches.

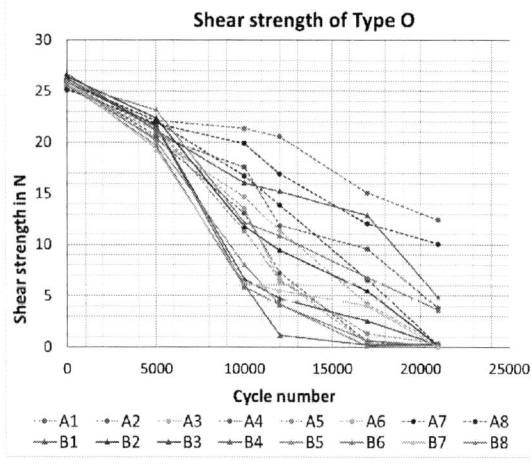

Fig. 3. Shear strengths of DUTs with no encapsulation.

The bond stitches of Type O were not protected by any encapsulations during the power cycling tests. Thus the shear strengths of DUTs from Type O, as illustrated in Fig. 3, can be considered as the shear strength reference to other types. As shown in Fig. 3, the degradation rates of the shear strength of the bond stitches at position B are comparably faster than the degradation rates at position A, which may due to the different strain distribution on the interface of the bond stitches. This phenomenon was also observed in authors' previous works [7] [8].

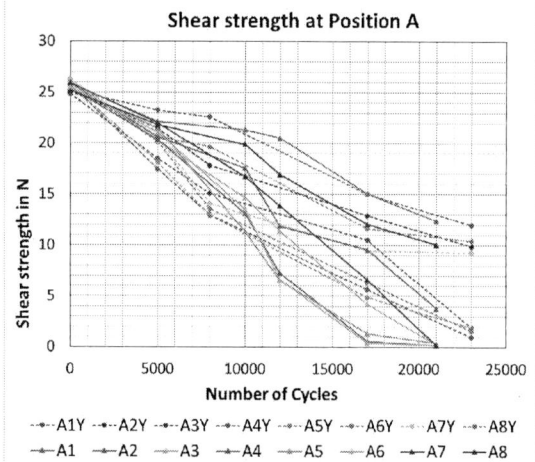

Fig. 4. The comparison of the shear strengths at position A (Type O and Type Y).

The shear strengths of the bond stitches at position A are compared in Fig. 4. It has to be mentioned, the shear tests can only be performed for the uncovered bond stitches, so the comparison here is mainly focused on the influence of the encapsulated regions on the uncovered bond stitches. As shown in Fig. 4, no significant improvement of the shear strength degradation at position A of DUTs from Type Y can be observed, which indicates the encapsulation at position B cannot increases the lifetime of bond stitches at position A.

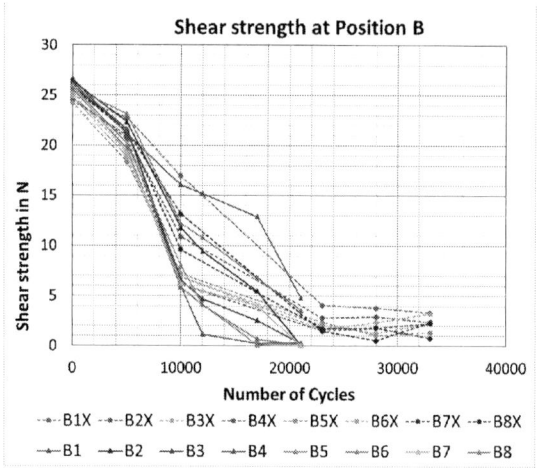

Fig. 5. The comparison of the shear strengths at position B (Type O and Type X).

The shear strengths of the bond stitches at position B are compared in Fig. 5: no significant improvement of the shear

strength degradation at position B of DUTs from Type X can be observed. In fact, all the bond stitches at position B of DUTs after 20000 cycles showed a weak bonding strength, but the DUTs did not reach the end of lifetime until 340000 cycles. It has to be mentioned, the measured forward voltage during the power cycling tests only jumps when both bond stitches at one bond wire lose their contacts. The comparisons of shear strengths depicted in Fig. 4 and Fig. 5 indicate the encapsulation at position A has a stronger influence on the reliability improvement of DUTs than the encapsulation at position B.

C. Simulation

In this study, a simplified structure related to the Al wire bonding on the top side of the chip metallization and the upper copper layer of the substrate was built in order to simulate the strain distribution in the bond wires. The cross-section view of one bond wire during the power cycle is illustrated in Fig. 6. It can be seen that the maximum strain locates on the bonding surface when no encapsulation is applied (Type O), which implies the bond wire lift-off can be considered as the main failure mechanism in this case. However, when the inorganic encapsulation is covered over the bond stitches, then the deformation of the bond wire is restricted by the encapsulation. The strain distribution is highly depending on the bonding loop and the shape of the encapsulated area. As shown in Fig. 6 (Type Z), a similar strain distribution is observed at the encapsulated bonding region of position A, while more strains are concentrated at the heel of the bond wires of position B. The simulation results implies the heel crack failure may have stronger influence on the encapsulated bond wires at position B than the bond wires at position A.

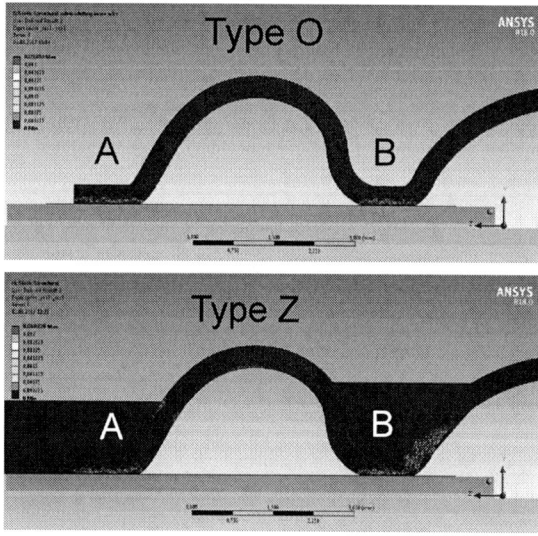

Fig. 6. Plastic strain in the bond wires of DUTs with or without encapsulation (Type O and Type Z).

The simulation results of the strain distributions on the bonding interface of Type O and Type Z are demonstrated in Fig. 7. The maximum strains of both cases are located at the bonding region of position B. But the value of the maximum strain reduced when the encapsulation is applied.

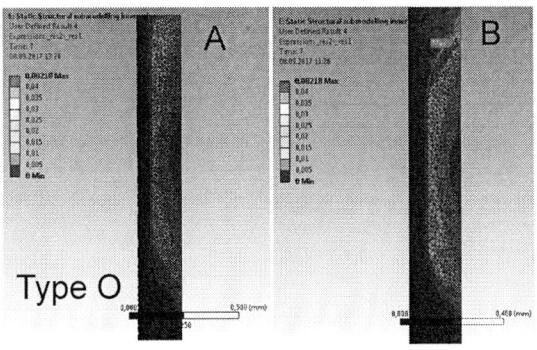

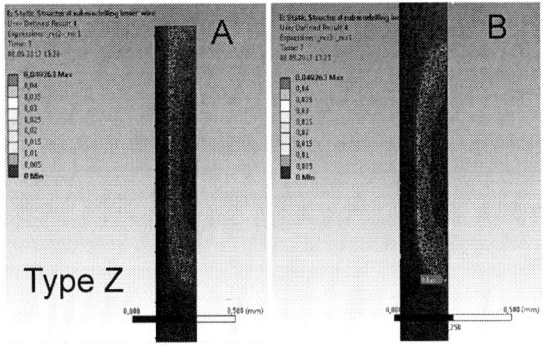

Fig. 7. Plastic strain on the bonding surface of DUTs with or without encapsulation (Type O and Type Z).

TABLE II.   MAXIMUM STRAIN AND LOCATION

Type	Max plastic strain		First failure location	
	Position A	*Position B*	*Simulation*	*Test*
Type O	1.20E-02	1.21E-02	B	B
Type X	4.55E-03	1.21E-02	B	B
Type Y	1.16E-02	4.61E-03	A	A
Type Z	4.43E-03	4.68E-03	A	unclear

The results of the maximum plastic strain and the locations are shown in TABLE II. The maximum strain reduces when the encapsulation is applied on the bond stitches. However, in the case of Type Y, the strongly reduced maximum strain on the bonding interface at position B does not result in a significant reliability improvement of DUTs. Furthermore, the similar reduced strain at position A of Type X resulted in the 1.5 times of lifetime (as shown in Fig. 2 and TABLE I). The less reliability improvement of the encapsulated bond wires at position B can be explained by the change of strain distribution inside the bond wires as shown in Fig. 6.

IV.   CONCLUSION AND OUTLOOK

In this work, power cycling tests were performed for aluminum wires bonded DUTs with inorganic encapsulation covered at different bond stitches. Shear tests were conducted in parallel with PC tests in order to monitor the degradation of the bond stitches.

Significant reliability improvements of DUTs encapsulated at position A were observed. However, the encapsulation of bond stitches at position B showed a weaker influence on the lifetime of DUTs. The comparison of shear strengths degradation of DUTs encapsulated at different positions also indicates the encapsulation of bond wires at position A has a stronger influence on the reliability improvement of the DUTs than the encapsulation of bond wires at position B.

However, the simulation results of the maximum strain on the bonding interface had a different result. The maximum strain on the both bonding interfaces reduced significantly when the encapsulation was applied, which was against the results of the performed power cycling tests. But the cross-section view of the bond wire in the simulation showed a variation of strain distributions from the bonding surface to the heels of the bond wires at position B when covered with inorganic encapsulations.

For further studies, more power cycling tests and simulations of DUTs with different encapsulation types and coverage are demanded in order to develop a most suitable encapsulation for the Al bond wires.

REFERENCES

[1] M. L. Locatelli, "Evaluation of encapsulation materials for high temperature power device packaging," IEEE Trans. Power Electr., vol. 29, pp. 2281 - 2288, August 2014.

[2] B. Boettge, "Material characterization of advanced cement-based encapsulation system for efficient power electronics with increased power density" , In: Pro. ECTC 2018, San Diego, USA, in press.

[3] R. Schmidt and U. Scheuermann, "Using the chip as a temperature sensor-The influence of steep lateral temperature gradients on the Vce(T)-measurement." In: Proc. EPE 2009, Barcelona, Spain.

[4] C. Herold, J. Sun, P. Seidel, L. Tinschert and J. Lutz, "Power cycling methods for SiC MOSFETs." In: Proc. ISPSD 2017, Sapporo, Japan.

[5] U. Scheuermann and S. Schuler, "Power cycling results for different control strategies." Microelectron. Rel., vol. 50, pp. 1203–1209, 2010.

[6] J. Göhre, M. Schneider-Ramelow, U. Geißler and K.-D. Lang, "Interface degradation of Al heavy wire bonds on power semiconductors during active power cycling measured by the shear test." In: Proc. CIPS 2010, Nuremberg, Germany.

[7] N. Jiang, M. Kalajica and J. Lutz, "On-time dependency on the power cycling capability of Al bond wires measured by shear test" , In: Pro. CIPS 2018, Stuttgart, Germany, in press.

[8] N. Jiang, B. Fabian, M. Kalajica and J. Lutz, "Investigation of ton dependency of Al-clad Cu bond wires under power cycling tests" , In: Pro. PCIM 2018, Nuremberg, Germany, in press.

Proceedings of the 30th International Symposium on Power Semiconductor Devices & ICs
May 13-17, 2018, Chicago, USA

Sn- and Cu-Oxide Reduction by Formic Acid and Its Application to Power Module Soldering

Naoto Ozawa, Tatsuo Okubo, Jun Matsuda and Tatsuo Sakai

R&D Headquarters, Origin Electric Co., Ltd., Saitama, Japan, t_sakai@origin.jp

Abstract— **In this paper, surface analysis of a copper substrate and solder foil, and real time measurement of oxide film thickness during reduction by formic acid with an ellipsometer are described. From the measurement results, the native oxide films of the copper substrate and the solder foil are presumed to be Cu_2O and SnO, respectively, and their thicknesses are confirmed as 4 nm and 5.3 nm, respectively. It is also found that SnO has a higher reduction rate by formic acid than Cu_2O. Furthermore, the contact angle of the melted solder ball becomes smaller as the copper oxide film becomes thinner, but complete oxide removal is necessary to obtain favorable solder wettability. On the basis of these formic acid reduction data, a void rate of 1% or less is achieved in a soldered sample processed with a formic acid reduction reflow machine.**

Keywords— *Formic Acid, Ellipsometer, Real-time, in-situ*

I. INTRODUCTION

Reduction in solder void for power device die attachment is important for good heat dissipation and therefore high reliability. In recent years, fluxless solder is sometimes used for die attachment, and a reducing agent, such as hydrogen and formic acid, is used for metal oxide reduction in a fluxless soldering process. Formic acid is a promising candidate as a reducing agent since it effectively reduces metal oxide film from low temperature [1] - [2].

The reduction of copper oxide film by formic acid has already been reported [3] - [4]. In addition to the copper oxide film, reduction properties of tin oxide film, which is the Pb-free solder surface oxide, are necessary to be clarified to design a favorable soldering process which realizes a low void rate.

In this paper, surface oxide films of a copper substrate and solder foil are identified by the surface analysis. Then, reduction rates of these oxide films by formic acid are obtained by the real time measurement with an ellipsometer. Finally, the soldering result of two solder layers with the optimized process is described.

II. EXPERIMENTAL METHOD

A. Materials used in experiments

Copper substrates (purity:> 99.96%, size: 20 × 20 × 2 mm) and solder foils (SnAg3.0Cu0.5, size 20 × 20 × 0.07 mm) were used for the oxide film thickness measurement, and solder foils (SnAg3.0Cu0.5, size 5 × 5 × 0.07 mm) and solder balls (SnAg3.0Cu0.5, size φ 1.5 mm) were used for evaluation of wettability. Formic acid (manufactured by Kanto Kagaku, special grade, formic acid content:> 98%) was used as the

reducing agent, and N_2 gas was supplied from an N_2 cylinder of 99.99% purity.

B. Experimental system

A high-speed spectroscopic ellipsometer (M-2000, manufactured by J. A. Woollam) is used to measure the thicknesses of surface oxide film on the copper substrate and the solder foil. The configuration of the measurement system is shown in Fig. 1. The system is constructed by combining a vacuum chamber (internal volume of 170 ml) capable of introducing formic acid and the ellipsometer. Since the ellipsometer can analyze oxide thickness with non-contact and non-destructive measurement, the reduction process of the oxide film by formic acid can be measured in real time from the outside of the chamber. The wavelength range of measurement light is 400 nm – 1,000 nm, the incident angle θ is 70 degrees, and the data acquisition interval during real time measurement is set to about 0.6 – 1sec which is short enough to check the oxide film thickness change.

C. Analysis of surface oxide film of copper substrate and solder foil

TEM observation (Field emission type transmission electron microscope JEM-2010F manufactured by JEOL Ltd.) and TEM-EELS analysis (863 GIF Tridiem manufactured by Gatan, beam diameter: about 1.0 nm) were carried out for the native oxide film of the copper substrate and the solder foil.

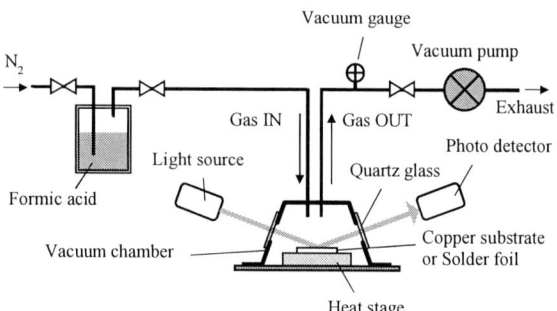

Fig.1   Schematic view of the experimental equipment

978-1-5386-2928-4/18 $31.00 © 2018 IEEE   248

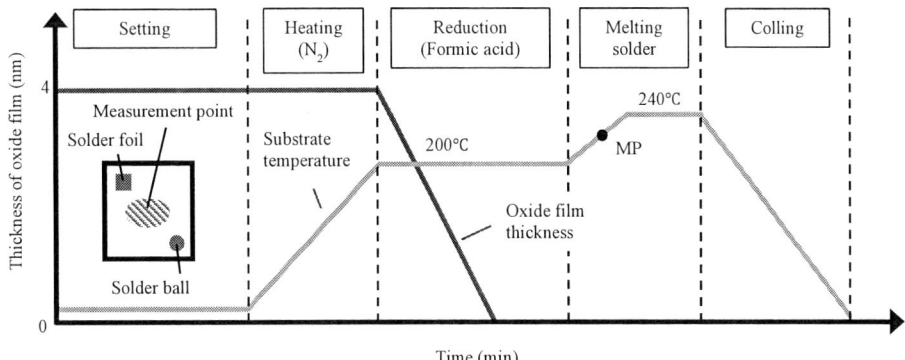

Fig.2 Experimental profile

D. Measurement procedure of reduction rate of oxide film by formic acid

The procedure for measuring the reduction rate of the oxide film on the copper substrate and the solder foil is described below.

i) Place the copper substrate or the solder foil with the surface native oxide film on the heat stage. Replace the chamber atmosphere to N_2 and then heat the copper substrate or the solder foil to 200°C.

ii) After the temperature is stabilized, reduce the pressure inside the chamber. Introduce formic acid gas vaporized by N_2 bubbling to atmospheric pressure while measuring the oxide film thickness in real time. The introduction time of the formic acid gas is 1 sec or less.

iii) After the chamber pressure reaches the atmospheric pressure, supply of the formic acid gas is stopped and held at that pressure until the oxide film is removed. The real time measurement of the oxide film continues during this period.

iv) After the oxide film is removed, the formic acid gas is exhausted, and the copper substrate or the solder foil is cooled to room temperature and taken out from the chamber.

E. Measurement of oxide thickness during formic acid reduction and evaluation of solder wettability

In order to observe the relationship between the oxide film thickness of the copper substrate and the solder wettability, samples with different oxide film thicknesses were prepared by adjusting the time for formic acid reduction, and the wettability was evaluated by melting the solder on the sample substrates. The schematic experimental profile is shown in Fig. 2, and the experimental procedure is described below.

i) Place the copper substrate on the heat stage inside the vacuum chamber and mount the solder foil and the solder ball in places other than the measuring part of the ellipsometer on the substrate.

ii) Heat the copper substrate to 200°C after replace the chamber atmosphere to N_2.

iii) Introduce formic acid gas vaporized by N_2 bubbling into the chamber and reduce the oxide film.

iv) Increase the temperature to 240°C in N_2 atmosphere to melt the solder, then cool the copper substrate and take it out from the chamber.

v) Observe the solder foil and the solder ball with an optical microscope, and evaluate the contact angle for the solder ball and the spread area ratio to the initial solder area for the solder foil.

In addition to the sample from which the copper oxide film was completely removed, the sample in which the reduction was stopped during the process and resulting in the remained oxide film, and the sample in which the copper oxide film was not reduced at all were also prepared, and the relationship between the copper oxide film thickness and the solder wettability was examined.

III. RESULTS AND DISCUSSION

A. Analysis of surface oxide film of copper substrate and solder foil

Fig. 3 shows a cross-sectional TEM image of the copper substrate. A portion with a different contrast in the image was defined as the copper oxide film, and its thickness was measured to be 4 nm. The results of TEM-EELS analysis at points 1 to 6 in Fig. 3 are shown in Fig. 4. Comparing this analysis result with CuO and Cu_2O standard samples, it was found that the oxide film of the copper substrate surface was Cu_2O [3].

Next, a cross-sectional TEM image of the solder foil is shown in Fig. 5. A portion with a different contrast in the image was defined as the tin oxide film, and its thickness was found to be 5.3 nm. Fig. 6 shows TEM-EELS analysis results at points 1 to 5 in Fig. 5. From references [4] - [5], it can be inferred that the surface of the solder foil is tin oxide film SnO.

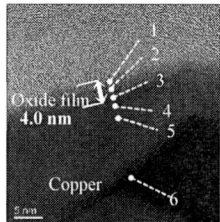

Fig.3 TEM image of copper oxide film section

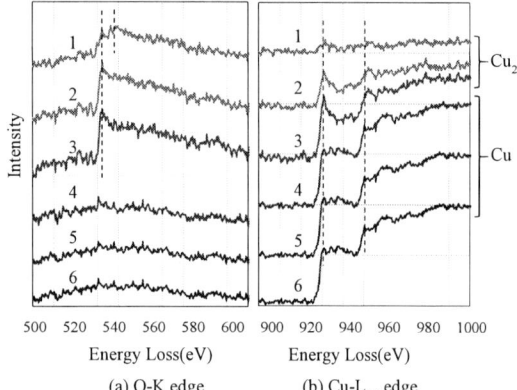

Energy Loss(eV)　　　　Energy Loss(eV)

(a) O-K edge　　　　(b) Cu-L$_{2,3}$ edge

Fig.4 EELS analysis of copper oxide film section

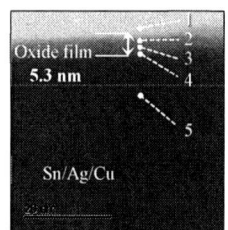

Fig.5 TEM image of Sn oxide film section

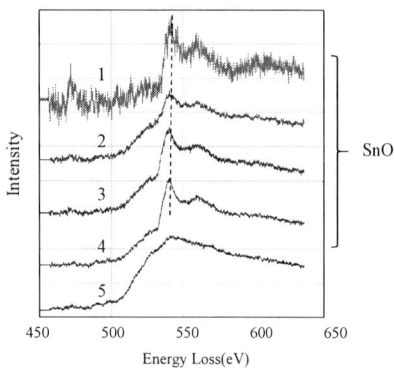

Fig.6 EELS analysis of Sn oxide film section

B. Measurement of reduction rate of copper oxide film and tin oxide film by formic acid

Fig. 7 shows real-time measurement results of the film thicknss of Cu$_2$O and SnO when reduced by formic acid. Formic acid gas was started to introduce at 0.1 min, and before that point, the film thicknesses of Cu$_2$O and SnO are their initial values. They are 4.2 nm and 6.2 nm, respectively, which roughly agree to the measurement results of the above-mentioned TEM analysis. After the introduction of formic acid, it was confirmed that the film thicknesses of Cu$_2$O and SnO both decreased as the reduction advanced. It was found that the reduction rate of SnO is higher than that of Cu$_2$O, and the SnO film is completely removed before the Cu$_2$O film is removed. From this result, data were acquired focusing on the relationship between the copper oxide film thickness and the solder wettability.

C. Relationship between copper oxide film thickness and solder wettability

Fig. 8 shows the temperature profile of the copper substrate and the measured film thickness by the ellipsometer for three samples. Fig. 8(a) is the case when the solder is melted after completely removing the copper oxide film, Fig. 8(b) shows the case when the reduction was stopped in the middle and solder was melted, and Fig. 8(c) shows the case where the solder was melted without reducing the copper oxide film at all.

Fig. 9 shows the relationship between the thickness of the copper oxide film and the contact angle of the solder ball. The contact angle of the solder ball starts to decrease from about 1 nm of the oxide thickness, and the contact angle decreases to about 40 degrees when the oxide film is completely removed.

Regarding the spreading area of the solder foil, it becomes 100% to the foil's original area when the oxide film is completely removed. However, when the oxide film thickness is even only 0.6 nm, the spreading area degrades to 18.1%. From this result, it turned out that it is necessary to completely remove the oxide film in order to obtain favorable solder wettability although the contact angle becomes small as the oxide film thickness approaches 0 nm.

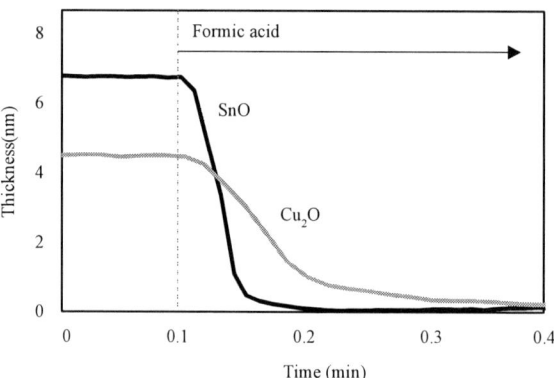

Fig.7 Reduction of Oxide film by Formic acid

978-1-5386-2928-4/18 $31.00 © 2018 IEEE

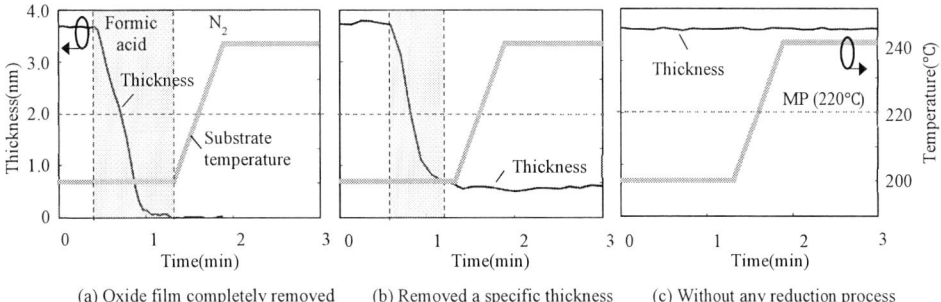

(a) Oxide film completely removed   (b) Removed a specific thickness   (c) Without any reduction process

Fig.8 Measured oxide film thickness by ellipsometer

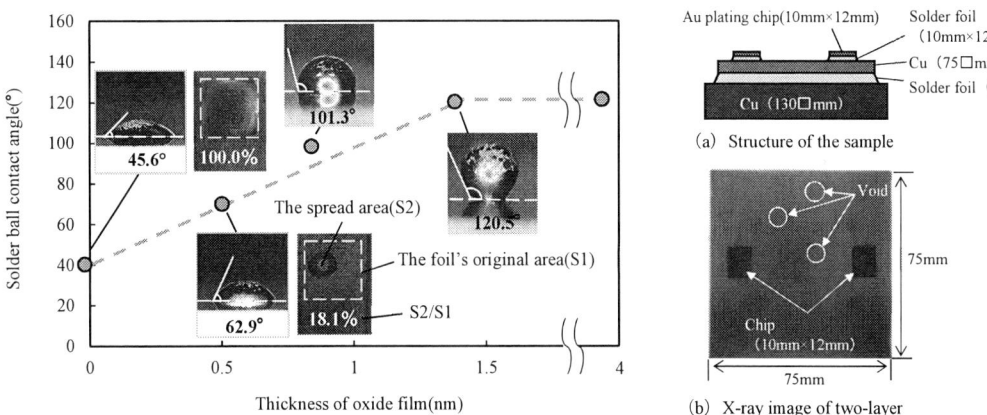

Fig.9 Interrelation between oxide film thickness and solder wettability

Fig.10 Power module soldering sample

D. Application to power device soldering

As shown in Fig. 10, a solder foil (SnAg3.0Cu0.5, size 75 × 75 × 0.07 mm) was placed on a copper base (130 mm × 130 mm × 3 mm) and a copper substrate (75 mm × 75 mm × 1 mm) was placed on the solder foil, and further a solder foil (SnAg3.0Cu0.5, size 10 × 12 × 0.07 mm) and a gold plated chip (10 mm × 12 mm × 0.4 mm) were mounted. This test piece was processed with a formic acid reduction reflow machine by the optimized process based on the obtained experimental data. An X-ray photograph of the sample is shown in Fig. 10 (b). A void rate of 1% or less are obtained both under the copper substrate and under the chip, confirming the effectiveness of oxide film removal by formic acid.

IV. CONCLUSIONS

Native surface oxides of a copper substrate and solder foil are presumed to be Cu_2O and SnO, respectively, by TEM and TEM-EELS analysis, and it is confirmed that their thicknesses are about 4 nm and 5.3 nm, respectively. From the real time measurement of the oxide film thickness with the ellipsometer, it is revealed that the reduction rate of SnO is higher than that of Cu_2O. Furthermore, the contact angle of the solder ball becomes smaller as the copper oxide film becomes thinner by formic acid

reduction, but the solder wettability remains poor even if a small amount of oxide film remains. It is found that it is necessary to completely remove the oxide film to 0 nm in order to obtain good wettability.

Based on the obtained reduction data, solder joint samples are processed with a formic acid reduction reflow machine, and a good result of a void rate of 1% or less is obtained.

REFERENCES

[1] W. Lin, "Study of Fluxless Soldering Using Formic Acid Vapor," IEEE Trans. Advanced Packaging, vol.22, pp.592-601, 1999.

[2] M. Akaike, "Copper Direct Joint by Formic Acid with Platinum Catalyst," in Proc. MES-2013, 2013, pp.277. (in Japanese)

[3] N. Ozawa, "Real-time Measurement and Study of Reduction Process of Copper Oxide Film by Formic Acid," Journal of The Japan Institute of Electronics Packaging vol.20, No,4, pp.219-227, 2017,

[4] N. Ozawa, "Observation and Analysis of Metal Oxide Reduction by Formic Acid for Soldering," IMPACT 2016,

[5] Q. Zhao, "Catalytic characterization of pure SnO2 and GeO2 in methanol steam reforming," Applied Catalysis A: General, Vol.375, Issue 2, 1 March 2010, pp.188-195,

[6] H. Lorenz, "Preparation and structural characterization of SnO2 and GeO2 methanol steam reforming thin film model catalysts by (HR)TEM," Materials Chemistry and Physics,vol.122,Issues2–3,1 August 2010, pp.623-629

Proceedings of the 30th International Symposium on Power Semiconductor Devices & ICs
May 13-17, 2018, Chicago, USA

Dynamic Characterisation and Optimisation of Multiply Contacted Power Busbars

Vanessa Basler, Andreas Wagner, Wolfgang Hoelzl, Gerhard Wachutka
Institute for Physics of Electrotechnology
Technical University of Munich
Arcisstrasse 21, 80333 Munich, Germany
Email: basler@tep.ei.tum.de

Abstract—Today's semiconductor power modules have become highly efficient thanks to fast switching modulation techniques, which are based on the safe control of high power densities in the commutation circuit. This requires, among others, that the structure and topography of the interconnects and busbars has to be optimised with a view to reducing current crowding and local eddy currents. For the analysis of multiply-contacted busbars, the concept of a generalised impedance operator proves to be the appropriate approach to setting up circuit-level models of entire power modules. Our paper is intended to show how the transient inductive behaviour of power busbars can be properly characterised and, then, optimised with a revisited formulation of the commonly used inductance and capacitance matrix.

I. INTRODUCTION

The progressively faster switching transients and high power densities of semiconductor power modules and applications tighten the requirements on the commutation circuit. Among other, low-inductive passive components with high current capability such as DC busbars have to be employed. Furthermore, these components have to be optimized with respect to current crowding and local eddy currents. So far, a number of tools have been developed for the numerical analysis of field and current distributions in the frequency domain (see [1], [2] and [3] for example). In today's high power technology, we have to cope with fast pulse-shaped voltage and current waveforms encountered in power modules, which contain fast switching semiconductor devices connected by complex busbar structures. Therefore, a tool for the accurate numerical simulation of waveforms in the time domain is needed as well. The inductance extraction programs available from commercial distributors mostly rely on "Neumann's formula" for a static version of the inductance matrix (see e.g. [4]). However, using the concept of time-invariant self- and mutual inductance coefficients is not appropriate for modeling and optimising the dynamic behavior of multiply-contacted busbars with complicated shape and composed of several electrically separated components as depicted in Figure 1. Instead, a generalized impedance operator has to be introduced, which is well-defined for transient waveforms and applicable to interconnects with more than two contact electrodes. For the proper

description of transient current crowding, skin effect and displacement current, this generalized impedance matrix is extracted from the space and time dependence of the electro-magnetic field distribution in quasi-stationary approximation. To this end, it is necessary to model and to numerically simulate these physical effects in the time domain and not in the frequency domain.

II. ELECTROMAGNETIC FIELD COMPUTATION

For a detailed description of the current distribution and the electric and magnetic fields inside and outside the busbar structure, a transient approach must be pursued. To this end, a formulation based on the scalar and vector potentials φ and $\mathbf{A}$ (4-potential), is appropriate. The vector potential $\mathbf{A}$ generates the magnetic field as $\mathbf{B} = \nabla \times \mathbf{A}$ inside and outside the busbars, and the scalar potential φ together with $\mathbf{A}$ generates the electric field as $\mathbf{E} = -\nabla \varphi - \frac{\partial \mathbf{A}}{\partial t}$ in the interior of the busbars. For the situation and configuration considered, the quasi-stationary approximation is applicable, in which the occurence of electromagnetic waves is suppressed [7]. Using Maxwell's equations the following governing equations can be deduced [5]:

$$\nabla \cdot (\varepsilon \nabla \varphi) + \frac{\partial}{\partial t} \left(\nabla \cdot (\varepsilon \mathbf{A}) \right) = 0 \tag{1}$$

$$\nabla \times \frac{1}{\mu} \nabla \times \mathbf{A} + \sigma \frac{\partial \mathbf{A}}{\partial t} = -\sigma \nabla \varphi. \tag{2}$$

Here, ε denotes the permittivity, μ the permeability, and σ the conductivity of the respective material. These equations are valid inside the conducting material Ω_c. $\mathbf{A}$ and φ are not uniquely determined by equations (1) and (2) due to the fact that they can be altered by a gauge transformation [5]. In our approach, the Coulomb gauge is used to obtain a unique solution. If ε is constant inside the conducting domain, equation (1) becomes a Laplace equation for the scalar potential φ: $\Delta \varphi = 0$. The Finite Element Method implemented in our simulation tool uses Nédélec's edge elements as basis functions for the representation of $\mathbf{A}$. These basis functions have their degrees of freedom in the middle of the edges and satisfy the condition $\nabla \cdot \mathbf{A} = 0$ in the strong sense [6]. A complete solution requires the

978-1-5386-2928-4/18 $31.00 © 2018 IEEE

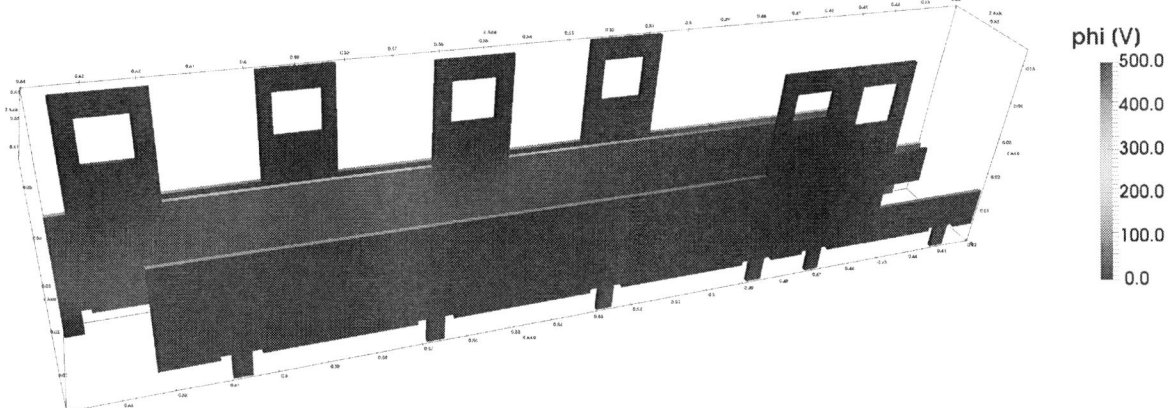

Fig. 1. Potential distribution in the copper parts of H-bridge at the end of a voltage ramp applied to the lower contacts

calculation of the magnetic field also outside the busbars in the non-conductive region Ω_n. For this purpose we use equation (1) with a formal conductivity σ_n set to zero, leading to the equation

$$\nabla \times \frac{1}{\mu} \nabla \times \mathbf{A} = 0. \tag{3}$$

This conforms with the approach reported in [7]. For a complete problem definition, boundary and interface conditions must be specified. These are:

$\nabla \varphi \cdot \mathbf{n}_c = 0$ on Γ_{cn} and

$\varphi(t)$ known at the contacts C_j for all $j = 1, \ldots, N$ (4)

$$\mathbf{A} \times \mathbf{n}_c = 0 \text{ on } \Gamma_n \tag{5}$$

$$\frac{1}{\mu} \nabla \times \mathbf{A} \times \mathbf{n}_c = \frac{1}{\mu} \nabla \times \mathbf{A} \times \mathbf{n}_n \text{ on } \Gamma_{nc} \tag{6}$$

$$\mathbf{A} \times \mathbf{n}_c - \mathbf{A} \times \mathbf{n}_n \text{ on } \Gamma_{nc} \tag{7}$$

The vector $\mathbf{n}_c$ denotes the outward unit normal vector along the boundary of domain Ω_c; evidently we have $\mathbf{n}_n = -\mathbf{n}_c$. Condition (4) implies that a voltage-controlled problem is considered. This means that the potential applied at the contacts is given and the potential-driven contribution to the current flow $\mathbf{j}_{\text{pot}} = \sigma \mathbf{E}_{\text{pot}} = -\sigma \nabla \varphi$ through the interface Γ_{cn} between the conducting domain Ω_c and the surrounding air is suppressed. This condition ensures the correct potential distribution and, hence, the correct electric field in the asymptotic stationary state attained for $t \to \infty$. The boundary condition (5) forces $\mathbf{A} \approx 0$ along Γ_n, the boundary of the domain Ω_n. The interface conditions (6) and (7) ensure continuity of the tangential component of $\mathbf{A}$ and $\mathbf{B}$ along the interface Γ_{cn}.

III. DYNAMIC IMPEDANCE OPERATOR

Having computed the field distributions as primary quantities, a generalized impedance matrix can be calculated as postprocessed step as follows. We consider a configuration of busbars consisting of M electrically separated components Ω_α, where each of them has N_α contacts $C_{k,\alpha}$ ($k = 1, \ldots, N_\alpha$). Hence, we set $\mathbf{B} = \nabla \times \mathbf{A}$ throughout the entire domain Ω, inside and outside the busbars, and $\vec{\mathbf{E}} = -\nabla \varphi - \frac{\partial \mathbf{A}}{\partial t}$ within the conducting subdomain $\Omega_c = \cup_{\alpha=1}^{M} \Omega_\alpha$. From now, φ_α calculated by the Laplace equation in the interior of busbar Ω_α denotes the appropriate scalar potential, while φ is the superposition of all the scalar potentials: $\varphi = \sum_{\alpha=1}^{M} \varphi_\alpha$. The boundary and interface conditions are as in section II. This notation can be found in Figure 2.

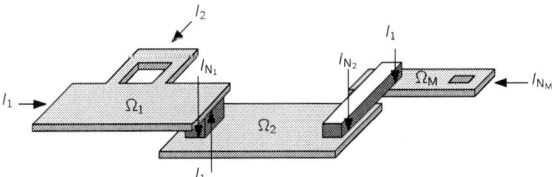

Fig. 2. Multicomponent busbar configuration

The concept and definition of a dynamic impedance operator has to be independent of the operating point considered; it has only to reflect the geometric configuration of the busbar components together with their physical properties. Considering the stored magnetic energy and introducing the form functions $\mathbf{j}_{k,\alpha}(\mathbf{r}) = -\sigma \nabla \varphi_{k,\alpha}(\mathbf{r})$ and $\mathbf{A}_{k,\alpha}(\mathbf{r}, t)$ for each structure α and contact k we get the following inductance matrices:

$$\mathbf{L}^{(1)}{}_{k\alpha,l\beta}(t) = -\int\limits_{\Omega_a} \mathbf{j}_{k,\alpha}(\mathbf{r}) \cdot \mathbf{A}_{l,\beta}(\mathbf{r},t)\,\mathrm{d}\mathbf{r} \quad \text{and}$$

$$\mathbf{L}^{(2)}{}_{k\alpha,l\beta}(t_1,t_2) = -\sigma(\mathbf{r})\int\limits_{\Omega_c} \frac{\partial \mathbf{A}_{k,\alpha}(\mathbf{r},t_1)}{\partial t} \cdot \mathbf{A}_{l,\beta}(\mathbf{r},t_2)\,\mathrm{d}\mathbf{r}.$$

In the same way, the stored electric energy can be expressed by three capacitance matrices. This system of five matrices provides a complete description of the transient dynamic behavior of a multiply-contacted, multi-component busbar structure. As it is required, these quantities are independent of the applied terminal voltages and currents. They can be employed for optimising the design of all kind of complex interconnects, in particular those encountered in high power modules.

IV. RESULTS

An exemplary busbar structure made of copper was considered to demonstrate the capability and practicality of the implemented method. A voltage ramp up to $10\,\mathrm{V}$ in $1\,\mu s$ is applied. Figure 3 shows the structure of the busbar, where the contacts are marked by labels. The resulting electric field distribution at the end of the switching process is displayed in Figure 4. Most of the electric field lines flow in the vicinity of the corners and edges of the conducting material. In the middle of the busbar, the potential-driven current is largely suppressed, and an induced current flowing in reverse direction prevails. Evidently, the skin effect dominates the current distribution inside the structure. In Figure 3 the electric field of the stationary state attained after $1\,\mathrm{ms}$ is depicted.

The potential-driven current and the induced current

$$i_{\mathrm{pot}}(t) = -\int\limits_C \sigma\,\nabla\varphi \cdot \mathrm{d}a, \qquad i_{\mathrm{rot}}(t) = -\int\limits_C \sigma\,\frac{\partial \mathbf{A}}{\partial t} \cdot \mathrm{d}a,$$

where C denotes the contact with the positive potential, are displayed in Figure 5 together with the total current $i(t) = i_{\mathrm{pot}}(t) + i_{\mathrm{rot}}(t)$. The potential-driven current $i_{\mathrm{pot}}(t)$ attains immediately to its final value, proportional to the waveform of the potential. The induced current, however, has opposite sign and rises as long as the potential at the contacts vary. When the switching process is finished and the potential attains its final value, the current density starts to redistribute and the induced current falls down until the stationary state is reached.

Figure 6 shows the apparent ohmic resistance of the busbar. In consequence of the skin effect the resistance is strongly enhanced at the beginning of the switching process. The transient inductance of the busbar structure is displayed in the same figure. At the beginning of the switching process the inductance is very small. With progressing time both the resistance and the inductance converge to their asymptotic stationary value. It is clearly visible that the time required for the busbar to reach its stationary state is much longer than the switching time of $10\,\mu s$.

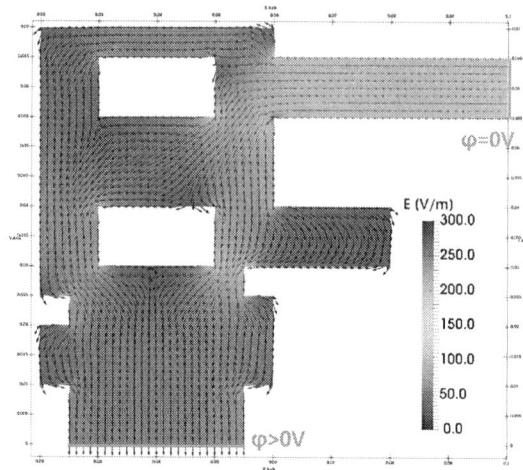

Fig. 3. Transient behaviour of a theoretical structure as response to a voltage ramp applied to the lower contact, Electric field at $t = 1\,\mathrm{ms}$, stationary state

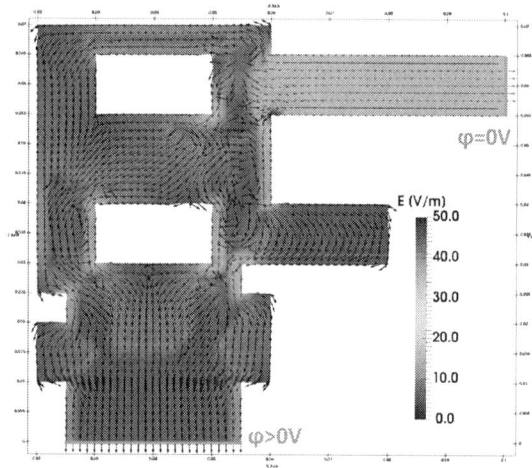

Fig. 4. Transient behaviour of a theoretical structure as response to a voltage ramp applied to the lower contact, Scalar potential at $t = 0.1\,\mu s$

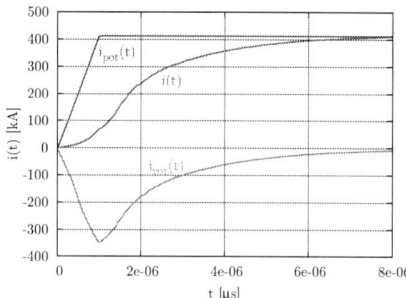

Fig. 5. Resulting current $i(t)$, potential driven current $i_{\mathrm{pot}}(t)$ and induced current $i_{\mathrm{rot}}(t)$ over time

978-1-5386-2928-4/18 $31.00 © 2018 IEEE

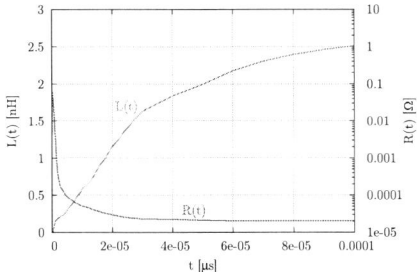

Fig. 6. Resistance and inductance over time

Using our method, structures with more than two contacts can be analyzed as well. An illustrative example is depicted in Figure 8 and 9. Figure 7 shows the scalar potential φ in a similar busbar was considered, but now four contacts are attached. Figure 9 shows the current distribution inside the conducting busbar during the switching process. Also in this structure the skin effect is visible. In Figure 8 the asymptotic stationary state is shown at a time long after the voltage reached its final

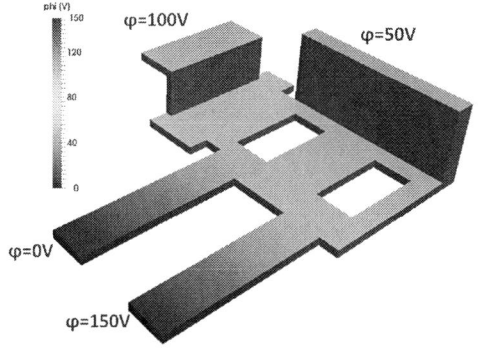

Fig. 7. Examplary structure with 4 contacts, Scalar potential at $t = 50\,\mathrm{ns}$

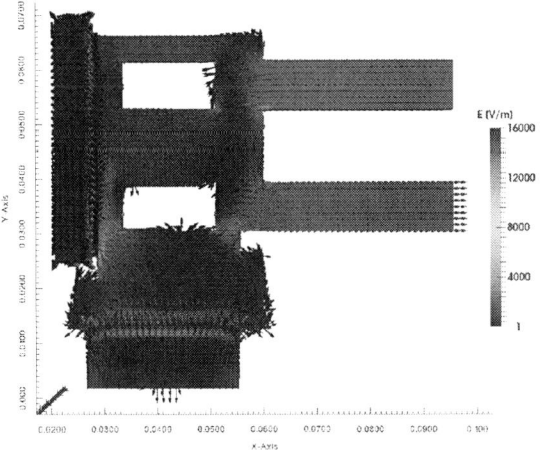

Fig. 8. Examplary structure with 4 contacts

V. CONCLUSION

For the future optimisation of power busbars it is decisive to gain a qualitative and quantitative insight into their transient behavior under realistic switching conditions. A complete simulation model using the $A - \varphi$-formulation has been presented. By using edge elements for the space discretisation, a fast and intelligible algorithm has been developed to solve the dynamic inductance problem. Our results show that the model is able to reproduce the involved physical effects and visualize the internal current density. The current distribution and the vector potential can be used to extract a generalised impedance matrix with a view to optimising the busbar structure.

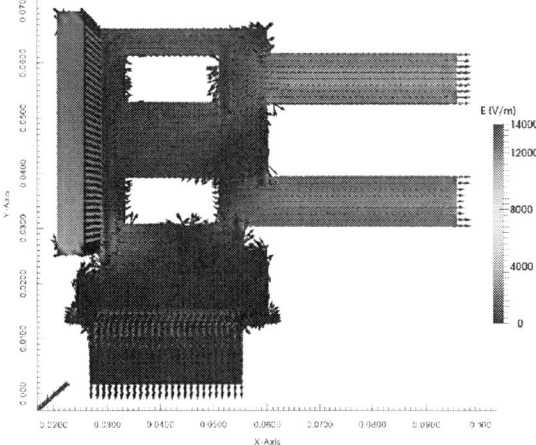

Fig. 9. Electric field in a busbar struture with 4 contacts at $t = 50\,\mathrm{ns}$; skin effect state

REFERENCES

[1] Hu Xin, "Full-wave Analysis of Large Conductor Systems over Substrate", *Massachusetts Institute of Technology*. 2006.

[2] Cangellaris, A.C. and Prince, J. and Vakanas, L.P., "Frequency-dependent inductance and resistance calculation for three-dimensional structures in high-speed interconnect systems", *IEEE Transactions on Components, Hybrids, and Manufacturing Technology*. 154-159, 1990.

[3] Hai Van Jorks and Erion Gjonaj and Thomas Weiland, "Eddy current analysis of a PWM controlled induction machine", *COMPEL - The international journal for computation and mathematics in electrical and electronic engineering*. 1609-1619, 2013.

[4] Kamon, M. and Tsuk, M.J. and White, J.K., "FASTHENRY: a multipole-accelerated 3-D inductance extraction program", *IEEE Transactions on Microwave Theory and Techniques*. 1750-1758, 1994.

[5] Biro, O. and Preis, K. and Richter, K.R., "On the use of the magnetic vector potential in the nodal and edge finite element analysis of 3D magnetostatic problems", *IEEE Transactions on Magnetics*. 651-654, 1989.

[6] Nedelec, J.C., "Mixed finite element in 3D in H(div) and H(curl)", *Lecture Notes in Mathematics, Springer Berlin Heidelberg*. 321-325, 1986.

[7] P. Böhm and G. Wachtuka, "Transient Electromagnetiv Behavior of Multiply Contacted Interconnects", *2nd Int. Conf. on Modeling and Simulation of Microsystems, Semiconductors, Sensors and Actuators*. 1999, pp. 301-304.

[8] V. Basler and H. Hölzl and G. Wachutka, "Physical Modeling and High-Fidelity Simulation of the Transient Behavior of Multiply-Contacted Power Busbars", *PCIM Europe 2016; International Exhibition and Conference for Power Electronics, Intelligent Motion, Renewable Energy and Energy Management*. 2016.

Proceedings of the 30th International Symposium on Power Semiconductor Devices & ICs
May 13-17, 2018, Chicago, USA

Development of a Highly Integrated 10 kV SiC MOSFET Power Module with a Direct Jet Impingement Cooling System

Bassem Mouawad, Robert Skuriat, Jianfeng Li,
C. Mark Johnson
Power Electronics, Machines and Control
University of Nottingham, University Park Campus
Nottingham, UK
bassem.mouawad@nottingham.ac.uk

Christina DiMarino
Center for Power Electronics Systems
Virginia Tech
Blacksburg, VA, USA
dimaricm@vt.edu

Abstract—High-density packaging of fast-switching power semiconductors typically requires low thermal resistance and low parasitic inductance. High-density packaging of high voltage semiconductors, such as 10kV SiC MOSFETs, has brought additional challenge. This work proposes a wire-bond-less, highly integrated planar SiC half-bridge module, with embedded decoupling capacitors and a high performance integrated thermal management system.

Keywords—*silicon carbide, packaging, high voltage, high density, SiC MOSFET, power module, direct jet impingement*

I. INTRODUCTION

High-voltage silicon carbide (SiC) MOSFETs, such as Wolfspeed's 3rd-generation 10 kV, 350 mΩ devices are capable of switching high voltages faster and with lower losses, than Si IGBTs [1]. However, it has been shown that the package can have profound impacts on the module performance. In particular, the high power density due to the application of high voltage devices requires high performance cooling system which can be well aligned with the electrical and electromagnetic parts. To reap the full benefits of these unique devices; therefore, an optimized power module package must be developed.

The aim of this work is to develop a high-density, high-speed and highly integrated power package for 10 kV SiC MOSFETs. It is based on a planar structure, wire-bond-less, sandwich structure with embedded decoupling capacitors and stacked ceramic substrates in order to realize a high-density module with low parasitic inductance and low thermal resistance. This work builds on work previously reported by the authors in [2][3] with the addition of new material on the practical realization and characterization of a prototype module and gives more insight on the thermal performance of the integrated direct jet impingement cooling system installed within the 10 kV SiC MOSFET package.

This work was supported by the Engineering and Physical Sciences Research Council (EPSRC) through research grant [EP/K035304/1] and the CPES High Density Integration Consortium.

II. MODULE DESIGN

Fig. 1 shows the designed half-bridge power module, which has one 10 kV, 350 mΩ, 20 A SiC MOSFET per switch position. The module has a planar structure using molybdenum (Mo) posts and a direct bonded aluminum (DBA) substrate for the interconnections instead of wire bonds. This type of structure allows for increased power density, and reduces the parasitic inductances and capacitances in the module, thereby improving the transient performance. The DBAs have special conducting vias for forming low-inductance electrical connections within the power module. Each MOSFET has its own decoupling capacitor placed directly over it. The module also features spring connectors for the terminations. The footprint of the designed module is 69.2 mm × 61.9 mm × 27.29 mm with the housing which gives a power density of 1.45 W/mm³. For reference, the power density of Wolfspeed's 10 kV, 240 A SiC MOSFET module is 4.2 W/mm3, not including the cooling system [4].

To address the enhanced electric field associated with a 10 kV, high-density design, two DBAs are stacked together, as reported in the previous work [2]. Furthermore, Mo bumps were used to provide sufficient insulation distance between the top and bottom DBAs. It should especially be noted that the filled vias in the DBAs have been to form the electrical connection between the MOSFETs and the top DBA. While planar structures with metal posts and spacers have been explored in the past, the voltages have been limited to below a few kilovolts [5].

Instead of mounting the module onto a baseplate, a direct-substrate, jet-impingement cooling system, was custom designed and fabricated. Each jet impingement cell is located under one SiC MOSFET die, and the coolant is impinged on the bottom side of the stacked DBAs to remove the heat generated by the SiC MOSFETs (Fig. 2). As far as we are aware, this is the first time a highly integrated design for 10 kV SiC MOSFETs with an integrated cooler is being reported.

III. CFD ANALYSIS

ANSYS software has been used to perform Computational Fluid Dynamic (CFD) analysis in order to investigate the thermal behaviour of the cooler and calculate the velocities and

978-1-5386-2928-4/18 $31.00 © 2018 IEEE

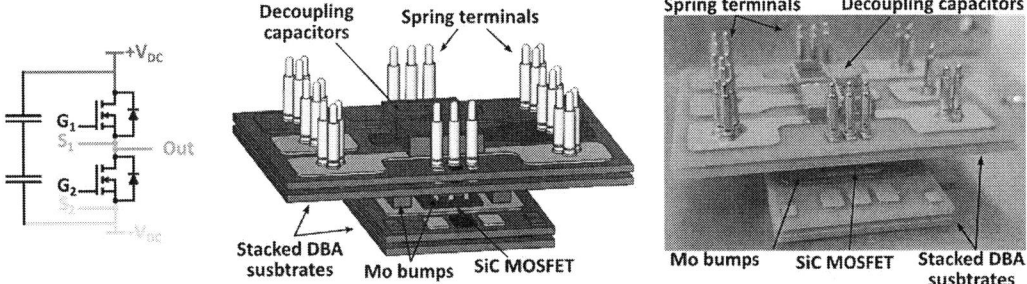

Fig. 1. Schematic and 3D model of the complete assembly. The color in the 3D model correspond to the node in the schematic with the same color. The right photo shows the assembled module with stacked DBAs, spring terminals and embedded capacitors.

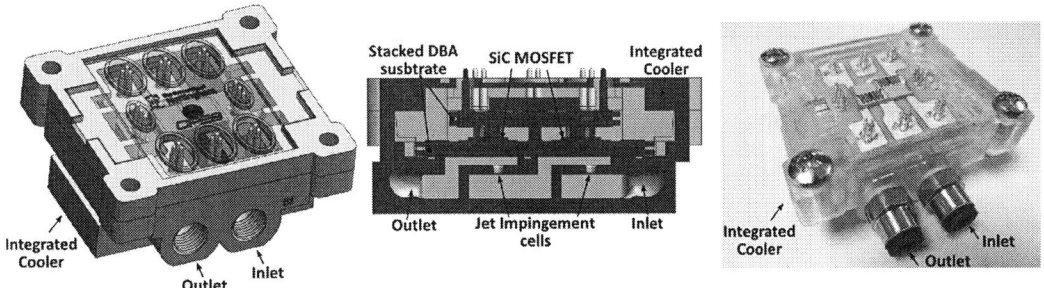

Fig. 2. 3D model of the assembly with its integrated cooler. A cross-section of the 3D model shows the jet impingement cells cooling directly each SiC MOSFET at the back of the stacked DBA substrates. The right photo shows the assembled module embedded with its integrated cooler and filled with silicone gel.

the pressure drop of the coolant at different flow rates. The internal heat generation was assumed to be on one SiC MOSFET only, with a loss generation of 139 W. The coolant temperature at the inlet was set to 20 °C and different flow rate conditions, varying from 0.47 l/min to 3 l/min. Using ANSYS Fluent, a steady-state thermal analysis was performed by employing the appropriate viscosity, energy and turbulence models. The results of interest simulated from these models were temperature and pressure fields under the different flow rates.

Fig. 3 shows the velocity field of the coolant and the temperature distribution at flow rate of 1.5 l/min, and the extracted maximum junction temperature and pressure drop with respect to the different flow rates. As expected, the pressure drop is almost proportional to while the maximum junction temperature is inversely proportional to the flow rate. This can be understood because increasing the flow rate would require increased driving force. On the other hand, this would also

increase the heat exchange between the cooling surface of the substrate and the coolant. Especially it should be noted that a flow rate of 1.5 l/min can be used to achieve the maximum junction temperature below 80 °C.

IV. MODULE FABRICATION

The designed module was fabricated using the following assembling process. First, two DBA substrates were stacked together using pressure-assisted silver sintering. Silver sintering was chosen because it has higher thermal conductivity and improved reliability than solder [6]. Furthermore, since the melting temperature after sintering (960 °C) is higher than the sintering temperature ($\leq$ 260 °C), multiple sintering processes can be done without affecting the previously-sintered joints. The thermal resistance of the sintered-Ag bondline between the two DBA substrate was measured at the locations where the dies would be placed on the bonded substrate. The measured thermal

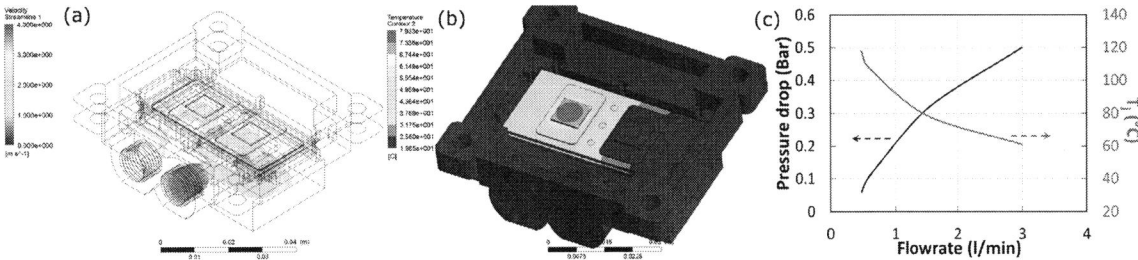

Fig. 3. CFD results showing (a) the velocity profile and (b) the temperature distribution for 139 W and 1.5 l/min. The maximum junction temperature at these conditions was calculated to be 79 °C. The pressure drop and the simulated junction temperature for different flow rates are shown in (c).

978-1-5386-2928-4/18 $31.00 © 2018 IEEE   257

resistances ranged from 0.11–0.14 K/W. These values indicate good quality and uniformity of the bond, and suggests that no significant voids exist in these locations.

Then the dies and some of the bumps were attached on the bottom DBA stack and subsequently the rest of the bumps were bonded on top sides of the dies. The Al top metal of the as-received dies was first coated with Nb and Au, using e-beam evaporation through a shadow mask, to provide a suitable surface for sintering. Die attach to the bottom DBA stack was accomplished with nano-particle Ag, using a die bonder for accurate positioning, followed by pressure-assisted sintering. The bumps used for the interconnection contains 74 % Mo in thickness in the middle and 13% Cu in thickness at the two ends. The surfaces of the Cu was Au-finished for the silver sintering. The bumps were also accurately placed on the corresponding positions and sintered using the die bonder.

The next step is the bonding of the top DBA stack (shown in Fig. 1) on the top sides of all the bumps by using solder joints. Solder joints were chosen instead of Ag joints because a thicker solder layer can be achieved to accommodate height variations and it is more compliant so that it can absorb more of the thermo-mechanical stresses than a rigid thinner Ag layer. It should be noted that the 10 kV SiC MOSFETs used in this module were semi-functional dies and were used to evaluate the fabrication processes for the final module containing more MOSFETs.

The spring pin terminals and the decoupling capacitors were then soldered on the top using another solder alloy with a melting temperature lower than the one used in the last step to top DBA stack on the bumps.

Finally, the assembly was placed in the cooler and sealed with silicone sealant to avoid water leaks. Then silicone gel was poured in the package for electrical isolation, within a vacuum

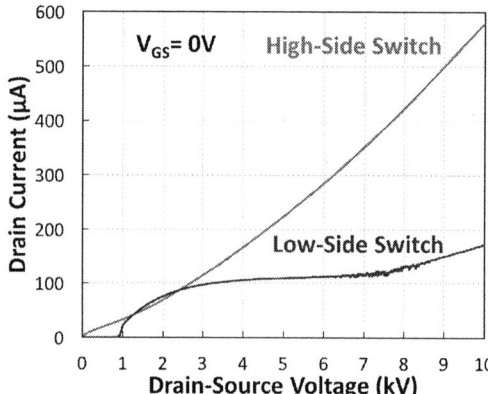

Fig. 4. Breakdown voltage test preformed on the assemble module. It should be noted that the devices used for the assembly were semi-functional. However, they were able to block the voltage up to 10 kV with a leakage current of 580 µA and 173 µA for top and bottom switch, respectively. No breakdown or partial discharge could be detected.

oven for removing the air bubbles before the curing which can be performed at room temperature for ~24 hours.

V. EXPERIMENTAL RESULTS

A. Static Characterisation

The breakdown voltages for the semi-functional SiC MOSFET dies after the module fabrication are shown in Fig. 4. The breakdown voltage of the module is greater than 10 kV, though the leakage current of the high-side MOSFET is 580 µA at 10 kV, which is higher than the fully-functional 10 kV SiC MOSFETs. This shows that the semi-functional dies used to fabricate this module were able to block up to 10 kV.

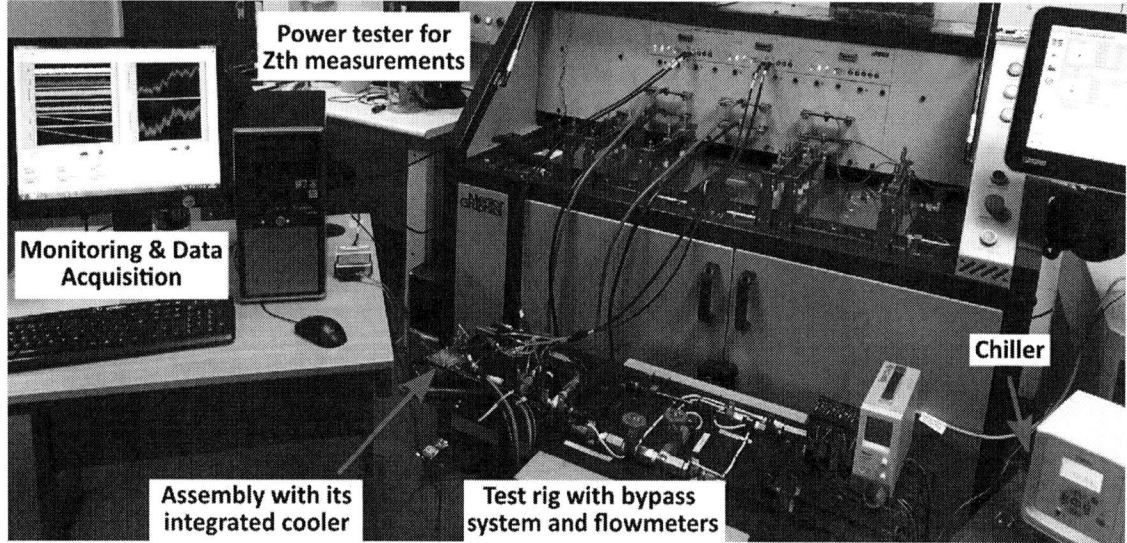

Fig. 5. Test setup for thermal impedance measurements showing Mentor Graphics power tester; a chiller with deionized water at 20 °C; test rig with bypass system to regulate the flow rates through the assembled module, accomodating flowmeters, thermocouples, and pressure sensors; and a PC for monitoring and data acquisition.

978-1-5386-2928-4/18 $31.00 © 2018 IEEE

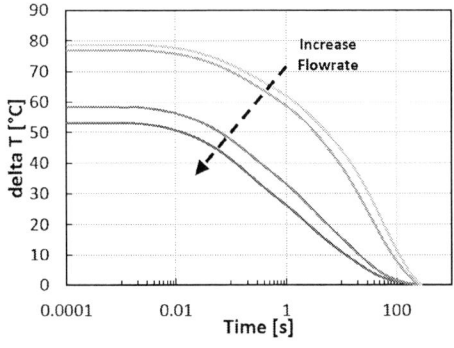

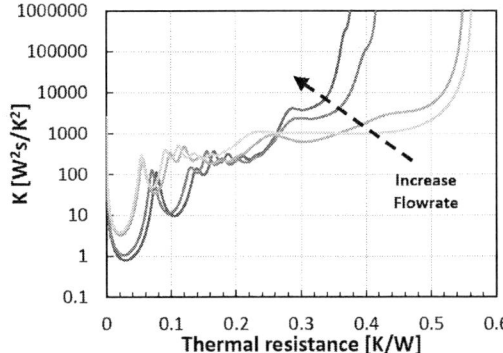

Fig. 6. Results of the thermal impedance measurement showing the ΔT at different flow rates (top) and the differential structure function of the power module with the stacked DBA substrate at different flow rates (bottom). It should be noted that the thermal impedance measurements were performed on individual devices using the bodydiode of the 10 kV SiC MOSFET as a heating source with a heating power of 139 W.

B. Thermal Impedance Measurements

To evaluate the performance of the integrated direct-substrate jet-impingement cooler, thermal impedance measurements were carried out on a Mentor Graphics Power Tester [7]. A test setup was built with a bypass system to regulate the flow rates through the assembled module. The test rig included flowmeters, thermocouples and pressure sensors at the inlet and outlet. Fig. 5 shows the test setup used to perform the measurements. In order to measure the transient temperature response, the body diodes of the 10 kV SiC MOSFETs were used as the heating devices. The forward voltages of the body diodes at a low measurement current of 100 mA were used as the temperature sensitive parameters.

The ΔT curves obtained during the transient tests and the differential structure functions obtained from the transient thermal analysis for the power module are shown in Fig. 6 for the different flow rates. It can be seen that the temperature decreases and hence the thermal resistance also decreases with increase in the flow rate of the coolant.

The lowest junction-to-ambient specific thermal resistance of the module was measured to be 26 mm²•K/W (0.38 K/W) for a flow rate of about 1.5 l/min. This value is lower than the results using a stacked substrate structure reported in the literature [8].

VI. CONCLUSION

The design, fabrication and testing of a wire-bond-less planar 10 kV SiC MOSFET module has been presented. This planar structure and the embedded decoupling capacitors offer very low parasitic inductances. The module includes a high-performance jet impingement cooling system, which offers very efficient cooling for a high-density application. Transient thermal analysis was used to evaluate the performance of the integrated cooler at different flow rates. It shows that the junction temperature and the thermal resistance decreases when the flow rate increases. The junction-to-ambient specific thermal resistance of the power module with an integrated direct-substrate, jet-impingement cooler is 26 mm²•K/W at a flow rate of about 1.5 l/min.

ACKNOWLEDGMENT

The authors acknowledge Dr. Robert Abebe for supporting the CFD analysis, DOWA for the donation of substrates and Wolfspeed (a Cree company) for the 10 kV SiC MOSFET dies.

REFERENCES

[1] V. Pala, E. V. Brunt, L. Cheng, M. O'Loughlin, J. Richmond, A. Burk, S. T. Allen, D. Grider, J. W. Palmour, and C. J. Scozzie, "10 kV and 15 kV silicon carbide power MOSFETs for next-generation energy conversion and transmission systems," in *2014 IEEE Energy Conversion Congress and Exposition, ECCE 2014*, 2014, pp. 449–454.

[2] C. DiMarino, M. Johnson, B. Mouawad, J. Li, D. Boroyevich, R. Burgos, G.-Q. Lu, and M. Wang, "Design of a novel, high-density, high-speed 10 kV SiC MOSFET module," in *IEEE Energy Conversion Congress and Exposition (ECCE'17)*, 2017, pp. 4003–4010.

[3] C. DiMarino, D. Boroyevich, R. Burgos, M. Johnson, and G.-Q. Lu, "Design and development of a high-density, high-speed 10 kV SiC MOSFET module," in *19th European Conference on Power Electronics and Applications (EPE'17 ECCE Europe)*, 2017.

[4] B. Passmore, Z. Cole, B. McGee, M. Wells, J. Stabach, J. Bradshaw, R. Shaw, D. Martin, and T. McNutt, "The next generation of high voltage (10 kV) silicon carbide power modules," in *IEEE 4th Workshop on Wide Bandgap Power Devices and Applications (WiPDA)*, 2016.

[5] B. Mouawad, J. Li, A. Castellazzi, and C. M. Johnson, "Hybrid half-bridge package for high voltage application," in *28th International Symposium on Power Semiconductor Devices and ICs (ISPSD'16)*, 2016, pp. 147–150.

[6] X. Cao, T. Wang, K. D. T. Ngo, and G. Q. Lu, "Characterization of lead-free solder and sintered nano-silver die-attach layers using thermal impedance," *IEEE Trans. Components, Packag. Manuf. Technol.*, vol. 1, no. 4, pp. 495–501, 2011.

[7] M. Graphics, "Power tester 1500A—Lifetime testing and failure diagnosis of high-power semiconductors.", 04-14 MGC, 1032500.

[8] F. Kato, H. Takahashi, H. Tanisawa, K. Koui, S. Sato, Y. Murakami, H. Nakagawa, H. Yamaguchi, and H. Sato, "Evaluation of Thermal Resistance Degradation of SiC Power Module Corresponding to Thermal Cycle Test," in *Conference & Exhibition on High Temperature Electronics Network (HiTEN 2017)*, 2017.

Proceedings of the 30th International Symposium on Power Semiconductor Devices & ICs
May 13-17, 2018, Chicago, USA

A More Accurate Electromagnetic Modeling of WBG Power Modules

Ivana Kovačević-Badstübner
and Ulrike Grossner
Advanced Power Semiconductor Laboratory
ETH Zurich
Zurich, Switzerland
Email: kovacevic@aps.ee.ethz.ch

Daniele Romano
and Giulio Antonini
University of L'Aquila
L'Aquila, Italy
Email: giulio.antonini@univaq.it

Jonas Ekman
Luleå University of Technology
Luleå, Sweden
Email: jonas.ekman@ltu.se

Abstract—A major requirement for further development of wide-band gap (WBG) power devices and their applications is the optimization of packages and PCB layouts to enable fast-switching capabilities. Electromagnetic modelling allows the prediction of parasitic inductances, capacitances, and resistances of the current paths within power modules, which cannot be easily approached in measurements. As a result, electromagnetic-circuit-coupled modeling enables the optimization of package layouts and interconnections before manufacturing actual power modules. The accuracy and limitations of present numerical techniques for three-dimensional (3D) electromagnetic modeling of power modules is still neither well understood nor verified. This paper presents the extraction of parasitics of power semiconductor packages using two electromagnetic modelling approaches. The first approach is based on a well-established 3D electromagnetic quasi-static solver, ANSYS Q3D Extractor. For the second approach, a numerical solver based on the Partial Element Equivalent Circuit (PEEC) method is developed and assessed in terms of modelling accuracy required by fast switching WBG-based power converters. The PEEC method is presented as a promising numerical technique, which can potentially be used to overcome the limitations of the EM modeling based on the ANSYS Q3D Extractor.

I. INTRODUCTION

Wide band-gap (WBG)-based power converters operate at higher switching frequencies and produce current and voltage waveforms with much faster slopes than Si-based power electronic systems. Therefore, small parasitic inductances and capacitances have a more severe impact on the electromagnetic (EM) behavior of WBG-based power converters. As the stray inductances seen from the device terminals have different impact on the switching properties of power converters, potentially hampering the utilization of WBG power semiconductor devices, it is highly useful to have the information on these inductances and minimize them in an optimal design. EM-circuit coupled, i.e. multi-physics, modeling enables the optimization of package layouts and interconnections before fabricating actual power modules. The circuit modelling is related to the development of compact device models, while the EM modelling of the power semiconductor packages is closely related to the procedures for the extraction of parasitics in electrical circuits. The accuracy and limitations of present numerical techniques for three-dimensional (3D) electromagnetic modeling of power modules are not well

explored in literature. First, the capability of the ANSYS Q3D Extractor, a well-established 3D EM modelling tool used to estimate the parasitics within power semiconductor packages, is comprehensively analyzed in terms of accuracy. Second, the Partial Element Equivalent Circuit (PEEC) method is presented as a promising numerical technique for 3D EM-circuit coupled modeling for future power electronics.

II. MODELLING FOR PACKAGE PARASITIC PREDICTION

Electromagnetic modelling relies on numerical techniques used to solve Maxwell's equations in terms of unknown electric and magnetic field distributions and/or current and charge distributions in space. The selection of a numerical technique mainly depends on the application and, hence, only relative (dis)advantages of the specific method can be discussed. In this paper, the limitations of the existing modeling approach implemented in ANSYS Q3D Extractor for an accurate extraction of parasitics within WBG power modules are described. Then, the Partial Element Equivalent Circuit (PEEC) method is introduced as a promising numerical technique that can be used to overcome these limitations. In this section, the physical background of the Q3D-based modeling and PEEC-based modelling is described in order to provide a better understanding of the modelling challenges coming along with these two numerical solvers.

A. ANSYS Q3D Extractor

The well-known ANSYS tool, Q3D Extractor, has frequently been used for the extraction of parasitics by power electronics engineers both in academia and industry. Additionally, ANSYS provides the EM-circuit-coupled modeling capability using Q3D and the ANSYS Simplorer circuit simulator. The power of the ANSYS Q3D Extractor in comparison to other commercial 3D quasi-static EM solvers used for EM compatibility analysis of power electronics systems is its capability to directly extract the stray inductances, parasitic capacitances and resistances. Furthermore, ANSYS Q3D calculates the mutual inductive couplings between current paths and the capacitive couplings of conductive areas inside of packages. The Q3D Extractor is based on two numerical techniques: the Finite Element Method (FEM) and the Method

978-1-5386-2928-4/18 $31.00 © 2018 IEEE

of Moments (MoM). The modelling is based on dividing the solution in two parts, the low frequency (dc, $f < f_{dc}$) and high frequency (ac, $f > f_{ac}$). For the dc solution, a uniform current distribution across the cross sections of conductors is assumed and modelled using the FEM, i.e. skin depth higher than the conductor thickness. For the ac solution, the assumption that the skin-effect is fully developed, i.e. skin depth $\approx$ three times smaller than the conductor thickness, and the currents are distributed only on the surface of conductors, is exploited using MoM, which leads to the ac resistances increasing as $\sqrt{f}$. In the mid-frequency range ($f_{dc} < f < f_{ac}$), the resistance and inductance are approximated based on the dc and ac solutions.

The parasitic extraction in Q3D is based on placing the equi-potential surfaces, referred to *source* and *sink* contacts, defining the current paths. When calculating the commutation loop inductance of power modules, $L_{\sigma,\mathrm{loop}}$ in Q3D, the semiconductor devices have to be modeled as conductive blocks, e.g. typically copper blocks, in order to simulate the current path. On the other hand, for the calculation of the distributed commutation loop inductance in Q3D, i.e. $L_{\sigma,\mathrm{loop}} = \sum_{ij} L_{\mathrm{p},ij}$, the actual current loop has to be divided into partial current segments ($L_{\mathrm{p},i}$) by removing the 3D models of the devices and setting up the corresponding *source* and *sink* contacts. As these contacts are equi-potential, the cut current path approximates the actual current path accurately to some extent, which mainly depends on the modelled geometry, as shown in [1].

B. A quasi-static PEEC solver

The PEEC method was introduced in the 1970s, and since then, different formulations have been developed: quasi-static and full-wave formulations, formulations for including electric and/or magnetic field effects, and dielectric and/or magnetic material properties [2]. The PEEC method provides a circuit interpretation of the Maxwell's equations in terms of partial elements, namely resistances, partial inductances and coefficients of potential. The resulting equivalent circuit can be then analyzed in both time and frequency domain in a circuit environment such as SPICE-like circuit solvers. In the frequency range of interest for modern PE applications (from kHz to GHz range), the current has to be represented as a 3D vector in order to accurately capture the skin and proximity effects, which can significantly increase the number of unknowns, and thus, the computational cost of the PEEC method. Exactly this has been an obstacle for exploiting the PEEC modeling in a wide range of PE applications. Therefore, most efforts today are directed towards an acceleration of the PEEC solvers in order to allow the analysis of more complex circuits in a wide-frequency range [3], [4]. In power electronic applications, there is a strong requirement to simultaneously take into account Ohmic losses (R matrix), as well as magnetic (L matrix) and electric (P matrix) field effects within a unique modeling environment. A PEEC-based tool for multi-physics modeling, which can take into account all design aspects (resistive, inductive, capacitive, and additionally thermal) is

still under research [5], [6]. The advantage of the PEEC method for 3D EM modeling of power modules has been described in literature [7], [8]; however, a detailed verification of PEEC solvers for 3D EM modeling of power modules in terms of accuracy is missing, particularly the impact of the PEEC mesh and modeling of non-orthogonal geometries, e.g. bond-wires. This paper summarizes for the first time the required conditions for accurate and computationally efficient PEEC modeling of power semiconductor packages using a (R, L, P) PEEC solver.

III. MODELING RESULTS AND VERIFICATION

The first results are demonstrated for a 1.2 kV 80 mΩ SiC power MOSFET in TO-247-3 package (C2M0080120D). Fig. 1 and 2 describe the corresponding 3D PEEC-modeling. A non-uniform PEEC mesh (number of unknowns $n_{\mathrm{sys}} = n_{\mathrm{edges}} + n_{\mathrm{nodes}} = 14896 + 2325 = 17221$) is applied to discretize the TO-247-3 package. For the modelling the drain-source (D-S) current path, the package can be represented with 4 pins corresponding to drain and source package terminals and the internal drain and source contacts for the die, as shown on Fig. 2. The size of the extracted 4-pins PEEC system (17221×17221) can be further reduced to e.g. a (576×576) system by applying a Model Order Reduction (MOR) technique [9], which enables to calculate both transient and frequency response of the package in a circuit domain at a lower computational cost. Further improvements of the PEEC-MOR solver with respect to the required memory storage and computational speed could be achieved as suggested in e.g. [10]. The modeling methods are verified by the D-S impedance, Z_{DS}-Θ_{DS}, measurements using a Keysight Impedance Analyzer E4990 (20 Hz-120 MHz). The high frequency (HF) D-S loop inductance $L_{\mathrm{DS,loop}}$ (relevant in the switching transients) is calculated from the measured Z_{DS}-Θ_{DS} for the MOSFET switched on and off, as described in [1]. In comparison to Q3D, the PEEC $L_{\mathrm{DS,loop}}$ perfectly matches the loop inductance calculated from the extracted partial inductances $\sum_{ij} L_{\mathrm{p},ij}$ as shown in Fig. 3. The internal node was placed at the central position of the die. As the modelled geometry cannot fully replicate the actual geometry since the shape of the bond wires is only approximated, a mismatch of less than 8% between the measured and modelled $L_{\mathrm{DS,loop}}$ is present. From Fig. 3 and Table I, it can be further concluded that the accuracy of Q3D is similar to the accuracy of the PEEC solver for modelling the D-S loop inductance of a TO-247-3 package in the frequency range up to 50 MHz. However, in comparison to the PEEC method, Q3D Extractor introduces an error for the estimation of partial inductances and does not include the parasitic self-capacitance of the current path, which can become influential at higher frequencies, i.e. above several hundred MHz.

A similar comparison was performed for a 3rd-Gen 1.2 kV 75 mΩ SiC power MOSFET in four-lead TO-247-4 package (C3M0075120K), as shown in Fig. 4. As previously described, the prediction of the partial package inductances is performed without and with cutting the current path using the PEEC

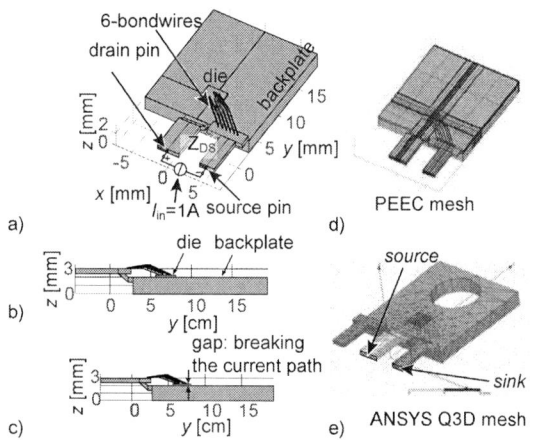

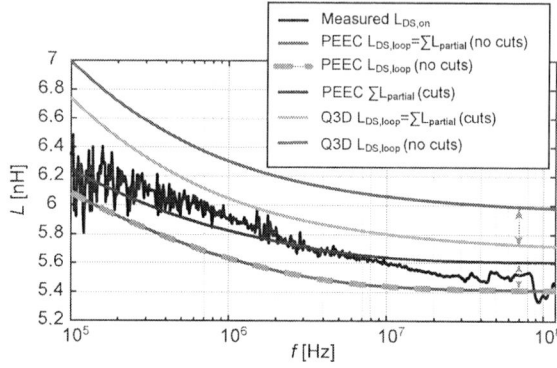

Fig. 3. Verification of PEEC and Q3D modelling of TO-247-3 package for the configurations with and without cutting the current path. The errors introduced by dividing the current path into sub-paths to extract partial inductances are marked.

Fig. 1. 3D PEEC model of a TO-247-3 package for the estimation of L_{DS}: a) 3D structure before meshing, b) a YZ view showing the PEEC modeling without, and c) with breaking the loop into two current paths to extract drain and source inductances, d) non-uniform PEEC mesh, and e) ANSYS Q3D mesh.

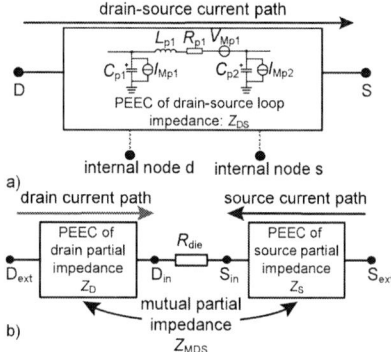

Fig. 2. PEEC modeling of Z_{DS} and constituting partial impedances: a) without, see Fig. 1b, and b) with, see Fig. 1c, breaking the current path. The PEEC method enables solving 3D PEEC models together with lumped circuit elements, which directly simplifies the coupling between the circuit and EM domains.

solver and Q3D Extractor, respectively. Two current loops, the D-S and D-Kelvin source (D-KS) loops are modelled and measured using the Keysight E4990 Impedance Analyzer. The results are summarized in Table II and illustrated for the D-S current loop example in Fig. 4. Here, it should be noted that the mismatch between the modelled and measured current loop inductances also comes from the limitation to fully accurately extract the actual shape and position of the

TABLE I
L_{DS} OF A TO-247-3 PACKAGE AT 50 MHz, WHERE THE MEASURED INDUCTANCE $L_{meas} = 5.54$ nH.

L [nH]	L_D	L_S	M_{DS}	$L_{tot,eq} = \sum L_{partial}$	L_{tot}
Q3D	3.53	5.39	-1.58	5.76	5.98
PEEC	4.06	5.37	-2.0	5.42	5.42

TABLE II
L_{D-S} AND L_{D-SK} OF A TO-247-4 PACKAGE AT 10 MHz.

L [nH]	L_S	L_D	M_{DS}	$L_{tot,eq} = \sum L_{partial}$	L_{tot}
Drain-Source current loop: $L_{D-S,meas} = 8.16$ nH					
Q3D	5.58	5.55	-1.56	8.00	8.45
PEEC	6.61	6.23	-2.43	7.98	7.98
Drain-Kelvin Source current loop: $L_{D-KS,meas} = 11.32$ nH					
Q3D	6.93	5.55	-1.08	10.33	11.47
PEEC	8.178	6.23	-1.18	10.86	10.86

bond wires. Therefore, in Q3D, the loop inductance calculated from the partial inductances, $L_{tot,eq} = \sum L_{partial}$, is different from the total loop inductance calculated without breaking the loop into the current segments L_{tot}. However, L_{tot} represents the actual current path more accurately. This difference depends on the loop geometry and the definition of Q3D *source* and *sink* equipotential ports. With certain surface areas of the equipotential ports, $L_{tot,eq}$ can be equal to L_{tot} at high frequencies ($f > f_{ac}$); however, finding such areas is not straightforward. This is further demonstrated on an *all-SiC* half-bridge(HB) power module with planar interconnections.

The *all-SiC* half-bridge power module based the planar interconnection technology and the corresponding PEEC model are shown in Fig. 5a-b. In order to measure the low-inductance loop between DC+ and DC- ($L_{DC+,DC-}$) without the contribution of the bus-bars, the Keysight E4990 Impedance analyzer with a 42941 impedance adapter and a pin probe was used. The comparison between the PEEC-based, Q3D results and the measurements of $L_{DC+,DC-}$ is shown in Fig. 5d and Table III and IV. Here, a difference between the modelled and measured $L_{DC+,DC-}$ is due to a difficulty to accurately represent the excitation points, DC+ and DC-, marked in Fig. 5a.

IV. CONCLUSION

This paper presents the EM modelling of power semi-conductor packages using the well-established ANSYS Q3D

978-1-5386-2928-4/18 $31.00 © 2018 IEEE

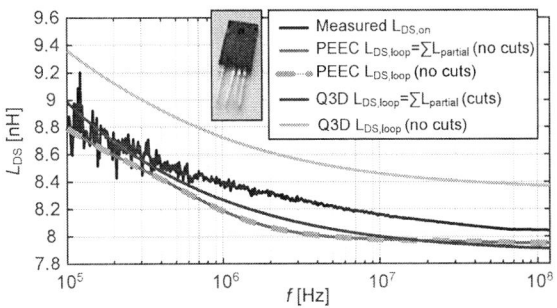

Fig. 4. Verification of PEEC and Q3D modelling of TO-247-4 package for the configurations with and without cutting the current path. The number of unknowns $n_{sys} = n_{edges} + n_{nodes} = 9960 + 1988 = 11948$.

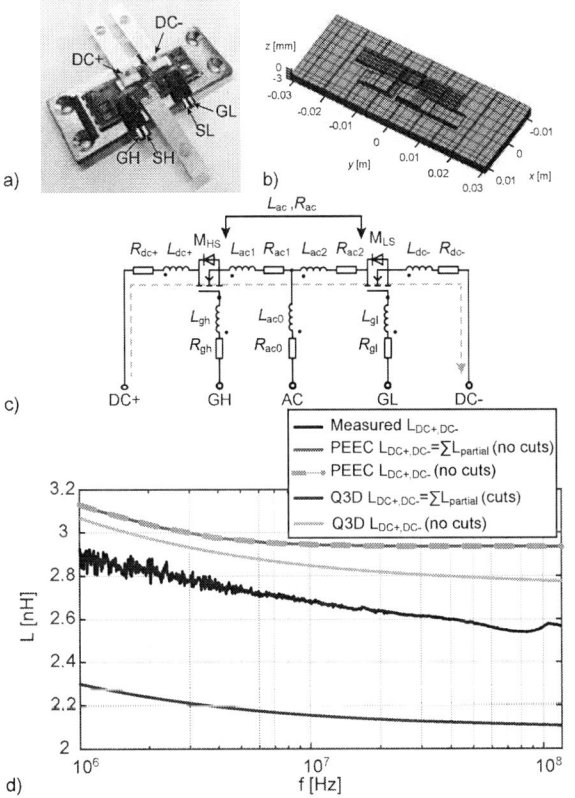

Fig. 5. *All-SiC* HB power module package: a) a photo, b) the PEEC mesh with $n_{sys} = n_{edges} + n_{nodes} = 26256 + 4411 = 30667$ unknowns, c) the equivalent circuit, and d) the verification of PEEC and Q3D modelling for the configurations with and without cutting the current path.

Extractor tool, and a self-developed quasi-static (R, L, P) PEEC-based solver. The verification of both modelling approaches are verified using the examples of two TO-247 packages and an *all-SiC* half-bridge power module. Analyzing the modelling challenges and limitations of two modelling tools, the PEEC method is shown to be a promising numerical technique enabling a more accurate prediction of package

TABLE III
$L_{DC+,DC-}$ OF THE *all-SiC* HB POWER MODULE AT 10 MHz, THE MEASURED INDUCTANCE $L_{DC+,DC-,meas} = 2.69$ nH.

L [nH]	$L_{tot,eq} = \sum L_{partial}$	L_{tot}	rel.diff
Q3D	2.12	2.84	-25 %
PEEC	2.94	2.94	0 %

TABLE IV
PARTIAL INDUCTANCES OF THE *all-SiC* HB POWER MODULE AT 10 MHz.

L [nH]	L_{dc+}	L_{ac}	L_{dc-}	$L_{dc+,ac}$	$L_{dc+,dc-}$	$L_{ac,dc-}$
Q3D	0.80	1.51	0.88	-0.219	-0.187	-0.127
PEEC	1.08	2.09	1.89	-0.207	-0.461	-0.397

parasitics, and additionally, allowing EM-circuit coupled modelling in a wide-frequency range. However, further improvements of the PEEC method with respect to computational speed and memory requirements are required in the future.

ACKNOWLEDGMENT

The *all-SiC* HB power module was designed in the course of the Horizon2020 European Project 636170 - Integrated, Intelligent Modular Power Electronic Converter (I2MPECT). The authors would like to thank the Power Electronic Systems Laboratory at ETH Zurich and *Siemens*^{TM} AG for providing the module sample.

REFERENCES

[1] I. Kovacevic-Badstuebner, R. Stark, M. Guacci, J. Kolar, and U. Grossner, "Parasitic extraction procedures for SiC power modules," in *Proc. of* 10th *Int. Conf. on Integrated Power Electronics (CIPS)*, 2018.

[2] A. E. Ruehli, G. Antonini, and L. Jiang, *Circuit Oriented Electromagnetic Modeling Using the PEEC Techniques*. John Wiley & Sons, Inc., Hoboken, New Jersey, 2017.

[3] D. Romano and G. Antonini, "Partitioned model order reduction of partial element equivalent circuit models," *IEEE Tran. on Components, Packaging and Manufacturing Technology*, vol. 4, no. 9, pp. 1503 – 1514, 2014.

[4] G. Antonini and D. Romano, "Adaptive-cross-approximation-based acceleration of transient analysis of quasistatic partial element equivalent circuits," *IET Microwaves, Antennas & Propagation*, vol. 9, no. 7, pp. 700–709, 2015.

[5] I. Lombardi, R. Raimondo, and G. Antonini, "Electrothermal formulation of the partial element equivalent circuit method," *Int. Journal of Numerical Modelling: Electronic Networks, Devices and Fields*, 2017.

[6] K. Li, P. Evans, and M. Johnson, "Using multi time-scale electro-thermal simulation approach to evaluate SiC-MOSFET power converter in virtual prototyping design tool," in *Proc. of 18th Workshop on Control and Modeling for Power Electronics (COMPEL)*, 2017, pp. 1–8.

[7] P. L. Evans, A. Castellazzi, and C. M. Johnson, "Design tools for rapid multidomain virtual prototyping of power electronic systems," *IEEE Tran. on Power Electronics*, vol. 31, no. 3, pp. 2443–2455, 2016.

[8] G. Regnat, P.-O. Jeannin, G. Lefevre, J. Ewanchuk, D. Frey, S. Mollov, and J.-P. Ferrieux, "Silicon carbide power chip on chip module based on embedded die technology with paralleled dies," in *Proc. of IEEE Energy Conversion Congress and Exposition (ECCE)*, 2015.

[9] A. Odabasioglu, M. Celik, and L. T. Pileggi, "PRIMA: passive reduced-order interconnect macromodeling algorithm," *IEEE Tran. on Computer-Aided Design of Integrated Circuits and Systems*, vol. 17, no. 8, pp. 645–654, Aug. 1998.

[10] G. Antonini and D. Romano, "Efficient frequency-domain analysis of PEEC circuits through multiscale compressed decomposition," *IEEE Tran. on Electromagnetic Compatibility*, vol. 56, no. 2, pp. 454–465, 2014.

Proceedings of the 30th International Symposium on Power Semiconductor Devices & ICs
May 13-17, 2018, Chicago, USA

Accelerated Thermal Fatigue Test of Metallized Ceramic Substrates for SiC Power Modules by Repeated Four-Point Bending

Hiroyuki Miyazaki, Hideki
Hyuga, Kiyoshi Hirao
Structural Materials Research
Institute
National Institute of Advanced
Industrial Science and Technology
(AIST)
Nagoya, Japan

Hiroshi Sato, Hiroshi Yamaguchi
Advanced Power Electronics
Research Center (ADPERC)
National Institute of Advanced
Industrial Science and Technology
(AIST)
Tsukuba, Japan

Shoji Iwakiri and Hideki
Hirotsuru
Denka Innovation Center
Denka Co., Ltd.
Tokyo, Japan

Abstract— Maximum tensile stress in the ceramics during thermal cycle test (from -40 to 250°C) of active metal brazing (AMB) substrate was estimated by the finite element method (FEM) analysis, because such a tensile stress is the driving force of Cu plate delamination from the ceramic plate. In order to accelerate thermal fatigue of the AMB substrate, tensile stress 1.5-2.1 times larger than the maximum thermal stress at -40°C was applied to ceramic plate by four-point bending the AMB substrate at 250°C repeatedly at a frequency of 1 Hz. The time to failure by repeated bending of the Si_3N_4-AMB substrate was less than 1/40 of the time to delamination of the Cu plate by the thermal cycling.

Keywords— *thermal fatigue; silicon nitrides; aluminum nitrides; AMB substrate;*

I. INTRODUCTION

Highly thermal conductive ceramic insulating substrates such as AlN or Si_3N_4 are employed in heat dissipating board for the power modules [1-4]. Mechanical reliability of the metallized substrates after thermal cycles becomes especially important when the operating temperature of SiC power module increases up to 250°C since delamination of Cu layers from ceramic substrates has been reported. 1000 thermal cycles test using the active metal brazing (AMB) substrates is the minimum requirement for the reliable assurance of the power modules, which takes more than a month to accomplish. Usually, longer thermal cycle tests are preferred for some specific products, which span more than several months. Such a long test time hinders quick development of reliable power modules, thus shortening of the test time is strongly required. In our previous studies of thermal cycle test from -40 to 250°C [5, 6], it was found that residual thermal stress caused by mismatch in thermal expansion of Cu and ceramics was the driving force of the delamination of Cu plate from AMB substrate. It is likely that the same order of tensile stress can be reproducible in the ceramic when the AMB substrate is bended in four-point flexural manner, where quick loading and

unloading is possible and the magnitude of tensile stress can be enlarged easily. Thus, it is rational to suppose that such an external loading of the AMB substrate can accelerate fatigue behavior of the AMB substrate and shorten the test time. In this study, variation of maximum tensile stress in the ceramics during thermal cycle (from -40 to 250°C) was estimated by the finite element method (FEM) analysis. Cyclic tensile stress swing was given to the ceramic substrates by four-point bending the AMB substrate repeatedly at a frequency of 1 Hz at 250°C. The magnitude of tensile stress during the cyclic external loading was varied by changing the maximum load. The dependence of the cycle to failure by the repeated external loading on the maximum load was studied for three kinds of AMB substrates in order to study the potential of such an accelerated fatigue test.

II. EXPERIMENTA PROCEDURE

Two kinds of Si_3N_4 and AlN were employed as ceramic substrates. Table I shows thermal and mechanical properties of ceramics substrates. Both conventional Si_3N_4 and AlN for electric substrates were fabricated by tape casting method, whereas the bulks of the commercially available Si_3N_4 for the application to molten metal were machined into thin plates with thickness of 0.32 mm. Four-point bending strength was measured using ceramic test plates with the dimension of 40 mm × 10 mm × ca. 0.32 mm. The outer and inner spans were 30 mm and 10 mm, respectively. The average strength was obtained with 10 test pieces. Fracture toughness was measured by the modified Single Edge-Precracked Plate (SEPP) method [7]. The shape and size of both Si_3N_4- and AlN-AMB substrates were determined according to International Standard ISO 17841 "Test method for thermal fatigue of fine ceramics substrate" (Fig. 1) [8]. The AMB method was employed to join Cu plates with ceramic substrates. The size of Si_3N_4 and AlN plates were 40 mm × 10 mm × ca. 0.32 mm. The dimension of four Cu layers was 17 mm × 8 mm × 0.3 mm. In order to prevent the oxidation of Cu during thermal cycling, 5 μm-thick

---

This work was supported by the Council for Science, Technology and Innovation (CSTI), the Cross-ministerial Strategic Innovation Promotion Program (SIP), "Next-generation power electronics/Consistent R&D of next-generation SiC power electronics" (funding agency: the New Energy and Industrial Technology Development Organization (NEDO)).

TABLE I THERMAL AND MECHANICAL PROPERTIES OF CERAMIC SUBSTRATE AND CYCLE TO DELAMINATION OF COPPER PLATE FROM CERAMIC SUBSTRATE

Ceramic substrate	Thermal conductivity (W/mK)	Fracture toughness (MPa·m$^{1/2}$)	Four-point bending strength (MPa)	Cycle to delamination of the copper plate from ceramic substrate
Conventional Si$_3$N$_4$ for electric substrates (SN-A)	90	8.0	600	2800-3000
Conventional Si$_3$N$_4$ for molten metal (SN-B)	21	5.2	910	20
Conventional AlN for electric substrates	180	3.2	460	7

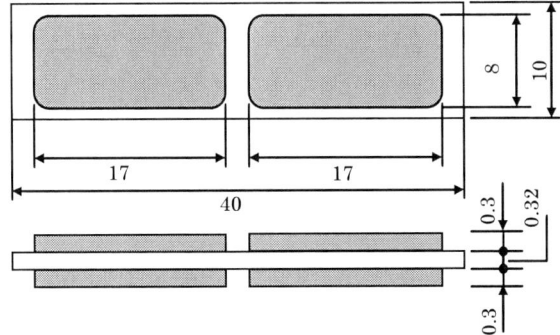

Fig. 1. Shape and dimension of metallized Si$_3$N$_4$ or AlN substrate [8]. The gray area corresponds to copper plate and the remainder is ceramic substrate. Dimension in millimeters.

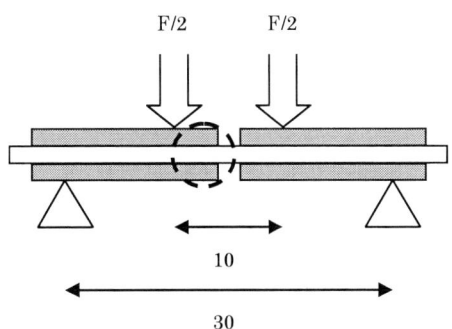

Fig. 2: Schematic illustration of the repeated four-point bending test at 250°C. The results of FEM analysis in the open dotted circle are shown in Fig. 4. Dimension in millimeters

nickel layer and 0.05 µm-thick Au film were formed subsequently on the Cu layer.

Both Si$_3$N$_4$- and AlN-AMB substrates were tested under thermal cycle condition from -40 to 250°C. The duration time for both top and bottom temperatures were ca. 16 min and time for heating and cooling stages were about 1 to 2 min. After certain number of thermal cycles, eight to ten samples were selected from the thermal shock chamber, followed by the visual inspection. Downward and upward four-point bending (Fig. 2) were repeated alternately at a frequency of 1 Hz in the hot chamber. In this preliminary experiment, the cyclic bending test was conducted at 250°C with the intention that the swing of thermal tensile stress in the ceramic/metal interface generated by temperature variation from 250 to -40°C was simulated by unloading and loading. In order to accelerate thermal fatigue of the AMB substrate, tensile stress 1.5-2.1 times larger than the maximum thermal stress at -40 °C was applied. A special bending fixture with outer and inner span of 30/10 mm was employed. The maximum load was varied from 13 N to 18 N depending on the ceramic substrates and the number of cycle was recorded.

The residual stress caused by cooling of the brazed substrate was estimated using the FEM with the Abaqus code. The same calculation procedures and material parameters as those in our previous report were used [5, 6], except that the yield stress of Cu annealed at 700°C was employed instead of

that of as-received Cu. The maximum tensile stress due to the four-point bending at 250°C was also analyzed with the same code and material parameters.

III. RESULTS AND DISCUSSION

Delamination of Cu layer from the AlN plate was observed after only 7 thermal cycles. The Si$_3$N$_4$-AMB substrates using Si$_3$N$_4$ for molten metal application (hereafter, SN-B) exhibited Cu delamination after 20 cycles. In contrast, delamination of Cu layer from Si$_3$N$_4$ for electric substrate (SN-A) occurred after 2800-3000 cycles. The results are summarized in Table I with the mechanical properties of the ceramic substrates. It is obvious that the resistance to thermal fatigue of the AMB substrate depended on the fracture toughness of ceramic substrate, but not on the bending strength.

Fig. 3 shows the contour plot of residual stress in the Si$_3$N$_4$ substrate at –40°C. Peak tensile stress appeared just below the corner of Cu plate and the stress level was estimated to be about 270 MPa. The tensile stress was decreased to about 80 MPa when the AMB substrate was heated up to 250°C. The similar results were also obtained for the AlN-AMB substrate. FEM analysis was also applied to the AMB substrate bended in four point manner at 250°C. The tensile stress distribution in the cross-section (dotted circle in Fig. 2) along the central longitudinal line in the Si$_3$N$_4$-AMB substrate is presented in Fig. 4. Maximum tensile stress in the Si$_3$N$_4$ substrate appeared

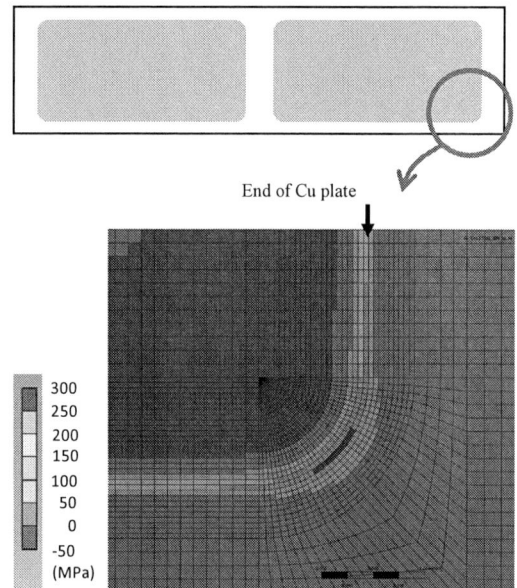

End of Cu plate

Fig. 3. FEM analysis of the maximum principle stress in the Si₃N₄ just below the corner of Cu plate at -40°C

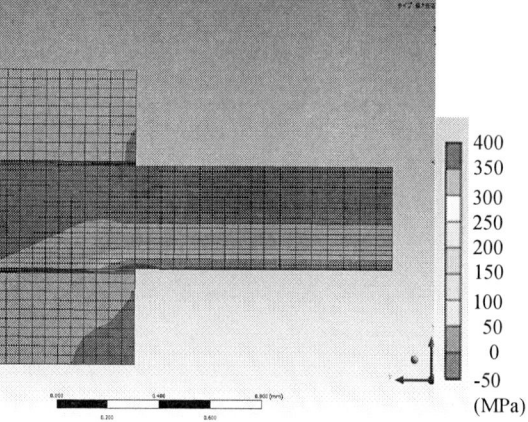

Fig. 4 Contour plot of maximum principle stress at the center part of the Si₃N₄-AMB substrate (open dotted circle in Fig. 2) when bended with a load of 13 N at 250 °C.

at the portion adjacent to the end of Cu plate on the tensile surface. The maximum tensile stress in Si_3N_4 was calculated as a function of load. Maximum tensile stresses at a load of 13, 15, 17 and 18 N were estimated to be 400, 480, 540 and 570 MPa, respectively. The maximum tensile stress in AlN was also analyzed and found to be almost the same as that in Si_3N_4.

Bending loads from 13 to 18 N were applied to both AlN- and Si_3N_4- AMB substrates repeatedly at 250°C and the cycle to failure was counted (Fig. 5). Maximum tensile stress caused by the external loading was 1.5-2.1 times larger than that of the peak residual stress under cooling at -40°C. When the maximum load of 13 N was applied to the AlN-AMB substrate repeatedly at 250°C, the AMB substrate was broken at only 20 cycles. Larger loads more than 13 N were not applied to the AlN-AMB substrate, because tensile stress due to higher load was expected to exceed the fracture strength of the AlN substrate. In the case of the Si_3N_4-AMB substrate (SN-B), cycle to failure at the maximum load of 15 N was 1277 cycles, whereas cycle to failure at the maximum load of 17 N was decreased significantly to 158 and 389 cycles. The Si_3N_4-AMB substrate (SN-A) survived after 10599 and 144000 cycles under the maximum load of 17 N, but failed at 35 cycles when the maximum load was increased up to 18 N. As compared with the fracture toughness of ceramic substrate listed in Table I, it is obvious that the data points of the AMB substrate using ceramic with high fracture toughness located in both higher-load side and longer-cycle side in Fig. 5. The correlation between the cycle to failure during repeated bending at a maximum load of 17 N and the cycle to delamination of Cu plate during thermal cycle for both SN-A- and SN-B- AMB

substrates is presented in Fig. 6. The cycle to failure during repeated loading increased with the cycle to delamination of Cu plate during thermal cycling. Time to failure during repeated loading for SN-A-AMB substrate was 40 h at the most, whereas the test time for thermal cycle test was 1680 h at the least, indicating that the repeated loading at constant temperature can be used as a quick screening method for resistance to thermal fatigue of the AMB substrate.

IV. CONCLUSION

In order to reproduce thermal stress swing in the AMB substrate during the thermal cycling (-40 to 250°C), tensile stress swing was given repeatedly to the AMB substrate by four-point bending the AMB substrate at a frequency of 1 Hz at 250°C. The magnitude of maximum tensile stress during the external loading test was varied from 1.5 to 2.1 times larger than the peak thermal residual stress at -40°C. When the AlN-AMB substrate with low resistance to thermal fatigue was bended to apply tensile stress 1.5 times larger than the stress at -40°C, it was broken at 20 cycles. In contrast, the Si_3N_4-AMB substrate with high resistance to thermal fatigue survived after more than 10000 cycles of external loading even when the tensile stress level was two times larger than the stress at -40°C. The cycle to failure by the repeated external loading of two kinds of Si_3N_4-AMB substrates showed correlation with the cycle to delamination of the Cu plate by the thermal cycling, indicating our new approach can reduce the time needed for test of resistance to thermal fatigue of AMB substrate significantly.

REFERENCES

[1] F. Lang, H. Yamaguchi, H. Nakagawa and H. Sato, "Cyclic thermal stress-induced degradation of Cu metallization on Si3N4 substrate at -40°C to 300°C," J. Electron. Mater., vol. 44 (1), pp. 482-489, 2015.

[2] A. Fukumoto, D. Berry, K. D. T. Ngo and G.-Q. Lu, "Effects of extreme temperature swings (−55°C to 250°C) on silicon nitride active metal

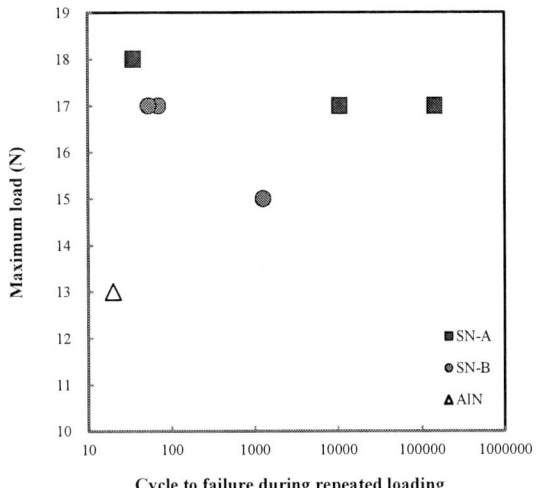

Fig. 5 Relationship between maximum load during cyclic loading and number of cycle to failure.

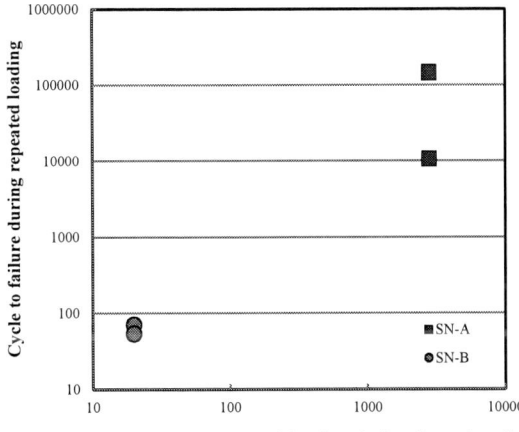

Fig. 6 Relationship between cycle to failure during repeated loading of a 17 N and cycle to delamination of copper plate during thermal cycling.

brazing substrates," IEEE Trans. Device Mater. Reliab., vol. 14 (2), pp. 751-756, 2014.

[3] L.Dupont, Z. Khatir, S. Lefebvre and S. Bontemps, "Effect of metallization thickness of ceramic substrates on the reliability of power assemblies under high temperature cycling," Microelectron. Reliab., vol. 46 (9-11), pp. 1766-1771, 2006.

[4] M. Goets, N. Kuhn, B. Lehmeier, A. Meyer and U. Voeller, "Comparison of silicon nitride DBC and AMB substrates for different applications in power electronics," PCIM Europe Conference Proceeding, Nuremberg, Germany, May. 2013, pp. 14-16..

[5] H. Miyazaki, et al., "Effect of high temperature cycling on both crack formation in ceramics and delamination of copper layeres in silicon nitride active metal brazing substrates," Ceram. Int., vol. 43, pp. 5080-5088, 2017.

[6] H. Miyazaki, et al., "Improved resistance to thermal fatigue of active metal brazing substrates for silicon carbide power modules using tough silicon nitrides with high thermal conductivity," Ceram. Int., in press.

[7] H. Miyazaki, Y. Yoshizawa, K. Hirao and T. Ohji, "Evaluation of fracture toughness of ceramic thin plates through modified single edge-precracked plate method," Scr. Mater., vol. 103, pp. 34-36, 2015.

[8] ISO 17841, Fine Ceramics (Advanced Ceramics, Advanced Technical Ceramics)—Test method for thermal fatigue of fine ceramics substrate, international organization for standards, 2015.

Proceedings of the 30th International Symposium on Power Semiconductor Devices & ICs
May 13-17, 2018, Chicago, USA

Dynamic Stability Analysis based on State-space Model and Lyapunov's Stability Criterion for SiC-MOS and Si-IGBT Switching

Xiao Zeng, Zehong Li, Yuzhou Wu, Wei Gao, Jinping Zhang, Min Ren, Bo Zhang

The State Key Laboratory of Electronic Thin Films and Integrated Devices
University of Electronics Science and Technology of China (UESTC)
Chengdu, China

Abstract—This paper proposes a new stability evaluation method for SiC-MOS and Si-IGBT in switching transient state. This method is based on Lyapunov's stability criterion and statistics. A state-space model is established based on SiC-MOS equivalent circuit with consideration of straw parameters. The system matrix in the state-space model can be used for stability analysis by Lyapunov's stability criterion for the samples that include the parameters in system matrix. The statistical method is implemented for the stability results of the samples in device switching transient state and the unstable level which is used to descript the stability performance during transient state has been defined. From the simulation and experimentation results, the unstable level has the capability to evaluate the EMI performance for comparing the impact of parasitic parameters in design of package, module and chip, even the suitability of the application circuits to suppress device self-excited oscillation.

Keywords—SiC-MOS; IGBT; stability; self-excited oscillation

I. INTRODUCTION

SiC-MOS has become a hotspot both in research and engineering application in recently years due to its higher power density, higher switching frequency and lower switching loss. It has the trend to replace Si Insulated Gate Bipolar Transistor (Si-IGBT) in design of middle power system in the near future, especially in automotive power solutions, such as the electrical drive in electrical automobile. In engineering application, self-excited oscillation is a common problem in the power system based on SiC-MOS or Si-IGBT. The self-excited oscillation is a significant contributor of the EMI issue during SiC-MOS and Si-IGBT switching. Many factors can be the candidates of the self-excited oscillation, such as the parasitic inductance, capacitance and resistance from external circuits, device package and chip. Moreover the working condition also impacts. The system level analysis should be performed to find the proper parameters or working condition to eliminate the EMI issue to improve the switching performance.

Many previous literatures are focusing on characteristic equation of device equivalent circuits with consideration of parasitic parameters to work out suitable range of parameters. In [1], a characteristic equation has been derived from differential mode circuit of a single IGBT in a parallel connected IGBT typology. All the parameters in this equation are parasitic parameters, including some variable parameters that highly depend on collector voltage, collector current and temperature, except for the gate resistor. The Routh-Hurwitz criterion is used to define the stability of the system based on the characteristic equation. Same analysis method can be found in [2] to guide the package design for SiC-MOS and Si-IGBT to suppress EMI issue. The only way to make the system stable is to find the minimum gate resistor after finding other parameters in whole range condition. Related analysis can also be found in [3] and so forth. Nowadays, from the application of power semiconductor device, the active gate control (AGC) technology has been adopted by more and more power system design to improve switching performance of device, such as [4]. So the AGC can be implemented during device switching to perform stability control based on feedback signal by measuring methods, such as [5]. So finding an effective way to do stability evaluation during device switching based on particular computational resource is absolutely necessary. This can also benefit to device design during simulation loop as well as the package and application circuit design.

In order to solve this problem, this paper proposed a dynamic stability analysis method based on state-space model and Lyapunov's stability criterion for SiC-MOS and Si-IGBT switching. Section II will explain the establishment of state-space model and the proposed method of stability analysis. Section III will discuss the simulation and experimentation. At last, section IV will summarize the whole study.

II. THE PRICIPLE OF PROPOSED DYNAMIC STABILITY EVALUATION

The switching of SiC-MOS or Si-IGBT is a significant non-linear process. So many physical parameters are variable during the switching. For example, the gate-drain capacitor C_{gd} of SiC-MOS and gate-collector capacitor C_{gc} of Si-IGBT are highly depended on drain voltage and collector voltage, respectively, as well as the drain-source capacitor C_{ds} of SiC-MOS and collector-emitter capacitor C_{ce} of Si-IGBT, and so on. These variable parameters and the parasitic parameters from package, external circuits make the switching system time-varied. The state-space model is suitable for describing this kind of system. Once the state-space model is established,

This work was supported in part by the National Natural Science Foundation of China (Grant No. 61474017; Grant No. 61404023) and the Project of National Energy Application Technology Research and Engineering Demonstration (Grant No. NY20150703).

978-1-5386-2928-4/18 $31.00 © 2018 IEEE

the system matrix is available for stability analysis by Lyapunov's stability criterion, but the state-space model establishment for time-varied system is complex, but the system for one of the samples from device switching process is time-invariant. So the state-space model establishment for one of the samples during device switching is easy, but this space-model cannot be suitable for all samples during device switching process due to mentioned above variable parameters. If the variation of these variable parameters is worked out, the space-model can be used for other samples with changes of the variable parameters, because the circuit is same for all samples in device switching process and this means the structure of state-space model is the same. The difference of state-space model is induced by these variable parameters. Based on mentioned above analysis, the time-varied state-space model for device switching process can be equivalent to time-invariant system space-model for one sample with the curves of variable parameters that in system matrix.

A. State-space Model Establishment

The establishment of state-space model for one sample is based on MOS equivalent circuit, as shown in Fig. 1, with consideration of parasitic parameters. In Fig.1, the R_G, R_S and R_D are parasitic resistance in gate, source and drain, respectively. The L_G, L_S and L_D are parasitic resistance in gate, source and drain, respectively. The device is equivalent to gate-source capacitance C_{gs}, C_{gd}, C_{ds}, on resistance R_{on} and controllable current source $g_m V_{GS}$.

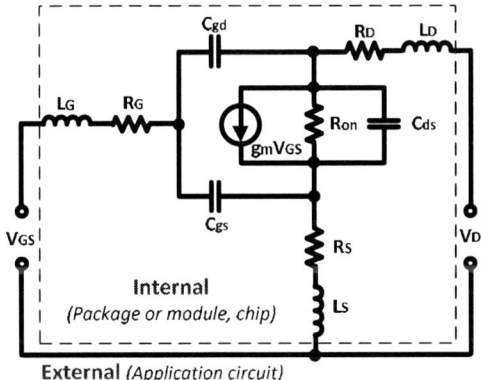

Fig. 1. MOS equivalent circuit.

The method in [1] can also be referred to get the transfer function of the circuit in Fig.1, as shown below:

$$G(s) = \frac{V_{DS}}{V_{GS}}$$

$$= \frac{C_{gd}R_{on}g_m(LL)s^3 + C_{gd}R_{on}g_m(LR)s^2 + R_{on}g_m(L_S + C_{gd}(RR))s + R_S}{R_{on}(LL)(CC)s^4 + [((LL)(C_{gd} + C_{gs}) + R_{on}(LR)(CC)]s^3 +}$$

$$[(C_{gd} + C_{gs})(LR) + R_{on}(C_{ds}(L_D + L_S) + C_{gd}(L_D + I_G) + C_{gs}(I_G + L_S) + (CC)(RR))]s^2 +$$

$$[L_D + L_S + (C_{gd} + C_{gs})(RR) + R_{on}(R_D(C_{ds} + C_{gd}) + R_G(C_{gd} + C_{gs}) + R_S(C_{ds} + C_{gs})]s +$$

$$R_D + R_S + R_{on}$$

(1)

where

$$LL = L_D L_G + L_D L_S + L_G L_S$$
$$RR = R_D R_G + R_D R_S + R_G R_S$$
$$CC = C_{ds} C_{gd} + C_{ds} C_{gs} + C_{gd} C_{gs}$$
$$LR = L_D R_G + L_G R_D + L_D R_S + L_S R_D + L_G R_S + L_S R_G$$

(2)

This transfer function can be converted to state-space model, as shown in (3).

$$\begin{cases} \dot{x} = Ax + Bu \\ y = Cx \end{cases}$$

(3)

where

$$A = \begin{pmatrix} 0 & 1 & 0 & 0 \\ 0 & 0 & 1 & 0 \\ 0 & 0 & 0 & 1 \\ \dfrac{R_D + R_S + R_{on}}{R_{on}(LL)(CC)} & \dfrac{R_S(C_{ds} + C_{gs}) +}{(LL)(CC)} & \dfrac{\substack{(C_{gd} + C_{gs})(LR) + \\ R_D(C_{ds} + C_{gd}) + \\ R_G(C_{gd} + C_{gs}) + \\ C_{gd}(L_D + I_G) + \\ C_{ds}(L_G + L_S)] + (CC)(RR)}}{R_{on}(LL)(CC)} & \left(\dfrac{C_{gd} + C_{gs}}{R_{on}(CC)} + \dfrac{(LR)}{(LL)}\right) \end{pmatrix}$$

(4)

$$B = \begin{pmatrix} 0 \\ 0 \\ 0 \\ 1 \end{pmatrix}$$

(5)

$$C = \begin{pmatrix} \dfrac{R_S R_{on} g_m}{R_{on}(LL)(CC)} & \dfrac{(L_S + C_{gd}(RR))R_{on}g_m}{R_{on}(LL)(CC)} & \dfrac{C_{gd}(LR)R_{on}g_m}{R_{on}(LL)(CC)} & \dfrac{C_{gd}}{(CC)} \end{pmatrix}$$

(6)

A is system matrix, B is input matrix and C is output matrix.

B. Stability Analysis by Lyapunov's Stability

The system matrix A can be used to solve the Lyapunov's equation, as shown in (7) to get the matrix P.

$$A^T P + PA + I = 0$$

(7)

where I is unit matrix. Then, judge P is positive definite or not by Sylvester criterion to know the SiC-MOS or Si-IGBT based system is stable or not.

Some of the parasitic parameters which mentioned above in matrix A are changing with external conditions as well as some electrical parameters, so the A is variable and it needs to perform above mentioned steps for all samples in device switching dynamically to get the binomial probability distribution of stability. The unstable level is defined by the percentage of unstable samples to indicate the overall stability level of device with different parasitic parameters in package and chip or external circuits.

C. Acquiring Parameters in System Matrix A

The time-invariant parameters, such as resistance and inductance can be obtained from device datasheet, SPICE model file or parameter extraction experiment. The time-varied parameter C_{gd}, C_{ds} can be obtain from the capacitance test curve from datasheet and perform curve fitting to get an equivalent equation. The R_{on} can be calculated by (8) or by curve fitting from I-V characteristics on device datasheet.

$$R_{on} = V_{DS} / I_{DS} \qquad (8)$$

III. SIMULATION AND EXPERIMENTATION

Before simulation and experimentation, a stability algorithm which runs in Mathworks/Matlab has been developed according to proposed method, which is used for stability evaluating in simulation and experimentation. The SiC-MOS device is implemented for simulation while the Si-IGBT device is implemented for experimentation.

A. Simulation

The SiC-MOS (CREE C2M0045170D) based BUCK DC-DC converter is implemented in simulation circuit, as shown in Fig.2. The acquired the time-invariant parameters in matrix A are shown in TABLE I. The curve fitting equations of C_{gd} and C_{ds} are shown in (9) and (10).

CREE C2M0045170D

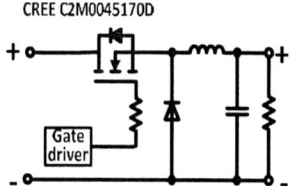

Fig. 2. SiC-MOS based BUCK DC-DC converter in simulation.

TABLE I. TIME-INVARIANT PARAMETERS (C2M0045170D)

Para.	Value (unit)
L_G	15e-9 H
L_S	9e-9 H
L_D	6e-9 H
R_G	1.3 Ω
R_S	2e-3 Ω
R_D	2e-3 Ω
Cgs	3.9e-9F

$$C_{gd} = (13.76e-12) - (0.1281e-9)/V_{DS} + (7.6648e-9)/V_{DS}^2 \qquad (9)$$

$$C_{ds} = (0.2209e-9) - (16.093e-9)/V_{DS} + (43.767e-9)V_{DS}^2 \qquad (10)$$

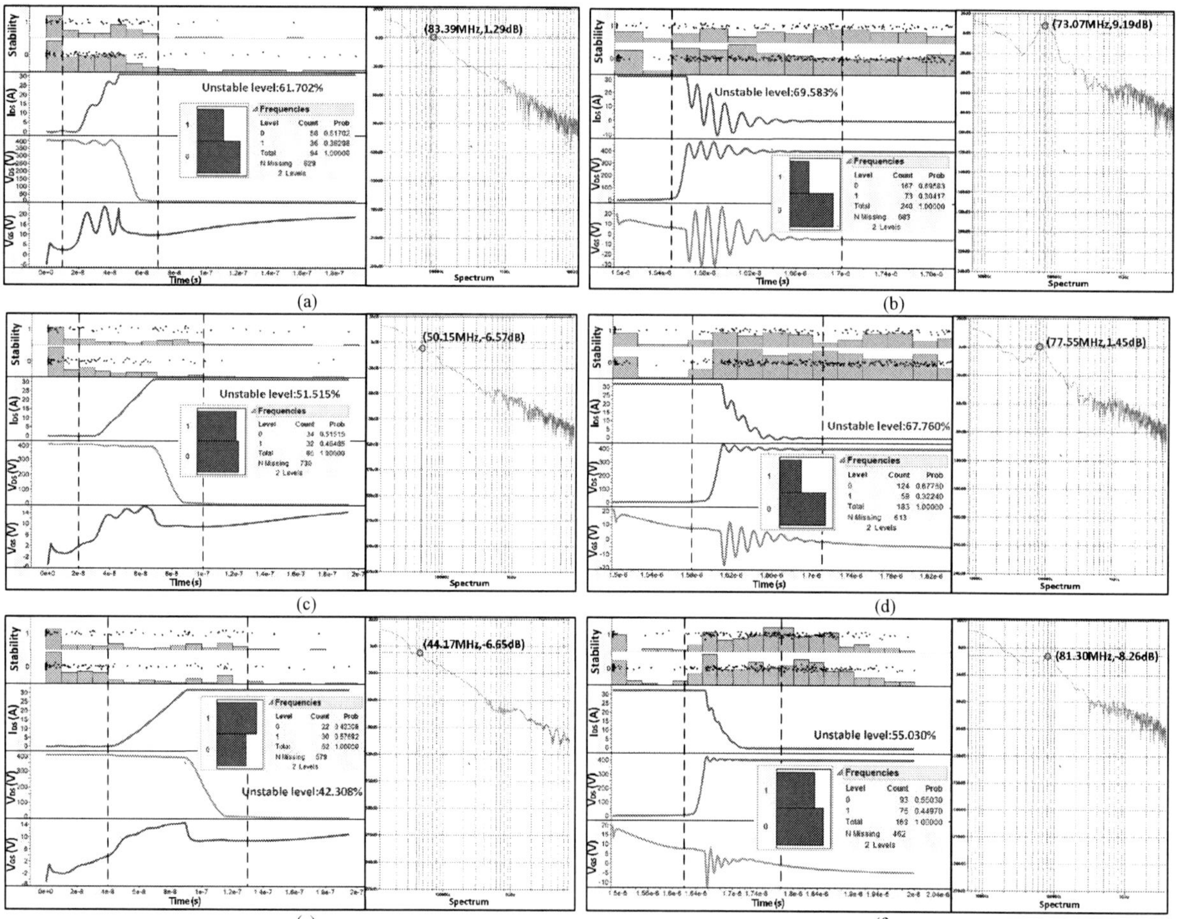

Fig. 3. SiC-MOS based BUCK DC-DC converter simulation results per round, (a) R_{G_EXT}=10Ω turn-on;(b) R_{G_EXT}=10Ω turn-off;(c) R_{G_EXT}=20Ω turn-on;(d) R_{G_EXT}=20Ω turn-off;(e) R_{G_EXT}=30Ω turn-on;(f) R_{G_EXT}=30Ω turn-off.

There are 3 rounds in the simulation with different external gate resistor R_{G_EXT}, as shown in Fig.3. The summary of the simulation can be found in Fig. 4 and it concludes that the proposed method is effective to evaluate the EMI level or stability during SiC-MOS switching.

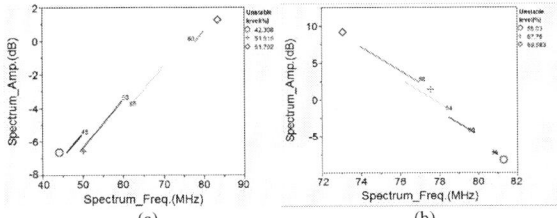

Fig. 4. The summary of relationship between unstable level and EMI performance, (a) relationship between turn-on unstable level and spectrum; (b) relationship between turn-off unstable level and spectrum.

B. Experimetation

The experimentation of stability evaluating for Si-IGBT (Infineon IKW40N65H5) is performed on power semiconductor device dynamic tester (ITC 57240 test head), as shown in Fig.5 and the time-invariant parameters are shown in TABLE II. The curve fitting equations of C_{gc}, C_{ce} and R_{on} are shown in (11), (12) and (13).

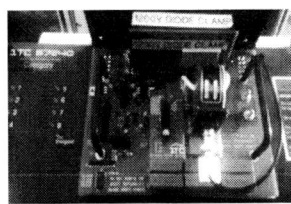

Fig. 5. Si-IGBT test platform (ITC57240).

TABLE II. TIME-INVARIANT PARAMETERS (C2M0045170D)

Para.	Value (unit)
L_G	5e-9 H
L_E	5e-9 H
L_C	3e-9 H
R_G	1.3 Ω
R_E	2e-3 Ω
R_C	2e-3 Ω
C_{ge}	2.5e-9F

$$C_{gc} = (8.539e-12) - (55.82e-12)/V_{CE} + (5.553e-14)/V_{CE}^2 \quad (11)$$

$$C_{ce} = (9.941e-12) - (0.3236e-9)/V_{CE} + (0.3225e-12)/V_{CE}^2 \quad (12)$$

$$R_{on} = 0.0234448 - 0.8637419/I_{DS} + 0.1843049/I_{DS}^2 \quad (13)$$

There are two rounds with different loads (load2>load1) in the experimentation, the results are shown in Fig. 6. The unstable level of load2 is higher than load1 and load2 deserves worse EMI performance. It proves that the proposed method can evaluated the EMI performance by unstable level for Si-IGBT switching.

IV. CONCULSION

A dynamic stability evaluation method based on Lyapunov's stability criterion is proposed for SiC-MOS and Si-IGBT. The simulation and experimentation results show that this method is capable to evaluate and compare the device EMI performance with different internal (package and chip) or external (application, drive circuits) parameters. It can be used for on-line stability investigation as well as device design and application circuit design to improve EMI performance.

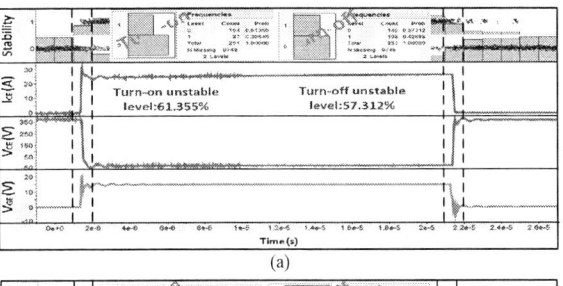

(a)

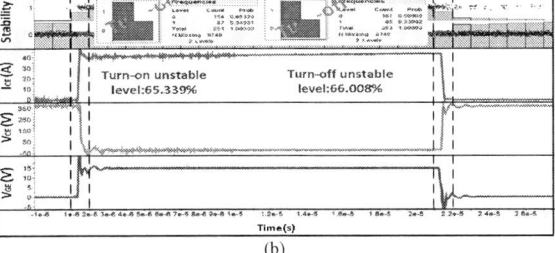

(b)

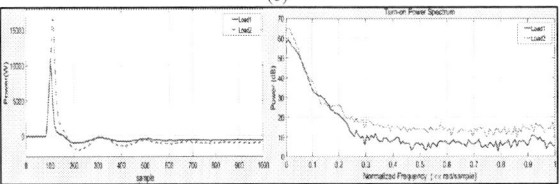

(c)

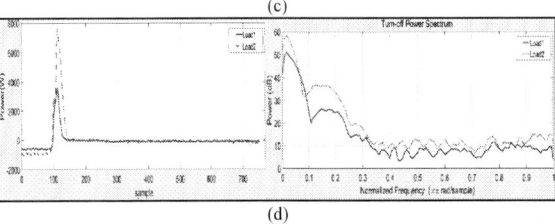

(d)

Fig. 6. Si-IGBT experimental results (load2>load1), (a) load1 switching waveform; (b) load2 switching waveform; (c) turn-on power loss and spectrum; (d) turn-off power loss and spectrum.

REFERENCES

[1] John C. Joyce, "Current sharing and redistribution in high power IGBT modules," University of Cambridge, pp. 53, May. 2001.

[2] K. Saito, T. Miyoshi, D. Kawase, S. Hayakawa, T. Masuda and Y. Sasajima, "Suppression of self-excited oscillation for common package of Si-IGBT and SiC-MOS," 2017 29th International Symposium on Power Semiconductor Devices and IC's (ISPSD), Sapporo, 2017, pp. 427-430

[3] M. Spang, S. Buetow and G. Katzenberger, "Differential-mode oscillations between parallel IGBTs in power modules," 2015 17th European Conference on Power Electronics and Applications (EPE'15 ECCE-Europe), Geneva, 2015, pp. 1-10.

[4] X. Yang, X. Zhang and P. R. Palmer, "IGBT converters conducted EMI analysis by controlled multiple-slope switching waveform approximation," 2013 IEEE International Symposium on Industrial Electronics, Taipei, Taiwan, 2013, DOI: 10.1109/ISIE.2013.6563663, pp. 1-6.

[5] X. Zeng et al., "A Novel Virtual Sensing with Artificial Neural Network and K-means Clustering for IGBT Current Measuring," in IEEE Transactions on Industrial Electronics, in press.

Proceedings of the 30th International Symposium on Power Semiconductor Devices & ICs
May 13-17, 2018, Chicago, USA

1 kV/1.3 mΩ·cm² Vertical GaN-on-GaN Schottky Barrier Diodes with High Switching Performance

Shu Yang[*], Shaowen Han, Rui Li, Kuang Sheng

College of Electrical Engineering, Zhejiang University, Hangzhou, China
[*]Email: eesyang@zju.edu.cn

Abstract—In this paper, we present vertical GaN Schottky barrier diodes implemented on bulk GaN substrates, delivering a breakdown voltage of ~1 kV, a specific ON-resistance of 1.3 mΩ·cm² with current spreading considered, a high current swing over 13 orders of magnitude and a low ideality factor of 1.04. The developed devices exhibit current-collapse-free operation under 400 V/500 ns switching condition as well as zero reverse recovery characteristics, showing great potential for high-power and high-frequency applications.

Keywords—*GaN, power semiconductor devices, Schottky barrier diode, breakdown voltage, switching performance, current-collapse-free*

I. INTRODUCTION

In addition to the tremendous progress in lateral GaN-on-Si power electronics [1], the recent availability of high-quality bulk GaN substrate has led to the development of vertical GaN-on-GaN power devices. With zero lattice and thermal expansion coefficient mismatch between the substrate and epitaxial layer, the GaN-on-GaN structure enables larger drift layer thickness (up to 40 μm) and lower dislocation density (10^3–10^6 cm^{-2}). The GaN-on-GaN vertical devices can reach high voltage blocking and current handling capability [2, 3]. Unlike the lateral GaN-on-Si devices that could be prone to surface/buffer trapping [4-7], vertical GaN-on-GaN devices are inherently less susceptible to dynamic ON-resistance (R_{ON}) degradation (or current collapse).

Among the vertical power rectifiers, GaN Schottky barrier diodes (SBDs), with low forward-voltage drop and absence of minority carrier storage effect, are capable of delivering high conduction/switching efficiency and high switching speed.

In this work, we demonstrate 1 kV/1.3 mΩ·cm² vertical GaN-on-GaN SBDs with nearly ideal Schottky contact and nitridation-based termination, showing excellent static and switching performance. The dynamic performance of the vertical GaN SBDs has been evaluated by using a high-speed double pulse tester (DPT) in conjunction with a clamping circuit.

II. DEVICE STRUCTURE AND FABRICATION

The GaN-on-GaN structure used in this work was grown on a commercially available free-standing n$^+$-GaN substrate. The 11-μm thick n$^-$-GaN homo-epitaxial layer has a net doping concentration of ~8×10^{15} cm^{-3}. The vertical GaN SBD in this work consists of Pt/Au anode metal on the pre-cleaned

surface, Ti/Al/Au cathode on the substrate backside, and nitridation-based termination structure around the device periphery (Fig. 1(b) inset). The termination structure, aiming at suppressing the edge leakage current, was formed by plasma nitridation [8]. Unterminated SBD was also fabricated for comparison. Thick pad metal was deposited on top of the anode for wire bonding, to facilitate switching and dynamic performance characterizations in a DPT circuit.

III. STATIC PERFORMANCE

As shown in Fig. 1, the vertical GaN SBDs exhibit well-behaved ON-state characteristics, including a high forward current density of 2000 A/cm² and current swing of 13 orders of magnitude. The near-unity ideality factor (η = 1.04) in the barrier-limited region suggests the high quality of the Pt/n-GaN Schottky contact and GaN homo-epitaxial layer [8]. The differential specific ON-resistance ($R_{ON,sp}$) is 1.3 mΩ·cm² with lateral current spreading taken into consideration.

By virtue of the plasma nitridation that can favorably modify GaN surface condition and suppress the excess leakage at the junction edge [8-10], the vertical GaN SBD with nitridation-based termination can achieve a *BV* of ~995 V (defined at a reverse current density of 0.1 A/cm²), in comparison with the unterminated control device with a *BV* ~ 335 V (Fig. 2). The terminated SBD can maintain a leakage current density of ~10^{-5} A/cm² at –600 V, resulting in a high ON/OFF current ratio (I_{ON}/I_{OFF} at –600 V) of ~10^8.

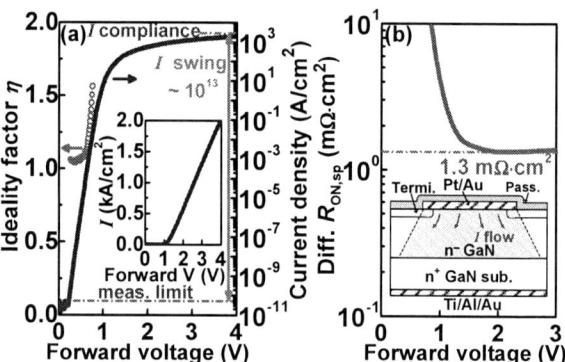

Fig. 1. (a) Forward *I–V* characteristics of the vertical GaN SBD in semi-log scale with ideality factor η. Inset: Forward *I–V* characteristics in linear scale. (b) Differential $R_{ON,sp}$ vs. forward bias voltage. Inset: Schematic cross section of the vertical GaN SBD with nitridation-based termination.

978-1-5386-2928-4/18 $31.00 © 2018 IEEE

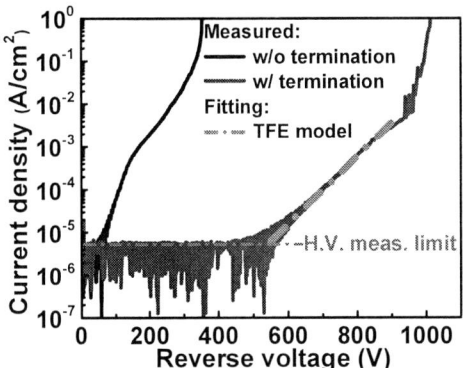

Fig. 2. Measured reverse *I–V* characteristics of the unterminated SBD and the SBD with nitridation-based termination. The calculated *I–V* based on the TFE model is also shown for reference.

The reverse leakage current is below the high-voltage measurement limit at a reverse bias lower than 550 V. Within the bias range of ~550–900 V, the measured reverse *I–V* characteristics primarily follow thermionic field emission (TFE) model [11], which is described by

$$J_{\text{TFE}} = \frac{A^* T q \hbar E}{k} \sqrt{\frac{\pi}{2m_{\text{n}}kT}} \exp[-\frac{1}{kT}(\phi_{\text{B}} - \frac{(q\hbar E)^2}{24m_{\text{n}}(kT)^2})] \quad (1)$$

$$E = \sqrt{\frac{2qN_{\text{D}}(V_{\text{d}} - V)}{\varepsilon_s \varepsilon_0}} \quad (2)$$

where A^*, m_{n}, E, ϕ_{B}, N_{D}, V_{d}, V are the effective Richardson constant, electron effective mass of GaN, electric field at the metal/semiconductor interface, Schottky barrier height, net donor concentration, diffusion voltage and applied voltage, respectively. ϕ_{B} is ~0.93 eV, as extracted from the temperature-dependent forward *I–V* characteristics. The reverse leakage beyond 900 V is possibly related to the edge effect [12]. The trap-related space-charge-limited current (SCLC) [13, 14], which could have adverse impact on the dynamic performance, is considered to be minimal in the vertical GaN SBDs in this work. In addition, the repeated sweeps between forward and reverse biases do not show any trapping-related dispersion/hysteresis (Fig. 3). No degradation in the forward *I–V* was observed after subjecting to a high

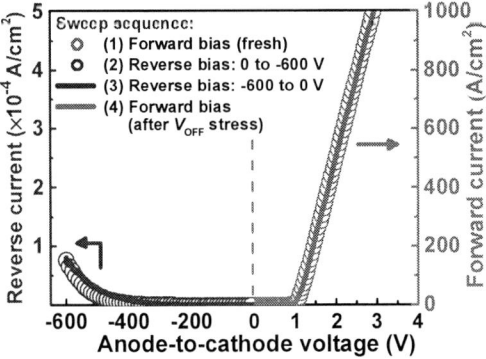

Fig. 3 Multiple sweeps of forward and reverse *I–V* measurements.

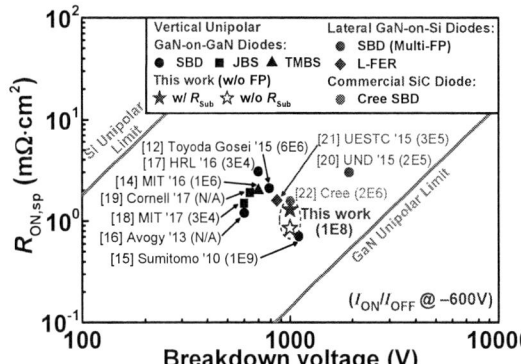

Fig. 4. Benchmark of *BV* vs. $R_{\text{ON,sp}}$ of the vertical GaN SBD in this work, the previously reported vertical *unipolar* GaN-on-GaN diodes, state-of-the-art lateral GaN diodes and commercial SiC SBD. $I_{\text{ON}}/I_{\text{OFF}}$ at –600 V is denoted. $R_{\text{ON,sp}}$ of the vertical GaN SBD in this work with R_{Sub} subtracted is also shown for reference.

Table I. Summary of $R_{\text{ON,sp}}$, $I_{\text{ON}}/I_{\text{OFF}}$, *BV* and *BFOM* of the vertical unipolar GaN-on-GaN diodes (including SBD, TMBS, JBS).

Vertical GaN-on-GaN diodes	$R_{\text{ON,sp}}$ (mΩ·cm²)	$I_{\text{ON}}/I_{\text{OFF}}$ @–600V	*BV* (V)	*BFOM* (MW/cm²)
[15] SBD	0.7	1×10⁹	1100	1700
[16] SBD	1.2	N/A	600	300
[12] SBD	2.1	6×10⁶	790	300
[17] SBD	3.1	3×10⁴	700	160
[14] TMBS	2.0	1×10⁶	700	250
[18] JBS	1.5	3×10⁴	600	240
[19] JBS	1.9	N/A	640	220
(This work) SBD	**1.3**	**1×10⁸**	**995**	**760**

reverse bias up to 600 V (overlapped (1) and (4)), further verifying the minimal trapping effect in these devices.

The vertical GaN-on-GaN SBD in this work leads to a Baliga's figure-of-merit ($BFOM = BV^2/R_{\text{ON,sp}}$) of 760 MW/cm², which is competitive among the best *unipolar* vertical GaN diodes [12, 14-19], and also compares favorably with the state-of-the-art lateral GaN diodes [20, 21] as well as the SiC diode [22] (Fig. 4 and Tab. I).

IV. SWITCHING/DYNAMIC PERFORMANCE

To accurately characterize the switching/dynamic performance of the power diodes, a DPT with an inductive load was designed and implemented, which can minimize the contact resistance and parasitics.

The reverse recovery characterizations of the vertical GaN SBD developed in this work, commercial SiC SBD (600 V/1 A) and Si FRD (700 V/1 A) were carried out using the DPT (Fig. 5). When switching from a forward current of 1 A to a reverse bias of 400 V, the vertical GaN SBD exhibits fast reverse recovery with reverse recovery time (t_{rr}) of ~17 ns and reverse recovery charge (Q_{rr}) of ~0.8 nC, outperforming the Si FRD and SiC SBD of similar ratings (Fig. 5(d)). t_{rr} is defined as the time in which the reverse current recovers to 10% of its peak value at the turn-off response. Furthermore, the vertical GaN SBD could also deliver higher switching

978-1-5386-2928-4/18 $31.00 © 2018 IEEE

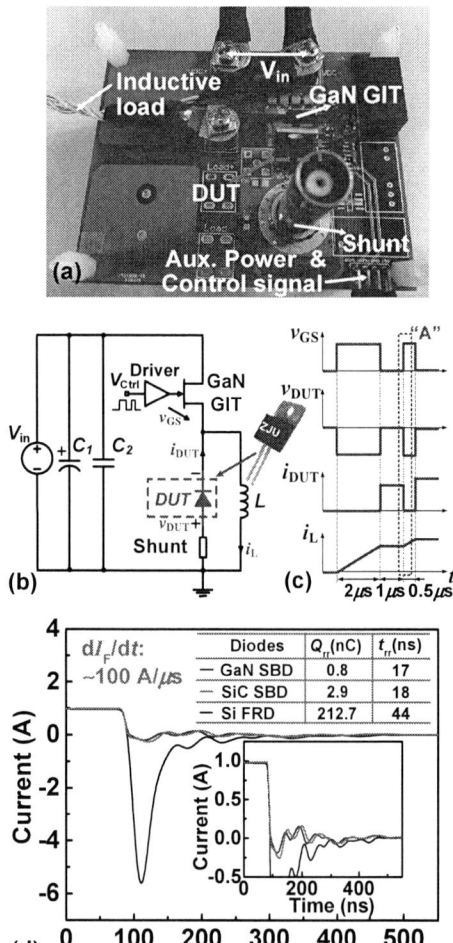

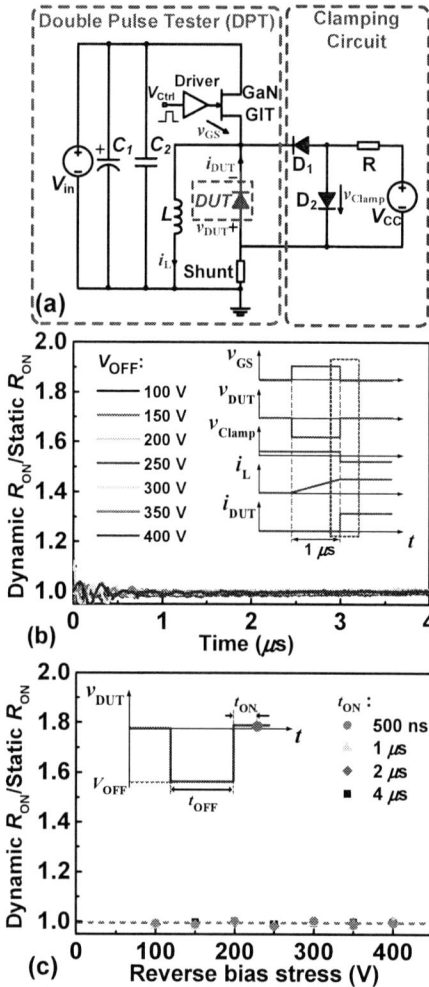

Fig. 5 (a) Photo and (b) schematic of the DPT which was specially designed for the evaluation of GaN transistors/diodes up to 400 V. (c) Schematic waveforms in the reverse recovery measurements. (d) Reverse recovery characteristics of the vertical GaN SBD developed in this work, SiC SBD and Si FRD when switching from a forward current of 1 A to a reverse bias of 400 V (corresponding to "A" in Fig. 5(c)).

Fig. 6 Dynamic R_{ON} characterization of the vertical GaN SBD by using a DPT with a clamping circuit: (a) Circuit schematic. (b) Time-resolved normalized dynamic R_{ON} after switching from various OFF-state biases (V_{OFF}: 100 V–400 V) to ON-state. Inset: Schematic waveforms with the measurement region highlighted. (c) Normalized dynamic R_{ON} with V_{OFF} stress up to 400 V extracted at t_{ON} varying from 500 ns to 4 μs.

performance than the lateral GaN diode, owing to the more uniform distribution of drift region charges and more efficient removal of these charges during turn-off [23].

In addition, the DPT in conjunction with a clamping circuit was developed and used to accurately characterize the dynamic R_{ON} of the vertical GaN SBD. During the power diode's switching from hundreds of volts at OFF-state to a small ON-state voltage, it is challenging to accurately measure the small ON-state bias directly by a conventional DPT. Therefore, a voltage clamping circuit was incorporated to facilitate the precise extraction of the dynamic R_{ON} during the high-voltage switching (Fig. 6 (a)). The schematic waveforms for dynamic R_{ON} extraction are shown in Fig. 6(b) inset. Right after the high-voltage OFF-state bias stress, the ON-state voltage and current of the vertical GaN SBD were recorded,

Table II. Summary of switching/dynamic performance of Si FRD, SiC SBD, lateral GaN-on-Si SBD and vertical GaN-on-GaN SBD developed in this work.

Diodes	Reverse recovery test condition	t_{rr} (ns)	Q_{rr} (nC)	Norm. dynamic R_{ON}	
Si FRD	1 A/400 V $di/dt = 100$ A/μs	44	212.7	N/A	
SiC SBD	1 A/400 V $di/dt = 100$ A/μs	18	2.9	N/A	
GaN-on-Si Lateral SBD*	4 A/400 V $di/dt = 300$ A/μs	31	30	>1.5	t_{ON}: 10 ms V_{OFF}: 400 V
GaN-on-GaN Vertical SBD	1 A/400 V $di/dt = 100$ A/μs	17	0.8	~1.0	t_{ON}: 500 ns V_{OFF}: 400 V

* t_{rr} and Q_{rr} of the GaN-on-Si lateral SBD were extracted from [23]. The normalized dynamic R_{ON} of the lateral GaN-on-Si SBD listed here was from [24], which could be epi-specific due to different buffer trapping.

from which the dynamic R_{ON} can be extracted. The ratios between dynamic R_{ON} and static R_{ON} with t_{ON} varying from 500 ns to 4 μs under different OFF-state biases are shown in Fig. 6(c). The vertical GaN SBD can maintain a *current-collapse-free* operation with a reverse bias stress up to 400 V and a minimum t_{ON} of 500 ns. As summarized in Tab. II, the vertical GaN SBD shows superior switching/dynamic performance compared with Si FRD, SiC SBD and lateral GaN-on-Si SBD.

V. CONCLUSIONS

1 kV/1.3 m$\Omega\cdot$cm^2 vertical GaN-on-GaN SBDs have been demonstrated in this work, yielding a Baliga's figure-of-merit of 760 MW/cm^2. The developed vertical GaN SBDs exhibit nearly ideal Schottky contact, large forward current density of 2000 A/cm^2, 8-order ON/OFF current ratio at −600 V, zero reverse recovery characteristics and *current-collapse-free* operation under sub-μs switching condition. The superior static and switching/dynamic performance shows great potential of vertical GaN-on-GaN Schottky power rectifiers for high-power and high-frequency applicaitons.

ACKNOWLEDGMENT

This work was supported by the National Key Research and Development Program of China (No. 2017YFB0404100).

REFERENCES

[1] K. J. Chen, O. Häberlen, A. Lidow, C. Tsai, T. Ueda, Y. Uemoto, and Y. Wu, "GaN-on-Si power technology: Devices and applications," *IEEE Trans. Electron Devices*, vol. 64, no. 3, pp. 779-795, Mar. 2017.

[2] H. Ohta, N. Kaneda, F. Horikiri, Y. Narita, T. Yoshida, T. Mishima, and T. Nakamura, "Vertical GaN p-n junction diodes with high breakdown voltages over 4 kV," *IEEE Electron Device Lett.*, vol. 36, no. 11, pp. 1180-1182, Nov. 2015.

[3] I. C. Kizilyalli, A. P. Edwards, H. Nie, P. Bui-Quang, D. Disney, and D. Bour, "400-A (pulsed) vertical GaN p-n diode with breakdown voltage of 700 V," *IEEE Electron Device Lett.*, vol. 35, no. 6, pp. 654-656, Jun. 2014.

[4] R. Vetury, N. Q. Zhang, S. Keller, and U. K. Mishra, "The impact of surface states on the DC and RF characteristics of AlGaN/GaN HFETs," *IEEE Trans. Electron Devices*, vol. 48, no. 3, pp. 560-566, Mar. 2001.

[5] S. Yang, C. Zhou, Q. Jiang, J. Lu, B. Huang, and K. J. Chen, "Investigation of buffer traps in AlGaN/GaN-on-Si devices by thermally stimulated current spectroscopy and back-gating measurement," *Appl. Phys. Lett.*, vol. 104, no. 1, p. 013504, Jan. 2014.

[6] P. Moens, M. J. Uren, A. Banerjee, M. Meneghini, B. Padmanabhan, W. Jeon, S. Karboyan, M. Kuball, G. Meneghesso, and E. Zanoni, "Negative dynamic Ron in AlGaN/GaN power devices," in *Proc. Int. Symp. on Power Semiconductor Devices and IC's (ISPSD)*, May/Jun. 2017, pp. 97-100.

[7] S. Yang, C. Zhou, S. Han, J. Wei, K. Sheng, and K. J. Chen, "Impact of substrate bias polarity on buffer-related current collapse in AlGaN/GaN-on-Si power devices," *IEEE Trans. Electron Devices*, vol. 64, no. 12, pp. 5048-5056, Dec. 2017.

[8] S. Han, S. Yang, and K. Sheng, "High-voltage and high-I_{ON}/I_{OFF} vertical GaN-on-GaN Schottky barrier diode with nitridation-based termination," *IEEE Electron Device Lett.*, 2018, in press.

[9] S. Yang, Z. Tang, K.-Y. Wong, Y.-S. Lin, C. Liu, Y. Lu, S. Huang, and K. J. Chen, "High-quality interface in Al$_2$O$_3$/GaN/AlGaN/GaN MIS structures with *in situ* pre-gate plasma nitridation," *IEEE Electron Device Lett.*, vol. 34, no. 12, pp. 1497-1499, Dec. 2013.

[10] Z. Zhang, B. Li, Q. Qian, X. Tang, M. Hua, B. Huang, and K. J. Chen, "Revealing the nitridation effects on GaN surface by first-principles calculation and X-Ray/ultraviolet photoemission spectroscopy," *IEEE Trans. Electron Devices*, vol. 64, no. 10, pp. 4036-4043, Oct. 2017.

[11] J. Suda, K. Yamaji, Y. Hayashi, T. Kimoto, K. Shimoyama, H. Namita, and S. Nagao, "Nearly ideal current–voltage characteristics of Schottky barrier diodes formed on hydride-vapor-phase-epitaxy-grown GaN free-standing substrates," *Appl. Phys. Express*, vol. 3, no. 10, p. 101003, Oct. 2010.

[12] N. Tanaka, K. Hasegawa, K. Yasunishi, N. Murakami, and T. Oka, "50 A vertical GaN Schottky barrier diode on a free-standing GaN substrate with blocking voltage of 790 V," *Appl. Phys. Express*, vol. 8, no. 7, pp. 071001-1-071001-3, Jun. 2015.

[13] C. Zhou, Q. Jiang, S. Huang, and K. J. Chen, "Vertical leakage/breakdown mechanisms in AlGaN/GaN-on-Si devices," *IEEE Electron Device Lett.*, vol. 33, no. 8, pp. 1132-1134, Aug. 2012.

[14] Y. Zhang, M. Sun, Z. Liu, D. Piedra, M. Pan, X. Gao, Y. Lin, A. Zubair, L. Yu, and T. Palacios, "Novel GaN trench MIS barrier Schottky rectifiers with implanted field rings," in *IEDM Tech. Dig.*, Dec. 2016, pp. 10.2.1-10.2.4.

[15] Y. Saitoh, K. Sumiyoshi, M. Okada, T. Horii, T. Miyazaki, H. Shiomi, M. Ueno, K. Katayama, M. Kiyama, and T. Nakamura, "Extremely low on-resistance and high breakdown voltage observed in vertical GaN Schottky barrier diodes with high-mobility drift layers on low-dislocation-density GaN substrates," *Appl. Phys. Express*, vol. 3, no. 8, p. 081001, Jul. 2010.

[16] D. Disney, H. Nie, A. Edwards, D. Bour, H. Shah, and I. C. Kizilyalli, "Vertical power diodes in bulk GaN," in *Proc. Int. Symp. on Power Semiconductor Devices and IC's (ISPSD)*, May 2013, pp. 59-62.

[17] Y. Cao, R. Chu, R. Li, M. Chen, and A. J. Williams, "Improved performance in vertical GaN Schottky diode assisted by AlGaN tunneling barrier," *Appl. Phys. Lett.*, vol. 108, no. 11, p. 112101, Mar. 2016.

[18] Y. Zhang, Z. Liu, M. J. Tadjer, M. Sun, D. Piedra, C. Hatem, T. J. Anderson, L. E. Luna, A. Nath, A. D. Koehler, H. Okumura, J. Hu, X. Zhang, X. Gao, B. N. Feigelson, K. D. Hobart, and T. Palacios, "Vertical GaN junction barrier Schottky rectifiers by selective ion implantation," *IEEE Electron Device Lett.*, vol. 38, no. 8, pp. 1097-1100, Aug. 2017.

[19] W. Li, K. Nomoto, M. Pilla, M. Pan, X. Gao, D. Jena, and H. G. Xing, "Design and realization of GaN trench junction-barrier-Schottky-diodes," *IEEE Trans. Electron Devices*, vol. 64, no. 4, pp. 1635-1641, Apr. 2017.

[20] M. Zhu, B. Song, M. Qi, Z. Hu, K. Nomoto, X. Yan, Y. Cao, W. Johnson, E. Kohn, D. Jena, and H. G. Xing, "1.9-kV AlGaN/GaN lateral Schottky barrier diodes on silicon," *IEEE Electron Device Lett.*, vol. 36, no. 4, pp. 375-377, Apr. 2015.

[21] Q. Zhou, Y. Jin, Y. Shi, J. Mou, X. Bao, B. Chen, and B. Zhang, "High reverse blocking and low onset voltage AlGaN/GaN-on-Si lateral power diode with MIS-gated hybrid anode," *IEEE Electron Device Lett.*, vol. 36, no. 7, pp. 660-662, Jul. 2015.

[22] http://www.wolfspeed.com/power/products/sic-schottky-diodes/table.

[23] L. Efthymiou, G. Camuso, G. Longobardi, F. Udrea, E. Lin, T. Chien, and M. Chen, "Zero reverse recovery in SiC and GaN Schottky diodes: A comparison," in *Proc. Int. Symp. on Power Semiconductor Devices and IC's (ISPSD)*, Jun. 2016, pp. 71-74.

[24] J. A. Croon, G. A. M. Hurkx, J. J. T. M. Donkers, and J. Šonský, "Impact of the backside potential on the current collapse of GaN SBDs and HEMTs," in *Proc. Int. Symp. on Power Semiconductor Devices and IC's (ISPSD)*, May 2015, pp. 365-368.

Proceedings of the 30th International Symposium on Power Semiconductor Devices & ICs
May 13-17, 2018, Chicago, USA

Reverse-Blocking AlGaN/GaN Normally-Off MIS-HEMT with Double-Recessed Gated Schottky Drain

Jiacheng Lei, Jin Wei, Gaofei Tang, and Kevin J. Chen
Department of Electronic and Computer Engineering
The Hong Kong University of Science and Technology
Hong Kong SAR, CHINA
E-mail: jleiaa@connect.ust.hk

Abstract—A reverse blocking AlGaN/GaN normally-Off MIS-HEMT featuring double-recessed gated Schottky drain was demonstrated on a double-channel HEMT platform. Two recess steps with robust recess depth tolerance are performed to form the MIS-gated Schottky drain. The shallow recess stops at the upper GaN channel layer where a MIS-gated section (i.e. MIS field plate) is formed to suppress the reverse leakage current. The deep recess cuts through the lower 2DEG channel where a metal-2DEG Schottky contact with low turn-on voltage is formed along the sidewall. Since the lower channel is separated from the surface of the shallow recess, the MIS-gated section in the drain maintains a high mobility channel to yield a sheet resistance of 806 Ω/Square. The device exhibits a threshold voltage of +0.6 V at a drain leakage current of 10 μA/mm and +1.9 V from linear extrapolation, and a low differential ON-resistance of ~11 Ω/mm. Owing to the presence of the metal-2DEG Schottky contact and the leakage-suppression MIS field plate in the drain, a low forward turn-on voltage of 0.5 V and a low reverse leakage current of 20 nA/mm (at −100 V) are achieved simultaneously. The device also exhibits a high forward and reverse breakdown voltage of 700 V and −600 V.

Keywords—AlGaN/GaN, double-channel, double-recessed, gated Schottky drain, high breakdown voltage, low reverse leakage, low turn-on voltage, MIS-HEMT, reverse blocking.

I. INTRODUCTION

GaN based power devices, especially with AlGaN/GaN heterojunction featuring high-mobility channel, high breakdown voltage and high switching frequency, have emerged as a promising candidate for next generation power conversion applications [1]. As a component of bi-directional switches which are highly desired in attractive AC-AC matrix converters [2], reverse blocking AlGaN/GaN HEMTs (RB-HEMT [3]) with a built-in rectifier in the drain terminal are preferred when the matrix converters operate with a two-step current-direction-based commutation [2]. In this commutation scheme, the switch pair usually do not turn on simultaneously. With bi-directional switch built by series connected AlGaN/GaN HEMTs (Fig. 1 (a)), switch G2 works as a lateral field-effect rectifier (LFER) when G1 turns on and G2 turns off, resulting in a trade-off between the threshold voltage (V_{th}) under the gate and the turn-on voltage (V_{on}) of the rectifier by $|V_{on}| = |V_{GS} - V_{th}|$ [4], [5]. The anti-parallel connected RB-MIS-HEMTs avoid this trade-off (Fig. 1 (b)) [6]. In RB-MIS-HEMT

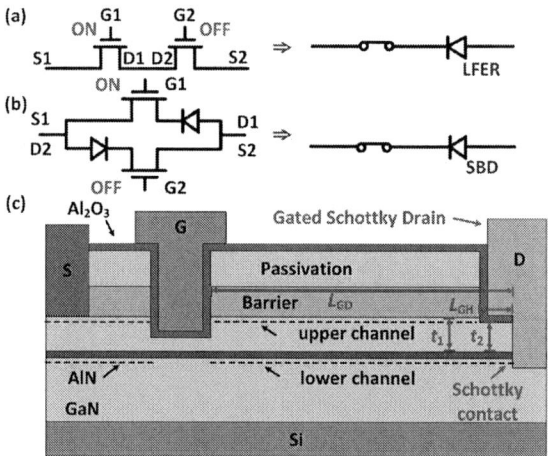

Fig. 1. (a) Bi-directional switch built by series connected AlGaN/GaN HEMT. When G1 turns on and G2 turns off, G2 works as a lateral field-effect rectifier, so V_{th} is coupled with reverse turn-on voltage by $|V_{on}| = |V_{GS} - V_{th}|$. (b) Bi-directional switch built by anti-parallel connected RB-MIS-HEMTs. When G2 is off, G1 works as a SBD. (c) Schematic cross-section of the normally-Off RB-MIS-HEMT with double-recessed gated Schottky drain. The Schottky drain contacts 2DEG directly from the sidewall. A leakage suppression MIS field plate is deployed adjacent to the Schottky contact.

design, a hybrid Schottky-Ohmic drain has been used to obtain a low V_{on} [7]–[9]. However, the reverse leakage current (I_R) through the Schottky-Ohmic drain is relatively large. A hybrid tri-anode Schottky drain structure with nanoscale fins has been demonstrated to achieve a low I_R, but the switch exhibits normally-on operation and relatively large differential R_{ON} [10].

Recently, a high performance double-channel Schottky barrier diode (SBD) featuring a dual-recess gated anode was demonstrated. The MIS field plate can effectively shield the Schottky contact from the high electric field, leading to a low reverse leakage current. Besides, a low turn-on voltage is achieved simultaneously by deploying a metal-2DEG Schottky contact [11].

In this work, we demonstrate a normally-Off GaN RB-MIS-HEMT (Fig.1 (c)) with the aforementioned double-recessed gated Schottky drain using a double-channel heterojunction grown on Si substrate. The Schottky drain features two recess

978-1-5386-2928-4/18 $31.00 © 2018 IEEE   276

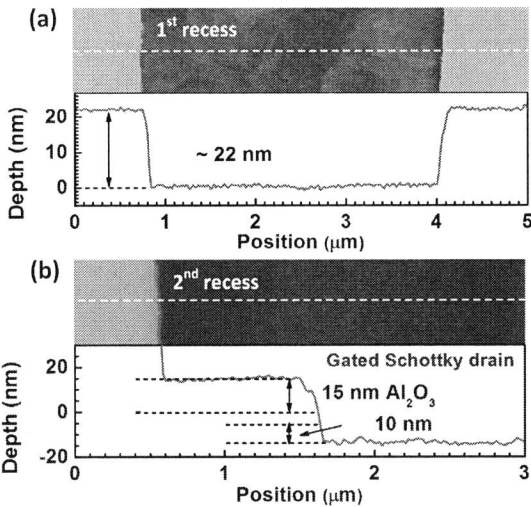

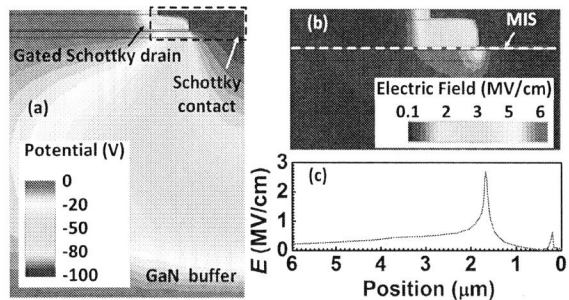

Fig. 3. (a), (b) Potential and Electric field distribution at the Schottky drain region at $V_{DS} = -100$ V. (c) The electric field along the dashed line. The high electric field is concentrated at the edge of MIS field plate, while the electric field at the Schottky contact is reduced.

Fig. 2. (a) AFM image of the MIS field plate region with a recess depth of 22 nm, indicating a 0.5-nm over-etch ($t_2 = 5.5$ nm) on upper channel GaN layer. (b) The deep recess in Schottky contact region shows a 10-nm over-etch on lower channel, indicating a metal-2DEG Schottky contact.

steps. The shallow recess terminates on the upper GaN channel layer to facilitate the formation of a leakage suppression MIS field plate (i.e. MIS-gated section in the drain), while the deep recess cuts through the two channels to allow the drain metal to contact the sidewall of the 2DEG channel directly, leading to a low V_{on}. In addition, the leakage suppression MIS field plate adds limited resistance to the device, which is indicated by the sheet resistance measured by transfer length method.

II. EXPERIMENT

The devices were fabricated on a GaN-on-Si double-channel heterostructure sample [12], [13]. The epi-structure consists of a barrier layer (including 3-nm GaN cap, 17-nm AlGaN and 1.5-nm AlN), a 6-nm upper GaN channel layer, a 1.5-nm AlN insertion layer and a 4-μm GaN buffer/transition layer. The upper channel is below the barrier layer and the lower channel is below the AlN insertion layer. The total carrier density and mobility is 8.6×10^{17} cm^{-2} and 2080 cm^2/(V·s), respectively, obtained from the hall measurement. The fabrication process began with AlN/SiN$_x$ (4/60 nm) stack deposition by PEALD/PECVD (plasma enhanced atomic layer deposition / plasma enhanced chemical vapor deposition) in sequence to serve as a passivation layer. After removing the passivation in source region, Ti/Al/Ni/Au ohmic contacts were formed, followed by planar isolation using ion implantation. The main process of the gated Schottky drain and MIS-gate are as follows.

(1) 1st recess on both MIS-gate region and gated Schottky drain region: A recess-window-1 was defined and opened at the MIS-gate region and Schottky-drain region simultaneously, followed by the 1st recess (barrier layer recess) using digital etch which featured plasma oxidation and oxide removal by

HCl solution [14]. Since the total barrier thickness of the heterojunction is 21.5 nm, the 22-nm recess depth of 1st recess obtained by AFM (Fig. 2 (a)) indicates a totally removal of the barrier layer and a 0.5-nm over-etch on the upper GaN channel layer (which is $t_1 = 6$ nm).

(2) Formation of Schottky-drain contact and MIS field plate: An AlN/Al$_2$O$_3$ (1/15 nm) dielectric stack was deposited in sequence by PEALD, followed by a post-deposition annealing with 500 °C oxygen ambient. This 1-nm AlN was deposited to provide a normally-ON channel below MIS field plate. Next, a recess-window-2 was defined and opened inside the recess-window-1 of Schottky-drain region by diluted KOH. The 2nd deep recess is performed at the Schottky contact region (recess-window-2) by Cl$_2$/BCl$_3$ mixture plasma using ICP-RIE (inductively coupled plasma reactive ion etching). The 2nd recess cut through the lower channel, showing a 10-nm over-etch of lower GaN channel layer (Fig. 2 (b)). Ni/Au metal stack was then deposited and patterned at Schottky-drain region (covering the entire recess-window-1 on the drain side), forming a metal-2DEG Schottky contact and a MIS field plate simultaneously. Since the lower heterojunction channel is separated from the etched interface, a high conductivity channel with high electron mobility under MIS field plate is achieved.

(3) Formation of MIS-gate and contacting pads: An AlN/Al$_2$O$_3$ dielectric stack in MIS-gate region was also removed to get a normally-Off device (1-nm AlN introduces a net positive charge). A channel cleaning was performed by digital etch at the MIS-gate region, and then surface plasma nitridation was performed, followed by 20-nm PEALD grown Al$_2$O$_3$ gate dielectric deposition. Finally, gate electrode and contacting pads were evaporated and patterned.

III. RESULT AND DISCUSSION

Device simulation (Fig. 3) reveals that the MIS field plate (on the 1st recess stair) can pinch off the channel and shield the metal-2DEG Schottky junction from high electric field at reverse drain bias. This effective technique of reverse leakage suppression is confirmed by the low I_R of -20 nA/mm measured at $V_{DS} = -100$ V (Fig. 5 (d)).

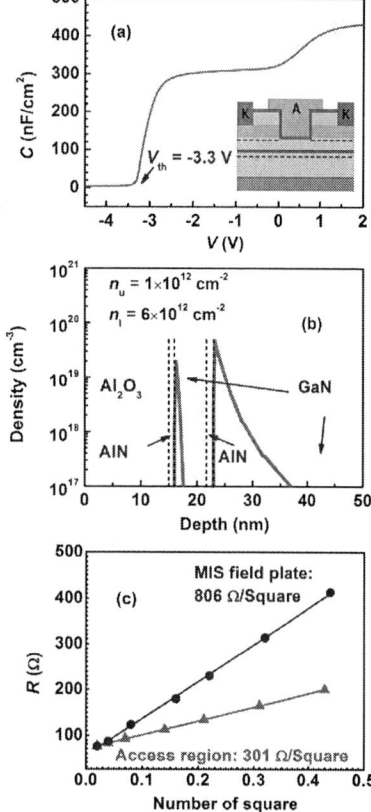

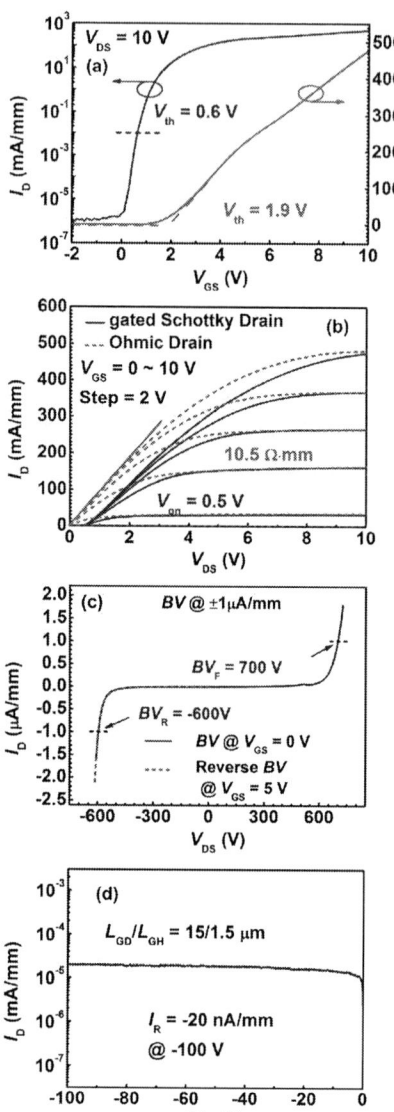

Fig. 4. (a) *C-V* characteristics of MIS field plate measured on a MIS-diode. (b) Simulated carrier distribution below MIS field plate when the device is at forward ON-state (V_{DS} = 5 V). (c) Sheet resistance beneath the field plate and gate to drain access region. The sheet resistance is measured by transfer length method.

The *C-V* characteristic of MIS field plate shows a V_{th} of −3.3 V and this negative threshold voltage is created by inserting a 1-nm PEALD-AlN with net positive charge under a 15-nm ALD-Al$_2$O$_3$ (Fig. 4 (a)). The simulated carrier distribution below the MIS field plate shows a high carrier density concentrating at the lower channel when the device is at forward ON-state (Fig. 4 (b)). Since the lower channel is spatially separated from the etched surface, a high-conductivity channel (below the MIS field plate) with a sheet resistance of 806 Ω/Square is achieved (Fig. 4 (c)). Although this resistance is larger than that of the gate-to-drain access region (~301 Ω/Square), the MIS-gated section still exhibits low resistance with a fully recessed AlGaN barrier layer. With a length of 1.5 μm, the MIS field plate and gate-to-drain access region presents a resistance of 1.2 Ω·mm and 0.45 Ω·mm respectively, indicating additional resistance of 0.75 Ω·mm is introduced at the Schottky drain region of the RB-MIS-HEMT.

The transfer, output (I_D-V_D), forward and reverse Off-state *I-V* curves of an RB-MIS-HEMT are plotted in Fig. 5. The forward I_D-V_D characteristic of a MIS-HEMT with a conventional Ohmic drain was also measured for comparison.

Fig. 5. *I-V* characteristics of a RB-MIS-HEMT with a gate-to-drain distance (L_{GD}) of 15 μm. (a) Transfer curve in linear and log scale. (b) Forward I_D-V_D curves. The device shows a low V_{on} of 0.5 V. The ON-resistance of the reference MIS-HEMT is 10.5 Ω·mm. The blocking curves in (c) show a BV_F (forward blocking) and BV_R (reverse blocking) of 700 V and −600 V with a grounded substrate. (d) Reverse leakage current in log scale.

The RB-MIS-HEMT exhibits a V_{th} of 0.6 V at a drain leakage current of 10 μA/mm and 1.9 V from linear exploration. A low V_{on} of 0.5 V (at 1 mA/mm) is achieved by employing a metal-2DEG contact. The device exhibits a forward (and reverse) blocking voltage of 700 V (and −600 V) at ±1 μA/mm with a grounded substrate. A low reverse Off-state leakage current of −20 nA/mm is achieved at −100 V. The RB-MIS-HEMT shows a slightly higher differential ON-resistance (~11 Ω·mm) than that of the conventional MIS-HEMT (~10.5 Ω·mm), as the MIS

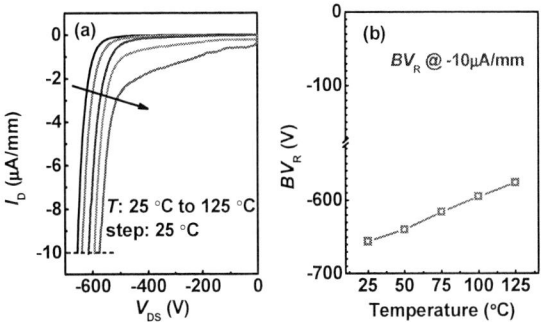

Fig. 6. (a) Temperature dependent reverse blocking characteristics. The reverse leakage current increases with temperature increasing. The reverse breakdown voltage decreases in magnitude from −656 V to −576 V with temperature increasing from 25 °C to 125 °C, determined at a drain current criterion of −10 μA/mm as shown in (b).

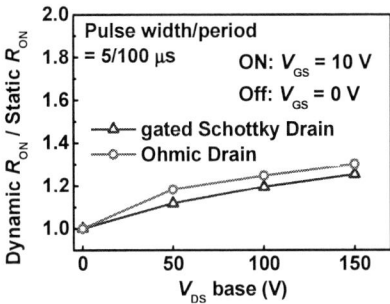

Fig. 7. Dynamic performance of MIS-HEMT with Schottky drain (RB-MIS-HEMT) and with Ohmic drain. The RB-MIS-HEMT shows smaller dynamic / static R_{ON} ratio. The R_{ON} of RB-MIS-HEMT is defined at I_D of 100 mA/mm.

field plate at Schottky drain region results in additional resistance.

The junction temperature of GaN power switches and rectifiers could easily reach 100 °C during circuit operation, thus it is important to characterize the devices at high temperature. The reverse blocking voltage at a leakage current of −10 μA/mm decreases in magnitude from −656 V to −576 V with the temperature rising from 25 °C to 125 °C , as shown in Fig. 6.

Pulsed I-V characteristics are measured with an AMCAD high-voltage high-speed system. With a pulse period of 100 μs and a pulse width of 5 μs, the dynamic R_{ON} of RB-MIS-HEMT shows a 25% increase compared to the static R_{ON} when the device is switched from an Off-state drain bias of 150 V. The dynamic performance of a MIS-HEMT is also shown for comparison and the RB-MIS-HEMT shows smaller dynamic/static R_{ON} ratio than MIS-HEMT (Fig. 7).

IV. CONCLUSIONS

A normally-Off reverse-blocking GaN MIS-HEMT with double-recessed gated Schottky drain has been successfully demonstrated on a double-channel heterojunction platform, enabling high-voltage blocking capability at both reverse and forward Off-state. The recessed metal-2DEG Schottky contact enables a low turn-on voltage. The MIS-gated section (i.e. MIS field plate) of the drain electrode plays the role of shielding the Schottky junction from large electric field and enabling low leakage current at reverse Off-state. In addition, as a high-mobility heterojunction lower channel is preserved under this MIS field plate, a low resistance can be obtained in the MIS-gated Schottky drain, leading to a low overall ON-resistance.

REFERENCES

[1] K. J. Chen, O. Häberlen, A. Lidow, C. l Tsai, T. Ueda, Y. Uemoto, Y. Wu, "GaN-on-Si Power Technology: Devices and Applications," *IEEE Trans. Electron Devices*, vol. 64, no. 3, pp. 779–795, Mar. 2017.

[2] P. W. Wheeler, J. Rodriguez, J. C. Clare, L. Empringham, and A. Weinstein, "Matrix converters: a technology review," *IEEE Trans. Ind. Electron.*, vol. 49, no. 2, pp. 276–288, Apr. 2002.

[3] E. Bahat-Treidel, R. Lossy, J. Wurfl, and G. Trankle, "AlGaN/GaN HEMT With Integrated Recessed Schottky-Drain Protection Diode," *IEEE Electron Device Lett.*, vol. 30, no. 9, pp. 901–903, Sep. 2009.

[4] T. Morita, M.Yanagihara, H. Ishida, M. Hikita, K. Kaibara, H. Matsuo, Y. Uemoto, T. Ueda, T. Tanaka, D. Ueda, "650 V 3.1 mohm·cm² GaN-based monolithic bidirectional switch using normally-off gate injection transistor," in *IEDM Tech. Dig., Dec. 2007*, pp. 865–868.

[5] T. Morita, S. Tamura, Y. Anda, M. Ishida, Y. Uemoto, T. Ueda, T. Tanaka, D. Ueda, "99.3% Efficiency of three-phase inverter for motor drive using GaN-based Gate Injection Transistors," in *Proc. IEEE Appl. Power Electron. Conf. Expo. (APEC)*, Mar. 2011, pp. 481–484.

[6] Y. Shi, W. Chen, C. Liu, G. Hu, J Liu, X. Cui, H. Tao, J. Zhang, Y. Shi, A. Zhang, Z. Li, Q. Zhou, B. Zhang, "A high-performance GaN E-mode reverse blocking MISHEMT with MIS field effect drain for bidirectional switch," in *Proc. IEEE Int. Symp. Power Semiconductor Device ICs (ISPSD)*, May 2017, pp. 207–210.

[7] C. Zhou, W. Chen, E. L. Piner, and K. J. Chen, "Schottky-Ohmic Drain AlGaN/GaN Normally Off HEMT With Reverse Drain Blocking Capability," *IEEE Electron Device Lett.*, vol. 31, no. 7, pp. 668–670, Jul. 2010.

[8] Y. W. Lian, Y. S. Lin, H. C. Lu, Y. C. Huang, and S. S. H. Hsu, "AlGaN/GaN HEMTs on Silicon With Hybrid Schottky-Ohmic Drain for High Breakdown Voltage and Low Leakage Current," *IEEE Electron Device Lett.*, vol. 33, no. 7, pp. 973–975, Jul. 2012.

[9] J. Lee, S. Han, B. Park, H. Cha, "Unidirectional AlGaN/GaN-on-Si HFETs with reverse blocking drain," *Appl. Phys. Express*, vol. 7, no. 1, p. 014101, Dec. 2013.

[10] J. Ma, M. Zhu, and E. Matioli, "900 V Reverse-Blocking GaN-on-Si MOSHEMTs with a Hybrid Tri-anode Schottky Drain," *IEEE Electron Device Lett.*, vol. PP, no. 99, pp. 1–1, 2017.

[11] J. Lei, J. Wei, G. Tang, Z. Zhang, Q. Qian, Z. Zheng, M. Hua, K. J. Chen, "650-V Double-Channel Lateral Schottky Barrier Diode With Dual-Recess Gated Anode," *IEEE Electron Device Lett.*, vol. 39, no. 2, pp. 260–263, Feb. 2018.

[12] J. Wei, S. Liu, B. Li, X. Tang, Y. Lu, C. Liu, M. Hua, Z. Zhang, G. Tang, K. J. Chen, "Enhancement-mode GaN double-channel MOS-HEMT with low on-resistance and robust gate recess," in *IEDM Tech. Dig., Dec. 2015*, p. 9.4.1-9.4.4.

[13] J. Wei, S. Liu, B. Li, X. Tang, Y. Lu, C. Liu, M. Hua, Z. Zhang, G. Tang, K. J. Chen, "Low On-Resistance Normally-Off GaN Double-Channel Metal-Oxide-Semiconductor High-Electron-Mobility Transistor," *IEEE Electron Device Lett.*, vol. 36, no. 12, pp. 1287–1290, Dec. 2015.

[14] S. Liu, S. Yang, Z. Tang, Q. Jiang, C. Liu, M. Wang, K. J. Chen, "Al₂O₃/AlN/GaN MOS-Channel-HEMTs With an AlN Interfacial Layer," *IEEE Electron Device Lett.*, vol. 35, no. 7, pp. 723–725, Jul. 2014.

978-1-5386-2928-4/18 $31.00 © 2018 IEEE

Proceedings of the 30th International Symposium on Power Semiconductor Devices & ICs
May 13-17, 2018, Chicago, USA

Recess-Free AlGaN/GaN Lateral Schottky Barrier Controlled Schottky Rectifier with Low Turn-on Voltage and High Reverse Blocking

Xuanwu Kang[1], Xinhua Wang[1], Sen Huang[1], Jinhan Zhang[1], Jie Fan[1], Shuo Yang[1], Yuankun Wang[1],
Yingkui Zheng[1], Ke Wei[1], Jin Zhi[1], Xinyu Liu[1]
[1]Key Laboratory of Microelectronic Devices & Integrated Technology
Institute of Microelectronics of Chinese Academy of Sciences, Beijing 100029, China
Email: kangxuanwu@ime.ac.cn; liuxinyu@ime.ac.cn

Abstract—High-performance AlGaN/GaN-on-Si diodes are fabricated with lateral Schottky barrier controlled Schottky rectifier (LSBS) on thin-barrier (5nm) AlGaN/GaN heterostructures, which features a recess-free process, enabling better electrostatic control to pinch off the channel under the anode region. In this way, low leakage currents (3-orders of magnitude lower than conventional recessed Schottky-barrier-diode (SBD)) and a high reverse breakdown voltage of 1700 V (@10μA/mm) are reached, together with a low onset voltage of only 0.37 V and a record low on-state resistance $R_{on,sp}$ of 1 mΩ·cm^2 for GaN-based SBDs with an Anode-to-Cathode distance (L_{AC}) of 10 μm. This is attributed to the combination of an effectively preserved 2DEG by LPCVD SiN$_x$ passivation and a hybrid Schottky /Ohmic anode. Thanks to the recess free technology and low damage in the Schottky region, stable preliminary HTRB performance are obtained. The proposed diode fabrication is compatible with GaN depletion/enhancement MIS-HEMTs process flow, enabling integration in the promising smart GaN platform.

Keywords—*AlGaN/GaN, Schottky-barrier-diode, SBD, Lateral, high breakdown voltage, low turn-on voltage*

I. INTRODUCTION

GaN-based Schottky diodes are of great importance in power switching applications, as they offer fast switching speed, high electric field breakdown strength, and good thermal properties [1]–[2]. The next-generation high-efficiency power systems require diodes with a low turn on voltage, a high breakdown voltage, and a fast switching speed. Various structures have been proposed to reduce the turn-on voltage of AlGaN/GaN SBDs, e.g. a recessed Schottky [2]-[3], a lateral field-effect rectifier employing fluorine plasma treatment [4]. In addition to that, a gated ohmic anode rectifier [5], a MIS-Gated hybrid anode rectifier [6] and gate edge terminated Schottky diodes [3][7] are designed to reduce the turn-on voltage and suppress the reverse leakage current. However, the solutions [2]-[6] proposed need extra Schottky barrier "treatment", e.g. recess or implantation, which might introduce

Schottky barrier damage and further result in potential reliability issues[8][9].

To get rid of extra Schottky barrier "treatment", an recess-free lateral Schottky barrier controlled Schottky rectifier (LSBS) has been proposed and fabricated on thin-barrier AlGaN/GaN heterostructures, with two-dimensional electron gas (2DEG) being effectively preserved by SiN$_x$ passivation grown by low-pressure chemical-vapor-deposition (LPCVD). The fabricated LSBSs exhibit low onset voltage, high reverse blocking and good stability.

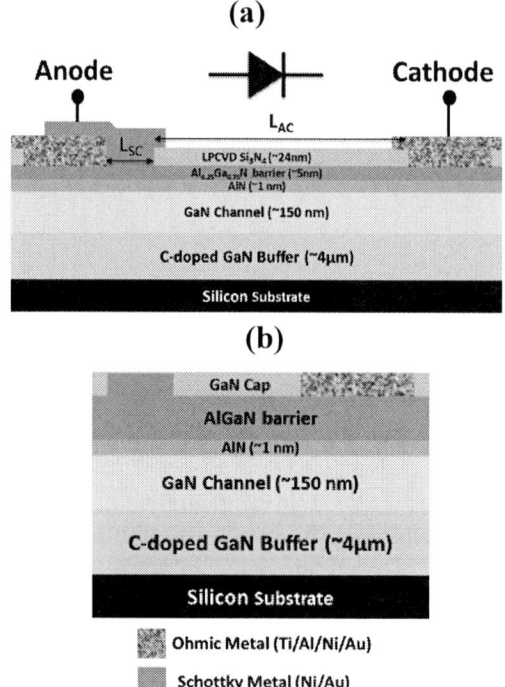

Fig. 1. The schematic cross-section of the recess free LSBSs and epi structure (Al$_{0.25}$Ga$_{0.75}$N barrier is 5nm), where, L_{SC} and L_{AC} are the length of the Schottky contact region and the drift length of the LSBS, respectively.

This work was supported in part by the National Key R&D Program of China (No. 2016YFB0400100, No. 2017YFB0403000), in part by Natural Science Foundation of China (No. 61334002, 61404163, 61474138, 61534007, 61527816, and 11634002), in part by the Key Project of Chinese Academy of Sciences (No. QYZDB-SSW-JSC012 and Y7YT024002).

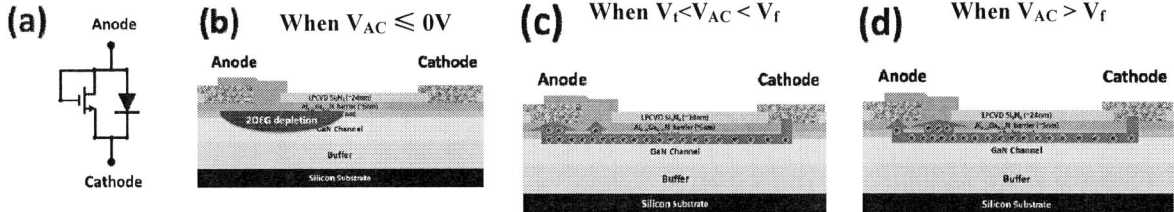

Fig. 2: (a) Equivalent circuit and operation mechanism of thin- barrier AlGaN/GaN recess-free LSBS. (b) Reverse bias ($V_{AC} \leq 0V$), (c) the dominant current path for small forward bias ($V_t < V_{AC} < V_f$), and (d) the dominant current path for large forward bias ($V_{AC} > V_f$) conditions.

II. TECHNOLOGY

The AlGaN/GaN heterostructure epitaxial wafer is a commercial product from Enkris Semiconductor, starting from a 4-inch Si<111> substrate, consists of a ~4 μm C-doped GaN buffer stack, a 150 nm GaN channel, 1-nm AlN interface enhancement layer and a 5-nm $Al_{0.25}Ga_{0.75}N$ barrier. With a 24-nm LPCVD-SiN_x passivation (780°C, with RCA pretreatment)[10], the sheet resistance is reduced from 1780 Ω/□ to 450 Ω/□. Meanwhile, conventional recessed barrier (with 5-nm $Al_{0.25}Ga_{0.75}N$ barrier remaining) SBDs on GaN-on-Si wafer were fabricated as reference.

The schematic cross-section of the recess-free LSBS and recessed SBD is shown in Fig. 1. (a) and Fig. 1. (b), respectively. The cathode electrode (C) of the LSBS is made of an electrode in Ohmic contact (Ti/Al/Ni/Au) with the 2DEG, while the anode electrode (A) is made of electrically shorted ohmic contact and Schottky contact, which is formed by opening the LPCVD SiN_x passivation with Fluorine-based ICP, followed by Ni/Au Schottky metal evaporation.

The equivalent circuit (Fig. 2. a) of the proposed LSBS is composed of a Schottky diode and a normally-off HEMT [5][11]. When the diode is under reverse biased condition ($V_{AC} \leq 0V$), the 2DEG channel under the Schottky contact region is effectively depleted, which is similar to an operation of normally-off thin-barrier HEMT fabricated in [11] (Fig. 2. b). When the diode is under forward biased condition ($V_{AC} \geq 0V$), the forward turn-on voltage (V_t, @1mA/mm) of the diode is determined by the pinching-off voltage of the channel instead of the on-voltage of Schottky (Fig. 2. c), while the forward on-state voltage (V_f, @100mA/mm) of the diode is determined by the combination of channel and Schottky (dual-conduction paths), which is the key feature of the LSBS to contribute to the low V_t and low V_f (Fig. 2. d).

III. DEVICE CHARACTERISTICS & DISCUSSION

On-wafer DC characterization is carried out on single-finger diodes with anode finger width of 50 μm and on power diodes with total anode finger width of 20 mm. The measured devices have anode-cathode length (L_{AC}) of 10 μm and Schottky junction length (L_{SC}) of 1 μm for both recess-free LSBS and recessed SBD. The measured I–V characteristics of small diodes, as a function of anode-to-cathode voltage V_{AC} is

shown in Fig. 3. The recessed SBD shows high reverse leakage, while the recess-free LSBS reverse leakage current is 3-order of magnitude lower (Fig. 3, left). This leakage reduction can be ascribed to the damage free and good pinch-off characteristics of the thin-barrier [11], which blocks the leakage current in reverse bias. Forward IV characteristics is shown in Fig. 3 (right), where the recess-free LSBS exhibits lower turn-on behavior than the SBD.

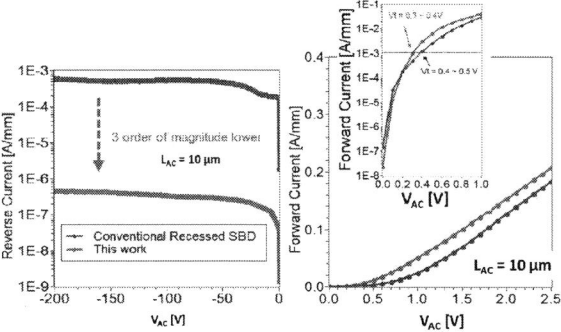

Fig. 2. Typical reverse (left) and forward (right) curves of the recess-free LSBS and the conventional recessed SBD, both with an anode finger width of 50 μm, $L_{SC} = 1$ μm and $L_{AC} = 10$ μm. Red and gray lines: characteristics of recess-free LSBSs and conventional recessed SBDs, respectively.

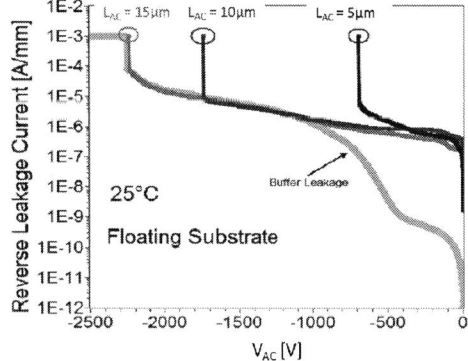

Fig. 3: Typical reverse leakage and breakdown characteristics of the recess-free LSBSs and buffer isolation structure (15 μm spacing) with a variation of anode-to-cathode spacing L_{AC}. The final hard breakdown of devices is limited by the buffer V_{bd}. The device dimensions are $L_{SC}/W=1/50$ μm.

Typical leakage and breakdown characteristics of recess-free LSBS (50 µm width) with varying L_{AC} and buffer isolation structure (15 µm space) is evaluated at room temperature. The results, from which low leakage is evident for the recess-free LSBS (~1 µA/mm at −600 V for L_{AC} = 5, 10 and 15 µm) are shown in Fig. 4. Hard breakdown of 1700 V is reached with L_{AC} = 10 µm.

The temperature dependence of reverse and forward characteristics is assessed. As shown in Fig. 5, a temperature increase from 25 °C to 150 °C results into an increase of leakage by about one order of magnitude. Meanwhile, a positive temperature coefficient is observed on V_f, and the "Thermal stable point" of forward current is at a very low current level, resulting in less "thermal runaway" problems [12].

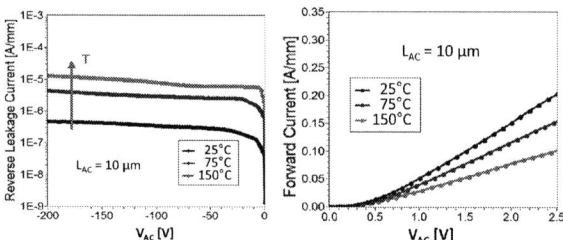

Fig. 5. Reverse leakage current (left) and forward (right) characteristics for the recess-free LSBS at different temperatures. The device dimensions are L_{AC} /L_{SC} /W=10/1/50 µm.

The turn-on voltage V_t (1 mA/mm) and on-state voltage V_f (100 mA/mm) with different L_{AC} are extracted by statistical measurement (Fig. 6 and Fig. 7) and resulted to be 0.37 V and 1.5 V for L_{AC} = 10 µm, respectively. The turn-on voltage (V_t) is comparable with [2][3][5], where a value of ~0.4 V has been demonstrated. Typical forward I–V characteristics after 100s reverse stress for different quasi-bias conditions of recess-free LSBS (L_{AC} = 10µm) is shown in Fig. 8. The device shows relatively stable forward characteristics without total current

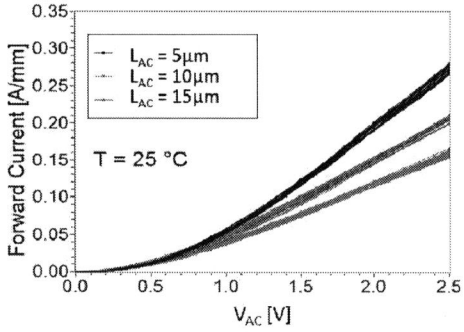

Fig. 6: Forward characteristics for the recess-free LSBSs with different L_{AC} at room temperature (20 recess-free LSBSs for each different L_{AC}.). The current is tightly distributed with varying L_{AC}.

collapse problem (within 30% dynamic R_{on} increase at 200V). Preliminary HTRB at -200V and 150 °C has been carried out on devices (50 µm width) with L_{AC} = 10 µm (Fig. 9). Recess-free LSBS at V_{AC} = -200 V are stable over time at 150 °C, yet the recessed SBDs show earlier devices failure.

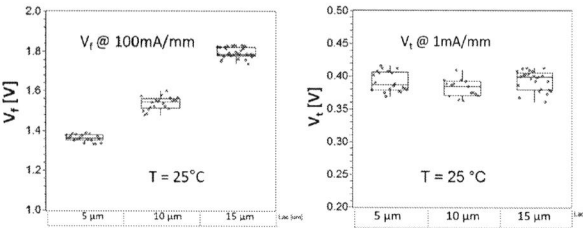

Fig. 7: Statistics on V_f and V_t of 20 recess-free LSBS for each different L_{AC} at room temperature (W = 50 µm). Both V_f and V_t are tightly distributed. V_f increases with L_{AC} and V_t is independent of L_{AC}.

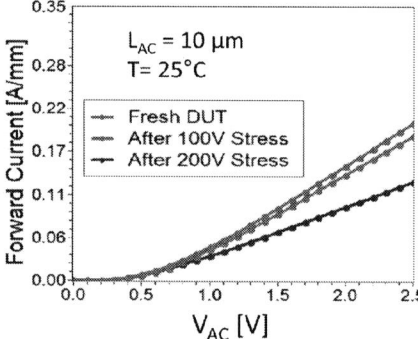

Fig. 8. Typical forward I–V characteristics after 100s reverse stress for different quasi-bias conditions of recess-free LSBS (L_{AC} = 10 µm, W =50 µm). The device shows relatively stable forward characteristics without total current collapse.

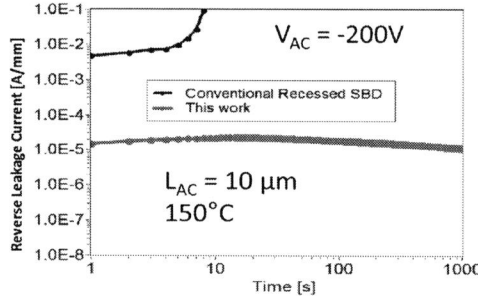

Fig. 9. HTRB performance of recess-free LSBS at V_{AC} = -200 V are stable over time at 150 °C, while the conventional recessed SBD shows much earlier devices failure. The device dimensions are L_{AC} /L_{SC} /W=10/1/50 µm.

978-1-5386-2928-4/18 $31.00 © 2018 IEEE

Multi-finger recess-free LSBS power diodes with 20-mm anode width are characterized at room temperature (Fig. 10). Low leakage (10 μA@-650V), hard breakdown of 800V (Fig. 10, left) and forward current of 2A@1.5V (Fig. 10, right) is obtained. The junction capacitance of the recess-free LSBS power diodes is 20% of commercial SiC JBS (C3D02065E [13]) with identical current and voltage rating, enabling fast switching and reduction of power loss for DC-DC converters.

preserved 2DEG by LPCVD Si_3N_4 passivation and a hybrid Schottky /Ohmic anode, a low onset voltage of only 0.37 V , a record low specific on-resistance $R_{on,sp}$ of 1 mΩ·cm², a high reverse breakdown voltage of 1700 V (@10μA/mm), and preliminary stable HTRB performance are obtained. The proposed diode fabrication is compatible with GaN depletion/enhancement MIS-HEMTs process flow, enabling integration in the promising smart GaN platform.

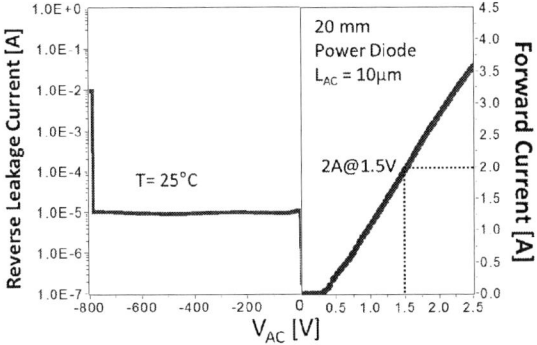

Fig. 10. Reverse leakage current and forward characteristics of power LSBS with total anode finger width of 20 mm and L_{AC} =10 um.

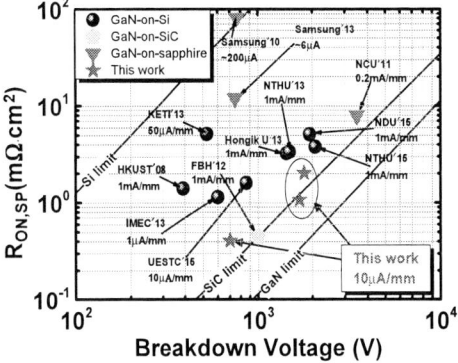

Fig. 12. Benchmark of $R_{ON,SP}$ vs. BV of GaN diode on SiC/sapphire/Si substrates. The recess-free LSBS show a record low on-state resistance $R_{on,sp}$ of 1 mΩ·cm² for GaN-based SBDs with the BV at 1700V with L_{AC} = 10 μm.

ACKNOWLEDGMENT

The authors would like to thank Dr. S. You from IMEC, Belgium, Dr. Q. Zhou from UESTC, Dr. M. Wang and M. Tao from PKU and Dr. K. Cheng from Enkris Semiconductor for the fundamental devices physics discussion.

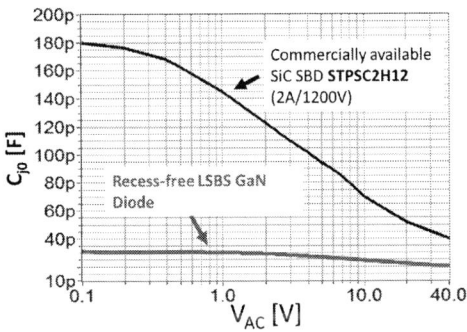

Fig. 11. The capacitances of power LSBS measured at 1MHz is significantly lower than the identical voltage and current rated commercial SiC SBD (Cree C3D02065E). The device dimensions are $L_{AC}/L_{SC}/W$=10μm /1μm /20mm.

Fig. 12 shows the $R_{ON,SP}$ vs. BV of the state-of-the-art GaN diode on different substrates. The recess-free LSBS with L_{AC} of 10 μm delivers a BV of 1700 V with corresponding $R_{ON,SP}$ of 1 mΩ·cm². This value is among the best results reported for GaN-on-Si power diode at reverse leakage as low as 10 μA/mm.

IV.   CONCLUSION

High-performance lateral Schottky barrier controlled Schottky rectifier (LSBS) is fabricated on thin-barrier AlGaN/GaN heterostructures. Combined with effectively

REFERENCES

[1] B. N. Shashikala et al., Int. J. Eng. Sci. Technol., vol. 2, no. 12, pp. 7586–7591, Feb. 2010.

[2] O. Hilt et al., *IEEE Electron Device Lett.* vol. 33, no. 3, pp. 2011–2013, 2012.

[3] S. Lenci et al., IEEE Electron Device Lett., vol. 34, no. 8, pp. 1035–1037, 2013.

[4] W. Chen, et al., *IEDM*, 2008, no. 852, pp. 1–4.

[5] J.-G. Lee et al., IEEE Electron Device Lett., vol. 34, no 2, pp 214–216, Feb. 2013.

[6] Q. Zhou et al., IEEE Electron Device Lett., vol. 36, no. 7, pp. 660–662, 2015.

[7] A. Kamada, K. Matsubayashi, A. Nakagawa, et al.,ISPSD 2008, pp. 225–228.

[8] S. K. Jha et al., Proc. COMMAD, 2004, pp. 33–36.

[9] C. Yi et al., IEDM 2007, no. 852, pp. 389–392.

[10] S. Huang et al., EDL, vol. 37, no. 13, pp. 1617-1620, Dec. 2016.

[11] J. Derluyn et al., Tech. Dig. – IEDM 2009, pp. 157–160, 2009.

[12] D. Wellekens et al., ESSDERC, 2012, pp. 302–305.

[13] http://www.wolfspeed.com/media/downloads/118/C3D02065E.pdf

Proceedings of the 30th International Symposium on Power Semiconductor Devices & ICs
May 13-17, 2018, Chicago, USA

An Industry-Ready 200 mm p-GaN E-mode GaN-on-Si power Technology

N.E. Posthuma, S. You, S. Stoffels, D. Wellekens, H. Liang, M. Zhao, B. De Jaeger, K. Geens, N. Ronchi and S. Decoutere
imec, Leuven, Belgium,
Email: niels.posthuma@imec.be

P. Moens, A. Banerjee, H. Ziad and M. Tack

ON Semiconductor, Oudenaarde, Belgium

Abstract— Enhancement mode 650V rated p-GaN gate HEMTs are fabricated on 200 mm p⁺ Si substrates by using an industrial, Au-free process. The devices show true e-mode performance, with a high V_t of 2.8 V, low off-state leakage current and are dynamic R_{DS-ON} free over the complete V_{DS} and temperature range. High temperature reverse bias (HTRB) testing is done on-wafer and after packaging. For the first time, 650 V e-mode power HEMTs realized on 200 mm Si substrates, show industry ready device performance and pass 1008 hour reliability testing, at V_{GS}=0 V, V_{DS}=650 V.

Keywords— *Power transistors, gallium-nitride, GaN-on-Si, HEMT, p-GaN gate, 650V, reliability, packaging, 200mm*

I. INTRODUCTION

GaN-on-Si technology is accepted as a breakthrough technology for power electronics [1]. The (Al)GaN material system has favorable material characteristics to obtain a high blocking voltage and a low on-resistance. Initial trapping effects and reliability issues have been solved and first commercial products have entered the market [2, 3, 4].

Normally-off power components, also called enhancement mode (e-mode) transistors, are preferred for fail-safe operation. To realize e-mode transistors, the gate region needs to be adapted to fully switch off the transistor at a gate voltage of 0 V. One of the approaches for e-mode operation is to use a Mg-doped p-type GaN (p-GaN) gate, lifting the conduction band in the channel under zero gate bias [5, 6, 7, 8].

This paper reports on an industrial process for 650 V rated p-GaN gate enhancement mode GaN-on-Si power devices on 200 mm substrates, see Fig. 1. To the authors knowledge this is the first time that true e-mode p-GaN 650 V power transistors are realized on 200 mm Si substrates, which are dynamic R_{DS-ON} free and pass 1008 hour reliability testing. Challenges of using 200 mm substrates are discussed in paragraph II. Device fabrication and the characterization method of the p-GaN HEMTs are described in the third section. Characterization results including static and dynamic measurements and reliability testing, both on wafer level and on packaged devices, will be discussed in detail in paragraph IV. The paper is concluded in the final section.

II. CHALLENGES FOR MANUFACTURING OF 200MM GAN-ON-SI SUBSTRATES

Although today's industry standard wafer size for GaN-on-Si is 150 mm, a transition to 200 mm is believed to happen in

the near future. With the newest generation MOCVD reactors entering the market, a substantial improvement in layer uniformity is expected, especially for 200 mm.

Fabrication of 650 V rated lateral GaN-on-Si power devices on 200 mm silicon substrates is challenging, especially due to the relatively thick GaN buffer stack that is needed for this voltage range. A first challenge is to obtain a wafer warp within ± 50 μm after deposition of the GaN epitaxial stack. By implementation of an innovative and proprietary stepped superlattice buffer concept, the wafer warp can be controlled within this limit for an epitaxial stack that also fulfills the breakdown voltage and dispersion requirements. A second challenge is to make the GaN-on-Si wafer mechanically strong. Optimization has been done by careful design of the epi stack and meticulous tuning of all growth parameters. Additionally a boron doped p⁺ Si (111) substrate (5 mΩ.cm) is used to provide a better mechanical strength [9] compared to typical medium doped substrates (1 to 10 Ω.cm). Prior to device fabrication a dedicated wafer screening procedure is done to remove the mechanically weak wafers and avoid unscheduled wafer breakage during processing. A third challenge is related to packaging of the devices. As final processing, prior to dicing and device packaging, back-side grinding is performed down to a wafer thickness of 200 μm followed by back-side metallization. Within the imec industrial affiliation program it is shown that, after process optimization, it is possible to perform back-side metallization on 200 μm thin 200 mm GaN-on-Si wafers.

Fig. 1. Photograph of a 200 mm GaN-on-Si wafer with p-GaN HEMT power transistors processed in imec's 200 mm pilot line using a Au-free process flow.

III. DEVICE FABRICATION AND CHARACTERIZATION METHOD

The GaN buffer is grown by metalorganic chemical vapor deposition (MOCVD) on 200 mm p⁺ boron-doped Si (111) substrates using a 3×200 mm Veeco Maxbright reactor. The total thickness of the GaN epitaxial layer, including the active

978-1-5386-2928-4/18 $31.00 © 2018 IEEE

layers, is limited to less than 5 μm. The p-GaN layer is grown in sequence with the GaN buffer, channel and barrier, and is Mg doped with a doping level of approximately 3×10^{19} at/cm^3. Mg activation is done in-situ during cool down after p-GaN layer growth.

Enhancement mode transistor processing is done in imec's 200 mm pilot line using a Au-free process flow. We consider a p-GaN gate with a Schottky contact realized by TiN metal deposited on the p-GaN layer, as illustrated in Fig 2, such to control the gate leakage current in on-state. First the TiN layer is etched, followed by a selective p-GaN to AlGaN dry etch using a BCl$_3$/SF$_6$ chemistry. The access area is passivated by a dielectric stack containing Al$_2$O$_3$ and SiO$_2$. A thick gate metal stack is used, patterned separately from the p-GaN layer, such that a low sheet-resistance of the gate metal is achieved and a gate metal field plate can be applied [10]. Source and drain ohmic contacts are realized by recess etching, cleaning, Ti/TiN/Al based metallization and low-temperature ohmic anneal [11]. The process flow is completed by back-end metallization modules, where SiO$_2$ is used for the inter-metal dielectric layers.

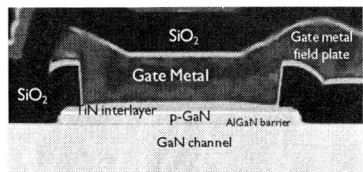

Fig. 2. TEM micrograph of the gate area of the p-GaN HEMT device, realized with separate patterning of the p-GaN layer and gate metal stack. In-between the p-GaN layer and the gate metal stack a TiN interlayer is applied.

In total 4 metal field plates can be applied from the source side to redistribute the electric field peaks, and improve the device breakdown voltage and the dynamic R$_{DS-ON}$. Various design variations, including variation of the field plate length, are implemented. Two different designs are presented in this work, namely a design with 3 metal field plate levels (C3) and a design with 4 metal field plate levels, but with slightly shorter field plates (C6). The gate drain spacing used for 650 V applications is 16 μm.

Electrical characterization of large-area power transistors with a gate width of W=36 mm is performed using a fully automated Eagle T200/FT test system, measuring the main static and dynamic device parameters on wafer-level. Additional static manual measurements are done using an Agilent B1505 analyzer and fast pulsed IV characterization is performed with an Auriga AU4850 system. After assessment of the main device parameters, initial wafer-level high temperature reverse bias (HTRB) reliability testing is done. Finally, devices are packaged in a surface mount QFN8x8 package with Kelvin connection, which are submitted to 1008 hour HTRB reliability testing.

IV. RESULTS AND DISCUSSION

A. Device performance

The device performance of the fabricated 36 mm enhancement mode power transistors is shown by the transfer characteristic and I$_D$V$_{GS}$ output characteristic in Fig. 3.

The AlGaN barrier layer thickness, Al percentage and p-GaN growth conditions are optimized to obtain the best trade-off between a high V$_t$, low on-resistance R$_{on}$ and stable off-state performance. For this condition the 2DEG sheet resistance (R$_{SH-2DEG}$) is around 650 Ohm/sq, see Fig 4(a). The devices are true enhancement mode power transistors with a threshold voltage V$_t$ = 2.8 V (extracted at maximum transconductance). The contact resistance of source and drain Ohmic contacts is as low as 0.35 Ohm.mm. The on-resistance R$_{on}$ at 25 °C is typically 17 Ohm.mm, as shown in Fig 4 (b).

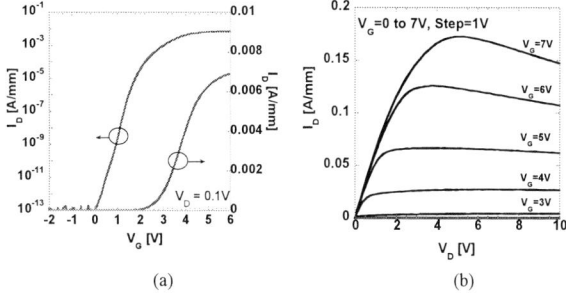

Fig. 3. (a) Transfer characteristic of a 36 mm power transistor at V$_{DS}$=0.1V, and I$_D$V$_{GS}$ output characteristic with V$_{GS}$ stepped from 0 to 7 V.

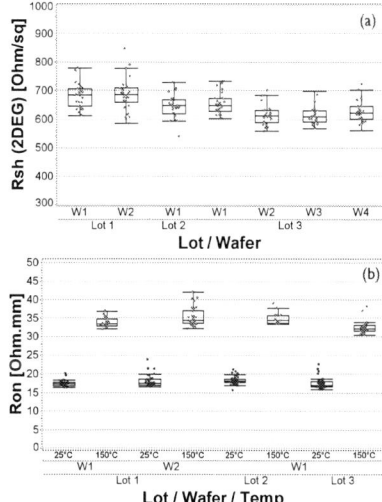

Fig. 4. Wafer-to-wafer variation of 2DEG R$_{SH}$ extracted from a Van der Pauw structure (a) and R$_{on}$ measured on 36 mm power transistors across different 200 mm GaN-on-Si wafers at 25 and 150 °C (b).

Off-state performance at V$_{GS}$ = 0 V is shown in Fig. 5. For enhancement mode devices, it is critical to have a sufficiently high threshold voltage, such that the channel is fully pinched-off and a low leakage level can be obtained at V$_{GS}$ = 0 V. The distribution of the off-state drain leakage current at V$_{DS}$ = 650 V (V$_{GS}$ = 0 V, 25 °C) across different wafers and field plate designs is shown in Fig. 5(a). The drain leakage current at room temperature is well below 100 nA/mm and shows a tight distribution. An example of the device breakdown characteristics for W=36 mm power transistors, measured until hard breakdown at both room temperature and 150 °C, is

978-1-5386-2928-4/18 $31.00 © 2018 IEEE

shown in Fig. 5(b). The hard breakdown voltage is above 1kV at 25 °C and close to 1.2kV at 150 °C. Device failure occurs between gate and drain.

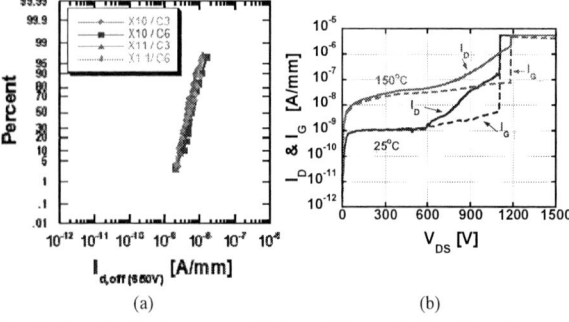

(a)   (b)

Fig. 5. $I_{D\text{-off}}$ distributions across different 200 mm GaN-on-Si wafers and designs for V_{DS}=650 V, V_{GS}=0 V at 25°C (a), and off-state drain and gate (I_D, I_G) leakage currents meausred until hard breakdown at 25 and 150 °C (b).

B. Pulsed IV characteristics

Two methods were applied to characterize the device dynamic $R_{DS\text{-}ON}$. An industrial tester with ms pulse width was used for fast screening of process conditions, design splits and verification of the distribution over the 200 mm wafer. A shorter pulse width in the micro-second range was applied using a manual tester. The results of both methods are presented below.

Industrial tester with ms pulse width

Automated dynamic $R_{DS\text{-}ON}$ measurements are done at 25 °C on an Eagle test system with V_{DS_q} varying from 0 to 650 V with a pulse duration of 47 ms off-state and 500 μs on-state. The dynamic $R_{DS\text{-}ON}$ of 2 wafers is presented in Fig. 6, showing a tight distribution over the 200 mm wafers. For both the field plate designs, the dynamic $R_{DS\text{-}ON}$ is less than ±5 %, showing the devices are dynamic $R_{DS\text{-}ON}$ free.

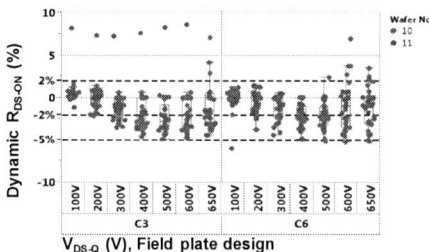

Fig. 6. Dynamic $R_{DS\text{-}ON}$ device dispersion of the p-GaN HEMTs for 2 field plate designs on 2 wafers, showing dynamic $R_{DS\text{-}ON}$ free devices thightly distributed over the 200 mm wafers. The data is on W=36 mm powerbars.

Manual tester with μs pulse width

Manual dynamic $R_{DS\text{-}ON}$ measurements were done using the Auriga AU4850 system with V_{DS_q} varying from 0 to 600 V with a pulse duration of 90 μs off-state and 10 μs on-state. Five devices were measured per wafer, distributed over the full wafer area. In Fig. 7, the normalized $R_{DS\text{-}ON}$ for two field plate configurations at 25 and 150 °C are given. The devices have a normalized dynamic $R_{DS\text{-}ON}$ level well below 20 percent, with a

tight distribution. Similarly as for the ms pulse condition, for the fast pulse condition the devices are dynamic $R_{DS\text{-}ON}$ free. The two field plate configurations perform very similar in terms of dynamic $R_{DS\text{-}ON}$ behavior.

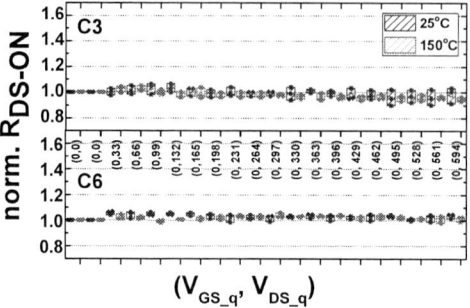

Fig. 7. Normalized dynamic $R_{DS\text{-}ON}$ for the voltage range of $V_{DS\text{-}q}$ between 0 and 600 V and field plate designs C3 and C6, measured with the Auriga AU4850. The data is on W=36 mm power transistors.

C. HTRB reliability

High temperature reverse bias (HTRB) reliability of the p-GaN HEMT devices was assessed first on wafer-level. Subsequently devices were packaged and subjected to a 1008 hour HTRB test. The results of both tests are described below.

HTRB results on wafer level

Wafer-level HTRB tests were done on a Keithley system at 80 percent of the rated voltage (V_{DS}=520 V), during 2 hours at 150 °C and V_{GS}=0 V, measuring 9 devices per wafer. The device parameters were measured before and after the stress at T=25 °C, and were continuously monitored during the stress at T=150 °C, see Fig. 8 for the off-state leakage at the stress condition (V_{GS}=0 V, V_{DS}=520 V). The drain leakage currents pre- and post-HTRB stress at 25 °C are indicated by "0" and "1", respectively. The drain leakage current during stress at T=150 °C is shown as a function of the stress time. The leakage current decreases during the stress, an effect that is retained at T=25 °C (distribution reduced from ~300nA to ~100nA). The distributions across the wafer and between wafers is tight.

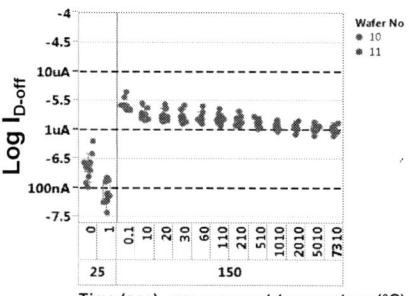

Fig. 8. Evolution of the drain leakage current during wafer-level HTRB stress testing at V_{DS} = 520 V and 150 °C during 2 hours, pre- ("0") and post ("1") measurements at 25 °C, indicated on the left-hand side of the figure, show stable drain leakage current before and after HTRB testing (devices with W=36 mm).

978-1-5386-2928-4/18 $31.00 © 2018 IEEE

HTRB results packaged devices

Long term HTRB testing at T=150 °C was done on packaged power transistors (W=36 mm), for devices with the two different field plate designs. The transistor parameters were measured at T=25 °C after 0, 24, 96, 168, 504 and 1008 hours of HTRB stress at T=150 °C. First a 1008 hour HTRB test was performed at 80 percent of the rated voltage (V_{DS}=520 V) and V_{GS}=0V. Since all devices passed the 1008 hours HTRB test at V_{DS} 520 V, the same set of packaged devices were also subjected to a second 1008 hours HTRB test at V_{DS} of 650 V.

Fig. 9 shows the $R_{on} \cdot W$ of the packaged devices during the HTRB stress, for a drain-to-source voltage of both 520 and 650 V. Fig. 9(a) shows the performance of the C3 design and Fig. 9(b) gives the data for the C6 design. These graphs indicate a very stable device performance during HTRB testing. All the devices with the C3 design passed the 1008 hour test for both a V_{DS} = 520 and 650 V. For the other design (C6) all the devices passed the test at 520 V, but several devices broke during the test after 168 hours at the increased drain-to-source voltage level of 650 V. These results indicate that the devices with the longer, 3-level field plate design are more reliable during the 1008 hour HTRB testing, compared to the design with shorter field plates.

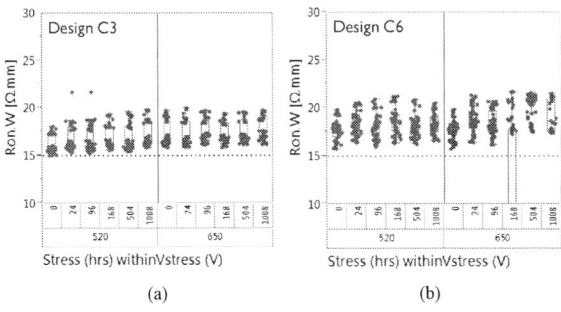

(a)　　　　　　　　(b)

Fig. 9. $R_{on}W$ as a function of V_{DS} measured at intermediate times during the 1008 hour HTRB tests performed at V_{DS}=520 V and V_{DS}=650 V. The left graph (a) shows the data for the C3 field plate design whereas the right graph (b) the C6 design. Note the dropout of the C6 devices during the 650V HTRB test, after 168 hours.

The pinch-off voltage, extracted at 1 µA/mm, measured during HTRB testing of the packaged devices is shown in Fig. 10. The different colors indicate the $V_{th\_1\mu A/mm}$ measured at various times during the HTRB stress. The threshold voltage of the devices is constant for both field plate configurations during the 1008 hour HTRB test.

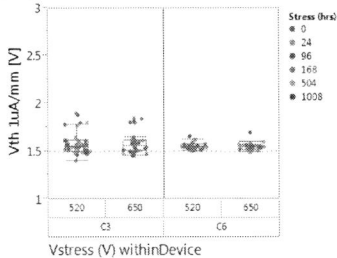

Fig. 10. Threshold voltage of the packaged devices during the 1008 hour HTRB test performed at V_{DS} of 520 and 650 V for 2 field plate designs.

V. CONCLUSION

650V rated enhancement mode p-GaN gate HEMTs are fabricated in imec's pilot line on 200 mm p$^+$ Si substrates by implementation of an industrial, Au-free process. Using 200 mm Si substrates is beneficial for cost reduction, but is challenging, as besides the electrical device performance and reliability, the wafer warp and mechanical wafer strength needs to be well controlled. Key aspects for device performance are the buffer and p-GaN layer growth, gate formation, passivation of the access area and field plate design.

True enhancement mode power transistors are fabricated with a low contact resistance of 0.35 Ω.mm and a V_t of 2.8 V. The drain leakage current at room temperature is well below 1 µA/mm and the device hard breakdown voltage is above 1kV. The devices are dynamic R_{DS-ON} free (less than 20% at T=25°C and 150 °C up to 600V). They show stable device performance without parametric shift during wafer-level HTRB testing. Devices assembled in QFN8x8 packages were subjected to 1008 hour HTRB testing at 520 V, followed by testing at a V_{DS} of 650 V. For the best field plate design all prime device parameters including dynamic R_{DS-ON} did not shift more than 10% after HTRB testing. To the author's knowledge, it is for the first time that 650 V, e-mode p-GaN power transistors are realized on 200 mm GaN-on-Si substrates, with excellent device performance and passing 1008 hour HTRB testing.

ACKNOWLEDGMENT

The authors thank the imec industrial affiliation program partners for their contribution, as well as Filip Bogman, Denise Thienpont and Mario Deblock (ON Semiconductor) for their assistance in the assembly and reliability testing.

REFERENCES

[1] K. Chen, "GaN-on-Si Power Technology: Devices and Applications", IEEE TED VOL. 64, NO. 3 (2017), pp. 779-795.

[2] GaN Systems (2017) GS66502B 650V [Online] http://www.gansystems.com/datasheets/GS66502B%20DS%20Rev%20 171101.pdf

[3] EPC (2017) EPC2019 200V [Online]: http://epc-co.com/epc/Portals/0/epc/documents/datasheets/EPC2019_datasheet.pdf

[4] Panasonic (2017) PGA26E07BA 600 V [Online] https://industrial.panasonic.com/ww/products/semiconductors/powerics/ganpower

[5] O Hilt, "Normally-off GaN Transistors for Power Applications" Journal of Physics: Conference Series 494 (2014), 012001

[6] J. Kim, "High Threshold Voltage p-GaN Gate Power Devices on 200 mm Si", ISPSD 2013, pp. 315

[7] N.E.. Posthuma, "Impact of Mg out-diffusion and activation on the p-GaN gate HEMT device performance" ISPSD 2016, pp. 95

[8] Uemoto, "Gate Injection Transistor (GIT) - A Normally-Off AlGaN/GaN Power Transistor Using Conductivity Modulation", IEEE TED (2007), 54(12), pp. 3393–3399.

[9] S. Stoffels, "Next generation 200mm substrates for GaN power devices", 41st Workshop on Compound Semiconductor Devices and Integrated Circuits Held in Europe: May 21-24, 2017, Gran Canaria, Spain.

[10] N.E. Posthuma, "Gate architecture design for enhancement mode p-GaN gate HEMTs for 200 and 650V applications", ISPSD 2018, this conference.

[11] Patent US 9,634,107 B2

Proceedings of the 30th International Symposium on Power Semiconductor Devices & ICs
May 13-17, 2018, Chicago, USA

Application-driven device/circuit co-simulation framework for power MOSFET design and technology development

Tirthajyoti Sarkar, Ashok Challa
Power Solutions Group
ON Semiconductor,
Sunnyvale, CA, USA

Kirk Huang, Prasad Venkatraman
Power Solutions Group
ON Semiconductor,
Phoenix, AZ, USA

Dean Probst
Corporate R&D, Process Technology
ON Semiconductor,
Gresham, OR, US

Abstract— **Physics-based device-circuit co-simulation turns out to be an extremely valuable tool to guide and optimize process technology development and device design for high-performance power MOSFETs. It is particularly well-suited to the need of emerging trend of integration of discrete power devices and analog ICs in the same package. In this article, we outline the key features and benefits of such an integrated design framework.**

Keywords— Power MOSFET, circuit simulation, device design, low-voltage, DrMOS.

I. INTRODUCTION

Power MOSFET technology [1] continues to be the key enabler for a multitude of high-performance switching power management systems, employed in diverse applications such as – mobile and ultra-portable gadgets, cloud and internet-of-things (IoT) infrastructure, personal computing, industrial automation, consumer appliances, and automotive electronics (Fig. 1). Ever increasing demand to deliver better energy efficiency and cost effectiveness depends critically on the progress of power MOSFET technology, which are capable of high-frequency switching, have low on-state resistance, and extends reliability [2]. However, traditional technology indicators are falling short of describing the dynamics of high-frequency, power-dense applications employing power MOSFETs.

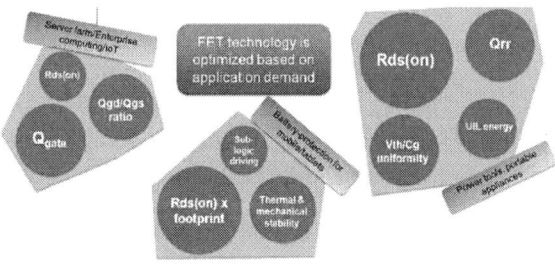

Fig. 1: Multi-domain need for Power MOSFET optimization.

On the other hand, customers are demanding more value/functions in a single package/foot-print. Single MOSFET products are giving way to duals, multi-chip-modules (MCMs), and smart-power-stages (SPS). In this respect, co-packaging of drivers, controllers, and passives is the next emerging trend. This necessitates FET technology design to be a synergistic effort along with other elements in the package/system. Therefore, a holistic simulation framework, which can address the challenges of real-world applications, must be established to develop MOSFET technology. MOSFET technology development, therefore, can no longer be accomplished without guidance from device-circuit co-simulation, which captures this complex interplay at multiple levels (Fig. 2), and optimizes the design towards a unified goal of higher efficiency/reliability.

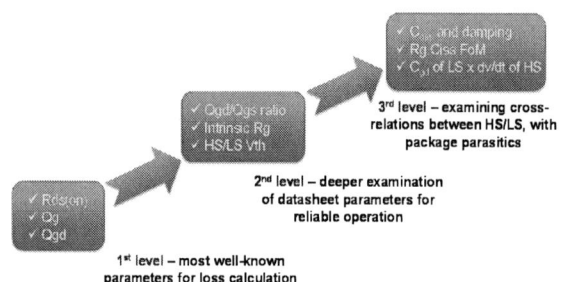

Fig. 2: Various levels of design optimizations for power MOSFET.

II. DEVICE-CIRCUIT CO-SIMULATION APPROACH

A. TCAD Physics-based Models at the Core

At its core, the mixed-mode approach relies on finite element TCAD models for the power device. Overall, the simulator solves fundamental semiconductor physics equations (carrier continuity equations in a drift-diffusion framework along with Poisson's equations and various mobility and recombination models) within a framework of circuit laws in a tightly coupled manner. This ensures a high degree of accuracy in situations where complex device dynamics impacts the circuit performance and where the usual approach of PSPICE modeling fails to capture that dynamics in sufficient detail. The idea is shown in Fig. 3.

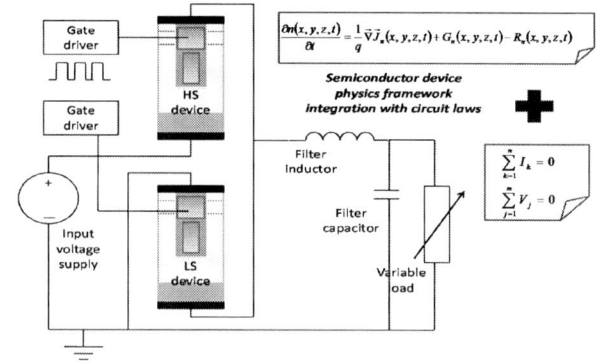

Fig. 3: Basic idea of Mixed-mode device/circuit co-simulation.

978-1-5386-2928-4/18 $31.00 © 2018 IEEE

Table I shows the relative merits and limitations of the Mixed Mode approach as compared to traditional PSPICE equivalent circuit modeling approach [3]. It is to be noted that the mixed-mode approach is particularly suited in the technology development stage, where small but significant changes in the process parameters yield different models that can impact circuit dynamics but which are hard to capture through tuning PSPICE parameters.

TABLE I: PSPICE AND MIXED MODE MODEL

Traditional PSPICE modeling	Mixed Mode Platform
Fast circuit simulation. Many switching cycles can be simulated i.e. circuit dynamics can be studied.	Slower simulation. Limited number of switching cycles can be simulated. Steady-state to be assumed.
Model accuracy often limited. Body-diode/ C-V models are generally inadequate.	No limitation on device model accuracy. Complete semiconductor physics is used. Body-diode, nonlinear C-V are modeled most accurately.
Need to generate fresh model for every iteration of die design/layout and process technology iteration	Process corner modeling is easy as every small change in device feature can be simulated with same level of accuracy.

B. Increasing Layers of Complexity added as Necessary

Even an apparently simple circuit topology like nonisolated synchronous buck converter embodies multi-level, complex interaction between MOSFETs, driver, and controller (Fig. 4). Therefore, it is necessary to capture as much details of package and associated active/passive elements as possible in the Mixed-mode simulation platform to model the device/circuit coupled dynamics accurately.

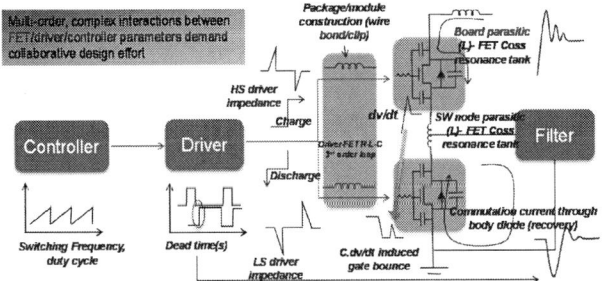

Fig. 4: Multi-dimensional interaction between FET, driver, controller in a typical DC-DC converter.

III. MIXED MODE PLATFORM FEATURES. CAPABILITES

A. Input Features

As discussed in the preceding section, it is crucial to develop the simulation platform with enough flexibility to adopt to various design and application scenarios. While it is possible to extend the circuit topology in many directions, here we focus on the non-isolated synchronous buck converter. In that context, HS and LS FET denotes the high-side (control FET) and low-side (synchronous FET) respectively. To model the complex interplay between MOSFET, package, driver, and

passive elements [4], it is necessary for the simulation platform to have following features as part of its 'input space'. The idea is also illustrated in Fig. 5.

- ✓ HS and LS FET technology, Integrated Schottky, and any
- ✓ non-active area factors
- ✓ HS and LS FET die size, Schottky %, any thickness of FET
- ✓ die including spreading resistance
- ✓ HS, LS FET, and driver junction temperature
- ✓ Intrinsic Rg of FETs (depending on layout)
- ✓ FET package resistance and parasitic inductances
- ✓ Gate driver to FET parasitics (co-packaged or discrete)
- ✓ Gate driver o/p impedances (driving strength)
- ✓ Controller dead times (Rising and Falling edges)
- ✓ Board parasitics, filter inductor DCR, capacitor ESR

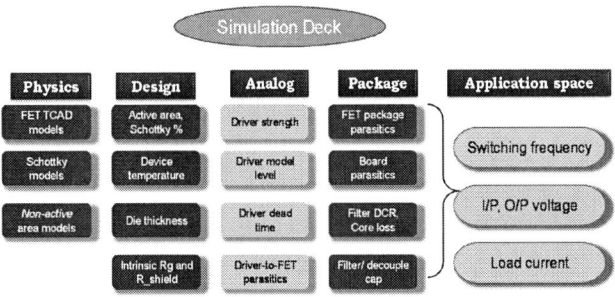

Fig. 5: Desired input space of Mixed-mode simulation framework.

Additional challenges include numerical stability of the simulation model to adopt to a wide range of circuit conditions – input and output voltages, switching frequency, filter inductor and capacitor, over a range of load current. This entails optimum choice of numerical solver algorithms and dynamic time stepping which can ensure high degree of accuracy of the resulting time-domain waveforms but, at the same time, does not make the computational load too high.

B. Outputs Needed for Optimum MOSFET Design Addressing the Efficiency-Reliability Trade-off

On the output side, we want the model to generate and store various parameters – both at device and circuit level. Typical outputs should include,

- ✓ Efficiency at SW node and at o/p load
- ✓ HS and LS silicon loss, Schottky loss
- ✓ Gate driver loss
- ✓ Package resistance loss, filter inductor loss
- ✓ Conduction and switching loss breakup at silicon level
- ✓ Key waveforms (VDS and VGS of FET, SW ringing, gate driver current pulses, etc.) and their 'quality'

The last point deserves special mention as it intrinsically couples the switching performance of the device with the aspect of reliability. Generally, fast switching dynamics is desired of the FETs for high-power-density converters to minimize switching losses. However, high dv/dt and di/dt, resulting from extremely short rise and fall time, lead to voltage overshoot and EMI noise issue in a high-frequency power application. This often poses significant reliability concern for system designers and end customers.

C. Multi-scale modeling

Careful optimization of process and device design parameters are necessary during MOSFET technology development to foresee the impact of such design choices on the system level. Time-domain waveform data must be analyzed together with power loss extractions to make process integration and design decisions. Mixed-mode simulation, as described in this article, is particularly well-suited to provide such outputs. In this capacity, Mixed-mode simulation serves as a multi-scale modeling framework which helps propagating changes at the semiconductor fabrication level all the way up to circuit dynamics. The idea is illustrated in the Fig. 6.

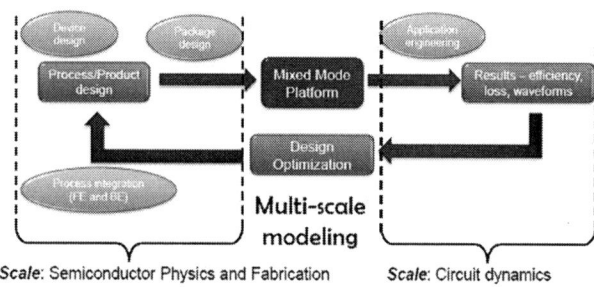

Fig 6: Idea of multi-scale modeling with Mixed-mode platform.

IV. REPRESENTATIVE RESULTS

Figs. 7 – 11 show key examples of typical outputs from this which govern the process of device design and technology development. Fig. 7 shows the efficiency plots (with load current) for various MOSFET technologies and switching frequencies. This helps use to judge the merits of one FET technology over another as a function of application domain. Fig. 8 shows the breakup of various types of losses at the silicon level. This kind of insight in relative loss percentages is almost impossible to get from standard application testing but it may have strong influence on the optimum die design for a particular application.

Going further on the trade-off between various types of losses, Fig. 9 shows how cross-over in efficiency performance happens based on either FET technology or die size. By analyzing this kind of result, one can target specific design or technology choice as optimally suited to the relevant application domain.

Fig. 10 illustrates how the choice of gate runner layout impacts the intrinsic gate impedance of the FET which determines the ultimate switching speed and also impacts reliability from a gate bounce immunity point of view. In the mixed-mode simulation, by dividing the die into small segments, we can study the gate bounce voltage at different locations of the die, which cannot be measured.

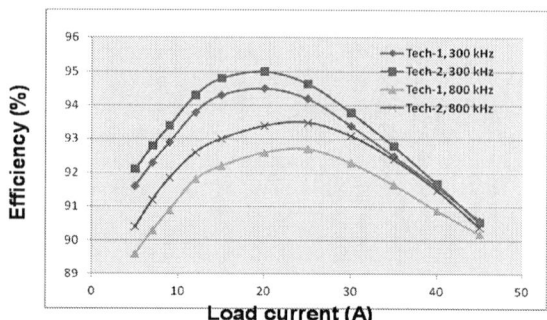

Fig. 7: Efficiency plots with FET tech and switching frequencies.

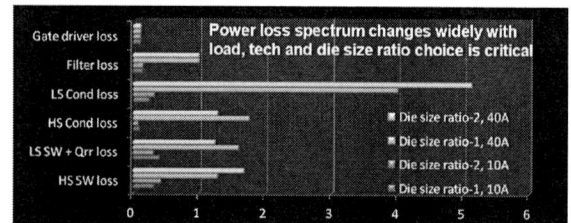

Fig. 8: Detailed loss breakdown insight to optimize the design/die ratio as per application (switching loss vs. conduction loss driven).

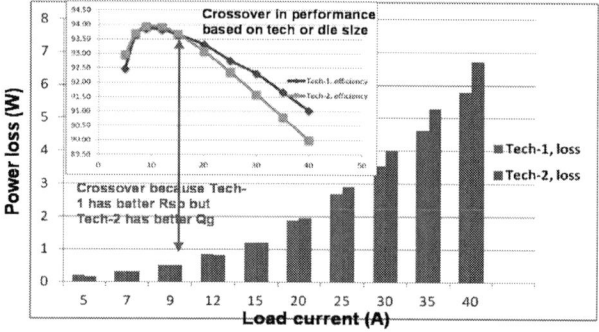

Fig. 9: Crossover in efficiency: optimization of die size/ technology is based on switching and conduction loss share in specific application.

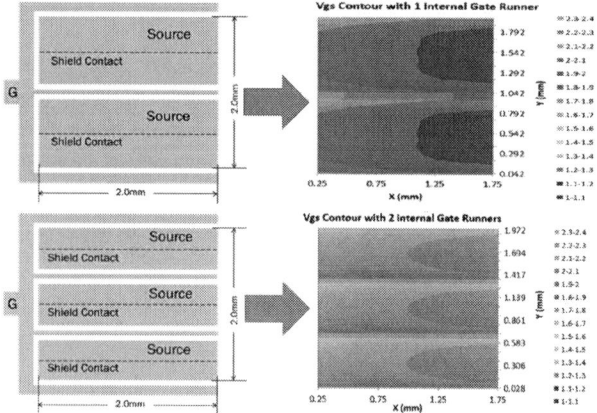

Fig. 10: (A) Low-side FET layout with 1 and 2 internal gate runners; (B) gate bounce V_{gs} contours (within the die) corresponding to die with 1 and 2 internal gate runners, during switching-off of the low-side FET.

Next, in Figs. 11-13, we show the typical waveform extracts from the Mixed-mode model. This shows the immense utility of this analysis where the impact of a set of process parameter (e.g. gate oxide, field plate design, or trench depth) on application level can be gauged without undergoing full application evaluation with experimental prototypes.

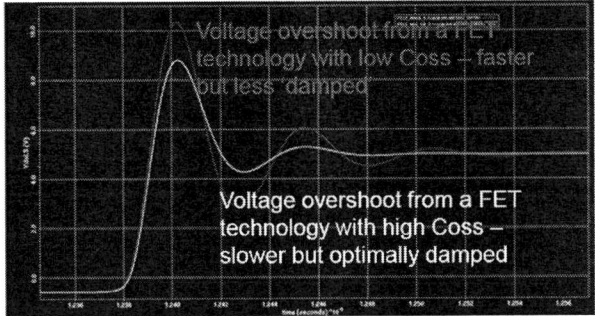

Fig. 11: VDS voltage waveforms across the FET with two different technology optimizations. Process and cell architecture of the FET can be tuned to trade-off between switching speed and overshoot voltage in an application circuit.

Accurate switching dynamics data can help the overall product design where the impact of package parasitics can also be inferred easily. In high-frequency converters, optimum setting of package and board level parasitics have a huge impact on system performance. Mixed-mode simulation can help uncover the trend of this impact without going through a slew of costly and time-consuming packaging/assembly iterations.

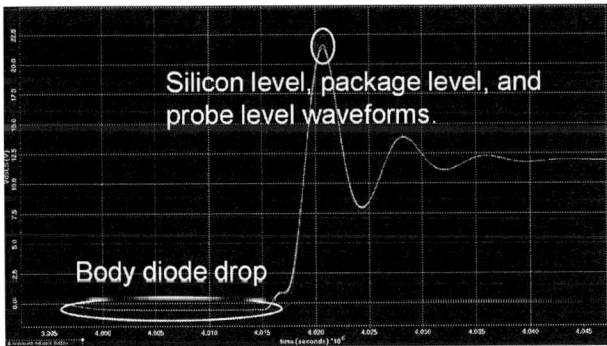

Fig. 12: Waveform observation capability at silicon, package, and board level from the Mixed-mode simulation data. This helps the package design optimization from inductive parasitic point of view.

Furthermore, this kind of rich time-domain data help analog designers to determine the boundaries of the driver IC which controls the power MOSFET. Dead time and driving strength (impedance of output stage transistors inside the IC) can be optimized from the point of view of safe operation – e.g. avoidance of cross-conduction power loss and dv/dt induced gate bounce and shoot-through. Consequently, Design and development of smart power-stage products (e.g. DrMOS) can be accomplished holistically by considering both power FET and analog IC features.

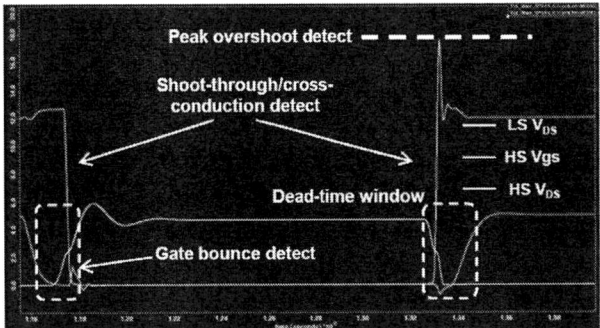

Fig. 13: Gate bounce and dead-time window detection from Mixed-mode time-domain data. This helps analog driver IC designers.

V. SUMMARY AND FUTURE EXPANSION

A. Key Benefits

Physics-based device-circuit co-simulation is a valuable tool to guide and optimize power MOSFET process technology development. It is well-suited to the need of emerging trend of integration of discrete and analog ICs in the same package. It has the potential of saving cost and engineering time by propelling the technology design faster towards the ultimate goal of higher energy efficiency and enhanced reliability. We outline the key features and benefits of such an integrated design framework.

B. Future Expansion with Machine Learning

It can also be argued that the spectrum of rich data, obtained from such Mixed-mode co-simulation, can be potentially analyzed with the help of advanced machine learning algorithms to uncover impactful functional mappings between process/design choices and system performance. In general, this kind of pattern discovery is hard for domains which are fundamentally different on scale and scope e.g. ion implantation physics and inductive energy transfer in a switching circuit. However, tools like Mixedmode simulation make it possible to bridge the gap between these two disparate domains and to generate sufficiently dense high-accuracy data set so that machine learning techniques (e.g. Nonlinear Regression, Random Forest or Deep Neural Network) can be applied to learn useful relationships.

REFERENCES

[1] R. Sodhi, A. Challa, J. Gladish, S. Sapp, and C. Rexer, "High cell-density, shielded-gate power MOSFET for improved DC-DC converter efficiency", *Proceedings of PCIM Europe*, 2010.

[2] T. Sarkar, A. Challa, and S. Sapp, "Enhanced Shielded-Gate Trench MOSFETs for High-frequency, High-Efficiency Computing Power Supply Applications", *Proceedings of IEEE Applied Power Electronics Conference*, 2013, pp. 507 -511.

[3] J. Victory, et al., "A Physically based scalable SPICE model for shielded-gate trench power MOSFETs", *Proceedings of International Symposium on Power Semiconductor Devices and ICs* (ISPSD), 2016, pp – 217.

[4] T. Sarkar, A.Upadhaya, S. Pearson, R. Sodhi, and S. Sapp., "Effect of inductive parasitics on the device loss and system Efficiency in a DCDC synchronous buck converter for computing applications", *Proceedings of PCIM Europe*, 2011.

Proceedings of the 30th International Symposium on Power Semiconductor Devices & ICs
May 13-17, 2018, Chicago, USA

A Novel High Performance Medium-Voltage DEnMOS in 40nm CMOS Technology

Lin Wei, Upinder Singh, Jeoung Mo Koo, Jiang Huihua
Technology Development Department, GLOBALFOUNDRIES Singapore Pte Ltd.
Phone: +65 93388716, mail: Wei.Lin@globalfoundries.com

Abstract—A new kind of symmetric medium-voltage DEnMOS embedded in 40nm CMOS process is developed out for parasitic capacitance improvement and specified on resistance decrease by adding one to two masks except for baseline set. Hot carrier induced degradation and DC measurement result are studied. Our results clearly show that the new concept gets tremendous improvement in both on-resistance and parasitic capacitance.

Keywords— Symmetric Drain Extended MOSFET, 40nm CMOS Technology

I. INTRODUCTION

Symmetric Medium-Voltage Drain-Extended-MOSFET (DEnMOS) embedded in CMOS process, is widely used in power management application as switches since DEnMOS provides several advantages in terms of large driving current and high breakdown voltage. Specific on-resistance (Rdson) is the key concern of power consumption and lot of works had been developed for on-resistance improvement in 0.13um and 0.18um platform [1] [2], parasitic capacitance (Cgd, Cgg and Cdb) is another key factor for power efficiency considering. As technology code migrates to 65nm and below, both gate oxide thickness and power supply decrease significantly, it brings the challenge to fabricate the compatible medium-voltage devices considering reliability issue. Hereby two new concepts based on 40nm low power technology are carried out for the first time, targeting at minimal Rdson and parasitic capacitance by architecture optimization (Figure 1).

II. EXPERIMENT AND RESULT

A. Device fabrication

Targeted devices of this letter are 5.0V n-type DEMOS transistors and fabricated by 40nm CMOS compatible technology. Figure 1 shows the schematic of different concepts. Operation voltage of Gate terminal is always 2.5V, Gate Oxide is as thick as 65A, and the width of each device is 20um. A lightly doped drift region, also called extension is fabricated to sustain the high voltage in drain/source terminal. Drift region is protected by silicide block (SBLK) to prevent silicide formation which reduces the high voltage capability.

The drawbacks of typical DEnMOS are long channel and large parasitic capacitance introduced by deep extension, IONW in this letter, not suitable for high performance switching application. While in "CONCEPT-1", two more

masks are added. They are HVPW and Extn. Lightly doped HVPW forms the channel and highly doped Boron (BN), retrograded doped Phosphor (PH) and Arsenic (AS), forms the extension(Extn). BN doping controls threshold voltage (Vth). PH/AS retrograded doping leads to high conductivity of the extension in favor of low on-resistance. In this way, minimum channel can be realized in favor of on-resistance together with lower parasitic capacitance. Short channel of the order of 0.2 um and lower is designed whereas shallow Extn junction is designed which contribute to less parasitic capacitance. The dimension and junction are well controlled due to better control capability of 40nm technology node.

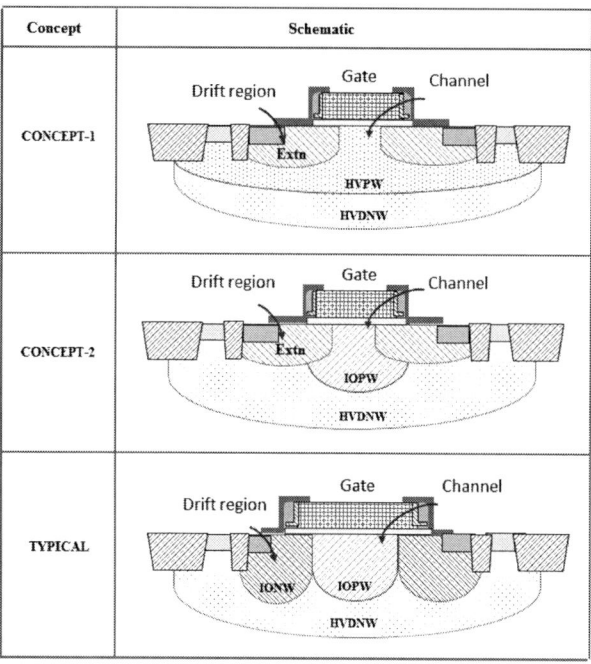

Figure 1 Simple Schematic of high performance 5.0V DEnMOS, Typical device as reference.

The other concept is "CONCEPT-2", which needs one extra mask "Extn". IOPW forms the channel, extra mask "Extn" forms extension. Different with CONCEPT-2, Vth is mainly controlled by IOPW. Extn implant contributes to higher

978-1-5386-2928-4/18 $31.00 © 2018 IEEE   292

conductivity of the extension. Compared with CONCEPT-1, CONCEPT-2 is cost-effective, and immune to Latchup due to high doping of body. The doping profile by TCAD simulation is shown in Fig. 2. Assumed benefits of new concepts are listed in Table. I.

TABLE .I Assumption of Devices Fabrication

Perspective	CONCEPT-1	CONCEPT-2	TYPICAL
Advantage	• Lightly Body • Heavy Extn doping • Shallow Extn junction • Minimized channel length	• IOPW Body • Heavy Extn doping • Shallow Extn junction	• Cost-free
Disadvantage	• Two extra masks;	• One extra mask;	• Short channel effect • Parasitic capacitance
Extn Scheme	• Dedicated implants	• Dedicated implants	• Reuse exiting well

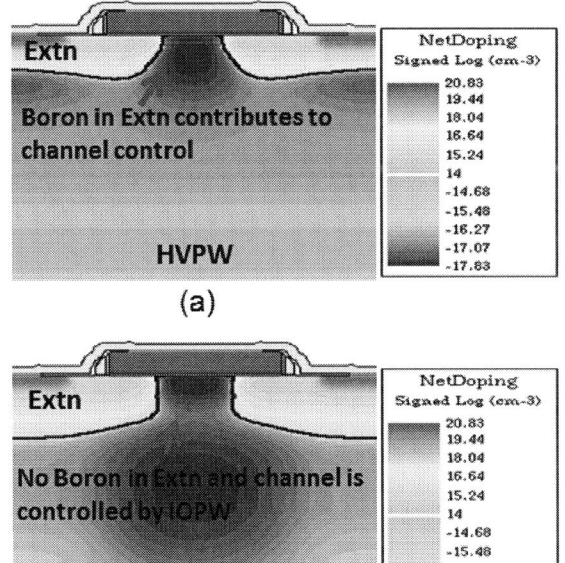

Figure 2 (a) CONCEPT-1 doping Profile (b) CONCEPT-2 doping Profile.

B. Device Characterization

On-state breakdown (Vgs=2.5V) and off-state breakdown are characterized in Figure.3, both concepts work well. The leakage at Vds=5.5V (1.1Vdd) is at ten Pico Ampere level, drive current is >300uA/um, both Bvoff and Bvon are above 7.0V which is acceptable for most switching applications. Extracted data is listed in Table. II including Rdson calculated by the linear current and Source-Drain pitch, Vth, Bvoff and Bvon. CONCEPT-1 shows the lowest Rdson as 2.1mohm.mm2 which is far better than Typical, while CONCEPT-2 has the highest Bvon, it is due to higher doping of IOPW body which prevents formation of parasitic bipolar in the bulk. The only drawback of CONCEPT-1 is Vth is as high as 0.82V due to high Boron doping in Extn.

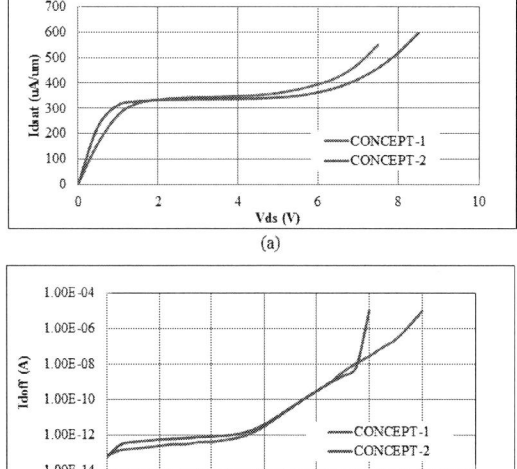

Figure 3 (a) On-state breakdown voltage measurement of "CONCEPT-1" and "CONCEPT-2", Vgs=2.5V; (b) Off-state breakdown voltage measurement, Vgs=0V.

The comparison of parasitic Capacitance between CONCEPT-1 and Typical is presented in Fig.4. 20% to 30% decrease of Gate-overlap-Drain Capacitance (Cgd) and Gate-overlap-Body Capacitance (Cgg) is observed since both channel and extension dimension shrunk tremendously for CONCEPT-1. Drain-overlap-Body capacitance (Cdb) is also decreased around 80% due to depth of extension junction significantly decreases compared with that of IONW in Typical.

Finally, Idsat degradation due to stressing at Isubmax condition is shown in Fig.5. There is almost no degradation of Idsat within 1000 second for both CONCEPT-1 and CONCEPT-2 devices.

In summary, both CONCEPT-1 and CONCEPT-2 concepts offers better performance in terms of lower on-resistance and less parasitic capacitance under acceptable breakdown voltage. CONCEPT-1 offers the lowest Rdson and main drawbacks are higher Vth and higher cost for extra mask.

TABLE .II Data Table of Devices

Parameters	CONCEPT-1	CONCEPT-2	TYPICAL
Vth (V)	0.82	0.69	0.67
Rdson (mohm.mm2)	2.1	2.4	>5.0
Bvoff (V)	12.0	10.0	>12.0
Bvon (V)	7.0	9.3	>8.0

III. CONCLUSION

In order to get the high performance symmetric MV DEnMOS compatible with advanced technology node, two new concepts so called CONCEPT-1 and CONCEPT-2 are carried out by adding one to two masks. Device dimension and junction can be well fabricated due to the control capability of advanced technology.

Device performance is cross checked. Rdson is decreased to 35% of that of typical device due to channel length decrease, while both Bvoff and Bvon are in the level of 1.5Vdd to 2.5Vdd. Parasitic capacitance is also checked including Cgd, Cgg and Cdb, 30% to 80% improvement is observed. Hot carrier induced degradation is checked and almost no change of Idsat post 1000 seconds stressing at Vgs@Isubmax. In sum, "Extn" is a promising method to fabricate high performance device in advanced technology node.

ACKNOWLEDGMENT

Authors would like to thanks for the measurement for excellent support of Pu Jinfei.

REFERENCES

[1] Lin.W, Cheng,C. Upinder.S, Ruchil J., L. L.Goh Purakh R. V. "A Novel Contact Field Plate Application in Drain-Extended-MOSFET Transistors", Pm. IEEE ISPSD, pp. 335- 337, 2017.
[2] A. Shibib, S. Xu, Z. Xie, P. Gammel, M. Mastrapasqua, I. Kizilyalli, "Control of Hot Carrier Degradation in LDMOS Devices by A Dummy Gate Field Plate", Pm . IEEE ISPSD, pp. 233- 235, 2004

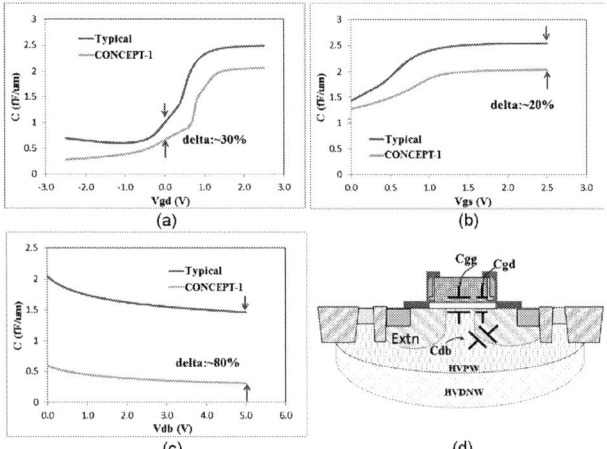

Figure 4 Parasitic capacitance characterization, a) Cgd, Gate-overlap-Drain Capacitance; b) Cgg, Gate-overlap-Body Capacitance; c) Cdb, Drain-overlap-Body capacitance; d) Model of parasitic capacitance of "CONCEPT-1". Measurement frequency is 100 KHz.

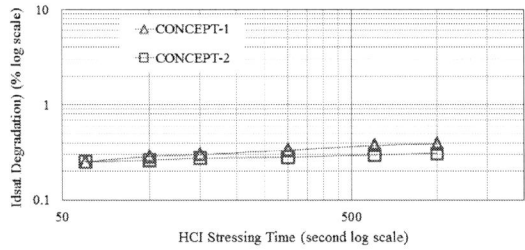

Figure 5 Idsat Degradation of DEnMOS up to 1000 seconds. Stressing condition is Vgs@Isubmax and 1.1Vdd.

Proceedings of the 30th International Symposium on Power Semiconductor Devices & ICs
May 13-17, 2018, Chicago, USA

Novel Current Re-Distribution Structure for Improved and Easy-to-Manufacturing 24V LDMOS

Cheng-Hua. Lin, Yan-Liang. Ji, C. H. Jan, C. W. Hu, Keven Chang, HW Kao

Product Engineering. MediaTek Inc. Hsinchu, Taiwan. R.O.C.

Email: cheng-hua.lin@mediatek.com, Tel:(+886)-3-5670766 ext. 23623

Fax: (+886)-3-567-0178

Abstract— This paper presents a novel structure combined POLY-CAP and P+ Cut Edge which are implemented in s.LDMOS (switching LDMOS) to avoid trapping at active region. s.LDMOS is widely used in SMPS (Switch Mode Power Supply) which requires reliable lifetime and stable turn-on performance. P+ Cut Edge could perfectly solve hump effect while turn-on; with Drain side POLY-CAP, lifetime could be drastically improved under NCS (Non-Conductive Stress) to sustain DC 10 years with comparable BV (Breakdown Voltage) and HCI (Hot Carrier Injection) lifetime.

I. INTRODUCTION

SMPS integrated in SOC for higher efficiency is required, and Switching LDMOS to sustain high voltage is implemented in SMPS output stage [1], which makes suitable BV of device is the first task to accomplish. Lots of experiments would be necessary in order to balance the trade-off between Rdson and BV; reliability assessment is also required at the same time to ensure the lifetime of product operating. It would be a huge impact once the fragile Reliability behavior and Hump issue are discovered when the process conditions are fixed.

0.18um technology is widely used as mixed-mode platform, and the process would be demonstrated in this paper. LDMOS as shown in Fig.1 is developed with ND IMP (N-type Drift Implant) which defines a light and deep doped region. Light-doped region is designed for fully-depleted concept and the deeper profile is also necessary for High-Side application, i.e. the source/bulk of a LDMOS are connected to a high voltage. Meanwhile, a thicker Oxide (StepOx) is inserted between Gate and Drain to sustain high voltage drop up to 24V. StepOx stage is right before CMOS is fabricated which keeps the baseline comparable [2]. However, there are 2 major challenges must be overcome during LDMOS development: Reliability and Hump; result in quality concern after the development is finished. Herein, POLY-CAP and P+ Cut Edge are introduced to provide reliable devices with Current Re-Distribution idea: POLY-CAP could detour the current path and improve Idlin lifetime more than 1000 times and P+ Cut Edge could eliminate Low Vt device and no parasitic corner device turns on. Both solutions are not extra mask or process change required and LDMOS performance would not be impacted, which are practical and easy-to-manufacturing.

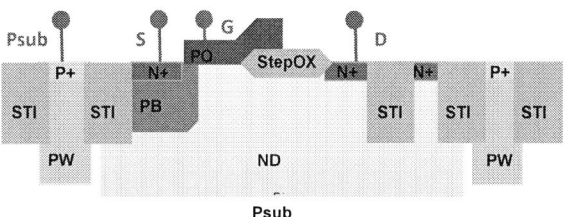

Fig. 1.   LDMOS structure.

II. MECHANISM AND DESIGN CONCEPT

A. POLY-CAP

Trapping is strongly related to Reliability which is always the challenge for LDMOS to meet 10-years lifetime. Especially, Idlin is the key parameter for switch application, so the lifetime assessment is started from Idlin. Poor Reliability of LDMOS is discovered when 24V s.LDMOS is stressed by NCS [3]. SiN liner in STI plays a role in trapping center to accumulate charge at active edge after NCS. In Fig.2, the Idlin degradation of conventional structure could not sustain more than 0.01 yrs which indicates the trapping of oxide between Drain to Gate is severe, and POLY-Cap structure could drastically improve the degradation and get saturated at <8%.

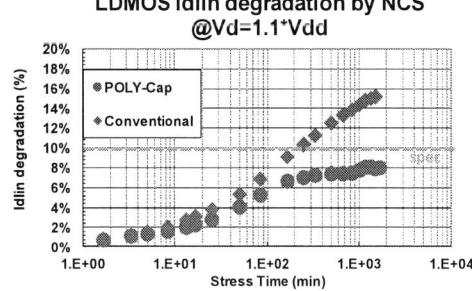

Fig. 2.   Idlin degradation is serious in traditional s.LDMOS by NCS. Improvement is obvious and s.LDMOS could pass 10 years lifetime while POLY-CAP is implemented.

978-1-5386-2928-4/18 $31.00 © 2018 IEEE   295

NCS (Non-Conductive Stress) or Drain Side TDDB (Off-state with Vd=1.1xVdd at ambient temperature 125°C) is one of key reliability test, and it is necessary for high voltage devices which is frequently implemented in Buck or Boost application. The oxide quality between Drain to Gate is important to LDMOS because LDMOS always handles high power which makes itself sustain high voltage drop at high temperature. Reliability fails DC 10yrs would cause circuit failure while handling high power, and therefore, stability and safety would be a risk. The NCS lifetime of LDMOS is key for SMPS, especially Buck/Boost Converters, because LDMOS would always sustain high drain voltage at off-state. [4]

In Fig.3, without extra mask, POLY-CAP constructs at drain side; blocking N+ implantation and Salicide formation to detour the current away from trapping center. The rule (Pov) of POLY capping on active is not critical as shown in Fig.4. That is to say, once the current is far away from trapping center, the degradation by NCS could be suppressed.

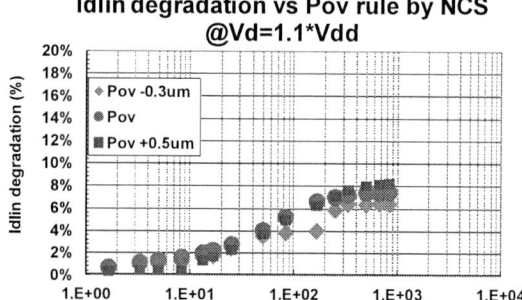

Fig. 4.   NCS Lifetime is not dependent on Pov rule.

B. P+ Cut Edge

Serious hump is discovered during characterization, and makes the modeling hard to fit well. SiN liner in STI generates parasitic devices at edge with lower Vt which turns on earlier and results in Hump and influences the quality of model fitting accordingly. Inaccurate fitting at subthreshold region would causes incorrect sensing circuit and invalid close-loop feedback. Structure of P+ Cut Edge as shown in Fig.5. is to block the parasitic device at edge and only Nominal device is able to turn on as shown in Fig.6. P+ on active edge could eliminate the source of parasitic device with Lower Vt and makes it unable to turn on [5]. The Current Re-Distribution still works without process change or extra mask.

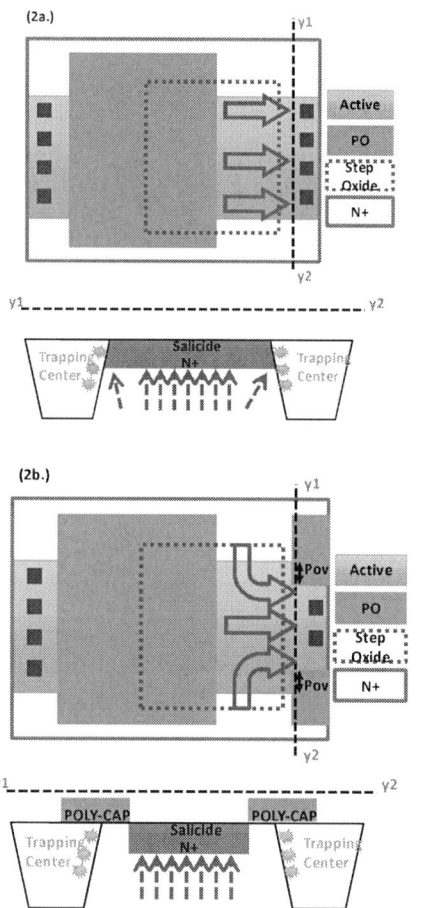

Fig. 3.   Schematic topview and cross-section comparison of s.LDMOS. POLY-CAP is utilized to detour current (2a) Conventional s.LDMOS, (2b) POLY-CAP s.LDMOS

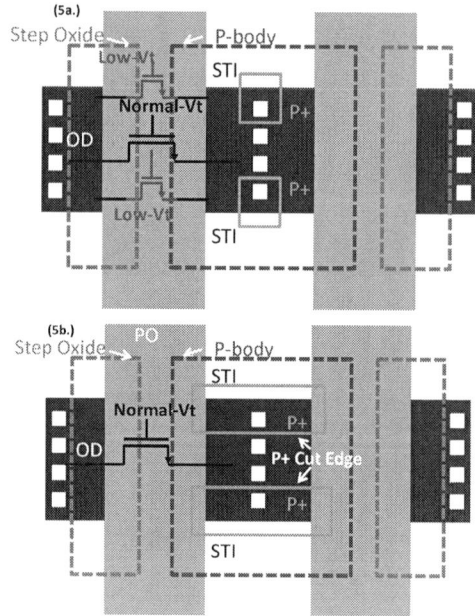

Fig. 5.   Schematic topview and cross-section comparison of n-type s.LDMOS. POLY-CAP is utilized to detour current (5a) Conventional s.LDMOS, (5b) P+ Cut Edge s.LDMOS

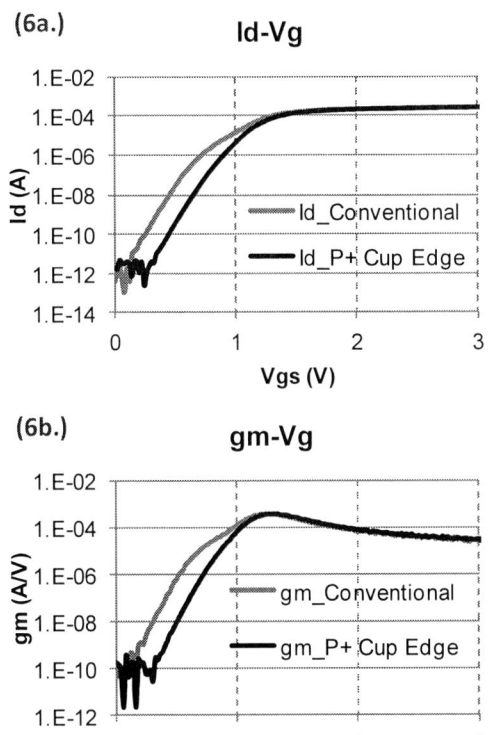

(6a.)

Id-Vg

(6b.)

gm-Vg

Fig. 6. Hump is serious in conventional LDMOS, and P+ Cut Edge is the solution to turn off the parasitic low Vt devices

III. EXPERIMENT RESULT

The Current Re-Distribution Structure is to detour the current away from active region which decreases the influence of SiN Liner. SiN Liner is the baseline which could releases the stress in STI and avoid unexpected dislocation. However, the trapping is enhanced accordingly, and even worse while VDD is up to 26.4V, and it is difficult to solve the trapping because the trapping center is at the edge of active region. Trapping center can be easily removed directly with one extra mask layer. The SiN Liner can be locally removed, but the side effects such as process complexity, defects, and even worse hump effect. Worse divot might induce stronger corner devices, and even more, the unexpected dislocation could happen located at SMPS output stage and this would result in more serious risk.

POLY-CAP is verified to suppress the trapping with sacrifice of active region; therefore, Rdson impact is expected owing to smaller effective width. Nevertheless, once the rule Pov is fixed, the Rdson impact is dependent on Single Width as Fig.7a. The utility of sLDMOS is mostly large width, in other words, if we focus on Width ≥35 um, the impact is less than

0.16%, and makes the Rdson comparable to conventional structure. Moreover, the HCI performances are comparable which implies the dominant location of HCI is not identical to NCS from Active Edge as Fig.7b, and reliability could refer to conventional structure, which could save resource, shorten the device learning cycle time, and keep the schedule for time to market.

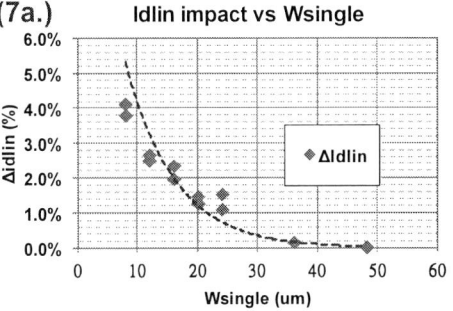

(7a.)

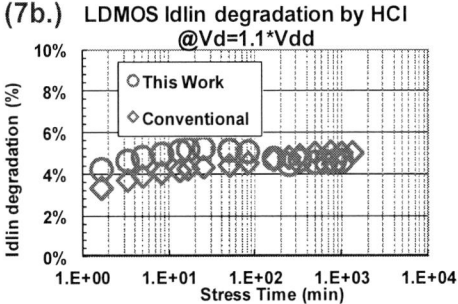

(7b.)

Fig. 7. Device impact by POLY-CAP implementation. (7a) Idlin impact is negligible when Single Width is larger than 35um. (7b) HCI performance is comparable between conventional structure and novel structure.

Before the novel structure are implemented, the experiment of process split includes StepOX width and physical thickness, and IMP energy and dopant concentration already show the improvement are not significant or the device performance is dropped by higher Rdson or lower BV. The experiment includes different structure, but no obvious impact of NCS lifetime. Since there is no good progress found in various experiments, in order to verify the root cause which induces the Idlin degradation by NCS, the SiN Liner split is required for the next fundamental study - trapping center. The result of process split in Fig.8. is demonstrated to verify the trapping center and the experiment shows the degradation of POLY-cap is comparable to the same structure without SiN liner, which indicates the trapping is from SiN liner. It would be a huge impact if the SiN Liner process is skipped. Not only the baseline would be changed, but the dislocation is hard to prevent. Nevertheless, not only POLY-CAP could detour the current away from trapping center by Current Re-Distribution concept but also keeps the same baseline and performance of LDMOS. If any process with SiN Liner requests High Voltage

978-1-5386-2928-4/18 $31.00 © 2018 IEEE   297

LDMOS, the concept is practical to implement once the electrical field is so high that could induce degradation. In this process, the voltage above 20V would suffer the NCS degradation. Although the initial shift would be higher depends on the stress voltage, it would always get saturated around 20%, which is to say, the trapping center is occupied, and the mechanism is different from the failure model results from oxide failure. [6]

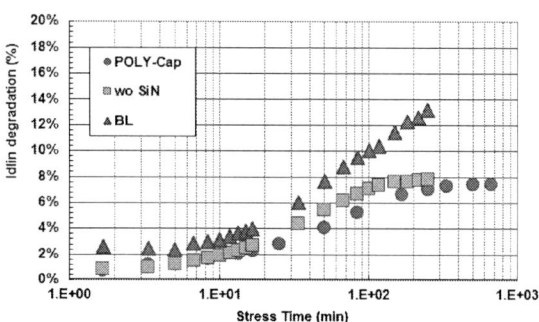

Fig. 8. NCS degradation comparison between BL & SiN liner remove & POLY-CAP.

IV. CONCLUSION

Integration of HV MOS and LV MOS is widely used for power management ICs. Conventionally, HV devices are integrated into a standard CMOS process by adding extra layers and process [7, 8]. A novel structure POLY-CAP and P+ Cut Edge are proposed and characterized to keep the LVCOMS process. Device performance BV/Rdson with 39.8V/14.1mOhm*mm2 and 44.8V/16.2mOhm*mm2 are maintained even they suffer NCS failure after the process and device rules are fixed. POLY-CAP and P+ Cut Edge could be implemented in all process without extra mask to accomplish the Current Re-Distribution concept and self-align to block N+ and P+ and Salicide which makes the concept easy-to-manufacturing. The result demonstrates that POLY-CAP and P+Cut Edge are promising and practical to develop reliable s.LDMOS.

REFERENCES

[1] Don Disney et. al., "180nm HVIC Technology for Digital AC/DC Power Conversion", ISPSD, pp. 287-290, 2017.

[2] Tsung-Yi Huang1 et. al., "Demonstration of a HV BCD Technology with LV CMOS Process", ISPSD, pp. 193-196, 2015.

[3] T. Cheng , Lee, M.Z. amd Yang, M.T., "A new device reliability evaluation method for overdrive voltage circuit application" in Reliability Physics Symposium, 2008. IRPS 2008. IEEE International, p.715 – 716

[4] Varghese. D, Reddy. V, Shichijo. H, et. al, "A comprehensive analysis of off-state stress in drain extended PMOS transistors: Theory and characterization of parametric degradation and dielectric failure." IEEE IRPS, pp. 566–574, 2008.

[5] Elizabeth Kho Ching Tee, Deb Kumar Pal, Tia Swee Hua, Hu Yong Hai, "High Voltage NMOS Double Hump Prevention by Using Baseline CMOS P-well Implant" ICEDSA.2010.5503059

[6] S.W. Chang, Chia Lin Chen, C. J. Wang, Kenneth Wu, "A New Off-state Drain-bias TDDB Lifetime Model for DENMOS Device", IRPS, pp 421-425, 2009

[7] Paul G.Y. Tsui et. al., "Integration of Power LDMOS into a Low-Voltage 0.5 um BiCMOS Technology ", IEDM Digest, pp.231-234, 1992.

[8] Kwang-Sik Ko et. al., "HB1340 Advanced 0.13um BCDMOS technology of complimentary LDMOS including fully isolated transistors", ISPSD, pp. 159-162, 2013.

Proceedings of the 30th International Symposium on Power Semiconductor Devices & ICs
May 13-17, 2018, Chicago, USA

A Novel Divided STI-based nLDMOSFET for Suppressing HCI Degradation under High Gate Bias Stress

Takahiro Mori, Shunji Kubo and Takashi Ipposhi

Technology Division, Renesas Semiconductor Manufacturing Co., Ltd.
751 Horiguchi, Hitachinaka, Ibaraki 312-8504, Japan
E-mail: takahiro.mori.cj@renesas.com

Abstract— Incorporating a P-type REduced SURface Field (RESURF) layer under an N-type drift region in an nLDMOSFET is a well-known means of improving the trade-off between the on-resistance (Rsp) and off-state breakdown voltage (BVoff), as well as reducing hot carrier injection (HCI) degradation. However, the N-type buffer layer under the drain N+ region must be eliminated to maintain a high BVoff in such structures. Generally, HCI degradation occurs near the channel-side STI edge at the maximum substrate current (Isubmax). In addition, new HCI degradation occurs in the vicinity of the drain-side STI edge under high gate bias and drain bias conditions because a high electric field is generated near this region. Herein, a new nLDMOSFET in which the STI is divided near the drain-side edge is proposed to suppress this HCI degradation while maintaining the Rsp-Bvoff trade-off.

I. INTRODUCTION

LDMOSFETs are widely used in power IC chips for automotive and industrial applications. Recently, the long-term reliability of these devices and the effects of hot carrier injection (HCI) degradation have become increasingly important. In addition, as a result of improvements in CMOS processing nodes, shallow trench isolation (STI) sections have been applied as insulating films in the drift regions of LDMOSFETs. HCI degradation at the maximum substrate current (Isubmax) tends to increase upon introducing an STI section into an nLDMOSFET to a greater extent than is observed for LOCal Oxidation of Silicon (LOCOS). Many approaches have been developed to suppress this HCI degradation [1] [2], including the use of a P-type REduced SURface Field (RESURF) layer under the N-type drift region [3] [4]. This structure both reduces HCI degradation and improves the trade-off between the on-resistance (Rsp) and off-state breakdown voltage (BVoff).

To maintain a high BVoff value when using a P-type RESURF layer, without increasing the number of processing steps, it is necessary to eliminate the N-type buffer layer under the drain N+ region of the nLDMOSFET. However, it has been determined that new HCI degradation also occurs under high gate bias and drain bias conditions in this structure. Herein, the mechanism of this new HCI degradation is explained based on technology computer-aided design (TCAD) simulations. Moreover, a novel structure is proposed to suppress this HCI degradation while maintaining the Rsp-BVoff trade-off.

II. DEVICES AND SIMULATIONS

A. Conventional structure and new HCI degradation

TCAD simulations were performed using an STI-based 65 V class nLDMOSFET with a gate voltage (Vg) rating of 3.3 V. Figure 1 shows cross-sections of several structures with P-type RESURF layers. Figure 1(a) is an nLDMOSFET with an N-type buffer layer, while Fig. 1(b) corresponds to a conventional nLDMOSFET without an N-type buffer layer but having the same P-type RESURF layer as in Fig. 1(a). The BVoff for the device in Fig. 1(a) is limited by the vertical electric field between the N-type buffer layer and the P-type RESURF layer even if the STI width is increased. Thus, to improve the BVoff, it is important to eliminate the N-type buffer layer. Figure 2 shows the results of TCAD simulations for these two devices at a Vg of 0 V and a drain voltage (Vd) of 65 V.

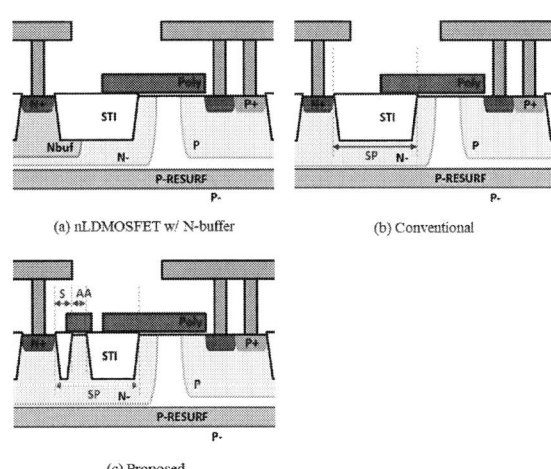

(a) nLDMOSFET w/ N-buffer   (b) Conventional

(c) Proposed

Fig. 1. Cross-sections of (a) an nLDMOSFET with an N-type buffer layer, (b) a conventional nLDMOSFET and (c) the proposed nLDMOSFET. In the latter, the STI in the drift region is divided into two parts near the drain-side region and the active area between these STI regions is covered with a floating polysilicon pattern.

978-1-5386-2928-4/18 $31.00 © 2018 IEEE   299

Figure 3(a) summarizes the impact ionization and electrostatic potential results at Isubmax, at a Vg equal to approximately half of Vgmax and a Vd of 65 V, for the 65 V nLDMOSFET in Fig. 1(b). It is evident that HCI degradation occurs near the channel-side STI edge and that elimination of the N-type buffer layer does not affect this degradation. Figure 3(b) shows the same results but under high gate bias and drain bias conditions, meaning that Vg is 3.3 V (the rated value) and Vd is 65 V. Here, it can be seen that removing the N-type buffer layer increases both the electric field and impact ionization values near the drain-side STI edge. These results suggest that new HCI degradation will occur in this location under these conditions. Section III provides experimental results for comparison purposes.

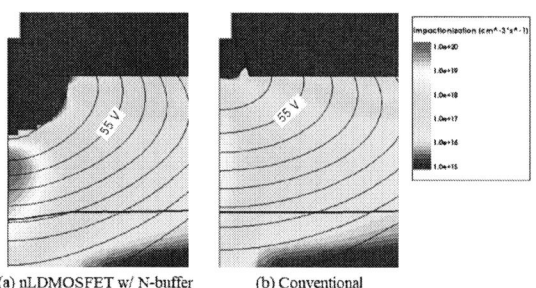

(a) nLDMOSFET w/ N-buffer   (b) Conventional

Fig. 2. TCAD simulations of the impact ionization (color contours) and electrostatic potential (line contour; 5 V steps) values for the 65 V nLDMOSFETs in Figs. 1(a) and (c), biased at Vg = 0 V and Vd = 65 V.

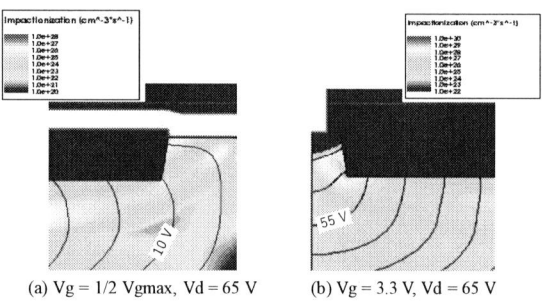

(a) Vg = 1/2 Vgmax, Vd = 65 V   (b) Vg = 3.3 V, Vd = 65 V

Fig. 3. TCAD simulations of the impact ionization (color contour) and electrostatic potential (line contour; 5 V steps) values for the conventional 65 V nLDMOSFET biased at (a) Vg = 1/2 Vgmax and Vd = 65 V, and (b) Vg = 3.3 V and Vd = 65 V.

B. Proposed structure

The simulations in the previous section demonstrate that a high electric field and increased impact ionization will occur near the drain-side STI edge under high gate bias and drain bias conditions. If the electric field near the drain-side STI edge can be reduced, it is predicted that the new HCI degradation will be suppressed. A new nLDMOSFET structure having a divided STI in the N-type drift region (Fig. 1(c)) is proposed for this purpose. A major feature of this structure is that the STI is split into two parts near the drain N+ region. The new active area

formed by this divided STI is also covered with a floating polysilicon pattern to prevent N-type ion implantation and thus to restrict N+ formation in this region and silicidation of its surface. The most important design parameters for this device are the drain-side STI width (S) and the new active area width (AA). TCAD simulations were carried out for S/SP = 0.062 and AA/SP = 0.055, where SP is the total STI width in the N-type drift region, while maintaining the other parameters (including the SP value and bias) the same as for the conventional structure modeled in Fig. 3. Figure 4(a) shows the impact ionization and electrostatic potential results obtained at Isubmax. Here, the proposed structure does not appear to affect the extent of HCI degradation, since the results are equivalent to those obtained for the conventional structure (Fig. 3(a)). In contrast, Fig. 4(b) presents the simulations for high gate bias and drain bias conditions. In this case, the impact ionization near the drain-side STI edge of the conventional structure can be shared between the two STI edges as a result of the divided structure. Figure 5 plots the impact ionization near the drain-side STI edge as a function of Vg for both structures. It is clear that the impact ionization for the proposed structure is less than that for the conventional structure, indicating significant immunity to HCI degradation under high gate bias and drain bias conditions.

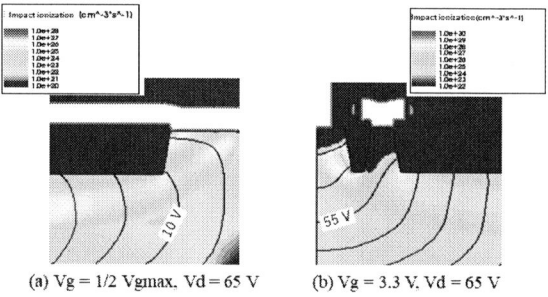

(a) Vg = 1/2 Vgmax, Vd = 65 V   (b) Vg = 3.3 V, Vd = 65 V

Fig. 4. TCAD simulations of the impact ionization (color contours) and electrostatic potential (line contour; 5 V steps) values for the proposed 65 V nLDMOSFET in Fig. 1(c), biased at (a) Vg = 1/2 Vgmax and Vd = 65 V, and (b) Vg = 3.3 V and Vd = 65 V.

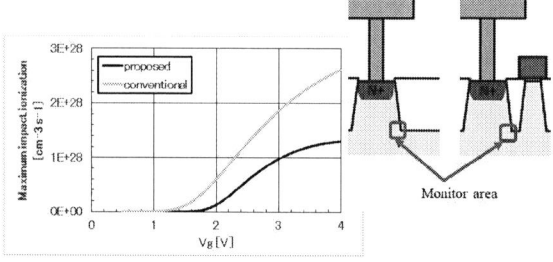

Fig. 5. Simulated plots of maximum impact ionization near the drain-side STI edges (the "monitor area" in each cross-section) as functions of Vg in the structures in Figs. 1(b) and (c) at Vd = 65 V.

III. MEASUREMENT RESULTS AND DISCUSSION

A 65 V nLDMOSFET incorporating a divided STI was fabricated by our STI-based bulk BiCDMOS process, having a gate voltage rating of 3.3 V, and employed for HCI evaluation. It should be noted that no extra processing steps were required to create this structure.

A. DC characteristics

The changes in BVoff and Rsp between the proposed and conventional structures are plotted against S/SP and AA/SP in Figs. 6 and 7. It is apparent that S/SP had little effect on either parameter within the measurement range (Fig. 6). However, AA/SP did modify both BVoff and Rsp in the proposed structure (Fig. 7). Specifically, BVoff decreases with increasing AA/SP, since it is difficult to deplete the larger AA region. This same effect can also be observed in the TCAD simulation results between AA/SP = 0.055 and 0.172 in Fig. 8. Rsp also decreases as AA/SP increases because the total area of the N-type drift region increases simultaneously. HCI stress tests of the proposed structure were performed primarily at S/SP = 0.062 and AA/SP = 0.055 because the BVoff and Rsp values were equal to those for the conventional structure at these dimensions.

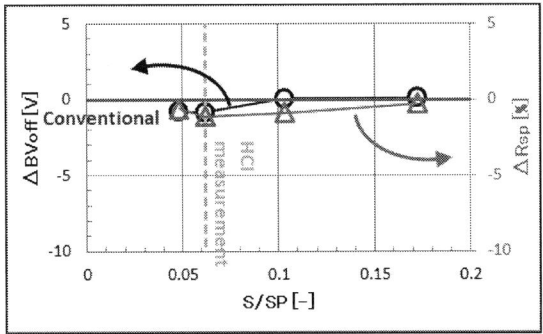

Fig. 6. Experimental BVoff and Rsp shifts as functions of S/SP in the proposed 65 V nLDMOSFET at room temperature. The value of AA/SP was set to 0.055 in this graph.

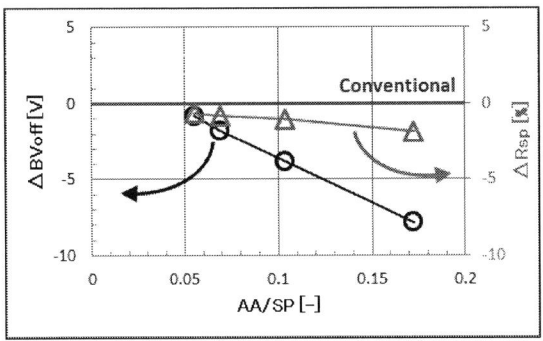

Fig. 7. Experimental BVoff and Rsp shifts as functions of AA/SP in the proposed 65 V nLDMOSFET at room temperature. The value of S/SP was set to 0.048 in this graph.

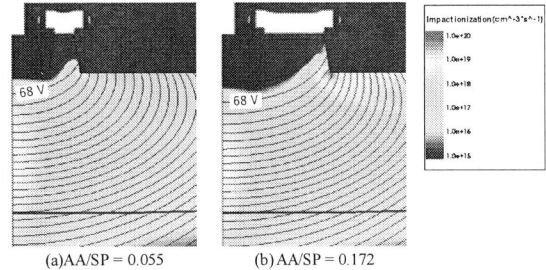

(a)AA/SP = 0.055   (b) AA/SP = 0.172

Fig. 8. TCAD simulations of the impact ionization (color contours) and electrostatic potential (line contours; 2 V steps) values for the proposed 65 V nLDMOSFET, biased at Vg = 0 V and Vd = 70 V (S/SP = 0.048).

B. HCI characteristics

HCI stress tests were performed at Ta = -40 °C at either Vd = 65 V at Isubmax or Vd = 65 V at Vg = 3.3 V, since it is well known that lower temperatures accelerate HCI degradation. In Figs. 9 to 12, ΔId is plotted against stress time, where ΔId is the maximum Id decrease under the monitor conditions (Vd = 0-8 V, Vg = 3.3 V). Initially, HCI degradation at Isubmax was compared between the conventional and proposed structures (Fig. 9). Since HCI degradation under these conditions occurred near the channel-side STI edge, there was no difference regardless of the shape near the drain edge.

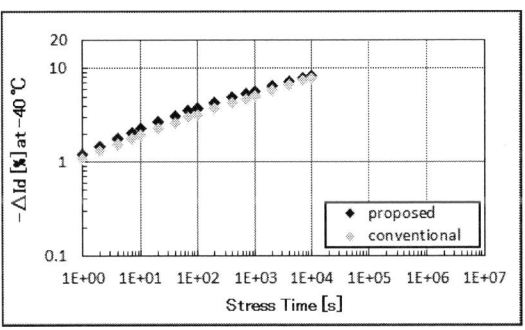

Fig. 9. Experimental ΔId values as functions of stress time at Vd = 65 V (Isubmax condition) and Ta = -40 °C. The measurements were performed at a monitor gate bias of 3.3 V, with the drain voltage swept from 0 to 8 V.

The difference in the extent of HCI degradation under high gate bias and drain bias conditions between the two structures was subsequently compared (Fig. 10). As predicted by the TCAD simulations, HCI degradation occurs under these conditions, but is greatly suppressed in the proposed structure at a relatively early stress time compared with the conventional structure. In this plot, the ΔId values for the proposed and conventional structures after 1×10^5 s were 5.8% and 7.9%, respectively.

Because HCI degradation can be suppressed by reducing the electric field near the drain-side STI edge (through applying the proposed structure), these results confirm that this degradation does in fact occur in this region. Furthermore, when considering actual operation in an IC chip, the duty cycle under high gate bias and drain bias conditions is very low, so it is highly advantageous to suppress HCI degradation within a relatively short stress time. Figures 11 and 12 present plots of ΔId against stress time for various values of S/SP and AA/SP in the proposed structure. The effect of S/SP on ΔId is lowest at a value of 0.062 and increases along with S/SP (Figure 11). This effect can presumably be attributed to reduced electric field relaxation near the drain-side STI edge as S/SP becomes larger. The effect of AA/SP on ΔId is smaller than that of S/SP (Figure 12). This is because AA/SP is presumed to have little effect on the electric field relaxation near the drain-side STI edge.

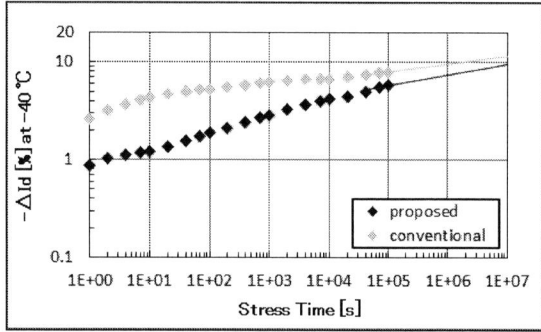

Fig. 10. Experimental ΔId values as functions of stress time at Vd = 65 V (Vg = 3.3 V) at Ta = -40 °C, with the same monitor conditions as in Fig. 9.

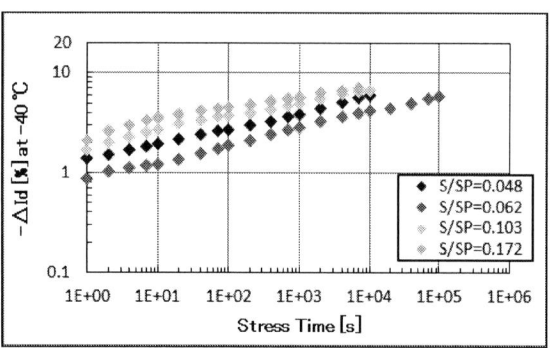

Fig. 11. Experimental ΔId values as functions of stress time and S/SP at Vd = 65 V (Vg = 3.3 V) and Ta = -40 °C, with AA/SP set to 0.055, and the same monitor conditions as in Fig. 9.

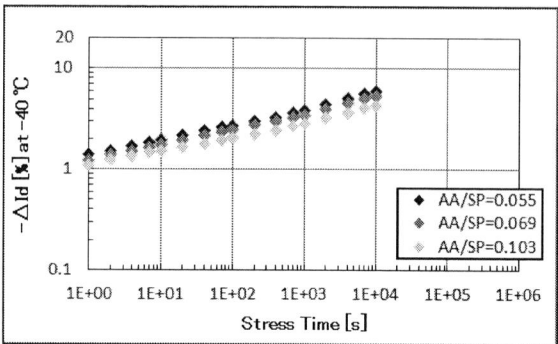

Fig. 12. Experimental ΔId values as functions of stress time and AA/SP at Vd = 65 V (Vg = 3.3 V) and Ta = -40 °C, with S/SP set to 0.048, and the same monitor conditions as in Fig. 9.

IV. CONCLUSION

New HCI degradation occurs in nLDMOSFETs having a P-type RESURF layer without an N-type buffer layer under the drain N+ region. This effect appears near the drain-side STI edge under high gate bias and drain bias conditions in both TCAD simulations and in experimental results. The present work proposed a new nLDMOSFET in which the STI is divided near the drain N+ region as an approach to reducing HCI degradation. The results demonstrate that this structure has the desired effect, especially within a relatively short stress time. The incorporation of this new structure into IC chips is anticipated to increase long-term reliability, since the duty cycle at a high gate bias and a high drain bias is very low in such devices.

ACKNOWLEDGMENTS

The authors would like to thank H. Sugiura, Y. Iwanaga, A. Komuro and S. Tokumitsu for BiCDMOS process integration.

REFERENCES

[1] D. Disney, W. Chan et al., "60V Lateral Trench MOSFET in 0.35 μm Technology," Proc. of ISPSD08, pp. 24-27.

[2] T. Mori, H. Fujii, S. Kubo and T. Ipposhi, "Investigation into HCI Improvement by a Split-Recessed-Gate Structure in an STI-based nLDMOSFET," Proc. of ISPSD17, pp.459-462.

[3] K.-Y. Ko, I.-Y. Park, Y.-K. Choi et al., "BD180LV 0.18 um BCD Technology with Best-in-Class LDMOS from 7V to 30V," Proc. of ISPSD10, pp. 71-74.

[4] H. Fujii, S. Tokumitsu et al., "A 90nm Bulk BiCDMOS Platform Technology with 15-80V LD-MOSFETs for Automotive Applications," Proc. of ISPSD17, pp.73-76.

Proceedings of the 30th International Symposium on Power Semiconductor Devices & ICs
May 13-17, 2018, Chicago, USA

Hot-carrier Induced Off-state Leakage Current Increase of LDMOS and Approach to Overcome the Phenomenon

Keita Takahashi, Kanako Komatsu, Toshihiro Sakamoto, Koji Kimura and Fumitomo Matsuoka

Toshiba Electronic Devices & Storage Corporation
580-1 Horikawa-cho, Saiwai-ku, Kawasaki-city, Kanagawa
212-8520, Japan
Phone: +81-44-548-2294
E-mail: keita2.takahashi@toshiba.co.jp

Yoshiaki Ishii, Katsumi Egashira and Masaki Sakai

Japan Semiconductor Corporation Oita Operations
3500 Oaza Matsuoka, Oita-city, Oita
870-0197, Japan
Phone: +81-97-524-6109
E-mail: yoshiaki2.ishi@toshiba.co.jp

Abstract—**We found and reported the unique drastic Ioff increase of LDMOS caused by HC induced trapped charge in the STI under the off-state condition. In this paper, we propose two approaches to overcome this phenomenon, one is a cost-oriented structure which can realize high BVdss and the other is low Ron characteristics suitable for high efficiency output circuit.**

Keywords—Hot-carrier; Off-state leakage current increase

I. INTRODUCTION

Shallow trench isolation (STI) -based LDMOS is widely used for Mixed-signal ICs, which are constituted with low-voltage CMOS devices as digital and high-voltage LDMOS devices as analog components [1]. There is an advantage that the manufacturing process can be shortened by using a STI, which is made for the low voltage CMOS devices and LDMOS devices at the same time. However, the STI-based LDMOS has a disadvantage that current flow concentrates at the bottom edge of STI where impact-ionization occurs, and generates hot-carrier (HC) [2]. It is known that the performance of the STI-based LDMOS devices is degraded by this HC, trapping charges in the STI and generating interface states at the STI/Si interface [3]. Many researchers had discussed on degradation of the STI-based LDMOS, such as a threshold voltage shift [4], a gate oxide film destruction [5], and a drain current and an on-resistance degradation [6]. However, there are few papers written on a relation between the off-state leakage current (Ioff) and HC stress [7]. We have found and reported a HC induced drastic Ioff increase phenomenon and investigated in detail [8]. Based on this study, we propose two N-ch LDMOS structures to overcome the problem of Ioff increase due to HC stress and discussed the results obtained from the experiments. The mechanisms of these experimental results were intensively studied utilizing TCAD simulation.

II. MECHANISM OF DRASTIC IOFF INCREASE

The phenomenon of HC induced drastic off-state leakage current increase for the N-ch LDMOS is shown in Fig. 1. According to Fig. 1, the Ioff of the N-ch LDMOS starts to increase drastically after 2×10^4 s, and it increases more than three decades at 1×10^5 s.

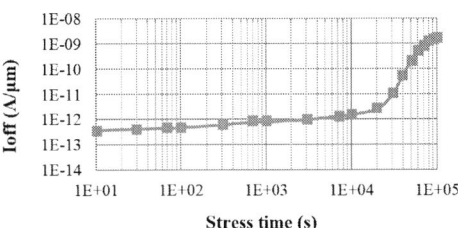

Fig. 1. The Ioff of N-ch LDMOS up to 1×10^5 s under HC stress condition (Vds=30V, Vgs=2V, T=27 ℃) and measurement condition of Ioff (Vds=30V, Vgs=0V, T=27℃).

In order to understand the phenomenon of the Ioff increase, TCAD simulation was utilized. Fig. 2 shows the result of impact-ionization distribution and the 1D-profile of impact-ionization rate along STI bottom (a - a') under HC stress (Vds=30V, Vgs=2V) by TCAD simulation.

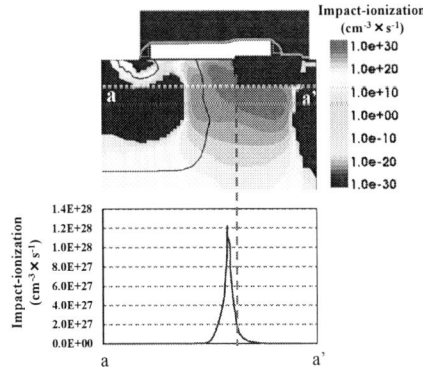

Fig. 2. Simulated results of impact-ionization rate 2D-distribution and its 1D-profile along STI bottom (a - a'), under HC stress condition of Vgs=2V, Vds=30V.

978-1-5386-2928-4/18 $31.00 © 2018 IEEE

According to Fig. 2, the impact-ionization rate under the HC stress condition is increasing at the STI edge. The electrons are accelerated by high Vds stress condition and flow toward the STI edge, which results in impact-ionization increase, and generates the fixed charge at the Si/SiO$_2$ interface of the STI. The mechanism of Ioff increase by forming the fixed charge has been verified by adding and increasing the amount of the fixed charge at the interface Si/SiO$_2$ of the STI as shown in Fig. 3.

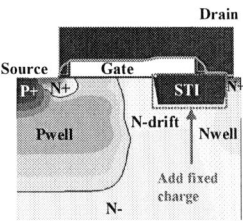

Fig. 3. TCAD simulated device structure with fixed charge around STI edge. The fixed charge amount was calculated under each condition of (1) none, (2) 6×10^{11} cm^{-2} and (3) 7×10^{11} cm^{-2}.

Fig. 4 shows the amount of fixed charge dependence on the Ioff. Fig. 5 is the result obtained from simulation of the impact-ionization at the off state condition (Vds=30V, Vgs=0V) with the fixed charge added along Si/SiO$_2$ interface at the STI. Comparing the fixed charge (1) none, (2) 6×10^{11} cm^{-2} and (3) 7×10^{11} cm^{-2}, despite the Vds conditions of (2) and (3) is same as (1), the impact-ionization increases as the fixed charge increases. The electric field at the STI becomes high due to the increase of the fixed charge along Si/SiO$_2$ interface at the STI and the impact-ionization increases. From these results, we considered that the fixed charge trapped at the STI edge is contributing to Ioff increase under the HC stress.

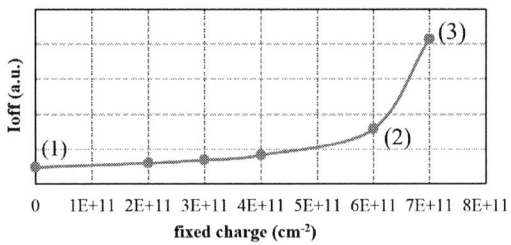

Fig. 4. Simulated Ioff dependence of the fixed charge along Si/SiO$_2$ interface at the STI. The numbers correspond to Fig. 5.

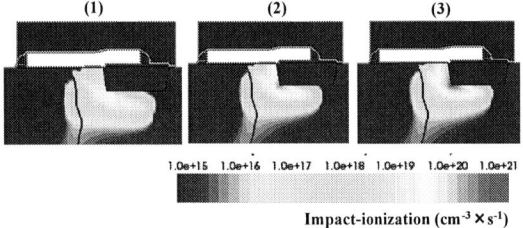

Fig. 5. TCAD simulation results of impact-ionization during off-state condition (Vgs=0V, Vds=30V) with the fixed charge along Si/SiO$_2$ interface at the STI (1) no charge, (2) 6×10^{11} cm^{-2} and (3) 7×10^{11} cm^{-2}.

III. DEVICE STRUCTURE AND EXPERIMENTS

To overcome the drastic Ioff increase under HC stress, two structures are proposed which have excellent tolerance. Since more hot-carriers are generated when the electric field between the drain and the source is larger, a structure with extended STI is proposed to relax the electric field in the first structure. In the second structure, a stepped-oxide structure [9] in which an oxide film is formed on the Si surface has been studied in order to remove the STI from drift region where hot-carriers are generated. These devices are fabricated with the process of 0.13 µm CMOS technology embedded with LDMOS. A cross-sectional view of a conventional STI-based N-ch LDMOS is shown in Fig. 6 (a). The channel length is 0.5 µm, the gate width is 80 µm, the gate oxide film thickness is 13 nm and the STI depth is 300 nm. These device parameters are common to all the devices studied in our work. Fig. 6 (b) shows the cross-sectional structure of N-ch LDMOS with extended STI. This N-ch LDMOS is formed by using the same N-drift diffusion layer as the conventional device and the STI length is stretched by 0.5 µm compared to the conventional N-ch LDMOS which STI length is 2.34 µm. This device can be associated without any additional manufacturing processes. Compared with the conventional structure, this N-ch LDMOS has BVdss increased by 34 % and on-resistance (Ron-sp) also increased by 30 %. Fig. 6 (c) shows the cross-sectional structure of N-ch LDMOS using a stepped-oxide without forming STI in N-drift region.

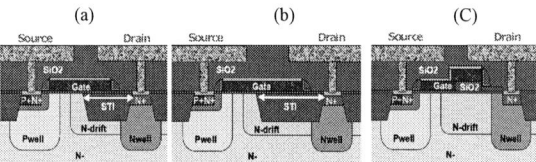

Fig. 6. Cross-sectional view of three N-ch LDMOS structures.
(a) Conventional STI-based N-ch LDMOS, (b) Extended STI-based N-ch LDMOS and (c) Stepped-oxide N-ch LDMOS

In order to obtain the same BVdss as the N-ch LDMOS of the conventional structure, the implant condition of the N-drift layer and the element design are adjusted for the stepped-oxide N-ch LDMOS. This N-ch LDMOS requires additional masks and manufacturing process steps. In the stepped-oxide N-ch LDMOS structure, Ron-sp is decreased by 22% comparing to the conventional N-ch LDMOS with attaining the same BVdss as the conventional one.

IV. RESULTS AND DISCUSSIONS

Fig. 7 shows the results of the Ioff increase under the HC stress of a conventional STI-based N-ch LDMOS (a), the extended STI-based N-ch LDMOS (b) and a stepped-oxide N-ch LDMOS (c). The Ioff of the extended STI-based N-ch LDMOS was half of the conventional structure and no drastic Ioff increase was observed up to 1×10^5 s. On the other hand, although Ioff value of the stepped-oxide N-ch LDMOS was as same as the conventional N-ch LDMOS at the beginning, a drastic Ioff increase was not observed under the continuous HC stress up to 1×10^5 s.

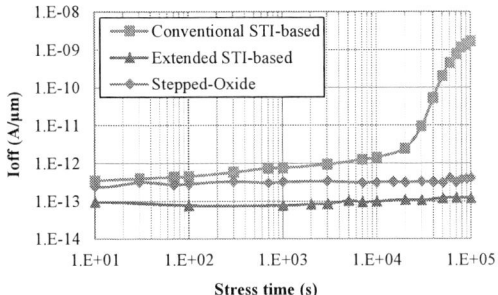

Fig. 7. Measured Ioff of LDMOS up to 1×10^5 s under HC stress conditions (Vds=30V, Vgs=2V and T=27 ℃). Ioff measurement conditions (Vds=30V, Vgs=0V and T=27℃)
(a) Conventional STI-based N-ch LDMOS, (b) Extended STI-based N-ch LDMOS and (c) Stepped-oxide N-ch LDMOS

We use Synopsys's TCAD simulation [10] to investigate the reason why the Ioff of extended STI-based N-ch LDMOS and the stepped-oxide LDMOS was not increased by HC stress. Fig. 8 shows the simulated results of the electron trapped charge after HC stress of 1×10^5 s. The trapped charge distributions along STI edge (b - b') are also shown. The peak value of the trapped charge density for the N-ch LDMOS of both structure (a) and (b) are about 5×10^{10} cm^{-3}. However, when compared the total amount of fixed charge existing along the Si/SiO$_2$ interface at the STI edge, that is reduced by one order of magnitude in structure (b) comparing with that of structure (a). To the contrary, it is noted that both the peak value and the total amount of trapped charge is extremely small and less than 1×10^{10} cm^{-3} in the N-ch LDMOS of structure (c).

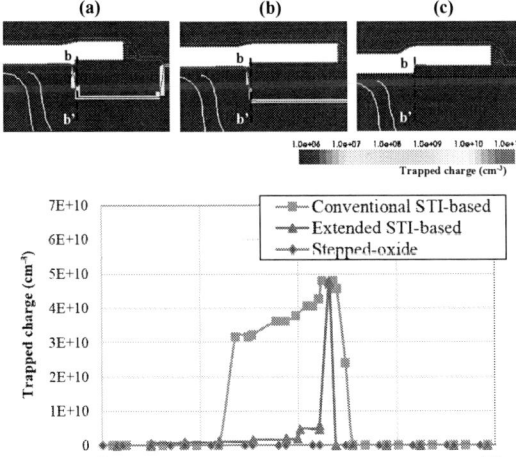

Fig. 8. Simulated results of trapped charge and its 1D-profile along STI edge (b - b'), after HC stress condition (Vgs=2V, Vds=30V and stress time=1×10^5 s). (a) Conventional STI-based N-ch LDMOS, (b) Extended STI-based N-ch LDMOS and (c) Stepped-oxide N-ch LDMOS

The results of simulated trapped charge well describe and correspond to the experimental results with considering the total amount of fixed charges existing along the Si/SiO$_2$ interface at the STI edge.

Fig. 9 shows the amount of impact-ionization distribution under HC stress condition of Vgs=2V and Vds=30V. The 1D impact-ionization profiles along STI edge (c - c') are also shown. The impact-ionization at the STI edge of extended STI-based N-ch LDMOS (b) is suppressed less than half in comparing with the conventional structure (a) due to the relaxation of electric field at STI edge. On the other hand, the drain current of the stepped-oxide N-ch LDMOS (c) flows detached from the oxide region, and thus the impact-ionization does not occur near the Si/SiO$_2$ interface. In structure (c), we considered that the amount of trapped charge at the STI/Si interface is smaller than other structures, since the location where impact-ionization occurs is away from the oxide film.

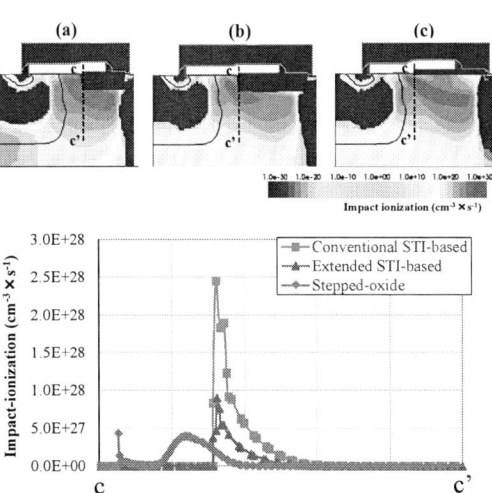

Fig. 9. Simulated results of impact-ionization under HC stress condition of Vgs=2V, Vds=30V and 1D-Profile of impact-ionization along the STI edge (c - c'). (a) Conventional STI-based N-ch LDMOS, (b) Extended STI-based N-ch LDMOS and (c) Stepped-oxide N-ch LDMOS

The total current density distribution under HC stress condition of Vgs=2V and Vds=30V are shown in Fig. 10. According to the simulation results, it was recognized that the STI-based N-ch LDMOS of (a) and (b) has high current density (7×10^5 A/cm^2) at STI edge, although the current density of the stepped-oxide N-ch LDMOS (c) is low (1×10^3 A/cm^2) near the stepped-oxide. As the results of this simulation, it was verified that the stepped-oxide N-ch LDMOS has low impact-ionization rate because the current path under the HC stress condition is far from the oxide film, therefore it is difficult to generate the trapped charge. Based on the simulation results, it was confirmed that the impact-ionization rate was suppressed by not placing the STI in the current path, and consequently the Ioff increase after the HC stress is suppressed.

978-1-5386-2928-4/18 $31.00 © 2018 IEEE

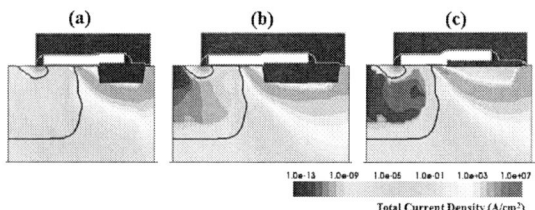

Fig. 10. Simulated results of total current density under HC stress condition of Vgs=2V and Vds=30V. (a) Conventional STI-based N-ch LDMOS, (b) Extended STI-based N-ch LDMOS and (c) Stepped-oxide N-ch LDMOS

Table I is the comparison of Ron-sp, BVdss, and Ioff after HC stress of 1×10^5 s of the studied three N-ch LDMOS structures: conventional, extended STI and the stepped-oxide structure. For the purpose of avoiding drastic Ioff increase under HC stress, both proposed structures (b) and (c) have excellent immunity. In case of extended STI LDMOS structure (b), the stretched drift region can realize high BVdss without additional mask. On the other hand, LDMOS structure with stepped-oxide (c) can achieve superior low Ron characteristics although it requires some additional masks compared to the conventional STI-based structure. Based on the results of this study, we can select the suitable LDMOS structure depending on the purpose such as chip design or applications.

The extended STI LDMOS can be easily used in analog circuit as well as conventional LDMOS without any process step increase. To the contrary, LDMOS for high efficiency output circuits requires the low Ron and the high drain current characteristics, and moreover it occupies a large portion of product chip's area. In this case, a stepped-oxide LDMOS structure is preferable, because the low Ron-sp characteristic can realize LDMOS size shrink, which directly affects the total chip size shrink, thus the structure is cost effective even taking into account the additional manufacturing steps.

TABLE I. The comparison of Ron-sp, BVdss and Ioff after HC stress of 1×10^5 s for Conventional STI-based N-ch LDMOS, Extended STI-based N-ch LDMOS and Stepped-oxide N-ch LDMOS.

Device	Ron-sp $(m\Omega \cdot mm^2)$	BVdss (V)	Ioff after HC stress of 1×10^5s $(A/\mu m)$
Conventional STI-based N-ch LDMOS	20.7	35.6	1.6×10^{-9}
Extended STI-based N-ch LDMOS	29.7	47.7	1.2×10^{-13}
Stepped-oxide N-ch LDMOS	16.1	35.1	3.0×10^{-13}

V. CONCLUSIONS

In this paper, we propose two approaches to overcome the hot-carrier induced drastic off-state leakage current increase phenomenon. The LDMOS structure with extended drift region can realize high BVdss without additional mask. On the other hand, LDMOS structure with stepped-oxide can achieve superior low Ron characteristics although it requires some additional masks. We can select the suitable LDMOS structure depending on the purpose such as chip design or applications.

ACKNOWLEDGMENT

The authors would like to thank Mr. Kenya Kobayashi for his continuous encouragement and valuable support.

REFERENCES

[1] K. Shirai, K. Yonemura, K. Watanabe, and K. Kimura, "Ultra-low On-Resistance LDMOS Implementation in 0.13µm CD and BiCD Process Technologies for Analog Power IC's" International Symposium on Power Semiconductor Devices & ICs (ISPSD), 2009, pp. 77-79.

[2] J. Roig, P. Moens, F. Bauwens, D. Medjahed, S. Mouhoubi, and P. Gassot, "Accumulation Region Length Impact on 0.18µm CMOS Fully-Compatible Lateral Power MOSFETs with Shallow Trench Isolation" ISPSD, 2009, pp. 88-91.

[3] S. Poli, S. Reggiani, G. Baccarani, E. Gnani, A. Gnudi, M. Denison, S. Pendharkar, and R. Wise. "Full Understanding of Hot-Carrier-Induced Degradation in STI-Based LDMOS Transistors in the Impact-Ionization Operating Regime", ISPSD, 2011, pp. 152-155.

[4] S. Poli, S. Reggiani, G. Baccarani, E. Gnani, A. Gnudi, G. Baccarani, M. Denison, S. Pendharkar, R. Wise, and S. Seetharaman, "Investigation on the temperature dependence of the HCI effects in the rugged STI-based LDMOS transistor", ISPSD, 2010, pp. 311–314.

[5] Y. Huang, J. Shih, C.C. Liu, Y.-H. Lee, R. Ranjan, P. Chiang, D.-C. Ho and K. Wu, "Investigation of Multistage Linear Region Drain Current Degradation and Gate-Oxide Breakdown Under Hot-Carrier Stress in BCD HV PMOS", in Proc. International Reliability Physics Symposium (IRPS), pp. 444-448, 2011.

[6] S. Reggiani, G. Barone, E. Gnani, A. Gnudi, G. Baccarani, S. Poli, M.-Y. Chuang, W. Tian, and R. Wise "TCAD Predictions of Linear and Saturation HCS Degradation in STI-based LDMOS Transistors Stressed in the Impact-Ionization Regime", ISPSD, 2013, pp. 375-378.

[7] H. Fujii, M. Ushiroda, K. Furuya, K. Onishi, Y. Yoshihisa and T. Ichikawa, "HCI-induced Off-state I-V Curve Shifting and Subsequent Destruction in an STI-based LD-PMOS Transistor", ISPSD, 2013 pp. 379-382.

[8] K. Takahashi, K, Komatsu, T. Sakamoto, K. Kimura and F. Matsuoka, "Hot-carrier Induced Drastic Off-state Leakage Current Degradation in STI-based N-channel LDMOS", Extended Abstracts of International Conference on Solid State Devices and Materials (SSDM), 2017, pp. 765-766.

[9] K. Hara, T. Kakegawa, S. Wada, T. Utsumi and T. Oda, "Low On-Resistance High Voltage Thin Layer SOI LDMOS Transistors with Stepped Field Plates", ISPSD, 2017, pp. 307-310.

[10] Synopsys TCAD Sentaurus Manuals, Version K-2015.06, pp. 503-534.

Proceedings of the 30th International Symposium on Power Semiconductor Devices & ICs
May 13-17, 2018, Chicago, USA

Novel Approach for NLDMOS Performance Enhancement by Critical Electric Field Engineering

Jaroslav Pjenčák[*], Moshe Agam[**], Ladislav Šeliga[*], Thierry Yao[**] and Agajan Suwhanov[**]

[*] ON Semiconductor, Rožnov pod Radhoštěm, Czech Republic, [**] ON Semiconductor, Gresham, US
jaroslav.pjencak@onsemi.com, moshe.agam@onsemi.com

Abstract— **Novel design approach for LDMOS device operating at voltages exceeding 100V is demonstrated and architecture to address all aspects of critical electric field is described. As a bonus, no changes in doping conditions required to achieve the device targets. This paper presents a design study for a resurf layer to engineer the critical electric field in LDMOS device. Results provide guidance on how to push high electric field lines away from the surface and how to successfully align them in the direction of the current. The new architecture is intended for small devices where area saving due to smaller isolation brings another benefit. This approach is compared with conventional single resurf design that normally unable to meet these conditions. LDMOS device breakdown is designed to predominantly occur underneath the drain of the device to avoid interaction with the gate oxide. Finally, results in this publication show how reversing source and drain can improve ESD ruggedness.**

Keywords— *sinker; NLDMOS; resurf; deep drain; low cost; DTI isolation; HV operation; lateral and vertical electric field; quasi-lateral field; bulk breakdown*

I. INTRODUCTION

For decades the bulk silicon device technology remains as a mainstream in the high voltage semiconductor industry and it is likely to continue for some time [1]. Sustaining high electric field in semiconductor materials typically requires creating a depletion layer between N and P type regions where no mobile charge carriers are present. This condition is created under reverse biased P-N junction. The maximum reverse bias voltage that can be applied is limited by the junction breakdown. It is characterized by rapid increase of current flow under certain reverse bias. There are two nondestructive mechanisms that can cause breakdown; avalanche multiplication and quantum mechanical tunneling of carriers through the bandgap [2]. For the LDMOS device, tunneling mechanism due to low doping and large dimensions in the silicon can be excluded. Doping level lowering for avoiding the exceeding of critical electric field is not usually preferred due to its direct impact on current transfer which increases the on-resistance. One way of addressing the trade-off is creation of a resurf layer under the drain to deplete the drain junction vertically [3]. Further improvement can be achieved by adding field plate to enhance equipotential lines of electric field [4]. However, even by use of advanced methods the maximum breakdown of conventional junction-isolated LDMOS is limited < 100V [5].

This paper demonstrates a new approach that enables significantly higher operating voltage range. This is achieved by managing the electric field at the depth in the silicon, promoting a lateral direction of the current, and eliminating the vertical field component.

II. DEVICE ARCHITECTURE

Multiple placements of LDMOS in the same pocket with multi-fingers width scaling need special consideration. For this reason the device is constructed with the specially placed drain isolation. The DTI (*Deep Trench Isolation*) spacing from the body of the device is dictated by operation voltage options [6]. However, this approach may not be optimal for small devices since it has isolation overhead.

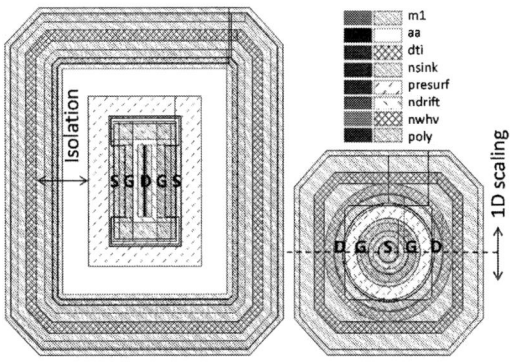

Fig. 1. Layout comparison between two NLDMOS transistors (wg = 25um): reference device (left), new device (right).

Authors propose a new NLDMOS device which is constructed as a single transistor inside of isolated DTI pocket. The drain of the transistor is connected to deep highly conducting sinker area adjacent to the DTI. The source/body of the device is placed in the middle of the pocket and poly Si gate is drawn around. Such architecture is beneficial for higher breakdown as well as for making gate area robust against the equipotential line contour abnormalities. It requires paying attention to proper termination of the device in all directions. The design with gate corners was investigated and results demonstrated how an inhomogeneity of charge balance can lead to lower breakdown. To address this issue circular shape design

978-1-5386-2928-4/18 $31.00 © 2018 IEEE

providing the same conditions in all directions was introduced. Device size scaling can be performed in one dimension (Fig. 1) only. New architecture is intended for small currents up to ~10mA to achieve beneficial footprint. Multi-finger architecture is better suited for high current regime. According to the device area calculations about 60% reduction can be achieved (Fig. 2).

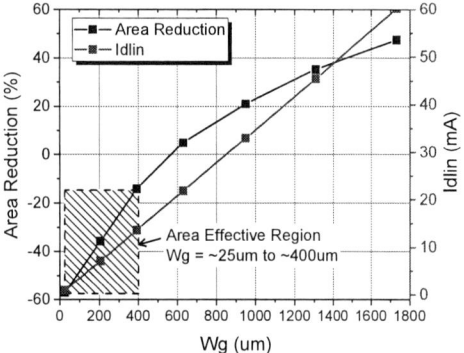

Fig. 2. Layout size comparison over device width (*multi-finger vs. new*).

III. DEVICE PHYSICS

When looking at the architecture of typical reference single resurf device (Fig. 3), the main limitation or drawback is in the vertical electric field distribution below the drain. The breakdown voltage is adjusted by setting appropriate charge balance between the drift and resurf layers. Exact setup of process conditions is needed to achieve acceptable trade-off between device breakdown (Bvdss) and device on-resistance (Rdson).

Fig. 3. Doping concentration of reference device on the left and impact ionization with equipotential lines of electrostatic potential on the right.

The physical principle of new resurf technique is demonstrated in Fig. 4. The sinker region at the drain improves the equipotential lines underneath of drain and

results in a quasi-lateral electric field that enhances the high voltage capabilities. BVdss no longer is limited by high electric field below the drain. One of the important parameters of the new architecture is spacing between the resurf layer and the new deep drain layer (N-sink). Zero spacing between N-sink and P-resurf was selected in our structure to achieve high breakdown located in the bulk far away from the channel. The next two important parameters are: Lsti which represents the drift length and Lpov which represents the length of the gate poly field plate (Fig. 4).

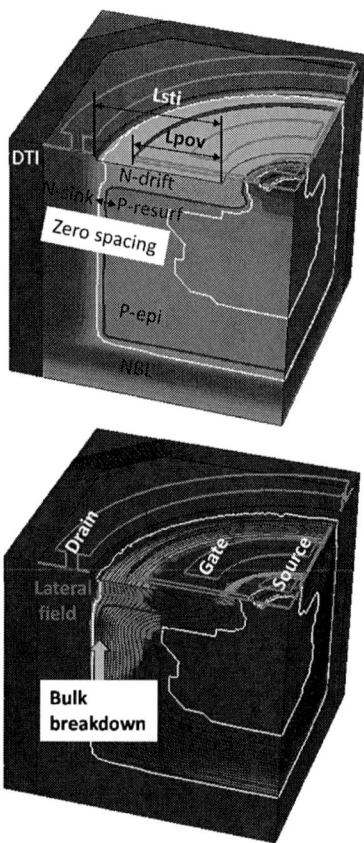

Fig. 4. Doping concentration of the new device on the top and impact ionization with equipotential lines of electrostatic potential on the bottom.

The trade-off between these two parameters is evaluated in Fig. 5 using a surface chart. Various combinations can be extracted for the same breakdown voltage. The selection is intentionally focused on a specific part of the chart. Lsti is tuned to be as short as possible for the lowest Rdson, while Lpov is tuned to be longer to achieve a stable process.

Three regions are recognized in Figure 5. Region A is characterized by lower breakdown voltage due to excessively long poly field plate electrode. It is too close to the drain terminal resulting in an increased electric field at the STI (*Shallow Trench Isolation*) edge close to the drain. Critical electric field can be reached at lower reverse bias of the drain. Region B corresponds to the highest breakdown due to the position of critical electric field in the bulk. The electric field is changing its orientation from vertical to quasi-lateral and is more capable for high breakdown. Region C is a region that is not suitable for the high breakdown. The drift region is too long which increases Rdson, and the poly field plate is too short therefore it is not able to push the electric field deeper into the bulk. In this case the device behavior is comparable with conventional single resurf architecture and the breakdown occurs between drift and resurf layer vertically (Fig. 6).

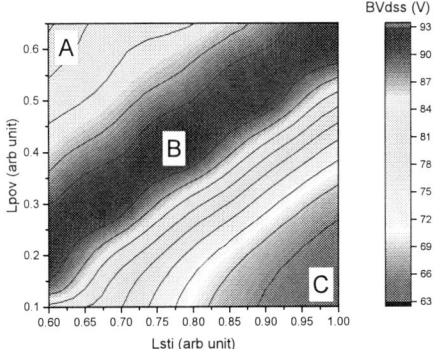

Fig. 5. Device breakdown in relation to Lsti and Lpov geometrical parameters.

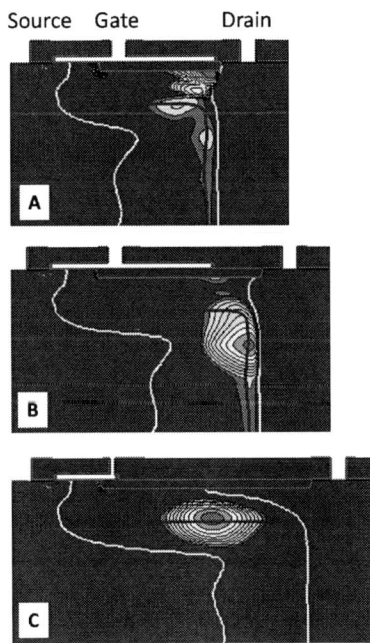

Fig. 6. Impact ionization of the new device at breakdown. The A, B, C symbols are the same as in Fig. 5 indicating on the different locations of device breakdown.

IV.   EXPERIMENTAL RESULTS

Silicon data corresponds to the trends evaluated in TCAD (*Technology Computer-Aided Design*). The highest breakdown ~107V was achieved by using of the 45V drift and resurf layers from the baseline 0.18um BCD technology. The new architecture doubles voltage capability of this device (Fig. 7). The predicted higher Rdson was confirmed and is demonstrated in output characteristics (Fig. 8). Ids current decrease over Vds is caused by Rdson increase due to self-heating of the device in the conventional voltage sweep test.

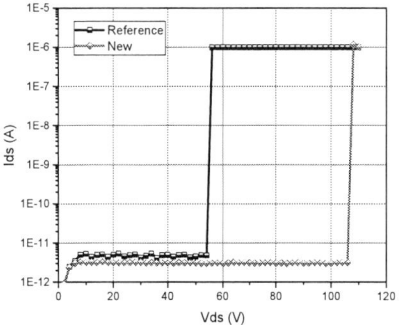

Fig. 7. Breakdown voltage (Bvdss) of the reference and new device.

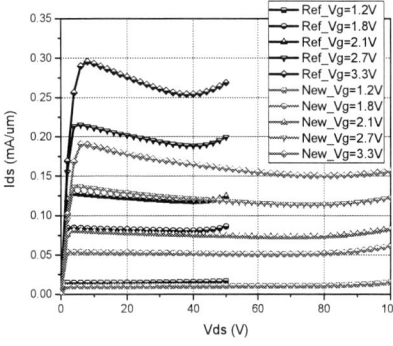

Fig. 8. Output characteristics of the reference and new device.

The concern related to the potential shift of threshold voltage (Vth) in the new architecture was not confirmed. It appears that additional mechanical stress originating from nearby DTI trench is negligible and the transfer characteristics for two drain voltages are identical (Fig. 9). The larger area of the drain which is extended by N-sink results in higher drain to body capacitance (Cdb), but also has a potential to be more rugged during ESD events. Therefore, both devices were evaluated with TLP (*Transition Line Pulse*) tester (Fig. 10). Each point of IdVd curve corresponds to one TLP pulse with the rise time of 10ns and the hold time of 100ns. The voltage and current are read in the hold time region of the pulse [7]. It is clear that the new device exhibits high voltage capability up to 120V due to higher Bvdss. The ESD ruggedness is usually classified by the use of extracted parameters from the TLP curves. One of the parameters is the threshold voltage Vt1 which defines the maximum voltage that is generally measured on the structure before snapback. Related to this is the maximum current before snapping It1. The second parameter is the holding voltage Vh1 which defines (a) the voltage drop after turning-

978-1-5386-2928-4/18 $31.00 © 2018 IEEE

on the parasitic components and (b) corresponding current Ih1 representing how much current can be handled by the device. Reaching the Ih1 point may be destructive for some devices. The TLP curve may contain other parameters for the complicated structures such as the second threshold, the holding voltage and current.

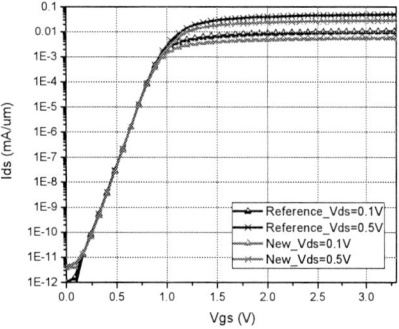

Fig. 9. Transfer characteristics of the reference and new device.

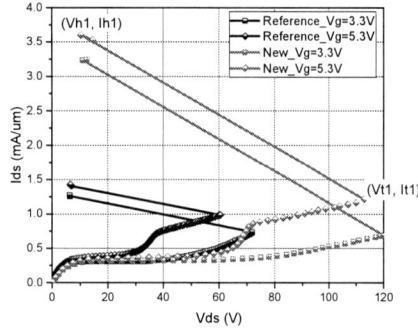

Fig. 10. TLP characteristics of the reference and new device.

The above described parameters of the TLP curves are extracted in the Table 1. Data demonstrates no difference in It1 for the new device at 3.3V on the gate. For the higher Vg = 5.3V which is out of the normal operation ~ 20% improvement is observed. Vt1 is ~ 10V lower for the higher Vg. The main benefit of new device is in ~ 2.5x higher Ih1 current. It results in higher current capability of the new device. The higher Vh1 about ~ 3.5V may also be beneficial.

TABLE 1. TLP DATA COMPARISON

Device	Vg (V)	Vt1 (V)	It1 (mA/um)	Vh1 (V)	Ih1 (mA/um)
Reference	3.3	72	0.74	6.37	1.26
	5.3	60.5	0.98	6.99	1.40
New	3.3	120	0.69	11	3.23
	5.3	113	1.21	10.4	3.59

V. CONCLUSION

Novel resurf technique was introduced with optimized electric field in the drain to increase operating voltage of NLDMOS device and still maintain rugged ESD performance. Rdson is increased in comparison to conventional design, but it is compensated by more efficient isolation for low current applications. There are few limitations and constrains which need to be taken into account; the isolation has limitation due to drain connection to buried layer. Hence, special considerations are needed not to forward bias the body diode for preventing substrate leakage. It also has about ~3.5 times higher Cdb capacitance which has disadvantages for high frequency applications like ultrasound drivers (~15 MHz). In conclusion, advantages of higher operating voltage range and rugged ESD performance will be further studied against the limitations related to the usability of this device. Authors observe a promising utility of the novel device design in automotive, medical and other segments where small isolated high voltage device is needed with no additional cost to fabricate it.

ACKNOWLEDGMENT

The authors would like to thank Mark Griswold and his team for providing feedback and electrical characterization of the device. This work has been partially supported by TH03010006 grant awarded by the Technology Agency of the Czech Republic.

REFERENCES

[1] Claudio Contiero, Antonio Andreini, Paola Galbiati, Roadmap Differentiation and Emerging Trends in BCD Technology, 32nd European Solid-State Device Research Conference (ESSDERC), 2002, ISBN 88-900847-8-2.

[2] Badih El-Kareh, Lou N. Hutter, Silicon Analog Components; Device Design, Process Intergration, Characterization, and Reliability, Springer Science+Business Media, New York 2015, ISBN: 978-1-4939-2750-0, pp. 88 – 96.

[3] Yue Fu, Zhanming Li, Wai Tung Ng, Johnny K.O. Sin, Integrated Power Devices and TCAD Simulation, Taylor & Francis Group, LLC 2014, ISBN: 978-1-4665-8381-8, pp. 206 – 220.

[4] Moshe Agam, Thierry Yao, Agajan Suwhanov, Tracy Myers, Yutaka Ota, Sallie Hose, Matt Comard, "New Modular High Voltage LDMOS Technology Based on Deep Trench Isolation and 0.18um CMOS Platform" , ASMC 2014.

[5] Ankit Kumar, Emita Yulia Hapsari, Vasanth Kumar, Design of a low on resistance high voltage (<100V) novel 3D NLDMOS with side STI and single P-top layer based on 0.18um BCD process technology, Nanotechnology Meterials and Devices Conference (NMDC), 2013, pp. 303 – 306, ISBN 978-1-4799-3387-7.

[6] Moshe Agam, Jaroslav Pjencak, Dusan Prejda, Agajan Suwhanov, Ladislav Seliga, Management of Parasitic Bipolars in Modular High Power LDMOS Technology, 46th European Solid-State Device Research Conference (ESSDERC), 2016, pp. 303 – 306, ISSN 2378-6558.

[7] O. Semenov, H. Sarbishaei and M. Sachdev, ESD protection device and circuit design for advanced CMOS technologies, Dordrecht: Springer, 2008. ISBN 1402083009, pp. 32 - 43.

Proceedings of the 30th International Symposium on Power Semiconductor Devices & ICs
May 13-17, 2018, Chicago, USA

A 0.35μm 600V Ultra-Thin Epitaxial BCD Technology for High Voltage Gate Driver IC

Huihui Wang[1*], Ming Qiao[2*], Feng Jin[1], Yang Yu[2], ZhangYi'an Yuan[2], Binbin Miao[1], Wenqing Yang[1], Jie Wu[1], Wensheng Qian[1], Tong Deng[1], Donghua Liu[1], Ziquan Fang[1], Wenting Duan[1], Jiye Yang[1], Weiran Kong[1] and Bo Zhang[2]

[1]Shanghai Huahong Grace Semiconductor Manufacturing Corporation, Shanghai, P.R. China,
[1*]Huihui.wang@hhgrace.com

[2]State Key Laboratory of Electronic Thin Films and Integrated Devices, University of Electronic Science and Technology of China, Chengdu, P.R. China, [2*]qiaoming@uestc.edu.cn

Abstract—A 0.35μm 600V ultra-thin epitaxial BCD technology for high voltage gate driver IC is developed in this work, including 600V LDMOS, 600V isolation ring, asymmetric 20V NMOS, asymmetric 20V PMOS, symmetric 20V NMOS, symmetric 20V PMOS and BJT. Only 15 photo layers are used for the proposed 1P2M technology. The experimental devices demonstrate that the BVs of divided RESURF structures are 770V, and the isolation BVs between V_D of the 600V LDMOS and high side V_B are above 30V. The off-state BVs of the NMOS and the PMOS structures are greater than 38V, respectively, while the current capacity in the same area is doubled, namely, the area of the CMOS driver is decreased by 50% at least at same drive capability, compared with the competitor A.

Keywords—Ultra-thin epitaxial layer; 600V BCD technology; high voltage gate driver; high current capability;

I. INTRODUCTION

High voltage gate driver ICs, which have the advantages of short response time, low power consumption, high integration and high reliability, are widely used in motor drives, automotive electronics, electronic ballasts, switching mode power supplies and other fields [1-11]. High voltage LDMOS is used as level shift transistors and high voltage isolation region from high side to low side devices which usually account for a small proportion of the total area of high voltage gate driver ICs. The conventional high voltage gate drive ICs are mostly based on the thick epitaxial technology and have slightly larger feature size, resulting large area for high side and low side CMOS circuits.

In this work, a 0.35μm 600V ultra-thin epitaxial BCD technology for high voltage gate driver IC is developed, which achieves smaller feature size and less thermal processes. The presented BCD technology enables monolithic solutions for high voltage gate driver IC, which features a smaller area of CMOS driver and high voltage application up to 600V. In addition, only 15 photo layers are used in this 1P2M BCD technology, resulting low manufacturing cost.

II. STRUCTURE AND PROCESS

Fig. 1 shows the cross-sectional views of main devices for the proposed 0.35μm 600V ultra-thin epitaxial BCD technology with only 15 photo layers for high voltage gate

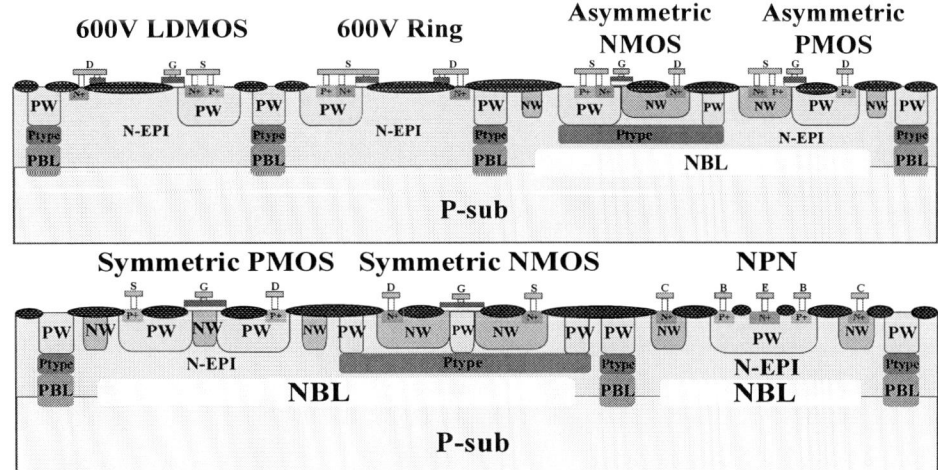

Fig. 1: Schematic cross-sectional views of main devices for the proposed 0.35μm 600V ultra-thin epitaxial BCD technology with only 15 photomasks for high voltage gate driver IC.

978-1-5386-2928-4/18 $31.00 © 2018 IEEE   311

driver IC, including 600V single RESURF LDMOS, 600V isolation ring, asymmetric 20V NMOS, asymmetric 20V PMOS, symmetric 20V NMOS, symmetric 20V PMOS and NPN type BJT. For this BCD technology, the thickness of N-epi layer is only 5µm and after the traditional formation of field oxide layer, the NW, PW and Ptype regions are all realized by multiple implantations with different implant energies. Due to thermal process minimized and using ultra-thin EPI process, the feature size of isolation area is extremely shrunk, resulting less device size. The lateral isolation from high side devices to low side devices is formed by a divided RESURF structure which consists of 600V single RESURF LDMOS and 600V isolation ring. While the vertical isolation is formed by N-epi, high doped NBL layer and high resistance of p-substrate. PW, Ptype and PBL form the isolation well for the target to 20V V_{DD} voltage applications. The technology provides a set of both asymmetric and symmetric 20V full isolated CMOS which are focused on high side gate driver IC application. The cross-section of them is also shown in Fig.1. Ptype layer is introduced in 20V device under NW region, and picked up by PW which is surrounding NW. Then NW, drain of 20V NMOS, would be isolated to P-substrate by vertical N(NW)-P(Ptype)-N(NBL)-P(P-sub) structure. This full isolated structure can be used in high side application to avoid noise leaking from drain to P-substrate. The capability of isolation is measured in silicon, and the sustained punch-through voltage is above 35V which is satisfied to 20V V_{DD} application. Furthermore, the NPN type BJT shown in Fig. 1 is also fabricated in the presented BCD technology.

Table I shows the main measured results of these devices performance in the proposed BCD technology, including the specific on-state resistance ($R_{on,sp}$), off-state breakdown voltage (BV_{off}), the on-state breakdown voltage (BV_{on}), the saturation current (I_{Dsat}) and the threshold voltage (V_{TH}).

Table I. ELECTRICAL PARAMETERS OF MAIN DEVICES FOR 0.35µm 600V ULTRA-THIN EPITAXIAL BCD TECHNOLOGY

Device	BV_{off}	BV_{on}	$R_{on.sp}$	I_{Dsat}	V_{TH}
	(V)	(V)	(mΩ·mm²)	(µA/µm)	(V)
600V isolation ring	770	/	/	/	/
600V LDMOS	770	>600	/	/	1.75
Asymmetric 20V NMOS	42	>20	37	405	1.75
Asymmetric 20V PMOS	38	>20	73	275	1.75
Symmetric 20V NMOS	42	>20	/	195	1.75
Symmetric 20V PMOS	38	>20	/	97	1.75

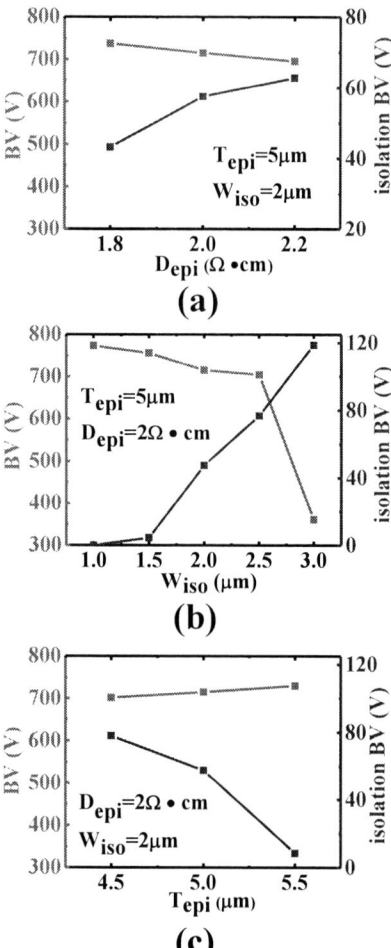

Fig. 2: Simulated BVs of the 600V divided RESURF structures and isolation BVs between V_D of the 600V LDMOS and high side V_B as functions of (a) Depi, (b) Wiso, and (c) Tepi.

III. EXPERTIMENTAL RESULTS AND DISCUSSIONS

A. *600V single RESURF LDMOS in the divided RESURF structure*

For the isolation of high-side region to low-side part, there are mainly two-part isolations should be taken into account: the one is LDMOS BV to GND (BV), the other is V_B to V_D (isolation BV). Considering the process tolerance, the first isolation of the LDMOS BV for 600V BCD technology should be higher than 650V. The second isolation BV value of V_B to V_D that is set to a target value of above 35V in this work which can cover the gate driver ICs with the V_{DD} of 20V.

To evaluate the electrical performance of 5μm ultra-thin epitaxial BCD technology with this novel nLDMOS, we split different EPI resistivity (D_{epi}) and the width dimensions of PW, Ptype and PBL combined junctions between V_B and V_D (W_{iso}). Fig. 2(a), (b) and (c) show simulated BVs of the 600V divided RESURF structures and isolation BVs between V_D of the 600V LDMOS and high side V_B as functions of D_{epi}, W_{iso}, and T_{epi}, respectively. The simulated results indicate that the two-part BV exhibit opponent trends when these variables increasing on the typical process condition nearby. To realize the isolation of high-side region to low-side part, the trade-off of BV of 600V divided RESURF structure and isolation BV should be considered cautiously.

Fig. 3 shows the measured performances of the 600V divided RESURF structures on the proposed BCD technology. Fig.3 (a) shows the measured BVs of the 600V divided RESURF structures and isolation BVs between V_D of the 600V LDMOS and high side V_B as functions of W_{iso}, in which we

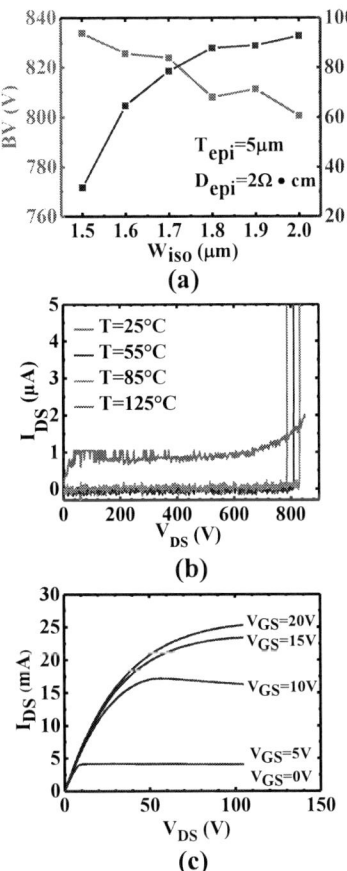

Fig. 3: Measured performances of the 600V divided RESURF structures: (a) *BV*s of the 600V divided RESURF structures and isolation *BV*s between V_D of the 600V LDMOS and high side V_B as functions of W_{iso}, (b) Off-state breakdown characteristic curves of the 600V divided RESURF structure at different temperatures and (c) Output characteristic curves of the 600V single RESURF LDMOS in the divided RESURF structure.

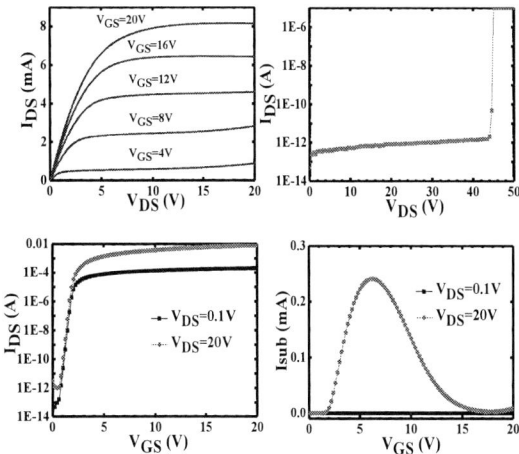

Fig. 4: Measured results of asymmetric NMOS: (a) Output characteristic curves, (b) Off-state breakdown characteristic curve, (c) Transfer characteristic curves and (d) I_{SUB}-V_{GS} curves.

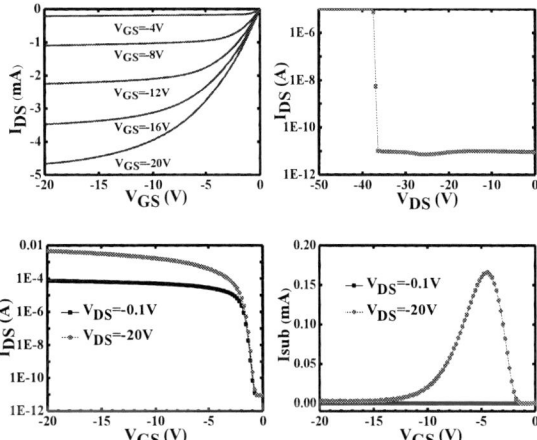

Fig. 5: Measured results of asymmetric PMOS: (a) Output characteristic curves, (b) Off-state breakdown characteristic curve, (c) Transfer characteristic curves and (d) I_{SUB}-V_{GS} curves.

can clearly see the correlation of W_{iso} to BV values. In order to ensure that the BV is higher than the target value of 650V while the isolation BV above 35V in this work, the 5μm thickness epitaxial layer, the 2μm width dimensions of PW, Ptype and PBL combined junctions and 2ohm·cm resistivity are selected as the typical process conditions. Fig. 3(b) shows the off-state breakdown characteristic curves of the 600V divided RESURF structure at different temperatures on the typical process condition, indicating that the BV_{off} is about 780V while leakage current before breakdown is smaller than 0.2μA at room temperature. At 125°C, the BV_{off} is above 700V while leakage current is around 1μA. Fig.3 (c) shows the output characteristic curves of the 600V single RESURF LDMOS in the divided RESURF structure. But the output characteristic is evaluated under 100V only, due to the test prober limited.

978-1-5386-2928-4/18 $31.00 © 2018 IEEE

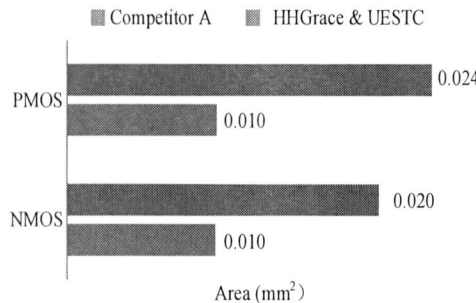

Fig. 6: Comparison of the area of CMOS driver between competitor A and HHGrace & UESTC at the same rated current.

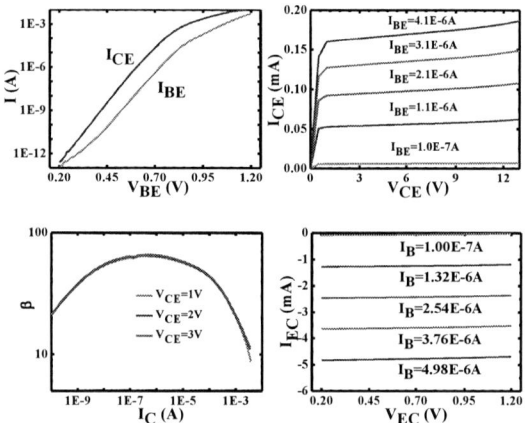

Fig. 7: Measured results of NPN type BJT: (a) I_{CE}, I_{BE}-V_{BE} curves, (b) I_{CE}-V_{CE} curves, and (c) β-I_C curves and (d) I_{EC}-V_{EC} curves.

B. 20V asymmetric CMOS

Fig. 4 and Fig. 5 show the measured asymmetric NMOS characteristics and the measured asymmetric PMOS characteristics, respectively. While the breakdown voltages are 42V for NMOS and -38V for PMOS, respectively. Fig. 4(a) and Fig. 5(a) show the measured output characteristic curves of 20 V asymmetric NMOS and PMOS, respectively. They indicate that the great SOA for the 20V asymmetric CMOS is achieved. Fig. 4(b) and Fig. 5(b) show the measured off-state breakdown characteristic curves of 20 V asymmetric NMOS and PMOS, respectively. Fig. 4(c) and Fig. 5(c) separately illustrate the measured transfer characteristic curves of 20 V asymmetric NMOS and PMOS. The experiment results indicate that the threshold voltages of 20V asymmetric CMOS are 1.75V for NMOS and -1.75V for PMOS, respectively. Fig. 4(d) and Fig. 5(d) show the measured I_{SUB}-V_{GS} curves of 20 V asymmetric NMOS and PMOS, respectively.

Fig.6 shows the comparison of the area of CMOS driver between competitor A and HHGrace & UESTC at the same rated current. Due to 0.35μm technology is adopted, which has the smaller feature size, the area of CMOS driver can be reduced significantly. As indicated from the measured results, the area of the CMOS driver is decreased by 50% at least at same drive capability, compared to the competitor A.

C. NPN type BJT

Fig.7 (a), (b), (c) and (d) show the measured I_{CE}, I_{BE}-V_{BE} curves, (b) I_{CE}-V_{CE} curves, (c) β-I_C curves and (d) I_{EC}-V_{EC} curves of NPN type BJT fabricated in the presented BCD technology, respectively. β of NPN BJT is stable with different V_{CE}, and it is above 20 when I_C from 1nA to 1mA. When I_C is 1μA nearby, the maximumβ is around 70.

IV. SUMMARY & CONCLUSION

A 0.35μm 600V ultra-thin epitaxial BCD technology for high voltage gate driver ICs has been developed and only 15 photo layers are used for the proposed 1P2M technology. By using divided RESURF structure on 5μm N-epi process, the breakdown voltage for 600V single RESURF LDMOS is up to 770V. While with the benefit of minimizing the thermal budget in the process, using ultra-thin epitaxial layer and 0.35um technology, a competitive CMOS area has been achieved. This presented BCD technology enables monolithic solutions for high voltage gate driver IC, which features a high voltage application and a small area of the CMOS driver is at same drive capability. The major performances of this BCD technology for high voltage gate driver IC have been clearly demonstrated.

REFERENCES

[1] Kim, Sunglyong, et al. "A new ESD self-protection structure for 700V high side gate drive IC." ISPSD, pp 467-470,2017.

[2] Qiao, Ming, et al. "A versatile 600V BCD process for high voltage applications." ICCCAS,pp1248-1251, 2007.

[3] Yu, J. S., W. J. Zhang, and W. T. Ng. "A segmented output stage H-bridge IC with tunable gate driver." ISPSD,pp205-208, 2014.

[4] Zhang, Yunwu, et al. "A capacitive-loaded level shift circuit for improving the noise immunity of high voltage gate drive IC." ISPSD, pp173-176,2015.

[5] To, Duc-Ngoc, et al. "Modeling and characterization of 0.35 μm CMOS coreless transformer for gate drivers." ISPSD, pp330-333,2014.

[6] Jonishi, Akihiro, et al. "1200V-Class HVIC technology with a divided high-side well structure for high-functionality and downsizing of circuits." ISPSD, pp422-425,2014.

[7] Nam-Chil Moon,et al."Design and Optimization of 700V HVIC Technology." ISPSD,pp151-154, 2013.

[8] Huang T Y, Huang C H, Huang C F, et al. "Demonstration of a HV BCD technology with LV CMOS proces", ISPSD, pp193-196,2015.

[9] Moscatelli A, Croce G, et al. "LDMOS implementation in a 0.35μm BCD technology (BCD6). " ISPSD, pp323-326, 2000

[10] Lee J H, Su H D, Chan C L, et al. "The influence of the layout on the ESD performance of HV-LDMOS." ISPSD,pp303-306,2010.

[11] Shimizu K, Terashima T. "The 2nd generation divided RESURF structure for high voltage ICs" ISPSD, pp311-314, 2008

Proceedings of the 30th International Symposium on Power Semiconductor Devices & ICs
May 13-17, 2018, Chicago, USA

Impact of Self-Heating Effect in Hot Carrier Injection Modeling

Dong Seup Lee, Dhanoop Varghese, Arif Sonnet, Jungwoo Joh, Archana Venugopal, Srikanth Krishnan
Analog Technology Development, Texas Instruments, USA (dongseup@ti.com)

Abstract— This paper studies the impact of self-heating effects in DC-based Hot Carrier Injection (HCI) modeling in power LDMOS devices. Continuous and large power consumption under the on-state DC stress can result in substantial increase in device temperature, which potentially causes non-negligible error in the HCI modeling. The issue is systematically investigated and verified through various approaches such as comparison of HCI degradation between the devices with different voltage ratings and finger widths, junction temperature estimation with 3D thermal simulation, and pulse-based stress modeling. In addition, it is shown that reliability projection methodology based on the actual circuit waveforms can be more immune to the potential errors caused by the self-heating effect in the conventional DC-based modeling.

Keywords— *hot carrier injection, self-heating, junction temperature, thermal simulation*

I. INTRODUCTION

In semiconductor technology, continuous pursuit for high performance often encounters limitations due to various reliability issues. In the recent power technology, Hot Carrier Injection (HCI) becomes a critical factor in the power device design [1]. In this aspect, accurate modeling and projection of HCI degradation are important to find the optimum device design to meet the reliability requirement and achieve the maximum performance. In the actual power switching applications, HCI events occur only during the hard switching transition whose stress duty cycle is typically less than 1% as shown in Fig. 1. On the other hand, DC-based stress setup with 100% duty cycle has been widely used to model HCI degradation due to several advantages such as setup simplicity and test time reduction. However, there is a risk to underestimate the degradation due to unintended self-heating [2, 3]. This paper studies the self-heating effect in the conventional DC-based modeling through various approaches. In addition, in order to assess its impact more realistically, projection methodology based on actual circuit waveforms is evaluated.

II. EXPERIMENTS AND DISCUSSION

In this study, HCI lifetime (τ) is defined as the time to reach a certain $\Delta I_{d.lin}$ limit. HCI scaling factor (S-factor), which is a relative ratio of inverse of HCI lifetime (S-factor $\sim \tau^{-1}$), is used as the measure of HCI degradation rate in each bias condition as shown in Fig.2 [4]. Fig. 3 shows the S-factors as a function of the gate voltage (V_{GS}) under the fixed drain voltage (V_{DS}),

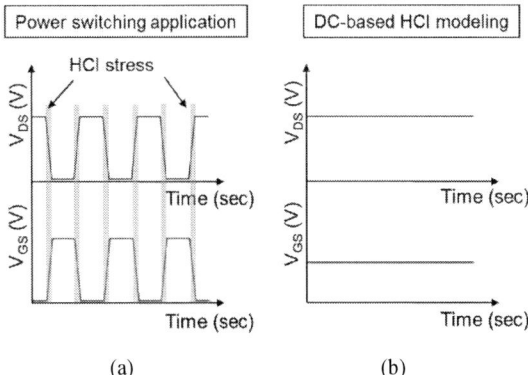

Fig. 1. Simplified gate and drain waveforms in (a) power switching application and (b) DC-based HCI modeling. Different from the actual application case, constant bias stress with 100% duty cycle is applied in the DC modeling setup.

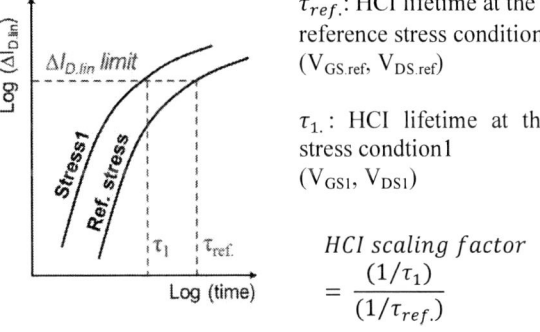

$\tau_{ref.}$: HCI lifetime at the reference stress condition ($V_{GS.ref}$, $V_{DS.ref}$)

$\tau_{1.}$: HCI lifetime at the stress condtion1 (V_{GS1}, V_{DS1})

$$HCI\ scaling\ factor = \frac{(1/\tau_1)}{(1/\tau_{ref.})}$$

Fig. 2. HCI-driven $I_{d.lin}$ degradation behavior and HCI scaling factor (S-factor) definition. The S-factor is the measure of relative speed of the HCI degradation at the given bias condition to the reference condition.

measured with the DC-based modeling setup in different voltage rating LDMOS devices. The devices are from the same process technology, so that their basic device structure and process are the same except the drift region length. All the plots follow the typical bell-shape curve, but two unique trends are observed as the device voltage rating increases. First, the S-factor in mid V_{GS} range (2-4V) becomes lower in the higher voltage rating devices. Next, the second peak in the high V_{GS} range (> 4V) becomes more prominent. As V_{GS} increases well beyond V_T, the self-heating under DC stress becomes non-negligible. It is more serious in the higher

978-1-5386-2928-4/18 $31.00 © 2018 IEEE   315

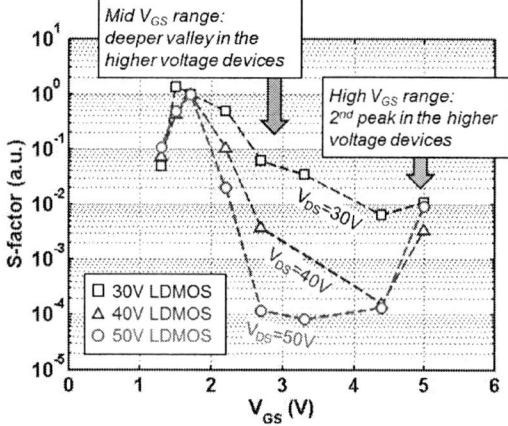

Fig. 3. S-factor as a function of V_{GS} under the fixed drain voltage in the different voltage rating devices in the same generation technology.

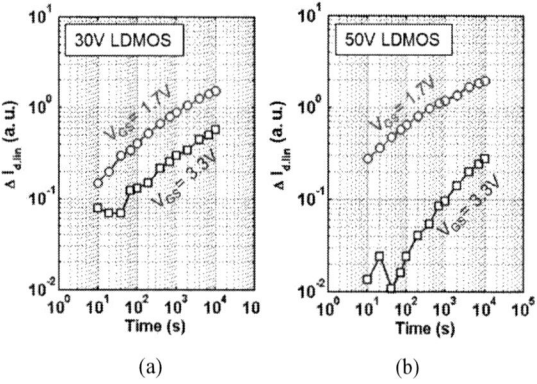

Fig. 4. Comparison of HCI-driven $\Delta I_{d\,lin}$ at V_{GS}=1.7V/3.3V between 30V and 50V LDMOS devices. The separation of curves between two gate voltages becomes larger in the 50V LDMOS device due to the larger self-heating.

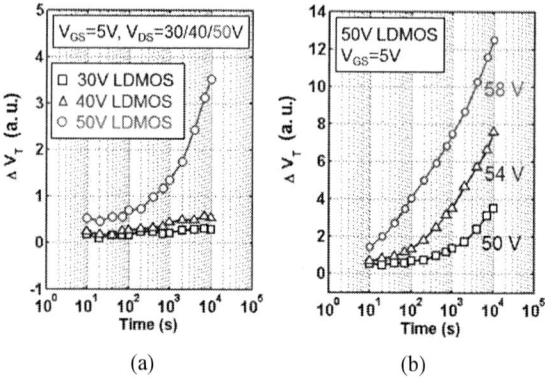

Fig. 5. (a) Comparison of ΔV_T at V_{GS}=5.0V in the different voltage rating LDMOS devices. The self-heating induced ΔV_T increases in the higher voltage rating devices. (b) ΔV_T at V_{GS}=5V depending on V_{DS} in 50V LDMOS.

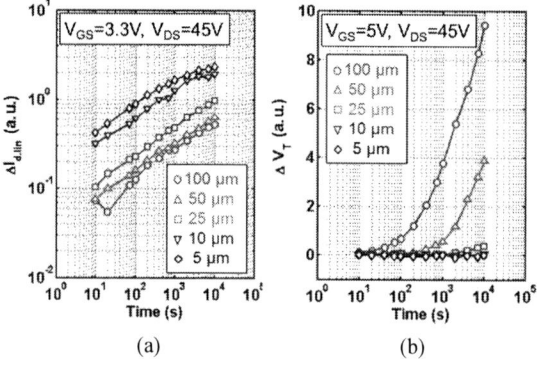

Fig. 6. (a) HCI-driven $\Delta I_{d\,lin}$ at V_{GS}=3.3V and (b) ΔV_T at V_{GS}=5.0 V depending on the finger width in 50V LDMOS devices. With increasing the finger width, HCI degradation becomes slower in the mid-V_{GS} range. In addition, more positive V_T shift is observed in the high-V_{GS} range in the wider finger width devices.

voltage rating devices due to the larger power consumption during the stress. Since the hot carrier generation rate decreases under the high junction temperature, the strong self-heating effect suppresses the HCI-driven degradation. This is why separation of $\Delta I_{d\,lin}$ curves between V_{GS}=1.7V and 3.3V becomes larger in the 50V LDMOS than 30V LDMOS as shown in Fig. 4. However, in the case of even higher V_{GS} range above 4V, the trend observed in the mid V_{GS} range becomes reversed. At this bias range, the extremely high device temperature and high electric field in the gate oxide result in the positive V_T shift as shown in Fig 5 [5]. The amount of the V_T shift shows the correlation to the self-heating. Under the same gate voltage (gate electric field), the high voltage rating device shows more shift due to the larger power consumption (Fig. 5(a)). Also, even in the same voltage rating device, the shift increases with increasing the stress drain voltage (Fig. 5(b)). The high device temperature suppresses the conventional HCI degradation mechanism, but

the positive V_T shift introduces another $I_{d\,lin}$ degradation mechanism by reducing the gate overdrive (V_{GS}-V_T). This is the reason why the second peak in the S-factor plot shown in Fig. 3 is generated. Similar trends are also observed with increasing the finger width (W_f) of the test module as shown in Fig. 6. In the wider finger width devices, the self-heating effect becomes stronger because the heat dissipation along the finger width direction is limited. This also suggests that the DC-based modeling can be sensitive to the device layouts.

To estimate the junction temperature under DC-based HCI tests, 3D ANSYS thermal simulation was conducted. As shown in Fig. 7, the simulation structure was designed to emulate wafer-level test environment. The height of the structure is comparable to the wafer thickness and the lateral size is also large enough to be regarded as the wafer size. The device active channel area was set as a heat source and uniform heat generation was assumed. The heat generation rate was calculated from the product of the voltage and current under the DC stress. In 50V LDMOS with 40 μm finger

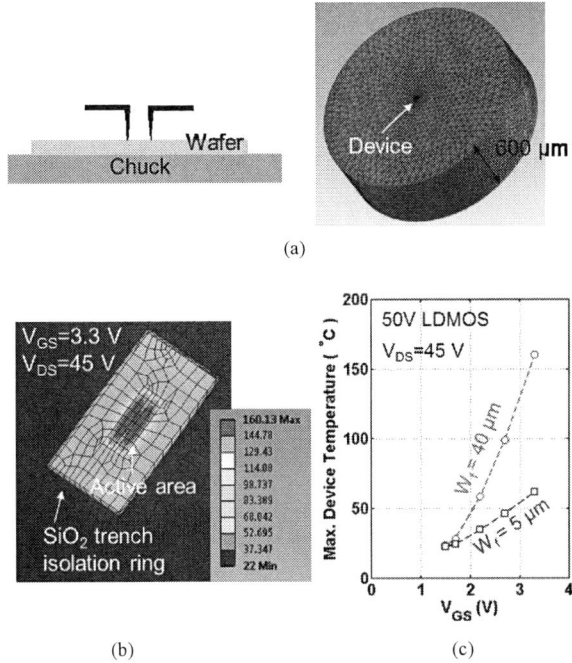

(a)

(b)   (c)

Fig. 7. (a) Simulation structure to emulate the wafer-level test. (b) Simulated temperature map under DC stress with V_{GS}=3.3V, V_{DS}=45V in 50V LDMOS. (c) Simulated maximum temperature as a function of V_{GS} (V_{DS}=45V) in 5 μm and 40 μm finger width devices.

width, the simulation predicts that the device temperature increases to above 100 °C with V_{GS} over 3V. In addition, the finger width dependence is also observed, showing more than 50% increase in the self-heating in the 40 μm width device compared to the 5 μm at V_{GS}=3.3V.

In addition to the simulation, a pulse-based HCI modeling setup was developed with double pulse system for verification as shown in Fig. 8. Instead of stressing the device with the constant bias, 1 μs stress pulses were applied to gate and drain with 1% duty cycle. The pulse width is not short enough to eliminate the self-heating completely, but its impact can be substantially reduced compared to the DC stress. In addition, in order for a fair comparison with DC stress, 50 ns delay was applied between drain and gate pulses. With the delay, the gate voltage transition first occurs under the low drain voltage and then the drain voltage changes. Since the load line follows path 1 instead of path 2 as shown in Fig. 8(a), the degradation during the gate voltage transition can be prevented. As shown in Fig 8(b), at V_{GS}=1.5V, the data from DC and pulse setups is well matched because of negligible self-heating in this bias condition. In contrast, at V_{GS}=3.3V, the degradation becomes a few hundred times faster with the pulse setup due to the reduction of the self-heating. This is consistent with the trends observed in S-factor measurement data and thermal simulation. Also, no positive ΔV_T was observed at the high V_{GS} with the pulse stress, which is another indication of the suppression of the self-heating effect.

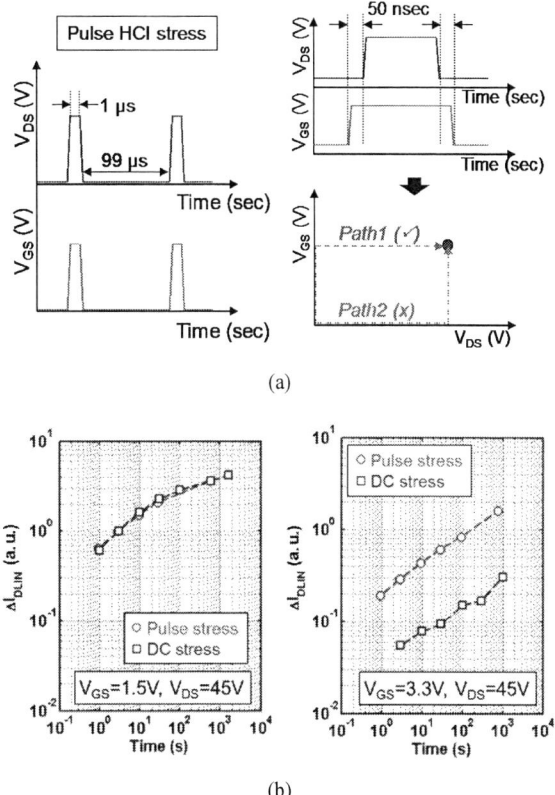

(a)

(b)

Fig.8. (a) Pulse stress setup with 1 μsec pulse width and 1% stress duty cycle. 50 ns delay is applied between drain and gate pulses to follow the path 1 (green line) instead of the path 2 (red line) during the on/off transition. (b) Comparison of HCI-driven $I_{d.lin}$ degradation at V_{GS}=1.5V/3.3V (V_{DS}= 45V) between DC-stress and pulse-stress setups in 50V LDMOS device.

TABLE I. COMPARISON OF HCI PROJECTION BETWEEN DC AND PULSE MODELING BASED ON THE FIXED STRESS CONDITION.

Stress condition	$\Delta I_{d.lin}$ (%) (DC)	$\Delta I_{d.lin}$ (%) (Pulse)	Difference
V_{GS}=1.7V, V_{DS}=40V	7.5 %	7.5 %	0.0 %
V_{GS}=2.2V, V_{DS}=40V	6.6 %	7.5 %	0.9 %
V_{GS}=2.7V, V_{DS}=40V	4.4 %	6.4 %	2.0 %
V_{GS}=3.3V, V_{DS}=40V	2.3 %	5.3 %	3.0 %

Based on the data collected with two different modeling setups (DC and pulse), potential risk for underestimation of HCI was analyzed. 100 k Power-On-Hour (POH) with 0.4% HCI stress duty cycle was assumed for the end-of-lifetime projection. Table I shows the difference in $\Delta I_{d.lin}$ projection between DC and pulse-based modeling with direct estimation at fixed bias conditions. The difference increases with raising the stress V_{GS} level and reaches about 3% level in mid V_{GS} range. Since the typical HCI lifetime limit in a qualified technology is generally set in the 10% range, the error is non-

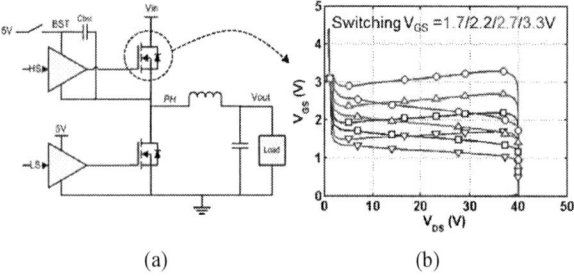

(a)   (b)

Fig. 9. Generation of hard switching stress waveforms based on the actual power switching circuit data. The switching waveforms extracted from the high-side FET in DC/DC converter were scaled to four different load lines to study the impact of the self-heating.

TABLE II.   COMPARISON OF HCI PROJECTION BETWEEN DC AND PULSE MODELING BASED ON THE SWITCHING WAVEFORMS.

Stress condition	$\Delta I_{d.lin}$ (%) (DC)	$\Delta I_{d.lin}$ (%) (Pulse)	Difference
$V_{GS}=1.7V$, $V_{DS}=40V$	7.5 %	7.5 %	0.0 %
$V_{GS}=2.2V$, $V_{DS}=40V$	7.6 %	7.9 %	0.3 %
$V_{GS}=2.7V$, $V_{DS}=40V$	7.7 %	7.9 %	0.2 %
$V_{GS}=3.3V$, $V_{DS}=40V$	7.9 %	8.1 %	0.2 %

negligible. Moreover, considering that the pulse setup does not remove the self-heating completely, the actual error is expected to be larger than the estimation.

Since gate and drain voltages continuously change during hard switching in real applications, the projection based on fixed conditions cannot capture the real nature of HCI stress. In order to solve the issue, projections based on the actual switching waveforms were conducted. The baseline stress waveform was extracted from a high-side FET in the half-bridge DC/DC converter circuit as shown in Fig. 9 (a). Then, the switching V_{GS} levels in the waveforms were scaled to generate the load lines covering the different self-heating regimes (Fig. 9(b)). Table II summarizes the projection result based on the switching stress waveforms. Unlike the projection with the fixed stress conditions, the maximum difference between DC and pulse-based modeling is reduced below 0.5%. In addition, regardless of the switching V_{GS} levels, the projected $\Delta I_{d.lin}$ becomes similar. During hard switching, V_{GS} first rises from 0V to the switching voltage level under high drain bias. The net HCI degradation is determined by the accumulated damage occurred in this transition and it is mainly dominated by the peak HCI point which is in the low V_{GS} range as shown in Fig. 10. In other words, in spite of the self-heating driven error in mid/high V_{GS} ranges during DC modeling, its impact is significantly lower in the projection based on the actual switching waveforms. In addition, since this is valid as long as the peak HCI point is formed in the low self-heating regime, it applies to the most power devices with the typical bell-shape S-factor curves.

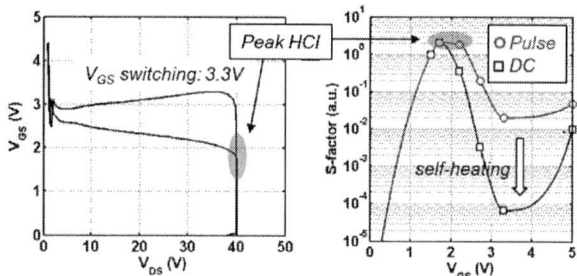

Fig.10. Cause of error reduction in the projection method based on the actual switching waveforms.

Thus, in general, the actual circuit-based HCI projection method can be more immune to the potential error caused by the self-heating effect in the conventional DC-modeling. Also, based on this, the combination of DC-based HCI test and circuit-based projection methodology can be used to achieve test efficiency and minimize potential errors.

III. SUMMARY

We investigated the impact of self-heating in conventional DC-based HCI modeling methods in power devices. We found that self-heating during DC stress can result in non-negligible error in the estimation of HCI degradation in the mid/high V_{GS} ranges. Also, for the same reason, the modeling data can be sensitive to the device layouts. In spite of these issues, the risk for lifetime projection error can be minimized by using the reliability projection methodology based on the actual switching waveforms. Since this can be applied to the most power devices with typical bell-shape HCI curves, it is expected to be a useful solution, providing both test efficiency and minimum error.

REFERENCES

[1] S. Pendharkar, "Technology requirements for automotive electronics," in Proc. 2005 *IEEE Conf. Vehicle Power and Propulsion*, 2005, pp. 834-835)

[2] C. Cheng, J. F. Lin, and T. Wang, "Impact of Self-Heating Effect on Hot Carrier Degradation in High-Voltage LDMOS," in Proc. *IEDM Symp. Tech. Dig.*, Dec. 2007, pp. 881-884.

[3] J. Hao, M. Pelletier, R. Murphy, and T. E. Kopley, "Self Heating Effect on Hot Carrier Degradation Characteristic in High Voltage n-channel LDMOS," in Proc. *Int. Rel. Phys. Symp.*, Apr. 2009, pp. 634-639.

[4] D. Varghese, P. Moen, and M. A. Alam, "ON-State Hot Carrier Degradation in Drain-Extended NMOS Transistors," *IEEE Trans. Elec. Dev.*, vol. 57, no. 10, pp. 2704-2710, Oct. 2010.

[5] P, Moens, J. F. Cano, and B. Desoete, "Anomalous BTI Effects in n-Type Integrated Power Transistor," *IEEE Elec. Trans. Dev. Lett.*, vol. 27, no. 6, pp. 502-504, Jun. 2006.

Proceedings of the 30th International Symposium on Power Semiconductor Devices & ICs
May 13-17, 2018, Chicago, USA

Duty-cycle-accelerated Hot-carrier Degradation and Lifetime Evaluation for 700V Lateral DMOS

Siyang Liu, Zhichao Li, Wangran Wu, Weifeng Sun*

National ASIC System Engineering Research Center
Southeast University, Nanjing, China
*E-mail: swffrog@seu.edu.cn

Shulang Ma, Yuwei Liu, Wei Su, Xiaohong Liu

CSMC Technologies Corporation
Wuxi, China

Abstract— Due to serious self-heating effect, traditional DC stress is hard to be used for evaluating the hot-carrier degradation of the LDMOS above 600V. In this work, the hot-carrier degradation for a 724V-breakdown LDMOS is studied by adopting gate duty-cycle-accelerated AC stress. It demonstrates that hot-electrons injection and donor-like interface states generation happen near the drain when the gate pulse is high. No obvious degradation and recovery can be observed when the gate pulse is zero. Moreover, the short pulse edges enhance the decrease of on-resistance (R_{on}) due to transient hot-holes injection into bird's beak. The device lifetime is also calculated according to the proposed models related to duty-cycle.

Keywords— *700V LDMOS, Hot-carrier degradation, Duty-cycle, AC stress, Lifetime model*

I. INTRODUCTION

Lateral DMOS has expanded to the application area above 600V to improve the integration level of high-voltage power chip [1-2]. Such high operation voltage and electrical field make the device suffer from serious hot-carrier degradation. Main existing literatures about hot-carrier degradations were focused on the below-200V LDMOS by using DC maximum substrate current stress or maximum gate voltage stress, and the presented lifetime models were essentially based on Hu-model or $1/V_d$-model [3-4]. However, due to the limitation of heat dissipation at high power density, high-voltage LDMOS above 600V usually owns very narrow DC thermal SOA. Thereby, traditional hot-carrier investigation methodology based on DC stress is unavailable, which might directly destroy the LDMOS.

By adopting the gate duty-cycle-accelerated stress, the self-heating can be restricted, as a result, the hot-carrier degradation mechanism and lifetime of the 700V-class device can be obtained. In this work, the hot-carrier degradation for a 724V-breakdown LDMOS is studied by adopting gate duty-cycle-accelerated AC stress, and the related lifetime evaluation models are proposed. Finally, the lifetime of investigated LDMOS under the normal operation condition can be calculated.

II. DEVICE STRUCTURE AND EXPERIMENTS

The studied LDMOS is processed in the 0.5μm BCD technology. Fig. 1 shows the schematic cross-section of the device. The buried-pwell and buried-nwell are set to get high breakdown voltage. The effective channel length and accumulation region length are 4.0μm and 1.5μm, respectively. The thick gate oxide (100nm) and long drift region (73μm) are needed for the high operational voltage. For the device, the off-state breakdown voltage is 724V and the threshold voltage is 1.7V. Moreover, the normal operation V_{ds} and V_{gs} are 600V and 12V with 2% gate duty-cycle (200ns) and 10ns edges.

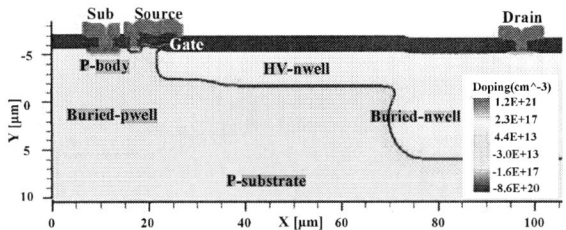

Fig.1: The schematic cross section of investigated 700V LDMOS

Fig.2 shows the substrate current (I_{sub}) variations with V_{gs} at different V_{ds} for the LDMOS (DC measurements). It can be seen that the device is destroyed when the DC stress is small (@V_{gs}=3.2V&V_{ds}=500V) due to the thermal accumulation. Thereby, the pulse stress is necessary. In this paper, the gate duty-cycle-accelerated method with constant operation V_{ds} will be adopted to investigate the degradation mechanism and lifetime. The stress tests are interrupted selectively to monitor the degradation of key electrical parameter, on-resistance (R_{on}).

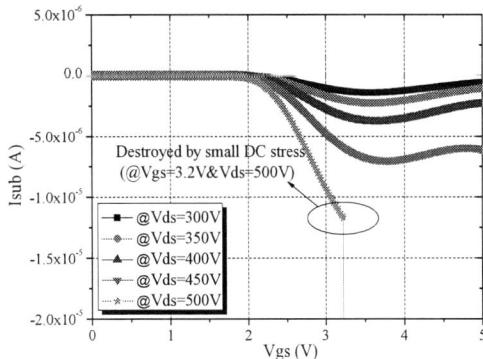

Fig.2: The substrate current (I_{sub}) variations with V_{gs} at different V_{ds} for 700V LDMOS (DC measurement)

The work is supported by CSMC Technologies Corporation.

978-1-5386-2928-4/18 $31.00 © 2018 IEEE

319

III. MEASUREMNTS AND DISCUSSIONS

Fig.3 shows the R_{on} degradations under different gate duty-cycle stresses with 12V pulse amplitude and constant 600V V_{ds}. The R_{on} decreases at the beginning of the stress but increases subsequently, and larger duty-cycle makes the device degrade more quickly and seriously. Moreover, it is noted that the self-heating effect can be ignored for below-30% duty cycle stress according to our temperature monitoring.

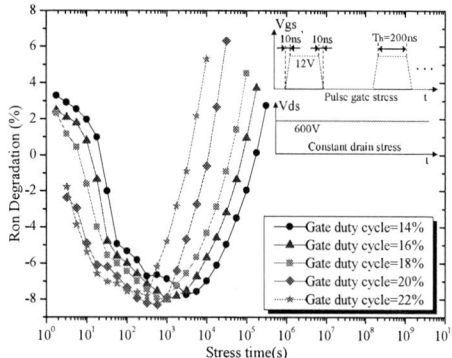

Fig.3: R_{on} degradations under different gate duty-cycle stresses

To understand the inner degradation mechanism, the T-CAD simulations have been performed. Firstly, the stage of high gate pulse level is concerned. As shown in Fig.4, at the stage of 12V high pulse level, the peak impact ionization is located near the drain due to Kirk effect, leading to the interface states generation there [5]. Moreover, the positive perpendicular electrical field near the drain is favor of active hot-electrons injection into field oxide, repelling the flowing electron current and increasing the R_{on} [6]. Furthermore, Fig.5 shows that the influences of interface states with different types upon R_{on} degradations. It can be inferred that the effective interface states may be donor-like types, making the R_{on} decrease. Meanwhile, the results in Fig.5 also verify the effect of the injected electrons, which brings the increase of R_{on}.

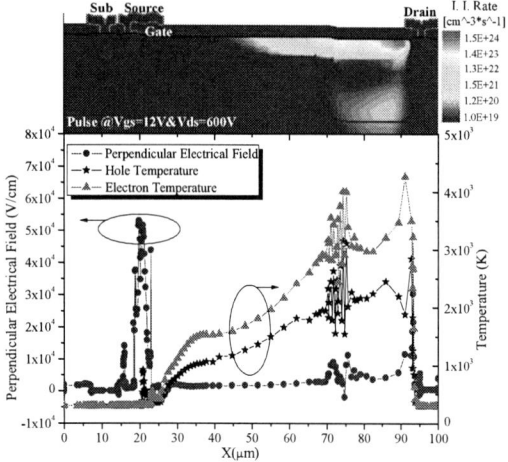

Fig.4: Whole 2D impact ionization distribution, perpendicular electrical field, hot hole and electron temperatures distributions along the device surface for the LDMOS at stage of 12V high pulse level

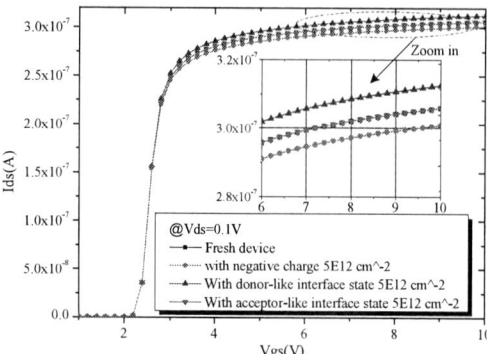

Fig.5: Transfer characteristic curves for fresh LDMOS and the devices with different damage defect types at the Si/SiO₂ interface near the drain (X=89-91μm)

Then, the stage of zero level is concerned. The R_{on} degradations at $V_{gs}=0V\&V_{ds}=600V$ and $V_{gs}=12V\&V_{ds}=0V$ are shown in Fig.6. It is clear that the degradations are very small and can be ignored. The degradation recovery observation (@$V_{gs}=0V\&V_{ds}=600V$) after 1000s 20% gate duty-cycle stress is also shown in Fig.6, indicating that the 0V level stage indeed doesn't contribute the degradation.

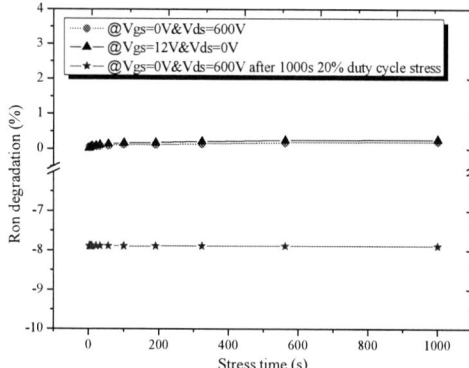

Fig.6: R_{on} degradations at $V_{gs}=0V\&V_{ds}=600V$ and $V_{gs}=12V\&V_{ds}=0V$, as well as the degradation recovery observations after 1000s 20% gate duty-cycle stress

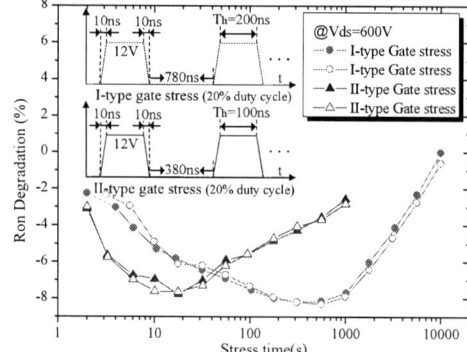

Fig.7: R_{on} degradations for the LDMOS at different 20% gate duty-cycle-accelerated stresses (I-type with 200ns high level holding time, II-type with 100ns high level holding time)

978-1-5386-2928-4/18 $31.00 © 2018 IEEE

Next, the gate pulse edges are concerned. As shown in Fig.7, when more pulse edges are inserted into gate stress sequence, the R_{on} decreases more obviously at the beginning of the stress. To understand the mechanism, Fig.8 simulates the substrate current and V_{gs} variations with the pulse time. It is interesting to observe that the substrate current peaks occur during the pulse edges, indicating more serious impact ionization rates at those moments. The 2D impact ionization distribution at the moment that peak substrate current appears during the edge is also shown in Fig.8. It can be seen that the peak impact ionization is located at the bird's beak.

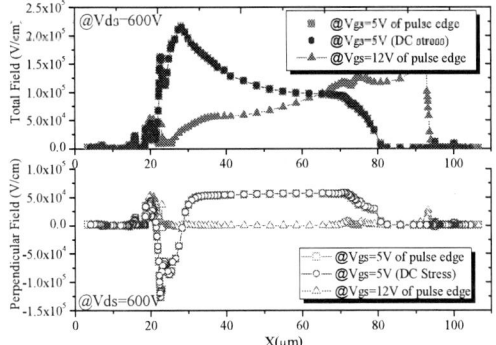

Fig.8: The substrate current and V_{gs} variations with the pulse time, as well as the 2D impact ionization distribution at the moment that peak substrate current appears during the edge

Fig.9 shows the total electrical field and perpendicular electrical field distributions along the device surface for the LDMOS at different moments of the gate pulse edge and DC stress, while Fig.10 indicates the corresponding total flowing current density distributions along the device surface. Clearly, the enhancement phenomenon of impact ionization during the pulse edge is due to the lagging behind of flowing current density change comparing to the electrical field change. The lagged carriers supply enough source to generate more impact ionization. Also, the negative perpendicular electrical field at the bird's beak during the pulse transient helps hot-holes injection into field oxide, leading to R_{on} decrease due to induced mirror negative charges in the HV-nwell. As a result, more pulse edges enhance the decrease of R_{on} at the beginning.

Fig.9: Total electrical field and perpendicular electrical field distributions along the device surface for the LDMOS at different moments of the gate pulse edge and DC stress

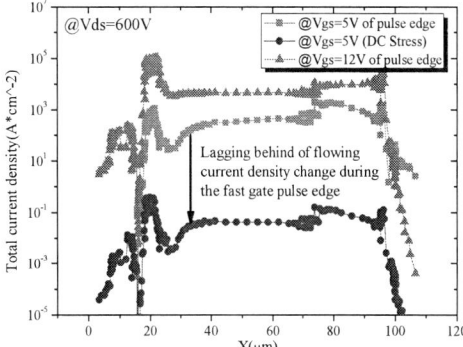

Fig.10: Total flowing current density distributions along the device surface for the LDMOS at different moments of the gate pulse edge and DC stress

To calculate the device lifetime at the duty-cycle-accelerated stress, the following model-I has been presented to fit the R_{on} degradation with the stress time.

$$\Delta R_{on} = A_1 * \exp\left(-\frac{t}{m}\right) + A_2 * \exp\left(-\frac{t}{n}\right) + B \qquad (\text{I})$$

in which, A_1, A_2, m, n and B are constants, t is stressing time. The fitting results with different gate duty-cycle stresses are shown in Fig.11. We define 10% R_{on} shift as the HCI failure of the LDMOS, in this way, the device lifetime at each duty-cycle-accelerated stress can be obtained.

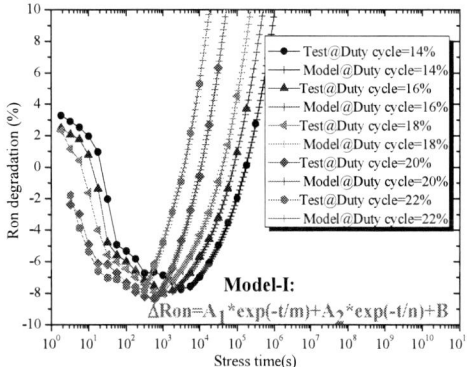

Fig.11: Data fitting of R_{on} degradations with the stress time based on the proposed model

Finally, the following model-II is also proposed to descript the relationship between duty-cycle and device lifetime.

$$D = k * lg(\tau) + b \qquad (\text{II})$$

in which, D is duty cycle, τ is device lifetime, k and b are constants. The fitting curve is shown in Fig.12. To improve the fitting accuracy, three groups results for each stress have been adopted. Moreover, to verify the model, the errors for the stresses with 16% and 30% duty-cycles have been considered. As shown in Fig.12, the errors between the test results and model predictions can be less than 6%. Therefore, the presented model is useful. According to this, the lifetime of 700V LDMOS under the normal operation condition with 2% duty cycle can be calculated as 17.2 years.

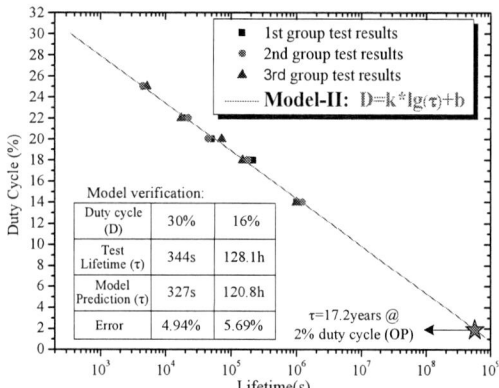

Fig.12: Data fitting based on the proposed model that links the duty-cycle and device lifetime, model verification, as well as the lifetime evaluation of the target LDMOS

IV. Conclusions

In this work, the hot-carrier degradation for 700V LDMOS is studied by adopting gate duty-cycle-accelerated AC stress. The hot-electrons injection and donor-like interface states generation happen near the drain when the gate pulse is high. The former one increases the R_{on}, but the latter one decreases the R_{on}. No obvious degradation and recovery can be observed when the gate pulse is zero. Moreover, more gate pulse edges enhance the decrease of R_{on} due to more transient hot-holes injection into bird's beak and induce mirror negative charges in the HV-nwell.

Finally, the device lifetime models related to duty-cycle are proposed. Based on these, the lifetime of investigated LDMOS under the normal operation condition is calculated as 17.2 years.

Acknowledgment

This work was supported by National Natural Science Foundation of China (61604038, 61674030), and Natural Science Foundation of Jiangsu Province (BK20160691).

References

[1] M. Qiao, Z. Wang, H. Wang, et al., "Edge Termination Design of 700V Triple RESURF LDMOS with N-type Top Layer", *International Symposium on Power Semiconductor Devices & IC's (ISPSD)*, 2017, p319-322.

[2] T. Okawa, H. Eguchi, M. Taki, et al., "2000V SOI-LDMOS with New Drift Structure for HVICs", *International Symposium on Power Semiconductor Devices & IC's (ISPSD)*, 2016, p435-438.

[3] T. Mori, H. Fujii, S. Kubo, et al., "Investigation into HCI Improvement by a Split-Recessed-Gate in an STI-based nLDMOSFET", *International Symposium on Power Semiconductor Devices & IC's (ISPSD)*, 2017, p459-462.

[4] K. Cho, S. Ko, F. Machida, et al., "Investigation of HCI Reliability in Interdigitated LDMOS", *International Symposium on Power Semiconductor Devices & IC's (ISPSD)*, 2015, p69-72.

[5] S. Y. Liu, X. F. Ren, Y. Fang, et al., "Hot-carrier-induced Degradation and Optimizations for Lateral DMOS Transistor with Multiple Floating Poly-gate Field Plates", *IEEE Trans. Electron Devices*, vol. 64, no. 8, pp. 3275-3281, 2017.

[6] P. Moens, F. Bauwens, M. Nelson, and M. Tack, "Electron trapping and interface trap generation in drain extended pMOS transistors," *in Proc. Annu. Int. Rel. Phys. Symp. (IRPS)*, 2005, pp. 555-559.

Proceedings of the 30th International Symposium on Power Semiconductor Devices & ICs
May 13-17, 2018, Chicago, USA

A High-Speed SOI-LIGBT with Electric Potential Modulation Trench and Low-Doped Buried Layer

Shaohong Li[1], Long Zhang[1], Jing Zhu[1], Weifeng Sun*[1], Qingxi Tang[1], Hao Wang[1], Ling Sun[1], Yan Gu[2], Shikang Cheng[2], Sen Zhang[2], Yangbo Yi[3]

[1] National ASIC System Engineering Research Center, Nanjing, China
[2] CSMC Technologies Corporation, Wuxi, China
[3] Chipown Micro-electronics limited Technologies Corporation, Wuxi, China
*swffrog@seu.edu.cn

Abstract—A high-voltage SOI-LIGBT with high turn-off speed and low turn-off loss (E_{OFF}) is proposed in this paper. The proposed SOI-LIGBT features a Low-doped Buried N-layer (LBN) region and an emitter-side Electric Potential Modulation Trench (EPMT) shorted with the P+ emitter. By employing the LBN and EPMT, fast extraction of the stored carrier and the high turn-off speed are realized due to the accelerated depletion of N-drift region. The simulated results show that the proposed SOI-LIGBT can achieve a 73% lower turn-off loss compared with the conventional SOI-LIGBT at the same V_{ON} of 1.52V. Moreover, the hole heat flux distribution in the proposed device predicts an improvement of ruggedness under high-voltage and high-current conditions.

Keywords—SOI-LIGBT; low-doped buried N-layer; electric potential modulation trench; turn-off speed; turn-off loss

I. INTRODUCTION

SOI-LIGBTs are one of the key components in high-voltage monolithic ICs such as the three-phase single chip inverter ICs [1-8]. In order to enhance the current density of the SOI-LIGBTs in monolithic ICs, many efforts, such as multi-channel [9] and U-shaped channel [10] structures have been investigated. For the high frequency switching applications, the power losses mainly come from the turn-off process and thus the SOI-LIGBTs with high turn-off speed are more preferred to the ones with high current density. Employing independent extraction path is the typical principle to improve the turn-off speed. Based on this principle, various approaches including shorted anode [11], segmented NPN controlled anode [12], Schottky anode [13] and multi-gate [14], have been reported. However, all of these approaches adopt the anode (collector) engineering, and inevitably effect the hole injection level from P+ anode (collector) and degrade the on-state voltage drop (V_{ON}).

In this paper, a high-speed SOI-LIGBT adopting the cathode (emitter) engineering is proposed and investigated by TCAD simulations. By employing the Electric Potential Modulation Trench (EPMT) and the Low-doped Buried N-layer (LBN), the high turn-off speed can be realized without sacrificing the injection level of P+ collector.

This work was supported by the National Natural Science Foundation of China (BK20150627 and 61674030), Natural Science Foundation of Jiangsu Province (61504025) and the National key research and development plan (2017YFB0402900).

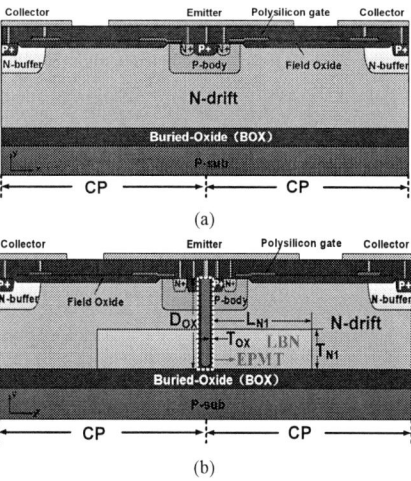

Fig. 1. The cross section views of (a) the conventional SOI-LIGBT and (b) the proposed SOI-LIGBT.

II. DEVICE CONCEPT

Fig. 1 shows the schematic cross-sections of the conventional and the proposed SOI-LIGBTs. The doping concentration of the N-drift region is 8.3×10^{14} cm^{-3}. The thickness of the top layer and the buried oxide (BOX) layer are 18μm and 3.5μm, respectively. In the proposed structure, the EPMT and the LBN region are employed. The EPMT consists of the grounded polysilicon and the surrounded oxide. The depth of the EPMT is D_{OX}, and the thickness of the sidewall oxide of the EPMT is T_{OX}. The length of and the thickness of the LBN are L_{N1} and T_{N1}, respectively. The doping concentration of the LBN region is lower than that of the N-drift region. By virtue of the low doping, the LBN can be depleted rapidly during the turn-off period. The electric potential distribution is laterally reshaped by the EPMT, accelerating the depletion of N-drift region. Combining LBN with EPMT, fast extraction of stored carrier and high turn-off speed are realized. Additionally, the hole inversion layer can be formed along the EPMT, which provides an extra heat flowing path under short-circuit conditions.

978-1-5386-2928-4/18 $31.00 © 2018 IEEE

III. Results and Disscussions

Fig. 2 (a) shows the equipotential contours of the conventional and the proposed SOI-LIGBTs at breakdown. The optimal cell pitch for the conventional and the proposed structures are 52μm and 48μm, respectively. The optimal parameters for the proposed structure are: $L_{N1} = 34$μm, $T_{N1} = 9$μm, $D_{OX} = 18$μm, $T_{OX} = 100$Å and $N_{N1} = 3 \times 10^{14}$ cm^{-3}. With the same BV of 594V, the N-drift length of the proposed structure can be shortened by 4μm compared with the conventional structure, which can be seen from Fig. 2 (b).

Fig. 3 shows the turn-off waveforms of the conventional and the proposed structures. The t_1 and T_1 are the moment that the load current I_L (0.5A) drops to 70%I_L in the conventional and the proposed structures, respectively. At $L_{N1} = 34$μm, $N_{N1} = 3e14$cm^{-3}, $T_{N1} = 9$μm, $D_{OX} = 18$μm and $T_{OX} = 100$ Å, the proposed structure exhibits a turn-off speed much faster than the conventional structure.

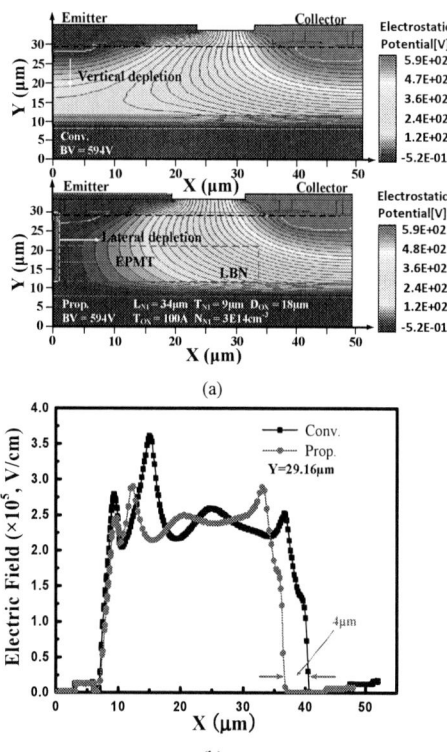

(a)

(b)

Fig. 2. (a) Potential contours of the conventional and the proposed SOI-LIGBTs at breakdown. (b) The surface electric field distribution of the both structures along line Y=29.16μm.

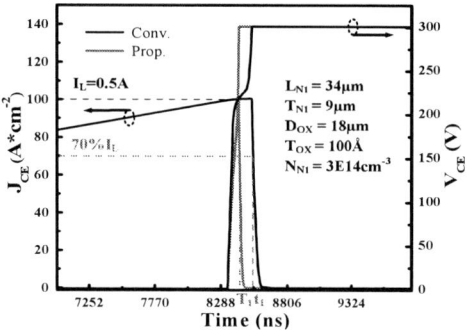

Fig. 3. Turn-off waveforms of the conventional and the proposed SOI-LIGBTs.

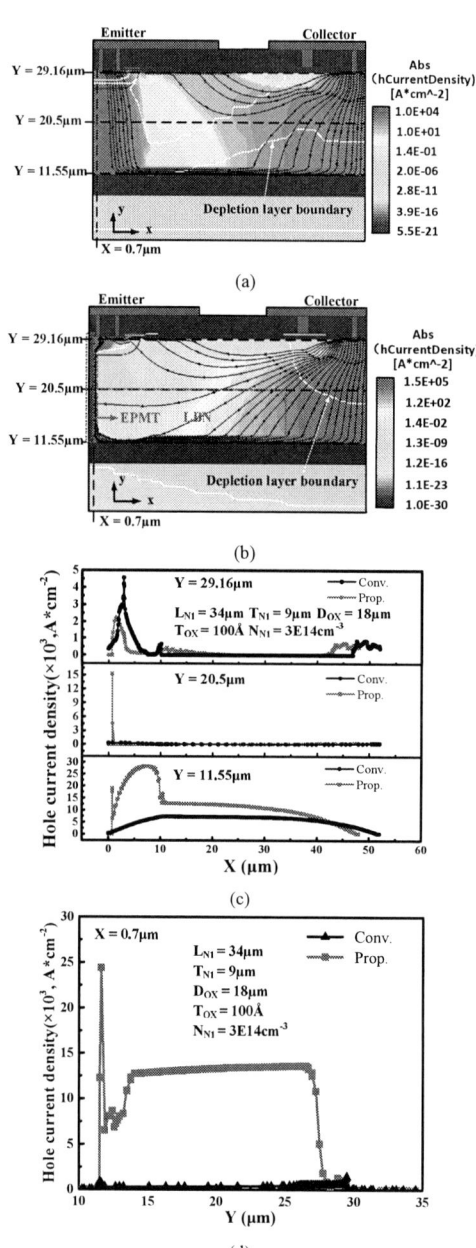

(a)

(b)

(c)

(d)

Fig. 4. The hole current density distribution of (a) the conventional and (b) the proposed structures at t_1 and T_1 moment, respectively, during the turn-off period. (c) The horizontal and (d) vertical profiles of the hole current density in both structures. T_1 and t_1 are labelled in Fig. 3.

978-1-5386-2928-4/18 $31.00 © 2018 IEEE   324

Fig. 4 (a) and (b) show the hole current density of the conventional and the proposed structures at t_1 and T_1 moments, respectively, during the turn-off period. The hole current in the proposed structure tends to flow laterally along the buried oxide layer and vertically along the EPMT, which can be revealed by the hole current density profiles in Fig. 4 (c) and (d). Compared with the conventional structure, the proposed structure shows a higher hole current density along the BOX layer (along the line Y=11.55μm) and a hole current path can be formed along the EPMT (along the line X=0.7μm).

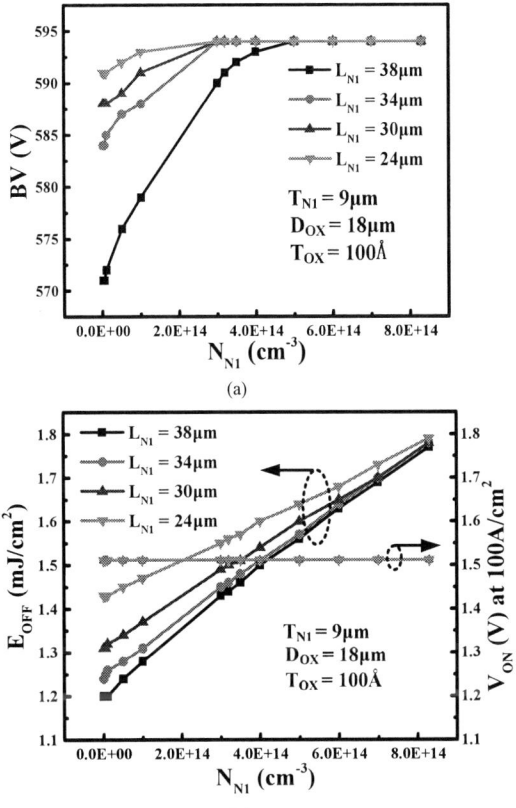

Fig. 5. The dependence of (a) BV, and (b) E_{OFF} and V_{ON} on N_{N1} at different L_{N1} of the proposed structure.

Fig. 5 (a) shows the dependence of BV on N_{N1} at different L_{N1}. BV increases firstly with the increase of N_{N1} and then keeps at 594V. Shorter LBN is beneficial to BV. Fig. 5 (b) shows the dependence of E_{OFF} and V_{ON} on N_{N1} at different L_{N1}. E_{OFF} increases with the increase of N_{N1}. The optimal L_{N1} and N_{N1} are 24μm and 3×10^{14} cm^{-3}, respectively. It should be pointed out that the introduction of LBN and EPMT has no effect on V_{ON}.

Fig. 6 shows the dependence of E_{OFF} and BV on D_{OX} at different T_{OX}. The deeper D_{OX} and the thinner T_{OX} are good choice for BV and E_{OFF}.

Fig. 7 shows the trade-off performances between V_{ON} and E_{OFF} of the conventional SOI-LIGBT and the proposed SOI-LIGBT. The proposed SOI-LIGBT exhibits a better trade-off performance, in which E_{OFF} of the proposed SOI-LIGBT (1.45mJ/cm^2) can be reduced by 73% compared with the conventional SOI-LIGBT (5.42mJ/cm^2) at the same V_{ON} of 1.52V.

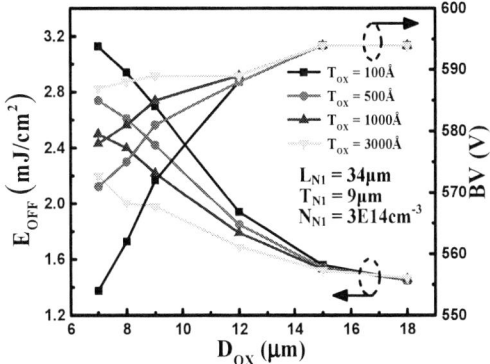

Fig. 6. The dependence of E_{OFF} and BV on D_{OX} at different T_{OX}. (b) Turn-off waveforms of the conventional and the proposed SOI-LIGBTs.

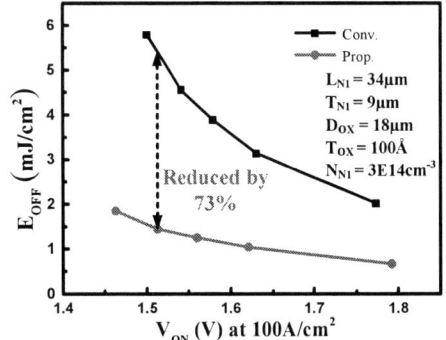

Fig. 7. Trade-off between V_{ON} and E_{OFF} of the conventional and the proposed structures at their optimal dimension parameters.

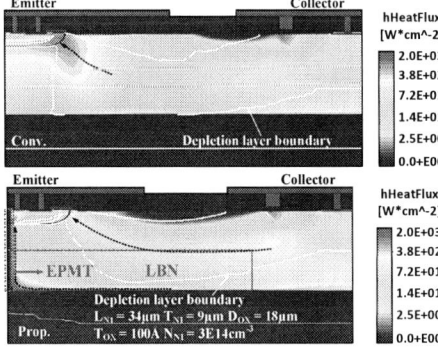

Fig. 8. The hole heat flux distribution of the conventional structure and the proposed structure under short-circuit condition with $V_{CE} = 450$V at short-circuit duration of 1μs.

Fig. 8 compares the hole heat flux of both structures under short-circuit conditions with $V_{CE} = 450V$ at short-circuit duration of 1µs. The proposed structure provides an extra path along the EPMT for the heat dissipation, which predicts an enhanced ruggedness.

IV. CONCLUSION

A 500V high speed SOI-LIGBT with the EPMT and LBN is proposed and investigated by simulations in this paper. Combining LBN with EPMT, the fast extraction of the stored carrier and the high turn-off speed are realized. The proposed SOI-LIGBT can achieve a 73% lower turn-off loss (E_{OFF}) compared with the conventional SOI-LIGBT at the same V_{ON} of 1.52V.

REFERENCES

[1] L. Zhang, J. Zhu, W. Sun, Y. Du, H. Yu, K. Huang and L. Shi, "A High Current Density SOI-LIGBT with Segmented Trenches in the Anode Region for Suppressing Negative Differential Resistance Regime," in Proc. 27th ISPSD, May 2015, pp. 49-52.

[2] J. Zhu, L. Zhang, W. Sun, M. Chen, F. Zhou, M. Zhao, L. Shi, Y. Gu and S. Zhang, "Electrical Characteristic Study of an SOI-LIGBT with Segmented Trenches in the Anode Region," IEEE Trans. Electron Devices, vol. 63, no. 5, pp. 2003-2008, May 2016.

[3] W. Sun, J. Zhu, L. Zhang, H. Yu, Y. Du, K. Huang, S. Lu, L. Shi and Y. Yi, "A Novel Silicon-on-Insulator Lateral Insulated Gate Bipolar Transistor with Dual Trenches for Three-Phase Single Chip Inverter ICs," IEEE Electron Device Lett., vol. 36, no. 7, pp. 693-695, Jul. 2015.

[4] L. Zhang, J. Zhu, W. Sun, M. Chen, C. Huang, F. Zhou, Y. Gu, S. Zhang and W. Su, "A novel high-voltage interconnection structure with dual trenches for 500V SOI-LIGBT," in Proc. 28th ISPSD, 2016, pp. 439–442.

[5] A. Nakagawa, H. Funaki, Y. Yamaguchi, and F. Suzuki, "Improvement in lateral IGBT design for 500 V 3 A one chip inverter ICs," in Proc. 11th ISPSD, May 1999, pp. 321-324.

[6] J. Sakano, S. Shirakawa, K. Hara, S. Yabuki, S. Wada, J. Noguchi, M. Wada, "Large current capability 270V lateral IGBT with multi-emitter," in Proc. 22th ISPSD, Jun. 2010, pp. 83-86.

[7] K. Hara, S. Wada, and J. Sakano,"600V single chip inverter IC with new SOI technology," in Proc. 26th ISPSD, Jun. 2014, pp. 418-421.

[8] L. Zhang, J. Zhu, W. Sun, M. Chen, M. Zhao, X. Huang, J. Chen, Y. Qian and L. Shi, "Novel Snapback-Free Reverse-Conducting SOI-LIGBT With Dual Embedded Diodes," IEEE Trans. Electron Devices, vol. 64, no. 3, pp. 1187-1192, March 2017.

[9] K. Hara, S. Wada, and J. Sakano, "600 V single chip inverter IC with new SOI technology," in Proc. 26th ISPSD, Jun 2014, pp. 418–421.

[10] J. Zhu, W. Sun, L. Zhang, Y. Du, H. Yu, K. Huang, Y. Gu, S. Zhang and W. Su, "High Voltage Thick SOI-LIGBT with High Current Density and Latch-up Immunity," in Proc. 27th ISPSD, May 2015, pp. 169-172.

[11] P.A. Gough, M.R. Simpson, and V. Rumennik, "Fast switching lateral insulated gate transistor," in IEEE IEDM Tech. Dig., pp. 218–221. 1986.

[12] S. Hardikar, R. Tadikonda, and M. Sweet, "A fast switching segmented anode NPN controlled LIGBT," IEEE Electron Device Letters, vol. 24, pp. 701–703, 2003.

[13] S. Takahashi, A. Akio, Y. Youichi, S. Satoshi, and N. Norihito, "Carrier-Storage Effect and Extraction-Enhanced Lateral IGBT (E²LIGBT): A Super-High Speed and Low On-state Voltage LIGBT Superior to LDMOSFET," in Proc. 24th ISPSD, Jun. 2012, pp. 393-396.

[14] N.K. Udugampola, R.A. Mcmahon, F. Udrea, and G.A.J. Amaratunga "Analysis and design of the dual-gate inversion layer emitter transistor," in IEEE Trans. Electron Devices, vol. 52, pp. 99-105, 2005.

Proceedings of the 30th International Symposium on Power Semiconductor Devices & ICs
May 13-17, 2018, Chicago, USA

A Comparison of Close-cell, Stripe-cell, and Orthogonal-cell Low Voltage Superjunction Trench Power MOSFETs for Linear Mode Application

Yi Su, Madhur Bobde, Sik Lui, and Hong Chang

Alpha and Omega Semiconductor, Inc.
Sunnyvale, CA, USA
Email: {ysu, mbobde, slui}@aosmd.com

Qinhai Jin, and Lei Zhang

Jireh Semiconductor, Inc.
Hillsboro, OR, USA
Email: {qinhai.jin, lei.zhang}@jfab.aosmd.com

Abstract— Trade-offs between the On Resistance and Linear Mode performance have been compared for different Low Voltage superjunction cell structures, including stripe-cell, closed-cell & orthogonal-cell. The closed-cell has measured R_{SP} of 4.4mΩ.mm², which is 20% lower than that of the stripe-cell for the same die size and same pitch at the same package. Despite higher trans-conductance, the closed-cell linear mode performance is the same as that of the stripe-cell. The orthogonal-cell has the best linear mode performance, but has high Rds(on) due to current conduction path partially blocked by p-type superjunction column. The tested maximum load current is 38.4A at a drain voltage Vds of 10V and turn-on time of 10 mSec from the orthogonal-cell, which is the highest among the competitors' parts for a power 5x6 clip package. The Rds(on) for orthogonal-cell can be further improved by device optimization.

Keywords— superjunction, trench power MOSFETs, low voltage, linear mode, FBSOA

I. INTRODUCTION

The advanced narrow pitch trench power MOSFETs targets low specific on-resistance and low gate charge for high frequency DC to DC power switching application. For the same on-resistance product, the die size becomes smaller with each generation. This is detrimental to the linear mode performance of the FET. During linear mode operation, device is electro thermally stressed by high drain voltage and high drain current simultaneously, resulting in high power dissipation. When electro-thermal stress exceeds a certain critical value, a hot spot occurs on a localized weak spot of active die area, due to thermal run away resulting in device destruction. In order to prevent device failure during the linear mode operation, better power dissipation capability from package design and better forward-bias safe operation area (FBSOA) from the device design are required. Linear mode robustness has been studied in terms of pitch size, channel length, drain current temperature coefficient, threshold voltage for trench power MOSFETs [1-6]. This paper studies three different device structures, close-cell, stripe-cell and orthogonal-cell low voltage superjunction devices for the first time for linear mode performance.

II. DEVICE DESIGN AND FABRICATION

A. Device Design Considerations

Low specific on-resistance Rsp requires narrow pitch cell, which is also valid for low voltage superjunction device. In order to have low on-resistance, close-cell and orthogonal-cell were designed to compare with stripe cell. A mask matrix with the same pitch size and the same die size is designed for strip-cell, close-cell and orthogonal-cell for device performance comparison. The orthogonal-cell is designed because trench pitch density can be increased to improve on-resistance regardless of superjunction pitch size. A schematic layout is shown in Fig.1. The termination regions were designed to have a higher breakdown voltage compared to the core. TCAD software Synopsys Taurus Workbench was used for simulating drain current temperature coefficient vs drain current Id for low voltage superjunction trench MOSFETs (SJ), standard trench MOSFETs (TrenchFET), and gate shielded trench MOSFETs (SGT). The simulation results show that positive temperature coefficient

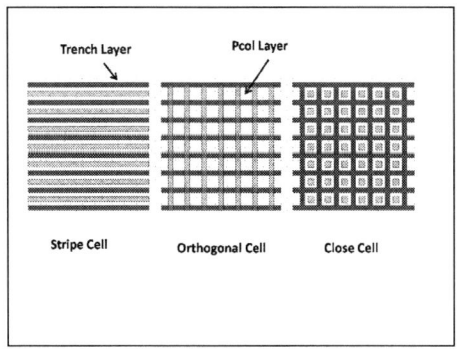

Fig.1 A schematic layout designs for low voltage superjunction trench MOSFETs

978-1-5386-2928-4/18 $31.00 © 2018 IEEE   327

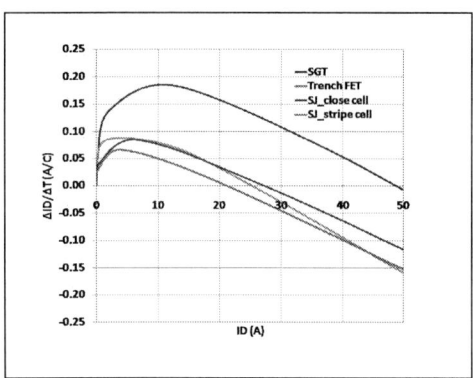

Fig.2 The simulated drain current temperature coefficient vs the drain current Id.

is lowest for superjunction device compared to other device structures. This indicates better linear mode device performance for the superjunction devices.

B. Device Fabrication

The device fabrication followed standard trench power MOSFETs with an added superjunction implant step. The drain breakdown voltage BVdss was determined by N-type epi layer doping level and its thickness, and P-type (Pcol) superjunction doping concentration. A SEM image from a fabricated close-cell superjunction trench MOSFET is shown in Fig. 3. The P-type doping profile can be observed from junction stain. The tested BVdss was 27.4V for the close-cell device, 28.7V for the stripe cell, and 29V for the orthogonal-cell. Since the stripe-cell and the orthogonal-cell had the same superjunction pitch size, the N-type epi layer and P-type implant doping were the same. For the close-cell, N-type epi doping concentration and P-type doping implant were adjusted for achieving the same drain breakdown voltage.

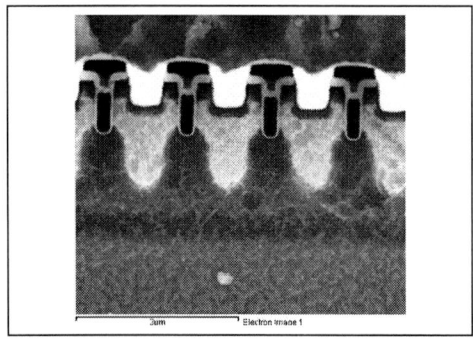

Fig.3 A SEM image from a fabricated close-cell superjunction trench MOSFET.

III. EXPERIMENTAL RESULTS AND DISCUSSIONS

Table I shows electrical performance comparison for device structures from different technologies including Trench FET, shielded gate trench MOSFETs (SGT), superjunction trench MOSFETs (close-cell, stripe-cell and orthogonal-cell SJ), and competitors' parts using power 5x6 clip package. The device performance is normalized to the close-cell SJ. The on-resistance Rds(on) at Vgs of 10V from the close-cell SJ is 0.50 mΩ, which is the lowest among the competitors' parts at power 5x6 package. The close-cell Rds(on) is 20% lower than that of the stripe-cell, 28% lower than that of the orthogonal-cell. The high Rds(on) from the orthogonal-cell is thought due to current path is partially blocked under the active trench by the P-type doped superjunction column region. The orthogonal-cell high Rdson can be possibly improved by device optimization.

All the devices listed in Table 1 were also tested for linear mode performance. There were some minor difference in packaging, especially the size of the clip used that could have a minor impact on the SOA performance The measured maximum load drain current Io,max from the linear mode evaluation are plotted in Fig. 4. The orthogonal-cell SJ trench MOSFETs shows the best linear mode performance among all tested devices. The maximum load drain current is 38.4A at a Vds of 10V and turn-on time of 10mSec from the orthogonal-cell.

The high linear mode performance from the orthogonal-cell is due to threshold voltage difference from the blocked and un-blocked channel region by the P-type column in the active cells. The close-cell has the same linear mode performance as the stripe-cell, indicating that transconductance is not the key factors in determining the liner mode performance for superjunction structure, which may be different from SGT and Trench structures. Since the close-cell and the stripe-cell have the same pith size, the same die size and the same clip package, the influence from package side can also be neglected.

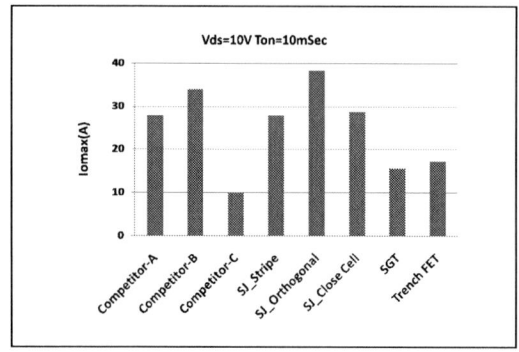

Fig.4 A comparison of the tested linear mode maximum drain current from different technologies.

978-1-5386-2928-4/18 $31.00 © 2018 IEEE

TABLE I

A COMPARISON OF DEVICE PERFORMANCE BETWEEN AOS PARTS AND THE COMPETITORS' PARTS

Product		SJ Closed-cell	Competitor A	Competitor B	Competitor C	SJ Stripe-cell	SJ Orthogonal-cell	Trench FET	SGT
Vds	V	27.4	>25	>30	>30	28.7	29	>30	>30
Vgs	V	20	20	20	20	20	20	20	20
Vth	V	2.0	1.6	1.4	1.9	2.0	2.0	1.7	1.8
Rds(on) (10V)	mΩ	100%	114%	190%	160%	120%	128%	230%	100%
Ciss (50% Vds)	pF	100%	73%	62%	99%	77%	113%	61%	101%
Coss (50% Vds)	pF	100%	129%	77%	132%	87%	84%	46%	147%
Crss (50%Vds)	pF	100%	35%	6%	16%	39%	40%	49%	207%
Qg(10V)	nC	100%	57%	N/A	78%	66%	87%	73%	120%
Qg(4.5V)	nC	100%	54%	42%	67%	62%	77%	75%	139%
Qgs	nC	100%	53%	40%	96%	81%	121%	56%	79%
Qgd	nC	100%	26%	20%	29%	41%	38%	70%	202%

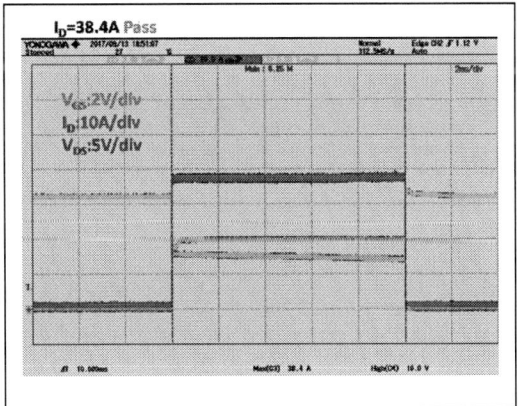

Fig.5 Waveforms recorded during the linear mode evaluation from a passed unit.

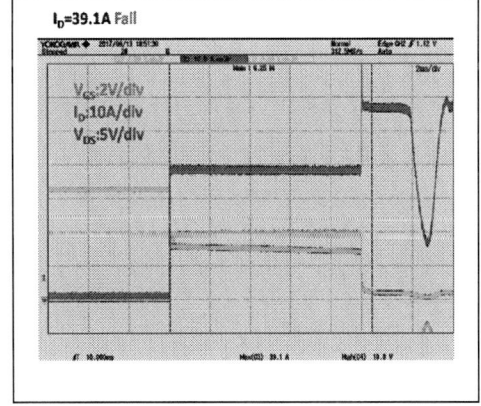

Fig. 6 Waveforms recorded during the linear mode evaluation from a failed unit.

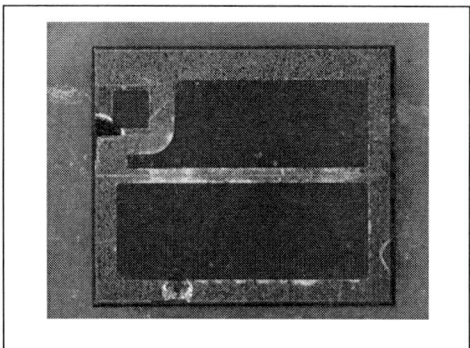

Fig.7 A top-view SEM image showing failure spots after the linear mode evaluation.

Super-Junction FETs can be designed to push the peak electric field down the drift region, thereby mitigating the short channel effects seem on default Trench-FETs. Reduction in peak electric field near body junction for superjunction devices as compared to SGT and trench MOSFETs is the reason why the superjunction trench MOSFETs provides superior linear mode performance even for high gain close-cell device.

Waveforms for Vgs, Vds and drain current Ids were recorded during the linear mode evaluation. The passed unit after the linear mode evaluation is shown in Fig. 5; and the failed unit is shown in Fig. 6. For the failed unit, the device loses its voltage blocking capability after the linear mode evaluation because of catastrophic destruction. Decap was done for the failed unit to check failure location. A top view SEM image is shown in Fig. 7. Two failure spots were observed at the die surface after test at Vds of 10V, turn-on time of 10mSec at the maximum of drain current of 39.1A. All of these failure spots are located outside of the pad opening. This indicates current crowding near the edge of the device, outside the clip region as the root cause of destructive failure. A larger clip or thicker top metal can be used for further improvement of SOA performance.

IV. CONCLUSIONS

A close-cell superjunction trench power MOSFET was reported for the first time for 25V product for linear mode application. A stripe-cell and an orthogonal-cell were designed for comparison. The orthogonal-cell has the highest linear mode device performance among the three superjunction designs and the competitors' parts available so far. The high Rds(on) from the orthogonal-cell can be possibly improved by device optimization.

ACKNOWLEDGMENT

We would like to thank Yohern Chow for taking SEM images, Hailin Zhou and Xuejun Wu for part assembling and device characterization, and Quan Li for linear mode evaluation.

REFERENCES

[1] P. Spirito, G. Breglio, V. d'Alessandro, N. Rinaldi, "Analytical model for thermal instability of low voltage power MOS and S.O.A in pulse operation", ISPSD 2002, pp. 269-272.

[2] O. M. Alatise, I. Kennedy, G. Pethkos, K. Khan, A. Koh, and P. Rutter, "Understanding linear mode robustness in low-voltage trench power MOSFETs," IEEE Trans. on Device and Materials Reliability, Vol. 10p, No. 1, pp. 123-129, 2010.

[3] M. Chang, and P. Rutter, "Optimizing the trade-off between the Rdson of power MOSFETs and linear mode performance by local modification of MOSFET gain," ISPSD2016, pp. 379-382.

[4] P. Spirito, G. Breglio, and V. d'Alessandro, "Modeling the onset of thermal instability in low voltage power MOS: an experimental validation," ISPSD 2005, pp. 159-162.

[5] G. Breglio, F. Frisina, A. Magri, and P. Spirito, "Electro-thermal instability in low voltage power MOS: experimental characterization, " ISPSD 1999, pp. 233-236.

[6] P. Spirito, G. Breglio, V d'Alessandro, N. Rinaldi, "Thermal instabilities in high current power MOS devices: experimental evidence, electro-thermal simulations and analytical modeling, " Proc. 23rd International Conference on Microelectronics, Vol. 1, pp.23-30, 2002.

Proceedings of the 30th International Symposium on Power Semiconductor Devices & ICs
May 13-17, 2018, Chicago, USA

A 150V Novel High-Voltage LDMOS in a 0.18um BCD Plug-In Process

Yen-Ming Chen, Chiu-Ling Lee, Min-Hsuan Tsai, Chiu-Te Lee, Chih-Chong Wang

Specialty Technology
United Microelectronics Corporation
Hsinchu, Taiwan
e-mail : jora_chen@umc.com

Abstract—**The feature of this BCD180 process is flexible to plug-in varied high voltage devices based on fundamental 1.8V/5V BCD process, and we present the plug-in LDMOS reach to 150V operation voltage in this paper.**

Keywords—Plug-In; LDMOS; BCD; 150V

I. INTRODUCTION

The current trend of PMIC's development is embedded high voltage power devices into CMOS platform. The planar-type devices are attracting a great deal with attention for highly integrated chip design. The most suitable candidate of high voltage device is the Laterally Diffused Metal Oxide Semiconductor (LDMOS). In order to provide SOC power units, the BCD platform is needed to have flexibility and different voltage rating power devices [1]. It is not easy to integrate power device with wide voltage rating using unit process. The concept of "plug-in" process will be the best solution for highly integration PMIC. The device performance tuning can be modified by the layout schematic or adding extra implant steps without any thermal budget change. The voltage rating lager than 100V will be the increasing demand for consumer, industrial and automotive application.

II. PROCESS FEATURE

The fundamental process uses the standard 0.18um 1.8/5V platform. All the high voltage devices are plugged in the basic flow. Up to now, the voltage rating range is from 7V to 150V. The key process flow is shown in Fig. 1. Starting material is a p-type silicon wafer. NBL (N+ buried layer) is formed. The P-type epitaxial layer is grown on the NBL. The high voltage N/P well implant steps set after epitaxial process. These wells achieve high breakdown voltage by proper implant condition and thermal drive-in anneal. The following steps are standard STI (shallow trench isolation) and CMOS twin well process. The body and drift formation steps put after CMOS twin well sequence narrow the influence of exist devices. The following process steps are back to standard BCD process flow. The device performance can be improved by inserted extra masks and schematic layout optimum.

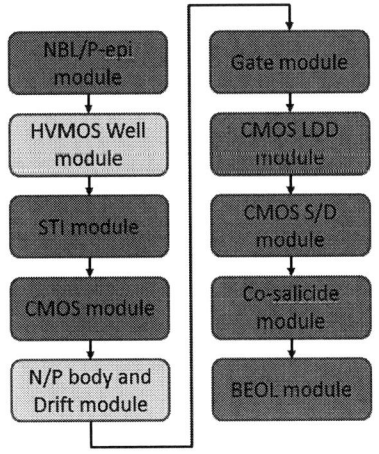

Fig. 1. Key Process Flow

TABLE I.   RON-SP AND BVD CHARACTERISTICS

DEVICE	Vt(V)	Ron-sp(mΩ*mm^2)
7VN	1.15	2.2
8VN	1.15	2.5
12VN	0.75	4.8
24VN	0.97	13.8
30VN	0.97	19.8
55VN	1.34	54
60VN	1.14	66.3
150VN	0.71	410
8VP	0.80	12
12VP	0.81	15.1
24VP	0.85	24.1
40VP	0.85	82
150VP	0.85	860

978-1-5386-2928-4/18 $31.00 © 2018 IEEE   331

III. HIGH VOLTAGE DEVICE PERFORMANCE

The proposed 150V LDMOS, and also medium voltage (7V~60V) NLDMOS could be "plug-in" as using same basic process (see Table. I). All high voltage transistors have same gate oxide thickness.

A. 150V NLDMOS

The Fig. 2 shows the top-view and cross-section of proposed 150V NLDMOS. The structure uses double RESURF [2] and doesn't have NBL (N+ buried layer) underneath. The threshold voltage is obtained 0.71V at Vd=0.1V shown in Fig. 3. The off-state breakdown voltage is larger than 170V. Fig. 4 shows the Ids-Vds characteristics for the proposed unit which is measured by Keysight-4156. The specific on resistance of Rdson=410 mΩ-mm².

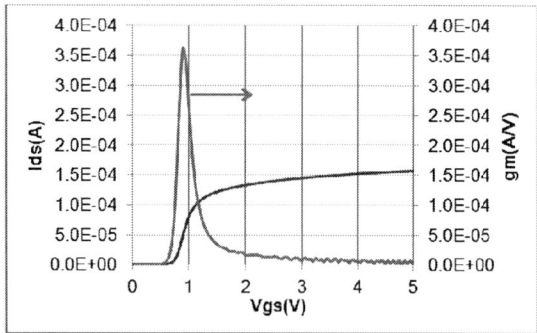

Fig. 3. Ids-Vgs characteristics of 150V NLDMOS

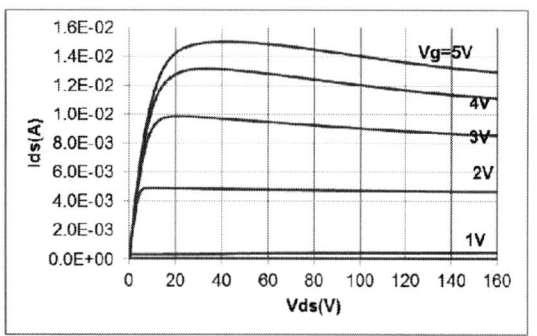

Fig. 4. Ids-Vds characteristics of 150V NLDMOS

As shown in Fig. 5 the SEM cross-section of 150VN device, N+ buried layer is not included in the structure so non-Epi basic BCD process is acceptable to plug-in this device, T-CAD simulation with doping profile (Fig. 6) and potential contour lines distribution of 150V NLNMOS at Vd=5V (gate, source/bulk and substrate are connected to ground) see in Fig. 7.

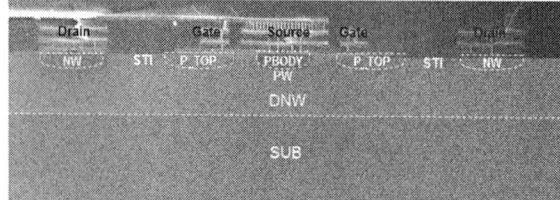

Fig. 5. SEM cross-section of 150V NLDMOS

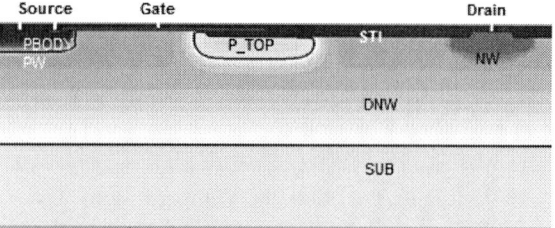

Fig. 6. T-CAD simulation doping profile of 150V NLDMOS

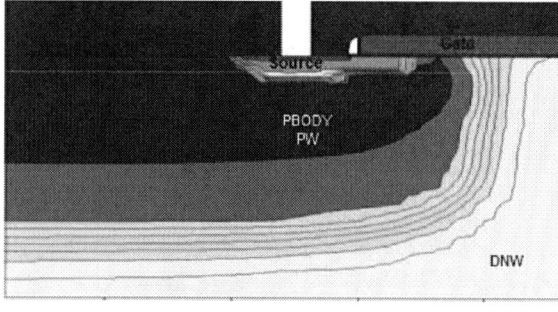

Fig. 7. T-CAD simulation potential contour of 150V NLDMOS

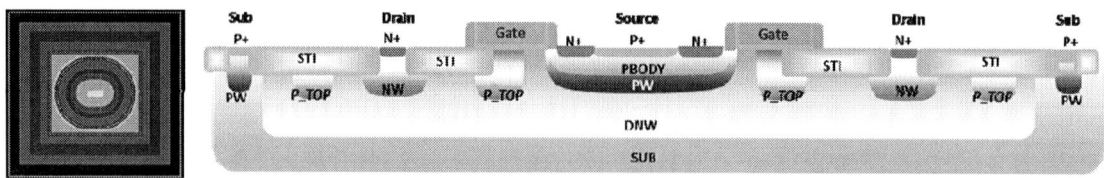

Fig. 2. 150V NLDMOS top view and cross-section

B. 150V PLDMOS

The Fig. 8 shows the top-view and cross-section of proposed 150V PLDMOS. The structure includes NBL (N+ buried layer) and DPW (Deep P Well) to ensure high breakdown voltage and low on-resistance. The threshold voltage is obtained 0.85V shown in Fig. 9 at Vd=0.1V. The "off-state" breakdown voltage is larger than 170V. Fig. 10 shows the Ids-Vds characteristics for the proposed unit which is measured by HP4156. The specific on resistance of Rdson=860 mΩ-mm².

As shown in Fig. 11 the SEM cross-section of 150VP device, N+ buried layer is necessary for this structure therefore

BCD process with Epi is required for plug-in this device, T-CAD simulation with doping profile (Fig. 12) and potential contour lines distribution of 150V PLDMOS at Vd=-5V (gate, source/bulk and substrate are connected to ground) see in Fig. 13.

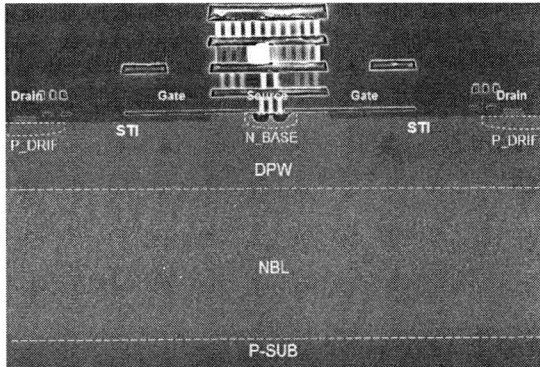

Fig. 11. SEM cross-section of 150V PLDMOS

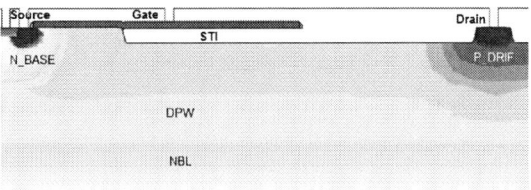

Fig. 12. T-CAD simulation doping profile of 150V PLDMOS

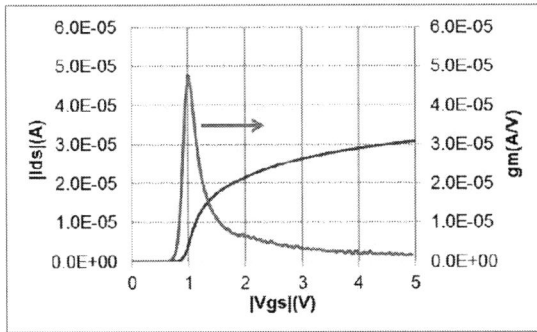

Fig. 9. Ids-Vgs characteristics of 150V PLDMOS

IV. CONCLUSION

The "plug-in" BCD process is a better choice for wide voltage rating application. It is very flexible and effective to combine the units which you need. It also suit for the complex and high integrated PMIC application.

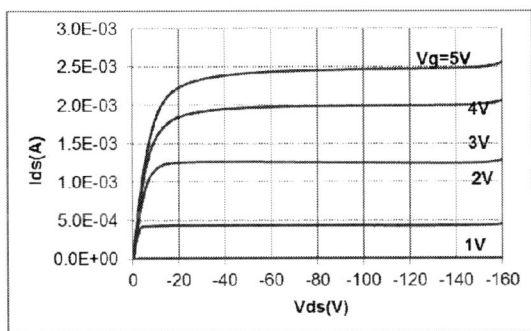

Fig. 10. Ids-Vds characteristics of 150V PLDMOS

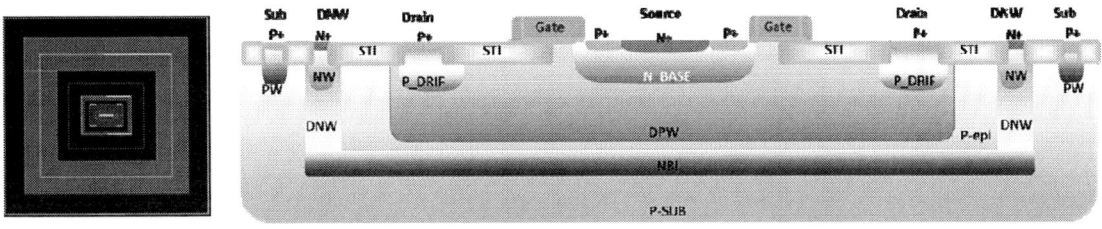

Fig. 8. 150V PLDMOS top view and cross-section

ACKNOWLEDGMENT

The authors would like to thank United Microelectronics Corporation BCD team members for process integration and device characterization support.

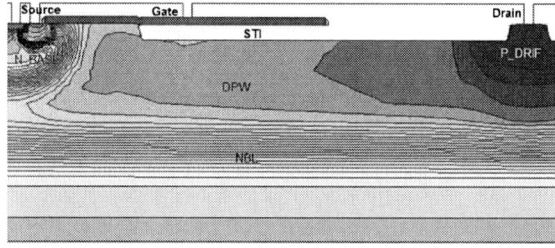

Fig. 13. T-CAD simulation potential contour of 150V PLDMOS

REFERENCES

[1] B. J. Baliga, "The Future of Power Semiconductor Device Technology", Proceedings of the IEEE, vol. 89, 2001, pp. 822-832.

[2] A. W. Ludikhuize, "A review of RESURF technology," Proc. of ISPSD, 2000, pp. 11–18.

Proceedings of the 30th International Symposium on Power Semiconductor Devices & ICs
May 13-17, 2018, Chicago, USA

Application of CS-MCT in DC Solid State Circuit Breaker (SSCB)

Wanjun Chen[1,2], Hong Tao[1], Chao Liu[1], Yawei Liu[1], Chengfang Liu[1], Jie Liu[1], Yijun Shi[1], Qi Zhou[1], Zhaoji Li[1] and Bo Zhang[1]

[1]State Key Laboratory of Electronic Thin Films and Integrated Devices, University of Electronic Science and Technology of China, Chengdu 610054, China (e-mail: wjchen@uestc.edu.cn)

[2]Institute of Electronic and Information Engineering of UESTC in Guangdong, Dongguan 523808, China

Abstract—In this work, the application of cathode-short MOS-controlled thyristor (CS-MCT) in DC solid state circuit breakers (SSCB) is evaluated. Firstly, static and dynamic characteristics of CS-MCT are presented, and a low on-state resistance combined with a high surge current capability makes CS-MCT a promising candidate for SSCB. Furthermore, fault interruption process of SSCB based on CS-MCT is demonstrated via TCAD simulation as well as experiment. According to the research, SSCB based on CS-MCT can achieve a high fault current interruption capability once the mis-triggering of internal thyristor is effectively suppressed. And a contradictory relationship between energy loss and interruption capability of the SSCB is also found, which implies a trade-off design should be taken into consideration.

Keywords—CS-MCT; SSCB; interruption capability; energy loss;

I. INTRODUCTION

The circuit breaker is a fault protection device for power systems. In the normal state, circuit breaker is connected in series with load and source, and a circuit breaker with low energy loss is essential to improve system efficiency. When fault happens, circuit breaker will be disconnected, and a short response time as well as a high fault interruption capability is required to effectively protect the power system. Traditionally, mechanical circuit breakers (MCBs) have been used to protect the system under fault condition, but due to the arc generated in tripping process, MCBs suffer from a short lifetime [1]. Moreover, MCBs based on thermal and electromagnetic principle have a long response time (tens of milliseconds [1]). However, the speed and reliability of MCBs cannot meet the protection requirements of emerging DC power systems such as DC microgrids [2], EV charging station. Specifically, there is no natural zero voltage cross point in DC power system [3] and the short-circuit capability of power semiconductor devices used in DC-DC or AC-DC conversion stages is much worse than transformers used in AC system[4]. Besides, a compact structure design in DC system causes a low system inductance [5], causing to a high current increasing rate when short-circuit happens. To effectively protect the DC system under fault condition, a variety of solid state circuit breakers using power semiconductors have been proposed [1, 3, 4, 6-10]. Compared with MCB, with the introduction of high speed fault current sensing and control circuits and removal of arc during the fault interruption process, SSCB features a short response time as well as a long-term reliability. But the application of SSCB is hindered by a large energy loss and a low fault interruption capability. With a cathode-short thyristor structure inside, CS-MCT possesses a low on-state resistance and a simplified gate

control strategy [11]. In this work, a 400V DC SSCB based on CS-MCT is presented via TCAD simulation and experiment. Corresponding results indicate that with optimizations of critical device and circuit parameters, CS-MCT can be a promising power semiconductor device in DC SSCB applications.

II. STRUCTURE AND CHARCTERISTICS OF CS-MCT

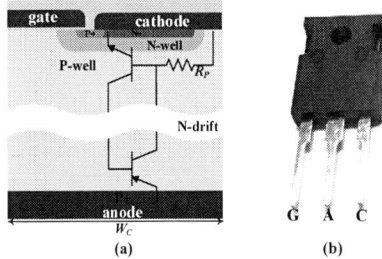

Fig. 1. (a)Cross section of CS-MCT (b) photo of TO-247 packaged CS-MCT

Fig.1 shows the cross section of CS-MCT and photo after package. Similarly to a conventional MCT, a PNPN thyristor structure formed by P+/N-drift/P-well/N-well layers appears in the CS-MCT, which is favorable to get a small on-state resistance. With a cathode-short structure, leakage current at forward blocking state can flow into cathode without a negative gate bias, which simplifies the gate driving circuit. At on-state, internal NMOS structure formed by N-well/P-well/N-drift region is turned on and supplies base current for internal PNP transistor. Holes, emitted from anode, will be collected by P-well and flow horizontally to cathode, which causing a voltage drop on P-well resistance (R_P). At low current level, CS-MCT works like an IGBT, where electrons are only

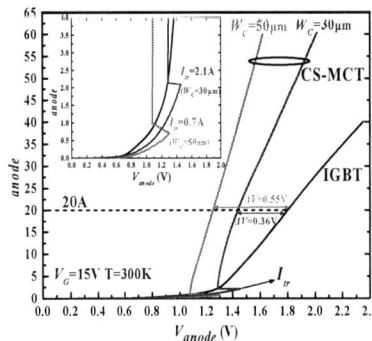

Fig. 2. Measured forward conduction characteristics of CS-MCTs with different values of W_C and a commercial IGBT.

978-1-5386-2928-4/18 $31.00 © 2018 IEEE   335

injected from NMOS channel. Once anode current reaches the triggering current (I_{tr}), voltage drop on R_p (i.e., forward voltage drop on emitter junction of internal NPN transistor) is large enough to turn on NPN transistor, which triggers the thyristor structure. Then, CS-MCT works in thyristor mode and features a low resistance. Fig.2 plots the forward I-V curves of CS-MCTs with different cell width (W_C). As shown in this figure, CS-MCT with a W_C of 30μm possesses a forward voltage drop (V_F) of 1.44V at 20A, with a larger cathode area of internal thyristor, V_F can be further reduced to 1.25V when W_C is increased to 50μm. To compare with the commonly used IGBTs, forward I-V curve of a 600V 20A commercial IGBT (HGTG20N60B3D) with similar chip area is also plotted in Fig.2. It can be seen, CS-MCTs show 0.55V (30%) and 0.36V (20%) reductions in V_F at 20A when W_C is 50μm and 30μm, respectively. Since most energy loss in SSCB is caused by

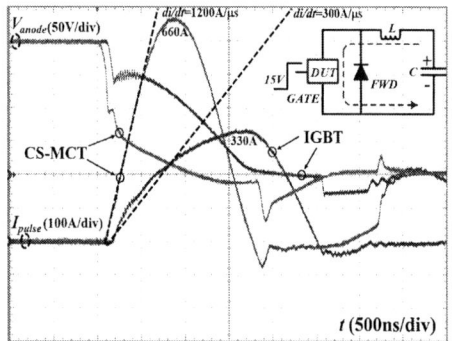

Fig. 3. Measured pulse discharge waveforms of CS-MCT and IGBT

internal power semiconductor devices, SSCB based on CS-MCT could attain a similar reduction in energy loss compared with that using IGBTs. As seen in the inset in Fig.2, CS-MCT has an I_{tr} of 2.2A when W_C is 30μm. When W_C is increased to 50μm, a larger R_p can be obtained, which causes a lower I_{tr}. Fig.3 shows the capacitive discharge behavior of the mentioned IGBT and CS-MCT with a W_C of 50μm under 200V, it can be seen CS-MCT shows a peak surge current of 660A, leading to a di/dt of 1200A/μs. By contrast, measured peak current and di/dt of IGBT is about 330A and 300 A/μs, respectively. And such a difference is actually due to the different conduction mechanisms. Specifically, without current saturation, CS-MCT possesses a low on-state resistance even under a large current condition. While the current of IGBT is controlled by gate voltage and saturation will appear at high current level, which may cause a burnt-out problem especially when the transient fault current in SSCB is large enough to drive IGBTs into deep saturation condition.

III. CHARCTERISTICS OF SSCB BASED ON CS-MCT

Considering the inherent low turn-off capability of MCT under hard-switching condition, a force commutation circuit is adopted to interrupt the fault current. Fig.4 shows the schematic of SSCB based on CS-MCTs. Principle of this circuit is explained as follows. Firstly, Capacitor C_a should be pre-charged to a given voltage (pre-charge circuit is ignored in Fig.4). Under normal condition, DC source supplies power for loads through main path, and CS-MCT M_1 is applied in this path to achieve a low energy loss, while commutation path is at

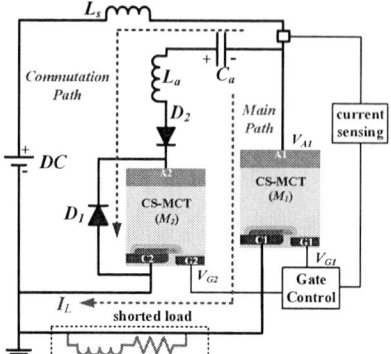

Fig. 4. Schematic of SSCB using CS-MCTs in mixed-model simulation

off-state. When short circuit happens, fault current in main path starts to rapidly increase, the rate of which is determined by the magnitude of DC source voltage and system stray inductance (L_s). Once the fault current reaches the overcurrent trip setting, device M_2 is turned on, triggering the current commutation process between main path and commutation path. Consequently, fault current in main path will be reduced to zero within the commutation process, and CS-MCT M_2 is applied in this path to obtain a high di/dt performance, which is benefit to accelerate the elimination of fault current through the load. After the commutation, current in commutation path will still exist until it decreases to zero in the subsequent resonant process. And energy stored in L_s will be transferred to C_a within the process.

Fig.5 (a) shows the simulated results of fault interruption processes with different magnitudes of peak fault current (I_{PEAK}). It can be seen the fault current through the shorted load (I_L) can be quickly eliminated within the current commutation process. However, with a larger I_{PEAK}, fault current may rise again at zero voltage cross (ZVC) point of V_{A1} (i.e., anode voltage of device M_1), causing the failure of the SSCB. Fig.5 (b) shows the average hole current density flowing towards cathode side (J_{PC}) near ZVC point, it can be seen the magnitude of J_{PC} gradually decays. But with a higher I_{PEAK}, peak hole current density at the ZVC point ($\triangle J_{PC}$) could be larger. As we have analyzed in Section II, a large hole current at cathode side can trigger the thyristor structure. And the mis-triggering of CS-MCT at ZVC point could cause the failure of

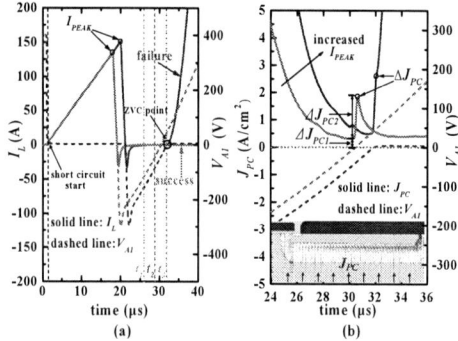

Fig. 5. (a)Simulated fault current interruption processes and (b) average hole current density at cathode side of device M_1 near the ZVC point of V_{A1} with different I_{PEAK}

SSCB. To clarify the origin of J_{PC}, distributions of hole concentration in N-drift region of device M_1 at time points of t_1, t_2 and t_3 in Fig.5 (a) are plotted in Fig.6. It can be seen that injected holes during the on-state still exist in N-drift region. Due to the gradient of carrier concentration, holes near P-well flow towards cathode side, which produces a hole diffusion current (i.e., J_{PC}). With continuous carrier extraction and recombination, the number of stored holes as well as the concentration gradient is getting smaller, causing a lower J_{PC}. At ZVC point, J_{PC} is reduced to ΔJ_{PC1}. Right after ZVC point, M_1 is forward biased, and P-well/N-drift junction will be

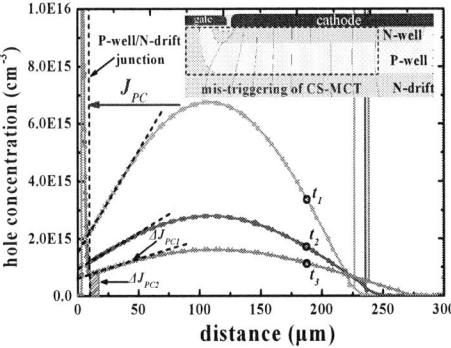

Fig. 6. Distribution of hole concentration of M_1 at time points of t_1, t_2 and t_3 with a larger I_{PEAK} shown in Fig.5 (a).

reverse biased. But firstly, excess holes near this junction should be eliminated to form depletion region, which causes another hole extraction current ΔJ_{PC2}. With a higher hole concentration or a larger dV/dt of V_{A1} at ZVC point, more holes will be extracted out from N-drift region per unit time, producing a larger ΔJ_{PC2}. Once the hole current density is large enough to trigger the thyristor, as shown in the inset of Fig.6, fault current will rise again, causing the failure of SSCB.

To enhance the fault interruption capability of the SSCB, mis-triggering of CS-MCT at ZVC point should be effectively suppressed. Thus, a lower ΔJ_{PC} is required. According to above analysis, a longer duration of hole current decay and a smaller dV/dt of V_{A1} at ZVC point contribute to a lower ΔJ_{PC1} and ΔJ_{PC2}, respectively. In fact, device M_1 is in semi-floating state after current commutation, variation of V_{A1} is mainly determined by the resonant process in commutation path. It is known a larger

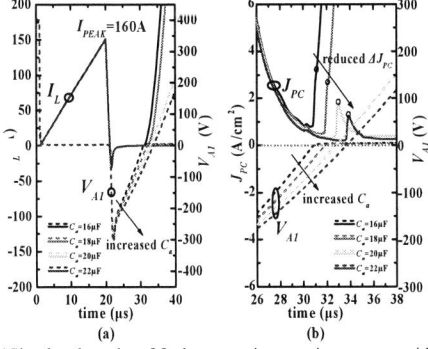

Fig. 7. (a)Simulated results of fault current interruption process with different C_a and (b) J_{PC} of device M_1 near the ZVC point.

capacitance brings a longer resonant cycle in a *LC* circuit. Therefore, a larger C_a in commutation path is preferred to obtain a smaller ΔJ_{PC} at ZVC point. Fig.7 (a) plots the fault interruption processes with different C_a, it can be seen with a larger C_a, mis-triggering at ZVC point is suppressed and failure of the SSCB can be avoided. Corresponding results of J_{PC} are given in Fig.7 (b), ΔJ_{PC} including carrier concentration gradient induced ΔJ_{PC1} and dV/dt induced ΔJ_{PC2} is indeed reduced with an increased C_a, which confirms previous analysis.

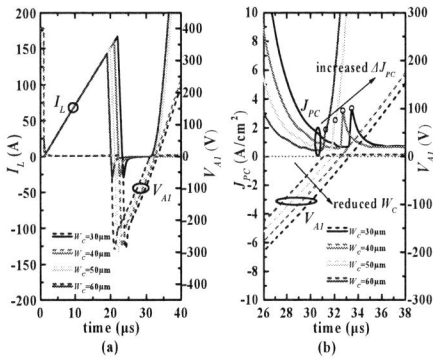

Fig. 8. (a)Simulated results of fault current interruption process with different W_C and (b) J_{PC} of device M_1 near the ZVC point where C_a=18μF.

Except for a larger C_a, fault interruption capability of the proposed SSCB can be enhanced via structure design of CS-MCT. As shown in Fig.8 (a), with a smaller cell width (W_C) of M_1, a higher fault current can be successfully eliminated. Corresponding results of J_{PC} are shown in Fig.8 (b), though a larger ΔJ_{PC} is observed with a smaller W_C, mis-triggering of internal thyristor is still suppressed. The reason behind is that a smaller W_C causes a higher triggering current (I_{tr}), as shown in Fig.2, which makes it more difficult for CS-MCT to be mis-triggered at ZVC point.

IV. EXPERIMENT RESULTS AND DISCUSSIONS

Fig. 9 show the schematic of SSCB test circuit, where an air core inductor and a pre-charged capacitor (C_s) are introduced to represent the equivalent stray inductance (L_s) and 400V DC voltage source, respectively. A hall current sensor combined with a voltage comparator forms a simplified current sensing and gate control unit. In the test, short-circuit condition is simulated by connecting the cathode of M_1 to ground and short-circuit starts when M_1 is turned on.

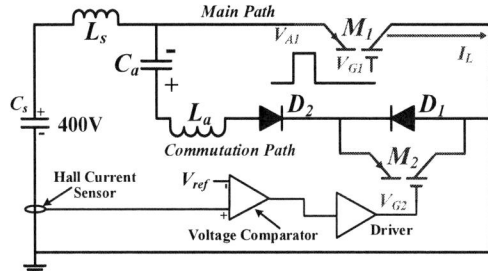

Fig. 9. Schematic of 400V SSCB test circuit

978-1-5386-2928-4/18 $31.00 © 2018 IEEE   337

Fig.10 (a) shows the measured results of fault current interruption processes under 400V DC condition with different values of C_a. It can be seen a larger C_a is indeed conducive to a SSCB with an enhanced fault current interruption capability.

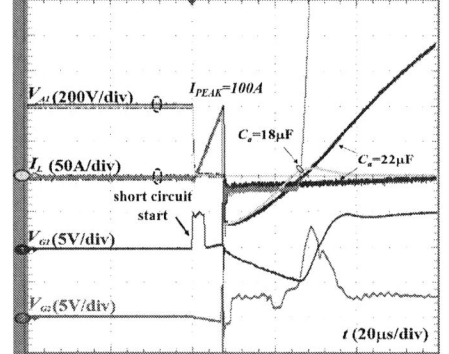

Fig. 10 (a). Fault current interruption processes with different C_a

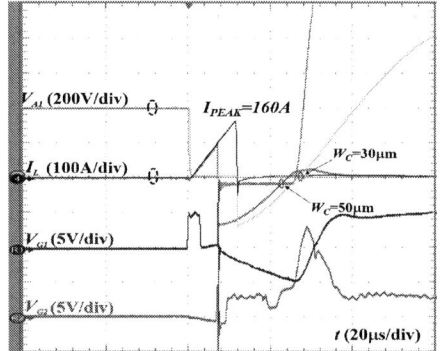

Fig. 10 (b). Fault current interruption processes with different W_C of M_1

Corresponding results with different values of W_C are plotted in Fig. 10 (b), with a smaller W_C, mis-triggering of CS-MCT at ZVC point is effectively suppressed. However, as shown in Fig.2, on-state voltage drop of CS-MCT is also increased with a larger W_C, causing a higher energy loss. Consequently, a trade-off relationship exists between the energy loss and fault current interruption capability of SSCB based on CS-MCT, as shown in Fig.11, which implies the balance of above two critical SSCB characteristics should be taken into consideration for the design of CS-MCT structure parameters.

V. CONCLUSIONS

With a low on-state resistance and a high surge current capability, CS-MCT shows the potential to be a solid state switch in SSCB. In this work, the behavior of SSCB based on CS-MCT is demonstrated, it is found that fault interruption capability of the SSCB is limited by the mis-triggering of CS-MCT, which can be effectively suppressed with a larger capacitance and a smaller cell width of CS-MCT. The trade-off relationship between the energy loss and fault interruption capability will guide the design of SSCB based on CS-MCT.

ACKNOWLEDGMENT

This research is funded in part by the Sichuan Youth Science and Technology Foundation (No. 2017JQ0020), the

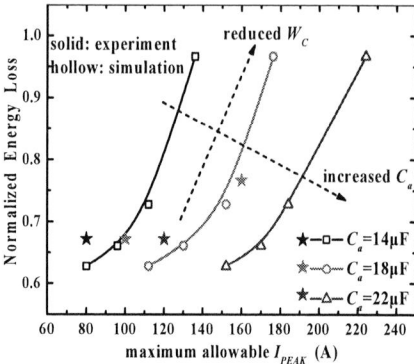

Fig. 11. Trade-off relationship between normalized energy loss and maximum allowable I_{PEAK} that can be successfully interrupted by the SSCB.

Fundamental Research Funds for the Central Universities (No. ZYGX2016Z006) and the Natural Science Foundation of Guangdong Province, China (Grant No. 2015A030311016).

REFERENCES

[1] S. Krstic, E. L. Wellner, A. R. Bendre and B. Semenov, "Circuit Breaker Technologies for Advanced Ship Power Systems," *2007 IEEE Electric Ship Technologies Symposium*, Arlington, VA, 2007, pp. 201-208.

[2] Elsayed, et al., "DC microgrids and distribution systems: An overview," *Electric Power Systems Research*, vol. 119, pp. 407-417, Feb. 2015.

[3] A. Würfel, J. Adler, A. Mauder and N. Kaminski, "Over current breaker based on the dual thyristor principle," *2016 28th International Symposium on Power Semiconductor Devices and ICs (ISPSD)*, Prague, 2016, pp. 143-146.

[4] Z. J. Shen, G. Sabui, Z. Miao and Z. Shuai, "Wide-Bandgap Solid-State Circuit Breakers for DC Power Systems: Device and Circuit Considerations," in *IEEE Transactions on Electron Devices*, vol. 62, no. 2, pp. 294-300, Feb. 2015.

[5] X. Feng, L. Qi and J. Pan, "Fault inductance based protection for DC distribution systems," *13th International Conference on Development in Power System Protection 2016 (DPSP)*, Edinburgh, 2016, pp. 1-6.

[6] Zyborski, J., Czucha, J., Sajnacki, M, "Thyristor circuit breaker for overcurrent protection of industrial dc power installations," *In Proceedings of the Institution of Electrical Engineers*, vol. 123, no. 7, pp. 685-688, July.1976.

[7] Naoaki Yamato, Akiyoshi Fukui and Keiichi Hirose, "Effect of breaking high voltage direct current (HVDC) circuit on demonstrative project on power supply systems by service level in Sendai," *INTELEC 07 - 29th International Telecommunications Energy Conference*, Rome, 2007, pp. 46-51.

[8] D. P. Urciuoli, V. Veliadis, H. C. Ha and V. Lubomirsky, "Demonstration of a 600-V, 60-A, bidirectional silicon carbide solid-state circuit breaker," *2011 Twenty-Sixth Annual IEEE Applied Power Electronics Conference and Exposition (APEC)*, Fort Worth, TX, 2011, pp. 354-358.The Technical Writer's Handbook. Mill Valley, CA: University Science. 1989.

[9] Z. J. Shen *et al.*, "First experimental demonstration of solid state circuit breaker (SSCB) using 650V GaN-based monolithic bidirectional switch," *2016 28th International Symposium on Power Semiconductor Devices and ICs (ISPSD)*, Prague, 2016, pp. 79-82.

[10] Y. Sato, Y. Tanaka, A. Fukui, M. Yamasaki and H. Ohashi, "SiC-SIT Circuit Breakers With Controllable Interruption Voltage for 400-V DC Distribution Systems," in *IEEE Transactions on Power Electronics*, vol. 29, no. 5, pp. 2597-2605, May 2014.

[11] Wanjun Chen *et al.*, "Experimentally demonstrate a Cathode-Short MOS-controlled thyristor (CS-MCT) for single or repetitive pulse applications," *2016 28th International Symposium on Power Semiconductor Devices and ICs (ISPSD)*, Prague, 2016, pp. 311-314.

Proceedings of the 30th International Symposium on Power Semiconductor Devices & ICs
May 13-17, 2018, Chicago, USA

ESD Failure Analysis and Robustness Improvement for Multi-STI-Finger LDMOS Used as Output Device

Ran Ye, Siyang Liu, Zhigang Dai, Hongting Chen,
Wangran Wu, Weifeng Sun*, Shengli Lu
National ASIC System Engineering Research Center
Southeast University, Nanjing, China
swffrog@seu.edu.cn

Wei Su, Feng Lin, Guipeng Sun
CSMC Technologies Corporation
Wuxi, China

Abstract—The MSF-LDMOS will suffer from high ESD failure risk when used as output device. In this work, it is found that the different ratios of STI width to silicon width have impact not only on breakdown voltage and $R_{dson.sp}$, but also on ESD robustness. Thus, the ESD failure mechanism of MSF-LDMOS is comprehensively investigated. Furthermore, a novel structure with multiple stair-STI fingers is proposed by alleviating the crowded current density. For the novel MSSF-LDMOS, the ESD robustness increases by 10% and the $R_{dson.sp}$ decreases by 2.5% with only 1.2% reduction of breakdown voltage comparing with the conventional MSF-LDMOS.

Keywords—MSF-LDMOS; ESD robustness; Output device; TLP; Failure analysis

I. INTRODUCTION

Lateral DMOS is more popular to be used as output device because it can be easily integrated in the power managements, automotive applications and display drivers[1-3]. Decreasing specific on-resistance($R_{dson.sp}$) which has positive effect on low

power dissipation and small chip size is one of the most challenge for LDMOS design. Multi-STI-Finger LDMOS (MSF-LDMOS) attracts more attention because it can maximize the reduction of $R_{dson.sp}$ with minimal sacrifice of the breakdown voltage (BV)[4]. Meanwhile, the earlier studies mainly focused on the BV and $R_{dson.sp}$ for the MSF-LDMOS[5-6]. However, the MSF-LDMOS will be up against the high electro-static discharge (ESD) failure risk when it is used as output device due to its direct exposure to the output stage without other ESD protection cells. Thus, it requires this device to possess ESD self-protection ability. There is less available information on the ESD reliability for this device. In this work, the ESD failure mechanism of the MSF-LDMOS is comprehensively investigated. A novel structure based on the failure mechanism is also proposed to improve its ESD robustness.

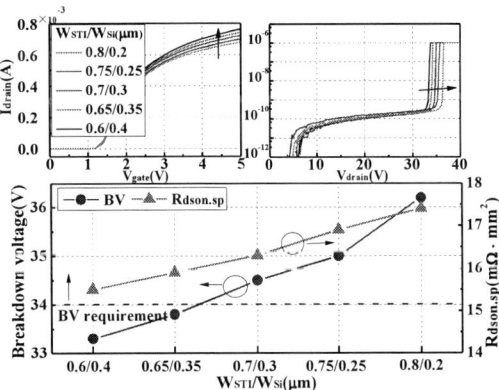

Fig. 2: The measured turn-on characteristic curves and breakdown voltage curves of the MSF-LDMOS for different ratios of W_{STI}/W_{Si}.

II. DEVICE STRUCTURE AND EXPERIMENTS

Fig.1 shows the studied MSF-LDMOS device. The effective channel length is 0.2µm and the thickness of gate oxide is 13nm. The STI is embedded in the drift region, presenting multi-finger distribution. The width of silicon is

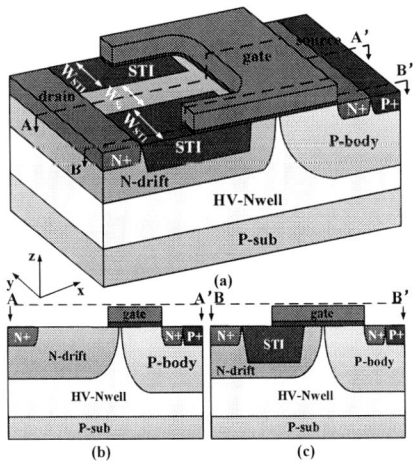

Fig. 1: (a) The 3D schematic diagram of MSF-LDMOS. (b) The cross section along line AA'. (c) the cross section along line BB'.

This work is supported by CSMC Technologies Corporation.

978-1-5386-2928-4/18 $31.00 © 2018 IEEE   339

labelled as W_{Si} and the width of STI is labelled as W_{STI}. Fig.2 shows the measured turn-on characteristic curves and breakdown voltage curves of the MSF-LDMOS for different ratios of W_{STI} to W_{Si}. It demonstrates that large ratio of W_{STI} to W_{Si} can not only make the BV of the MSF-LDMOS increase but also bring in large $R_{dson.sp}$. Our application requires the minimum BV must be larger than 34V. As a result, according to the measured results, the best ratio of W_{STI} to W_{Si} is 0.7/0.3.

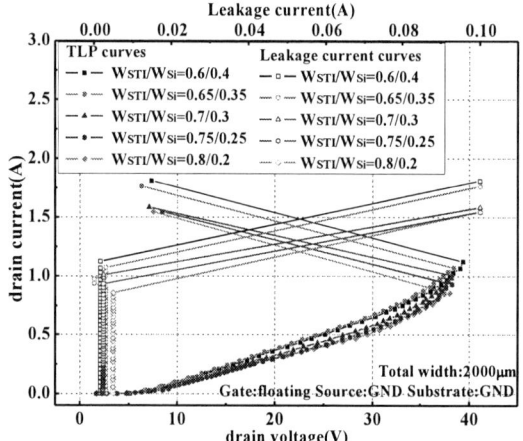

Fig. 3: The measured TLP characteristic curves of the MSF-LDMOS for different ratios of W_{STI}/W_{Si}.

Furthermore, the Transmission Line Pulse (TLP) with 100ns pulse width has been carried out to evaluate these devices' ESD robustness[7]. During the measurement, the TLP stress is applied on the drain while the gate stays floating and the source is grounded because the MSF-LDMOS's gate is connected with pre-stage drive circuit rather than ground in practical application. When the ESD stress comes, its gate stays at uncertain floating state. The measured results are displayed in Fig.3. We find that the secondary breakdown current (It_2) decreases with the increasing ratio of W_{STI} to W_{Si}. That is to say large ratio of W_{STI} to W_{Si} can weaken the ESD robustness of the MSF-LDMOS.

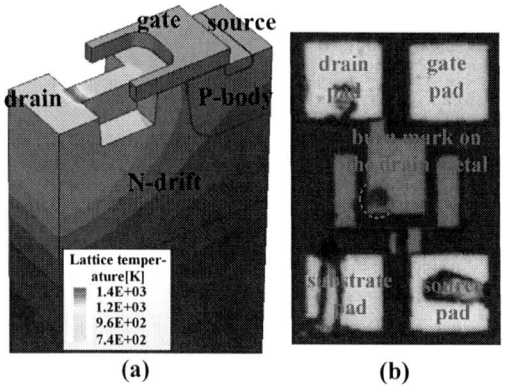

Fig. 4: (a) The simulated lattice temperature distribution of MSF-LDMOS with W_{STI}/W_{Si}=0.7/0.3 under the 1.0A TLP current stress. (b) The top view of MSF-LDMOS with W_{STI}/W_{Si}=0.7/0.3 after TLP test.

III. ANALYSIS AND DISCUSSIONS

TCAD simulator Sentaurus has been used to understand the ESD failure mechanism of the MSF-LDMOS. The best ratio of W_{STI} to W_{Si} equaling 0.7/0.3 is chosen as the research object. Based on the measured results in Fig.3, the It_2 of the device with W_{STI}/W_{Si}=0.7/0.3 is 1A. So the 1A TLP stress is applied on its drain during the simulation. The lattice temperature distribution of the MSF-LDMOS with W_{STI}/W_{Si}=0.7/0.3 is shown in Fig.4(a). The maximum lattice temperature appears at the silicon zone between two STI fingers near the drain side. Fig.4(b) shows that the drain-side metal has burn mark after TLP test, which is coincident with the simulation results.

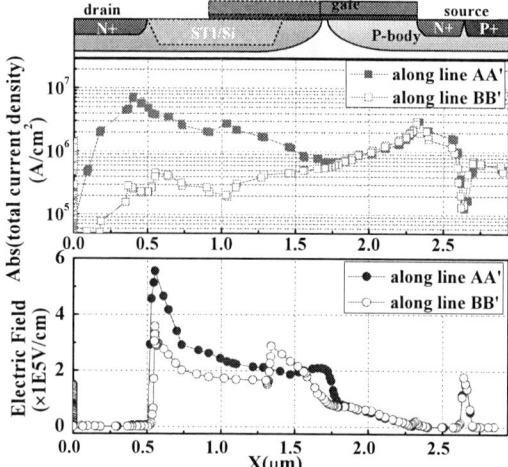

Fig. 5: The total current density and electric field distributions at the Si/SiO₂ interface of MSF-LDMOS along line AA' and line BB' respectively under 1.0A TLP current stress.

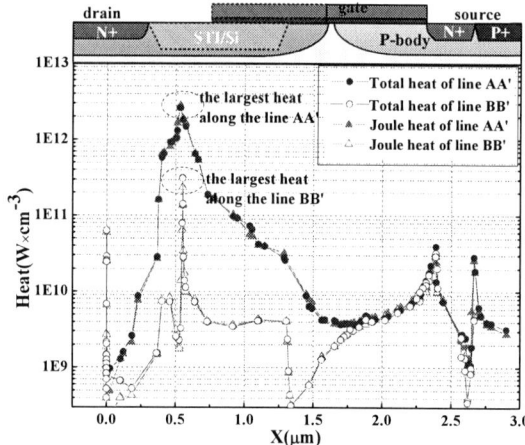

Fig. 6: The Joule heat and total heat distributions at the Si/SiO₂ interface of MSF-LDMOS along line AA' and line BB' respectively under 1.0A TLP current stress.

The main heat source making lattice temperature rise comes from the Joule heat that is influenced by total current density and electric field[8]. Thus, Fig.5 depicts the peaks of total current density and electric field along line AA' and along line BB' under the 1A TLP stress. Along line AA', most of current tends to flow through the silicon zone between two STI fingers attributed to its smaller drift resistance. As a result, the total current density along line AA' is larger than that along line BB'. Not only that, the serious Kirk effect caused by the larger current density also makes the peak electric field higher at the drain side along line AA'. In addition, the potential lines are easier to concentrate near the drain side along line AA' without the pushing effect of STI. Thus, as shown in Fig.6, the generation of Joule heat at silicon zone is more than that at STI zone. The lattice temperature is also increased by the much more Joule heat and it finally fails at the silicon zone between two STI fingers at the drain side.

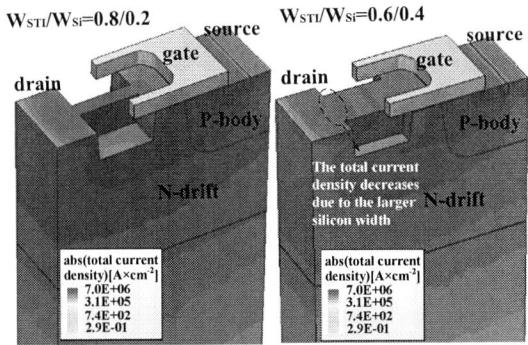

Fig. 7: The simulated total current density distributions of MSF-LDMOS with W_{STI}/W_{Si}=0.8/0.2 and W_{STI}/W_{Si}=0.6/0.4 respectively under the 0.85A TLP current stress.

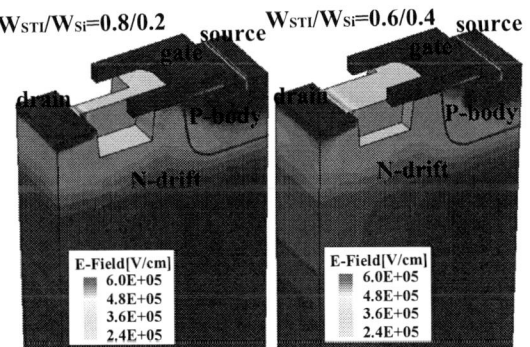

Fig. 8: The simulated electric field distributions of MSF-LDMOS with W_{STI}/W_{Si}=0.8/0.2 and W_{STI}/W_{Si}=0.6/0.4 respectively under the 0.85A TLP current stress.

The ESD failure mechanism of MSF-LDMOS has been clarified above. The reason why larger ratio W_{STI}/W_{Si} can weaken the ESD robustness of MSF-LDMOS will be further discussed. In order to be more obvious for the results, the largest ratio of W_{STI}/W_{Si}=0.8/0.2 and the smallest ratio of W_{STI}/W_{Si}=0.6/0.4 are selected to compare with each other. According to the above-mentioned ESD failure mechanism of MSF-LDMOS, the total current density and the electric field distributions are the direct factors to cause the failure of MSF-LDMOS. Thus they are also given in Fig.7 and Fig. 8 respectively for W_{STI}/W_{Si}=0.8/0.2 and W_{STI}/W_{Si}=0.6/0.4 under

the 0.85A TLP stress (the It_2 of the device with W_{STI}/W_{Si}=0.8/0.2). In Fig.7, the simulation results illustrate that the total current density for W_{STI}/W_{Si}=0.8/0.2 is more crowded than that for W_{STI}/W_{Si}=0.6/0.4. The wider STI fingers compress the current path in the silicon zone and the total current density can concentrate in silicon zone easily. Meanwhile, the Kirk effect is more serious for W_{STI}/W_{Si}=0.8/0.2 because of the large current in the silicon zone. As a result, its electric field near the drain side is larger the than that for W_{STI}/W_{Si}=0.6/0.4, which has already reflected in Fig.8 clearly. Therefore, the larger W_{STI}/W_{Si} ratio can induce the weaker ESD robustness.

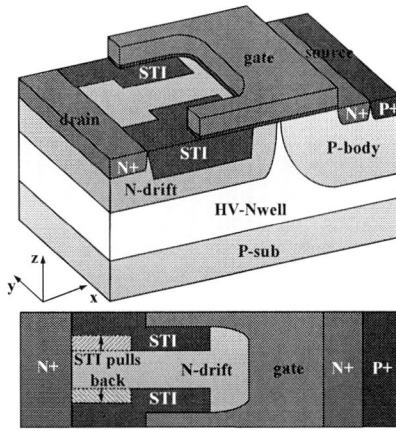

Fig. 9: The 3D schematic diagram of the MSSF-LDMOS.

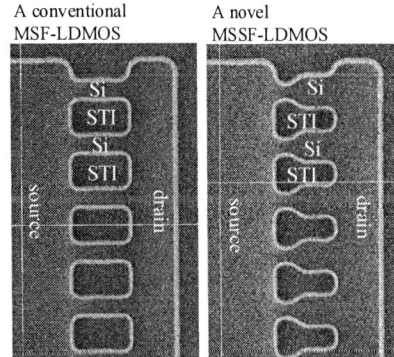

Fig. 10: The top view of active areas of MSF-LDMOS and MSSF-LDMOS.

IV. OPTIMIZATION AND VERIFICATION

Based on the explanation above, the widened silicon width is helpful to improve device's ESD robustness. However, if the silicon width is enlarged, the breakdown voltage also decreases and may be out of the BV requirement. The relationship between ESD robustness and breakdown voltage seems to be contradictory. To solve this problem, a novel structure with multiple stair-STI fingers LDMOS (MSSF-LDMOS) is proposed, as shown in Fig.9 and Fig.10. Based on the best ratio of W_{STI}/W_{Si} =0.7/0.3, the STI of the novel device near the drain side pulls back about 0.1μm.

978-1-5386-2928-4/18 $31.00 © 2018 IEEE

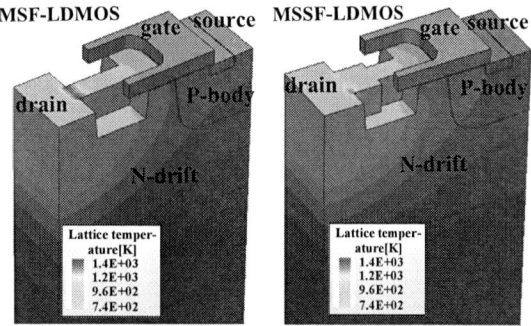

Fig. 11: The lattice temperature distributions of MSF-LDMOS (W_{STI}/W_{Si}=0.7/0.3) and the MSSF-LDMOS under the 1.0A TLP current stress.

Fig.11 depicts the lattice temperature distributions of the novel MSSF-LDMOS and the conventional MSF-LDMOS with W_{STI}/W_{Si}=0.7/0.3 under the 1A TLP stress (the It_2 of the device with W_{STI}/W_{Si}=0.7/0.3). It is obvious that the maximum lattice temperature of the novel one is suppressed effectively since the broadened silicon width at drain side can alleviate the crowded current density at vulnerable region.

In addition, the measured results of the novel MSSF-LDMOS and the conventional MSF-LDMOS, including breakdown voltages, turn-on characteristics and TLP measurements, are shown in Fig.12. Compared with the conventional MSF-LDMOS with W_{STI}/W_{Si}=0.7/0.3, the ESD robustness of the novel device increases by 10%, the $R_{dson.sp}$ decreases by 2.5% due to the smaller drift resistance resulting from the wider silicon zone near the drain side, the BV only reduces by 1.2%.

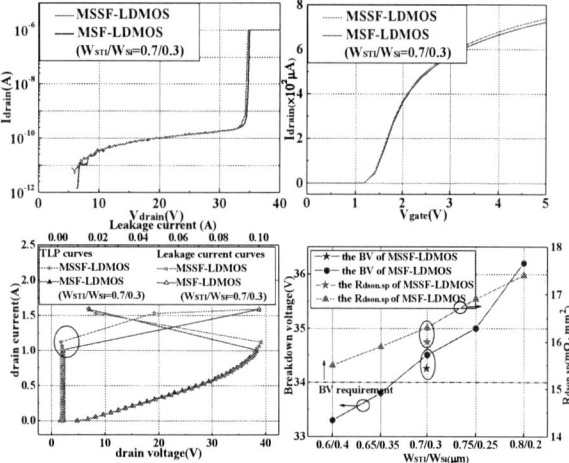

Fig.12: The measured electric characteristics of MSSF-LDMOS and MSF-LDMOS (W_{STI}/W_{Si}=0.7/0.3) including breakdown voltage, turn-on characteristic and TLP curves.

V.   CONCLUSION

In this work, it is found that the different ratios of STI width to silicon width not only influence the BV and $R_{dson.sp}$, but also have distinct effects on the ESD robustness. The ESD failure mechanism demonstrates that the MSF-LDMOS is damaged by accumulated Joule heat at the silicon zone between two STI fingers near the drain as the crowded current density and high peak electric field. Furthermore, a novel structure with multiple stair-STI fingers is proposed by alleviating the crowded current density. The experiment results demonstrate that, for MSSF-LDMOS, the ESD robustness increases by 10% and the $R_{dson.sp}$ decreases by 2.5% with only 1.2% reduction of BV comparing with the conventional MSF-LDMOS, meanwhile almost without sacrificing other main electric characteristics.

ACKNOWLEDGMENT

This work was supported by Nation Natural Science Foundation of China (61604038, 61674030), and Natural Science Foundation of Jiangsu Province (BK20160691).

REFERENCES

[1] S. L. Chen, Y. T. Huang, "Design and Layout Strategy in the 60-V Power pLDMOS with Drain-End Modulated Engineering of Reliability Considerations", *IEEE Transactions on Power Electronics*, vol.7, no.31, pp.5113-5121, 2016.

[2] H. L. Liu, Z.W. Jhou, S.-T. Huang, et al., "A novel high-voltage LDMOS with shielding-contact structure for HCl SOA enhancement", *International Symposium on Power Semiconductor Devices and IC's(ISPSD)*, 2017, pp.311-314.

[3] C. Schmidt, G. Spitzlsperger, "Increasing Breakdown Voltage of p-Channel LDMOS in BCD Technology with Novel Backside Process. *International Symposium on Power Semiconductor Devices and IC's(ISPSD)*, 2017, pp. 339-342.

[4] S. Poli, S. Reggiani, G. Baccarani, et al., "Hot-carrier stress induced degradation in Multi-STI-Finger LDMOS: An experimental and numerical insight", *Solid-State Electronics*, vol.65–66, no.1, pp.57-63, 2011.

[5] J. Jang, K. H. Cho, D. Jang, et al., "Interdigitated LDMOS", *International Symposium on Power Semiconductor Devices and IC's(ISPSD)*, 2013, pp.245-248.

[6] H.-C. Tsai, R.-H. Liou, C. Lien, "Resurf model and electrical characteristics of finger-type STI DEMOS," *IEEE Transcation on Electron Devices*, vol.63, no.12, pp.4603-4609, 2016.

[7] S. Malobabic, J. A. Salcedo, J.J Hajjar, et al., "NLDMOS ESD Scaling Under Human Metal", *IEEE Electron Device Letters*, vol.33, no.11, pp. 1595-1597, 2012.

[8] R. Ye, S. Liu, W. Sun, et al., "ESD failure mechanism and optimization for the LDMOS with low on-resistance and large geometric array used as output device", *International Symposium on Power Semiconductor Devices and IC's(ISPSD)*, 2016, pp.239-242.

978-1-5386-2928-4/18 $31.00 © 2018 IEEE

Proceedings of the 30th International Symposium on Power Semiconductor Devices & ICs
May 13-17, 2018, Chicago, USA

Integrated Symmetrical High-voltage Inverter for the Excitation of Touch Sensitive Electroluminescent Devices

Katrin Hirmer, Muhammad Bilal Saif, Klaus Hofmann

Integrated Electronic Systems Lab
Technische Universtität Darmstadt, Germany
Email: Katrin.Hirmer@ies.tu-darmstadt.de

Abstract—**Smart Power Integrated Circuits are commonly used where different operating voltages are required since it can enable miniaturization of the system electronics. For the combination of an electroluminescent device with a capacitive touch sensor, a 1 µm SOI integrated circuit is implemented with a high-voltage inverter and a low-voltage spread spectrum clock generator. The transient signals with up to $\pm$ 300 V$_\mathrm{p}$ and 5 kHz show interferences on the digital sensor excitation signal.**

Keywords—**Smart Power Integrated Circuits, SPIC, High-voltage Inverter, Spread Spectrum Clock Generator, High-voltage interferences**

I. INTRODUCTION

Smart Power Integrated Circuits (SPIC) are commonly used where different operating voltages are required [1]. Due to the high level of integration, SPICs enable effective miniaturization. Hence, these devices become more and more relevant where construction size is limited.

Electroluminescent (EL) devices are one example of components that require high voltage excitation to emit light. Although EL devices have been investigated for many years, they are still getting more important for applications where very thin, bendable or glare-free light sources are required. Quite often, only a limited area for the electronic excitation is available. If the lighted device can also be used as a control sensor, many applications such as lighted push buttons or hand rails can arise in future.

SPICs enable the requirement of high voltage excitation and low voltage sensing at limited area. However, due to the integration of high voltages as well as low voltage signals on one single die, interferences can arise. The influence of disturbing signals on power lines or other sensitive nodes within integrated circuits was the subject of intense research for a very long time (e.g. [2], [3], [4]). The main noise sources that have been investigated are switching noises in RF circuits as well as electromagnetic interferences.

There is only little research on the influence of high voltage signals on low voltage nodes. In the automotive sector, smart power ICs are used to combine signals with up to 20 V with low voltage analog signals [5]. Investigations show that the influence in CMOS technologies is mainly due to substrate coupling within the die [6]. Parasitic PNP and NPN transistors are activated by voltages above and below the supply voltages,

respectively. For switching devices with ultra-high voltage signals more expensive silicon-on-insulator (SOI) processes are preferred since the different power domains can have a separate substrate.

This paper focuses on the influences of alternating high voltage signals on sensitive low voltage sensor nodes. It is organized as follows: Section II gives an overview of the implemented high-voltage and low-voltage functions within the IC. Section III discusses the measurement results. Section IV concludes the paper.

II. SYSTEM OVERVIEW

Apart from the electronics, the main component is an electroluminescent device which can emit light when exposed to an alternating high voltage. The devices are mostly screen printed due to the flexibility and economy of the manufacturing process. Therefor, first a well-conducting rear electrode is deposited on a substrate. Afterwards, an insulation layer prevents short circuits. Then the luminescence layer, often called phosphor, is deposited followed by a transparent front electrode and an encapsulation layer. Due to its structure shown in Fig. 1, the EL device, representing the load of the inverter and providing the sensor electrode, can be regarded as a lossy parallel plate capacitor C_EL with a series resistor R_Pcdot due to the finite conductivity of the transparent electrode. The capacitance C_EL is depending on the printing parameters, the used materials as well as on the size of the EL. The samples available have a capacitance of up to 10 nF with a series resistance R_Pedot of $\leq$ 100 Ω which are the important parameters for the inverter design.

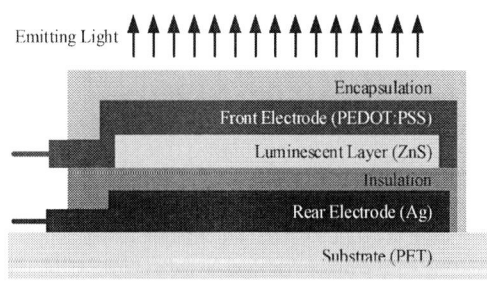

Fig. 1. Structure of a printed electroluminescence device.

978-1-5386-2928-4/18 $31.00 © 2018 IEEE   343

The overall system is composed of the EL device, a high voltage inverter and a sensor part, consisting of a spread spectrum clock (SSC) generator for the excitation and a sensor evaluation as shown in Fig. 2 to indicate the touch event.

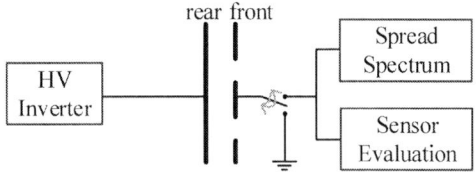

Fig. 2. Block diagram of the different system components.

Notwithstanding an encapsulation layer added during the printing process of the ELs, the front electrode is always connected to ground while the EL is lighted to ensure the safety of the user of the touch sensor. This results in the need of a symmetrical $\pm$ alternating high voltage signal on the rear electrode. Hence, most commonly used H-bridges are excluded in this design [7], [8]. It also follows that the electronic components of the output stage of the inverter have to withstand twice the peak high voltage which is required to be more than $200\,V_p$. For the integration of the inverter a technology allowing at least $400\,V$ is needed. This is why a $1\,\mu m$ SOI technology with up to $700\,V$ is chosen to ensure a feasible lifetime of the integrated circuit (IC). Since the luminence of the EL is proportional to the clock frequency of the inverter, up to $5\,kHz$ are required to enable a sufficient brightness of the printed EL devices in research. Hence, in addition to the two half bridges instead of a single H-bridge, the implemented inverter has to deliver higher frequencies than previous inverters [9],[10].

The sensor part needs high noise immunity for reliable operation in robust environments. This is why, a spread spectrum (SSC) technique is chosen to excite the capacitive sensor. The sensor evaluation is realized off chip to maximize flexibility. For the investigations within this work, the evaluation is not significant and is therefore not described further.

The logic within the IC enables an independent control of the inverter and the SSC generator. Hence, the two functionalities can be performed one after the other or at the same time. The following subsections explain the architectures of the integrated high voltage inverter and the sensor excitation.

A. High Voltage Inverter

The high voltage is generated by a boost and a buck-boost converter for a positive (+HV) and a negative (-HV) high voltage, respectively. The output amplitudes of both converters can be controlled by the IC via an SPI interface followed by an ADC which creates the reference voltage. Fig. 3 shows the block diagram of the implemented inverter. The high voltage generation needs external components (shown in gray) such as an inductor, a diode and a capacitor to enable a stable and efficient high voltage generation.

The inverter itself is fully integrated requiring a gate driver for the positive HV switch as well as for the negative HV

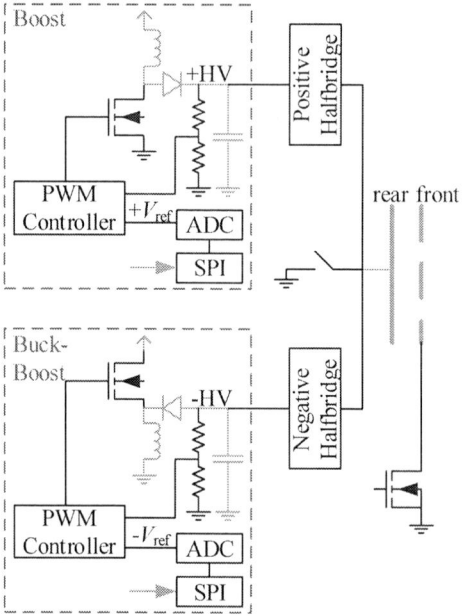

Fig. 3. Block diagram of the high voltage inverter with internal components (in black) and external components (in gray).

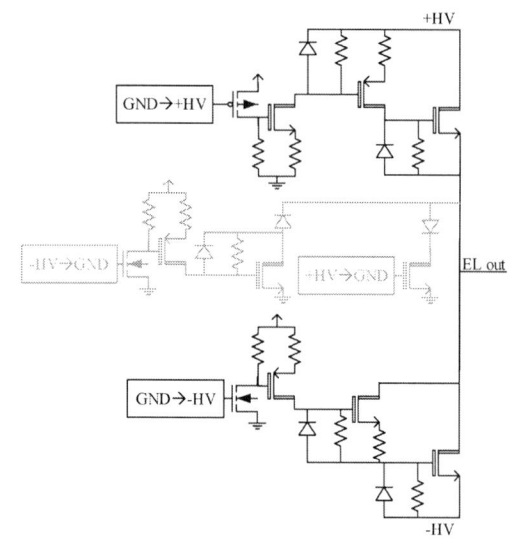

Fig. 4. Schematic of the fully integrated output stage (switching to ground colored in gray for better visibility).

switch as shown in Fig. 4. It can switch from ground (GND) to the negative high voltage (-HV) and back to ground. Afterwards it switches to the positive high voltage (+HV) and again back to ground. With this procedure it is ensured that the EL lighting always stops at the low level voltage before the sensor starts its turn.

978-1-5386-2928-4/18 $31.00 © 2018 IEEE

B. Low Voltage Spread Spectrum Clock Generator

On-chip a 9 bit spread spectrum clock generator is implemented which can vary its frequencies randomly between 1.6 MHz and 8 MHz. Compared to many other concepts (e.g. [11], [12]), this concept relies more on analog circuit components since the benefits of all-digital architectures are more relevant for technologies with smaller feature sizes where digital standard cell sizes are smaller. The principle, shown in Fig. 5, is based on a current adding ADC which is connected to a pseudo-random number (PRN) generator. The current from the ADC is used to charge an integrated capacitor up to a reference voltage. Once the level is reached, the capacitor is discharged by a second ADC drawing the same current. A subsequent logic then transfers the voltage level into the output frequency (f_{DCO}) and delivers enough fan out for subsequent use.

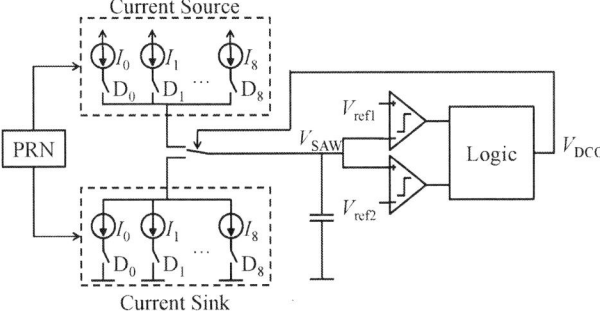

Fig. 5. Block diagram of the integrated spread spectrum clock generator.

III. MEASUREMENT RESULTS

The electronics for high voltage EL and low voltage sensor excitation are integrated in a 1 μm high-voltage SOI technology (see Fig. 6) enabling the miniaturization of the system components. The overall chip size is 6570 μm x 5865 μm.

The integrated inverter can drive loads of up to 10 nF providing a maximum peak current of 263 mA at + 300 V at the switching point. The SSC generator with a resolution of 9 bit can vary its frequency between 1.6 MHz and 8 MHz showing good correspondence between post-layout simulation and measurement (see Fig. 7).

The chip can control the output voltage of up to ± 300 V, the EL frequency up to 5 kHz, the modulation frequency of the sensor, as well as the on and off times for the two components. This allows measuring the influence of the HV signal on the low voltage digital excitation signal of the sensor (see Fig. 8).

Measurements show that the frequency of the digital output is changing significantly when the EL is switching at the same time. The magnitude of change is depending on the amplitude of the HV signal as well as on the load. Hence, it can be concluded that the interference of the high voltage signal is proportional to the output current of the high voltage stage. Furthermore it can be assumed that the coupling effect is not at the final digital output but at the sensible analog parts such

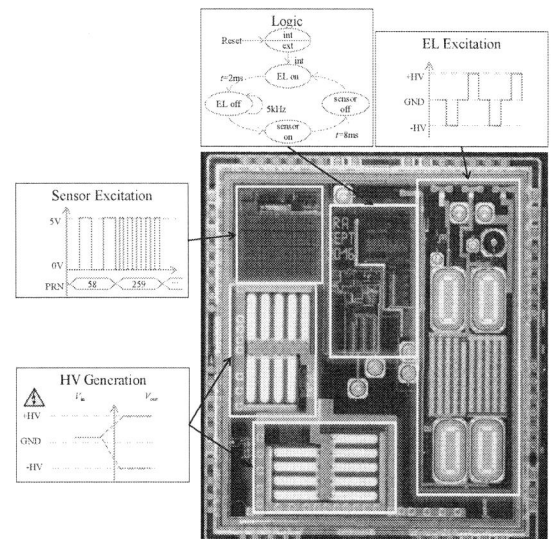

Fig. 6. Chip photograph with circuit block explanations.

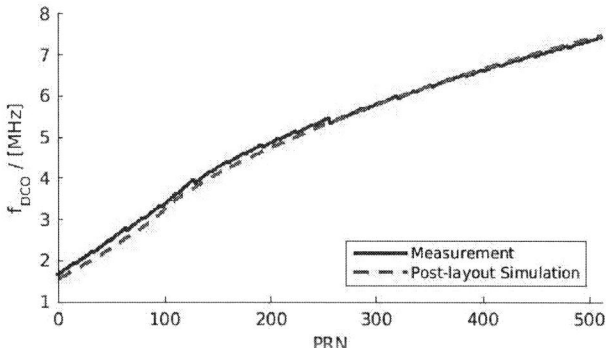

Fig. 7. Output frequency f_{DCO} of the spread spectrum clock generator with respect to the pseudo random number (PRN) showing good conformity of measurement and post-layout simulation.

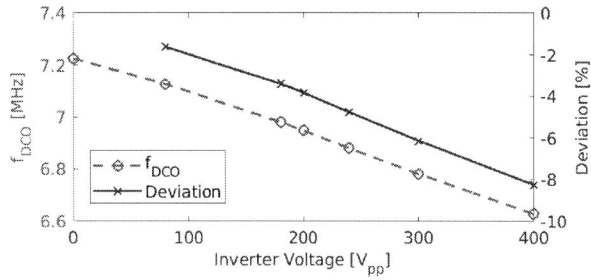

Fig. 8. Measurement results of the influence of the high-voltage on the output frequency of a spread spectrum clock generator f_{DCO} over the peak-to-peak voltage of the inverter and its percentage deviation from the frequency without EL switching.

as supply voltage or any reference node within the spread spectrum clock generator. Due to the SOI technology, coupling effect through the substrate can be excluded.

Fig. 9 shows the time resolved influence of the HV signal on the output frequency of the DCO for the most significant bit, and thus the highest frequency. It can be seen that the frequency is clearly changing at the transitions of the HV signal. Due to the ringing within the frequency, inductive coupling has to be taken into account. Depending on the maximum HV and the load of the inverter the output frequency of the SSC generator can vary up to 14 %. Further investigations are necessary to gain comprehensive results.

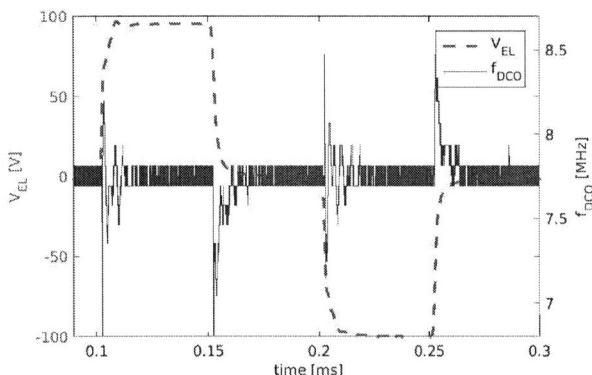

Fig. 9. Measurement results of the influence of the high-voltage switching on the output frequency of a spread spectrum clock generator.

To reduce this influence and to enable the sharing of the front electrode, a so called time multiplexing is proposed alternating between sensor and EL. To avoid flicker, the EL is switched off for up to 8 ms while the sensor passes randomly through all its 512 different frequencies. Afterwards, the EL is switched on for 2 ms, making ten transitions at 5 kHz before the sensor is switched on again.

IV. CONCLUSION

A smart power integrated circuit is presented combining an ultra-high voltage inverter with up to ± 300 V and a low voltage sensible capacitive sensor excitation. The chip shows excellent conformity of measurement and post-layout simulation. The two components can be controlled independent of each other, which enables the measurements of influences from the high voltage part on low voltage nodes. The influence on the digital output of a spread spectrum clock generator has been investigated showing frequency changes of up to 14 %. The influences are depending on the load as well as on the magnitude of the high voltage signal which indicates that the output current of the inverter is relevant for the interferences.

For the first time, the presented chip provides a highly monolithic integrated solution for high voltage inverters with sensible sensor technology.

ACKNOWLEDGMENTS

The authors would like to thank the German Federal Ministry of Education and Research for the funding within the Project ELSE under grant no. 13N13079.

REFERENCES

[1] B. Murari, F. Bertotti, and G. A. Vignola, *Smart power ICs: technologies and applications*. Springer Science & Business Media, 2002, vol. 6.

[2] K. T. Tang and E. G. Friedman, "Simultaneous switching noise in on-chip cmos power distribution networks," *IEEE Transactions on Very Large Scale Integration (VLSI) Systems*, vol. 10, no. 4, pp. 487–493, 2002.

[3] K. Yoshikawa, Y. Harada, N. Miura, N. Takeda, Y. Saito, and M. Nagata, "Immunity evaluation of inverter chains against rf power on power delivery network," in *Electromagnetic Compatibility of Integrated Circuits (EMC Compo), 2013 9th Intl Workshop on*. IEEE, 2013, pp. 232–237.

[4] H. A. Huynh and S. Kim, "Design of on-chip power noise sensing circuit and its applications," in *Electrical Design of Advanced Packaging and Systems Symposium (EDAPS), 2015 IEEE*. IEEE, 2015, pp. 209–212.

[5] European FP7 - AUTOMICS Project web site. [accessed 15-December-2017]. [Online]. Available: https://www.automics.eu/

[6] V. Tomasevic, A. Boyer, and S. B. Dhia, "Development of an on-chip sensor for substrate coupling study in smart power mixed ics," in *Electromagnetic Compatibility (APEMC), 2015 Asia-Pacific Symposium on*. IEEE, 2015, pp. 67–70.

[7] I. Sidiropoulos and S. Siskos, "An integrated dc-ac inverter for electroluminescent lamps," in *Design of Circuits and Integrated Circuits (DCIS), 2014 Conference on*. IEEE, 2014, pp. 1–6.

[8] I. Goncharov, A. Kabyshev, E. Kozyrev, and A. Maldzigaty, "Power source for electroluminescent panels," *Journal of Communications Technology and Electronics*, vol. 62, no. 6, pp. 634–637, 2017.

[9] T. Ge, L. Guo, Y. Kang, J. Zhou, H. He, P. Ng, E. Fitzgerald, K. Lee, and J. Chang, "A driver circuit based on the emerging gan-on-cmos process for the emerging electroluminescent panels," in *Circuits and Systems (MWSCAS), 2015 IEEE 58th International Midwest Symposium on*. IEEE, 2015, pp. 1–4.

[10] Microchip Technology Inc., "Datasheet HV830," Tech. Rep., 2013, [accessed 18-October-2017]. [Online]. Available: http://www.microchip.com/wwwproducts/en/HV830

[11] N. Da Dalt, P. Pridnig, and W. Grollitsch, "An all-digital pll using random modulation for ssc generation in 65nm cmos," in *Solid-State Circuits Conference Digest of Technical Papers (ISSCC), 2013 IEEE International*. IEEE, 2013, pp. 252–253.

[12] C.-C. Chung, D. Sheng, and W.-D. Ho, "A low-cost low-power all-digital spread-spectrum clock generator," *IEEE Transactions on Very Large Scale Integration (VLSI) Systems*, vol. 23, no. 5, pp. 983–987, 2015.

Proceedings of the 30th International Symposium on Power Semiconductor Devices & ICs
May 13-17, 2018, Chicago, USA

A Power Inductor Integration Technology Using a Silicon Interposer for DC-DC Converter Applications

Yixiao Ding[*], Xiangming Fang[†], Yuan Gao[*], Yuefei Cai[*], Xing Qiu[‡], Philip K.T. Mok[*], S. W. Ricky Lee[‡], Kei May Lau[*], and Johnny K. O. Sin[*]

[*]Department of Electronic and Computer Engineering, Hong Kong University of Science and Technology, Hong Kong, China
[†]Shenzhen CoilEasy Technologies Limited, Shenzhen, China
[‡]Department of Mechanical and Aerospace Engineering, Hong Kong University of Science and Technology, Hong Kong, China
Email: ydingah@connect.ust.hk; eesin@ust.hk

Abstract— In this paper, a new power inductor integration technology using a silicon interposer for DC-DC converter applications is proposed and experimentally demonstrated. Using this technology, coreless spiral inductors can be embedded from the back of a silicon interposer and connected with the front-side metal routing through TSVs (through-silicon vias). Trenches for the inductor windings and TSVs are etched from the back of the silicon interposer and filled with copper by using a two-step electroplating process. In this way, process difficulties and limitations in current TSV based interposer technologies for embedded power inductor applications can be overcome. Using this embedded power inductor integration technology, a 4.2 µH power inductor is embedded at the backside of a silicon interposer and integrated with a buck converter IC and LED chips to form a compact LED system. The buck converter in the LED system achieves an overall power efficiency of 69% while the size of the system is reduced by 5 times compared to the discrete counterparts. The silicon interposer embedded power inductor shows a good power efficiency of 89%.

Keywords—integrated inductor; silicon interposer; DC-DC converter

I. INTRODUCTION

Inductor integration is of vital importance to DC-DC converter miniaturization [1, 2]. Using a silicon interposer to integrate power inductors is a promising approach for achieving a compact power-supply-in-package solution [3 - 5]. Compared with monolithic integration [6], power inductor integration using a silicon interposer does not only enable the use of optimal processes for fabricating the power inductor and IC chips, but also provides heterogeneous integration capability for IC chips and load chips.

In current integration schemes which adopt a silicon interposer, thin-film power inductors are fabricated on the silicon interposer, and DC-DC converter chips are flip-chip bonded on the silicon interposer [3 - 5]. However, those power inductors are usually built on the surface of the silicon interposer with a winding thickness up to tens of microns only. The limited winding thickness leads to a relatively large DC resistance and power loss.

In order to increase the winding thickness and achieve a higher power efficiency, a promising approach is to embed the

inductor in the silicon substrate [7 - 9]. Embedded coreless power inductor can efficiently utilize the silicon substrate by using deep trenches (a few hundred microns) filled with copper as very thick windings, which has the advantages of small dc resistance, large current-carrying capability, and no core saturation issue.

Previously reported silicon-embedded inductors are designed to be embedded in the silicon substrate where the power IC will also be fabricated. To suppress the substrate loss of the inductor operating at frequencies up to several MHz, a high-resistivity substrate must be used [8]. However, the high-resistivity substrate is not compatible with the low-resistivity substrate used in standard CMOS or Bipolar-CMOS-DMOS (BCD) technology for power IC fabrication. Thus, the previously reported silicon-embedded inductor cannot be practically integrated with power ICs to achieve a functional system unless a more complicated isolation scheme is used. Instead of integrating the silicon-embedded inductor with the power IC on the same substrate, embedded inductor integration using a silicon interposer allows the use of high-resistivity silicon substrate without the restriction from the standard CMOS or BCD technology.

In this paper, a new power inductor integration technology using a silicon interposer is proposed and demonstrated for buck converter applications.

II. POWER INDUCTOR INTEGRATION TECHNOLOGY

The proposed power inductor integration technology can be applied to a 2.5D integrated power management system. Fig. 1 shows the integration of the embedded power inductor with other chips using a silicon interposer for a 2.5D integrated

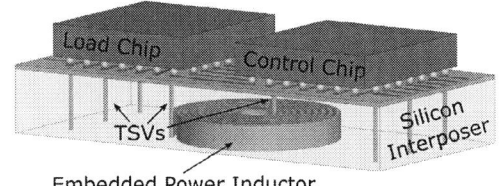

Fig. 1: Integration of the embedded power inductor with other chips using a silicon interposer.

This work was supported by the Research Grants Council of the Hong Kong Special Administrative Region under Grant T23-612/12-R.

978-1-5386-2928-4/18 $31.00 © 2018 IEEE   347

power management system. A control chip for the power management system and a load chip are flip-chip bonded on the front-side of the silicon interposer. The power inductor is embedded into the back of the interposer and has deep trenches as thick windings. The electrical connections between the front-side chips and the embedded power inductor are realized through TSVs, since TSVs can shorten the interconnection paths and offer better electrical performance than wirebonding.

The power inductor integration technology is described below. The major fabrication steps are shown in Fig. 2. The process begins with a double-side polished high-resistivity (~3000 Ω·cm) silicon wafer. Low-temperature oxide (LTO) is first deposited on both sides of the silicon wafer. The oxide layer on the back of the wafer is patterned with the shape of the trenches and vias, and will be used as a hard mask for both the trench and via etching later on. Then, a photoresist layer is spin coated on the oxide hard mask and is exposed to reveal only the via region. After this, the via region is first etched to a certain depth by deep-silicon reactive-ion-etching (DRIE) (Fig. 2 (a)). Then the photoresist is removed and the oxide hard mask is used for the subsequent etching. The etching of the trenches is then begun while the via etching is continued using the oxide hard mask. The etching of the trenches and vias is finished when the vias are etched all the way to the front-side oxide layer. Fig. 2 (b) shows the trenches and vias after DRIE. After trench and via etching, the oxide hard mask is removed by wet etching. An insulation oxide layer is then deposited on the sidewall and bottom of the trenches and vias as well as on the wafer surface (Fig. 2 (c)). Then, seed layers of first TiW and then Cu is sputtered on the front-side. Electroplating is used to seal the mouth and the front portion of the vias (Fig. 2 (d)). The excess electroplated copper and the sputtered seed layer on the front surface of the silicon wafer are removed by blanket wet etching. During the wet etching, the back of the wafer is protected by photoresist. After the wet etching, the copper surface on the via region will be approximately at the same level as the wafer surface (Fig. 2 (e)). The first electroplating seals the vias and minimizes the surface topology variation on the via region. Then, metal routing can be easily made on the relatively smooth surface (Fig. 2 (f)). After this, the metal routing layer on the front-side is protected by photoresist, and a seed layer is sputtered on the back of the wafer. Another electroplating is used to fill the trenches and the unfilled portion of the TSVs (Fig. 2 (g)). Since the front portion of the TSVs has already been filled with copper, the depth of the unfilled part of the TSVs is close to the depth of the trenches. Bottom-up electroplating can be used to minimize the electroplating time and overburden. Finally, the excess copper is removed by chemical mechanical polishing (CMP) (Fig. 2 (h)).

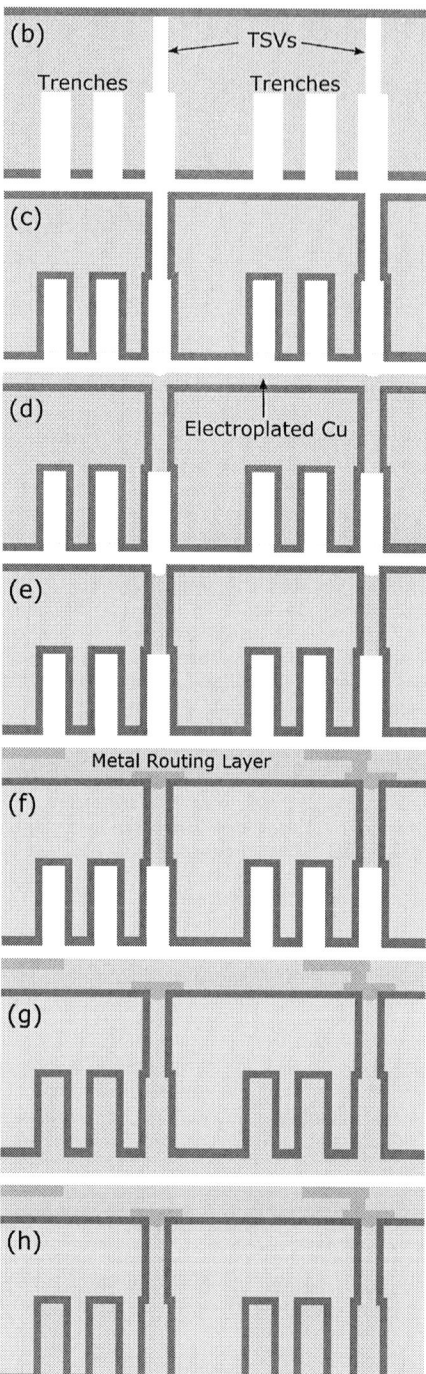

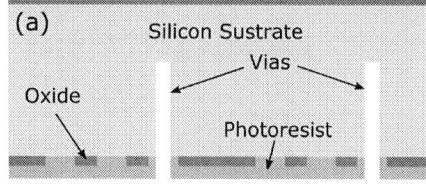

Fig. 2: Schematic of major fabrication steps: (a) via etching; (b) via and trench etching; (c) original oxide mask removal and oxide insulation layer deposition; (d) first copper electroplating to seal the via mouth; (e) overplated copper removal; (f) front-side metal routing formation; (g) second copper electroplating to fill the trench and via; (h) overplated copper removal.

III. EXPERIMENTAL RESULTS

Using the proposed power inductor integration technology, a power inductor is embedded from the back of a silicon interposer. Fig. 3 (a) shows the front-side of the fabricated silicon interposer with two metal routing layers on it. The white layer is an Al layer and the brown layer is a Cu layer. Fig. 3 (b) shows the back of the silicon interposer with a power inductor embedded inside. The inductor has an area of 5 mm × 5 mm, and the interposer has a total thickness of 0.35 mm. The inductor winding width, spacing and thickness are 20 µm, 25µm and 180µm, respectively. Two TSV arrays (12 TSVs per array) are used to connect the inductor to the front-side.

Fig. 4 (a) shows the cross-sectional view of the trenches for the inductor windings and TSVs after the front-side metal routing formation corresponding to Fig. 2 (f). Fig. 4 (b) shows the cross-sectional view of the trenches and TSVs after the electroplating and CMP corresponding to Fig. 2 (h). The diameter and depth of the TSVs are 44 µm and 350 µm, respectively. After the two-step electroplating and final polishing, the TSVs and trenches are filled with copper without any voids as previously shown in Fig. 4 (b). A good electrical connection between the front-side devices and the backside embedded inductor is achieved by using the proposed technology.

The electrical performance of the backside embedded power inductor is measured using two pads on the front-side of the silicon interposer. The DC resistance of the inductor is measured to be 1.86 Ω using four-point probe method. The measured AC performance of the embedded inductor by R&S ZVB 8 network analyzer is plotted in Fig. 5. A peak quality factor of 12.6 is achieved at 2 MHz. An inductance value of 4.2 µH is maintained up to 10 MHz.

To show the usefulness of the proposed integration technology, the silicon interposer embedded power inductor was integrated with a buck converter IC and LED chips (as the load for the buck converter) to form a compact LED system. Fig. 6 shows the LED system under testing. The LED chips were flip-chip bonded on the Cu layer on the silicon interposer. The buck converter control IC could also be flip-chip bonded on the Al layer on the silicon interposer. But at the current stage of testing, the control IC chip was attached to the silicon interposer and wire bonded to a testing PCB board instead. Other circuit components, including the capacitors and power diodes, were also placed on the PCB board.

For 130 V DC input, the overall efficiency of the buck converter is approximately 69%. To estimate the power efficiency of the inductor in the buck converter, the power loss on the inductor is calculated. The current flowing through the inductor was measured. It is represented as a sum of the DC and sinusoidal AC currents by Fourier analysis. The amplitude of the DC and major sinusoidal AC currents are plotted in Fig. 7. The power loss on the inductor is calculated by summing up the DC and AC power losses using the following equation [10]:

$$P_{loss} = I_{DC}^2 R_{DC} + \frac{1}{2}\sum I_{AC}^2 R_{AC} \qquad (1)$$

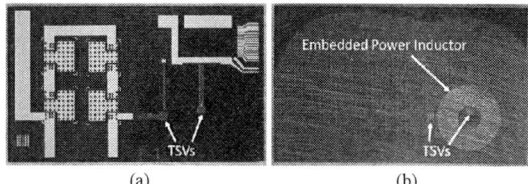

Fig. 3: Top view of silicon interposer: (a) front-side and (b) backside.

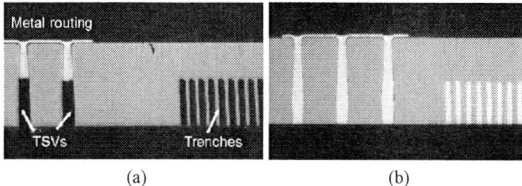

Fig. 4: Optical images of cross-sections of trenches and TSVs: (a) after first electroplating and metal routing; (b) after second electroplating and final polishing.

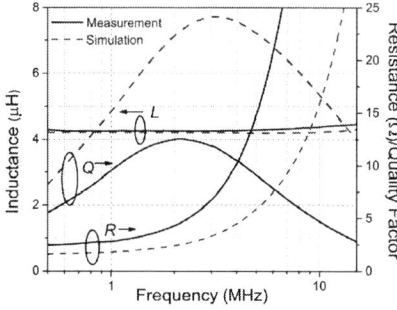

Fig. 5: Measured and simulated AC performance of the embedded power inductor.

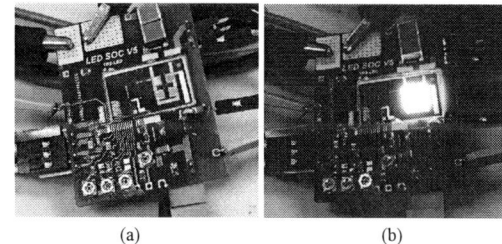

Fig.6: Integrated LED system under testing: (a) off and (b) on.

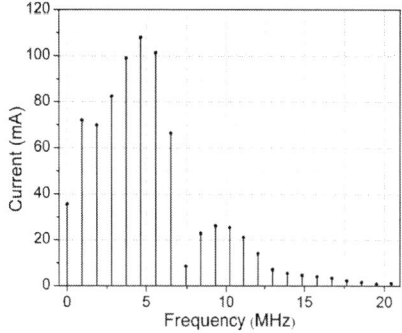

Fig. 7: DC and major sinusoidal AC components of measured inductor current.

I_{DC} and R_{DC} are the amplitude of the DC current and DC resistance of the inductor, respectively. I_{AC} and R_{AC} are the amplitude of the AC current and AC resistance, respectively, of the inductor at the corresponding frequency. The power loss on the inductor is 0.39 W using the data obtained in Fig. 5 and Fig. 7. The measured output power for the LED is around 3.15 W. So the power efficiency of the fabricated integrated inductor is:

$$\eta_{inductor} = \frac{P_{output}}{P_{output} + P_{inductor}} = 89\% \qquad (2)$$

IV. Discussion

The simulated performance obtained by using HFSS [11] of the fabricated inductor is also plotted in Fig. 5 for comparison with the measured performance. The measured inductance value agrees well with the simulated value. The discrepancies between the simulated and measured values of the resistance and Q-factor come from the higher electroplated copper resistivity and the higher substrate loss caused by the non-optimized CVD deposited insulation layer in the trenches. With advanced electroplating and CVD technologies, the DC and AC resistance can be further reduced and the simulated performance shown in Fig. 5 can be achieved.

To estimate the power efficiency of a well-fabricated embedded inductor using an advanced process, the HFSS simulations are also used. With the same current flow, the power loss on the inductor can be reduced to 0.15 W, and the inductor power efficiency can be improved to 95%.

The thickness of the silicon interposer embedded inductor is around one tenth of a typical surface mounted power inductor. Using the silicon interposer embedded inductor, the thickness of the LED system is reduced to 1mm, and the system size is reduced by approximately 5 times compared to the discrete implementation of placing the control IC, LED chips and power inductor side-by-side on a PCB board [12].

V. Conclusion

In this paper, a power inductor integration technology using a silicon interposer for DC-DC converter applications was proposed and experimentally demonstrated. Using the proposed technology, a 4.2 µH power inductor was integrated with a buck converter on the silicon interposer. A compact LED system was finally implemented on the fabricated silicon interposer. With the help of the integration technology, a 5 times reduction in the LED system size compared to the discrete counterparts was obtained. The fabricated embedded inductor has demonstrated a power efficiency of 89%, and the power efficiency can be further improved to 95% using a more optimized fabrication process. This embedded inductor integration technology showed significant potential for integrated DC-DC converter applications.

Acknowledgment

The authors would like to thank the staff of the Nanosystem Fabrication Facility and the Semiconductor Product Analysis and Design Enhancement Center at the Hong Kong University of Science and Technology for their excellent support.

References

[1] Q. Li et al., "Technology road map for high frequency integrated DC-DC converter," in 2010 Twenty-Fifth Annual IEEE Applied Power Electronics Conference and Exposition, Palm Springs, CA, USA, pp. 533–539, 2010.

[2] C. Ó. Mathúna, N. Wang, S. Kulkarni, and S. Roy, "Review of integrated magnetics for Power Supply on Chip (PwrSoC)," IEEE Trans. Power Electron., vol. 27, no. 11, pp. 4799–4816, 2012.

[3] N. Sturcken et al., "A 2.5D integrated voltage regulator using coupled-magnetic-core inductors on silicon interposer," IEEE J. Solid-State Circuits, vol. 48, no. 1, pp. 244–254, 2013.

[4] H. J. Bergveld, R. Karadi, and K. Nowak, "An inductive down converter system-in-package for integrated power management in battery-powered applications," in Proc. IEEE Power Electron. Spec. Conf., Rhodes, Greece, pp. 3335–3341, 2008.

[5] K. Ishida, K. Takemura, K. Baba, M. Takamiya, and T. Sakurai, "3D stacked buck converter with 15µm thick spiral inductor on silicon interposer for fine-grain power-supply voltage control in SiP's," in 3D Systems Integration Conference (3DIC), Tokyo, Japan, pp. 1–4, 2010.

[6] J. Wibben and R. Harjani, "A High-Efficiency DC–DC Converter Using 2 nH Integrated Inductors," IEEE J. Solid-State Circuits, vol. 43, no. 4, pp. 844–854, Apr. 2008.

[7] R. Wu and J. K. O. Sin, "A novel silicon-embedded coreless inductor for high-frequency power management applications," IEEE Electron Device Lett., vol. 32, no. 1, pp. 60–62, 2011.

[8] R. Wu, J. K. O. Sin, and C. P. Yue, "High-Q Backside silicon-embedded inductor for power applications in µh and MHz range," IEEE Trans. Electron Devices, vol. 60, no. 1, pp. 339–345, 2013.

[9] X. Fang, R. Wu, L. Peng, and J. K. O. Sin, "A new embedded inductor for ZVS DC-DC converter applications," in 2012 24th International Symposium on Power Semiconductor Devices and ICs, Bruges, Belgium, pp. 53–56, 2012.

[10] R. P. Wojda and M.K. Kazimierczuk, "Analytical Winding Foil Thickness Optimization of Inductors Conducting Harmonic Currents", IET Power Electronics, vol. 6, no. 5, pp. 963-973, 2013.

[11] HFSS, ANSYS Inc., Canonsburg, PA, USA, 2005.

[12] T.H. Mak, Z. Liu, W.C. Chong, Y. Gao, X. Fang, J.K.O. Sin, P.K.T. Mok, and K.M. Lau, "Integration Scheme Toward LED System-on-a-Chip (SoC)", Solid-State and Organic Lighting, Canberra, Australia, Dec 2014.

Proceedings of the 30th International Symposium on Power Semiconductor Devices & ICs
May 13-17, 2018, Chicago, USA

A New 1200 V HVIC with High Side Edge Trigger in order to Solve the Latch on Failure by the Negative VS Surge

Kinam Song, Wonhi Oh, Jinkyu Choi, Seunghyun Hong and Sangmin Park
High Power Division, Power Solutions Group, ON Semiconductor
Bucheon, Republic of Korea
kinam.song@onsemi.com

Abstract—This paper investigates the root cause of the latch on failure by a short turn-on input signal with a negative VS surge and proposes a new 1200 V HVIC with a high side edge trigger in order to solve the latch on failure. The proposed HVIC is fabricated using a 1.2 μm 1200 V BCDMOS process. The experimental results show the latch on failure no longer occurs on the new 1200 V HVIC because it doesn't overlap the VS recovery period and the RESET pulse period. The proposed HVIC can be applied to IPM modules (intelligent power modules) and APM modules (automotive power modules) which require a more robust HVIC solution.

Keywords—HVIC, negative voltage surge, latch on failure

I. INTRODUCTION

High voltage integrated circuits (HVICs) based on high voltage level shifter are widely used in power electrical systems, such as motor drives, automotive electronics, home appliances and light-emitting diode lighting etc. which require high integration density and efficiency [1]. Generally, the conventional HVICs are vulnerable to a negative VS surge, where VS is a high side floating supply ground that would drop several tens of volts below the ground reference value for the duration of several hundreds of nanoseconds. The negative VS surge leads to the HVIC destruction or the HVIC malfunction [2-6]. Especially, an unwanted turn-on phenomenon of the output signal of HVIC, we call the latch on failure.

In particular, the latch on failure has occurred more frequently when the signal having a short turn-on pulse width is applied to the input terminal of HVIC. Because the short turn-on input signal causes reducing the turn-off propagation delay of the output signal and it also causes an overlapping between the negative VS recovery period and the turn-on period of RESET LDMOS in the high voltage level shifter. Generally, the HVIC needs the pulse generator that produces two pulses, respectively the "SET pulse" at the occurrence of the rising edge and the "RESET pulse" at the occurrence of the falling edge of the input signal for low power dissipation in performing the level shifting function [7]. Fig. 1 shows the HVIC latch on failure by the short turn-on input signal. We can see that the overlapping of the VS recovery period and the RESET pulse period occurs by decreasing the input pulse width. As seen in Fig. 1(a) (HIN=1μs), if the VS recovery period and the RESET pulse period are overlapped, the latch on failure happens.

This paper is organized as follow. Section II covers the root cause of the latch on failure by the short turn-on input signal. We describe the effect of the short turn-on input signal, and the relation of the VS recovery period and the RESET pulse period. From the root cause analysis, the new high side driving method is presented in Section III. The experimental results of the proposed HVIC are shown in Section IV. Finally, conclusions are discussed in Section V.

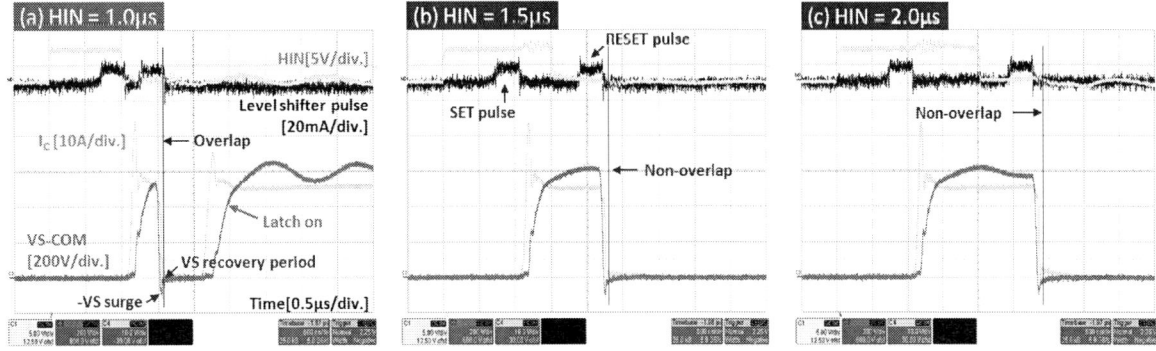

Fig. 1. HVIC latch on failure by the short turn-on input signal

978-1-5386-2928-4/18 $31.00 © 2018 IEEE   351

II. THE ROOT CAUSE OF LATCH ON FAILURE

A. The effect of the short turn-on input signal

Fig. 2 shows the switching diagram regarding the pulse widths of the input signal. If the turn-off signal is triggered before the gate-emitter voltage is fully charged up to VBS that is the high side power supply voltage by the short turn-on input signal, the turn-off propagation delay time is reduced since the gate-emitter voltage is low compared with a longer input pulse width. As a result, the overlapping between the VS recovery period and the RESET pulse period is happening by the short turn-on input signal. In addition, the overlapping is frequently able to happen when the operating conditions are high current switching, high line voltage, low VBS voltage, and low junction temperature.

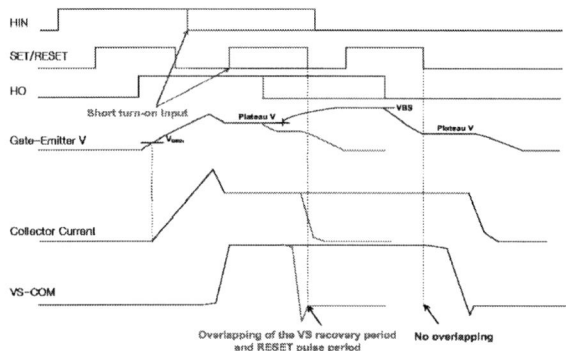

Fig. 2. Switching diagram regarding the pulse width of the input signal

B. Relation between the latch on failure and the each period

The relation between the latch on failure and the each period are shown in Fig. 3. Where, I_C and I_B indicate a collector current of a high side IGBT and a VB current of the HVIC, respectively. When the negative VS happens, the body diodes of LDMOS in HVIC turn on because the COM node that is an anode of the body diodes is 0 V and the VB ($\approx$15 V + VS) node that is a cathode of the body diodes is the negative voltage. When passing over from the conductive state to the blocking state, the storage charge of the body diodes has to be discharged. This results in a current flowing in the reverse direction in the body diodes. The I_B current shows this current

flowing. The latch on failure doesn't occur although the falling dVS/dt and the negative VS surge are overlapped with the RESET pulse period as seen in Fig. 3(b). Eventually, the latch on failure occurs when the falling dVS/dt, the negative VS surge, and the VS recovery period (reverse recovery time) are overlapped with the RESET pulse period as seen in Fig. 3(c). It indicates the latch on failure is deeply related to the VS recovery period and the RESET pulse period.

C. The root cause of the latch on failure by the overlapping of the VS recovery period and the RESET pulse period

The high voltage level shifter and the operation of the LDMOS's body diodes are shown in Fig. 4. The high voltage level shifter transfers and isolates between the low side signal and the high side signal. It is basically composed of the pull-up resistors, the Zener diodes and the high voltage LDMOSs with the body diodes. If VB voltage drops under 0V by the freewheeling operation on the low side, the body diodes turn on and the forward currents flow from COM to VB node as seen in Fig. 4(a). When VB voltage recovers over 0V, the body diodes turn off and the reverse recovery currents flow VB node to COM at the same time. However, Fig. 4(b) shows that if the reverse recovery phenomenon occurs when RESET LDMOS remains turn-on state by RESET pulse signal and SET LDMOS is a turn-off state, each body diode has a different reverse recovery characteristic. It is because the body diode's storage charge decreases along with the increase of the LDMOS gate-source voltage [8]. Therefore, RESET body diode reverse recovery current (I_{RR}) and the reverse recovery time (t_{rr}) are declined by higher gate-source voltage compared with the SET LDMOS. This difference of the reverse recovery behavior between SET and RESET LDMOS causes malfunction.

Fig. 5 show the switching diagrams when is overlapping between the RESET pulse period and the VB recovery period. Where, SET and RESET are the gate-source voltage of the SET and the RESET LDMOS, respectively. VB represents a transient of the VB node voltage. I_{BODY} indicates a current flowing through the SET and the RESET body diode. The RS latch set and reset signal are the high side RS latch inputs. HO shows the high side switching device's gate-source voltage or gate-emitter voltage. Since the RESET LDMOS remains turn-on state by the RESET pulse and the SET LDMOS is turn-off state, the RESET body diode has a small reverse recovery

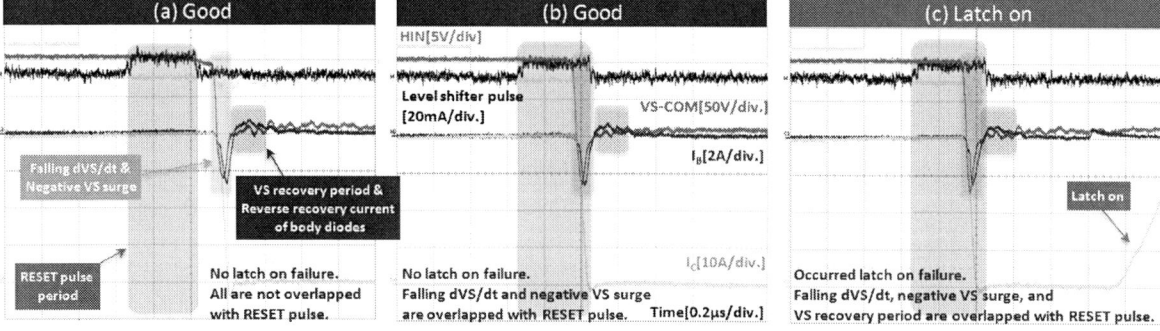

Fig. 3. Relation between the latch on failure and the each period

978-1-5386-2928-4/18 $31.00 © 2018 IEEE   352

charge in comparison with the SET body diode. As a result, an unwanted RS latch set signal is generated by the mismatch of the reverse recovery charge between the SET and the RESET LDMOS. The unwanted HO signal is generated by the RS latch set signal, the latch on failure eventually occurs.

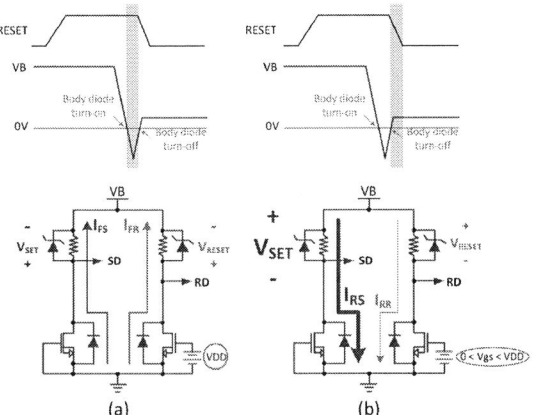

Fig. 4. High voltage level shifter and the operations of the LDMOS's body diodes:
(a) On state of body diodes during the negative VB surge period
(b) Off state and the reverse recovery behavior of the body diodes during the VB recovery period

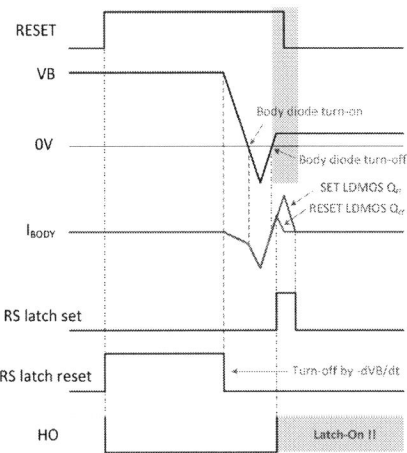

Fig. 5. Overlapping of the RESET pulse period and the VB recovery period

On the other hand, if the VB recovery period occurs when the SET and the RESET LDMOS have a same gate-source voltage (turn-off state), the SET and the RESET body diode have a same reverse recovery charge. The reverse recovery charges become the common mode noise and it is completely rejected by the common mode noise canceller in HVIC. Therefore, the latch on failure never happens on this condition.

In summary, if the RESET pulse period and the VB recovery period are overlapped by the short turn-on input

signal, the reverse recovery charge of the RESET body diode is decreased by the higher gate-source voltage in comparison with the SET LDMOS. The drop voltage of the SET LDMOS drain becomes higher than RESET LDMOS. As a result, the unwanted RS latch set signal is generated by the mismatch, and HO becomes high and eventually occurs the latch on failure.

III. PROPOSED NEW HIGH SIDE DRIVING METHOD

In order to solve the latch on failure, this paper proposes new high side driving method which never overlaps between the VS recovery period and the RESET pulse period of the high voltage level shifter. The method is inserting the high side edge trigger between the noise canceller and the high side RS latch. The block diagram of the proposed HVIC that is implemented the high side edge trigger is shown in Fig. 6.

Fig. 7 shows the timing diagram illustrating operation of the conventional HVIC and the proposed HVIC in the short turn-on input signal condition. The high side edge trigger generates two pulses corresponding to the rising edge of the noise canceller's output signals (SD' and RD'). The circuit prevents timing overlap between the VS recovery period and the RESET pulse period. Because always the HO signal goes in a low state after the end of the RESET pulse period.

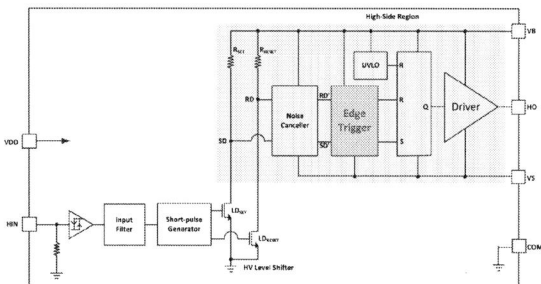

Fig. 6. Block diagram of new high side driving method that is implemented the high side edge trigger

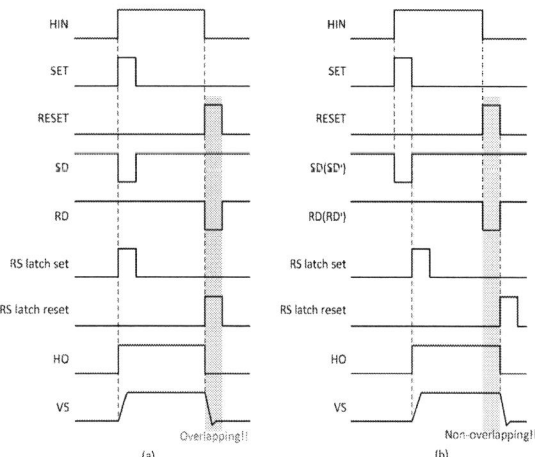

Fig. 7. Timing diagrams of the conventional HVIC (a) vs. the proposed HVIC (b) in the short turn-on input signal condition

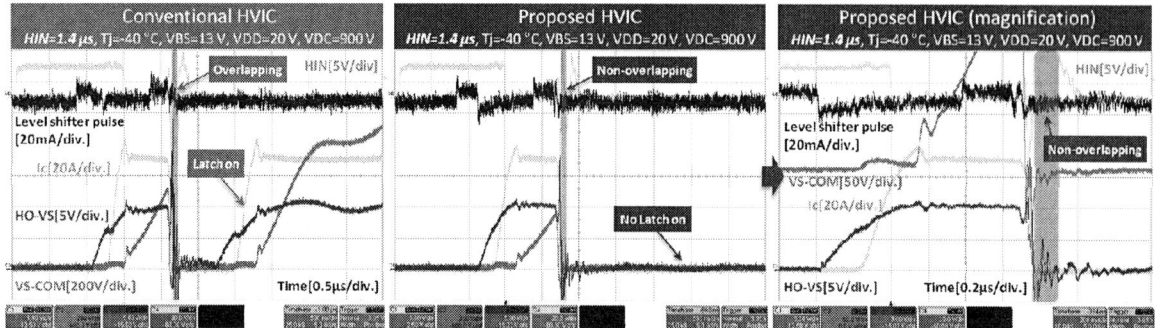

Fig. 8. Experimental results: The conventional HVIC vs. the proposed HVIC

TABLE I. EXOPERIMENTAL RESULTS REGARDING THE TURN-ON PULSE WIDTH OF THE INPTU SIGNAL: THE CONVENTIONAL HVIC VS. THE PROPOSED HVIC

VBS [V]	VDD [V]	T_J [°C]	VDC [V]	I_C [A]	DUT	HIN [μs]												
						1.0	1.1	1.2	1.3	1.4	1.5	1.6	1.7	1.8	1.9	2.0	2.6	3.0
13	20	25	900	70	Conv.	Latch On	Latch On	Latch On	Latch On	Latch On	Latch On	Latch On	Latch On	Latch On	Good	Good	Good	Good
		25	900	70	Proposed	Good	Good	Good	Good	Good	Good	Good	Good	Good	Good	Good	Good	Good
13	20	-40	900	70	Conv.	Latch On	Latch On	Latch On	Latch On	Latch On	Latch On	Latch On	Latch On	Latch On	Latch On	Latch On	Latch On	Good
		-40	900	70	Proposed	Good	Good	Good	Good	Good	Good	Good	Good	Good	Good	Good	Good	Good

IV. EXPERIMENTAL RESULTS

The proposed HVIC that is implemented the high side edge trigger is fabricated using a 1.2 μm 1200 V BCD (Bipolar-CMOS-DMOS) process technology. The experimental results show the latch on failure no longer occurs on the proposed HVIC because it doesn't overlap the VS recovery period and the RESET pulse period.

Fig. 8 and Table I show the experimental results. Note that VBS, VDD, VDC, and I_C indicate the high side and the low side supply voltage, the DC line voltage and the high side IGBT current, respectively. In the case of the conventional HVIC, the latch on failure occurs up to HIN is 2.6 μs at low temperature. However, the proposed HVIC doesn't happen the latch on failure at all test conditions. Fig. 8 is well shown the non-overlapping between the VS recovery period and the RESET pulse period on the proposed HVIC.

V. CONCLUSION

This paper presented the root cause of the latch on failure by the short turn-on input signal by describing the switching characteristics regarding the input pulse width, and the relation of the VS recovery period and the RESET pulse period. To prevent the latch on failure, we propose the new high side driving method that includes the high side edge trigger. The experimental results show that the latch on failure no longer occurs on the proposed HVIC. The proposed HVIC can be applied to IPM modules (intelligent power modules) and APM modules (automotive power modules) which require a more robust HVIC solution.

REFERENCES

[1] Sungpah Lee, Kunhee Cho, Minwoo Lee and Wookang Jin, "A leakage reduced HVIC with coarse-fine UVLO," IEEE International SoC Design Conference, pp. 408-411, 2012.

[2] S. Pawel, M. Rossberg and R. Herzer, "600V SOI gate drive HVIC for medium power applications operating up to 200°C," IEEE International Symposium on Power Semiconductor Devices and ICs, pp. 55-58, 2005.

[3] M. Rossberg, B. Vogler and R. Herzer, "600V SOI Gate Driver IC with advanced level shifter concepts for medium and high power applications," IEEE European Conference on Power Electronics and Applications, pp. 1-8, 2007.

[4] Jing Zhu, Guodong Sun, Weifeng Sun and Yunwu Zhang, "Negative voltage surge resistant circuit design in HVIC," IEEE Electronics Letters, 7th, vol. 49, no. 23, pp. 1476-1477, November 2013.

[5] Masaharu Yamaji, Akihiro Jonishi, Takahide Tanaka, Hitoshi Sumida and Yoshio Hashimoto, "A New Enhanced Noise Tolerance Technique for a 600V High Voltage IC," IEEE International Conference on Power Electronics and Drive Systems, pp. 108-111, 2015.

[6] Akihiro Jonishi, Masashi Akahane, Masaharu Yamaji, Hiroshi Kanno, Takahide Tanaka, Noriyuki Tochinai and Hitoshi Sumida, "A breakthrough concept of HVICs for high negative surge immunity," IEEE International Symposium on Power Semiconductor Devices and ICs, pp. 57-60, 2015.

[7] Aldo Novelli, Luca Giussani and Ignazio Bellomo, "New generation of half bridge gate driver ICs for use with low power 3.3 V control applications," IEEE Power Electronics Specialists Conference, pp. 3237-3242, 2004.

[8] Arash Elhami Khorasani, Mark Griswold and T. L. Alford, "Gate-Controlled Reverse Recovery for Characterization of LDMOS Body Diode," IEEE Electron Device Letters, vol. 35, no. 11, pp. 1079-1081, November 2014.

Proceedings of the 30th International Symposium on Power Semiconductor Devices & ICs
May 13-17, 2018, Chicago, USA

A High-Voltage Half-Bridge Gate Drive Circuit for GaN Devices with High-Speed Low-Power and High-Noise-Immunity Level Shifter

Xin Ming, Xuan Zhang, Zhi-wen Zhang, Xu-dong Feng, Li Hu, Xia Wang, Gang Wu and Bo Zhang

State key Laboratory of Electronic Thin Films and Integrated Devices
University of Electronic Science and Technology of China
Chengdu, China
Email: mingxin@uestc.edu.cn

Abstract—A high-voltage half-bridge gate drive circuit for E-mode GaN device with high-speed low-power and high-noise-immunity level shifter is proposed in this paper. It adopts digital-level detection concept by combining fast-slewing and output-maintain circuits in high noise environments to achieve both small response time and high dV_{SW}/dt noise immunity. The proposed gate driver is fabricated in 0.5μm 80V HV CMOS process where the active area is $1699 \times 1522 \mu m^2$. Simulation and experimental results of the level shifter demonstrate that propagation delay is only about 1.618ns and dV_{SW}/dt noise immunity is up to 50V/ns with a low FOM equal to 0.04ns/μm·V.

Keywords—level shifter; GaN gate drive; fast slewing circuit; output maintain circuit; digital-level detection.

I. INTRODUCTION

High-voltage half-bridge gate drive techniques for GaN devices become more critical in high-frequency high-power-density power conversion recently, such as half and full-bridge converters as well as synchronous buck converters. The high-speed energy-efficient level shifter is a key block in high-side gate driver [1]-[6], since it mainly determines high-side channel's propagation delay (i.e. ~25ns for 100V half-bridge architecture). Moreover, negative V_{SW} bias for floating ground and dV_{SW}/dt noise are two critical issues for the traditional resistive-loaded level shifter design, which is based on detection of voltage drop of resistors and may false trigger RS flip-flop at high-voltage and heavy-load conditions [7]-[8].

Compared to Si power device, the E-mode GaN gains a more negative reverse body diode voltage (~1.5V-2.5V) during dead time and higher dV_{SW}/dt when turning on high-side switch (i.e. ~50V/ns for 100V half-bridge application) [9], which will even worsen the problem. Different methods have been reported to resolve this trouble, including adding a buffer with filtering function [10], adopting common mode noise cancellation circuit [11] or capacitive-loaded level shifter [7]. However, the previous ones may sacrifice other switching performance or need complex circuit structure and layout design. In addition, the transmission delay of high-side channel and dV/dt noise immunity are hard to reach the requirements of high-frequency and fast-switching application by GaN devices.

This work was supported by the National Key R&D Program of China under Grant 2017YFB0402800.

In this work, a novel digital-level detection concept for high-voltage level shifter is proposed to substitute for traditional analog voltage detection and resolve design challenges mentioned above. Concept of the proposed technique is discussed in Section II. Circuit design and implementation are shown in Section III. Experimental results and conclusions are given in Sections IV and V, respectively.

II. DESIGN OF PROPOSED LEVEL SHIFTER

In this section, the different level shifting strategies are explained in detail to introduce the trade-off for circuit design. Then system theory of the proposed high-speed low-power and high-noise-immunity level shifter is given, showing the advancement compared to traditional ones.

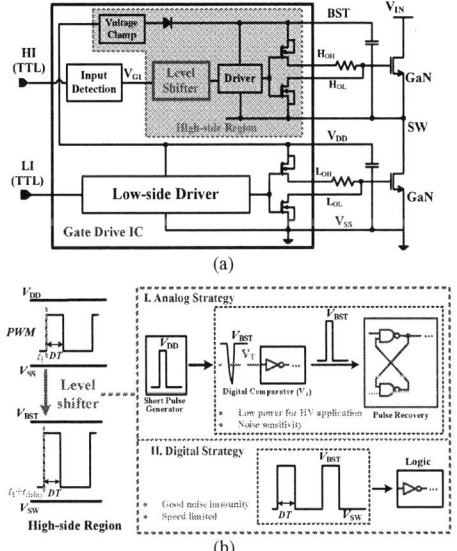

Fig. 1 Half-bridge gate drive for E-mode GaN (a) system architecture (b) different level shifting strategies

Fig. 1(a) shows the block diagram of the high-voltage (HV) half-bridge gate drive for E-mode GaN. Level shifter with LDMOS in the signal path is a key block, since it determines propagation delay of the high-side channel and will restrict high-frequency operation of gate drive (i.e. minimum on-time for Buck converter). It also should be insensitive to noise,

978-1-5386-2928-4/18 $31.00 © 2018 IEEE   355

preventing misoperation during fast switching process. Fig. 1(b) gives a comparison of different level shifting strategies for HV application. Traditional analog detection method compares voltage drop and threshold of the inverter during short-pulse time and recovers the PWM signal with the help of R-S latch. The propagation delay and power consumption are small, however, it is sensitive to noises and will introduce reliability problems [7]. Conventional digital method with latch circuit adopts actual input PWM signal for control in the signal conversion process. The response speed is slow although it gains a good noise margin [2]. None of the architectures mentioned above are well suited for high-frequency high-power-density half-bridge gate drive for GaN devices, which is the motivation of the paper.

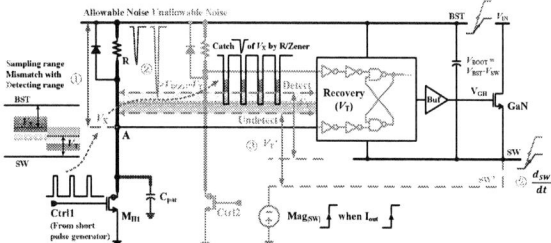

Fig. 2 Reliability problems of the conventional resistive-loaded level shifter

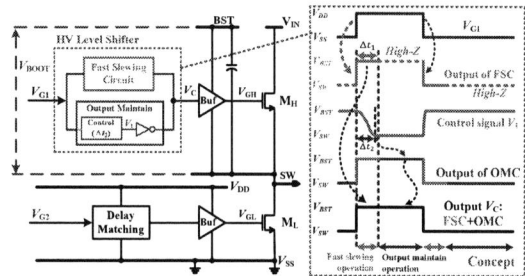

Fig. 3 System theory of the proposed high-speed low-power and high-noise-immunity level shifter

In order to further observe the potential reliability issues of conventional circuits for GaN application, the problems of the resistive-loaded level shifter, which is based on voltage-undershoot detection, is shown in Fig. 2. Signal mismatch often occurs to introduce misoperation [7]-[8]. For example, referring to problem ①, each time when the LDMOS is turned on, output current I_{MH1} is small when V_{BST} is small. Also considering temperature coefficient and process variation of R_1, V_X may not cross V_T when $I_{MH1}R_1$ is small and is not detected by the recovery circuit. Problem ② shows dV/dt issue for fast switching operation. As can be seen, V_{BST} always follows with V_{SW} based on the bootstrap technique and dV_{BST}/dt is approximately equal to dV_{SW}/dt. However, due to a large parasitic capacitance C_{par} at node A of M_{H1}, voltage slope dV_A/dt can not be as fast as dV_{BST}/dt. The noise amplitude ($\sim RC_{par}$dV_{SW}/dt) might be caught falsely by the recovery circuit when its amplitude is larger than ($V_{BOOT}-V_T$), where V_{BOOT} is equal to $V_{BST}-V_{SW}$. Therefore, dV_{SW}/dt immunity capability is limited by parasitic capacitance of HV devices (i.e. dV_{SW}/dt$_{max}$=($V_{BOOT}-V_T$)/RC_{par}). For problem ③ with negative V_{SW} noise issue, when V_{SW} becomes negative during

dead time, V_T will be decreased by βV_{SW}, where β is aspect ratio of the first inverter in recovery circuit. The output signal of level shifter is not caught by the next stage if the threshold is decreased below GND (i.e. the allowable negative voltage $V_{SW}>-V_T/\beta$). GaN devices operate with a larger negative reverse body diode voltage, higher dV/dt and wider input voltage range for half-bridge application, which deteriorates the reliability of traditional gate drive circuit mentioned above.

Fig. 3 shows the control theory of the proposed high-speed low-power and high-noise-immunity level shifter. The concept is to split the signal conversion into two stages. Fast-slewing circuit (FSC) provides fast response for PWM signal V_{G1} during short-pulse time Δt_1 and then remains in high-impedance state to save power. It is a high-power block in Δt_1 and mainly determines propagation delay of the level shifter. Low-power output-maintain circuit (OMC) starts to respond when the logic level of V_{G1} changes and is prepared to sustain output state at Δt_2 (i.e. control signal V_1 should be pulled down to V_{SW} during Δt_2). If $\Delta t_1>\Delta t_2$ is always satisfied by circuit design, then output signal V_C can be maintained after Δt_1 is over and is the same as PWM signal V_{G1}. The input logic control signal is then fully recovered by the new level shifter at fast switching condition. Therefore, this co-working mode uses digital-level detection concept to substitute for traditional analog voltage detection without sacrificing speed and power. Moreover, it provides an important reliability enhancement: after the short-pulse time, the node V_C does not stay in high-impedance state [7]-[8], which is latched to V_{SW} or V_{BST} by OMC instead. So V_C is less sensitive to process, temperature and noises, being especially suited for high-frequency half-bridge GaN gate drive application.

III. CIRCUIT REALIZATION

A. Level Shifter Design

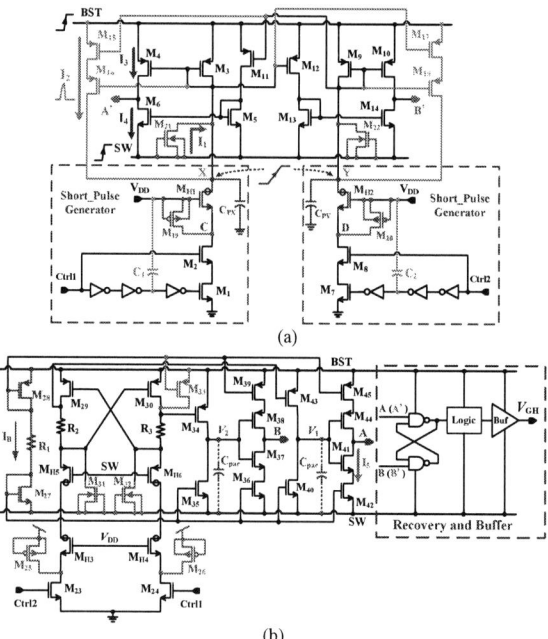

(a)

(b)

Fig. 4 Circuit realizations of level shifter (a) FSC (b) OMC

978-1-5386-2928-4/18 $31.00 © 2018 IEEE

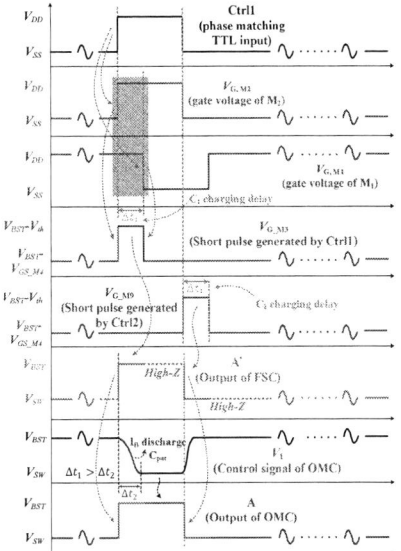

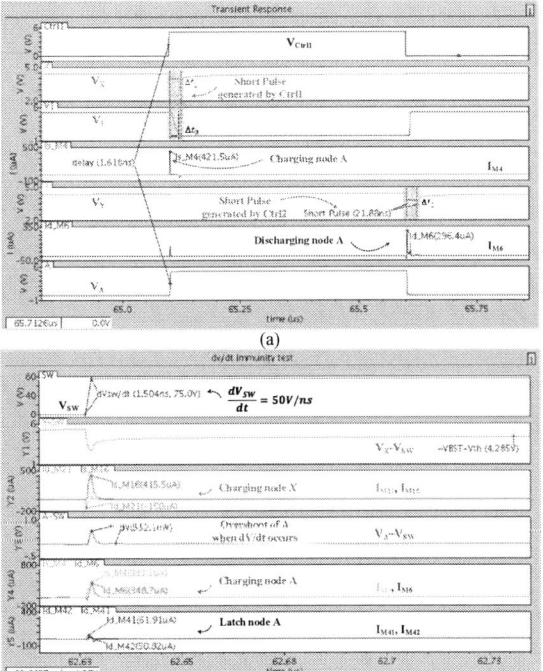

Fig. 5 Detailed working process of the proposed level shifter

Fig. 6 Simulated transient response of the proposed level shifter (a) working process (b) dV_{SW}/dt noise immune capability

Fig. 4 shows the circuit realizations of level shifter. OMC's response speed is controlled by I_B, where the conversion of control signals V_1 and V_2 ($\Delta t_2 = C_{par}(V_{BST} - V_{SW})/I_B$) should be finished before short pulse time Δt_1 ($\approx$20ns). Voltage clamp diodes (green parts) are adopted to clamp maximum operating voltage of low-voltage devices. Here, three parts contribute to dV/dt noise rejection, including dynamic charging, common-

mode noise cancellation and output latch by OMC. For example, If positive dV_{SW}/dt issue occurs at Ctrl1=0, dynamic charging paths (M_{15}-M_{16}) and clamp diode M_{21} are turned on to charge C_{PX}, increasing voltage slope dV_X/dt quickly and reducing false turn-on currents I_3. In the level-shifting core of FSC, the mis-triggered currents I_3 and I_4 during fast switching appear as common-mode noises and the net output noise current is reduced greatly to $I_3 - I_4$, which can also be rejected by the output current I_5 of OMC and is a design tradeoff between response speed and reliability.

The detailed working process is shown in Fig. 5. Signal conversion is finished in short-pulse time Δt_1, which is realized by charging capacitor C_1/C_2 in Fig. 4(a) and is a fast and low-power level-shifting strategy. The simulated transient waveforms are shown in Fig. 6 to prove fast response speed and high noise immunity. For voltage rising edge of node A in Fig. 6(a), the output charging current I_{M4} reaches 421.5μA, establishing a logic high level immediately. The input signal V_1 of the inverter is pulled down to SW before Δt_1 is up, which is controlled by $C_{par}V_{BOOT}/I_{M40}$. The total transmission time from the level shifter is only 1.618ns due to the use of high-bandwidth current mirrors. During dV/dt test in Fig. 6(b), V_{SW} pin is increased with 50V/ns after a delay (~5ns) of short pulse time, considering transmission time of high-side channel. Due to the three contributions for noise rejection as mentioned in Fig. 4, the ground bounce with respect to V_{SW} is only 532mV, which can hardly false trigger subsequent logic circuits, where the threshold voltage is about 2.5V. FOM (=delay time/node L·max V_{DD}) is equal to 0.04ns/μm·V and the current consumption is about 175μA with 1MHz logic input signal.

B. High-speed PCB Design

As can be seen, the printed circuit board (PCB) layout technique is significant to high-speed GaN-based converter performance, which is mainly due to parasitic inductances [12]. The high-frequency high-power-density PCB design concern for the proposed GaN gate driver is shown in Fig. 7. Power loop and gate loop (① ~ ③) parasitic inductances will introduce power loss and degrade efficiency. Bootstrap charging loop (④) inductance affects amplitude of V_{BOOT} at MHz switching frequency. All of these loops with high di/dt current characteristic should be minimized.

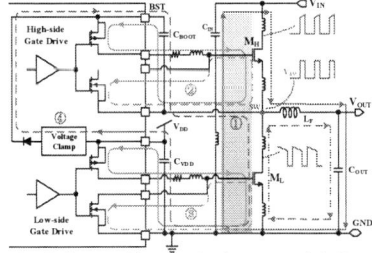

Fig. 7 PCB design consideration for the proposed GaN gate driver

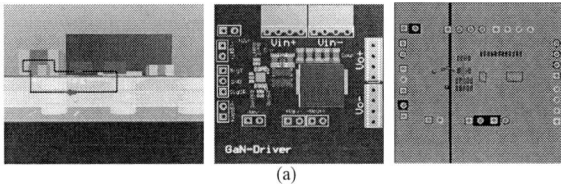

(a)

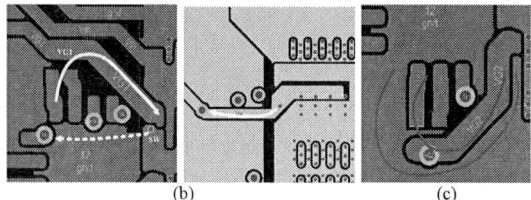

(b)   (c)

Fig. 8 Optimal PCB layout techniques for power/drive loops (a) Hybrid layout for power loop (b) Gate drive loop for top switch: Top layer (red) and Inner layer2 (blue) (c) Gate drive loop for bottom switch

For example, as shown in Fig. 8, in the PCB design for a multilayer structure, we try to minimize all these loops with the smallest area and utilize field self-cancelling effect in power loop to reduce parasitic inductance.

IV. EXPERIMENTAL RESULTS AND DISCUSSION

This circuit has been fabricated in 0.5μm 80V HV CMOS process where the active area is $1699 \times 1522 \mu m^2$. Chip micrograph and test bench of the gate drive is shown in Fig. 9. The tested E-mode GaN device is EPC2015. Symmetric layout design has been utilized for high-side and low-side channels to reduce propagation delay mismatching.

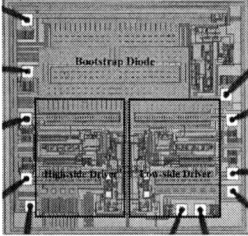

Fig. 9 Chip micrograph and test bench of the gate drive

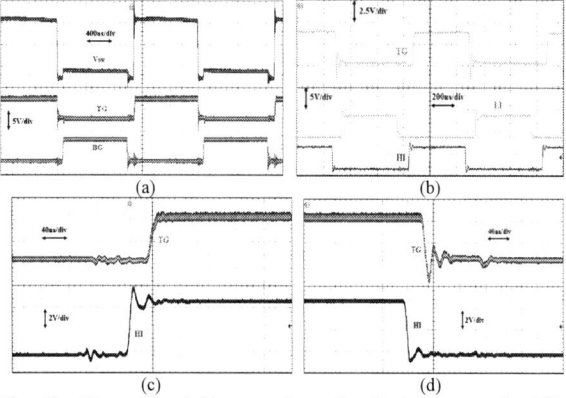

(a)   (b)

(c)   (d)

Fig. 10 Measured switching waveform of a Buck converter for POL application at V_{IN}=12V. (a) f_{sw}=500kHz (b) f_{sw}=1MHz (c) $\Delta t_{delay,rise}$=28ns (d) $\Delta t_{delay,fall}$=24ns

Fig. 10 shows the measured switching waveform of a Buck converter for POL application at V_{IN}=12V. The half-bridge gate drive can operate normally up to MHz switching frequency with the level shifter and small propagation delays for the high-side driver are observed (i.e. 28ns delay for the rising edge and 24ns delay for the falling edge).

Table I Performance Comparison With Previous Published Work

Process	[2]	[4]	[5]	This Work
	HV CMOS	CMOS	HV CMOS	HV CMOS
Node L(μm)	0.5	1.5	0.35	0.5
Supply voltage V_{DD}(V)	40	12	10	80
Delay time D(ns)	2	3	3	1.618
dV/dt noise immunity(V/ns)	40	N.A.	N.A.	50
Current consumption at 1MHz(μA)	12.3	N.A.	15.2	175
FOM=$D/L \cdot V_{DD}$ (ns/μm·V)	0.1	0.1	0.86	0.04

Table I shows performance comparison with some previously reported level shifters. Here, the figure of merit (FoM) [2] is just utilized to evaluate design superiority, considering input supply, propagation time and process technology. The proposed circuit achieves the lowest FOM and high noise immunity as well.

V. CONCLUSION

A high-speed low-power and high-noise-immunity level shifter is proposed in this paper, which detects logic high/low level instead of voltage undershoot. The response time is only 1.68ns and when dV_{SW}/dt noise is increased to as large as 50V/ns, there are no logic errors or latch-up issues in the gate driver. This circuit has also been applied in a Buck converter to evaluate switching performance at MHz frequency. The co-working mode with FSC and OMC is promising to meet the particular applications for GaN devices very well.

REFERENCES

[1] Z. Liu and H. Lee, "A 100 V gate driver with sub-nanosecond-delay capacitive-coupled level shifting and dynamic timing control for ZVS based synchronous power converters," in *Proc. IEEE Custom Integrated Circuits Conf.*, 2013, pp. 1–4.

[2] Z. Liu, L. Cong and H. Lee, "Design of on-chip gate drivers with power-efficient high-speed level shifting and dynamic timing control for high-voltage synchronous switching power converters," *IEEE J. Solid-State Circuits*, vol. 50, no. 6, pp. 1463–1477, Jun. 2015.

[3] F. Li *et al.*, "A low loss high-frequency half-bridge driver with integrated power devices using EZ-HV SOI technology," in *Proc. IEEE Applied Power Electronics Conf. Expo.*, 2002, pp. 1127–1132.

[4] S. C. Tan and X. W. Sun, "Low power CMOS level shifters by bootstrapping technique," *IEEE Electron. Lett.*, vol. 38, no. 16, pp. 876–878, Aug. 2002.

[5] Y. Moghe, T. Lehmann, and T. Piessens, "Nanosecond delay floating high voltage level shifters in a 0.35 m HV-CMOS technology," *IEEE J. Solid-State Circuits*, vol. 46, no. 2, pp. 485–497, Feb. 2011.

[6] J. T. Hwang, M. S. Jung, S. K. Jin, and H. K. Dong, "Noise immunity enhanced 625V high-side driver," in *Proc. 32nd Eur. Conf. IEEE Solid-State Circuits*, Montreux, Switzerland, Sep. 2006. pp. 572–575.

[7] Y. Zhang, *et al.*, "A capacitive-loaded level shift circuit for improving the noise immunity of high voltage gate drive IC," in *Proc. Int. Symp. IEEE Power Semicond. Devices*, Hong Kong, May 2015, pp. 173–176

[8] J. Zhu, *et al.*, "Noise immunity and its temperature characteristics study of the capacitive-loaded level shift circuit," *IEEE Trans. Ind. Electron.*, vol. 65, no. 4, pp. 3027–3034, Apr. 2018.

[9] E. A. Jones *et al.*, "Application-based review of GaN HFETs," in *Proc. IEEE Workshop on Wide Bandgap Power Devices and Applications*, 2014, pp. 24–29.

[10] M. Akahane *et al.*, "A new level up shifter for HVICs with high noise tolerance," in *Proc. Int. Conf. IEEE Power Electron.*, Hiroshima, Japan, May 2014, pp. 2302–2309.

[11] P. Tseng *et al.*, "Configuration and method for improving noise immunity of a floating gate driver circuit," U.S. Patent 8 723 552, May 2014

[12] David Reusch and Johan Strydom, "Understanding the effect of PCB layout on circuit performance in a high-frequency gallium-nitride-based point of load converter," *IEEE Trans. Power Electron.*, vol. 29, no. 4, pp. 2008–2015, Apr. 2014.

978-1-5386-2928-4/18 $31.00 © 2018 IEEE

Proceedings of the 30th International Symposium on Power Semiconductor Devices & ICs
May 13-17, 2018, Chicago, USA

AC/DC flyback controller with UHV integrated start-up current source in 180nm HVIC Technology

Hing Kit Kwan*, Bai Yen Nguyen*, Wen-Cheng Lin[+], Xiaoxin Liu[+], Swapnil Pandey[+], Saikat Chakraborty[+], Jongjib Kim[+], and Don Disney[+]

IP Engineering* and Technology Development[+] Departments

GLOBALFOUNDRIES

Singapore

hingkit.kwan@globalfoundries.com

Abstract—This work demonstrates an AC/DC flyback converter controller fabricated in an 180nm HVIC technology that enables reduced layout size and increased digital content. It features a monolithically integrated ultra-high-voltage (UHV) start-up circuit. An integrated UHV depletion-mode transistor is used to draw current directly from the ~400V DC bus and charge up VDD to ~13V. Controlling the gate of this depletion transistor allows the start-up current to be completely cut off after the VDD is charged and the controller starts normal operation, which cannot be done using the resistor start-up approach. The start-up circuit includes built-in protection features and enables tight tolerances for key design parameters.

Keywords—power integrated circuits; high-voltage integrated circuits (HVIC); ultra-high-voltage (UHV); AC/DC power converter; switching converters; UHV start-up; start-up current source.

I. INTRODUCTION

AC/DC switch-mode power supplies (SMPS) have their presence in most of the AC-powered electronics. Offline flyback converter dominates in the area of low power AC/DC power converters. The core of these converters is a small transformer switched by a 600V (or higher) power MOSFET controlled by an SMPS controller IC. The simplified schematic is shown in Fig. 1.

The controller IC obtains power from an auxiliary winding of the transformer after the converter output has settled into the steady state. However, it needs to obtain power from the high-voltage DC bus to power up the controller chip VDD before the converter output settles. The current path is highlighted in Fig. 1. In many cases, this power start-up is done by a lossy RC circuit. The disadvantage of such a simple solution is that this leaky resistor start-up path cannot be cut off after the converter powers up. With more stringent standby regulation of AC/DC converters, the use of HV resistors to provide start-up power becomes infeasible. The use of UHV depletion device as start-up device eliminates the power loss in the start-up resistor and provides fast-startup at low-line. Some products make use of discrete UHV depletion MOSFET or dual-die methodology [1], which co-package the controller IC and UHV depletion MOSFET in the same package.

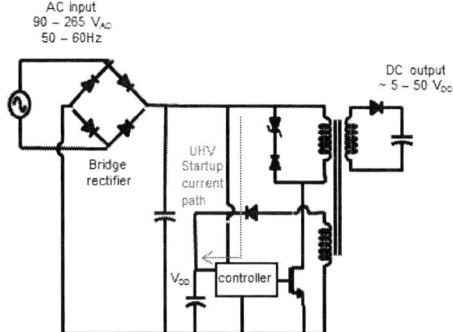

Fig. 1. The simplified circuit schematic of an AC/DC flyback converter and its start-up current path.

AC/DC flyback controller ICs have traditionally been fabricated using technologies with feature sizes of 350nm or larger. However, a new high-voltage process technology on the 180nm node was recently introduced and is optimized for increasing digital control content [2].

This paper demonstrates an AC/DC flyback controller fabricated in the GLOBALFOUNDRIES 180nm UHV process. Using 3.3V CMOS for the digital and analog circuitry enabled a significant reduction in the layout area, compared to conventional 5V CMOS designs. A monolithically integrated UHV start-up current source is incorporated to provide improved performance compared to using a HV start-up resistor and to eliminate the need for a separate UHV depletion MOSFET.

The organization of the paper is as follows. Section II introduces the architecture of the flyback controller implemented and the design of the UHV current source. Silicon characterization of the controller is in Section III. Conclusions are drawn in Section IV.

II. DESIGN OF THE AC/DC CONTROLLER

The block diagram of the AC/DC controller is shown in Fig. 2.

978-1-5386-2928-4/18 $31.00 © 2018 IEEE   359

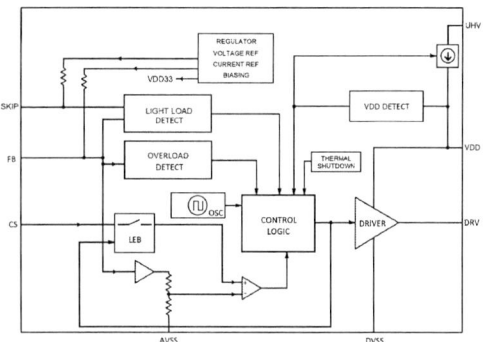

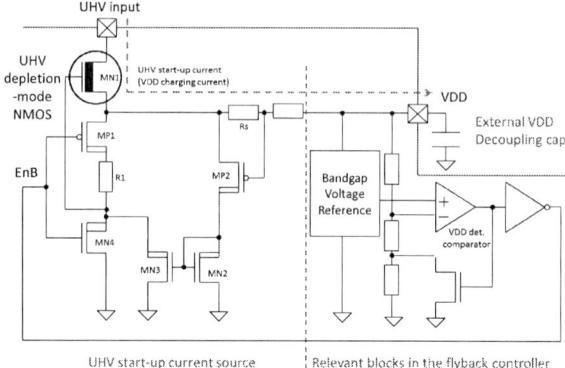

Fig. 2. Block diagram of the AC/DC flyback controller with UHV current source.

Fig. 3. Schematic of the UHV start-up current source and its connection with other relevant blocks in the flyback controller.

A. Design of the flyback controller

This design implemented a secondary-side-sensing peak-current-mode flyback control with cycle-by-cycle current limit. Internal leading-edge blanking (L.E.B.) has been implemented in current-sensing. The internal oscillator is running at 65kHz with frequency dithering for EMI reduction. It will run into skip-cycle operation at output light-load condition. Output feedback and overload detection are monitored at the feedback (FB) pin through an optical coupler. VDD is constantly monitored for VDD ready, over-voltage protection and under-voltage lock-out. Operation is auto-recovering from protection mode upon under-voltage lock-out. This provides a hiccup mode operation under fault condition.

30V HV CMOS devices were used to provide robust operation across a wide range of VDD voltage. The controller is equipped with HV bandgap voltage reference, HV internal 3.3V-output cap-less regulator and HV output driver to enable the operation with a single HV supply. Thermal shutdown with hysteresis has also been designed to stop the operation of the controller at temperature out of the specified operating range.

B. UHV start-up current source

An important feature of the present work is the UHV normally-ON (depletion-mode) transistor integrated to provide a built-in start-up function. Previous literature [3]-[3] on UHV start-up current source either suffer from the lack of current regulation, or the use of process and temperature dependent device parameters (e.g. VBE of a BJT and Vt of MOSFET) to control the shut-down and restart threshold voltages of the current source. Fig. 3 shows the schematic of the UHV current source implemented. Besides the UHV depletion-mode transistor, 30V asymmetric devices were also used.

The shut-down and restart of the current source is controlled by the bandgap reference, VDD sensing resistor and the detection comparator in the flyback controller, which provides robust, trimmable, process-invariant implementation without consuming extra die area.

With reference to Fig. 3, the operation is as follows: When the VDD is below the VDD-ON threshold of the controller (also known as the shutdown threshold of the UHV current source, 13.1V in this design), the VDD detection comparator generates a logic low for the EnB signal through an inverter. This turns on MP1 which connects the gate of the UHV depletion-mode NMOS NM1 to its source. As the Vt of the depletion NMOS is less than zero, its turns on even with zero Vgs. This delivers a start-up current through the path shown by the dotted line in Fig. 3, from the UHV input to the VDD of the controller, charging up the off-chip capacitor at VDD. This corresponds to the "Current Source ON" state marked in Fig. 4.

The start-up circuit incorporates a current regulation via current sensing and feedback to generate negative Vgs voltage on the depletion-mode transistor to limit the power dissipation and keep this transistor within its safe operating area (SOA). As current flows through the sensing resistor Rs and voltage drop across Rs is higher the threshold voltage of the MP2, current starts to flow into MP2 and MN2. MN2 and MN3 form a current mirror which copies this current signal and forms a voltage drop across R1. This voltage drop generates a negative Vgs on the UHV depletion NMOS MN1 and reduces its current. Hence, a negative feedback loop is formed to regulate the start-up current to be ~ $V_{gs,MP2}$/Rs. Rs is implemented using a doped poly resistor, and the current regulation point is easy to change using only a metal mask change.

This start-up current regulation is important under fault conditions, such as secondary-side output shorts, in which the auxiliary winding of the flyback converter fails to deliver power and the UHV start-up current source is being turned on with high duty cycle in hiccup mode as in Fig. 4. Regulation of this current also makes the start-up current value less dependent on process variation of the UHV depletion-mode device, making the design more robust.

978-1-5386-2928-4/18 $31.00 © 2018 IEEE

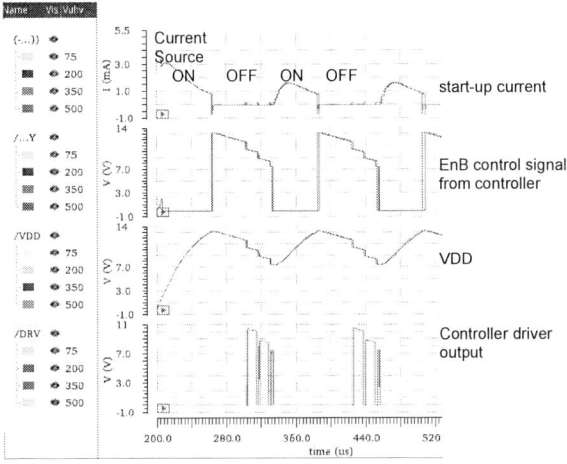

Fig. 4. Simulation waveform of the AC/DC controller with UHV current source in hiccup protection mode under different UHV DC bus voltage.

When the VDD capacitor is charged up to 13.1V and the flyback controller starts normal operation, the VDD detection comparator toggles and the circuit generates a logic high on the EnB signal. MN4 pulls the gate of the UHV depletion NMOS MN1 to ground. The body of the MN1 is connected to ground. As long as the VDD is above under-voltage lockout threshold of the flyback controller (also the restart threshold of the UHV current source, 7.5V in this design), this negative Vgs and Vbs cause MN1 to turn off, thereby reducing power loss from UHV DC bus during normal operation of the SMPS.

III. EXPERIMENTAL RESULTS

The AC/DC flyback controller IC with integrated UHV start-up has been fabricated using GLOBALFOUNDRIES 180nm HVIC process [2]. Die photos of the UHV start-up current source and flyback controller are shown in Fig. 5. The compact design rules of this advanced 180nm process allow the entire control circuit (without bond pads) to have an area less than 0.35mm². This demonstration makes use of a racetrack shape UHV depletion-mode NMOS, with the drain connection made by a bond wire and the source region completely surrounding the drain region, to avoid routing high-voltage metal over the drift region. Smaller circular UHV NMOS layouts are also available in the process.

TABLE I. shows the measured untrimmed oscillator frequency and the current-sense peak-current threshold (Vcs_max) of the AC/DC controller in a lot of 10 wafers. Among the 50 dies measured on each wafer, the untrimmed peak-current threshold can demonstrate a tolerance of about +/-1.1% in different process corners.

Fig. 6 shows the measured line regulation of the UHV start-up current source. It demonstrates a good line regulation up to 500V. For example, with VDD=0V, the start-up current varies only +/- 3.4% from 75V to 500V.

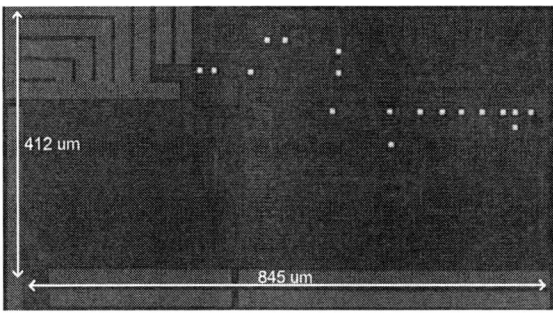

(a)

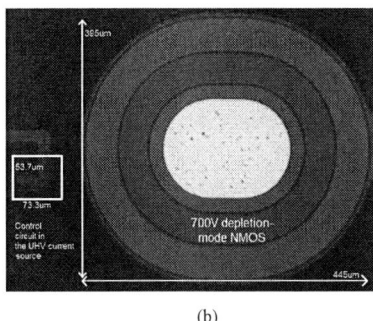

(b)

Fig. 5. Die photo of (a) the AC/DC flyback controller, (b) integrated UHV current source.

TABLE I.   MEASURED DISTRIBUTION OF TWO KEY PARAMETERS OF THE CONTROLLERS IN A LOT OF 10 WAFERS OF DIFFERENT SKEWED CORNERS (50 DIES PER WAFER).

Test items	Wafer No.	i	ii	iii	Iv	v	vi	vii	viii	ix	x
	Corner	TT	TT	TT	TT	FF	SS	FS	SF	TT +resS	TT +resF
Oscillator frequency - untrimmed (kHz)	Min	68.0	66.6	66.9	68.6	68.8	69.2	69.1	69.5	66.0	67.0
	Med	69.6	67.8	68.6	69.8	70.2	70.8	70.3	71.1	67.2	67.9
	Max	72.2	69.0	69.8	72.1	72.7	72.9	72.4	73.3	68.6	69.3
Current-sense peak-current threshold, Vcs_max – untrimmed (V)	Min	0.89	0.89	0.89	0.89	0.89	0.89	0.89	0.89	0.89	0.9
	Med	0.9	0.9	0.9	0.9	0.9	0.9	0.9	0.9	0.9	0.9
	Max	0.91	0.91	0.91	0.91	0.91	0.91	0.91	0.91	0.9	0.91

The application waveform of the AC/DC flyback controller under 500V DC bus voltage is shown in Fig. 7. A 0.14uF capacitor is connected to VDD. Upon the application of the 500V power source, the start-up current source turns on and charges the VDD pin up. After VDD reach 13.1V, it turns off the current source and the controller starts normal operation delivering driving pulse at ~65kHz. The auxiliary winding is intentionally not connected to VDD to emulate a fault condition. The current consumption of the controller discharges the VDD capacitor until its voltage falls to 7.5V which stops the flyback operation and driving pulse. The UHV start-up current source is then turned on again to charge the VDD capacitor again forming a hiccup operation.

978-1-5386-2928-4/18 $31.00 © 2018 IEEE

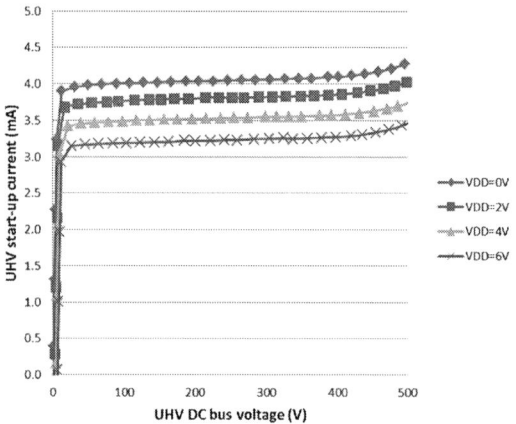

Fig. 6. Meased line regulation of the UHV current source at different output VDD voltage.

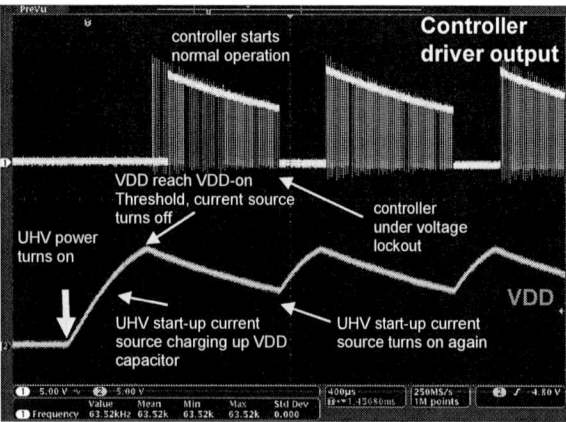

Fig. 7. Waveform of the AC/DC controller in hiccup operation mode at 500V DC bus voltage.

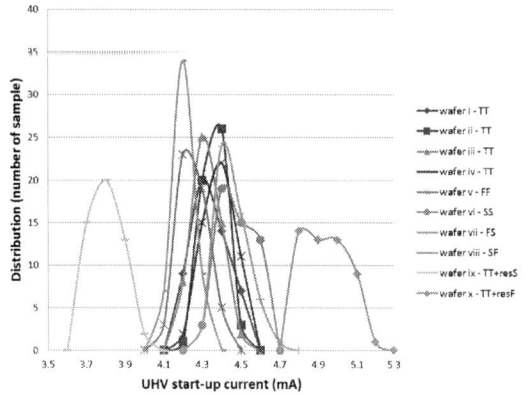

Fig. 8. Measured start-up current (UHV DC bus = 500V, VDD=0V) in a lot of 10 wafers of different process corners at 25°C.

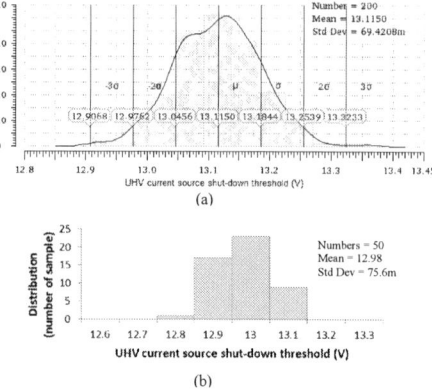

Fig. 9. UHV current source shut-down threshold at 25°C: (a) local Monte Carlo simulation result with typical process parameters and (b) measurement result on a typical wafer.

Fig. 8 shows a distribution of the current sourced at 500V DC bus voltage in wafers of different process corners, demonstrating tight tolerance. The wafers of skewed resistor corners demonstrate the use of resistor value to control the start-up current value. The mean off-state leakage of the current source is measured to be far smaller than 10nA at 25°C (less than equipment measurement resolution). The simulated and measured distributions of shut-down threshold of the current source on a typical process wafer are shown in Fig. 9 demonstrating a very tight distribution and a good match between measurement and simulation.

IV. CONCLUSION

An AC/DC flyback SMPS controller IC, including a monolithically integrated ultra-high-voltage (UHV) start-up circuit, has been designed and fabricated in an 180nm HVIC technology. Test results confirm that the controller IC is fully functional and demonstrates excellent start-up current regulation, robust protection features, and very tight tolerances.

REFERENCES

[1] W. Langeslag, R. Pagano, K. Schetters, A. Strijker and A. van Zoest, "VLSI Design and Application of a High-Voltage-Compatible SoC–ASIC in Bipolar CMOS/DMOS Technology for AC–DC Rectifiers," in IEEE Transactions on Industrial Electronics, vol. 54, no. 5, pp. 2626-2641, Oct. 2007.

[2] D. Disney, W. C. Lin, X. Liu, S. Pandey and J. Kim, "180nm HVIC technology for digital AC/DC power conversion," 2017 29th International Symposium on Power Semiconductor Devices and IC's (ISPSD), Sapporo, 2017, pp. 287-290.

[3] C. L. Chen, J. C. Tsai, Y. T. Chen, C. L. Ni, C. Y. Chen and K. H. Chen, "800V ultra-high-voltage start-up mechanism for pre-regulator in power factor correction (PFC) controller," 2011 IEEE 54th International Midwest Symposium on Circuits and Systems (MWSCAS), Seoul, 2011, pp. 1-4.

[4] Y. C. Kang, C. C. Chiu, M. Lin, C. P. Yeh, J. M. Lin and K. H. Chen, "Quasiresonant Control With a Dynamic Frequency Selector and Constant Current Startup Technique for 92% Peak Efficiency and 85% Light-Load Efficiency Flyback Converter," in IEEE Transactions on Power Electronics, vol. 29, no. 9, pp. 4959-4969, Sept. 2014.

978-1-5386-2928-4/18 $31.00 © 2018 IEEE

Proceedings of the 30th International Symposium on Power Semiconductor Devices & ICs
May 13-17, 2018, Chicago, USA

Evaluation of Gate Oxide Reliability in 3.3 kV 4H-SiC DMOSFET with J-Ramp TDDB Methods

Masakazu Sagawa, Hiroshi Miki, Yuki Mori, Haruka Shimizu, and Akio Shima
Center for Technology Innovation-Electronics, Research and Development Group
Hitachi, Ltd.
1-280 Higashi-Koigakubo, Kokubunji, Tokyo, 185-8601 Japan
masakazu.sagawa.xa@hitachi.com

Abstract—In order to verify a gate oxide reliability of 3.3 kV 4H-SiC DMOS for rail car application, we developed a J-Ramp TDDB and a constant current stress screening method. We examined a conventional gate stack structure device with a single layer gate electrode and an improved one with double layered gate electrode; the latter one reveals a low hazard rate less than 1 FIT under a gate operation voltage of ±15 V at 150 °C.

Keywords—4H-SiC; DMOS; Gate oxide; J-Ramp; TDDB

I. INTRODUCTION

Invertor modules are commonly installed in the power units of electric train cars because of their low energy consumption, and the recognition they receive is now spreading worldwide. Main invertor modules are overwhelmingly occupied by Si-based IGBTs and diodes, but new emerging devices based on a wide band gap semiconductor SiC are being introduced into commercial markets [1]. Hybrid and full SiC invertor modules are already being used by many railway companies in Japan in particular. SiC-based devices (Schottky barrier diode and MOSFET) are considered effective for reducing an energy conversion loss at high switching frequencies and for reducing the mass and volume of power units. However, there is a lot of work to be done in order to make the modules cost effective and reliable in the long-term.

II. GATE OXIDE RELIABILTY

We deal with the later reliability issues, and focus on the gate oxide reliability of SiC power MOS devices. For the train car applications, the MOS switching device must tolerate a minus gate bias voltage—e.g., −15 V to avoid a false turn-on caused by a switching noise under a high applied voltage 3.3 kV and a high current over 1000 A. But SiC power MOS-FETs are vulnerable to negative bias temperature instability (NBTI) and Vth hysteresis. Many researchers have reported that the poor reliability is due to many traps in the interface of SiO₂/SiC and remaining bulk defects in the SiC crystals. To satisfy the above criteria, we have to find out a proper way to estimate a lifetime distribution and determine a screening method and the conditions. However, for power devices with a large chip size and an expensive SiC substrate, time-dependent dielectric breakdown (TDDB) tests—very common examinations—are terribly time and cost consuming. So we have developed a new prediction method of gate oxide reliability for the SiC power MOS and have successfully obtained a preferably low hazard rate.

III. EXPERIMENTS

A. Samples

The sample die we used here were 6.8 × 6.8 mm² 4H-SiC DMOSFET for 3.3 kV / 375 A invertor modules [2]. The gate insulator was made of oxi-nitrides SiO₂ fabricated by CVD and NO annealing 50 nm thick. Two types of devices were prepared: type-A was a conventional DMOS with a single layer poly-Si gate electrode (Fig. 1) and type-B had a double-layered gate electrode.

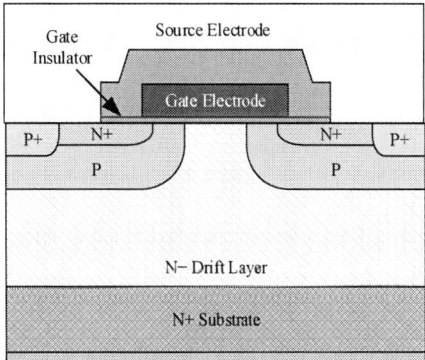

Fig. 1 Schematic cross section view of 4-H SiC DMOSFET

The differences in V_{GS}-I_G characteristics for negative gate bias between both versions clearly appear in Fig. 2. In the type-A version, the F-N tunneling leakage current started to appear at the gate voltage over −15 V. On the other hand, the type-B version showed that the leakage current flowed at the relatively higher voltage over −25V. The resulting difference between two types was almost 10 V. The details of the

978-1-5386-2928-4/18 $31.00 © 2018 IEEE

physical mechanism for this phenomenon will be discussed at another time.

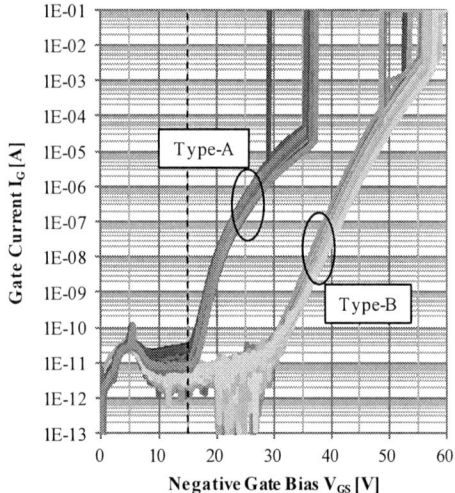

Fig. 2 V_{GS}-I_G characteristics of type-A and type-B devices in negative gate bias at 150 °C

B. J-ramp TDDB and contant current stress screening

As described above, the reliability of gate oxide in SiC MOS is mainly dominated by extrinsic defects, which results in A-mode and the B-mode distributions in TDDB Weibull plots. Conventional constant voltage stress (CVS) TDDB tests are not suitable because their relatively high applied voltage makes it impossible to distinguish both modes correctly. We applied a J-Ramp TDDB method [3] to get precise information on both modes. We also combined the J-Ramp TDDB method with a constant current stress (CCS) screening method to eliminate a lower portion of a Q_{BD} distribution of the J-Ramp TDDB test. The two measurements were done with the apparatus of source measurement units, Keithley 2636A, and a semi-auto prober, Vector Semiconductor Summit 20000.

IV. RESULTS

A. Results of type-A device

We started a J-Ramp TDDB test with type-A dies. In the J-Ramp measurements, the stress current I was set as a function of the stress time t, $I = I_0 * r^t$ up to an upper limit J_{MAX}. In our experiment, we set $I_0 = 10$ nA, r = 1.002868, and $I_{MAX} = 200$ nA, respectively. Figure 3 is a typical waveform of J-Ramp TDDB of a die in $V_{DS} = 0$ V at 150 °C. The Q_{BD} Weibull plots presented in Fig. 4 indicate a Weibull scale parameter η is 3×10^{-3} C. As previously stated, the V_{GS}-I_G characteristics of the type-A dies in Fig. 2 show a high gate leakage current almost 1×10^{-11} A at the gate bias of $V_{GS} = -15$ V. This means the total gate charge during 30 years of continuous operation reaches the scale parameter η in which almost 63% of all devices have worn out. When taking reliability into account, we ultimately doubt the practicality of using the type-A device for our application.

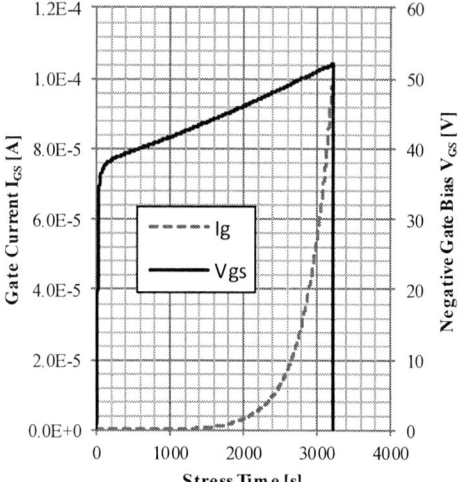

Fig. 3 Typical waveform of J-Ramp TDDB test

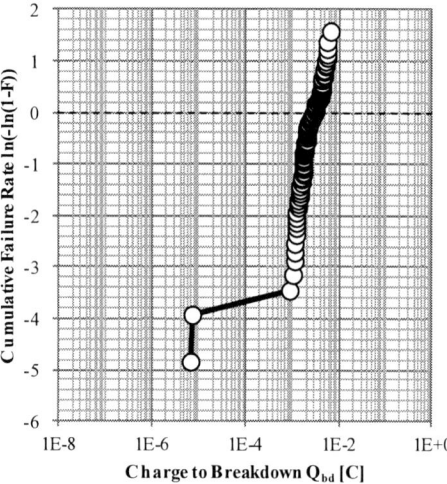

Fig. 4 Q_{BD} distribution of type-A device in Weibull chart

B. Results of type-B device

The charge to breakdown Q_{BD} vs the cumulative failure rate F in the Weibull chart is presented in Fig. 5. In this chart, the dashed blue line stands for raw data and the bold black line stands the data after screening. The vertical dashed black line near $Q_{BD} = 1 \times 10^{-3}$ C is the boundary between C-mode and B-mode. In the B-mode region, three low Q_{BD} chips—B1, B2, and B3—were observed. These dropouts show relatively lower breakdown voltages compared to the C-mode chips, as we can see in the locus curves during the J-Ramp TDDB test in Fig. 6. In order to guarantee a low hazard rate, we have to eliminate them with an appropriate screening method.

978-1-5386-2928-4/18 $31.00 © 2018 IEEE   364

Figure 7 denotes the waveforms of the constant current screening. We applied a constant current of 100 nA at 150 °C for 7 sec. At the first stage, the gate voltage V_{GS} went up in accordance with a displacement current charging an input gate capacitance of C_{ISS}. V_{GS} then reached a constant voltage level of near −40 V for a period of almost 2 sec in which the gate current was actually injected into the gate oxide.

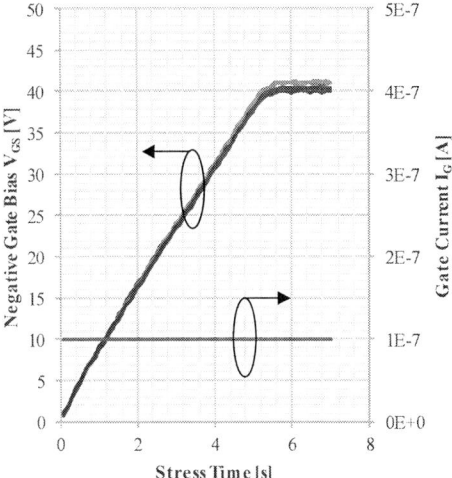

Fig. 7 Typical waveforms of constant current stress screening at 150 °C

This screening is capable of removing high leakage current chips that show lower constant gate voltage levels less than −40 V and short life-time chips under the gate stress of −40 V. Therefore, as shown in Fig. 6, since low-Q_{BD} chips B1, B2, and B3 showed low breakdown voltage less than −40 V, we assumed that the above three dropouts would have been eliminated by adopting the CCS screening for them. Finally, the black bold line showed the B-mode data after screening. To calculate a hazard rate, we used a Weibull distribution. In (1), F is a cumulative failure rate, η is a Weibull structure parameter, and m is a scale parameter.

$$F = 1 - exp[-(Q/\eta)^m] \qquad (1)$$

In (1), a curve fitting was carried out to the B-mode data after screening. We obtained Weibull parameters of $m = 9 \times 10^{-1}$ and $\eta = 8 \times 10^{-3}$ C. The hazard rate of the DMOS could be calculated by equations (2) and (3) [4]. First, we had to transform a charge to breakdown Q_{BD} into a time to breakdown t_{BD} with (2).

$$Q_{BD} - I_{GL} \times t_{BD} \qquad (2)$$

$$\lambda = I_{GL} \times (dF/dQ) / (1-F)$$
$$= I_{GL} \times (m/\eta) \times (Q/\eta)^{m-1} \qquad (3)$$

In (2), I_{GL} means a gate leakage current under the rated condition—e.g., $V_{GS} = -15$ V at 150 °C. However, I_{GL} could not be monitored directly because of the high level of background leakage current, order of 1×10^{-11} A, due to the large chip size. We had to rely on an extrapolation of the V_{GS}-I_G curve alone. Figure 8 is the enlarged V_{GS}-I_G characteristic in the negative gate bias at 150 °C. Generally speaking, t_{BD} is proportional to exp[−E] under a low electric field strength E in the SiO₂ gate insulator. It is referred to as "E-model" in a

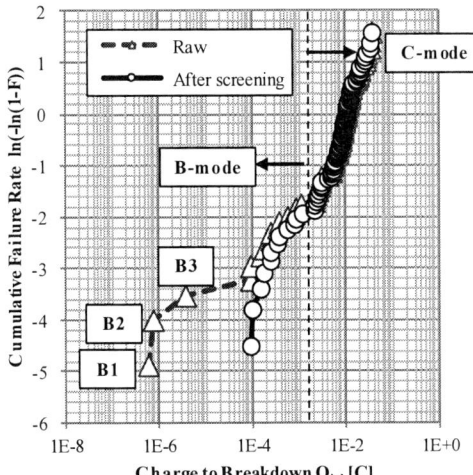

Fig. 5 Q_{BD} distribution of type-B device in Weibull chart

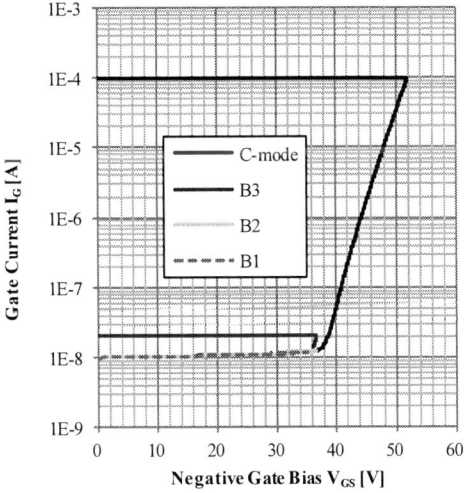

Fig. 6 Locus of J-Ramp TDDB test at 150 °C

978-1-5386-2928-4/18 $31.00 © 2018 IEEE

TDDB theory [5]. Taking this model into account, we could transform (2) into (4) as shown in below.

$$log[I_{GL}] = log[Q_{BD}/t_{BD}] \sim -E \qquad (4)$$

In the last transformation, we assumed Q_{BD} was independent of CCS stress conditions (i.e., E). Consequently, we employed a semi-log extrapolation and obtained I_{GL} values as 1×10^{-18} A for $V_{GS} = -8$ V and 1×10^{-16} A for -15 V.

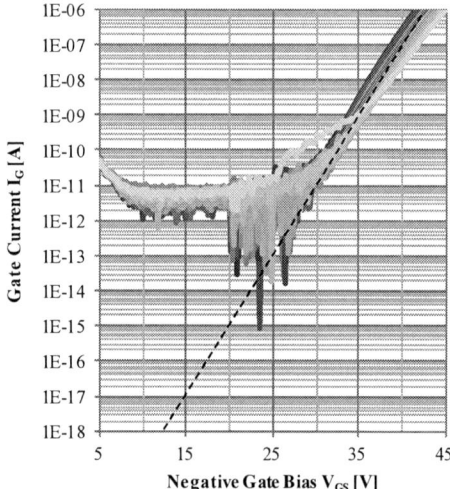

Fig. 8 V_{GS}-I_G characteristic of type-B devices in negative gate bias

at 150 °C

We could calculate hazard rates by (3) as the function of the injected charge during the CCS screening in Fig. 9. We assumed the injected charge was 200 nC as shown in Fig. 7.

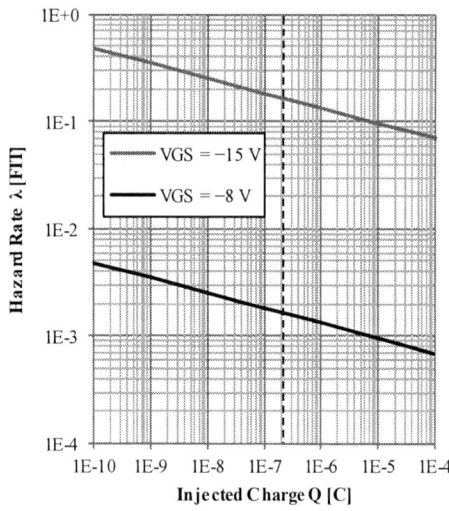

Fig. 9 Injected charge dependency of hazard rate in negative gate bias at 150 °C

The estimated hazard rate was 2×10^{-3} FIT in $V_{GS} = -8$ V at 150 °C, which is equivalent to the cumulative failure rate of 0.3 ppm after 20 years of continuous operation. As you can see in (3), the hazard rate is proportional to the gate leakage current I_{GL}. In cases of $V_{GS} = -15$ V, the hazard rate increases to 0.2 FIT by two orders of the previous one. In the same way, we evaluated the hazard rate of the positive gate bias. We summarized the results in Table 1. These values are almost as same as the commercially available SiC trench gate MOSFET [6] for electric vehicle and hybrid electric vehicle applications.

TABLE 1. Estimated hazard rate

	Hazard Rate λ
$V_{GS} = -8$ V	2E–3 FIT
$V_{GS} = -15$ V	2E–1 FIT
$V_{GS} = +15$ V	3E–1 FIT

V. CONCLUSIONS

We have developed J-Ramp TDDB and CCS screening methods suitable for SiC power MOS-FETs that are still suffering from the negative influences of several interface states and bulk crystal defects. With the goal of adopting these methods, we confirmed that our SiC DMOS has a low hazard rate even under the gate bias of −15 V at the junction temperature of 150 °C, which is essential for electric train car applications.

REFERENCES

[1] T. Ishigaki, et al., "3.3kV/450A A Full-SiC nHPD² (next High Density Dual) with smooth swtching," Proceedings of PCIM Europe 2017, Nuremberg, Germany, p. 33.

[2] A. Shima, et al., "3.3kV 4H-SiC DMOSFET with Higly Reliable Gate Insulator and Body Diode," Abstract of 11th ECSCRM 2016, Halkidiki, Greece, p. 137.

[3] JEDEC STANDERD, JESD35-A, https://www.jedec.org/standards-documents.

[4] Slimane Oussalah and Boualem Djezzar, "Field Acceleration Model for TDDB; Still a Valid Tool to Study the Reliability of Thick SiO²-Based Dielectric Layers?" *IEEE Transcntions on Electron Devices*, Vol. 54, No. 7, 2007, p. 1713.

[5] J. W. MacPherson, "Reliability Physics and Engineering Time-to-Failure Modeling," Springer, p. 158.

[6] D. Peters, et al., "The new CoolSiC™ Trench MOSFET Thecnology for Low Gate Oxide Stress and High Performance," Proceedings of PCIM Europe 2017, Nuremberg, Germany, p. 168.

Proceedings of the 30th International Symposium on Power Semiconductor Devices & ICs
May 13-17, 2018, Chicago, USA

Repetitive surge current test of SiC MPS diode with load in bipolar regime

Shanmuganathan Palanisamy[a,*], Jens Kowalsky[a], Josef Lutz[a], Thomas Basler[b], Roland Rupp[b], Jasmin Moazzami-Fallah[b]

[a] Technische Universität Chemnitz, Chair of Power Electronics and EMC, Chemnitz, Germany,
[b] Infineon Technologies AG, Neubiberg, Germany
Email-Address: shanmuganathan.palanisamy@etit.tu-chemnitz.de

Abstract—**The reliability of power diodes under surge current is an important factor that has to be taken into account for power electronic applications. In this work, latest generation SiC MPS (Merged Pin Schottky) diodes (650V, 1200V, 1700V) with different current classes from Infineon are exposed to repetitive high surge current stress. Furthermore, the temperature was estimated using different methods including direct measurement, Sentaurus TCAD simulation, Cauer network and an analytical model using temperature dependent mobility. It was found that the diodes could withstand a high number of high-current pulses. However, before reaching final destruction, different ageing phenomenon were observed at the unipolar and bipolar regime of the MPS diode. A detailed investigation on the aging mechanisms including failure analysis was performed.**

Keywords— *4H-SiC; Merged Pin Schottky diode; Repetitive surge current behaviour; Surge current; Degradation mechanisms; Bipolar and Unipolar degradation; Schottky barrier*

I. INTRODUCTION

Silicon carbide is one of the promising semiconductor materials with superior physical properties such as wide band gap, high critical field strength and high thermal conductivity. Such favorable physical properties make SiC a reliable material for power devices. SiC MPS diodes are commercially available with various voltage classes since 2005 [1, 2]. The optimized hexagonal p$^+$ cell structure of the used diode helps to withstand higher surge current stress. Despite the fact that the device concept, basic working principle and the strong surge current ruggedness of SiC MPS diode are well known so far [3, 4], the repetitive surge current stress and the related degradation mechanism of SiC MPS diodes are still unevidenced. In this work, SiC MPS diodes (650V, 1200V and 1700V) with different current classes are exposed to high repetitive surge current stress reaching the bipolar (resistivity modulation) mode. The corresponding degradation mechanisms were investigated.

II. EXPERIMENTAL DETAILS

An automated test bench was built to investigate ageing mechanisms during repetitive surge current event as shown in Fig. 1. A half sinusoidal signal is applied by the control system to an IGBT, working in a linear amplifier mode which results in half sine surge current pulse. To determine the junction temperature (T_{vj}) before and after the pulse as shown in Fig. 2, a measurement current (10mA) flows continuously through the test device. In this case, each device is calibrated in advance with respect to the temperature dependent forward voltage drop

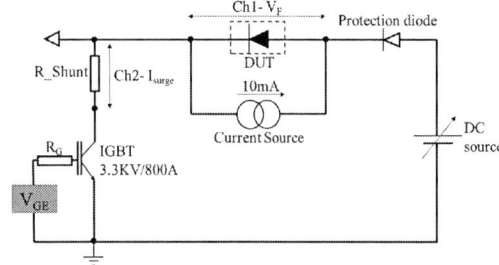

Fig. 1. A simplified circuit for repetitive surge current test

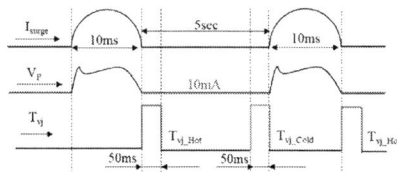

Fig. 2. Pulse pattern for surge current and the temperature measurement

at the given measurement current (V_j (T) method, similarly used in power cycling). Several *End-of-lifetime* (EOL) criteria were considered for the test device such as 20% increase in the forward voltage drop (V_F), 5% increase in the maximum deposited energy (E_{max}), loss in the blocking or 100% increase in the leakage current.

III. MEASUREMENT RESULTS AND DISCUSSION

A. Surge current limits

The single-event surge current limit for 650V, 1200V and 1700V diode samples with various nominal current was tested first at 10ms surge current pulse. The voltage waveform *v(t)* and the corresponding I-V curve of a 20A/1200V rated diode at different single-event surge current pulses is shown in Fig. 3. The diode remains unipolar up to 130A, after which minority charge carriers start to trigger from merged p$^+$ cells [5] resulting in a lower voltage drop due to resistivity modulation. The diode reaches bipolar regime at ~160A. The diode survived up to 285A (~14x the rated current) which is significantly above the limit given in datasheet at T_c=100°C. For the sudden destruction, the failure can be linked to high dissipated energy corresponding to a molten front side metallization [6]. Similar behavior was seen for all diode samples with different voltage and current classes.

978-1-5386-2928-4/18 $31.00 © 2018 IEEE   367

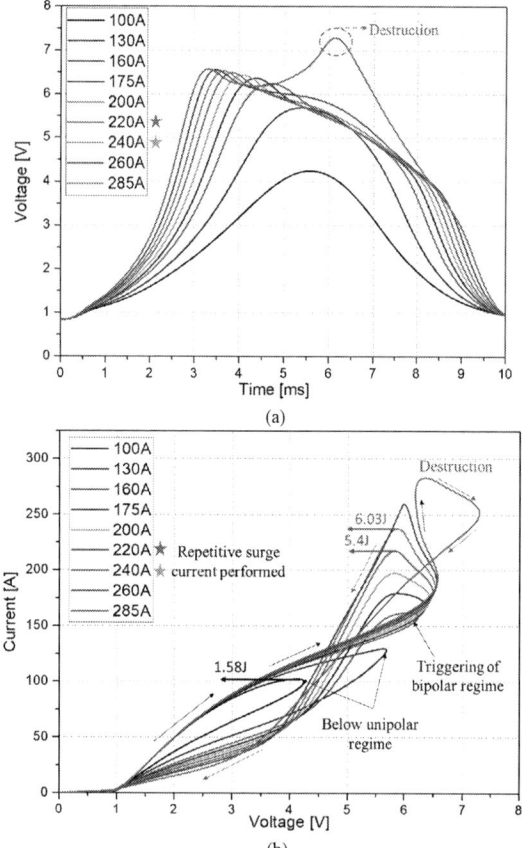

Fig. 3. (a) Single-event surge current limit: Voltage waveform *v(t)* of 10ms half sine current pulses of latest generation 1200V 20A rated SiC MPS diode, (b) Corresponding I-V curve

The single-event surge current limit at 10ms pulse and the corresponding deposited maximum energy for different diode samples are given below in TABLE I.

TABLE I: 10ms single-event surge current limit for 650V, 1200V and 1700V diodes at different nominal current

Diode rating	Single-event surge current limit at 10ms pulse at *T=300K* [A]	E_{max} [J]
20A/650V	225	8
20A/1200V	285	8
10A/1700V	155	4.2

B. *Repetitive surge current test*

The following results belong to 20A/1200V rated diodes (from single diode of IDW40G120C5B). In order to avoid early destruction without clear sings of aging, repetitive surge current measurement was carried out on several test devices at the peak values of 220A and 240A. At this current, the diodes are clearly in the bipolar regime, as well as ~20% lower than the destructive surge current limit. Fig. 4 shows one of the test device's I-V curves of repetitive surge current measurement at

240A. This test device results in a strong forward voltage drift at the (ascending branch) unipolar regime up to ~500+ pulses. However, there is no degradation visible at the bipolar branch and unipolar descending branch. A possible explanation is that, higher acceptor ionization rate at higher temperature leads to improved emitter efficiency at the bipolar branch and compensates the higher losses in the unipolar branch. The strong degradation in the unipolar regime could be caused by several reasons. Due to higher temperature swing in the surge current event, the bond wire foot is degraded and/or a strong change in the front side metallization can be observed as shown in Fig. 5.

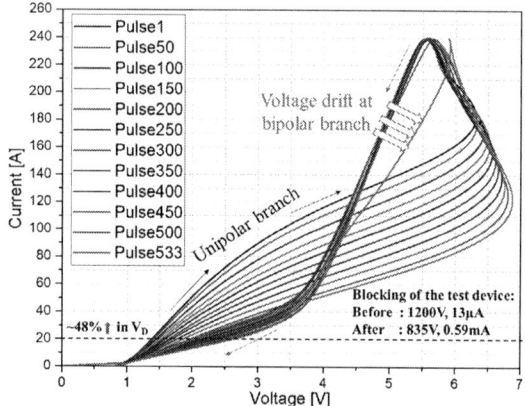

Fig. 4. I-V curve of repetitive surge current measurement; ϑ_{case}= 25°C, I_{max} = 240A, P_{length} = 10ms, P_{break} = 5sec. Lower blocking voltage with increase in leakage current after the repetitive surge current measurement

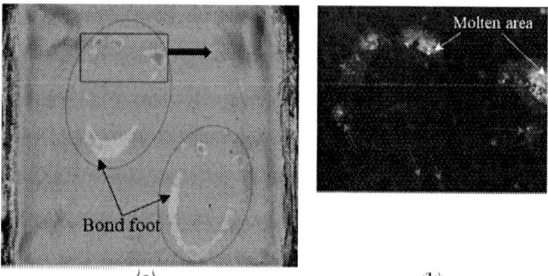

Fig. 5. Optical microscope image of test device of Fig. 4. Decapsulation of the backside metallization of the chip (view from backside). (a) Irregularities around the bond wedges marked in red. (b) Zoom area from (a): Molten areas are visible and the red arrows shows some irregularities probably due to missing front side metallization

On the other hand, diodes were tested with 220A, out of which one of the test device I-V curves are shown in Fig. 6. This diode survived up to 2569 pulses, which should be sufficient for the application requirement. In this case, a clear degradation at the unipolar and bipolar branch was visible. The voltage drop at the bipolar branch starts to increase from 1310th pulse onwards. Fig. 7 shows the change in V_F and maximum deposited energy over each pulse. At the end, this test device shows 23% increase in the V_F at the nominal current. Besides, a Schottky junction lowering was observed which impacts the temperature measurement T_{vj} as mentioned before with 10mA constant current. An increase in temperature (Fig. 8a) T_{vj_Hot} (measurement after 250µs measurement delay) and T_{vj_Cold} (measured shortly before next surge current pulse) after 1310th pulse was detected. The junction voltage was lowered by ~9mV after the repetitive surge current measurement as shown in Fig.8b. However, the barrier lowering was seen only for the devices stressed more than 2000 pulses. At this point, the used temperature measurement leads to error. Therefore, different methods were used to estimate the temperature during the surge current event.

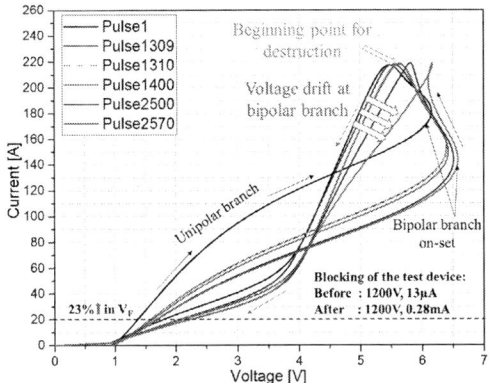

Fig. 6. I-V curve of repetitive surge current measurement; ϑ_{case}= 25°C, I_{max} = 220A, P_{length} = 10ms, P_{break} = 10sec. Leakage current increased after the repetitive surge current measurement

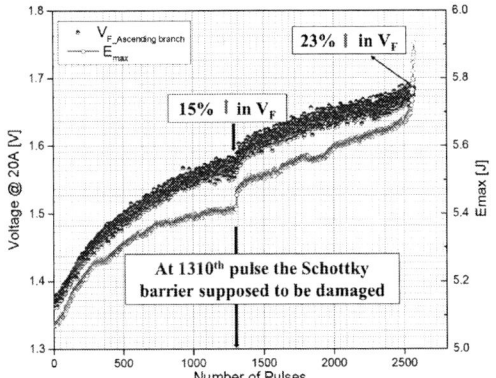

Fig. 7. Corresponding voltage drift at 20A and maximum deposited energy of each surge current pulse of Fig. 6

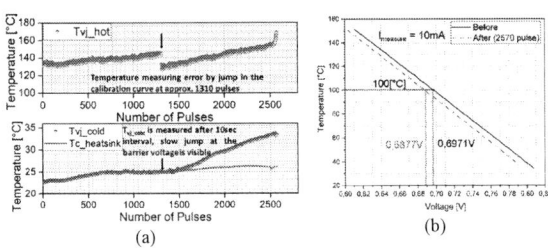

Fig. 8. (a) Junction (T_{vj}) and case temperature (T_C) after each surge current pulse (T_{vj_Hot} & T_{C_Hot}) and before next surge current pulse (T_{vj_Cold} & T_{C_Cold}) of Fig. 6. (b) Corresponding temperature calibration (T_{vj}) of Fig. 6, before and after 2570 pulses

All the test devices show similar failure at ~>5% increase in the maximum deposited energy (E_{max}) as given below in the TABLE II.

TABLE II: Average maximum deposited energy out of 20 test devices rated 20A/1200V (IDW40G120C5B) at each surge current peak

Surge current event	Average E_{max} at 1st pulse	Average E_{max} at the last pulse before destruction
240A	5.7J	6.1J
220A	5.2J	5.6J

C. Temperature estimation

The temperature was estimated by different methods like, simplified 2D Cauer model simulation, a direct temperature measurement during the surge current event, Sentaurus simulation using diode model from the previous work [5] and a model using temperature dependent mobility. Temperature comparison of all the models is shown in Fig. 10.

The analytical model is used for estimating the temperature dependent mobility of 4H-SiC [7]. The mobility of each surge current pulse is calculated from the on-state resistance at the rated current with equation (1). Then the temperature is calculated out of the mobility equation (3). The 2D Cauer model simulation is performed with the chip dimension from the previous work [5].

The temperature dependent mobility as well as the Cauer model show reasonable agreement with the measurement, while a Sentaurus simulation resulting in a higher temperature which exceeds the melting temperature of Al. This is also seen from the measured device after several thousand pulses in the failure analysis. However, temperature extracted from Sentaurus simulation is maximum temperature in the device structure during the single-surge current event.

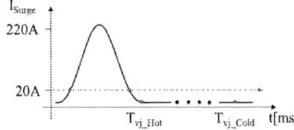

Fig. 9. Schematic of the surge current event where the temperature estimated

→ Temperature estimation from on-state resistance (R_Ω: V_F=V_S+R_Ω*I_F)

$$R_\Omega = \frac{W_B}{q \cdot \mu_e \cdot N_D \cdot A} \quad \Longrightarrow \quad \mu_e = \frac{W_B}{q \cdot R_\Omega \cdot N_D \cdot A} \quad (1)$$

→ Active area loss calculated from on-state resistance

$$A = \frac{W_B}{q \cdot R_\Omega \cdot N_D \cdot \mu_e} \quad (2)$$

→ Temperature dependent mobility of 4H-SiC at 300K

$$\mu_e = \mu_\infty + \frac{\mu_0 \cdot (T/300)^\alpha - \mu_\infty}{1 + \left(N/N_{ref}\right)^\gamma} \quad (3)$$

978-1-5386-2928-4/18 $31.00 © 2018 IEEE   369

$$T = 300 \cdot \left[\frac{(\mu_e - \mu_\infty) \cdot (1 + (N/N_{ref})^\gamma)}{\mu_0} \right]^{\frac{1}{\alpha}} \quad (4)$$

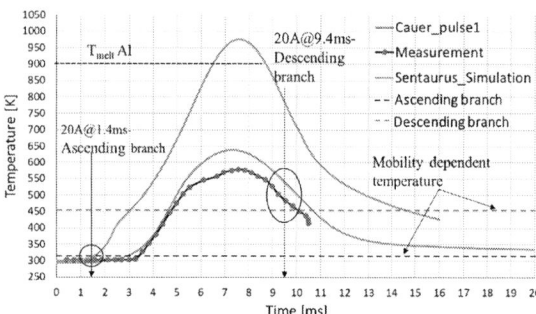

Fig. 10. Temperature comparison of the 2D Cauer model simulation, temperature measurement by $V_j(T)$ method during the surge current event, analytical model by on-state resistance method (mobility dependent temperature - this result corresponds to test device shown in Fig. 6) and maximum temperature is extracted from Sentaurus TCAD simulation

D. *Aging mechanisms*

During repetitive very high current and high temperature stress, different aging mechanisms were identified. A crack was found in the test device from the backside exactly in the middle of the chip (Fig. 11a). Due to very high thermal cycling, there is a thermo-mechanical mismatch between the solder and the chip which results in crack formation at the back side of the chip. But this crack did not propagate into the epi-layer. However this test device was able to block 1200V with higher leakage current (Fig. 6). Hotspot was found from the back side EMMI (Emission microscopy) during blocking (Fig. 11b).

At 1ms high surge current stress (340A) the test device survived ~10000 pulse with strong top side metallization (Aluminum) degradation is visible and the bond wire is partially detached (Fig. 12).

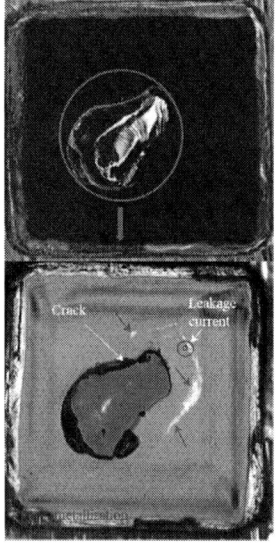

Fig. 11a. Optical microscope picture of test device (Fig. 6). After repetitive surge current stress. Crack at back side of the chip after decapsulation of the metallization

Fig. 11b. Emission spot with in the active region encircled in red (3µA/775V/10sec). The red arrow marks shows the abnormalities in the front-side metal

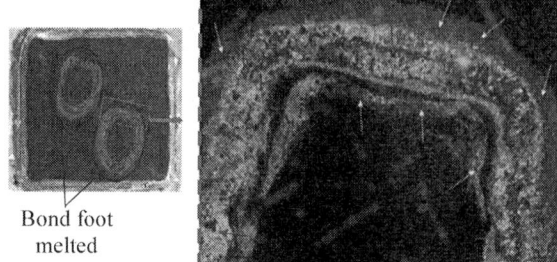

Bond foot melted

Fig. 12. Optical microscope image of test device after back side metallization decapsulation. Strong degradation at the front side metallization and the area under bond foot is found melted away

IV. CONCLUSION

Under very high repetitive surge current stress and high temperatures, almost all the test devices show similar failures. Voltage drift at the bipolar falling branch leads to destruction, and the maximum deposited energy (E_{max}) was increased up to ~5% after the last survived pulse.

The following failure mechanisms were observed during the repetitive surge current stress: A change in the Schottky barrier/power metal, chip crack at the middle of the chip, partial bond wire lift-off and front side metallization delamination. It seems that in tested devices under same condition, different competing failure mechanisms are triggered in parallel. Apparently, at this very high current stress, stacking faults could also be unavoidable. However, the SiC MPS diodes are quite robust against very high repetitive surge current stress which could be quite sufficient for application related surge current events.

REFERENCES

[1] R. Rupp, M. Treu, S. Voss, F. Björk, T. Reimann, " '2nd Generation' SiC Schottky diodes: A new benchmark in SiC device ruggedness.", Proceedings of the 18th International Symposium on Power Semiconductor Devices & IC's (ISPSD), pp. 1-4, 2006.

[2] M. Draghici, R. Rupp, R. Gerlach, B. Zippelius, "A new 1200V SiC MPS Diode with improved performance and ruggedness", Materials Science Forum, vols. 821-823, pp. 608-611, 2015.

[3] S. Fichtner, J. Lutz, T. Basler, R. Rupp, and R. Gerlach, "Electro-Thermal Simulations and Experimental Results on the Surge Current Capability of 1200V SiC MPS Diodes", Proceedings of the 8th International Conference on Integrated Power Systems (CIPS), pp. 1–6, 2014.

[4] R. Rupp, R. Gerlach, and A. Kabakow, "Current Distribution in the Various Functional Areas of a 600V SiC MPS Diode in Forward Operation", Materials Science Forum, vol. 717, pp. 929–932, 2012.

[5] S.Palanisamy, S.Fichtner, J.Lutz, T.Basler and R.Rupp, "Various structure of 1200V SiC MPS diode models and their simulated surge current behavior in comparision to measurement", Proceedings of the 28th International Symposium on Power Semiconductor Devices & IC's (ISPSD), pp. 235-238, 2016.

[6] B. Heinze, J. Lutz, R. Rupp, M. Holz, and M. Neumeister, "Surge current ruggedness of silicon carbide Schottky-and merged-PiN-Schottky diodes", Proceedings of the 20th International Symposium on Power Semiconductor Devices & IC's (ISPSD), pp. 245–248, 2008.

[7] J. Lutz, H. Schlangenotto, U. Scheuermann, R. De Doncker, Semiconductor Power Devices-Physics, Characteristic, Reliability. Springer, 2011

Proceedings of the 30th International Symposium on Power Semiconductor Devices & ICs
May 13-17, 2018, Chicago, USA

Accumulation channel vs. Inversion channel 1.2 kV rated 4H-SiC Buffered-Gate (BG) MOSFETs: Analysis and Experimental Results

Kijeong Han, B. Jayant Baliga
Electrical and Computer Engineering
PowerAmerica Institute, North Carolina State University
Raleigh, NC 27695, USA
khan5@ncsu.edu

Woongje Sung
Colleges of Nanoscale Science and Engineering
State University of New York Polytechnic Institute
Albany, NY 12203, USA

Abstract—SiC power MOSFETs with improved high frequency figures-of-merit (HF-FOMs) are needed for various applications to reduce switching energy losses. This paper compares the electrical performance of novel 1.2 kV-rated 4H-SiC Accumulation channel (Accu) and Inversion channel (Inv) Buffered-Gate (BG) MOSFETs with conventional (C) and Split-Gate (SG) devices. It is demonstrated that Accu BG-MOSFETs exhibit 4.0× and 2.6× smaller HF-FOM [$R_{on} \times Q_{gd}$], and 3.6× and 2.1× smaller HF-FOM [$R_{on} \times C_{gd}$], when compared to the Accu C-MOSFET and Accu SG-MOSFET, respectively.

Keywords—4H-SiC; MOSFET; Split gate; Buffered gate; C_{gd}; Q_{gd}; HF-FOMs; Accumulation channel; Inversion channel

I. INTRODUCTION

In the past decades, there has been significant progress in the electrical performance of 4H-SiC MOSFETs such as high speed switching and reduced specific on-resistance [1]. One major on-resistance performance limiting factor in SiC MOSFETs is the poor channel mobility due to high interface trap densities [2]. The accumulation mode channel has been demonstrated [3] to achieve improved channel mobility. It is also well known that small reverse transfer capacitance (C_{gd}) and gate-to-drain charge (Q_{gd}) are required to achieve fast switching speed and low switching power losses [4]. Therefore, many research groups have tried to obtain smaller high-frequency figures-of-merit (HF-FOMs), defined as [$R_{on} \times C_{gd}$] and [$R_{on} \times Q_{gd}$] [5], for high frequency applications.

Central implant (CI) MOSFET have been reported by Wolfspeed to reduce C_{gd}, Q_{gd}, and HF-FOMs [6]. An extra p-region is implanted in the center of the JFET region so that the electrical characteristics are improved. It has been recently demonstrated that 4H-SiC Split-Gate (SG) MOSFET structures have superior HF-FOMs using the same fabrication steps as conventional MOSFETs [7]. An even superior 4H-SiC Buffered-Gate (BG) MOSFET structure, with the split-gate electrode buffered from the drain by extending the P⁺ shielding region, has also been recently reported [8]. In this paper, experimental results for both Accumulation channel (Accu) and Inversion channel (Inv) BG-MOSFETs are compared in detail with Accu conventional (C) and Accu Split-Gate (SG) MOSFETs.

The information, data, or work presented herein was funded in part by the Office of Energy Efficiency and Renewable Energy (EERE), U.S. Department of Energy, under Award Number DE-EE0006521 with North Carolina State University, PowerAmerica Institute.

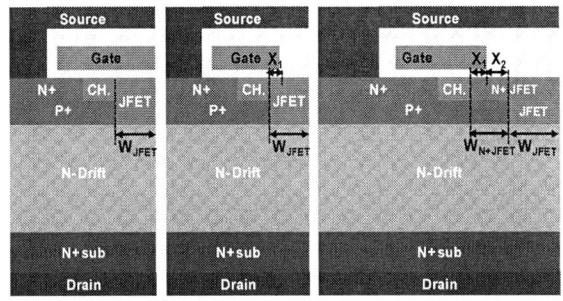

Fig. 1: Cross-sectional views of fabricated 4H-SiC 1.2kV conventional (C), Split-Gate (SG), and Buffered-Gate (BG) MOSFETs. All structures have the same channel length (0.5μm) and W_{JFET} (0.7μm).

II. DEVICE STRUCTURE AND ANALYSIS

Fig. 1 shows cross-sectional views of the fabricated Accu C-MOSFET, Accu SG-MOSFET, and Accu or Inv BG-MOSFET structures. All the structures have the enhanced JFET doping of 3×10^{16} cm⁻³, JFET width (W_{JFET}) of 0.7 μm, and channel length of 0.5 μm. An additional N⁺ JFET region is included in the BG-MOSFET structure as shown in Fig. 1 to prevent complete depletion above the P⁺ shielding region beyond the gate edge. The N⁺ JFET region doping concentration was carefully studied by 2-D TCAD numerical simulations in order to optimize the device operation.

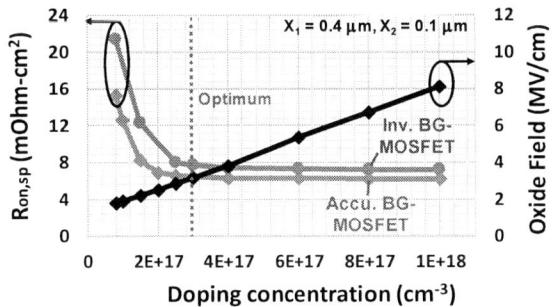

Fig. 2: Simulation results for optimization of the N⁺ JFET doping concentration in Inv and Accu BG-MOSFETs to minimize the Oxide electric field and on-resistance. Optimum value is 3×10^{17}cm⁻³.

Fig. 2 shows the simulation results for specific on-resistance ($R_{on,sp}$) and maximum gate oxide electric field at V_d=1600 V as a function of N⁺ JFET doping concentration. An N⁺ JFET region thickness of 0.2 μm, poly-Si gate overhang

978-1-5386-2928-4/18 $31.00 © 2018 IEEE   371

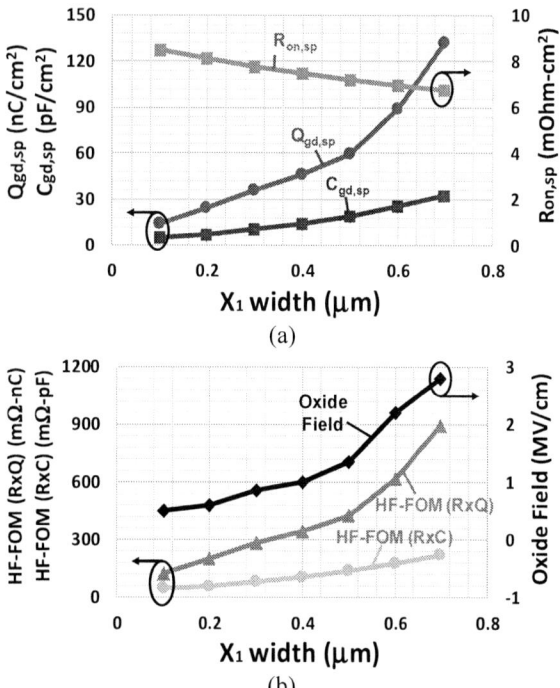

Fig. 3: Simulation results of (a) $Q_{gd,sp}$, $C_{gd,sp}$, and $R_{on,sp}$, (b) Oxide electric field, HF-FOM [$R_{on} \times Q_{gd}$], and HF-FOM [$R_{on} \times C_{gd}$] as a function of X_1 width for BG-MOSFETs. W_{N+JFET} was fixed to 0.7μm for the simulations. From the simulation results and alignment tolerances, optimum X_1 of 0.4μm is chosen.

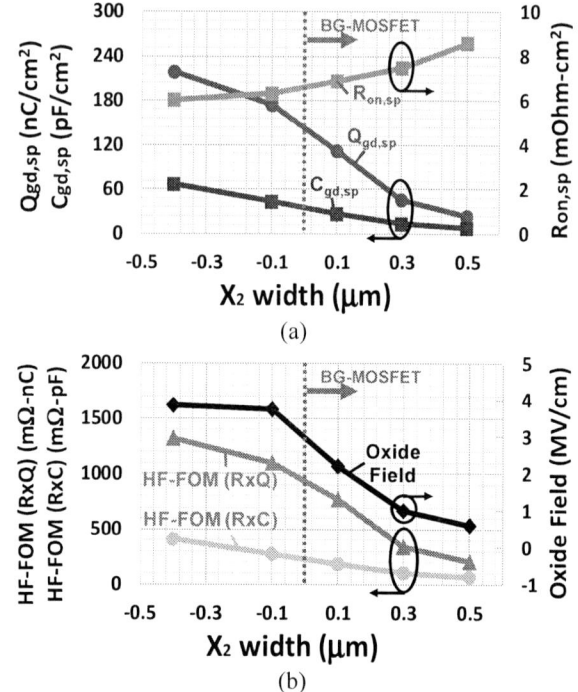

Fig. 4: Simulation results of (a) $Q_{gd,sp}$, $C_{gd,sp}$, and $R_{on,sp}$, (b) Oxide electric field, HF-FOM [$R_{on} \times Q_{gd}$], and HF-FOM [$R_{on} \times C_{gd}$] as a function of X_2 width for BG-MOSFETs. X_1 of 0.4μm was used for the simulations. From the simulation results, optimum X_2 is determined to be 0.3μm.

into the JFET region (X_1) of 0.4 μm, and P^+ shielding region width beyond the poly-Si gate edge (X_2) of 0.1 μm were used in the simulations. The poly-Si gate edge is buffered by the P^+ shielding region in the BG-MOSFETs. However, the oxide electric field exceeds 4 MV/cm with high N^+ JFET doping concentration (>4×10^{17} cm^{-3}) without any benefits in $R_{on,sp}$ as shown in Fig. 2. This indicates an optimum N^+ JFET doping concentration of 3×10^{17} cm^{-3} for both Accu and Inv BG-MOSFETs.

Fig. 3 and Fig. 4 compare the simulation results of $Q_{gd,sp}$, $C_{gd,sp}$, $R_{on,sp}$, maximum oxide electric field (at V_d=1200 V and V_g=0 V), HF-FOM [$R_{on} \times Q_{gd}$], and HF-FOM [$R_{on} \times C_{gd}$] for the Accu BG-MOSFETs as a function of X_1 and X_2 widths, respectively. The N^+ JFET doping of 3×10^{17} cm^{-3} was used for all the simulations. The oxide electric field (see Fig. 3(b) and Fig. 4(b)) is well below 3 MV/cm in the BG-MOSFET structures demonstrating that the split poly-Si gate edges are buffered by the P^+ shielding regions as previously discussed.

Significantly reduced $Q_{gd,sp}$ and $C_{gd,sp}$ with gradually increased $R_{on,sp}$ are observed in Fig. 3(a) when X_1 decreases from 0.7 to 0.1 μm. Consequently, the HF-FOMs are significantly reduced with decrease of X_1 as shown in Fig. 3(b). Fig. 3(a) and Fig. 3(b) indicate X_1 width shorter than 0.5 μm are necessary due to the rapid increases in $Q_{gd,sp}$, oxide electric field and HF-FOM [$R_{on} \times Q_{gd}$] for larger values. Based on the simulation results and process alignment tolerances, an X_1 value of 0.4μm was chosen for the fabricated SG- and BG-MOSFETs.

Fig. 4(a) shows significantly reduced $Q_{gd,sp}$ and $C_{gd,sp}$ when X_2 increases from 0 to 0.5 μm due to enhanced screening of

the gate by the P^+ shielding region. The HF-FOMs are greatly reduced with increase of X_2 as shown in Fig. 4(b) in spite of a gradual increase in $R_{on,sp}$. Although the best HF-FOMs occur with X_2 of 0.5μm, the small improvement of the HF-FOMs with the increased $R_{on,sp}$ is not desirable when compared to the structure with X_2 of 0.3μm. Consequently, the optimum X_2 was determined to be 0.3μm. Numerical simulations of the BG-MOSFETs, with X_1 of 0.4 μm and X_2 of 0.3 μm, demonstrate that about 3× reduced HF-FOMs can be expected compared with the SG-MOSFET.

III. FABRICATION TECHNOLOGY AND EXPERIMENTAL RESULTS

Accu and Inv BG-MOSFETs with X_1 of 0.4μm and X_2 of 0.3μm were fabricated at X-FAB, TX, on a 6-inch, N^+ 4H-SiC wafer with an n-type epi-layer (Thickness: 10 μm, Doping: 8×10^{15} cm^{-3}) at X-FAB, TX. using a 12 mask process. Accu C-MOSFET and Accu SG-MOSFET with X_1 of 0.4μm were also fabricated at the same time for comparison. The basic fabrication steps to create the BG-MOSFET are shown in Fig. 5.: (a) The enhanced JFET, P^+ shielding region, accumulation or inversion channel, N^+ source, P^+ contact, and P^- JTE regions were formed by a series of implants of N and Al; (b) N^+ JFET region was formed by additional implantation step, followed by an implant activation anneal at 1650°C for 10-min with a carbon cap; (c) Dry oxidation at 1175 °C was used to form the 50 nm thick gate oxide, followed by a post oxidation anneal in nitric oxide ambient and N-type poly-Si gate deposition; (d) The poly-Si gate and gate oxide were patterned to define the buffered gates; (e) Thick oxide interlayer dielectric was deposited,

978-1-5386-2928-4/18 $31.00 © 2018 IEEE

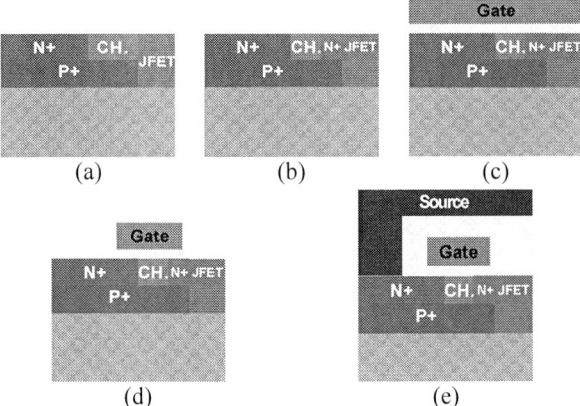

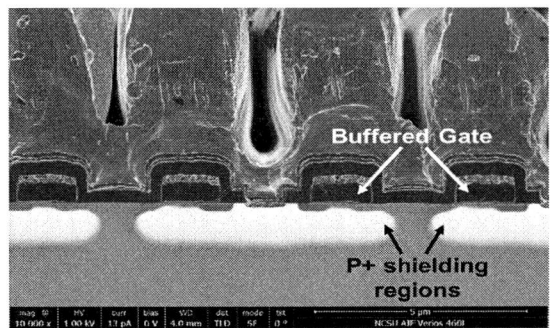

Fig. 5: Fabrication steps to create BG-MOSFETs: (a) Implantations for JFET, P$^+$ shielding with Channel, N$^+$, and P$^+$, (b) Implantation for N$^+$ JFET, (c) Gate oxide growth and Poly-Si deposition, (d) Patterning of the Poly-Si and Gate oxide, (e) Deposition of ILD, Ohmic contact, and Deposition of Source Metal.

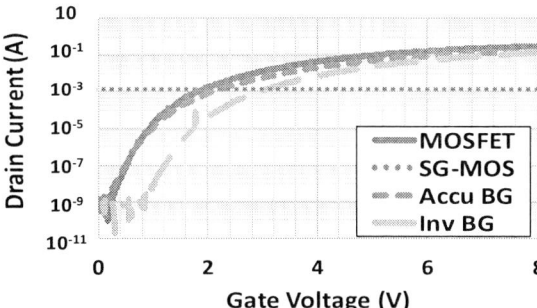

Fig. 7: Measured transfer characteristics for fabricated SiC Accu C-, SG-, and BG-MOSFETs. V$_{th}$ was extracted at I$_d$=1mA.

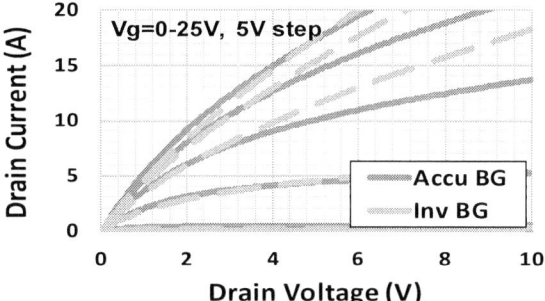

Fig. 8: Typical measured output characteristics of fabricated 1.2 kV rated Accu and Inv channel BG-MOSFETs. Active area is 4.5mm^2.

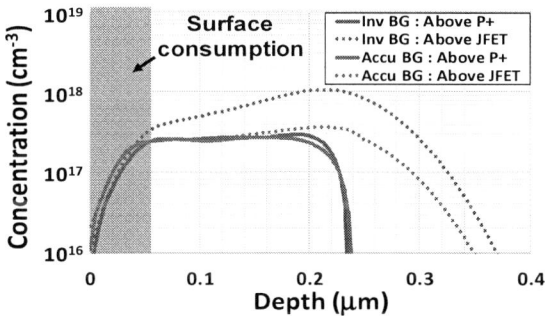

Fig. 9: N$^+$ JFET implantation profiles for Accu and Inv channel BG-MOSFETs. Higher N$^+$ JFET doping is required to compensate the P-base in Inv BG-MOSFET, resulting in the reduced BV (Fig. 9).

Fig. 6: SEM image for the fabricated Accu BG-MOSFET. It is clearly seen that the split poly-Si gates are buffered by the P$^+$ shielding regions.

followed by Ni ohmic contact process with an RTA on the front and back sides; (f) The source and gate pads were formed with 4 µm thick Al-based metal; (g) Nitride and polyimide passivation layers were deposited on the front side and patterned; (h) Finally, the back side was covered by a solderable metal stack. The SEM image of the fabricated Accu BG-MOSFET in Fig. 6 demonstrates that the buffered gate structure was very well defined through the fabrication process.

Fig. 7 compares the transfer characteristics at room temperature for the fabricated MOSFETs measured at V$_d$=0.1V. The threshold voltages (V$_{th}$) for Accu C-, Accu SG-, Accu BG-, and Inv BG-MOSFETs are 1.84, 1.80, 2.04, and 2.98 V, respectively. The threshold voltages of the Accu structures are about 1 V lower than that of the Inv BG-MOSFET.

Fig. 8 shows the measured output characteristics of the fabricated Accu and Inv BG-MOSFETs. The I-V characteristics were measured at gate biases of 0 to 25 V with 5 V steps. R$_{on,sp}$ at V$_g$=25V and I$_d$=1A are 8.39, and 8.14 mΩ·cm^2 for the Accu and Inv BG-MOSFETs, respectively. Although the Accu structure has higher channel mobility [3], the Inv BG-MOSFET shows lower R$_{on,sp}$. This is due to smaller N$^+$ JFET region resistance determined by higher N$^+$ JFET doping concentration above the JFET region as shown in Fig. 9, where the N$^+$ JFET implantation profiles for Accu and Inv channel

BG-MOSFETs are shown. Higher N$^+$ JFET implantation profile is required to compensate the P-base region in the Inv BG-MOSFET. A N$^+$ JFET doping concentration of 3×10^{17} cm^{-3} above the P$^+$ shielding region is obtained with very high doping concentration of 1×10^{18} cm^{-3} above the JFET region as shown in Fig. 9.

The excellent measured blocking characteristics of the C-, SG-, and Accu BG-MOSFETs with the same Hybrid-JTE edge termination are shown in Fig. 10. In contrast, the breakdown voltage for the Inv BG-MOSFET is significantly reduced because the high N$^+$ JFET doping concentration leads to much higher electric field at the edge of the P$^+$ shielding region.

Fig. 11 shows the measured C$_{gd}$ up to 1000 V for all the structures. The C$_{gd}$ extracted at V$_d$=1000V (500V for the Inv BG-MOSFET) are 5.43, 3.07, 1.04, and 2.60 pF for the Accu C-, SG-, and BG-MOSFETs, respectively. The higher C$_{gd}$ value in the Inv BG-MOFSET compared with the Accu BG-MOSFET

978-1-5386-2928-4/18 $31.00 © 2018 IEEE

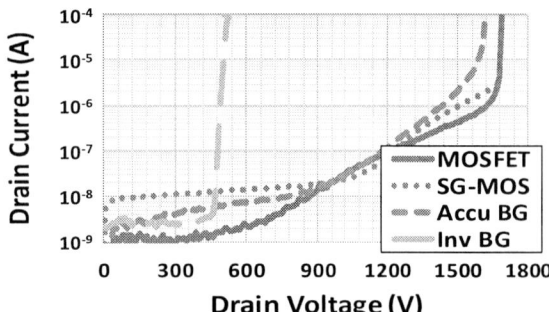

Fig. 10: Typical room temperature blocking characteristics (V_{gs} = 0V) of fabricated SiC Accu C-, Accu SG-, Accu and Inv channel BG-MOSFETs. BV is defined at I_d = 1mA.

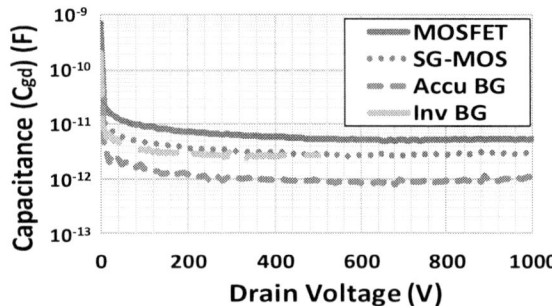

Fig. 11: Typical measured reverse transfer capacitance of fabricated SiC C-, SG-, and BG-MOSFETs. The Accu BG-MOSFET has much smaller C_{gd}.

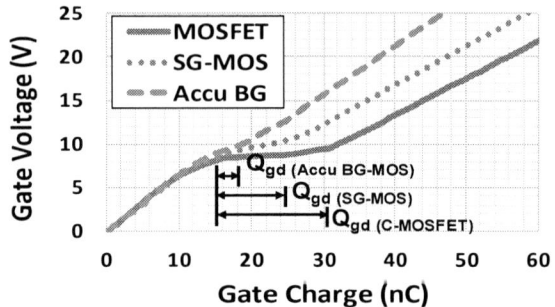

Fig. 12: Measured gate charge at V_d=800V and I_d=10A of fabricated SiC Accu C-, SG-, and BG-MOSFETs. Much smaller plateau (Q_{gd}) is observed in the Accu BG-MOSFET.

TABLE I
SUMMARY OF EXPERIMENTAL RESULTS FOR C-MOSFET, SG-MOSFET, AND BG-MOSFETs

	Accu C-MOS	Accu SG-MOS	Accu BG-MOS	Inv BG-MOS
Cell pitch [μm]	5.6	5.6	7.0	7.0
BV [V]	1689	1688	1617	520
V_{th} [V]	1.84	1.80	2.04	2.98
$R_{on,sp}$ [mΩ-cm²]	5.78	5.96	8.39	8.14
$C_{gd,sp}$ [pF/cm²]	121	68	23	57.8*
$Q_{gd,sp}$ [nC/cm²]	347	216	60	-
FOM <$R_{on} \times C_{gd}$> [mΩ-pF]	698	406	194	407*
FOM <$R_{on} \times Q_{gd}$> [mΩ-nC]	2006	1287	503	-

*$C_{gd,sp}$ at V_d=500V

[$R_{on} \times Q_{gd}$] is 4.0× and 2.6× lower than the Accu C- and SG-MOSFETs.

IV. CONCLUSION

Novel 1.2 kV-rated 4H-SiC both Accu and Inv BG-MOSFET structures have been successfully fabricated at the 150mm commercial X-Fab foundry and experimentally demonstrated to have superior C_{gd}, Q_{gd}, and HF-FOMs characteristics. The Inv BG-MOSFET shows a little bit lower on-resistance due to the inevitable higher N$^+$ JFET doping concentration above the JFET region. However, the higher doping concentration results in the poor breakdown voltage, which makes the Inv BG-MOSFET unsuitable for the 1.2 kV device.

From the measured electrical characteristics, it has been experimentally confirmed that the Accu BG-MOSFET has reduced HF-FOM [$R_{on} \times C_{gd}$] by 3.6× and 2.1×, and HF-FOM [$R_{on} \times Q_{gd}$] by 4.0× and 2.6× compared with the Accu C-MOSFET and SG-MOSFET, respectively due to the significantly improved C_{gd} and Q_{gd}.

REFERENCES

[1] J. W. Palmour, "Silicon Carbide Power Device Development for Industrial Markets," in IEEE Int. Electron Devices Meeting (IEDM), Dec. 2014, pp. 79-82

[2] T. Zheleva, A. Lelis, G. Duscher, F. Lui, I. Levin and M. Das, "Transition layers at the SiO2/SiC interface," Applied Physics Letters, vol. 93, no. 2, 2008.

[3] W. Sung, K. Han, and B. J. Baliga, "A comparative study of channel designs for SiC MOSFETs: accumulation mode channel vs. inversion mode channel," in Proc. 29th ISPSD 2017, May 2017, pp. 375-378.

[4] T. Sakai and N. Murakami, "A New VDMOSFET Structure with Reduced Reverse Transfer Capacitance," IEEE Trans. Electron Devices, vol. 36, no. 7, pp. 1381-1386, Jul. 1989.

[5] B. J. Baliga, Fundamentals of Power Semiconductor Devices. New York, NY, USA: Springer, 2008, ch. 6, pp. 279-434

[6] Q. C. J. Zhang, G. Wang, H. Doan, S.-H. Ryu, B. Hull, J. Young, S. Allen, and J. Palmour, "Latest Results on 1200V 4H-SiC CIMOSFETs with Ron,sp of 3.9mΩ·cm2 at 150°C," in Proc. 27th Int. Symp. Power Semiconductor Devices ICs, May 2015, pp. 89-92.

[7] K. Han, B. J. Baliga, and W. Sung, "Split-Gate 1.2-kV 4H-SiC MOSFET: Analysis and Experimental Validation," IEEE Electron Device Letters, vol. 38, no. 10, Oct. 2017.

[8] K. Han, B. J. Baliga, and W. Sung, "A Novel 1.2 kV 4H-SiC Buffered-Gate (BG) MOSFET: Analysis and Experimental Results," IEEE Electron Device Letters, vol. 39, no. 2, Feb. 2018.

is due to the higher N$^+$ JFET doping concentration. Fig. 12 compares the measured gate charge at V_d=800V and I_d=10A of the fabricated MOSFETs. A much smaller plateau (Q_{gd}) can be observed in the Accu BG-MOSFET structure when compared with the other devices. The Q_{gd} values extracted from the gate voltage plateaus are 15.6, 9.7, and 2.7 nC for the Accu C-, SG-, and BG-MOSFETs, respectively.

All the experimental results are summarized in Table I. The Inv BG-MOSFET has a lower $R_{on,sp}$ due to the higher N$^+$ JFET doping concentration as discussed above, but this results in lower breakdown voltage.

It is experimentally demonstrated that the Accu BG-MOSFET has reduced $C_{gd,sp}$ by 5.2× and 3×, and $Q_{gd,sp}$ by 5.8× and 3.6× when compared with the Accu C- and SG-MOSFETs. The HF-FOM [$R_{on} \times C_{gd}$] of the Accu BG-MOSFET is 3.6× and 2.1× lower than the Accu C- and SG-MOSFETs. The HF-FOM

978-1-5386-2928-4/18 $31.00 © 2018 IEEE

Proceedings of the 30th International Symposium on Power Semiconductor Devices & ICs
May 13-17, 2018, Chicago, USA

Characterization of 1.2 kV SiC Super-Junction SBD Implemented by Trench and Implantation Technique

Baozhu Wang, Hengyu Wang, Xueqian Zhong, Shu Yang, Qing Guo and Kuang Sheng*

College of Electrical Engineering, Zhejiang University, Hangzhou, China

*shengk@zju.edu.cn

Abstract—This paper presents the characterizations of SiC Super-Junction SBD, which can break the theoretical one dimensional limit of the SiC unipolar devices. The Super-Junction structure is implemented by trench and implantation technique. The mesa width of Super-Junction device is a key structure parameter, therefore its impacts on device forward and reverse characteristics are analyzed by measurement and numeric simulations. The temperature dependence of the specific on-resistance ($R_{sp,on}$) of this device is measured and compared with that of a conventional SiC SBD. SiC Super-Junction SBD shows a slower increase in $R_{sp,on}$ with temperature. The switching capability of SiC Super-Junction device is demonstrated for the first time by double pulse tester.

Keywords—silicon carbide; Super-Junction; characteristics

I. INTRODUCTION

SiC power devices are promising for high-voltage and high-temperature power electronics applications, owing to the superior physical properties. Recent years have witnessed tremendous progress in power MOSFETs [1], JFETs [2] and JBSs [3,4]. SiC unipolar devices can realize higher switching speed and lower switching loss than Si bipolar devices within the voltage range of 300 V~6500 V [5]. However, without conductivity modulation, it's challenging to further reduce the on-resistance in the conventional SiC unipolar devices. SiC Super-Junction devices can improve the trade-off between the breakdown voltage and on-resistance of SiC unipolar devices.

SiC Super-Junction devices have been demonstrated recently [6,7,8]. Our group has demonstrated SiC Super-Junction SBD which successfully breaks the theoretical one dimensional limit of the SiC unipolar devices [7]. However, comprehensive investigations of SiC Super-Junction devices, which are important for the further optimization of SiC Super-Junction devices, are still lacking to date.

In this work, systematic electrical characterizations of the SiC Super-Junction SBD are carried out. Forward and reverse performance, capacitance-voltage characteristics and dynamic performance are reported and analyzed.

II. DEVICE STRUCTURE AND FABRICATION

Fig. 1 shows the schematic cross section of the SiC Super-Junction device, which was implemented by trench and implantation technique. The 12 μm-thick n-type epi-layer has a doping concentration of 7×10^{16} cm^{-3}. The trench depth and width are 6 μm and 3.2 μm, respectively. The mesa width varies from 1.4 μm to 2.2 μm. The width of p pillar is designed to be 0.3 μm. A 1.5 μm-thick SiO$_2$ layer was firstly deposited by PECVD and deep trenches were formed by ICP technique. After etching, the SiO$_2$ layer was used as self-aligned implantation mask. Al was selected as the p-type dopant and there were two-step vertical/tilted implantations. The p-type dopants were activated at the temperature of 1350°C, after which the deep trenches were filled with SiO$_2$. Ni was used to form Schottky contact and ohmic contact. The detailed description can be found in our previous paper [7].

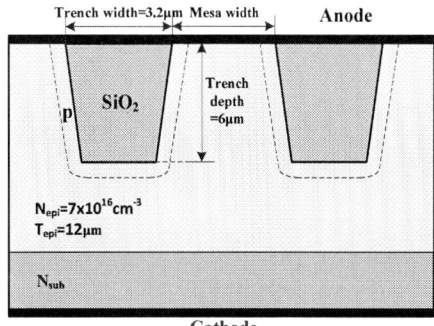

Fig. 1. Schematic cross section and key parameters of SiC Super-Junction SBD

III. EXPERIMENTAL RESULTS AND DISCUSSIONS

Forward and reverse performance and capacitance-voltage characteristics of SiC Super-Junction SBD are evaluated by Keysight B1505A, and its switching capability is characterized by double pulse tester.

A. Static Characteristics

In Fig. 2, the forward voltage drops of the SiC Super-Junction SBDs at 500 A/cm^2 increase from 2.39 V to 2.68 V when the mesa widths decrease from 2.2 μm to 1.4 μm. The $R_{sp,on}$ of SiC Super-Junction SBDs at various temperatures are plotted and compared with that of conventional SiC SBD in Fig. 3. At room temperature, the $R_{sp,on}$ of SiC Super-Junction SBDs increase from 1.05 mΩ·cm^2 to 1.3 mΩ·cm^2 when the mesa widths decrease from 2.2 μm to 1.4 μm. The $R_{sp,on}$ of 1.4 μm structure is still nearly 50% less than that of the conventional SiC SBD. With temperature increasing from 25°C to 150°C,

978-1-5386-2928-4/18 $31.00 © 2018 IEEE   375

the SiC Super-Junction SBDs exhibit ×1.8 higher $R_{sp,on}$, while the $R_{sp,on}$ of conventional SiC SBD increases by more than 2.2 times. Such difference is due to the more dominating substrate resistance (or mobility) in the Super-Junction device which is less temperature dependent.

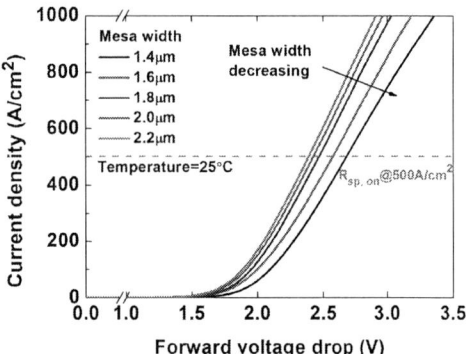

Fig. 2. Forward characteristics of SiC Super-Junction SBDs with various mesa widths

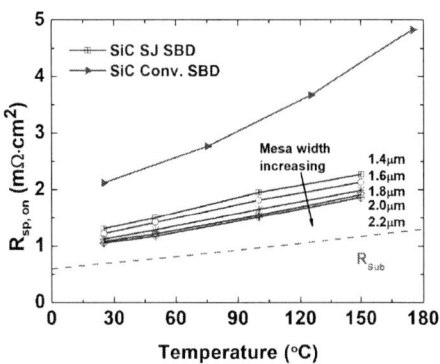

Fig. 3. Specific on-resistance of SiC Super-Junction SBDs and SiC conventional SBD (1200 V rated commercial product) at various temperatures

In Fig. 4, the leakage currents of SiC Super-Junction SBDs are lower and the reverse blocking voltages of SiC Super-Junction SBDs are higher when the mesa widths decrease from 2.2 μm to 1.4 μm. The trade-off between $R_{sp,on}$ and breakdown voltage with various mesa widths (in Fig. 5) shows that when the mesa widths decrease from 2.2 μm to 1.4 μm, the $R_{sp,on}$ merely increases by 25% but the breakdown voltage increases from 400 V to 1150 V. The $R_{sp,on}$ and breakdown voltage of these devices are plotted in Fig. 6. With the device mesa width decreased to 1.4 μm, the device performance is approaching the 4H-SiC unipolar limit. This result also suggests that higher breakdown voltage can be achieved by forming narrower mesa. The impact of mesa width on breakdown voltage and $R_{sp,on}$ is also investigated through numerical simulation. As shown in Fig. 7, although a narrow mesa (1.4 μm) increases current crowding in the mesa region, it can mitigate the peak electric field at the mesa top,

which leads to a more uniform electric field distribution along the Super-Junction pillar and thus a higher breakdown voltage.

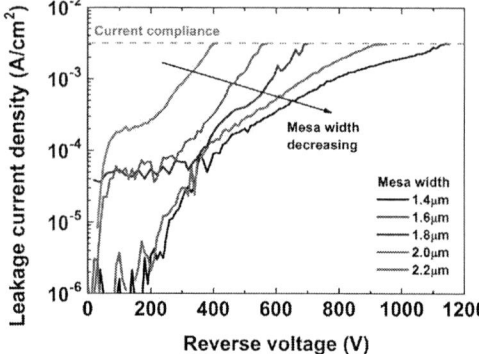

Fig. 4. Reverse characteristics of SiC Super-Junction SBDs with various mesa widths

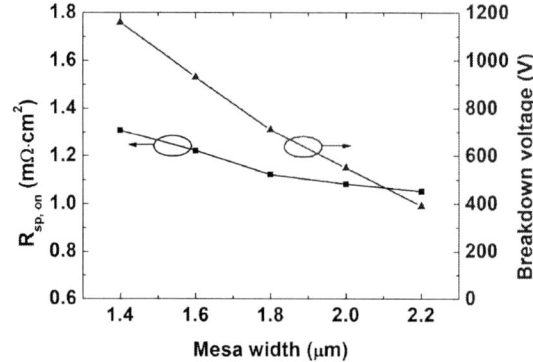

Fig. 5. Trade-off between $R_{sp,on}$ and breakdown voltage of SiC Super-Junction SBDs with various mesa widths

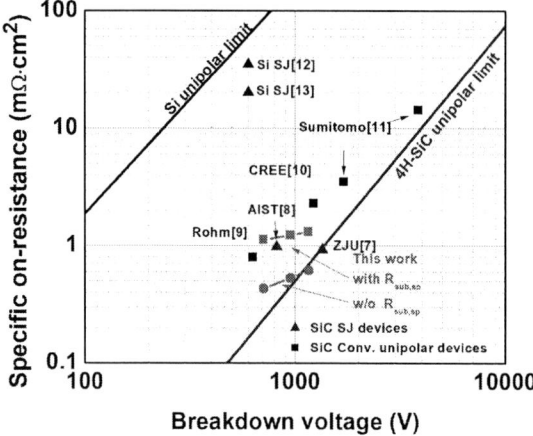

Fig. 6. Benchmark of BV versus $R_{sp,on}$ of this work, previously reported Si and SiC Super-Junction devices, and other representative SiC power devices

978-1-5386-2928-4/18 $31.00 © 2018 IEEE

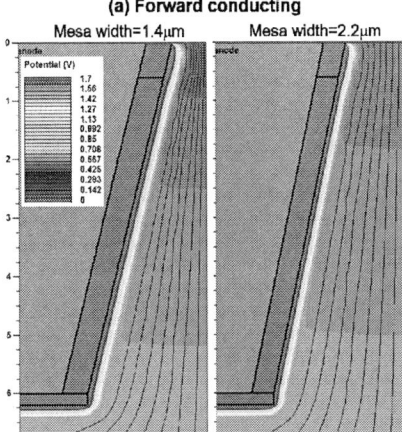

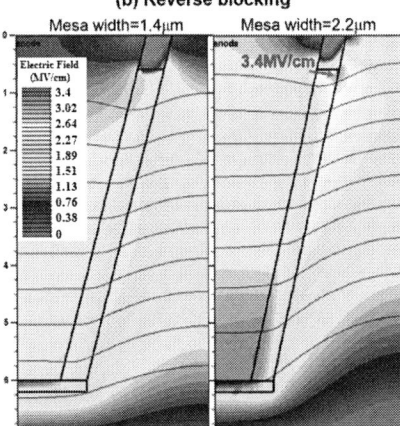

Fig. 7. (a) Simulated forward-conduction-state potential and current flowline distribution in devices with mesa widths of 1.4 μm and 2.2 μm. The Simulated $R_{sp,on}$ of devices with mesa widths of 1.4 μm and 2.2 μm are 0.63 mΩ·cm² and 0.37 mΩ·cm², respectively. The simulated results are in good agreement with measurements as shown in Fig. 3. (b) Simulated reverse-blocking-state electric field and equal-potential lines

B. Capacitance-Voltage Characteristics

C-V characteristics of SiC Super-Junction SBDs with various mesa widths are measured and analyzed by numeric simulations (Fig. 8). Compared with the conventional SiC SBD, the capacitance of the device decreases more quickly with reverse bias, due to the lateral depletion of Super-Junction pillar. On the other hand, no abrupt drop of capacitance is observed in this Super-Junction device because the Super-Junction pillar is gradually depleted, as verified in simulation as shown in Fig. 9. This gradual depletion may be

due to the tilted P pillar and the non-uniform doping profile of the pillar from top to bottom.

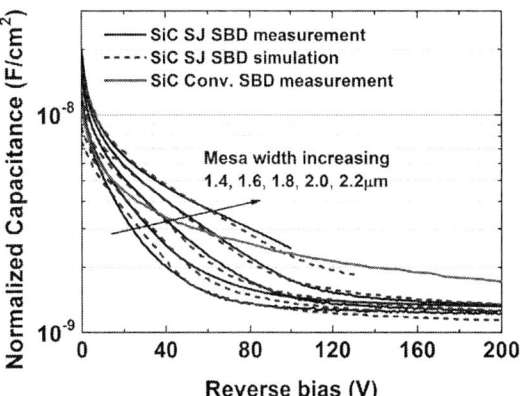

Fig. 8. Measured and simulated C-V characteristics of SiC Super-Junction SBDs with various mesa widths and measured C-V characteristics of SiC conventional SBD

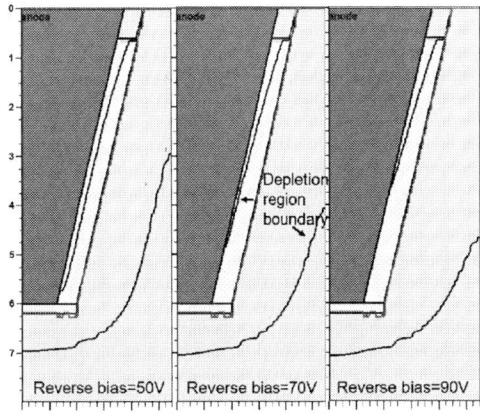

Fig. 9. The simulated depletion region boundary of SiC Super-Junction SBD as reverse bias increased from 50 V to 90 V

C. Switching Characteristics

The switching capability of SiC Super-Junction device is demonstrated by double pulse tester (shown in Fig. 10 (a)). The DC voltage is 400 V and the load current is 1 A. The reverse recovery waveform is shown and compared with that of the conventional SiC SBD and Si FRD, as shown in Fig. 10(b). Significant reduction in maximum reverse recovery current is observed. The maximum reverse recovery currents of SiC super-Junction SBD, conventional SiC SBD and Si FRD are 0.3 A, 0.9 A and 6.5 A, respectively.

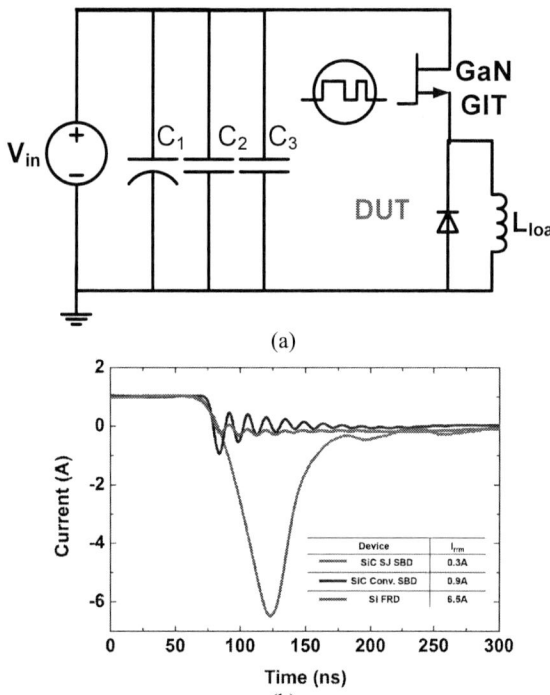

(a)

(b)

Fig. 10. (a) Schematic of double pulse tester. (b) Reverse recovery waveforms of SiC Super-Junction SBD, conventional SiC SBD and Si FRD. Double pulse test condition: DC voltage=400 V, load current=1 A

Device	I_{rm}
SiC SJ SBD	0.3A
SiC Conv. SBD	0.9A
Si FRD	6.5A

IV. CONCLUSION

SiC Super-Junction SBDs implemented by trench and implantation technique are fabricated and characterized systematically in this paper. The measured $R_{sp,on}$ of SiC Super-Junction SBD increases slower with increasing temperature compared with the conventional SiC SBD. This is caused by the fact that the more dominating substrate resistance in the Super-Junction device has weaker temperature dependence due to higher doping concentration. Forward and reverse characteristics are investigated by measurement and numeric simulations. It is shown that structures with narrower mesa widths (within 1.4 µm~2.2 µm) can achieve higher breakdown voltage due to more uniform electric field distribution along the Super-Junction pillar, while $R_{sp,on}$ will be increased due to the current crowding at Schottky contact surface. The performance of the device with 1.4 µm-wide mesa approaches the 4H-SiC unipolar limit. The measured device capacitance shows a faster decrease rate with increasing reverse voltage than that of the conventional SiC SBD. The switching capability of SiC Super-Junction device is characterized and the maximum reverse recovery current of the device is relatively low. The proposed SiC Super-Junction device not only improves the trade-off between breakdown voltage and specific on-resistance but also offers better dynamic performance. These advantages make it a promising candidate for the medium to high voltage rating applications.

ACKNOWLEDGMENT

This work was supported by the National Key Research and Development Program of China (No. 2016YFB0400502) & National Natural Science Fundation of China (No. U51777187).

REFERENCES

[1] J. Wei, M. Zhang, H. Jiang, H. Wang, and K. J. Chen, "Charge storage effect in SiC trench MOSFET with a floating p-shield and its impact on dynamic performances," in *Proceedings of 29th International Symposium on Power Semiconductor Devices and ICs*, 2017, pp. 387-390.

[2] S. Chen, A. Liu, J. He, S. Bai, and K, Sheng, "Design and Application of High-Voltage SiC JFET and Its Power Modules," *IEEE Journal of Emerging & Selected Topics in Power Electronics*, vol. 4, no. 3, pp. 780-789, September 2016.

[3] N. Ren, J. Wang, and K. Sheng, "Design and Experimental Study of 4H-SiC Trenched Junction Barrier Schottky Diodes," *IEEE Transactions on Electron Devices*, vol. 61, no. 7, pp. 2459-2465, July 2014.

[4] N. Ren, and K. Sheng, "An Analytical Model With 2-D Effects for 4H-SiC Trenched Junction Barrier Schottky Diodes," *IEEE Transactions on Electron Devices*, vol. 61, no. 12, pp. 4158-4165, December 2014.

[5] T. Kimoto, "Material science and device physics in SiC technology for high-voltage power devices," *Japanese Journal of Applied Physics*, vol. 54, 040103, 2015.

[6] R. Kosugi, Y. Sakuma, K. Kojima, S. Itoh, A. Nagata, T. Yatsuo, Y. Tanaka and H. Okumura, "First experimental demonstration of SiC super-junction (SJ) structure by multi-epitaxial growth method", in *Proceedings of 26th International Symposium on Power Semiconductor Devices and ICs*, June 2014, pp. 346-349.

[7] X. Zhong, B. Wang, and K. Sheng, "Design and experimental demonstration of 1.35 kV SiC super junction Schottky diode," in *Proceedings of 28th International Symposium on Power Semiconductor Devices and ICs*, June 2016, pp. 231-234.

[8] T. Masuda, R. Kosugi, and T. Hiyoshi, "0.97 mΩcm2/820 V 4H-SiC Super Junction V-Groove Trench MOSFET," *Materials Science Forum. Trans Tech Publications*, vol. 897, pp. 483-488, 2017.

[9] T. Nakamura, Y. Nakano, M. Aketa, and Y. Yokotsuji, "High performance SiC trench devices with ultra-low ron", in *Proceedings of Electron Devices Meeting*, 2011, pp. 26.5.1-26.5.3.

[10] J. W. Palmour, L. Cheng, V. Pala, E. V. Brunt, D. J. Lichtenwalner, G-Y Wang, J. Richmond, M. O'Loughlin, S. Ryu, S. T. Allen, A. A. Burk, and C. Scozzie, "Silicon carbide power MOSFETs: breakthrough performance from 900V to 15kV," in *Proceedings of 26th International Symposium on Power Semiconductor Devices and ICs*, June 2014, pp. 79-82.

[11] Y. Mikamura, K. Hiratsuka, T. Tsuno, H. Michikoshi, S. Tanaka, T. Masuda, K. Wada, T. Horii, J. Genba, T. Hiyoshi, and T. Sekiguchi, "Novel designed SiC devices for high power and high efficiency systems," *IEEE Transactions on Electron Devices*, pp. 382-389. 2015.

[12] G. Deboy, M. Marz, J. P. Stengl, H. Strack, J. Tihanyi, and H. Weber, "A new generation of high voltage MOSFETs breaks the limit line of silion," in *Proceedings of Electron Devices Meeting*, 1998, pp. 783-685.

[13] W. Saito, I. Omura, S. Aida, M. Yamaguchi, and T. Ogura, "An 20mΩ·cm² 600V-class Superjunction MOSFET," in *Proceedings of 16th International Symposium on Power Semiconductor Devices and ICs*, June 2004, pp. 459-462.

Proceedings of the 30th International Symposium on Power Semiconductor Devices & ICs
May 13-17, 2018, Chicago, USA

Normally-OFF Dual-gate Ga$_2$O$_3$ Planar MOSFET and FinFET with High I$_{ON}$ and BV

H. Y. Wong[*], N. Braga and R. V. Mickevicius

Synopsys, Inc.,
Mountain View, CA, USA
[*]hiuyung.wong@ieee.org

F. Ding

Dept. of Electrical Engineering and Computer Sciences
University of California, Berkeley
Berkeley, CA, USA

Abstract— Ga$_2$O$_3$ is a promising Wide-Band-Gap material for power electronics due to its large bandgap and inexpensive native substrate. However, due to technological difficulties, only normally-ON n-type junctionless MOSFET (V$_{TH}$ < 0V) can be made easily. We propose using dual-gate configuration to achieve normally-OFF device for both Ga$_2$O$_3$ planar MOSFET and FinFET. Through TCAD simulations with calibrated parameters, it is found that normally-OFF dual-gate planar device and FinFET can be achieved with 6X and 1X enhancement in ON-current (I$_{ON}$), respectively, as higher doping is allowed, while breakdown voltage is not sacrificed.

Keywords—Ga$_2$O$_3$; Normally-OFF; Enhancement Mode; Recessed; Cascode; FinFET

I. INTRODUCTION

β-Ga$_2$O$_3$ is a promising material for the next generation power device due to its wide bandgap (4.5eV-4.9eV) and low cost native substrate [1][2]. However, planar Ga$_2$O$_3$ MOSFET is usually normally-ON (i.e. Threshold Voltage, V$_{TH}$ < 0V) which requires complex gate drive circuit for fail-safe operation. It also usually delivers lower drive current (I$_{ON}$) than its competitor such as GaN HEMT as the epitaxial layer cannot be too thick or too heavily doped in order to be turned on at reasonable gate voltage. To make it normally-OFF (V$_{TH}$ > 0V), even lower epitaxial doping or thinner epitaxial layer (t$_{epi}$) is needed but I$_{ON}$ will be sacrificed. Chabak et al. have demonstrated normally-OFF Ga$_2$O$_3$ device by using FinFET with narrow fins [1] and gate recessed planar technologies [2]. In this paper, we further propose 1) gate recessed (dubbed as "Recessed-gate" in the following discussion) FinFET and 2) dual-gate (dubbed as "Recess-dual" in the following discussion) planar MOSFET and FinFET to achieve high I$_{ON}$, high breakdown voltage (BV) and normally-OFF characteristics. The "Recess-dual" idea is developed from the dual gate idea in GaN HEMT [3][6].

Figures 1 and 2 show the structures of Ga$_2$O$_3$ MOSFET and FinFET studied, respectively. "Recess-gate" (e.g. L$_{G,re}$=2µm, L$_{G,long}$=0µm) means recessing the epaxial layer under the whole gate region. "Recess-dual" (e.g. L$_{G,re}$=0.5µm, L$_{G,long}$=1.5µm) means recessing the epitaxial layer under only a smaller part of the gate at the source side. Essentially, the two parts of the gate form a cascode configuration. "Recess-dual" is expected to have better performance than "Recess-gate" due to its cascode nature [3][6].

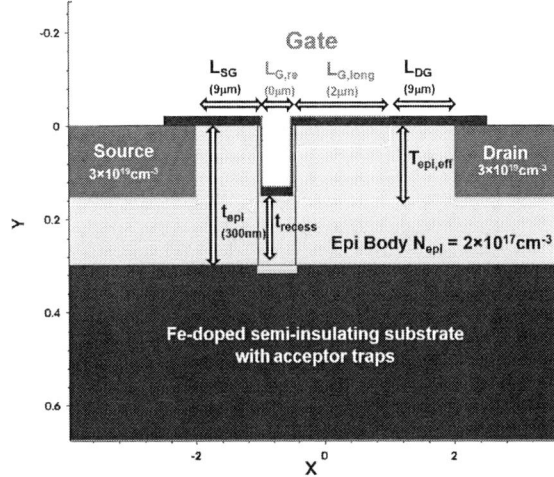

Fig. 1: Cross section of the planar Ga$_2$O$_3$ MOSFET in this study. Parameters in parenthesis are the values of norminal device in Ref. [4] (i.e. with no gate recess, L$_{G,re}$=0). White lines are the depletion layer boundaries at zero bias. Legend is the same as that in Fig. 2. Device is not drawn to scale for clarity. Gate oxide is 20nm Al$_2$O$_3$.

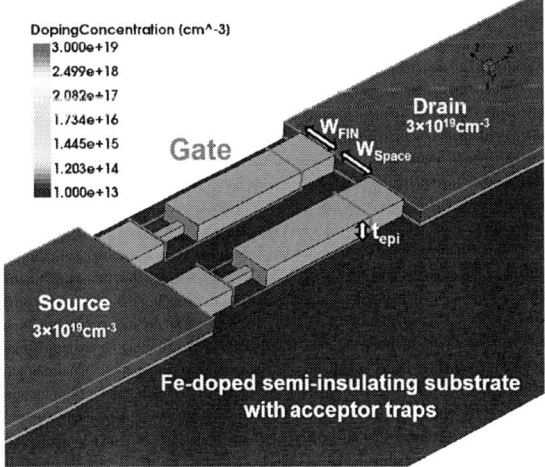

Fig. 2: Ga$_2$O$_3$ FinFET used in this study. Gate Al$_2$O$_3$ (20nm) is not shown for clarity. It shares the same symbols as the planar device in Fig. 1. Additionally, Fin width is W$_{FIN}$ and Fin spacing is W$_{Space}$. The recessed part of the Fin has width = W$_{Fin,recess}$ and height = t$_{recess}$.

978-1-5386-2928-4/18 $31.00 © 2018 IEEE   379

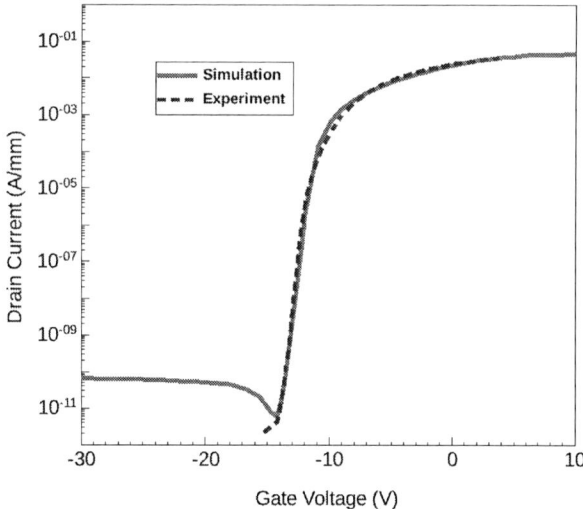

Fig. 3: I_DV_G of experimental data in Ref. [4] and simulated nominal planar device in Fig. 1 with $V_D = 25V$, $\varepsilon_{Al2O3} = 8.9$ and mobility $\mu = 180cm^2/Vs$.

II. SIMULATION SETUP AND CALIBRATION

TCAD Sentaurus [7] is used to study Ga_2O_3 planar device and FinFET physics, their design space and to verify the proposed structures, which are difficult to obtain experimentally. Firstly, TCAD parameters are calibrated to experimental I_DV_G, CV and BV curves of planar Ga_2O_3 MOSFET in Ref. [4] and [5] as shown in Figure 3, 4 and 5, respectively. $N_{epi} = 2\times10^{17}cm^{-3}$ is found to best match the experimental data. Electron low field mobility is set to be $180cm^2/Vs$, which gives the best fitting. Since recessed epitaxial layer is expected to have worse quality due to the damage of etching, ten times lower electron low field mobility (i.e. $\mu = 18cm^2/Vs$) is used in the recessed region in the

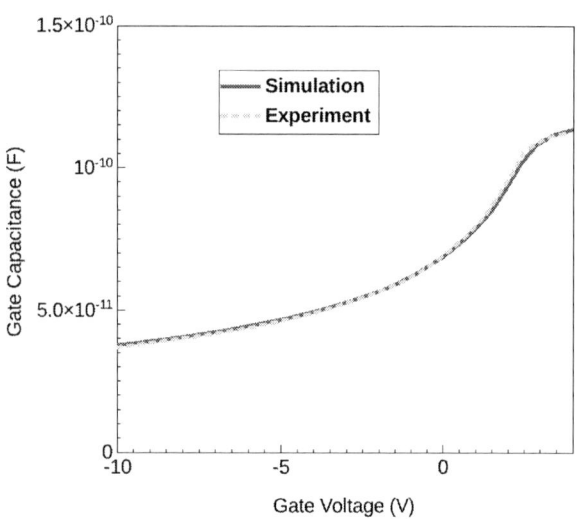

Fig. 4: Simulated and experimental (Ref. [5]) gate CV curves of Ga_2O_3 MOS capacitor. $\varepsilon_{Al2O3} = 8.6$ is used.

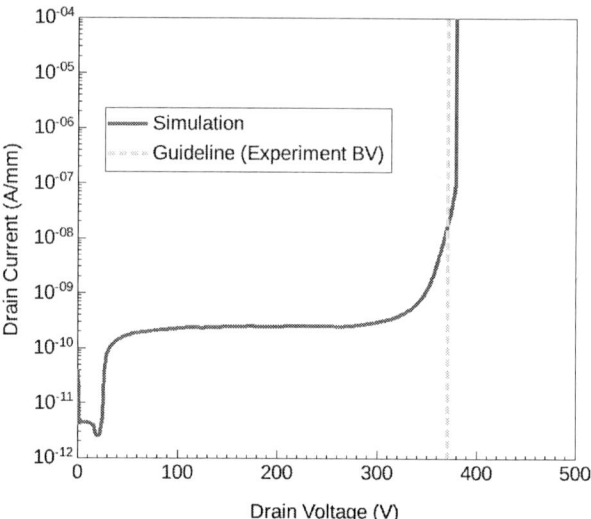

Fig. 5: BV curve of nominal planar device in Fig. 1 at $V_G = -20V = V_{TH,(VD=0.01V)}$ -16.5V. Experimental BV in Ref. [4] is drawn as guidance.

simulations. High field saturation model for mobility is turned on. The substrate is assumed to be semi-insulating with 10^{20} cm^{-3} acceptor traps at valence band edge due to Fe compensation doping. Same set of parameters are used in all simulations. Self-heating is not included in the simulations for simplicity.

III. SIMULATION RESULTS

Ga_2O_3 MOSFET is a junctionless device with switching behavior based on depletion layer modulation under the gate. Therefore, its I_{ON} and V_{TH} depend on the effective epitaxial doping (N_{epi}) and effective epitaxial thickness ($t_{epi,eff}$). As shown in Fig. 1, in general, there is a depletion region at the bottom of the epitaxial layer due to the existence of semi-

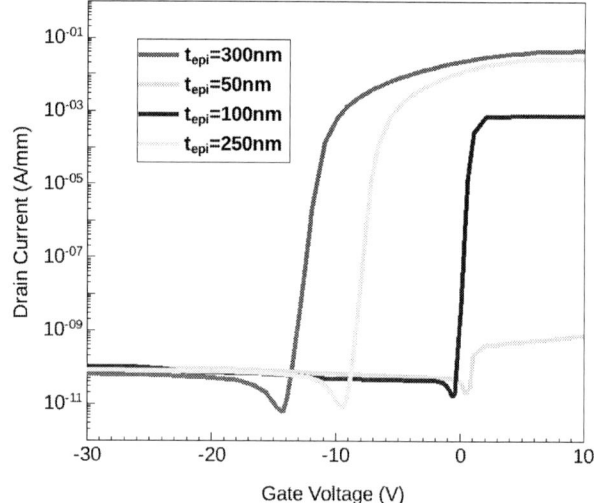

Fig. 6: Simulated I_DV_G curves as a function of t_{epi} for the nominal device in Fig. 1 (i.e. $L_{G,re}=0\mu m$ / no recess). V_D is 25V.

Table 1: Device characteristics of planar Ga$_2$O$_3$ devices. Current enhancement is compared to the "Nominal" device.

Device Type	Nepi	tepi	trecess	LG,long	LG,re	VTH (Vd=25V)	Max gm (mS/mm)	Sub-Threshold Slope (mv/dec)	Ion (Vg=Vth+10V) (mA/mm)	Current Enhancement (%)
Nominal	2.00E+17	300	150	2	0	-11.1	4	515	20	0
Recessed-gate	2.00E+17	300	50	0	2	1.9	3	234	24	23
Recessed-dual	2.00E+17	300	50	1.5	0.5	1.5	3	199	27	39
Recessed-gate	1.00E+18	300	50	0	2	1.0	16	232	86	337
Recessed-dual	1.00E+18	300	50	1.5	0.5	0.2	18	211	143	631

insulating substrate in Ga$_2$O$_3$ junctionless device at zero gate bias. Therefore, $t_{epi,eff}$ in general is less than t_{epi} (the physical epitaxial thickness), resulting in thinner conduction channel.

We first study the impact of t_{epi} on $I_D V_G$. To achieve positive V_{TH}, for $N_{epi} = 2 \times 10^{17} cm^{-3}$, t_{epi} needs to be less than 100nm but I_{ON} will be reduced by ~97% (Fig. 6). The I_{ON} reduction in thinner epitaxial layer is more than simple t_{epi} scaling because $t_{epi,eff} < t_{epi}$. But one can achieve positive V_{TH} by only recessing the epi under the gate region instead of the whole epitaxial layer. For "recess-gate" device (i.e. $L_{G,re}$=2μm, $L_{G,long}$=0μm) with t_{recess} = 50nm, V_{TH} becomes 1.9V at V_D=25V and $I_{ON,(VG=VTH+10V)}$ is 23% larger than the nominal device because of better gate control and similar access region resistance (Table 1). The better gate control in "recess-gate" device can be confirmed from the reduction of sub-threshold slope. Since recessed epitaxial layer allows higher N_{epi} for normally-OFF device, one can use higher doping to further enhance I_{ON}. As shown in Table 1, by using $N_{epi} = 10^{18} cm^{-3}$, I_{ON} can be increased by more than 300% while V_{TH} is still positive (i.e. 1V).

To further enhance the device performance, "recess-dual" is used. As shown in Figure 1, in a "recess-dual" device, a small portion of the gate at the source side (e.g. $L_{G,re}$=0.5μm) is recessed to give positive V_{TH} and a larger part of the gate at the drain side (e.g. $L_{G,long}$=1.5μm) is not recessed (so with negative

sV_{TH}). This forms a cascode like structure. When device is turned on, the long gate region has very large gate overdrive (V_G-V_{TH}) due to its negative V_{TH}. Therefore, the long gate part is very conductive and the device behaves as if it is a short gate device (with $L_{gate} = L_{G,re}$), resulting in very high I_{ON}. On the other hand, when the device is turned off, both the short and long gate regions are turned off due to the cascode nature. As a result, it has large effective gate length (with $L_{gate} = L_{G,re} + L_{G,long}$) and can block very high drain voltage like a regular long gate MOSFET. Therefore, its BV and punchthrough voltage are not degraded.

Figure 7 shows the simulated $I_D V_G$ curves of nominal, "recessed-gate" and "recessed-dual" devices. It clearly shows that "recessed-dual" device performs better than "recessed-gate".

Table 1 shows that by using "recess-dual", I_{ON} can further be increased by 39% for $N_{epi} = 2 \times 10^{17} cm^{-3}$. One can again use higher $N_{epi} = 10^{18} cm^{-3}$ and obtain more than 600% enhancement in I_{ON} in "recess-dual" while $V_{TH} > 0$. It should be noted that "recessed-dual" results in higher I_{ON} than "recessed-gate" because of the smaller effective gate length in ON state.

Depending on the design of the transistor, the breakdown may occur at the gate-edge at drain side or at the drain region.

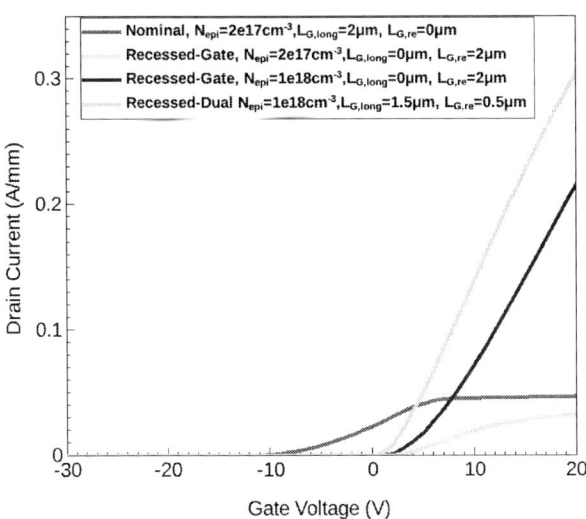

Fig. 7: Simulated $I_D V_G$ curves of various devices with t_{epi}=300nm and t_{recess}=50nm. μ is 180cm^2/Vs and 18cm^2/Vs for unrecessed and recessed regions, respectively.

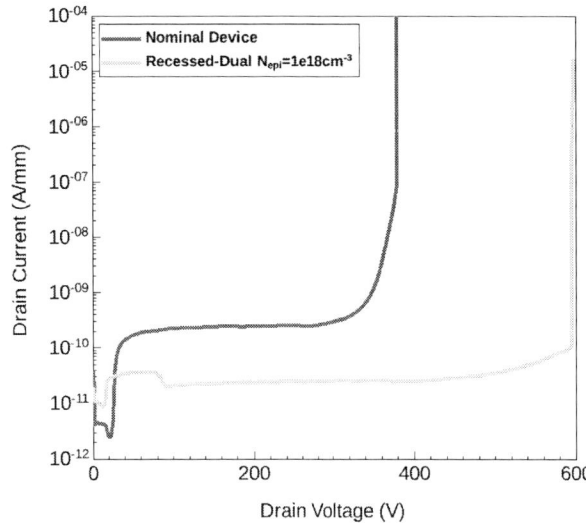

Fig. 8: BV curves of nominal and Recessed-dual devices with N_{epi}=10^{18}cm^{-3}, $L_{G,re}$ = 0.5μm, $L_{G,long}$ = 1.5μm at $V_G = V_{TH,(VD=0.01V)}$ -16.5V.

978-1-5386-2928-4/18 $31.00 © 2018 IEEE   381

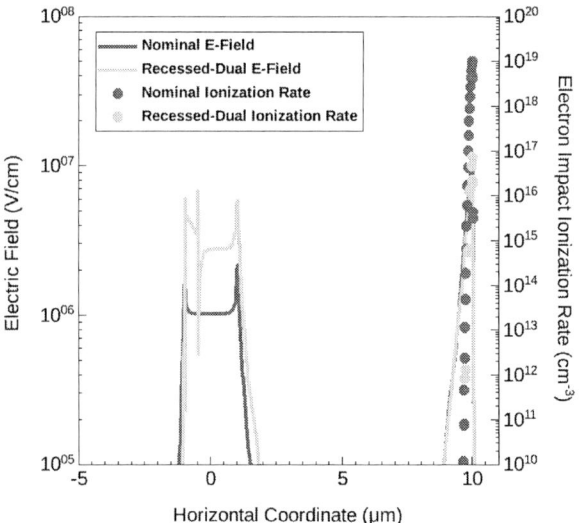

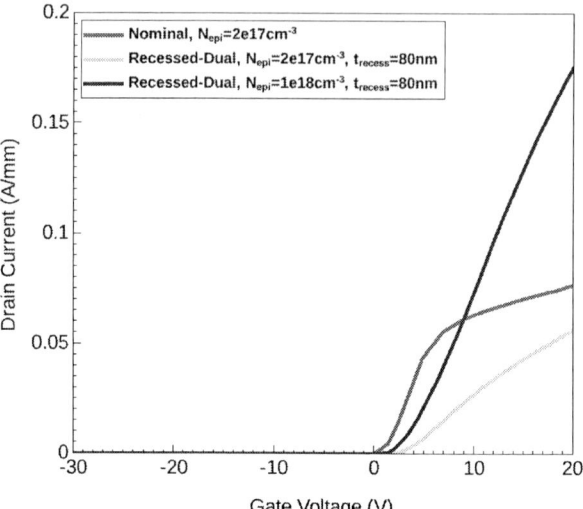

Fig. 9: Electric field and electron impact ionization rate along the horizontal Al_2O_3/Ga_2O_3 interface in Fig. 1 for nominal and recessed-dual devices with $N_{epi}=10^{18}cm^{-3}$, $L_{G,re} = 0.5\mu m$, $L_{G,long} = 1.5\mu m$ at $V_{G = VTH,(VD=0.01V)}$ -16.5V when $V_D=300V$.

Fig. 10: Simulated $I_D V_G$ curves of Ga_2O_3 FinFET. $t_{epi} = 200nm$ for all devices (same as Ref. 1). $W_{FIN}/W_{SPACE} = 0.2\mu m/0.7\mu m$ for nominal device and $0.5\mu m/0.4\mu m$ for Recessed-Dual devices. $V_D=10V$ and $L_{SG} = L_{DG} = 0.5\mu m$ are used as in Ref. 2. $W_{Fin,recess} = t_{recess} = 80nm$.

If N_{epi} is the same, for both cases, "recess-dual" device is expected to have similar BV and punch-through voltage compared to nominal device because of the similar electric field and carrier distributions at those regions in both devices at off-state [6]. Since "recess-dual" allows higher N_{Epi}, the situation becomes different. Fig. 8 shows the BV of "recess-dual" device ($N_{epi} = 10^{18}cm^{-3}$) and the BV is increased compared to the nominal device ($N_{epi} = 2\times10^{17}cm^{-3}$). This is because of the change of electric field due to higher N_{Epi} (Fig. 9). As the epitaxial layer is more heavily doped, there is smaller doping gradient at the drain region, resulting in smaller electric field at the drain region. As a result, the breakdown voltage is increased. On the other hand, the peak electric field at the gate edge at drain side becomes larger. Therefore, N_{epi} cannot be indefinitely increased without reducing the BV eventually.

Similar idea is then applied to design Ga_2O_3 FinFET (Fig. 2). Fig. 10 shows the drive current can be doubled by using "recessed-dual" with $N_{epi}=10^{18}cm^{-3}$ per unit layout area with more positive V_{TH} than the nominal device. The nominal device is similar to the experimental device in [1] with $W_{FIN}/W_{SPACE} = 0.2\mu m/0.7\mu m$. The "recessed-dual" one has $W_{FIN}/W_{SPACE} = 0.5\mu m/0.4\mu m$. W_{FIN} can be larger in "recessed-dual" while maintaining positive V_{TH} because of the recessed fin under the gate.

IV. CONCLUSIONS

Novel normally-OFF dual-gate Ga_2O_3 MOSFET and FinFET are proposed and validated in TCAD simulations with experimentally calibrated parameters. Due to the cascode-like configuration in "recessed-dual" device, the device behaves as if it has short gate length in ON-state and long gate length in OFF-state. As a result, it has higher I_{ON} and similar or even

higher BV and punch-through voltage compared to nominal device. The novel device also allows higher epitaxial doping resulting in higher I_{ON} while it is still normally-OFF. As a result, 6 times higher current is observed in dual-gate Ga_2O_3 MOSFET or one time higher current can be obtained in dual-gate Ga_2O_3 FinFET. Due to thinner epitaxial layer under the gate, the sub-threshold slope is also significantly improved in the proposed device.

REFERENCES

[1] K. D. Chabak, N. Moser, A. J. Green, D. E. Walker Jr., S. E. Tetlak, E. Heller, A. Crespo, R. Fitch, J. P. McCandless, K. Leedy, M. Baldini, G. Wagner, Z. Galazka, X. Li, and G. Jessen, "Enhancement-mode Ga_2O_3 wrap-gate fin field-effect transistors on native (100) β-Ga_2O_3 substrate with high breakdown voltage," Applied Physics Letters, 109, 213501 (2016).

[2] K. D. Chabak, J. P. McCandless, N. A. Moser, A. J. Green, K. Mahalingam, A. Crespo, N. Hendricks, B. M. Howe, S. E. Tetlak, K. Leedy, R. C. Fitch, D. Wakimoto, K. Sasaki, A. Kuramata and G. H. Jessen, "Gate-Recessed, Laterally-Scaled β-Ga_2O_3 MOSFETs with High-Voltage Enhancement-Mode Operation," 75th Device Research Conference, 2017.

[3] B. Lu, O. I. Saadat and T. Palacios, "High-Performance Integrated Dual-Gate AlGaN/GaN Enhancement-Mode Transistor," Electron Device Letters, vol. 31, no. 9, pp. 990-992, Sept. 2010.

[4] M. Higashiwaki, K. Sasaki, T. Kamimura, M. H. Wong, D. Krishnamurthy, A. Kuramata, T. Masui, and S. Yamakoshi, "Depletion-mode Ga_2O_3 MOSFETs," 71st Device Research Conference, Notre Dame, IN, 2013.

[5] T. Kamimura, K. Sasaki, M. H. Wong, D. Krishnamurthy, A. Kuramata, T. Masui, S. Yamakoshi, and M. Higashiwaki, "Band alignment and electrical properties of $Al_2O_3/β$−Ga_2O_3 heterojunctions," Applied Physics Letters, 104, 192104 (2014).

[6] H. Y. Wong, N. Braga and R. V. Mickevicius, "Normally-Off GaN HFET Based on Layout and Stress Engineering," Electron Device Letters, vol. 37, no. 12, pp. 1621-1624, Dec. 2016.

[7] Synopsys Inc., CA, USA, Sentaurus™ Device User Guide (2017).

Proceedings of the 30th International Symposium on Power Semiconductor Devices & ICs
May 13-17, 2018, Chicago, USA

Analysis of Short-Circuit Break-Down Point in 3.3 kV SiC-MOSFETs

Kazuki Tani, Jun-ichi Sakano, and Akio Shima
Center for Technology Innovation
Hitachi, Ltd.
Hitachi, Ibaraki, Japan
E-mail: kazuki.tani.ru@hitachi.com

Abstract—**To identify short-circuit (SC) failure mechanism in 3.3 kV silicon carbide metal-oxide-semiconductor field-effect transistors (SiC- MOSFETs), we analyzed the break-down point and device degradation of the 3.3 kV SiC-MOSFETs using our proposed SC test method. The proposed test method enabled the application of a SC stress larger than the SC capability of the device without break-down, which facilitates the analysis of SC degradation. Degradation of the gate oxide in the junction field-effect transistor (jFET) area in the SiC-MOSFETs resulted in deterioration of the break-down characteristics and was the starting point of the SC failure. Therefore, improved reliability of the gate oxide can increase the SC capability of the SiC-MOSFETs.**

Keywords—SiC, MOSFET, Short-circuit

I. INTRODUCTION

Silicon carbide metal-oxide-semiconductor field-effect transistors (SiC-MOSFETs) have gained much attention due to their wide gap, improved temperature capability, and high thermal conductivity. However, one of the issues in the SiC-MOSFETs is that their short-circuit (SC) capability is smaller than that in silicon insulated gate bipolar transistors (Si-IGBTs). Identification of the SC failure mechanism helps to improve SC robustness of the devices. There have been many studies about the SC failure mechanism based on the SC waveforms and simulation results [1]-[5]. Several SC failure mechanisms, such as gate structure degradation and thermal runaway, have been proposed based on destructive SC tests with 600 V and 1200 V SiC MOSFETs [3]-[5]. Meanwhile, the analysis of the degraded characteristics and the starting point of the SC break-down is also helpful to identify the SC failure mechanism in terms of improvement of the device design. However, large scale SC destruction makes it difficult to identify the starting point of the SC break-down and limits SC stress for degradation analysis.

In this paper, we propose a SC test method which prevents break-down during the SC tests and enables the application of a SC stress larger than the SC capability of the device. The tested device was analyzed in terms of electrical characteristics and identification of the starting point of the SC break-down to identify the SC failure mechanism of the 3.3 kV SiC-MOSFETs.

II. APPROACH

A. Circuit configuration and SC test method

Figure 1 shows a schematic setup of the SC tests. A 3.3 kV SiC-MOSFET and a protective Si-IGBT module were connected in series to decrease the SC break-down scale. The SiC-MOSFET was heated to 150°C by hotplate and the DC bus voltage (Vdc) was set to 1800 V prior to the SC tests. The drain-source current (Id) was measured by a current transformer. The gate-emitter voltage of the protective Si-IGBT (VgeH), gate-source voltage of the SiC-MOSFET (VgsL), and drain-source voltage of the SiC-MOSFET (Vds) were also measured to evaluate the SC waveforms. Pulse patterns used in this study are shown in table 2. Two types of protection methods were demonstrated for each purpose. A small destruction test was demonstrated to evaluate the SC capability of the SiC-MOSFETs. The SC duration was gradually increased by 200 ns from 2000 ns up to the failure point. The maximum SC duration the device was able to sustain was recorded as a SC capability. A non-destructive degradation test was demonstrated to evaluate degradation of the SiC-MOSFET by SC stress. In this test, the SC current was cut by the protective Si-IGBT, so the SiC-MOSFET remained on-state after the SC event.

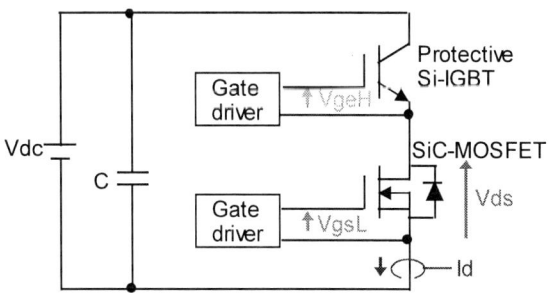

Fig. 1: Schematic circuit model of short-circuit test.

978-1-5386-2928-4/18 $31.00 © 2018 IEEE

Tab. 1: Pulse patterns of small destruction and non-destructive degradation test.

Method	Pulse pattern	Vds after SC	Break-down mode
Small destruction test		1800 V	Break-down during off-state
Non-destructive degradation test		0 V	Break-down during on-state

B. SC capability

Figures 2 (a) and (b) show SC waveforms and energy loss during the SC break-down events obtained through a small destruction tests with various Vdc conditions. Both devices broke down during the off-state after the SC cutoff, therefore we estimated that the SC capability of the 3.3 kV SiC-MOSFETs is limited by degradation of the break-down voltage characterisitcs. The SC duration was shortened with the increase of Vdc, but the energy loss was almost the same. These results suggest that the determining factor of the degradation of the break-down characteristics may be the temperature increasing with SC joule heating rather than the electrical field applied to the devices. We tested the electrical characteristics of a 3.3 kV SiC-MOSFET after the SC test with a SC duration shorter than the capability of the device. However, no charecteristic variations such as break-down voltage degradation were found. Therefore, we thought that perhaps the acceleration of degradation by application of a SC stress larger than the capability of the device without destruction would help to analyze the degraded electrical characteristics.

C. Non-destructive degradation test

Figures 3 (a) and (b) show the SC break-down waveforms obtained by both small destruction test and non-destructive degradation test. In the small destruction test, Vds remained at 1800 V after the SC events, and the SiC-MOSFET broke down during off-state after the SC cutoff. In the non-destructive degradation test, the SC stress was limited by protection time because SC current was cut by the protective Si-IGBT. Therefore, the SiC-MOSFET broke down during the SC events. As a result, the duration of the SC stress increased to 8.5 μs and energy increased to 14.0 J/cm^2 in the non-destructive degradation test, compared to 5.9 μs duration and 10.7 J/cm^2 of critical energy of the SC capability of the SiC-MOSFET. For the purpose of analyzing the degradation and starting point of the break-down, a non-destructive degradation test with the SC durationof 8.1 μs was demonstrated. The waveforms are shown in Fig. 4. Although the SC energy of the test was 13.8 J/cm^2 larger than the critical energy of 10.7 J/cm^2, the device did not break because the Vds was reduced to 0 V after the SC cutoff.

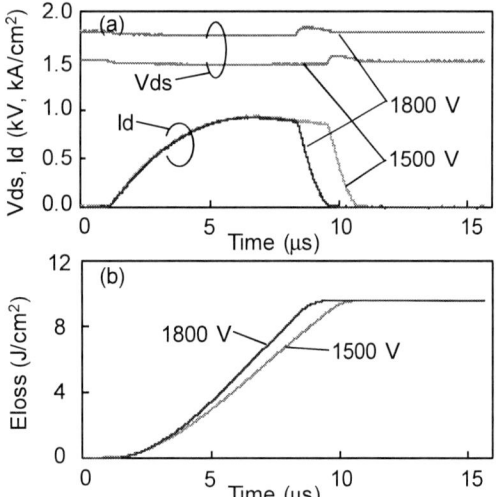

Fig. 2: (a) Waveforms and (b) energy loss in the short-circuit break-down.

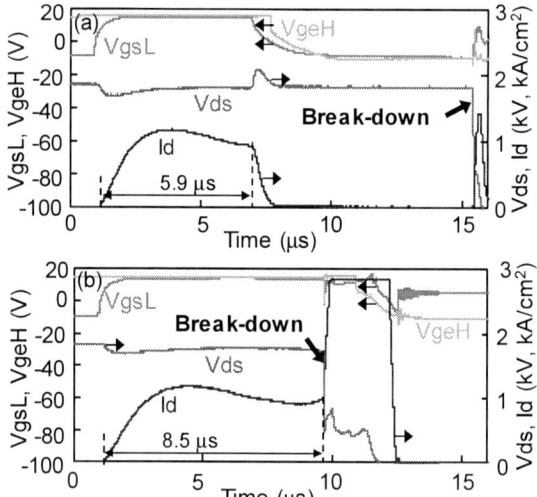

Fig. 3: Short-circuit break-down waveforms of (a) small destruction test and (b) non-destructive degradation test.

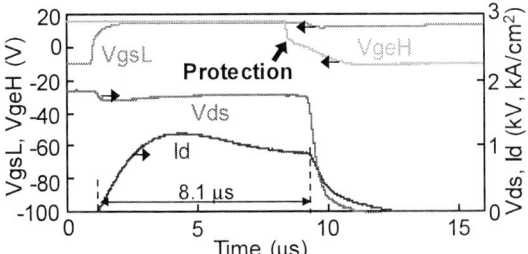

Fig. 4: (a) Short-circutit waveforms of non-destructive degradation test.

III. RESULTS AND DISCUSSION

To identify the SC failure mechanism of the 3.3 kV SiC-MOSFETs, we analyzed the device after the non-destructive degradation test with three types of evaluation composed of electrical characteristic evaluation, identification of degraded area in the device, and identification of damaged materials. First, we analyzed the electrical characteristic variations by the SC stress. Fig. 5 shows the break-down voltage characteristics of the 3.3 kV SiC-MOSFET before and after the non-destructive degradation test. The Vgs was set to 0 V and Vds was increased from 0 V to 3 kV in the break-down voltage characteristic test. The maximum current in this test was limited to less than 100 µA by the measurement system to limit the damage of the device. Both drain-gate leakage current (Ig) and drain-source leakage current (Id) of the device after the non-destructive degradation test were slightly larger than before the SC test, and the device after the test broke down at 2.8 kV, unlike the device before the SC test. These results proved that the SC stress larger than the capability of the device applied in the non-destructive degradation test accelerated the SC degradation of the device. The break-down voltage characteristic degradation caused the SC break-down during off-state after the SC events. It means that the break-down at 2.8 kV in this test was equivalent to the reprodction test of the SC break-down during off-state after SC events with

a 100 µA current limitation. Also, the increased gate leakage current suggested that the degraded point was related to the gate oxide.

Second, we identified the area with maximum degradation by the SC stress, which would be the starting point of the SC break-down, in the device after the break-down voltage characeristics test with emission microscopy (EMS) observation. The Vds was set to 0 V and the drain-gate voltage (Vdg) and the drain-gate (or source) leakage current increased to 15 V and 800 nA in the EMS observation. Fig. 6 shows an EMS image of the maximum drain-gate leakage area of the device. The maximum leakage area was found in the junction field-effect transistor (jFET) region in the active area. It was not the maximum electrical field region, which is the region farthest from the source contact, but the region where the two channels faced each other and where the SC current concentration was the highest. This result suggests that the break-down voltage characteristics were degraded by joule heating during the SC events rather than by electrical field applied to the device, and this is consistent with the SC break-down characteristics shown in Fig. 2 (a), (b).

Finally, we identified the materials damaged in the SC stress by cross-sectional observation. The most degraded area found by the EMS observation is shown in Fig. 6. The cross section of the degraded point was identified with slice-and-view using a focused-ion-beam (FIB) system and a scanning electron microscope system. A cross-sectional scanning transmission electron microscope (STEM) image of the most degraded point in the device is shown in Fig. 7. Voids in the gate oxide layer and in the gate poly-Si, as well as on the surface of SiC layer were found. The gate oxide degradation which resulted in voids in the gate oxide layer should be critical to the degraded break-down voltage characteristics. These voids and damages may have been generated in the break-down voltage characteristics test, which was equivalent to the reproduction test of the SC break-down during off-state after the SC events with 100 µA current limitation. Therefore the broken gate oxide shown in Fig. 7 should be the starting point of the SC break-down during off- state after the SC

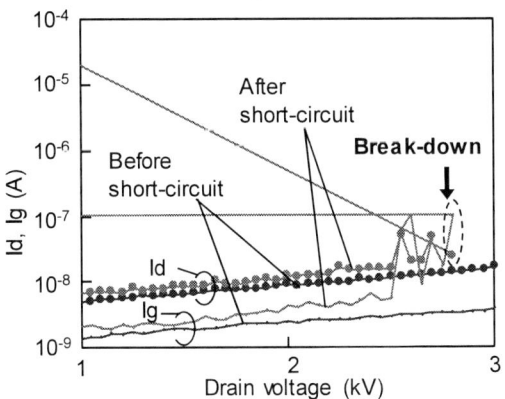

Fig. 5: Break-down voltage characteristics before and after short-circuit test.

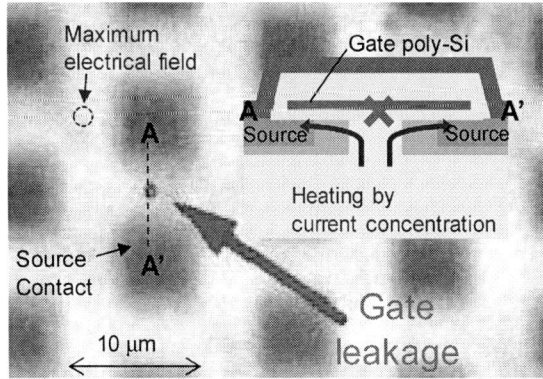

Fig. 6: An emission microscopy image of break-down point.

978-1-5386-2928-4/18 $31.00 © 2018 IEEE

events. In conclusion, joule heating during the SC degrades the gate oxide in the jFET region and the degraded break-down voltage characteristics causes the SC break-down.

IV. CONCLUSION

We proposed a non-destructive degradation test method, which enables the application of a SC stress larger than the capability of the device without destruction, to analyze the SC failure mechanism of 3.3 kV SiC-MOSFETs. The tested device was evaluated in terms of degraded electrical characteristics, identification of degraded area in the device, and identification of damaged materials. We concluded that the degradation of the gate oxide, which was caused by the joule heating during the SC, is critical to the SC capability of the SiC-MOSFETs and that the thermal management of the gate oxide can increase the SC capability of the SiC-MOSFETs.

REFERENCES

[1] G. Romano, L. Maresca, M. Riccio, V. d'Alessandro, G. Breglio, A. Irace, A. Fayyaz, and A. Castellazzi, "Short-circuit Failure Mechanism of SiC Power MOSFETs," in proceedings of the 27th International Symposium on Power Semiconductor Devices (ISPSD), pp. 345-348, 2015.

[2] A. März, T. Bertelshofer, R. Horff, M. Helsper, and M. M. Bakran, "Explaining the short-circuit capability of SiC MOSFETs by using a simple thermal transmission-line model," in proceedings of the 18th European Conference on Power Electronics and Application (EPE'16 ECCE Europe), pp. 1-10, 2016.

[3] G. Romano, A. Fayyaz, M. Eiccio, L. Maresca, G. Breglio, A. Castellazzi, and A. Irace, "A Comprehensive Study of Short-Circuit Ruggedness of Silicon Carbide Power MOSFETs," IEEE J. Emerging and Selected Topics in Power Electron., vol. 4, no. 3, pp. 978-987, 2016.

[4] Z. Wang, X. Shi, L. M. Tolbert, F. Wang, Z. Liang, D. Costinet, and B. J. Blalock, "Temperature-Dependent Short-Circuit Capability of Silicon Carbide Power MOSFETs," IEEE Trans. Power Electron. Vol. 31, no. 2, pp. 1555-1566, 2016..

[5] R. Tanaka, Y. Kagawa, N. Fujiwara, K. Sugawara, Y. Fukui, N. Miura, M. Imaizumi, and S. Yamakawa, "Impact of Groundign the Bottom Oxide Protection Layer on the Short-Circuit Ruggedness of 4H-SiC Trench MOSFETs", in proceedings of the 26th International Symposium on Power Semiconductor Devices (ISPSD), pp.75-78, 2014.

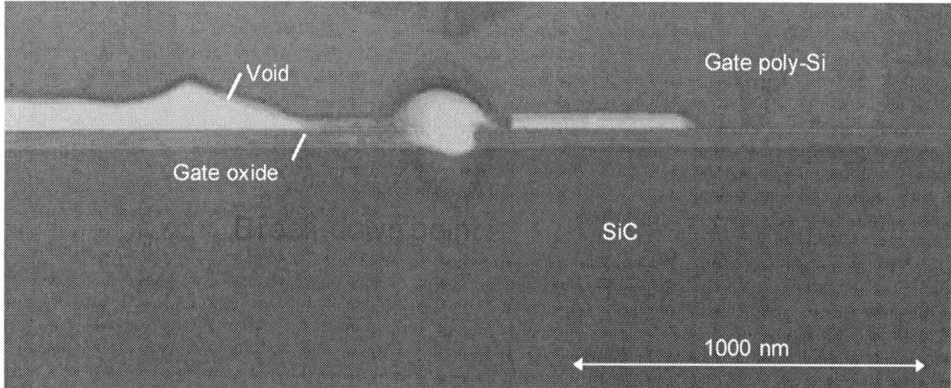

Fig. 7: A cross sectional scanning transmission electron microscope image of break-down point.

Proceedings of the 30th International Symposium on Power Semiconductor Devices & ICs
May 13-17, 2018, Chicago, USA

Electrical Characterization of 1.2kV SiC MOSFET at Extremely High Junction Temperature

Jiahui Sun, Hongyi Xu, Shu Yang and Kuang Sheng[*]

College of Electrical Engineering, Zhejiang University, Hangzhou, China

[*]shengk@zju.edu.cn

Abstract— **Threshold voltage and channel mobility of 1.2kV SiC MOSFET at high junction temperature up to 700°C have been extracted and analyzed for the first time, by virtue of a specially designed short-circuit measurement technique we developed. During the short-circuit operation, the junction temperature of the SiC MOSFET can rise significantly within a few microseconds, which can be extracted based on the short-circuit waveforms and electro-thermal calculations. The SiC MOSFET investigated in this work can maintain normally-off operation at a junction temperature up to 700°C. Furthermore, the underlying mechanisms of the temperature dependence of the threshold voltage and channel mobility are also discussed.**

Keywords— High temperature characterization, SiC power MOSFET, threshold voltage, channel mobility

I. Introduction

Owing to the wide bandgap, high electron saturation velocity and thermal conductivity of SiC, SiC MOSFET can deliver superior performance for high power and high temperature applications[1][2]. However, the robustness issues have been concerns toward the widespread adoption of SiC MOSFETs[3][4]. A SC across the load can result in a direct connection of the power device to the high bus bar voltage which could cause device failure within several microseconds. Our previous work reveals that the SC capability of SiC MOSFET is significantly weaker than that of Si IGBT with 80% shorter short circuit withstand time (SCWT) under single pulse SC test condition[3]. Such short SCWT requires much shorter response time of SC protection circuits[5]. Hence, it is utmost significant to investigate the SC failure mechanisms, which could guide the optimization of device structure and process to enhance the SC capability of SiC MOSFET. The SC failure mechanisms of SiC MOSFTEs have been studied using TCAD simulations[6]. However, most device models and parameters within high temperature range (>500°C) are still missing to date. Given the fact that the junction temperatures (T_j) of SiC MOSFET could exceed 1000°C under the SC condition[3][7], it is essential to characterize SiC MOSFET systematically at high T_j. However, the characterization temperature for most of the commercially available SiC MOSFETs is limited by the packaging materials (i.e., resin and lead-free solder) with a melting point of ~220°C[8]. With high temperature packaging materials (i.e., sintered silver and high temperature resin), the characteristics of a 1.2 kV SiC MOSFET at T_j ~ 380°C have been investigated[9]. So far, the electrical characteristics of SiC vertical power devices have been extracted at a T_j up to 407°C[10].

In this work, a new SC characterization technique is

developed so that the SiC MOSFET can be locally heated up to 700°C within several microseconds. The electrical characteristics of the device under test (DUT) can be extracted in a wide T_j range. Moreover, the temperature-dependence of threshold voltage (V_{th}) and channel mobility are extracted and analyzed.

II. Characterization Platform and Electro-Thermal Simulation

A. Experimental setup

A 1.2kV/80mΩ SiC MOSFET[11] in standard TO-247 is used as DUT which can be self-heated under SC condition. As

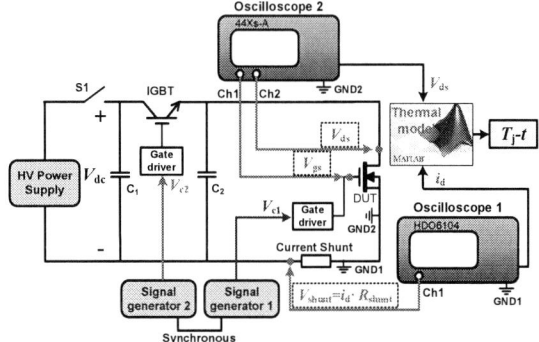

Fig. 1. Schematic diagram of dynamic characterization platform and data analysis, and the current shunt is used to measure SC drain current.

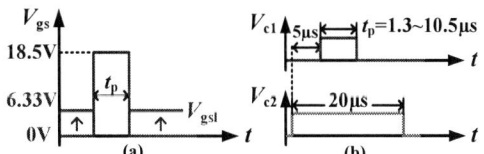

Fig. 2. (a) Gate voltage waveforms of DUT at different low-level drive voltages. (b) Gate control signal of DUT and IGBT.

shown in Fig. 1, the characterization platform is primarily composed of a SC test circuit, a high voltage (HV) power supply, two synchronized signal generators and two oscilloscopes. The SC test circuit is composed of a capacitor bank (C_1 and C_2), a current shunt, an IGBT and DUT and their gate drivers. C_1 is a group of energy storage capacitors and C_2 provides rapidly changing current in the main circuit. The IGBT functions as a switch which controls the power supply of the DUT. As shown in Fig. 2, the gate driver of the DUT can provide positive V_{gsl} varying from 0 to 6.33V when the gate control signal (V_{c1}) generated from the signal generator 1 is

978-1-5386-2928-4/18 $31.00 © 2018 IEEE   387

low. SC waveforms of the SiC MOSFET were captured and recorded by two oscilloscopes (Fig. 1), which are isolated by transformers (Fig. 3) to avoid the interference among the three captured signals arising from parasitic inductance and resistance between the ground nodes of the shunt and DUT. One is used to measure the voltage of the shunt ($V_{shunt}=i_d \cdot R_{shunt}$) while the other monitors V_{ds} and V_{gs}. Time-resolved temperature distribution mapping can be realized based on the experimental data in conjunction with the electro-thermal simulation, as shown in Fig. 1.

At the beginning of each measurement, the high voltage switch S1 is turned on and C_1 is charged to ~300V. After the voltage on C_1 is stable, S1 is turned off. Then, the IGBT is turned on to charge C_2. After 5 μs, the DUT is switched on

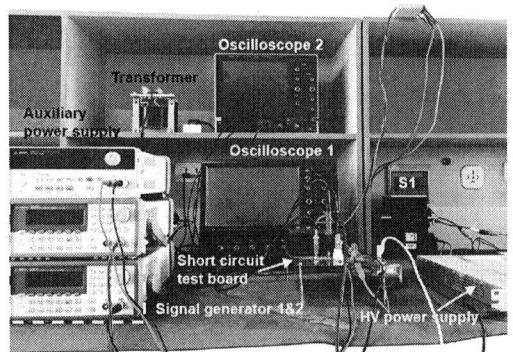

Fig. 3. Experimental setup of the dynamic characterization platform.

(Fig. 2). The gate pulse width (t_p) of the DUT is set between 1.3 μs and 10.5 μs in order to reach different maximum junction temperatures (T_{jmax}'s) at the end of each SC pulse. After the gate control voltage V_{c1} falls down, i_d at V_{gs1} and a certain T_j range can be obtained.

B. Electro-thermal simulation

Heat will be transferred in both lateral and vertical directions in the DUT during SC operation. In this work, lateral

heat conduction is neglected because vertical heat transfer dominates. Heat transfer in the solder and copper is also

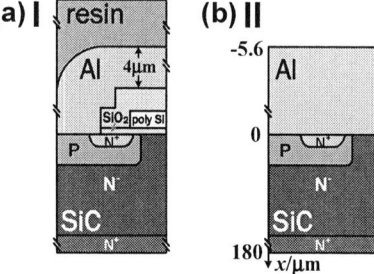

Fig. 4. Cross-sectional view of the (a) real device structure and (b) simplified structure used in the 1D (vertical) electro-thermal simulation.

neglected as heat will not be transferred to them within 15 μs. 1D transient electro-thermal simulations have been carried out in the vertical direction. Fig. 4 shows the cross-sectional view of the real structure of DUT and the simplified one used in the electro-thermal simulation. Based on the simplified structure II and the SC experimental data, the heat diffusion equation is solved using finite difference method in Matlab[3].

Since the temperature of Al above SiC can reach its melting point (T_{melt}), the enthalpy of fusion (ΔH_{fus}) at the melting point has been considered. All the structure parameters and thermal properties used in the simulation are listed in Table I.

The waveforms and T_j-t plot in Fig. 5 illustrate the drain current extraction at a certain T_j. During the SC pulse, T_j rises up to ~750°C due to the excessive heat generated at high i_d ~ 215A and high V_{ds} ~ 290V. When V_{gs} falls down from 18V and stabilizes at 3.5V, the drain current of 9.2A at T_j=700°C and V_{ds}=290V can be captured. Fig. 5(c) demonstrates that heat is only transferred in a depth of ~120μm in SiC within 15μs, which confirms that solder and copper can be neglected in the thermal simulation. By changing V_{gs1} and t_p, transfer IV curves of the DUT at different T_j can be obtained.

TABLE I. STRUCTURE PARAMETERS AND THERMAL PROPERTIES USED IN THE ELECTRO-THERMAL SIMULATIONS

Symbol	Parameters	Values or models	Unit
λ_{SiC}	Thermal conductivity of SiC[12]	$(-3\times10^{-4}+1.05\times10^{-5}T)^{-1}$ (0<T<1800K)	W/m·K
c_{SiC}	Specific heat of SiC[12]	$925.65+0.3772 \cdot T - 7.9259\times10^{-5}T^2 - 3.1946\times10^7/T^2$ (200<T<2400K)	J/kg·K
ρ_{SiC}	SiC density	3.21	g/cm³
A	DUT active area	6.344 (Area of all cells in DUT)	mm²
x_j	Junction depth	0.6	μm
N_d	Doping concentration of N⁻ epilayer	6.74×10^{15} (Extracted from C_{ds}-V_{ds} curve)	cm⁻³
t_{SiC}	Thickness of SiC	180	μm
t_{Al}	Thickness of Al above SiC	4	μm
$k_{s, SiC}$	Relative dielectric constant of SiC	9.7	-
λ_{Al}	Thermal conductivity of Al[13]	$2.943\times e^{(-0.0003691 \cdot T)} - 1.379\times e^{(-0.005605 \cdot T)}$ $200K < T < 933.32K$	W/m·K
		$1.148\times sin(5.072\times10^{-4} \cdot T + 0.438)$ $933.32K < T < 2200K$	
c_{Al}	Specific heat of Al[14]	$1.041\times10^3 - 0.2 \cdot T + 3.173\times10^{-4} \cdot T^2 + 1.270\times10^{-7} \cdot T^3 - 1.028\times10^7 T^{-2}$ (298K<T<933.32K)	J/kg·K
		1.177 (933.32K <T<2790K)	
ρ_{Al}	Al density[15]	2.70	g/cm³
T_{melt}	Melting point of Al[15]	933.32	K
ΔH_{fus}	Enthalpy of fusion at the melting point[15]	3.97×10^5	J/kg

978-1-5386-2928-4/18 $31.00 © 2018 IEEE

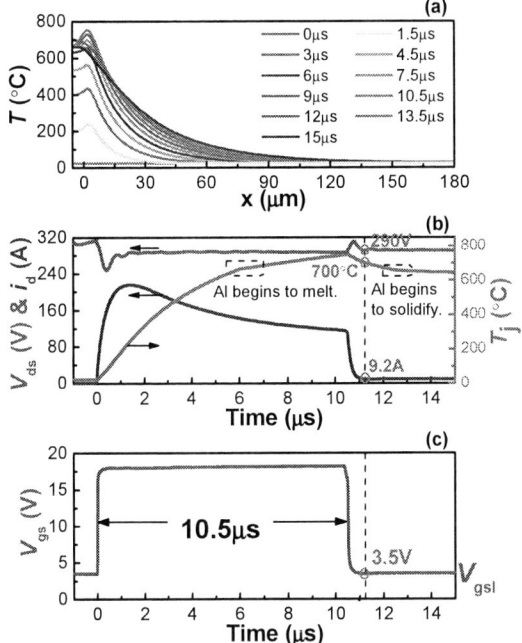

Fig. 5. Extraction of drain current at high T_j: (a) i_d-t, V_{ds}-t waveforms and simulated T_j-t, (b) V_{gs}-t waveform and (c) temperature distribution in x direction. Test conditions: t_p=10.5μs, V_{gsl}=3.5V, V_{dc}=300V and T_c=25°C. V_{ds} is less than 300V after DUT is turned off as SC energy dissipation causes voltage drop of C_1 while part of the voltage in C_1 drops across the IGBT.

III. EXPERIMENTAL RESULTS AND ANALYSIS

By implementing the characterization technique presented in Section II, drain currents at T_j=120°C~700°C and V_{gs}=0.91~6.33V can be extracted, as shown in Fig. 6. The transfer curves shift toward the negative direction with increasing T_j, suggesting a negative temperature coefficient (NTC) of V_{th}. Please note that V_{th} shows less temperature dependence at T_j>340°C. The saturation current is given by

$$I_{dsat} = \frac{Z\mu_n(T_j)\varepsilon_{ox}}{2Lt_{ox}}(V_{gs}-V_{th}(T_j))^2 \quad (1)$$

where Z is the channel width, μ_n is the channel mobility, ε_{ox} is dielectric constant, L is the channel length, and t_{ox} is the oxide thickness. In (1), only the channel mobility and the threshold voltage are temperature-dependent. As shown in Fig. 6(b), when $V_{gs}\leq3$V, the drain current rises with higher T_j, which is dominated by the NTC of V_{th}. When $V_{gs}\geq3.5$V, the drain current shows a peak value with increasing T_j, arising from the temperature-dependence of the channel mobility (Fig. 7). The average channel mobility ($\mu_{ch,av}$) in V_{gs} range of 1V~6.33V is extracted by using (1). $\mu_{ch,av}$ shows a peak value of 22.8cm²/Vs at 190°C and subsequently decreases to 12.6cm²/Vs at 700°C.

Using (1), V_{th} can also be extracted as a function of T_j (120°C~700°C), as shown in Fig. 8. The SiC MOSFET can maintain normally-off operation with positive V_{th} even at T_j up to 700°C. The V_{th}-T_j curve measured by a High Power Curve

Tracer (Tektronix 371A) (-160°C~200°C) is also shown for reference. When measured by Tektronix 371A, the DUT was heated in a hot oven and cooled down using liquid nitrogen.

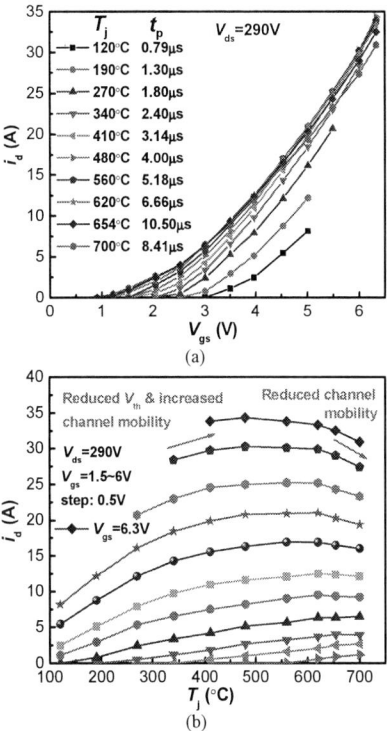

Fig. 6. (a) Transfer IV characteristics at different junction temperatures and (b) temperature-dependent drain current at different V_{gs}. Different $V_{gs,max}$ at T_j of 140°C, 290°C and 410°C is limited by self-heating. The maximum T_j is constrained by current degradation of DUT.

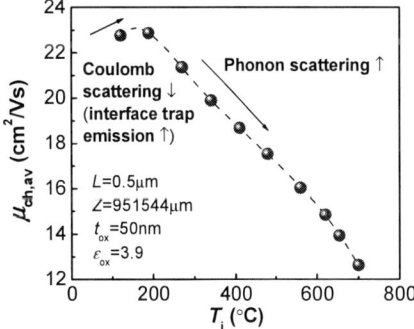

Fig. 7. Temperature dependent average channel mobility in V_{gs} range of 0.91V~6.33V (solid circles in the figure).

According to the difference between the decreasing rate of the measured V_{th}, the V_{th}-T_j curve can be divided into two regions. In region I, the decrease of the measured V_{th} is more significant due to thermally enhanced electron emission from interface traps near interface near the conduction band. In region II, the extracted V_{th} decreases with temperature at a much lower rate which is dominated by the reduction of bulk potential.

978-1-5386-2928-4/18 $31.00 © 2018 IEEE

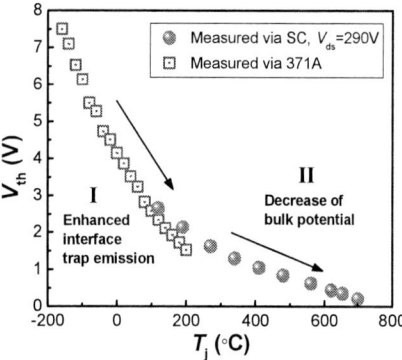

Fig. 8. Temperature dependent threshold voltage extracted via SC characterization technique (120°C~700°C) and threshold voltage measured using Tektronix 371A (-160°C~200°C).

IV. CONCLUSION

Threshold voltage and channel mobility of the SiC MOSFET at high junction temperature up to 700°C have been extracted for the first time, by using a dynamic characterization technique combined with electro-thermal simulation developed in this work. The SiC MOSFET can maintain normally-off operation even at a junction temperature up to 700°C. The V_{th}-T_j characteristics in SiC MOSFETs have been analyzed and attributed to the distinct emission dynamics of interface traps within different T_j ranges.

ACKNOWLEDGMENT

This work was supported by the National Key Research and Development Program of China (No. 2016YFB0100603) & National Natural Science Foundation of China (No. 51577169).

REFERENCES

[1] X. Wu, S. Cheng, Q. Xiao, and K. Sheng, "A 3600 V/80 A series-parallel connected silicon carbide MOSFETs module with single external gate driver," *IEEE Trans. Power Electronics*, vol. 29, no. 5, pp. 2296–2306, May 2014.

[2] Z. Wang, X. Shi, L. Tolbert, F. Wang, Z. Liang, D. Costinett, and B. Blalock, "A high temperature silicon carbide MOSFET power module with integrated silicon-on-insulator based gate drive," *IEEE Trans. Power Electron.*, vol. 30, no. 3, pp. 1432–1445, Mar. 2015.

[3] J. Sun, H. Xu, X. Wu, S. Yang, Q. Guo and K. Sheng, "Short circuit capability and high temperature channel mobility of SiC MOSFETs," in *Proc. IEEE 29th Int. Symp. Power Semicond. Devices IC's (ISPSD)*, May 2017, pp. 399-402.

[4] G. Romano, A. Fayyaz, M. Riccio, L. Maresca, G. Breglio, A. Castellazzi, and A. Irace, "A comprehensive study of short-circuit ruggedness of silicon carbide power MOSFETs," *IEEE J. Emerging and Selected Topics in Power Electronics*, vol. 4, no. 3, pp. 978-987, Sept. 2016.

[5] Z. Wang, X. Shi, Y. Xue, L. M. Tolbert, F. Wang and B. J. Blalock, "Design and performance evaluation of overcurrent protection schemes for silicon carbide (SiC) power MOSFETs," *IEEE Trans. Industrial Electronics*, vol. 61, no. 10, pp. 5570 – 5581, Oct. 2014.

[6] G. Romano, L. Maresca, M. Riccio, V. d'Alessandro, G. Breglio, A. Irace, A. Fayyaz, and A. Castellazzi, "Short-circuit failure mechanism of SiC power MOSFETs," in *Proc. IEEE 27th Int. Symp. Power Semicond. Devices IC's (ISPSD)*, May 2015, pp. 345-348.

[7] M. Namai, J. An, H. Yano and N. Iwamuro, "Experimental and numerical demonstration and optimized methods for SiC trench MOSFET short-circuit capability," in *Proc. IEEE 29th Int. Symp. Power Semicond. Devices IC's (ISPSD)*, 2017, pp. 363-366.

[8] I. E. Anderson, F. G. Yost, J. F. Smith, C. M. Miller and R. L. Terpstra, "Pb-free Sn-Ag-Cu ternary eutectic solder," U.S. Patent 5 527 628, Jun. 18, 1996.

[9] Y. Nanen, M. Aketa, Y. Nakano, H. Asahara and T. Nakamura1. "Electrical characterization of 1.2 kV-class SiC MOSFET at high temperature up to 380°C," *Materials Science Forum*, vol. 858, pp. 885-888, 2016.

[10] K.V. Vassilevski, I.P. Nikitina, N.G. Wright, A.B. Horsfall, A.G. O'Neill, C.M. Johnson, "Device processing and characterisation of high temperature silicon carbide Schottky diodes," *Microelectronic Engineering*, vol. 83, no. 1, pp. 150-154, Jan. 2006.

[11] *C2M0080120D*, Cree Inc., Durham, NC, USA, 2014. [Online]. Available: http://www.wolfspeed.com/media/downloads/167/C2M0080120D.pdf

[12] L. L. Snead, T. Nozawa, Y. Katoh, T. Byun, S. Kondo, D. A. Petti, "Handbook of SiC properties for fuel performance modeling," *Journal of Nuclear Materials*, vol. 371, no. 1-3, pp. 329-377, Sep. 2007.

[13] C. Y. Ho, R. W. Powell, and P. E. Liley, "Thermal conductivity of the elements," *J. Phys. Chem. Ref. Data*, vol. 1, no. pp. 279-421, 1972.

[14] M. W. Chase Jr, "NIST-JANAF themochemical tables", 4th ed., *J. Phys. Chem. Ref. Data*, Monograph 9, pp. 1-1951, 1998.

[15] W. M. Haynes, David R. Lide, Thomas J. Bruno, "Properties of solids," in *CRC Handbook of Chemistry and Physics*. 95th ed. Boca Raton, FL, USA: Taylor & Francis Group, 2014, pp. 12-1-12-236.

Proceedings of the 30th International Symposium on Power Semiconductor Devices & ICs
May 13-17, 2018, Chicago, USA

Methodology for Enhanced Short-Circuit Capability of SiC MOSFETs

Junjie An, Masaki Namai, Hiroshi Yano,
Noriyuki Iwamuro

Graduate School of Pure and Applied Sciences
University of Tsukuba
Tsukuba, Japan
e-mail: iwamuro.noriyuki.fb@u.tsukuba.ac.jp

Yusuke Kobayashi*, Shinsuke Harada

National Institute of Advanced Industrial Science and
Technology (AIST)
Tsukuba, Japan

Abstract—It has been demonstrated in many power transform systems that silicon carbide (SiC) MOSFETs with low power loss, high thermal capability are superior candidate to replace the status of silicon power devices. Short-circuit capability is one of most essential robustness issues for the reliability of SiC MOSFETs, which should be carefully considered. In this study, the special failure mechanisms of SiC MOSFETs have been investigated and analyzed by the experimental and numerical analyses during the short-circuit transient. In addition, a robustness enhanced SiC MOSFET with the thick gate oxide of 85nm has been fabricated and been showed the high short-circuit capability even with higher short-circuit energy and longer short-circuit transient.

Keywords—SiC MOSFETs; Enhanced short-circuit capability; Gate switch off voltage; Thick gate oxide layer

I. INTRODUCTION

Owing to the excellent properties of the wide bandgap, the high breakdown electric filed, and the high thermal capability, SiC can be the superior semiconductor material for the high power and reliability applications [1]. Nowadays, with rapid development of fabrication process and device structure technology, the SiC MOSFETs including planar and trench structures are commercial available. Short-circuit ruggedness is one of the most attractive issue for SiC MOSFET due to high requirement of stability for the whole application system.

Recently, based on the experimental analysis and simulation method, various papers related to short-circuit capability have been reported for the SiC MOSFETs [2]-[5]. Thermal runaway is a well-known mechanism of the short-circuit failure for the SiC MOSFETs as well as the Si IGBTs. In addition, the gate voltage degradation and gate oxide layer damage always occur during the short-circuit transient or after device switch-off. However, little of them exhibit the gate switch-off voltage dependence of the short-circuit capability. There are still few research works which report the methodology for the enhanced short-circuit capability of SiC MOSFET.

An investigation of short-circuit failure with different DC voltages are very important because the DC input voltage would be frequently changed to control the output AC motor

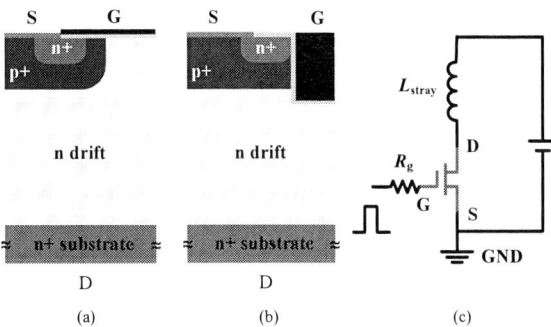

(a)　　　　(b)　　　　(c)

Fig. 1. (a) The cross section of the planar SiC MOSFET, (b) the trench SiC MOSFET, and (c) the test circuit of short-circuit for measurement and simulation.

speed. In this study, the special failure mechanisms of the SiC MOSFETs during the short-circuit state and robustness-enhanced structure/operation are investigated and demonstrated by the experimental and numerical analyses.

II. EXPERIMENTAL AND SIMULATION RESULTS

A. Gate voltage dependence of short-ciruit capability

The devices under test used at first experiment are 1200 V SiC planar MOSFETs. Fig. 1 shows the structures and the test circuit for the tested SiC MOSFET, respectively.

Fig.2 gives the short-circuit waveforms for the SiC MOSFET with the DC voltage of 600V. In order to investigate the gate switch-off voltages (V_{g_off}) dependence of short-circuit capability, the values are set to be -5V and -10V, respectively. It is obvious that the drain currents in Fig. 2(a) increases promptly and both two devices enter from the linear region to the active region until they achieve their saturation current at the initial pulse time. After that, the drain current decreases from a peak current of 110 A to 54.6 A due to the degradation of carrier mobility. The slope of drain current become horizontal at the end of short-circuit transient. According to the experimental results, the device fails at 9.0μs with the V_{g_off} of -5 V in spite of the lower drain surge voltage of 734 V. On the contrary, the device with the V_{g_off} of -10 V can be safely

978-1-5386-2928-4/18 $31.00 © 2018 IEEE　　　391

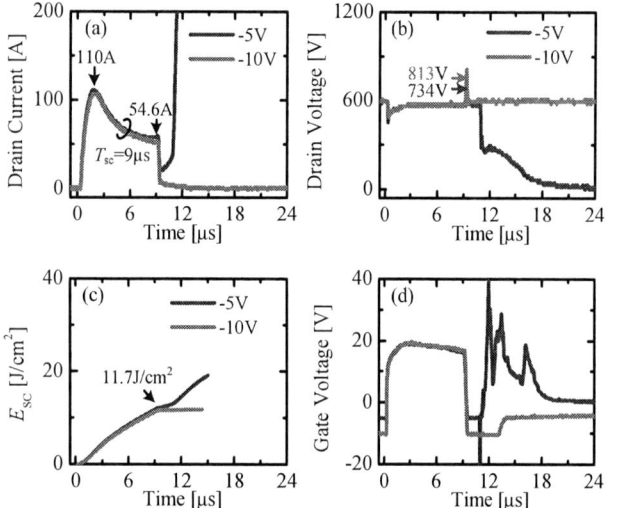

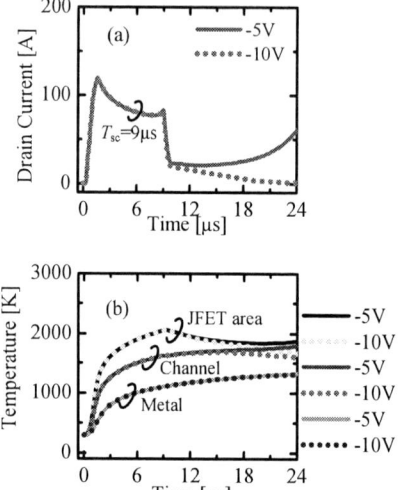

Fig. 2. Experimental results of short-circuit waveforms for the 1200V SiC MOSFET at room temperature. (a) the drain current, (b) the drain voltage, (c) the short-circuit energy, and (d) the gate voltage as functions of the short-circuit transient with the DC voltage of 600V and the gate turn-off voltage of -5V and -10V.

Fig. 3. Simulated drain current and lattice temperature as functions of the short-circuit transient with the DC voltage of 600V and the gate turn-off voltage of -5V and -10V at room temperature.

switched off. It is obvious that the larger surge voltage does not affect short-circuit capability if this voltage is not beyond the breakdown voltage of the device. The calculated short-circuit energy are around 11.7 J/cm² during the short-circuit period for both devices. Unfortunately, despite the thermal runaway can be avoided for the SiC MOSFET with the V_{g_off} of -10 V, the high lattice temperature associated with high electric field leads to the degradation of gate oxide layer. Therefore, the reliability issue for gate oxide layer during the short-circuit transient should be carefully considered.

As reported in [6], two failure mechanisms are recognized for the short-circuit failure. One is that increasing of leakage current produced by the high lattice temperature triggers the on-state of the parasitic of *npn* transistor during the relatively transitory short-circuit period. The second one is the melting of metal or the degradation of SiO₂ during the relative long short-circuit transient, resulting in the gate shorted between the gate and source terminals. The lattice temperature is the main point

that takes responsibility for those failures.

In order to further investigate the relationship between the gate switch-off voltages and short-circuit capability, the systematic numerical simulations by using Sentaurus TCAD are implemented. Fig. 3 exhibits the simulation results of the drain current and the lattice temperature as functions of simulation time for the SiC MOSFET with the V_{g_off} of -5 V and -10 V, respectively. It should be noted that the simulation waveforms are in good agreement with the experimental ones. According to the simulation results, a larger negative gate voltage can effectively reduce the tail current for the SiC MOSFET after the device switches off. The maximum lattice temperature locates at the JFET area and is over 2000 K. In addition, the maximum lattice temperature nearby channel region increases to 1750 K at the device switch-off point. However, the lattice temperature in metal is near to about 1000 K, which can still permit the safe operation for the devices.

Fig.4 shows the calculated results of the electron current distribution during different short-circuit point. It is clear that

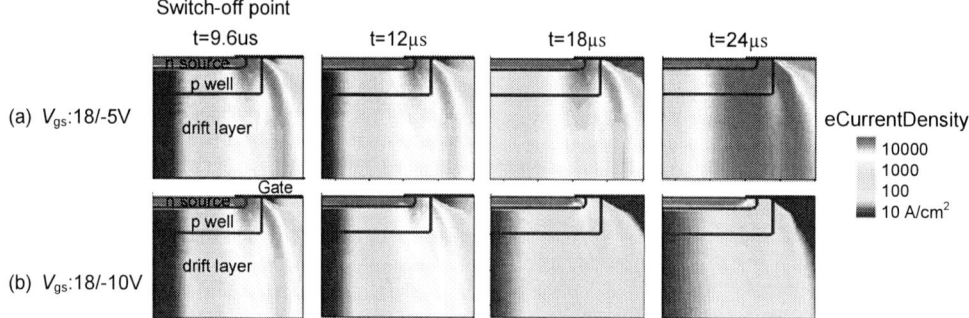

Fig. 4. Simulation results of the electron current density distribution with different short-circuit period.

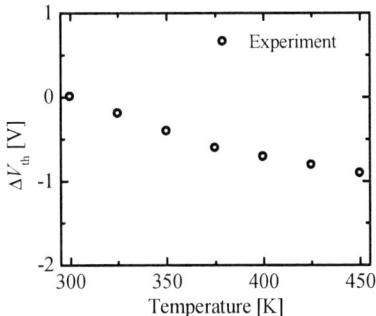

Fig. 5. Experimental results of temperature dependence of threshold voltage decrease. The minimum threshold voltage of the device is 2 V at room temperature. ΔV_{th} is equal to -7 V when the lattice temperature increases to 1750 K.

the electron current keeps flowing through the MOS channel after the device switches off for both devices with different gate switch-off voltages. It can be found that the current gradually decreases when the V_{g_off} of -10 V is implied in the SiC MOSFET. However, the tail current still exists in the MOS channel and does not change anymore for the device with the V_{g_off} of -5 V. Finally, the larger tail current triggers on the parasitic *npn* transistor for the device with the V_{g_off} of -5 V.

Furthermore, Fig. 5 shows the experimental and fitting results of the temperature dependence of the threshold voltage decrease (ΔV_{th}). The minimum threshold voltage of the tested device is 2 V at 298 K. The ΔV_{th} is equal to -7 V when the temperature increases to 1750 K. Combined with these results, it can be concluded that the SiC MOSFET is more likely to be

destroyed by the uncontrollable gate switch-off. Therefore, the larger V_{g_off} can effectively switch off the device and thus improve the short-circuit capability by reducing the probability of channel normally-on induced by the higher lattice temperature.

B. Dependence of short-ciruit capability on gate oxide layer thickness

It is reported that not only the planar MOSFET but also the trench MOSFET with thin gate oxide layer suffers from both high electric field and temperature during the short-circuit transient, leading to the gate voltage degradation during short-circuit transient and the gate oxide layer rapture after the device switches off [6]-[7].

Fig.6 exhibits the comparison of short-circuit waveforms between the 1200V SiC trench MOSFETs with the thin (50 nm) and thick (85 nm) gate oxide layers. The electric field relaxation structures at the trench bottom are applied to the MOSFETs. $R_{on,sp}$ are 2.9 mΩcm^2 (50nm) and 3.88 mΩcm^2 (85nm) at the gate voltage (V_g) of 20 V, respectively. In order to get a long short-circuit transient, the DC voltage used in experiment is only 400 V. As shown in Fig. 6, the short-circuit withstand time (T_{SC}) for two devices are 20.4 and 31.1 μs, respectively. It is easy to accumulate the Joule heat inner the device during a long T_{SC}, resulting in the damage of the device. From the experiment results, the trench MOSFET with the thin gate oxide layer is easier to breakdown after the device switches off. Currently, the short-circuit energy is 14.6 J/cm^2 for the enhanced SiC MOSFET with the thick gate oxide layer, which is 17.1% higher than that of the one with the thin gate oxide layer. It can be concluded that the robustness enhanced MOSFET with the thick gate oxide of 85nm shows the high short-circuit capability even with the higher short-circuit

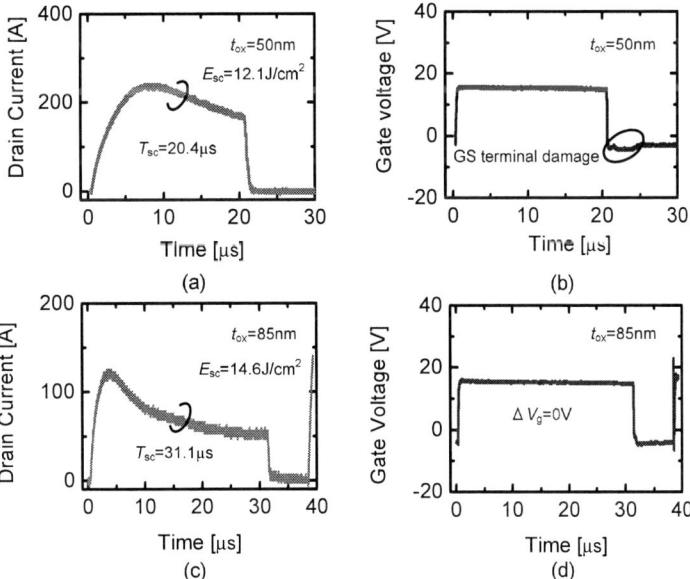

Fig. 6. The short-circuit waveforms with the DC voltage of 400V for the SiC trench MOSFETs with different SiO$_2$ thickness at room temperature. (a) the drain current, and (b) the gate voltage with SiO$_2$ thickness of 50nm; (c) the drain current, and (d) the gate voltage with SiO$_2$ thickness of 85nm. The electric field relaxation structures at the trench bottom are applied to the MOSFETs. $R_{on,sp}$ are 2.9 mΩcm^2 (50nm) and 3.88 mΩcm^2 (85nm) at V_g=20V, respectively.

978-1-5386-2928-4/18 $31.00 © 2018 IEEE

TABLE I
IMPEDANCE TEST AFTER DEVICE FAILURE

Symbol	SiO₂:50nm	SiO₂:85nm
R_{gs}	167 Ω (Shorted)	3.4 Ω (Shorted)
R_{ds}	∞ (Blocking)	9.8 Ω (Shorted)
R_{gd}	23 MΩ (Blocking)	10.6 Ω (Shorted)

R_{gs}: Resistance between gate and source,
R_{gd}: Resistance between gate and drain,
R_{ds}: Resistance between drain and source.

energy and the longer short-circuit transient.

From the measured resistances between all electrodes after the device failure as shown in Table.I, it is clear that the failure modes of two devices are totally different. The gate voltage degradation or gate shorted can be eliminated and the short-circuit capability is successfully improved by thickening gate oxide layer. Therefore, moderately thick gate oxide layer is required to avoid short-circuit failure related to the gate oxide breakdown in the SiC MOSFETs before the thermal runaway occurs.

III. CONCLUSION

In this work, the gate voltage dependence of short-circuit capability for the SiC MOSFETs is presented. It is verified by experiment and numerical simulation that the large gate switch off voltage can effectively improve the short-circuit capability by reducing the probability of channel normally-on induced by the higher lattice temperature.

In addition, a robustness enhanced SiC MOSFET with the thick gate oxide of 85 nm and the electric field relaxation structures at the trench bottom is compared with the SiC trench MOSFET with the thin gate oxide of 50 nm, which shows the high short-circuit capability even with the higher short-circuit energy and longer short-circuit transient. Thus, the moderately thick gate oxide layer is necessary to satisfy the requirement of high short-circuit capability.

ACKNOWLEDGMENT

A part of this work has been implemented under a joint research project of Tsukuba Power Electronics Constellations (TPEC).

*Assignor company of co-author Y. Kobayashi is Fuji Electric Co., Ltd.

REFERENCES

[1] T. Kimoto and J. A. Cooper, *Fundamentals of Silicon Carbide Technology: Growth, Characterization, Devices and Applications*: John Wiley & Sons, 2014.

[2] G. Romano, L. Maresca, M. Riccio, V. d'Alessandro, G. Breglio, A. Irace, *et al.*, "Short-circuit failure mechanism of SiC power MOSFETs," in *Proceedings of the International Symposium on Power Semiconductor Devices and ICs*, 2015, pp. 345-348.

[3] A. Castellazzi, A. Fayyaz, L. Yang, M. Riccio, and A. Irace, "Short-circuit robustness of SiC Power MOSFETs: experimental analysis," in *Proceedings of the International Symposium on Power Semiconductor Devices and ICs*, 2014, pp. 71-74.

[4] X. Huang, G. Wang, Y. Li, A. Q. Huang, and B. J. Baliga, "Short-circuit capability of 1200V SiC MOSFET and JFET for fault protection," in *Applied Power Electronics Conference and Exposition (APEC), 2013 Twenty-Eighth Annual IEEE*, 2013, pp. 197-200.

[5] T.-T. Nguyen, A. Ahmed, T. Thang, and J.-H. Park, "Gate oxide reliability issues of SiC MOSFETs under short-circuit operation," *IEEE Transactions on Power Electronics*, vol. 30, pp. 2445-2455, 2015.

[6] J. An, M. Namai, and N. Iwamuro, "Experimental and theoretical analyses of gate oxide and junction reliability for 4H-SiC MOSFET under short-circuit operation," *Japanese Journal of Applied Physics*, vol.55, 124102-1~4, 2016

[7] M. Namai, J. An, H. Yano and N. Iwamuro, "Experimental and Numerical Demonstration and Optimized Methods for SiC Trench MOSFET Short-Circuit Capability," in *Proceedings of the International Symposium on Power Semiconductor Devices and ICs*, 2017, pp. 363-366.

Proceedings of the 30th International Symposium on Power Semiconductor Devices & ICs
May 13-17, 2018, Chicago, USA

27.5 kV 4H-SiC PiN Diode with Space-Modulated JTE and Carrier Injection Control

Koji Nakayama[1], Tomonori Mizushima[1,2], Kensuke Takenaka[1,2], Akihiro Koyama[1,3], Yuji Kiuchi[1,4],
Shinichiro Matsunaga[1,2], Hiroyuki Fujisawa[1,2], Tetsuo Hatakeyama[1], Manabu Takei[1,2],
Yoshiyuki Yonezawa[1], Tsunenobu Kimoto[5] and Hajime Okumura[1]

[1]Advanced Power Electronics Center, National Institute of Advanced Industrial Science and Technology,
16-1 Onogawa, Tsukuba, Ibaraki, Japan
[2]Electronic Devices Business Group, Fuji Electric Co., Ltd., 4-18-1 Tsukama, Matsumoto, Nagano, Japan
[3]Advanced Technology R&D Center, Mitsubishi Electric Corporation, 8-1-1 Tsukaguchi-Honmachi, Amagasaki, Hyogo, Japan
[4] Power Device Development Department, New Japan Radio Co., Ltd., 2-1-1 Fukuoka, Fujimino, Saitama, Japan
[5]Department of Electronic Science & Engineering, Kyoto University, Katsura, Nishikyo, Kyoto, Japan
E-mail: koji.nakayama@aist.go.jp

Abstract—Ultra-high-voltage 4H-SiC PiN diode with a space-modulated junction termination extension and carrier injection control has been investigated. The introduction of a space-modulated region results in a high breakdown voltage of 27.5 kV, that is the highest among the values reported for 20 A class 4H-SiC PiN diodes. The simulated and measured forward characteristics of the 4H-SiC PiN diode with the carrier injection control are also reported. Forward voltage and on-resistance decrease as carrier lifetime increases. The introduction of carrier injection control at the anode and cathode sides results in reduction in carrier concentration. The measured characteristics exhibit good correlation with simulated results. Based on these results, we can confirm the effect of carrier lifetime on electrical characteristics.

Keywords—SiC; Ultra High Voltage; PiN Diode; space-modulated JTE; Carrier Lifetime Control; Carrier Injection Control

I. INTRODUCTION

Ultra-high-voltage 4H-SiC devices provide the possibility of significantly changing the design concept of existing power electronics components for high-voltage inverters and transmission and distribution systems. This includes the introduction of ultra-high-voltage DC power systems. As a result, the developments of more than 20 kV class 4H-SiC bipolar power devices for ultra-high-voltage AC-DC converters and inverters are expected. Until now, 20 kV class 4H-SiC PiN diodes [1,2], and insulated gate bipolar transistors [3] have been developed. To produce these devices, it is highly significant to optimize the junction termination extension (JTE) or the thickness and doping concentration of drift layers. Moreover, carrier lifetime control and carrier injection control techniques are required to achieve low on-state loss and switching loss. Carbon implantation [4] and thermal oxidation [5] have been reported to increase the carrier lifetime of 4H-SiC. When carrier lifetime becomes sufficiently long to reduce the on-state resistance of a drift layer (R_{ond}), the charge stored

This work was supported by Council for Science, Technology and Innovation (CSTI), Cross-ministerial Strategic Innovation Promotion Program (SIP), "Next-generation power electronics" (funding agency: NEDO).

in the drift layer (Q_s) and switching loss increase. It is crucial to concurrently realize three features, i.e., high breakdown voltage, low R_{ond}, and small Q_s.

In this study, we investigate 4H-SiC PiN diodes with a space-modulated JTE and carrier injection control. We compare experimental results with numerical simulation results and demonstrate that the proposed PiN diode exhibits high breakdown voltage, relatively low R_{ond}, and small Q_s.

II. EXPERIMENTAL METHOD

A. Device Structure

Figure 1 shows the device structure of the developed ultra-high-voltage 4H-SiC PiN diode with the following three features:

1) Two-zone and space-modulated JTE

We adopted a two-zone JTE and a space-modulated JTE [2] because the dose of a JTE should gradually decrease with the distance from the edge of an active region.

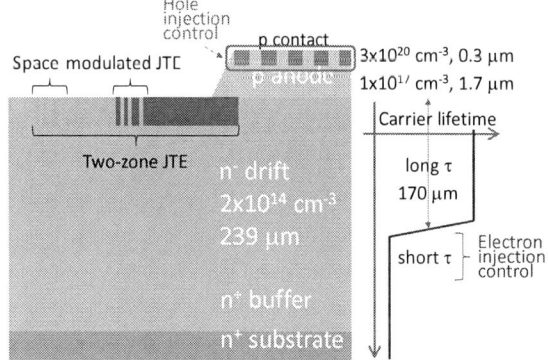

Fig. 1. Device structure of developed UHV 4H-SiC PiN diode with three features, i.e., two-zone and space-modulated JTE, electron injection control from the cathode, and hole injection control from the anode. Active area of fabricated diodes is 5.75 mm².

2) Electron injection control from a cathode

Carbon implantation [4] and thermal oxidation [5] have been used to increase the carrier lifetime of 4H-SiC. Carbon diffusion from a surface is used in both of these techniques. By controlling annealing temperature and time, the carrier lifetime of the anode side in a drift layer is increased while the carrier lifetime close to the cathode layer is maintained at a low level. The electron injection from the cathode can be controlled.

3) Hole injection control from an anode

Contact resistance increases when the doping concentration of an anode layer decreases to reduce hole injection. Using the line and space pattern of an anode contact layer, the total hole injection from the anode can be controlled while maintaining low contact resistance.

B. Device Fabrication

An n$^+$ buffer layer, an n$^-$ drift layer, and a p$^+$ anode layer were grown on an n$^+$ substrate. The substrates employed for this work were n-type (0001) 4H-SiC wafers off-angled by 4° toward <11-20>. The three layers were formed through epitaxial growth. Table I shows the thickness and doping concentration of each layer. The breakdown voltage calculated using the thickness and doping concentration of the drift layer is 31 kV.

TABLE I.   THICKNESS AND DOPING CONCENTRATION OF EACH LAYER

	n$^+$ Buffer	n$^-$ Drift	p$^+$ Anode
Thickness	15 μm	239 μm	2 μm
Doping Concentration	1×10^{18} cm^{-3}	2×10^{14} cm^{-3}	1×10^{17} cm^{-3}

Carbon atoms were implanted after p$^+$ epitaxial growth. The activation annealing process was performed at 1600 °C for 30 min. [4] Figure 2 shows the estimated depth profile of diffused carbon concentration and carrier lifetime. The implanted carbon atoms were diffused to depths of up to 170 μm. The carrier lifetime measured using μ-PCD was improved from 0.99 μs to 3.64 μs by the carbon implantation and annealing process.

Al atoms were implanted for the line and space pattern of the anode contact layer. The concentration and depth of the box sharp profile were 3×10^{20} cm^{-3} and 200 nm, respectively. A two-zone and space-modulated JTE were formed using Al ion implantation. Figure 3 shows the top view of the fabricated 4H-SiC PiN diodes. The PiN diode active areas were 5.75 mm^2.

C. Measurement and Calculation

Carrier lifetime was measured by Lifetime Measuring System (WAFER-τLTA-1800SP/01, KOBELCO research institute) at room temperature. WAFER-τLTA-1800SP/01 measured the carrier lifetime by reflection of a 26 GHz microwave with differential system. We used the 349 nm line of a frequency-tripled Nd:YLF pulse laser for carrier excitation. The pulse width was 5 ns, and the laser spot size was 2.0 mmφ. Penetration depth in 4H-SiC at 349 nm is estimated about 25-30 um. [6] The excitation intensity of Nd:YLF pulse laser was fixed at 3.9×10^{13} Photons/cm^2. The excess carrier concentration was about 1.6×10^{15} cm^{-3}, corresponding to a high injection level. We measured the $1/e^2$ lifetime map of 4 inches thick epitaxial 4H-SiC wafer at 2.0mm pitch. The median value from the histogram of $1/e^2$ lifetime map was showed for τ_a.

Forward characteristics were measured in the pulse mode of Power Device Analyzer and Curve Tracer (B1505A, KEYSIGHT Technologies). Reverse characteristics were measured using a high-voltage system. Voltage (high voltage power supply: HAR-30, Matsusada Precision, voltage measurement: Keithley 2000, current measurement: Keithley 2636A)

2D numerical device simulation was performed using DESSIS (Synopsis) to investigate the electrical performance of 4H-SiC PiN diodes. DESSIS calculates the electrical response of a device by solving the transport equation, continuity equation, and Poisson equation. The models and parameters were modified for 4H-SiC bipolar device simulation. [7,8]

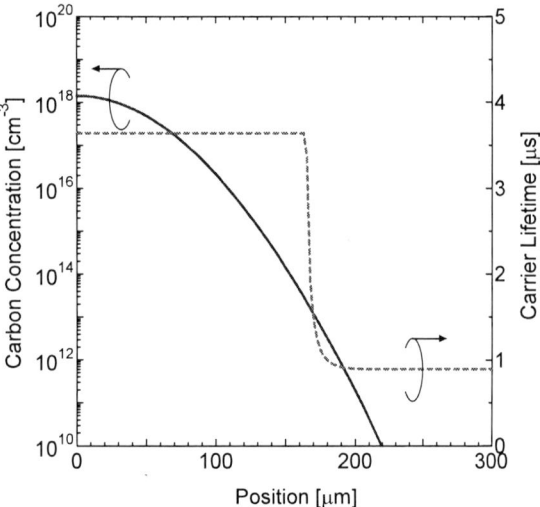

Fig. 2. Estimated depth profile of diffused carbon concentration and carrier lifetime. When the activation annealing process were performed at 1600 °C for 30 min, the implanted carbons were diffused to depths of up to 170 μm.

Fig. 3. Top view of fabricated 4H-SiC PiN diodes. Active area of fabricated diodes is 5.75 mm^2.

III. RESULTS AND DISCUSSION

A. Reverse Characteristics

Figure 4 shows the simulated breakdown voltage of the two-zone and space-modulated JTE. In the simulation, the doping concentration and thickness of drift layer are 3×10^{14} cm^{-3} and 229 μm, respectively. The maximum breakdown voltage without the space-modulated JTE was 23.1 kV, which is approximately 85% of the theoretical value calculated using the doping concentration and thickness of the drift layer, at a

JTE dose of 1.45×10^{12} cm^{-2}. By introducing the space-modulated JTE, the maximum breakdown voltage increased to 25.5 kV, which is approximately 94% of the theoretical value, at a JTE dose of 1.50×10^{12} cm^{-2}. The introduction of the space-modulated region reduces the electric field at the end of the JTE; thus, results in relatively high breakdown voltage.

Figure 5 shows the reverse characteristics of the fabricated ultra-high-voltage 4H-SiC PiN diode. The measured breakdown voltage is 27.5 kV. This voltage is approximately 89% of the theoretical value calculated using the doping concentration and thickness of the drift layer, and it is the highest among the values reported for 20 A class 4H-SiC PiN diodes. The experimental value is higher than the calculated value because the doping concentration and thickness of the fabricated drift layer are lower and thicker than the value for the simulation.

B. Forward Characteristics

Figure 6 shows the simulated forward characteristics of the ultra-high-voltage 4H-SiC PiN diode with carrier injection control. We considered the following 3 types of PiN diodes:

1) Fully diffused 4H-SiC PiN diode

We assume that the carbon atoms diffuse all over the drift layer. The carrier lifetime of the drift layer is 3.64 μs. The contact area defines the entire surface area of the anode layer.

2) Partially diffused 4H-SiC PiN diode

We assume that the carbon atoms diffuse to depths of up to 170 μm. The penetration depth of μ-PCD is approximately 25-30 μm. The carrier lifetime measured through μ-PCD is the value at the surface of the drift layer. The carrier lifetimes of the drift layer are 3.64 μs (the interface between the anode layer and the drift layer ~ 170 μm) and 0.99 μs (170 μm ~ the interface between the drift layer and buffer layer). The area ratio of the anode contact is reduced to 40% to simulate the line and space pattern of the anode contact.

3) 4H-SiC PiN diode without carbon diffusion

We assume that the carbon diffusion process does not apply to the drift layer. The carrier lifetime is 0.99 μs. Similar to the first type of diode, the contact area defines the entire surface area of the anode layer.

Forward voltage and on-resistance decrease from 16 V and 68 mΩcm^2 to 7.6 V and 28 mΩcm^2, respectively, as carrier lifetime increases.

Figure 7 shows the simulated carrier distribution of the ultra-high-voltage 4H-SiC PiN diode with carrier injection control. The reduction in anode contact area results in decrease in the carrier concentration at the anode side in the drift layer. This indicates that carrier injection can be controlled. The carrier concentration at the cathode side in the drift layer of the PiN diode with partial or no diffusion decreases by approximately half compared to that with full diffusion. The introduction of carrier injection control at the anode and cathode sides results in reduction in carrier concentration. The carrier stored in the drift layer at a current density of 100 A/cm^2 decreases form 1.36 μC to 0.48 μC. It is assumed that switching loss decreases.

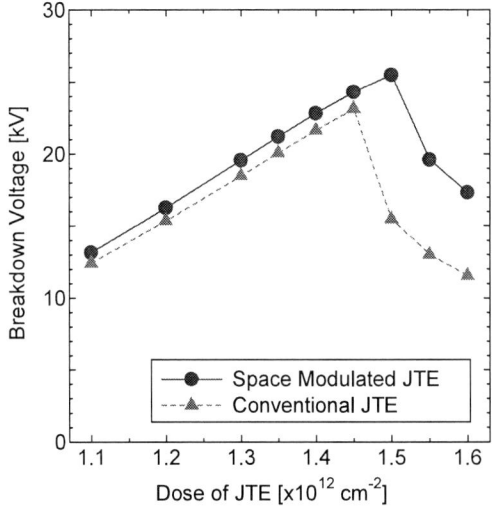

Fig. 4. Simulated breakdown voltage of PiN diodes with two-zone JTE and two-zone and space-modulated JTE. Doping concentration and thickness of drift layer are 3×10^{14} cm^{-3} and 229 μm, respectively. The introduction of space-modulated JTE results in high breakdown voltage.

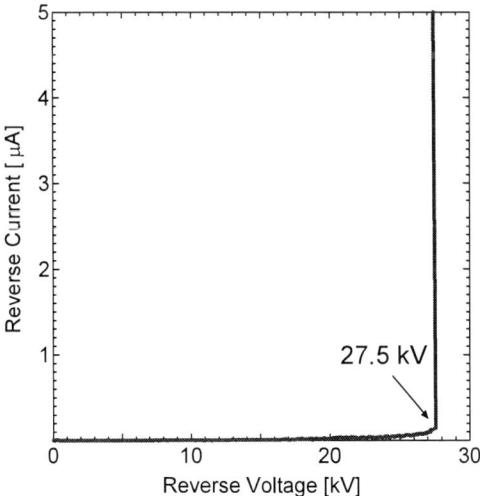

Fig. 5. Reverse characteristics of fabricated UHV 4H-SiC PiN diode. The measured breakdown voltage is 27.5 kV. This voltage is approximately 89% of the theoretical value calculated using the doping concentration and thickness of the drift layer, and it is the highest among the values reported for 20 A class 4H-SiC PiN diodes.

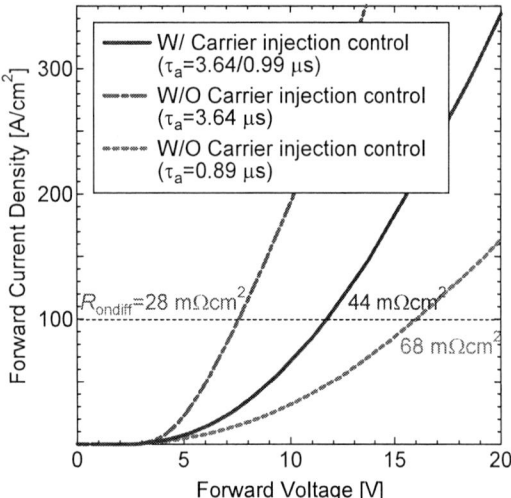

Fig. 6. Simulated forward characteristics of UHV 4H-SiC PiN diodes with carrier injection control. Forward voltage and differential on-resistance (R_{ondiff}) decrease as carrier lifetime increases.

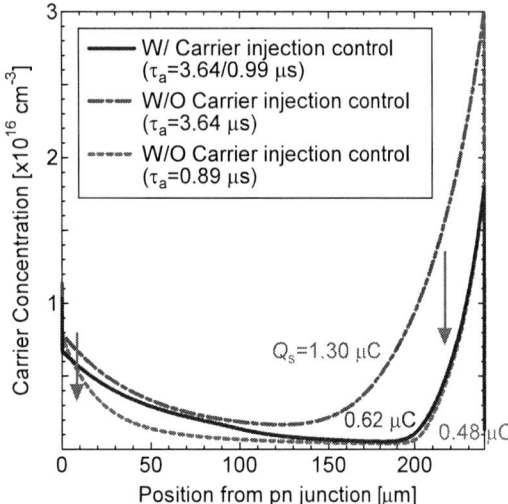

Fig. 7. Simulated carrier distribution at 100 A/cm² of UHV 4H-SiC PiN diodes with carrier injection control. The introduction of carrier injection control at the anode and cathode sides results in reduction in carrier concentration.

Figure 8 shows the forward characteristics of the fabricated ultra-high-voltage 4H-SiC PiN diode. The carrier lifetime of the epi-layer without the carbon diffusion was 0.89 μs. The carrier lifetime of the epi-layer with carbon diffusion was improved from 0.99 μs to 3.64 μs by the carbon implantation and annealing process. The forward voltage and R_{ond} decrease by the carbon diffusion. The measured characteristics exhibit good correlation with the simulated results. Based on these results, we can confirm the effect of carrier lifetime on electrical characteristics.

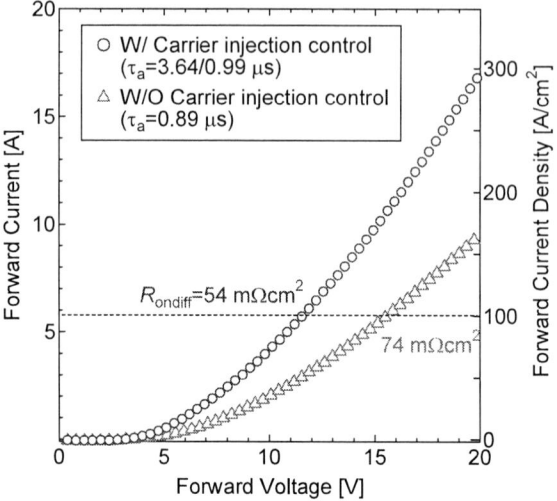

Fig. 8. Forward characteristics of fabricated UHV 4H-SiC PiN diodes. The measured characteristics exhibit good correlation with the simulated results. The effect of carrier lifetime on electrical characteristics is confirmed.

IV. CONCLUSION

The introduction of a space-modulated region results in a high measured breakdown voltage of 27.5 kV. This voltage is approximately 83% of the theoretical value calculated using the doping concentration and thickness of the drift layer, and it is the highest among the values reported for 20 A class 4H-SiC PiN diodes.

Forward voltage and on-resistance decrease as carrier lifetime increases. The introduction of carrier injection control at the anode and cathode sides results in reduction in carrier concentration. The measured characteristics exhibit good correlation with the simulated results. The effect of carrier lifetime on electrical characteristics is confirmed based on these results.

REFERENCES

[1] M. K. Das, J. J. Sumakeris, S. Krishnaswami, M. J. Paisley, A. K. Agarwal and A. Powell, International Semiconductor Device Research Symposium 2003, pp.364-365 (2003).

[2] H. Niwa, J. Suda and T. Kimoto, App. Phys. Express, 5, 064001 (2012).

[3] E. V. Brunt, L. Cheng, M. O'Loughlin, C. Capell, C. Jonas, K. Lam, J. Richmond, V. Pala, S. Ryu, S. T. Allen, A. A. Burk, J. W. Palmour and C. Scozzie, 2014 IEEE 26th International Symposium on Power Semiconductor Devices & IC's (ISPSD), pp.358-361 (2014).

[4] L. Storasta and H. Tsuchida, Appl. Phys. Lett, 90, 062116 (2007).

[5] T. Hiyoshi and T. Kimoto, App. Phys. Express, 2, 041101 (2009).

[6] S.G. Sridhara, T.J. Eperjesi, R.P. Devaty, W.J. Choyke, Materials Science and Engineering B61–62, pp.229–233 (1999).

[7] T. Hatakeyama, K. Fukuda and H. Okumura, IEEE Trans. on Electron Devices, 60, 2, pp.613-621 (2013).

[8] T. Hatakeyama, K. Nakayama, Y. Yonezawa and H. Okumura1, Proceedings of International Conference on Silicon Carbaide and Related Materials, MO.BP.2.

Proceedings of the 30th International Symposium on Power Semiconductor Devices & ICs
May 13-17, 2018, Chicago, USA

Investigation on Degradation Mechanism and Optimization for SiC power MOSFETs under Long-Term Short-Circuit Stress

Jiaxing Wei, Siyang Liu, Jiong Fang, Sheng Li, Ting Li, Weifeng Sun*
National ASIC System Engineering Research Center
Southeast University
Nanjing, China
*E-mail: swffrog@seu.edu.cn

Abstract—In this paper, the degradation mechanism of silicon carbide (SiC) power metal-oxide-semiconductor field-effect transistors (MOSFETs) under long-term short-circuit (SC) stress is investigated. With the help of Silvaco TCAD simulations and measurements on degraded parameters, the injection of electrons into gate oxide above channel region of the device is demonstrated to be the dominant degradation mechanism. It results in the positive shift of threshold voltage (V_{th}) and the increase of on-state resistance (R_{dson}) under low gate voltage bias condition. Simulated electrical properties of the device with electrons trapped into gate oxide above channel region share similar degradation trend with measured ones, proving the correctness of our analysis. Furthermore, an improved device structure with an additional shallow inverted-doping p-well, which can effectively lower the impact ionization rate (I.I.) along the SiC/SiO₂ interface above channel region during SC process, is proposed to restrict the degradations under long-term SC stress.

Keywords—SiC power MOSFET; short-circuit; degradation mechanism; optimization

I. INTRODUCTION

The silicon carbide (SiC) power metal-oxide-semiconductor field-effect transistor (MOSFET) is considered to be one of the most promising candidates in power electronic fields for superior electrical and thermal properties [1]-[3]. However, due to the poor quality of SiC/SiO₂ interface, it has been suffering from reliability issues working under abominable conditions including overcurrent, unclamped-inductive-switching (UIS) and short-circuit (SC) conditions [3]-[5]. When connected to resistance or inductance loads, such as being employed in converter systems, frequent SC situations happen to SiC power MOSFETs unexpectedly. During SC procedure, saturated current (I_{dsat}) and high drain source voltage (V_{ds}) are added to the device at a same time, which may result in failure or degradations with high probability.

Short-term SC robustness and failure mechanisms of SiC power MOSFETs have been investigated in details [5]-[7]. However, few studies have focused on the hot-carrier-induced (HCI) long-term SC damage of the device, not to

mention revealing the accurate degradation mechanism. In this work, with the help of experiments on the degradations of electrical parameters and Silvaco TCAD simulations, the degradation mechanism of SiC power MOSFETs under repetitive SC stress is verified. Moreover, an improved structure with a shallow inverted-doping p-well adjacent to original p-body, which can decrease the impact ionization rate (I.I.) along the SiC/SiO₂ interface above channel region effectively, is adopted to enhance the long-term SC endurance capacity.

II. DEVICE STRUCTURE AND STRESS CONDITIONS

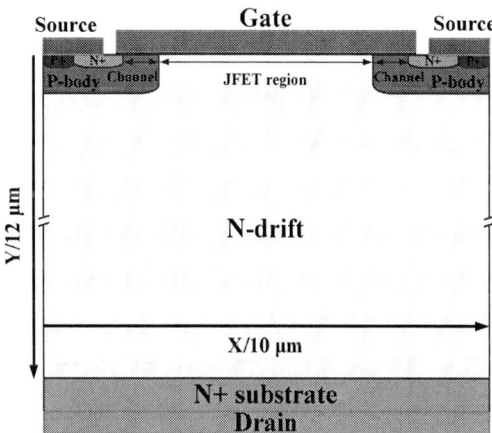

Fig.1 Schematic cross section of the SiC power MOSFET investigated in this work.

The schematic cross section of the SiC MOSFET investigated in this work is shown in Fig.1. It is fabricated on a 12μm-thick n-type epitaxial layer to sustain a more than 1200V breakdown voltage (BV). The channel length, the gate length and the thickness of the gate oxide are 0.5μm, 5μm and 50nm respectively. The total length of a symmetric cell is 10μm. Simulations presented in this work are based on this device structure.

978-1-5386-2928-4/18 $31.00 © 2018 IEEE   399

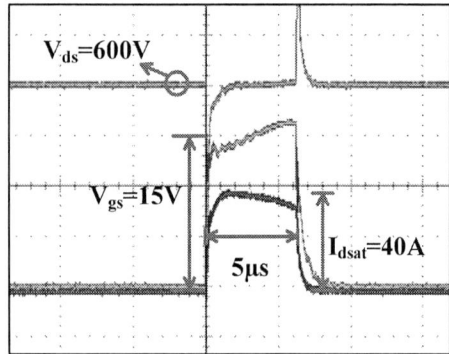

Fig.2 V_{gs}, V_{ds} and I_{ds} waveforms of the SiC power MOSFET under SC condition.

Fig.2 shows the current and voltage waveforms of the device under SC stress conditions at room temperature. During the stress process, gate source voltage (V_{gs}) is set to be 15V while V_{ds} is 600V. At the very beginning of gate pulse signal, drain source current (I_{ds}) rises sharply to the value of I_{dsat}, which is nearly 40A in this case. Then it drops smoothly due to thermal accumulation. In order to suppress the impact of thermal on the SiC power MOSFET, the width of repetitive gate signal is set to be 5µs, which is far less than the single-pulse SC capability of the device and does not result in a critical damage on gate oxide. Moreover, a 0.1% duty for each signal pulse is adopted to further lower the junction temperature.

III. RESULTS AND DISCUSSIONS

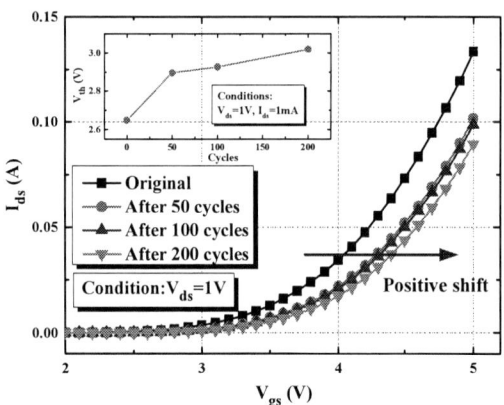

Fig.3 transfer characteristics and extracted V_{th} of the SiC power MOSFET under different SC cycles.

Transfer characteristics and extracted V_{th} of the SiC power MOSFET under different SC cycles are plotted in Fig.3. An obvious positive shift in I_d-V_g curve is observed. The V_{th} increases along with the increase of SC cycles, implying that repetitive SC stress might lead to injection of electrons into gate oxide of the device.

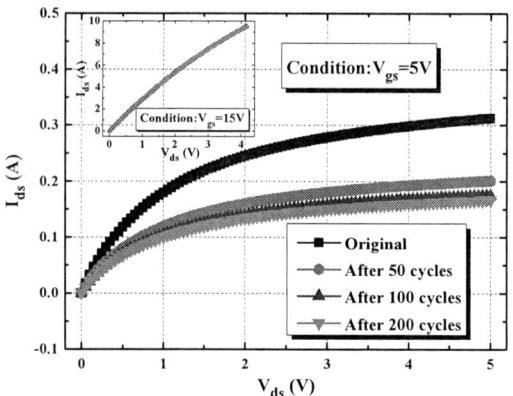

Fig.4 I_d-V_d characteristics under V_{gs}=5V&15V conditions of the SiC power MOSFET after being stressed by different SC cycles.

Fig.4 shows the I_d-V_d characteristics under different bias conditions of the stressed device. As can be seen, I_{ds} under low bias condition (V_{gs}=5V) degrades a lot, which is resulted from the increased V_{th} shown in Fig.3. Therefore, on-state resistance (R_{dson}) under V_{gs}=5V condition rises versus the increasing SC cycles. Meanwhile, I_d-V_d characteristic under high bias condition (V_{gs}=15V) almost keeps unchanged, illustrating that JFET region and drift region of the device are not affected by the stress. The blocking characteristic of the SiC power MOSFET also remain stable during SC procedure.

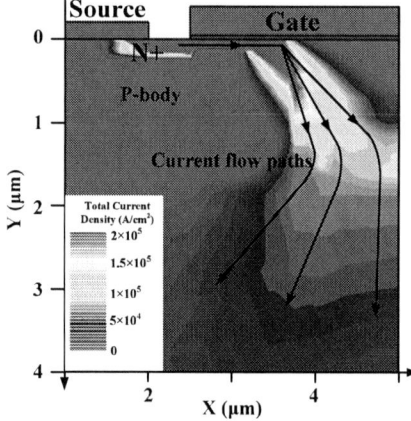

Fig.5 Current flow paths of the device under SC condition.

In order to verify the accurate HCI long-term SC degradation mechanism of SiC power MOSFETs, Silvaco TCAD simulations are performed. Fig.5 presents the simulated current flow paths of the device during SC procedure. The current flows concentrately through the interface of channel, which should be the main damaged region, and then spreads into JFET region and drift region.

I.I. and perpendicular electrical field (PEF) along the gate oxide interface are then extracted, just as presented in Fig.6.

978-1-5386-2928-4/18 $31.00 © 2018 IEEE   400

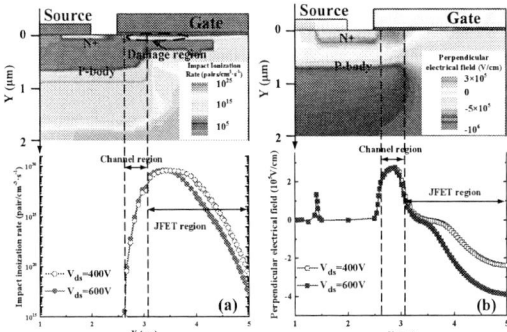

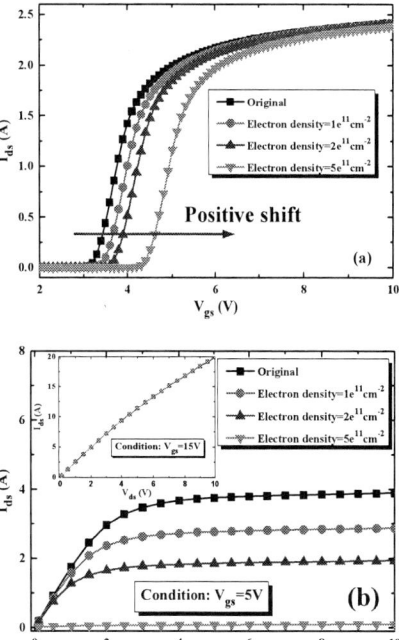

Fig.6 (a) Impact ionization rate (I.I.) and (b) Perpendicular electrical field (PEF) along the SiC/SiO$_2$ interface under SC stresses with V_{gs}=15V, Vds=400V&600V respectively.

Even though the peak value of I.I. occurs in the JEFT region, the PEF at the same location is extremely low, meaning that it rarely results in severe interfacial damage in JFET region. Meanwhile, the positive peak PEF and high I.I. values appear in channel region simultaneously, leading to the injection of electrons into the gate oxide above channel, which takes the responsibility for the increase of V_{th} and the rise of R_{dson} under low V_{gs} condition. When it is under high V_{gs} condition, the channel resistance occupies little portion of the total resistance, weakening the influence of trapped electrons on the R_{dson}. Hence the I_d-V_d characteristic under V_{gs}=15V bias condition remain unchanged.

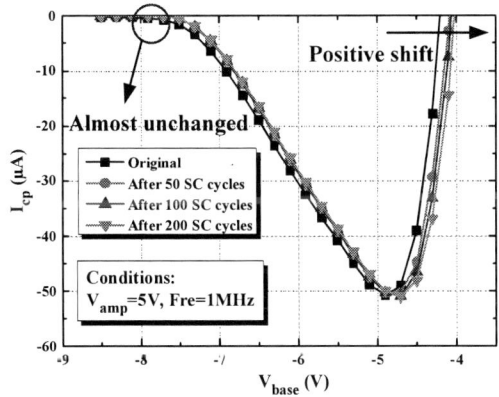

Fig.7 Charge pumping (CP) experiment variations of the device after different SC cycles.

To further verify the exact degradation mechanism, three-terminal charge pumping (CP) measurements with constant amplitude voltage (V_{amp}) CP signals are performed [8]. Fig.7 shows that during long-term SC stress process, left edge of CP current (I_{cp}) curve is almost unchanged while the right edge shifts to the positive direction. It indicates that there are electrons injected into the oxide above channel region but the stress rarely damage the JFET region, which are consistent with the conclusions made from previous simulation analysis.

Fig.8 Simulated variations of (a) transfer characteristics and (b) I_d-V_d characteristics of the SiC power MOSFET with different densities of trapped electrons in the oxide above channel region.

Fig.8 shows the simulated variations of transfer and I_d-V_d characteristics of the SiC power MOSFET with different densities of electrons trapped only into gate oxide above channel region. All the curves exhibit similar degradation trends as that of the experiment ones, proving the correctness of our investigation.

IV. OPTIMIZATION AND VERIFICATION

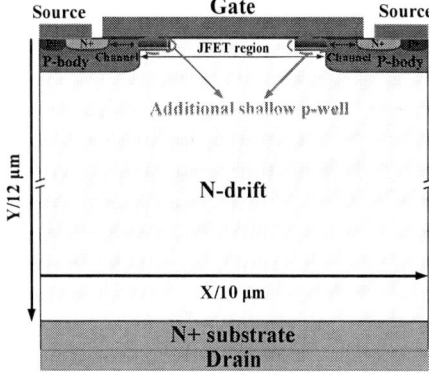

Fig.9 Schematic cross section of the improved device with a shallow inverted-doping p-well assistant to original p-body.

For the sake of restraining the degradations of SiC MOSFETs under repetitive SC stress, an improved device

978-1-5386-2928-4/18 $31.00 © 2018 IEEE   401

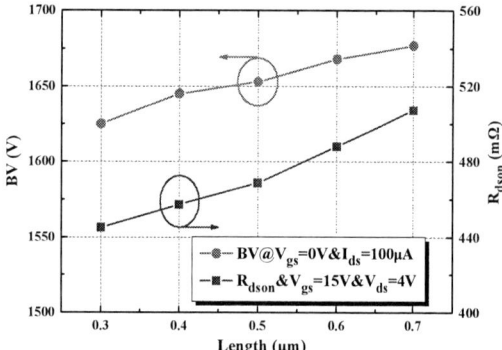

Fig.10 Simulated BVs and R_{dsons} of the improved SiC power MOSFET with different length of the shallow p-well.

structure is proposed, just as presented in Fig.9. A shallow p-well with light-doping surface, which can be fabricated by multi-step doping process with different implantation energies, is added to original p-body region to suppress the impact of the stress on SiC/SiO$_2$ interface. Fig.10 shows that BV and R_{dson} of the improved device increase with the increasing length of the shallow p-well region. To keep the BV stable and obtain a lower R_{dson}, a device structure with a 0.3μm-long additional shallow p-well is adopted.

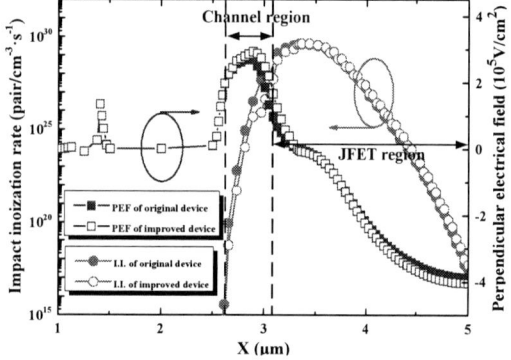

Fig.11 I.I. and PEF along the SiC/SiO$_2$ interface of the conventional and the improved device structures.

Fig.11 shows the comparisons of both interfacial I.I. and PEF between improved and original structures. According to simulated results, even though PEF increases slightly, the I.I. of the novel structure in channel region decreases by orders when compared with that of the original one, meaning the proposed structure can suppress the degradations under repetitive SC stress effectively.

V. Conclusions

The dominant HCI degradation mechanism of SiC power MOSFETs under long-term SC stress is investigated. It demonstrates that the repetitive SC stress results in the injection of electrons into the gate oxide above channel region, but rarely leads to damage in JFET region. It is the presence of those injected electrons that increases the V_{th} and the R_{dson}

under low V_{gs} bias condition of the device. Based on the degradation mechanism, an improved device structure with a shallow inverted-doping p-well assistant to original p-body region is proposed to reduce the I.I. along the interface above channel region, so as to suppress the degradations under long-term SC stress.

Acknowledgements

This work was supported by National Natural Science Foundation of China (61604038), Natural Science Foundation of Jiangsu Province (BK20160691, BK20171156), the Scientific Research Foundation of Graduate School of Southeast University (YBJJ1764) and the Fundamental Research Funds for the Central Universities.

Reference

[1] Y. Tanimoto, A. Saito, K. Matsuura, H. Kikuchihara, H. J. Mattausch. M. M. Mattausch, and N. Kawamoto, "Power-Loss Prediction of High-Voltage SiC-MOSFET Circuits With Compact Model Including Carrier-Trap Influences", *IEEE Trans. on Power Electronics*, vol. 31, no. 6, pp. 4509-4516, Jun. 2016.

[2] J. Wei, S. Liu, R. Ye, X. Chen, H. Song, W. Sun, W. Su, S. Ma, Y. Liu, and F. Lin, "Interfacial Damage Extraction Method for SiC Power MOSFETs Based on C-V Characteristics", in *Proc. ISPSD*, Sapporo, Japan, May. 2017, pp. 359-362.

[3] M. D. Kelley, B. N. Pushpakaran, and S. B. Bayne, "Single-Pulse Avalanche Mode Robustness of Commercial 1200 V/80 mΩ SiC MOSFETs", *IEEE Trans. on Power Electronics*, vol. 32, no. 8, pp. 6405-6415, Aug. 2017.

[4] J. A. Schrock, B. N. Pushpakaran, A. V. Bilbao, W. B. Ray II, E. A. Hirsch, M. D. Kelley, S. L. Holt and S. B. Bayne, "Failure Analysis of 1200-V/150-A SiC MOSFET Under Repetitive Pulsed Overcurrent Conditions", *IEEE Trans. Power Electronics*, vol. 31, no.3, pp. 1816-1821, Mar. 2016.

[5] G. Romano, M. Riccio, L. Mareca, G. Breglio, A. Irace, A. Fayyaz, and A. Castellazzi, "Influence of Design Parameters on the Short-Circuit Ruggedness of SiC Power MOSFETs", in *Proc. ISPSD*, Prague, Czech Republic, Jun. 2016, pp. 47-50.

[6] Z. Wang, X. Shi, L. M. Tolbert, F. Wang, Z. Liang, D. Costinett, and B. J. Blalock, "Temperature-Dependent Short-Circuit Capability of Silicon Carbide Power MOSFETs", *IEEE Trans. on Power Electronics*, vol. 31, no. 2, pp. 1555-1566, Feb. 2016.

[7] J. Sun, H. Xu, X. Wu, S. Yang, Q. Guo, and K. Sheng, "Short Circuit Capability and High Temperature Channel Mobility of SiC MOSFETs", in *Proc. ISPSD*, Sapporo, Japan, May. 2017, pp. 399-403.

[8] S. Liu, C. Gu, J. Wei, Q. Qian, W. Sun and A. Q. Huang, "Repetitive Unclamped-Inductive-Switching-Induced Electrical Parameters Degradations and Simulation Optimizations for 4H-SiC MOSFETs", *IEEE Trans. Electron Devices*, vol. 63, no.11, pp. 4331-4338, Nov. 2016.

Proceedings of the 30th International Symposium on Power Semiconductor Devices & ICs
May 13-17, 2018, Chicago, USA

High Accuracy Large-signal SPICE Model for Silicon Carbide MOSFET

Fu-Jen Hsu*, Cheng-Tyng Yen, Chien-Chung Hung, Chwan-Ying Lee, Lurng-Shehng Lee, Kuo-Ting Chu, Ya-Fang Li

Hestia Power Incorporated (HPI)
Hsinchu, Taiwan R.O.C.
Frane.Hsu@outlook.com

Abstract—**This paper provides a method to match characteristics of SiC MOSFET by a simple SPICE model. Besides, this method not only reaches highly approximate results in an accuracy of characteristics compared to commercial SiC SPICE model but also reduces lots of quantities of parameters from modified BSIM model. This method expresses the 1st quadrant I_D-V_{DS} and I_D-V_{GS} curve well by some additional modified equations. Also, the model development of the 3rd quadrant characteristics, which combines a diode with a JFET model, obtains a good fitting result. Finally, compared to conventional models, the R-square value and normalized RMSD value are significantly improved.**

Keywords- SPICE, SiC MOSFET, empirical model, body diode

I. INTRODUCTION

Silicon Carbide (SiC) MOSFET have been applied in many high-end fields due to its short switching duration, high breakdown voltage, and high power density [1]. With the usage of SiC MOSFET increases, the requirement of SPICE models is significant for circuit simulation. However, the unique phenomenon such as the short channel effect when applying a high drain-to-source voltage (V_{DS}), the 3rd quadrant performances of various gate-to-source voltage (V_{GS}), and the different output characteristic in high junction temperature (T_j) couldn't be fit very well. The output characteristics in the saturation region or lower V_{GS} level simulated by conventional models usually induce enormous errors where researchers have proposed many methods to reduce them [2][3][4]. Inaccurate models could cause the wrong judgment at operation points, and this will induce some condition over devices' guarantee margins and may trigger unexpected failures.

Thus, to gain more precision simulation result, researchers developed modified BSIM (Berkeley Short-channel Insulated-Gate FET Model) to fit the characteristics of SiC MOSFET [5]. Although the fitting result of using modified BSIM is pretty well, the BSIM is so complicated that users can't build it easily. This paper provides another empirical method to build SiC MOSFET SPICE model. By reducing the quantities of parameters and simplifying the equivalence equations, a simple and reliable SiC MOSFET SPICE model could be established.

II. BEHAVIORAL SPICE MODEL DESCRIPTION

A. Structure of Model and 1st quadrant characteristics

A general SPICE model of silicon power MOSFET has been established and well known for decades. Therefore, a packaged silicon carbide power MOSFET device model could be developed as shown in Fig.1. This empirical model consists 3 parts of circuits: (i.) thermal grid based on the package, (ii.)

parasitic parameters of wire bonding and die soldering, and (iii.) bare die characteristics model. To be specific, the thermal grid, which phrases the heat conductive path could be extracted from the structure-function in Fig.2. We could divide the thermal impedance, including the thermal resistance and the thermal capacitance, to multi-stage structures by this figure. According to the practical performance of SiC MOSFET, transfer characteristics saturate at a high-level V_{GS}, and these curves might be approximated by a hyperbolic tangent function [6].

Hence, we could express the output current I_D with modified equations as shown in Table I. By modify the method in [7], it could achieve a good accuracy to fit the unique I_D-V_{DS} characteristics of SiC MOSFET. The crucial equations to fit that characteristic is the modified equations $\psi(V_{GS})$, $\alpha(V_{GS})$, $\beta(V_{GS})$, and $\lambda(V_{GS})$.

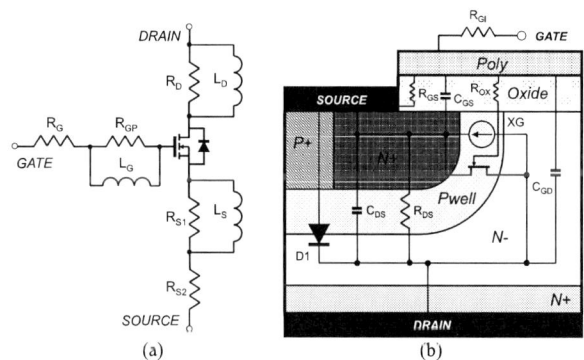

Fig. 1. The (a) parasitic components (b) structure and an equivalent circuit of a packaged SiC power D-MOSFET in this work.

TABLE I

SIGNIFICANT EQUATIONS OF SPICE MODEL IN THIS PAPER

Item	Equation
1st quadrant I-V curve	$I_{DS} = \beta \cdot (1 + \tanh(\psi))(1 + \lambda V_{DS}) \tanh(\alpha V_{DS})$
	$\psi = P_1(V_{GS} - V_{pk}) + P_2(V_{GS} - V_{pk})^2 + P_3(V_{GS} - V_{pk})^3$
	α, β, λ as modified functions, reference from Fig.3.
3rd quadrant I-V curve	Composed of a PN-diode and an N-type JFET model.
Thermal dependency	$P_{nTj} = P_{nT0} \times \left(1 + P_{nT}(T_J - T_{NOM})\right)$

978-1-5386-2928-4/18 $31.00 © 2018 IEEE

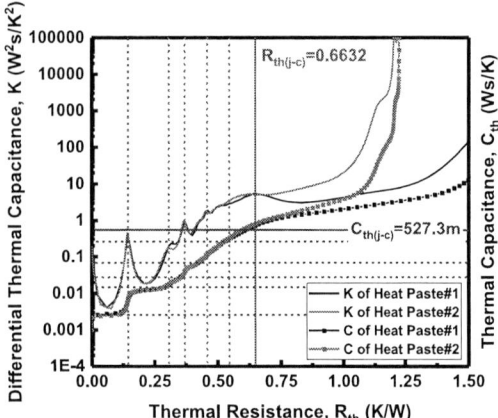

Fig. 2. The thermal impedance structure function of a packaged SiC MOSFET, which contains thermal resistance and thermal capacitance.

Due to the strong V_{GS} dependency to the saturation voltage parameter α, the drain current gain β, and the channel length modulation parameter λ, these parameters should not be constant values during various V_{GS} bias be applied. Thus, this work fit the equation of each parameter by various V_{GS} and the result is shown in Fig.3. In short, $\alpha(V_{GS})$, $\beta(V_{GS})$, and $\lambda(V_{GS})$ are modified functions of V_{GS} due to the channel mobility would change by various V_{GS}. Moreover, V_{pk} is defined by V_{GS} of the maximum transconductance point, which is the inflection V_{GS} point on the I_D-V_{GS} curve. These modified equations could help people to fit the characteristics more precise and easier.

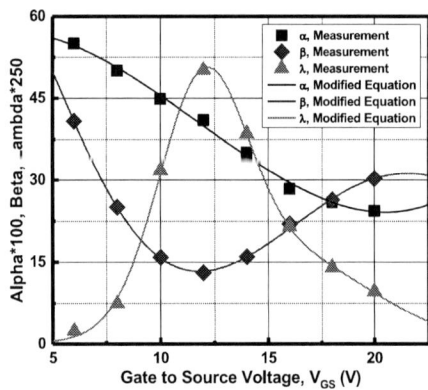

Fig. 3. α, β, λ which are functions of V_{GS}. In this work, we could find that the stationary points of β and λ and the inflection point occur at $V_{pk} \approx 11$V.

B. The 3rd quadrant and thermal characteristics

In this work, a modified diode model combined with an N-type JFET model is established to fit the 3rd quadrant output characteristics which are changing in different V_{GS} bias. Owing to the light doping concentration of P-well in a planner SiC MOSFET, the "N+ to P-well to N-" area forms a parasitically partial pinch-off JFET and induces some leakage current at V_{GS}=0V. By applying a negative bias on V_{GS}, the field effect finally pinches-off the channel and shows the pure body diode

characteristics. From this discussion, combining a P-N diode model and a JFET model together will be helpful to present the operation mode of the 3rd quadrant characteristics on SiC MOSFET.

C. The thermal dependency and CV characteristics

In order to consider the thermal dependency of electrical characteristics, it could be easily fitting by linear equations due to the little variation in a narrow range of temperature. Although the overall temperature dependence of electrical characteristics and parameters such as V_{th}, V_{pk}, $\alpha(V_{GS})$, $\beta(V_{GS})$, and $\lambda(V_{GS})$ are quite nonlinear, by narrowing down the operation region into 25°C to 150°C, we could find the performances are much linear and could be fit by simple equations showed in Table. I.

For the CV characteristics, since the mechanism of parasitic capacitance forming is the same as conventional silicon devices, equations which mentioned in [3] could be used to calculate the result of CV measurement in SiC MOSFET. For example, when fitting a 1200V/60mΩ SiC MOSFET device H1M120F060, we can approximate its C_{GS} by equation (1).

$$C_{GS} \approx \frac{\epsilon_{SiC} \cdot A_{GS}}{t_{ox}} = const. = C_{GS0} \tag{1}$$

In this equation, t_{OX} means the gate oxide thickness of devices. By the measurement result, we can simply fit C_{GS} by a constant parameter C_{GS0}. Also, because C_{DS} is decided by p-well to n-drift layer and the p-well region is always shorted to the source region through the P+ contact, the C_{DS} capacitance could be described precisely by equation (2). In (2), V_{bi} is the built-in voltage across the P/N-drift region which could be shown as equation (3).

$$C_{DS} = C_{DS0} \cdot \left(\frac{V_{bi}}{V_{bi} + V_{DS}} \right)^M \tag{2}$$

$$V_{bi} = \frac{kT}{q} ln \left(\frac{n_{dope} \times 1 \times 10^{16}}{n_i^2} \right) \tag{3}$$

Considering the structure of SiC MOSFET, equation (4) is a good method to express the C_{GD} capacitance. In equation (4), t_{GDdep} could be described as equation (5), and a transition voltage parameter V_{CGDT} has been defined to model the transition of the gate-drain capacitance.

$$C_{GD} = \frac{\epsilon_{SiC} \cdot A_{GD}}{t_{ox} + t_{GDdep}} \tag{4}$$

$$t_{GDdep} = \frac{1}{2} \{ 1 - tanh(V_{GD} - V_{CGDT}) \} \sqrt{\frac{2\epsilon_{SiC}|V_{CGDT} - V_{GD}|}{q \cdot n_{dope}}} \tag{5}$$

III. SIMULATION RESULTS AND DISCUSSIONS

A. The Simulation Result of I-V Characteristics

This paper provides a better method to match the I_D-V_{DS} characteristics of SiC MOSFET by a simple SPICE model. Besides, this method reduces lots of parameters from the modified BSIM, the complicated model, and expresses the 1st quadrant I_D-V_{DS} curve well (see Fig.4). Comparing to the commercial SiC MOSFET model, the R^2 value (coefficient of determination) and normalized RMSD (Root-Mean-Square Deviation, calculated by (6)) value are significantly improved.

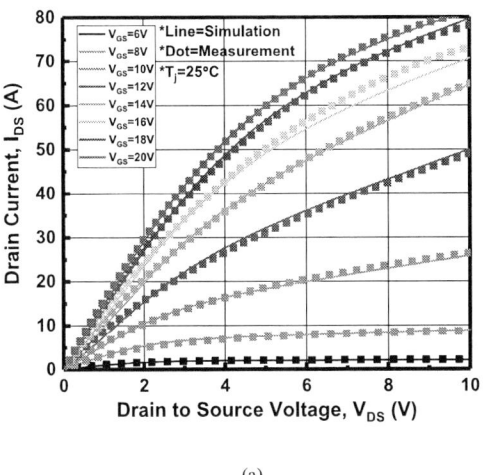 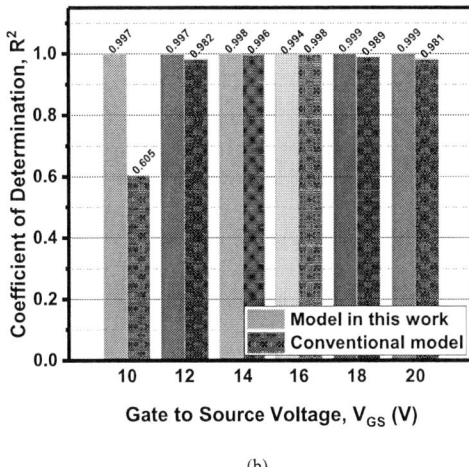

(a)　　　　　　　　　　　　　　　　　　　(b)

Fig. 4. (a) Comparison of I_D-V_{DS} curves from SPICE simulation and practical measurement results of a SiC power MOSFET. (b) The R^2 value (coefficient of determination) of I_D-V_{DS} curves from simulation and measurement of a SiC power MOSFET at T_j=25°C, the model in this work is more approximate than the conventional model given by the manufacturer.

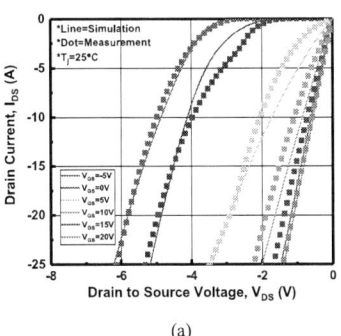 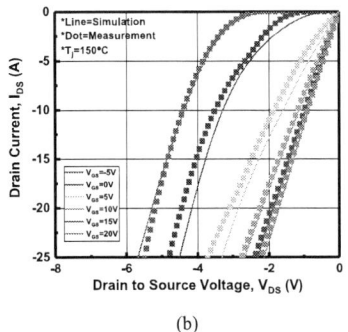 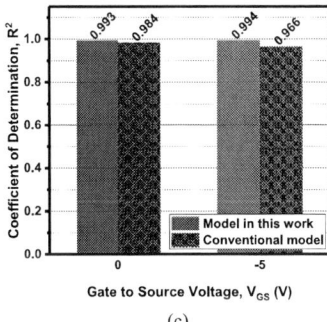

(a)　　　　　　　　　　　　(b)　　　　　　　　　　　　(c)

Fig. 5. Comparison of 3rd quadrant I_D-V_{DS} curves from SPICE simulation and practical measurement of a SiC power MOSFET at (a) T_j=25°C and (b) T_j=150°C. (c) The R^2 value of 3rd quadrant I_D-V_{DS} curves from simulation and measurement of a SiC power MOSFET at T_j=25°C.

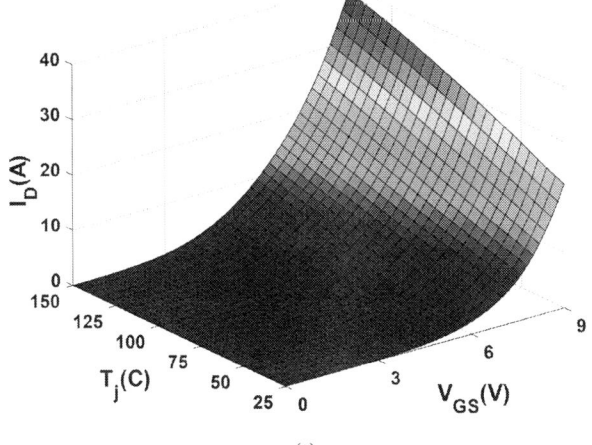 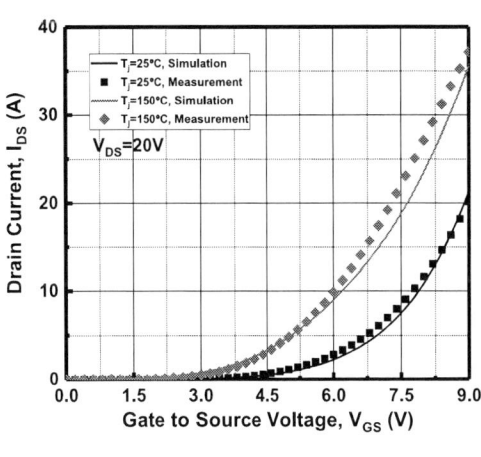

(a)　　　　　　　　　　　　　　　　　　　(b)

Fig. 6. The (a) I_D-V_{GS}-T_j surface plot from SPICE simulation. (b) Comparison of I_D-V_{GS} curves from SPICE simulation and practical measurement of a SiC planner MOSFET.

978-1-5386-2928-4/18 $31.00 © 2018 IEEE　　　　405

TABLE II

NORMALIZED RMSD OF THE 1ST QUADRANT BEHAVIOR

V_{GS} (V)	20	18	16	14	12	10
This work	9.6E-3	1.5E-2	4.0E-2	1.7E-2	2.9E-2	2.5E-2
Conventional	8.3E-1	8.0E-1	7.7E-1	7.1E-1	6.9E-1	4.8E0

TABLE III

NORMALIZED RMSD OF THE 3RD QUADRANT BEHAVIOR

V_{GS} (V)	-5	0
This work	0.095	0.127
Conventional	7.403	0.008

$$Normalized\ RMSD = \frac{\sqrt{\frac{\sum_i^n (y_{i,model} - y_{i,meas.})^2}{n}}}{\overline{y}} \quad (6)$$

Also, the model of the 3rd quadrant characteristics, which combines a diode with a JFET model, obtains a good fitting result in Fig.5. Currently, many commercial SiC MOSFET models do not fit the 3rd quadrant I-V curve respect to the various V_{GS} especially in V_{GS}=0V to -5V. Therefore, end users hardly simulate the behavior precisely in the reverse conducting mode. This will confuse users and cause some unpredictable fail. In contrast to commercial SiC MOSFET model, using a typical P-N diode model to simulate the 3rd quadrant behavior of SiC MOSFET, this work combining a JFET and diode model together to approximating the practical characteristics of SiC MOSFET. By this modified method, the R^2 value and the normalized RMSD value of 3rd quadrant I_D-V_{DS} are significantly improved especially at V_{GS}= -5V.

Finally, the I_D-V_{GS} fitting result could be found in Fig.6. In Fig.6, it also presents a high accuracy result which is acceptable for applications.

B. The Simulation Result of Thermal and Dynamic Behavior

We could detect the linear approximation of the thermal dependency of I_D-V_{GS} in Fig.6 (a) and $R_{DS(on)}$ in Fig.7, these results are quite close and also show a high coefficient of determination. As a result, a linear approximation by the equation in Table I could satisfy users' requirement.

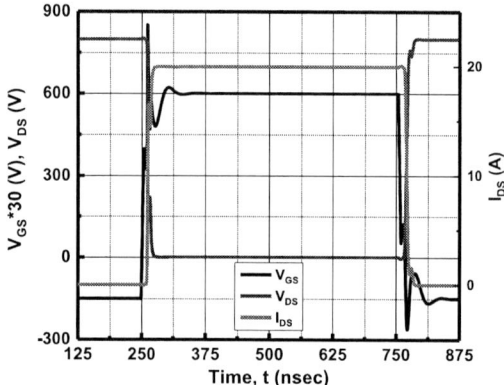

Fig. 8. The 800V/20A resistive switching waveform of the SiC MOSFET SPICE simulation result.

IV. CONCLUSION

According to this work, a simple and reliable SPICE model has been established. Compared to conventional models, the R^2 values and the NRMSD values gain significant improvements in many characteristics such as the 1st and 3rd quadrant I_D-V_{DS}, the I_D-V_{GS}-T_j, and $R_{DS(on)}$ -T_j. Also, the dynamic performance of this model exhibits a great simulation result as Fig.8. Instead of a complicated physical model, an empirical SPICE model in this work could get more useful in practical requirements for system designers.

REFERENCES

[1] B. J. Baliga, *Silicon Carbide Power Devices.*, Singapore: World Scientific Pub. Co. Inc., 2005.
[2] T. R. McNutt *et al.*, "Silicon carbide power MOSFET model and parameter extraction sequence" *IEEE Trans. Power Electron.*, vol. 22, no. 2, pp. 353–363, Mar. 2007.
[3] M. Mudholkar *et al.*, "Datasheet Driven Silicon Carbide Power MOSFET Model" *IEEE Trans. Power Electronics*, vol. 29, no. 5, pp. 2220–2228, May 2014.
[4] J. Wang *et al.*, "Characterization, Modeling, and Application of 10-kV SiC MOSFET" *IEEE Trans. Electron Devices*, vol. 55, no. 8, pp. 1798–1806, Aug. 2008.
[5] Z. Dilli *et al.*, "An Enhanced Specialized SiC Power MOSFET Simulation System" *Proc. IEEE SISPAD '15*, pp. 463–466, Sep. 2015.
[6] I. Angelov *et al.*, "An Empirical Table-Based FET Model" *IEEE Trans. Microwave Theory and Techniques*, vol. 47, no. 12, pp. 2350–2357, Dec. 1999.
[7] I. Angelov *et al.*, "A new empirical nonlinear model for HEMT and MESFET devices" *IEEE Trans. Microwave Theory and Techniques*, vol. 40, no. 12, pp. 2258–2266, Dec. 1992.

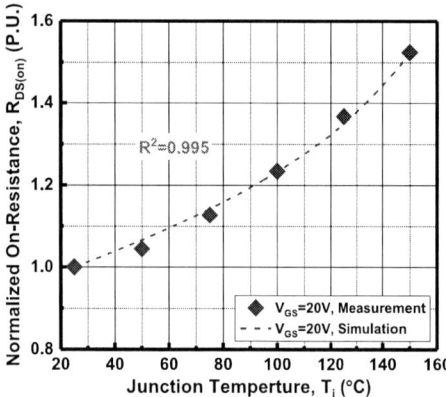

Fig. 7. The simulation result and measurement data of $R_{DS(on)}$, it shows that a linear approximation satisfies the trend of $R_{DS(on)}$ when T_j is ranging from 25°C to 150°C.

Proceedings of the 30th International Symposium on Power Semiconductor Devices & ICs
May 13-17, 2018, Chicago, USA

Analysis of Parameters Determining Nominal Dynamic Performance of 1.2 kV SiC Power MOSFETs

Roger Stark, Ivana Kovačević-Badstübner, Alexander Tsibizov, Bhagyalakshmi Kakarla,
Yanrui Ju, Beat Jaeger, Thomas Ziemann and Ulrike Grossner
Advanced Power Semiconductor Laboratory, ETH Zurich
Physikstrasse 3, 8092 Zurich, Switzerland
Phone/Fax : +41 44 632 99 76 / +41 44 632 12 02
Email: kovacevic@aps.ee.ethz.ch

Abstract—A good understanding of internal MOSFET capacitances is required in order to accurately model the dynamic characteristics of SiC MOSFETs. MOSFET compact models used to simulate and optimize power converter systems have to take into account the effects of non-linear voltage-dependent internal MOSFET capacitances correctly. In this paper, the individual influence of the voltage-dependent drain-source capacitance C_{ds} and the drain-gate capacitance C_{dg} on the MOSFET dynamics is investigated in detail. A comprehensive analysis of the switching performance of 1.2 kV SiC power MOSFETs with respect to C_{ds} and C_{dg} by means of a physics-based compact model, TCAD modeling, and the switching measurements of a 1.2 kV 80 mΩ SiC power MOSFET is presented. It is demonstrated that the non-linearity of C_{gd} impacts the turn-off delay time $t_{d,OFF}$, while the non-linearity of C_{ds} does not have a significant impact on the switching transients.

I. INTRODUCTION

Virtual prototyping based on electromagnetic circuit coupling can be used as a tool to provide a deep insight into the fast switching behavior of SiC power MOSFETs as shown in [1]. Development of accurate compact models suitable for virtual prototyping requires the knowledge of the MOSFET design and fabrication parameters and their impact on the fast switching capabilities. Nowadays, a great effort is being made in the direction of developing compact models for SiC MOSFETs, which are able to take into account temperature, non-linear behavior of capacitances [2]–[5] and voltage-dependent mobility [6]. However, the previous publications do not clarify which effects are dominant and to which extent each parameter influences the dynamic characteristics of SiC MOSFETs. This paper presents a comprehensive analysis of the switching performance of 1.2kV SiC power MOSFETs with respect to different MOSFET design parameters by means of a physics-based compact model, TCAD modeling, and the switching measurements of a 1.2 kV 80 mΩ SiC power MOSFET. Identifying the dominant effects based on the semiconductor physics will allow the optimization of SiC MOSFET models for computationally extensive multi-physics simulations.

II. LITERATURE SURVEY

The MOSFET switching transients have been analyzed extensively in literature, observing the impact of the stray inductances seen from the drain, source and gate terminals [7], the threshold voltage, and the MOSFET transconductance [8]. However, the individual influence of the non-linear voltage-dependent drain-source capacitances C_{ds} and drain-gate capacitance C_{dg} is still unclear. In [9], a MOSFET analytical model assuming simplified piece-wise constant voltage-dependency of MOSFET internal capacitances was used to show the influence of parasitic effects on the MOSFET switching performance. However, such a simplified model does not include the non-linear voltage dependence of MOSFET capacitances, and hence, does not represent the effect of this non-linearity on the switching waveforms. Recent publications [4], [10]–[12] clearly demonstrate the importance of modeling the non-linear behavior of MOSFET capacitances, especially the drain-gate capacitance. In [4], [10], [11], behavioral MOSFET models were developed, and the corresponding analytical models describing C_{dg} and C_{ds} as the functions of the MOSFET terminal voltages were parametrized based on either S-parameter measurements of MOSFET capacitances or using the slow switching measured current and voltage waveforms. These modelling approaches do not provide any physical understanding of the MOSFET capacitances. On the other hand, in [12] a physics-based model of C_{dg} was developed based on the surface-potential approach. This model, however, does not address the modelling of the so called kink point in the non-linear C_{dg}-V_{ds} curve as described in [5] and also observed in the C_{dg} measurements of the modelled 1.2 kV 80 mΩ SiC power MOSFET. Therefore, in this paper we analyze the importance of modelling C_{dg} with two different decay rates for the increase of the drain voltage around a kink in the measured C_{dg}-V_{ds} curve. Additionally, this paper presents also an analysis of switching waveforms with respect to C_{ds}.

978-1-5386-2928-4/18 $31.00 © 2018 IEEE

III. PROPOSED MODELLING APPROACH

As the numerical modeling of semiconductor physics in TCAD tools is computationally expensive, behavioral and/or physics-based compact models are typically used for the device and system analysis. In this paper, a physics-based compact model for vertical SiC MOSFETs is used for a parametric study on dynamic behavior of a $1.2\,\mathrm{kV}$ $80\,\mathrm{m\Omega}$ SiC MOSFET (C2M0080120D) from Cree. This model accurately simulates the device characteristics by iteratively calculating the surface potential and considering important material and geometry effects. The physical background and consideration of mobility degradation effects due to interface traps and surface roughness distinguishes this model from other physics-based compact models of SiC power MOSFETs proposed in literature [6]. The model is established using the Spectre circuit simulator from Cadence. It allows evaluating and visualizing the switching waveforms with much less computational resources compared to TCAD mixed-mode simulations while still providing the link between semiconductor device physics and MOSFET dynamics. As a result, it enables an efficient evaluation of the impact of different MOSFET design parameters on both static and dynamic MOSFET behaviors. TCAD modeling is used to derive the physics-based models of MOSFET parameters, e.g. channel mobility as function of V_{gs}, $C_{\mathrm{ds}}(V_{\mathrm{ds}})$ and $C_{\mathrm{gd}}(V_{\mathrm{gd}}, V_{\mathrm{ds}})$.

IV. DYNAMIC PERFORMANCE ANALYSIS

The unknown parameters of the compact model were extracted from the I-V and C-V measurements of a C2M0080120D sample as described in [13]. The developed compact model was validated by the double-pulse test (DPT) measurements using a C2M0080120D MOSFET as the low-side switch and a Cree C4D10120D SiC diode as the free-wheeling diode (see Fig. 1), while considering the parasitics of the measurement setup [1]. The verification of the compact model in the switching transient simulation is shown in Fig. 2. The difference between the measured and simulated switching energy losses calculated from $I_{\mathrm{S}}(t)$ and $V_{\mathrm{DS}}(t)$ is $\approx 5\%$, which indicates that the model can be used to predict accurately the switching losses. In comparison to the behavioral compact model for C2M0080120D MOSFET provided by the device manufacturer, the developed compact model provides a physical insight into the dynamic MOSFET characteristics, however, by its nature it is more sensitive to convergence problems. As a result, the oscillations observed in the simulated $V_{\mathrm{DS}}(t)$ at turn on (Fig. 2b) are more damped in the measurements, which will be further investigated in the course of future research.

In the next step, the analysis of switching waveforms is performed to identify the impact of MOSFET capacitances on its dynamic performance under nominal operation. TCAD modeling based on Synopsys Sentaurus is used to better understand the physics of the nonlinear voltage dependency of the capacitance.

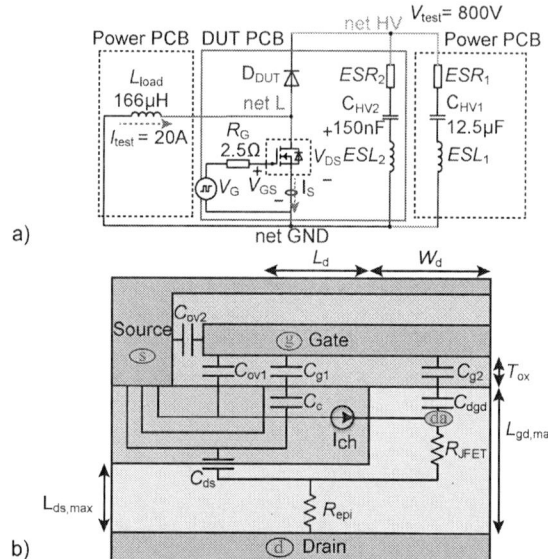

Fig. 1. Simplified equivalent circuits of: a) the DPT measurement setup, and b) the implemented compact model of the $1.2\,\mathrm{kV}$ SiC MOSFET.

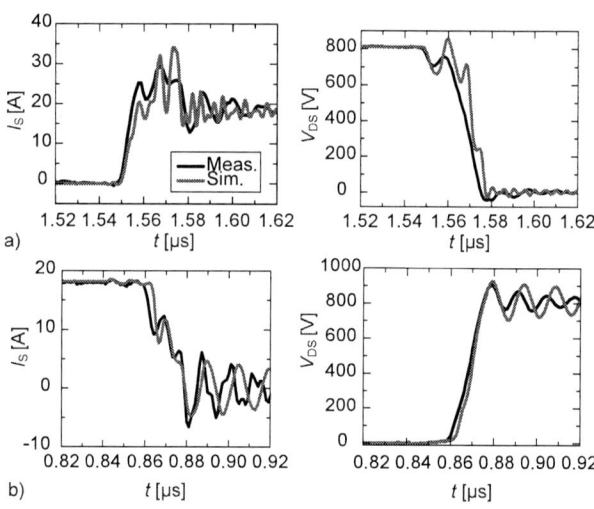

Fig. 2. Comparison between the measured and simulated switching waveforms: a) turn on ($t_{\mathrm{start,on}} = 11.541\,\mu\mathrm{s}$), and b) turn off ($t_{\mathrm{start,off}} = 0.841\,\mu\mathrm{s}$).

A. Effect of C_{gd}

Using the Sentaurus device simulator, a half pitch cell of the $1.2\,\mathrm{kV}$ Cree SiC MOSFET is designed with two different configurations having the same threshold voltage of $V_{\mathrm{th}} = 3.0\,\mathrm{V}$. The first MOSFET configuration (DUT0) is designed in TCAD with constant doping of the JFET and the drift region ($10^{16}\,\mathrm{cm}^{-3}$) over a total epi-layer thickness of $L_{\mathrm{gd_{max}}} = 10\,\mu\mathrm{m}$. In the second configuration (DUT1), the JFET doping was increased up to $3 \cdot 10^{16}\,\mathrm{cm}^{-3}$ in the top $0.6\,\mu\mathrm{m}$ of the drain region under the gate oxide, which resulted in a non-linear

C_{gd}-V_{ds} curve with a kink point around $V_{DS1} = 10\,V$ and the same voltage-dependency as DUT0 for $V_{ds} > 20\,V$, see Fig. 3. In comparison to DUT0, the on-state performance of DUT1 is improved in terms of on-state resistance, i.e. $R_{ds,on,DUT1} \approx 0.71 \cdot R_{ds,on,DUT0}$, determined at $V_{ds} = 20\,V$ and $I_d = 20\,A$ as shown in Fig. 4. A higher C_{gd} of DUT1 is due to the larger doping concentration, whereas the decay of C_{gd} around the kick-point $V_{ds} = 10\,V$ can be explained by a combination of two effects: an extension of the depletion in the JFET region in the vertical direction from the oxide surface with an increase of V_{gd}, and an extension of the depletion in lateral direction from the base-drain p-n junction with an increase of V_{ds} leading to the pinch-off of the JFET region, see also Fig. 1b.

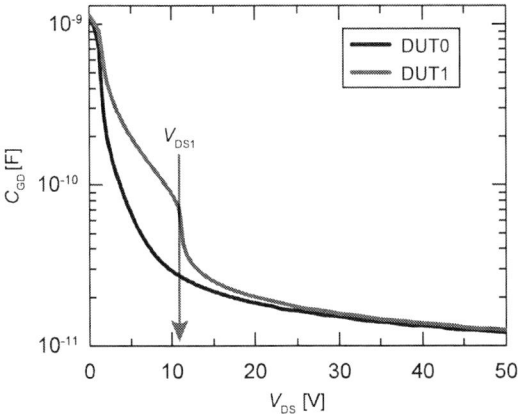

Fig. 3. C_{gd} characteristics of the two MOSFET designs DUT0 and DUT1 at 1 MHz.

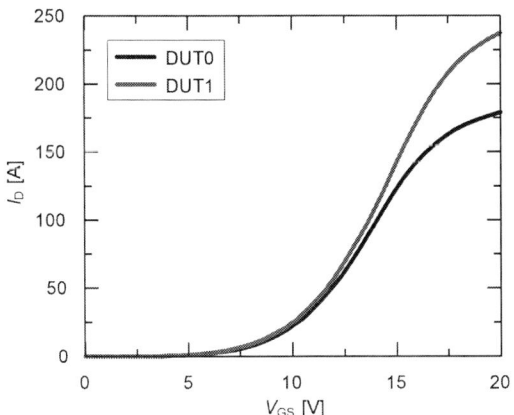

Fig. 4. Transfer characteristics of the two MOSFET designs DUT0 and DUT1 at 300 K. DUT1 has a smaller on-state resistance than DUT0 while having the same threshold voltage $V_{th} = 3.0\,V$.

A TCAD simulation was used to extract the non-linear C_{gd} as a function of both the gate-source, V_{GS}, and drain-source, V_{GS}, voltages. A TCAD small-signal simulation of $C_{gd,DUT1}$ at 1 MHz during the switch off transient is shown

in Fig. 6 together with the $C_{gd,DUT1}(V_{DS}, V_{GS})$ under a quasi-stationary variation of V_{DS} for different gate-source voltages $V_{GS} = 0 - 20\,V$. For $V_{GS} < V_{th}$, the C_{GD} dependence on V_{GS} mostly leads to a parallel shift of $C_{gd}(V_{DS})$ in comparison to $C_{gd}(V_{DS})$ at $V_{GS} = 0\,V$. For $V_{GS} > V_{th}$, increasing V_{GS} leads to a smaller channel resistance, and lower C_{gd} magnitude. Additionally, the voltage drop across the drain resistance (R_{JFET}) increases with the drain current $I_D(V_{GS})$, which leads to a $C_{gd}(V_{DS})$ dependency different than $C_{gd}(V_{DS})$ at $V_{GS} = 0\,V$. Analyzing the dynamic response of two designs DUT0 and DUT1, we observed that the non-linear C_{gd} of DUT1 influences only the turn-off transient introducing a time delay, Δt_{shift}, of the current and voltage waveforms as shown in Fig. 5. This impact increases if the kick-point V_{DS1} is at higher V_{DS}. Accordingly, the non-linearity of C_{gd} does not affect the turn-on transient significantly, and introduces a time-delay of current and voltage curves at turn-off.

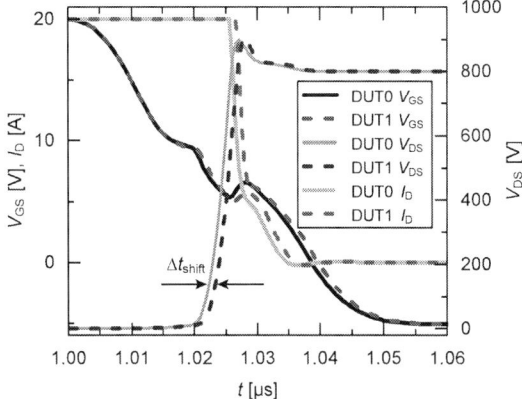

Fig. 5. Comparison between turn off switching waveforms for the two MOSFET designs, DUT0 and DUT1.

B. Effect of C_{ds}

The effect of the non-linear drain-source capacitance C_{ds} is investigated for three different C_{ds} models: a p-n junction capacitance model fitted to match the measured $C_{ds}(V_{DS})$, a $C_{ds}(V_{DS,approx1})$ model approximated with values $C_{ds,high}$ and $C_{ds,low}$ at low and high V_{DS} voltages respectively, and for $C_{ds}(V_{DS,approx2}) = \text{const.}$, equal to the measured C_{ds} at high V_{DS} voltages, as shown in Fig. 7.

The simulations of the turn-on and turn-off transients, including the parasitics of measurement setup and the device packages, are shown in Fig. 8a) and b) for three different C_{DS} models. The comparison of the I_S and V_{DS} curves in Fig. 8a) and b) points out that an assumption of $C_{DS} = \text{const.}$ can be used to model the MOSFET dynamic performance with an acceptable accuracy. Similarly, the constant value of $C_{GS} = \text{const.}$ can be adopted to accurately reproduce the switching transients.

978-1-5386-2928-4/18 $31.00 © 2018 IEEE

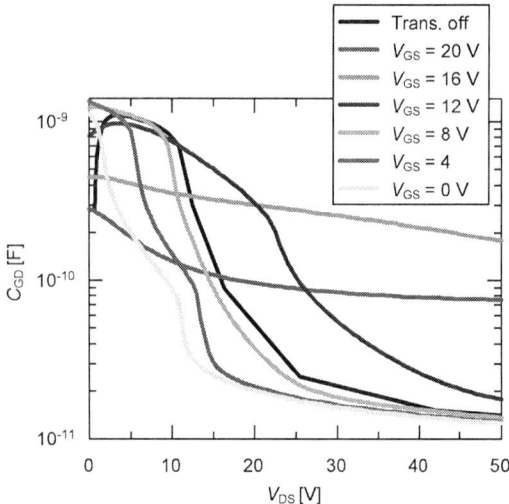

Fig. 6. TCAD small-signal simulation of $C_{gd,DUT1}$ at 1 MHz during switch off (black curve, V_{GS} changing from 20 V to -5 V), and for the quasi-stationary dc variation of V_{DS} for different $V_{GS} = 0 - 20$ V.

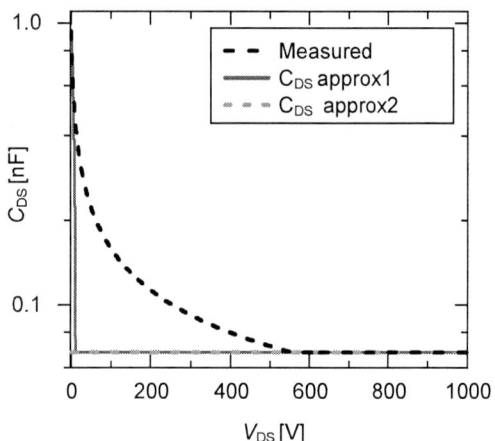

Fig. 7. Three models of C_{ds}: $C_{ds}(V_{DS})$ dependency.

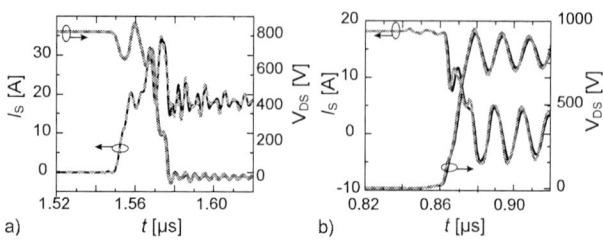

Fig. 8. Three models of C_{ds}: a) turn-on and b) turn-off switching transients.

V. CONCLUSION

This paper presents a detailed analysis of dynamic performance of a 1.2 kV SiC MOSFET with respect to non-linear drain-source C_{ds} and gate-drain C_{gd} semiconductor capacitances. It is demonstrated that the non-linearity of C_{gd} impacts the turn-off delay time $t_{d,OFF}$, while the non-linearity of C_{ds} does not have a significant impact on the switching transients. These results can help to develop a simplified MOSFET model, which can be used for electromagnetic-circuit coupled, i.e. multiphysics, simulations.

REFERENCES

[1] I. Kovacevic-Badstübner, T. Ziemann, B. Kakarla, and U. Grossner, "Highly accurate virtual dynamic characterization of discrete SiC power devices," in *Proc. of 29th Int. Symp. on Power Semiconductor Devices and IC's (ISPSD)*, 2017, pp. 383–386.

[2] K. Chen, Z. Zhao, L. Yuan, T. Lu, and F. He, "The impact of nonlinear junction capacitance on switching transient and its modeling for SiC MOSFET," *IEEE Transactions on Electron Devices*, vol. 62, no. 2, pp. 333–338, 2015.

[3] Z. Duan, T. Fan, X. Wen, and D. Zhang, "Improved SiC Power MOSFET model considering nonlinear junction capacitances," *IEEE Transactions on Power Electronics*, vol. 33, no. 3, pp. 2509–2517, 2018.

[4] Y. Mukunoki, Y. Nakamura, K. Konno, T. Horiguchi, Y. Nakayama, A. Nishizawa, M. Kuzumoto, and H. Akagi, "Modeling of a silicon-carbide MOSFET with focus on internal stray capacitances and inductances, and its verification," *IEEE Transactions on Industry Applications*, 2018.

[5] J. Wang, T. Zhao, J. Li, A. Q. Huang, R. Callanan, F. Husna, and A. Agarwal, "Characterization, modeling, and application of 10-kV SiC MOSFET," *IEEE Transactions on Electron Devices*, vol. 55, no. 8, pp. 1798–1806, 2008.

[6] R. Kraus and A. Castellazzi, "A physics-based compact model of SiC power MOSFETs," *IEEE Transactions on Power Electronics*, vol. 31, no. 8, pp. 5863–5870, 2016.

[7] Z. Chen, "Characterization and modelling of high-switching-speed behaviour of SiC active devices," Master's thesis, The Faculty of the Virginia Polytechnic Institute and State University, 2009.

[8] S. Y. Liu, Y. F. Jiang, W. J. Sung, X. Q. Song, B. Baliga, W. F. Sun, and A. Q. Huang, "Understanding high temperature static and dynamic characteristics of 1.2 kV SiC power MOSFETs," in *Materials Science Forum*, vol. 897, 2017, pp. 501–504.

[9] J. Wang, H. S.-h. Chung, and R. T.-h. Li, "Characterization and experimental assessment of the effects of parasitic elements on the MOSFET switching performance," *IEEE Transactions on Power Electronics*, vol. 28, no. 1, pp. 573–590, 2013.

[10] K. Oishi, M. Shintani, M. Hiromoto, and T. Sato, "Input capacitance determination of power mosfets from switching trajectories," in *Proc. of Int. Conf. of Microelectronic Test Structures (ICMTS)*. IEEE, 2017, pp. 1–6.

[11] H. Sakairi, T. Yanagi, H. Otake, N. Kuroda, H. Tanigawa, and K. Nakahara, "Measurement methodology for accurate modeling of SiC MOSFET switching behavior over wide voltage and current ranges," *IEEE Transactions on Power Electronics*, vol. PP, no. 99, 2017.

[12] M. Shintani, Y. Nakamura, M. Hiromoto, T. Hikihara, and T. Sato, "Measurement and modeling of gate–drain capacitance of silicon carbide vertical double-diffused mosfet," *Japanese Journal of Applied Physics*, vol. 56, no. 4S, p. 04CR07, 2017.

[13] B. Jaeger, Y. Ju, R. Stark, and U. Grossner, "Continuous compact model of a SiC VDMOSFET based on surface potential theory," *Materials Science Forum*, (accepted for publication) 2018.

Proceedings of the 30th International Symposium on Power Semiconductor Devices & ICs
May 13-17, 2018, Chicago, USA

SiC Trench IGBT with Diode-Clamped p-Shield for Oxide Protection and Enhanced Conductivity Modulation

Jin Wei[1], Meng Zhang[2], Huaping Jiang[3], Suet To[2], SungHan Kim[1], Jun-Youn Kim[1] and Kevin J. Chen[4]

[1]Innoscience Technology Co., Ltd, Zhuhai, China   [2]Dept. of ISE, The Hong Kong Polytechnic University, Hong Kong
[3]Dynex Semiconductor Ltd, Lincoln, UK   [4]Dept. of ECE, The Hong Kong University of Science and Technology, Hong Kong
Email: jinwei@innoscience.com; jweiaf@connect.ust.hk

Abstract—In this paper, a diode-clamped p-shield is proposed as a feasible approach for SiC trench IGBT. The introduction of the p-shield effectively suppresses the high electric field in the gate oxide around the trench corner, which is a notorious feature for SiC trench gate devices. Additionally, the p-shield also results in a reduced C_{rss} and thus better switching characteristics. The traditional grounded p-shield (widely adopted in SiC trench MOSFETs), however, significantly reduces the electron/hole density near the emitter side of the SiC trench IGBT. The diode-clamped p-shield structure proposed in this paper successfully solves this issue, since the potential of the p-shield in the on-state is elevated to near the turn-on voltage of the diode, and thus the extraction of holes from the p-shield is suppressed. Therefore, the proposed SiC trench IGBT with a diode-clamped p-shield reduces the high oxide field, improves the switching characteristics and maintains a low V_{ON} simultaneously.

Keywords—*SiC trench IGBT; diode-clamped; p-shield; gate oxide field; conductivity modulation; switching characteristics*

I. INTRODUCTION

SiC power devices have attracted great interest because of its fundamentally superior material property [1-4]. For > 9 kV power applications, SiC IGBT is preferred over SiC MOSFET since the IGBT utilizes conductivity modulation to reduce the on-state voltage drop [5]. SiC planar IGBTs have already been experimentally demonstrated [5-8]. For silicon IGBT, the trench IGBT is the state-of-the-art architecture since the trench structure enhances the conductivity modulation at the emitter side of the device [9-12]. However, the device structure for the SiC trench IGBT is not established. In this paper, we discuss the approaches to realize high-performance SiC trench IGBTs. It is well-known that SiC trench gate devices suffer from high oxide field [13-15]. Thus, the gate oxide reliability would be of great concern if the conventional trench IGBT (C-IGBT) structure (Fig. 1(a)) is adopted for SiC. In SiC MOSFET, a grounded p-shield region is typically adopted to protect the trench gate [16-18]. If this technology is adopted for SiC IGBT (GS-IGBT), as in Fig. 1(b), the gate oxide field could be effectively reduced as well. However, unlike MOSFET, the IGBT requires a high minority carrier concentration at the emitter side of the device to reduce the conduction loss. The adoption of a grounded p-shield for SiC trench IGBT provides a quick path to extract the holes at the emitter side of the device, and degrade the conductivity modulation, resulting in a higher on-state voltage drop (V_{ON}) [9, 19].

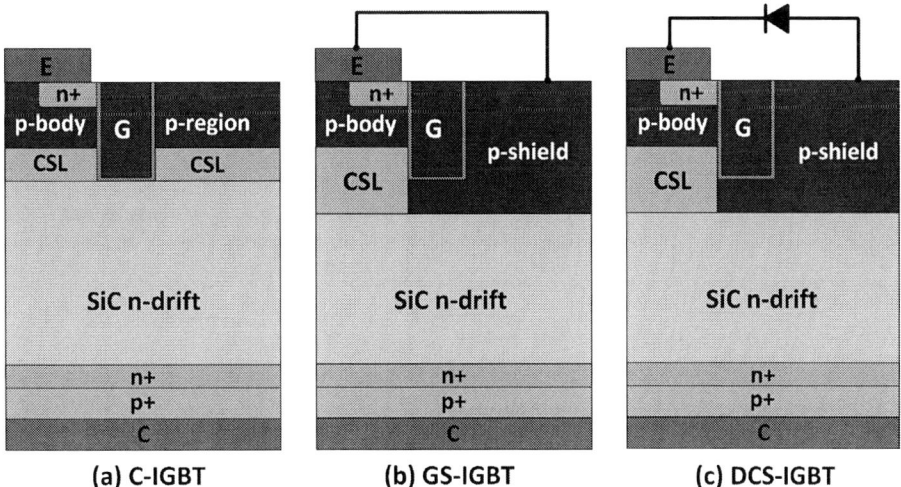

(a) C-IGBT   (b) GS-IGBT   (c) DCS-IGBT

Fig. 1. Cross-sectional structures of (a) Conventional SiC IGBT with a floating p-region (C-IGBT), (b) SiC trench IGBT with a grounded p-shield (GS-IGBT), and (c) SiC trench IGBT with a diode-clamped p-shield (DCS IGBT).

978-1-5386-2928-4/18 $31.00 © 2018 IEEE   411

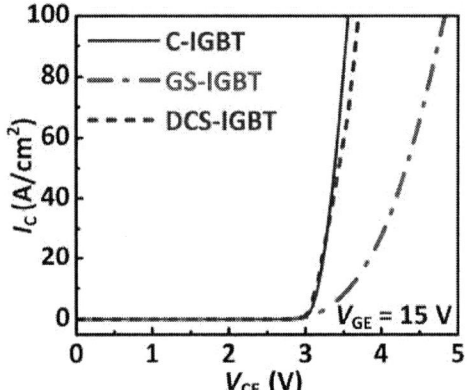

Fig. 2. *I-V* characteristics of the SiC IGBTs.

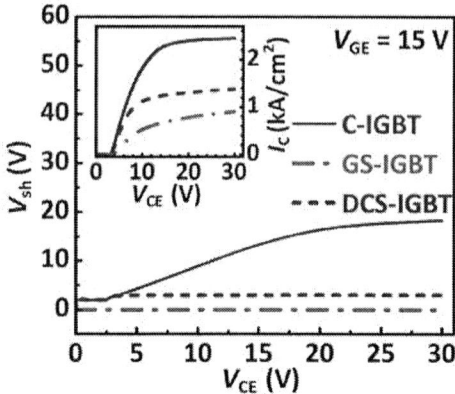

Fig. 4. V_{sh} as a function of V_{CE} in the on-state. The *I-V* characteristics are shown in the inset.

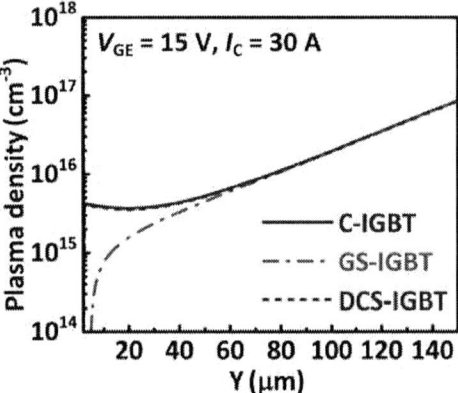

Fig. 3. Plasma distribution along the depth of the drift region.

In this work, a SiC trench IGBT with a diode-clamped p-shield (DCS-IGBT) is proposed, as shown in Fig. 1(c). Compared to SiC trench IGBT without a p-shield, the DCS-IGBT boasts a much lower gate oxide field in the off-state, and an improved switching performance. Furthermore, the potential of the p-shield (V_{sh}) in the DCS-IGBT is elevated to the turn-on voltage of the diode, which suppresses the extraction of minority carriers so a strong conductivity modulation is maintained near the emitter side of the device. A comparison of the DCS-IGBT, the C-IGBT and the GS-IGBT is made through comprehensive numerical simulations. The proposed DCS-IGBT achieves a low gate oxide field, a fast switching speed, and a low V_{ON} simultaneously.

II. DEVICE PERFORMANCES AND DISCUSSION

Figure 1 plots the schematic structures of the C-IGBT, the GS-IGBT and the proposed DCS-IGBT. The clamping diode in the DCS-IGBT could be a SiC PN diode, or a series of other diodes to make the turn-on voltage equal or slightly larger than that of the SiC PN diode. In this study, a SiC PN diode is adopted. The devices are designed for 15-kV level, with a drift region doping concentration of 2×10^{14} cm^{-3} and a thickness of 150 μm. The cell pitch is 4 μm. The thickness of gate oxide is 50 nm. The acceptor concentration for p-body is 2×10^{17} cm^{-3},

and the channel length is 0.5 μm. The mobility of the channel is 25 cm^2/V-s. The gate in the studied devices protrudes over the p-body by 0.8 μm. In the GS-IGBT and DCS-IGBT, the p-shield is deeper than the trench gate by 0.6 μm. The doping concentration of the n-type carrier storage layer (CSL) is 2×10^{16} cm^{-3}. The doping density of the p+ collector region is 5×10^{19} cm^{-3}.

The simulated *I-V* characteristics of the studied SiC IGBTs are plotted in Fig. 2. The V_{ON} (at I_C = 30 A/cm^2) of the GS-IGBT is much larger than the C-IGBT and DCS-IGBT, due to the lower plasma concentration near the emitter side of the GS-IGBT. Figure 3 presents the plasma concentration through a cutline across the middle of the gate towards the collector. The extraction of holes by the grounded p-shield results in very low plasma concentration near the emitter side of the GS-IGBT. For the C-IGBT and DCS-IGBT, the plasma near the emitter side is high owing to the IE effect. Although the DCS-IGBT has a p-shield, the p-shield potential (V_{sh}) is elevated to near the turn-on voltage of the clamping diode (~2.7 V) once the IGBT is in on-state, as shown in Fig. 4. Therefore, the extraction of holes through p-shield is suppressed in the DCS-IGBT.

Figure 4 also presents the saturation current of the studied IGBTs. The GS-IGBT and the DCS-IGBT have low saturation currents than the C-IGBT owing to the JFET effect introduced by the p-shields. As a result, the GS-IGBT and DCS-IGBT are expected to have better short circuit capabilities than the C-IGBT [20, 21].

Figure 5 plots the *BV* characteristics of the studied SiC IGBTs. All of them boast a *BV* larger than the designed 15 kV. The off-state electric field distributions under the condition of V_{GE} = −5 V and V_{CE} = 15 kV are exhibited in Fig. 6. Without the screening effect of a p-shield, the C-IGBT suffers from an extremely high oxide electric field (E_{ox}) of 9.64 MV/cm. An off-state oxide field below 3 MV/cm is commonly desired for long time reliability. With the protection of the p-shield, most of the electric field lines are terminated into the p-shields instead of the gate. Therefore, a much lower E_{ox} of 2.78 MV/cm is achieved in the GS-IGBT and DCS-IGBT. Here, the potential of the diode-clamped p-shield is clamped at

978-1-5386-2928-4/18 $31.00 © 2018 IEEE

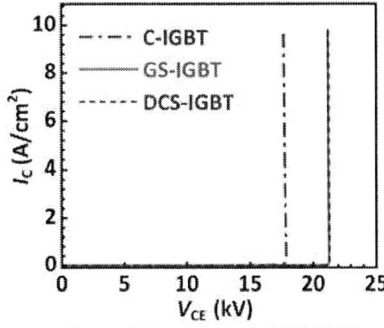

Fig. 5. *BV* characteristics of the studied SiC IGBTs.

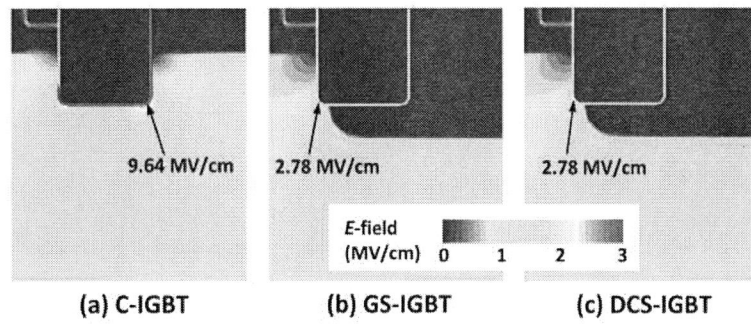

Fig. 6. Off-state electric field distribution in the SiC IGBTs with V_{CE} = 15 kV and V_{GE} = -5 V.

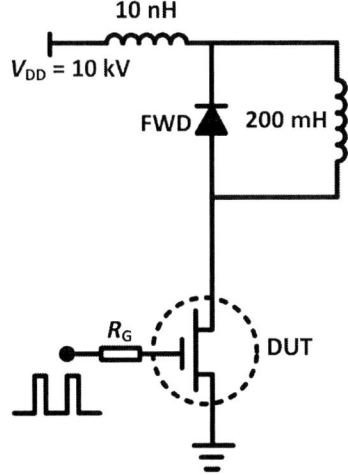

Fig. 7. Test circuit for switching performance of the SiC IGBTs.

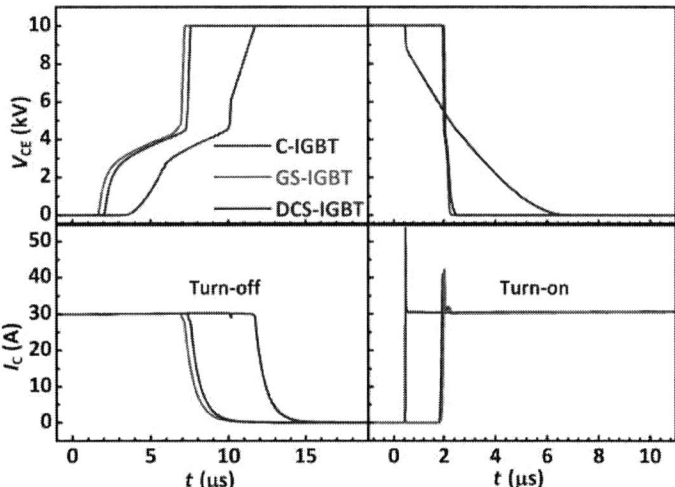

Fig. 8. Switching waveforms of the SiC IGBTs.

~2.7 V, and thus it plays a similar role as the grounded p-shield.

The test circuit for switching performances of the studied SiC IGBTs is shown in Fig. 7. The supply voltage is 10 kV. The load current is 30 A. The areas of the device under test (DUT) and the freewheeling diode (FWD) are 1 cm². A parasitic inductance of 10 nH is assumed for the power loop. A 50-Ω gate resistance (R_G) is used. The gate voltage is switched between −5 V and 15 V to set the studied IGBT off and on, respectively. The switching waveforms are shown in Fig. 8. The p-shields in the GS-IGBT and the DCS-IGBT screens the capacitive coupling between the collector and the gate, resulting in faster switching speed than the C-IGBT. The C-IGBT further suffers from a large current overshoot (I_{OS}) during turn-on transient because of the floating p-region [22-25].

The trade-off relationship between the switching energy (E_{ON}) and I_{OS} during turn-on transient is presented in Fig. 9 for the studied SiC IGBTs by tuning R_G. A larger R_G reduces the I_{OS} by slowing down the switching speed, but the E_{ON} is increased accordingly. The DCS-IGBT and GS-IGBT achieve a much better E_{ON}-I_{OS} trade-off than the C-IGBT.

The trade-off between turn-off loss (E_{OFF}) and the V_{ON} is presented in Fig. 10 for the studied SiC IGBTs by tuning the doping density of the p+ collector region. A higher collector doping concentration facilitates hole injection into the n-drift region, and thus enhances the conductivity modulation and leads to lower V_{ON}. However, a higher hole injection efficiency makes it more difficult to remove the minority carriers during turn-off transient, resulting in higher E_{OFF}. The GS-IGBT suffers from a poorer E_{OFF}-V_{ON} trade-off compared to the C-IGBT and the DCS-IGBT.

III. CONCLUSION

Design optimization of SiC trench IGBTs has been investigated by collectively considering the gate oxide field, on-state conduction loss, and switching performance based on numerical simulation. The conventional trench IGBT structure established for silicon is not desired for SiC IGBT, due to the high gate oxide field. The grounded p-shield concept developed for SiC trench MOSFET cannot be readily transferred to SiC trench IGBT, because of weakened conductivity modulation and the consequent higher V_{ON}. We propose to use a diode-clamped p-shield for the SiC trench IGBT, in which the off-state oxide field can be effectively

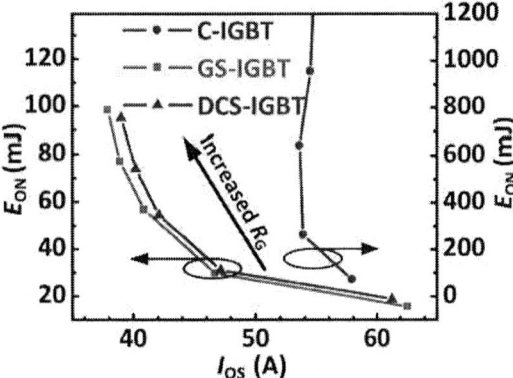

Fig. 9. The trade-off curve between E_{ON} and I_{OS} under tunable R_G.

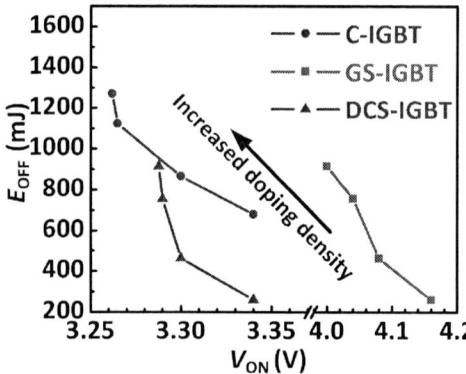

Fig. 10. The trade-off curve between E_{OFF} and V_{ON} under tunable collector injection efficiency.

suppressed while maintaining a low V_{ON} in the on-state. At on-state, the positively biased p-shield enhances the conductivity modulation. Meanwhile, at off-state, the p-shield helps to screen the gate oxide from high electric field. The diode clamped p-shield also improves the dynamic characteristics of the DCS-IGBT owing to reduced C_{rss}. Therefore, the proposed DCS-IGBT has achieved low oxide field, fast switching speed and low V_{ON} simultaneously.

REFERENCES

[1] T. Kimoto and J. A. Cooper, *Fundamentals of Silicon Carbide Technology.* Singapore: Wiley, 2014.

[2] J. W. Palmour, "Silicon carbide power device development for industrial markets," in *IEDM Tech. Dig.*, San Francisco, CA, USA, Dec. 2014, pp. 1-8.

[3] J. A. Cooper and A. Agarwal, "SiC power-switching devices—the second electronics revolution?" *Proc. IEEE*, vol. 90, no. 6, pp. 956-968, Jun. 2002.

[4] J. Wei, H. Jiang, Q. Jiang, and K. J. Chen, "Proposal of a GaN/SiC hybrid field-effect transistor for power switching applications," *IEEE Trans. Electron Devices*, vol. 63, no. 6, pp. 2469-2473, Jun. 2016.

[5] S. Ryu, C. Capell, C. Jonas, L. Cheng, M. O'Loughlin, A. Burk, A. Agarwal, J. Palmour, and A. Hefner, "Ultra high voltage (>12 kV), high performance 4H-SiC IGBTs," in *Proc. ISPSD*, Bruges, Belgium, Jun. 2012, pp. 257-260.

[6] A. Kadavelugu, S. Bhattacharya, S. Ryu, E. Van Brunt, D. Grider, A. Agarwal, and S. Leslie, "Characterization of 15 kV SiC n-IGBT and its application considerations for high power converters," in *Proc. ECCE*, Denver, CO, USA, Sep. 2013.

[7] K. Chu, W. Lee, C. Cheng, C. Huang, F. Zhao, L. Lee, Y. Chen, C. Lee, and M. Tsai, "Demonstration of lateral IGBTs in 4H-SiC," *IEEE Electron Device Lett.*, vol. 34, no. 2, pp. 286-288, Feb. 2013.

[8] Y. Yonezawa, T. Mizushima, K. Takenaka, *et al.*, "Low Vf and highly reliable 16 kV ultrahigh voltage SiC flip-type n-channel implantation and epitaxial IGBT," in *IEDM Tech. Dig.*, Washington, DC, USA, Dec. 2013, pp. 164-167.

[9] M. Kitagawa, I. Omura, S. Hasegawa, T. Inoue, and A. Nakagawa, "A 4500 V injection enhanced insulated gate bipolar transistor (IEGT) operating in a mode similar to a thyristor," in *IEDM Tech. Dig.*, Washington, DC, USA, Dec. 1993, pp. 679-682.

[10] M. Mori, K. Oyama, Y. Kohno, J. Sakano, J. Uruno, K. Ishizaka, and D. Kawase, "A trench-gate high-conductivity IGBT (HiGT) with short-circuit capability," *IEEE Trans. Electron Devices*, vol. 54, no. 8, pp. 2011-2016, Aug. 2007.

[11] S. Linder, *Power Semiconductors.* Lausanne, Switzerland: EPFL Press, 2006.

[12] H. Jiang, J. Wei, B. Zhang, W. Chen, M. Qiao, and Z. Li, "Band-to-band tunneling injection insulated-gate bipolar transistor with a soft reverse-recovery built-in diode," *IEEE Electron Device Lett.*, vol. 33, no. 12, pp. 1684-1686, Dec. 2012.

[13] R. Singh, "Reliability and performance limitations in SiC power devices," *Microelectron. Reliab.*, vol. 46, no. 5-6, pp. 713-730, Dec. 2006.

[14] J. Wei, M. Zhang, H. Jiang, C. Cheng, and K. J. Chen, "Low ON-resistance SiC trench/planar MOSFET with reduced OFF-state oxide field and low gate charges," *IEEE Electron Device Lett.*, vol. 37, no. 11, pp. 1458-1461, Nov. 2016.

[15] M. Zhang, J. Wei, H. Jiang, K. J. Chen, and C. Cheng, "A new SiC trench MOSFET structure with protruded p-base for low oxide field and enhanced switching performance," *IEEE Trans. Device and Mater. Reliab.*, pp. 2592-2598, Jun. 2017.

[16] D. Peters, R. Siemieniec, T. Aichinger, and T. Basler, "Performance and ruggedness of 1200V SiC - trench - MOSFET," in *Proc. ISPSD*, Sapporo, Japan, May 2017, pp. 239-242.

[17] H. Jiang, J. Wei, X. Dai, M. Ke, I. Deviny, and P. Mawby, "SiC trench MOSFET with shielded fin-shaped gate to reduce oxide field and switching loss," *IEEE Electron Device Lett.*, vol. 37, no. 10, pp. 1324-1327, Oct. 2016.

[18] J. Wei, M. Zhang, H. Jiang, H. Wang, and K. J. Chen, "Dynamic degradation in SiC trench MOSFET with a floating p-shield revealed with numerical simulations," *IEEE Trans. Electron Devices*, vol. 64, no. 6, pp. 2592-2598, Jun. 2017.

[19] M. Sumitomo, J. Asai, H. Sakane, K. Arakawa, Y. Higuchi, and M. Matsui, "Low loss IGBT with partially narrow mesa structure (PNM-IGBT)," in *Proc. ISPSD*, Bruges, Belgium, Jun. 2012, pp. 17-20.

[20] R. Tanaka, Y. Kagawa, N. Fujiwara, K. Sugawara, Y. Fukui, N. Miura, M. Imaizumi, and S. Yamakawa, "Impact of grounding the bottom oxide protection layer on the short-circuit ruggedness of 4H-SiC trench MOSFETs," in *Proc. ISPSD*, Waikoloa, Hawaii, USA, Jun. 2014, pp. 75-78.

[21] W. Sung, A. Q. Huang and B. J. Baliga, "A novel 4H-SiC IGBT structure with improved trade-off between short circuit capability and on-state voltage drop," in *Proc. ISPSD*, Hiroshima, Japan, Jun. 2010, pp. 217-220.

[22] M. Yamaguchi, I. Omura, S. Urano, S. Umekawa, M. Tanaka, T. Okuno, T. Tsunoda, and T. Ogura, "IEGT design criterion for reducing EMI noise," in *Proc. ISPSD*, Kitakyushu, Japan, May 2004, pp. 115-118.

[23] M. Shiraishi, T. Furukawa, S. Watanabe, T. Arai, and M. Mori, "Side gate HiGT with low dv/dt noise and low loss," in *Proc. ISPSD*, Prague, Czech Republic, Jun. 2016, pp. 199-202.

[24] Y. Onozawa, H. Nakano, M. Otsuki, K. Yoshikawa, T. Miyasaka, and Y. Seki, "Development of the next generation 1200V trench-gate FS- IGBT featuring lower EMI noise and lower switching loss," in *Proc. ISPSD*, Jeju, Korea, May 2007, pp. 13-16.

[25] H. Feng, W. Yang, Y. Onozawa, T. Yoshimura, A. Tamenori, and J. K. O. Sin, "A new fin p-body insulated gate bipolar transistor with low miller capacitance," *IEEE Electron Device Lett.*, vol. 36, no. 6, pp. 591-593, Jun. 2015.

Proceedings of the 30th International Symposium on Power Semiconductor Devices & ICs
May 13-17, 2018, Chicago, USA

Surge Current Failure Mechanisms in 4H-SiC JBS Rectifiers

Edward Van Brunt, Thomas Barbieri, Adam Barkley, Jim Solovey, Jim Richmond, and Brett Hull

Wolfspeed, A Cree Company
Research Triangle Park, North Carolina/USA
Edward.vanbrunt@wolfspeed.com

Abstract— **4H-SiC Junction Barrier Schottky (JBS) rectifiers can operate in both a purely unipolar mode during normal operation, and a bipolar mode during surge conditions. The transition between operation modes is not only a function of the static device physics, but also the dynamic heating that occurs during a surge transient. For long surge transients on the order of 1ms, heat diffusion into the package allows for 4H-SiC JBS diodes to absorb up to an order of magnitude more energy than surge transients that occur in the 10 µs time scale. Data from surge operation of 4H-SiC diodes at varying time scales was used to map surge currents to an allowed operation space.**

Keywords—4H-SiC, JBS Diode, Surge Operation

I. INTRODUCTION

There are at least two standards that relate to the surge-current testing of power semiconductor rectifiers: the so-called peak non-repetitive surge test [1], which consists of short, high current pulses in the range of 10 µs, and the forward non-repetitive surge test, which consists of longer, lower current pulses in the range of 10 ms [2]. These two operating regimes are an oversimplification of the continuum of possible surge conditions that a diode can experience in the field.

In addition to the varying surge timescales that are possible in real-world applications, 4H-SiC JBS rectifiers exhibit unique operating regimes corresponding to unipolar and bipolar injection that may differ from manufacturer to manufacturer. The current-voltage relationship of a 4H-SiC JBS rectifier is a strong function of the temperature and may also exhibit a distinct bipolar turn-on as shown in the I-V characteristic of a Wolfspeed C4D10120A 10 A, 1200 V JBS diode in Fig. 1.
In a surge transient, the temperature in a 4H-SiC power rectifier may rise rapidly due to joule heating within the device, and, coupled with the device's current-voltage behavior at extreme currents, can give rise to dynamic behavior patterns and negative differential resistance. This work explores the failure mechanisms and transient operation modes of 4H-SiC JBS rectifiers as relevant to surge current stress, with particular emphasis on pulse durations between the extremes described in [1] and [2]. A volume of transient surge data was taken at fine timescales spanning the regime of 10 µs to 10 ms. This data was then used to determine the allowable operation space for an example application from calculated surge current/time/energy characteristic.

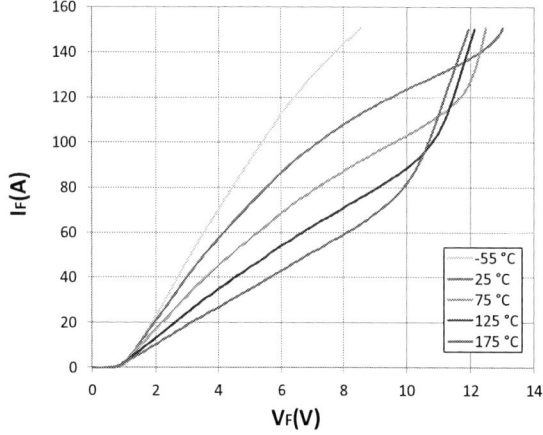

Fig. 1. I-V characteristic of Wolfspeed C4D10120A 10 A, 1200 V JBS diode, showing bipolar behavior at high currents.

II. SURGE CURRENT CHARACTERIZATION

The Wolfspeed C4D10120A 10 A, 1200 V JBS diode was selected as a vehicle to explore surge current behavior under a variety of conditions. Although surge current is typically characterized using a square current pulse for short time scale (e.g. 10 µs) conditions and a half-sine current pulse for longer time scales, square voltage pulses were used to characterize diode surge behavior due to the ease of implementation of a square pulse. All tests were performed at room temperature.

A large 264 mF capacitor bank was used to ensure that the device voltage remained constant during the surge testing. A silicon power switch in series with the capacitor bank was used to realize surge current pulses of lengths ranging from 10 µs to 10 ms with device voltage amplitudes of up to 25 V. The timescales used were 10 µs, 20 µs, 50 µs, 100 µs, 200 µs, 500 µs, 1 ms, 2 ms, 5 ms, and 10 ms. An oscilloscope with a differential voltage probe was used to observe the device under test (DUT) voltage, and a current viewing shunt resistor was used to characterize the DUT current. The realized surge current waveforms for a pulse duration of 10 µs are shown in Fig. 2. The peak DUT voltage is inset into the graph.

978-1-5386-2928-4/18 $31.00 © 2018 IEEE   415

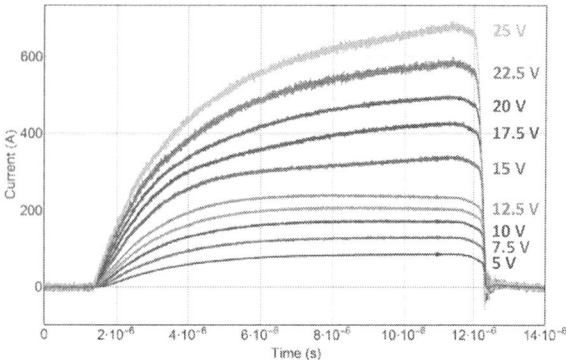

Fig. 2. Surge current waveforms for pulses 10 μs in duration. The DUT peak voltage extracted from the voltage waveform (not shown) is shown to the right of each current waveform.

For each timescale, the DUT peak voltage was stepped in 2.5 V increments until device failure was observed, indicated by a loss in blocking capability. Although each waveform represents a single pulse, to ensure reliable operation of the device as well as consistency with the non-repetitive surge test described in [1], 100 pulses were used for each operation point. Custom gel-filled packages were employed to aid in visual analysis of the failed parts to determine the failure mode.

III. DIODE SURGE ANALYSIS UNDER REPETITIVE CONDITIONS

A number of operational regimes were observed during the surge current testing. Short-time pulses in the regime of 10 μs (as shown in Fig. 2) have a negative differential resistance (increasing current for a constant voltage), consistent with turn-on of the P+/N junction at high currents and high temperatures. This is consistent with an adiabatic heating model that confines heat entirely to the device active region, resulting in increased minority carrier lifetime, increased ionization of P type dopants resulting in increased injection efficiency, and decreased built-in potential of the P+/N junction, all of which contribute to increased injection from the JBS grid in the diode.

For longer timescales in the range of 500 μs, bipolar operation can be observed at much lower currents than in the short pulse regime. This behavior is shown in Fig. 3. The surge current waveforms show two distinct operational modes. For very low current, purely unipolar operation is observed. For higher currents, the device begins operation in unipolar mode, however, eventually enough energy is evolved to heat the device sufficiently to allow for injection from the P+/N junction, giving rise to the increase in current with time observed for the 10V condition.

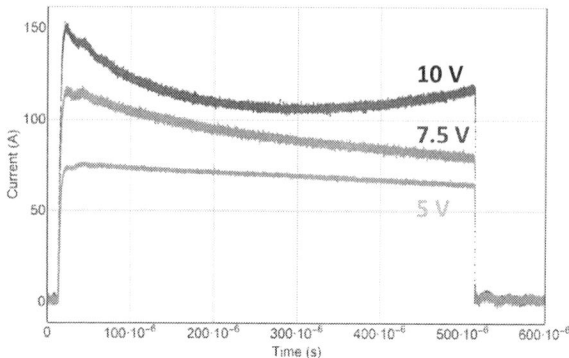

Fig. 3. Surge current waveforms for pulses 500 μs in duration. Bipolar injection is observed in the 10 V waveform.

For each timescale, the voltage and current waveforms captured by the oscilloscope were integrated to obtain surge energy. The resultant plane of surge current, time, and energy can be depicted using a contour plot, as shown in Fig. 4. The uncolored areas represent operational conditions that result in device failure. The graph is largely consistent with the data presented in the device datasheet [3].

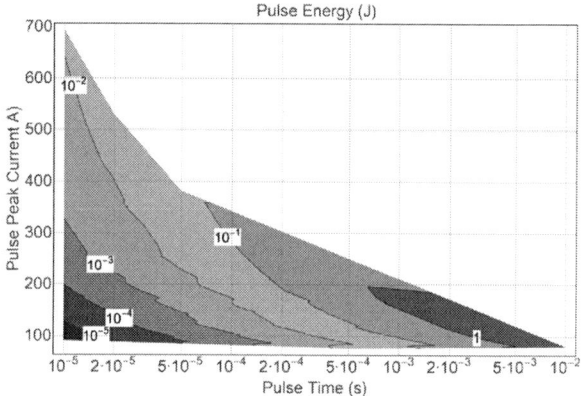

Fig. 4. Contour plot of surge current energy as a function of pulse time and pulse peak current. Points outside of the envelope of color on the graph result in failure in less than 100 surge pulses.

Although peak surge current is an easy-to-define quantity, it is perhaps of less meaning to a circuit designer than pulse mean current, taken by averaging the current in a surge current waveform. This results in an abstraction of different pulse shapes into a single metric, and results in a smooth log/linear boundary between viable operation modes and failure. The resultant contour plot is shown in Fig. 5. The jagged appearance of the plot at lower current is a result of more samples being taken at lower currents (due to the averaging of the current waveforms). The mean current represents a quantity not captured in standard surge current testing, however, it is defined when assuming either a purely square current pulse or purely sinusoidal current pulse. Given the appearance of the surge

978-1-5386-2928-4/18 $31.00 © 2018 IEEE

current waveforms shown in Fig. 2 for 10 µs "square" pulses, this simplification may be inaccurate, and the mean current represents a more general metric of device performance.

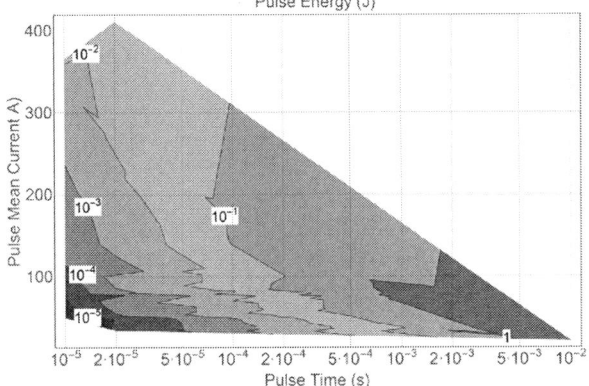

Fig. 5. Contour plot of surge current energy as a function of pulse time and pulse mean current.

The surge current energy data shown in Fig. 5 can be fit to an empirical formula that is valid for positive values of the output function. The functional form is given for the observed data for the C4D12010A diode as

$$E_{surge} = -\frac{2.04}{I_{surge}^{0.2}} - \frac{7.51}{\ln\left[7.72\ t_{surge}\right]}$$

The generated surface is plotted below in Fig. 6 and is shown to capture the behavior that is present in Fig. 5. To avoid non-physical negative energy values, the function should be bounded at 10 µJ on the lower end. It is apparent that even for repetitive conditions at this small energy value, non-negligible self-heating can result in the device when operated at high frequency.

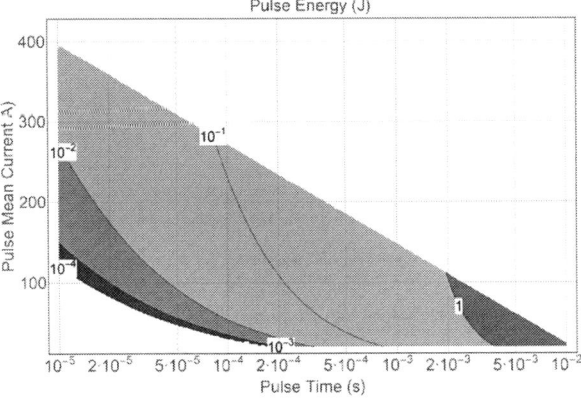

Fig. 6. Surge energy as a function of current duration/mean current generated by equation fit as shown above. Negative values of the function are non-physical and not plotted. The functional form was chosen to capture the behavior of the empirical data.

IV. DIODE SURGE FAILURE

Of note in the plot shown in Fig. 4 is the far higher energies that can be sustained before failure at long pulse times: the surge energy-to-fail at 10 ms is over 10 times the energy required to induce failure at 10 µs. The surge current failure energies are shown, along with surge current and pulse duration in Fig. 7.

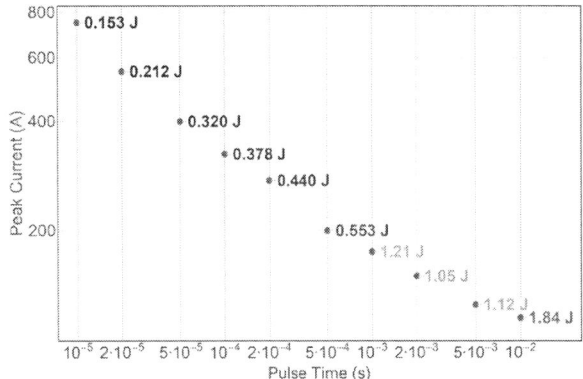

Fig. 7. Plot of peak surge current conditions that lead to device failure in less than 100 pulses, as a function of surge current pulse duration. To the right of each point is the computed energy of each pulse.

Light microscope images of the failed device die for each surge pulse duration as observed through the package gel are shown in Fig. 8. The observed failure characteristic offers some explanation behind the increased failure energy at long pulse times that is depicted in Fig. 7.

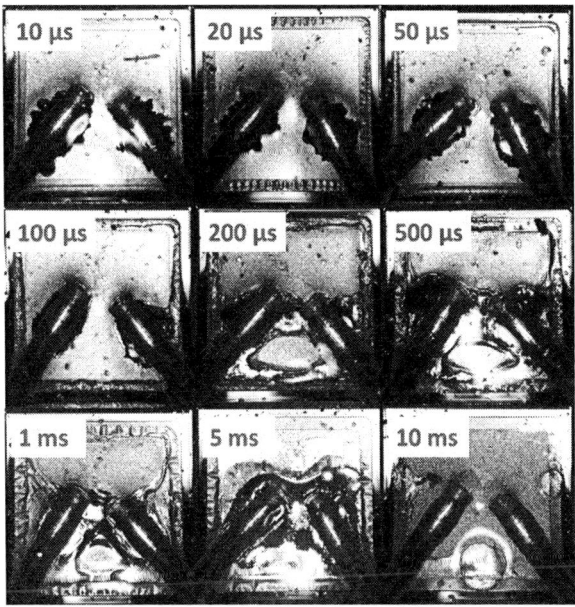

978-1-5386-2928-4/18 $31.00 © 2018 IEEE

Fig. 8. Images of die subjected to surge conditions resulting in failure for operational points shown in Fig. 7.

For short duration pulses such as 10 µs, the observed damage is confined entirely to the wirebond pads, even constrained to near the boundaries of the wires themselves, indicating that heat diffusion was confined not only to a limited vertical direction, but also laterally. Limited melting of the aluminum anode metallization at the device edges is observed at the 20 µs condition; this melting increases at 50 µs and 100 µs and begins to spread throughout the anode metallization at 200 µs. These increases in melted area correspond with a steady increase in the energy required to lead to failure and is indicative of increased lateral and vertical heat diffusion. This trend continues to the maximum current pulse duration of 10 ms, which results in complete reconstruction of the bond pad metal and a substantial increase in the maximum tolerable non-repetitive surge energy. The complete reconstruction of the bond pad indicates that sufficient time has been provided for thermal equilibration within the device structure.

V. DISCUSSION AND CONCLUSION

The increase in tolerable non-repetitive surge energy with increasing pulse duration as shown Fig. 7, when coupled with the localized bond pad damage that is observed for short timescales in Fig. 8 indicates that a simple 1-dimensional thermal model is insufficient to capture the dynamic behavior of the device for ultra-short current pulses. 3-dimensional electrothermal TCAD modeling should be employed to accurately capture these details. Short circuit failures in active switch devices such as MOSFETs occur on even shorter timescales and have similar values of current and could also potentially benefit from a 3-dimensional simulation approach.

The surge current, time, and energy relationship shown in Fig 5 can be used to generate a more general SOA curve for surge operation of 4H-SiC JBS rectifiers, however, this SOA curve will depend on the operation conditions for a given application including converter operation frequency and device current, as well as any switching losses incurred during ordinary, non-surge operation of the 4H-SiC power rectifier. Thus, the surge energy is a more general device characteristic represents a sufficient abstraction to allow for use in a loss model to determine steady-state device heating that can occur during repetitive surge operation. This additional surge power loss can then be added to the other device losses to determine an application-specific device surge-SOA that does not exceed maximum rated junction temperature in steady state device operation.

REFERENCES

[1] IEC 60747-2, "Semiconductor Devices- Discrete Devices and Integrated Circuits Part 2: Rectifier Diodes," 2nd Edition, 2000.

[2] JEDEC JSD282-B Rev. .01, "Silicon Rectifier Diodes," Section 4.2, 2002.

[3] C4D10120A Datasheet, http://www.wolfspeed.com/downloads/dl/file/id/82/product/56/c4d10120a.pdf

Proceedings of the 30th International Symposium on Power Semiconductor Devices & ICs
May 13-17, 2018, Chicago, USA

Surge Capability of 1.2kV SiC Diodes with High-Temperature Implantation

Hongyi Xu, Jiahui Sun, Jingjing Cui, Jiupeng Wu, Hengyu Wang, Shu Yang, Na Ren, Kuang Sheng*

College of Electrical Engineering
Zhejiang University, Hangzhou, China
*shengk@zju.edu.cn

Abstract—**This paper presents a high-temperature implanted 4H-SiC JBS diode with improved surge capability. The fabrication of the P+ region is implemented with 500 °C implantation. It was found that Ti can form ohmic contact on high temperature implanted P+ region without any additional annealing. This could simplify the ohmic contact process for MPS fabrication. In this work, a wide transition P+ region between cell and termination is designed, which can alleviate snapback phenomenon and improve the surge capability.**

Keywords—*SiC diode, high temperature implantation, surge capability*

I. INTRODUCTION

Silicon carbide (SiC) diodes have shown great advantages in high temperature, high frequency and high voltage applications[1]. Power diodes would undergo a high surge current in some situations, which can result in thermal runaway. Therefore, surge capability has been regarded as one of the most important parameters for power diodes. However, the enhancement of surge capability usually requires additional fabrication processes. For instance, additional metallization and annealing process are usually implemented to form ohmic contact to the P+ region in the merged PN structure (MPS) to enhance the surge capability.

Additionally, snapback phenomenon usually occurs in conventional JBS diodes, which is caused by the short P+ region length. The snapback could cause unbalanced current for paralleled devices, which eventually result in thermal runaway[2].

In this paper, a fabrication technology featuring high temperature ion implanted P+ region is developed, which can form ohmic contact to the as-deposited Titanium and realize improved surge capability without any additional process step. Furthermore, the large P+ transition region could alleviate the snapback phenomenon.

II. DEVICE FABRICATION

JBS diodes were fabricated on n-type 4° off-axis 4H-SiC. The epi-layer has a doping concentration of $8 \times 10^{15} cm^{-3}$ and thickness of 12μm. Fig. 1 shows process flow of the JBS. Al with a dose of 2.12×10^{15} cm^{-2} was implanted at room temperature (RT) or 500 °C. As shown in Fig. 2, the implanted area at 500 °C shows a lighter color compared with that at room

temperature, implying less damage from the high temperature implantation.

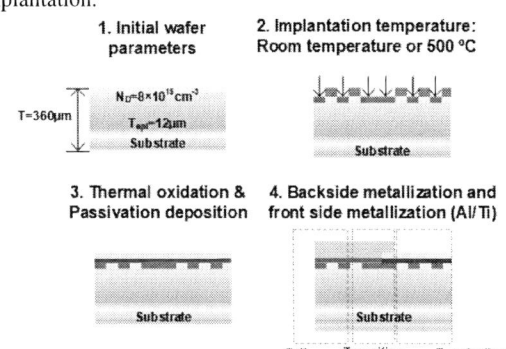

Fig. 1. JBS Process flow in this work.

After RCA cleaning, post-implantation annealing was carried out at 1550 °C in Ar ambient. Device passivation was implemented by wet thermal oxidation and SiO₂ deposition. The backside metal is sputtered and annealed to form ohmic contact. After passivation via opening, Al/Ti was sputtered on the front side without any additional annealing.

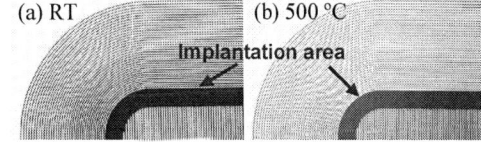

Fig. 2. Different surface colors after implantation at (a) RT and (b) 500 °C.

III. EXPERIMENTAL RESULTS OF HIGH TEMPERATURE IMPLANTED JBS DIODES

To compare the electrical parameters of P+ region implanted at room temperature and 500 °C, four probe test and hall effect test were conducted on Van Der Pauw structures. The test results of the sheet resistance, mobility and doping concentration are summarized in Table I. Compared to Device A with implantation at RT, Device B with implantation at 500 °C has a higher activation rate, leading to lower sheet resistance. Lower mobility is caused by higher coulomb scattering due to higher hole concentration. Higher activation rate with

978-1-5386-2928-4/18 $31.00 © 2018 IEEE   419

implantation at higher temperature was also previously reported[3].

TABLE I. Hall effect test results of P+ doped area formed by Al implantation at RT and 500 °C.

Parameters	Device A (w/ implantation at RT)	Device B (w/ implantation at 500 °C)
Sheet resistance (Ω/sq)	40200	20400
Mobility (cm²V⁻¹s⁻¹)	14.2	7.96
Ns (cm⁻²)	1.1×10^{13}	3.85×10^{13}

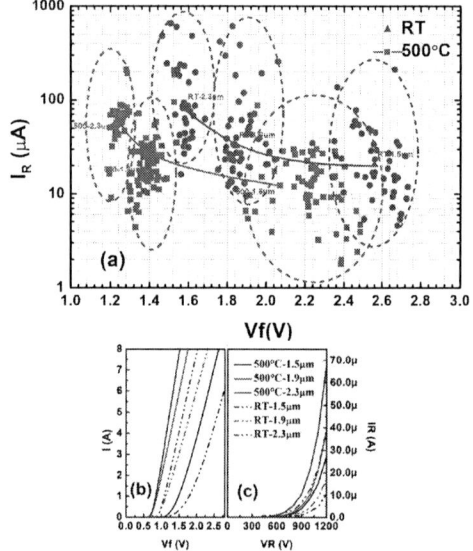

Fig. 3. (a) Comparison of forward voltage (VF@IF=5A) and leakage current (IR@V=1200V) of diodes. The devices were fabricated with implantation at room temperature (RT) and 500 °C, respectively. (b) Typical forward characteristics of these devices. (c) The typical blocking characteristics of these devices.

The two types of devices were packaged into TO220, and static performance results are shown in Fig. 3. Forward voltage drop is extracted at a forward current of 5A. The leakage current is extracted at reverse bias of 1200V. The spacing between adjacent P+ regions was varied as 1.5μm, 1.9μm and 2.3μm. In comparison to Device A with implantation at RT, Device B with high temperature implantation could lower the leakage current, as a result of less defect density from ion implantation.

The static performance of two devices was measured by Tektronix 371A, which is shown in Fig. 4(a). Device B implanted at 500 °C converts to bipolar mode at a lower voltage, which is nearly 3V lower than that of Device A implanted at RT. The I-V characteristics on Van Der Pauw structure are also measured before ohmic annealing shown in Fig. 4(b). Device B shows ohmic contact behavior, whereas the Device A shows rectifying characteristic. 500 °C implantation results in higher activation rate and less lattice damage in the P+ implanted region which contributes to ohmic contact characteristic with Ti without annealing.

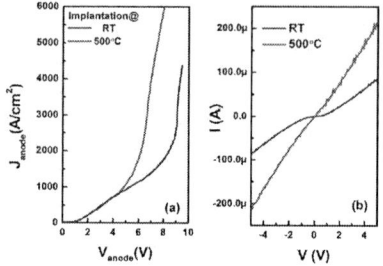

Fig. 4. (a) Forward I-V characteristics of the JBS diodes. (b) I-V characteristics of the Van Der Pauw structure.

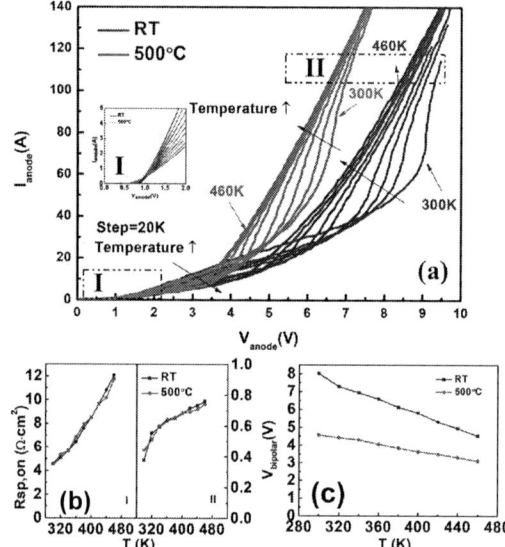

Fig. 5. (a) Comparison of the forward characteristics of devices A and B at different measurement temperatures. The inset shows the static performance in region I. (b) Comparison of the specific on-resistance extracted from region I and II individually. (c) Comparison of the bipolar turn-on voltage of device A and B at different temperature.

The static performance of two devices was measured by Tektronix 371A, which is shown in Fig. 4(a). Device B implanted at 500 °C converts to bipolar mode at a lower voltage, which is nearly 3V lower than that of Device A implanted at RT. The I-V characteristics on Van Der Pauw structure are also measured before ohmic annealing shown in Fig. 4(b). Device B shows ohmic contact behavior, whereas the Device A shows rectifying characteristic. 500 °C implantation results in higher activation rate and less lattice damage in the P+ implanted region which contributes to ohmic contact characteristic with Ti without annealing.

Both devices were characterized at elevated temperature. The devices could reach above 140A with pulse width 250 μs without failure, as shown in Fig. 5(a). The differential on-resistances in the unipolar mode (region I) and bipolar mode (region II) of the two devices are similar at various temperatures, as shown in Fig. 5(b). The only difference is the bipolar turn-on voltage, as shown in Fig. 5(c). Device B with 500 °C implantation shows lower bipolar turn-on voltage than Device A with RT-implantation.

To confirm the different bipolar turn-on voltage is caused by different implantation temperature, PN diodes fabricated with different implantation temperature are measured under 300K to 380K shown in Fig. 6. The turn on voltage is ~ 2.5V in Device B, whereas in Device A the turn-on voltage is above 3V.

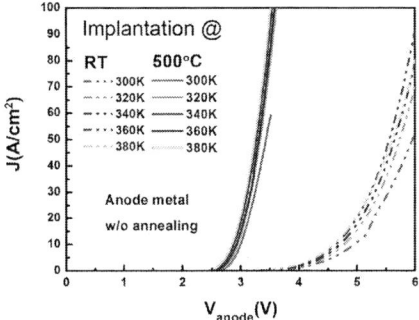

Fig. 6. I-V characteristic of the PN diodes with elevated temperature.

IV. SURGE CURRENT CAPABILITY

The surge capability test has been carried out to evaluate the two types of devices. After each surge test, the breakdown voltage is measured to check if the device is degraded. The last non-degradation test waveforms is shown in Fig. 7. At about 1ms, the change of slope shows the devices enter into bipolar mode. Device B shows good surge capability due to lower power dissipation and heat generated within the device.

The forward I-V characteristics of the device was simulated by Silvaco, with the line width determined by SEM measurement. The total cell pitch is 4.9μm and ±0.1μm error due to measurement is also considered. In the simulation result in Fig. 8, distinct snapback phenomena are observed, whereas the snapback is insignificant in the measurement. The main reason may be ascribed to a wide P+ transition region, as shown in Fig. 9.(b).

In the JBS diode, the on state resistance could be separated into channel resistance (R_{ch}), spreading resistance (R_{spread}), drift resistance (R_{drift}), substrate resistance (R_{sub}) and ohmic contact resistance (R_c) (marked in fig. 9.(a))[4]. The voltage between P+ region and the n-epi region beneath is defined as V_{pn}. V_{pn} is determined by the diode current, R_{ch} and R_{spread}. If the resistance of the two parts is relatively high, the V_{pn} could easily reach the on state voltage. Otherwise it needs higher current to turn on the pn junction and trigger the bipolar mode, which is undesirable to surge capability as more heat would be generated. Channel resistance should be relatively small. The conventional way is to design a wide P+ region to increase the spreading resistance. In this work, although the P+ region in the cell is small, the transition region is wide enough to drive the device into bipolar mode at a lower current.

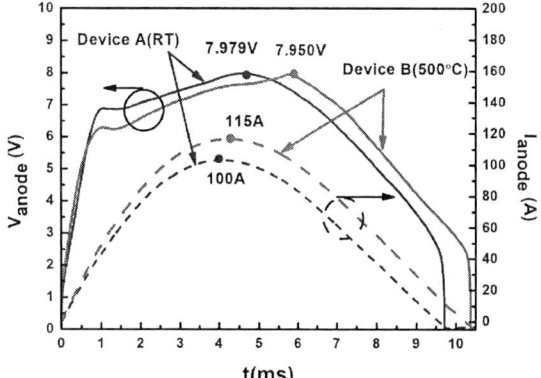

Fig. 7. The last non-degradation surge capability test waveforms. Devices enter bipolar mode during surge operation. Due to the ohmic contact with P+ region, Device B (500 °C) shows better surge capability.

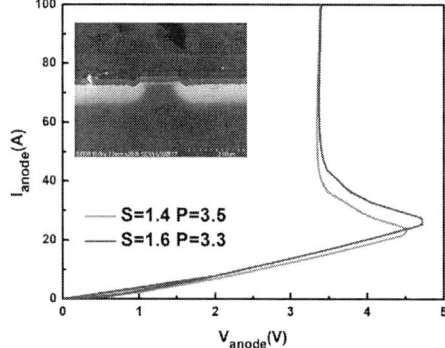

Fig. 8. Simulation results of active region is based on structure parameters measured by the SEM. The I-V characteristic evidences obvious snap back point which is not in accordance with the measured curve.

To investigate the impact of the P+ region on the snapback phenomenon, the forward I-V characteristics of the devices were simulated. Fig. 10 shows the simulated electron and hole current density, which shows the P+ transition region enters bipolar mode first. And the turn-on of the P+ transition region could lead to more holes injecting into the drift layer and could assist the P+ regions nearby to be turned on.

To investigate the influence of the higher hole concentration in the wider P+ region at the beginning of the bipolar operation, the surge test was used to evidence if the P+ region was destructed. As shown in the inset of Fig. 11, at the front side metal has melted after surge capability test. The location is in the middle of the chip due to localized high current density and heat concentration. The surge current waveform is plotted in Fig. 11. Both devices do not turn back to Schottky mode until the anode voltage drops to 2.5V, which is caused by the increased resistance with high temperature during the surge process. At point A, PN structure in the device is turned on and a slight snap back phenomenon is observed. After surge current reaches the peak, negative differential resistance occurred[5].

978-1-5386-2928-4/18 $31.00 © 2018 IEEE   421

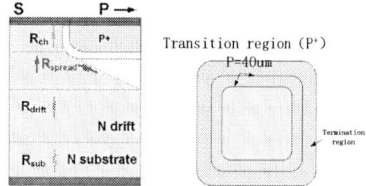

Fig. 9. Schematic diagrams of device (a) structure and (b) layout. The wider P+ region would increase R_{spread}, which result in the turn-on of the PN junction at a lower voltage.

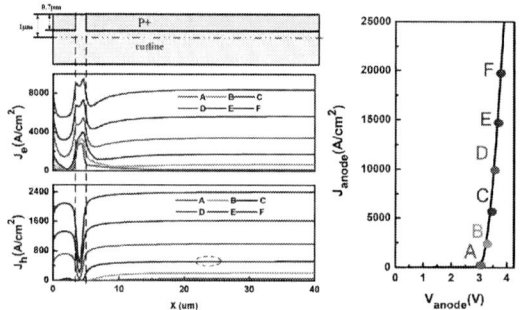

Fig. 10. The turn-on behavior in transition region. The transition region is firstly turned on and could assist the P+ region nearby to be turned on, and thus, the snap back phenomenon is alleviated.

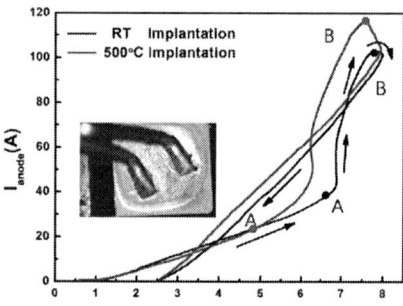

Fig. 11. I-V characteristics of both devices in the last non-degradation surge capability test.

In Fig. 7, the comparison of the last non-destruction waveform of two devices is illustrated. After the surge test, the blocking voltage and forward characteristic are monitored. The destruction of the diode can be predicted by blocking voltage. The results are shown in Fig. 12. The active region area of Device A and B is 2.62mm², while the active region area of Device C (C2D10120A[6]) is 7.29mm². The Device A reaches 100A with 10ms half sine wave at 25 °C. After that, its breakdown voltage is reduced from 1500V to 900V. Device B with the same measurement condition could reach 115A with breakdown voltage reduced from 1550V to 1000V. In addition, Device C reaches 110A with breakdown voltage decreased

from 2200V to 1150V. Device C has high voltage at peak surge current because it does not enter bipolar mode. The larger forward voltage drop would generate more heat leading to Device C destructed at lower current compared with Device A and B.

Fig. 12. The degradation (last) point denotes the surge current capability for each diode. The device fabricated via 500 °C implantation demonstrates the best surge capability among the three devices.

V. CONCLUSION

In this paper, high-temperature (500 °C) implanted P+ region is implemented in the JBS diode fabrication. The high temperature implantation is shown to be effective in reducing lattice damage and increasing the activation rate. As a result, the implanted P+ region can form ohmic contact directly to the as-deposited Ti. In addition, with the wide P+ transition region, snap-back phenomenon is alleviated, compared with the conventional JBS diode with narrower P+ regions. The surge current experiments verified that an improved surge capability is achieved in the JBS diodes with high-temperature implantation.

ACKNOWLEDGMENT

This work was supported by the National Key Research and Development Program of China (No.2016YFB0400502) & National Natural Science Foundation of China (No. 51577169)

REFERENCE

[1] N. Ren and K. Sheng, "An analytical model with 2-D effects for 4H-SiC trenched junction barrier Schottky diodes," IEEE Transactions on Electron Devices, vol. 61, no. 12, pp. 4158-4165, December. 2014.

[2] H. Niwa, J. Suda and T. Kimoto, "Ultrahigh-voltage SiC MPS diodes with hybrid unipolar/bipolar operation," IEEE Transactions on Electron Devices, vol. 64, no. 3, pp. 874-881, March 2017.

[3] N. S. Saks, A. V. Suvorov, D. C. Capell. "High temperature high-dose implantation of aluminum in 4H-SiC," Applied Physics Letters, vol. 84, no. 25, pp. 5195-5197, June. 2004.

[4] N. Ren, J. Wang and K. Sheng, "Design and experimental study of 4H-SiC trenched junction barrier Schottky diodes," IEEE Transactions on Electron Devices, vol. 61, no. 7, pp. 2459-2465, July 2014.

[5] S. Fichtner, S. Frankeser, J. Lutz, R. Rupp, T. Basler, R. Gerlach. "Ruggedness of 1200 V SiC MPS diodes," Microelectronics Reliability, vol. 55, no. 9–10,pp. 1677-1681, June. 2015.

[6] C2D10120A, Cree Inc., Durham, NC, USA, 2016. [Online]. Available: http://www.wolfspeed.com /c2d10120a.pdf.

978-1-5386-2928-4/18 $31.00 © 2018 IEEE

Proceedings of the 30th International Symposium on Power Semiconductor Devices & ICs
May 13-17, 2018, Chicago, USA

Ruggedness of 6.5 kV, 30 A SiC MOSFETs in Extreme Transient Conditions

Ashish Kumar*, Sanket Parashar*, Shadi Sabri[†], Edward Van Brunt[†], Subhashish Bhattacharya* and Victor Veliadis*

* Department of Electrical and Computer Engineering, North Carolina State University, Raleigh, NC, USA.
[†]Wolfspeed, A Cree Company, Research Triangle Park, Raleigh, NC, USA.
Email address: akumar19@ncsu.edu

Abstract— **6.5 kV silicon IGBTs are used in multi-megawatt medium voltage power converters for rail traction applications, AC drives and grid-connected power converters. The 6.5 kV, 30 A SiC MOSFETs, recently launched by Wolfspeed, have the potential to replace Si IGBTs. Static and dynamic characteristics of the 6.5 kV SiC MOSFETs have been reported earlier. During the extreme transient conditions, the SiC MOSFETs can be subjected to high voltage overshoot and high short circuit current. In this work, ruggedness of these MOSFETs is established by short-circuit tests and single shot avalanche tests at room temperature. The 6.5 kV SiC MOSFET is observed to withstand 7.75 J of short-circuit energy at 20 V gate voltage with a stress withstanding time of 6 μs. The single shot avalanche test result shows an avalanche energy of 5.0 J. The increase in junction temperature of the SiC MOSFET at the avalanche failure is estimated from the experimental results.**

Keywords—6.5 kV SiC MOSFET, avalanche ruggedness, short circuit, unclamped inductive switching.

I. INTRODUCTION

6.5 kV silicon IGBTs are widely used in medium voltage power converters in rail traction, ship propulsion systems and medium voltage drives, where 3.6 kV dc bus voltage level is required to source high current. Due to the MOS-gated structure and ease of controllability, IGBTs are preferred in the lower range of the medium voltage applications [1]. Short circuit capability of Si IGBTs adds to the reliability of power converters. However, ruggedness limitations of 6.5 kV silicon IGBTs have been reported in [2], [3]. Moreover, 6.5 kV silicon IGBTs used in MW converters can only be switched at less than 1 kHz, typically around 300 Hz to 500 Hz due to high losses. Given their inherent unipolar nature, 6.5 kV silicon carbide MOSFETs have a significantly higher switching frequency limit, making them a potential replacement to 6.5 kV silicon IGBTs. Recently, Wolfspeed launched a new generation 6.5 kV/30 A SiC MOSFET for medium voltage applications [4]. Their ruggedness in extreme transient conditions must be evaluated to ensure reliability of the converters. Short-circuit fault and unclamped inductive switching (UIS) are two major extreme transient conditions in which these MOSFETs are expected to absorb the fault energy, and still survive after the fault clearance. These limits are important parameters required for both the converter design and the fault protection circuit design to ensure the safety of the devices and converter operation.

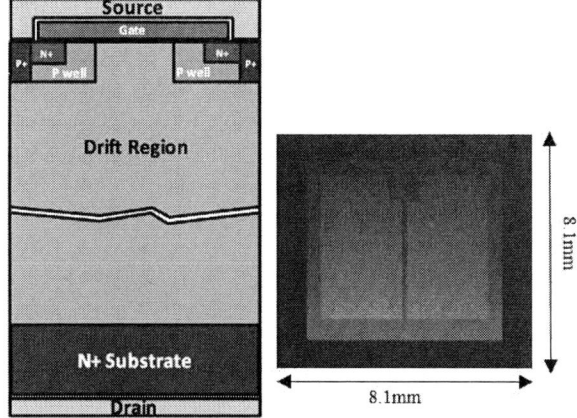

Fig 1: Device structure and die layout of the 6.5 kV, 30 A SiC MOSFET.

There have been many recent reports on the avalanche ruggedness and short-circuit characterization of commercial 1200 V SiC MOSFETs and 10 kV SiC MOSFETs in [5], [6], [7] and [8]. Short-circuit and UIS failure mechanism of SiC MOSFETs are proposed in [9]-[13]. In this paper, both the short circuit and the avalanche ruggedness of 6.5 kV, 30 A SiC MOSFETs are reported, for the first time. The minimum energy dissipated into the die for the short circuit and the avalanche failures are determined experimentally.

II. DEVICE STRUCTURE OF 6.5 KV SIC MOSFET

Device structure and die layout of the 6.5 kV SiC MOSFET are shown in Fig 1. It has a planar structure with n-doped drift region. The die has an active area of $0.41\,cm^2$. Forward blocking characteristics of the MOSFET at room temperature is shown in Fig 2. The blocking voltage is observed to be higher than 7.25 kV. Output I-V characteristics of these MOSFETs are reported in [4].

III. SHORT CIRCUIT RUGGEDNESS

A. Test Methodology

Short circuit performance of the MOSFET is evaluated under a hard switch fault condition at room temperature [14]. Schematic of the short-circuit test bench is shown in Fig 3. The 6.5 kV MOSFET is connected across a dc bus of 3500 V. Initially, the MOSFET is turned off by applying -5 V to the gate. A single shot gate pulse of +20 V is applied to emulate the short-circuit fault. Turn-on duration of the gate pulse is varied from

978-1-5386-2928-4/18 $31.00 © 2018 IEEE   423

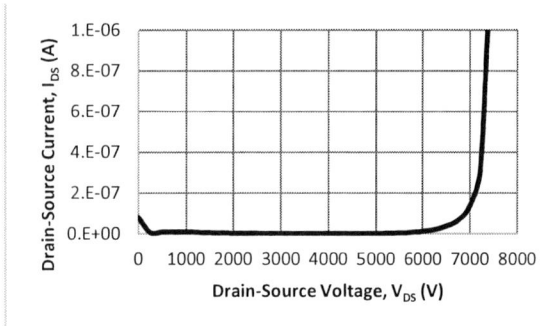

Fig 2: Forward blocking characteristic of the 6.5 kV/30 A SiC MOSFET at V_{GS}=0 V and 25°C.

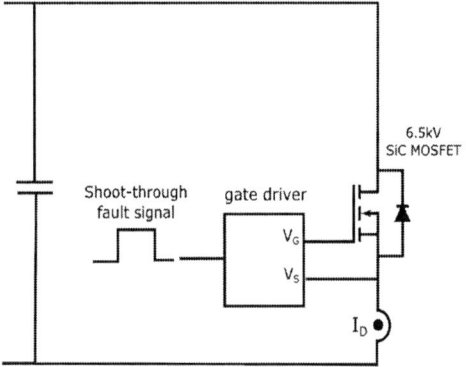

Fig 3: Schematic of the experimental set-up for the short circuit fault test

Fig 4: Photograph of the experimental test setup for the short circuit characterization of the 6.5 kV SiC MOSFET.

2 µs to 6 µs in 1 µs steps. This gradual increase is required to approach the failure condition cautiously considering this experiment is a destructive test. Photograph of the experimental set-up is shown in Fig 4. The setup is enclosed in an air-tight safety box. A high bandwidth current monitor Pearson-3972 is employed to detect the short-circuit switch current.

B. Results

The short-circuit experimental results at different fault durations are shown in Fig 5. The device does not fail until the ON pulse duration of 5 µs. At 6 µs, the short circuit energy

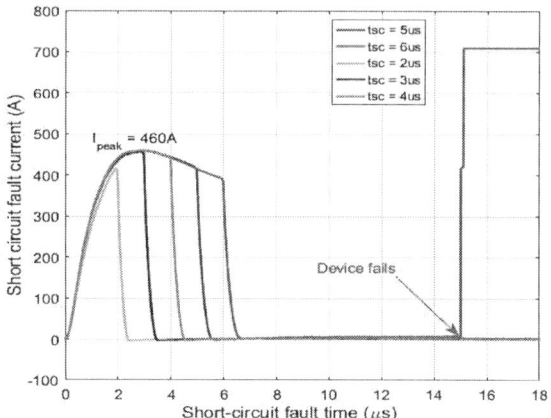

Fig 5: Experimental results from the short circuit fault test of the 6.5 kV SiC MOSFET. The fault duration is increased from 2 µs to 6 µs, and the device eventually fails at 6 µs with 7.75 J of the total fault energy. The peak current is 460 A at VDS = 3500 V and V_{GS} = 20 V.

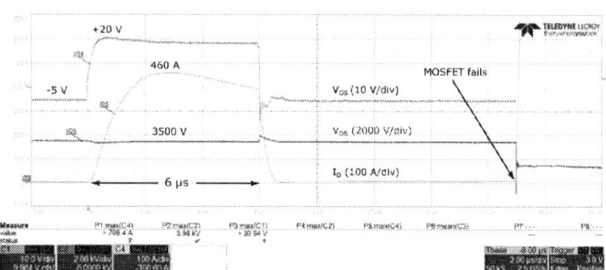

Fig 6: Experimental results of short circuit characterization of 6.5 kV SiC MOSFET at V_{DS}= 3500 V and V_{GS} = 20 V at room temperature.

becomes sufficient to cause a failure with a total fault energy of 7.75 J. The peak current is measured to be 460 A. Experimental waveforms at the failure with the drain voltage V_{DS} and the gate voltage V_{GS} are shown in Fig 6. Owing to the small inductance in the dc bus path, there is a small fluctuation in the drain voltage when the fault occurs.

IV. AVALANCHE RUGGEDNESS

A. Approach

Single shot avalanche test provides a good measure of ruggedness in transient conditions. A schematic of the test bench is shown in Fig 7. The avalanche current and the time at the failure are observed while varying the inductor, L, and keeping the dc bus voltage constant. Three different inductance values are explored.

A typical waveform during a single shot avalanche test is shown in Fig 8. The avalanche energy E_{AV} for the device failure is calculated when the device voltage shoots up to the avalanche voltage V_{AV} to the point it eventually fails at time T_{AV}. The method to estimate the increase in the device junction temperature, T_J, during the avalanche condition is discussed in [6].

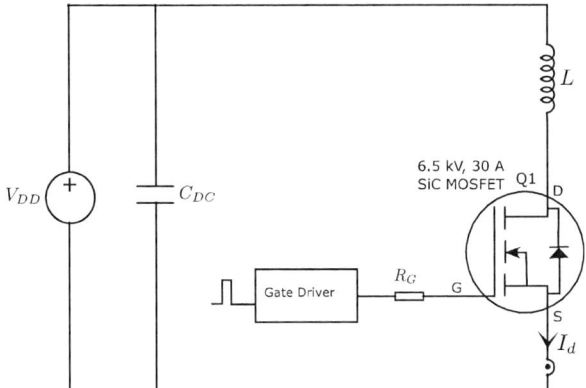

Fig 7: Schematic of the test bench for single pulse avalanche test.

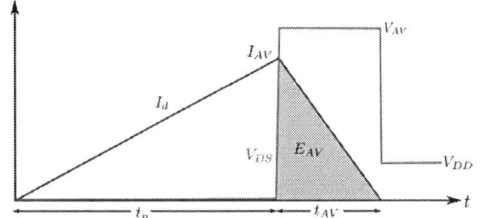

Fig 8: Waveform of the single pulse avalanche test [6].

Fig 9: Photograph of the experimental set-up for the single pulse avalanche test.

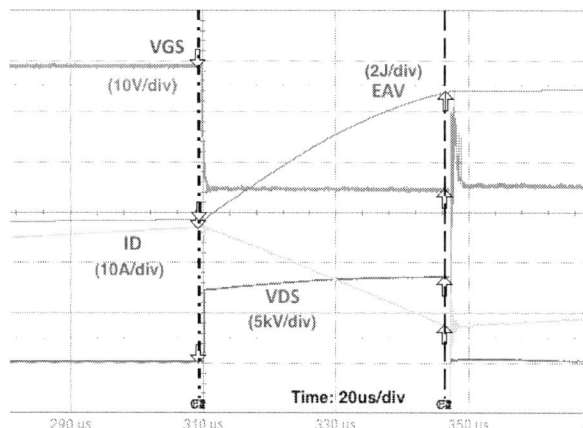

Fig 10: Experimental results for the single pulse avalanche test of the 6.5 kV SiC MOSFET at room temperature. E_{AV} = 5.0 J, I_{AV} = 26.9 A, T_{AV} = 37 μs, V_{AV} = 8.51 kV, V_{DD} = 1300 V, L = 14 mH.

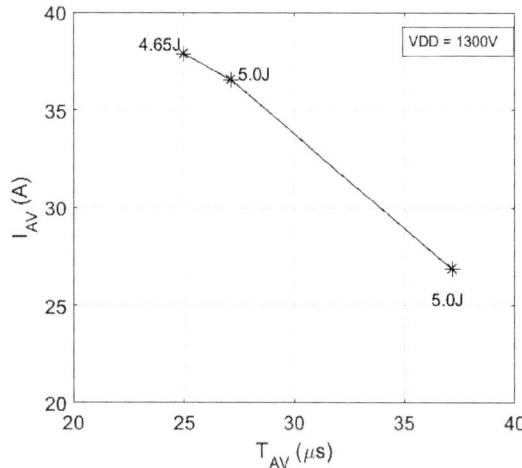

Fig 11: Experimental results from the single pulse avalanche test of the 6.5 kV SiC MOSFET at various current level and at room temperature. The avalanche energy E_{AV} varies from 4.65 J to 5.0 J.

B. Results

The experimental waveform at the single pulse avalanche failure is shown in Fig 10 at a peak current of 26.9 A, resulting in an avalanche energy, E_{AV}, of 5.0 J, meanwhile pushing the avalanche voltage to 8.51 kV. The avalanche time T_{AV} of 37 μs is sufficient to clear the avalanche fault using modern protection circuitry. Data points are obtained at three current levels, and are plotted in Fig 11. Using the relation $E_{AV}=0.5*L*I^2_{AV}$, an avalanche energy of 5.0 J sets the limit of maximum allowable inductance of 11.1 mH that can be connected to the 6.5 kV SiC MOSFET carrying the rated current of 30 A at room temperature.

V. DISCUSSION

A. Short-circuit Behavior

As shown in Fig 6, It is worth noting that the device fails at 9 μs after the short circuit fault pulse is cleared. This delayed failure could be attributed to the increase in the leakage current in the blocking state [9]. The leakage current increases as a result of the increase in temperature during the fault. The temperature becomes sufficiently high to trigger the latch up, and results in the failure.

The short-circuit energy required for failure of the 6.5 kV SiC MOSFET is observed to be 7.75 J at V_{DS} of 3500 V and V_{GS} of 20 V, resulting in the short-circuit withstand time T_{SC} of 6 μs. Power MOSFETs are typically operated at around 80% of their rated voltage. Assuming the same short-circuit fault energy and the current behavior at 5000V, the withstand capability T_{SC} gets reduced to 4.2 µs, which is larger than the SC trip time demonstrated by modern gate driver protection circuitry, making it practically employable in medium voltage power converters [15]. The ruggedness parameter T_{SC} can be further investigated for the MOSFET at elevated junction temperature up to 175˚C as it will provide more reliable ruggedness characteristics for the MOSFET in actual practical applications. The SiC MOSFET can be further characterized for the short circuit behavior under load fault condition [14].

B. Avalanche failure

An insight into the junction temperature of the MOSFET during the avalanche failure is discussed here based on the analysis given in [6]. The junction temperature at the failure is estimated by the equation:

$$T_{jmax} = T_0 + \frac{I_{AV} T_{AV}}{2} \frac{E_c}{A_c C_v}$$

where T_0 is the initial junction temperature, A_C is the active area of the device, C_V is the volumetric heat capacity, and E_C is the critical electric field for the failure. The junction temperature based on the experimental results shown in Fig 10 is estimated using the values of C_V and E_C given in [16], and the device parameters. The increase in the junction temperature is plotted in Fig 12. The junction reaches the maximum temperature of 980˚C at the failure.

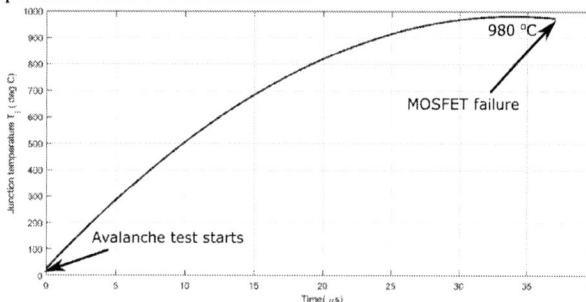

Fig 12: Estimation of the junction temperature T_J of the MOSFET during the avalanche test. The MOSFET fails after 37 µs with estimated T_J of 980˚C. E_{AV} = 5.0 J, I_{AV} = 26.9 A, T_{AV} = 37 µs, V_{AV} = 8.51 kV, V_{DD} = 1300 V, L = 14 mH.

The maximum avalanche energy required for the MOSFET failure indicates the safety margin to the circuit designer. Fig 11 shows the maximum current before the failure at the various avalanche times. At the rated current of 30 A, the allowable avalanche time T_{AV} is estimated to be 34 µs, which is larger than the typical avalanche time in most power converters. Assuming the same avalanche energy of 5.0 J at 30 A, the maximum inductance comes out to be 11.1 mH that can be connected to the MOSFET without any additional protection circuitry for avalanche failure. This ruggedness can be further evaluated at higher operating temperatures.

VI. CONCLUSION

The ruggedness of the 6.5 kV SiC MOSFET under extreme transient conditions is characterized by measuring the energy required for the failure to occur during short-circuit and avalanche condition. At 3500 V dc bus voltage, the short circuit energy is observed to be 7.75 J resulting in a short-circuit

withstand time of 6 µs. At a typical operating voltage of 5000 V, the SC stress withstand time comes down to 4.2 µs, which is larger than the protection time offered by modern gate drivers. The single shot avalanche needs 5.0 J of energy for the device to fail. The junction temperature of the MOSFET is estimated to reach up to 980˚C at the avalanche failure.

REFERENCES

[1] M. Hiller, R. Sommer and M. Beuermann, "Converter Topologies and Power Semiconductors for Industrial Medium Voltage Converters," *2008 IEEE Industry Applications Society Annual Meeting*, Edmonton, Alta., 2008, pp. 1-8.

[2] J. G. Bauer, O. Schilling, C. Schaeffer and F. Hille, "Investigations on the ruggedness limit of 6.5 kV IGBT," *Proceedings. ISPSD '05. The 17th International Symposium on Power Semiconductor Devices and ICs, 2005.*, 2005, pp. 71-74.

[3] A. Kopta, M. Rahimo, U. Schlapbach, N. Kaminski and D. Silber, "Limitation of the short-circuit ruggedness of high-voltage IGBTs," *2009 21st International Symposium on Power Semiconductor Devices & IC's*, Barcelona, 2009, pp. 33-36.

[4] S. Sabri *et al.*, "New generation 6.5 kV SiC power MOSFET," *2017 IEEE 5th Workshop on Wide Bandgap Power Devices and Applications (WiPDA)*, Albuquerque, NM, 2017, pp. 246-250

[5] M. D. Kelley, B. N. Pushpakaran and S. B. Bayne, "Single-Pulse Avalanche Mode Robustness of Commercial 1200 V/80 m SiC MOSFETs," in IEEE Transactions on Power Electronics, vol. 32, no. 8, pp. 6405-6415, Aug. 2017

[6] A. Kumar et al., "Single Shot Avalanche Energy Characterization of 10kV, 10A 4H-SiC MOSFETs, 2018 Applied Power Electronics Conference (APEC), San Antonio, TX, Mar. 2018.

[7] C. Ionita, M. Nawaz, K. Ilves and F. Iannuzzo, "Short-circuit ruggedness assessment of a 1.2 kV/180 A SiC MOSFET power module," *2017 IEEE Energy Conversion Congress and Exposition (ECCE)*, Cincinnati, OH, 2017, pp. 1982-1987.

[8] E. P. Eni, S. Bęczkowski, S. Munk-Nielsen, T. Kerekes and R. Teodorescu, "Short-circuit characterization of 10 kV 10A 4H-SiC MOSFET," *2016 IEEE Applied Power Electronics Conference and Exposition (APEC)*, Long Beach, CA, 2016, pp. 974-978.

[9] G. Romano *et al.*, "Short-circuit failure mechanism of SiC power MOSFETs," *2015 IEEE 27th International Symposium on Power Semiconductor Devices & IC's (ISPSD)*, Hong Kong, 2015, pp. 345-348..

[10] G. Romano *et al.*, "A Comprehensive Study of Short-Circuit Ruggedness of Silicon Carbide Power MOSFETs," in *IEEE Journal of Emerging and Selected Topics in Power Electronics*, vol. 4, no. 3, pp. 978-987, Sept. 2016.

[11] A. Fayyaz *et al.*, "UIS failure mechanism of SiC power MOSFETs," *2016 IEEE 4th Workshop on Wide Bandgap Power Devices and Applications (WiPDA)*, Fayetteville, AR, 2016, pp. 118-122.

[12] Tomoyuki Shoji *et al.*, "Theoretical analysis of short-circuit capability of SiC power MOSFETs", *Japanese Journal of Applied Physics*, Jan 2015.

[13] A. Fayyaz *et al.*, "UIS failure mechanism of SiC power MOSFETs," *2016 IEEE 4th Workshop on Wide Bandgap Power Devices and Applications (WiPDA)*, Fayetteville, AR, 2016, pp. 118-122.

[14] R. Chokhawala, J. Catt and L. Kiraly, "A discussion on IGBT short circuit behavior and fault protection schemes," *Proceedings Eighth Annual Applied Power Electronics Conference and Exposition,*, San Diego, CA, 1993, pp. 393-401.

[15] A. Kumar, A. Ravichandran, S. Singh, S. Shah and S. Bhattacharya, "An intelligent medium voltage gate driver with enhanced short circuit protection scheme for 10kV 4H-SiC MOSFETs," *2017 IEEE Energy Conversion Congress and Exposition (ECCE)*, Cincinnati, OH, 2017, pp. 2560-2566.

[16] V. Pala, B. Hull, J. Richmond, P. Butler, S. Allen and J. Palmour, "Methodology to qualify silicon carbide MOSFETs for single shot avalanche events," *2015 IEEE 3rd Workshop on Wide Bandgap Power Devices and Applications (WiPDA)*, Blacksburg, VA, 2015, pp. 56-59.

Proceedings of the 30th International Symposium on Power Semiconductor Devices & ICs
May 13-17, 2018, Chicago, USA

Next Generation 1200V, 3.5mΩ.cm² SiC Planar Gate MOSFET with Excellent HTRB Reliability

Sauvik Chowdhury, Kevin Matocha, Blake Powell, Gin Sheh and Sujit Banerjee

Monolith Semiconductor, Inc.
Round Rock, TX, U.S.A.
schowdhury@monolithsemi.com

Abstract—**In this paper, we report on 1200V SiC planar gate MOSFETs with an improved tradeoff between on-resistance and reverse bias gate oxide electric field. The improved tradeoff was obtained by optimizing the JFET doping profile and unit cell design. These MOSFETs showed a specific on-resistance of 3.5mOhm.cm² at room temperature, increasing to 5.9mOhm.cm² at 175°C, along with excellent High Temperature Reverse Bias (HTRB) reliability as shown by no failures after stressing at 1440V, 175°C for 1000hours.**

Keywords—silicon carbide; reliability; power MOSFET; high voltage)

I. INTRODUCTION

For high voltage power electronics applications, silicon carbide (SiC) power MOSFETs offer several advantages over silicon bipolar IGBTs, such as significantly lower switching losses due to no minority carrier storage, a fast intrinsic body diode with negligible reverse recovery loss, and a resistive forward characteristic without a diode voltage drop which helps to improve efficiency under light load conditions. To reduce the cost of SiC devices, we have adopted the approach of processing SiC MOSFETs on 150mm substrates, fabricated in a high volume silicon foundry [1]. Integration of the SiC MOSFET process flow in a silicon production line allows efficient usage of available fab capacity, and hence minimizes the processing cost of SiC power devices.

Our first generation (Gen-1) commercial SiC MOSFETs devices show a specific on-resistance ($R_{on,sp}$) of about 5mΩ.cm², with excellent reliability and ruggedness characteristics, as shown by negligible threshold voltage drift and high gate oxide reliability [2]. To achieve further cost reduction, it is necessary to further reduce specific on-resistance, which will result in a smaller die area and improved manufacturing yield. In this paper, we present detailed characteristics of our next generation (Gen-2) 1200V SiC MOSFET technology platform, with an $R_{on,sp}$ of 3.5mΩ.cm². This state-of-the-art performance is enabled by optimizing the cell design for lowest $R_{on,sp}$, while maintaining a low gate oxide electric field (E_{ox}) in the blocking state, resulting in excellent HTRB reliability.

This work was supported by funding from ARPAe (DE-AR0000442) and the U.S. Army Research Laboratory (W911NF-15-2-0088).

II. DEVICE STRUCTURE AND PERFORMANCE

A. Device structure and static performance

1200V SiC planar gate MOSFETs with a die area of

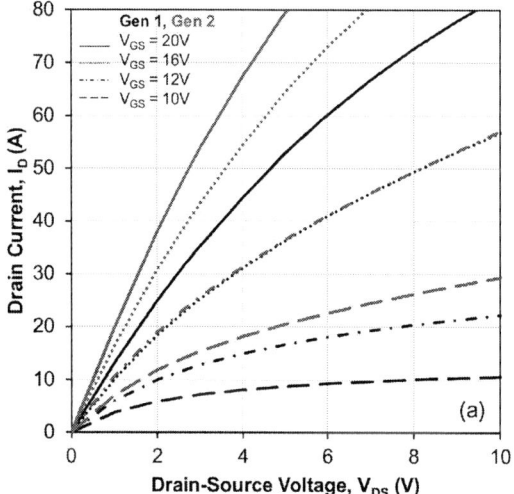

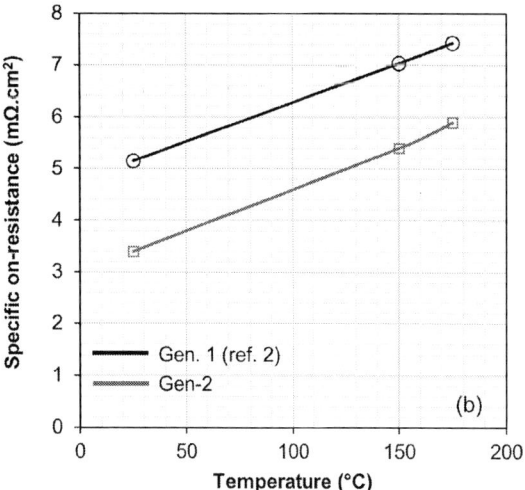

Fig. 1 (a) Room temperature output characteristics of Gen-1 and Gen-2 MOSFETs, and (b) normalized $R_{on,sp}$ variation with junction temperature.

978-1-5386-2928-4/18 $31.00 © 2018 IEEE   427

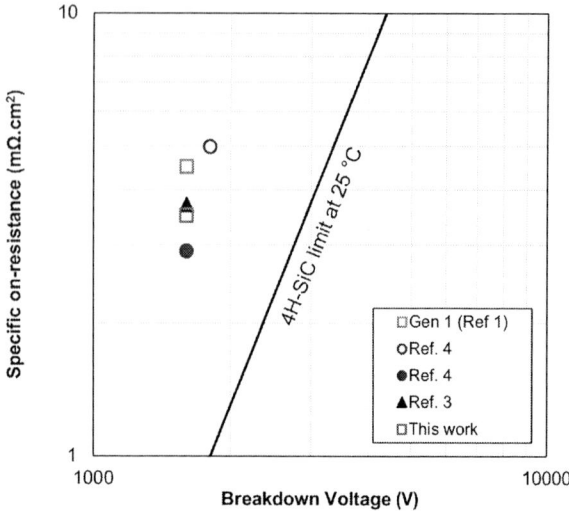

Fig. 2 $R_{on,sp}$ – BV tradeoff for state of the art 1200V SiC MOSFETs.

approximately $10mm^2$ were fabricated. Drift layer parameters were chosen for an avalanche breakdown voltage of approximately 1600V, for a breakdown voltage rating of 1200V. One of the key design elements for SiC planar gate MOSFETs is the constituent junction field effect transistor (JFET) formed between adjacent P-wells. The gap between adjacent P-wells (W_{JFET}), unit cell layout, and the JFET doping profile have a significant effect on both on-state performance and off-state reliability. Gen-2 MOSFETs utilize a heavily doped JFET region formed by nitrogen ion implantation, which has been optimized for the best tradeoff between $R_{on,sp}$ and reliability.

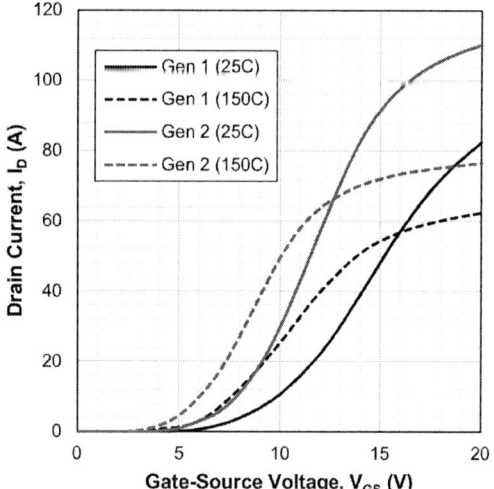

Fig. 3 Transfer characteristics (V_{DS} = 10V) of Gen-1 and Gen-2 SiC MOSFETs at 25 °C and 150 °C.

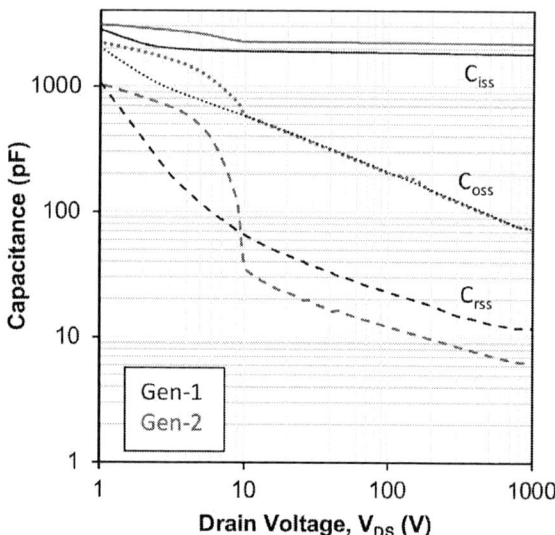

Fig. 4 Comparison of device capacitances for Gen-1 and Gen-2 MOSFETs.

Fig. 1 shows compares the output characteristics of Gen-2 MOSFET with a commercial Gen-1 MOSFET (LSIC1MO120E0080). Both devices showed a threshold voltage of 2.8V at room temperature (measured at I_D = 10mA, V_{DS} = V_{GS}). The specific on-resistance of the Gen-2 MOSFET was about $3.5m\Omega.cm^2$ (measured at V_{GS} = 20V, I_D = 20A). The thick N^+ substrate contributes $0.7m\Omega.cm^2$ to the total on-resistance, hence the device $R_{on,sp}$ can be further reduced to approximately $3.0m\Omega.cm^2$ by thinning of the N^+ substrate. The on-resistance of Gen-2 MOSFETs is comparable to that of other state of the art planar and trench gate 1200V SiC MOSFETs [3], [4], as shown in Fig. 2. The temperature dependence of specific on-resistance for MOSFETs with and without JFET doping is compared in Fig. 1 (b). In SiC MOSFETs, the channel resistance component has a negative temperature coefficient due to reduction in threshold voltage and increase in inversion channel mobility, whereas other resistance components (drift, JFET, source etc.) have a positive temperature coefficient. In Gen-2 MOSFETs, the channel resistance component is a smaller part of the total on-resistance as a result of the higher channel and JFET packing density. This leads to a slightly faster increase in on-resistance with temperature for Gen-2 MOSFETs, however the on-resistance is still only $5.9m\Omega.cm^2$ at a junction temperature of 175 °C.

Typical transfer characteristics for both types of devices are compared in Fig. 3. Gen-2 MOSFETs show significantly improved transconductance as a result of the higher channel density in these devices. This also results in a wider gate voltage region over which the drain current shows a negative temperature coefficient. This behavior is an especially attractive feature for uniform current sharing across multiple paralleled devices, especially under dynamic conditions.

978-1-5386-2928-4/18 $31.00 © 2018 IEEE

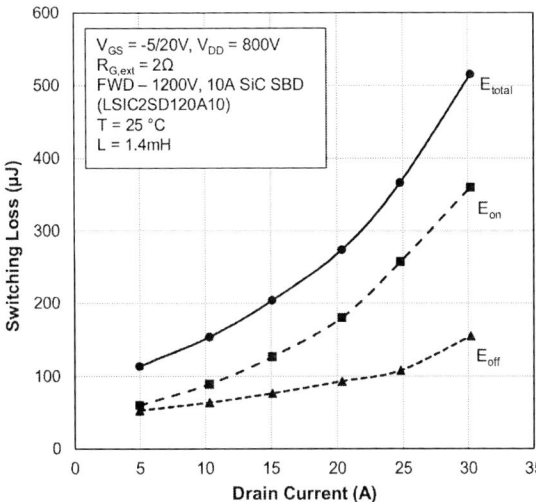

Fig. 5 Inductive load switching losses as a function of drain current for Gen-2 MOSFETs ($R_{G,ext}$ = 2Ω, V_{DD} = 800V).

B. Dynamic Performance

The devices were assembled in standard TO-247 3-lead package for dynamic and reliability tests. Fig. 4 compares the standard capacitance-voltage characteristics of Gen-1 and Gen-2 MOSFETs. The Gen-2 MOSFET shows lower C_{gd} at high drain voltages due to the smaller gate-drain overlap, with a slight overall increase in C_{iss} due to higher channel density. Inductive load switching measurements were performed with an external gate resistance of 2Ω and a 1200V, 10A SiC Schottky diode (LSIC2SD120A10) as the freewheeling device. Fig. 5 shows the switching losses as a function of drain current. At a drain bias of 800V and drain current of 20A, total switching losses are less than 300μJ with an external gate resistance of 2Ω. Despite the increase in channel density, Gen-

2 MOSFETs still showed good short circuit capability, as shown in Fig. 6. The device was able to turn off safely after a short circuit pulse of 4μs at a DC bus voltage of 600V.

III. IMPACT ON DEVICE RELIABILITY

The high critical electric field of SiC which results in superior performance characteristics, also necessitates proper field management to prevent degradation of the gate oxide or other passivation layers when the device is operated in the off-state. In the active region of a planar MOSFET, the peak electric field in the gate oxide (E_{ox}) under reverse bias occurs in the middle of the JFET region, and an excessively high electric field can lead to gate oxide degradation and early device failures [5], [6]. This field can be reduced by shrinking the JFET width or modifying the JFET doping profile, which however also increases the on-resistance. Thus there is an inherent trade-off between on-resistance and oxide electric field. Fig. 7 compares this $R_{on,sp} - E_{ox}$ tradeoff for Gen-1 and Gen-2 MOSFETs, which clearly demonstrates the improved tradeoff for Gen-2 MOSFETs. For a given oxide electric field, Gen-2 MOSFETs show about 30% lower $R_{on,sp}$. In addition, due to the heavily doped JFET region in Gen-2 MOSFETs, $R_{on,sp}$ is nearly independent of oxide field. This allows a dramatic reduction in E_{ox}, with only a small penalty in $R_{on,sp}$. The heavily implanted JFET design also eliminates the variability in oxide electric field arising out of epi doping variations, resulting in improved process margin and manufacturability.

To study the effect of E_{ox} on device reliability, two Gen-2 MOSFET designs (Designs A and B, as indicated in Fig. 7) were chosen for a stepped voltage, High Temperature Reverse Bias (HTRB) stress test (20 devices per design). The drain voltage was stepped from 960V (80% of rated BV) to 1440V at

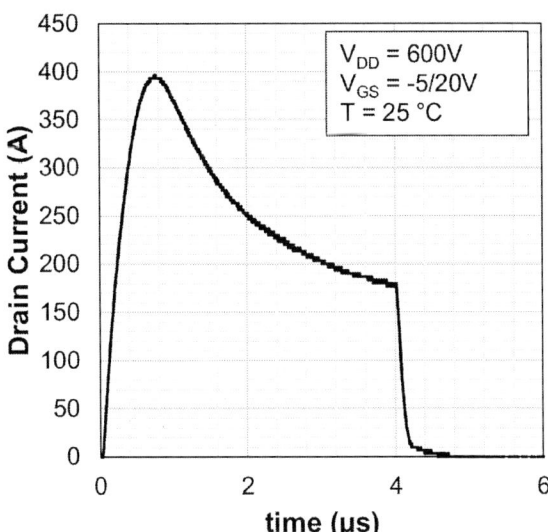

Fig. 6 Short circuit capability of Gen-2 MOSFETs (V_{DD} = 600V).

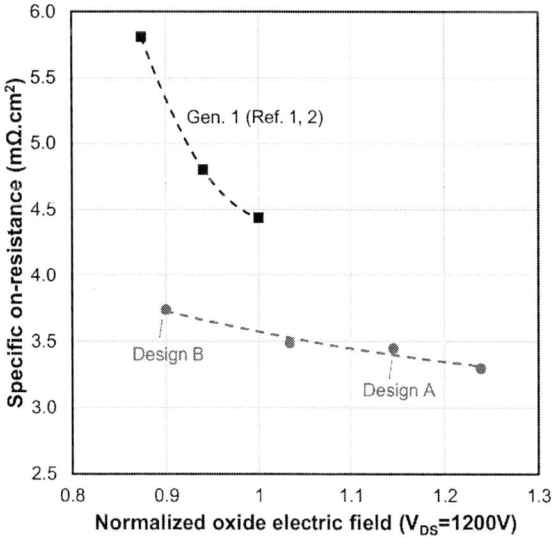

Fig. 7 Specific on-resistance as a function of peak oxide electric field (at V_{DS} = 1200V) for Gen-1 and Gen-2 SiC MOSFETs.

978-1-5386-2928-4/18 $31.00 © 2018 IEEE   429

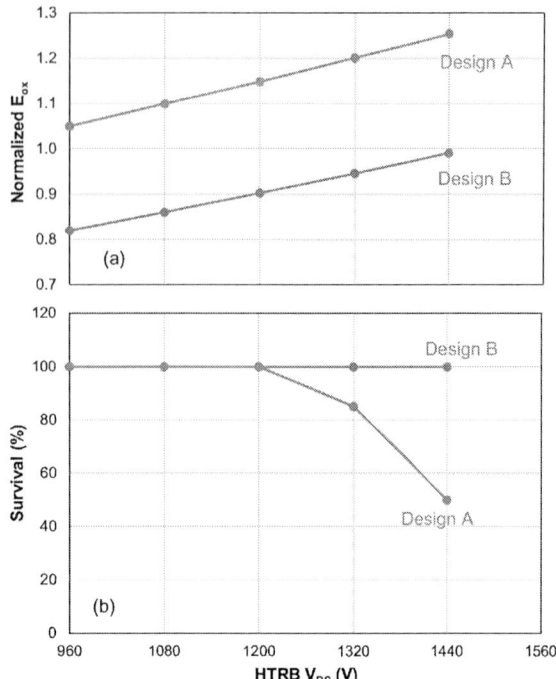

Fig. 8 (a) Peak oxide electric field (normalized to Gen-1 MOSFET) at V_{DS} = 1200V and (b) survival rate for designs A and B as a function of HTRB stress voltage. HTRB voltage was increased from 960V to 1440V at 168 hours per step, T_J = 175°C. (N = 20 devices per design)

120V steps, with 168 hours per step, at a junction temperature of 175 °C, with gate and source connected to ground. Fig. 6 shows the estimated E_{ox} and survival rates after stressing at different V_{DS} bias conditions for designs A and B. Design B did not show any failures throughout the test, whereas for design A, failures were observed at 1320V and 1440V. It should be noted that all failed parts showed high gate-drain leakage, indicating that failure occurred due to rupture of the gate oxide. No termination related (drain-source leakage type)

failures were observed. Parts from design B were subsequently stressed for an additional 1000 hours at 1440V, with all devices passing. These results clearly indicate that HTRB failure in planar SiC MOSFETs is driven by E_{ox}, and the improved $R_{on,sp}$ − E_{ox} tradeoff in Gen-2 MOSFETs allows a highly HTRB reliable design, while still maintaining very low $R_{on,sp}$.

IV. SUMMARY

In this study, we have studied the impact of JFET design on on-resistance and HTRB reliability of 1200V SiC planar gate MOSFETs. By utilizing an optimized JFET design, Gen-2 MOSFETs achieved a significantly improved $R_{on,sp}$ − E_{ox} tradeoff. As a result, Gen-2 MOSFETs can achieve a low $R_{on,sp}$ of 3.5 mΩ.cm^2, while maintaining excellent HTRB reliability.

REFERENCES

[1] S. Banerjee, K. Matocha, K. Chatty, J. Nowak, B. Powell, D. Gutierrez, and C. Hundley, "Manufacturable and rugged 1.2 KV SiC MOSFETs fabricated in high-volume 150mm CMOS fab," in *Proc. Int. Symp. Power Semicond. Devices ICs*, pp. 279–282, 2016.

[2] S. Chowdhury, L. Gant, and B. Powell, "Reliability and Ruggedness of 1200V SiC Planar Gate MOSFETs Fabricated in a High Volume CMOS Foundry," to be published in *Proc. ICSCRM*, 2018.

[3] D. Heer, D. Domes and D. Peters, "Switching Performance of a 1200V SiC-Trench-MOSFET in a Low-Power Module", *Proc. PCIM*, p. 53 – 59, 2016

[4] J. W. Palmour, L. Cheng, V. Pala, E. V. Brunt, D. J. Lichtenwalner, G. Y. Wang, J. Richmond, M. O'Loughlin, S. Ryu, S. T. Allen, a. a. Burk, and C. Scozzie, "Silicon carbide power MOSFETs: Breakthrough performance from 900 v up to 15 kV," *Proc. Int. Symp. Power Semicond. Devices ICs*, pp. 79–82, 2014.

[5] P. A. Losee, A. Bolotnikov, L. C. Yu, G. Dunne, D. Esler, J. Erlbaum, B. Rowden, A. Gowda, A. Halverson, R. Ghandi, P. Sandvik, L. Stevanovic, and R. Hristov, "SiC MOSFET design considerations for reliable high voltage operation," in *2017 IEEE International Reliability Physics Symposium (IRPS)*, 2017, p. 2A–2.1–2A–2.8.

[6] D. A. Gajewski, B. Hull, D. J. Lichtenwalner, S. Ryu, E. Bonelli, H. Mustain, G. Wang, S. T. Allen, and J. W. Palmour, "SiC Power Device Reliability," in *Proc. International Integrated Reliability Workshop (IIRW)*, pp. 29–34, 2016.

Proceedings of the 30th International Symposium on Power Semiconductor Devices & ICs
May 13-17, 2018, Chicago, USA

Investigation on Single Pulse Avalanche Failure of 900V SiC MOSFETs

Na Ren, Hao Hu, Kang L. Wang
Electrical Engineering Department
UCLA
Los Angeles, USA

Zheng Zuo, Ruigang Li
AZ Power Inc.
Los Angeles, USA

Kuang Sheng
Electrical Engineering Department
Zhejiang University
Hangzhou, China

Abstract—**In this work, avalanche ruggedness and failure mechanisms of 900V SiC MOSFETs under single-pulse Unclamped Inductive Switching (UIS) test are investigated and compared with Si counterparts. It was found in this work that, due to the higher resistance to BJT latch-up, only uniform heating related device temperature limit failure exists in SiC MOSFETs. Experimental results also show that, SiC MOSFETs have 9 times higher avalanche energy per area and 50% higher avalanche current than Si MOSFETs in low inductance/short pulse condition. In large inductance/long pulse condition, SiC MOSFETs have shorter avalanche duration, lower avalanche current and only similar avalanche energy per area compared to Si, due to the much smaller (15×) chip size, thinner active layer thickness and higher power density.**

Keywords—SiC; MOSFET; avalanche ruggedness

I. INTRODUCTION

Silicon Carbide has been identified as a material with the potential to replace Si devices in the near term because of its superior properties, such as high thermal conductivity and high critical breakdown field strength [1, 2]. These advantages could make SiC devices attractive in high-voltage, high frequency, and high-temperature applications [3, 4].

As SiC MOSFET products are becoming technologically mature, the assessment of stability and reliability are essential for the development of these devices [5]. MOSFETs are often used in high speed switching applications. Electromagnetic force produced during device turn-off due to the abrupt change of drain current as a result of inductive loads may force the MOSFET into avalanche and damage the device. Therefore, understanding the cause of the failure under avalanche condition is essential [6].

For traditional Si power MOSFETs, there are two known avalanche failure modes, i.e. (1) BJT latch-up and localized heating caused by high avalanche current dissipated over short avalanche durations (low inductance/short pulse) and (2) intrinsic temperature limit associated with low avalanche current dissipated over longer avalanche durations (large inductance/long pulse). However, SiC MOSFETs should have higher resistance to BJT latch-up due to its higher pn junction voltage drop and less temperature dependence of resistance. On the other hand, although SiC has higher intrinsic temperature limit, this benefit may be offset by the smaller chip size and higher electric field/ current density during avalanche events. Therefore, failure mechanisms of SiC MOSFETs may be

different from that of Si and avalanche robustness could depend on inductance (i.e. avalanche duration). The purpose of this work is to investigate these problems.

II. UNCLAMPED INDUCTIVE SWITCHING TEST

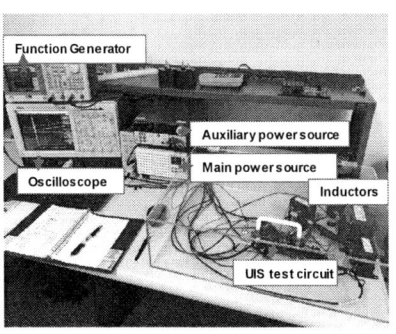

(a)

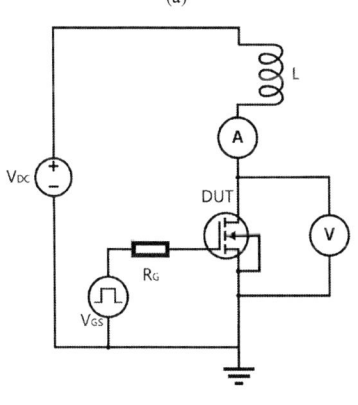

(b)

Fig.1 (a) UIS test bench. Test conditions: 25°C temperature, 80V DC bus voltage, (b) schematic view of test circuit.

Fig. 1 shows the UIS test bench and circuit diagram that includes power supplies, test circuit, function generator, oscilloscope, inductive load. In this circuit, the device under test (DUT) itself is used as the turn-on switch to charge the inductor. When the DUT is switched ON, the inductor is charged to a peak current that is proportional to the duration of the gate pulse. When the DUT is switched OFF, the drain-source voltage immediately increases up to breakdown voltage

978-1-5386-2928-4/18 $31.00 © 2018 IEEE

and the device enters the avalanche breakdown mode. The avalanche breakdown mode lasts till the device consumes all the energy stored in the inductive load. In this study, the turn on duration (t_{on}) was incremented stepwise until avalanche failure occurred. The peak inductor current in the last test before failure is regarded as the maximum avalanche current.

With the UIS test circuits, 900V/11.5A SiC MOSFETs from vendor 1, 650V/18A Si MOSFETs from vendor 2 and 3 were tested. Active areas of these three devices are: 1.8mm², 27mm² and 27mm². The load inductors (L) are selected to be 0.5mH, 1mH, 5mH, and 10mH. DUT devices were tested to their failure point at each inductance. The typical single-pulse UIS test results, including the last test before failure and the failure test are shown in Fig. 2. Fig. 2(a) is the results of SiC MOSFET from vendor 1, and Fig. 2(b) and (c) are the results of Si MOSFET from vendor 2 and 3, respectively. Two Si devices show similar behaviors but SiC MOSFET shows big differences.

III. AVALANCHE ROBUSTNESS AND FAILURE MECHANISM

A. Avalanche Robustness

The experimental results of avalanche current, avalanche duration and avalanche energy per area for SiC MOSFET and two Si MOSFETs at different inductances are summarized and plotted in Fig. 3. It is shown that, with an increasing inductance, the avalanche current is decreased for both SiC and Si MOSFETs due to an increased avalanche time (longer pulse). The avalanche current capability of SiC MOSFET at

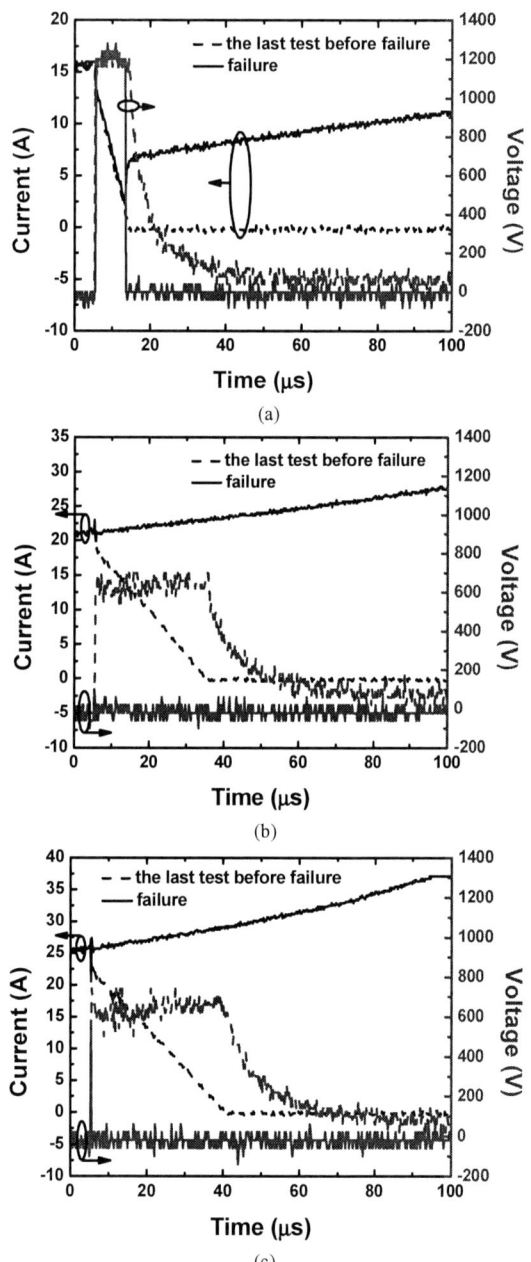

Fig. 2 The typical measured drain current and drain-source voltage waveforms of three devices under 1mH inductance condition. (a) SiC MOSFET (vendor 1), (b) Si MOSFET (vendor 2), (c) Si MOSFET (vendor 3).

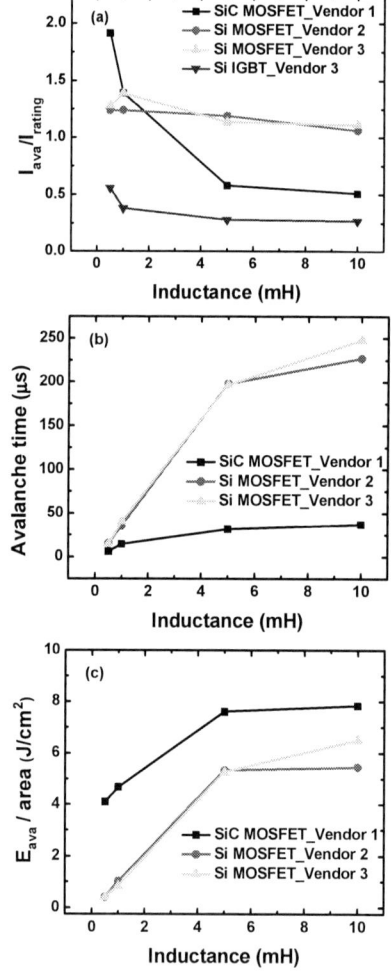

Fig. 3 (a) Avalanche current (I_{ava}) normalized with rated current (I_{rating}), (b) avalanche withstand time comparison and (c) avalanche energy (E_{ava}) per area comparison between different test devices with varied inductances.

low inductances (L<1mH) is 1.9 times current rating, which is 50% higher than Si MOSFETs and then decreases rapidly at higher inductances, which results into a 55% lower value than that of Si MOSFETs at $L\geq$5mH. On the other hand, SiC MOSFETs have shorter avalanche time than Si MOSFET, especially at higher inductances. Avalanche time can be calculated using (1), where the avalanche time is shown to be proportional to the load inductance (L), avalanche current (I_{av}) and inversely proportional to avalanche breakdown voltage (V_{BR}). The 900V rated SiC MOSFET has higher avalanche breakdown voltage (~1200V) than 650V rated Si MOSFETs (~700V), which contributes to a shorter avalanche time.

$$t_{av}=L\cdot I_{av}/V_{BR} \qquad (1)$$

$$E_{av}\approx 1/2\cdot V_{BR}\cdot I_{av}\cdot t_{av} \qquad (2)$$

According to (2), avalanche energy can be estimated by avalanche current, avalanche time and avalanche breakdown voltage. Although SiC MOSFET has lower avalanche energy than Si MOSFETs, its device active area is 15 times smaller than Si devices as well. When comparing the avalanche energy per area, SiC MOSFET is advantageous and has 9 times higher value than Si MOSFET at low inductances.

B. Failure Mechanisms

It is well known that, there are two kinds of avalanche failure modes for Si MOSFETs (as shown in Fig. 4), i.e., (1) BJT latch-up and localized heating caused by high avalanche current dissipated over short avalanche durations (low inductance/short pulse) and (2) intrinsic temperature limit associated with low avalanche current dissipated over longer avalanche durations (large inductance/long pulse).

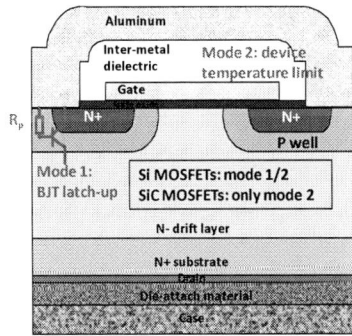

Fig. 4 Cross-sectional view of MOSFET showing avalanche failure modes.

In Mode 1, BJT latch-up secondary breakdown is caused by the positive feedback of temperature and current. The resulted localized heating can lead to the ultimately destruction of the semiconductor. In mode 2, avalanche duration is long enough to allow uniform temperature rises across the chip. Without localized heating, device will be failed when intrinsic temperature limit of the device is reached. For SiC MOSFETs, the avalanche failure mode could be different from that of Si MOSFETs. Since the SiC pn junction built-in voltage is higher (3V) than Si (~0.7V) and the p-base resistance is less temperature dependent, the BJT latch-up in SiC MOSFETs is less likely to be happened.

To get insight into the avalanche failure mechanisms of SiC MOSFETs, an electro-thermal modeling based on actual device structures was carried out. The measured current/voltage waveforms at four different inductances are set as the input to calculate the dynamic energy dissipation and temperature distribution in SiC and Si MOSFETs. The heat distribution is assumed to be uniform in the simulation. For SiC MOSFETs, the simulated junction temperature during avalanche and the temperature distribution along vertical axis from device surface to bottom copper layer (when peak temperature is happened)

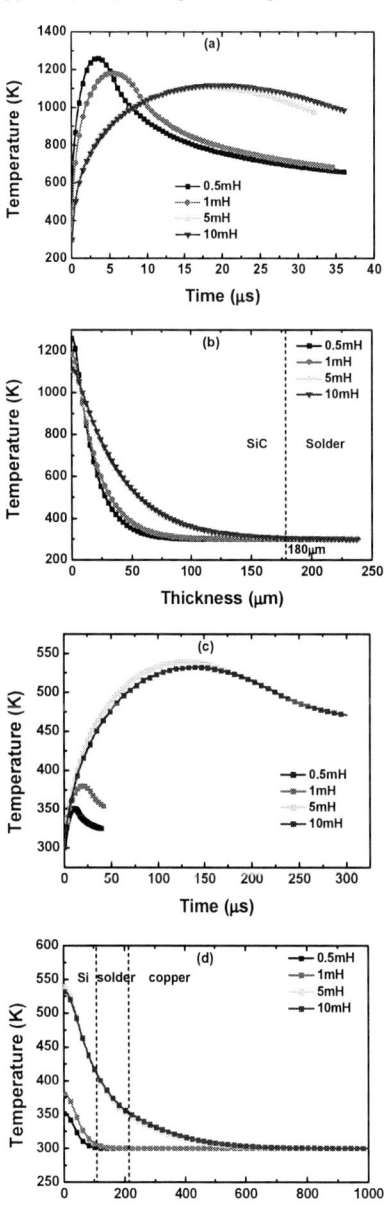

Fig. 5 (a) Calculated junction temperature and (b) temperature distribution along vertical path in SiC MOSFETs with electro-thermal model. (c) Calculated junction temperature and (d) temperature distribution along vertical path in Si MOSFETs.

are shown in Fig. 5 (a) and (b), respectively. And those simulated results for Si MOSFETs are shown in Fig. 5(c) and (d). Fig. 5(a) shows that the junction temperature of SiC MOSFETs could reach >1100K at both low and high inductances. Since this high temperature can already destroy a device, localized heating can be ruled out and only mode 2 exists in SiC MOSFETs. On the contrary, in Fig. 5(c), Si MOSFETs can reach ~540K at high inductances but only ~380K at low inductances, which confirms that localized heating must be happened in Si at low inductances.

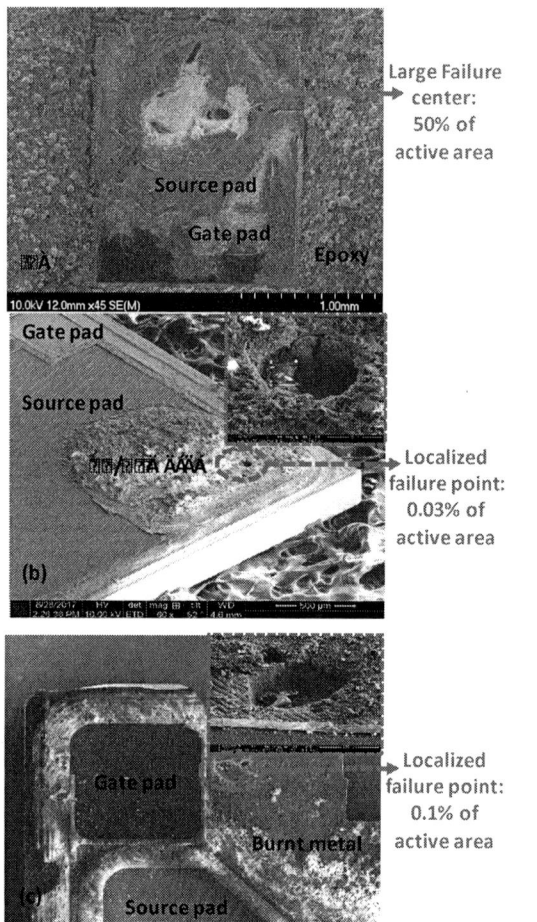

Fig. 6 SEM photographs of failed devices in low inductance tests. (a) failed SiC MOSFET device with large failure center (in red dash line). (b) failed Si MOSFET device (vendor 2), (c) failed Si MOSFET device (vendor 3) with localized failure point.

Although the temperature values calculated from the electro-thermal model need further experimental validation, the contrastive analysis is useful for the failure mechanism analysis. Based on this, it can be concluded that SiC MOSFETs have more uniform heat distribution at low inductances, which become the reason for the 9 times higher avalanche energy per area and 50% higher avalanche current as illustrated in Fig. 3. In high inductance tests, failure mechanisms are device temperature limits for both Si and SiC. Although SiC have higher intrinsic temperature (~1550K) than Si (~550K), some other device parts can degrade first at lower temperature and SiC device is operated with much higher power density. As a result, in high inductance tests, SiC MOSFETs have shorter avalanche duration, lower avalanche current and only similar avalanche energy per area when compared to Si (Fig. 3).

The failed devices were observed with Focused Ion Beam (FIB) and the images of failure location at low inductance are shown in Fig. 6. It is obvious that SiC MOSFETs have large failure center, which is about 50% of the total active area. It should be noted that, although the device is not shown 100% uniformly destroyed, other device parts should be heated up and degraded too. In contrast to SiC MOSFETs, Si MOSFETs have localized failure points, which is only 0.03% or 0.1% of the total active area. These results confirm that BJT latch-up related localized failure is suppressed in SiC MOSFETs, while it is happened in Si MOSFETs at low inductance.

IV. CONCLUSION

Single pulse UIS tests at different inductances were conducted on 900V SiC MOSFETs. The avalanche robustness and failure mechanisms are investigated and compared with Si counterparts. Based on the temperature calculation and avalanche mechanism analysis, it is found that BJT latch-up related localized failure is suppressed in SiC MOSFETs at low inductances. Therefore, with a more uniform heat distribution, SiC MOSFETs have 9 times higher avalanche energy per area and 50% higher avalanche current as compared to Si MOSFETs at low inductances. On the other hand, in high inductance tests, failure mechanisms are device temperature limits for both Si and SiC. Although SiC has higher intrinsic temperature limit, this benefit is offset by the smaller chip volume and higher electric field/ current density during avalanche events. The experimental results show that SiC MOSFETs have shorter avalanche time, lower avalanche current and only similar avalanche energy per area when compared to Si at high inductances.

REFERENCES

[1] N. Ren, J. Wang and K. Sheng, "Design and Experimental Study of 4H-SiC Trenched Junction Barrier Schottky Diodes," in *IEEE Transactions on Electron Devices*, vol. 61, no. 7, pp. 2459-2465, July 2014.

[2] N. Ren, K. L. Wang, Z. Zuo, R. Li and K. Sheng, "A novel 4H-SiC pinched barrier rectifier," *2017 IEEE Applied Power Electronics Conference and Exposition (APEC)*, Tampa, FL, 2017, pp. 1950-1957.

[3] X. Lyu, N. Ren, Y. Li and D. Cao, "A SiC-Based High Power Density Single-Phase Inverter With In-Series and In-Parallel Power Decoupling Method," in *IEEE Journal of Emerging and Selected Topics in Power Electronics*, vol. 4, no. 3, pp. 893-901, Sept. 2016.

[4] X. Lyu, Y. Li and D. Cao, "DC-Link RMS Current Reduction by Increasing Paralleled Three-Phase Inverter Module Number for Segmented Traction Drive," in *IEEE Journal of Emerging and Selected Topics in Power Electronics*, vol. 5, no. 1, pp. 171-181, March 2017.

[5] M. Treu, R. Rupp and G. Sölkner, "Reliability of SiC power devices and its influence on their commercialization - review, status, and remaining issues," *2010 IEEE International Reliability Physics Symposium*, Anaheim, CA, 2010, pp. 156-161.

[6] L. Yang, A. Fayyaz and A. Castellazzi, "Characterization of high-voltage SiC MOSFETs under UIS avalanche stress," *7th IET International Conference on Power Electronics, Machines and Drives (PEMD 2014)*, Manchester, 2014, pp. 1-5.

978-1-5386-2928-4/18 $31.00 © 2018 IEEE

Proceedings of the 30th International Symposium on Power Semiconductor Devices & ICs
May 13-17, 2018, Chicago, USA

Long Term High Temperature Reverse Bias (HTRB) Test on High Voltage SiC-JBS-Diodes

Felix Hoffmann*, Andrei Mihaila+, Lukas Kranz+, Philippe Godignon[1], Nando Kaminski*

*Institute for Electrical Drives, Power Electronics, and Devices, University of Bremen, Germany
+ABB Switzerland Ltd, Corporate Research Center, Baden-Dättwil, Switzerland
[1]Centre Nacional de Microelectrònica, CNM-CSIC, Barcelona, Spain
felix.hoffmann@uni-bremen.de

Abstract—In this paper, the result of a long term HTRB Test with a test time over 5000 hours on novel 3.3kV SiC-JBS-Diodes is presented. The diodes under test have an area of 5x5mm² with a p-stripe design and a JTE-based edge-termination. Two splits were tested, one with globtop cover underneath silicone gel potting and the other with gel potting only. Both splits show a significant decrease in leakage current over the course of the test. A different behavior of leakage current is observed when test conditions are reapplied after an interruption for intermediate measurements. It is evident that the leakage current is not only affected by the chip but also by the surrounding materials. Both splits passed over 5000 hours without failure.

Keywords—SiC, Schottky, Diode ,JBS ,Reliability, High Temperature Reverse Bias, HTRB, Passivation, Globtop

I. INTRODUCTION

Today, SiC-JBS-Diodes are well-established in commercial applications up to 1.7kV and devices for higher voltages are under development [1], [2]. However, higher voltage also increases electrical stress on the edge-termination. The High Temperature Reverse Bias (HTRB) is the standard test procedure to survey the integrity of a power electronic device under combined thermal and electrical stress [3]. For this paper, an HTRB test is conducted with novel 3.3kV SiC-JBS-Diode chips mounted on a test substrate and potted with silicone gel. Whereas reliability results on 1.7kV JBS-Diodes and 3.3kV SiC-MOSFETs have been reported previously e.g. in [4], [5], to our knowledge this is the first report on HTRB with packaged 3.3kV SiC-JBS-Diodes.

II. TEST DEVICES

The structure of the diodes under test is shown in Fig. 1.

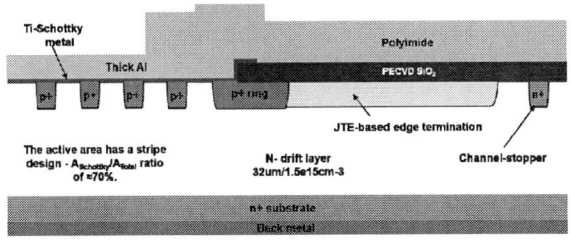

Fig. 1. Schematic cross-section of a 3.3kV SiC-JBS-Diode (not drawn to scale) showing the active area with p+ stripes and a JTE-based edge-termination.

It consists of a p-stripe design with a relatively large area of 5x5mm². The chip design features a JTE-based edge-termination with an n+ doped channel stopper similar to the design described in [1]. The surface of the edge termination is covered with an SiO_2/polyimide stack. A total of 8 substrates with 4 chips each had been packaged for testing. Fig. 2 shows a test substrate populated with 4 diode chips covered with globtop. The chips are soldered to a DBC substrate and mounted on a copper baseplate to increase thermal capacitance and avoid thermal runaway during testing. Two different types of package insulation were tested. Whereas all substrates received a silicone gel potting, the chips on substrates S1 to S4 had been coated with an additional globtop covering before the silicone gel potting was applied. The globtop covering used for this test is an epoxy-based compound. It is applied to avoid direct contact of the silicone gel with the chips' surface and prevent interference of possible charges in the silicone gel with the electric field applied during HTRB. As a reference, the substrates S5 to S8 were covered with silicone gel potting only.

III. SAMPLE PREPARATION

Prior to testing, the blocking characteristics of all substrates were measured. During those initial measurements (IM), some chips showed premature breakdown. Fig. 3 (a) shows an overlay image of a chip with the breakdown spot in the lower right corner of the junction termination close to the point of

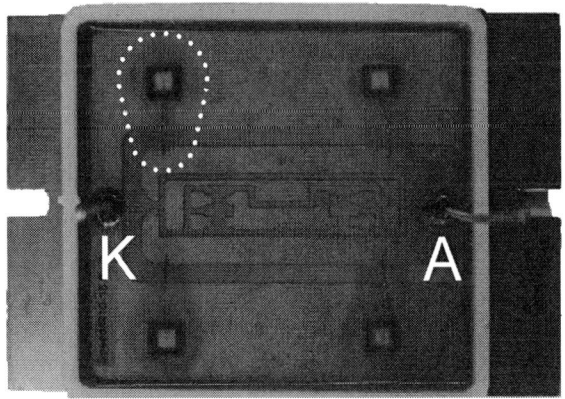

Fig. 2. Substrate for HTRB test populated with 4 chips with globtop covering. Deficient chips have been insulated by cutting the bond wires. The globtop covering (indicated with a dotted white circle on the upper left chip) is visible through the silicone gel as a blurred circle around the chip

978-1-5386-2928-4/18 $31.00 © 2018 IEEE

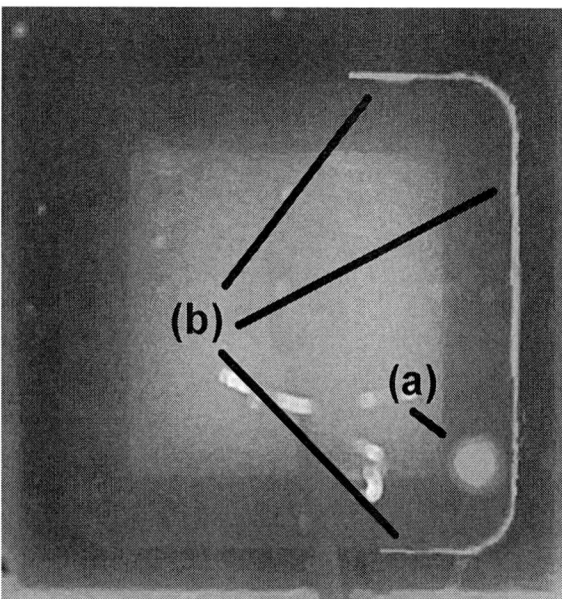

Fig. 3. Microscope image of a deficient chip graphically enhanced via overlay of single video frames. (a) Premature breakdown at the bright spot leading to catastrophic failure. (b) Avalanche luminescence on the right side

highest field strength. Those chips were removed from the test batch by cutting the bond wires of the affected chips. During this test preparation, substrate S3 was removed completely since all chips showed insufficient blocking behavior, and all other substrates except S6 were tested with a reduced chip count. Table I shows the number of chips per substrate after test preparation. Hence, the HTRB has been performed on a total of 17 chips on 7 substrates. 5 of those chips were part of the first split with globtop covering and 12 chips were part of the second split with gel potting only. All these chips showed the asymmetric avalanche inception, which is also shown in overlay image in Fig. 3 (b). It is shown that avalanche luminescence is visible only on the right side of the chip and propagating on the top and bottom edge to approximately one third of the chip's dimension. This is caused by anisotropic impact ionization, well-known from other silicon carbide devices with similar edge-termination [6], [7].

TABLE I.   CHIPS PER SUBSTRATE

Substrate	S1	S2	S4	S5	S6	S7	S8
Globtop cover	Yes	Yes	Yes	No	No	No	No
Chip count	1	2	2	3	4	3	2

IV.   TEST EXECUTION AND RESULTS

The test setup consists of a heat chamber where the test substrates are stored and a high voltage supply to apply the desired reverse bias. A high-performance measurement system, capable of measuring down to three-digit nano-ampere range is used to measure and log the leakage current for each substrate over the course of the test. In case of a failure, the device is disconnected immediately from the power supply by means of a high-voltage reed relay.

The test is conducted with an ambient temperature of 125°C and a reverse bias of 2640V, corresponding to 80% of the diode's nominal voltage ($V_{AK,max}$) [3]. The leakage current for each substrate logged over the course of the test is shown in Fig. 4. The upper graph shows the current of the substrates with additional globtop covering, whereas the lower graph shows the current of the substrates with silicone gel only. To verify thermal stability, the test was initialized with a slow increase in temperature from 85°C to 125°C for 500 hours. Thus, the leakage current also increases during that phase. This initialization phase is shaded in yellow in Fig. 4. One chip on substrate S2 (red line, upper graph), which was already suspicious during initial static blocking measurement, failed during this test initialization and was removed from the test batch. After the initial phase, the test was conducted for more than 5000 hours under steady state reverse bias of 2640V and constant temperature of 125°C.

A. Online Monitoring

The current log indicates a different leakage current characteristic over the course of the test for substrates with globtop versus silicone gel potting only. For the devices with gel potting only, the leakage current is not falling back to its previous value immediately when resuming the test after an interruption for intermediate measurements, whereas for the ones with globtop the current is stable after the test conditions are reapplied. Therefore, it is evident that the leakage current is not only affected by the chip itself but also by the insulation material surrounding it. This effect is associated with charges from the silicone gel accumulating on the chip surface. For the substrates with globtop, where the silicone gel is not in direct contact with the chip, those surface charges are kept away and hence no impact on the leakage current can be observed. Although this effect is not necessarily affecting the device reliability, surface charges can distort the field distribution, leading to inhomogeneities. This is critical particularly on the edge of the junction termination were those distortions can even lead to device failure. Furthermore, the leakage current curves show a significant decrease over time for both test splits. This "burn-in"-behavior has already been noticed for SiC-MOSFETs, however it has not been discussed in detail [5].

B. Blocking Characteristics

Reverse-blocking characteristics at room temperature were measured initially before the test was performed and at several intermediate measurements over the course of the test. Fig. 5 shows the blocking curves for all substrates under test. The substrates with globtop are labeled in red (S1, S2 and S4) whereas the ones with gel potting only are labeled in blue (S5, S6, S7 and S8). For the sake of clarity, just one intermediate measurement after 2500 hours is shown in addition to initial and final measurements. Substrate S6 shows a significantly higher leakage current below 2000V. Even though this substrate is populated with at least one more chip compared to the other substrates, this cannot explain that the leakage current is higher by an order of magnitude. This effect can be

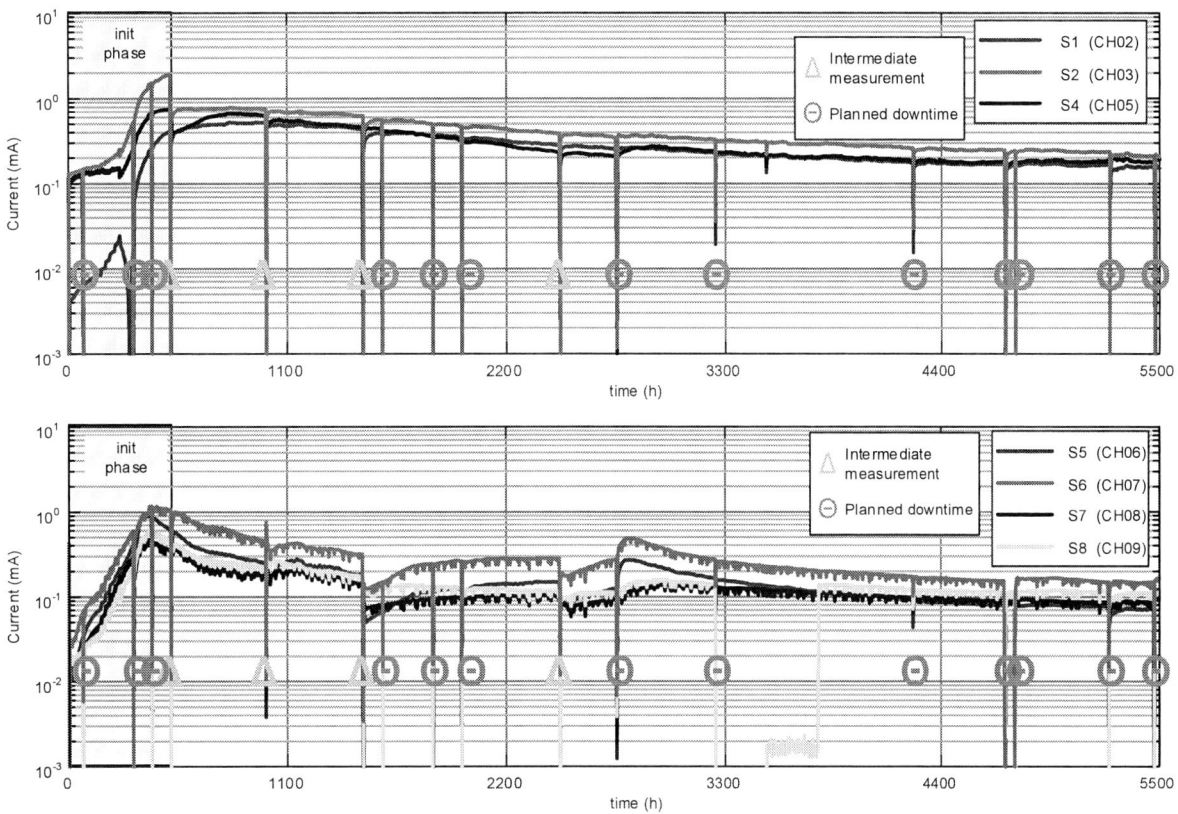

Fig. 4. Leakage current log during HTRB for substrates with globtop covering (upper graph) and substrates with silicone gel only (lower graph). The spikes, pointing downwards are caused by voltage outages e.g. due to intermediate measurements or planned downtime and maintenance interruptions. Substrate S8 (green line, lower graph) was disconnected from the voltage for some 500 hours at approx. 3200 hours due to a wiring problem. After the initialization phase, the leakage current was generally decreasing (apart from the recovery phases after IM for substrates with gel potting only).

attributed to a local lowering of the Schottky barrier height on the diodes which populate S6. However, this effect levels out at higher voltages and did not have any effect on the diodes' HTRB performance. Additionally, it can be observed that the curves at room temperature are not identical throughout the measurements. All substrates except S8 show increased leakage currents between 2200V and 3400V after 5500 total test hours including the initial phase. For S5, this behavior is already visible after 2500 hours and for S6 and S8 the leakage current is elevated after 2500 hours but lowers back to the initial value after 5500 hours. However, at higher blocking voltages this effect is no longer visible due to generally higher leakage currents and the blocking curves are aligning with the initial curve for all test substrates. Considering the development of blocking characteristics over the course of the test, no difference between substrates with globtop and substrates with silicone gel only is visible.

V. CONCLUSION

Some of the tested diodes showed intrinsic deficiencies in blocking capability. It shows that an initial screening is necessary to remove infant mortality sensible diodes. Those

chips were not tested and also disregarded for further test analysis. The blocking curves indicate that almost all substrates exhibit a change in their reverse blocking characteristics i.e. a leakage current increase in a certain voltage range. Both test splits trending towards the same direction concerning these changes. Hence it can be concluded that there is no significant impact of the insulation material, or respectively its embedded charge, on HTRB performance and surface charges from the silicone gel are not emerging as an important factor influencing the lifetime of a device under HTRB stress. Additionally, the current logs clearly indicate that both test splits show a significant general decrease in leakage current over the course of the test at high temperature, which is consistent with other publications. However, the root cause of this behavior could not be unveiled during this test.

The test was terminated after more than 5000 hours of constant HTRB stress. During our test none of the substrates had a significant increase in leakage current, which would have been an indicator for degradation or failure of a chip. Also, the intermediate measurements of reverse blocking characteristics showed that no significant degradation of blocking capability could be observed. Hence, all substrates passed the HTRB test

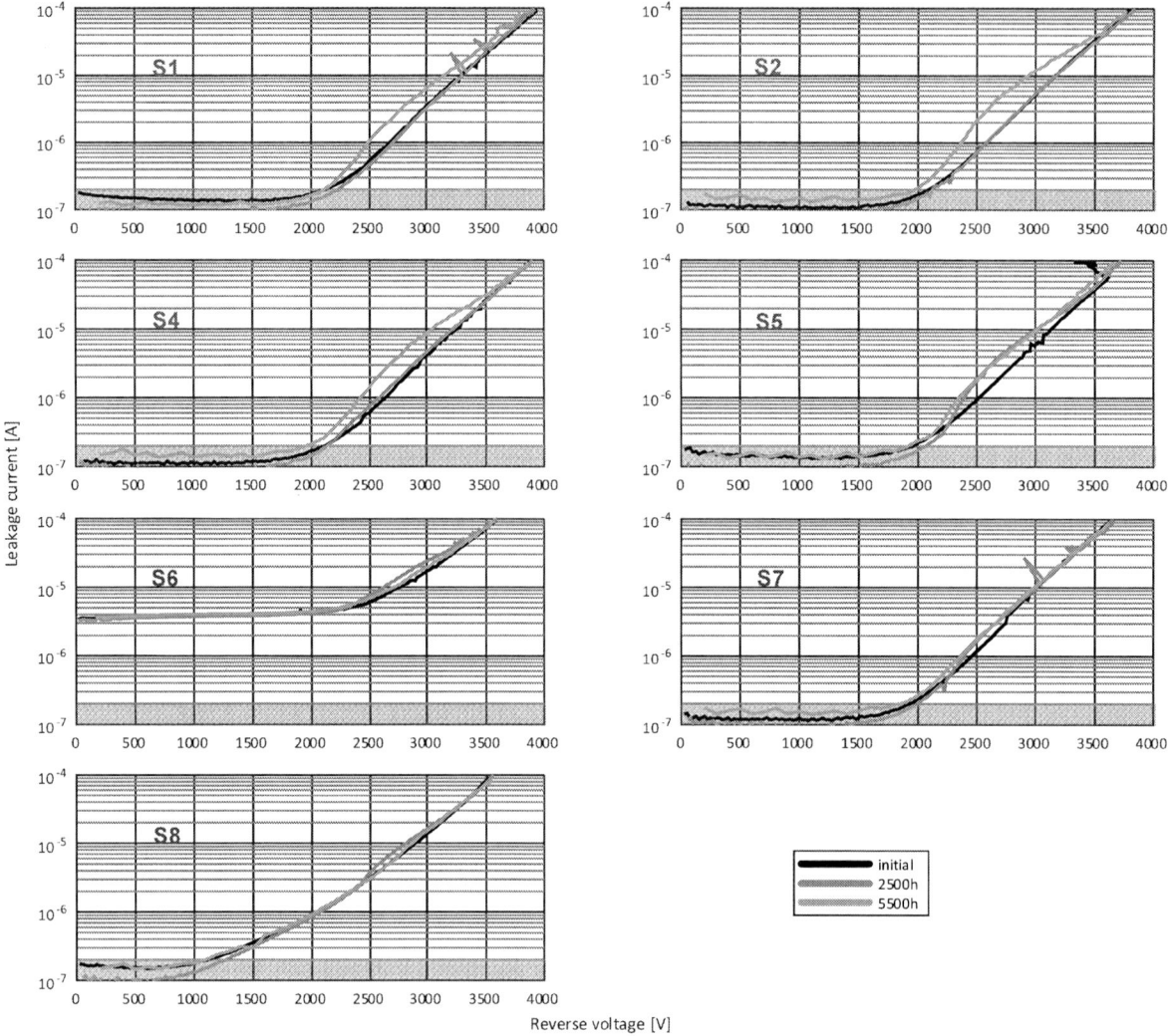

Fig. 5. Room temperature blocking curves of all substrates during initial measurement, after 2500 hours and after 5500 hours for globtop substrates (numbered in red) and substrates with gel potting only (numbered in blue). The initial leakage current is below the measurement range, hence differences in leakage current due to the substrates' chip count cannot be resolved

without any failures. This result suggests that the quality of silicon carbide high voltage epi material approaches the maturity level required for commercial application. For stable and robust devices, a major challenge of silicon carbide material is the edge-termination due to its almost ten times higher critical field strength compared to silicon, and the results of this long term HTRB test show that an appropriate edge-termination design can meet the necessary stability requirements and can achieve reliable devices.

ACKNOWLEDGMENT

This work was supported by the European Seventh Framework Programme (FP7) project SPEED (Silicon Carbide Power Technology for Energy Efficient Devices - NMP3-LA-2013-604057).

REFERENCES

[1] A. Mihaila *et al.*, 'Experimental investigation of SiC 6.5kV JBS diodes safe operating area', in *ISPSD 2017*, pp. 53–56.

[2] J. Millán *et al.*, 'High-voltage SiC devices: Diodes and MOSFETs', in *CAS 2015*, pp. 11–18.

[3] cf. JEDEC JESD22-A108

[4] L. Kranz *et al.*, 'Robust SiC JBS Diodes for the Application in Hybrid Modules', in *PCIM Europe 2017*; pp. 1–6.

[5] E. v. Brunt *et al.*, 'Reliability assessment of a large population of 3.3 kV, 45 A 4H-SiC MOSFETs', in *ISPSD 2017*, pp. 251–254.

[6] R. Rupp *et al.*, 'Avalanche behaviour and its temperature dependence of commercial SiC MPS diodes: Influence of design and voltage class', in *ISPSD 2014*, pp. 67–70.

[7] C. Zorn *et al.*, 'H³TRB Test on 650V SiC JBS Diodes', in *ICSCRM 2017*, in press.

Proceedings of the 30th International Symposium on Power Semiconductor Devices & ICs
May 13-17, 2018, Chicago, USA

Robustness improvement of short-circuit capability by SiC trench-etched double-diffused MOS (TED MOS)

Naoki Tega, Kazuki Tani, Digh Hisamoto, and Akio Shima

Research & Development Group
Hitachi, Ltd.
Kokubunji-shi, Tokyo, 185-8601, Japan
E-mail: naoki.tega.ub@hitachi.com

Abstract—A 3.3-kV SiC trench-etched double-diffused MOS (TED MOS) is designed and fabricated for robust short-circuit (SC) capability. Because of its low-V_{over} (V_g - V_{th}) operation, the TED MOS successfully reduces the drain current in saturation region to less than 700 A/cm^2 at SC tests. The low drain current in a saturation region enhances the SC capability of the TED MOS. As a result, the SC endurance time of the TED MOS is 2.8 times longer than that of the conventional SiC DMOS.

Keywords— SiC, MOS, short-circuit capability, and trench

I. INTRODUCTION

Silicon carbide (SiC) trench MOSs have gained much attention as a replacement for SiC double-diffused MOSs (DMOS) in power systems because of their low loss. Nevertheless, their insufficient short-circuit (SC) capability remains as a robustness issue [1].

Recently, three dimensional SiC MOSs have been reported as providing extremely low on-resistance ($R_{on}A$) [2, 3]. At the ISPSD 2015, we proposed a novel trench-etched double-diffused MOS (TED MOS) that can provide wider channel width (W_g) and higher channel mobility (μ_{eff}) than conventional SiC MOSs so that the TED MOS can improve its channel resistance [2]. The present work demonstrates that a TED MOS

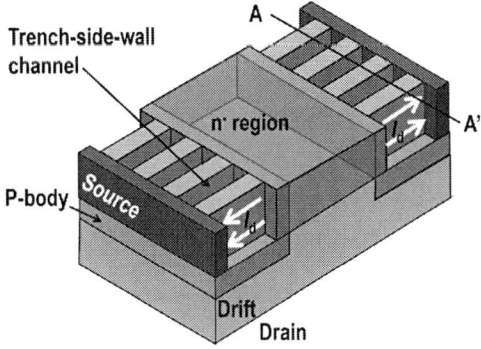

Fig. 1: Bird's eye view of 3.3-kV SiC trench-etched double-diffused MOS (TED MOS) with narrow trench pitch. Gate electrode is not shown.

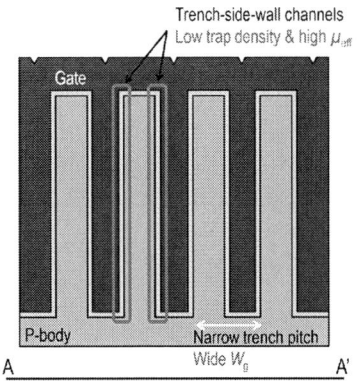

Fig. 2: Cross sectional image between A and A' as illustrated in Fig. 1.

TABLE I. SAMPLE INFORMATION

Sample name	TED1	TED2	DMOS
Structure	TED MOS	TED MOS	DMOS
Ratio of trench pitch	1(Base)	0.3	-
Ratio of max μ_{eff}	3	3	1 (Base)

with the low channel resistance also has an extraordinary advantage in improving the robustness of the SC capability in comparison with conventional SiC power MOSs.

II. DESIGN OF TED MOS FOR ROBUST SC CAPABILITY

The TED MOS is based on the double-diffused MOS (DMOS) structure as illustrated in Fig. 1 (the gate electrode is not shown in this figure). The key feature is that both trenches and gates are formed solely on the P-body to improve the reliability of the gate oxide on the trenches at a high drain-source blocking voltage. Moreover, the designed 3.3-kV SiC TED MOS, the TED2, provides wider W_g than the conventional SiC DMOS as a reference device because TED2 has the narrow trench pitch as illustrated in Fig. 2. Table I shows that the trench pitch of TED2 is about one third that of TED1. The other device parameters of TED2 are the same as those of TED1. Maximum

978-1-5386-2928-4/18 $31.00 © 2018 IEEE

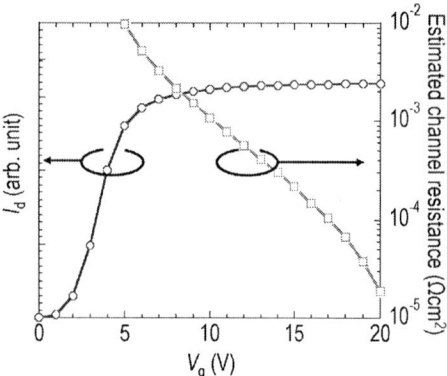

Fig. 3: Extremely low channel resistance. Channel resistance of TED2 is negligible, and drift resistance determines I_d. $V_d = 10$ V, $T = 150°C$.

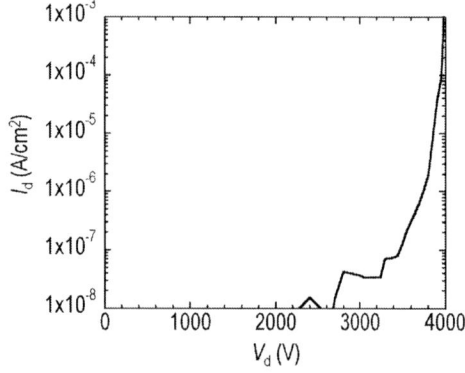

Fig. 4: Breakdown test of TED2.

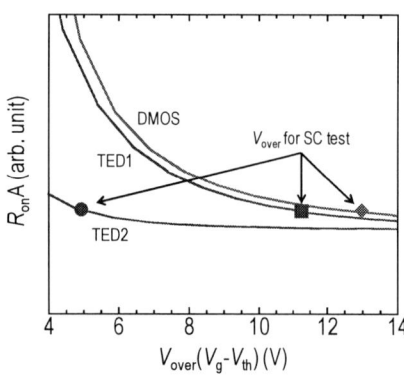

Fig. 5: Dependence of $R_{on}A$ on V_{over}. $T = 150°C$.

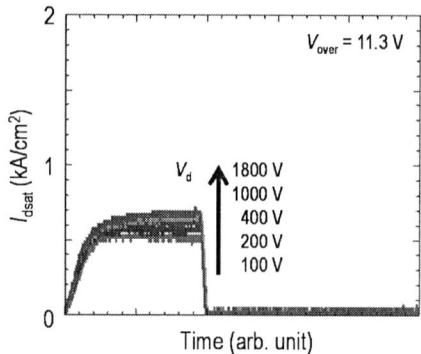

(a) I_{dsat} of TED1. $V_{over} = 11.3$ V.

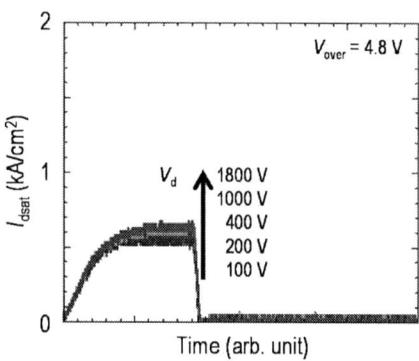

(b) I_{dsat} of TED2. $V_{over} = 4.8$ V.

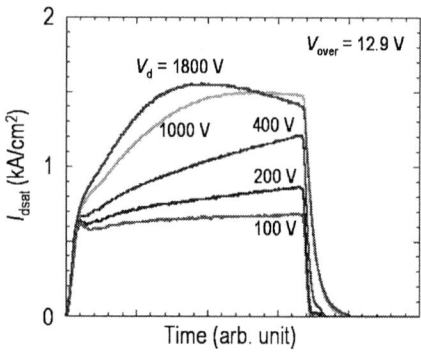

(c) I_{dsat} of DMOS. $V_{over} = 12.9$ V.

Fig. 6: Short-circuit tests. $T = 150°C$.

μ_{eff} of a trench-side-wall channel in the TED MOS is approximately three times higher than that of a planer channel in DMOS because the interface trap density of the trench-side-wall channel is lower than that of the DMOS, as clarified by using a charge pumping method [4]. The wide W_g and high μ_{eff} of the TED2 achieves extremely low channel resistance, as shown in Fig. 3. The TED2 achieves less than an estimated 0.2 m ·cm² channel resistance when the gate voltage (V_g) is 15 V. Therefore, the TED MOS can be operated by low V_{over}, which is defined as a difference voltage between V_g and the threshold voltage (V_{th}). It means that the TED MOS can give the same $R_{on}A$ as the DMOS, even though V_{over} of the TED MOS is lower than that of the DMOS. The low V_{over} has a certain advantage in reducing I_{dsat} because I_{dsat} is more dependent on V_{over} than the other device

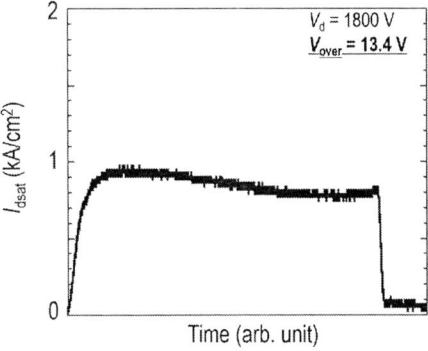

(a) Behavior of I_{dsat} immediately before breakdown.

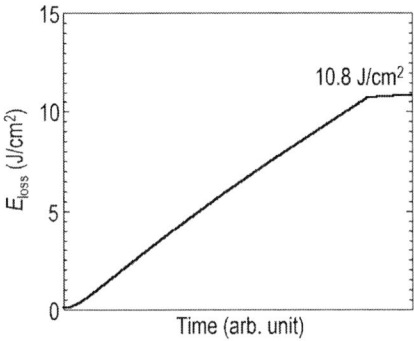

(b) Estimation of energy loss.

Fig. 7: SC capability of TED1. $T = 150°C$.

TABLE II. ENERGY LOSS

Sample name	TED1	TED2	DMOS
E_{loss} (J/cm^2)	10.8	12.6	10.5

parameters according to the gradual channel approximation [5]. We demonstrates that the TED MOS achieves robust SC capability with a small I_{dsat} in addition to its good performance.

Almost all SiC trench MOSs have been used for a low voltage devices because their gate oxide at the bottom of trenches is weak at a high blocking voltage. In fact, a proposed high-voltage SiC trench MOS needed a very complex structure to prevent the gate oxide from breakdown [6]. In contrast, the TED MOS structure is a simple, but successfully reached approximately 4 kV during a breakdown test, as shown in Fig. 4. The TED MOS structure certainly provides a wide product line from low-voltage to high-voltage SiC MOSs.

III. DRAIN CURRENT AT SC TEST

It is important to determine the V_{over} of each sample at constant $R_{on}A$ before evaluating the SC capabilities. Dependence

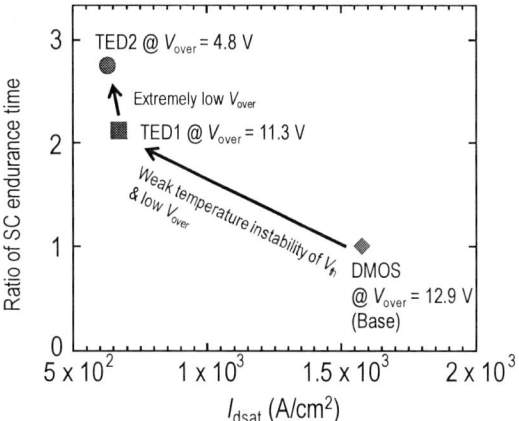

Fig. 8: Dependence of SC endurance time on I_{dsat}. TED MOSs can freely set low V_g and high V_t because low V_{over} is efficient to operate them. $V_d = 1800$ V, $T = 150°C$.

of $R_{on}A$ on V_{over} is shown in Fig. 5. An occupied diamond indicates typical V_{over} of the DMOS (12.9 V), and thereby, targeting $R_{on}A$ at the typical V_{over} of DMOS. An optimized V_{over} of TED1 (11.3 V) is relatively lower than that of the DMOS due to the high μ_{eff} of the TED MOS. Furthermore, an optimized V_{over} of TED2 (4.9 V) is extremely lower than that of TED1 because of the widest W_g of TED2 in the three samples. The I_{dsat}s at the SC test are shown in Fig. 6. The I_{dsat}s of TED1 and TED2 (500 ~ 700 A/cm^2) are smaller than that of the DMOS (over 1500 A/cm^2), and they are weakly dependent on time. One of the reasons for the low I_{dsat}s of TED1 and TED2 is the low V_{over}. It is notable that the I_{dsat} of the DMOS increases as the V_{over} rises and is not saturated. Moreover, it is strongly dependent on time. A. Castellazzi et al. also found the unique behaviors [7].

Previous work indicated that the temperature instability of V_{th} caused by interface traps results in this behavior in the conventional SiC DMOS [8]. In the DMOS, high I_d generates even more heat. The increased heat decreases V_{th} because of electron emissions from interface traps. The increased V_{over} results in the increased I_d, and I_d eventually becomes unstable and increases until drift and JFET resistances becomes relatively large. On the other hand, the TED MOS can provide low interface traps and temperature stability for V_{th} [4]. The TED MOS is likely to suppress the temperature instability of V_{th} due to the low interface traps and control the increase in I_{dsat} during the SC test compared to a conventional SiC DMOS [8].

IV. SC CAPABILITY OF TED MOS

The SC endurance time just before breakdown of TED1 is demonstrated in Fig. 7. Applied V_{over}, 13.4 V, is relatively higher than optimized V_{over}, 11.3 V, because I_{dsat} must be maintained at a control level in wide-range V_{over} for practical use. I_{dsat} of TED1 truly remains stable over time. An integrated energy loss (E_{loss}) of TED1 is 10.8 J/cm^2. All E_{loss}s are summarized in Table II. E_{loss}s of the TED MOSs are the same or slightly higher than that of the DMOS. A breakdown mechanism of the TED MOS at the short circuit appears similar to that of the DMOS.

978-1-5386-2928-4/18 $31.00 © 2018 IEEE

There is a clear relationship between the SC endurance time and I_{dsat} as shown in Fig. 8. TED1 provides less than half I_{dsat} of the DMOS, and the SC endurance time of TED1 corresponds by being over two times longer than that of the DMOS by being operated by a lower V_{over} than that of the DMOS. Its V_{th} is stabilized against temperature compared to the DMOS. Likewise, TED2 achieves 2.8 times longer SC endurance time than the DMOS because W_g of TED2 is far wider than those of TED1 and the DMOS. The V_{over} of TED2 is much lower than that of TED1, and its I_{dsat} is even smaller. Furthermore, the SC endurance time of the TED MOS is longer than that of the SiC DMOSs. We confidently conclude that the 3.3-kV SiC TED MOS achieves an incredibly robust SC capability compared to the all conventional SiC MOSs.

V. CONCLUSION

We demonstrated that the 3.3-kV SiC TED MOS achieves an incredibly robust SC capability compared to the conventional SiC MOSs because of the low-V_{over} operation and weak temperature instability of V_{th}. The TED MOS reduces I_{dsat} to less than 700 A/cm^2. The SC endurance time of the TED MOS is 2.8 times longer than that of the DMOS. We are absolutely convinced that the SiC TED MOS raises the SC capability of SiC MOS to a whole new level, in addition to low loss.

REFERENCES

[1] R. Tanaka, Y. Kagawa, N. Fujiwara, K. Sugawara, Y. Fukui, N. Miura, M. Imaizumi, and S. Yamakawa, "Impact of Groundign the Bottom Oxide Protection Layer on the Short-Circuit Ruggedness of 4H-SiC Trench MOSFETs," in Proceedings of the 26th International Symposium on Power Semiconductor Devices (ISPSD), pp.75–78, 2014.

[2] N. Tega, H. Yoshimoto, D. Hisamoto, N. Watanabe, H. Shimizu, S. Sato, Y. Mori, T. Ishigaki, M. Matsumura, K. Konishi, K. Kobayashi, T. Mine, S. Akiyama, R. Fujita, A. Shima, and Y. Shimamoto, "Novel trench-etched double-diffused SiC MOS (TED MOS) for overcoming tradeoff between $R_{on}A$ and Q_{gd}," in Proceedings of the 27th International Symposium on Power Semiconductor Devices (ISPSD), pp. 81–85, 2015.

[3] J. A. Cooper, N. Islam, R. Ramamurthy, M. Sampath, and D. T. Morisette, "Vertical Tri-Gate Power MOSFETs in 4H-SiC," in Proceedings of the International Conference on Silicon Carbide and Related Materials(ICSCRM), 2758009, 2017.

[4] N. Tega, D. Hisamoto, A. Shima, and Y. Shimamoto, "Channel Properties of SiC Trench-Etched Double-Implanted MOS (TED MOS)," IEEE Trans. Electron Devices, vol. 63, no. 9, pp. 3439–3444, 2016.

[5] B. Jayant Baliga, Fundamentals of Power Semiconductor Devices, New York: Springer, 2008, pp.325–326.

[6] T. Kojima, S. Harada, Y. Kobayashi, M. Sometani, K. Ariyoshi, J. Senzaki, M. Takei, Y. Tanaka, and H. Okumura, "Self-Aligned Formation of the Trench Bottom Shielding region in 4H-SiC UMOSFETs," in Proceedings of the International Conference on Solid State Devices and Materials, pp. 948–949, 2015.

[7] A. Castellazzi, T. Funaki, T. Kimoto, and, T. Hikihara, "Thermal instability effects in SiC power MOSFETs," Micro. Rel., vol. 52, pp.2414–2419, 2012.

[8] C. Unger, and M. Pfost, "Energy Capability of SiC MOSFETs," in Proceedings of the 28th International Symposium on Power Semiconductor Devices (ISPSD), pp. 275–278, 2016.

Proceedings of the 30th International Symposium on Power Semiconductor Devices & ICs
May 13-17, 2018, Chicago, USA

High-temperature validated SiC power MOSFET model for flexible robustness analysis of multi-chip structures

M. Riccio, V. d'Alessandro, G. Romano, L. Maresca,
G. Breglio, A. Irace

Dept. of Electrical Eng. and Information Technology
University of Naples Federico II
Naples, Italy
Email: michele.riccio@unina.it

A. Castellazzi

PEMC Group
University of Nottingham
Nottingham, UK
Email: Alberto.Castellazzi@nottingham.ac.uk

Abstract—**This paper presents a statistical analysis on the effect of parallel connection of SiC power MOSFETs in high current applications. To this purpose, a reliable temperature-dependent SPICE model is calibrated on static and dynamic experimental curves of 1.2kV-36A commercial SiC MOSFET. The statistical fluctuation of threshold voltage and on-resistance is evaluated on 20 device samples and modeled with Gaussian functions. The proposed analysis, based on SPICE electro-thermal Monte Carlo simulations, is then aimed to improve the design of high current systems with multi-chip devices. Therefore, the study is focused on the evaluation of current and energy unbalance during device switching under inductive load. Results achieved for nominal switching condition and out-of-SOA current levels are discussed.**

Keywords—*Electro-thermal modeling; Monte Carlo simulation; power MOSFET; silicon carbide (SiC); SPICE.*

I. INTRODUCTION

Silicon carbide (SiC) power MOSFETs are finding widespread adoption in many application areas, such as energy distribution, automotive and aircraft, thanks to many excellent features. In several high power applications there is the need to use parallel devices, since commercial SiC MOSFETs are mostly available for low current ratings. Although for Si MOSFETs and IGBTs paralleling is well known and commonly used in many applications [1], [2], poor information is available in literature for SiC MOSFETs. Compared to commercial SiC modules [3], the use of discrete devices in parallel has some benefits: *(i)* the generated heat could be more evenly distributed over the heatsink (thus reducing temperature peaks); *(ii)* during the design process, flexibility is gained in terms of number of devices to use; *(iii)* lower cost is obtained thanks to the high volume production of discrete parts.

However, when paralleling two or more SiC MOSFETs, their currents may not be balanced due to the statistical fluctuations of the on-state resistance (R_{on}) and threshold voltage (V_{TH}) from sample to sample. This phenomenon can drastically reduce the reliability of the entire power system [4]. Previous works [5], [6] have suggested different feedback techniques for balancing drain currents during switching transients. However, those methods could be applied just for two parallel devices, and increase the overall system cost. In [7] an experimental study on the SiC MOSFET's self-balancing capability without adding any sensing or control

circuit is presented. The parameters used to balance paralleled devices are the gate drive voltage and resistance. However, this solution requires a tailored gate drive circuit for each device and could not be applied to modules. Another way to address this problem could be the adoption of design rules to determine the maximum allowed device parameters dispersion and parasitic circuit elements unbalance [8]. To optimize these rules, a valid approach involves a simulation analysis that quantifies the impact of device and circuit mismatches on the application performances. Since SiC MOSFETs often operate under harsh conditions, reliable electro-thermal (ET) simulations are mandatory for design optimization. In the last years, several papers have focused on the modeling of SiC MOSFETs [9]; some of them rely on empirical functions [10], while others are based on physics-based descriptions [11].

In this paper, a temperature-dependent SPICE model for SiC MOSFETs [12] is exploited for dynamic ET simulations of single and paralleled devices, both in SOA and out-of-SOA conditions. An experimental measurement campaign on 20 virtually identical devices is used to characterize the statistical distribution of R_{ON} and V_{TH}. Finally, an analysis based on Monte-Carlo ET simulations of 4 paralleled devices during switching is presented. The impact on the statistical energy dissipation is evaluated during turn-on and turn-off transients. Lastly, the potential unbalance in temperature is discussed.

II. SiC MOSFET COMPACT MODEL

In this paper, an extended formulation of a previously presented SPICE model [13] is reported, which has been experimentally verified on a broad range of operating conditions. The schematic representation of a planar SiC MOSFET structure is depicted in Fig. 1.

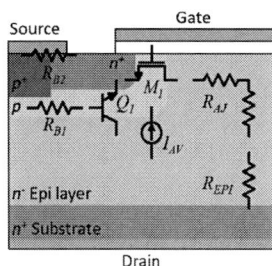

Fig. 1 Structure of a planar SiC power MOSFET and main equivalent circuit components.

978-1-5386-2928-4/18 $31.00 © 2018 IEEE

$$R_D\left(V_{GS}, V_{drift}, T\right) = R_{AJ}\left(V_{GS}, V_{drift}, T\right) + R_{EPI}\left(T\right)$$

$$R_{AJ}\left(V_{GS}, V_{drift}, T\right) = \frac{V_{drift}}{V_1 + V_{drift}} \cdot \left[R_{AJ1}\left(T\right) + R_{AJ2}\left(T\right)\left(1 + \frac{V_{GS}}{V_2}\right)^{-\eta}\right]$$

$$V_{TH}\left(T\right) = \left[V_{TH}\left(T_0\right) - \beta_{TH}\right]e^{-\varphi_{TH}\left(T - T_0\right)} + \beta_{TH}$$

$$M = 1 + a_{II}\tan\left[f_i\left(I_D\right)\frac{\pi}{2}\left(\frac{V_{DS}}{BV_{DS}\left(T\right)}\right)^{b_{II}}\right]$$

$$I_{Therm} = A_{Therm}n_i\left(T\right)^{\alpha_{Therm}}V_{DS}^{\gamma}$$

$$I_{AV} = \left(M - 1\right)I_{D\left(M_i\right)} + M \cdot I_{Therm}$$

$$\mu_n\left(T\right) = \mu_n\left(T_0\right)\left(\frac{T}{T_0}\right)^{-m\left(T\right)}$$

$$m\left(T\right) = -a_m + \left(a_m + b_m\right)\left[1 - c_m \exp\left(-d_m\frac{T}{T_0}\right)\right]$$

$$I_{MOS} = \frac{I\left(V_{SENS}\right) \cdot f_\mu\left(T\right)}{\left[1 + \theta_1\left(V_{GS} - V_{TH}\left(T\right)\right)\right]\left(1 + \theta_2 \cdot V_{DS}\right)}\left(1 + \lambda \cdot V_{DS}\right)$$

$$C_{DS}\left(V_{ds}\right) = \frac{C_{DS0}\left[\frac{\pi}{2} + \arctan\left(-\frac{V_{ds}}{V_{ds}^*}\right)\right]}{\pi/2} + C_{DSMIN}$$

$$C_{GD}\left(V_{gd}\right) = \left(C_{GD0} - C_{GDMIN}\right)\left[1 + \frac{2}{\pi}\arctan\left(\frac{V_{gd}}{V_{gd}^*}\right)\right]$$

Fig. 2 Most relevant equations for compact elements in the SPICE sub-circuit of Fig. 3.

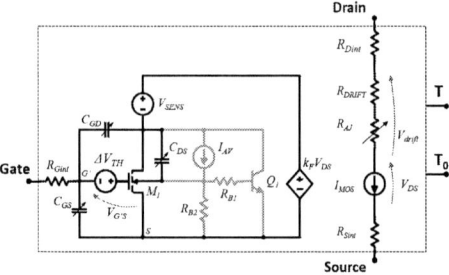

Fig. 3 Developed SPICE sub-circuit, with electrical and thermal nodes. Elements in gray model the out-of-SOA operation.

The model is based on the partitioning of the device into an 'intrinsic' (channel) MOSFET, a bias-dependent resistance for the accumulation and JFET regions, and a constant resistance for the epitaxial drift region. The influence of SiO$_2$/SiC interface traps on threshold voltage and channel mobility, impact ionization and capacitance nonlinearity are accounted for. The most relevant model equations are reported in Fig. 2, while the related SPICE sub-circuit is depicted in Fig. 3.

The parameters require a simple optimization procedure based on experimental data; the details are given in [12].

A. Model calibration and verification

The device under test (DUT) selected as a case-study is an 80 mΩ 1.2kV–36A 4H-SiC power MOSFET. The model parameters were calibrated on measurements performed at different baseplate temperatures under pulsed (isothermal) conditions. Fig. 4 and Fig. 5 show the transfer and output characteristics at 300 and 400 K.

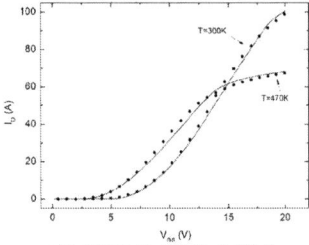

Fig. 4 I$_D$-V$_{GS}$ curves at T=300 K, T=470 K. Solid lines are SPICE results; symbols are measurements.

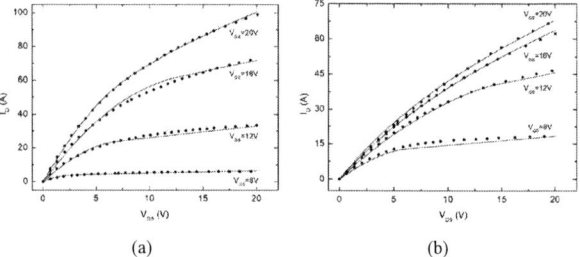

Fig. 5 I$_D$-V$_{DS}$ curves at (a) T=300K and (b) T=470K. Solid lines are SPICE results; symbols are measurements.

Excellent agreement was obtained with the experimental dc characteristics despite the smooth triode-saturation transition occurring in SiC transistors. A double-pulse test was used to verify the model accuracy under dynamic conditions. Both turn-on and turn-off current and voltage evolutions are well predicted, as reported in Fig. 6.

III. STATISTICAL ANALYSIS

In order to perform the Monte Carlo analysis, a statistical description of the device parameters fluctuation (R$_{on}$ and V$_{TH}$) is needed. To this purpose, the MOSFET current factor K and the threshold voltage V$_{TH}$ for all the 20 devices were directly extracted from the highest-slope portion (medium V$_{GS}$) of the isothermal I$_D$-V$_{GS}$ transfer characteristics measured at various baseplate temperatures using the quadratic extrapolation method (QEM). The on-state resistance was evaluated on the output I$_D$-V$_{DS}$ curve for V$_{GS}$=20 V.

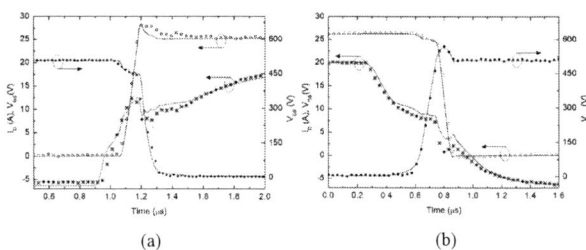

Fig. 6 Inductive switching waveforms (I$_D$, V$_{DS}$, V$_{GS}$): (a) turn-on; (b) turn-off. Solid lines are SPICE numerical results; symbols refer to the experiment. V$_{BATT}$=500 V, L$_{LOAD}$=1.9 mH, R$_G$=50 Ω.

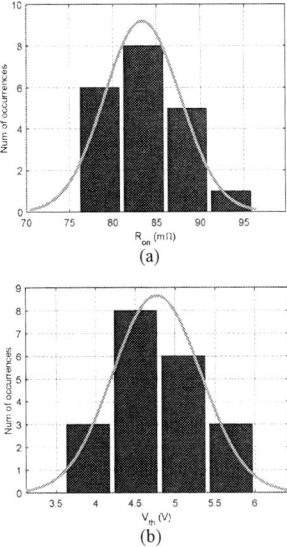

(a)

(b)

Fig. 7 Histograms of statistical distribution for both V_{TH} and R_{on}, measured on 20 DUT samples at T=27°C, and fitting Gaussian functions (light blue lines).

Fig. 7 shows the histograms of statistical distribution for both V_{TH} and R_{ON} at room temperature. The on-resistance exhibits a spread of about 20 mΩ for the analyzed devices, while the spread for the threshold voltage (evaluated with the QEM) is ≤3.5 V. Optimized Gaussian functions are used to describe the parameters variation in the following Monte Carlo analysis. In particular, the expected value µ and variance σ of the current factor and threshold voltage are properly included in the SPICE sub-circuit.

A. Monte Carlo ET simulations

The proposed Monte Carlo analysis was based on the paralleling of 4 SiC MOSFETs in a double-pulse test. The circuit is shown in Fig. 8 along with parasitic elements and test parameters. An example of the impact of devices mismatches is reported in Fig. 9, which confirms that the individual transistor with lower R_{on} (MOSFET 4) conducts an higher current, while the current sharing during turn-off transient is affected by the V_{TH} unbalance. In particular, the MOSFET with the lower V_{TH} will switch-on earlier and switch-off later than the higher V_{TH} others. As a consequence, a nonuniform temperature increase takes place. As a relevant result, the MC analysis also allowed quantifying the statistical distribution of the energy dissipation during turn-on and turn-off transients as dictated by the V_{TH} and R_{on} variation. The first analyzed case was for nominal device current I_D=20 A, involving a total load current of 80 A. In this analysis, 1500 MC ET simulation were carried out and the histograms of the dissipated energy evaluated during turn-off and turn-on for each of the 4 MOSFETs are reported in Figs. 10 and 11, respectively. The resulting shape of statistical energy distributions were almost Gaussian with expected E_{off} and E_{on} values of about 580 µJ and 1.3 mJ, respectively. The most useful finding is the considerable switching loss spreading in real applications. For example, derating rules for SiC MOSFETs can be extracted using these data. Another critical

point regards the energy unbalance within the same multi-chip structure. Fig. 12 reports the maximum energy unbalances for each Monte Carlo simulation run, evaluated as:

$$\Delta E_{off} = \max\left(E_{off1}, E_{off2}, E_{off3}, E_{off4}\right) - \min\left(E_{off1}, E_{off2}, E_{off3}, E_{off4}\right) \quad (1)$$

$$\Delta E_{on} = \max\left(E_{on1}, E_{on2}, E_{on3}, E_{on4}\right) - \min\left(E_{on1}, E_{on2}, E_{on3}, E_{on4}\right) \quad (2)$$

Pointing the attention only on the switching power loss, the unbalance in temperature rise ΔT can be estimated as:

$$\Delta P_{sw} = \left(\Delta E_{on} + \Delta E_{off}\right) \cdot f_{sw} \quad (3)$$

$$\Delta T_{sw} = \Delta P_{sw} \cdot R_{th} \quad (4)$$

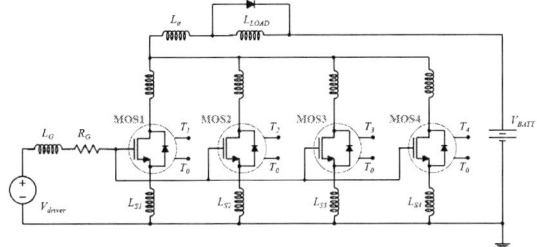

Fig. 8 Circuit model of the four paralleled SiC MOSFET. For each MOSFET, an equivalent thermal network (provided by manufacturer) was used to enable fully coupled ET simulations. V_{BATT}=800 V, I_{LOAD}=80 A, L_{LOAD}=142 µH, R_G=2.5Ω.

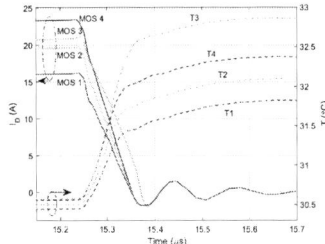

Fig. 9 Paralleled 4 SiC MOSFETs inductive turn-off: currents and device temperatures.

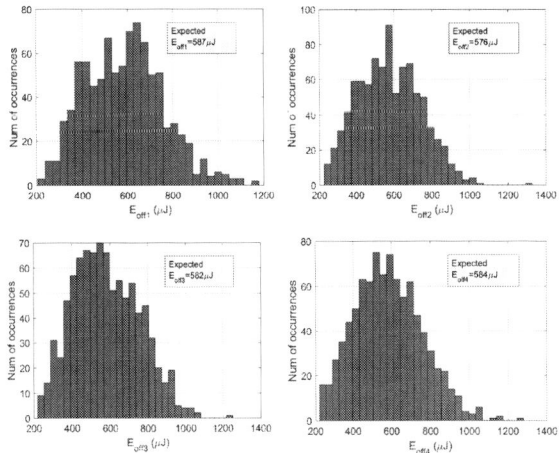

Fig. 10 Histograms of the turn-off dissipated energy evaluated over 1500 MC ET simulations for the 4 MOSFETs at I_{LOAD}=80 A.

978-1-5386-2928-4/18 $31.00 © 2018 IEEE

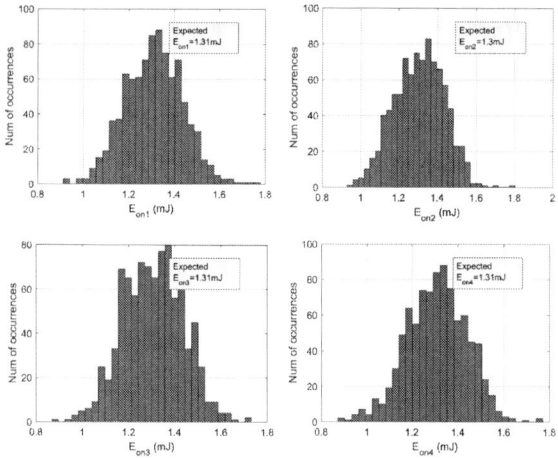

Fig. 11 Histograms of the turn-on dissipated energy evaluated over 1500 MC ET simulations for the 4 MOSFETs at I_{LOAD}=80 A.

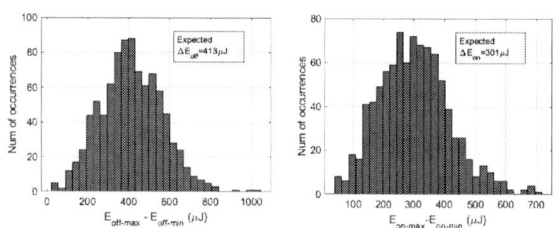

Fig. 12 Histograms of the difference between the maximum and minimum dissipated energy evaluated during (a) turn-off and (b) turn-on over 1500 MC ET simulations at I_{LOAD}=80 A.

Considering the thermal resistance R_{th} of the SiC MOSFET used in this paper (R_{th}=0.6 K/W), with a switching frequency of 80 kHz the above expressions provide a maximum temperature difference $\Delta T \approx 34$ K. However, (4) just gives an indication for the design of a power system. In fact, due to the negative temperature coefficient of V_{TH} [12], the switching loss difference will increase due to ET effects, and then a positive feedback can arise. On the other hand, the R_{on} increase with temperature is expected to partially compensate such temperature difference. It is desired and important to minimize the switching loss difference caused by threshold voltage variance, since the switching loss difference turns into difference in temperatures of the paralleled devices.

Lastly, in Fig. 13 a first attempt to the out-of-SOA analysis is given with the evaluation of switching loss unbalance for a total load current of 160 A (40 A × 4 MOSFETs). The results in terms of ΔE_{off} and ΔE_{on} can be used to analyze the impact of device mismatches on the reliability of the entire power multichip module.

IV. CONCLUSION

In this work, a statistical analysis on the effect of parallel connection of SiC power MOSFETs in high-current applications has been presented. The analysis relies on a temperature-dependent SPICE model calibrated on dc and transient experimental curves of a 1.2 kV–36 A commercial SiC MOSFET. The statistical fluctuations of threshold voltage and on-resistance have been evaluated on 20 devices, and a Gaussian fitting was used to perform SPICE electro-thermal Monte Carlo simulations. The study, focused on the evaluation of current and energy unbalance during switching under inductive load, gives helpful indication on the impact of V_{TH} and R_{on} statistical fluctuation in real applications.

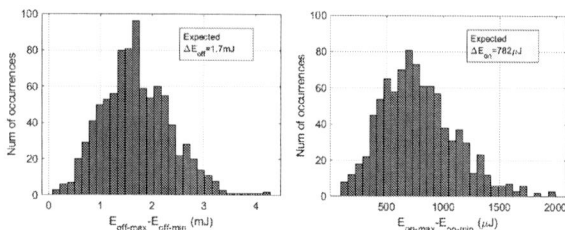

Fig. 13 Histograms of the difference between the maximum and minimum dissipated energy evaluated during (a) turn-off and (b) turn-on over 1500 MC ET simulations at nominal I_{LOAD}=160 A.

V. REFERENCES

[1] J. B. Forsythe, "Paralleling of Power MOSFETs For Higher Power Output," (http://www.irf.com/technical-info/appnotes/para.pdf).

[2] P. R. Palmer and J. C. Joyce, "Current redistribution in multi-chip IGBT modules under various gate drive conditions," in *Proc. IEE Power Electronics and Variable Speed Drives*, 1998, pp. 246-251.

[3] G. Wang *et al.*, "Performance comparison of 1200V 100A SiC MOSFET and 1200V 100A silicon IGBT," in *Proc. IEEE ECCE*, 2013, pp. 3230-3234.

[4] R. Horff *et al.*, "Current mismatch in paralleled phases of high power SiC modules due to threshold voltage unsymmetry and different gate-driver concepts," in *Proc. IEEE ECCE*, 2016, pp. 1-9.

[5] F. Chimento, A. Raciti, A. Cannone, S. Musumeci, and A. Gaito, "Parallel connection of super-junction MOSFETs in a PFC application," in *Proc. IEEE ECCE*, 2009, pp. 3776-3783.

[6] Y. Xue *et al.*, "Active current balancing for parallel-connected silicon carbide MOSFETs," in *Proc. IEEE ECCE*, 2013, pp. 1563-1569.

[7] G. Wang, J. Mookken, J. Rice, and M. Schupbach, "Dynamic and static behavior of packaged silicon carbide MOSFETs in paralleled applications," in *Proc. IEEE APEC*, 2014, pp. 1478-1483.

[8] H. Li *et al.*, "Influences of device and circuit mismatches on paralleling silicon carbide MOSFETs," *IEEE Trans. Power Electron.*, vol. 31, no. 1, pp. 621-634, 2016.

[9] H. A. Mantooth *et al.*, "Modeling of Wide Bandgap Power Semiconductor Devices—Part I," *IEEE Trans. Electron Devices*, vol. 62, no. 2, pp. 423-433, 2015.

[10] K. Sun *et al.*, "Improved modeling of medium voltage SiC MOSFET within wide temperature range," *IEEE Trans. Power Electron.*, vol. 29, no. 5, pp. 2229-2237, 2014.

[11] R. Kraus and A. Castellazzi, "A physics-based compact model of SiC power MOSFETs," *IEEE Trans. Power Electron.*, vol. 31, no. 8, pp. 5863-5870, 2016.

[12] M. Riccio, V. d'Alessandro, G. Romano, L. Maresca, G. Breglio and A. Irace, "A temperature-dependent SPICE model of SiC power MOSFETs for within and out-of-SOA simulations," *IEEE Trans. Power Electron.*, doi: 10.1109/TPEL.2017.2774764.

[13] V. d'Alessandro, A. Magnani, M. Riccio, G. Breglio, A. Irace, N. Rinaldi, A. Castellazzi, "SPICE modeling and dynamic electrothermal simulation of SiC power MOSFETs," in *Proc. IEEE ISPSD*, 2014, pp. 285-288.

Proceedings of the 30th International Symposium on Power Semiconductor Devices & ICs
May 13-17, 2018, Chicago, USA

Reliability Investigation with Accelerated Body Diode Current Stress for 3.3 kV 4H-SiC MOSFETs with Various Buffer Epilayer Thickness

Yuji Ebiike, Takeshi Murakami, Eisuke Suekawa, Shigehisa Yamamoto, Hiroaki Sumitani,
Masayuki Imaizumi and Masayoshi Tarutani
Power Device Works, Mitsubishi Electric Corporation
1-1-1 Imajukuhigashi Nishi-ku Fukuoka 819-0192/Japan
Ebiike.Yuji@aj.Mitsubishi Electric.co.jp

Abstract—**3.3 kV 4H-SiC MOSFETs with various buffer layer thickness has been fabricated in order to investigate the bipolar degradation associated with the expansion of stacking faults (SFs). The body diode stress tests under DC current of 240 A/cm² were performed at 200 °C. Shifts in specific on-resistance ($R_{on,sp}$) and forward voltage (V_f) of body diode were markedly reduced for the MOSFETs with thick buffer layer of 30 μm. This result indicates that the body diode reliability may be improved by suitably designed buffer layer. The photoluminescence (PL) image of these SiC epilayer after the stress tests revealed the SF expansion due to the bipolar current stress was suppressed by the thick buffer layer.**

I. INTRODUCTION

R&D on SiC-MOSFETs have been conducted to improve energy-saving performance in power electronics systems [1-3]. We have reported 3.3 kV class SiC power module for latest traction inverter systems and demonstrated superiority in power loss reduction [4]. For the purpose of reducing the number of parallel-connected power chips in power modules, integrating SBDs and MOSFETs [5], or elimination or reduction in number of SBD chips is proposed. To realize the SiC module without SBDs, we have to utilize the parasitic body diodes in SiC-MOSFETs. However, body diode current may cause the bipolar degradation by the stacking fault (SF) expansion located in an SiC drift layer, leading to an increase in resistance [6]. To investigate the body diode stability, we fabricated 3.3 kV-class 4H-SiC MOSFETs with various thickness of the buffer layer, and the accelerated body diode current test for the MOSFETs has been conducted.

II. DEVICE FABRICATION

Figure 1 shows the cross sectional structure of the drift layer and buffer layer on SiC substrate. In this experiment, we formed n-type buffer layer, heavily doped to 8×10^{18} cm^{-3}, and an n-type drift layer, lightly doped to 3×10^{15} cm^{-3}, on an n-type 4 ° off-axis 4H-SiC (0001) commercially available substrates. The thickness of the buffer layer was widely varied to be (a) 2 μm, (b) 10 μm, (c) 20 μm and (d) 30 μm, respectively in order to investigate the merit and demerit of the buffer layer. Figure 2 shows the schematic cross sectional structure of the fabricated 3.3 kV SiC-MOSFETs. The thickness of the epilayer was designed as 30 μm which was determined to suppress

leakage current appropriately at high drain voltage (V_D) of 3.3 kV.

Device fabrication processes are as follows. Ion implantation was performed to form the p-well regions, the n-source regions and the n-type JFET doping (JD) regions. The doping condition of JD was carefully selected in order to realize both a low on-resistance and a high breakdown voltage.

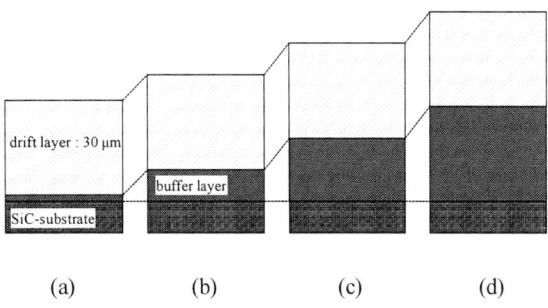

Fig.1: The drift layer and buffer on the SiC substrate. The thickness of the buffer layer was designed to be (a) 2 μm, (b) 10 μm, (c) 20 μm and (d) 30 μm, respectively.

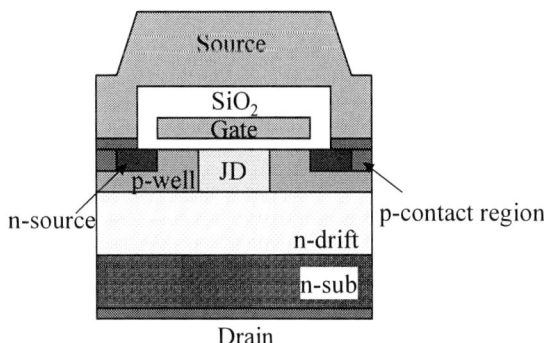

Fig.2: Schematic cross sectional structure of the SiC-MOSFET.

978-1-5386-2928-4/18 $31.00 © 2018 IEEE   447

We formed the p-contact regions by a high-temperature ion implantation. After thermal annealing over 1600 °C to activate the implanted impurities, a gate oxide layer was formed by thermal oxidation of the epilayer. Poly-Si was deposited onto the gate oxide layer as a gate electrode. A thermally grown oxide was deposited to isolate the gate electrode from the SiC surface. Finally, metallization was performed for a drain and a source electrode. A more detailed fabrication process is reported in the literature [7].

III. RESULTS AND DISCUSSION

Static on-state resistance of the MOSFETs were measured at room temperature. Figure 3 shows the $R_{on,sp}$ of fabricated MOSFETs as a function of buffer layer thickness. We confirmed that the increase in $R_{on,sp}$ is negligible for the high voltage class MOSFETs due to the low resistivity of heavily-doped buffer layer. No negative impact was observed on the characteristics of the MOSFETs by utilizing thick buffer layer.

Body diode current stress test with forward current (I_f) of 240 A/cm² was carried out for these devices at 200 °C to accelerate the degradation. Figure 4 shows the circuit diagram of the body diode current stress test.

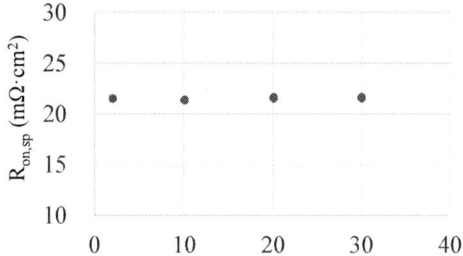

Fig.3: $R_{on,sp}$ of fabricated MOSFET as a function of buffer layer thickness. The increase in $R_{on,sp}$ is negligible for the MOSFETs due to the low resistivity of heavily-doped buffer layer.

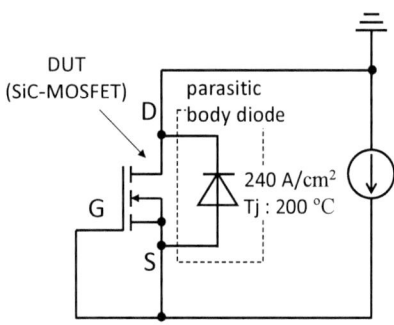

Fig.4: Circuit diagram of the body diode current stress test.

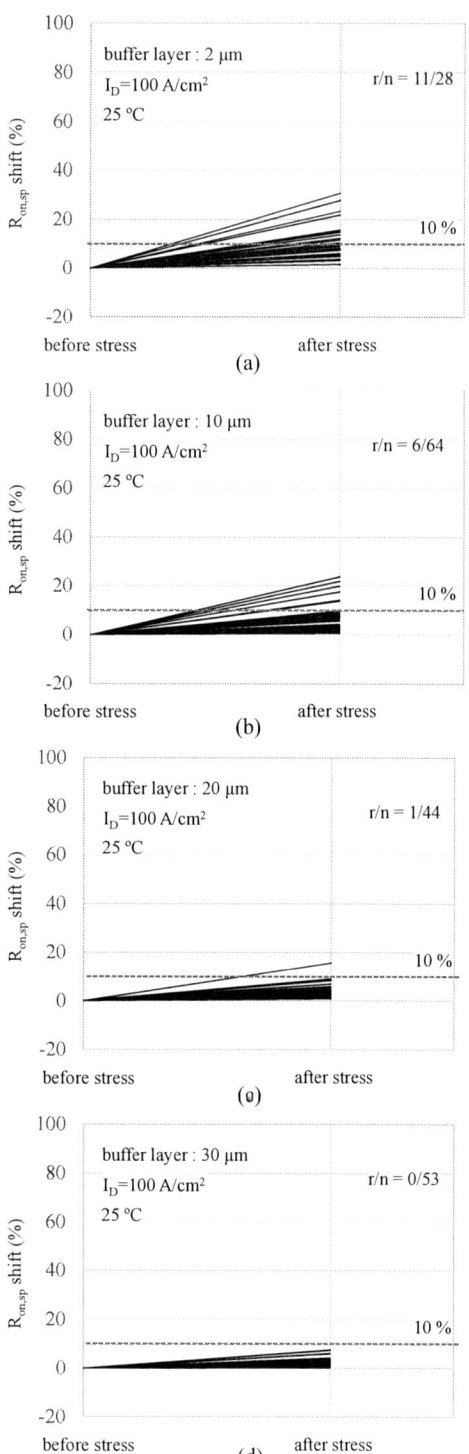

Fig.5: Results of stress test for the MOSFETs with the buffer layer thickness of 2 μm (a), 10 μm (b), 20 μm (c) and 30 μm (d).

Length of stress time was fixed at the value where the body diode degradation fully saturate. $R_{on,sp}$ of the MOSFETs were measured at room temperature and the shifts were calculated by comparison of the measurement before and after current stress. Figure 5 shows the results of the stress test for the MOSFETs with the buffer layer thickness of (a) 2 μm, (b) 10 μm, (c) 20 μm and (d) 30 μm, respectively. Increase in $R_{on,sp}$ due to the degradation of the device with the buffer layer of 30 μm exhibits at most below 10 % at I_D of 100 A/cm², 25 °C.

We also measured forward voltage (V_f) of the body diodes. Figure 6 shows the relationship between $R_{on,sp}$ shift and body diode V_f shift before and after current stress. Kimoto et al, reported that V_f shift of an SiC pin diode can be reduced by utilizing a recombination-enhancing layer on an SiC substrate [8]. However, $R_{on,sp}$ shifts of the MOSFETs are much more severe than V_f shifts the body diodes because of the linear I_D-V_D characteristic from the coordinate origin. Therefore, to obtain the stable characteristics of the SiC-MOSFET by the bipolar current stress is much more difficult than SiC-PiN diode.

Figure 7 shows the typical photoluminescence (PL) image of the of the MOSFET which has a buffer layer of 2 μm (a), 10 μm (b) and 20 μm (c) after the body diode current stress. From the PL image (a) We confirmed that SF expansion of the MOSFET which have thin buffer layer converted from the interface between the substrate and the epilayer. While in the case of thicker buffer layer over 10 μm, SF expansion from the interface between the substrate and the epilayer was well suppressed; however, SF expansion from the midst of buffer layer was confirmed. Figure 5 and 7 indicate that to improve the stability of the MOSFET from the body diode current stress, the buffer layer of 10 μm is effective but insufficient.

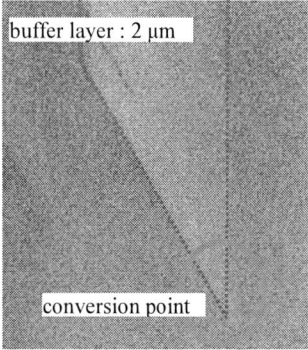

conversion point : interface between the substrate and the epilayer

(a)

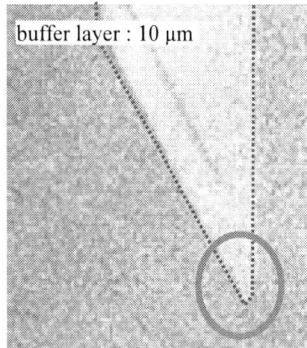

conversion point : inside the buffer layer

(b)

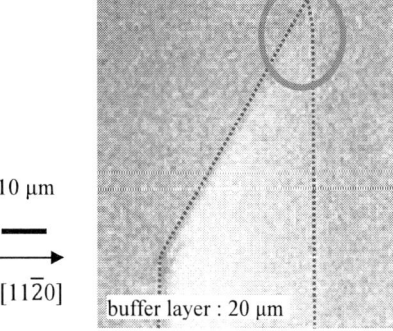

conversion point : inside the buffer layer

(c)

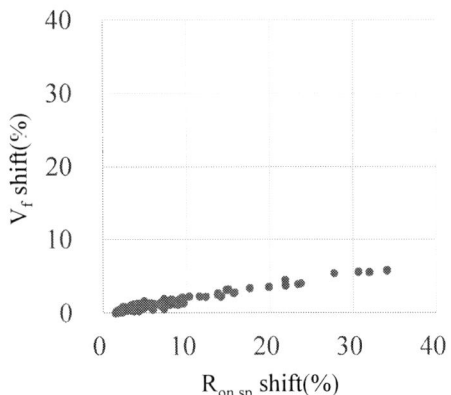

Fig.6: Relationship between $R_{on,sp}$ shift and V_f shift.

Fig.7: Typical PL image of MOSFET which has thick buffer layer of 2 μm (a), 10 μm (b) and 20μm (c) after the body diode current stress.

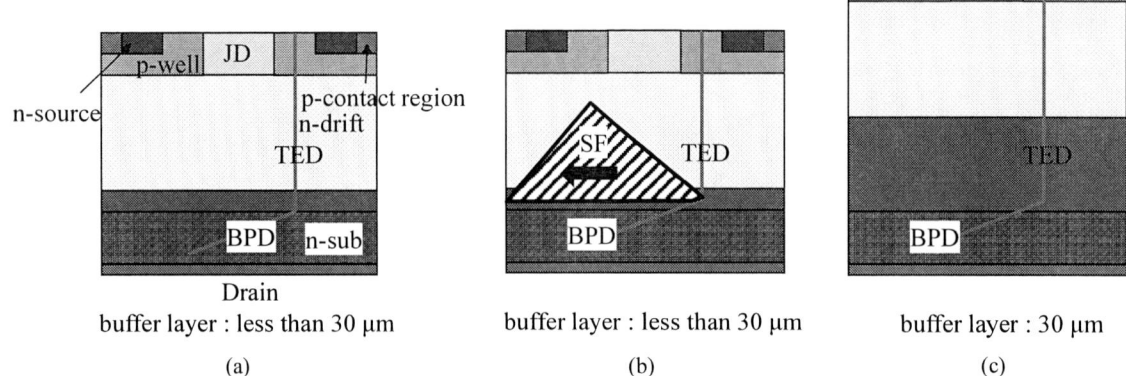

Fig.8: Schematic cross sectional image of the suppression of degradation owing to the thick buffer layer. Suppression of SF expansion from BPD converted at the interface of the substrate (a), SF expansion from BPD converted in the buffer layer (b) and suppression of SF expansion from BPD converted in the thick buffer layer (c).

Figure 8 shows the schematics of the suppression of degradation by utilizing the thick buffer layer. Figure 8 shows the suppression of SF expansion from BPD converted at the interface of the substrate (a), SF expansion from BPD converted in the buffer layer (b) and suppression of SF expansion from BPD converted in the thick buffer layer (c). We estimate that the thick buffer layer contribute to reducing minority carriers reaching the SF expansion original points, such as BPD-TED conversion points, in the drift layer.

IV. CONCLUSION

We investigate the effect of the reliability of the body diode in the SiC-MOSFET by the buffer layer. From the investigation by the body diode stress test of the SiC-MOSFETs with various buffer layer thickness, we confirmed the degradation of the MOSFETs is supressed by thick buffer layer. $R_{on,sp}$ shift of the MOSFET with 30 μm buffer layer of exhibited below 10% (I_D=100 A/cm^2, 25 °C) after the body diode current stress test (I_f=240 A/cm^2). From this result, we confirmed that the body diode reliability is significantly improved by the suitably designed buffer layer. By using SiC-MOSFETs with improved body diode reliability, we believe the number of parallel-connected power chips in power modules are drastically reduced.

REFERENCES

[1] N. Miura, K. Fujihira, Y. Nakao, T. Watanabe, Y. Tarui, S. Kinouchi, M. Imaizumi and T. Oomori, Proc. ISPSD2006, pp 251-264.

[2] A. Furukawa, S. Kinouchi, H. Nakatake, Y. Ebiike, Y. Kagawa, N. Miura, Y. Nakao, M. Imaizumi, H. Sumitani and T. Oomori, ISPSD2011, pp 288-291.

[3] M. Imaizumi and N. Miura, Electron devices, 62 (2015) pp 390-395.

[4] K. Hamada, S. Hino, N. Miura, H. Watanabe, S. Nakata, E. Suekawa, Y. Ebiike, M. Imaizumi, I. Umezaki and S. Yamakawa, Jpn. J. Appl Phys. 54 (2015) 04DP07G.

[5] K. Kawahara, S. Hino, K. Sadamatsu, Y. Nakao, Y. Yamashiro, Y. Yamamoto, T. Iwamatsu, S. Nakata, S. Tomohisa and S. Yamakawa, Proc. ISPSD2017, pp 41-44.

[6] M.Skowronski and S. Ha, J. Appl Phys. 99 011101 (2006).

[7] K. Hamada, N. Miura, S. Hino, T. Kawakami, M. Imaizumi, H. Sumitani and T. Oomori, Jpn. J. Appl Phys. 52 (2013) 04CP03.

[8] T. Kimoto, A. Iijima, T. Tawara, A. Otsuki, H. Tsuchida, T. Miyazawa, T. Kato and Y. Yonezawa, RIPS2017, pp 2A-1.1-2A-1-7.

Proceedings of the 30th International Symposium on Power Semiconductor Devices & ICs
May 13-17, 2018, Chicago, USA

Dynamic Switching and Short Circuit Capability of 6.5kV Silicon Carbide MOSFETs

L. Knoll, A. Mihaila, L. Kranz, M. Bellini, S. Wirths, E. Bianda

ABB Switzerland
Corporate Research
CH-5405 Baden-Dättwil, Switzerland
lars.knoll@ch.abb.com

C. Papadopoulos, M. Rahimo

ABB Switzerland
Semiconductors
CH-5600 Lenzburg, Switzerland

Abstract— **Electrically robust 6.5kV SiC MOSFETs are investigated for the static and dynamic performance, short circuit capability and safe operation area (SOA). SiC MOSFETs rated at 6.5kV were fabricated with different cell pitches from 12μm to 26μm that are able to withstand short circuit pulses of up to 8μs and have a turn-off SOA at 4400V up to twice the nominal current I_{NOM}. The paralleling of four MOSFETs was tested to represent a realistic setup while showing a substantial reduction in the switching loss by more than 80% compared to a silicon IGBT and Diode.**

Keywords—Wide band-gap; SiC; high voltage; power MOSFET; short circuit; SOA; reliability

I. Introduction

Over the last few years, silicon carbide MOSFETs have provided a power device technology alternative to silicon modules in various low voltage applications. In comparison to commercially available low-voltage components, the medium to high voltage SiC-based technology is still not mature enough to guarantee reliable operation. The focus of most publications in the field of medium to high voltage Silicon Carbide MOSFETs has been on improving static and dynamic performance [1,2]. However, a key requirement for reliable operation in many applications, especially regarding traction and grid systems is the electrical robustness of the devices [3,4]. Whereas 6.5kV MOSFETs are designed for power conversion applications at DC link voltage of 3.6kV, the margin of over-voltage spikes during switching is required to be large enough to ensure safe operation even in case of DC-link voltage fluctuations of up to 4.4kV and large circuit stray inductances. Moreover, the modules have to be rugged against short circuit currents. Thus, it is essential to examine the short circuit current capability as well as to outline the safe operation principles. In this work, we present a pitch-dependent design trade-off between on state, switching speed and short-circuit current capability of 6.5kV SiC MOSFETs. Furthermore, the aspect of parallelizability is investigated.

II. Concept

Vertical planar 6.5kV power MOSFETs were fabricated on 4H-SiC (0001) wafers with a lowly nitrogen doped epitaxial layer deposited on a high conductivity n-type substrate. Several cell pitches such as 26μm (p26), 14μm (p14) and 12μm (p12) were processed using different cell-to-cell distances while keeping the source contact dimensions fixed. In this approach, at smaller pitches, the current limiting effect is driven by an increased JFET effect, which is counter balancing a larger channel width in conjunction with a higher cell density, since the active area was kept constant. Static on-state measurements were performed with a pulsed I-V curve tracer with a duty cycle of 1% and pulse duration of 200μs to avoid self-heating. For dynamic characterization the devices were soldered on test substrates, wire bonded and mounted in a double pulse tester with a system inductance ($L\sigma$) of 480nH and variable gate resistors. For paralleling experiments four MOSFETs were bonded on the same substrate, thus sharing the same gate resistor and system stray inductance. For the short circuit analysis the drain potential (V_D) was set up to nominal voltage of 3600V and the device was switched on with the gate potential (V_G) of 15V for up to 8μs.

III. Results and Discussion

A. Static characterization

Figure 1 shows the static I_D-V_D characteristics of the fabricated MOSFETs comparing p26, p14 and p12 designs with a conventional state of the art 6.5kV Si IGBT at a junction temperature (Tj) of 125°C and a gate voltage of V_G=15V. A horizontal line indicates the rated current of 37.5 A/cm² (I_{NOM}) of the IGBT. As can be seen, the SiC MOSFETs provide comparable (p26) or even higher (p14, p12) normalized current capability compared to the Si IGBT. At I_{NOM} of the IGBT the p12 and p14 MOSFETs reduce on state voltage by 1V. At similar power density of ~130W/cm², the nominal current for the MOSFET is 45A/cm², thus improving current density by 20% at lower losses.

978-1-5386-2928-4/18 $31.00 © 2018 IEEE

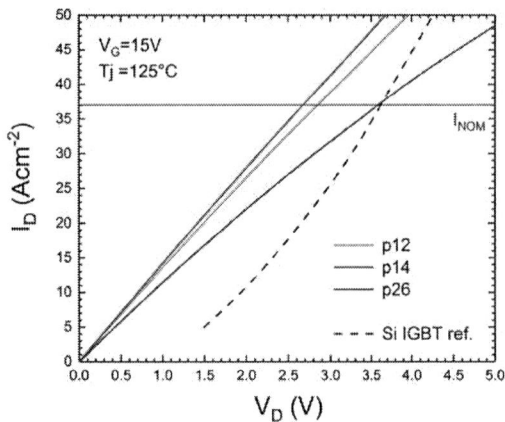

Figure 1 Static ID-VD of 6.5kV SiC MOSFETs with cell pitch 12, 14 and 26 μm and of a Si IGBT reference. At the IGBT nominal current density the on state loss can be reduced by up to 1V.

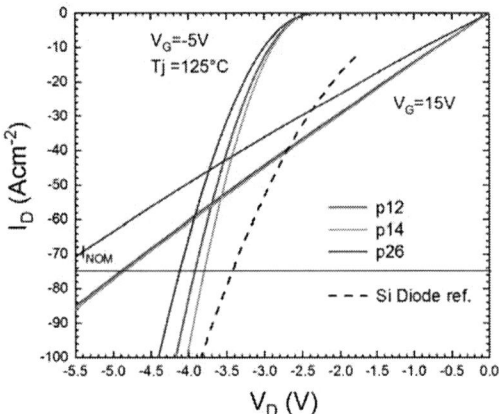

Figure 2 Static I_D-V_D of the body diode of 6.5kV SiC MOSFETs with cell pitch 12, 14 and 26 μm and of a Si diode reference.

In Figure 2, the diode conduction in the third quadrant is demonstrated for different V_G. The body diode of the MOSFET has a turn on threshold of -2.5V for negative V_G, but shows linear conduction for V_G=15V. At the rated diode current of 75Acm^{-2} a comparable voltage drop can be observed for all diodes, the silicon diode however, has the lowest voltage drop of 3.5V followed by p12 (3.8V), p14 (4.0V) and p26 (4.3V). A smaller cell pitch significantly reduces the body diode on state by 0.5V. For a fair comparison to the Si diode, we have to point out that by using the bidirectional functionality of the MOSFETs more diode area (~3x) is available in the module than it would be the case with separate diodes. The temperature dependence of the specific source drain resistance ($R_{DS,ON}$) is given in Figure 3. Both, the p12 and p14 MOSFETs show similarly low $R_{DS,ON}$ over the full temperature range having only minor differences. The p12 devices tend to have a slightly larger JFET resistance at high temperatures. Remarkably, the p26 MOSFET does not show a large difference in $R_{DS\ ON}$ as seen in case of higher currents in

Figure 1. A low ratio of $R_{DS\ ON}^{25C}$ / $R_{DS\ ON}^{200C}$=2.5 is observed for both, p12 and p14 pointing out the high temperature capability of SiC MOSFETs.

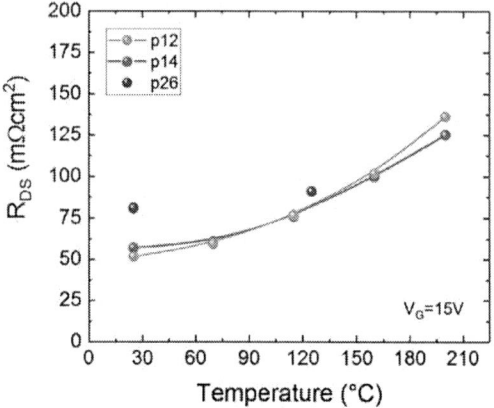

Figure 3 Extracted specific source drain resistance with cell pitch p12, p14 and p26 μm for a temperature range from T_j=25°C to T_j=200°C.

B. Dynamic switching

Figure 4 shows the turn on waveforms of a single MOSFET with pitches p26, p14 and p12 under nominal conditions of V_D=3600V, V_G=±15V and I_D=8A at T_j=125°C. The measurements were performed using R_G=33Ω and a stray inductance of 480nH. Devices with smaller pitches demonstrate faster switching and larger overshoot of I_D. In p12 we observe a total switching time of 300ns, and ~1μs in p26, respectively. The overshoot is present during the dV/dt swing, indicating a purely capacitive nature. Consequently, higher dV/dt results in a higher peak and a shorter duration.

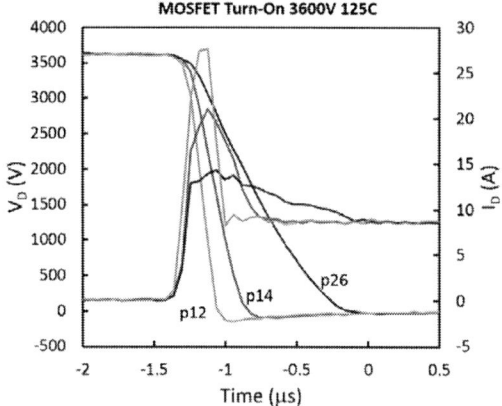

Figure 4 Turn on waveform of p12 p14 and p26 MOSFETs with R_G=33Ω, L_σ=480nH at Tj=125°C.

The turn-off characteristics for a single MOSFET at V_D=3600V, V_G=±15V and I_D=8A at T_j=125°C using R_G=33Ω and Lσ=480nH are plotted in Figure 5. The switching time is less sensitive on the design and measures approx. 500ns.

978-1-5386-2928-4/18 $31.00 © 2018 IEEE   452

During this nominal turn-off event the drain potential peaks at around 3800V for all pitches. The current waveform shows the biggest difference between the pitch designs where p12 devices have the highest dI/dt while p26 devices have the lowest.

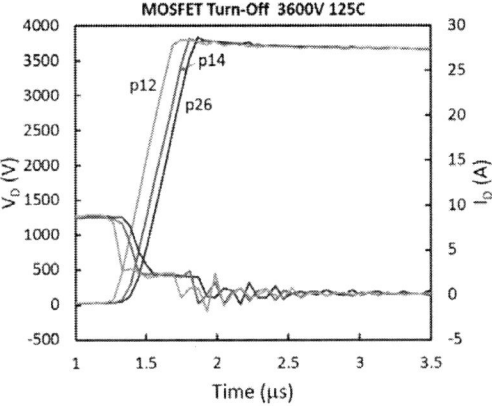

Figure 5 Turn off waveform of p12 p14 and p26 MOSFETs with R_G=33Ω, L_σ=480nH at Tj=125°C.

Turn off characteristics of p14 and p26 MOSFETs for over current and over voltage conditions are indicated in Figure 6. The waveforms are measured at V_G= ±15V, V_D=4400V and I_D=16A, which is twice I_{NOM}. Large oscillations are observed in the current waveform after turn-off since the MOSFET was switched fast with R_G=10Ω and a rather large inductance of L_σ=480nH. The voltage peak reaches 4700V and shows minor oscillations at high voltage. However, no failures were observed during this test.

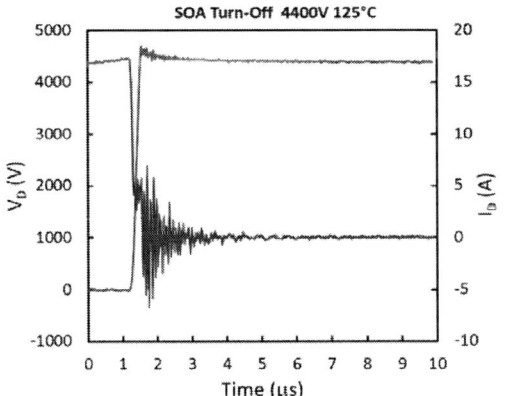

Figure 6 Turn off waveform of p14 and p26 MOSFETs with R_G=10Ω, L_σ=480nH at Tj=125°C under overcurrent and overvoltage conditions of V_D=4400V and I_D=16A (2x I_{NOM}).

Figure 7 depicts the short circuit (SC) waveforms of pitches p14 and p26 at T_j=125°C. The lower on state of p26 results in about 50% lower saturation current (I_D^{sat}) yielding longer SC capability of up to 8µs. Remarkably, the maximum SC energy of both pitches is very similar and measures about 2.3J at

125°C. Therefore, the saturation current mainly defines the SC capability. P14 devices have a saturation current of >200A, which is more than double of what is observed for p26. Thus, the maximum duration until failure reduces to less than half. Another important aspect is the fact that the saturation current does not depend on the applied drain potential. Therefore, we can show that no short channel effects at high V_D further reduce the SC capability in these devices. It depends essentially on the threshold voltage V_{th}.

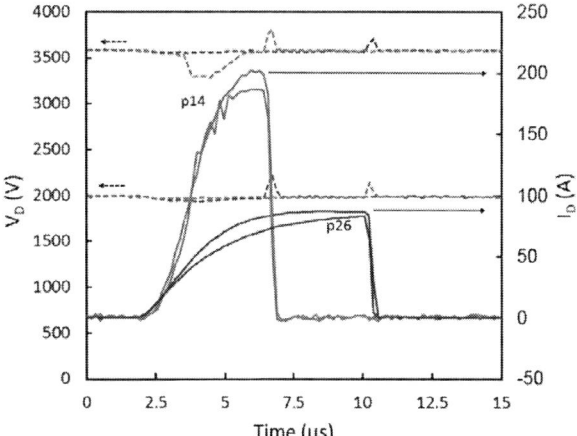

Figure 7 Short circuit waveforms of p14 and p26 MOSFETs up to nominal conditions of V_D=3600V, V_G=15V at 125°C.

C. Paralleling and dynamic characterization

Up to this point, performance and electrical robustness have been successfully demonstrated. Another important aspect is the possibility to connect MOSFETs in parallel. Especially for small components, paralleling is of high importance. The fast switching of large currents at high voltages can cause undesirable effects in the switching process, which can directly result in a fault. Important characteristic is the even distribution of current during operation. The positive temperature coefficient as observed in Figure 3 is very beneficial in this case. In addition to the temperature coefficient, it is also crucial whether the gate signal can be supplied to the MOS cell without disturbances. Oscillations generated in one MOSFET can increase in interaction with other components and have destructive effects, which again lead to a failure. It has been shown that additional resistors could suppress these oscillations for each MOSFET. In this work we demonstrate paralleling capability of four MOSFETs without using any additional gate resistors. Figure 8 shows the turn on characteristics of parallel p12 and p14 MOSFETs at V_G= ±15V, V_D=3600V and I_D=40A at T_j=125°C with an external gate resistor of R_G=18Ohm. As seen in Figure 4 an overshoot in the current signal is observed during the dV/dt swing. The total switching time measures 600ns and is thus reasonably fast and shows no critical oscillations neither in the current nor in the voltage signal.

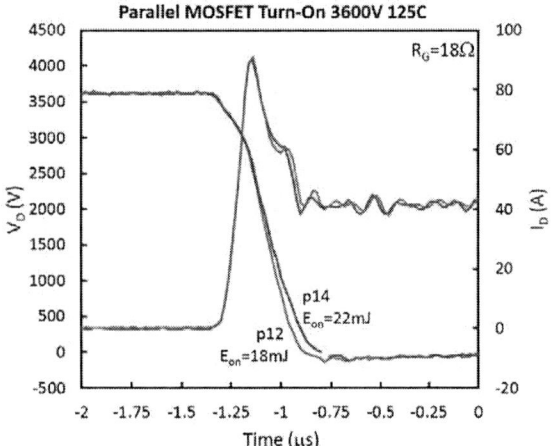

Figure 8 Turn on waveform of four parallel p12 and p14 MOSFETs with R_G=18Ω, L_σ=480nH at Tj=125°C.

Figure 9 shows the turn off characteristics of parallel p12 and p14 MOSFETs at V_G= ±15V, V_D=3600V and I_D=40A at T_J=125°C with an external gate resistor of R_G=33Ohm. As in the turn on event above, no critical oscillations can be observed here as well. At the end of the dV/dt swing the drain voltage peaks at 4000V. Similar to the switching of single MOSFETs in Figure 5 the smaller p12 device depicts a faster turn off in the current signal. Due to the lower output capacitance a significantly lower switching energy (E_{off}) is observed.

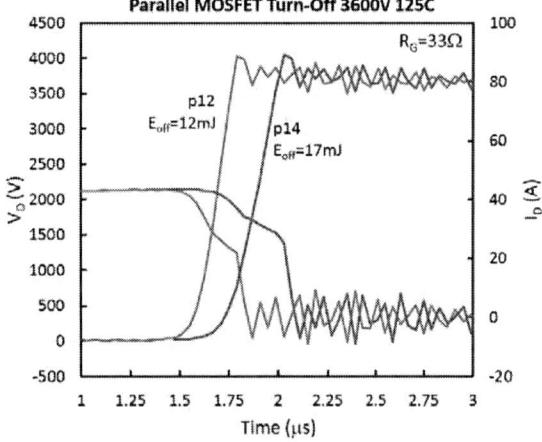

Figure 9 Turn off waveform of four parallel p12 and p14 MOSFETs with R_G=18Ω, L_σ=480nH at Tj=125°C.

In Figure 10 we show a comparison of the normalized switching loss per 1A of p14 and p26 MOSFETs to a 1000A HiPaK2 IGBT module [5]. The MOSFET losses were measured with a single device at the same power density of 130Wcm^{-2} and a comparatively large inductance. With both pitches the switching loss can be tremendously reduced by

>80%. A relatively slow switching has been applied using a gate resistor of 33Ohm for turn-on and turn-off. Significant improvements can be expected if either higher current densities are applied or by turning on with a smaller gate resistor.

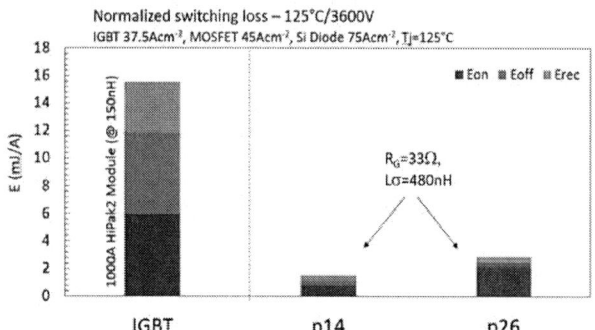

Figure 10 Comparison of the switching loss per ampere of a 1000A Si IGBT HiPak2 module and p14 and p26 MOSFETs for the same power density of 130Wcm^{-2}.

IV. CONCLUSIONS

The excellent features of the wide band gap material SiC for 6.5kV MOSFETs is demonstrated with clear improvements in the static and dynamic characteristic when compared to conventional Si technology. In addition, switching events with over voltage and over current conditions show robust behavior and a wide SOA. However, SC capability depends very much on the design with clear a performance trade-off with respect to conduction losses. If large SC times are required, compromises have to be made with respect to the pitch design and hence on the on state performance. Another interesting aspect is that no additional diode will be required, which saves space and semiconductor area in the module. Finally, we conclude from the presented results that for a well-designed module layout, the paralleling of HV SiC MOSFETs on a single substrate will be of little concern for high current applications. Low switching losses and fast switching times will enable the use of such devices in high power and high frequency applications, which has been a major limitation for silicon devices in the past.

REFERENCES

[1] K. Kawahara *et al.*, "6.5 kV schottky-barrier-diode-embedded SiC-MOSFET for compact full-unipolar module," ISPSD Sapporo, 2017, pp. 41-44. doi: 10.23919/ISPSD.2017.7988888

[2] T. Sakaguchi, M. Aketa, T. Nakamura, M. Nakanishi and M. Rahimo, "Characterization of 3.3 kV and 6.5 kV SiC MOSFETs," PCIM Europe 2017; pp. 1-5.

[3] M. Rahimo, "Performance Evaluation and Expected Challenges of Silicon Carbide Power MOSFETs for High Voltage Applications", Materials Science Forum, Vol. 897, pp. 649-654, 2017.

[4] L. Knoll *et al.*, "Robust 3.3kV Silicon Carbide MOSFETs with Surge and Short Circuit Capability", Proceedings of ISPSD 2018, Sapporo, Japan

[5] C. Papadopoulos *et al.*, "The third generation 6.5kV HiPak2 module rated at 900A and 150°C", PCIM Europe Proceedings 2018 (in press)

Proceedings of the 30th International Symposium on Power Semiconductor Devices & ICs
May 13-17, 2018, Chicago, USA

Improvement of Power Cycling Reliability of 3.3kV Full-SiC Power Modules with Sintered Copper Technology for $T_{j,max}$=175°C

Kan Yasui, Seiichi Hayakawa, Masato Nakamura,
Daisuke Kawase, Takashi Ishigaki, Koji Sasaki,
Toshihito Tabata, and Toshiaki Morita
Hitachi Power Semiconductor Device Ltd.
Hitachi-shi, Ibaraki-ken, Japan
E-mail: kan.yasui.ky@hitachi.com

Masakazu Sagawa, Hiroyuki Matsushima, and
Toshiyuki Kobayashi
Research & Development Group
Hitachi Ltd.
Hitachi-shi, Ibaraki, Japan

Abstract— **Higher maximum junction temperature operation requires higher power cycling reliability especially for silicon carbide power modules. In this work, with the help of a novel sintered copper die attach technology, 3.3kV/450A full-SiC power modules for up to 175°C maximum junction temperature were developed. Demonstrated power cycling lifetime shows an improvement of six times conventional Pb-rich solder die attach modules.**

Keywords— SiC MOS, silicon carbide power module, power cycling, sintered copper, reliability

I. INTRODUCTION

Recent progress in the development of silicon carbide power devices has facilitated the widening of the operational temperature range of power modules. Higher maximum junction temperature ($T_{j,max}$) is required for silicon carbide power modules to take advantage of its material properties. Higher $T_{j,max}$ durability allows for lower cooling capacity, making it possible to reduce the dimensions of power conversion equipment. An increase of $T_{j,max}$ leads to wider temperature swing and correspondingly requires more power cycling reliability. However, recent investigations have clarified the power cycling reliability of silicon carbide power modules were inferior to that of silicon power modules [1,2]. In this work, the power cycling capability of 3.3kV silicon carbide power modules was investigated and an innovative sintered copper die attach technology, to improve the power cycling reliability for $T_{j,max}$=175°C operation, is proposed.

II. TEST AND PROCESS METHODOLOGIES

A. Power Cycling Test

To clarify the failure mechanism and lifetime, simple power cycling experiments with multiple single-die substrates as well as standard modules were conducted. Table 1 shows power cycling test conditions. 3.3kV SiC MOSFET (4H-SiC) and 3.3kV silicon P-N diode were chosen for comparison purposes. For the single-die substrates power cycling tests, four to eight samples were connected in series. Among several test

options for power cycling using SiC MOSFET, constant source to drain current heating with shorted gate to source configuration was applied. End of life was defined by a 5% increase in output voltage or 20% increase in the junction temperature swing (ΔT_j). The junction temperature is represented by virtual junction temperature, estimated by temperature sensitive electrical parameters [2].

TABLE I. POWER CYCLING TEST CONDITIONS

Sample	$T_{j,min}$	$T_{j,max}$	ΔT_{j0}	t_{on}	t_{off}
Single-die Substrate	50°C	175°C	125K	2s	6s
Module ($_n$HPD^2)	50°C	175°C	125K	1-2s	$>t_{on} \times 3$

B. Sintered Copper Die Attach Process

Among the promising candidates to shift from conventional solder die attach technology, sintered metal technologies were desirable for their high temperature reliability [3-5]. Silver and Copper both have sufficiently high melting points for $T_{j,max}$=175°C operation. Sintered silver has been widely investigated and reported for improved cycling reliability. But higher material cost and the requirement apply a noble metal underlayer like Silver plating, fabrication costs increase. In terms of chemical stability, copper is deemed more desirable to achieve a strong bond. However, one drawback of sintered Copper is that copper nanoparticles may oxidize easily. A copper oxide (CuO) reduction technique was effective to overcome this potential issue [6]. In Fig. 1, dispersed nanoparticles for copper sintering and bonded structure are

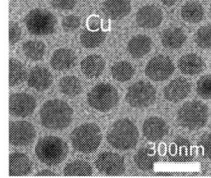 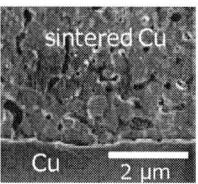

Fig. 1. Electron microscope images of copper nanoparticles (left) and bonded structure of sintered copper (right).

978-1-5386-2928-4/18 $31.00 © 2018 IEEE   455

shown. A reduction of the copper oxide in the sintering process was carried out by a carefully controlled hydrogen gas mixture. The interface has a hetero-epitaxial crystal orientation, which means strong bond has been obtained [6,7]. Furthermore, sintered Copper has better fabrication process compatibility than sintered silver because it does not need noble metal underlayer unlike sintered silver. Conventional nickel plating, the copper metallization layer of the substrates or a nickel film on a die backside are adequate for the sintered Copper.

III. COMPARISON OF DIE ATTACH TECHNOLOGY FOR SILICON CARBIDE POWER DEVICES

Since silicon carbide has a Young's modulus around three times larger than that of silicon, the stress around silicon carbide die during thermal cycles is assumed to be much larger. Maximum strain energy density has been previously reported to be larger in the case of silicon carbide than in that of silicon, with power cycling lifetime shown to be one third compared to silicon [1,2]. To confirm reported results and to clarify weakest package elements, power cycling tests were conducted. Fig. 2 shows a comparison of power cycling tests in relation to T_j evolution, comparing SiC MOSFET and silicon P-N diode, both adopting conventional Pb-rich solder die attach. The experiments were carried out by single-die substrates without a baseplate. Power cycling lifetimes of the SiC MOS dies were inferior by two to three times compared to that of the silicon P-N diode.

SEM cross sectional images of SiC MOS die after the tests are arranged in Fig. 3. Primary failure points of both the SiC MOSFET and the silicon P-N diode were assessed to be edges in the solder layer, where large cracks inside the solder are observed. These results indicated that a greater stress around silicon carbide dies shortened the power cycling lifetime. Reinforcement of the die attach is required especially in silicon carbide power modules.

As an alternative to conventional solder die attach technology, sintered Copper and sintered Silver were chosen and compared in power cycling tests. 3.3kV SiC MOSFET was assembled on single-die substrates. Power cycling test results in Fig. 4 show that a junction temperature swing in the sintered Copper die attached samples were more stable than the sintered

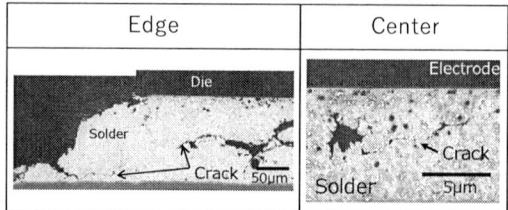
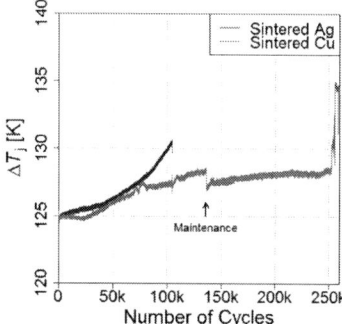

Fig. 3. SEM cross sectional images of SiC MOS die after power cycling.

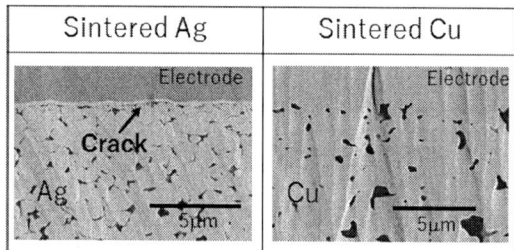

Fig. 4. ΔT_j during power cycling of SiC MOSFET with sintered silver and with sintered copper.

silver die attached samples, showing a lifetime improvement of approximately 2.5 times longer versus the sintered Copper. SEM cross sectional images after the tests are compared in Fig. 5. Thin cracks are observed near the interlayer between die backside electrode and the sintered Silver layer. Assumed root cause: temperature swing increase by the degradation of thermal impedance. In case of sintered Copper, crack propagation was limited to the perimeter of the die. The associated thermal impedance degradation was smaller and a longer lifetime was achieved.

IV. POWER CYCLING EXPERIMENTS USING SINTERED COPPER TECHNOLOGY

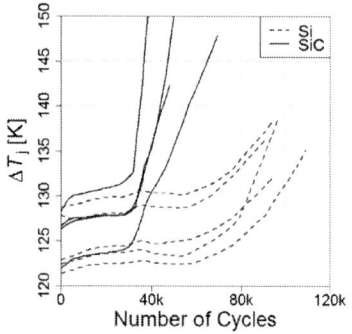

Fig. 2. Comparison of power cycling tests for SiC MOSFET and silicon P-N diode with conventional solder die attach.

Fig. 5. SEM cross sectional images after power cycling with sintered silver and with sintered copper, both at the die centers.

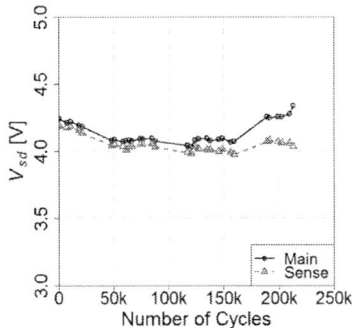

Fig. 6. Output voltages at both main source terminal and sense terminal in a power cycling.

As sintered Copper technology greatly enhanced the durability of die attach, the limiting factor of power cycling lifetime shifts from the die attach to another part of the interconnection mechanics. In the power cycling profile, end of life estimations by abrupt output voltage change were a common characteristic in case of sintered Copper die attach. The gradual temperature swing increase indicated delamination of the die attach layer. On the other hand, the abrupt output voltage fluctuation suggested breakdown of wire bonds and top electrode interconnections. In situ monitoring of wire bond degradation was possible by output voltage measurement both at the source main current terminal and source sense terminal [8]. The output voltages at both main source and sense terminals in a power cycling test using single-die substrates are plotted in Fig. 6. The voltage at the main source terminal is generally higher than that at the sense terminal due to the voltage drop by the main current. As expected, voltages by the end of the test had become larger. It indicated that wire bonds connected to the main terminal had been degraded gradually and the abrupt output voltage change was the result of complete wire bond lift off. Because the sintered Copper die attach layer degradation is small, detection of wire bond degradations independent from the die attach degradation is possible.

To reinforce the bond wire and the top electrode connections, a hard film onto a conventional AlSi top electrode was introduced. The reinforcement process was applied to both SiC MOSFET and silicon P-N diode, and power cycling tests with single-die substrate experiments were carried out. These results are summarized in Fig. 7. The lifetime, defined by the number of cycles until end of life, of the sintered Copper die attach without the hard film is plotted in center for both the SiC MOSFET and the silicon P-N diode test results. The corresponding lifetime of both the silicon carbide and the silicon adopting the hard electrode film is increased about three times than that without the hard film. In case of the silicon carbide, improved lifetime is 10 times better than that of the conventional solder die attach samples.

The difference of the lifetime between silicon carbide and silicon almost disappeared after introducing the sintered Copper die attach. This is understood to be the lifetime of the

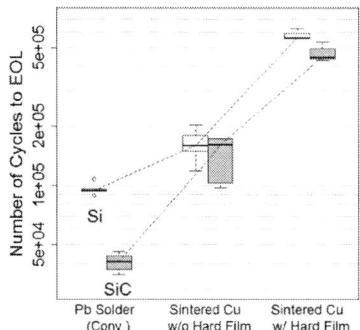

Fig. 7. Box plot of power cycling lifetimes in case of Pb-rich solder and sintered Copper with and without hard film for each SiC and Si by single-die substrate experiments. Definition of box plot follows standard way, bottom and top of the box are the first quartile to third quartile, a segment inside the box means median.

sintered Copper being dominated by the wire bond and the electrode connection, whose stress is assumed to be less affected by the material properties of the semiconductor than at the die attach. Sintered Copper is a valuable and effective in mitigating the larger Young's modulus of silicon carbide to improve the high temperature-reliability.

V. IMPROVEMENT OF POWER CYCLING IN POWER MODULES

To verify the superiority of the sintered Copper technology, a 3.3kV/450A Full-SiC $_n$HPD^2 (next-High-Power-Density-Dual) power modules[9] with sintered Copper die attach were fabricated for the first time. Fundamental electric characteristics are shown in Table 2. Output voltage $V_{ds, HT}$ was 3.9 V at rated current 450A, and equal to conventional solder die attached modules. Threshold voltage and other electric characteristics did not show a significant change in the case of the sintered Copper, demonstrating that the new die attach process did not adversely affect the electrical properties of the modules.

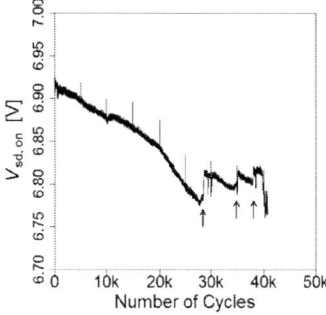

Fig. 8. $V_{sd,on}$ fluctuation during power cycling of a module with sintered Copper without hard film. Arrows indicate suspected wire bond lift off.

TABLE II.   ELECTIRICAL CHARACTERISTICS OF MODULES WITH SINTERED COPPER

Sample Name	V_{th}	$V_{ds,RT}$	$V_{ds,HT}$	$V_{sd,HT}$	$I_{dss,HT}$
MSM450GS33A (Sintered Copper)	2.9V	2.1V	3.9V	3.5V	<1μA (N.D.)
Measured Conditon	R.T. V_{ds}=10V	R.T. V_{gs}=15V I_{ds}=450A	150°C V_{gs}=15V I_{ds}=450A	150°C V_{gs}=15V I_{ds}=450A	V_{gs}=0V V_{ds}=3.3kV

In a power cycling test of the module with the sintered Copper without the hard film process, abrupt output voltage V_{sd} fluctuations were observed, indicating the moment when the bond wire lift off occurred. Fig. 8 highlights these occurrences, with arrow indicators. Confirmed was made by physical observation of the module under test. Fig. 9 shows a top view of a delaminated wire on the SiC MOS die after the power cycling test. Wire bonds presenting an undamaged appearance revealed, under cross sectional SEM analysis, that the bond wire and the top electrode connection was broken, and also shown in Fig. 9.

Power cycling test results of the modules with sintered Copper, with and without the hard film process applied to the top electrode, are plotted in Fig. 10. Test conditions listed in Table 1. It should be noted that the test condition was a highly accelerated and harsh for $T_{j,max}$=175°C operation. Standard Pb-rich solder die attached module without the hard film reached end of life at 18k cycles, whilst the sintered Copper without the hard film module endured until 29k cycles. A reason for the lower lifetime enhancement effect versus the single-die substrate is assumed to be the wider distributions of the

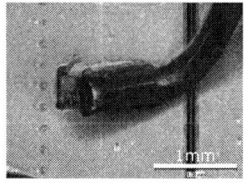

Fig. 9. Delaminated wire after power cycling test (left) and cross sectional SEM image of another wire bond (right).

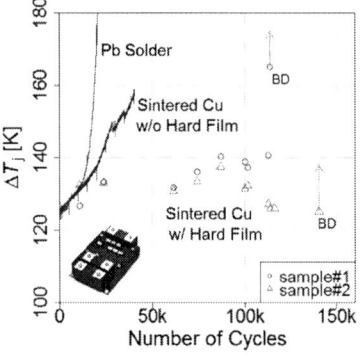

Fig. 10. ΔT_j evolution during highly accelerated power cycling of full-SiC $_n$HPD^2 modules with sintered copper technology compared to conventional module.

lifetime, seen in Fig. 7. In Fig. 10, circular and triangular symbols indicate power cycling results of the modules with the sintered Copper die attach and the hard film process to the top electrode. The number of cycles to reach end of life surpassed more than 115k cycles, four times greater than without the hard film application and corresponds to six times enhanced performance versus conventional solder die attached module. Even in the module with longest lifetime, reliability was limited by wire bonding and the top electrode connection. The sintered Copper die attach showed superior reliability versus conventional die attach technology. Further reliability improvement of the wire and the top electrode connection is on-going.

VI.   CONCLUSION

Sintered Copper die attach technology was applied to SiC MOSFET. Power cycling lifetimes in single-die substrate experiments were enhanced by 10 times those of a conventional solder die attach by the adoption of a sintered Copper die attach and hard film on the top electrode. Since the failure point shifts, from the die attach to the bond wire and the top electrode connections, the inherently inferior power cycling reliability of silicon carbide was improved to that of silicon. 3.3kV/450A full-SiC power modules for $T_{j,max}$=175°C operation with sintered copper technology were fabricated for the first time. Power cycling tests showed that lifetime was enhanced to six times that of conventional solder die attach modules.

REFERENCES

[1] Ch. Herold, M. Schäfer, F. Sauerland, T.Poller, J. Lutz, O. Schilling, "Power cycling capability of Modules with SiC-Diodes", Proc. CIPS, pp. 25-27, 2014.

[2] C. Herold, J. Sun, P. Seidel, L. Tinschert, J. Lutz, "Power Cycling Methods for SiC MOSFETs", Proc. 29th ISPSD, pp. 367-370, 2017.

[3] Akitoyo Konno, Takaaki Miyazaki, Yuusuke Yasuda, Osamu Ikeda, Hioshi Nakano, Toshiaki Morita, Hiroshi Houzouji, Mutsuhiro Mori, Masato Nakamura and Yoshihiko Koike, "Highly Reliable and Lead-Free High Power IGBT Modules Using Novel Copper Sintering Die Attachment", Proc. PCIM Europe, pp. 78-83, 2016.

[4] A. Uhlemann, K. Weidner, G. Mitic, S. Stegmeier, "Reliability study of SiC-JFET including new copper, planar and silver based interconnection and joining technologies", Proc. CIPS 2016.

[5] T. Furukawa, et al., "High power density Side-gate HiGT Modules with Sintered Cu Having Superior High-temperature Reliability to Sintered Ag", Proc. 29th ISPSD, pp. 263-266, 2017.

[6] Toshiaki Morita and Yusuke Yasuda, "New Bonding Technique Using Copper Oxide Materials", Materials Transactions, Vol. 56, No. 6, pp. 878-882, 2015.

[7] Tomohisa Suzuki, Yusuke Yasuda, Takeshi Terasaki, Toshiaki Morita, Yuki Kawana, Dai Ishikawa, Masato Nishimura, Hideo Nakako and Kazuhiko Kurafuchi, "Thermal Cycling Lifetime Estimation ofSintered Metal Die Attachment", Proc. ICEP, pp. 400-404, 2016.

[8] Haoze Luo, Nick Baker, Francesco Iannuzzo, Frede Blaabjerg, "Die degradation effect on aging rate in accelerated cycling tests of SiC power MOSFET modules", Microelectronics Reliability, Vol.76–77, p.p.415–419, 2017.

[9] T. Ishigaki, et al., "3.3 kV/450 A Full-SiC nHPD^2 (next High Power Density Dual) with smooth switching", Proc. PCIM Europe, pp. 33-38, 2017.

Proceedings of the 30th International Symposium on Power Semiconductor Devices & ICs
May 13-17, 2018, Chicago, USA

Enhanced Breakdown Voltage and Low Inductance of All-SiC Module

Motohito Hori, Yuichiro Hinata, Katsumi Taniguchi, Yoshinari Ikeda and Tomoyuki Yamazaki

Electric Devices Business Group, Fuji Electric Co., Ltd., Matsumoto, Nagano, Japan

Email: hori-motohito@fujielectric.com

Abstract— **SiC devices are expected to be used in fields that require in high voltage fields from 3kV to 10kV such as railways, and high reliability such as hybrid vehicles and electric vehicles. And it is also expected to realize high current to expand the application range. This paper presents the packaging technologies for enhanced breakdown voltage and low inductance corresponding to high current of All-SiC modules.**

I. INTRODUCTION

As interest in environmental issues including global warming is increasing, reduction of emissions of greenhouse gases such as $CO2$ is called for, and it is expected that high efficient power conversion technologies realize energy saving. Power semiconductors play a major role in power conversion equipment. Silicon (Si) semiconductor devices, which have been the mainstream, have improved over many years and their performance is approaching their theoretical limits based on their physical properties. Accordingly, wide band gap semiconductor devices such as silicon carbide (SiC) and gallium nitride (GaN) have been developed vigorously. In particular, SiC devices are capable of dramatically reducing the loss and expected to contribute to energy saving by decreasing the losses of power electronics products. At present, they are used in fields that require the rated voltage of approximately 1 kV such as power conditioning subsystems of solar power, and air conditioners. In the future, it is expected that SiC devices will be employed in fields that require high reliability such as hybrid electric vehicles and electric vehicles, and in high voltage fields from 3kV to 10 kV such as railways. And it is also expected to realize high current to expand the application range.

II. BASIC MODULE STRUCTURE AND ISSUES TO BE RESOLVED FOR ENHANCED BREAKDOWN VOLTAGE AND HIGH CURRENT

As shown in Fig. 1, the structure of All-SiC module is significantly different from that of conventional silicon insulated gate bipolar transistor (Si-IGBT) module [1]-[4]. For developed All-SiC module, copper pins formed on the power substrate are used as joint parts instead of conventional aluminum wires. This structure enables high current and high density packaging of SiC devices. As the ceramic insulating substrate to mount semiconductor chips, silicon nitride (Si3N4) insulating substrate with a thicker copper plate compared with conventional substrate has been used to reduce the thermal resistance. In addition, application of epoxy resin instead of the conventional silicone gel as a sealing material prevents degradation of the solder layer and deterioration of the insulation performance in high temperature.

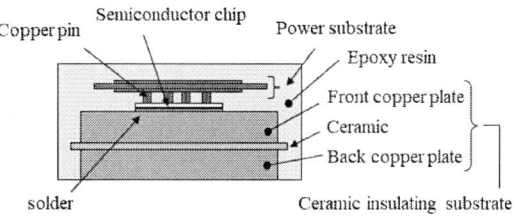

(a) Developed structure (All-SiC module)

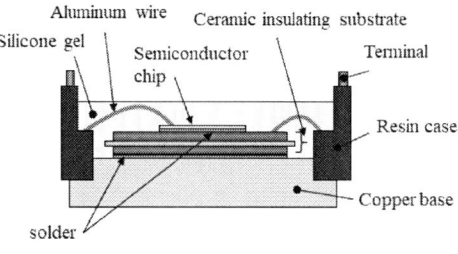

(b) Conventional structure (Si-IGBT module)

Fig.1 Module structure

Next, the issue to be resolved for enhanced breakdown voltage and high current will be described. In order to realize high current and enhanced breakdown voltage, it is necessary to increase the package size. However as a result, as shown in Fig. 2, the distance of the main current path extends and the stray inductance increases. As the stray inductance increases, the turn off surge and the switching loss increase. The greatest feature of SiC devices are high speed switching and low switching loss. In order to make use of SiC device characteristic, suppressing increase in stray inductance and achieving high current become a challenge.

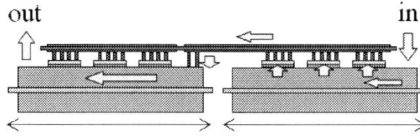

Fig.2 Main current path

978-1-5386-2928-4/18 $31.00 © 2018 IEEE   459

III. PACKAGE DESIGN TECHNOLOGY FOR HIGH BREAKDOWN VOLTAGE

For the insulation design of power semiconductor modules, the electric field strength is one of the important factor. The electric field strength is greatly affected by the voltage applied to the materials, the shapes of the constituent materials and permittivity. In addition, the electric field strength generally increases at the defects of the sealing material, such as voids and peeled parts, and the edge of copper electrodes on ceramics insulating substrate. A silicone gel used as a sealing material for conventional structure tends to generate voids and cracks during operation at high temperature of 175 degrees C or higher, possibly increasing the electric field strength. For that reason, determination of appropriate resin molding and ceramic is important in order to develop All-SiC packages with high breakdown voltage capability for high temperature operation. Furthermore, it is necessary to develop the structure that enables the electric field relaxation at the boundary region of each parts. As mentioned above, the regions with high electric field strength in power module tend to be located in the insulators, such as epoxy resin and ceramics, at the edge of copper plate or at the edge of semiconductor chip surface. Breakdown modes of power modules are classified into ceramic penetration breakdown originating from high electric field strength point and creeping breakdown along the joint region between the epoxy resin and the surface of the insulators, such as ceramics.

Focusing on the triple junction of copper plate, epoxy resin, and ceramics, which will be a high electric field, and electric field simulation was conducted. Fig. 3 shows the electric field strength distribution of the power module. Fig. 3(a) shows the result when the length from the edge of the ceramic to the both copper plates are same. Fig. 3(b) shows the result when the length from the edge of the ceramic to the both copper plates are different. In both cases, the electric field simulation were conducted under the same condition for the thickness and type of ceramic, the thickness of copper plate, and the type of epoxy resin. The results of the simulation indicate that the highest electric field strength point is located at the triple junction. The electric field strength distributions shown in Fig. 3(a) and Fig. 3(b) are different. This reason is that the total amount of electric potential lines is equal, but the electric field strength locally increases by changing the position of the front copper plate according to the Gauss' law. Fig. 4 shows the maximum electric field strength change in case of changing the ceramic thickness and the position of front copper plate. Fig. 4(a) shows the results in the case of changing the ceramic thickness. Fig. 4(b) shows the results in the case of changing the position of front copper plate. From these simulation results, increasing the thickness of the ceramic and equalizing the distances between the edge of the ceramic and both side of copper plates lead to the relaxation of the electric field strength. In this way, by relaxing the electric field strength, the required thickness for ensuring the withstand voltage can be minimized, and the package size can be optimized.

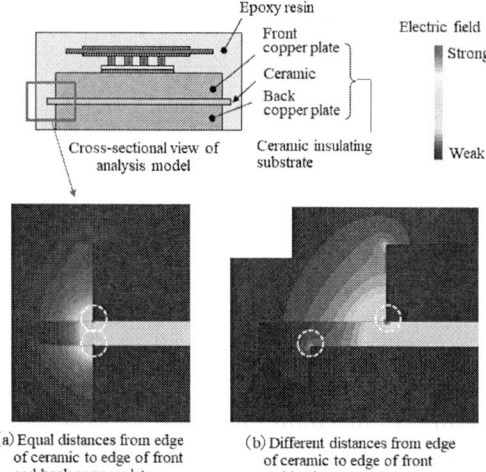

(a) Equal distances from edge of ceramic to edge of front and back copper plates

(b) Different distances from edge of ceramic to edge of front and back copper plates

Fig. 3 Results of electric field simulation (electric field strength distributions)

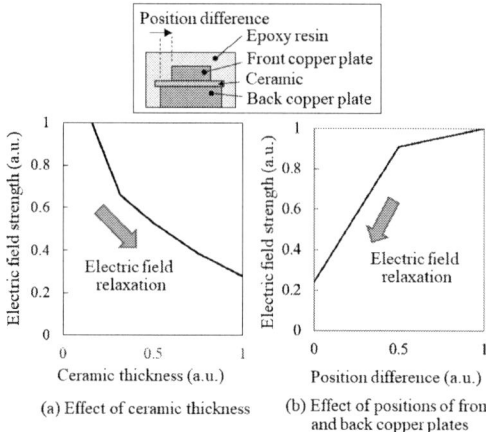

(a) Effect of ceramic thickness

(b) Effect of positions of front and back copper plates

Fig. 4 Results of electric field simulation (changes in electric field strength)

IV. EVALUATION OF RESIN MOLDING FOR ENHANCED BREAKDOWN VOLTAGE

A. Breakdown voltage test of resin molding

We compared the insulation performance of silicone gel used for the conventional structures and epoxy resin used for the developed structure. Fig. 5 shows schematic structure of test sample, that is filled with silicon gel or epoxy resin. Voltage was applied between the front electrode and the back electrode, and the breakdown voltage was measured. Fig. 6 shows the relationship between the breakdown voltage and cumulative breakdown rate, and Fig. 7 shows the breakdown points after experiments. At the cumulative breakdown rate of 1%, the breakdown voltage of epoxy resin is 16.3kV, which is approximately 2 times higher than that of silicone gel 8.8kV. The breakdown path at silicone gel proceed in silicone gel from the triple junction of front copper plate, ceramics and silicone gel to the back side. On the other hand, for the epoxy resin, it is penetration destruction of ceramic. This reason is presumed that epoxy resin has good adhesion to ceramic and copper, so voids as a starting point of fracture hardly occur. This indicates that the insulation performance of epoxy resin is determined by the breakdown capability of the ceramic insulating substrate itself, and by improving the thickness and withstand voltage to be further enhanced. As a result, it is possible to shorten the length of the ceramic to the limit, thereby realizing miniaturization of the package.

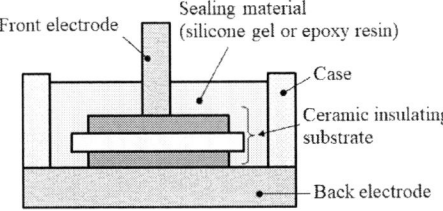

Fig. 5 Schematic structure of test sample

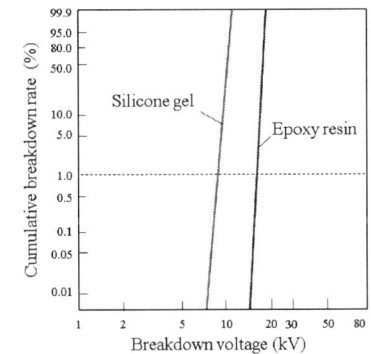

Fig. 6 Relationship between breakdown voltage and cumulative breakdown rate

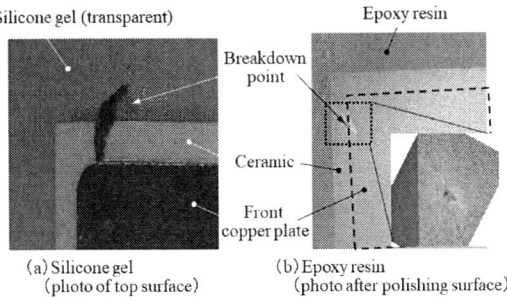

Fig. 7 Breakdown points

B. Partial discharge test of resin molding

As a method of evaluating the long term product lifetime based on an initial product evaluation, it is effective to investigate the existence of partial discharge. If partial discharge is generated, the degradation of resin molding originated from the discharge point is propagated, and that is likely to result in a breakdown after the long term operation. Therefore, how to suppress the partial discharge is important [5]-[8].

Fig. 8 shows the results of partial discharge test. The voltage at which the electric charges start to increase as the voltage rises is defined as the partial discharge inception voltage (PDIV), and the voltage at which the electric charges decrease to zero as the voltage drops is defined as the partial discharge extinction voltage (PDEV). For silicone gel, the PDIV was 7kV. Meanwhile, with epoxy resin, no partial discharge occurred even at 10 kV, indicating it is less likely to generate a partial discharge compared with silicone gel. Therefore, we conclude that the degradation due to partial discharge is not likely to occur in long time operation, and the resin molding is promising technology to enhance breakdown voltage of All-SiC modules.

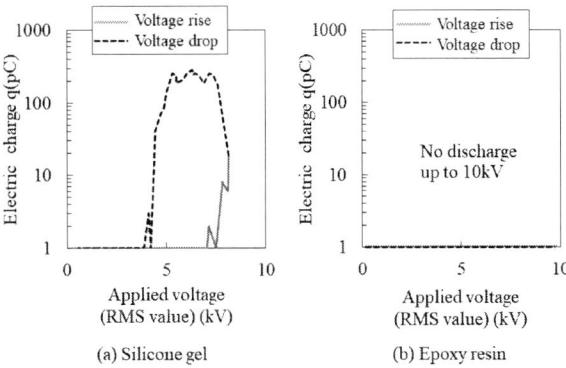

Fig. 8 Results of partial discharge test

V. PACKAGE DESIGN TECHNOLOGY FOR LOW INDUCTANCE CORRESPONDING TO HIGH CURRENT OF ALL-SIC MODULES

The method for reducing the stray inductance of All-SiC modules will be described. Fig. 2 shows the main current path and Fig. 9 shows the classification of the stray inductance. The terminal inductance occupies more than half of the total. Then, we designed to achieve high current by reducing the terminal inductance with thinning the package, and extending the width of package to satisfy the creepage and clearance from the terminal to grand, and keeping the chip mounting area by increasing the footprint size. Fig. 10 shows the relationship between the stray inductance and the footprint size when the terminal height is used as a parameter. There is a trade-off relationship between stray inductance and footprint size. We selected the condition that the stray inductance is low and the footprint size is as small as possible. As a result, a lineup idea of All-SiC modules ranging from small capacity to large capacity as shown in Fig. 11 could be realized.

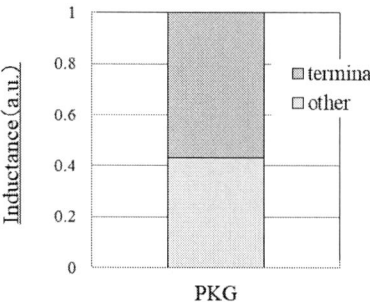

Fig. 9 Classification of stray inductance

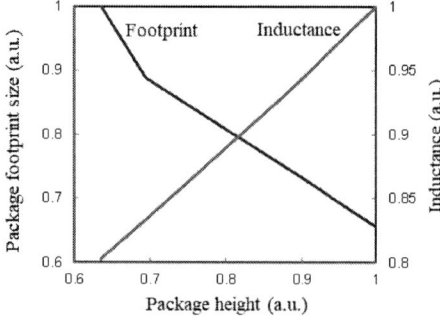

Fig. 10 Relationship between stray inductance and footprint size

Item		Type 1	Type 2	Type 3L
external appearance				
Dimensions (mm)		W62 x D20 x H12	W68 x D26 x H13	W126 x D45 x H13
Rating	voltage(V)	1200		
	current(A)	25, 50	75, 100	200, 300. 400

Fig. 11 Lineup idea of All-SiC modules

VI. CONCLUTION

This paper has described the package design technologies for enhanced breakdown voltage and low inductance corresponding to high current of All-SiC modules. The effect of the structure for the power module on the relaxation of electric field strength has been studied based on simulation. Furthermore, we investigated the difference in insulation performance depending on sealing material. The relaxation of electric field strength and choosing epoxy resin realized both of miniaturization of package satisfying insulation performance and low stray inductance. Furthermore, by reducing the package height, the terminal inductance is reduced, and achieved high current without increasing the stray inductance.

In the future, by expanding the application area of All-SiC modules with enhanced breakdown voltage by further improving their reliability, we will contribute to the development of power electronics technology and the realization of a low carbon society.

VII. REFERENCES

[1] Horio, M. et al. "New Power Module Structure with Low Thermal Impedance and High Reliability for SiC Device", Proceedings of PCIM, 2011, p.229-234.

[2] Ikeda, Y. et al. "Investigation on Wirebond-less Power Module Structure with High-density Packaging and High Reliability", Proceedings of ISPSD, 2011, p.272-275.

[3] Hinata, Y. et al. "Full SiC Power Module with Advanced Structure and its Solar Inverter Application", APEC , 2012.

[4] Hori, M. et al. "Compact, Low Loss and High Reliable Next Generation Si-IGBT Module with Advanced Structure", Proceedings of PCIM , 2014, p.472-476.

[5] Mitsudome, H. et al. "High Accuracy Examination on Idenfitication of Partial Discharge Location in Power Module using Waveform Information", The papers of technical meeting on electric discharges, IEE Japan, 2017, p.59-63.

[6] Mitsudome, H. et al. "High Accuracy Examination on Location Idenfitication of Partial Discharge Location in Power Module", 23rd Symposium on "Microjoining and Assembly Technology in Electronics", 2017.

[7] Taniguchi, K. et al. "3.3kV All-SiC Module for Power Distribution Apparatus", Proceedings of ICSJ, 2017, p.199-202.

[8] Hori, M. et al. "Enhanced Breakdown Voltage for All-SiC Modules", Proceedings of ICSJ, 2017, p.127-130.

Proceedings of the 30th International Symposium on Power Semiconductor Devices & ICs
May 13-17, 2018, Chicago, USA

Dynamic performance analysis of a 3.3 kV SiC MOSFET half-bridge module with parallel chips and body-diode freewheeling

Abdallah Hussein, Bassem Mouawad, Alberto Castellazzi
Power Electronics Machines and Control Group
University of Nottingham
Nottingham, UK
abdallah.hussein@nottingham.ac.uk; bassem.mouawad@nottingham.ac.uk ; alberto.castellazzi@nottingham.ac.uk

Abstract— Recently, 3.3 and 6.5 kV power MOSFETs have been introduced . Based on the 3.3 kV device, a 100 A half-bridge power module has been developed , using parallel chips for current scaling and relying exclusively on the use of the transistors body-diode for current free-wheeling (i.e., no anti-parallel external diode chips are used). This paper presents a thorough parametric characterization of the module switching performance. Single-chip and parallel-chip operation are investigated in both double-pulsea type tests and realistic single-phase inverter operation.

Keywords—Silicon carbide; SiC MOSFETs; power modules; inverter.

I. INTRODUCTION

Fig. 1 shows the chips used in this study and Fig. 2 reports its output and body-diode forward characteristics [1]. As can be seen, the devices, with a nominal rating of 50 A, exhibit nearly a 4 to 5 times bigger cross-section than counterpart lower voltage rated SiC MOSFETs. Moreover, in view of the body-diode characteristics, they are clearly designed with synchronous rectification in mind for optimum performance.

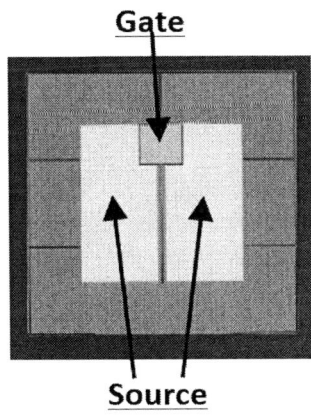

Fig. 1. 3.3 kV-50 A SiC MOSFET chip. The die lateral dimensions are 7.2 x 7.2 mm².

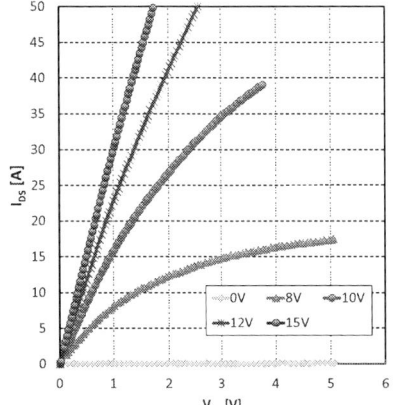

a)

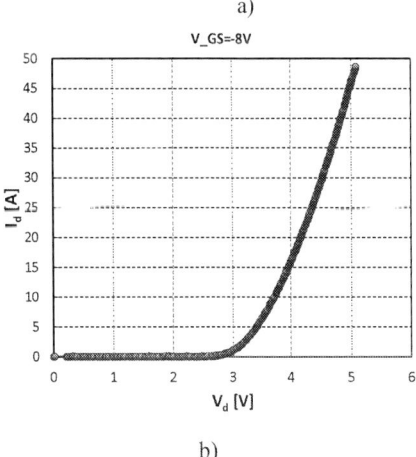

b)

Fig. 2. Measured MOSFET output characteristics, a), and body-diode forward characteristics, b), at ambient temperature.

Fig. 3 shows the measured temperature dependence of tehse characteristics: although, as expected, the increase of on-state

978-1-5386-2928-4/18 $31.00 © 2018 IEEE   463

resistance is more pronounced than the decrease of the diode forward voltage as the temperature increases, the advantages of adopting synchronous rectification for free-wheeling are manifest up to considerable operational temperature values.

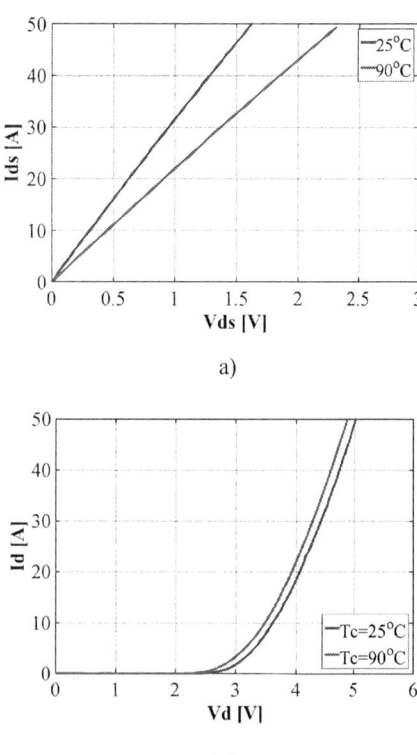

Fig. 3. Tempearture dependence of transistor output characteristics, a), and body-diode forward characteristics, b).

Finally. Fig. 4 shows the measured spread among 4 devices. The observed differences in the output characteristics could be related to a difference in threshold voltage of only some tens of mV, which is considered a very tight distribution in light of comparison with commercial products at lower voltage ratings. The spread in the body-diode on-state voltage is also very satisfactory in view of the high voltage technology maturity level. These results reasonably motivate the development of a multi-chip module.

II. MODULE DESIGN AND PROTOTYPE DEVELOPMENT

The chips are assembled in a half-bridge switch architecture, as depicted in Fig. 5 a). Two parallel transistors are used for both the high-side and the low-side switch implementation to yield the target 100 A nominal current rating. As indicated in Fig. 5 a), in the case of high voltage transistors, both in reason of higher dV/dt capabilities and because of larger dies cross-sectional area for a given current rating, parasitic capacitance of the terminals switching at high frequencies, such as the middle point in the half-bridge architecture, become more relevant and require dedicated

attention in the development of a power module to avoid the circulation of excess common mode currents, which may severely impair the maximum achievable switching frequencies and transition speeds. This is true in particular of the more traditional assembly approaches, of the kind pursued here, which rely on the use of ceramic substrates soldered onto a baseplate, typically grounded.

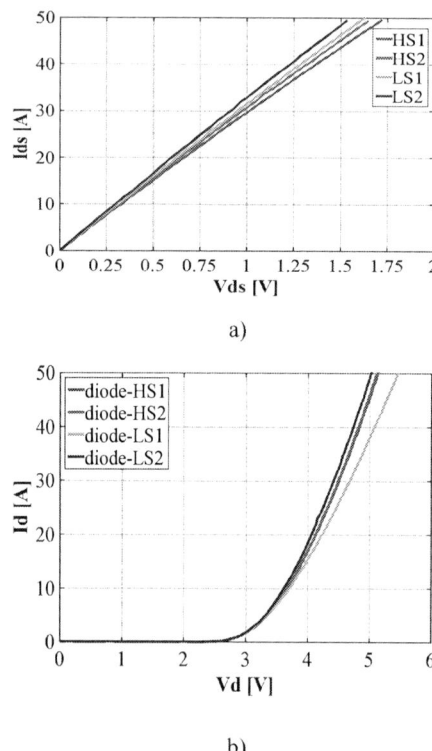

Fig. 4. Measured spread in transistor output characteristics, a), and body-diode forward characteristics, b), for 4 devices.

So the module design here is compatible with standard printed circuit board (PCB) mounting, but its design targets minimum and well balanced values of parasitic inductance and common mode capacitance towards the baseplate. The design formulated to jointly meet these design intentions is shown in Fig. 5 b): salient features of the module are:

- full separation of drive and power loops, to avoid interference of switching load currents with gate-loop parasitic inductance; the importance of the adoption of Kelvin-source type interconnects in the drive loop have been recently pointed out to be key in the case of SiC MOSFETs [2];

- asymmetry in the design of the substrate metal tracks for the high-side and low-side switch, respectively, so as to minimize the parasitic capacitance created by the drain terminal connection of the low-side chip with the base-plate, through the ceramic substrate dielectric constant. With the device dimensions given above, assuming aluminum nitride (AlN) is used for the ceramic substrates, then the parasitic capacitance value is estimated at at least 5 pF per chip, when only the chip cross-section is considered. However, the minimum acceptable track size needs to be decided on the basis of the module thermal performance and, more precisely, needs to ensure uniform and symmetric electro-thermal performance among the chips. The dimensions shown in Fig. 3 are the final ones, deriving from thorough finite element based analysis of the thermal response of the module, with results as shown in Fig. 5 c):

- symmetrical parasitic inductance for the high-side and low-side switches;

- use of two separate smaller substrates for mounting of the transistors, so as to optimize ease of manufacturing and ensure good control of the quality of the solder underneath the substrate (that is, between substrate bottom-side copper track and baseplate).

Fig. 6 a) shows the assembled power module, fully covered in dielectric gel and ready for final encapsulation. The test setup is a single-phase inverter leg, designed so as to be easily modified into a double-pulse tester for initial experiments. One connoting feature of the test setup, visible in Fig. 4 b), is a heatsink equipped with both cooling fans and heating resistors, so as to decouple the switched power level from the baseplate temperature during the device and module characterization.

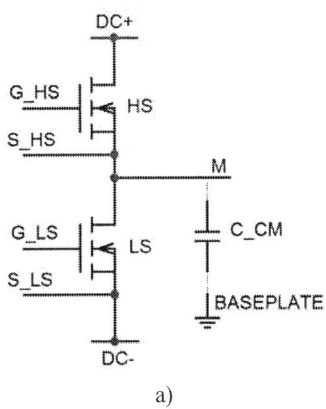

a)

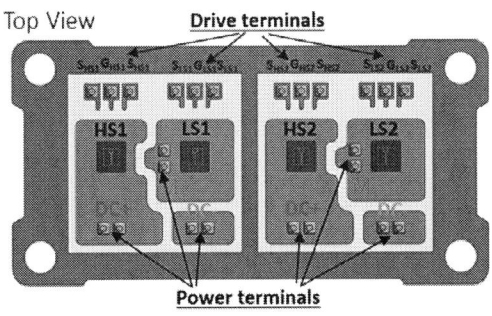

b)

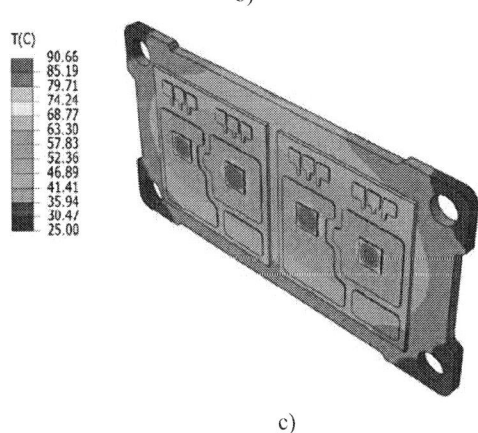

c)

Fig. 5: a), schematic of a half-bridge switch with indication of common-mode capacitance, a major limiting factor for high dV/dt's and switching frequencies; b), module layout, with fully symmetrical electro-magnetic paths, separation of drive and power loops and reduced drain-side mounting area for the low-side switch to reduce common mode capacitance; the devices are paralleled outside of the module; c), module thermal analysis showing uniform temperature distribution notwithstanding different track sizes.

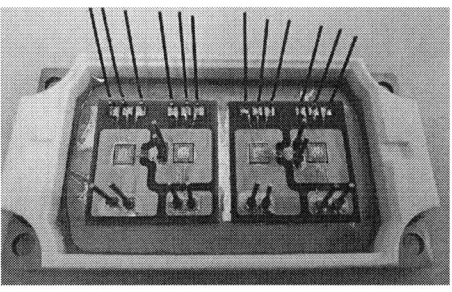

a)

b)

Fig. 6: a), assembled module with insulating gel and high-temperature plastic frame; b), detail of test-setup showing heatsink with fans and heating resistors mounted to fully decouple point of load from heatsink temperature.

978-1-5386-2928-4/18 $31.00 © 2018 IEEE   465

III. EXPERIMENTAL TESTING

A parametric characterization of the switching performance of the module has been presented in [3]. In particular, is has been shown that the switching energy is virtually independent of the baseplate temperature over a broad range of temperature values. This finding confirms previous observations on lower voltage SiC MOSFETs, for which this particular feature has already been proven to be associated with disruptive possibilities to optimize system level power densities by joint optimization of the switching frequency and the heat-sink [4]. Here, results for the parallel operation of the transistors and their body-diodes are presented. Fig. 7 shows turn-on and turn-off switching current transient waveforms for the two transistors in the low-side switch; Fig. 8 shows the reverse recovery current in the body-diodes of the high-side-switch transistors, in a), and its stability over temperature, in b). The devices are being subjected to extensive testing pushing towards their actual limits in terms of both current and temperature. The results gathered so far indicate that the devices can be used well in parallel and achieve good power sharing even during fast transient events. Also, the body-diode works reliably and no signs of aging or any degradation have been detected.

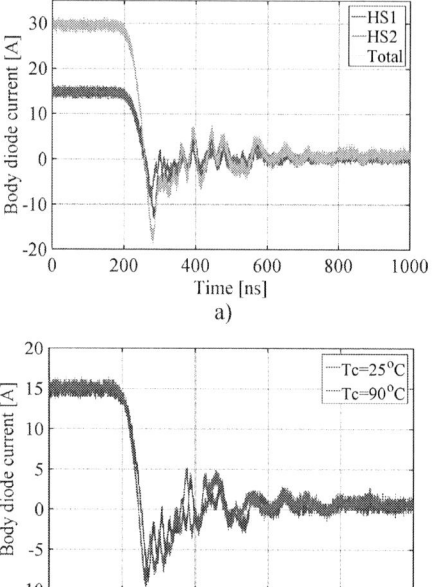

a)

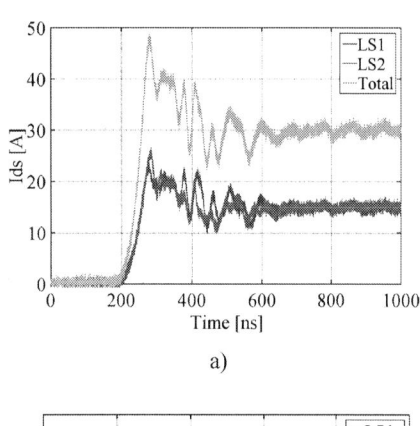

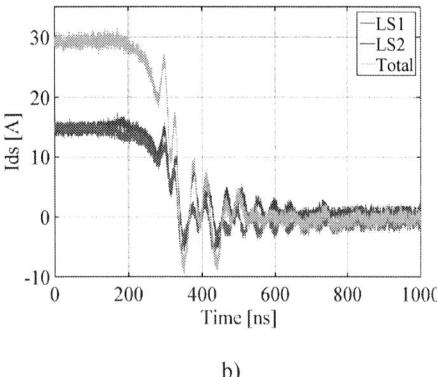

b)

Fig. 8: Transient switching body-diode turn-off current for two parallel transistors in the high-side switch, a); temperature stability of body-diode turn-off performance, b).

IV. CONCLUSION

High voltage SiC MOSFETs are rapidly making their way towards becoming an industrial reality. While the technology maturity still needs further development to access a number of application domains, first results on the device performance considering both single and parallel chip operation are very encouraging, even with conventional packaging solutions.

ACKNOWLEDGMENT

The authors gratefully acknowledge the support of Mr. Takui Sakaguchi, Dr. Takashi Nakamura and Mr. Masaharu Nakanishi of ROHM Semiconductors, Japan, and Dr. Prasad Bhalerao of ROHM Semiconductors GmbH, Germany.

REFERENCES

[1] T. Sakaguchi, M. Aketa, T. Nakamura, M. Nakanishi, M. Rahimo, *Characterization of 3.3kV and 6.5kV SiC MOSFETs*, in Proc. PCIM2017, Nuremberg, Germany, 2017.

[2] C. Boedeker, E. Ayerbe, N. Kaminski, *Impact of a Kelvin Source Connection on Discrete High Power SiC-MOSFETs*, in Proc. ICSCRM2017, Washington D.C., USA, Sep. 2017.

[3] B. Mouawad, A. Hussein, A. Castellazzi, *A 3.3 kV SiC MOSFET Half-Bridge Power Module*, In Proc. CIPS2018, Stuttgart, Germany, Mar. 2018.

[4] E. Gurpinar, A. Castellazzi, *Tradeoff Study of Heat Sink and Output Filter Volume in a GaN HEMT Based Single-Phase Inverter*, IEEE Transactions on Power Electronics (Volume: 33, Issue: 6, June 2018), 5226 - 5239 .

Fig. 7: Transient switching drain current waveforms for two parallel transistors in the low-side switch: a), turn-on; b), turn-off.

Proceedings of the 30th International Symposium on Power Semiconductor Devices & ICs
May 13-17, 2018, Chicago, USA

Power cycling reliability results of GaN HEMT devices

Jörg Franke, Guang Zeng, Tom Winkler and Josef Lutz
Chemnitz University of Technology
Chair of Power Electronics and Electromagnetic Compatibility
D-09126 Chemnitz, Germany

Abstract— **The GaN HEMT is a novel wide bandgap device which could improve the overall efficiency and at the same time shrink the system size. In order to verify the reliability of this promising semiconductor device, new measurement and testing methods have to be developed. In this work, as a general basis for performing reliability tests junction temperature measurement methods for GaN HEMT were investigated. By using suitable temperature measurement method, several power cycling tests were performed on GaN HEMT from three different manufacturers. The main failure mechanism of the GaN HEMT from two manufacturers under power cycling tests is the degradation of solder layer between device and printed circuit board. The main failure mechanism of the devices from the third manufacturer is bond wire lift-off. In GaN the piezoelectric effect is involved in the formation of the 2DEG, and electrical characteristics are sensitive to compressive and tensile stress. The question is whether repetitive deformation leads to new failure mechanisms compared to Si devices.**

Keywords—GaN HEMT; reliability; temperature measurement; power cycling test; lifetime estimation; failure mechanisms

I. INTRODUCTION

Temperature measurement in GaN HEMT devices is challenging due to the special structure of these devices with a two dimensional electron gas and therefore the impossibility to use the $V_{CE}(T)$ method [1]. The common and widely applicated technology for temperature measurement in GaN HEMT devices is the use of temperature dependency of drain-source on-state resistance. However, precise estimation of temperature by control of $R_{DS,on}$ is unsafe due to its very low temperature resolution. At 25°C the voltage drop of GaN HEMT devices is about 40 mV (at a measurement current of 500 mA) and 100 mV for 150°C respectively. The resulting resolution is in the range of 0.9-1.1 mOhm/K, therefore the temperature measurement can be strongly influenced by the measurement noise. A poor precision means strong deviations in lifetime estimation. An error of 5 K in power cycling test for a temperature swing of 80 K for example can cause a relative error in estimated lifetime of about 31% [2]. Temperature estimation by using infrared camera is also unsafe due to the housing materials and the varying emission factors. After presentation of gate structures for normally-off condition in [3] and [4], another promising way for temperature measurement is the use of the pn-junction of enhancement mode GIT HEMTs with p-gate [5]. Conventional components have a specified gate to source voltage where a remarkable gate current can flow. By

implementation of a suitable measurement current, a similar resolution like in silicon devices is obtainable. In power cycling tests, devices of three manufacturers were investigated under comparable conditions. It is possible to make statements for the most likely reasons for the failures addressed.

II. TEMPERATURE MEASUREMENT - SAMPLES AND SETUP

The investigations focused on three GaN HEMT devices of different manufacturers-(i) TPH3212PS (Transphorm), (ii) PGA26E19BA (Panasonic) and (iii) GS66504B (GaNSystems). The cascode device of Transphorm was not considered in the comparison due to the limitation to the $R_{DS,on}$ temperature measurement. In the first assessment, $R_{DS,on}$ vs. temperature was measured at measurement current of 500 mA to enable a remarkable temperature resolution which is shown in TABLE 1 as well as the Residual Sum of Squares (RSS). Second, temperature calibration by using V_{SD} of the reverse potential barrier was conducted at measurement current of 100 mA and gate voltage of 0 V. However, a relatively high measurement error due to the poor temperature resolution makes this measurement method inaccurate. Another promising approach is the temperature measurement by using pn-junction of the p-gate of enhancement mode GaN HEMTs. Fig. 1 depicts the gate to source output characteristics of PGA26E19BA. Datasheet information allows a gate to source voltage of maximum 4.5 V. Between approximately 3.0 V and 4.5 V a significant current flow can be observed and enables a calibration of temperature. The later on performed power cycling tests utilized a measurement current of 30 mA where a ΔV of approximately 270 mV between 25°C and 150°C was observed.

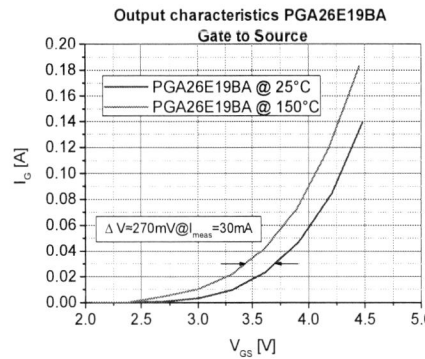

Fig. 1. Output characteristics gate to source PGA26E19BA @ 25°C and 150°C (dev.A).

978-1-5386-2928-4/18 $31.00 © 2018 IEEE

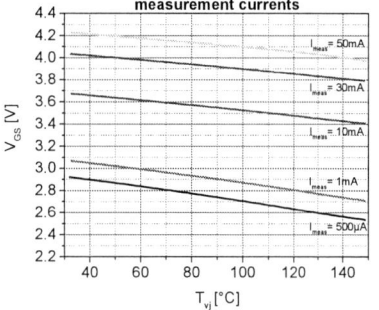

Fig. 2. Calibration curves Gate to Source at different measurement currents (dev. B).

Fig. 2 shows the calibration characteristics of the p-gate of PGA26E19BA for different measurement currents. Similarly to silicon devices a higher slope can be observed for even lower I_{meas}.

TABLE 1. CALIBRATION ACCURACY FOR DEVICES OF TWO MANUFACTURERS FOR DIFFERENT TEMPERATURE SENSITIVE ELECTRICAL PARAMETERS (TSEP).

TSEP	GS66504B	
	Temperature resolution	Residual Sum of Squares (RSS)
$R_{DS,on}$	0.9 mΩ/K[a]	4.924E-6 (pol. fit 2nd order)
V_{SD}	1.7 mV/K[b]	1.457E-4 (pol. fit 2nd order)
V_{GS}	-	-

TSEP	PGA26E19BA	
	Calibration resolution	Residual Sum of Squares (RSS)
$R_{DS,on}$	1.1 mΩ/K[a]	4.507E-6 (pol. fit 2nd order)
V_{SD}	0.6 mV/K[b]	1.021E-5 (lin. fit)
V_{GS}	2.3 mV/K[c]	2.760E-6 (pol. fit 2nd order)

[a] I_{meas}: 500 mA.
[b] I_{meas}: 100 mA.
[c] I_{meas}: 10 mA.

III. POWER CYCLING TEST

A. Cascode device Transphorm TPH3212PS

In the first test series, GaN HEMT cascode devices (TPH3212PS) in TO-220 package were used (see Fig. 3). All devices were calibrated individually by applying a measurement current of 500 mA and the voltage drop V_{DS} was measured between drain of HEMT and source of MOSFET. Gate voltage was set to 15 V. Relation of $R_{DS,on}$ values of GaN HEMT and MOSFET in cascode circuit is about 30:1, an influence due to increase of $R_{DS,on}$ of the MOSFET due to degradation is negligible. Test conditions are show in TABLE 2.

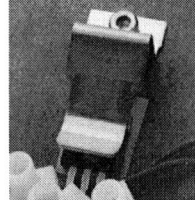

Fig. 3. Test setup TPH3212PS mounted on a copper adapter plate.

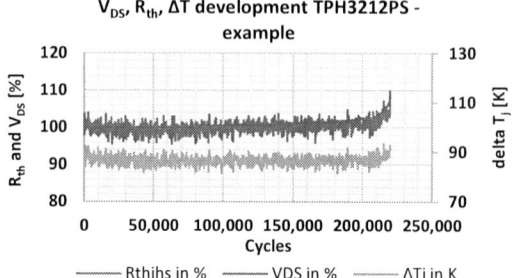

Fig. 4. V_{DS}, $R_{th,js}$ and ΔT-development in power cycling test of example TPH3212PS (Transphorm).

TABLE 2. TEST CONDITIONS POWER CYCLING TEST TPH3212PS.

I_{load} [A]	ΔT [K]	$T_{j,max}$ [°C]	t_{on} [s]
21.8	80-100	150	1

All investigated devices reached the End of Life criteria (EoL) by exceedance of 105% V_{DS}. A remarkable increase of $R_{th,js}$ was not ascertainable. The sharp increase of V_{DS}-voltage indicates bond wire lift-off (see Fig. 4). In Fig. 5 all performed power cycling test results of the Transphorm cascode are shown. Based on the Coffin-Manson law for low cycle fatigue, an exponent of -8.112 can be derived. It is much higher than the value shown in the CIPS 2008 lifetime model for IGBT power modules (-4.416) [6]. Therefore a stronger dependence on temperature swing ΔT seems obvious. It is however to be noted, that discrete devices may have a stronger current dependency of lifetime than predicted in CIPS 2008 lifetime model, which is not considered in this discussion.

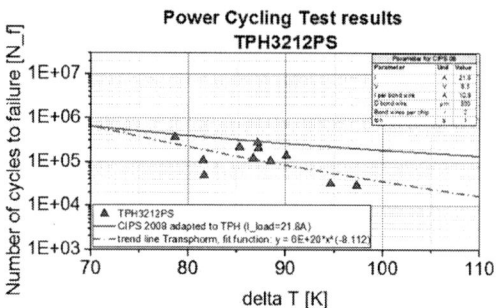

Fig. 5. Test results power cycling test TPH3212PS. Values normalized to $T_{j,max}$=150°C.

B. Enhancement GaN HEMT devices PGA26E19BA (Panasonic) and GS66504B (GaNSystems)

For testing the E-HEMT, GaN devices of Panasonic and GaNSystems were soldered on adapter PCBs due to their SMD packages. In Fig. 6 the PCBs are shown. The PCBs were mounted on water cooling system connected aluminium plates. Thermal foil at the bottom side of the PCB works as an electrical isolation but it nevertheless enables a proper thermal connection to the aluminium plate. Additionally the thermal vias of the adapter PCBs (main source pad) were filled with solder to provide equal thermal conditions for all samples. As the devices show a remarkable dispersion of measurement values, all test devices were calibrated individually by using

978-1-5386-2928-4/18 $31.00 © 2018 IEEE

$R_{DS,on}$ as TSEP. In TABLE 3 the test conditions are depicted. Fig. 7 shows a typical curve of degradation sensitive parameters of one Panasonic device under power cycling test, which is also similar to the curves of the GaNSystems devices.

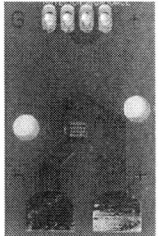

Fig. 6. Adapter PCBs for testing of SMD GaN HEMT devices. Left picture: GaNSystems. Right picture: Panasonic.

TABLE 3. TEST CONDITIONS OF POWER CYCLING TEST.

Test conditions	PGA26E19BA	GS66504B
I_{load} [A]	7	7
ΔT [K]	80-100	80-100
$T_{j,max}$ [°C]	150	150
t_{on} [s]	1	1

Main failure mode of the devices is a sharp increase of V_{DS}, $R_{th,js}$ and undefined measurement values of the controlling system. This makes a loose contact or a lost connection from the device to the adapter PCB obvious and confirms investigations in [7]. Fig. 8 provides an overview of the power cycling test results (GS66504B and PGA26E19BA). Similarly to the cascode devices of Transphorm an increased dependence on temperature swing ΔT is visible. In a second power cycling test series the feasibility of temperature measurement by using the temperature dependent voltage drop of the p-gate of PGA26E19BA was investigated. All tested devices were individually calibrated by applying a measurement current of 30 mA into the gate. During power cycling test the p-gate measurement method was established at two devices and in parallel to this, one sample was tested by using $R_{DS,on}$ as TSEP. To be comparable, the measurement current of 500 mA necessary for the $R_{DS,on}$ measurement was applied to the devices with p-gate measurement from drain to source too. Additionally, for better clarification of the measured temperature values I-V-characteristics for selected samples

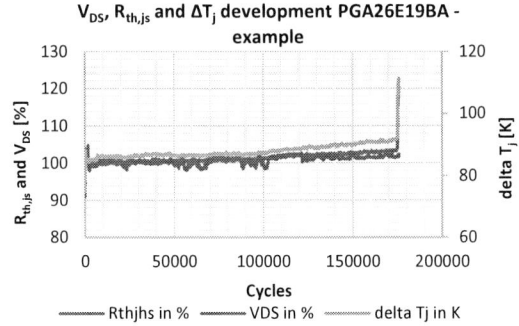

Fig. 7. V_{DS}, $R_{th,js}$ and ΔT-development in power cycling test of example PGA26E19BA (Panasonic).

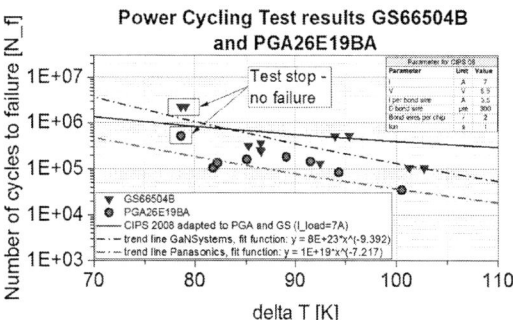

Fig. 8. Test results power cycling test GS6504B and PGA26E19BA. Values normalized to $T_{j,max}$=150°C:

were used to compare voltage drop at the applied load current for several temperature points. A curve fitting of these points served as plausibility assessment for the temperature estimation. Power cycling test results by using voltage drop of p-gate confirm the results of $R_{DS,on}$ measurement. To benefit from the better temperature resolution of the p-gate measurement method, an application in power cycling tests of GIT HEMTs is recommended. Further and extensive investigation to confirm the plausibility is useful.

IV. FAILURE ANALYSIS

The achieved results in the power cycling tests of the different devices diverge slightly. However, the application of different load currents (Transphorm: 127% of $I_{DS(max)}$, GaNSystems 55% and Panasonic 85%, respectively) due to the strongly different thermal resistance R_{thjh} has to be considered. Analysis of end-of-life cause showed bond wire lift-off for the cascode device at the drain side. For the SMD devices of Panasonic and GaNSystems, a different end-of-life cause was obvious: All devices failed due to cracked solder connections to the connecting printed board circuit adapter (see Fig. 9 and Fig. 10) [8]. The bow of the device during load pulse reaches its maximal level at the edges, therefore the outer contacts, particularly the gate contact of the device, are affected most.

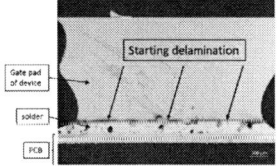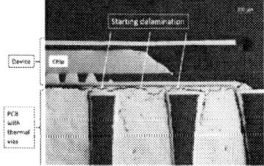

Fig. 9. Cross section Gate pad - solder - PCB (partial view optical microscopy). Delamination between Gate pad of PGA26E19BA and solder is visible. 100-fold.

Fig. 10. Cross section (partial view optical microscopy) source pad - solder attach – PCB of GS66504B. 50-fold.

A starting loose gate contact (possibly only temporarily during load pulse) is more critical for device with p-gate, as delivery of a continuous current into the gate has to be ensured. For devices with isolated gate a loose gate contact is not immediately noticeable when contact is established from time to time and the slightly larger number of cycles to failure in Fig. 8 is explainable. TABLE 4 provides an overview of the main failure modes and the number of devices analyzed.

TABLE 4. FAILURE MODE OF TESTED DEVICES.

Failure	TPH3212PS	PGA26E19BA	GS66504B
EoL (V_DS)	11	4	6
EoL (Rth)	-	3	-
Gate current increased	-	-	4
Bond wire lift-off (analysed)	3	-	-
Delamination (analysed)	-	3	2

V. FEM SIMULATION

In order to verify the failure mechanism of GaN devices from GaNSystems and Panasonic under power cycling tests, thermal-mechanical simulations were performed. The definition of boundary conditions of FEM simulation for PGA26E19BA is shown in Fig. 11.

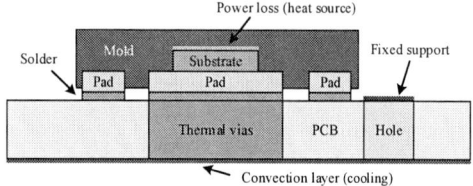

Fig. 11. Boundary conditions of the FEM simulation (PGA26E19BA).

The simulation was divided into two parts. The first part was a thermal simulation, which calculated the transient thermal behavior of the device under power cycling conditions. In total, 10 power cycles were simulated in order to heat up the whole system to the thermal stationary state. In the simulation, power loss was defined at the surface of silicon substrate assuming the complete power loss was generated in the 2DEG. The cooling was modelled by a convection layer at the bottom of the PCB. In the second part, the temperature result of the last power cycle was given to the thermal-mechanical simulation and the corresponding mechanical stress and deformation were calculated. Similar FEM-simulation was also performed on GS66504B from GaNSystems. The simulation results are shown in Fig. 12 and Fig. 13 respectively. It is clear to see from the simulation results shown in Fig. 13 that the solder joints between device and PCB suffer from a strong mechanical stress under power cycling tests. For gate pad solder, which is usually on the edge of the device, a higher deformation is to be observed. The loss of gate connection could explain the sudden failure of the devices under power cycling tests.

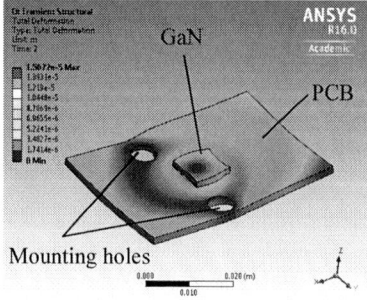

Fig. 12. Deformation of one PGA26E19BA under power cycling test at time point of turning-off load current (scaling factor: 200).

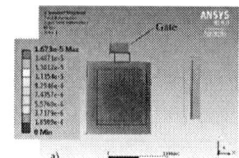 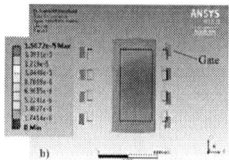

Fig. 13. Deformation of PCB solder layers under pc test at time point of turning-off I_load. a): GS66504B, b): PGA26E19BA (scaling factor: 200).

VI. CONCLUSION

Temperature measurement strategies for GaN HEMT devices were investigated to enable a precise lifetime prediction by using power cycling test. For GIT HEMT devices, the use of the temperature dependent voltage drop of the p-gate was successfully tested and its higher temperature resolution can guaranty the accuracy of measurement. Power cycling tests showed comparable results to IGBT devices of similar power ranges, however different failure modes occurred. TPH cascode devices failed due to bond wire lift-off whereas SMD devices of PGA and GS lost their function by cracked solder connections to the adapter PCBs. A more sophisticated device attach (sintering eg.) could possibly solve this issue. Threshold voltage measurement before and after power cycling test showed a drift in the range of maximum 200 mV. No hints on specific new failure mechanisms due to the piezoelectric effect were found.

ACKNOWLEDGMENT

This work is supported by the German Federal Ministry of Education and Research (BMBF) in the frame of the project GaNMOBIL.

REFERENCES

[1] R. Schmidt and U. Scheuermann: Using the chip as a temperature sensor - The influence of steep lateral temperature gradients on the Vce(T)-measurement, 2009 13th European Conference on Power Electronics and Applications, Barcelona, 2009, pp. 1-9.

[2] C. Herold, J. Franke, R. Bhojani, A. Schleicher and J. Lutz: Methods for virtual junction temperature measurement respecting internal semiconductor processes, 2015 IEEE 27th International Symposium on Power Semiconductor Devices IC's (ISPSD), pp. 325-328.

[3] Saito W et al. Recessed-gate structure approach towards normally off highvoltage AlGaN/GaN HEMT for power electronics applications. IEEE Trans Electron Dev 2006

[4] Cai Y et al. High performance enhancement-mode AlGaN/GaN HEMTs using fluoride-based plasma treatment. IEEE Electron Dev Lett 2005

[5] Uemoto Y et al. Gate injection transistor (GIT), a normally-off AlGaN/GaN power transistor using conductivity modulation. IEEE Trans Electron Dev 2007

[6] R. Bayerer, T. Licht, T. Herrmann, J. Lutz and M. Feller: Model for Power Cycling lifetime of IGBT Modules - various factors influencing lifetime. In: Proceedings of CIPS, 2008

[7] S. Song, S. Munk-Nielsen, C. Uhrenfeldt and I. Trintis: Failure mechanism analysis of a discrete 650 V enhancement mode GaN-on-Si power device with reverse conduction accelerated power cycling test, 2017 IEEE Applied Power Electronics Conference and Exposition (APEC), pp. 756-760

[8] K. Hofmann, C. Herold, M. Beier, J. Lutz and J. Friebe: Reliability of discrete power semiconductor packages and systems - D²Pak and CanPAK in comparison, 15th European Conference on Power Electronics and Applications (EPE), Lille, 2013

978-1-5386-2928-4/18 $31.00 © 2018 IEEE

Proceedings of the 30th International Symposium on Power Semiconductor Devices & ICs
May 13-17, 2018, Chicago, USA

Individual device active cooling for enhanced system-level power density and more uniform temperature distribution

Y. Zeng, A. Hussein, A. Castellazzi

Power Electronics Machines and Control Group
Nottingham University
Nottingham, United Kingdom
eeyyqz@nottingham.ac.uk; alberto.castellazzi@nottingham.ac.uk

Abstract—**This paper provides a method of individual device active cooling system to balance the temperature distribution of system-level power density. 3L-ANPC GaN inverter was used to test and prove the feasibility of it in using multi-level systems.**

Keywords—Individual active cooling system, GaN HEMT, 3L-ANPC inverter

I. INTRODUCTION

Progress in semiconductor device technology evolution has not been matched by corresponding advancements in packaging, passives and cooling. The result is that system integration levels and performance optimization (efficiency and reliability) cannot quite realize the full potential offered by the active device technology and characteristics in terms of heat-generation rate and temperature withstand capability. The issue is particularly sensitive in the case of wide-band-gap devices (SiC, GaN), which find great favor in the implementation of multi-level inverter topologies, in which non-uniform power dissipation conditions among the chips are common. While bespoke packaging solutions are being looked at and filter size reduction is made possible by higher switching frequencies, deeper studies into cooling options are still wanting. This paper presents an original approach, based on the replacement of a common multi-chip cooler with an individual one, to yield: flexible cooling parameter setting based on individual chip dissipation; elimination of thermal cross-coupling effects; reduced overall volume.

II. CHARACTERISTICS OF THE SYSTEM

2.1 Individual coolers

As a case study, reference is made to a 2 kW 3-level GaN inverter intended for photo-voltaic applications [1]. The inverter uses 6 x 600V rated transistors, in TO220 discrete package shown in Fig.1; its power density is mainly determined by input/output filter elements and heatsink, which is shared by all devices, according to common thermal management design practice. In this work we replaced the single heat-sink with individual ones for each transistor (Fig.2): the design involves the use of a square ceramic heatsink interposed between device and fan. The HEMT is mounted onto the heatsink by a built-in bi-adhesive tape. The

fan with the same as size as the heatsink is mounted onto the ceramic heatsink using adhesive tape wound around, so that no drilling or other mechanical work is needed. It is very compact and easy to mount solution.

Fig. 1: 3-Level ANPC inverter assembly with indication of functional parts.

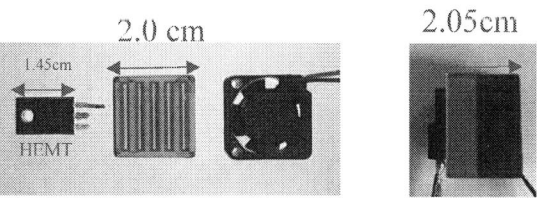

Fig. 2: Individual transistor cooling approach pursued in this study: transistor mounted on ceramic heatsink, mounted on mini-fan.

The Parameters of ceramic heatsink and fans are summarized in Table I. It is clear that the whole volume and weight significantly decrease.

978-1-5386-2928-4/18 $31.00 © 2018 IEEE

TABLE I. HEAT-SINK AND FAN PARAMETERS

Ceramic Heatsink	Fan
Weight < 8.784g	Rated Voltage = 5V Rated Current = 135mA
Thermal Conductivity = 41-50 W/m.k.	Operating range = 2.5-6VDC Operating Temp. = -10-70deg Air flow = 1.6CFM Weight = 4.64g Acoustic Noise = 23.0dB (A)

Fig. 3 shows the 3L-ANPC inverter after mounting with individual coolers

Fig. 3 The back view of 3L-ANPC inverter after mounting with new coolers.

2.2 Thermal analysis

The thermal network for one of devices of this ANPC inverter are illustrated in Fig. 4 where T_j is junction temperature, T_h is heat sink temperature, T_a is room temperature (22℃), P_loss is power loss across a single device, R_{jc} is junction-to-case thermal resistance, R_{ch} is case-to-heat sink thermal resistance and R_{hr} is required thermal resistance of the heatsink. R_{fan} is the thermal resistance of fan. X represents one of six switches. Table 2 shows known parameters of this model.

2.3 Control method of 3L-ANPC inverter

The topology of the above 3L-ANPC GaN inverter is shown in Fig. 5

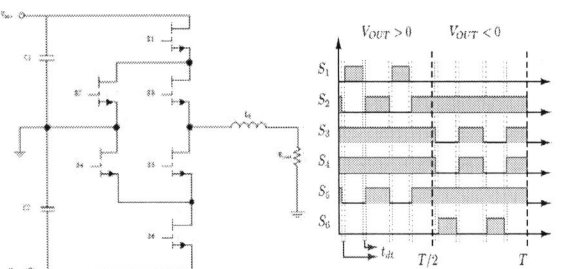

Fig. 5 The topology of 3L-ANPC GaN inverter (left); PWM control method in this case study [2] (right).

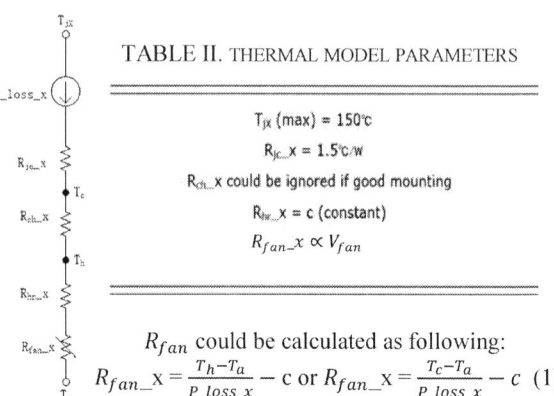

TABLE II. THERMAL MODEL PARAMETERS

T_{jx} (max) = 150℃
R_{jc}_x = 1.5℃/w
R_{ch}_x could be ignored if good mounting
R_{hr}_x = c (constant)
R_{fan}_x $\propto V_{fan}$

R_{fan} could be calculated as following:

$$R_{fan}\_X = \frac{T_h - T_a}{P\_loss\_x} - c \text{ or } R_{fan}\_X = \frac{T_c - T_a}{P\_loss\_x} - c \quad (1)$$

Fig. 4 The thermal model for individual device of 3L-ANPC inverter

Because GaN HEMT has capability of reverse conduction, no parallel diodes needed. One PWM control method described in [2] was used to control this 3L-ANPC inverter. Synchronous rectification applied to decrease the conduction loss. The PWM sequence is shown in Fig. 6.

III. EXPERIMENTAL RESULTS AND DISCUSSIONS

The aim of the experiment is to prove that this individual cooler is able to balance the temperature without at the expense of volume and weight.

3.1 Thermal resistance of the fan

The case temperature at the steady state of GaN HEMT was measured by thermal couple during the experiment and thermal resistance of fan could be calculated using equation (1). The measured temperature and thermal resistance are summarized in Table 3.

TABLE III. MEASURED TEMPERATURES

P_loss (W)	Fan Voltage (V)				
	No fan	3V	4.5V	5V	6V
0.05	25.6	24.8	24.6	24.6	24.7
0.3	27.1	26.3	26.2	26.2	26.1
0.6	30.2	28.6	28.2	28.1	28
1.2	35.3	32.4	31.6	31.4	31.2
2	44.4	38.5	36.5	36.4	35.6
2.7		45.5	42.1	42	41.4
3.85		55.6	51.3	50.2	49.2
5.2		66.2	61.3	60.6	59.1

When power loss is greater than 2W, the temperature will increase dramatically, the device will face with more possibility of breaking down. Therefore, if only heatsink works, we stopped at when the power loss is 2W. It should also be noted that the case temperature varies a lot at low power. Also, fig. 7 shows the temperature changes due to the power loss under different biased voltage of the fan.

978-1-5386-2928-4/18 $31.00 © 2018 IEEE

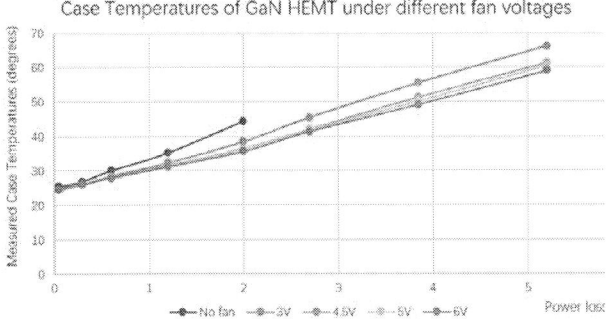

Fig. 7 The trend of the change of case temperatures due to power loss

From the above figure, these trends could be linearly interpolated. The interpolated equations are summarized in Table 4.

TABLE IV. INTERPOLATED EQUATIONS

Fan Voltage (V)	Equations
No fan (0V)	9.6944x + 24.474
3	8.1682x + 23.503
4.5	7.125x + 23.564
5	6.6592x + 23.677
6	6.6592x + 23.677

When fan voltage is 5V or 6V, the interpolated equations are the same because the air flow (speed of the fan) are almost the same even if the biased voltage increases. However, the measured case temperature shows the possibility of decreasing by at least 0.5 °C when increasing the biased voltage to 5V to 6V if the power loss is greater than 2W.

The thermal resistance could be calculated using Table 3 and equation (1). Assume R_{hr} is a small value, by approximately calculating the surface area of the heatsink, and apply equation $R_{hr} = \frac{L}{\sigma A}$ (2), where σ is thermal conductivity, the value of R_{hr} is 0.35 °C/W. Therefore, thermal resistances of the fan are summarized in Table 5 and the curve (Fig. 8) based on this shows the trend of change

TABLE V. FAN THERMAL RESISTANCE

P_loss (W)	Fan Voltage (V)			
	No fan	3V	4.5V	5V	6V
0.05	72	55.65	51.65	51.65	53.65
0.3	17	13.98	13.65	13.65	13.32
0.6	13.67	10.65	9.98	9.82	9.65
1.2	11.08	8.32	7.65	7.48	7.32
2	11.2	7.90	6.90	6.85	6.45
2.7		8.35	7.09	7.06	6.84
3.85		8.37	7.26	6.97	6.71
5.2		8.15	7.21	7.07	6.78

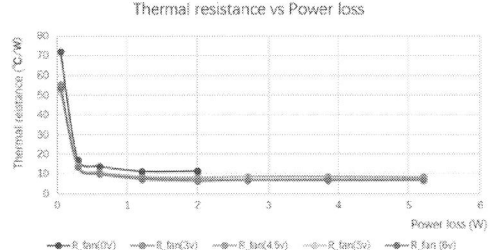

Fig. 8 Thermal resistances changes with fan voltage

It is clear that the thermal resistance decreases with the increase of the fan voltage. At low power, the temperature measurement might cause more error in calculating the thermal resistance which explain the significant thermal resistance at low power. When power loss is greater 2W, the value of fan thermal resistance is more accurate.

Also, the thermal resistance of fan does not change remarkably due to the limitation of the fan when increase the voltage which affect the effectiveness of this individual cooler.

3.2 Case study

The power loss of each device in this 3L-ANPC inverter was simulated using Plecs. Simulation parameters are summarized in Table 6.

TABLE VI. SIMULATION PARAMETERS

DC link Voltage (V)	700
Output RMS Voltages(V)	220
Output Power (W)	1K, 1.25K, 1.5K
Carrier Frequency (KHz)	100
Output Filter Inductance (mH)	1.6
Dead time (ns)	500
Modulation Index	0.95

Simulation results were illustrated in Table 7. Due to the symmetry of ANPC inverter, only power loss of the upper side switches loss.

TABLE VII. POWER LOSS OF HIGH-SIDE DEVICES

Output Power (W)	Power loss on each device (W)		
	S1	S2	S3
1000	1.67	1.61	2.15
1250	2.43	1.99	2.92
1500	3.28	2.48	3.9

From Table 7, S3 and S5 are the most stressed by power loss.

Therefore, we attempt to control the temperature of S3 by changing the fan voltage to make it more balanced. Under normal condition, we used rated 5V. At this time, the corresponding temperature of S3 could be interpolated using the equation in Table 4. To balance the temperature, we could turn off the fan when P_loss is 2.15W use 3V when P_loss is 2.92W and use 6V. The results could be seen in Table 8 and Fig. 9.

TABLE VIII. CASE TEMP. FOR 5V FAN BIAS

| Power Loss (W) | Case Temperature (degrees) | |
	Rated Voltage 5V	Varied fan voltage
2.15	38	45 (0V)
2.92	43	47 (3V)
3.9	50	49 (6V)

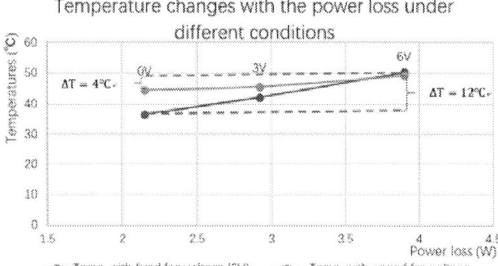

Fig. 9 The case temperatures when keeping the rated voltage and changing the fan voltage

It is clear that the temperature difference when the power loss increase from 2.15W to 3.9W decrease 8°C using varying fan voltage method rather than keeping at 5V.

Furthermore, we could balance the temperatures between two devices. From Table 7, the power loss difference between S2 and S3 are the maximum. (Fig. 10)

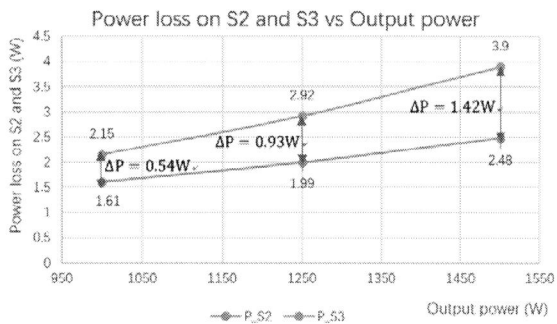

Fig. 10 Power loss and its difference between S2 and S3

The case temperatures for both S2 and S3 could be interpolated using equations in Table 4. Assume the constant fan voltage (5V). The temperatures for fixed fan voltage are summarized in Table 9. Then control the temperature by adjusting the fan voltage, the results are summarized in Table 10.

TABLE IX. CASE TEMPERATURE OF S2 AND S3 AT 5V FAN BIAS

| Power Loss (W) | Case Temperature (degrees) | | |
	S3	S2	Difference
2.15	38	34	4
2.92	43	37	6
3.9	50	40	10

From Table 9, the temperature difference will increase with the power loss increase.

TABLE X. CASE TEMPERATURE OF S2 AND S3 WITH REGULATED FAN BIAS

| Power Loss (W) | Case Temperature (degrees) | | |
	S3	S2	Difference
2.15	38 (5V)	37 (3V)	1
2.92	43 (5V)	44 (0V)	1
3.9	49 (6V)	48 (0V)	1

Table 10 shows the possibility that the time difference could be the same even the power loss increases.

IV. CONCLUSION

These individual active cooling systems have capability of balancing the temperatures of the fan when increasing or decreasing the power loss of each device. Also, the noise of the system is acceptable. However, the effectiveness is restricted by the characteristics of the fan and the ceramic heatsink. Firstly, the speed of the fan does not change significantly when we increase the biased voltage which causes the small change of thermal resistances. This kind of small fan is the minority number of fan which are available and suitable for this case study in the market even though they have some disadvantages.

Secondly, there are some limitations of the ceramic heatsink. We expect heatsinks with higher thermal conductivity (smaller thermal resistivity) and as the similar size as the device itself. The heat conductivity of ceramic material is worse than metal but it has the advantage of light weight lower thermal expansion coefficient. To improve the feasibility of the individual coolers, new material heatsink could be attempted.

Thirdly, the mounting method, in the case study, the ceramic heatsink has built-in thermal tape and we could use it to mount HEMT onto the heatsink. In the case study illustrated above, we were faced with a problem of not good stickiness for one or two heatsinks, which fell off during the experiment. This proves using thermal tape is not perfect. Furthermore, for the heatsink we used in the experiment, although it is very thin, it has very high thermal resistance which will decrease the effectiveness of The heatsink. To improve this, new method of mounting should be used.

REFERENCES

[1] E. Gurpinar, A. Castellazzi, 600 V normally-off p-gate GaN HEMT based 3-level inverter, in Proc. 2017 IEEE 3rd International Future Energy Electronics Conference and ECCE Asia (IFEEC 2017 - ECCE Asia), Kaohsiung, Taiwan, 3-7 June 2017.

[2] E. Gurpinar, A. Castellazzi, Trade-off Study of Heat Sink and Output Filter Volume in a GaN HEMT Based Single Phase Inverter, IEEE Transactions on Power Electronics, July 2017

Proceedings of the 30th International Symposium on Power Semiconductor Devices & ICs
May 13-17, 2018, Chicago, USA

Non-full Depletion Mode and its Experimental Realization of the Lateral Superjunction

Wentong Zhang[1,2], Song Pu[1], Chunlan Lai[1], Li Ye[1], Shikang Cheng[2], Sen Zhang[2], Boyong He[2],
Zhuo Wang[1], Xiaorong Luo[1], Ming Qiao[1], Zhaoji Li[1] and Bo Zhang[1]

[1]State Key Laboratory of Electronic Thin Films and Integrated Devices, University of Electronic Science and Technology of China, Chengdu, P.R.China

[2]CSMC Technologies Corporation, Wuxi, P.R.China

E-mail: zhwt@uestc.edu.cn

Abstract—To realize the minimum specific on-resistance $R_{on,sp}$ of the lateral superjunction (SJ) devices, the low resistance characteristic of the SJ should be adequately used and the adverse influence of substrate-assisted depletion (SAD) effect on the breakdown voltage V_B should be eliminated. From our previous equivalent substrate (ES) model, the SAD effect is completely suppressed if the ES is optimized. In this paper, the balanced symmetric SJ satisfying the optimized ES condition is defined as the ES-SJ. Based on the ES-SJ concept, the non-full depletion (NFD) mode of the lateral SJ is proposed and experimentally realized for the first time. In the experiments, the optimized ES is obtained by a linearly doped charge compensation layer (CCL) with a field plate covering the full drift region and the NFD SJ is implemented with a narrow width of 0.8 μm by implanting the SJ region thrice. The measured results exhibit a $R_{on,sp}$ of 30.9 mΩ·cm² with a V_B of 477 V, which obtains a reduction in $R_{on,sp}$ by 67.8% when compared with other SJ devices under the similar V_B.

Keywords—non-full depletion (NFD) mode; ES-SJ; LDMOS; the minimum $R_{on,sp}$; breakdown voltage V_B

I. INTRODUCTION

The lateral devices, as the heart of the smart power ICs, show the wide applications in the fields of automotive electronics, medical devices, and so on. The super junction (SJ) is the most important innovation concept in the area of the power MOSFET, which has been introduced into the lateral devices because of its low specific on-resistance $R_{on,sp}$ and high breakdown voltage V_B demonstrated by both the analytic models [1]-[4] and the experiments [5]-[12]. However, $R_{on,sp}$ of the lateral SJ device is relatively high if the SJ is designed with a wide SJ pillars and low doping concentration [6]. Besides, the best V_B performance of the lateral SJ can be achieved only when the substrate-assisted depletion (SAD) effect is completely suppressed [3]. A non-full depletion (NFD) mode of the vertical SJ has been proposed to obtain the minimum specific on-resistance $R_{on,min}$ [1]-[2]. Up to now, the NFD mode of the lateral SJ device has not been researched.

Based on the equivalent substrate (ES) model [3] and the global optimization [4], this paper proposes the NFD mode of the lateral SJ and gives its experimental realization in the SOI devices. Section II defines the ES-SJ concept and proposes the NFD mode of the lateral SJ. The optimized ES and NFD mode are used to gain the $R_{on,min}$ in Section III and the experiments of the NFD ES-SJ lateral double-diffused MOSFET (LDMOS) are discussed in Section IV.

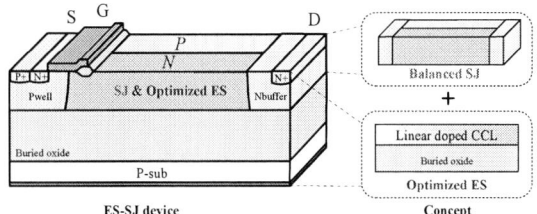

Fig. 1 The definition of the ES-SJ and the structure of the lateral ES-SJ MOSFET. The balanced symmetric SJ satisfying the optimized ES condition [3] is defined as the ES-SJ, which is a composite structure consisting of the balanced symmetric SJ and the optimized ES with a linearly doped CCL. The NFD mode and R-well of the vertical SJ can be extended to the lateral SJ devices if the SAD effect of the lateral SJ is completely suppressed.

II. NFD MODE OF ES-SJ

The full-depletion of the drift region is always a basic condition at breakdown in the theoretical analysis of the power MOSFET. With this condition, the well know "silicon limit" relationship of $R_{on,sp} \propto V_B^{2.5}$ [13] was found for the conventional vertical MOSFET and X. B. Chen et al gave the $R_{on,sp} \propto V_B^{1.32}$ [14] relationship for the vertical SJ devices, which breaks the conventional relationship. However, $R_{on,min}$ of the vertical SJ device is obtained with a non-full depleted drift region at breakdown. The NFD mode of the vertical SJ was thus proposed based on the lateral electric field modulation [1]. It was found from the mode analysis that the variation of $R_{on,sp}$ with the doping concentration of the SJ has a typical U-shape well distribution named the specific on-resistance well (R-well). The $R_{on,min}$ was searched along the R-well with the global optimization and a quasi-linear relationship of $R_{on,sp} \propto V_B^{1.03}$ [1] was developed for the vertical SJ. The NFD mode and R-well can be extended to the lateral SJ devices if the adverse influence of the SAD effect on V_B is eliminated.

The ES model in [3] reveals that the SAD effect can be completely suppressed if the ES satisfies the optimized ES conditions: electrical neutrality and uniform surface electric field "E_{ES} = constant". If the optimized ES condition is satisfied, the breakdown is only determined by the SJ region. The lateral balanced symmetric SJ satisfying the optimized ES condition is defined as the ES-SJ, which can keep the same V_B with the vertical SJ at the same pillar length. Fig. 1 presents the concept of the ES-SJ and the structure of the ES-SJ LDMOS, in which the N and P pillars have the same length L_d, width W and doping concentration N_{SJ}.

978-1-5386-2928-4/18 $31.00 © 2018 IEEE

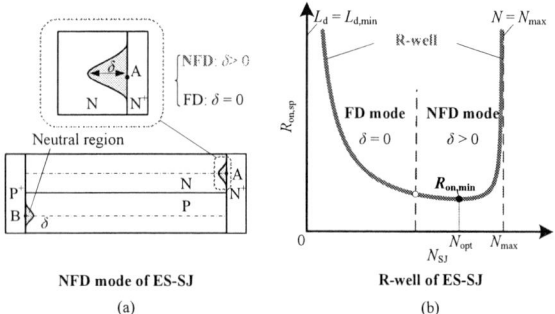

Fig. 2 (a) NFD mode of the ES-SJ. $R_{on,min}$ of the ES-SJ is obtained in the NFD mode with a neutral region at each pillar for the simplified diode structure, which can be realized by selecting the proper parameters of the SJ from (1); (b) R-well distribution of the ES-SJ. The unique $R_{on,min}$ can be searched along the R-well. For the NFD mode, the length of the neutral region $\delta > 0$ and for the FD mode, $\delta = 0$.

Because the ES-SJ shows the similar electric modulation characteristic as that of the vertical SJ, the NFD mode of the ES-SJ is proposed and shown with the simple structure in Fig. 2(a), in which δ is the length of the NFD region. For the NFD mode, $\delta > 0$ and for the FD mode, $\delta = 0$. The similar R-well is satisfied for the ES-SJ and shown in Fig. 2(b). $R_{on,sp}$ is reduced with N_{SJ} firstly before the optimized doping point N_{opt} because L_d is almost a constant, and then increased with N_{SJ} after N_{opt} because N_{SJ} is restricted by the breakdown of the lateral PN junctions and L_d is strongly increased to sustain the given V_B. The unique $R_{on,min}$ can be searched along the R-well.

In summary, the best optimization between $R_{on,sp}$ and V_B of the lateral SJ devices is realized by the NFD ES-SJ, the ES-SJ provides the high V_B and the NFD mode realizes the $R_{on,min}$.

III. $R_{ON,MIN}$ AND DESIGN

To realize $R_{on,min}$ of the ES-SJ, the optimized ES and NFD mode are needed in the ES-SJ device. Firstly, the optimized ES is realized by a linearly doped charge compensation layer (CCL) with a field plate covering the full drift region. Secondly, the NFD mode is obtained by the analytical design.

A. The optimized ES of ES-SJ

Fig. 3 shows the V_B of the ES optimized by the linearly doped CCL with a source field plate. The insert is the device structure with a field oxide thickness of 3.25 μm under the field plate. Fig. 3(a) gives the impacts of the implantation dose D_N of the CCL on V_B for the devices using the linearly doped CCL with field plate, using only the linearly doped CCL and using the uniform doped CCL with the field plate. Compared with the device with the uniform doped CCL, V_B of the new device is increased by 57%. D_N of the new device is increased by 61% when comparing with the device with only the linearly doped CCL because of the assisted-depletion effect from the field plate. Fig. 3(b) depicts V_B versus the lengths of the two field plates and the insert is the equal-potential line distribution. The optimized ES shows a short field plate 1 and a long field plate 2 because the long field plate 1 results in a low vertical V_B and the short field plate 2 leads to a weak depletion effect.

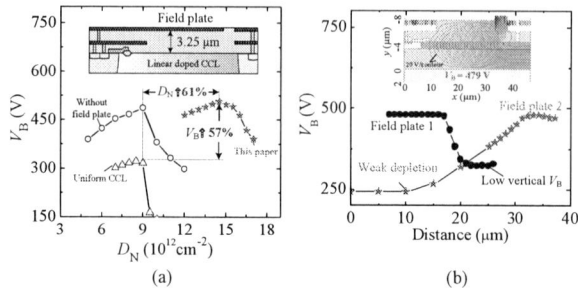

Fig. 3 V_B optimized by the linearly doped CCL with source field plates covering the drift region. (a) Impact of the implantation dose D_N of the CCL on V_B; (b) V_B versus the lengths of the two field plates.

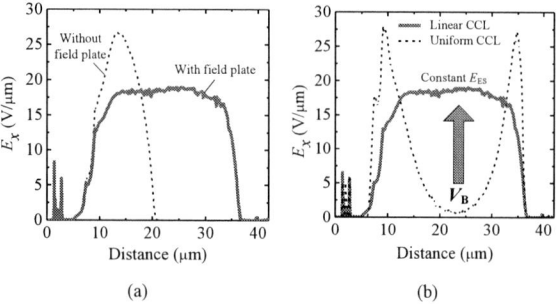

Fig. 4 Comparisons of the surface electric fields of the ES layers (a) with or without the field plate and (b) with the linearly or uniform doped CCL.

As shown in Fig. 4(a), only half of the drift region is depleted if the field plate is removed from the ES layer. The "E_{ES} = constant" distribution is also shown in Fig. 4(b). The high electric field peaks near the source and the drain of the device with the uniform doped CCL are alleviated in the new device. With the optimized ES, the ES-SJ structure is optimized to further reduce $R_{on,sp}$ with the NFD mode.

It is worthwhile to point out that this part only gives one method to realize the optimized ES and the ES-SJ can also be obtained by other technologies.

B. The NFD mode of ES-SJ

The main difference between the NFD mode and the FD mode of the ES-SJ is the NFD mode has a higher N_{SJ} to realize $R_{on,min}$ and longer L_d to sustain the same V_B. The optimized parameters of the ES-SJ corresponding to $R_{on,min}$ can be calculated with the design formula from [4]:

$$\begin{cases} N_{SJ} = 3.08 \times 10^{16} W^{-1.28} V_B^{0.07} & (cm^{-3}) \\ L_{SJ} = 2.38 \times 10^{-2} W^{0.006} V_B^{1.14} & (\mu m) \end{cases} \quad (1)$$

where W is in micrometer.

Therefore, the NFD mode of the ES-SJ can be simply realized by selecting the proper parameters of the SJ region from (1). In the experiments, W of 0.8 μm is implemented based on the 0.5 μm CSMC process platform and V_B is estimated to be 450 V for the SOI with 1 μm silicon layer and 3 μm buried oxide layer.

978-1-5386-2928-4/18 $31.00 © 2018 IEEE

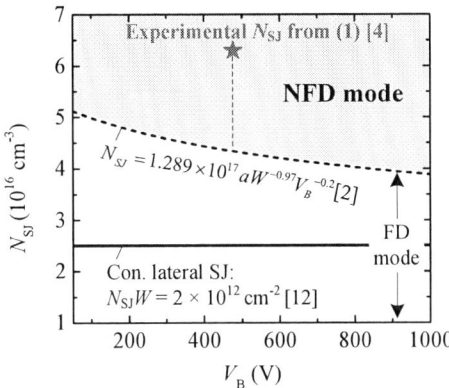

Fig. 5 The doping concentration N_{SJ} of the ES-SJ for difference designs. The NFD ES-SJ shows the higher N_{SJ} than that of the FD ES-SJ designed by the simple analogy to the RESURF condition.

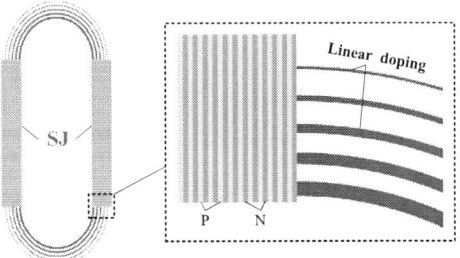

Fig. 6 The mask layouts of the SJ and the linearly doped CCL regions.

TABLE I.   KEY PARAMETERS IN EXPERIMENTS

Symbols	Descriptions	Values
D_{ES}	Linear drift dose (cm^{-2})	1.5×10^{13}
D_{SJ}	SJ pillar dose (cm^{-2})	8×10^{11}
D_{pwell}	Pwell region dose (cm^{-2})	1.2×10^{13}
$D_{Nbuffer}$	Nbuffer region dose (cm^{-2})	6×10^{12}
W	SJ pillar width (μm)	0.8
t_S	Silicon layer thickness(μm)	1
t_I	Buried oxide layer thickness (μm)	3
L_d	Drift length (μm)	30
L_{SJ}	SJ length (μm)	25
L_{f1}	First Source field plate length (μm)	6,7
L_{f2}	Second Source field plate length (μm)	27

Theoretically, the two modes can be separated by the theoretical boundary from [2], which is deduced from the condition that the electric fields at point A and B in Fig. 2(a) being zero. The boundary is given in Fig. 5. Obviously, the lateral SJ with the conventional dose of $N_{SJ}W = 2 \times 10^{12}$ cm^{-2} [12] works in the FD mode and the ES-SJ with the $R_{on,min}$ designed with (1) is in the NFD mode.

IV.   EXPERIMENTS AND ANALYSIS

The NFD ES-SJ LDMOS is experimentally realized by the following two key process steps: (a) the linearly doped CCL is obtained by a single N-type implantation with a dose of

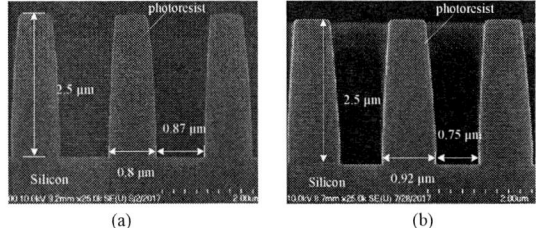

Fig. 7 The thick and high aspect ratio photoresists to realize the (a) N pillars and (b) P pillars of the ES-SJ.

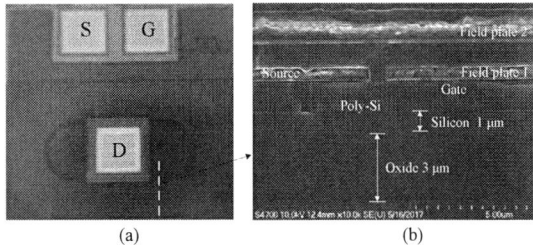

Fig. 8 Experimental results of the NFD ES-SJ LDMOS. (a) The top view of the NFD ES-SJ LDMOS; (b) The partial cross section view from the SEM image, in which the silicon and oxide layers are 1 and 3 μm, respectively.

1.5×10^{13} cm^{-2} followed by a 12.67 hours' diffusion, which enables the linear doping profile in the CCL; (b) after all the thermal process, the SJ region is generated by implanting N-type and P-type dopings thrice with different implantation energies to realize the depths of 0.25, 0.5 and 0.75 μm in the 1 μm thick silicon layer, respectively. By considering the process tolerance, a D_{SJ} of 8×10^{11} cm^{-2} is chosen to ensure the N_{opt} as the peak doping concentration. Fig. 6 gives the mask layouts of the SJ and the linearly doped CCL regions. The key parameters used in the experiments are given in Table I.

Fig. 7 shows the SEM of the photoresists to realize the N-type and P-type dopings of the SJ. The thick photoresists with high aspect ratio of 2.5/0.8 are observed, with which the narrow SJ pillar width of 0.8 μm is realized on the SOI layer. Experimental results of the NFD ES-SJ LDMOS are shown in Fig. 8 including (a) the top view of the NFD ES-SJ LDMOS and (b) the partial cross-sectional view of SEM, in which the silicon layer and buried oxide layer are 1 and 3 μm, respectively.

The measured V_B and output characteristic of the NFD ES-SJ LDMOS and the device without SJ are shown in Fig. 9. The test results ($T = 25$ °C) indicate that the SJ has a negligible impact on V_B. The saturation current of the NFD ES-SJ LDMOS is remarkably increased by 66.9% from 18.7 mA to 31.2 mA at $V_g = 10$ V. The measured results reinforce that the NFD ES-SJ can achieve a substantial reduction in $R_{on,sp}$ without the obvious V_B degradation. Fig. 10 illustrates the relationship of $R_{on,sp}$ versus V_B for different SJ LDMOS devices. The NFD ES-SJ LDMOS achieves a $R_{on,sp}$ of 30.9 mΩ·cm^2 ($V_g = 10$ V, $T = 25$ °C) that is 67.8% lower than those of the lateral SJ devices under the similar V_B [5]-[6]. $R_{on,sp}$ values of the new device are also lower than the theoretical results of the triple RESURF device [15]. The ES-SJ concept and NFD mode can also be

978-1-5386-2928-4/18 $31.00 © 2018 IEEE   477

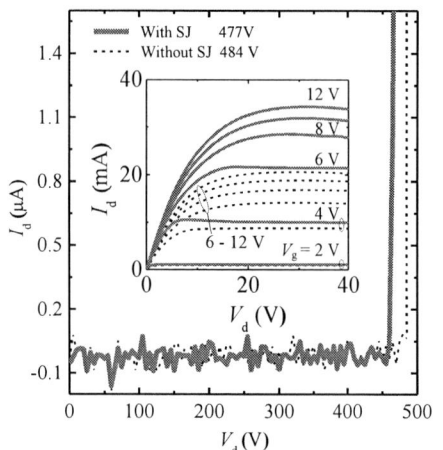

Fig. 9 Measured V_B and output characteristics of the NFD ES-SJ LDMOS and the device without SJ.

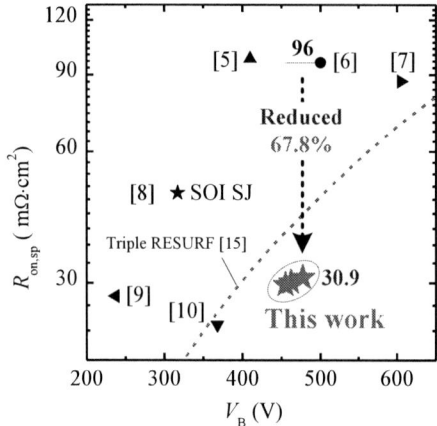

Fig. 10 Comparisons of $R_{on,sp}$ among different SJ LDMOS devices. The experiments of the NFD ES-SJ LDMOS achieves a $R_{on,sp}$ reduced by 67.8% lower than those of other SJ devices with the similar V_B [5]-[6].

introduced into the high voltage (> 950 V) device using the similar process [16].

V. CONCLUSION

In this paper, the ES-SJ concept and NFD mode are proposed for the first time to realize the minimum $R_{on,sp}$ of the lateral SJ devices. The NFD ES-SJ LDMOS is fabricated based on an optimized ES and the SJ region is realized by three-times implantations with a narrow width of 0.8 μm. The measured results of the new device exhibit a $R_{on,sp}$ of 30.9 mΩ·cm^2 with a V_B of 477 V, which shows superior $R_{on,sp}$ than the triple RESURF device and represents a reduction in $R_{on,sp}$ by 67.8% when compared with other SJ devices under the similar V_B.

ACKNOWLEDGMENT

This work was supported in part by the National Natural Science Foundation of China (grants 61704020 and 61474017), the 13th Five-year Plan for Microelectronics Advanced Research Program under project 31513030201-2 and the Fundamental Research Funds for the Central Universities (grants ZYGX2017KYQD159).

REFERENCES

[1] W. T. Zhang, B. Zhang, and M. Qiao, Z. J. Li "Theory of Superjunction With NFD and FD Modes Based on Normalized Breakdown Voltage," IEEE Trans. Electron Devices, vol. 62, pp. 4114-4120, December 2015.

[2] W. T. Zhang, B. Zhang, M. Qiao, Z. H. Li, X. R. Luo, and Z. J. Li, "The $R_{ON,min}$ of Balanced Symmetric Vertical Super Junction Based on R-Well Model" IEEE Trans. Electron Devices, vol. 64, pp. 224-230, January 2017.

[3] B. Zhang, W. T. Zhang, M. Qiao, and Z. J. Li, "Z JI Li, Equivalent Substrate Model for Lateral Super Junction Device," IEEE Trans. Electron Devices, vol. 61, pp. 525-532, February 2014.

[4] W. T. Zhang, B. Zhang, M. Qiao, Z. H. Li, X. R. Luo, and Z. J. Li, "Optimization of Lateral Superjunction Based on the Minimum Specific ON-Resistance" IEEE Trans. Electron Devices, vol. 63, pp. 1984-1990, May 2016

[5] B. X. Duan, Y. T. Yang, B. Zhang, "New Superjunction LDMOS With N-Type Charges' Compensation Layer," IEEE Electron Device Lett, vol. 30, pp. 305-307, March 2009.

[6] B Zhang, W. L. Wang, W. J. Chen, Z. H. Li, and Z. J. Li, "High-Voltage LDMOS With Charge-Balanced Surface Low On-Resistance Path Layer," IEEE Electron Device Lett, vol. 30, pp. 849-851, August 2009.

[7] M. Rub, M. Bar, G. Deml, H. Kapels, M. Schmitt, S. Sedlmaier, et al, "A 600V 8.7 Ohmmm2 Lateral Superjunction Transistor," ISPSD 2006, pp. 1–4, Naples, Italy.

[8] S. Honarkhah, S. Nassif-Khalil, C. A. T. Salama, "Back-etched super-junction LDMOST on SOI," ESSDERC 2004, pp. 117–120, Leuven, Belgium.

[9] M. J. Lin, T. H. Lee, F. L. Chang, C. W. Liaw and H. C. Cheng, "Lateral Superjunction Reduced Surface Field Structure for the Optimization of Breakdown and Conduction Characteristics in a High-Voltage Lateral Double Diffused Metal Oxide Field Effect Transistor," Jpn. J. Appl. Phys, vol. 42, pp. 7227–7231 December 2003.

[10] B. X. Duan, Z. Cao, X. N. Yuan, S. Yuan, and Y. T. Yang, "New superjunction LDMOS Breaking Silicon Limit by Electric Field Modulation of Buffered Step Doping," IEEE Electron Device Lett, vol. 36, pp. 47-49, March 2015.

[11] R. Zhu, V. Khemka, T. Khan, W. Huang, X. Cheng, P. Hui, et al, "A High Voltage Super Junction NLDMOS Device Implemented in 0.13μm SOI Based Smart Power IC Technology," ISPSD 2010, pp.79–82,Hiroshima, Japan.

[12] S. G. Nassif-Khalil, and C. A. T. Salama, "Super-Junction LDMOST on a Silicon-on-Sapphire Substrate", IEEE Trans. Electron Devices, vol. 50, May 2003

[13] C. M. Hu, "Optimum Doping Profile for Minimum Ohmic Resistance and High-Breakdown Voltage," IEEE Trans. Electron Devices, vol. 26, pp. 243-244, March 1979.

[14] X. B. Chen, J. K. O. Sin, "Optimization of the specific on-resistance of the COOLMOSTM," IEEE Trans. Electron Devices, vol. 48, pp. 344-348, February 2001.

[15] M. M-H. Iqbal, F. Udrea, and E. Napoli, "On the static performance of the RESURF LDMOSFETs for power ICs," ISPSD 2009, pp. 247–250, Barcelona, Spain.

[16] W. T. Zhang, Z. Y. Zhan, Y. Yu, S. K. Cheng, Y. Gu, S Zhang, et al, "Novel Superjunction LDMOS (>950 V) with a Thin Layer SOI", IEEE Electron Device Lett, vol. 38, pp. 1555-1558, November 2017.

978-1-5386-2928-4/18 $31.00 © 2018 IEEE

Proceedings of the 30th International Symposium on Power Semiconductor Devices & ICs
May 13-17, 2018, Chicago, USA

Cathode short structure to enhance the robustness of bidirectional power MOSFETs

Tanuj Saxena, Vishnu Khemka, Moaniss Zitouni, Raghu Gupta, Ganming Qin, Philippe Dupuy, Mark Gibson

NXP Semiconductors Inc., 1300 N Alma School Rd
Chandler, AZ, USA
tanuj.saxena@nxp.com

Abstract—**Bidirectional power MOSFETs can block voltages of either polarity between the drain and the source. This makes them attractive for many applications as they offer a distinct cost advantage over the conventional power MOSFETs. However, bidirectional power MOSFETs suffer from poor robustness as the parasitic bipolar is easier to trigger when the body is routed out separately from the source. In this paper, we propose and demonstrate the concept of a cathode short which reduces the parasitic bipolar gain by degrading the injection efficiency of the base-emitter junction and consequently improves the robustness of the bidirectional power MOSFET. The parasitic bipolar gain is shown to reduce by more than a factor of 3 and the current switching capability is enhanced by a factor of ~4.**

Keywords— *trench power MOSFET; SOA; bidirectional power MOSFET; body resistance; parasitic bipolar; cathode short; energy capability*

I. INTRODUCTION

Conventional power MOSFETs are unidirectional and can control current / block voltage of only one polarity between the source and the drain. This stems from the body and the source being "hard" tied on the chip, resulting in a body diode structure between the source and the drain terminals. Many applications such as solid state relays and reverse polarity protection switches [1,2] require voltage of both polarities to be blocked and usually employ back to back unidirectional MOSFETs to realize this function. A single bidirectional MOSFET can offer the same functionality replacing up to four conventional MOSFETs in such applications, hence providing significant cost savings.

While the advantages of a bidirectional power MOSFET are attractive, the poor robustness of the device compared to the unidirectional counterparts is the biggest challenge. The weakness in bidirectional MOSFETs arises from the body being routed out as a separate electrode. The additional resistance in the body path can prove detrimental for the device under harsh conditions as it can facilitate the turn-on of the parasitic bipolar through the I×R drop in the body, which may eventually lead to the second breakdown. To delay the onset of parasitic bipolar effects, two approaches can be adopted: the body resistance can be reduced by optimizing the layout [3] and/or the gain of the parasitic bipolar can be reduced by killing the minority carrier lifetime [4] in the base or degrading the injection efficiency of the base-emitter junction by engineering the doping. In this paper, we propose and demonstrate a cathode short structure in a bidirectional trench power MOSFET to reduce the parasitic

bipolar gain and improve the device robustness. The cathode short is shown to significantly reduce the parasitic bipolar gain and allows the device to switch more than 4x current compared to a conventional bidirectional device.

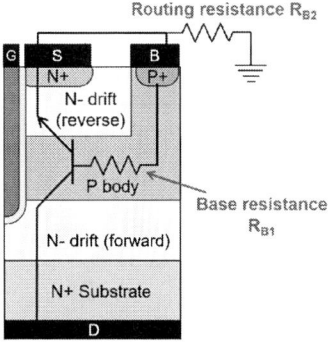

Fig. 1. A trench bidirectional power n-channel MOSFET showing the additional body resistance components: the base resistance (R_{B1}) and the external routing resistance (R_{B2}). The terminals are marked as G (gate), S (source), B (body) and D (drain).

II. DEVICE STRUCTURE AND OPERATION

The basic structure of a conventional bidirectional n-channel trench power MOSFET is shown in Fig. 1. The device has a deep implanted P body region connected to the body contact through a sinker implant region. The device includes two epitaxial N- drift regions to block voltage in both the forward ($V_D>V_S$) and the reverse ($V_S>V_D$) operation. The trench is lined with the gate oxide and the poly deposited inside the trench serves as the gate. The body of the device is connected to the circuit ground while the drain and source can have positive voltage swings.

The cathode short is formed by a replacing a part of the N+ region with a P+. This P+ region is shorted to the remaining source N+ through metallization as shown in Fig. 2(a). The P+ region is enclosed by the top N- drift region and is thus isolated from the body P region. However, the cathode short must be carefully designed to avoid any punch-through between this P+ and the body, which can affect the reverse blocking capability. The inclusion of this P+ region leads to a reduction in the parasitic bipolar gain and consequently improves the robustness

978-1-5386-2928-4/18 $31.00 © 2018 IEEE

of the device under forward operation. To understand the role of the P+ better, an equivalent circuit representation of the device under forward operation is shown in Fig. 2(b). The cathode short region introduces a PNP bipolar structure with its base and collector shorted. The P+ region acting as the collector of the PNP, creates a depletion region in the N- drift region. Because of the presence of the depletion region under the cathode short, the excess hole concentration injected in the N-drift region under the P+ drops more rapidly than under the N+. Since the diffusion current (low level injection) is proportional to the gradient of hole concentration (D_p being the diffusion coefficient),

$$J_{p,diffusion} = -qD_p \frac{dp}{dx} \qquad (1)$$

the hole diffusion current under the P+ is higher than under the N+. Simulated hole concentration profiles under N+ and P+ regions of a representational cathode short structure are shown in Fig. 3. While the excess hole concentrations at the depletion edge are similar, the gradient of the hole concentration is higher under the P+ than under the N+ leading to a significantly larger (~4x) hole current density in this case. The concentration gradient of holes under N+ is also adversely affected as the holes have lower diffusion coefficient in the heavily doped N+ region [5]. Thus, a cathode short structure with a part of the N+ region replaced by a P+ sinks a higher hole current.

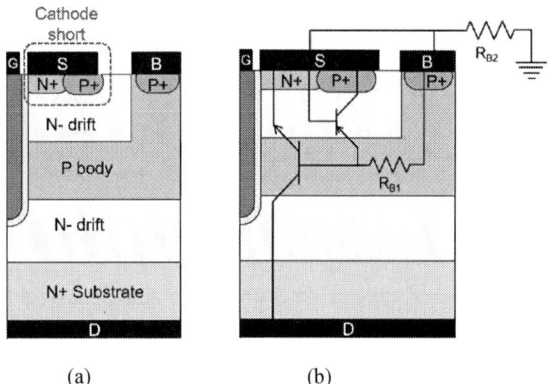

(a)   (b)

Fig. 2. (a) A trench bidirectional power MOSFET incorporating a cathode short, and (b) Equivalent circuit diagram of the device showing the parasitic NPN bipolar and the PNP bipolar introduced by the cathode short.

For any given forward bias of the body-source drift junction, the hole and electron concentrations at the edge of the junction are set up according to the law of the junction [5]. For the same forward bias (V_{BE}), the hole current is higher in the cathode short structure as compared to the conventional structure, as discussed above. The electron current back injected into the body, however, isn't affected and is only a function of the forward bias across the junction. The injection efficiency (γ) of the junction degrades as the hole component of the current (I_p) increases without disturbing the electron current (I_n),

$$\gamma = \frac{I_n}{I_n + I_p} \qquad (2)$$

which results in a decrease in the parasitic bipolar gain, β, (where α_T is the base transport factor):

$$\beta = \frac{1}{\frac{1}{\alpha_T \gamma} - 1} \approx \frac{I_n}{I_p} \qquad (\alpha_T \sim 1) \qquad (3)$$

When the MOSFET is in high voltage – high current operation or a high dV/dt condition exists at the drain, hole current going into the body creates an I×R drop, (R≈R_{B1}+R_{B2}). With sufficient hole current, the I×R drop may be high enough to forward bias the body-source drift junction and excess holes can start injecting into the source drift region. The electrons from the source region are back injected into the body and they diffuse to the drain drift region and undergo multiplication. This process can become self-sustaining if the hole current going into the body (I_p) is replenished by the generation of holes at the drain drift region. If the multiplication factor is M, and $I_{n,MOS}$ and $I_{n,BJT}$ are the electron currents from the channel and the parasitic bipolar respectively, the following condition emerges,

$$(M-1)(I_{n,MOS} + I_{n,BJT}) = I_p + \frac{V_{BE}}{R} \qquad (4)$$

Heat generated during this process leads to an increase in the base resistance, increasing the hole current contributing to the parasitic bipolar. This positive feedback can take the device into second breakdown. In case the channel current is small and the base resistance is huge, the condition transforms to that of the open base breakdown of a BJT [5]:

$$(M-1)I_{n,BJT} \approx I_p$$
$$\Rightarrow M = \frac{1}{\beta} + 1 \qquad (5)$$

When β is reduced, the multiplication factor needed at the drain-body junction to sustain the parasitic bipolar increases, which requires a higher drain voltage. The onset of the parasitic bipolar is thus delayed and the device robustness is improved.

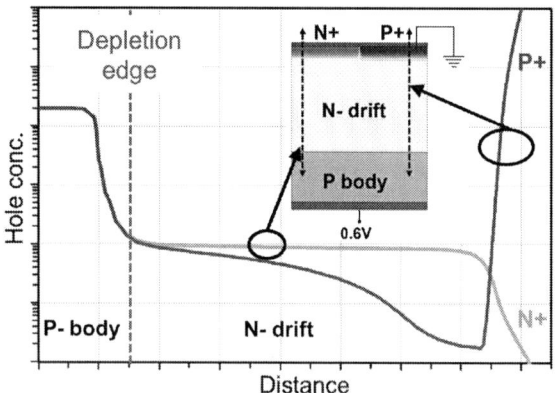

Fig. 3. Simulated hole concentration profiles for a representational cathode short structure under the N+ and P+ regions for a bias of 0.6V. A steeper gradient in hole concentration under the P+ leads to a higher hole current going into the P+ region.

978-1-5386-2928-4/18 $31.00 © 2018 IEEE

III. RESULTS

This section includes the electrical characterization results obtained from the devices including a cathode short (normalized cathode short size~4, unless stated otherwise) and their comparison with the conventional devices. Typical blocking characteristics of the devices are shown in Fig. 4. Both devices, with and without the cathode short, show identical blocking characteristics. No significant increase in the on resistance was observed with the inclusion of the cathode short, as most of the resistance is contributed by the channel and drift regions. The I_D-V_{DS} characteristics of a device with the cathode short in the first and third quadrant are shown in Fig. 5, demonstrating bidirectional conduction.

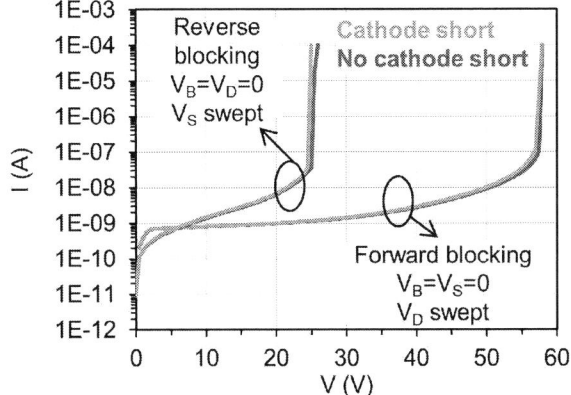

Fig. 4. Typical forward and reverse blocking characteristics of the devices with and without the cathode short.

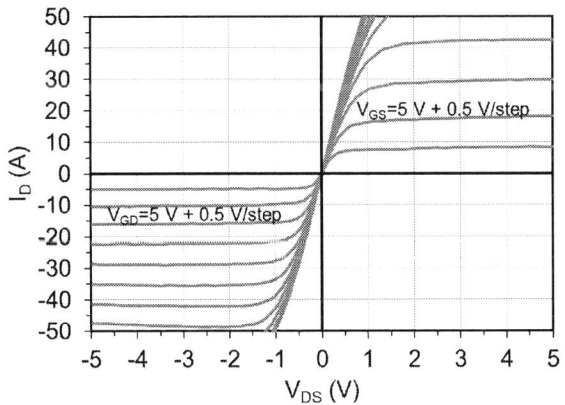

Fig. 5. I_D-V_{DS} characteristics of the device with the cathode short in the first and third quadrant demonstrating bidirectional conduction.

Fig. 6(a) shows the parasitic bipolar characteristics measured from the devices with and without the cathode short. The collector current of the parasitic bipolar is nearly identical for the two cases, however, the parasitic bipolar gain (β) shows a considerable difference. The dependence of peak β on the normalized cathode short size is shown in Fig. 6(b). The peak β is reduced as the cathode short size increases. This is a direct consequence of a higher fraction of hole current funneling into the cathode short P+ region, as opposed to the N+ region.

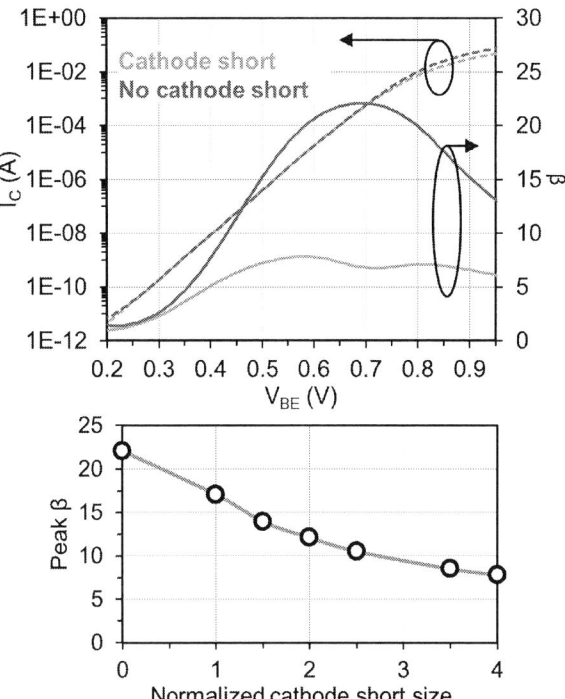

Fig. 6. (a) Parasitic bipolar characteristics of the devices with and without the cathode short, (b) Dependence of peak bipolar gain (β) on the size of the cathode short.

The strength of the parasitic bipolar was also assessed using a low-side "self-clamp" voltage test. In this configuration, the body of the device is left floating, the source is grounded and the drain is connected to an inductive load. The device is turned on with a gate pulse and the current is allowed to build. When the channel turns off, the drain voltage increases and latches up to a voltage that allows for the conduction of the current through the parasitic bipolar. This self-clamp drain voltage is akin to the open base breakdown voltage (BV_{CEO}) of the parasitic bipolar. The drain voltage and current waveforms for the devices with and without the cathode short are shown in Fig. 7. The inclusion of the cathode short leads to a significant increase (~7 V) in the self-clamp voltage or the BV_{CEO} of the parasitic bipolar.

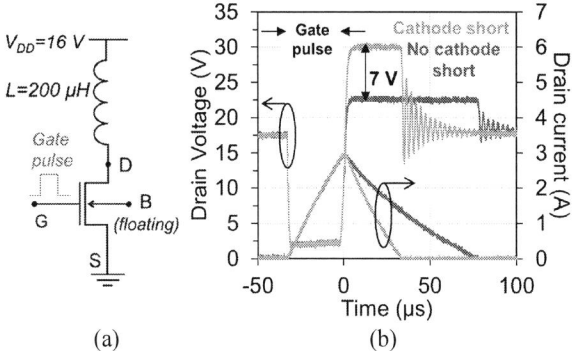

Fig. 7. (a) Schematic of the self-clamp test setup. (b) Waveforms obtained from the self-clamp test to assess the strength of the parasitic bipolar.

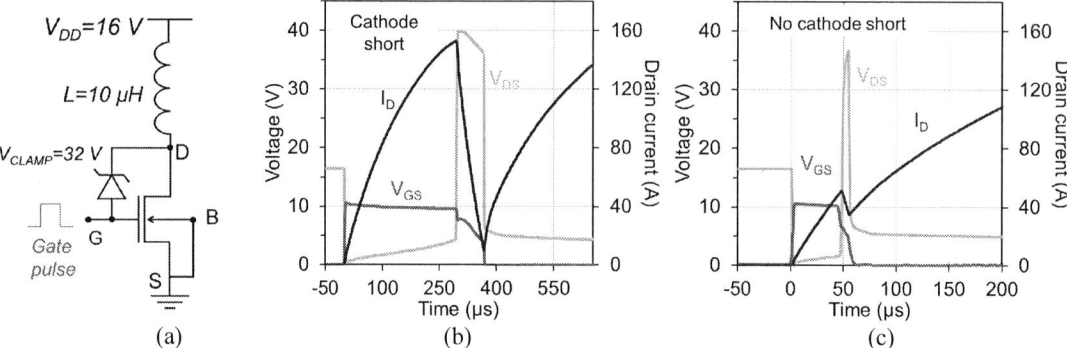

Fig. 8. (a) Schematic of the low side CIS test setup. Typical CIS (low side) waveforms at failure from the devices with (b) and without the cathode short (c).

The energy capability of the devices was tested in forward operation using low side clamped inductive switching (CIS), which represents a possible application scenario. A schematic of this measurement setup is shown in Fig. 8(a). In this mode, the body and source are tied externally to ground and the drain is tied to the power supply (V_{DD}=16 V) through an inductor with value 10 µH. To prevent the device from entering avalanche during switching, the drain to gate voltage is clamped using a Zener diode with breakdown ~32 V. The device is switched on with a gate pulse and the current is allowed to build. At gate turn off, the drain voltage increases and the drain pulls up the gate voltage through the clamp, and the device channel turns on weakly to allow the energy stored in the inductor to dissipate. Although far from avalanche, significant impact ionization may start to occur at the drain side leading to a significant hole current flowing through the body. Typical switching characteristics at failure for the two devices are shown in Fig. 8(b) and 8(c). The device without the cathode short goes into second breakdown during the rising phase of V_{DS}. Since the device is damaged during the rising phase, the overall energy dissipated is very small and the heating effects on the device are expected to be weak. This behavior suggests a very strong electrical stimulus for the damage, i.e. the parasitic bipolar effect. The device with the cathode short handles much higher current before getting destroyed, which is expected as the parasitic bipolar gets weakened. The maximum current switched by these two devices and the corresponding energy before failure during the described CIS is shown in Fig. 9. The device with the cathode short, on average, switches ~4x higher current and a correspondingly ~20x higher energy than the device without the cathode short. These results conclusively show that a significant improvement in the device robustness can be brought about by employing the cathode short structure.

IV. CONCLUSION

A new concept of a cathode short in a bidirectional trench power MOSFET to improve the robustness of the device was proposed and demonstrated. The cathode short P+ region in an n-channel MOSFET acts as a more efficient sink for the holes injected into the source drift region leading to a reduction in the parasitic bipolar gain. The weakened parasitic bipolar consequently leads to a significant enhancement in the current and energy switching capability of the device when operated in

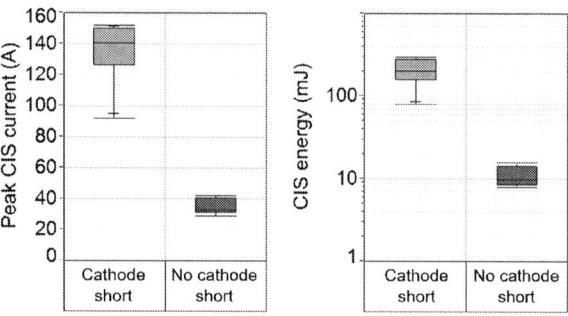

Fig. 9. Distribution graphs for peak CIS current and CIS energy handled by the devices with and without the cathode short.

the forward direction. The cathode short could be included in the device in addition to the other optimizations made to reduce the body resistance to further enhance the device robustness. A properly designed cathode short does not affect other electrical parameters of the device including the device leakage and breakdown voltages.

REFERENCES

[1] F. Robb, A. Ball and K. Huang, "A new p-channel bidirectional trench power MOSFET for battery charging and protection," *2010 22nd International Symposium on Power Semiconductor Devices and ICs (ISPSD)*, Hiroshima, 2010, pp. 405-408.

[2] D. H. Lu *et al.*, "Integrated Bi-directional Trench Lateral Power MOSFETs for One Chip Lithium-ion Battery Protection ICs," *2005 17th International Symposium on Power Semiconductor Devices and ICs (ISPSD)*, Santa Barbara, CA, 2005, pp. 355-358.

[3] P. Boos, A. Mels and S. Sque, "A 30V bidirectional power switch in a CMOS technology using standard gate oxide," *2016 28th International Symposium on Power Semiconductor Devices and ICs (ISPSD)*, Prague, 2016, pp. 247-250.

[4] M. S. Adler and P. V. Gray. "Bidirectional, high-speed power MOSFET devices with deep level recombination centers in base region," U.S. Patent 4 656 493, Apr. 7, 1987.

[5] B. J. Baliga, Fundamentals of Power Semiconductor Devices., New York:Springer, 2008.

Proceedings of the 30th International Symposium on Power Semiconductor Devices & ICs
May 13-17, 2018, Chicago, USA

40V to 100V NLDMOS built on Thin BOX SOI with High Energy Capability, State of the Art Rdson/BVdss and Robust Performance

Yang Hao, Sim Poh Ching, Madelyn Liew,
Alexander Hoelke, Uwe Eckoldt*

X-FAB Sarawak Sdn. Bhd., Kuching, Sarawak, Malaysia
*X-FAB Semiconductor Foundries, Erfurt, Germany
yang.hao@xfab.com

Martin Pfost

Chair of Energy Conversion
TU Dortmund University, Dortmund, Germany
martin.pfost@tu-dortmund.de

Abstract—We report in this paper for the first time that power LDMOS transistors integrated in thin BOX SOI can achieve high energy capabilities comparable to bulk BCD technologies and almost identical breakdown voltages for both high and low side applications. With proper top silicon thickness consideration, NLDMOS transistors for a voltage range from 40V to 100V with state of the art Rdson/BVdss trade-off and robust device performance are successfully constructed via a simple design concept.

I. INTRODUCTION

SOI BCD technologies have gained popularity in various application areas due to their superior electrical isolation capabilities, good EMC/EMI performance, as well as significant fewer redesign cycles and better likelihood of first time right success [1]. However, SOI technologies tend to have higher thermal impedances because of the lower thermal conductivity of the buried oxide layer (BOX) and are therefore often considered inferior for applications with high power dissipation requirements.

Based on the commercial 0.18µm SOI BCD technology XT018 described in [2], this new investigation aims to explore the way to advance the 40V, 60V and 100V power MOSFETs for X-FAB next generation smart power SOI platform in a smaller logic node. The NLDMOS transistors were engineered on a modified SOI substrate (thinner BOX and thicker top silicon) with further enhancement of thermal and electrical performance.

In this work, we have demonstrated for the first time that power MOSFETs built on SOI with thinner BOX can achieve both power dissipation capability similar to bulk BCD counterparts and almost identical breakdown voltage for both high and low side applications. With reasonable top silicon thickness, a range from 40V to 100V NLDMOS is built via a simple design concept, obtaining state of the art Rdson/BVdss and robust performance.

II. LDMOS DESIGN CONCEPT

A typical NLDMOS is shown in Fig. 1. The HV LDMOS relevant implant layers (NDF, HV pwell) have been

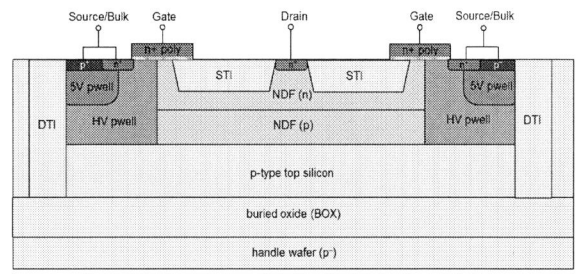

Fig. 1. Schematic cross-section of an NLDMOS used in this work. BOX thickness reduction is the main factor for higher energy capability. Top silicon thickness is adjusted to achieve around 140V BVdss for 100V LDMOS. It is noted that p-type top silicon allows formation of an accumulation layer at negative handle wafer bias (high side), therefore leads to identical Rdson between low side and high side.

introduced, following the same modular integration concept in [2], with purely high energy chain implant after baseline CMOS process thermal budget (both STI and DTI integration). Only one single mask is required for this drift engineering work.

The BOX thickness reduction is identified as the main factor for the thermal improvement. The 100V LDMOS on SOI substrate with a thinner BOX layer (30% of the thickness used in [2]) has to be realizable not only with the LDMOS drift, but also needs a sufficient dielectric isolation capability of the reduced BOX layer [3], which sets the minimum BOX thickness limit. In this work, the tunneling current of the BOX layer starts after 170V, which is sufficient for 100V operation.

Most of the vertical voltage drop has to be sustained inside the top silicon rather than in the thinner BOX, which only accommodates less than 10% of total BVdss. Therefore, the top silicon thickness is increased to guarantee around 140V BVdss for the 100V LDMOS. Fig. 2 shows the comparison of electrostatic potential distributions in 40V and 100V LDMOS transistors at breakdown between the original [2] and new modified SOI substrate.

Double RESURF drift engineering is obtained by integrating about 2e12cm^{-2} n-type phosphorus dose with approximately 1e12cm^{-2} boron dose, sharing the same NDF

978-1-5386-2928-4/18 $31.00 © 2018 IEEE

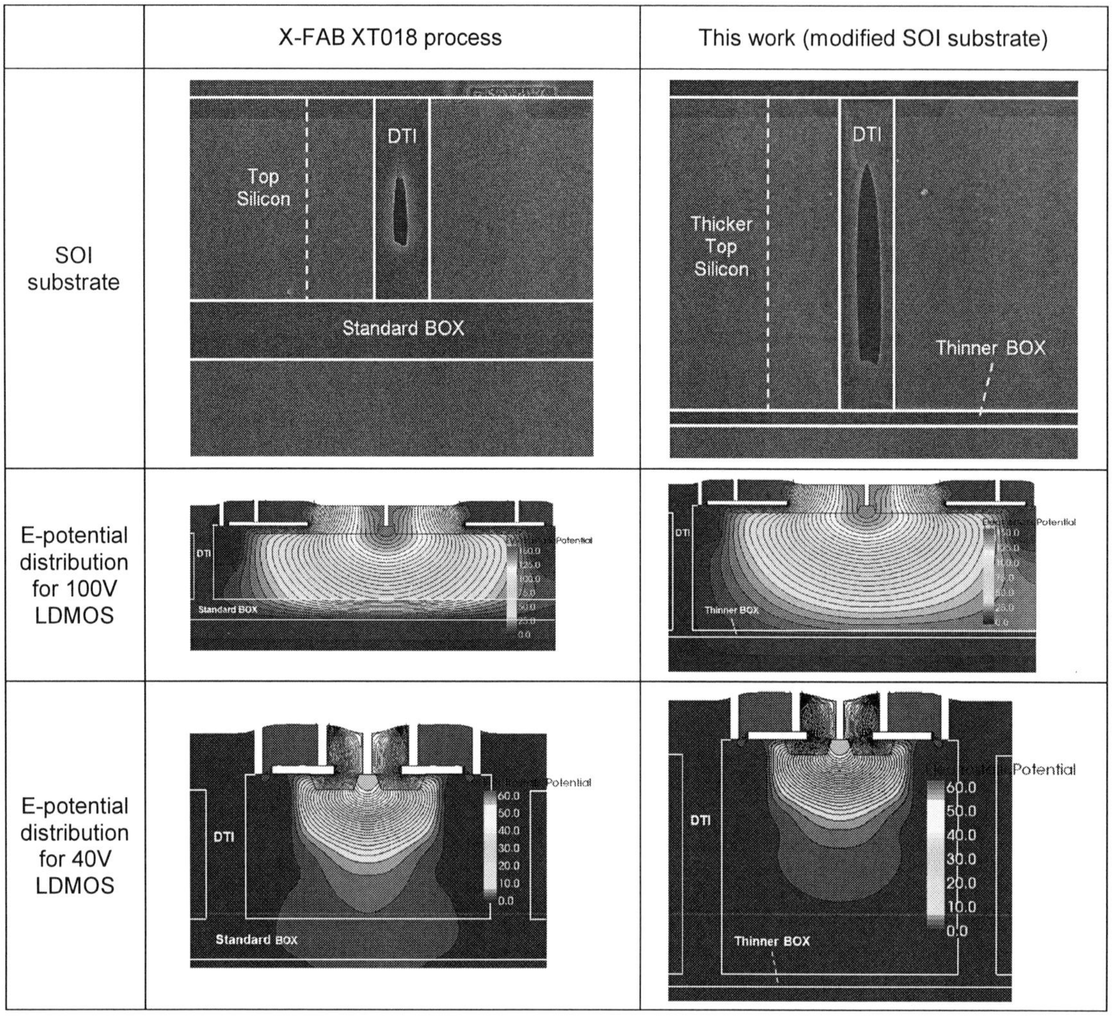

	X-FAB XT018 process	This work (modified SOI substrate)
SOI substrate	Top Silicon / DTI / Standard BOX	Thicker Top Silicon / DTI / Thinner BOX
E-potential distribution for 100V LDMOS	DTI / Standard BOX / Potential	DTI / Thinner BOX / Potential
E-potential distribution for 40V LDMOS	DTI / Standard BOX / Potential	DTI / Thinner BOX / Potential

Fig. 2: Compare to XT018 process, a thinner BOX (of 30% original thickness) and thicker top silicon is used in this work. DTI width remains the same. It is clearly shown that the whole top silicon and BOX is used to sustain the voltage in 100V LDMOS, but only few μm are needed in top silicon for 40V LDMOS.

implant mask. Local phosphorus doping concentration is further refined by different energy setting at the active/accumulation region and under the STI part. For the 100V LDMOS, a lower boron dose has to be chosen in order to construct a vertically uniform electrical field distribution for maximizing the breakdown voltage. As a result, for an Rdson/BVdss optimized drift design, two separated NDF masks are required for 40V/60V LDMOS and 100V LDMOS respectively.

III. SILICON RESULTS AND DISCUSSION

A. Energy Capability

The energy capability is an important figure of merit for many applications. Because of this, we investigated the performance of XT018 standard BOX SOI [2], thin BOX SOI, and bulk substrate. For this, special test structures containing

an NLDMOS with an active area of 880 μm x 600 μm were used. Results for the energy capabilities measured on-wafer for constant power pulses are shown in Fig. 3. A significant

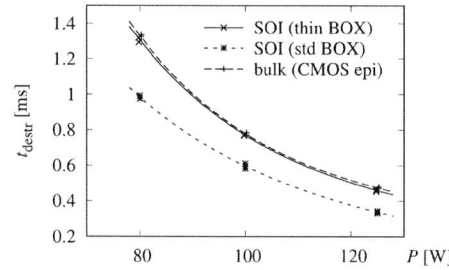

Fig. 3: Measured destruction time for the three investigated substrate types. The destruction times of the samples are denoted by symbols, obtained by applying constant power pulses of 80 W, 100 W, and 125 W. The lines are guides for the eye.

improvement achieved by thin BOX SOI can clearly be seen. Its energy capability is 30% larger than that of standard BOX SOI and even comparable to the bulk substrates with CMOS epitaxy used in many other BCD technologies – contrary to the common expectation that the energy capability of SOI technologies should be inferior to that of bulk processes.

The main reason for the excellent performance of the thin BOX turned out to be the better thermal conductivity of the lowly-doped SOI handle wafer which is 148 W/(mK). Contrary to that, CMOS substrates commonly have a highly doped p^+ bulk substrate with a lower thermal conductivity (here only 136 W/(mK) for 15 mΩcm, see [4]). Hence, comparable destruction times are observed for thin BOX SOI and CMOS bulk substrates with similar temperature distributions just before thermal runaway. The thermal resistance of the thin BOX is almost compensated by the higher thermal resistance of the p^+ bulk of the CMOS substrate as illustrated in Fig. 4.

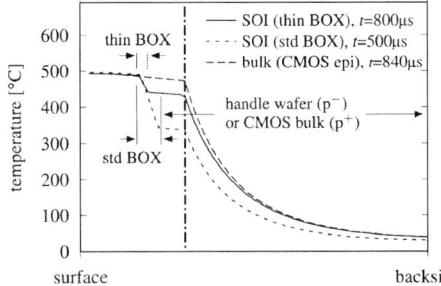

Fig. 4: Peak junction temperature in the LDMOS plotted over depth for a constant 100W pulse. Shown is the temperature distribution for different substrates for the point in time when the junction temperature reaches 500°C, which is close to destruction. (This point in time depends on the substrate type; see the legend for each graph.) Left is the surface, right the wafer backside. Note that the depth is piecewise linear, with a change of the scale at the dot and dash line, to enhance the region containing the BOX.

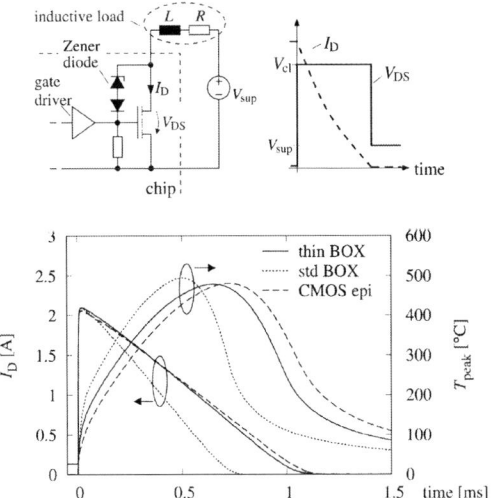

Fig. 5: Test circuit and typical waveforms as well as measured drain currents and simulated peak temperatures just before destruction for pulses under inductive clamping with a constant drain-source voltage of 50 V. The initial drain current was kept constant while the pulse length was increased until destruction occurs. Shown are the longest pulses just before device destruction.

The excellent performance of thin BOX is mostly independent of the operating condition. It also applies to triangular pulses that arise in clamped inductive switching, cf. [5], a common automotive application. Similar results as for rectangular pulses are observed, see Fig. 5. Again, bulk and thin BOX SOI perform similarly well while standard BOX SOI has a significantly lower energy capability. This remains valid even at elevated ambient temperatures of 150°C (not shown).

Moreover, Fig. 5 shows simulation results for the peak temperatures determined using the approach of [6], which has been successfully verified also for this technology. Temperatures approaching 500°C can be reached without destruction. However, a further increase of the pulse length causes thermal runaway in both measurements and simulations.

B. Electrical Performance

In contrast to other commercial SOI BCD technologies [1], [7], the off-state breakdown characteristics of the 100V NLDMOS at both 0V and -100V handle wafer bias is kept unchanged for the thin BOX variant as described in Fig. 6 as well as by TCAD in Fig. 7. In fact, the breakdown voltage of an LDMOS built on thin BOX is not influenced by handle wafer biasing even from 0V to -150V, see the TCAD result in Fig. 8. This identical breakdown voltage between high and low side bias configuration, is a desirable feature for circuit design.

The main reason for this is that the drift region is engineered in such way that most of the voltage drop is now sustained in the top silicon and not in the thin BOX. The negative voltage at the handle wafer (as in a high side configuration) drops in the BOX only, the LDMOS drift

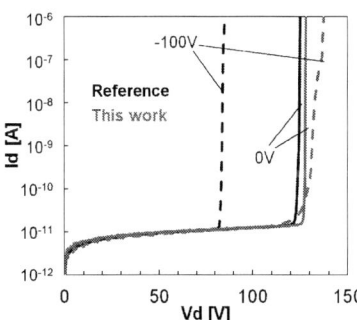

Fig. 6: Large reduction of 100V LDMOS breakdown at -100V handle wafer bias in other commercial SOI technologies (Reference) can be improved with thin BOX to almost identical off-state IdVd at both 0V and -100V handle wafer bias.

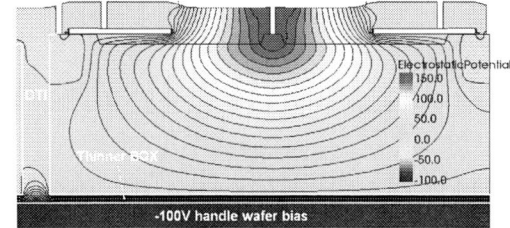

Fig. 7: TCAD simulation of the E-potential for a 100V NLDMOS at -100V handle wafer bias. It is in good agreement with the measurement result shown in Fig. 6. The E-potential distribution at 0V handle wafer is available in Fig. 2.

978-1-5386-2928-4/18 $31.00 © 2018 IEEE   485

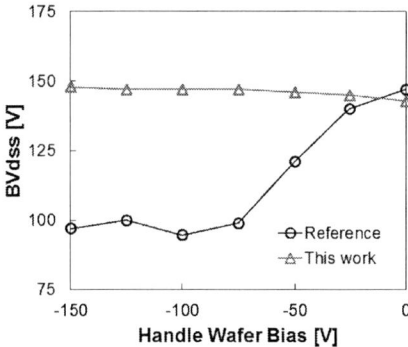

Fig. 8: TCAD simulations show that the LDMOS breakdown built on thin BOX has is not influenced by the handle wafer biasing (from 0V to -150V). Measurement condition: body, source, gate grounded, drain swept until breakdown (BVdss) at constant handle wafer bias.

RESURF effect is not disturbed. Note that the breakdown voltage obtained in silicon is around 20V lower than TCAD predicted, most likely due to an un-optimized edge termination.

Fig. 9 illustrates the use of separated NDF masks to achieve a low Rdson with sufficiently high BVdss for both 40V/60V and 100V voltage classes respectively. A drift engineering concept similar to [2] has been reused in this work with further optimization of local doping concentration and length in the accumulation region. As result, for the 40V LDMOS (which is often used in 12V automotive applications) an excellent value of 57V/23mΩ.mm² is obtained, to our knowledge one of the best-in-class values. In Fig. 10, state of the art Rdson/BVdss results for the whole voltage range from 40V to 100V are compared with other commercial technologies.

Lastly, a square electrical SOA is demonstrated by both high drift doping (for lower Rdson) and high HV pwell doping (to suppress parasitic BJT) engineering. Contrary to the thermal SOA, the electrical SOA does not have a BOX thickness dependency because the very short pulse that is imposed during TLP measurement can only create a very local heat generation. Fig. 11 shows that a square electrical SOA is obtained for both original and thinner BOX thickness variants. No dependency with BOX thickness is observed

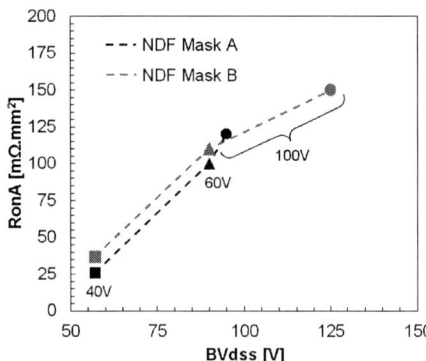

Fig. 9: Rdson/BVdss optimized NDF implant for 40V and 60V NLDMOSs restricts the maximum BVdss of 100V NLDMOS. A separated NDF mask is needed to achieve sufficient high BVdss of 100V NLDMOS, while only slightly degrading Rdson of 40V/60V NLDMOSs.

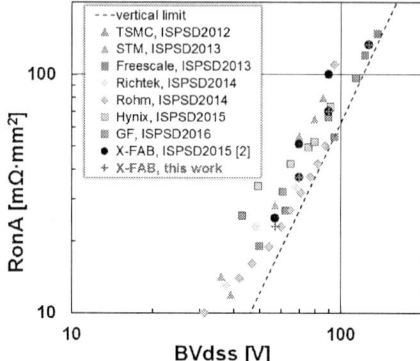

Fig. 10: State of the art Rdson/ BVdss is achieved for 40V to 100V NLDMOS.

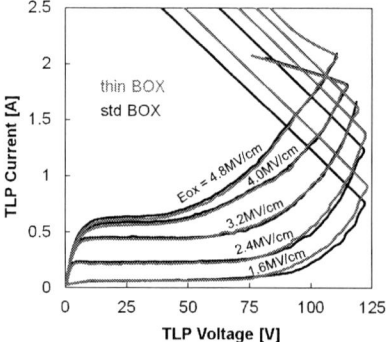

Fig. 11: Wafer level TLP characterization result of a 60V NLDMOS, width 32x80µm, 100ns TLP. Eox is the electric field in the 5V gate oxide.

IV. SUMMARY

In short, the authors believe that thin BOX enables robust and low-cost BCD SOI technologies as very attractive candidates for high power applications. State of the art Rdson/BVdss trade-off can be achieved in conjunction with a high energy capability comparable to CMOS substrate bulk processes.

REFERENCES

[1] P. Wessels, et al., "Advanced BCD technology for automotive, audio and power applications", *Solid-State Electronics*, Vol. 51, Issue 2, Feb. 2007, Pages 195-211

[2] Yang Hao, et al., "A 0.18µm SOI BCD technology for automotive application", *ISPSD2015*.

[3] S. Merchant, et al., "Realization of high breakdown voltage (> 700 V) in thin SOI devices", *ISPSD1991*.

[4] R. Krautbauer, C. Frey, S. Zitzelsberger, and L. Lehmann, "Process for producing doped semiconductor wafers from silicon, and the wafers produced thereby," *U.S. Patent 7*,202,146, Aug. 2015.

[5] D. Farenc, et al., "Clamped inductive switching of LDMOST for smart power IC's", *ISPSD1998*.

[6] M. Pfost, C. Boianceanu, H. Lohmeyer, and M. Stecher, "Electrothermal simulation of self-heating in DMOS transistors up to thermal runaway," *IEEE Trans. Electron Devices*, vol. 60, pp. 699–707, Feb. 2013.

[7] G. Toulon, et al., "Analysis of technological concerns on electrical characteristics of SOI power LUDMOS transistors", *ISPSD2010*.

Proceedings of the 30th International Symposium on Power Semiconductor Devices & ICs
May 13-17, 2018, Chicago, USA

Novel Integration Techniques of "Recessed" High Voltage Field-Drift MOSFET with HK/MG RMG Technology

C.P. Hsiung, P.H. Chiang, S.C. Pu, C.L. Wang, C.W. Lu, K.L. Liu, K.K. Chang, C.C. Yang, N.C. Lee, S.Y. Hsiao, W.F. Lee, C.C. Wang

Specialty Technology II Division, United Microelectronics Corporation (UMC), Taiwan

Email: Chang_Po_Hsiung@umc.com

Abstract— **We report herein on the first commercial production of 32V high voltage Field-Drift MOSFET (FDMOS) embedded in HK/MG with RMG technology. To overcome the challenges encountered during the RMG process, a recessed structure with a specifically designed poly gate patterning was developed. By using the new scheme, the devices can prevent metal gate disruptions. By comparison with the traditional devices in poly/SiON technology, the recessed devices can maintain good performance without degradation. Some extra benefits are obtained with the scheme by 4V improvement of gate dielectric breakdown. This approach can provide technological mass production techniques and reliable device characteristics for future PMIC and display driver ICs applications.**

Keywords—FDMOS; PMIC; display driver IC (DDI); HK/MG; RMG; recessed MOSFET

I. INTRODUCTION

Today, the high-voltage (HV) devices have been worldwide used as power management and display driver ICs [1]. Integrating HV devices and advanced low-voltage CMOS core devices on the same chip can provide the benefits of miniaturization, low power consumption, and cost effectiveness [2]. In HK/MG with the replacement metal gate (RMG) techniques, embedding of the HV devices will face significant challenges during the RMG process, as illustrated in Fig. 1. Because the traditional HV structures have abrupt difference of poly gate dimension and structural topography compared with the core devices, considerable risks of metal gate dishing or disruption in HV metal gates will arise during the CMP process of the RMG.

In this paper, we provide a new approach of "recessed" HV structure and a new design of poly gate patterning to overcome the challenges, and successfully demonstrate the first commercial production of FDMOS integration based on HK first and gate last HK/MG technology. In comparison with the traditional HV devices in the last generation of poly/SiON technology, the new HV device shows reliable and improved characteristics. This approach can not only provide suitable mass production techniques, but also reliable device performance for future applications.

II. DEVICE FABRICATION

The HV device used in the fabrication is a 32V field drift MOSFET (FDMOS) with an STI structure at each drift region with a depth of 250 nm and an HV gate oxide (HVGOX) with a thickness of 90 nm. Fig. 2 illustrates a brief process flow. After the HV well is implanted, and the thermal annealing and STI formation have been completed, the SIN hard mask (SIN-HM) is deposited and patterned to open the HVGOX growth area. In contrast to traditional flows, the HV device experiences an additional dry etching of Si recess, and the HVGOX is then grown by thermal oxidization. To ensure Si step height prior to the Poly deposition is on the same level between the HV and core devices, the Si recess depth should be carefully estimated. In this fabrication, recess depth is about 40 nm. After the HV fabrication has been completed, the core device fabrication starts, followed by the Poly deposition (50 nm thick), S/D formation, and RMG process.

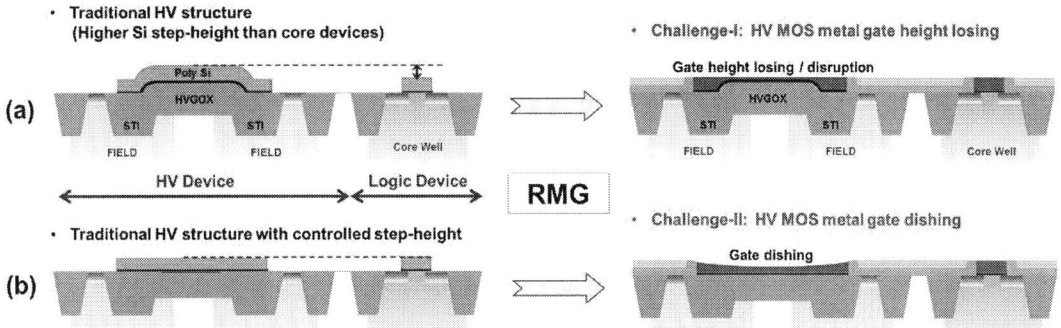

Fig. 1. Illustration of the process challenges during the RMG process. (a) Large Si step-height difference between HV and core devices cause metal gate disruption in HV gates, (b) For controlled HV structures (the same step-height level between HV and core devices), the abruptly large HV poly area will cause gate dishing.

978-1-5386-2928-4/18 $31.00 © 2018 IEEE
487

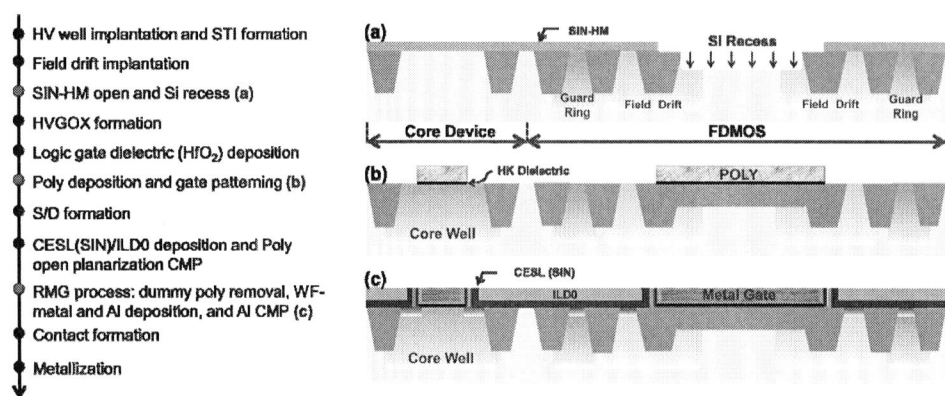

- HV well implantation and STI formation
- Field drift implantation
- SIN-HM open and Si recess (a)
- HVGOX formation
- Logic gate dielectric (HfO₂) deposition
- Poly deposition and gate patterning (b)
- S/D formation
- CESL(SIN)/ILD0 deposition and Poly open planarization CMP
- RMG process: dummy poly removal, WF-metal and Al deposition, and Al CMP (c)
- Contact formation
- Metallization

Fig. 2. Process flow of the HV device embedded in the HK first and gate last HK/MG technology platform. (a) SIN-HM open and Si recess of HV structures, (b) after poly deposition and gate patterning, and (c) after RMG process

This recessed structure approach benefits the CMP during the RMG process with lower control loading and without suffering gate height degradation of the HV structures. Figure 3 shows Si-recessed and HVGOX cross-sectional profiles. The corner ratio (HVGOX thickness ratio at the corner region to channel region) is ~82%, which is greatly improved compared to a traditional structure without Si recess (~70%). Since the HVGOX in the corner region is in a weaker position of gate oxide breakdown, improved corner ratio is expected to show better resistance to HVGOX breakdown and enhance TDDB performance.

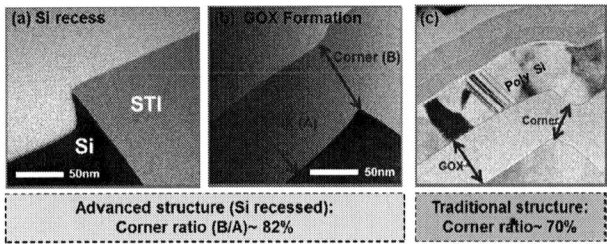

Fig. 3. TEM cross-sectional view of HV recessed structures and traditional structure; (a) Si recess profile with recess depth of ~40 nm, (b) after HVGOX growth with thickness of ~90 nm, and (c) traditional HV structure (without Si recess) which is demonstrated on Poly/SiON platform

III. DESIGN OF GATE PATTERNING

A. Effect of poly area on metal gate height (G_HT)

In the advanced CMOS with HK first and gate last integration techniques, there are two CMP involved during the RMG process: "poly open planarization (POP)" and "Al CMP". Extreme control of the CMP is needed to ensure proper device function, since variation in gate height leads to measureable device performance variability [3-4]. Here, we investigated the G_HT as a function of the poly area in a series of W/L, as shown in Fig. 4. An optimized G_HT is around 43 nm appearing at 10 um² of the poly area regardless of the poly W/L. With an increase in the poly area, the G_HT was reduced and suffered 3 nm of reduction as the area reached 100 um².

The Al CMP is suspected as the root cause since a higher Al gate area has a higher gate polish rate compared with a smaller one due to its higher component ratio of Al to the surrounding matrix (ILD0). Since the maximum dimension of our HV gate is respectively 50/10 um for the gate W/L, such a high area will entail considerable risks of metal gate dishing and gate height degradation; as a result, a considerable impact on device performance variability is expected. Therefore, it was necessary to redesign a new poly gate patterning to minimize the risks in the RMG process, as described in the following section.

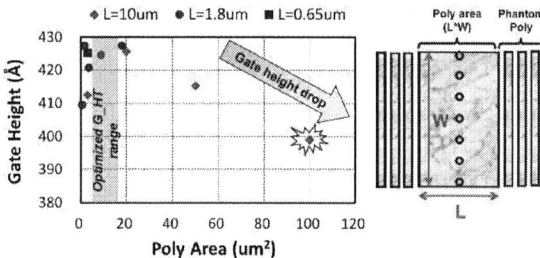

Fig. 4. Investigation of metal gate height (G_HT) as a function of the poly area (L*W); large poly area suffers G_HT degradation predominately caused by the ALCMP process. The right figure illustrates the patterns to be investigated. Each G_HT is calculated on average by measuring several positions through poly width direction (as marked circle points in the figure).

B. Design of poly gate patterning

A new gate patterning and the design procedure are illustrated in Fig. 5. To reduce the risk of a large poly area effect on the RMG process, the original poly layout was cut with a slot width of 0.1 um through the channel length direction per distance of 1 um. The slot patterns were designed via a specifically treated Boolean operation prior to mask making, and were used to exclude the original poly layout. Fig. 6 shows the plane- and cross-sectional views of the final poly patterns. With the use of the treated patterns, the Al metal gate can effectively minimize the gate height degradation and prevent possible gate dishing (as shown in Fig. 6). Figures 7 and 8 show the electric potential contour and Si surface potential as a function of slot widths by TCAD simulation under Vg=32V, while others are grounded. An increase in the slot width shows a reduction in surface potential underneath the slot regions.

978-1-5386-2928-4/18 $31.00 © 2018 IEEE

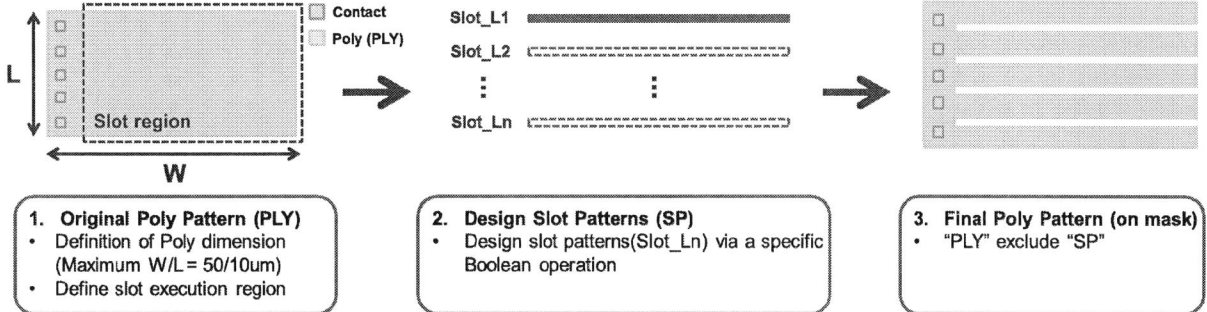

Fig. 5. Schematic illustration of a new HV poly patterning design procedure. The slot patterns are designed and patterned via a specifically treated Boolean operation before mask making, and are used to exclude the original poly patterns.

The reduction of the surface potential is attributed to the reduced electric fringe field due to wider spacing between the two poly edges (Fig. 8). Therefore, it suffers a loss in gate controlling ability under slot regions, and results in the degradation of the driving current. With a slot width of 0.1 um, the Si surface potential appears comparable with the traditional poly layout (without slots); as a result, the device Id-Vd characteristics can behave almost comparable between the two without suffering a loss in electric performance (Fig. 7).

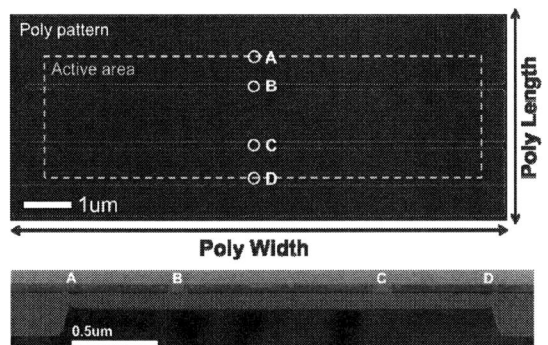

Fig. 6. SEM top-view of the HV poly patterns (upper), and TEM cross-sectional view of HV structure

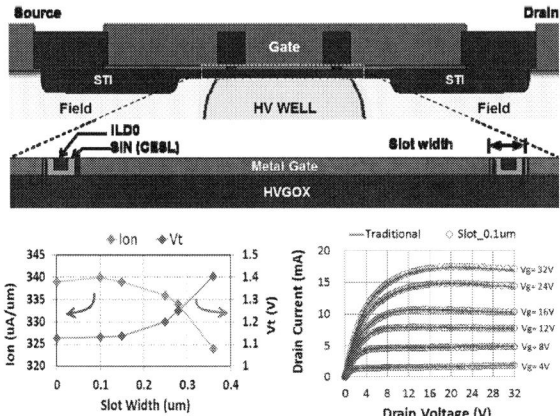

Fig. 7. Investigation of the device characteristics as a function of slot width, and the comparison of Id-Vd characteristics between a new designed poly pattern with slot width of 0.1 um and a traditional poly pattern by TCAD simulation.

IV. DEVICE PERFORMANCE

We describe the device performance of the recessed HV device and compare them with a traditional device fabricated by our previous techniques based on the poly/SiON technology in 40 nm generation. Figure 9 shows the Id-Vd characteristics of N/P MOS (W/L=50/2.5) in a series of Vg and normal characteristics with a saturation current of 375uA/um and 265uA/um under Vg=+/-32V for N/P MOS, respectively, without showing abnormal current humping characterized by Kirk effect [5]. In addition, the drain current is limited to a certain extent at high gate voltages. This is referred to as the "quasi-saturation effect" which is commonly shown in such kinds of high voltage devices [6]. Figure 10 shows the HVGOX breakdown characteristics of the devices. For the recessed HV structures, the average breakdown voltage is around 94V, which is higher than the traditional one by about 4V. This improved characteristic is possibly attributed to its improved corner ratio of the recessed HVGOX structure, as mentioned in the previous section. Figure 11 shows the device breakdown characteristics by applying drain voltage, while the others are grounded. The breakdown voltage of the recessed devices is about 40V on average, and is comparable to the traditional device. From our early investigation, the device breakdown is predominately caused by an avalanche breakdown occurring at the junction between the field drift region and the outer guard-ring region. This phenomenon seems not to be influenced by the recessed gate structure, but depends on the bulk dopant concentration and device layout. The recessed structure can truly maintain the breakdown performance without degradation.

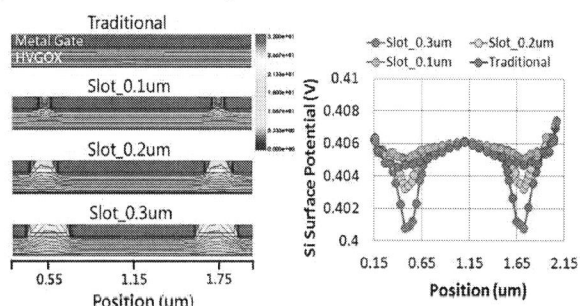

Fig. 8. Electric potential contour of the HV gate structures, and the Si surface potential of the channel for various slot widths under Vg=32V; others are grounded for NMOS device

978-1-5386-2928-4/18 $31.00 © 2018 IEEE

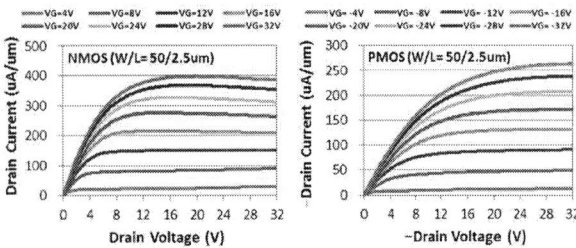

Fig. 9. HV N/PMOS Id-Vd characteristics of Si data with slot width of 0.1 um in a series of different gate voltages

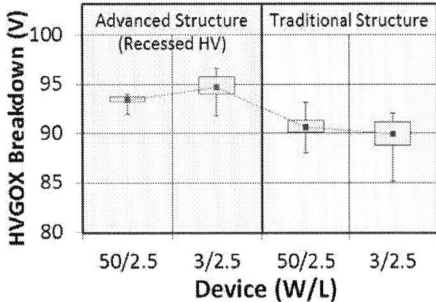

Fig. 10. Comparison of HVGOX breakdown voltage (extrapolated at 1uA) between advanced structures and traditional structures in a series of device W/L

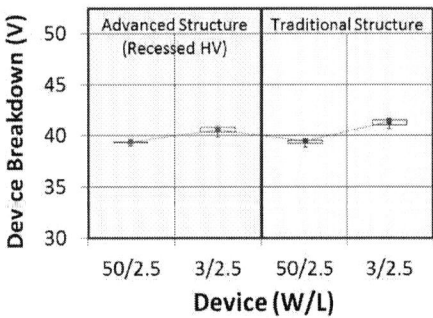

Fig. 11. Comparison of device breakdown voltage (extrapolated at 1uA) between advanced structures and traditional structures in a series of device W/L

Figure 12 shows the hot carrier effect on the device degradation of the package level reliability characteristic at room temperature. The stress is under the worst case condition with the drain voltage of 35.2V (1.1*Vcc) and the gate voltage under the maximum of bulk current. It shows a comparable degradation rate between the recessed and traditional devices; the AC lifetime for suffering 10% of degradation is about 145 years. The recessed device would not impact the hot carrier degradation effect at all.

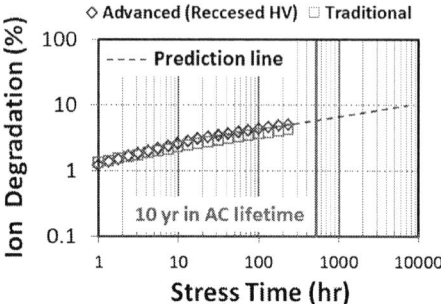

Fig. 12. Comparison of hot carrier effect on NMOS Ion degradation of advanced structure and traditional structure

V. CONCLUSIONS

We have demonstrated a new technological solution to the HV FDMOS integration in HK/MG with RMG technology. With recessed gate structure and a specific design of gate patterning, the novel HV device is suitable for technological mass production. In addition, the electric performance also shows reliable characteristics without degradation compared to the traditional HV devices in poly/SiON technology. Moreover, some improvements were found in the gate dielectric performance, which is the extra benefit of the recessed structures in the future applications.

REFERENCES

[1]   M. Annese, S. Bertaiola, G. Croce, A.Milani, R.Roggero, P. Galbiati, and C. Contiero, "0.18 μm BCD -high voltage gate (HVG) process to address advanced display drivers roadmap," IEEE ISPSD, 2005

[2]   D. Kim, J. Kim, K. Bae, H. Kim, L. Hwang, S. Shin, H. N. Park, I. T. Ku, J. Park, S. Pae, and H. Lee, "Reliability characterizations of display driver IC on high-k / metal-gate technology," IEEE IRPS, 2016, pp. 7C.4.1-7C.4.3.

[3]   R. Ghulghazaryan, J. Wilson, and A. Abouzeid, "FEOL CMP Modeling: Progress and Challenges," IEEE ICPT, 2015.

[4]   P. Feeney, "CMP for metal-gate integration in advanced CMOS transistors," Solid State Technology, p. 14, Nov. 2010.

[5]   A.W. Ludikhuize, "Kirk effect limitations in high voltage IC's" IEEE ISPSD, 1994

[6]   L. Wang, J. Wang, C. Gao, J. Hu, P. Li, W. Li, and Steve H. Y. Yang, "Physical description of quasi-saturation and impact-ionization effects in high-voltage drain-extended MOSFETs," IEEE Trans. Electron Devices, vol. 56, no. 3, pp. 492-498, Mar. 2009.

Proceedings of the 30th International Symposium on Power Semiconductor Devices & ICs
May 13-17, 2018, Chicago, USA

A novel carrier accumulating structure for 1200V IGBTs without negative capacitance and decreasing breakdown-voltage

MD TASBIR RAHMAN, KEISUKE KIMURA, TAKESHI FUKAMI, MASAKI KONISHI,
TSUYOSHI NISHIWAKI, JUN SAITO, KIMIMORI HAMADA
EHV Electronics Design Division, Toyota Motor Corporation
543, Kirigahora, Nishihirose-cho, Toyota, Aichi, Japan
Email: rahman_tasbir@mail.toyota.co.jp

Abstract—In order to improve the fuel efficiency of Hybrid Vehicles (HVs), it is very important to reduce the power losses of the power devices (such as IGBTs and FWDs) which account for about 20% of the total electric power losses of an HV system. To reduce this power losses, a new carrier accumulating structure was developed for the IGBTs called the Super Body Layer (SBL) structure, which improved the trade-off between Von and switching loss by -16% compared to its previous generation. Furthermore, this improvement was possible without the fundamental drawbacks of typical carrier accumulating structures, such as decrease in breakdown-voltage, latch-up capability and negative capacitance. This new IGBT contributed to improve the fuel efficiency of the new Toyota HVs by 2.4% in the JC08 test cycle compared to its previous generation. Thus, the new SBL structure for low loss and high current tolerant IGBT made HVs more fuel efficient than before.

Keywords— carrier accumulating structure; negative capacitance; breakdown-voltage, latch-up capability

I. INTRODUCTION

After the most recent wave of global motorization, vehicles have become a key part of the foundation supporting societies and economies around the world. However, vehicles are also a reason behind environmental issues such as global warming due to $CO2$ emissions and air pollution. For this reason, automakers regard the development of Hybrid Vehicles (HVs) and other fuel-efficient, low-emission environmentally friendly vehicles as an urgent task. Toyota Motor Corporation has a long history of developing and popularizing fuel efficient HVs as one measure to help resolve these issues. It is also making progress in the development of Electric Vehicles (EVs), Plug-in Hybrid Vehicles (PHVs), and Fuel Cell Vehicles (FCVs). All of these vehicles are powered by motors that are controlled by a Power Control Unit (PCU).

Power losses of the IGBTs and FWDs used in a PCU accounts for about 20% of the electric power losses of an HV (Fig.1). Therefore, higher PCU efficiency can be achieved by reducing power device losses. To reduce the power losses, a new carrier accumulating structure was developed for the IGBTs called the Super Body Layer (SBL) structure, which significantly improved the trade-off between Von and switching loss.

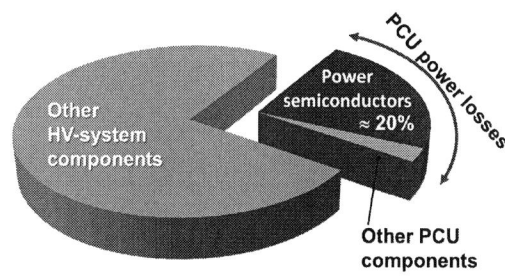

Fig.1: Breakdown of electrical power losses of an HV

Furthermore, the SBL structure was designed carefully to eliminate the following fundamental drawbacks of a typical carrier accumulating structure.

1. Decrease in breakdown-voltage

2. Negative capacitance

3. Decrease of latch-up capability

II. DEVICE STRUCTURE AND SIMULATION

A. Device Structure

The concept of the SBL structure is to introduce a hole barrier layer (a shallow N-layer) to accumulate large number of hole carriers in the drift region during ON state. Comparing with a conventional IGBT (structure A in Fig.2), the approach was to insert a hole barrier layer under the P-body region (structure B in Fig.2) or inside the P-body region (structure C in Fig.2, SBL structure). It was expected that both these carrier accumulating structures (structure B and C in Fig.2) have the effect of accumulating hole carriers from the collector side. As well as, both these structures would enhance the injection of electron carriers from the device surface. Therefore, the internal structures of these IGBTs are capable of accumulating more carriers (electrons and holes) than the conventional IGBT devices.

978-1-5386-2928-4/18 $31.00 © 2018 IEEE

• Structure A   • Structure B   • Structure C (SBL)

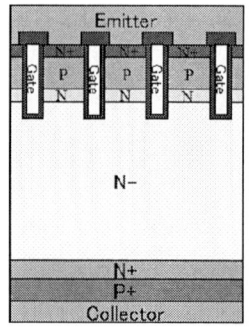

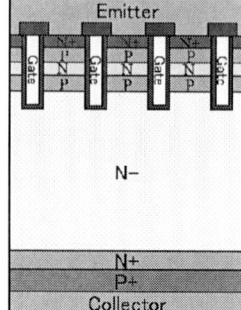

(a) Conventional IGBT   (b) Hole barrier under P-Body   (c) Hole barrier inside P-Body

Fig.2: Introduction of the carrier accumulating structure

B. Simulation Results

Both the hole barrier structures (structures B and C in Fig.2) have the same hole accumulating effect compared to a conventional IGBT during ON state (Fig.3).

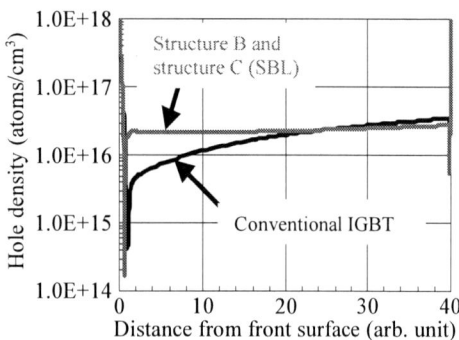

Fig.3: Simulation result of hole accumulation for structures shown in Fig.2

But, the hole barrier structure under the P-body (structure B) causes a significant decrease in the breakdown-voltage for its high N-doping concentration at the P-body and N-drift junction region.

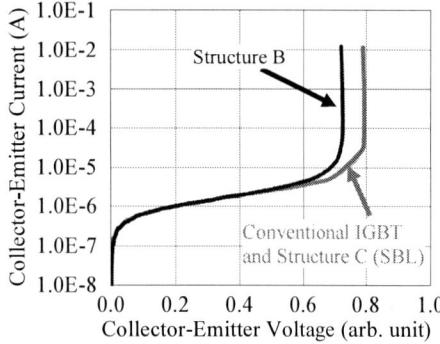

Fig.4: Simulation result of breakdown-voltage for structures shown in Fig.2

On the other hand, as the hole barrier is separated from the drift region by P-body for the SBL structure (Structure C), there is no decrease in breakdown-voltage compared to a conventional IGBT (Fig.4).

III. FABRICATION METHOD

To stabilize the hole accumulating effect of the proposed SBL structure, two P-body layers were implanted above and under the hole barrier layer called Double-peak SBL structure (Fig.5). This approach decreased the dispersion of major IGBT characteristics (such as threshold-voltage, on-voltage, etc) compared to a Single-peak SBL structure.

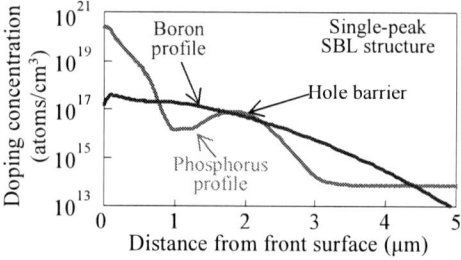

(a) Hole barrier layer inserted inside a deep P-body

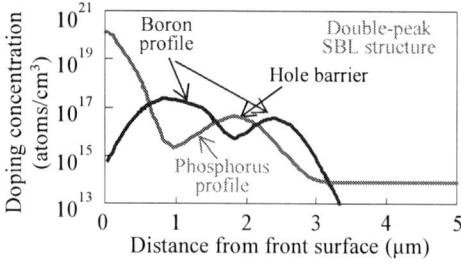

(b) Two P-body layers are implanted above and below the hole barrier to stabilize its peak concentration

Fig.5: Doping profile optimization for SBL structure

978-1-5386-2928-4/18 $31.00 © 2018 IEEE

IV. RESULTS AND DISCUSSION

A. On-voltage

The on-voltage of the fabricted device with a double-peak SBL structure is shown in Fig.6. When the IGBT is at on state, the high accumulation of carriers inside the device predicted from numerous simulatons shows the effect of lowering on-resistance of the device significantly. The result was a 19% decrease in on-voltage for the optimized SBL structure compared to a conventoinal IGBT.

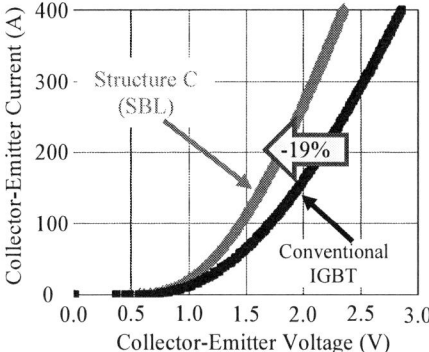

Fig.6: Measured curve of ON-voltage for the SBL structure compared to conventional IGBT

B. Negative capacitance

During switching, the accumulated carriers of the carrier accumulating structures (Structure B and C in Fig.2) concentrate around the bottom region of the trench gate. In case of structure C (SBL), as the hole barrier layer is buried in P-body, the hole accumulation around the bottom region of the trench gate is significantly lower compared to structure B (Fig.7).

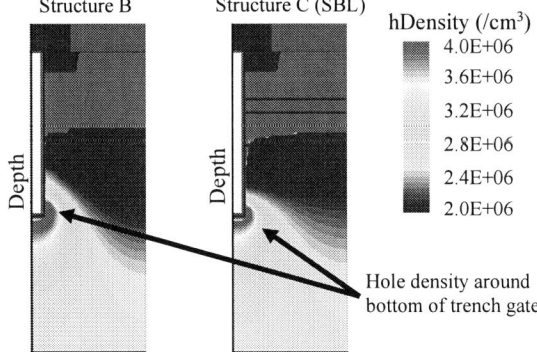

Fig.7: Hole density at the bottom of trench gate for structure C (SBL) compared to structure B

As a result, The SBL structure does not show any steep drop of gate capacitance around the threshold voltage during turn on unlike structure B (Fig.8).

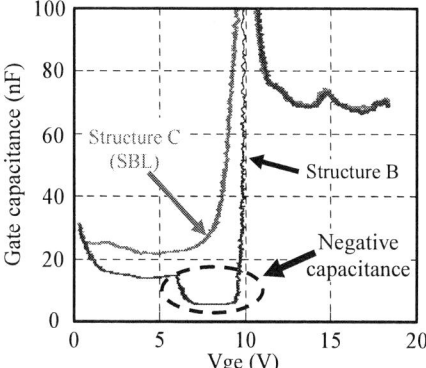

Fig.8: Solving the negative capacitance issue by adopting optimized SBL structure

As shown in Fig.8, structure C (SBL)is free from any negative capacitance and it is expected that the new SBL structure should not show any gate-voltage surge during turn on. The measured waveforms of Rg=150ohm for structure B and structure C (SBL) are shown in Fig.9. In case of Structure B, the current rise cannot be suppressed by the gate resistance and a gate-Volatge surge also with some voltage oscillation around the threshold volatge occurs due to its negative capacitance during turn on. On the other hand, the current rises smoothly for structure C (SBL) and does not show any gate-voltage surge or oscillation during turn on .

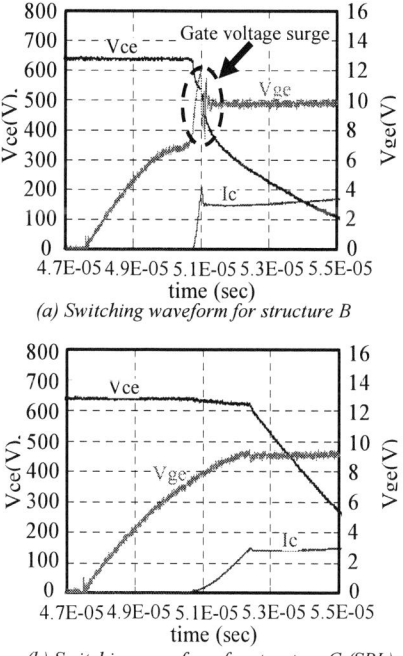

(a) Switching waveform for structure B

(b) Switching waveform for structure C (SBL)

Fig.9: Eliminating gate voltage surge by optimizing SBL structure

(Rg=150Ω ,Vcc=650V , Ic=150A , Vge=13.5V , L=100μH)

C. Latch up capability

To optimize the SBL structure, numerous simulations and prototypes were tested to identify the optimum hole barrier concentration range capable of accumulating a large number of carriers to reduce on-voltage without any negative capacitance during turn on. Fig.10 shows the dependence of on-voltage, gate voltage surge and reverse blocking safe operating area (RBSOA) for the hole barrier concentration of SBL structure. The lines and dots correspond to the simulated and measured results, respectively. As expected, SBL structure can offer an optimum hole barrier concentration to reduce on-voltage and avoid gate voltage surge and latch up phenomena due to a large number accumulated of carriers at the same time.

Experiment results confirmed that introducing the SBL structure improved the trade-off relation between on-voltage and switching loss of the IGBT by -16% compared to its previous generation (Fig.11).

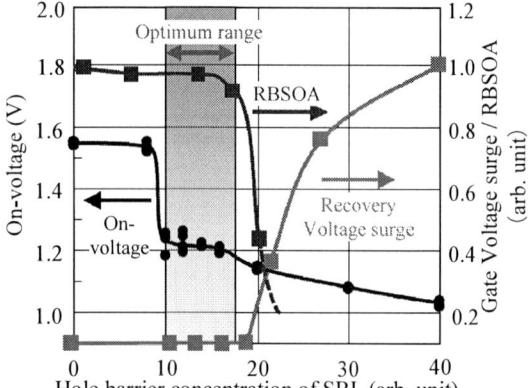

Fig.10: Optimum hole barrier concentration range for SBL structure

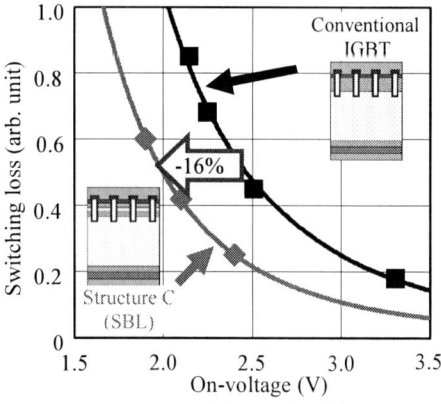

Fig.11: Improvement of the trade-off by SBL structure

D. Fuel efficiency improvement of HVs

The SBL structure was developed for the IGBTs that contributed to the low-loss and downsizing of the PCU. Thanks to its low-loss characteristics, the fuel efficiency of the new Toyota HVs has improved by 2.4% in the JC08 test cycle compared to its previous generation.

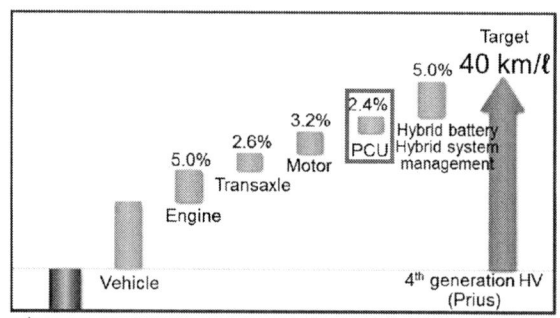

Fig.12: Fuel efficiency improvement of TOYOTA's new HV (Prius)

V. CONCLUSION

A new carrier accumulating structure called the SBL structure is developed for HVs. The new 1200V IGBTs with the SBL structure is capable of low on-voltage without any negative capacitance or decrease in latch-up capability. As a result, this new IGBTs have made the HVs more fuel efficient than before.

REFERENCES

[1] M. Kitagawa, I. Omura, S. Hasegawa, T. Inoue and A. Nakagawa, "A 4500 V injection enhanced insulated gate bipolar transistor (IEGT) operating in a mode similar to a thyristor" , IEDM Technical Digest, pp 679-682,1993

[2] I. Omura and W. Fichtner, "IGBT instability due to negative gate capacitance", Proceedings of EPE'97, pp.2.006-2.069, 1997

[3] M. Yamaguchi, I. Omura, S. Urano, S. Umekawa, M. Tanaka, T. Okuno, T. Tsunoda and T. Ogura, "IEGT Design Criterion for Reducing EMI Noise", Proc.ISPSD'04, pp 115-118, 2004

[4] H. Takahashi, H. Haruguchi, H. Hagino, and T. Yamada, "Carrier Stored Trench-Gate Bipolar Transistor (CSTBT) – A Novel Power Device for High Voltage Application –", Proc.ISPSD'96, pp 349-352, 1996

[5] T. Takahashi, Y. Tomomatsu, K. Sato, "CSTBTTM(III) as the next generation IGBT", Proc.ISPSD'08, pp 72-75, 2008

[6] K. Oyama, Y. Kohno, J. Sakano, J. Uruno, K. Ishizuka, D. Kawase, and M. Mori, "Novel 600-V trench high-conductivity IGBT (trench HiGT) with short-circuit capability," Proc.ISPSD'01, pp. 417–420, 2001.

[7] S. Machida, T Sugiyama, M. Ishiko, S. Yasuda, J. Saito, K. Hamada, "Invesigation of Correlation Between Device Structures and Switching losses of IGBTs", Proc.ISPSD'09, pp 136-139, 2009

[8] K. Hamada, "Power Semiconductor Devices and Modules for Hybrid Vehicles", Short Course ISPSD'14, pp 129-172, 2014

[9] K. Kimura, T. Rahman, T. Misumi, T. Fukami, "Development of New IGBT to Reduce Electrical Power Losses and Size of Power Control Unit for Hybrid Vehicles," SAE Int. J. Alt. Power. 6(2):2017, doi:10.4271/2017-01-1244

Proceedings of the 30th International Symposium on Power Semiconductor Devices & ICs
May 13-17, 2018, Chicago, USA

Study on the Improved Short-Circuit Behavior of Narrow Mesa Si-IGBTs with Emitter Connected Trenches

K. Eikyu, A. Sakai, †H. Matsuura, †Y. Nakazawa, Y. Akiyama and Y. Yamaguchi

Renesas Electronics Corp., Hitachinaka, Ibaraki, Japan

†Renesas Semiconductor Manufacturing Co. Ltd., Hitachinaka, Ibaraki, Japan

Tel: +81-29-272-3111, Fax: +81-29-270-2430, Email: katsumi.eikyu.ud@renesas.com

Abstract—The impacts of the self-heating and autonomous hole supply adjustment on the short-circuit (SC) behavior of narrow mesa Si-IGBTs are investigated. As reported previously, non-saturated output characteristics of very narrow mesa IGBT, which is originated from so-called CIBL effect, leads to larger current peak and lattice temperature rising during SC pulse. Newly introduced "GGEE" structure does not show non-saturated output curves thanks to autonomous hole supply adjustment by parasitic pMOS operation. 3D TCAD simulation shows MOS channel mobility degradation besides the parasitic pMOS plays an important role to establish the built-in negative feed-back loop against the thermally induced nearly instantaneous failure.

1. INTRODUCTION

Si IGBT is expected as key parts for controlling motors in the near coming ZEV era. The device will have been required for smaller power loss during conduction/switching operations and a large safe operation area simultaneously. Scaling the mesa width is proposed as a solution candidate for further performance improvement [1-3]. However, we have shown the very narrow mesa IGBTs, which were fabricated based on our 7th generation process and have excellent on-state performance, cannot be the mainstream for the future solution as it is because of its poor short-circuit (SC) behavior [4].

Recently a consideration for further cell size scaling is proposed [5] which is similar to the well-known CMOS scaling [6]. However, gate oxide scaling is inevitably accompanied by the driving voltage reduction, which means the necessity of redesign of the system surrounding switching power devices, such as gate driver circuits.

We propose a much simpler approach for avoiding such a large-scale replacement and physical mechanism is investigated in depth with focus on the self-heating effect and autonomous hole supply adjustment mechanism during short-circuit test operation.

2. SC BEHAVIOR OF NARROW MESA IGBTS

In previous report [4], we have shown the poor SC behavior of the narrow mesa devices which were fabricated

based on our 7th generation process. Such devices feature non-saturated output characteristics which were originated from CIBL (collector bias induced barrier lowering) effect and leads to much larger self-heating during SC pulse. Suppressing these non-saturated output characteristics is the key design factor of the narrow mesa devices.

3. ANOTHER DEVICE OPTION, EXPERIMENTS

As reported recently, introduction of emitter connected trenches into the advanced IGBT cell is found effective for better switching performance [7][8]. These structures have been adopted and fabricated using our 7th generation process. As shown in Fig. 1, cross section of the newly developed "GGEE" cell has 2 times wider cell pitch. This structure is found to show much better SC behavior than conventional "GG" structure with same mesa width, and more likely to be equivalent to wider mesa devices (Fig. 2). This "GGEE" device shows almost the same on-state voltage but slightly different saturation behavior in output characteristics. (Fig.3, 4)

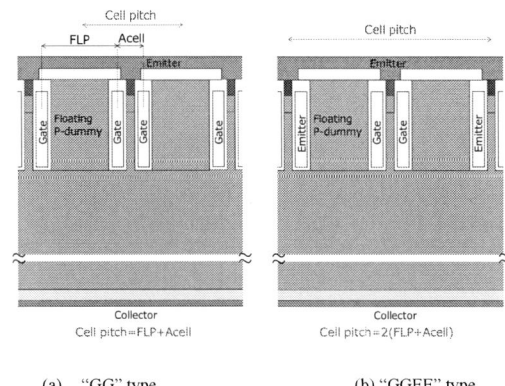

(a) "GG" type   (b) "GGEE" type

Fig. 1 2-dimentional cross section image of fabricated 7th generation (7G) based IGBT devices. (a) conventional "GG" type, (b) proposed "GGEE" type. *FLP/acell* ratio is kept to 3. Gate voltage is supplied to "G"-trench poly-Si and emitter voltage to "E"-trench. Unit cell pitch of "GGEE" device is ×2 larger than "GG" device.

978-1-5386-2928-4/18 $31.00 © 2018 IEEE   495

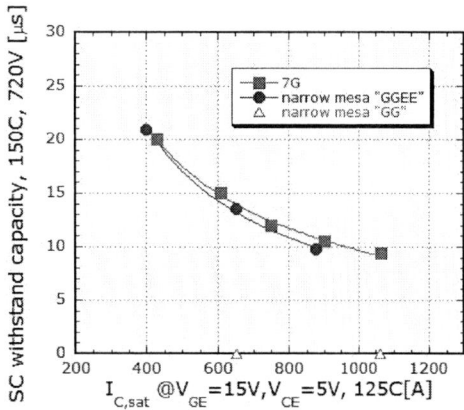

Fig. 2 Measured SC withstand capacity plot against saturation current level. SC test is performed under 150 °C and V_{CC} =720V and saturation current level is measured at 125 °C. SC capacity decreases with increasing saturation level. "*GGEE*" device shows roughly equivalent capacity as wider mesa 7G device while "*GG*" device shows immediate failure (t_{ON} ~0).

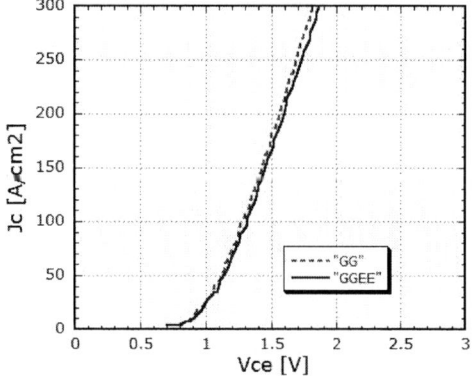

Fig. 3 Measured output curve J_C-V_{CE} of the device with the narrow mesa (*acell*=1.0μm). "*GGEE*" device shows almost the same on-state voltage V_{sat} (@J_C=200A/cm²) as "*GG*".

4. SIMULATION ANALYSIS

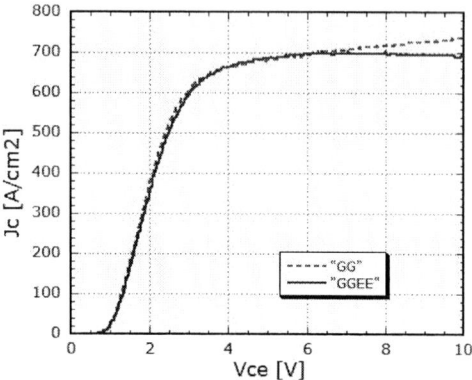

Fig. 4 Measured output curves J_C-V_{CE} in wider range. "GGEE" device shows saturated behavior with slight NDR at higher V_{CE}, while "GG" does non-saturated.

3D TCAD simulation of ½×½ cell is used for investigating physical mechanism. Detailed information on the simulation is described elsewhere [4]. Emitter diffusion size per unit cell is adjusted for saturation current level control. It should be noted "*GGEE*" device has denser emitter diffusion on the mesas between "*G*"-trenches than "*GG*" device because the mesas between "*E*"-trenches do not contribute for on-current. In Fig. 5, simulated output characteristics of "*GGEE*" device and "*GG*" device with the same mesa width are compared. 3D TCAD simulation without considering self-heating effect has been able to capture the apparent difference between these devices. "*GGEE*" shows nearly saturated curve while "*GG*", which suffers large CIBL [4], shows non-saturated curve. Electrostatic potential at the center of the mesa region between "*G*"-trenches is plotted in Fig. 6. "*GGEE*" is proved to be more robust against CIBL. Owing to parasitic p-channel MOSFET operation along "*E*"-trench, hole concentration is autonomously adjusted and conduction modulation around "*GG*" mesa is lowered after current saturation, and potential drop in the hole depleted region weaken the barrier lowering effect at mesa between "*G*"-trenches (Fig. 7). On the contrary, "*GG*" device does not show any hole supply adjustment operation after current saturation (Fig. 8).

Furthermore, observed negative differential resistance (NDR) of "*GGEE*" device is well reproduced by including self-heating effect and temperature dependent channel mobility (Fig. 5(b)). Using the calibrated model, SC waveform has been simulated. As shown in Fig. 9, "*GGEE*" shows much smaller current peak in 1st oscillation and this

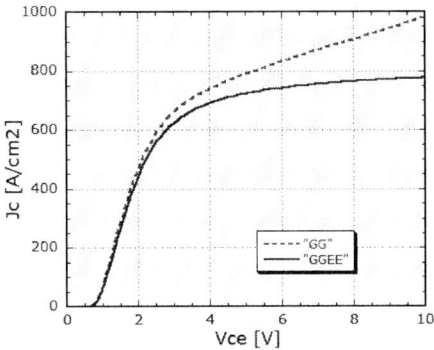

(a) w/o self-heating effect

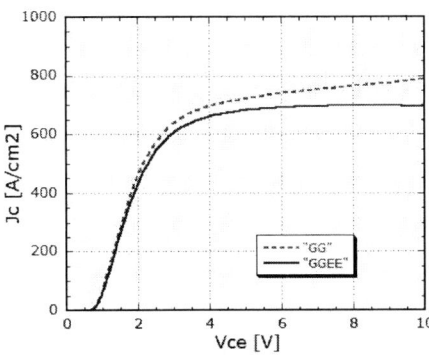

(b) with self-heating and mobility degradation

Fig. 5 Comparison between simulated output curves of the almost same saturation level devices, (a) without self-heating effect, (b) with self-heating effect and MOS channel mobility degradation.

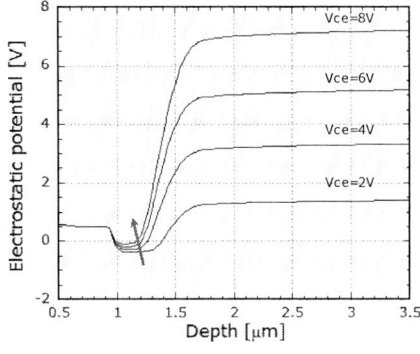

(a) GG-type

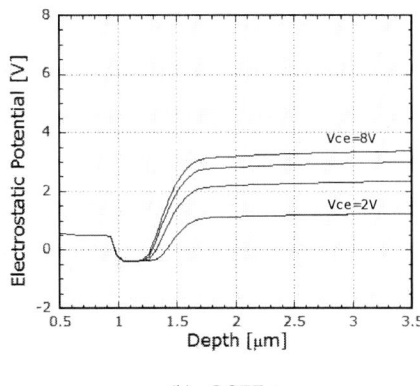

(b) GGEE-type

Fig. 6 Electrostatic potential in the middle of the Si mesa corresponding to Fig. 5(a), (a) "GG", (b) "GGEE". "GGEE" device maintains potential barrier at high V_C while "GG" suffers severe CIBL with same mesa width.

leads to moderate lattice temperature increase due to built-in negative feedback loop.

5. CONCLUSION

We have investigated the SC behavior of the narrow mesa IGBTs. Newly introduced "GGEE" structure is proved to have much higher SC capacity compared with conventional "GG" structure with same mesa width and saturation current level. 3D TCAD simulation elucidated autonomous hole supply adjustment by parasitic pMOS operation and temperature dependent MOS channel mobility degradation work as the built-in negative feed-back loop so as to circumvent thermal breakdown.

REFERENCES

[1] A. Nakagawa, "Theoretical Investigation of Silicon Limit Characteristics of IGBT," in Proc. ISPSD 2006, pp. 5-8.

[2] M. Takei et al., "DB (Dielectric Barrier) IGBT with Extreme Injection Enhancement," in Proc. ISPSD 2010, pp. 383-386.

[3] M. Sumitomo et al., "Low loss IGBT with Partially Narrow Mesa Structure (PNM-IGBT)," in Proc. ISPSD 2012, pp. 17-20.

[4] K. Eikyu et al., "On the scaling limit of the Si-IGBTs with very narrow mesa structure," in Proc. ISPSD 2016, pp. 211-214.

[5] M. Tanaka et al., "Conductivity modulation in the channel inversion layer of very narrow mesa IGBT," *in Proc. ISPSD 2017*, pp. 61-64.

[6] R. H. Dennard et al., "Design of Ion-Implanted MOSFET's with Very Small Physical Dimensions," *IEEE J. Solid-State Circuits*, vol. SC9, no.5, pp.256-267, 1974

[7] K. Konishi et al., "Experimental Demonstration of the Active Trench Layout Tuned 1200V CSTBT™ for Lower dV/dt Surge and Turn-on Switching Loss," *in Proc. ISPSD 2016*, pp. 363-366.

[8] M. Sawada et al., "Hole Path Concept for Low Switching Loss and Low EMI Noise with High IE-effect," *in Proc. ISPSD 2017*, pp. 65-68.

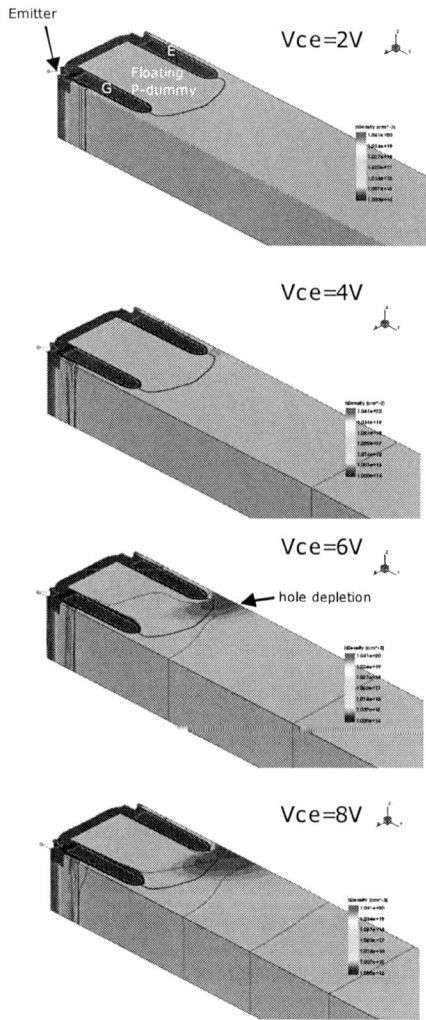

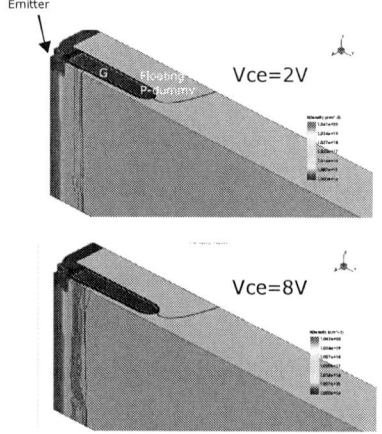

Fig. 8   Hole concentration of "*GG*" device corresponding to Fig. 5(a), 6(a) (Vce=2, 8V). High hole concentration due to conduction modulation is kept after current saturation.

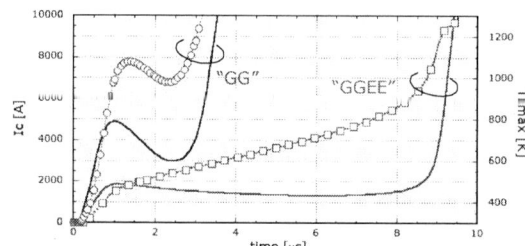

Fig. 9   Simulated SC waveform of 3D ½×½ cell at 300K, V_{CC}=400V. Self-heating effect is considered (solid line: collector current, open symbol: maximum lattice temperature). The "*GGEE*" device (red symbol/line) shows much higher SC capacity against the "*GG*" devices (blue symbol/line).

Fig. 7   Hole concentration of "*GGEE*" device corresponding to Fig. 5(b), 6(b) (Vce=2, 4, 6, 8V). Solid contour line shows electrostatic potential. After current saturation, hole supply to the "*GG*" mesa is autonomously suppressed by parasitic p-channel MOS operation. Voltage drop under the floating P-dummy region is observed due to hole depletion.

Proceedings of the 30th International Symposium on Power Semiconductor Devices & ICs
May 13-17, 2018, Chicago, USA

An Advanced Soft Punch Through Buffer Design for Thin Wafer IGBTs Targeting Lower Losses and Higher Operating Temperatures up to 200 °C

Elizabeth Buitrago, Athanasios Mesemanolis, Charalampos Papadopoulos, Chiara Corvasce, Jan Vobecky, Munaf Rahimo

ABB Switzerland Ltd, Semiconductors
Fabrikstrasse 3, CH-5600, Lenzburg, Switzerland
elizabeth.buitrago@ch.abb.com

Abstract—A shallow phosphorus buffer peak has been added to state-of-the-art planar soft punch through IGBT buffer (1200 V and 1700 V, 150 A, 13.6 x 13.6 mm²) to lower the leakage current and expand the temperature operation range up to 200°C. The new buffer design is experimentally demonstrated to provide rugged switching (RBSOA) up to 200°C without compromising other high performance characteristics like soft switching and short circuit capability.

I. INTRODUCTION

Advanced thin wafer processing technologies with different buffer design concepts s are necessary to produce high performing IGBTs with blocking capabilities rated below 2000 V. The careful design of the N+ field stop layer within the N-drift bulk region of an IGBT enables designers to produce devices with a higher forward blocking capability and in turn reduce the on-state voltage drop by thinning the device. Different field-stop concepts such as the soft [1] and controlled [2] punch-through (SPT and CPT) buffers have been designed to enable particular device characteristics. The SPT concept in particular (Fig. 1a), makes use of wafers with a predefined, low doped and deep buffer layers produced by deep diffusion (DD) [3, 4, 5]. Although the use of DD wafers allows for full manufacturing of the cathode front-side without any fabrication constraints, further processing is restricted to temperatures below 450°C. In addition, the phosphorus depth and doping profile are fixed. This does not allow for buffer design freedom and complete utilization of the device thickness. Wafer grinding also consumes the field-stop layer (Fig. 1a) further resulting in higher leakage current and reduced blocking. In order to overcome these limitations, the controlled punch through (CPT) concept was proposed [2] consisting of a double buffer profile: a phosphorous layer next to the anode to reduce the leakage current and a detached proton peak to stop the electric field farther away from it (Fig. 1b). To increase the design and processing freedom, we propose here an advanced SPT design that can be achieved by adding a single shallow (< 1 µm) N+ leakage buffer peak and reduce the device thickness (Fig. 1) for improved losses and 200°C operation.

With the implementation of the phosphorous shallow peak, the hole concentration in the drift region decreases as holes have to diffuse across the interface between this high phosphorous concentration area, the buffer layer and the n-drift region. This translates into a leakage reduction, increase of the on-state and a reduction in the turn-off losses. As the anode is also being activated in the same step, the bipolar gain can again be lifted so that the device has similar losses on the technology curve. The high doping concentration of the leakage buffer peak (at least one order of magnitude higher than the DD buffer) may restrict the electric field and prevent it from reaching the collector during blocking even when the DD profile is reduced by thinning.

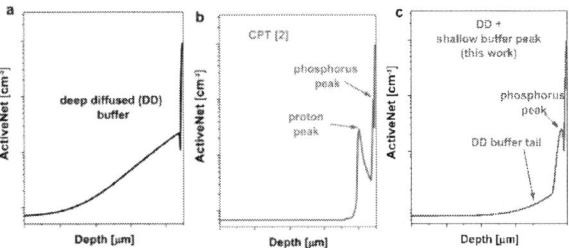

Fig. 1. Simulated buffer and anode profiles: a) reference SPT (DD buffer – black), b) CPT buffer (proton + phosphorous peaks [2] - green) and c) current work (DD buffer thinned + phosphorus peak by LA - red). Wafer thinning for the SPT buffer consumes the higher doped regions produced by deep diffusion and hence the leakage current increases and the blocking capability is reduced.

From the process point of view, the buffer and collector layers can be produced by the simple serial ion implantation of phosphorus and boron and the simultaneous dopant activation by a laser anneal process. Laser annealing allows for the low thermal budget dopant activation in very thin devices making it possible to produce IGBTs with a wide range of trade-off performances. In here we have used a laser system (Sumitomo Heavy Industries) for the activation of both phosphorous and boron in a single step or for anode alone. The reference device has a DD buffer and an anode that has been activated by a low temperature sintering step below 450°C.

II. EXPERIMENTAL RESULTS

A. Static performance at high temperatures

The simple and straightforward addition of an ultra-shallow phosphorus peak within the standard SPT buffer design and anode activation by a single laser anneal step enables a leakage current reduction of up to 50% at 175°C (Fig.2). We were able to measure up to maximum 193°C (Fig.3) due to the temperature limitation of the test system. The same leakage current reduction trend has been observed for both voltage classes. It is worth

978-1-5386-2928-4/18 $31.00 © 2018 IEEE

mentioning that the standard aluminum metallization was used and no special measures were implemented for better heat spreading.

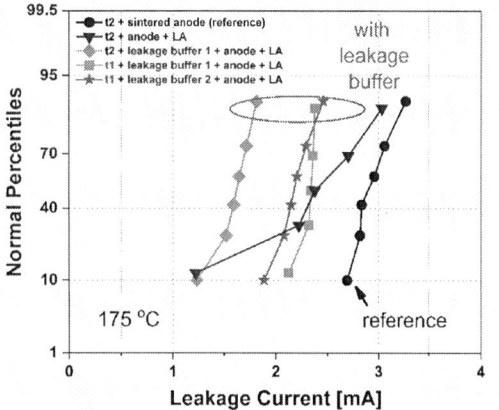

Fig. 2. 1200 V / 150 A IGBT leakage current measured at 175 °C. All devices with a leakage buffer show reduced leakage currents up to 50% vs. ref. The leakage current is reduced by the simple addition of a shallow leakage buffer peak even on thinned (t1) devices. Devices in overlapping technology points with and without a leakage buffer can be compared (Fig. 4).

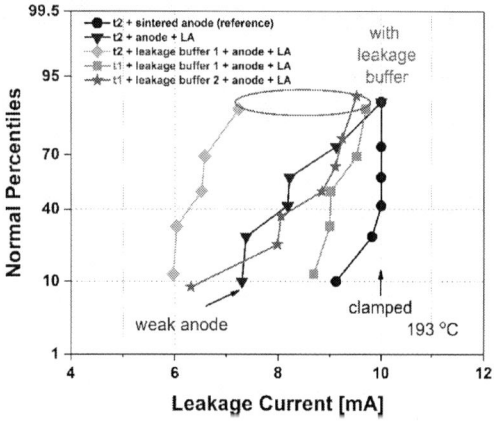

Fig. 3. 1200 V / 150 A IGBT leakage current measured at 193 °C. All devices with a leakage buffer show reduced leakage currents. Measurements are clamped at 10 mA. The leakage current is reduced by the simple addition of a shallow leakage buffer peak even on thinned (t1) devices. Devices in overlapping technology points with and without a leakage buffer can be compared (Fig. 4).

As can be seen in Fig. 4, the anode efficiency and bipolar gain can be optimized by tailoring the buffer and anode peak depth and height to produce a device with the same technology point (for the same thickness t2). With this approach there is no compromise in turn off losses or softness while targeting an application that requires an improved leakage current for higher operation temperature. This is particularly important at high temperatures when hole mobility is reduced and anode efficiency is reduced. Furthermore, on-state and turn-off switching loses are improved by thinning these advanced SPT devices while still maintaining leakage current levels below or equal to a device without the leakage buffer and therefore to allow for an improved performance at the same rated temperature of 175°C

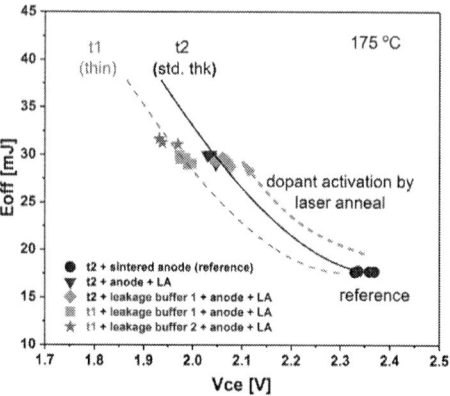

Fig. 4. Technology trade-off curves measured at 175 °C (L_s = 200 nH, 150 A) for IGBTs with two different device thicknesses. It is possible to move along technology curves by laser anneal dopant activation (anode + buffer and anode alone). Devices with and without the leakage buffer at overlapping technology points are shown.

Fig. 5 shows the leakage current values measured at reverse voltages up to 1200 V for different temperatures up to 193 °C. The leakage current improvement by the addition of a leakage buffer when comparing devices at the same technology point but realized by anode activation only is evident.

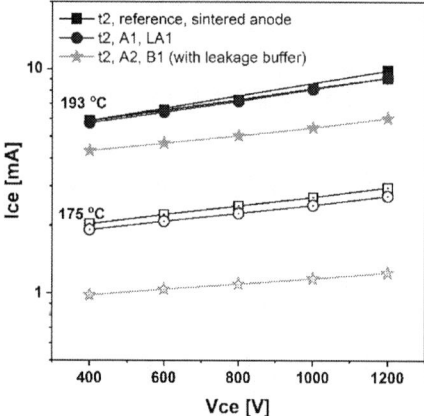

Fig. 5. Leakage current as a function of Vce measured up to 193 °C. Two IGBTs with and without a leakage buffer at overlapping technology points (std. thickness t2) are shown. The leakage current can be improved while maintaining the same switching losses and on-state characteristics by the addition of a shallow buffer phosphorous peak.

Fig. 6 provides the spreading resistance profile (SRP) which shows the activated carrier concentration in the backside buffer/anode side for a reference IGBT (std. DD buffer) and with the addition of a phosphorous leakage buffer peak at different depths respectively.

978-1-5386-2928-4/18 $31.00 © 2018 IEEE

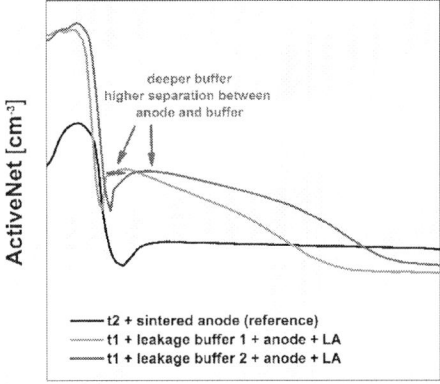

Fig. 6. Spreading resistance profile (SRP) for backside IGBTs with a std. SPT buffer (black - reference) and with the addition of an ultra-shallow leakage buffer produced by phosphorus implantation and laser anneal at a reduced wafer thickness (t1). Red curve shows device with deeper buffer peak and higher separation between anode and buffer.

B. RBSOA Capability up to 200°C

For the elevated temperature of 200°C measurements, the IGBT/diode substrate is mounted on a heating plate to maintain the device within the temperature range of interest (measured with thermocouples). A thermally conductive foil was used in order to minimize the thermal resistance of the interface between the substrate and the heat sink.

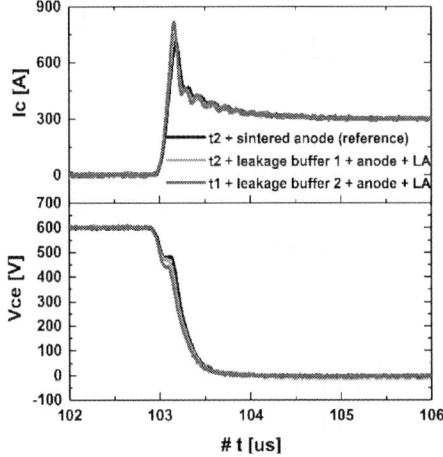

Fig. 7. Turn-on waveforms measured under double nominal conditions (600 V, 300 A, Ls = 70 nH, Rg = 8.2 ohm) at 200 °C for reference device, device with highest leakage reduction (green) and reduced deeper buffer peak (red).

Ruggedness under hard switching conditions and high operation temperatures was possible by the simple addition of the leakage buffer peak. Fig. 7 and 8 show the turn-on and turn-off waveforms for a reference device, an IGBT with the highest leakage reduction (green) and with deeper buffer peak (red) at twice nominal current and RBSOA conditions (200°C), respectively. Only devices with a leakage buffer allow for soft-turn off switching up to 200°C for the standard RBSOA conditions. Furthermore, the design with the highest leakage

reduction can be turned-off at 6X nominal current (Fig. 9) without leading to destruction.

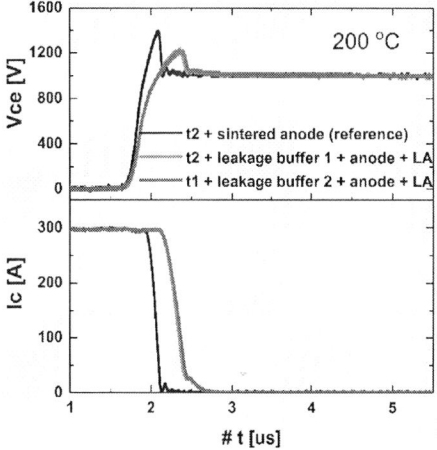

Fig. 8. Turn-off waveforms measured under RBSOA conditions (1000 V, 300 A, Ls = 70 nH, Rg = 8.2 ohm) at 200 °C for reference device, device with highest leakage reduction (green) and deeper buffer peak (red).

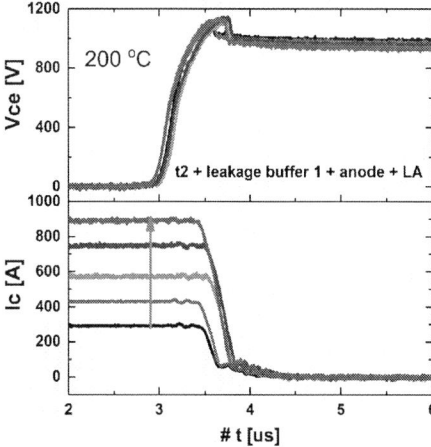

Fig. 9. Extended turn-off RBSOA waveforms up to 6X nominal conditions (1000 V, 900 A, Ls = 70 nH, Rg = 8.2 ohm) measured at 200 °C for IGBT with leakage buffer and highest leakage reduction (std. thickness t2). Reference SPT IGBT fails due to thermal runaway (not plotted here).

C. Short Circuit capability up to 200°C

We have investigated the IGBT's thermal destruction under short circuit operation where the temperature increase during testing is considerable. During short-circuit experiments, the power device is subjected to a high current and subsequently much higher temperatures due to self-heating. The dissipated power is continuous and is created by conduction and switching losses and undoubtedly causes the junction and overall temperature of the device to rise above 250 °C. Temperature stabilization is therefore difficult due to the slow internal control feedback system and non-existing effective cooling (Fig. 10). Increasing short circuit pulse cycles have been applied until the IGBT is destroyed. Destruction is induced by a very high instantaneous power loss and increased device temperature.

The short circuit SOA test is also a way to prove stability at high temperatures. Fig. 10 shows the characteristic waveforms of an increased short circuit pulse duration in which a 1200V and 1700V device were able to withstand at 200°C (a, b). The device sustains short-circuit pulses with a maximum time of 11 μs for 1200 V and 1700 V rated IGBTs. It is known in literature that the destruction of an IGBT under short circuit is induced by thermal runaway [6]. The capability of the devices with a leakage buffer to withstand long short-circuit current pulses as long as 11 μs applied at 200 °C demonstrate the higher operation temperature capability.

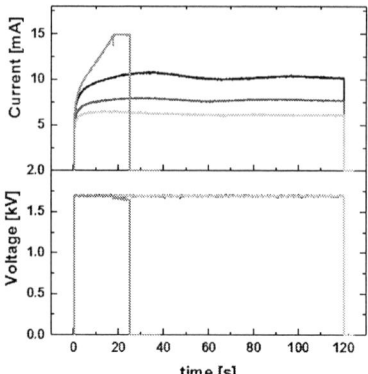

Fig. 11. Leakage current stability measurements (1700 V) at increasing heating plate temperatures from 182 °C. The device stabilizes over time as long as the leakage current stays below 10 mA where the device could fail due to thermal runaway.

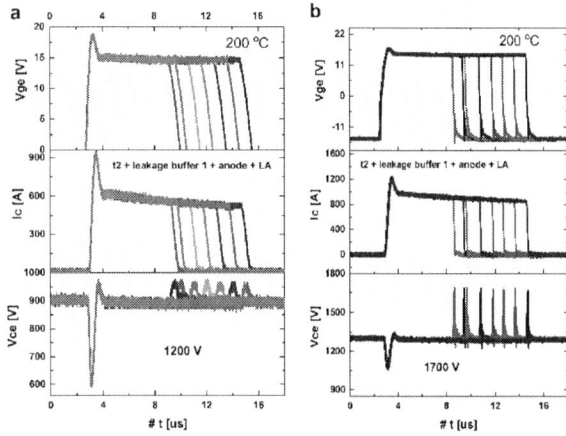

Fig. 10. Short-circuit waveforms (900 V, Ls = 70 nH, Rg = 8.2 ohm, 1200V (a) and 1300 V, Ls = 70 nH, Rg = 8.2 ohm, 1700V (a) at increasing pulses measured at 200 °C for IGBT with leakage buffer and highest leakage reduction (std. thickness t2). These device survives short-circuit pulses up to 11 μs at 200 °C.

D. HTRB stability

Stability measurements were also performed at 200°C (Fig. 11 for 1700 V device) for the High Temperature Reverse Bias HTRB tests. The leakage current is measured while the heating plate temperature is increased from 182 °C to 195 °C and applying a blocking voltage potential of 1700 V. The voltage is applied for 2 minutes to see if the device is thermally running away or manages to self-stabilize together with the system heating. The device can be stabilized as long as the leakage current stays below 10 mA (black curve, Fig. 11). Nevertheless, this value is more a test system limitation than a device limitation. The test system cannot cope with the increase of leakage current and therefore the temperature increases because its reaction in comparison is too slow. During the applied voltage, the temperature increase can be monitored through thermocouples. This increase is visible in the first step-up of the measurement duration, where the leakage current increases and after self-stabilization it drops back to its normal leakage value. In Fig. 11, it can be seen also that the leakage can rise by 30% in comparison to a short pulse value. The measurement of the temperature is monitored at the heating plate which is normally different from the real chip temperature. Therefore when the heating plate monitors a temperature increase of 0.5°C it can be assumed that a rise above 2°C happened on the device due to the leakage current.

I. Conclusions

The new buffer design has been demonstrated to fulfill the existing demands for 1200 and 1700 V IGBTs to extend their maximum operating temperature up to 200°C. A shallow phosphorus buffer peak added to a state-of-the-art planar SPT IGBT can lower the leakage current and expand the operation range up to 200°C. The new buffer design was experimentally demonstrated to provide rugged switching (RBSOA and SCSOA) up to 200°C. The power semiconductor's maximum operating temperature is important in circuit designs as thermal cycling affects the device performance and long term reliability. Nonetheless, high temperature operation capability of an IGBT is but one component that is necessary for the proper thermal management of power electronics converters and it is therefore critical that the system is considered as a whole. Therefore, research in packaging solutions for improved power dissipation must be conducted in parallel.

Acknowledgments

The authors wish to thank Sumitomo Heavy Industries for laser anneal support, J. Lauri, B. Karadaghi, W. Janisch and R. Jabrany for the frontend process support.

References

[1] M. Rahimo et al., " Novel Soft-Punch-Through (SPT) 1700V IGBT Sets Benchmark on Technology Curve," PCIM, Nuremberg, 2001.

[2] J. Vobecky et al., "Exploring the Silicon Design Limits of Thin Wafer IGBT Technology: The Controlled Punch Through (CPT) IGBT," ISPSD, Orlando, 2008.

[3] M. Rahimo et al., "Extending the boundary limits of high voltage IGBTs and diodes to above 8 kV," ISPSD, Santa Fe, 2002.

[4] K. Nakamura et al., "Advanced wide cell pitch CSTBTs having light punch-through (LPT) structures," ISPSD, Santa Fe, 2002.

[5] U. Schlapbach, et al., "1200V IGBTs operating at 200°C? An investigation on the potentials and the design constraints," ISPSD, Jeju Island, 2007.

[6] A. Ammous et al., "Transient temperature measurements and modeling of IGBT's under short circuit," in IEEE Transactions on Power Electronics, vol. 13, no. 1, pp. 12-25, Jan 1998.

Proceedings of the 30th International Symposium on Power Semiconductor Devices & ICs
May 13-17, 2018, Chicago, USA

Investigation of the mechanism of gate voltage oscillation in 1.2kV IGBT under short circuit condition

Takuo Kikuchi

Corporate Manufacturing Engineering
Center, Toshiba Corporation
33, Shin-Isogo-cho, Isogo-ku,
Yokohama 235-0017, Japan
takuo1.kikuchi@toshiba.co.jp

Kazutoshi Nakamura

Discrete Semiconductor Div., Toshiba
Electronic Devices & Storage Corporation
1-1, Iwauchi-machi, Nomi, Ishikawa
923-1293, Japan

Kazuto Takao

Corporate Research & Development
Center, Toshiba Corporation
31, Komukai-Toshiba-cho, Saiwai-
ku, Kawasaki 212-8582, Japan

Abstract—**The mechanism of the gate oscillation under short circuit condition was studied by experiment and simulation. In the both ways, dependence of input DC-link voltage V_{cc} and stray inductance L_e on the gate oscillation was examined to clarify the cause. It has been found that under short circuit a high electric field formed in the collector side and hole transit time delay plays a critical role for the onset of oscillation. Based on the model, its parametric dependence can be clearly interpreted. In addition, the design to suppress the oscillation has been discussed.**

Keywords—IGBT, TCAD, short circuit, oscillation, transit time

I. INTRODUCTION

The gate voltage oscillation in Insulated Gate Bipolar Transistor (IGBT) under short circuit condition is one of the instability phenomena which limit device operation. This problem causes harmful effects like EMI noise on adjacent circuits and even device destruction [1]. Although various models on this failure mode have been proposed such as IMPATT oscillation [2, 3], negative capacitance [4], PETT oscillation [5] and parasitic inductance in module [6], the detailed mechanism, especially its dependence of operating condition and device design is yet to be well understood.

In this paper, the mechanism of the gate oscillation has been investigated both experimentally and in simulation. In measurement, a 1.2kV trench gate IGBT chip was tested under

short circuit with different operating conditions. In simulation, TCAD model coupled with the circuit simulation was used to reproduce the measured oscillation and to investigate the behavior of the electric field and carrier distribution in the device during the oscillation. These results clarified the detailed condition for the occurance of the gate oscillation.

II. METHODOLOGY

A. Experimental setup

Figure1 shows the system and the equivalent circuit for the short circuit measurement. The system consists of a 1.2kV IGBT chip, a gate drive circuit, a condenser and CT. The device is a field-stop IGBT with a trench gate and n-buffer in the collector side. In measurement, 15V gate pulse with duration of 10µs was given to the gate electrode. The gate voltage V_g, the collector voltage V_{ce} and the collector current I_c were measured. To investigate the effect of operating condition, the dependence of input DC-link voltage V_{cc} and emitter stray inductance were obtained, changing the length of the emitter wire. Furthermore, the oscillation frequency and the current at the onset of the gate oscillation were well examined.

B. Simulation model

Figure2(a) shows the device simulation model of the 1.2kV

(a)   (b)

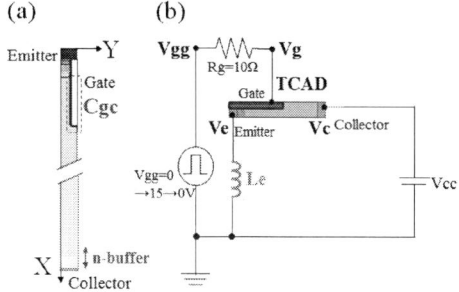

Figure 1: Gate oscillation measurement under short circuit. (a) System and (b) the equivalent circuit.

(a)   (b)

Figure 2: Simulation model.
(a) Device structure and (b) short circuit model.

978-1-5386-2928-4/18 $31.00 © 2018 IEEE   503

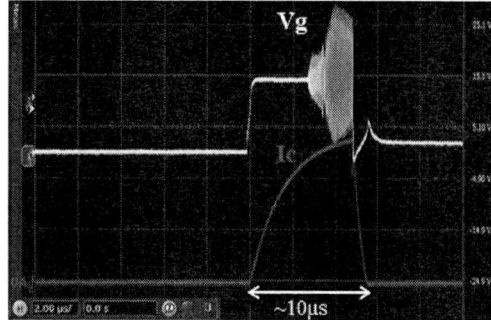

Figure 3: Measured gate oscillation at V_{cc}=500V.

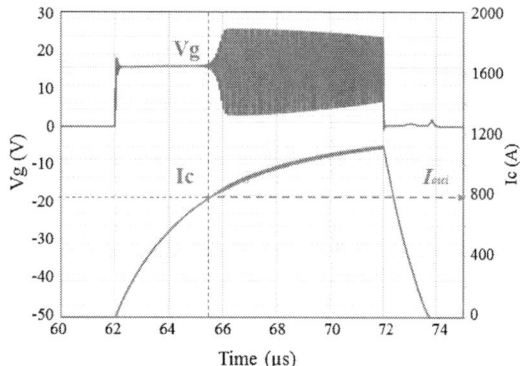

Figure 4: Simulated gate oscillation at V_{cc}=500V.

IGBT, including its vertical carrier profile. The model was calibrated to fit the static characteristics of the tested device. Figure2(b) shows circuit-coupled simulation model, including the external circuit shown in Fig. 1. In this model, the gate resistance R_g=10 Ω and the emitter inductance L_e were also considered as additional circuit elements. As for the physical model used in TCAD, a field saturation velocity model was considered whereas an impact ionization model was switched-off to see whether IMPATT mode is related to the oscillation.

III. EXPERIMENTAL AND SIMULATION RESULTS

Figure3 shows the experimental results at DC-link voltage V_{cc}=500V. When the gate pulse is applied to the gate, I_c starts to increase. Once I_c reaches a threshold current, the gate oscillation with increasing amplitude starts while the oscillation of I_c is invisibly small.

Figure4 shows the simulated gate oscillation at V_{cc}=500V. The oscillation starts around 65μs with the amplitude of 12V. The experimental results were qualitatively well reproduced in simulation without consideration of Impact Ionization, indicating IMPATT oscillation is not the root cause of the observed oscillation. Here the collector current at the time when the gate oscillation starts was defined as an oscillation onset current I_{osci} and plotted as a function of V_{cc} in Figure5. I_{osci} parabolically changes and has the minimum around 300V

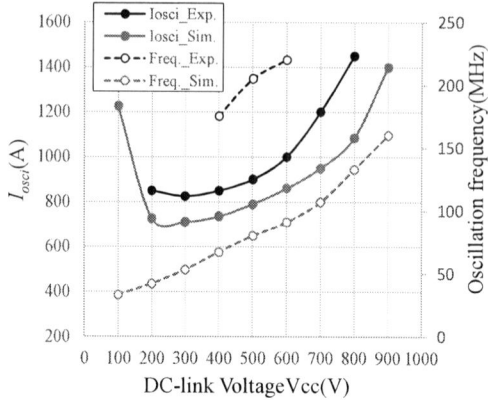

Figure 5: The oscillation onset current I_{osci} and the frequency as a function of V_{cc} in experiment and simulation.

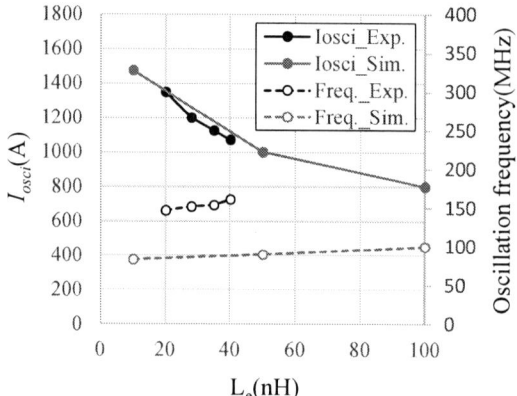

Figure 6: The gate oscillation frequency as a function of emitter parasitic inductance Le in experiment and simulation. Le was adjusted by changing the emitter wire length in experiment.

in both experiment and simulation. It should be noted that I_{osci} exhibits a sharp increase below 200V in simulation although such low electric field conditions have been reported to be weak to the gate oscillation [7]. The reason for this result is described later. The oscillation frequency increases with V_{cc} in both experiment and simulation.

Figure6 shows the I_{osci} and the oscillation frequency as a function of the emitter inductance L_e. I_{osci} decreases with increasing L_e, showing L_e has a significant influence on the onset of the oscillation. On the other hand, the frequency only slightly increases with larger L_e, indicating the oscillation frequency is not determined by LC resonant frequency since it should decrease with larger inductance. In order to understand the oscillation phenomena and its dependence of V_{cc} and L_e, the electric field and carrier distribution inside the device just before and during the oscillation were investigated in simulation.

978-1-5386-2928-4/18 $31.00 © 2018 IEEE   504

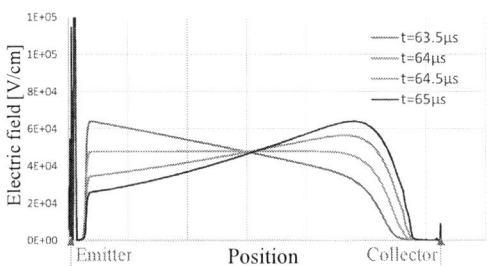

Figure 7: Time evolution of electric field distribution along device depth direction at V_{cc}=500V.

Figure7 shows the time evolution of the electric field distribution along the depth direction of the device at V_{cc}=500V until the oscillation started at 65μs in Fig. 4. A high electric field initially appears in the p-base and n-base junction or the emitter side. As I_c increases, a high field appears in the collector side until it finally reaches I_{osci} as shown in Fig. 4, indicating that the change of the electric field shape is linked to the onset of the gate oscillation.

Figure8 shows the collector current I_c at which the electric field peak in the collector side E_c became the same or twice as high as that in the emitter side E_e with different V_{cc}, which was compared with I_{osci} shown in Fig. 5. The obtained I_c and I_{osci} have a good agreement with each other and both have the minimum value at V_{cc}=300V. This result can be qualitatively interpreted by the following equation with the net charge ρ in the n-base and the anode efficiency γ ($= J/J_p$) [8].

$$p = J_p/qv_h, \quad n = J_n/qv_e \tag{1}$$

$$\rho = N_D + p\text{-}n = N_D + (\gamma/v_h + (\gamma\text{-}1)/v_e)J/q \tag{2}$$

where p and n denote hole and electron densities, v_h and v_e denote hole and electron velocities, respectively. In the following discussion, it has been assumed that the second term in Eq.(2) is negative and that γ is constant. At higher voltages (more than 300V), the carrier velocity v_h and v_e increases with increasing V_{cc} and it takes more current density J in the second term in Eq.(2) to turn the net charge ρ in the n-base to negative. On the other hand, at low voltages (100~300V), both hole and electron velocity don't saturate and are rather determined by these mobility. Since the mobility of electron is about three times higher than hole, hole positive charge prevails and the current J to turn the n-base to negative becomes higher, resulting in the parabolic dependence. The derived I_c from the electric field shape has a good agreement with I_{osci}, indicating that the high electric field in the collector side is associated with the onset of the oscillation.

Figure 9(a) is the simulated electric field along the X direction of the device (in Fig. 2) just before oscillation starts at different V_{cc}. In all the electric field distribution, high electric fields were formed in the collector side. Figure 9(b) shows the electron and hole velocity. At V_{cc}= 400V and more voltages, the both electron and hole velocities were close to the saturation velocities, 1×10^7 and 7×10^6 cm/s respectively. Below 200V, however, the velocities were less saturated and the electron velocity was about two times higher than the hole

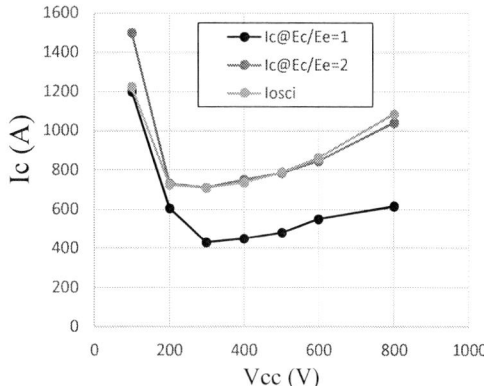

Figure 8: The relationship between electric field shape and the current at which oscillation starts.

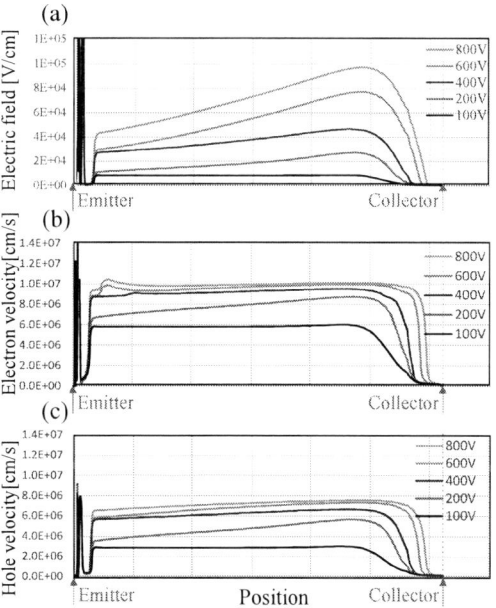

Figure 9: Simulated results. (a)Electric field, (b) electron velocity and (c) hole velocity along the device depth direction just before the gate oscillation starts.

velocity, validating the previous discussion in Fig. 8. In order to see the effect of the electric field shape, especially on capacitive effect, the carrier distribution in the device was examined. Figure10 shows the time evolution of hole and electron distribution at the gate-collector overlapping area (C_{gc}) shown in Fig. 2. The hole density increases with time as a result of the high electric field in the collector side which in turn lowers the electric field in the emitter side. On the other hand, the change of electron density is much smaller than that of holes, indicating that in terms of the impact on the gate capacitance, hole plays more significant role.

978-1-5386-2928-4/18 $31.00 © 2018 IEEE

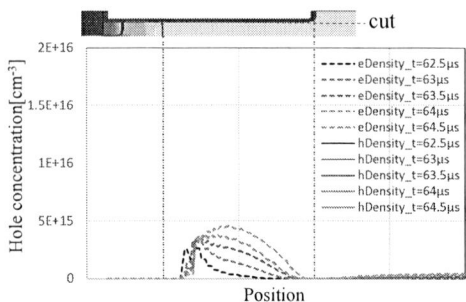

Figure 10: Time evolution of electron and hole distribution at the C_{gc} area in Fig. 2 along the device depth direction at V_{cc}=500V before the gate oscillation.

Figure11 shows the hole distribution and input capacitance C_i during the gate oscillation. It has been known that hole accumulation at C_{gc} under high V_g and V_{cg} gives rise to the reduction of gate input capacitance C_i (= C_{ge} + C_{gc}) [4]. Figure11(b) shows the C_i transient during the oscillation. The rise and fall of the C_i correspond to the decrease and increase of hole density at C_{gc} region in Fig. 11(b).

Figure12(a) shows V_{cg}, C_i, I_g and V_g during the oscillation respectively, focusing on the current loop in Fig. 12(b). Since C_i decreases with increasing V_{cg} with $\pi/2$ phase delay, the total phase difference is $3\pi/2$. The C_i induces the gate current I_g derived in the following equation, assuming C_{ge} is constant.

$$I_g = dQ/dt = d(CV)/dt$$
$$= C_{ge} \times dV_{ge}/dt + C_{gc} \times dV_{gc}/dt + V_{gc} \times dC_{gc}/dt \quad (3)$$

When dV_{gc}/dt (= $-dV_{cg}/dt$) is positive, dC_{gc}/dt becomes positive[4]. Since V_{gc} is always negative, the last term always has an opposite sign to the rest of the terms. If the current in the last term is dominant like in this case, the resulting gate current Ig goes ahead of C_i by $\pi/2$. It should be noted that Ig with this phase can flow even if C_{gc} is NOT negative. The resulting

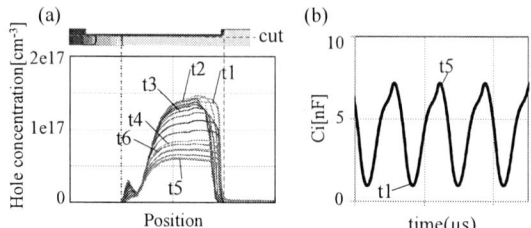

Figure 11: Time evolution of simulated hole concentration during the oscillation (a) at the gate and (b) the corresponding input capacitance C_i change.

phase difference between V_{cg} and I_g is π, indicating the presence of negative resistance which causes instability. The V_g oscillation is the product $R_g \times I_g$ and changes V_{cg} again to keep undamped oscillation. Therefore, the $\pi/2$ phase delay between V_{cg} and C_i leads to the negative resistance and the resulting gate oscillation. The phase delay can be estimated from the hole transit time (device thickness divided by carrier velocity) using the hole velocity in Fig. 9(c) and the resulting frequency was compared with the simulated oscillation frequency in Figure13. The both frequencies increase with higher V_{cc} and have an good agreement with each other, confirming the delay is attributed to the hole transit time.

IV. DISCUSSION

Based on the results shown above, the mechanism of the gate oscillation, its dependence on operating condition is summarized here. In addition, the device design to suppress the gate oscillation are discussed in this section.

A. The mechanism of the gate oscillation

- Under short circuit condition, as the collector current increases, a high electric field appears in the collector side as a result of electron build-up in n-base region.

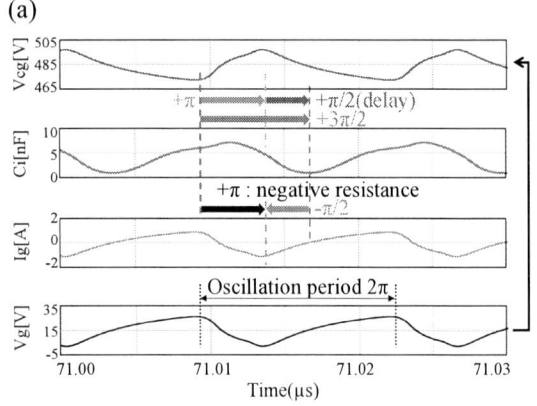

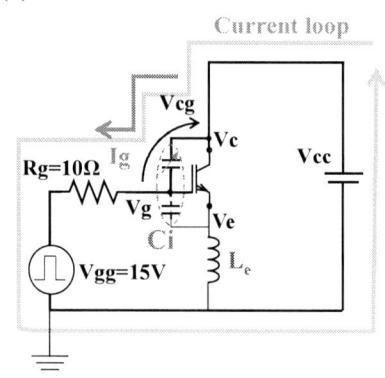

Figure 12: Simulated waveforms when V_{cc}=500V. (a)V_{cg}, C_i, I_g and V_g during oscillation and those phase relationships and (b) the corresponding schematic circuit diagram with terminal voltages and current flows. The product $R_g \times I_g$ was observed as the gate oscillation.

978-1-5386-2928-4/18 $31.00 © 2018 IEEE

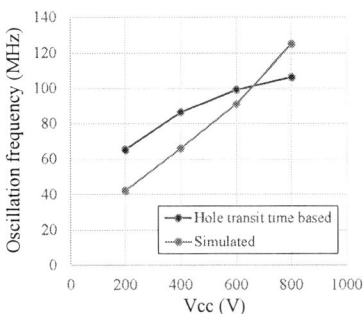

Figure 13: The relationship between the simulated gate oscillation frequency and the oscillation frequency derived by the hole transit time delay (device thickness divided by hole velocity shown in Fig. 9(c)).

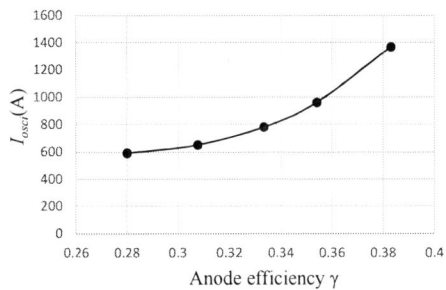

Figure 14: The simulated oscillation onset current I_{osci} as a function of anode efficiency γ.

- The high electric field in the collector side lowers the electric field in the emitter side, which results in the increase of the hole density at C_{gc} region.

- Once some V_{cg} fluctuation occurs in the device, it changes the hole density at the gate and induces the change of C_{gc} with the phase delay of $\pi/2$ due to the hole transit time. The resulting I_g with the phase difference π to V_{cg} causes the gate oscillation.

B. Dependence of V_{cc} and L_e on I_{osci}

As shown in Fig. 7, a high electric field in the collector side is strongly associated with the onset of the gate oscillation and I_{osci}. The collector current I_c for forming the electric field shape has a parabolic dependence on V_{cc} as shown in Fig. 8 and the reason was already discussed based on Eq(2). Therefore, the same dependency can also hold between I_{osci} and V_{cc}. Regarding the L_e dependence, it seems to have nothing to do with the phenomena inside the device. However, a larger L_e impedes the high frequency portion of the displacement current, $C_{ge} \times dV_{ge}/dt$ in the Eq.(3) to flow into the emitter in Fig. 12(b). As a result, the last term $V_{gc} \times dC_{gc}/dt$ with an opposite sign to the rest becomes more dominant, which leads to the onset of the gate oscillation with smaller I_{osci} as shown in Fig. 6.

C. Suppresing the gate oscillation

There are two ways to tackle the oscillation problem, i.e. at the circuit level and at the device level. At the circuit level, reducing L_e is an effective countermeasure as previously discussed. At the device level, it is essential to avoid a high field in the collector side. For this purpose, increasing hole or decreasing electron injection during short circuit is effective. Figure14 shows the result of simulated I_{osci} in which the anode efficiency γ was varied by changing the device design. I_{osci} increases with larger γ. The higher is the hole injection, the less electrons prevail in n-base, which prevents the n-base region from being negatively charged and leads to increasing I_{osci}.

V. CONCLUSION

We investigate the gate oscillation under short circuit with a 1.2kV trench gate IGBT chip with both experimental and simulation approach. It has been found that I_{osci}, the collector current at which the gate oscillation starts had a parabolic dependence on V_{cc} and that a high emitter stray inductance L_e lowered the current. A high electric field in the collector side as the result of electron building up in n-base has been strongly associated with the onset of the gate oscillation. The hole accumulation in C_{gc} region and the resulting C_{gc} change causes the gate oscillation. The oscillation frequency was determined by the hole transit time across the device. The dependence of V_{cc} and emitter stray inductance L_e on the oscillation can be clearly interpreted by the proposed model. Furthermore, it has been demonstrated that the larger anode efficiency γ effectively increases I_{osci} and could be an effective countermeasure to suppress the gate oscillation.

REFERENCES

[1] R. Siemieniec, J. Lutz, M. Netzel, and P. Mourick, "Transit time oscillations as a source of EMC problems in bipolar power devices", EPE'03, 2003.

[2] R. Siemienies, J. Lutz and R. Herzer "Analysis of Dynamic Impatt Oscillations caused by Radiation Induced Deep Centers" , ISPSD'03, pp.283-286, 2003.

[3] T. Hong, F. Pfirsch, B. Reinhold, J. Lutz, and D. Silber, "Transient avalanche oscillation of IGBTs under high current", ISPSD'14, pp.43-46, 2014.

[4] I. Omura, W. Fichtner and H. Ohashi, "Oscillation Effects in IGBT's Related to Negative Capacitance Phenomena", IEEE transaction on electron devices, vol.46, No.1(1999).

[5] R. Siemieniec, P. Mourick, M. Netzel, and Josef Lutz, "The Plasma Extraction Transit-Time Oscillation in Bipolar Power Devices-Mechanism, EMC Effects, and Prevention", IEEE TRANSACTIONS ON ELECTRON DEVICES, vol. 53, No.2, pp.369-379, 2006.

[6] T. Ohi, A. Iwata, K. Arai, "Investigation of gate voltage oscillations in an IGBT module under short circuit conditions" , PESC'02, pp.1758-1763, 2002.

[7] P.Diaz, F. Iannuzzo and M. Rahimo, "TCAD Analysis of Short-Circuit Oscillations in IGBTs", ISPSD'17, pp.151-154, 2017.

[8] A. Nakagawa, T. Matsudai, T. Matsuda, M. Yamaguchi and T. Ogura, "MOSFET-mode Ultra-Thin Wafer PTIGBTs for Soft Switching Application --- Theory and Experiments" , ISPSD'03, pp.103-106, 2003.

Proceedings of the 30th International Symposium on Power Semiconductor Devices & ICs
May 13-17, 2018, Chicago, USA

Design of LED Driver ICs for High-Performance Miniaturized Lighting Systems

Yuan Gao, Lisong Li, and Philip K.T. Mok

Department of Electronic and Computer Engineering, The Hong Kong University of Science of Technology
Clear Water Bay, Kowloon, Hong Kong
Emails: ygaoad@connect.ust.hk, lliar@connect.ust.hk and eemok@ust.hk

Abstract—This paper presents several high-performance LED driver ICs for AC-powered miniaturized lighting systems. With the help of presented techniques, the value of bulky inductors can be effectively reduced for the switching-converter-based LED drivers, while the flicker reduction and visible-light communication can be integrated into the compact inductor-less drivers. In addition, a hybrid driver with improved efficiency is achieved by combing two conventional driving approaches.

Keywords—LED driver, AC-DC, inductance reduction, inductor-less driver, hybrid driver.

I. INTRODUCTION

The past decade witnessed the unprecedented growth of the light-emitting diode (LED) lighting market. Besides the advantages of the long lifespan and supreme energy to light efficiency, the continuous reduction of the LED cost becomes the main driving force for this rapid market growth. Because LED shows distinct electrical characteristics from its conventional counterparts, driver circuits are needed to match the LEDs with the AC mains. A good LED driver should not only have little impact on the lighting quality of LEDs but also be with minimized hardware complexity and cost. Therefore, it is always desired to miniaturize the lighting system by reducing the sizes, the form factor or even ruling out the off-chip components inside the drivers.

There are two main AC input LED driving approaches. The first approach (switching-converter-based driver) scales down the rectified input with switching converters, as the example of the buck converter shown in Fig. 1(a), and then powers up the low voltage LEDs with the output of the converter [1], [2]. The second approach (inductor-less driver), as shown in Fig. 1(b), stacks a group of LEDs to get a high voltage, so the LEDs can be directly powered by the rectified AC input [3], [4]. Meanwhile, power transistors are used to segment and bypass some of the LEDs according to the change of the input voltage. In this paper, design considerations and developed techniques for both driving approaches will be discussed. In Section II, the main focus is to reduce the inductance of bulky off-chip magnetics as well as the system size of the switching-converter-based drivers. In Section III, besides the system integration, techniques of flicker reduction and visible-light communication will be introduced for high performance inductor-less drivers.

This work was partially supported by a grant from the Research Grants Council of the Hong Kong SAR, China (Project No. T23-612/12-R).

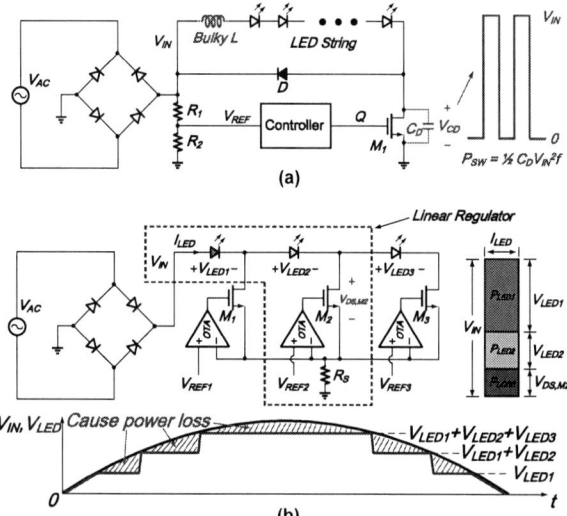

Fig. 1. Two types of AC input LED drivers. (a) Switching-converter-based driver. (b) Inductor-less driver.

Finally, a hybrid driver, which combines the merits of both two conventional driving approaches will be introduced.

II. INDUCTANCE REDUCTION FOR DRIVING TOPOLOGIES WITH SWITCHING CONVERTER

The power inductors, as essential off-chip components in the switching converter-based LED drivers, are not only expensive but also take up a lot of PCB space, especially in high voltage applications (usually in the hundreds of µH range), making the system bulky and less cost-effective. Increasing the switching frequency could reduce inductor values, but high switching frequency also leads to large switching loss because of the parasitic capacitor of the power switches.

A. Buck LED Driver in DCM Operation

Fig. 2 illustrates one way to achieve inductance reduction by operating the switching converter in the deep discontinuous conduction-mode (DCM). The single-stage buck LED driver, integrating both driving controller and power switch on a monolithic IC, has been fabricated in a 0.35-µm 700-V BCD process. The required inductance is reduced to µH range for high voltage AC power supply with moderate frequency of 1-MHz. Meanwhile, as shown in Fig. 2, the driver can find the optimal point to switch on the on-chip power transistor such that the wasted energy on the parasitic capacitor can be minimized. The DCM LED driver is measured with a commercial 4.7-µH inductor to power up 28 commercial LEDs

978-1-5386-2928-4/18 $31.00 © 2018 IEEE

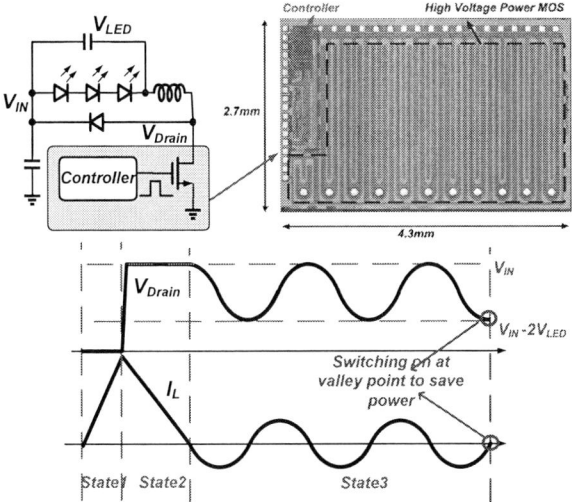

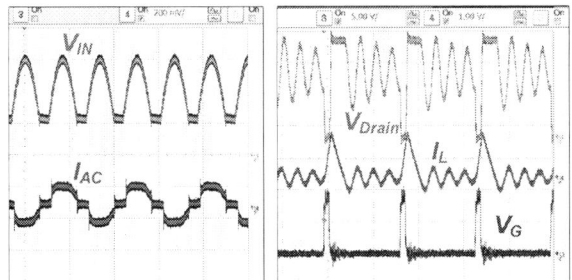

Fig. 2. Schematic, chip photo and key waveforms of the DCM driver.

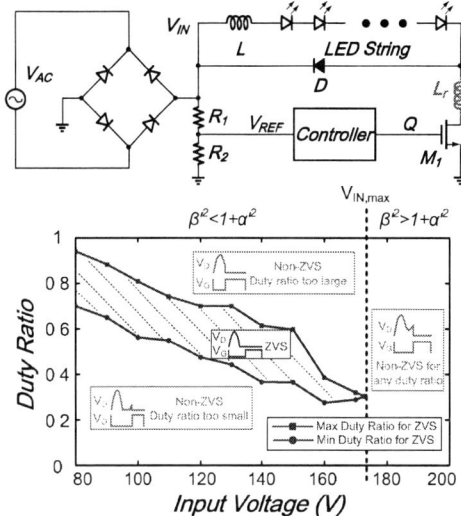

Fig. 3. Measured waveforms of the single-stage DCM driver.

in series and Fig. 3 shows some typical waveforms of the driver. Experimental results show that for the 110 V_{AC} input voltage and a 10-W output power, the driver achieves over 0.97 power factor and the overall efficiency are over 80%. With this inductance reduction method, it is possible to use the µH range silicon embedded inductors [5] in the design.

B. Quasi-Resonant LED Driver

Quasi-resonant converter, as shown in Fig. 4, is another possible method to achieve a small system form factor. Although an extra inductor L_r is needed in this topology, the total required inductance can be greatly reduced by pushing the operation frequency. Meanwhile, a good efficiency can be maintained with the help of zero-voltage switching (ZVS). However, as shown in Fig. 4, the ZVS condition cannot always be effective for varied input voltage with a fixed duty ratio, which poses challenges for the design of AC input quasi-resonant (QR) driver. To target for this issue, an auto-ZVS QR LED driver is developed, as illustrated in Fig. 5, in which the controller adaptively adjusts the duty ratio to ensure the zero voltage on the drain of power switch before it turns on [2]. Thus, the switching loss due to parasitic drain capacitance C_d can be eliminated in the entire operation range. Besides this, a 600V GaN HEMT is employed as the power switch to further improve the system efficiency. In addition, a fully integrated shunt protector is added to bypass the failed LEDs in series-connected LED strings to prolong the overall lifetime of the LED string. The driver IC, fabricated with a 0.35-µm 120-V

high-voltage process, can provide up to 25 W power to the LED with 2×3.3 µH inductors and achieves 91.4% peak efficiency and 0.973 peak power factor from 60 Hz 100–120 V_{AC} input. The chip photo is also included in Fig. 5 and the key measured waveforms are shown in Fig. 6.

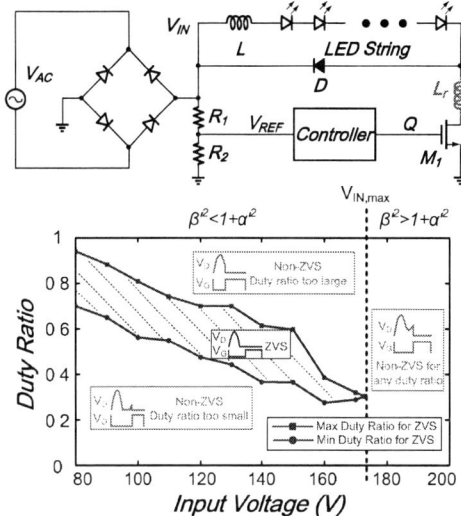

Fig. 4. Quasi-resonant buck converter and the conditions for ZVS (I_{LED} = 350 mA, L = L_r = 3.3µH).

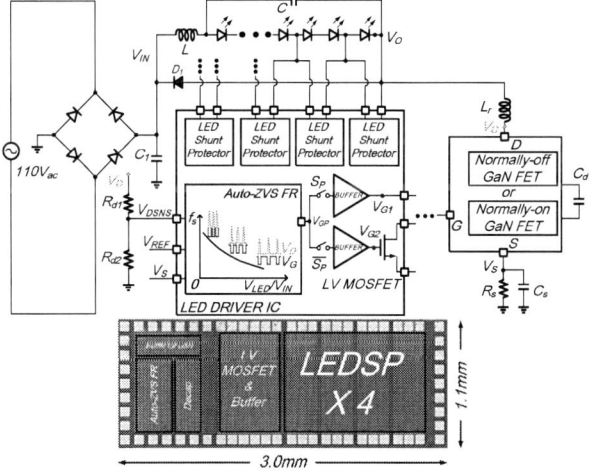

Fig. 5. System architecture and chip photo of the auto-ZVS quasi-resonant LED driver.

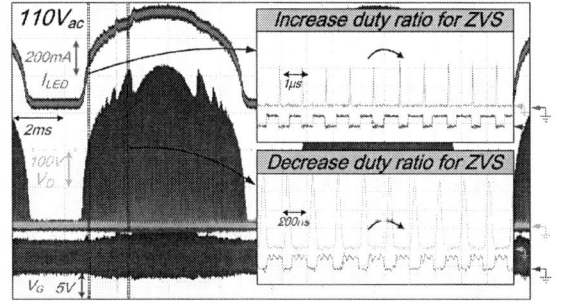

Fig. 6. Measured waveforms of the auto-ZVS quasi-resonant LED driver.

978-1-5386-2928-4/18 $31.00 © 2018 IEEE   509

III. Performance Improvement for Inductor-Less Driving Topologies

The inductor-less LED driver shows great potential to achieve high-level integration because it rules out the bulky magnetics. In addition, it does not require high-frequency switching, so the switching related losses are eliminated. As a result, for the low to moderate power applications, the inductor-less driver can achieve comparable efficiency to the switching-converter-based drivers.

A. Inductor-Less LED Driver with Low Flicker

The flicker is one of the major issue for the conventional inductor-less drivers. The customers will suffer from harmful 100% lighting flicker at the double-line-frequency (100 Hz or 120 Hz) with these designs in their LED lamps [6]. A novel inductor-less LED driving solution with high power efficiency, good power factor and small flicker is developed and the concept is illustrated in Fig. 7 [3]. The key idea for this approach is to regulate the total LED power instead of the LED current. When the voltage on the LED string changes, a variable number of LEDs are powered up accordingly to ensure efficient power delivery. Meanwhile, the controller will push the LED current in the opposite direction to keep the total LED power almost constant to minimize the flicker. Fig. 8 shows the chip micrograph and the test PCB. Fig. 9 illustrates the measured flicker and efficiency at different conditions. Experimental results show that the developed driver can achieve an over 0.9 power factor, 17.3% flicker and 87.6% peak efficiency. An over 80% reduction of flicker is achieved compared to the conventional inductor-less LED driver.

B. Inductor-Less VLC LED Driver

In addition to providing efficient illumination, the fast switching capability of LED can be utilized for high-speed

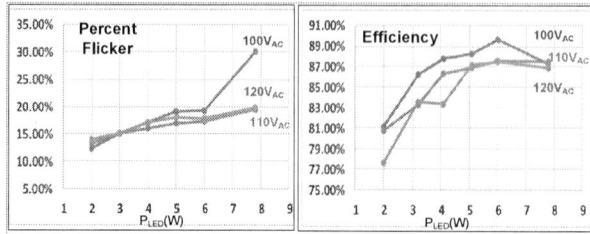

Fig. 9. Measurement results of percent flicker and efficiency at different input voltages and output powers.

visible light communication (VLC). Because of the LED current slew rate limitation, the conventional drivers can hardly support the high frequency modulation. In order to solve this issue, an inductor-less LED driver with fast dimming capability and low optical flicker is developed to simultaneously provide efficient lighting and data transmission [4]. As shown in Fig. 10, the driving topology is similar to the low-flicker inductor-less solution while the LED current regulation is accomplished by adjusting the digitalized power stage with a digital controller. In addition, the digital controller can achieve smooth transitions among the multiple current paths, while the keep-and-restore technique and auxiliary turn-on switch are included to further increase the switching speed. The VLC input DATA, modulated with on-off keying non-return-to-zero code, enables the controller when it is "1", and deactivates the power switches when it is "0".

The developed driver IC was fabricated in a 0.35-μm 120-V CMOS process and the test setup is shown in Fig. 11. The

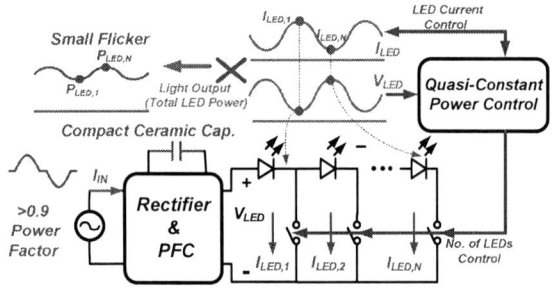

Fig. 7. Design concept of the inductor-less driver with low flicker.

Fig. 8. Test PCB and chip micrograph of the inductor-less LED driver with low flicker.

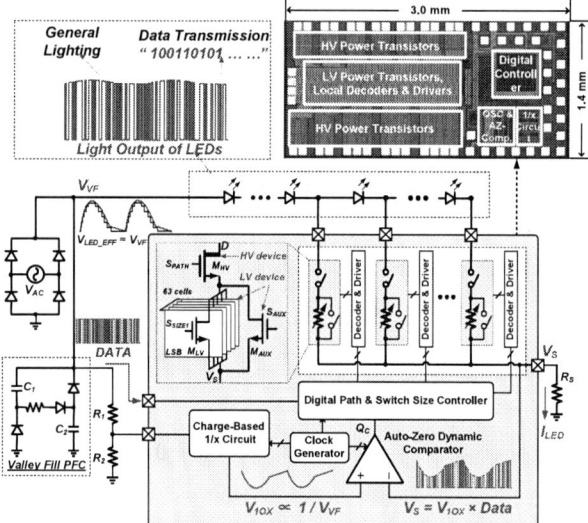

Fig. 10. System architecture and chip photo of the inductor-less VLC LED driver.

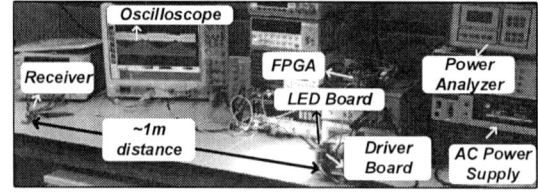

Fig. 11. Test setup of the VLC LED driver.

978-1-5386-2928-4/18 $31.00 © 2018 IEEE

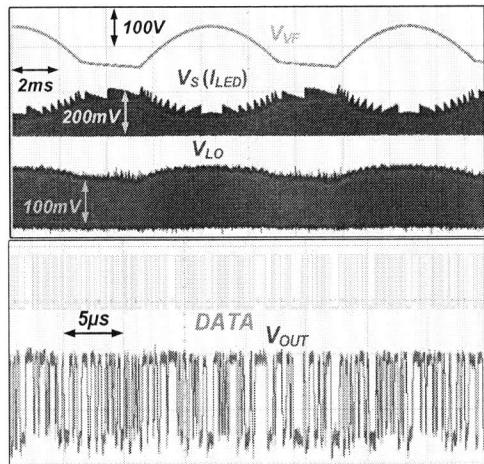

Fig. 12. Measured key waveforms for lighting and data transmission.

measurement results under 8 Mb/s data rate are given in Fig. 12, showing a good match between input data and received signal.

IV. HYBRID LED DRIVER

As shown in Fig. 1, the switching-converter-based LED driver is able to provide high efficiency for driving flexible numbers of LEDs, but it requires bulky and expensive off-chip inductors of hundreds of μH to ensure a high efficiency. Meanwhile, the inductor-less LED driver is with small form factor, but the maximum power efficiency is limited by the ratio of the LED voltage to the input voltage. If there is a large difference between the LED voltage and input voltage, significant power will be dissipated on the regulators, resulting in a low power efficiency.

To fully utilize the input power without causing large switching loss or excessive current and voltage stress, a hybrid LED driver is developed, as shown in Fig. 13, which combines the merits of the inductor-less and switching-converter-based LED drivers and adaptively switches between linear mode and switching mode according to the input voltage to enhance the power efficiency [7]. Measurement results show that the hybrid driver can achieve 97% efficiency and 0.996 power factor at 120 V_{AC} 60Hz input with a small inductor of 6.8 μH. The chip photo and measured waveforms are shown in Fig. 14 and Fig. 15, respectively.

V. SUMMARY

Five miniaturized LED driver ICs for high-performance AC input applications are reviewed in this paper. Several techniques are introduced to effectively improve the LED driving systems in different aspects, including inductance reduction, flicker elimination, data communication and efficiency enhancement.

REFERENCES

[1] J. T. Hwang, et al., "A simple LED lamp driver IC with intelligent power-factor correction," in IEEE Int. Solid-State Circuits Conf. (ISSCC) Dig. Tech. Papers, Feb. 2011, pp. 236–238.

[2] L. Li, et al., "An auto-zero-voltage-switching quasi-resonant LED driver with GaN FETs and fully integrated LED shunt protectors,"

IEEE J. of Solid-State Circuits, vol. 53, no. 3, pp. 913–923, Mar. 2018.

[3] Y. Gao, et al., "An AC input switching-converter-free LED driver with low-frequency-flicker reduction," IEEE J. of Solid-State Circuits, vol. 52, no. 5, pp. 1424–1434, May 2017.

[4] Y. Gao, et al., "An AC-input inductorless LED driver for visible-light-communication applications with 8Mb/s data-rate and 6.4% low-frequency flicker," in IEEE Int. Solid-State Circuits Conf. (ISSCC) Dig. Tech. Papers, Feb. 2017. pp. 384–385.

[5] X. Fang, et al., "A low substrate loss, monolithically integrated power inductor for compact LED drivers," in Proc. IEEE Int. Symp. Power Semi. Devices & IC's (ISPSD), May 2015, pp. 53–56.

[6] B. Lehman, et al., "Proposing measures of flicker in the low frequencies for lighting applications," in Proc. IEEE Energy Convers. Congr. Expo., Sep. 2011, pp. 2865–2872.

[7] L. Li, et al., "A multiple-string hybrid LED driver with 97% power efficiency and 0.996 power factor," in Proc. IEEE Symp. on VLSI Technology, Jun. 2016, pp. 106–107.

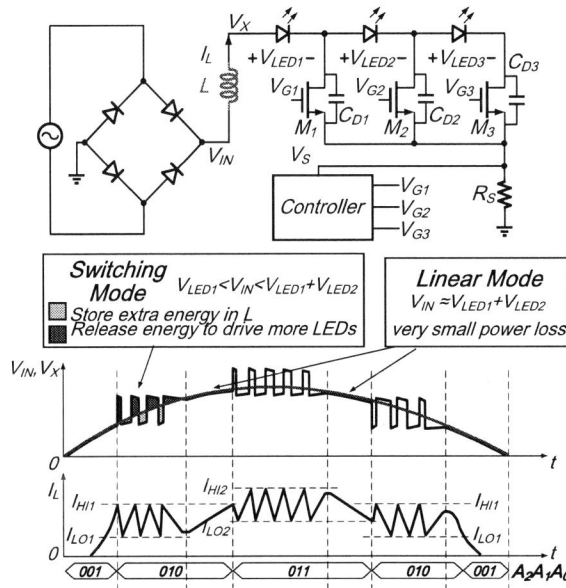

Fig. 13. Concept and waveforms of hybrid LED driver.

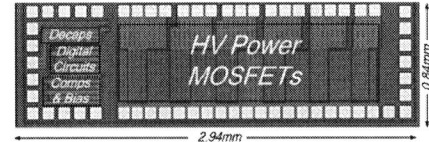

Fig. 14. Chip photo of the hybrid LED driver.

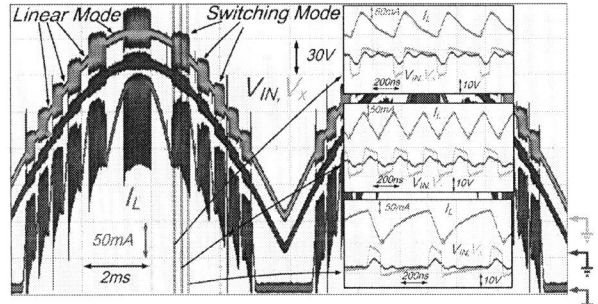

Fig. 15. Measured waveforms of the hybrid driver.

Proceedings of the 30th International Symposium on Power Semiconductor Devices & ICs
May 13-17, 2018, Chicago, USA

High Voltage Capacitive Voltage Conversion

Randall L. Sandusky
Development Engineering
Helix Semiconductors
Irvine, California
randy.sandusky@helixsemiconductors.com

Alexander Hölke
Technology
X-FAB Sarawak Sdn. Bhd.
Kuching, Sarawak, Malaysia
alexander.hoelke@xfab.com

Abstract— A novel high-voltage, ultra-high efficiency monolithic DC-DC Converter is presented. The disclosed architecture was integrated in the X-Fab XT018 SOI process using lateral Super Junction NMOS devices for up to 400V operation. Buck or boost voltage conversion is realized using high-voltage switched-capacitor and gate-driver techniques enabling efficiencies of > 98% peak efficiency, > 97% from 5% to 100% load. The topology is suitable for many applications including charging systems for cell-phones, tablets or other handheld devices, USB power conversion, AC-DC Adapters, DC-DC Point-of-Load (POL) and IoT applications. Block diagrams, simulated and measured results are presented showing the key features of this architecture, areas of interest and the performance summary.

Keywords—MuxCapacitor; Converter; SOI; Efficiency; Power Density; NLDMOS; Capacitive Conversion

I. INTRODUCTION

Power supply architecture has remained virtually unchanged over the last several decades, while at the same time the capability and function of the electronic devices powered by these power supplies have increased in accordance with Moore's law. The small evolutionary changes in traditional power supply technology resulting in only modest improvements in efficiency and power density are no longer acceptable for the continued proliferation of ever more capable electronic devices [1]. Recent advances in process technology allow for higher levels of integration of high voltage, high power transistors, passive components and low voltage logic elements which enable disruptive improvements in power density and conversion efficiency [2]. Power supply form factors will shrink along with wasted energy. This paper presents a novel approach to high voltage capacitive voltage conversion using an advanced CMOS HV SOI process.

A. Capacitive Voltage Conversion

MuxCapacitor IP enables higher levels of integration in AC-DC converters using high-voltage switched-capacitor technology to perform voltage conversion and power transfer. A novel use of the process and IP are shown in the high level block diagram of Fig. 1. The diagram proposes a solution for an AC-DC converter using the two integrated circuits developed in a 180nm 400V SOI process with a capacitive isolation barrier implemented in a multi-chip module. The IP design architecture utilized in the two integrated circuits has been successfully fabricated and tested in a 1.0um SOI high voltage process, a 180nm 60V SOI process and a 180nm 400v SOI process [3].

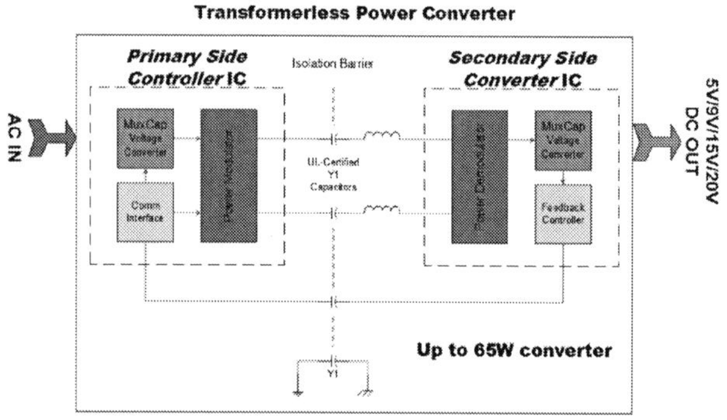

Fig. 1-MuxCapacitor Transformerless Power Converter

The primary and secondary side integrated circuits are both capable of operating with input voltages up to $400V_{DC}$. The power modulation and demodulation circuitry, enabling power transfer across the capacitive isolation barrier also works with voltages up to 400V. The MuxCapacitor circuits in each IC utilize high voltage devices to provide primary side buck and boost functions as needed to allow optimal power transfer, up to 65W, across the capacitive isolation barrier and secondary side buck functions to provide a regulated output voltage to the point of load.

In the following sections of this paper, details on the CMOS process and design of a 7.5W peak power primary side MuxCapacitor circuit are provided along with actual silicon test results.

II. PROCESS

A. Requirements

The MuxCapacitor requires the integration of multiple configurable switches and hence, an SOI technology is required. The switch network must be flexible to allow for various input voltages from 100V-400V.

B. Process Selection

To demonstrate feasibility, early versions of MuxCapacitor circuits were implemented in a 1.0um High Voltage SOI CMOS process, XDM10.

For the next generation IC a 180nm HV SOI process XT018 [4] was chosen due to the required higher integration density. The Partial SOI device architecture of the 200V process option is inherently voltage scalable and is practically free of substrate effects and thus, allows the realization of higher voltage rating as well as floatable switches. Fig. 2 shows a schematic cross-section of a NLDMOS and the Handle Wafer Diode.

978-1-5386-2928-4/18 $31.00 © 2018 IEEE

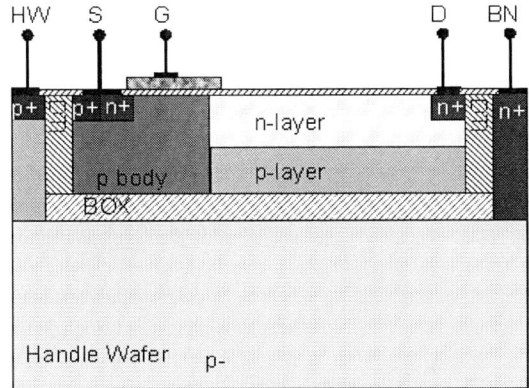

Fig. 2-Schematic cross-section of a 200V SJ NLDMOS on Partial SOI.

The layout of the NLDMOS as well as that of the associated Handle Wafer Diode was optimized for the higher voltage requirements.

C. Results

The measured results for the new 300V and 400V NLDMOS are summarized in Table 1. The achieved R_{dson} is low as is depicted in Fig. 3.

III. MuxCapacitor

A MuxCapacitor block is implemented using the XT018 NLDMOS devices using both uni-directional and bi-directional switches to control the flow of current in one or both directions [5]. A uni-directional switch is a single NLDMOS device with the bulk and source terminals shorted together.

Table 1 Parameters of selected devices

Device	NLDMOS		
	200V	300V	400V
Current Status	Existing	This work	This work
V_T (V)	1.15	1.13	1.12
ID_{sat} (µA/µm)	90	70	72
Rdson (mΩ·mm²)	1270	2230	3620
BV (V)	270	355	450

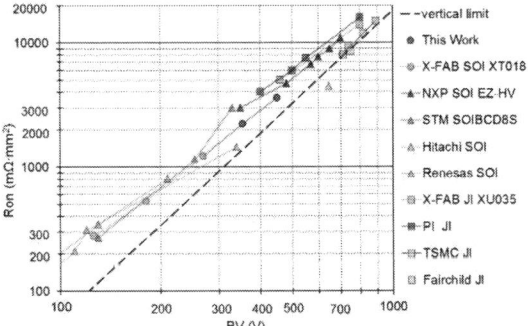

Fig. 3-Ron-BV Benchmarking against HV-IC technologies.

A. Bi-directional Switch

A bi-directional switch, comprised of two NLDMOS, is shown in Fig. 4.

The switch in Fig. 4 allows current to flow in either direction as long as the bulk terminal is shorted to the drain terminal at the lower voltage of the two devices. For current to flow from left to right in Fig. 4, the bulk terminal must be shorted to the drain of device M1B. This causes the bulk diode of M1B to be connected such that it cannot conduct current. Similarly, for current to flow from right to left, the bulk terminal must be shorted to the drain of device M1A shorting the bulk diode of M1A. This method ensures that the bulk diodes do not exceed their break-down voltage, and also prevents the conduction of current.

B. Switched Capacitor Network

When used in a switch-capacitor network, control of the bulk terminal for each of the bi-directional switches is provided which can short the bulk terminal to either device drain terminal as required by the switch-capacitor network. Fig. 5 and Table 2 present a sample switch-capacitor network with required bulk terminal settings to enable a gain setting of 0.66. C_1 and C_2 are flying capacitors external to the integrated circuit.

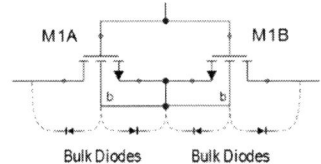

Fig. 4-MuxCapacitor Bi-Directional Switch

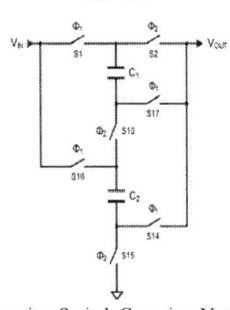

Fig. 5-MuxCapacitor Switch Capacitor Network – 0.66 Gain

Table 2 – Switch Configuration Settings Gain 0.66

Gain	Switch	Switch Nodes	Bulk	Bulk Control	
				Φ_1	Φ_2
0.66	S1	V_{IN} / C1T	Fixed	C1T	C1T
0.66	S2	C1T / V_{OUT}	Driven	V_{OUT}	V_{OUT}
0.66	S10	C1B / C2T	Driven	C1B	C2T
0.66	S17	C1B / V_{OUT}	Driven	C1B	C1B
0.66	S14	C2B / V_{OUT}	Driven	C2B	C2B
0.66	S15	C2B / GND	Fixed	GND	GND
0.66	S16	V_{IN} / C2T	Fixed	C2T	C2T

C. MuxCapacitor Dickson Gate Drivers

MuxCapacitor switches use Dickson charge pumps to develop the gate voltage required for each high-voltage switch. Since N-channel devices are used as the switch elements, a gate voltage is required that is higher than the source terminal of the device, which in some cases may be higher than any input voltage available to the IC.

To minimize the circuit area while achieving the highest overall efficiency, two Dickson circuit topologies are used. A simplified block diagram of the 'floating' Dickson topology is shown in Fig. 6 and is comprised of an oscillator and Dickson charge pump that float at the same potential as the high-voltage switch, which can be as high as the IC input voltage.

Only the switch control signals are level-shifted from the ground level to the high-voltage domain.

The oscillator and pump clock signals are generated in the high-voltage domain referenced to the N-channel switch source terminal. As a result, the Dickson pump capacitors can be low-voltage and have the benefit of being smaller in area.

A ground-based Dickson topology is also used (not shown) which provides the benefit of improved current efficiency at the expense of larger capacitor area. A complete MuxCapacitor converter block uses a hybrid implementation of the ground-based Dickson topology and the floating Dickson topology to achieve the best power transfer efficiency in the smallest physical area.

D. The Complete MuxCapacitor Block

A complete MuxCapacitor can be statically programmed to one of eight possible gain settings, each of which is designed to transfer charge through a network of switches and capacitors to obtain the specified voltage conversion. Both buck and boost operations are possible with the use of 18 integrated N-Channel switches and 3 external flying capacitors. The final charge is delivered to an external hold capacitor for use by the application design. The gain of the MuxCapacitor is controlled using the g<2:0> digital inputs as shown in Table 3.

Table 3 – MuxCapacitor Block Gain Settings

g<2:0> Inputs			MuxCapacitor Gain
g<2>	g<1>	g<0>	
0	0	0	0.25
0	0	1	0.33
0	1	0	0.50
0	1	1	0.66
1	0	0	0.75
1	0	1	1.0
1	1	0	1.5
1	1	1	2.0

In addition to the high voltage switches and gate drivers described above, the primary side integrated circuit includes low voltage control circuitries which implement proprietary algorithms to monitor and control start-up, in-rush current, over current and overvoltage conditions and secondary side feedback signaling. These functions are not included in this paper.

E. Results

Results for the MuxCapacitor block configured with a gain setting of 0.66 and optimized for 7.5W peak output power are shown below. Fig. 7 shows a picture of the die tested for this paper.

Fig. 8 shows the time response of the circuit with an input voltage of 400Vdc targeted for an output power of 7.5W. The MuxCapacitor switching frequency ("FCP") is 2kHz and the Dickson charge pump clock ("FDCK") is 5MHz.

Fig. 9 – Fig. 11 summarize the measured conversion efficiency of a MuxCapacitor stage such as could be used in the AC-DC Converter of Fig. 1. Configured for typical European voltage conditions, the MuxCapacitor stage receives a DC input voltage of 320V following rectification of the AC Line voltage. The MuxCapacitor then converts the input voltage to ~210Vdc for optimum power transfer across the isolation barrier.

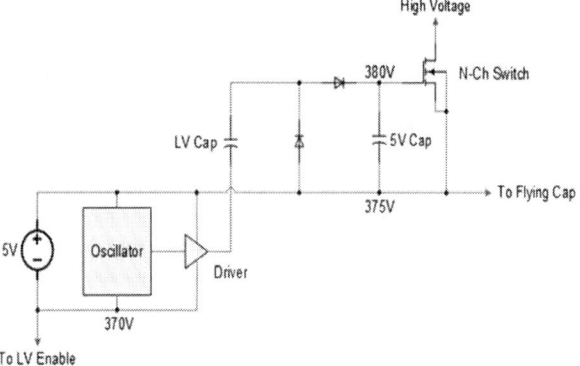

Fig. 6-MuxCapacitor /Floating Dickson Charge Pump Gate Driver

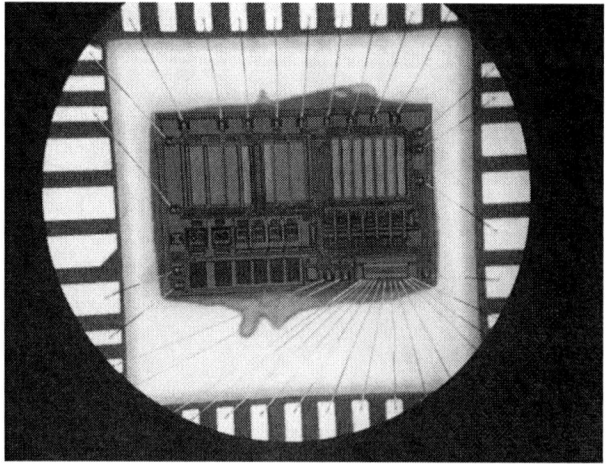

Fig. 7- MuxCapacitor Die Photo

FCP=2 kHz, FDCK=5MHz
Line Voltage=400V, Load Current=20mA

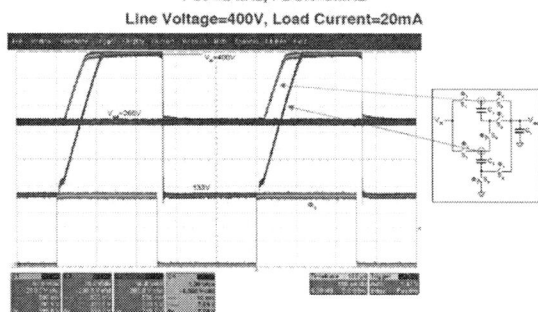

Fig. 8-Time Domain Response, MuxCapacitor Internal Nodes

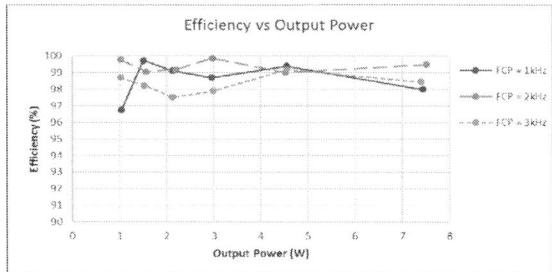

Fig. 9- Measured Efficiency, Multiple FCP values

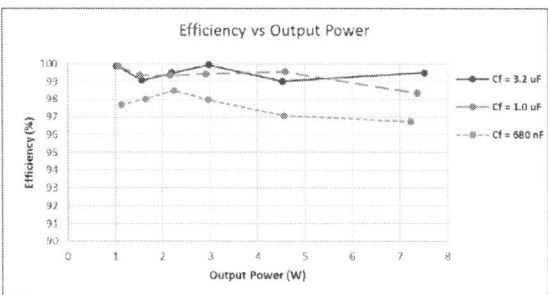

Fig. 10- Measured Efficiency, Multiple Fly Capacitor values

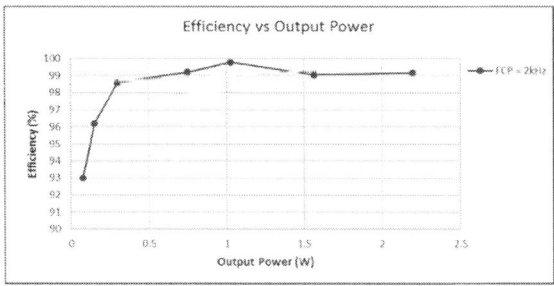

Fig. 11- Measured Efficiency, Low Load Condition

Fig. 8 presents the measured efficiency of the MuxCapacitor stage for output loads from 1.07W (5mA) to 7.5W (35mA), with an input voltage of 320Vdc and output voltage of ~210Vdc for 3 different FCP settings; 1kHz, 2kHz and 3kHz. The chart demonstrates the impact of FCP on MuxCapacitor conversion efficiency, showing that peak load efficiency can be improved with higher clock rates. In practice, the optimum clock rate may be chosen to minimize

the fCV current from switching the flying capacitors so that the light load efficiency is not compromised.

Fig. 9 presents measured conversion efficiency for output loads from 1.07W (5mA) to 7.5W (35mA), with an input voltage of 320Vdc and output voltage of ~210Vdc for 3 different values of flying capacitors; 3.2uF, 1.0uF and 680nF. The chart demonstrates the impact of flying capacitor value on MuxCapacitor conversion efficiency. Larger values of flying capacitance can improve efficiency due to their ability to hold larger quantities of charge. The optimum capacitance value can be chosen based upon the switching frequency and the power delivery requirements.

Finally, for the FCP = 2 KHz condition and Cf = 1.0uF, Fig. 10 presents measured data for low load (<1%) conditions, efficiency remains above 90% for 1% load condition.

In practice, the switching clock rate, flying and output hold capacitor values are optimized to obtain the best efficiency over a wide load range. The output voltage ripple for a MuxCapacitor stage can vary from a few millivolts to several volts, depending upon the capacitance value, switching frequency and load current. Output ripple can be minimized with a larger hold capacitance on the output of the MuxCapacitor, or by the regulation of the output by a subsequent conversion stage.

IV. SUMMARY

A novel approach to implement high voltage capacitive voltage conversion (MuxCapacitor) in a 180nm HV SOI process (XT018) is presented. The design details and results shown are for a circuit optimized for a maximum 7.5W output power and V_{IN} up to 400Vdc. The results show >97% efficiency from 10% to 100% output load and >90% for loads as low as 1% of maximum load. Next generation circuits currently being developed based on the technology presented in this paper will operate up to 65W maximum load and input voltages up to 400VDC with peak efficiency > 98%, >97% from 10% to 100% load and >90% for loads as low as 1% of maximum load.

ACKNOWLEDGEMENT

The primary authors of this paper acknowledge the following contributing authors of this paper: Bud Courville, Neaz Farooqi, Ken Harada, Brian Stevenson, Elizabeth Kho Ching Tee.

REFERENCES

[1] K. Shirriff, "Apple didn't revolutionize power supplies; new transistors did", Ken Shirrff's blog, February 2012.

[2] M. D. Seeman and S. Sanders, "Analysis and Optimization of Switched-Capacitor DC–DC Converters", IEEE Transactions on Power Electronics March 2008, pp. 841 – 851.

[3] Yang Hao, et al., "A 0.18µm SOI BCD Technology for Automotive Application", ISPSD 2015, Hong Kong

[4] A. Hölke, D. K. Pal, Yang H., Kee K. Y., Elizabeth K., G.Kittler, U. Kuniss and J. Gessner, "A 200V partial SOI 0.18µm CMOS technology," ISPSD 2010, Hiroshima, pp. 257–260.

[5] M.Kline, I Izyumin, B. Boser and S. Sanders, "Capacitive power transfer for contactless charging", APEC 2011, Fort Worth, TX, USA.

Proceedings of the 30th International Symposium on Power Semiconductor Devices & ICs
May 13-17, 2018, Chicago, USA

Chip-Scale Cooling of Power Semiconductor Devices

Fabrication of Jet Impingement Design

Feng Zhou[*], Ki Wook Jung[*], Yuji Fukuoka[†] and Ercan M. Dede[*]
Email: feng.zhou@toyota.com yuji_fukuoka@mail.toyota.co.jp eric.dede@toyota.com
[*]Electronics Research Department, Toyota Research Institute of North America, Ann Arbor, MI, USA
[†] EHV Electronics Design Div., Toyota Motor Corporation, Japan

Abstract—Chip-scale cooling is proposed for future wide band-gap (WBG) power semiconductor devices to overcome challenges associated with device power dissipation, high heat fluxes up to 1 kW/cm², and traditional package thermal resistance. In the current paper, fabrication of a chip-scale cooler that utilizes fluid jet impingement plus flow through an optimized branching microchannel topology is described. This chip-scale cooling structure is expected to provide an estimated 70% higher cooling performance for the same pumping power and more uniform cooling relative to a straight microchannel design based on prior numerical and experimental studies. The proposed embedded flow structure is defined by three layers, and each layer is fabricated by double-sided etching of a six-inch silicon wafer. The final device is obtained by bonding the three etched layers together and dicing the three-wafer stack into individual chips. The design and fabrication of the cooling chip, including process challenges and solutions, is the focus of the current paper. Discussion of the current technology and a vision of its future application is also provided.

Keywords—embedded cooling; topology optimization; MEMS; heat exchanger

I. INTRODUCTION

Power electronics packaging has traditionally been targeted toward the use of silicon-based (Si) semiconductor devices such as insulated gate bipolar transistors (IGBTs). As described in [1], standard high-power electronics packaging technology typically comprises a bare die soldered to a substrate, e.g. direct bond copper (DBC), which is then soldered to a heat spreader. A cold plate is then mechanically attached to the electronics package with an intermediary thermal interface material (TIM). This longstanding method provides means for proper electrical isolation and connection to the power device plus thermal pathways for dissipation of waste heat. Variations of this "remote" cooling strategy, Fig. 1(a), where the cold plate is connected to an external surface of the power module, often require size reduction efforts and experience coefficient of thermal expansion mismatch in the package leading to thermally induced stress. Furthermore, conductive thermal resistance between the cold plate and power device is a common roadblock to achieving lower device temperatures as heat flux increases. Thus, while effective, the established packaging approach for Si devices presents challenges for next-generation higher power density wide band-gap (WBG) devices such as silicon carbide (SiC) or gallium nitride (GaN).

Wide band-gap devices have lower losses and enable higher switching frequencies, which permits the reduction of the size of the inductors and capacitors in a typical power control unit boost converter, as explained by Hamada et al. [2].

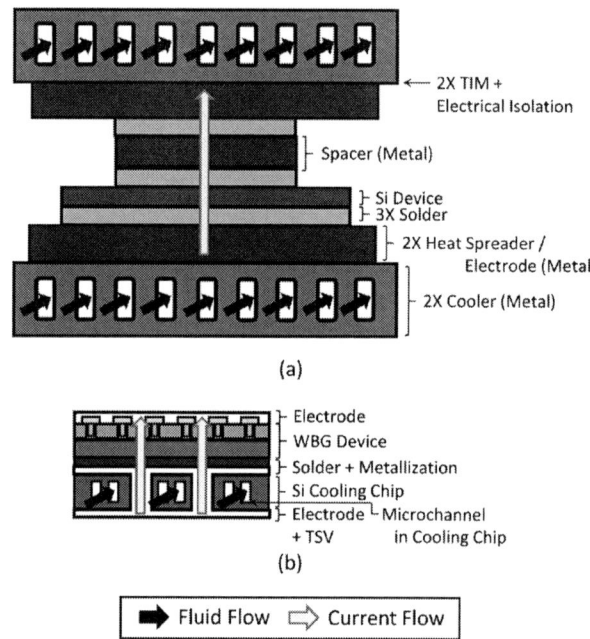

Fig. 1: Power electronics packaging: (a) double-side remote cooling using power card structure; (b) chip-scale embedded cooling. Note: images not to scale.

Up to 80% downsizing of a power control unit is anticipated in the future representing a significant increase in system power density. At the power semiconductor die level, concomitant heat fluxes of 1 kW/cm², up from current Si IGBT levels of 100~250 W/cm², must be handled for WBG devices, per Bar-Cohen et al. [3]. Additionally, from an electrical performance perspective, novel low parasitic inductance packaging schemes should be investigated to maximize the benefits of WBG power devices; see for example Huang et al. [4]. Accordingly, pioneering strategies for power-dense device packaging are required to circumvent the size, electrical, mechanical, and thermal bottlenecks of standard Si-based high-power electronics packaging technology.

One innovation for compact packaging and advanced thermal management of WBG power devices is "embedded" cooling, where the coolant liquid is brought near to the point of heat generation, i.e. the device junction. In [3], impressive examples of embedded cooling are presented for extreme heat flux, 30 kW/cm², lateral current GaN high electron-mobility transistors (HEMTs) for RF power amplifiers. In [5], Vladimirova, et al. investigated drift region integrated parallel microchannels for the direct cooling of vertical power PIN diodes, although concerns were identified in terms of modification of the device electrical performance, e.g. current distribution. More recently, Dede, et al. [6] proposed three concepts for embedded cooling of vertical current WBG semiconductor devices. These concepts include: 1) embedding the microchannels directly into the bottom region of the power device, 2) embedding the microchannels into the lower electrode of the device, or 3) bonding the vertical current power device directly to a Si cooling chip with embedded microchannels plus thru silicon vias (TSVs) for electrical function, as illustrated in Fig. 1(b). Here, we select Si for the cooling chip due to its good thermal-mechanical properties plus ease of manufacturing. In addition, non-doped Si may serve as an insulator between the electrical path and coolant flow path. In [7], the first concept from [6] was explored in depth by Jung et al. using a one-dimensional thermal-fluidic modeling approach, and analytical estimations indicated that ~1 kW/cm² heat fluxes may be cooled with WBG device temperatures less than 200 °C. As part of DARPA's ICECool program described in [8], Drummond et al. [9] recently demonstrated embedded two-phase cooling of heat fluxes up to 910 W/cm² using a Si-based hierarchical manifold microchannel heat sink, although the reported pressure drop was noticeable at 162 kPa and five stacked and bonded wafers were required for the complex heat sink assembly. Through this brief survey, we can observe that the field of embedded cooling is gaining momentum.

While truly embedding the cooling structure within the device itself likely provides the greatest compactness and highest thermal performance, there are significant challenges associated with fabricating microchannels in extremely thin (i.e. much less than 250 μm) next-generation WBG power devices. Additionally, as identified in [5], the embedded cooling structure may have unintended consequences for the electrical performance of the device. Thus, this manuscript builds off the work in [6] and the concept for a separately fabricated and bonded Si cooling chip, as shown in Fig. 1(b).

The paper is organized as follows. In Sec. II., an overview is provided of the straight microchannel chip-scale cooler from [6], and the new jet impingement design is introduced. The layer-by-layer fabrication of the jet impingement chip-scale cooler is described in Sec. III. Conclusions are provided in Sec. IV along with a brief discussion of ongoing work.

II. CHIP-SCALE COOLER OVERVIEW

In this section, the prior straight microchannel chip-scale cooler is briefly reviewed. This is followed by the introduction of the new jet impingement chip-scale cooler design. A thermal-fluid performance estimation based on prior numerical studies is summarized.

A. Straight Microchannel Design

A cross-section schematic of a two-layer straight microchannel chip-scale cooler is shown in Fig. 2(a). Layer 2 has two holes for both inlet and outlet and is bonded to Layer 1

to provide a leak-tight microchannel seal. After entering the inlet hole, fluid proceeds through an inlet manifold and travels through an array of straight microchannels positioned below a 1 cm² heat source region. Fluid then passes through an outlet manifold prior to exiting the device. Previously, the device was tested up to heat fluxes of 130 W/cm², as reported in [6]. Performance characteristics associated with the straight microchannel design were identified to include large device pressure drop (ΔP~40 kPa) and temperature gradients (ΔT~30 °C) as the fluid is pumped through the 12-mm microchannel length. This relatively large temperature gradient may cause non-uniform operation of the power device degrading performance under high current density conditions. Thus, the fabrication of a new chip-scale cooler design is considered to improve on these points by utilizing distributed fluid jet impingement plus flow through optimized branching microchannels. Additional details follow.

B. Jet Impingement Design

The cross-section schematic of the three-layer jet impingement chip scale cooler design is shown in Fig. 2(b). The key concept for this new design is to distribute the cool fluid via the jet array. So, the orientation of the flow path is changed from lateral straight channels to vertical jets. This design follows the micro heat exchanger design principles set forth by Harpole and Eninger [10], and builds off original studies in [11]. Specifically, fluid enters the inlet hole and travels through a fluid manifold in Layer 3, where it is distributed to a 12 × 12 array of jet orifices. From there, the fluid jets emitting from Layer 2 impinge 144 microchannel unit cells in Layer 1 for a total cooled area of 6 mm × 6 mm.

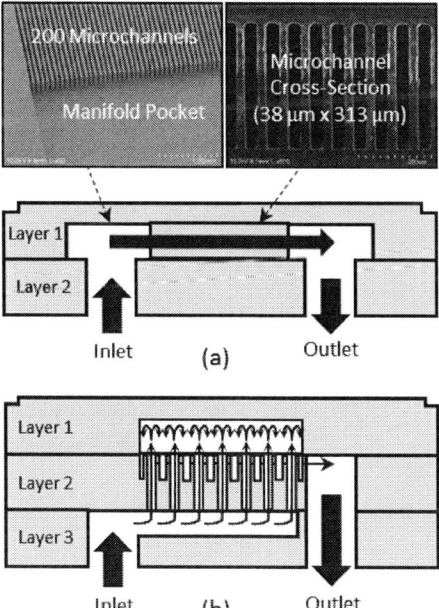

Fig. 2: Chip-scale cooler cross-section schematics: (a) prior two-layer 20 mm x 40 mm x 1 mm straight microchannel chip design with optical microscope images; (b) new three-layer jet impingement design. Note: images not to scale; solid arrows indicate fluid flow direction.

At each Layer 1 unit cell, fluid flows radially outward from the jet impingement location (i.e. stagnation point) to the cell edges. Finally, fluid empties from the cell edges into an outlet manifold in Layer 2 and flows to the outlet hole of the device.

Expected benefits of this chip-scale cooler include a more uniform chip temperature due to the distributed array of fluid jets, an approximate order-of-magnitude lower pressure drop for the same fluid flow rate due to the manifold and unit cell architecture, and an estimated 70% higher heat transfer coefficient for the same pumping power when compared with the straight microchannel design. The reader is referred to [6,11] for additional flow structure design and performance details.

III. FABRICATION OF JET IMPINGEMENT DESIGN

Six-inch double-side polished silicon wafers of 490 μm thickness are used for fabrication. Approximately 2 μm thick silicon oxide (SiO_2) is grown on the surfaces of the wafers, which is patterned and used as the mask for the Si etching. The cooling structure in each wafer is dry etched using the Bosch process with a customized recipe. The etched wafers are directly bonded together to create a monolithic structure. After that, the bonded wafer stacks are diced and metallized for testing. Additional process details are provided below.

A. Mask Layout & Layer Fabrication

The cooler comprises three layers, and each layer requires two masks for the double-side etching, as seen in the process flow in Fig. 3. The first layer has a relatively simple structure and no through holes. The front surface only has alignment marks and dicing lines etched. The optimized cooling structure is then etched on the back side of Layer 1. The front surface is etched first followed by the back surface. An optical microscope image of the cooling structure on the back surface is shown in Fig. 4.

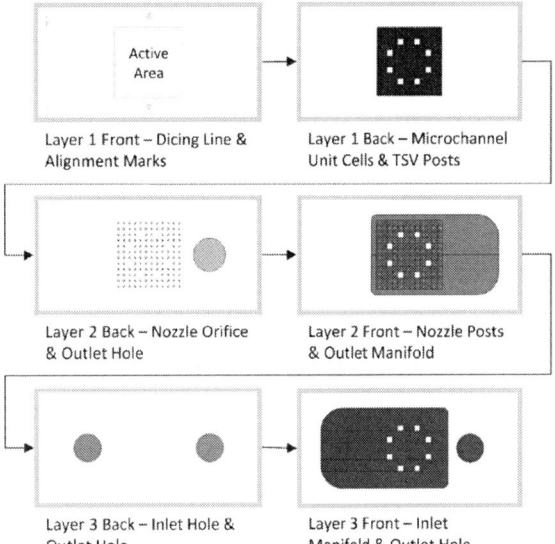

Fig. 3: Process flow for double-side etching of three wafers for the jet impingement chip-scale cooler design.

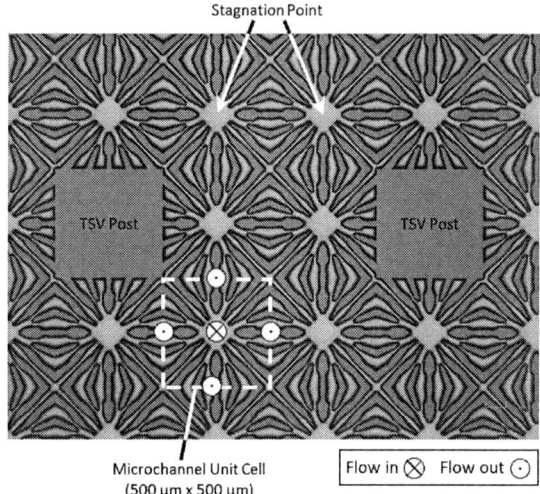

Fig. 4: Optical microscope image of the etched cooling structure on back side of Layer 1.

The unique microchannel design is obtained using topology optimization for conjugate heat transfer, as explained in [11], to minimize both the average temperature of the cooled region and the flow resistance through the microchannels. The minimum spacing between the fins is within the range of $w{\sim}20{-}30$ μm, and the fin height is about $H{\sim}250{-}300$ μm for a high channel aspect ratio, H/w, ranging from 8.3:1 to 15:1.

Layer 2 of the chip-scale cooler is the most challenging to fabricate. It has small etched features, i.e. square nozzle posts, residing in a larger etched feature, i.e. the fluid outlet manifold. Also, there are small nozzle orifice through holes in the posts plus a larger outlet through hole in the manifold. The ratio of the diameter of the nozzle orifices to the fluid outlet hole is 12.5:1. Due to the existence of through holes, the wafers were mounted on supporting wafers with mounting oil, Santovac 5, to prevent damage to the etching tool. The wide range of feature size leads to a nonuniform etching rate. On top of that, the etching depth of the orifices is the thickness of the wafer, 490 μm, while the etching depth of the outlet is the thickness of the wafer less the 275 μm depth of the outlet manifold. Therefore, if these through holes are etched from one side, say the back side, then the outlet is etched through at a much faster rate than the nozzle orifices, exposing the Si surface within the outlet manifold to the etching gas for the remainder of the etching period. This causes over-etching of the square nozzle post features in the manifold pocket and also generates a difficult to clean byproduct due to the reaction between etchant and the mounting oil. Hence, due to the unique design characteristics associated with the chip-scale cooling device, there are two critical points for the fabrication of the second layer: 1. The orifices should be etched from both sides instead of a single side; 2. The back-side etching should be performed first followed by the etching of the front side. As a result, the uniformly etched nozzle orifices are shown in the optical microscope image in Fig. 5. Note that Layer 3 of the chip-scale cooler design has a similar structure to Layer 2 but with less complicated features, and thus it was etched in similar manner.

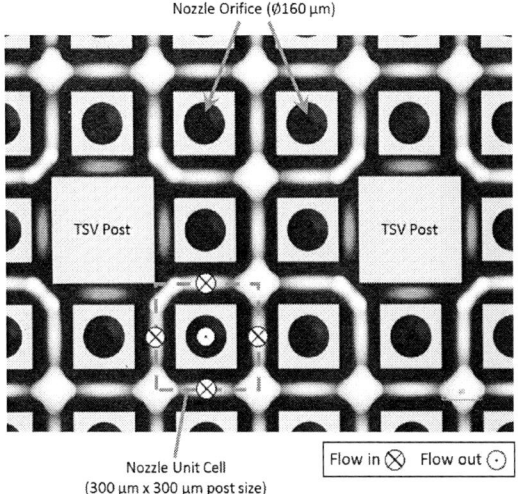

Fig. 5: Optical microscope image of the Layer 2 etched nozzle orifices and outlet manifold nozzle posts.

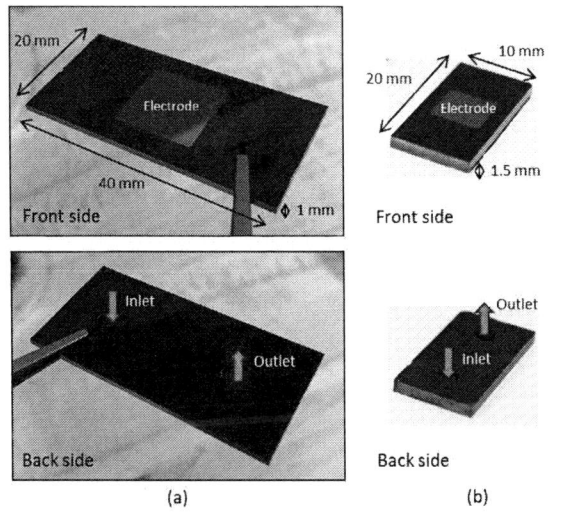

Fig. 6: Fabricated chip-scale Si cooling devices: (a) two-layer straight microchannel design; (b) new three-layer jet impingement design with an estimated 70% higher heat transfer coefficient for the same pump power.

B. Wafer Bonding and Dicing

After the three individual wafers are processed, they then go through the necessary pre-bond cleaning. The wafers are aligned and bonded using direct Si-to-Si fusion bonding followed by annealing at 1050 °C in a nitrogen, N_2, atmosphere for two hours in an Angstrom Engineering Furnace. The bonded wafer stack is diced into individual cooling chips. After that, a 5 mm × 5 mm titanium, nickel, gold (Ti/Ni/Au) metallization layer, for eventual power device or thermal test chip die attachment, is deposited on the center region of the chip. The as-fabricated jet impingement chip-scale cooling device is 1/4th the size of the straight channel device, as shown in Fig. 6.

IV. CONCLUSIONS

The concept of chip-scale cooling for future-generation power semiconductor devices was introduced. A previous straight microchannel device was reviewed, and performance challenges were identified. A new jet impingement chip-scale cooling structure was then described, and expected benefits of the design were discussed. An overview of the fabrication process was provided including a unique double-side etching technique required for realization of the application specific cooling structure topology. Continuing work includes process development to realize TSV electrical functionality. Integration of the existing prototype device into a test bench for experimental thermal-fluid performance evaluation is ongoing.

ACKNOWLEDGMENT

The authors thank Professor Mehdi Asheghi and Professor Kenneth E. Goodson from Stanford University for productive discussions related to this work.

REFERENCES

[1] A.B. Lostetter, F. Barlow, and A. Elshabini, "An overview to integrated power module design for high power electronics packaging," *Microelectronics Reliability*, vol. 40, no. 3, pp. 365-379, 1999.

[2] K. Hamada, M. Nagao, M. Ajioka, and F. Kawai, "SiC-Emerging power device technology for next-generation electrically powered environmentally friendly vehicles," *IEEE Transactions on Electron Devices*, vol. 62, no. 2, pp. 278-285, 2015.

[3] A. Bar-Cohen, J.J. Maurer, and A. Sivananthan, "Near-junction microfluidic cooling for wide bandgap devices," *MRS Advances*, vol. 1, no. 2, pp. 181-195, 2016.

[4] Z. Huang, Y. Li, L. Chen, Y. Tan, C. Chen, Y. Kang, and F. Luo, "Novel low inductive 3D SiC power module based on hybrid packaging and integration method," in *Proceedings of the IEEE Energy Conversion Congress and Exposition (ECCE)*, Nov. 2017, pp. 3995-4002.

[5] K. Vladimirova, J.-C. Crebier, Y. Avenas, and C. Schaeffer., "Drift region integrated microchannels for direct cooling of power electronic devices: advantages and limitations," *IEEE Transactions on Power Electronics*, vol. 28, no. 5, pp. 2576-2586, 2013.

[6] E.M. Dede, F. Zhou, and S.N. Joshi, "Concepts for embedded cooling of vertical current wide band-gap semiconductor devices," in *Proceedings of the 16th IEEE Intersociety Conference on Thermal and Thermomechanical Phenomena in Electronic Systems (ITherm)*, May-June 2017, 8 pages.

[7] K.W. Jung, C.R. Kharangate, H. Lee, J. Palko, F. Zhou, M. Asheghi, E.M. Dede, and K.E. Goodson, "Microchannel cooling strategies for high heat flux (1 kW/cm²) power electronic applications," in *Proceedings of the 16th IEEE Intersociety Conference on Thermal and Thermomechanical Phenomena in Electronic Systems (ITherm)*, May-June 2017, 7 pages.

[8] A. Bar-Cohen, J.J. Maurer, J.G. Felbinger, "DARPA's intra/interchip enhanced cooling (ICECool) program," in *Proceedings of the Compound Semiconductor Manufacturing Technology Conference (CS MANTECH)*, May 2013, pp. 171–174.

[9] K.P. Drummond, D. Back, M.D. Sinanis, D.B. Janes, D. Peroulis, J.A. Weibel, S.V. Garimella, "A hierarchical manifold microchannel heat sink array for high-heat-flux two-phase cooling of electronics," *International Journal of Heat and Mass Transfer*, vol. 117, pp. 319-330, 2018.

[10] G.M. Harpole and J.E. Eninger, "Micro-channel heat exchanger optimization," in *Proceedings of the 7th IEEE SEMI-THERM Symposium*, 1991, pp. 59–63.

[11] E.M. Dede and Y. Liu, "Scale effects on thermal-fluid performance of optimized hierarchical structures," in *Proceedings of the ASME/JSME 8th Thermal Engineering Joint Conference (AJTEC)*, Mar. 2011, 8 pages.

Proceedings of the 30th International Symposium on Power Semiconductor Devices & ICs
May 13-17, 2018, Chicago, USA

An Innovative Silicon Power Device (i-Si) through Time and Space Control of a Stored Carrier (TASC)

Mutsuhiro Mori, Tomoyuki Miyoshi, Tomoyasu Furukawa,
Yujiro Takeuchi, Yusuke Hotta, and Masaki Shiraishi
Hitachi, Ltd.
Ibaraki-ken, Japan
E-mail: mutsuhiro.mori.hw@hitachi.com

Abstract—An innovative silicon power device (i-Si) made entirely from silicon, which has an extremely low inverter loss beyond that of a conventional IGBT and a pn diode, is proposed. The i-Si is composed of a dual side-gate high-conductivity IGBT (HiGT) through time and space control of a stored carrier (TASC) and a novel MOS controllable stored-carrier diode (MOSD) by additional MOS gates. It is demonstrated that a 6.5 kV i-Si has a fabricated dual side-gate HiGT through TASC with a -36% collector-emitter saturation voltage and -27% turn-off loss of the conventional planar-gate IGBT and has a simulated MOSD with a -32% forward voltage drop and -64% reverse recovery loss of the conventional pn diode. As a result, the 6.5 kV i-Si has shown -40% inverter loss, similar to that of SiC-MOSFET.

Keywords— i-Si; dual side-gate HiGT; MOSD; TASC; time and space control; stored carrier; SiC; inverter loss

I. INTRODUCTION

For the last four decades, IGBTs have been improved in terms of reduced power dissipation with lower conductive and switching losses by various technologies such as an IGBT scaling rule [1] and a high-conductivity IGBT (HiGT) concept [2], [3], [4]. The leading-edge side-gate HiGT [5] with a low feedback gate capacitance (C_{res}) has showed the limitations of further lowering loss because of a trade-off relationship between a collector-emitter saturation voltage (V_{CEsat}) and a turn-off loss (E_{off}). A dual side-gate HiGT [6], [7] with an additional gate, which can dynamically control a stored carrier concentration right before turn-off, has broken through the limitations and demonstrated great potential for further improvement on power dissipation for a MOS controlled bipolar device made from silicon.

On the other hand, an anti-parallel diode for recent high-speed IGBT is also important for reducing the power dissipation and switching noise. An ultra-soft & fast recovery diode (U-SFD) [1], [8], [9], which has a composite structure of deep pn junctions and shallow p-type Schottky junctions, has improved the trade-off relationship between forward voltage drop (V_F) and reverse recovery loss (E_{rr}). A pn diode with an additional MOS gate has been suggested to decrease the power loss [10], [11], [12].

MOSFETs and Schottky barrier diodes (SBD) made from SiC are currently attracting considerable interest because of

their superior electrical characteristics of low loss. However, the SiC material requires high temperature, low speed crystal growth, and an epitaxial layer with a high crystal defect density compared to silicon and is thus costly.

The purpose of this paper is to propose innovative silicon power devices with further loss reduction beyond a conventional IGBT and a pn diode.

II. SIDE-GATE HiGT THROUGH TASC

Figure 1 shows the configuration of a side-gate HiGT with single- [5] and dual-gate [6], [7] control and a novel TASC between two different electrical characteristics chips, which are a single side-gate HiGT chip of high conductivity (Hc) with low V_{CEsat} and a dual side-gate HiGT chip of high speed (Hs) with low E_{off}. The side-gate HiGT through TASC has a control gate (Gc) that controls the stored carrier density and a switching gate (Gs) that turns on or off, as does the dual side-gate HiGT. However, Gc of TASC connects the gate of the Hc type single-gate chip and one gate Gc of the Hs type dual-gate chip, and Gs of TASC connects another gate Gs of the Hs type dual-gate chip.

During the on-state period for the side-gate HiGT through TASC, the current flows mainly through the Hc chip because of low V_{CEsat}. In the turn-off period, Gc turns off first, and the current of the Hc chip moves to the Hs chip in space during the timing delay (t_{d_off}) in the switching mode right before turn-off.

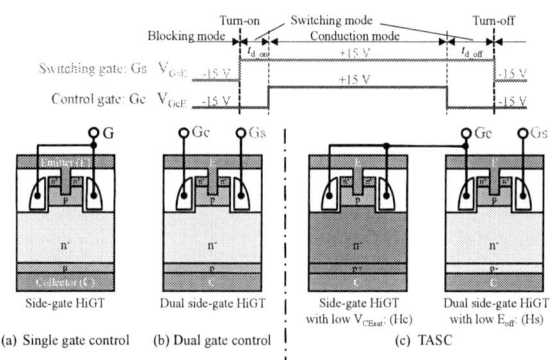

Fig. 1. Configuration of side-gate HiGT through time and space control of stored carrier (TASC).

978-1-5386-2928-4/18 $31.00 © 2018 IEEE

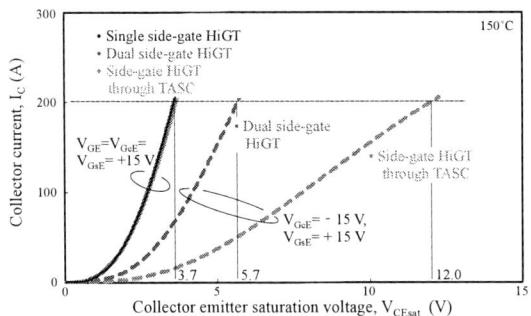

Fig. 2. Experimental current-voltage characteristics of 6.5 kV side-gate HiGTs.

At the same time, the stored carrier density around the emitter side of the Hs chip is reduced with time. The turn-off loss for the Hs chip is dramatically decreased because of a low injection from the p⁻-layer in the collector side and a low stored carrier in the emitter side. As a result, the side-gate HiGT through TASC has a low conduction loss and a low turn-off loss. When turning on, if Gs and Gc turn on at the same time, the turn-on loss will also be reduced because of the high conductivity characteristics of the Hc chip.

Figure 2 demonstrates an example of the experimental current-voltage characteristics for 6.5 kV side-gate HiGTs with single gates (2 chip parallel), dual gates (2 chip parallel), and TASC (1 Hc chip +1 Hs chip). The rating current for each chip is 100 A. At all gate voltages of +15 V, each side-gate HiGT has a similar low V_{CEsat} of 3.7 V. The dual side-gate HiGT with Gc of -15 V and Gs of +15 V has about half the output current of that with Gc and Gs of +15 V, which can control electron injection from MOS gates. On the other hand, the side-gate HiGT through TASC can reduce electron injection to a quarter and also hole injection from the p⁻-layer of the collector side. As a result, a high V_{CEsat} of 12.0 V is realized with a low stored carrier.

Figure 3 shows simulated current and voltage waveforms of Hc and Hs chips for the side-gate HiGT through TASC during turn-off. In the conduction mode, the current mainly flows in the Hc chip. In the switching mode, the Hc chip turns off, and the main current (stored carrier) moves to the Hs chip from the Hc chip in space with no change of total current (Hc + Hs). At the same time, the stored carrier in the emitter side of the n⁻-layer for the Hs chip is decreased with time. When turn-off

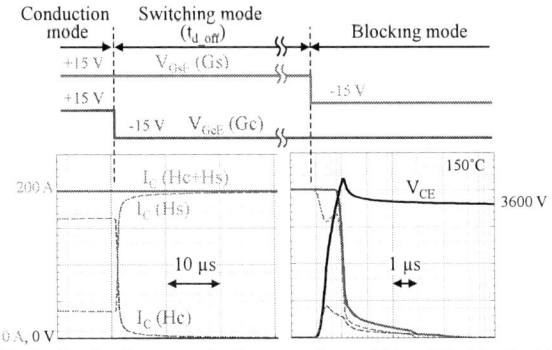

Fig. 3. Simulated current and voltage waveforms of Hc and Hs chips for side-gate HiGT through TASC.

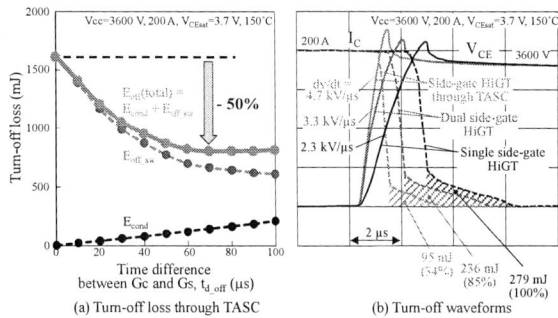

Fig. 4. Experimental turn-off characteristics for 6.5 kV side-gate HiGTs.

occurs, the main current flowing in the Hs chip is rapidly cut off. For the Hc chip the displacement and re-injection current caused by dv/dt during turn-off flows.

Figure 4 (a) shows the experimental turn-off losses regarding the time difference (t_{d_off}) between Gc and Gs for the side-gate HiGT through TASC. The switching loss (E_{off_sw}) in turn-off during the switching mode can be reduced in step with t_{d_off} because of the decreasing stored carrier. However, a large t_{d_off} leads to an increase in conduction loss (E_{cond}) during t_{d_off} because V_{CEsat} in the switching mode is higher than that in the conduction mode. As a result, the optimum t_{d_off} to get the lowest $E_{off(total)}$ exists at about 70 μs. For the side-gate HiGT through TASC, the Hs chip with a low injection p⁻-layer in the collector side creates a low tail current and E_{off_sw} compared to the single and dual side-gate HiGTs with the higher injection p-layer as illustrated in Fig. 4 (b).

Figure 5 presents the trade-off relationship between E_{off} and V_{CEsat}. For the side-gate HiGT through TASC, the trade-off relationship in which lowering the V_{CEsat} increases the E_{off} is attributed to a displacement and re-injection current caused by dv/dt for the Hc chip with the high injection efficiency of the p⁺-layer in the collector side. The side-gate HiGT through TASC has loss reductions with V_{CEsat} of -36% and E_{off} of -27% compared to the conventional planar-gate IGBT. It also has -40% lower E_{off} compared to the single side-gate HiGT, and further -24% lower E_{off} than the dual side-gate HiGT at the similar V_{CEsat} of 3.7 V. The simulated results for the 6.5 kV SiC-MOSFET indicate that the E_{off} is very low; however, the on-state voltage (=V_{CEsat}) is about 1.4 times higher than the 3.7 V of the side-gate HiGTs at the same current density.

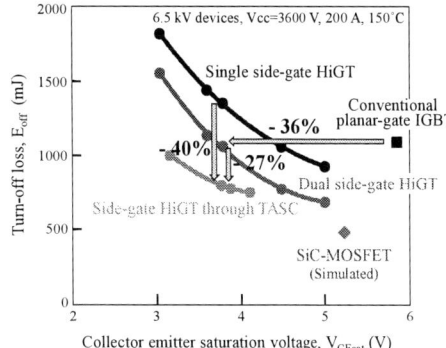

Fig. 5 Trade-off relationships between E_{off} and V_{CEsat} for side-gate HiGTs.

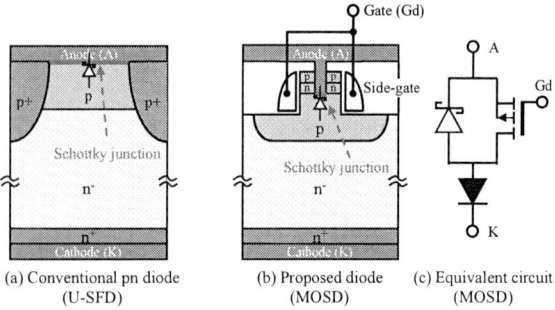

(a) Conventional pn diode (U-SFD)   (b) Proposed diode (MOSD)   (c) Equivalent circuit (MOSD)

Fig. 6 Structures and equivalent circuit of diodes.

III. MOS Controllable Stored-carrier Diode

Figure 6 illustrates the structures of a conventional pn diode (U-SFD [9]), the proposed novel <u>MOS</u> controllable stored-carrier <u>d</u>iode (MOSD), and the equivalent circuit.

Figure 7 shows the operating principle for MOSD. In forward bias (a) in Fig.6, the gate-anode voltage (V_{GdA}) of -15 V is applied to the p-channel MOSFET, and the hole is injected to the n⁻-layer from the p-layer via the p-channel MOSFET. In the operation (b) the V_{GdA} is changed to 0 V, and the hole injection current stops. A Schottky junction is reverse-biased, and the direct hole injection from the anode electrode is also refused. As a result, the stored carrier density in the n⁻-layer is reduced as it is in a dual side-gate HiGT. After the timing delay (t_{d_rr}), the IGBT of the opposite arm turns on, and the MOSD is in reverse recovery mode (c) while keeping the V_{GdA} of 0 V. The hole in the n⁻-layer flows to the anode electrode via the Schottky junction, and the MOSD moves to the reverse bias mode (d). Whenever V_{KA} is reverse-biased, the V_{GdA} returns to -15 V in the blocking state again and prepares for the turning-off of the opposite arm IGBT.

Figure 8 demonstrates the simulated current-voltage characteristics in forward (a) and reverse (b) bias for the MOSD. It has a low forward voltage drop of 3.3 V and a similar leakage current compared to the conventional actual pn diode. Furthermore, the MOSD possesses tough blocking characteristics even at V_{GdA}=+15 V as well as at 0 V and -15 V by the p-layer surrounding the MOS gate.

The simulated MOSD has an optimum timing delay (t_{d_rr})

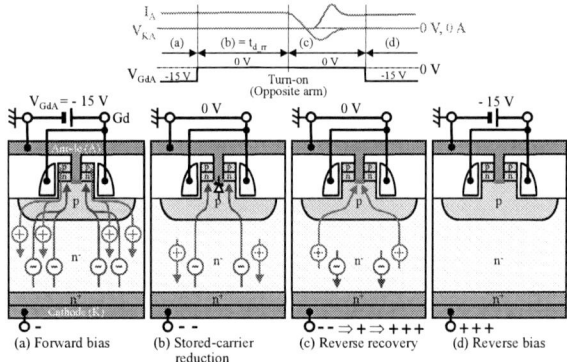

Fig. 7. Operating principle for MOSD.

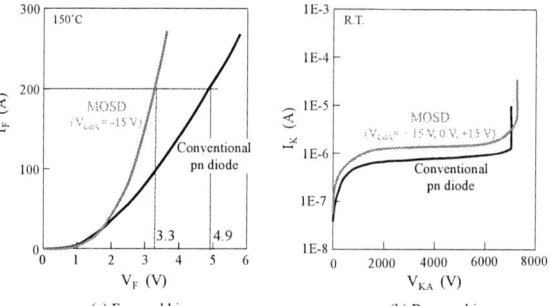

(a) Forward bias   (b) Reverse bias

Fig. 8. Simulated current-voltage characteristics of 6.5 kV MOSD.

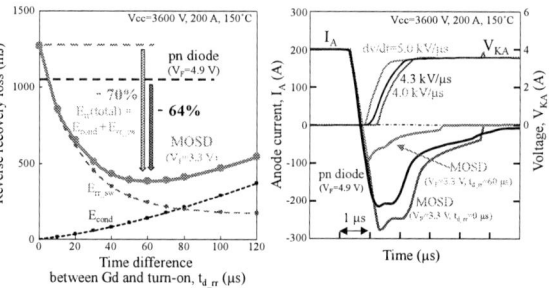

(a) Reverse recovery loss   (b) Reverse recovery waveforms

Fig. 9. Simulated reverse recovery characteristics of 6.5 kV MOSD.

of 60 μs right before the turning-on of the opposite arm IGBT as shown in Fig. 9 (a). The reverse recovery loss (E_{rr}(total)) for MOSD with a t_{d_rr} of 60 μs is dramatically reduced up to -70% compared to the t_{d_rr} of 0 μs while keeping a low dv/dt of 4 to 5 kV/μs as illustrated in Fig. 9 (b).

Figure 10 presents the trade-off relationship between V_F and $E_{rr(total)}$. The simulated 6.5 kV MOSD has -32% lower V_F and -64% lower $E_{rr(total)}$ compared to the conventional actual pn diode. The simulated results for the 6.5 kV SiC-MOSFET indicates that the $E_{rr(total)}$ is very low; however, even V_F of the SiC-MOSFET with synchronous rectification is about 1.6 times higher than the 3.3 V of the MOSD.

IV. Inverter Loss Calculation

Figure 11 shows the results of an inverter loss calculation of various combinations of 6.5 kV IGBTs and diodes at a

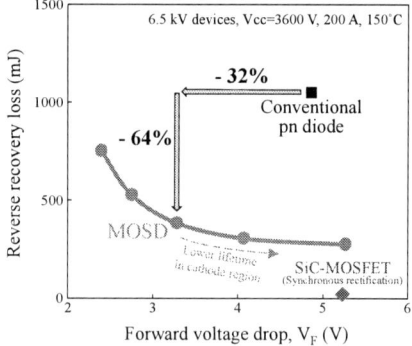

Fig. 10. Trade-off relationships between reverse recovery loss and V_F.

978-1-5386-2928-4/18 $31.00 © 2018 IEEE

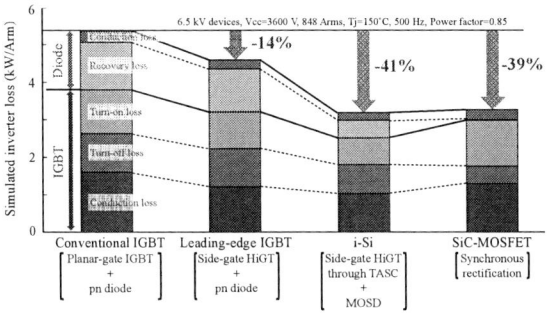

(a) Motoring mode

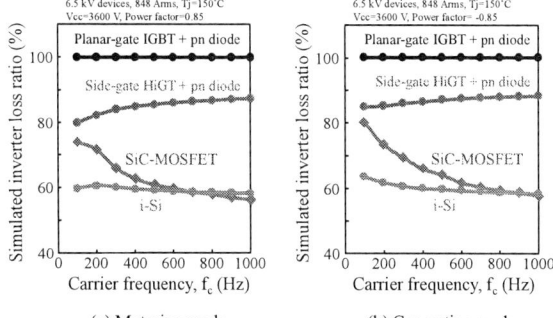

(a) Motoring mode   (b) Generating mode

Fig. 12. Relationships between inverter loss (ratio to conventional planar-gate IGBT + pn diode) and carrier frequency f_c.

(b) Generating mode

Fig. 11. Inverter losses calculation in motoring and generation modes.

carrier frequency (f_c) of 500 Hz in the motoring and generating modes. The calculation was based on the loss tables obtained by measured I-V characteristics and switching losses for the conventional planar-gate IGBT, pn diode, and the side-gate HiGTs and simulated values for the MOSD and SiC-MOSFET. The inverter loss of the i-Si that integrates the side-gate HiGT through TASC and the MOSD could be reduced by 40% during both motoring and generating modes compared to that of the conventional planar-gate IGBT and pn diode. This result indicates that the i-Si can achieve a lower inverter loss than that of the SiC-MOSFET.

Figure 12 shows the relationships between f_c and inverter losses (ratio to conventional planar-gate IGBT and pn diode). The i-Si can reduce inverter loss over a wide range of f_c in motoring and generation modes, and can achieve a low inverter loss of -40% even at high f_c, similar to that of the SiC-MOSFET.

V. CONCLUSION

In this paper, an innovative silicon power device (i-Si) through time and space control of a stored carrier (TASC) is proposed that exceeds the limitations of a conventional IGBT and pn diode. The i-Si is composed of a side-gate high-conductivity IGBT (HiGT) that combines a single-gate chip of high conductivity with a dual-gate chip of high speed and a novel MOS controllable stored-carrier diode (MOSD). The 6.5 kV i-Si achieves dramatic loss reductions at a collector-emitter saturation voltage of -36%, a turn-off loss of -27%, a forward voltage drop of -32%, and a reverse recovery loss of -64%

compared to a conventional planar-gate IGBT and pn diode. As a result, the i-Si made entirely from silicon has an extremely low inverter loss of -40%, similar to that of SiC-MOSFET.

REFERENCES

[1] M. Mori, N. Sakurai, K. Hanaoka, Y. Nakazawa, T. Kurosu, T. Shigemura, and H. Ohkubo, "A new generation IGBT module with low loss, soft switch and small package," in Proc. 1995 Int. Power Electronics Conf., pp. 916–920, 1995.

[2] M. Mori, K. Oyama, T. Arai, J. Sakano, Y. Nishimura, K. Masuda, K. Saito, Y. Uchino, and H. Homma, "A planar-gate high-conductivity IGBT (HiGT) with hole-barrier layer," IEEE Trans. Electron Devices, Vol. 54, pp. 1515–1520, 2007.

[3] M. Mori, K. Oyama, Y. Kohno, J. Sakano, J. Uruno, K. Ishizuka, and D. Kawase, "A trench-gate high-conductivity IGBT (HiGT) with shortcircuit capability," IEEE Trans. Electron Devices, Vol. 54, pp. 2011–2016, 2007.

[4] K. Oyama, Y. Kohno, J. Sakano, J. Uruno, K. Ishizuka, D. Kawase, and M. Mori, "Novel 600-V trench high-conductivity IGBT (trench HiGT) with short-circuit capability," in Proc. 13th Int. Symp. on Power Semiconductor Devices & ICs, pp. 417–420, 2001.

[5] M. Shiraishi, T. Furukawa, S. Watanabe, T. Arai, and M. Mori, "Side gate HiGT with low dv/dt noise and low loss," in Proc. 28th Int. Symp. on Power Semiconductor Devices & ICs, pp. 199–202, 2016.

[6] T. Miyoshi, Y. Takeuchi, T. Furukawa, M. Shiraishi, and M. Mori, "Dual side-gate HiGT breaking through the limitation of IGBT loss reduction," in Proc. PCIM Europe, pp. 315–322, 2017.

[7] Y. Takeuchi, T. Miyoshi, T. Furukawa, M. Shiraishi, and M. Mori, "A Novel Hybrid Power Module with Dual Side-gate HiGT and SiC-SBD," in Proc. 29th Int. Symp. on Power Semiconductor Devices & ICs, pp. 57–60, 2017.

[8] M. Mori, Y. Yasuda, N. Sakurai, and Y. Sugawara, "A novel soft and fast recovery diode (SFD) with thin p-layer formed by Al-Si electrode," in Proc. 3rd Int. Symp. on Power Semiconductor Devices & ICs, pp. 113–117, 1991.

[9] M. Mori, H. Kobayashi, and Y. Yasuda, "6.5 kV Ultra Soft & Fast Recovery Diode (U-SFD) with High Reverse Recovery Capability," in Proc. 12th Int. Symp. on Power Semiconductor Devices & ICs, pp. 115–118, 2000.

[10] Q. Huang and C. A. J. Amaratunga, "MOS Controlled Diodes-A New Power Diode," Solid-State Electronics Vol. 38, No. 5, pp. 977–980, 1995

[11] J.G. Bauer, T. Duetemeyer, F. Hille, and O. Humbel, "Experimental Study of a 6.5kV MOS Controllable Freewheeling Diode," in Proc. 20th Int. Symp. on Power Semiconductor Devices & ICs, pp. 40–43, 2008.

[12] M. Mori, "Semiconductor Device and Power Converter Using It," US Patent 8853736

978-1-5386-2928-4/18 $31.00 © 2018 IEEE

AUTHOR INDEX

A

Abdul Khadar, Riyaz	240
Agam, Moshe	307
Aichinger, Thomas	40
Akao, Shinya	132
Akiyama, Y.	495
An, Junjie	391
Antonini, Giulio	260
Arai, Daisuke	180
Ata, Yasuo	116

B

Baburske, Roman	108, 164
Bagatin, M.	92
Baines, Y.	200
Bakeroot, B.	92
Baliga, B. Jayant	371
Banerjee, A.	92, 284
Banerjee, Sujit	427
Barbieri, Thomas	415
Barkley, Adam	415
Basler, Thomas	40, 367
Basler, Vanessa	252
Basset, O.	140
Bellini, M.	156, 451
Bhalla, Anup	152
Bhattacharya, Subhashish	423
Bhojani, Riteshkumar	164
Bianda, E.	451
Biscarrat, J.	200
Bobde, Madhur	152, 327
Boksteen, B.K.	28
Botan, V.	156
Braga, N.	379
Brandt, P.	24
Breglio, G.	104, 144, 443
Briggs, M.	140
Buckley, J.	200

Buetow, Sven 196
Buitrago, Elizabeth 499
Burton, Richard 88

C

Cai, Yuefei 347
Camuso, Gianluca 220
Castellazzi, A. 443, 471
Castellazzi, Alberto 232, 463
Chakraborty, Saikat 359
Challa, Ashok 288
Chang, Hong 327
Chang, K.K. 487
Chang, Keven 295
Charles, M. 200
Chen, Hongting 339
Chen, Jia 224
Chen, Kevin J. 76, 276, 411
Chen, Max 220
Chen, Wanjun 96, 204, 212, 335
Chen, Xing Bi 148
Chen, Yen-Ming 331
Cheng, Junji 148
Cheng, Qian 96
Cheng, Shikang 80, 323, 475
Chern, Chan-Hong 76
Chiang, P.H. 487
Chien, Terry 220
Choi, Jinkyu 351
Chowdhury, Sauvik 427
Chowdhury, Srabanti 208
Christensen, Kim 68
Chu, Kuo-Ting 56, 403
Chung, Jayhoon 68
Corvasce, C. 28, 499
Cui, Jingjing 419

D

Dai, Xiaoping 112
Dai, Zhigang 339
d'Alessandro, V. 443
De Colvenaer, Bert 15
De Jaeger, B. 284
Decoutere, S. 188, 284
Dede, Ercan M. 516
Deng, Tong 311
Deviny, I. 112, 140

DiMarino, Christina	256
Ding, F.	379
Ding, Yixiao	347
Disney, Don	359
Dong, Changxu	204
Duan, Wenting	311
Dupuy, Philippe	479

E

Ebihara, Yasuhiro	44
Ebiike, Yuji	447
Eckoldt, Uwe	483
Efthymiou, Loizos	220
Egashira, Katsumi	303
Eikyu, K.	495
Ekman, Jonas	260

F

Fabian, Benjamin	244
Fan, Jie	280
Fang, Jiong	399
Fang, Xiangming	172, 347
Fang, Ziquan	311
Fayyaz, Asad	232
Felsl, H.-P.	24
Feng, Hao	172, 216
Feng, Xu-dong	355
Franke, Jörg	467
Fujisawa, Hiroyuki	395
Fujita, Y.	184
Fukami, Takeshi	491
Fukuoka, Yuji	516
Furukawa, Akihiko	128, 132, 160
Furukawa, Tomoyasu	520

G

Gao, Wei	268
Gao, Xiang	228
Gao, Yuan	347, 508
Garnier, G.	200
Geens, K.	284
Gerardin, S.	92
Gibson, Mark	479
Gillot, C.	200
Godignon, Philippe	435
Gossner, Harald	72

Grossner, Ulrike 260, 407
Gu, Xin 224
Gu, Yan 323
Guan, Lingpeng 152
Guo, Qing 375
Gupta, Raghu 479
Gwoziecki, R. 200

H

Hagihara, T. 184
Hamada, Kimimori 491
Han, Kijeong 371
Han, Shaowen 272
Hao, Yang 483
Harada, Shinsuke 391
Haruguchi, Hideki 116
Hatakeyama, Tetsuo 395
Hauf, Moritz 120
Hayakawa, Seiichi 455
He, Boyong 475
He, Jiabei 76
He, Lei 224
He, Liang 224
Herzer, Reinhard 196
Hinata, Yuichiro 459
Hirao, Kiyoshi 264
Hirmer, Katrin 343
Hirotsuru, Hideki 264
Hisamoto, Digh 439
Hoffmann, Felix 435
Hofmann, Klaus 343
Hölke, Alexander 483, 512
Hölzl, Wolfgang 252
Honda, Masaaki 180
Hong, Seunghyun 351
Hori, Motohito 459
Horikawa, Nobuyuki 52
Hotta, Yusuke 520
Hsiao, S.Y. 487
Hsiung, C.P. 487
Hsu, Fu-Jen 56, 403
Hu, C.W. 295
Hu, Hao 431
Hu, Kai 212
Hu, Kongsheng 80
Hu, Li 355
Hu, Zongyang 228

Huang, Kirk 288
Huang, Sen 280
Huang, Yanfen 224
Hull, Brett 415
Hung, Chien-Chung 56, 403
Hussein, A. 463, 471
Hutchings, J. 140
Hyuga, Hideki 264

I

Ichijo, Hisao 32
Ichimura, Aiko 44
Ide, Keiichiro 116
Ikeda, Yoshinari 459
Ikura, Yoshihiro 168
Imaizumi, Masayuki 447
Ipposhi, Takashi 299
Irace, A. 104, 144, 443
Ishigaki, Takashi 455
Ishii, Yoshiaki 303
Islam, A. 140
Iwakiri, Shoji 264
Iwamuro, Noriyuki 391
Izumisawa, Masaru 32

J

Jaeger, Beat 407
Jan, C.H. 295
Javid, Mahdi 88
Jena, Debdeep 228
Jeon, Jiho 176
Ji, Dong 208
Ji, Yan-Liang 295
Jiang, Huaping 411
Jiang, Huaxing 216
Jiang, Huihua 292
Jiang, Nan 244
Jin, Feng 311
Jin, Qinhai 327
Joh, Jungwoo 68, 315
Johnson, C. Mark 256
Ju, Yanrui 407
Jung, Jinyoung 176
Jung, Ki Wook 516

K

Kakarla, Bhagyalakshmi	407
Kalajica, Marko	244
Kalnitsky, Alexander	76
Kamibaba, Ryu	160
Kaminski, Nando	435
Kaneda, Mitsuru	128, 160
Kang, Xuanwu	280
Kanzawa, Yoshihiko	52
Kao, H.W.	295
Kawase, Daisuke	455
Kendig, Dustin	164
Khemka, Vishnu	479
Kikuchi, Takuo	503
Kim, Euntaek	176
Kim, Jongjib	359
Kim, Jun-Youn	411
Kim, Soo-Seong	176
Kim, So-Youn	176
Kim, SungHan	411
Kimoto, Tsunenobu	395
Kimura, Keisuke	491
Kimura, Koji	303
Kitchen, Jennifer	88
Kiuchi, Yuji	395
Kiyosawa, Tsutomu	52
Knoll, L.	451
Kobayashi, Toshiyuki	455
Kobayashi, Yusuke	391
Komatsu, Kanako	303
Kong, C.	140
Kong, Moufu	148
Kong, Weiran	311
Konishi, Masaki	491
Koo, Jeoung Mo	292
Kopta, A.	28
Kovačević-Badstübner, I.	260, 407
Kowalsky, Jens	164, 367
Koyama, Akihiro	395
Kranz, L.	435, 451
Krishnan, Srikanth	68, 315
Kubo, Shunji	299
Kudo, Tomohito	116
Kumar, Ashish	423
Kunori, Shinji	180
Kusakabe, Y.	184
Kwan, Hing Kit	359
Kwan, M.-H.	76

L

Lai, Chunlan	475
Lau, Kei May	216, 347
Lee, Chiu-Ling	331
Lee, Chiu-Te	331
Lee, Chwan-Ying	56, 403
Lee, Dong Seup	315
Lee, Lurng-Shehng	56, 403
Lee, N.C.	487
Lee, S.W. Ricky	347
Lee, W.F.	487
Lei, Jiacheng	76, 276
Leng, Jing	80
Li, Jianfeng	256
Li, Lisong	508
Li, Liuan	224
Li, Rophina	84
Li, Rui	272
Li, Ruigang	431
Li, Shaohong	323
Li, Sheng	399
Li, Ting	399
Li, Wenshen	228
Li, Wenwen	208
Li, Ya-Fang	56, 403
Li, Zehong	268
Li, Zhaoji	64, 204, 335, 475
Li, Zhichao	319
Liang, H.	188, 284
Liew, Madelyn	483
Lin, Cheng-Hua	295
Lin, Feng	339
Lin, Wei	292
Lin, Wen-Cheng	359
Lin, Y.-M.	76
Liu, Chao	240, 335
Liu, Chengfang	335
Liu, Donghua	311
Liu, Jie	335
Liu, K.L.	487
Liu, Siyang	60, 319, 339, 399
Liu, Tong	204
Liu, Xiaohong	319
Liu, Xiaoxin	359
Liu, Xinyu	280
Liu, Yang	224
Liu, Yawei	335
Liu, Yong	172

Liu, Yuwei	319
Liu, Zhenxin	224
Longobardi, G.	104, 220
Lorenz, Leo	1
Lu, C.W.	487
Lu, Lucas	8
Lu, Shengli	339
Lu, Yangyang	80
Lui, Sik	152, 327
Luo, H.	112, 140
Luo, Ping	64
Luo, Xiaorong	212, 475
Lutz, Josef	164, 244, 367, 467

M

Ma, Jun	192
Ma, Shulang	319
Machida, Norihisa	12
Maresca, L.	104, 144, 443
Matioli, Elison	192, 236, 240
Matocha, Kevin	427
Matsuda, Jun	248
Matsunaga, Shinichiro	395
Matsuoka, Fumitomo	303
Matsushima, Hiroyuki	455
Matsushita, Yukio	116
Matsuura, H.	495
Meier, K.U.	156
Meneghesso, G.	92
Meneghini, M.	92
Merlin, L.	144
Mesemanolis, Athanassios	499
Miao, Binbin	311
Mickevicius, R.V.	379
Mihaila, A.	435, 451
Miki, Hiroshi	363
Minato, T.	116, 184
Ming, Xin	355
Miric, Anton-Zoran	244
Mirone, P.	144
Mitani, Shuhei	44
Mitsuzuka, Kaname	172
Miyazaki, Hiroyuki	264
Miyazaki, Y.	184
Miyoshi, Tomoyuki	520
Mizuno, Shoji	44
Mizushima, Tomonori	395

Moazzami-Fallah, Jasmin ... 367
Moens, P. ... 92, 284
Mok, Philip K.T. ... 347, 508
Momii, S. ... 184
Mori, Mutsuhiro ... 520
Mori, Takahiro ... 299
Mori, Yuki ... 363
Morita, Toshiaki ... 455
Mouawad, Bassem ... 256, 463
Mouhoubi, S. ... 92
Murakami, Koichi ... 180
Murakami, Takeshi ... 447

N

Nakagawa, Akio ... 124, 168
Nakamura, Kazutoshi ... 503
Nakamura, M. ... 184, 455
Nakata, Yoji ... 116
Nakayama, Koji ... 395
Nakazawa, Y. ... 495
Namai, Masaki ... 391
Ng, Wai Tung ... 84
Nguyen, Bai Yen ... 359
Ngwendson, L. ... 140
Ngwendson, Luther-King ... 112
Ni, Yiqiang ... 224
Niedernostheide, F.-J. ... 24, 108, 120, 164
Ning, Xubin ... 112
Nishi, Koichi ... 128
Nishiwaki, Tsuyoshi ... 491
Noborio, Masato ... 44
Nomoto, Kazuki ... 228

O

Oh, Kwang-Hoon ... 176
Oh, Wonhi ... 351
Ohashi, Hiromichi ... 1
Ohoka, Atsushi ... 52
Okubo, Tatsuo ... 248
Okumura, Hajime ... 395
Ono, Syotaro ... 32
Onozawa, Yuichi ... 168, 172
Ozawa, Naoto ... 248

P

Paccagnella, A. 92
Padmanabhan, Karthik 152
Palanisamy, Shanmuganathan 367
Pandey, Swapnil 359
Papadopoulos, C. 28, 451, 499
Parashar, Sanket 423
Park, Sangmin 351
Park, Young-Joon 68
Paul, Milova 72
Peters, Dethard 40
Pfirsch, F. 24, 108
Pfost, Martin 48, 483
Pilla, Manyam 228
Pjenčák, Jaroslav 307
Plissonnier, M. 200
Poh Ching, Sim 483
Posthuma, N.E. 188, 284
Powell, Blake 427
Prindle, D. 28
Probst, Dean 288
Ptacek, Karel 88
Pu, S.C. 487
Pu, Song 475
Puschkarsky, Katja 40

Q

Qian, Wensheng 311
Qiao, Ming 64, 311, 475
Qin, Ganming 479
Qin, Rongzhen 112
Qiu, Xing 347
Que, Taotao 224

R

Rahimo, M. 451, 499
Rahman, Md Tasbir 491
Reisinger, Hans 40
Ren, Jie 216
Ren, Min 268
Ren, Na 419, 431
Rescher, Gerald 40
Riccio, M. 104, 144, 443
Richmond, Jim 415
Rivas-Davila, Juan M. 136
Romano, Daniele 260

Romano, G.	104, 443
Ronchi, N.	284
Rupp, Roland	367

S

Sabri, Shadi	423
Saddiqui, I.	140
Sagawa, Masakazu	363, 455
Saif, Muhammad Bilal	343
Saito, Jun	491
Saito, Wataru	32, 36
Saitou, Kouichi	52
Sakai, A.	495
Sakai, Masaki	303
Sakai, Tatsuo	248
Sakamoto, Toshihiro	303
Sakano, Jun-ichi	383
Sampath Kumar, B.	72
Sandow, C.	24, 108, 120
Sandusky, Randall L.	512
Sanfilippo, C.	144
Sano, Kazuya	116
Santos, F.	24
Sarkar, Tirthajyoti	288
Sasaki, Koji	455
Sato, Hiroshi	264
Sawada, Kazuyuki	52
Saxena, Tanuj	479
Scheibel, Markus G.	244
Schmidt, Gerhard	120
Schulze, H.-J.	24, 108, 164
Šeliga, Ladislav	307
Sheh, Gin	427
Sheng, Kuang	100, 272, 375, 387, 419, 431
Shi, Yijun	204, 335
Shi, Yu	212
Shi, Yuanyuan	96, 204
Shibib, Ayman	1, 220
Shima, Akio	363, 383, 439
Shimizu, Haruka	363
Shiraishi, Masaki	520
Shorten, Andrew	84
Shrivastava, Mayank	72
Shu, Lei	64
Sin, Johnny K.O.	172, 216, 347
Singh, Upinder	292
Skuriat, Robert	256

Solovey, James **415**

Soneda, Shinya **116, 132**

Song, Kinam **351**

Sonnet, Arif **315**

Sorada, Haruyuki **52**

Spaziani, Larry **8**

Stark, Roger **407**

Stegner, A. **24**

Stockman, A. **92**

Stoffels, S. **188, 284**

Su, R.-Y. **76**

Su, Wei **319, 339**

Su, Yi **327**

Suekawa, Eisuke **447**

Sumitani, Hiroaki **447**

Sun, Guipeng **339**

Sun, Jiahui **387, 419**

Sun, Ling **323**

Sun, Weifeng **60, 80, 319, 323, 339, 399**

Sung, Woongje **371**

Suwhanov, Agajan **307**

Suzuki, K. **128, 184**

T

Tabata, Toshihito **455**

Tack, M. **92, 284**

Tajalli, A. **92**

Takahashi, Keita **303**

Takahashi, Tetsuo **132, 160**

Takakura, K. **184**

Takao, Kazuto **503**

Takei, Manabu **395**

Takenaka, Kensuke **395**

Takeuchi, Yuichi **44**

Takeuchi, Yujiro **520**

Tan, Canjian **112**

Tanaka, Hiroyuki **172**

Tanaka, Masahiro **124**

Tang, Chak Wah **216**

Tang, Gaofei **76, 276**

Tang, Qingxi **323**

Tani, Kazuki **383, 439**

Taniguchi, Katsumi **459**

Tao, Hong **212, 335**

Tarutani, M. **184, 447**

Tega, Naoki **439**

Terrill, Kyle **220**

Thompson, J.	140
To, Suet	411
Tsai, Min-Hsuan	331
Tsai, Tom	76
Tsibizov, Alexander	407
Tsuji, Masataka	32
Tsuruta, Kazuhiro	44
Tuan, H.C.	76
Tugan, K.	156

U

Uchida, Masao	52
Udrea, Florin	220
Ueda, Tetsuzo	52
Uemura, H.	184
Umbach, F.	24
Unger, Christian	48
Uryu, K.	184

V

Van Brunt, Edward	415, 423
van Treek, Vera	108
Vandendaele, W.	200
Varghese, Dhanoop	315
Veliadis, Victor	423
Venkatraman, Prasad	288
Venugopal, Archana	315
Vobecky, J.	156, 499

W

Wachutka, Gerhard	252
Wagner, Andreas	252
Wagner, W.	24
Wang, Baozhu	375
Wang, C.C.	487
Wang, C.L.	487
Wang, Chih-Chong	331
Wang, Hao	323
Wang, Hengyu	375, 419
Wang, Huihui	311
Wang, Kang L.	431
Wang, Wenjing	224
Wang, Xia	355
Wang, Xinhua	280
Wang, Y.	140
Wang, Yuankun	280

Wang, Zhuo 475
Wei, D. 96, 212
Wei, Jiaxing 399
Wei, Jin 276, 411
Wei, Ke 280
Wei, P. 96
Wellekens, D. 284
Winkler, Tom 467
Wirths, S. 451
Wong, H.Y. 379
Wu, Gang 355
Wu, Han 232
Wu, Jie 311
Wu, Jiupeng 419
Wu, Wangran 60, 319, 339
Wu, Yuzhou 268
Wu, Zhisheng 224

X

Xiao, Chao 172
Xiao, Haibo 112
Xiao, Qiang 112
Xing, Huili Grace 228
Xu, Hongyi 387, 419

Y

Yachi, Megumi 116
Yamaguchi, Hiroshi 264
Yamaguchi, Y. 495
Yamaji, Mizue 180
Yamamoto, Shigehisa 447
Yamamoto, Toshimasa 44
Yamashita, Hiroaki 32
Yamazaki, Tomoyuki 459
Yang, C.C. 487
Yang, Chao 204
Yang, Jiye 311
Yang, Shu 100, 272, 375, 387, 419
Yang, Shuo 280
Yang, Thomas 76
Yang, Wenqing 311
Yang, Wentao 172, 216
Yang, Yi 212
Yano, Hiroshi 391
Yao, F.-W. 76
Yao, Thierry 307
Yao, Y. 112, 140

Yasui, Kan 455
Yasuzumi, Takenori 32
Ye, Li 475
Ye, Ran 339
Yen, Cheng-Tyng 56, 403
Yi, Bo 148
Yi, Yangbo 323
Yilmaz, Hamza 152
Yonezawa, Yoshiyuki 395
Yoshida, Daisaku 116
Yoshiura, Y. 184
You, S. 188, 284
Yu, J.-L. 76
Yu, Jingshu 84
Yu, Yang 311
Yu, Zhicheng 80
Yuan, ZhangYi'an 64, 311
Yun, Chongman 176

Z

Zanoni, E. 92
Zeng, Guang 467
Zeng, Xiao 268
Zeng, Y. 471
Zhang, A. 96, 204
Zhang, Bingke 148
Zhang, Bo 64, 96, 204, 212, 268, 311, 335, 355, 475
Zhang, Hanyuan 100
Zhang, Jialin 224
Zhang, Jinhan 280
Zhang, Jinping 268
Zhang, Lei 152, 327
Zhang, Lingfang 64
Zhang, Long 323
Zhang, Meng 411
Zhang, Sen 80, 323, 475
Zhang, Wei Jia 84
Zhang, Wentong 475
Zhang, Xuan 355
Zhang, Yunwu 80
Zhang, Zhaofu 76
Zhang, Zhi-wen 355
Zhao, M. 188, 284
Zhao, Yawen 224
Zheng, Jiexin 224
Zheng, Yingkui 280
Zhi, Jin 280

Zhong, Xueqian 375
Zhou, Feng 516
Zhou, Qi 96, 204, 212, 335
Zhou, Xianda 172, 216
Zhou, Xin 64
Zhu, C. 140
Zhu, Chunlin 112
Zhu, Jing 60, 80, 323
Zhu, L. 96, 212
Zhu, Mingda 228
Zhu, Minghua 236
Zhu, Ruopu 212
Ziad, H. 284
Ziemann, Thomas 407
Zitouni, Moaniss 479
Zulauf, Grayson 136
Zuo, Zheng 431

IEEE
445 Hoes Lane
Piscataway, NJ 08854-4141

ISBN 978-1-5386-2928-4

2019 31st International Symposium on Power Semiconductor Devices and ICs (ISPSD 2019)

Shanghai, China
19 – 23 May 2019

IEEE Catalog Number: CFP19ISP-POD
ISBN: 978-1-7281-0582-6